Practice
"Work the Problems"

 W9-BXW-537

Feature	Description	Benefit	Page
Concepts & Vocabulary	These Fill-in-the-Blank, True/False, and open-ended questions assess your understanding of vocabulary and statistical concepts.	Helps you to review the language of statistics. Plus, you are encouraged to explain statistics concepts in your own words.	116
Skill Building	"Drill and Practice" exercises that develop computational skills.	Working these exercises will increase your understanding of formulas and concepts.	116-117
Applying the Concepts	Exercises based in real data that ask a variety of questions, many of which require you to write a few sentences of explanation.	These exercises require a deeper level of thinking and analysis which, in turn, promotes statistical thinking.	117-123
Consumer Reports	A project that requires you to go through the same type of analysis that was used by researchers at Consumer Reports to help forms the basis for Product Ratings.	Gives you the chance to practice and apply the skills and concepts learned within the chapter and, also, to see how these skills are applied in every day life.	141
Activities	Short in-class projects that get students involved in collecting and analyzing their own data.	Promotes active learning in the classroom. Helps to generate enthusiasm and understanding of statistical concepts.	109

Review
"Study for Quizzes and Tests"

Feature	Description	Benefit	Page
Chapter Reviews at the end of each chapter contain...			
Summary, Formulas, and Vocabulary.	Provides an overview of important concepts and terms, and a listing of the formulas and vocabulary introduced in the chapter.	Review these and you'll know the most important material in the chapter!	168-170
Objectives	List of chapter objectives along with the corresponding review exercises that test your understanding of the objective.	Helps to get a "big picture" overview of the material presented in the chapter. Plus, makes review for the exam easier by organizing review problems by objective.	170
Review Exercises	These provide comprehensive review and practice of key skills, matched to the Learning Objectives for each section.	Practice makes perfect. These problems combine exercises from all sections, giving you a comprehensive review in one place.	170-176
Case Study	Each chapter concludes with a case study that requires you to use skills learned within the section to perform statistical analysis.	Brings together the chapter concepts within one case study.	On CD that accompanies the text

ANNOTATED INSTRUCTOR'S EDITION

Fundamentals of
STATISTICS

SECOND EDITION

Michael Sullivan, III
Joliet Junior College

PEARSON

Prentice
Hall

Upper Saddle River, New Jersey 07458

This work is protected by United States copyright laws and is provided solely for the use of instructors in teaching their courses and assessing student learning. Dissemination or sale of any part of this work (including on the World Wide Web) will destroy the integrity of the work and is not permitted. The work and materials from it should never be made available to students except by instructors using the accompanying text in their classes. All recipients of this work are expected to abide by these restrictions and to honor the intended pedagogical purposes and the needs of other instructors who rely on these materials.

Editor-in-Chief: Sally Yagan
Executive Acquisitions Editor: Petra Recter
Project Manager: Michael Bell
Executive Managing Editor: Kathleen Schiaparelli
Senior Managing Editor: Linda Mihatov Behrens
Production Editor: Bob Walters, Prepress Management, Inc.
*Assistant Managing Editor, Science
  and Math Supplements:* Karen Bosch
Media Production Editor: Traci Douglas
Editorial Assistant / Print Supplements: Joanne Wendelken
Manufacturing Buyer: Maura Zaldivar
Manufacturing Manager: Alexis Heydt-Long
Director of Marketing: Patrice Jones
Marketing Manager: Wayne Parkins
Marketing Assistant: Jennifer de Leeuwerk
Development Editor: Don Gecewicz
Editor-in-Chief, Development: Carol Trueheart
Art Director: Jonathan Boylan
Interior Designers: Dina Curro, Jonathan Boylan
Cover Designers: Jonathan Boylan, Michael Fruhbeis
Creative Director: Juan R. López
Director of Creative Services: Paul Belfanti

Composition: ESM Inhouse Formatting
  Manager, Composition: Allyson Graesser
  Assistant Manager, Formatting: Clara Bartunek
  Electronic Page Makeup: Karen Stephens, Julita Navarro, Lissette Quinones
Director, Image Resource Center: Melinda Reo
Manager, Rights and Permissions: Zina Arabia
Manager, Visual Research: Beth Brenzel
Cover Image Specialist: Karen Sanatar
Image Permission Coordinator: Joanne Dippel
Photo Researcher: Melinda Alexander
Cover Photo: David Mager; ©2007 Pearson Education
Art Studio: Artworks
  Senior Managing Editor, Art Production and Management: Patricia Burns
  Manager, Production Technologies: Matthew Haas
  Manager, Illustration Production: Sean Hogan
  Assistant Manager, Illustration Production: Ronda C. Whitson
  Art Editor: Tom Benfatti
  Illustrations: Stacy Smith, Scott Wieber, Nathan Storck, Ryan Currier, Mark Landis, Audrey Simonetti, David Lynch
  Quality Assurance: Pamela Taylor, Timothy Nguyen, Cathy Shelley

© 2008, 2005 Pearson Education, Inc.
Pearson Prentice Hall
Pearson Education, Inc.
Upper Saddle River, New Jersey 07458

All rights reserved. No part of this book may be reproduced, in any form or by any means, without permission in writing from the publisher.

Pearson Prentice Hall™ is a trademark of Pearson Education, Inc.

Printed in the United States of America
10 9 8 7 6 5 4 3 2 1

ISBN: 0-13-156991-0 (Annotated Instructor's Edition)

ISBN: 0-13-156987-2 (Student Edition)

Pearson Education LTD., *London*
Pearson Education Australia PTY, Limited, *Sydney*
Pearson Education Singapore, Pte. Ltd
Pearson Education North Asia Ltd. *Hong Kong*
Pearson Education Canada, Ltd., *Toronto*
Pearson Educación de Mexico, S.A. de C.V.
Pearson Education – Japan, *Tokyo*
Pearson Education Malaysia, Pte. Ltd

To My Wife Yolanda

and

My Children
Michael, Kevin, and Marissa

CONTENTS

Preface to the Instructor   xi
Student and Instructor Resources   xxi
List of Applications   xxiv
Photo Credits   xxx

Part I: Getting the Information You Need   1

Chapter 1   Data Collection   2

1.1   Introduction to the Practice of Statistics   3
1.2   Observational Studies, Experiments, and Simple Random Sampling   13
1.3   Other Effective Sampling Methods   23
1.4   Sources of Error in Sampling   33
1.5   The Design of Experiments   40
   Chapter Review   48
   Case Study: Chrysalises for Cash (on CD)

Part II: Descriptive Statistics   53

Chapter 2   Organizing and Summarizing Data   54

2.1   Organizing Qualitative Data   55
2.2   Organizing Quantitative Data   71
2.3   Graphical Misrepresentations of Data   93
   Chapter Review   98
   Case Study: The Day the Sky Roared (on CD)

Chapter 3   Numerically Summarizing Data   106

3.1   Measures of Central Tendency   107
3.2   Measures of Dispersion   123
3.3   Measures of Central Tendency and Dispersion from Grouped Data   142
3.4   Measures of Position   149
3.5   The Five-Number Summary and Boxplots   159
   Chapter Review   168
   Case Study: Who Was 'A Mourner'? (on CD)

Chapter 4 Describing the Relation between Two Variables 176

4.1 Scatter Diagrams and Correlation 177
4.2 Least-Squares Regression 195
4.3 The Coefficient of Determination 209
Chapter Review 215
Case Study: Thomas Malthus, Population,
and Subsistence (on CD)

Part III: Probability and Probability Distributions 221

Chapter 5 Probability 222

5.1 Probability Rules 223
5.2 The Addition Rule and Complements 238
5.3 Independence and the Multiplication Rule 249
5.4 Conditional Probability and the General Multiplication Rule 255
5.5 Counting Techniques 265
5.6 Bayes's Rule (on CD)
Chapter Review 279
Case Study: The Case of the Body in the Bag (on CD)

Chapter 6 Discrete Probability Distributions 284

6.1 Discrete Random Variables 285
6.2 The Binomial Probability Distribution 298
Chapter Review 314
Case Study: The Voyage of the *St. Andrew* (on CD)

Chapter 7 The Normal Probability Distribution 318

7.1 Properties of the Normal Distribution 319
7.2 The Standard Normal Distribution 331
7.3 Applications of the Normal Distribution 345
7.4 Assessing Normality 354
7.5 The Normal Approximation to the Binomial Probability Distribution 362
Chapter Review 367
Case Study: A Tale of Blood, Chemistry, and Health (on CD)

Part IV: Inference: From Samples to Population   373

Chapter 8   Sampling Distributions   374

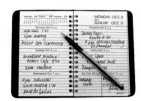

8.1 Distribution of the Sample Mean   375
8.2 Distribution of the Sample Proportion   392
Chapter Review   400
Case Study: Sampling Distribution of the Median (on CD)

Chapter 9   Estimating the Value of a Parameter Using Confidence Intervals   403

9.1 The Logic in Constructing Confidence Intervals about a Population Mean where the Population Standard Deviation is Known   404
9.2 Confidence Intervals about a Population Mean in Practice where the Population Standard Deviation is Unknown   423
9.3 Confidence Intervals about a Population Proportion   435
9.4 Putting It All Together: Which Method Do I Use?   443
Chapter Review   447
Case Study: The Search for a Fire-Safe Cigarette (on CD)

Chapter 10   Hypothesis Tests Regarding a Parameter   453

10.1 The Language of Hypothesis Testing   454
10.2 Hypothesis Tests for a Population Mean Assuming the Population Standard Deviation is Known   462
10.3 Hypothesis Tests for a Population Mean in Practice   480
10.4 Hypothesis Tests for a Population Proportion   493
10.5 Putting It All Together: Which Method Do I Use?   500
Chapter Review   503
Case Study: How Old Is Stonehenge? (on CD)

Chapter 11   Inference on Two Samples   507

11.1 Inference about Two Means: Dependent Samples   508
11.2 Inference about Two Means: Independent Samples   521
11.3 Inference about Two Population Proportions   534
Chapter Review   545
Case Study: Control in the Design of an Experiment (on CD)

Chapter 12 **Additional Inferential Procedures   550**

12.1 Goodness of Fit Test   551

12.2 Tests for Independence and the Homogeneity of Proportions   562

12.3 Testing the Significance of the Least-Squares Regression Model   580

12.4 Confidence and Prediction Intervals   594

Chapter Review   600

Case Study: Feeling Lucky? Well, Are You? (on CD)

Additional Topics on CD

C.1 Lines   C-1

C.2 Confidence Intervals about a Population Standard Deviation   C-7

C.3 Hypothesis Tests for a Population Standard Deviation   C-13

C.4 Comparing Three or More Means (One Way Analysis of Variance)   C-19

Appendix A   Tables   A-1

Appendix B Lines (on CD)   B-1

Answers   AN-1

Index   I-1

PREFACE TO THE INSTRUCTOR

Fundamentals of Statistics is a brief version of *Statistics: Informed Decisions Using Data*. This shorter text is intended to meet the needs of those schools that offer a one-semester introductory statistics course that do not wish or do not have the time to cover some of the more advanced topics found in *Statistics: Informed Decisions Using Data*. Some of these more advanced topics that appear in *Statistics: Informed Decisions Using Data*, such as residual analysis, nonlinear regression, the Poisson distribution, and nonparametric statistics, have been removed from this briefer text. The coverage of topics has been adapted to make it more suitable for usage in a one-semester course, but this does not compromise the depth of coverage.

Fundamentals of Statistics retains the strong pedagogy, features, and writing style that have made *Statistics: Informed Decisions Using Data* so successful. In addition, it retains the popular problems and discussion that encourage problem solving and statistical thinking.

Capturing a Powerful and Exciting Discipline in a Textbook

Statistics is a powerful subject. It is one of my passions. Bringing my passion for the subject together with my desire to create a text that would work for me, my students, and my school led me to write the first edition of this textbook.

I have taught introductory statistics for the past thirteen years at both 2-year and 4-year institutions, and during that time I have learned about the basic needs and issues confronting the introductory statistics course. The diversity of both the students taking the course and the instructors teaching the course heightens many of the challenges in presenting statistics at this introductory level. I brought these experiences and goals to the writing of the first edition.

The second edition of *Fundamentals of Statistics* builds on the success of the first edition as well as on my continuing experience in teaching and shaping the course. Furthermore, as I was writing the second edition, I made a special effort to abide by the *Guidelines for Assessment and Instruction in Statistics Education* (GAISE) for the college introductory course endorsed by the American Statistical Association (ASA). The GAISE Report gives six recommendations for the college introductory statistics course:

1. Emphasize statistical literacy and develop statistical thinking
2. Use real data in teaching statistics
3. Stress conceptual understanding
4. Foster active learning
5. Use technology for developing conceptual understanding
6. Use assessments to improve and evaluate student learning

Emphasize Statistical Literacy and Develop Statistical Thinking

The definition of *statistical thinking* presented in the GAISE Report is "understanding the basic language of statistics and understanding some fundamental ideas of statistics." Among these fundamental ideas are the need to work with data, the importance of data production and collection, the ability to read statistical graphs, and the capacity to select an appropriate technique to answer a research question. *Fundamentals of Statistics* addresses this recommendation in the following ways:

- **Extensive Discussion of Data Production**—Sections 1.2 through 1.4 discuss sampling techniques and pitfalls of sampling. Section 1.5 has a thorough discussion of the design of experiments.

- **Reading and Interpretation of Statistical Graphs**—The emphasis in Chapter 2 is on constructing and interpreting statistical graphs. In addition, we present Section 2.3, which illustrates the idea of misrepresenting data through graphs. The remainder of the text requires students to use graphs as a preliminary step in any statistical analysis. This continuing reliance on graphs is built into the examples and exercises.

- **Decision Making about Appropriate Techniques**—When a research question is posed, the researcher must decide on the type of analysis that is needed to answer the question. Different models may be used to answer the question. For example, in Section 1.5, students are often asked to design an experiment to answer a research question. Certainly, different designs may all be legitimate choices. The student is asked to justify his or her choice of design. Also, to help students in determining which technique to use, I have written new **Putting It All Together** sections in Chapters 9 and 10. The problems in these sections require that students decide which technique to use when performing inference, a key part of statistical thinking.

- **Decisions Projects**—Each chapter opens with a Decisions Project that requires students to collect, summarize, and interpret data to help make decisions we all must make, and would like to make intelligently, in our everyday lives. This edition highlights decision making by placing the Decisions Project on the chapter opener and weaving its resolution into the chapter itself.

- **Concepts and Vocabulary**—Each section exercise set has a list of problems that require students to learn the language of statistics and explain statistical concepts in their own words.

Use Real Data in Teaching Statistics

The examples and exercises both use real, sourced data to illustrate statistical concepts and procedures. The data come from a wide variety of disciplines such as psychology, sociology, finance, economics, biology, environmental studies, business, and chemistry so that all students can see the importance of data analysis in their field of study. A complete list of applications can be found in the List of Applications, pages xxiv-xxix.

Among the many interesting real data sets that I have chosen are from my own students. Data sets from students serve several purposes: First, I like to show students that thinking statistically applies to ordinary situations. (Instructors may recall the famous story of Fisher and the lady tasting tea.) I also like to show students that they can act like a statistician and gather real and meaningful information about the environment, consumer products, and even golf. Demonstrating to students at my campus, and in your class, that real data are everywhere helps to validate the course and statistics as a discipline.

I have also learned that many instructors rely on classroom activities to teach statistical concepts and make data more vivid. Each chapter has at least one *optional* classroom activity that fosters active learning by allowing students to collect and analyze their own data.

Stress Conceptual Understanding

It is important that students understand statistical concepts because facility with concepts will contribute to their understanding of statistical procedures. Conceptual understanding will provide them with a strong problem-solving mindset that will help them to succeed in future statistics courses. In addition, conceptual understanding develops a student's ability to make decisions and think critically. *Fundamentals of Statistics* helps students develop a conceptual understanding in the following ways:

- **Applying the Concepts Problems**—The Applying the Concepts problems in the exercise set not only ask students to analyze real data by drawing graphs or going through procedures, they also help to develop a student's ability to think statistically. Some examples of exercises that develop statistical thinking are: Problem 14 on page 98, Problem 25 on page 118, Problems 21 and 22 on page 135, Problem 31 on page 190, and the problems in Sections 9.4 and 10.5 (the problems in these sections require students first to determine the appropriate procedure to use for a given scenario).

- **Simulations and Applets**—Throughout the text, you will find exercises that require students to perform simulations or use applets that will contribute to their conceptual understanding. For example, in Problem 33 on page 391, the students use simulation to enhance their understanding of the Central Limit Theorem. In Problems 50-52 on page 422, the students use an applet to enhance their understanding of confidence intervals. Simulations and applets are powerful because they let students work actively with data. In addition, many students who are visual learners find that applets give them the reinforcement that they need to master a concept.

- **The Presentation of Formulas**—When formulas are presented, we offer both a "computational formula" and a "definition formula." For example, see the definition for variance on page 125. The student develops an understanding that the variance is a mean of squared deviations from the definition formula. Or consider the definition for the correlation coefficient on page 180. The students develop a conceptual understanding of what correlation measures. It is important for students to know a formula, but students should also be able to explain a concept and the workings of a formula in their own words.

Foster Active Learning in the Classroom

To promote active learning, we have included numerous in-class activities that encourage students to collect, summarize, and analyze data. Doing these activities promotes conceptual understanding of statistical concepts. In addition to the classroom activities, we have incorporated applets that can be used in the classroom or as a group project to illustrate and explore statistical ideas. Plus, each chapter has a case study located on the CD that accompanies the text that can be assigned as a small-group project. All these projects will help to foster active learning by encouraging students to discuss statistics among themselves and challenge each other's ideas.

Use Technology for Developing Concepts and Analyzing Data

We live in a digital world, and technology allows us to explore statistical ideas in new and exciting ways. Easily accessible technology now includes graphing calculators, statistical software packages, educational software, applets, and Web-based materials. For example, we can use graphing calculators or statistical software packages to explore "what if" scenarios. Doing so helps to develop the student's understanding of statistical concepts. The exercise sets in the text are written with "what if" scenarios built in. For example, Problem 24 in Section 4.1 asks the student to determine the impact of removing an observation from a data set. This problem foreshadows the idea of influential observations and is done easily with the help of technology.

Recognizing that various learning styles exist and that conceptual understanding can arise by working with formulas, each example is presented using computation "by hand." When appropriate, the results of the same example are obtained using a graphing calculator or software (MINITAB or Excel). The steps for obtaining the output using the TI-83/84 Plus graphing calculators, MINITAB, and Excel are included after the end-of-section exercise set.

To supplement the textbook, technology-specific manuals that contain tutorial instructions and worked examples and exercises are available (TI-83/84 Plus, MINITAB, Excel, and SPSS).

Use Assessments to Improve and Evaluate Student Learning

The exercises at the end of each section are divided into three parts: (1) Concepts and Vocabulary, (2) Skill Building, and (3) Applying the Concepts.

- The **Concepts and Vocabulary** exercises are fill-in-the-blank, true/false, and open-ended questions that assess students' understanding of vocabulary and statistical concepts. Formats such as true/false may be simple, but the Concepts and Vocabulary exercises have been written to review important concepts and to be thought provoking.

- The **Skill Building** exercises can be thought of as "drill and practice" problems that allow students to develop a level of comfort with the formulas and procedures in the section. These problems also serve as confidence-building problems, so that students experience early success and become more proficient with the quantitative aspects of a statistical analysis.

- The **Applying the Concepts** exercises are problems based in real data that ask a variety of questions to help develop statistical thinking. While these problems do ask the standard questions such as "find the mean" or "compute a 95% confidence interval," they also ask for students to explain the results or to respond to a variety of "what if" questions that broaden the scope of the concept and the analysis. For instance, the exercise may ask, "What is the impact of an outlier in the construction of a confidence interval?" Other exercises may highlight decision making about concepts by asking, "What is the best measure of central tendency for the given data set?" These "higher level of thinking" problems develop a student's statistical aptitude and teach the students to begin to frame their own questions and investigate problems on their own.

New to This Edition

Content Changes

- **Over 45% new and updated exercises,** with an emphasis on adding even greater variety, to give instructors an even larger range of exercises to choose from in selecting homework. The array of exercises includes many intriguing and meaningful topics, from M&Ms (always popular in statistics books) to the meaning of cholesterol counts, from identity theft to gas mileage, and from election results to cosmetic surgery. An interesting new exercise in Chapter 3 based on an experiment by Trina McNamara of Joliet

Junior College compares a name-brand chocolate-chip cookie with a store-brand chocolate-chip cookie—one of the many interesting data sets contributed by students to the textbook.

- **Over 60% new examples**, most of which contain real world data, acting to further illustrate and reinforce concepts and techniques. Like the exercises, the choice of topics is practical, ranges widely, and will hold students' attention.

- **All sourced data have been updated** and carefully selected to engage the students.

- **Each chapter opens with a scenario,** *Making an Informed Decision,* that poses a question to the student and then, within the chapter, presents the content that is necessary to use statistics to make the decision. This feature engages the student in the statistical-thinking process and highlights the practicality of statistics.

- **Chapter 5 on probability has been revamped** to ease the introduction of topics in probability. The early sections take greater care in delineating disjoint events and the complement rule. The concept of independence gets fuller coverage in this new edition, too.

- **Updated and rewritten Chapter 6** on discrete probability distributions reflects reviewers' concerns. I focused especially on enhancing coverage of the binomial probability distribution. At the request of reviewers, binomial tables have been added to the appendices.

- **New Chapter 8 on sampling distributions** was added to strengthen the coverage of this important topic that tends to cause students to stumble. The decision to give more play to sampling distributions by placing them in their own chapter was heartily endorsed by the reviewers of this new edition.

- **Chapters 9 and 10** continue to discuss confidence intervals and hypothesis testing about a mean by presenting the "sigma known use z, sigma unknown use t" approach. I find that students benefit from some discussion of confidence intervals and hypothesis testing when the population standard deviation is known, even if such situations seldom arise in practice. In both of these chapters, I present the sigma-known case as a model or template, followed by sections on tests more likely to be used in practice.

- **Putting It All Together** sections have been added to Chapters 9 and 10. These sections require that students first determine the appropriate confidence interval or hypothesis test to conduct. This forces students to think about when each technique should be used, which is typically a stumbling block on exams.

Pedagogical Changes

- **New side-by-side presentation of hypothesis testing**, which allows instructors the flexibility for presenting the classical approach, *P*-value approach, or both approaches. Recognizing that there are advocates of each of these two approaches, this format allows an instructor to choose the method that suits the philosophy of the course.

- **Activities have been added to each chapter,** giving instructors the flexibility to engage the class in an activity or assign them for homework. Use of activities promotes participation in the course by the students, makes concepts more vivid, and underscores the value of learning and using statistical techniques.

- **More emphasis on statistical thinking and interpretation of results** appears throughout the chapters. I added more follow-up questions to exercises that require consideration and interpretation of what the students have calculated. Furthermore, examples and exercises encourage the use of boxplots and normal probability plots so that students can use descriptive statistics throughout the text to interpret data and understand the workings of techniques. I also redoubled my effort to include screen captures and computer printouts so that students will learn to "eyeball" and interpret results when using technology.

- **Using Technology examples show the way that software and calculators assist in analysis**, once students have mastered "by hand" calculations. The *Using Technology* examples also give the students a feeling for statistics in practice, demonstrate when to rely on different devices and software, and show the output or screens to expect.

- **A division into parts makes the structure of the book more transparent** to instructors and students. The parts show the student that the order of topics is logical and practical by following the statistical process.

- **MyMathLab**, offering course management, online homework and tutorial and assessment, is fully integrated with the text. Since 2001, over one million students have done better in math and statistics with MyStatLab and MyMathLab!

Flexibility

Just as there are many different learning styles, there are many different teaching styles. One challenge in writing a text is to create a book that appeals to a broad audience. As I survey the halls of the mathematics department at Joliet Junior College, I see many high-quality instructors who present the same material in many different ways. Some like to incorporate as much technology as possible into their classroom, yet some prefer to minimize the technology. Some instructors prefer to use collaborative learning to present material, and others rely on lecture. For the introductory statistics course, some instructors are trained in the discipline of statistics, whereas others are trained in mathematics but teach statistics. Taking into consideration all these varied backgrounds, it is clear to me that a text should meet the needs of these diverse backgrounds and teaching philosophies.

"By Hand" Computation versus Technology Solutions

Let's consider how graphing calculators and statistical software are used in the course. Every concept in the text is first presented using the "by hand" approach. The reason for this is twofold. First, and probably most importantly, when a student learns to arrive at a solution by working a problem by hand, the student's ability to understand the concept is enhanced. How else can a student understand the concept of linear correlation, except by seeing it as the product of z-scores? Second, it allows for flexibility in philosophies. If your particular philosophy is to present statistics using "by hand" solutions, it is there for you.

If you are more apt to use technology, then you can use the *Examples Using Technology* feature. Following most "by hand" examples is a *Using Technology* example, which provides the output from a TI-83/84 Plus graphing calculator, MINITAB, or Excel.

In addition, any problem that has 15 or more observations in the data set has a CD icon, which indicates to the instructor and student that the data set is available on the data CD.

Chapter-by-Chapter Content

Chapter 1 Data Collection

This chapter deals with the methods of obtaining data. There is a careful presentation of the various sampling techniques along with circumstances under which each is used. In addition, an entire section is dedicated to nonsampling errors and how to control them. The chapter ends with a detailed discussion of experimental design.

Chapter 2 Organizing and Summarizing Data

This chapter addresses methods for summarizing qualitative data (Section 2.1) and quantitative data (Section 2.2). The chapter ends with a discussion of graphical misrepresentations of data. This section can be covered as a reading assignment.

Chapter 3 Numerically Describing Data

Sections 3.1 and 3.2 present numerical measures of central tendency and dispersion. Section 3.3 is optional and can be skipped without loss of continuity. However, if it is skipped and Section 6.1 is covered, the instructor should proceed slowly through the mean and standard deviation of a discrete random variable. Section 3.4 discusses measures of position, including the z-score, percentiles, and outliers. Section 3.5 presents exploratory data analysis, especially the boxplot, which will get much use in later chapters.

Chapter 4 Describing the Relation between Two Variables

Section 4.1 introduces scatter diagrams and correlation. Section 4.2 presents the least-squares regression line. Section 4.3 presents the coefficient of determination. The material in this chapter can be postponed until after Section 12.2.

Chapter 5 Probability

Section 5.1 introduces the basic concepts of probability and unusual events. Section 5.2 presents the Addition Rule and complements. Section 5.3 presents the Multiplication Rule for Independent Events. Section 5.4 presents conditional probability and the General Multiplication Rule and can be skipped without loss of continuity. Section 5.5 presents counting techniques and can be skipped without loss of continuity with the exception of Combinations.

Chapter 6 Discrete Probability Distributions

The entire chapter is optional. Section 6.1 introduces the concept of a random variable and discrete probability distributions, along with expected value. Section 6.2 presents the Binomial Probability Distribution. If you intend to cover Section 6.2, it is a good idea to cover Section 5.3 and Combinations in Section 5.5 also.

Chapter 7 The Normal Probability Distribution

Sections 7.1 through 7.3 introduce the normal probability distribution. Section 7.4 presents normal probability plots as a means for assessing normality and is required to cover topics presented in Chapters 8 through 12. Section 7.4 does not require the use of technology because the output generated by MINITAB is included. This section is necessary to help students see that verifying normality is necessary before proceeding with inference for small samples. Section 7.5 discusses the normal approximation to the binomial and is optional.

Chapter 8   Sampling Distributions (New)

A new stand-alone chapter on sampling distributions has been added to place added emphasis on the importance of this concept. Section 8.1 introduces the sampling distribution of the sample mean. Section 8.2 discusses the sampling distribution of the sample proportion.

Chapter 9   Estimating the Value of a Parameter Using Confidence Intervals

Section 9.1 introduces the construction of a confidence interval when the population standard deviation is known, and Section 9.2 constructs confidence intervals when the population standard deviation is unknown. This approach is different from that in some other texts, but it is logical. It's as simple as "sigma known, use z; sigma unknown, use t." In both cases, small samples require that the population from which the sample was drawn be normal, so we check this requirement with a normal probability plot. Section 9.3 covers confidence intervals about a population proportion. A new Section 9.4 has been added, entitled, "Putting It All Together: Which Method Do I Use?" When students are asked to construct a confidence interval in Section 9.2, they are fairly certain that they have to construct a confidence interval about a mean using the t-distribution. The material in this section does not provide hints, which play down the role of decision making; instead, it requires the student to choose the correct model first before employing any methods.

Chapter 10   Hypothesis Tests Regarding a Parameter

Section 10.1 provides an introduction to the language of hypothesis testing. Sections 10.2 and 10.3 test hypotheses regarding a population mean, again segmented by "sigma known, use z; sigma unknown, use t." New to this edition is the side-by-side "Classical" versus "P-value" approach to hypothesis testing. This allows instructors to choose the method that suits their philosophy. The side-by-side presentation continues for the rest of the text. Section 10.4 presents hypothesis testing about a population proportion. A new Section 10.5 has been added, entitled, "Putting It All Together: Which Method Do I Use?" Similar to the Putting It All Together from Chapter 9, students must first decide on the appropriate model to use before testing hypotheses.

Chapter 11   Inferences on Two Samples

Section 11.1 presents the analysis required for matched-pairs design. Section 11.2 presents the analysis for comparing two means from independent samples. Notice that the discussion regarding pooled estimates is absent. This discussion is not included because the pooled estimate approach requires that the two populations have a common variance, which is an extremely difficult requirement to test in practice. Because "pooling" versus "not pooling" often provides the same results, the pooled estimate approach is not necessary at this level to cloud the students' decision-making process. Section 11.3 discusses comparing two population proportions.

Chapter 12   Additional Inferential Procedures

Section 12.1 presents the chi-square test for goodness of fit. Section 12.2 discusses chi-square tests for independence and homogeneity. Again, if Sections 5.3 and/or

5.4 were skipped, proceed slowly. Sections 12.3 and 12.4 cover inference on the least squares regression line and require that Sections 4.1 and 4.2 be covered.

Using the Text Efficiently with Your Syllabus

To meet the varied needs of diverse syllabi, this book has been organized with flexibility of use in mind. When structuring your syllabus, you will notice the topics listed in the "Preparing for This Section" material at the beginning of each section, which will tip you off to dependencies within the course.

The two most common variations within an introductory statistics course are the treatment of regression analysis and the treatment of probability.

- **Coverage of Correlation and Regression** The text was written with the descriptive portion of bivariate data (Chapter 4) presented after the descriptive portion of univariate data (Chapter 3). Instructors who prefer to postpone the discussion of bivariate data until later in the course can skip Chapter 4 and return to it before covering Chapter 12.

- **Coverage of Probability** The text allows for a course to present either an extensive introduction to probability or a light coverage of probability. Instructors wishing to present light coverage of probability may cover Section 5.1 and skip the remaining sections. A mid-level treatment of probability can be accomplished by covering Sections 5.1 through 5.3. Instructors who will be covering the chi-square test for independence will want to cover sections 5.1 through 5.3. In addition, any instructor who will be covering binomial probabilities will want to cover independence in Section 5.3 and combinations in Section 5.5.

Acknowledgments

Textbooks evolve into their final form through the efforts and contributions of many people. First and foremost, I would like to thank my family, whose dedication to this project was just as much as mine: my wife, Yolanda, whose words of encouragement and support were unabashed, and my children, Michael, Kevin, and Marissa, who would come and greet me every morning with smiles that only children can deliver. I owe each of them my sincerest gratitude.

I would also like to thank the entire Mathematics Department at Joliet Junior College, my colleagues who provided support, ideas, and encouragement to help me complete this project. From Prentice Hall: Petra Recter, who has worked so hard to help bring this text into its second edition; Joanne Wendelken for her talent and skill in coordinating the manuscript and reviews; Sally Yagan for her support, encouragement, and words of wisdom; Patrice Jones, and Wayne Parkins, for their marketing savvy and dedication to getting the word out; Michael Bell, for his hard work in coordinating all the supplements; and the Prentice Hall sales team, for their confidence and support of this book. I thank Bob Walters, my Production Editor, for his dedication, enthusiasm, pride, and uncanny knack for attention to detail. I also extend many thanks to Don Gecewicz for his ability to see the forest for the trees and his talent getting me thinking about each sentence; Kathleen Miranda for her insight, suggestions, and in creation of answers (I owe you one). Craig Johnson and Sudhir Goel for their dedication to reading pages and verifying the accuracy in the answers. I also would like to extend my thanks to Kevin Bodden and Randy Gallaher, whose ability to write solutions, check answers, and provide insightful suggestions is amazing.

A very special thank you is extended to the following Consumer Reports staff for their effort in making the Consumer Reports Projects in each chapter a reality: Michael Saccucci, Director, Statistics & Quality Management; Frank Iacopelli, Director, Strategic Technology Development; Martin Romm, Statistical

Program Leader, Statistics & Quality Management; Keith Newsom-Stewart, Statistical Program Leader, Statistics & Quality Management; Malcolm Hazel, Statistician, Statistics & Quality Management; John Lipari, Project Leader, Statistics & Quality Management; Edward Kippel, Senior Program Leader, Health and Recreation Richard Handel, Project Leader, Health and Recreation.

We offer a very sincere thank you to all the reviewers, whose insights and ideas form the backbone of this text. We apologize for any omissions.

CALIFORNIA Charles Biles, *Humboldt State University* • Carol Curtis, *Fresno City College* • Freida Ganter, *California State University-Fresno* • Mary S. Martin, *Cosumnes River College* • Tamsen McGinley, *Santa Clara University* • Craig Nance, *Santiago Canyon College* • Irene Palacios, *Grossmont College* **COLORADO** Roxanne Byrne, *University of Colorado-Denver* **CONNECTICUT** Joseph Karnowski, *Norwalk Community College* • Sally A. Lesik, *Central Connecticut State University* • Kathleen McLaughlin, *Manchester Community College* • Dorothy Wakefield, *University of Connecticut* • Cathleen M. Zucco Teveloff, *Trinity College* **DISTRICT OF COLUMBIA** Jill McGowan, *Howard University* **FLORIDA** Randall Allbritton, *Daytona Beach Community College* • Abraham Biggs, *Broward Community College South* • Franco Fedele, *University of West Florida* • Laura Heath, *Palm Beach Community College* • Perrian Herring, *Okaloosa Walton College* • Marilyn Hixson, *Brevard Community College* • Philip Pina, *Florida Atlantic University* • Mike Rosenthal, *Florida International University* • Pali Sen, *University of North Florida* • James Smart, *Tallahassee Community College* **GEORGIA** Donna Brouillette, Todd Hendricks, *Georgia Perimeter College*, Virginia Parks, *Georgia Perimeter College* • Chandler Pike, *University of Georgia* • Jill Smith, *University of Georgia* **IDAHO** K. Shane Goodwin, *Brigham Young University-Idaho* • Craig Johnson, *Brigham Young University-Idaho* • Brent Timothy, *Brigham Young University-Idaho* • Kirk Trigsted, *University of Idaho* **ILLINOIS** Grant Alexander, *Joliet Junior College* • Kevin Bodden, *Lewis & Clark Community College* • Joanne Brunner, *Joliet Junior College* • James Butterbach, *Joliet Junior College* • Elena Catoiu, *Joliet Junior College* • Faye Dane, *Joliet Junior College* • Jason Eltrevoog, *Joliet Junior College* • Erica Egizio, *Joliet Junior College* • Randy Gallaher, *Lewis & Clark Community College* • Iraj Kalantari, *Western Illinois University* • Donna Katula, *Joliet Junior College* • Diane Long, *College of DuPage* • Jean McArthur, *Joliet Junior College* • Angela McNulty, *Joliet Junior College* • Judy Meckley, *Joliet Junior College* • Linda Padilla, *Joliet Junior College* • David Ruffato, *Joliet Junior College* • Patrick Stevens, *Joliet Junior College* • Robert Tuskey, *Joliet Junior College* • Stephen Zuro, *Joliet Junior College* **INDIANA** Jason Parcon, *Indiana University–Purdue University-Ft. Wayne* **IOWA** • Constance Marberry, *Kirkwood Community College* **KANSAS** Ingrid Peterson, *University of Kansas* **MARYLAND** John Climent, *Cecil Community College* • Rita Kolb, *The Community College of Baltimore County* **MASSACHUSETTS** Daniel Weiner, *Boston University* **MICHIGAN** Margaret M. Balachowski, *Michigan Technological University* • Susan Lenker, *Central Michigan University* • Timothy D. Stebbins, *Kalamazoo Valley Community College* • Sharon Stokero, *Michigan Technological University* • Alana Tuckey, *Jackson Community College* **MINNESOTA** Mezbhur Rahman, *Minnesota State University* **MISSOURI** Farroll Tim Wright, *University of Missouri-Columbia* **NEBRASKA** Jane Keller, *Metropolitan Community College* **NEW JERSEY** Jeffrey C. Jones, *County College of Morris* **NEW YORK** Jacob Amidon, *Finger Lakes Community College* • Stella Aminova, *Hunter College* • Pinyuen Chen, *Syracuse University* • Julie Hanson, *Clinton Community College* • Bryan Ingham, *Finger Lakes Community College* • Anne M. Jowsey, *Niagara County Community College* • Maryann E. Justinger, *Erie Community College-South Campus* • Kathleen Miranda, *SUNY at Old Westbury* • Robert Sackett, *Erie Community College-North Campus* **NORTH CAROLINA** Fusan Akman, *Coastal Carolina Community College* • Robert J. Bass, *Gardner-Webb University* • Mohammad Kazemi, *University of North Carolina-Charlotte* • Jacquelyn MacLaughlin, *Central Piedmont Community College* • Janet Mays, *Elon University* • Marilyn McCollum, *North Carolina State University* • Claudia McKenzie, *Central Piedmont Community College* • Said E. Said, *East Carolina University* • Karen Spike, *University of North Carolina-Wilmington* • Jeanette Szwec, *Cape Fear Community College* **OHIO** Richard Einsporn, *The University of Akron* • Michael McCraith, *Cuyaghoga Community College* • Mahbobeh Vezvaei, *Kent State University* **OREGON** Daniel Kim, *Southern Oregon University* **SOUTH CAROLINA** Diana Asmus, *Greenville Technical College* • Dr. William P. Fox, *Francis Marion University* • Cheryl Hawkins, *Greenville Technical College* • Rose Jenkins, *Midlands Technical College* • Lindsay Packer, *College of Charleston* **TENNESSEE** Nancy Pevey, *Pellissippi State Technical Community College* • David Ray, *University of Tennessee-Martin* **TEXAS** Jada Hill, *Richland College* • David Lane, *Rice University* **UTAH** Joe Gallegos, *Salt Lake City Community College* **VIRGINIA** Kim Jones, *Virginia Commonwealth University* • Vasanth Solomon, *Old Dominion University* **WEST VIRGINIA** Mike Mays, *West Virginia University* **WISCONSIN** William Applebaugh, *University of Wisconsin–Eau Claire* • Carolyn Chapel, *Western Wisconsin Technical College* • Beverly Dretzke, *University of Wisconsin-Eau Claire* • Jolene Hartwick, *Western Wisconsin Technical College* • Thomas Pomykalski, *Madison Area Technical College*

Finally, please do not hesitate to contact me, through Prentice Hall, with any suggestions or comments that would improve the text. I look forward to hearing from you.

Good luck with your studies.

Michael Sullivan, III
Joliet Junior College

STUDENT AND INSTRUCTOR RESOURCES

STUDENT RESOURCES

Data CD

Text comes with CD containing data sets (ASCII, SPSS, Excel, MINITAB, TI, and JMP), statistical applets, which are useful for illustrating statistical concepts, and PDFs of case studies, additional topics, formulas, and tables. Included topics are lines, confidence intervals about a population standard deviation, hypothesis tests for a population standard deviation, and comparing three or more means (one-way analysis of variance).

PRINTED RESOURCES

Student Study Pack

- Text and Student Study Pack bundle: 0-13-613730-X
- Stand-alone ISBN: 0-13-205051-X

Everything a student needs to succeed in one place. The Student Study Pack contains:

Student Solutions Manual

Fully worked solutions to odd-numbered exercises

Technology Manual

Contains detailed tutorial instructions and worked out examples & exercises for:

- TI-83/84 Plus Graphing Calculator
- Excel (including PHStat, an Excel plug-in)
- SPSS
- MINITAB

CD Lecture Videos

A comprehensive set of videos, tied to the textbook, in which examples from each chapter are worked out by a statistics instructor. The videos provide excellent support for students who require additional assistance, for distance learning and self-paced programs, or for students who missed class.

INTERNET RESOURCES

MyStatLab

- Text and MyStatLab Bundle: 0-13-613689-3
- Stand-alone ISBN: 0-13-147894-X

MyStatLab is a text-specific, online course that provides unlimited tutorial exercises closely correlated to the exercises in the text, a multimedia textbook with links to learning aids, such as animations and videos, student supplements in electronic format—the student solutions manual, PowerPoint Presentations and much more! Since 2001, over one million students have done better in math and statistics with MyStatLab. Now including StatCrunch.

StatCrunch

The award winning StatCrunch is now available in MyStatLab. This web-based statistical analysis tool offers such features as summary statistics, Z & T statistics, regression, proportions, variance, ANOVA, nonparametrics, as well as the ability to load text or Excel files from your computer or the Web, analyze data with a full range of numerical and graphical methods, create interactive graphics for exploratory analysis, subset data and grouping results, compute mathematical expressions, and generate HTML output that can be easily copied and saved. All datasets from the textbook are contained within StatCrunch for MyStatlab. Now you can spend more time analyzing your data and less time keying in datasets.

MathXL®

MathXL offers unlimited (algorithmically generated) practice exercises that are correlated directly to the exercises in the textbook.

- 12-month MathXL and Text Bundle: 0-13-205057-9
- 24-month MathXL and Text Bundle: 0-13-205054-4
- 12-month Stand-alone ISBN: 0-13-147892-3
- 24-month Stand-alone ISBN: 0-13-147897-4

INSTRUCTOR RESOURCES

INSTRUCTOR RESOURCES
Data CD

Text comes with CD containing data sets (ASCII, SPSS, Excel, MINITAB, TI, and JMP), statistical applets, which are useful for illustrating statistical concepts, and PDFs of case studies, additional topics, formulas, and tables. Included topics are Lines, Confidence Intervals About a Population Standard Deviation, Hypothesis Tests for a Population Standard Deviation, and Comparing Three or More Means (One-Way Analysis of Variance)

INSTRUCTOR RESOURCE CENTER

All instructor resources can be downloaded from http://www.prenhall.com (then search for Sullivan, *Fundamentals of Statistics*, and click on "Instructor Resources" in left column). This is a password protected site that requires instructors to set up an account or, alternatively, instructor resources can be ordered from your Prentice Hall sales representative.

Annotated Instructor's Edition (AIE)

(ISBN: 0-13-156991-0) Marginal notes include helpful teaching tips that suggest alternative presentations or point out common student errors. Also answers to all the exercises throughout the text are placed in the exercise set when space allows.

Mini Lectures

Mini-Lectures are provided for the Teaching Assistant, adjunct, part-time instructor, or even full-time instructor for lecture preparation by providing learning objectives, examples not found in the text, teaching notes, and the answers to said examples. Mini-Lectures are available from the Instructor's Resource Center and MyStatLab.

Instructor's Solutions Manual (ISBN: 0-13-157002-1)

Fully worked solutions to every textbook exercise.

TestGen (ISBN: 0-13-158828-1)

Test-generating software—create tests from textbook section objectives. Questions are algorithmically generated, allowing for unlimited test versions. Instructors may also edit problems or create their own.

Test Item File (ISBN: 0-13-234335-5)

A printed test bank derived from TestGen.

PowerPoint Lecture Slides

Fully editable and printable slides that follow the textbook. Use during lecture or post to a Web site in an online course. For download from the Instructor Resource Center and available within MyStatLab.

Technology Manual

Available only for download (PDF) from the Instructor Resource Center. Contains detailed tutorial instructions and worked out examples & exercises for:

- TI-83/84 Plus Graphing Calculator
- Excel
- SPSS
- MINITAB

Statistical Software Packages

We have bundling options available, at deep discounted prices for:

- MINITAB, Student Edition
- SPSS, Student Version
- JMP® INTRO

Ask your local Prentice Hall sales representative for more information.

INTERNET RESOURCES

MyStatLab

Powered by CourseCompass, the course management system, and MathXL, the online homework, tutorial and assessment system, MyStatLab is a text-specific, customizable course that allows you to deliver all or a portion of your course online. You can create and assign online homework and tests that are automatically graded and tightly correlated to the text. Manage your students' results in the online gradebook—designed specifically for mathematics and statistics—which automatically tracks students' homework and test results and gives you control over how to calculate final grades. You can customize your course to meet your course objectives. In addition to the e-book, you gain access to all instructor and student supplements electronically—PowerPoint Presentations, solutions manuals, videos, applets, etc. For added flexibility, instructors can import TestGen tests.

StatCrunch

The award winning StatCrunch is now available in MyStatLab. This web-based statistical analysis tool offers such features as summary statistics, Z & T statistics, regression, proportions, variance, ANOVA, nonparametrics, as well as the ability to load text or Excel files from your computer or the Web, analyze data with a full range of numerical and graphical methods, create interactive graphics for exploratory analysis, subset data and grouping results, compute mathematical expressions, and generate HTML output that can be easily copied and saved. StatCrunch is linked to all the datasets found within the textbook, now you can spend more time analyzing your data and less time keying in datasets.

MathXL®

Offered as a standalone product or as a component of MyStatLab, MathXL is the powerful online homework, tutorial and assessment system designed to make teaching your course easier and help your students succeed! With MathXL, instructors can create, edit, and assign online homework and tests and track all student work in the online class gradebook.

Request an access code from your local PH Sales Representative.

Statistical Applets

Built from the ground up by Webster West (Texas A&M) and David Lane (Rice University) these applets convey the key statistical concepts in a graphical manner. Instructions are included to assist with usage and understanding. The applets are integrated into some key textbook Activities and Exercises. Available for download from the Instructor Resource Center, on the CD within the AIE, and also available through MyStatLab.

Applets available are:

- Sampling
- Correlation by Eye
- Sampling Distribution
- Regression by Eye
- Random Number
- Mean and Median
- Long Run Probability
- Standard Deviation
- Confidence Intervals on a Mean
- Confidence Intervals on a Proportion
- Proportion Hypothesis Test
- Mean Hypothesis Test
- Comparing Distributions
- Contingency Tables
- Introduction to Sampling Distributions
- Binomial Distribution
- Characteristics of Estimators
- Components of Pearson's r
- Central Limit Theorem
- Regression Toward the Mean
- Normal Approximation to
- Effects of Violating the Binomial Assumptions of a t Test
- Reliability and Regression
- Two-way ANOVA
- Effect Size
- Confidence Intervals
- Chi Square Test of Deviations from Theoretically Expected Frequencies

LIST OF APPLICATIONS

Note: pages beginning with "C-" locate entries in the CD

Accounting
client satisfaction, 17-19
Aeronautics
bone length and space travel, 592
O-ring failures on Columbia, 112
Skylab, 51
Spacelab, 163-64
Agriculture
corn production, C-28
farm size, 478
farm subsidies, 31
growing season, 432-33, C-12
optimal level of fertilizer, 42-43
orchard damage, 51
yield
of orchard, 30
plowing method vs., 515
soybean, 491, C-28-29
Airline travel. *See under Travel*
Appliances
refrigerator life span, 329
Arboriculture
white oak tree diameter, 427-28
Astronomy
life on Mars, 543

Biology
alcohol effects, 47
blood classification, 567-70
blood plasma, 451
blood type, 68, 234, 255, 300, 364-65
bone length, 188, 206, 214, 592, 598-99
cholesterol level, 32, 316, 325, 369, 370, 390, 393-95, 441, 499, 505, 595-98
age and, 581-89
fertility rate, 104
gestation period, 330, 351, 389, 447
life expectancy vs., 188, 191, 205-6, 214
weight vs., 282
hemoglobin, 505
in cats, 157
in rats, 517
journal costs, 119
kidney weight, 140
lymphocyte count, 51
pulse rates, 118, 136, 158, 170
reaction time, 46, 117, 136
red blood cell mass, 523-25
resting pulse rate, 81
serum HDL, 119, 137, 148, 166
sickle-cell anemia, 260
step pulses, 532
weight and kidney stones, 20
Biomechanics
grip strength, 606
Business. *See also* **Work**
acceptance of shipment, 259-60, 278, 282
advertising
campaign, 30
humor in, 52

skin cream, 105
airline customer opinion, 30
Better Business Bureau, 282
billing process, 418
bolts manufacturing, 51, 140
car dealer profits, 120
car rentals, 518
car sales, 83
civilian labor force, 248
conference members, 263
customer satisfaction, 25-26
customer service, 361
defective products, 259, 263, 264, 275, 278
Disney World statistics conference, 236
drive-through rate, 489
fast-food, 71-73, 102, 390, 391, 420, 433, 450-51, 548
employee morale, 31, 32
entrepreneurship, 402
filling machine, 450
food court expansion, 38
food inspection, 38
Internet, 122, 140
Louisville workers, 281
marketing research, 32
minimum wage rates, 104
new store opening, 38
oil change time, 390
packaging error, 263, 278, 283
quality-control, 30, 255
restaurant
meat loaf recipe, 173
tipping in, 431
waiting time for table, 85-86, 549
salaries of principals, 402
Speedy Lube, 352
stocks on the NASDAQ, 276
stocks on the NYSE, 276
Target demographic information gathering, 32
textbook publishing, 38
traveling salesperson, 268, 276
union membership, 96
waiting in line, 295, C-18-19
worker injury, 97
worker morale, 20

Chemistry
acid rain, 477, 505, 548, 549, C-18
dissolved organic carbon
concentration of, 432
in soil, 157-58
gasoline additives, 462
pH in rain, C-32
pH meter calibration, 489
reaction time, 328
Combinatorics
arranging flags, 274, 283
combination locks, 277
committee formation, 277, 278
committee structures, 267, 283
forming words, 277

license plate numbers, 277, 283
Communication(s)
Americans online, 502
cell phone, 146, 398-99, 461
bill, 468-69, 471-74
brain tumors and, 20
rates, 329
talk time on, 369, 449
use of, 116, 134
fraud detection, 158
language spoken at home, 248
National Do Not Call Registry, 51
newspaper article analysis, 218-19
phone call length, 328
phone purchases, 506
text messaging, 398
Computer(s). *See also* **Internet**
downloading music, 504
download time, 30
DSL Internet connection speed, 31
high-speed Internet, 366-67
high-speed Internet connection, 311, 312
passwords, 277
phishing, 399
toner cartridges, 172, 502
user names, 277
Construction
basement waterproofing, 141
community center renovations, 32
of commuter rail station, 32
new homes, 96-97
Consumers
his and hers razors, 265
Crime(s)
aggravated assault, 446
conviction decision, 309
delinquency offense, 576
driving under the influence, 255
fraud, 560
identity, 63
larceny theft, 235
murder/homicide
by firearms, 311, 312, 367
victims of, 66
weapons used in, 100, 246
property, 315-16
rate of, 89-90, 175
by state, 102
robbery, 65, 120
speeding, 32
stolen credit card, 158
violent, 87, 151-53, 157, 166
Criminology
fraud detection, 158

Demographics
age of death-row inmates, 488
births
distribution of months of, 561
live, 100, 296
proportion born each day of week, 556-58
deaths. *See also* Mortality

due to cancer, 262
distribution of residents in United
    States, 553-56
family size, 101, 172, 450, 502
females living at home, 367
fertility rate, 104
household size, 81
life expectancy, 252, 254, 371
living alone, 499, 562
males living at home, 367
marital status, 256-58, 263
    gender and, 603
population
    of 5- to 9-year-old girls, 295
    of Florida, 237
    foreign-born, 65, 120
    of smokers, 263
    of United States residents under
        100, 148
southpaws, 254
teenage mothers, 460
Drugs. *See also* **Pharmaceuticals**
alcohol effects on reaction times, 515
marijuana legalization, 576
marijuana use, 461

Ecology
parasites in large-mouth bass, 20
Economics
abolishing the penny, 442
defense spending, 441
Economy
consumption versus income, 213
Dow Jones Industrial Average
    (DJIA), 94-95
health care expenditures, 98
home sales, 103, 171
housing prices, 89
income and birthrate, 187
national debt, 97
poverty, 62, 76-77, 86
Education
advanced degrees, 402
annual salary and, 504-5
area of country vs., 575
bachelor's degree, 261
    income and, 186-87
board work, 263
class attendance, 189, 206
class average, 148
classroom seating, 283
college
    algebra courses in self-study
        format, 505
    benefits of, 573
    campus safety, 24
    dropping a course, 578
    enrollment in, 91
    freshmen, 311
    grade-point average in, 194
    graduation rate, 316
    health-risk behaviors among
        students in, 65
    proportion of males and females
        completing, 544
    saving for, 105
confidence in schools, 499
developmental math, 45
distance learning, 98
doctorates, 64, 246-47
engineering degrees, 97
faculty opinion poll, 21

foreign language study, 68
GPA, 144, 148, 173
grade school enrollment, 295
grading time, 31
high school dropouts, 262
high school enrollment, 296
home schooling, 561
level of, 101, 312
    of adults 25 years and older, 59, 60-
        61
    age vs., 602
    distribution of, 602
    income and, 105, 261
    of males and females, 66
    marital status and, 264
    smoking and, 549
math education, 20, 450, 451
missing exam grade, 122
music's impact on learning, 44
parental involvement in, 294
political affiliation and, 578-79
poverty and school academic
    performance, 218
salaries of principals, 402
SAT scores, 94, 148
    combined, 97
    Math, 139, 461, 505
school enrollment, 84
school loans, 361
school violence, 602
standardized test score performance,
    515
student arrangement, 276
student government satisfaction
    survey, 248-49
student opinion poll/survey, 21, 30, 95
student services fees, 32
teaching reading, 45
teaching supplies, 402
visual vs. textual learners, 531
Electricity
battery life, 401, 432, C-12
light bulbs, 278, 351, 370
    halogen, 370
    intensity, 189
    mean life of, 172
Electronics
integrated circuits, 51
Employment. *See* **Work**
Energy
from coal, 88
consumption of, 99
home heating fuel, 63
household expenditure for, 488
oil output, 478
Engineering
batteries and temperature, 47
bearing failures, 174
bolts manufacturing, 51
concrete, 213
    mix, 116-17, 134
    strength, 530, 592, 599, C-32
    engine additives, 462
engine displacement versus fuel
    economy, 604-5
filling machines, 401-2, 456-57, 477, C-
    12-13, C-18
hardness testing equipment, 518
linear rotary bearing, 504
O-ring-thickness, 361
pump design, C-18
tensile strength, 431

tire design, 47
tire wear, 449
valve pressure, 461
**Entertainment. See also Leisure
    and recreation**Academy
    Award winners, **90, 391**
cable television subscription, 302-3,
    306, 308-9, 365
CD
    random playback, 263
    songs on, 276, 278
controversial film, 31
dramas, 419
movie attendance, 11
MP3 players, 12
musical play attendance, 303-5
neighborhood party, 263
People Meter measurement, 28
spending on, 158
student survey of spending on, 158
television
    government rules on content of,
        436, 437-38
    viewing by teenagers, 431
    watching hours, 360
    widescreen, 11
tickets to concert, 229
Environment
acid rain, 477, 505, 548, 549, C-18
carbon monoxide, 451-52
pH in rain, C-32
Secchi disk, 516-17
U.S. greenhouse emissions, 100
water clarity, 451
Epidemiology
SARS epidemic, 431
tuberculosis, 506
West Nile virus, 433
Exercise
by females, 454
Parkinson's disease and, 12
proportion of men and women after,
    544
pulse rate during, 447
treadmills, 69

Farming. *See also* **Agriculture**
incubation times for hen eggs, 328,
    350, 351-52
Finance. *See also* **Investment(s)**
ATM fees, 117, 136
ATM withdrawals, 476
billing process, 418
car buying decision, 138, 164
charitable contributions, 460
cost of kids, 97
credit card debt, 442, 504
credit card ownership, 398
domestic vacation costs, 391
Dow Jones Industrial Average
    (DJIA), 94
earnings and educational attainment,
    105
estate tax returns, 446
gasoline expenditures, 165
health care expenditures, 98
home sales, 103, 171
housing prices, 89
income
    annual salaries of married couples,
        515
    average, 86

birthrate and, 187
consumption vs., 213
household, 26-27, 237
median, 96, 97, 237
by region, 262
student survey of, 158
IRS audits, 92
law grads' pay, 447
mortgage rates, 502
national debt, 97
price of gasoline, 140
retirement funding, 446
school loans, 361
tips, 431
Firearms
gun ownership, 442, 544
muzzle velocity, 170, 446
Food. *See also* **Nutrition**
candy
M&Ms, 119, 138, 438, 447, 560
Skittles® Brand, 498
Snickers, 484-86, C-15-16
cookies
Chips Ahoy!, 350, 352, 360
chips per cookie, 167
diameter of, 103
fruits and vegetables, 461
insect fragments in, 391
meat loaf contents, 173
mixed chocolates, 148
number of possible meals, 265-67
nut mix, 148
peanuts, C-12-13
salads, 499
time spent eating or drinking, 416
vegetable consumption, 103

Gambling. *See also* **Game(s)**
craps, 281
Little Lotto, 277
lotteries, 275, 277, 282
Arizona's Fantasy 5, 283
Cash Five Lottery, 297
Pennsylvania's Cash 5, 283
Powerball, 297
pathological, 499
perfecta, 277
riverboat casino losses, 122
roulette, 234, 247, 251-52, 281, 283, 601
trifecta, 270
Game(s). *See also* **Gambling**
Big Game, 277
card drawing, 262, 283
carnival, 316
die/dice, 222
fair, 601
loaded, 236, 561
rolling, 83, 236
Jumble, 283
poker
flush, 264
royal flush, 264
stud, 233, 283
three of kind, 278
two of kind, 278
seven-card stud poker{delete: already under poker}, 233
World Championship Punkin Chunkin contest, C-29-30
Gardening, 398
planting tulips, 263
tulip planting, 234

Genetics
Huntington's disease, 235
sickle-cell anemia, 234-35
Geography
random sample of states, 21
Geology
density of Earth, 371
earthquakes, 85, 147, 350-51, 352
Old Faithful geyser (Calistoga, California), 213
Old Faithful geyser (Yellowstone Park), 120, 138, 390
Government
defense spending, 441
expenditures, 64-65
highway funds, 99
income, 64
IRS audits, 92
minimum wage rates, 104
national debt, 97
Senate committees, 283
Social Security numbers, 277
Social Security reform, 399

Health. *See also* **Exercise;**
   Medicine; Mental health
alcohol effects on the brain, 488
allergies, 311, 312, 367
to peanuts and tree nuts, 402
asthma control, 312
atopic dermatitis, 51
brain tumors and cell phones, 20
cancer, 262
cigar smoking and, 248
death due to, 262
prostate, tomatoes and, 47
skin, 353
smoking and, 264
cholesterol, 32, 316, 325, 369, 370, 393-95, 499, 505, 595-98
age and, 189
dieting and, C-32
in men vs. women, 549
serum HDL, 87, 119, 148, 166, 418, 420
cold, 604
college students' health-risk behaviors, 235
doctor visits, 248
fertility rate, 104
fitness club member satisfaction, 31
ginkgo and memory, 46
health care expenditures, 98
health-risk behaviors among college students, 65
hearing problems, 247-48, 396-97
HIV test false positives, 254
hospital admissions, 67
hypertension, 451
salt and, 47
indoor tanning, 400
injuries requiring rehabilitation, 112
insomnia, 45
insurance, 85, 147, 262, 264
life expectancy, 254
live births, 100
Lyme disease versus drownings, 193
migraine sufferers, 311, 312, 366
music and sleep, 51
obesity, 194
overweight, 98
in children, 461

para-nonylphenol effects, 488
physical rehabilitation of patient, 55, 57-58
pregnancy
E.P.T. Pregnancy Tests, 255
energy needs during, 401
psychic or spiritual healing, 544
St. John's wort and depression, 46
salmonella, 579
smoking, 311, 312, 366, 402
by age group, 576-77
cancer and, 264
cigar, 248, 262
cigarettes smoked per day, 482-84
education level and, 549
percentage among Americans, 579
population of smokers, 263
profile of smokers, 576
quitting, 316
rates of, 543
tar and nicotine levels in cigarettes, 592, 599
television stations and life expectancy, 193
tooth whitener, 45
tuberculosis, 506
Tylenol tablets, 165
vision problems, 247-48
water quality, 214
weight and kidney stones, 20
weight loss, 61
weights of males versus females, 175
West Nile virus, 433
wine and the heart, 218
Height(s)
arm span vs., 548, 549
average
of men vs. women, 157
of women, 476
of baseball players, 593-94, C-19
of boys, 606
girls vs., 217
sons vs. fathers, 517
ten-year-old, 329
of females, 351
five-year-old, 331
three-year-old, 323-24, 326-27, 345-49, 378-81, 383
head circumference vs., 187, 205, 214, 598
of males, 446
Houses and housing
apartments, 217
rents on, 605
construction of new homes, 96-97
garage door code, 276-77
haunted houses, 499
home ownership, 442
home sales, 103, 171, 433
mortgage rates, 502
number of rooms per unit, 247
paint comparisons, 47
paper towels, 533
price of, 89
rental units, 296
second homes, 399, 417
single-family home price, 460
vacation homes, 418
water filtration, 214

Insurance
  auto, 502
  health, 96
  life, 296, 316
  term life policy, 291
Intelligence
  brain size and, 190, 207
  IQ scores, 83, 121, 139, 391, 407-8, 420, 433
  Mozart effect, 10-11
Internet, 66, 461-62
  Americans online, 502
  connection time to ISP, 118-19
  high-speed access to, 311, 312, 366-67, 442
  phishing, 399
  time online, 420
  users of, 62
Investment(s)
  diversification, 192
  dividend yield, 87, 166, 431
  expected profit from, 296
  of inheritance, 528
  in mutual funds
    risk of, C-17-18
    selection of, 138
  P/E ratio, 490
  rate of return on, 81, 139, 145-46, 390, C-30
    mean, 547
  in real estate, 296
  risk of, C-13
  in Roth IRA, 73-76
  in second homes, 417
  stock price, 91
    appreciation of, 254
    over time, 91
  United Technologies vs. S&P 500, 592-93, 599
  in vacation home, 418
  volume of stock
    Altria Group, 87, 119, 137, 148, 166
    Dell Computer, 478
    Google, 419
    Harley-Davidson, 418-19
    Motorola, 478

Kids and leisure, 532

Landscaping
  golf course, 277
Language
  spoken at home, 248
Law(s)
  chief justices, 171
  death penalty, 442
  driver's license, 11
  fair packaging and labeling, 460
  jury selection, 277-78
Law enforcement
  age of death-row inmates, 488
  catching speeders, 447
  sentence length for men vs. women, 547
Leisure and recreation. See also Entertainment
  clubs, 359
  domestic vacation costs, 391
  fishing, 137
  kids and, 532
  movie attendance, 11, 312

  Nielsen ratings, 316
  Six Flags over Mid-America, 236
  theme park spending, 446
  time for, 442
  TV watching, 449
  waiting in line, 517-18
    at amusement ride, 328, 358, 428-29
Literacy. See Reading

Manufacturing
  ball bearings, 351
  copper tubing, 401
  steel rods, 351
Measurement. See also Height(s); Weight(s)
  pH meter calibration, 489
Media
  death penalty survey by, 500
  president's job approval rating, 500
Medicine. See also Drugs; Health; Pharmaceuticals
  antibiotics, 189, 207
    for children, 456-57, 458, 459-60
  apnea treatment, 544
  benign prostatic hyperplasia, 490
  blood plasma, 451
  blood type, 68, 255, 300
  blows to the head, 391
  cancer
    high-tension wires and, 19
    in rats, 19
  carpal tunnel syndrome, 20
  cholesterol level, 32, 316, 325, 369, 370, 390, 393-95, 441, 499, 505, 595-98
    HDL, 87, 119, 137, 148, 166, 418, 420
  clinical trials for treatment of skin disorder, 544
  cosmetic surgery, 62
  dentistry, 431
  depression therapy, 46
  diabetic rats, C-21-23
  epidurals, 10
  folate and hypertension, 10
  gastric bypass, 549
  hearing trouble, 396-97
  heart disease predictor, 12
  hip arthroplasty, C-33
  hormone replacement therapy, 529
  hospitals
    admissions, 67, 85, 121
    bacteria in, 531
    emergency room visits, 576
    mean length of stay in for-profit, 547
  Lipitor, 499
  live births by a mother 50 to 54 years old, 296
  medical residents' mean number of hours worked, 350, 352
  medical residents' weekly work hours, 390
  multiple births, 85
  osteoporosis treatment, 549
  Parkinson's disease, 12
  partial birth abortions, 441
  placebo effect, 249
  Prevnar vaccine, 496-97
  Salk vaccine, 543-44
  SARS epidemic, 431
  sickle-cell anemia, 260
  systolic blood pressure of surgical patients, 490

  temperature of surgical patients, 491
  ulcers, 441
  wart cure, 10
  West Nile virus, 433
Mental health
  attention deficit-hyperactivity disorder, 399
  bipolar mania, 529
  suicides, 100
Meteorology. See Weather
Military
  active-duty personnel, 249
  night vision goggles, 51
  peacekeeping missions, 30
  satellite defense system, 255
Miscellaneous
  aluminum bottles, 532
  average height of men vs. women, 157
  birth and gender orders, 277
  birthdays, 234, 247, 264
  Christmas lights, 254
  clothing options, 276
  gender preference for only child, 367
  lying, 367
  New Year's Holiday, 281
  sleeping, 416, 604
  tattoos, 502
  toilet flushing habits, 93
Money. See also Finance
  abolishing the penny, 442
Morality
  baby outside of marriage, 11
Mortality
  bicycle deaths, 561
  death due to cancer, 262
  pedestrian deaths, 561
  vehicle-related fatalities. See under Motor vehicle(s)
Motor vehicle(s). See also Transportation
  accidents, 98
    male vs. female rates, 190-91, 207-8
  age of, 478
  annual mileage, 467-68, 470-71
  BMWs, 11
  braking distance, 519
  car buying, 164, 204
  car color, 66
  car rentals, 518
  crash data, 116, 134, 417, C-31-32
  dependability survey of, 89
  driving age, 316
  driving under the influence, 255
    simulator for, 518-19
  emission inspection, 502
  engine additives, 462
  fatalities, 172
    alcohol-related, 83-84
    driver, 249, 262, 264
  fuel catalysts, 477
  gas mileage/fuel economy, 88, 165, 330, 390, 444-45, C-18
    engine displacement vs., 216, 604-5
    in Honda Accord, 431
    weight vs., 188, 191, 206, 214
  gasoline additives, 462
  gasoline expenditures, 165
  hybrid cars, 351, 352
  SUV, 371
  miles on a Cavalier, 420
  miles on a Saturn, 419-20
  octane in fuel, 519-20

oil change, 390
price of Cavaliers, 171
SUV towing capacity, 102-3
3-year-old Chevy Corvette, 456-57, C-11
tires
  testing, 434
  wear on, 369, 449
Music
  effect on learning, 44
  rhythm & blues vs. alternative, 532
  sleep and, 51
Neurology
  brain abnormalities and schizophrenia, 515
Nutrition. *See also* **Food**
  calcium consumption, 488, 541-42
  calories
    burning of, 96
    mean intake of, 387-88
    sugar vs., 213, 593, 599
  carbohydrates in smoked turkey, C-18
  chicken consumption, 579
  cholesterol in diet, 543
    dieting and, C-32
    men vs. women, 549
  daily dietary supplement, 479
  diets
    low-carbohydrate, 393
  eating together, 499
  energy needs during pregnancy, 401
  fat
    in fast food, 88, 90, 165
    saturated, 543
  fruits and vegetables, 461
  Girl Scout cookies, 281-82
  overweight, 98
  sodium consumption, 488, 505
  vegetable consumption, 103
  weight and kidney stones, 20

Obstetrics. *See also* **Pediatrics**
  birth(s)
    by day of week, 556-58, C-29
    by month, 561
    multiple, 148, 235, 247, 352
    weight at, 115-16, 156, 258, 328, 329, 505, 562
  caesarean sections, 603
  prenatal care, 575
Optometry
  eyeglasses, 492

Pediatrics. *See also* **Obstetrics**
  age of mother at childbirth, 149
  breast-feeding and obesity, 194
  caffeine and SIDS, 193
  crawling babies, 432, C-12
  heights
    of five-year-old females, 331
    head circumference vs., 187, 205, 214, 591, 598
    of ten-year-old males, 329
    of three-year-old females, 323-24, 326-27, 345-49, 378-81, 383
Pets
  talking to, 499
Pharmaceuticals. *See also* **Drugs;** **Medicine**
  Accupril, 544
  antihypertensive, 5-6

Celebrex, 577-78
cholesterol research, 32
Clarinex, 544
cold medication, 45, 508
drug effectiveness, 47
Lipitor, 441, 499
memory drug, 46
Nasonex, 537-38, 540-41
Nexium, 499
Pepcid, 441
Prevnar, 543
Tylenol tablets, 165
Zocor, 571-73
Zoloft, 549
Physics
  catching falling meter stick, 508, 510-12, 513
  muzzle velocity, 170, 516
Physiology
  fat-free mass vs. energy expenditure, 593, 599
  flexibility, 532
  reaction time, 46
  step pulses, 532
Politics
  community center renovations, 32
  elections
    Bosnian, 38
    county, 32
    presidential, 62, 66, 120, 442
  estate taxes, 30
  exit polls, 39
  farm subsidies, 31
  Future Government Club, 20, 31
  mayor and small business owners, 50
  party affiliation, 101
    education and, 578-79
  presidents, 68
    age at inauguration, 88, 165, 419
    favorite, 11
    inauguration costs, 96
    job approval rating, 500
    random sample of, 21
  public policy survey, 51
  student government satisfaction survey, 248-49
  village poll, 21
Polls and surveys, 264
  election, 20
  exit, 39
  of foreign language speakers, 38
  of gender of children in family, 229-30
  of illicit drug use, 38
  of infidelity, 38
  of police department, 38
  taste tests, 20
  village, 21
Psychiatry
  attention deficit-hyperactivity disorder, 399
  bipolar mania, 529
Psychology
  depression, 46
  insomnia relief, 45
Psychometrics
  IQ scores, 83, 121, 139, 391, 407-8, 420, 433
    of ninth-graders, 30
  stimulus response, 515
Publishing
  journal costs, 119

Reading
  amount of time devoted to, 420, 431
  habits of, 499
  number of books read, 370
Recreation. *See* **Leisure and recreation**
Residential. *See* **Houses and housing**

Security
  combination locks, 277
Seismology
  earthquakes, 350-51, 352
Society
  abortion issue, 576
  affirmative action, 442
  babies outside of marriage, 11
  charitable contributions, 460
  day care enrollment, 234
  death penalty, 442
  divorce rate, 88
  dog ownership, 234
  evolution vs. creation issue, 602-3
  family structure and sexual activity, 574-75
  housing prices, 89
  marijuana use, 461
  poverty, 62, 76-77, 86, 440
  race vs. region of United States, 603
  satisfaction of Americans, 577
  Social Security, 277, 399
  volunteerism, 234
Sports
  baseball
    American League vs. National League, 139
    baseballs manufacturing, 370
    batting averages, 140
    batting champions, 157
    cold streaks, 255
    designated hitter, 531
    ERA, 157
    heights and weights of players, 593-94, C-19
    home runs, 91, 167, 171, 236, 283
    Mets vs. Yankees, 173
    most hits in season, 295
    New York Yankees payroll, 114-15
    run production, 150-52
    St. Louis Cardinals, 282
    salaries by position, 67
    Simpson's Paradox in, 604
    starting lineup, 277
    winning percentage, 192
    World Series, 295
  basketball, 122, 234
    free throws, 84, 286, 290-91
    hot streaks, 255
    performance consistency, C-18
    points per game, 32
  bowling, 255
  disc golf, 433, C-12
  5-kilometer race, 160
  football
    blows to the head, 391
    bowl games, 147
    NFC Champions, 137
    Packer fans, 441-42
    Super Bowl, 116, 121-22, 165
  golf, 61, 177-79, 182-83, 184, 196-97, 199-203, 209-10, 234, 263, 271
    balls, 488-89, C-18

club comparisons, 46
  pitching wedge, 331
greyhound races, 355-57
high school students participating in, 234
hockey, 315
Indy 500, 277
NASCAR, 174-75, 255
Olympic medals, 282
soccer, 118, 136
softball, C-19
sports drinks, 52
Tour de France, 118, 137
200-meter dash, 218
winning predictions, 261
Surveys. *See* **Polls and surveys**

Temperature
cricket chirps vs., 216, 605
of humans, 488
summer of 2000, 478
of surgical patients, 491
Test(s)
IQ scores, 83, 121, 140, 391, 407-8, 420, 433
multiple-choice, 51
preparation for, 461
SAT
  math portion, 461, 479
  penalty, 297
  reasoning portion, 477-78
  scores, 94, 97, 461, 476-77
Wechsler Intelligence Scale, 369
  for Children (WISC), 433
Time
on cell phone, 449
to change oil, 390

on eating or drinking, 416
flight, 117, 134, 417
with friends, 417
to grade, 31
incubation, 328, 350, 351-52
online, 420
  connecting to ISP, 118-19
  downloading, 30
reaction, 46, 117, 136, 328, 516, C-30
  alcohol effects on, 515
  to catch falling meter stick, 508, 510-12, 513
  male vs. female, 530
reading, 420
travel, 118, 158
  to school, 136
  to work, 140, 227, 248, 417, 474-75, 547
waiting, 85, 172
watching TV, 449
Transportation. *See also* **Motor vehicle(s)**
alcohol-related traffic fatalities, 83-84
flight time, 117, 134, 417
highway funds, 99
highway repair, 86, 91
highway speeds, 405, 410-13, 414
male versus female accident rate, 190-91, 207-8
on-time flights, 366
seat belt use, 446
streetlight replacement, 502
students driving to school, 237
Travel
airline
  airport codes, 267
  flight time, 134, 417

on-time flights, 311
  satisfaction with, 311
  walking in airport, 530
creative thinking during, 371
driving age, 316
miles driven annually, 467-68, 470-71
time for, 118, 158
  commuting to work, 140, 227, 248, 417, 474-75, 547
  to school, 136

Weapons. *See* **Firearms**
Weather
in Chicago, 352
cyclones, 281
Memphis snowfall, 361
rainfall, 147-48
  April showers, 157
  days of, 262
summer of 2000, 478
temperature, 147
tornadoes, 167
Weight(s)
net, 486
Work. *See also* **Business**
civilian labor force, 248
commuting to, 140, 227, 248, 417, 474-75, 547
critical job skills, 499
employee morale, 20, 31, 32
hours worked per week, 450, 451
Louisville workers, 281
multiple jobs, 264
promotion chances, 499
willingness to work hard, 604
working students, 120, 147

PHOTO CREDITS

Chapter 1 Pages 1, 2, 37, Burke/Triolo Productions/Getty Images / Getty Images, Inc.-Photodisc; Page 8, PhotoEdit; Page 11, Corbis/Bettmann; Page 20, Bob Torrez / Workbook Co/Op Stock; Page 34, Kari Erik Marttila Photography; Page 40, A. Barrington Brown/Photo Researchers,; Inc. Page 44, Lonnie Duka/Getty Images Inc.-Stone Allstock; Page 46, Sergio Piumatti/Sergio Piumatti; Page 50, Don Smetzer/PhotoEdit; CD Case Study, Fritz Polking/Bruce Coleman Inc.

Chapter 2 Pages 53, 54, 95, Ryan McVay/Getty Images / Getty Images, Inc.- Photodisc; Page 71, PhotoEdit; Page 74, The Granger Collections; Page 90, IPOL-GLOBE PHOTOS / Globe Photos, Inc.; Page 91, Google Inc.; CD Case Study, Phil Degginger / Picturesque Stock Photo.

Chapter 3 Pages 53, 106, 164, Ryan McVay/Getty Images / Getty Images, Inc.-Photodisc; Page133, Sovfoto/Eastfoto; Pages 139, 167, 171, Major League Baseball Properties, Inc; Page 141, Consumer Reports; Page 163, Alfred Eisenstaedt / Getty Images/Time Life Pictures; CD Case Study, Art Resource, N.Y.

Chapter 4 Pages 53, 176, 204, Sergio Piumatti / Sergio Piumatti; Page180, Photo Researchers, Inc.; Pages 198, 199, CD Case Study, The Granger Collection.

Chapter 5 Pages 221, 222, 261, Jonathan Nourok / PhotoEdit; Page 225, Sergio Piumatti / Sergio Piumatti; Page 229, Mary Evans Picture Library / Image Works/Mary Evans Picture Library; Pages 230 and 231, Photo Researchers, Inc.; Page 234 SuperStock, Inc.; Page 260, Sovfoto/Eastfoto; Pages 270 and 275, AP Wide World Photos; Page 282, Darling Kindersley Media Library; CD Case Study, Peter Beck / Corbis/Bettmann.

Chapter 6 Pages 221, 284, 309, Steve Gorton / Darling Kindersley Media Library; Page 292, Photo Researchers, Inc.; Page 299, Corbis/Bettmann; Page 300, AP Wide World Photos; CD Case Study, Schwenkfelder Library and Heritage Center.

Chapter 7 Pages 221, 318, Burke Triolo Productions/Brand X Pictures/Getty Images; Page 322, Photo Researchers, Inc.; Page 322, Northwind Picture Archives; Page 325 Corbis/Bettmann; Page 362, North Wind Picture Archives; CD Case Study; eStock Photography LLC.

Chapter 8 Pages 373, 374, 388, Lawrence Manning / Corbis Royalty Free; Page 387, Corbis/ Bettmann; CD Case Study, Lester Lefkowitz/Getty Images, Inc.-Taxi.

Chapter 9 Pages 373, 403, 440, Michael Matisse / Getty Images, Inc.-Photodisc; Page 419, Google, Inc.; Page 423, The Granger Collection; Page 477, Consumer Reports; CD Case Study, Peter Skinner / Photo Researchers, Inc.

Chapter 10 Pages 373, 453, 486, Brian Hagiwara/Getty Images / Getty Images, Inc.-Photodisc; Pages 454, D. Graham / Getty Images/Retrofile; Page 458, J. Neubauer / Getty Images/Retrofile; Page 463, Paul Bishop / University of California-Berkeley; Page 492, Consumer Reports; CD Case Study, Jan Stromme / Bruce Coleman Inc.

Chapter 11 Pages 373, 507, 528, EMG Education Management Group; CD Case Study, Comstock Images.

Chapter 12 Pages 373, 550, 573, C Squared Studios / Getty Images, Inc.-Photodisc; Page 551, Richard B. Levine/Richard B. Levine/Franco; Page 553, Photo Researchers, Inc.; CD Case Study, Reza Estakhrian / Getty Images Inc.-Stone Allstock.

PART **1**

Getting the Information You Need

Statistics is a process—a series of steps that lead to a goal. This text is divided into four parts to help the reader see the process of statistics.

The first step in the process is to determine the research objective or question to be answered. Then, information is obtained to answer the questions stated in the research objective.

Data Collection

Outline

1.1   Introduction to the Practice of Statistics

1.2   Observational Studies, Experiments, and Simple Random Sampling

1.3   Other Effective Sampling Methods

1.4   Sources of Errors in Sampling

1.5   The Design of Experiments
- Chapter Review
- Case Study: Chrysalises for Cash (on CD)

DECISIONS

It is Monday morning and already you are thinking about Friday night—movie night. You don't trust the movie reviews published by professional critics, so you decide to survey "regular" people yourself. You need to design a questionnaire that can be used to help you make an informed decision about whether to attend a particular movie. See the Decision Activity on page 37.

Note to Instructor

Due to the tremendous number of terms in this chapter, you may want students to create a list of terms, definitions, and their own examples for each term. This will allow students to make a personal connection to the concepts and will lead to a greater understanding of the vocabulary. Remind students that many of the terms presented in this chapter will be presented again in upcoming chapters, so they should not feel overwhelmed by the number of definitions.

 Putting It All Together

For most of you, this is your first statistics course. Taking a statistics course is different from taking a mathematics course. While there are formulas and mathematical symbols in the course, statistics is not mathematics. So, whether you have struggled or had success in prior mathematics courses, you can succeed in statistics.

Before you begin the course, read "How to Use This Book" on the inside front cover of the text.

1.1 Introduction to the Practice of Statistics

Objectives

1 **Define statistics and statistical thinking**

2 **Understand the process of statistics**

3 **Distinguish between qualitative and quantitative variables**

4 **Distinguish between discrete and continuous variables**

1 Define Statistics and Statistical Thinking

What is statistics? When asked this question, many people respond that statistics is numbers. This response is only partially correct.

Definition

> **Statistics** is the science of collecting, organizing, summarizing, and analyzing information to draw conclusions or answer questions.

Note to the Instructor

This first objective defines what statistics is, but also lists some key elements to solid statistical thinking.

1. Anecdotal claims can be refuted with statistical analysis.
2. Poorly collected data are not useful.
3. Watch out for lurking variables in a study.
4. Results in statistics are not certain.

The Guidelines for Assessment and Instruction in Statistics Education (GAISE) report suggests these as major components of statistical thinking.

In Other Words

Anecdotal means that the information being conveyed is based on casual observation, not scientific research.

It is helpful to consider this definition in three parts. The first part of the definition states that statistics involves the collection of information. The second refers to the organization and summarization of information. Finally, the third states that the information is analyzed to draw conclusions or answer specific questions.

What is the information referred to in the definition? The information is *data*. According to the *American Heritage Dictionary*, **data** are "a fact or proposition used to draw a conclusion or make a decision." Data can be numerical, as in height, or they can be nonnumerical, as in gender. In either case, data describe characteristics of an individual. The reason that data are important in statistics can be seen in this definition: data are used to draw a conclusion or make a decision.

Analysis of data can lead to powerful results. Data can be used to offset anecdotal claims, such as the suggestion that cellular telephones cause brain cancer. After carefully collecting, summarizing, and analyzing data regarding this phenomenon, it was determined that there is no link between cell phone usage and brain cancer.

Because data are powerful, they can be dangerous when misused. The misuse of data usually occurs when data are incorrectly obtained or analyzed. For example, radio or television talk shows regularly ask poll questions in which respondents must call in or use the Internet to supply their vote. The only individuals who are going to call in are those that have a strong opinion about the topic. This group is not likely to be representative of people in general, so the results of the poll are not meaningful. Whenever we look at data, we should be mindful of where the data come from.

Even when data tell us that a relation exists, we need to investigate. For example, a study showed that breast-fed children have higher IQs than those who were not breast-fed. Does this study mean that mothers should breast-feed their children? Not necessarily. It may be that some other factor contributes to the IQ of the children. In this case, it turns out that mothers who breast-feed generally have higher IQs than those who do not. Therefore, it may be genetics that leads to the higher IQ, not breast-feeding. This illustrates an idea in statistics known as the *lurking variable*. In statistics, we must consider the lurking variables because two variables most often are influenced by a third variable. A good statistical study will have a way of dealing with the lurking variable.

Another key aspect of data is that they vary. To help understand this variability, consider the students in your classroom. Is everyone the same height? No. Does everyone have the same color hair? No. So, among a group of individuals there is variation. Now consider yourself. Do you eat the same amount of food each day? No. Do you sleep the same number of hours each day? No. So, even looking at an individual there is variation. Data vary. The goal of statistics is to describe and understand the sources of variation.

Because of this variability in data, the results that we obtain using data can vary. This is a very different idea than what you may be used to from your mathematics classes. In mathematics, if Bob and Jane are asked to solve $3x + 5 = 11$, they will both obtain $x = 2$ as the solution, if they use the correct procedures. In statistics, if Bob and Jane are asked to estimate the average commute time for workers in Dallas, Texas, they will likely get different answers, even though they both use the correct procedure. The different answers occur because they likely surveyed different individuals, and these individuals have different commute times. Note: The only way Bob and Jane would get the same result is if they both asked *all* commuters or the same commuters how long it takes to get to work, but how likely is this?

So, in mathematics when a problem is solved correctly, the results can be reported with 100% certainty. In statistics, when a problem is solved, the results do not have 100% certainty. In statistics, we might say that we are 95% confident that the average commute time in Dallas, Texas is 21.5 minutes. While uncertain results may sound disturbing now, it will become more apparent what this means as we proceed through the course.

Without certainty, how can statistics be useful? Statistics can provide an understanding of the world around us because recognizing where variability in data comes from can help us to control it. Understanding the techniques presented in this text will provide you with powerful tools that will give you the ability to analyze and critique media reports, make investment decisions (such as what mutual fund to invest in), or conduct research on major purchases (such as what type of car you should buy). This will help to make you an informed consumer of information and guide you in becoming a critical and statistical thinker.

② Understand the Process of Statistics

The definition of statistics implies that the methods of statistics follow a process.

Note to Instructor
Discuss how the process of statistics parallels the scientific method.

The Process of Statistics

1. *Identify the research objective.* A researcher must determine the question(s) he or she wants answered. The question(s) must be detailed so that it identifies a group that is to be studied and the questions that are to be answered. The group to be studied is called the **population**. An **individual** is a person or object that is a member of the population being studied. For example, a researcher may want to study the population of all 2005 model-year automobiles. The individuals in this study would be the cars.

2. *Collect the information needed to answer the questions posed in (1).* Gaining access to an entire population is often difficult and expensive. In conducting research, we typically look at a subset of the population, called a **sample**. For example, the U.S. population of people 18 years or older is about 218 million. Many national studies consist of samples of size 1,100. The collection-of-information step is vital to the statistical process, because if the information is not collected correctly, the conclusions drawn are meaningless. Do not overlook the importance of appropriate data-collection processes.

3. *Organize and summarize the information.* This step in the process is referred to as *descriptive statistics*.

! CAUTION

Many nonscientific studies are based on *convenience samples*, such as Internet surveys or phone-in polls. The results of any study performed using this type of sampling method are not reliable.

Definition   **Descriptive statistics** consists of organizing and summarizing the information collected.

Descriptive statistics describe the information collected through numerical measurements, charts, graphs, and tables. The main purpose of descriptive statistics is to provide an overview of the information collected.

4. *Draw conclusions from the information.* In this step the information collected from the sample is generalized to the population.

Definition

Inferential statistics uses methods that take results obtained from a sample, extends them to the population, and measures the reliability of the result.

For example, if a researcher is conducting a study based on the population of Americans aged 18 years or older, she might obtain a sample of 1,100 Americans aged 18 years or older. The results obtained from the sample would be generalized to the population. There is always uncertainty when using samples to draw conclusions regarding a population because we can't learn everything about a population by looking at a sample. Therefore, statisticians will report a level of confidence in their conclusions. This level of confidence is a way of representing the reliability of results. If the entire population is studied, then inferential statistics is not necessary, because descriptive statistics will provide all the information that we need regarding the population.

The following example will illustrate the process of a statistical study.

EXAMPLE 1   **Effectiveness of Antihypertensive Drugs**

According to researchers, little information exists on the effects that antihypertensive drugs have on patients who have heart disease and normal blood pressure.* Blood pressure is the force of blood against the walls of arteries and is presented as two numbers: the systolic pressure (as the heart beats) over the diastolic pressure (as the heart relaxes between beats). A blood pressure measurement of 120/80 mm Hg (millimeters of mercury) is normal. Hypertension or high blood pressure exists in individuals with a systolic blood pressure above 160 mm Hg or a diastolic blood pressure above 100 mm Hg. Researchers Steven E. Nissan and his associates wanted to determine the effectiveness of an antihypertensive drug** on preventing cardiovascular events such as congestive heart failure, stroke, or other heart-related problems. The following statistical process allowed the researchers to measure the effectiveness of the drug:

Group 1   Group 2

1. *Identify the research objective.* Researchers wished to determine the effectiveness of the drug on preventing cardiovascular events in patients who have heart disease and normal blood pressure.

2. *Collect the information needed to answer the questions.* The researchers divided 1,317 patients with heart disease and diastolic blood pressure less than 100 mm Hg into two groups. Group 1 had 663 patients and group 2 had 654 patients. The patients in group 1 received 10 mg daily of the antihypertensive drug. The patients in group 2 received a *placebo*. A **placebo** is an innocuous drug such as a sugar tablet. Group 1 is called the **experimental group**. Group 2 is called the **control group**. Neither the doctor administering the drug nor the patient knew whether he or she was in the experimental or control group. This is referred to as a **double-blind** experiment. After 24 months of treatment, each patient's blood pressure was recorded. In addition, the number of patients in each group who experienced a cardiovascular event was counted.

3. *Organize and summarize the information.* Before administering any drugs, it was determined that both groups had similar blood pressure. After the

Note to Instructor

What is a single-blind experiment? Are there situations where this type of experiment would be preferred over one that is double-blind?

---

*The discussion is based on a study done by Steven E. Nissan, E. Murat Tuzcu, Peter Libby, Paul D. Thompson, Magdi Ghali, Dahlia Garza, Lance Berman, Harry Shi, Ethel Buebendorf, and Eric Topol published in the *Journal of the American Medical Association*, Vol. 292, No. 18.

**The drug used in the study was 10 mg of amlodipine.

24-month period ended, the experimental group's blood pressure decreased by 4.8/2.5 mm Hg, whereas the placebo group's blood pressure increased 0.7/0.6 mm Hg. In addition, 16.6% of patients in the experimental group experienced a cardiovascular event, while 23.1% of patients in the control (placebo) group experienced a cardiovascular event.

4. *Draw conclusions from the data.* We extend the results from the sample of 1,317 patients to all individuals who have heart disease and normal blood pressure. That is, the antihypertensive drug appears to decrease blood pressure and seems effective in reducing the likelihood of experiencing a cardiovascular event such as a stroke. ██

In the study presented in Example 1, notice that the population is clearly identified in the research objective as patients who have heart disease and normal blood pressure. However, the researchers collected the information that they needed by looking at a subset of this population, the 1,317 patients.

Now Work Problem 45.

③ **Distinguish between Qualitative and Quantitative Variables**

Once a research objective is stated, a list of the information the researcher desires about the individual must be created. **Variables** are the characteristics of the individuals within the population. For example, this past spring my son and I planted a tomato plant in our backyard. We decided to collect some information about the tomatoes harvested from the plant. The individuals we studied were the tomatoes. The variable that interested us was the weight of the tomatoes. My son noted that the tomatoes had different weights even though they all came from the same plant. He discovered that variables such as weight vary.

If variables did not vary, they would be constants, and statistical inference would not be necessary. Think about it this way: If all the tomatoes had the same weight, then knowing the weight of one tomato would be sufficient to determine the weights of all tomatoes. However, the weights of tomatoes vary from one tomato to the next. One goal of research is to learn the causes of the variability so that we can learn to grow plants that yield the best tomatoes.

Variables can be classified into two groups: *qualitative* or *quantitative*.

Note to Instructor
Emphasize the point that variables vary. This is a key to helping students to develop statistical thinking.

Definition

Qualitative or categorical variables allow for classification of individuals based on some attribute or characteristic.

Quantitative variables provide numerical measures of individuals. Arithmetic operations such as addition and subtraction can be performed on the values of a quantitative variable and will provide meaningful results.

Many examples in this text will include a suggested **approach**, or a way to look at and organize a problem so that it can be solved. The approach will be a suggested method of *attack* toward solving the problem. This does not mean that the approach given is the only way to solve the problem, because many problems have more than one approach leading to a correct solution. For example, if you turn the key on your car's ignition and it doesn't start, one approach would be to look under the hood and try to determine what is wrong. (Of course, this approach would work only if you know how to fix cars.) A second, equally valid approach would be to call an automobile mechanic to service the car.

In Other Words
Typically, there is more than one correct approach to solving a problem.

EXAMPLE 2

Distinguishing between Qualitative and Quantitative Variables

Problem: Determine whether the following variables are qualitative or quantitative.

(a) Gender

(b) Temperature

(c) Number of days during the past week a college student aged 21 years or older has had at least one drink

(d) Zip code

Approach: Quantitative variables are numerical measures such that arithmetic operations can be performed on the values of the variable. Qualitative variables describe an attribute or characteristic of the individual that allows researchers to categorize the individual.

Solution

(a) Gender is a qualitative variable because it allows a researcher to categorize the individual as male or female. Notice that arithmetic operations cannot be performed on these attributes.

(b) Temperature is a quantitative variable because it is numeric, and operations such as addition and subtraction provide meaningful results. For example, 70°F is 10°F warmer than 60°F.

(c) Number of days during the past week that a college student aged 21 years or older had at least one drink is a quantitative variable because it is numeric, and operations such as addition and subtraction provide meaningful results.

(d) Zip code is a qualitative variable because it categorizes a location. Notice that the addition or subtraction of zip codes does not provide meaningful results.

Now Work Problem 15.

On the basis of the result of Example 2(d), we conclude that a variable may be qualitative while having values that are numeric. Just because the value of a variable is numeric does not mean that the variable is quantitative.

④ ### Distinguish between Discrete and Continuous Variables

We can further classify quantitative variables into two types.

Definition

A **discrete variable** is a quantitative variable that has either a finite number of possible values or a countable number of possible values. The term *countable* means that the values result from counting, such as 0, 1, 2, 3, and so on.

A **continuous variable** is a quantitative variable that has an infinite number of possible values that are not countable.

In Other Words
If you count to get the value of a variable, it is discrete. If you measure to get the value of the variable, it is continuous. When deciding whether a variable is discrete or continuous, ask yourself if it is counted or measured.

Figure 1 illustrates the relationship among qualitative, quantitative, discrete, and continuous variables.

Figure 1

Recognizing the type of variable being studied is important because it dictates the type of analysis that can be performed. An example should help to clarify the definitions.

EXAMPLE 3

Distinguishing between Discrete and Continuous Variables

Problem: Determine whether the following quantitative variables are discrete or continuous.

(a) The number of heads obtained after flipping a coin five times.
(b) The number of cars that arrive at a McDonald's drive-through between 12:00 P.M and 1:00 P.M.
(c) The distance a 2005 Toyota Prius can travel in city driving conditions with a full tank of gas.

Approach: A variable is discrete if its value results from counting. A variable is continuous if its value is measured.

Solution

(a) The number of heads obtained by flipping a coin five times would be a discrete variable because we would count the number of heads obtained. The possible values of the discrete variable are 0, 1, 2, 3, 4, 5.
(b) The number of cars that arrive at McDonald's drive-through between 12:00 P.M. and 1:00 P.M. is a discrete variable because its value would result from counting the cars. The possible values of the discrete variable are 0, 1, 2, 3, 4, and so on. Notice that there is no predetermined upper limit to the number of cars that may arrive.
(c) The distance traveled is a continuous variable because we measure the distance.

Continuous variables are often rounded. For example, when the miles per gallon (mpg) of gasoline for a certain make of car is given as 24 mpg, it means that the miles per gallon is greater than or equal to 23.5 and less than 24.5, or $23.5 \leq \text{mpg} < 24.5$.

> Now Work Problem 27.

The list of observed values for a variable is **data**. Gender is a variable, the observations male or female are data. **Qualitative data** are observations corresponding to a qualitative variable. **Quantitative data** are observations corresponding to a quantitative variable. **Discrete data** are observations corresponding to a discrete variable, and **continuous data** are observations corresponding to a continuous variable.

In Other Words
The singular of data is datum.

EXAMPLE 4

Distinguishing between Variables and Data

Problem: Table 1 presents a group of selected countries and information regarding these countries as of July, 2004. Identify the individuals, variables, and data in Table 1.

Approach: An individual is an object or person for whom we wish to obtain data. The variables are the characteristics of the individuals, and the data are the specific values of the variables.

Solution: The individuals in the study are the countries: Australia, Canada, and so on (in red ink). The variables measured for each country are *government type*, *life expectancy*, and *population* (in blue ink). The variable government type is qualitative because it categorizes the individual. The quantitative variables are life expectancy and population.

Country	Government Type	Life Expectancy (years)	Population (in millions)
Australia	Democratic	80.26	19.9
Canada	Confederation	79.96	32.5
France	Republic	79.44	60.4
Morocco	Constitutional Monarchy	70.35	32.2
Poland	Republic	74.16	38.6
Sri Lanka	Republic	72.89	19.9
United States	Federal Republic	77.43	293.0

Table 1

Source: CIA *World Factbook*

The quantitative variable life expectancy is continuous because it is measured. The quantitative variable population is discrete because we count people. The observations are the data (in green ink). For example, the data corresponding to the variable life expectancy are 80.26, 79.96, 79.44, 70.35, 74.16, 72.89, and 77.43. The following data correspond to the individual Poland: a *republic* government with residents whose *life expectancy* is 74.16 years and where *population* is 38.6 million people. Republic is an instance of qualitative data that results from observing the value of the qualitative variable government type. The life expectancy of 74.16 years is an instance of quantitative data that results from observing the value of the quantitative variable life expectancy.

Now Work Problem 51.

1.1 ASSESS YOUR UNDERSTANDING

Concepts and Vocabulary

1. Define statistics.
2. Explain the difference between a population and a sample.
3. Statistics is the science of collecting, organizing, summarizing, and analyzing _____ to answer questions or draw conclusions. information
4. A(n) _____ is a person or object that is a member of the population being studied. individual
5. _____ statistics consists of organizing and summarizing information collected, while _____ statistics uses methods that generalize results obtained from a sample to the population and measure the reliability of the results.
6. What does it mean when an experiment is double-blind?
7. Discuss the differences between an *experimental* group and a *control* group.

8. What are placebos? Why do you think they are needed?
9. _____ are the characteristics of the individuals of the population being studied. Variables
10. Contrast the differences between qualitative and quantitative variables.
11. Discuss the difference between discrete and continuous variables.
12. *True or False*: Both discrete and continuous variables are quantitative variables. True
13. Explain the process of statistics.
14. The age of a person is commonly considered to be a continuous random variable. Could it be considered a discrete random variable instead? Explain.

5. Descriptive; inferential

Skill Building

In Problems 15–26, classify the variable as qualitative or quantitative.

15. Nation of origin Qualitative
NW

16. Number of siblings Quantitative

17. Eye color Qualitative

18. Number on a football player's jersey Qualitative

19. Grams of carbohydrates in a doughnut Quantitative

20. Number of unpopped kernels in a bag of ACT microwave popcorn Quantitative

21. Assessed value of a house Quantitative

22. Phone number Qualitative

23. Population of a state Quantitative

24. Cost (in dollars) to fill up a 2005 Chevrolet Corvette

25. Student ID number Qualitative

24. Quantitative

26. Marital status Qualitative

In Problems 27–38, determine whether the quantitative variable is discrete or continuous.

27. Runs scored in a season by Albert Pujols Discrete

NW

28. Volume of water lost each day through a leaky faucet Continuous

29. Length of a country song Continuous

30. At rest pulse rate of a 20-year-old college student Discrete

31. Number of sequoia trees in a randomly selected acre of Yosemite National Park Discrete

32. Weight of a randomly selected hog Continuous

33. Temperature on a randomly selected day in Memphis, Tennessee Continuous

34. Internet connection speed in kilobytes per second Continuous

35. Number of tornadoes in a year in the United States Discrete

36. Points scored in an NCAA basketball game Discrete

37. Number of donors at a blood drive Discrete

38. Air pressure in pounds per square inch in an automobile tire Continuous

In Problems 39–44, a research objective is presented. For each research objective, identify the population and sample in the study.

39. The Gallup Organization contacts 1,028 teenagers who are 13 to 17 years of age and live in the United States and asked whether or not they had been prescribed medications for any mental disorders, such as depression or anxiety.

40. A quality-control manager randomly selects 50 bottles of Coca-Cola that were filled on October 15 to assess the calibration of the filling machine.

41. A farmer wanted to learn about the weight of his soybean crop. He randomly sampled 100 plants and weighed the soybeans on each plant.

42. Every year the U.S. Census Bureau releases the Current Population Report based on a survey of 50,000 households. The goal of this report is to learn the demographic characteristics of all households within the United States, such as income.

43. **Folate and Hypertension** Researcher John P. Forman and co-workers wanted to determine whether or not higher fo-

late intake is associated with a lower risk of hypertension (high blood pressure) in younger women (27 to 44 years of age). To make this determination, they looked at 7,373 cases of hypertension in younger women and found that younger women who consumed at least 1,000 micrograms per day (μg/d) of total folate (dietary plus supplemental) had a decreased risk of hypertension compared with those who consumed less than 200 μg/d.

(*Source*: "Folate intake and the risk of incident hypertension among US Women," John P. Forman, MD; Eric B. Rimm, ScD; Meir J. Stampfer, MD; Gary C. Curhan, MD, ScD; *Journal of the American Medical Association* 293(2005):320–329)

44. A large community college has noticed that an increasing number of full-time students are working while attending the school. The administration randomly selects 128 students and asks this question: How many hours per week do you work?

Applying the Concepts

For the studies in Problems 45–50, (a) identify the research objective, (b) identify the sample, (c) list the descriptive statistics, and (d) state the conclusions made in the study.

45. **A Cure for the Common Wart** A study conducted by re-

NW searchers was designed "to determine if application of duct tape is as effective as cryotherapy in the treatment of common warts." The researchers randomly divided 51 patients into two groups. The 26 patients in group 1 had their warts treated by applying duct tape to the wart for 6.5 days and then removing the tape for 12 hours, at which point the cycle was repeated for a maximum of 2 months. The 25 patients in group 2 had their warts treated by cryotherapy (liquid nitrogen applied to the wart for 10 seconds every 2 to 3 weeks) for a maximum of 6 treatments. Once the treatments were complete, it was determined that 85% of the patients in group 1 and 60% of the patients in group 2 had complete resolution of their warts. The researchers concluded that duct tape is significantly more effective in treating warts than cryotherapy.

(*Source*: "The Efficacy of Duct Tape vs. Cryotherapy in the Treatment of Verruca Vulgaris (The Common Wart)," Dean R. Focht III, Carole Spicer, Mary P. Fairchok; *Archives of Pediatrics and Adolescent Medicine*, Vol. 156, No. 10, October 2002)

46. **Early Epidurals** A study was conducted at Northwestern University in Chicago to determine if pregnant women in first-time labor could receive low-dose epidurals, an anesthesis to control pain during childbirth early without raising their chances of a Caesarean section. In the study, reported in the *New England Journal of Medicine*, "728 women in first-time labor were divided into two groups. One group received the spinal shot and then got epidurals when the cervix dilated to about 2 centimeters. The other group initially received pain-relieving medicine directly into their bloodstreams, and put off epidurals until 4 centimeters if they could tolerate the pain." In the end, the C-section rate was 18% in the early epidural group and 21% in the delayed group. The researchers concluded that pregnant women in first-time labor can be given a low-dose epidural early without raising their chances of a C-section.

(*Source*: *Associated Press*, Feb. 22, 2005)

47. **The Mozart Effect** Researchers at the University of California, Irvine, wished to determine whether "music cognition and cognitions pertaining to abstract operations such as mathematical or spatial reasoning" were related. To test the research question, 36 college students listened to Mozart's Sonata for Two Pianos in D Major, K488, for 10 minutes and then took a spatial reasoning test using the Stanford–Binet intelligence scale. The same students also took the test after sitting in a room for 10 minutes in complete silence. Whether the student experienced Mozart first or silence first was randomly determined. The average score on the test following the Mozart piece was 119, while the average test score following the silence was 110. The researchers concluded that subjects performed better on abstract and spatial reasoning tests after listening to Mozart.

(*Source*: "Music and Spatial Performance," Frances H. Rauscher et al.; *Nature* 365, 14 October 1993:611)

48. **Favorite Presidents** A poll commissioned by Washington College and conducted by Schulman, Ronca, and Bucuvalas, February 7–10, 2005, surveyed 800 U.S. adults. Asked who was the greatest president, 20% of respondents chose Abraham Lincoln, 15% Ronald Reagan, 12% Franklin Roosevelt, 11% John F. Kennedy, 10% Bill Clinton, 8% George W. Bush, and 6% George Washington. The remaining 18% selected other presidents. Washington College concluded that Abraham Lincoln ranks first among U.S. adults as the greatest president.

(*Source*: *Associated Press*, Feb. 22, 2005)

49. **Go to the Movies?** Gallup News Service conducted a survey of 1,003 American adults aged 18 years or older, December 5–8, 2004. The respondents were asked, "How many movies, if any, have you attended in a movie theater in the past 12 months?" Of the 1,003 adults surveyed, 65% said they attended at least one movie in a movie theater in the past 12 months. Gallup News Service concluded that 65% of all Americans aged 18 years or older attended at least one movie in a movie theater in the past 12 months.

50. **Morally Acceptable?** Gallup News Service conducted a survey of 1,000 U.S. adults aged 18 years or older, May 2–4, 2004. The respondents were asked, "Do you believe that having a baby outside of marriage is morally acceptable or morally wrong?" Of the 1,000 adults surveyed, 49% said that having a baby outside of marriage is morally acceptable. Gallup News Service concluded that 49% of all U.S. adults 18 years or older believe that having a baby outside of marriage is morally acceptable.

In Problems 51–54, identify the individuals, variables, and data corresponding to the variables. Determine whether each variable is qualitative, continuous, or discrete.

51. **Widescreen TVs** The following data relate to widescreen high-definition televisions.

Model	Size (in.)	Screen Type	Price ($)
Sanyo #PDP42H2W	42	Plasma	2994
Panasonic #PT-47WX54	47	Projection	1072
Tatung #P50BSAT	50	Plasma	4248
RCA #HD50LPW42	50	Projection	2696
RCA #D52W19	52	Projection	1194
JVC #HD52Z575	52	Projection	2850
Sony #KDF-60XS955	60	Projection	4100

Sources: walmart.com and sears.com

52. **BMW Cars** The following information relates to the entire product line of BMW automobiles.

Model	Body Style	Weight (pounds)	Number of Seats
M/Z3 Coupe	Coupe	2945	2
M/Z3 Roadster	Convertible	2690	2
3 Series	Coupe	2780	5
5 Series	Sedan	3450	5
7 Series	Sedan	4255	5
Z8	Convertible	3600	2

Source: *Car and Driver* magazine

53. **Driver's License Laws** The following data represent driver's license laws for various states.

State	Minimum Age for Driver's License (Unrestricted)	Blood Alcohol Concentration Limit	Mandatory Belt-Use Law Seating Positions	Maximum Allowable Speed Limit, 2003
Colorado	17	0.10	Front	75
Missouri	18	0.08	Front	70
Montana	15	0.08	All	75
New York	17	0.08	All	65
Texas	16.5	0.08	Front	75

Source: *Time Almanac*, 2005

54. **MP3 Players** The following information concerns various MP3 players that can be purchased online at circuitcity.com:

Product	Memory Size	Weight (oz)	Price (S)
RCA Lyra™ 256 MB MP3 Player	256 MB	2 oz	$149.99
iRiver 256 MB MP3 Player	256 MB	2.1 oz	$99.99
Samsung 256 MB MP3 Player	256 MB	0.85 oz	$129.99
Creative® 256 MuVo® Micro N200	256 MB	0.8 oz	$99.99
iRiver 512 MB MP3 Player	512 MB	2.1 oz	$149.99
SanDisk 512 MB MP3 Player	512 MB	1.4 oz	$139.99
Creative® 512 MuVo® Micro N200	512 MB	0.8 oz	$139.99
SanDisk 1 GB MP3 Player	1 GB	1.4 oz	$169.99

Source: circuitcity.com

For Problems 55 and 56, read the newspaper article and identify (a) the research question the study addresses, (b) the population, (c) the sample, (d) the descriptive statistics, and (e) the inferences of the study.

55. By Jeff Donn
Associated Press
February 20, 2005

BOSTON–Levels of a stress-related protein in the blood could give doctors a powerful new tool for deciding which patients with clogged heart arteries are most in danger and need aggressive treatment, a study found.

This protein is yet another predictor of heart trouble, in addition to such substances as cholesterol and C-reactive protein.

The Danish study, published in Thursday's *New England Journal of Medicine*, focused on a substance called B-natriuretic peptide, or BNP, released into the blood when the heart is stressed. Some doctors already test for this protein to help evaluate patients with shortness of breath who may be suffering from congestive heart disease.

The researchers measured the protein levels in 1,034 patients and followed their health for nine years. People with the highest protein levels were $2\frac{1}{2}$ times more likely to die from any cause than those with the lowest.

This protein could help doctors decide which patients need angioplasty or heart-bypass surgery.

"It sort of tells you that no matter what condition you have, higher levels of these peptides are associated with worse outcomes," said cardiologist Dr. James de Lemos, at the University of Texas Southwestern Medical Center.

56. **Study: Exercise May Prevent Parkinson's** By Kathleen Fackelmann, *USA Today*

Men who engaged in regular, vigorous exercise as teens and young adults drastically cut their risk of developing Parkinson's disease later in life, a study reports Tuesday.

As many as 1 million people in the USA, including actor Michael J. Fox, have this progressive neurological disease, which commonly strikes after age 50.

In addition to suggesting that exercise could ward off the disease, the findings also raise the hope that physical activity might help hold the line on brain cell destruction in people who already have it.

Robin Elliott, executive director of the Parkinson's Disease Foundation in New York, said the study is promising because there is no cure for Parkinson's. Doctors today have no way to stop or delay the progression of the disease, which affects the brain region that controls movement. Common symptoms of Parkinson's are tremors and a shuffling gait.

Researcher Alberto Ascherio of the Harvard School of Public Health and his colleagues studied 48,000 men and 77,000 women who were relatively healthy and middle-aged or older at the study's start. Over the course of the study, 387 people developed the disease.

The team did a statistical analysis to look for a link between physical activity and the risk of Parkinson's.

Men who said they jogged, played basketball or participated in some other vigorous activity at least twice a week in high school, college and up to age 40 had a 60% reduced risk of getting Parkinson's, says the study, which was published today in *Neurology*.

The team found no such protection for women. But the women in this study came of age in the '50s and '60s, an era when young women did not have as many opportunities to participate in sports. Ascherio says. The study had so few female athletes that any protection may have been missed, he says.

This study's findings are supported by animal research: Michael Zigmond, a researcher at the University of Pittsburgh and his colleagues have reported that exercise protects rats from developing a Parkinson's-like disease.

No one knows for certain whether exercise can prevent or delay the symptoms of Parkinson's in humans. But there is plenty of evidence suggesting that a regular fitness regimen might ward off a host of ailments as diverse as heart disease and Alzheimer's.

That leads experts such as Zigmond to recommend a vigorous workout on most days of the week.

"We know that exercise is good for you in ways that go beyond any one condition."

Find this article at:
http://www.usatoday.com/news/health/
2005-02-21-parkinsons-exercise_x.htm

57. Nominal versus Ordinal Just as a quantitative variable can be further classified as discrete or continuous, we can classify a qualitative (or categorical) variable as *nominal* or *ordinal*. A **nominal variable** is a qualitative variable that describes an attribute of an individual. An **ordinal vari-**

able is a qualitative variable that has all the properties of a nominal variable, but also has observations that can be ranked or put in order. For example, gender is a nominal variable, whereas a response to a customer satisfaction survey of poor, fair, good, or outstanding is considered ordinal. For each of the following qualitative variables, determine whether the variable is nominal or ordinal.

(a) Hair color Nominal
(b) Letter grade in a statistics class Ordinal
(c) Make of a television set Nominal
(d) Rank of a military officer Ordinal
(e) Method of payment (cash, check, debit card, credit card) Nominal

1.2 Observational Studies, Experiments, and Simple Random Sampling

Objectives

1 **Distinguish between an observational study and an experiment**

2 **Obtain a simple random sample**

We are now familiar with some of the terminology used in describing data. Now, we need to determine how to obtain data. When we defined the word *statistics*, we said it is a science that involves the collection of data. Data can be obtained from four sources:

1. A census
2. Existing sources
3. Survey sampling
4. Designed experiments

We start by defining a census.

Definition A **census** is a list of all individuals in a population along with certain characteristics of each individual.

If a census is available and the census data can be used to answer the questions posed in the research objective, then the census data should be used. This is because the results will answer the questions with 100% certainty.

The United States conducts a census every 10 years to learn the demographic makeup of the United States. Everyone whose usual residence is within the borders of the United States must fill out a questionnaire packet. The cost of obtaining the census in 2000 was approximately $6 billion. The census data provide information such as the number of members in a household, number of years at present address, household income, and more. Because of the cost of obtaining census data, most researchers obtain data through existing sources, survey samples, or designed experiments.

Have you ever heard this saying? *There is no point in reinventing the wheel.* Well, there is no point in spending energy obtaining data that already exist either. If a researcher wishes to conduct a study and a data set exists that can be used to answer the researcher's questions, then it would be silly to collect the data from scratch. For example, in the August 22, 2001, issue of the *Journal of the American Medical Association* ["Physical Activity, Obesity, Height, and the Risk of Pancreatic Cancer," Dominique S. Michaud, ScD, et al., Vol. 286, No. 8], researchers did a study in which they attempted to identify factors that increase the likelihood of an individual getting pancreatic cancer. Rather than conducting their own survey, they used data from two existing surveys: the Health Professionals Follow-Up Study and the Nurses' Health Study. By doing this, they saved time and money. The moral of the story: **Don't collect data that have already been collected!**

 ## Distinguish between an Observational Study and an Experiment

Survey sampling is used in research when there is no attempt to influence the value of the variable of interest. For example, we may want to identify the "normal" systolic blood pressure of U.S. men aged 40 to 44. The researcher would obtain a sample of men aged 40 to 44 and determine their systolic blood pressure. No attempt is made to influence the systolic blood pressure of the men surveyed. Polling data are another example of data obtained from a survey sample because the respondent is asked his or her opinion. No attempt is made to influence this opinion. Data obtained from a survey sample lead to an *observational study*.

Definition

> An **observational study** measures the characteristics of a population by studying individuals in a sample, but does not attempt to manipulate or influence the variable(s) of interest.

Observational studies are sometimes referred to as *ex post facto* (after the fact) studies because the value of the variable of interest has already been established. We distinguish an observational study with a *designed experiment*.

Definition

> A **designed experiment** applies a treatment to individuals (referred to as **experimental units** or **subjects**) and attempts to isolate the effects of the treatment on a **response variable**.

Data obtained through experimentation will be thoroughly discussed in Section 1.5, but the main idea is that the researcher is able to control factors that influence the experimental units. For example, suppose my son has two types of fertilizer and wants to determine which results in better tomatoes. He might conduct an experiment in which he divides 20 tomato plants into two groups. Group 1 receives the recommended amount of the first fertilizer, and group 2 receives the recommended amount of the second fertilizer. All other factors that affect plant growth (amount of sunlight, water, soil condition, and so on) are kept the same for the two groups. The two fertilizers are the two treatments, and the tomatoes are the experimental units. The weight of the tomatoes is the response variable. The observed weights are the data.

An example should clarify the difference between an observational study and a designed experiment.

EXAMPLE 1   ## Observational Study versus Designed Experiment

Note to Instructor
Have students brainstorm advantages and disadvantages of an observational study. What are some advantages and disadvantages of an experiment?

In most types of research, the goal is to determine the relation, if any, that may exist between two or more variables. For example, a researcher may want to determine whether there is a connection between smoking and lung cancer.* This type of study is performed using an observational study because it is *ex post facto* (after the fact) research. The individuals in the study are examined after they have been smoking for some period of time. The individuals are not controlled in terms of the number of cigarettes smoked per day, eating habits, and

*The interested reader may wish to read Chapter 18, "Does Smoking Cause Cancer?" in David Salsburg's book *The Lady Tasting Tea*. W. H. Freeman and Co., 2001.

so on. A researcher simply interviews a sample of smokers and monitors their rate of cancer and compares it with a sample of nonsmokers. The nonsmokers serve as a *control group;* that is, the nonsmokers serve as the benchmark upon which the smokers' rate of cancer is judged. If a significant difference between the two groups' rates of cancer exists, the researcher might want to claim smoking *causes* cancer.

In actuality, the researcher determined that smoking is *associated* with cancer. It may be that the population of smokers has a higher rate of cancer, but this higher incidence rate is not necessarily a direct result of smoking. It is possible that smokers have some characteristic that differs from the nonsmoking group, other than smoking, that is the *cause* of cancer. The characteristics that may be related to cancer, but that have not been identified in the study, are referred to as **lurking variables**. For example, a lurking variable might be the amount of exercise. Maybe smokers generally exercise less than nonsmokers, and the lack of exercise is the cause of cancer. In this observational study, we might be able to state that smokers have a higher rate of cancer than nonsmokers and that smoking is *associated* with cancer, but we would not be able to definitively state that smoking causes cancer.

Obtaining this type of data through experimentation requires randomly dividing a sample of people into two groups. We then require one group to smoke a pack of cigarettes each day for the next 20 years while the other group does not smoke. We then compare the incidence rate of lung cancer (the response variable) in the smoking group to the nonsmoking group. If the two cancer rates differ significantly, we could say that smoking causes cancer. By approaching the study in this way, we are able to control many of the factors that were beyond our control in the observational study. For example, we could make sure each group had the same diet and exercise regiment. This would allow us to determine whether smoking is a cause of cancer. Of course, moral issues preclude conducting this type of experiment. ▬▬▬

It is vital that we understand that observational studies do not allow a researcher to claim causation, only association.

> Observational studies are very useful tools for determining whether there is a relation between two variables, but it requires a designed experiment to isolate the cause of the relation.

Many observational studies, such as polls, are set up to learn the characteristics of a population. For example, the Gallup Organization routinely surveys U.S. residents in an attempt to identify opinion.

Observational studies are performed for two reasons:

1. To learn the characteristics of a population
2. To determine whether there is an association between two or more variables where the values of the variables have already been determined (again, this is often referred to as *ex post facto* research)

Ex post facto research is common when control of certain variables is impossible or unethical. For example, economists often perform research using observational studies because they do not have the ability to control many of the factors that affect our purchasing decisions. Some medical research requires that observational studies be conducted because of the risks thought to be associated with experimental study, such as the link between smoking and lung cancer.

Experiments, on the other hand, are used whenever control of certain variables is desired. This type of research allows the researcher to identify certain cause and effect relationships among the variables in the study.

The bottom line to consider is control. If control is possible, an experiment should be performed. However, if control is not possible or necessary, then observational studies are appropriate.

! **CAUTION**
Beware of observational studies that claim causation!

Now Work Problem 13.

Sampling

The goal in sampling is to obtain individuals for a study in such a way that accurate information about the population can be obtained. For example, the Gallup Organization typically polls a sample of about 1,000 adults from the population of adults aged 18 years or older. We want the sample to provide as much information as possible, but each additional piece of information has a price. So the question is this "How can the researcher obtain accurate information about the population through the sample while minimizing the costs in terms of money, time, personnel, and so on?" There is a balance between information and cost. An appropriate sample design can maximize the amount of information obtained about the population for a given cost.

We will discuss four basic sampling techniques: *simple random sampling*, *stratified sampling*, *systematic sampling*, and *cluster sampling*. These sampling methods are designed so that any biases introduced (knowingly and unknowingly) by the surveyor during the selection process are eliminated. In other words, the surveyor does not have a choice as to who is in the study. We discuss simple random sampling now and discuss the remaining three types of sampling in the next section.

Obtain a Simple Random Sample

The most basic sample survey design is **simple random sampling**, which is often abbreviated as **random sampling**.

Definition

> A sample of size n from a population of size N is obtained through **simple random sampling** if every possible sample of size n has an equally likely chance of occurring. The sample is then called a **simple random sample**.

In Other Words
Simple random sampling is like selecting names from a hat.

The sample is always a subset of the population, meaning that the number of individuals in the sample is less than the number of individuals in the population.

In-Class Activity: Illustrating Simple Random Sampling

This activity illustrates the idea of simple random sampling.

(a) Choose 5 students in the class to represent a population. Number the students 1 through 5.

(b) Form all possible samples of size $n = 2$ from the population of size $N = 5$. How many different simple random samples are possible?

(c) Write the numbers 1 through 5 on five pieces of paper and place the paper in a hat. Select two of the numbers. The two individuals corresponding to these numbers are in the sample.

(d) Put the two numbers back in the hat. Select two of the numbers. The two individuals corresponding to these numbers are in the sample. Are the individuals in the second sample the same as the individuals in the first sample?

Note to Instructor
The purpose of this activity is to allow students to visualize simple random sampling. It also helps to foreshadow the idea of a sampling distribution by showing that different samples lead to different individuals in the sample. It should take about 15–20 minutes.

Obtaining a Simple Random Sample

How do we select the individuals in a simple random sample? To obtain a simple random sample from a population, we could write the names of the individuals in the population on different sheets of paper and then select names from the hat.

Often, however, the size of the population is so large that performing simple random sampling in this fashion is not practical. Typically, random numbers are used by assigning each individual in the population a unique number between 1 and N, where N is the size of the population. Then n random numbers from this list are selected. Because we must number the individuals in the population, we must have a list of all the individuals within the population, called a **frame**.

In Other Words
A frame lists all the individuals in a population. For example, a list of all registered voters in a particular precinct might be a frame.

| EXAMPLE 2 | **Obtaining a Simple Random Sample** |

Problem: Senese and Associates has increased their accounting business. To make sure their clients are still satisfied with the services they are receiving, Senese and Associates decides to send a survey out to a simple random sample of 5 of its 30 clients.

Approach

Step 1: A list of the 30 clients must be obtained (the frame). Each client is then assigned a unique number from 01 to 30.

Note to Instructor

In-class discussion: What is meant by sampling "with replacement"? What is meant by sampling "without replacement"?

Step 2: Five unique numbers will be randomly selected. The client corresponding to the number is given a survey. This is called **sampling without replacement**. We sample without replacement so that we don't select the same client twice.

Solution

Step 1: Table 2 shows the list of clients. We arrange the clients in alphabetic order (although this is not necessary). Because there are 30 clients, we number the clients from 01 to 30.

Table 2		
01. ABC Electric	11. Fox Studios	21. R&Q Realty
02. Brassil Construction	12. Haynes Hauling	22. Ritter Engineering
03. Bridal Zone	13. House of Hair	23. Simplex Forms
04. Casey's Glass House	14. John's Bakery	24. Spruce Landscaping
05. Chicago Locksmith	15. Logistics Management, Inc.	25. Thors, Robert DDS
06. DeSoto Painting	16. Lucky Larry's Bistro	26. Travel Zone
07. Dino Jump	17. Moe's Exterminating	27. Ultimate Electric
08. Euro Car Care	18. Nick's Tavern	28. Venetian Gardens Restaraunt
09. Farrell's Antiques	19. Orion Bowling	29. Walker Insurance
10. First Fifth Bank	20. Precise Plumbing	30. Worldwide Wireless

Step 2: A table of random numbers can be used to select the individuals to be in the sample. See Table 3.* We select a starting place in the table of random numbers. This can be done by closing our eyes and placing a finger on the table. This may sound haphazard, but it accomplishes the goal of being random. Suppose we start in column 4, row 13. Because our data have two digits, we select two-digit numbers from the table using columns 4 and 5. We only select numbers greater than or equal to 01 or less than or equal to 30. Anytime we encounter 00, a number greater than 30, or a number already selected, we skip it and continue to the next number.

The first number in the list is 01, so the client corresponding to 01 will receive a survey. Moving down the list, the next number is 52. Because 52 is greater than 30, we skip it. Continuing down the list, the following numbers are selected from the list:

$$01, 07, 26, 11, 23$$

---

*Each digit is in its own column. The digits are displayed in groups of five for ease of reading. The digits in row 1 are 893922321274483 and so on. The first digit, 8, is in column 1; the second digit, 9, is in column 2; the ninth digit, 1, is in column 9.

Column 4

Table 3

Row Number	Column Number									
	01–05	06–10	11–15	16–20	21–25	26–30	31–35	36–40	41–45	46–50
01	89392	**23**212	74483	36590	25956	36544	68518	40805	09980	00467
02	61458	17639	96252	95649	73727	33912	72896	66218	52341	97141
03	11452	74197	81962	48433	90360	26480	73231	37740	26628	44690
04	27575	04429	31308	02241	01698	19191	18948	78871	36030	23980
05	36829	59109	88976	46845	28329	47460	88944	08264	00843	84592
06	81902	93458	42161	26099	09419	89073	82849	09160	61845	40906
07	59761	55212	33360	68751	86737	79743	85262	31887	37879	17525
08	46827	25906	64708	20307	78423	15910	86548	08763	47050	18513
09	24040	66449	32353	83668	13874	86741	81312	54185	78824	00718
10	98144	96372	50277	15571	82261	66628	31457	00377	63423	55141
11	14228	17930	30118	00438	49666	65189	62869	31304	17117	71489
12	55366	51057	90065	14791	62426	02957	85518	28822	30588	32798
13	96**101**	30646	35526	90389	73634	79304	96635	6626	94683	16696
14	381**52**	55474	30153	26525	83647	31988	82182	98377	33802	80471
15	850**07**	18416	24661	95581	45868	15662	28906	36392	07617	50248
16	855**44**	15890	80011	18160	33468	84106	40603	01315	74664	20553
17	104**46**	20699	98370	17684	16932	80449	92654	02084	19985	59321
18	672**37**	45509	17638	65115	29757	80705	82686	48565	72612	61760
19	230**26**	89817	05403	82209	30573	47501	00135	33955	50250	72592
20	674**11**	58542	18678	46491	13219	84084	27783	34508	55158	78742

Row 13 — 13

We skip 52 because it is larger than 30.

The clients corresponding to these numbers are

ABC Electric, Dino Jump, Travel Zone, Fox Studios, Simplex Forms

Each individual selected in the sample is set in boldface type in Table 3 to help you to understand where the numbers come from.

EXAMPLE 3   Obtaining a Simple Random Sample Using Technology

Note to Instructor
Many types of statistical software packages and calculators can be used in statistics. Because of the number of statistical packages available, it is impossible for a text to present each of them. Therefore, the three most popular (MINITAB, Excel, and TI83/TI84), as learned from our surveys, are presented in this text. The technology used will vary from example to example. However, the purpose of the examples that use technology is to illustrate how it can obtain the results we need to solve a problem. Once we have the results, work still must be done to determine its meaning. So, just because the software or calculator used in a technology example is not the one used by you, do not feel obligated to skip it.

Problem: Find a simple random sample of five clients for the problem presented in Example 2.

Approach: The approach is similar to that given in Example 2.

Step 1: A list of the 30 clients must be obtained (the frame). The clients are then assigned a number from 01 to 30.

Step 2: Five numbers are randomly selected using a random number generator. The client corresponding to the number is given a survey. We sample without replacement so that we don't select the same client twice. To use a random-number generator using technology, we must first set the *seed*. The **seed** in a random-number generator provides an initial point for the generator to start creating random numbers. It is just like selecting the initial point in the table of random numbers. The seed can be any nonzero number. Statistical spreadsheets such as MINITAB or Excel can be used to generate random numbers, but we will use a TI-84 Plus graphing calculator. The steps for obtaining random numbers using MINITAB, Excel, or the TI-83/84 graphing calculator can be found in the Technology Step by Step on page 22.

Solution

Step 1: Table 2 on page 17 shows the list of clients and numbers corresponding to the clients.

Step 2: See Figure 2(a) for an illustration of setting the seed using a TI-84 Plus graphing calculator where the seed is set at 34. We are now ready to obtain the list of random numbers. Figure 2(b) shows the results obtained from a TI-84 Plus graphing calculator. If you are using a TI-83, your results will differ.

USING TECHNOLOGY

If you are using a different statistical package or type of calculator, the random numbers generated will likely be different. This does not mean you are wrong. There is no such thing as a wrong random sample as long as the correct procedures are followed.

Figure 2

(a)   (b)

The following numbers are generated by the calculator:

$$4, 5, 6, 22, 15$$

The clients corresponding to these numbers are the clients to be surveyed: Casey's Glass House, Chicago Locksmith, DeSoto Painting, Ritter Engineering, and Logistics Management, Inc.

Now Work Problem 21.

! CAUTION

Random-number generators are not truly random, because they are programs and programs do not act "randomly." The seed dictates the "random numbers" that are generated.

There is a very important consequence when comparing the by hand and technology solutions from Examples 2 and 3. Because both samples were obtained randomly, they resulted in different individuals in the sample! For this reason, each sample will likely result in different descriptive statistics. Any inference based on each sample *may* result in different conclusions regarding the population. This is the nature of statistics. Inferences based on samples will vary because the individuals in different samples vary.

1.2 ASSESS YOUR UNDERSTANDING

Concepts and Vocabulary

1. Explain the difference between an observational study and an experiment. In your explanation, be sure to discuss the circumstances in which each is appropriate.
2. Explain why a frame is necessary to obtain a simple random sample.
3. Discuss why sampling is used in statistics.
4. What does it mean when sampling is done without replacement?
5. A(n) _____ is a list of all individuals in a population along with certain characteristics of each individual. *frame*

6. *True or False*: An observational study measures the characteristics of a population by attempting to manipulate or influence the variables of interest. *False*
7. *True or False*: Observational studies do not allow a researcher to claim causation. *True*
8. When using a random-number generator to generate a simple random sample, why is it necessary to set a randomly selected seed?

Skill Building

In Problems 9–20, determine whether the study depicts an observational study or an experiment.

9. A study to determine whether there is a relation between the rate of cancer and an individual's proximity to high-tension wires. *Observational study*

10. Rats with cancer are divided into two groups. One group receives 5 mg of a medication that is thought to fight cancer, and the other receives 10 mg. After 2 years, the spread of the cancer is measured. *Experiment*

11. Seventh-grade students are randomly divided into two groups. One group is taught math using traditional techniques; the other is taught math using a reform method. After 1 year, each group is given an achievement test to compare its proficiency with that of the other group.

12. A poll is conducted in which 500 people are asked whom they plan to vote for in the upcoming election.

13. A survey is conducted asking 400 people, "Do you prefer
NW Coke or Pepsi?" *Observational study*

14. While shopping, 200 people are asked to perform a taste test in which they drink from two unmarked cups. They are then asked which drink they prefer. *Experiment*

15. Sixty patients with carpal tunnel syndrome are randomly divided into two groups. One group is treated weekly with both acupuncture and an exercise regimen. The other is treated weekly with the exact same exercise regimen, but no acupuncture. After 1 year, both groups are questioned about the pain of carpal tunnel syndrome.

16. Conservation agents netted 250 large-mouth bass in a lake and determined how many were carrying parasites.

17. **Got Kidney Stones?** Researchers Eric N. Taylor and others wanted to determine if weight, weight gain, body mass index (BMI), and waist circumference are associated with kidney stone formation. To make this determination, they looked at 4,827 cases of kidney stones covering 46 years. Based on an analysis of the data, they concluded that obesity and weight gain increase the risk of kidney stone formation.
(*Source*: "Obesity, Weight Gain, and the Risk of Kidney Stones," Eric N. Taylor, MD; Meir J. Stampfer, MD, DrPH; and Gary C. Curhan, MD, ScD; *Journal of the American Medical Association* 293(2005):455–462)

18. **Cell Phones and Tumors** Researchers Helle Collatz Christensen and co-workers examined the possible association between the use of cellular telephones and the development of acoustic neuroma (a noncancerous growth in the inner ear) by comparing 106 cases of individuals with acoustic neuroma to 212 cases of individuals who did not have acoustic neuroma. The data obtained included information on the use of cellular telephones from personal interviews, data from medical records, and the results of radiologic examinations.
(*Source*: "Cellular Telephone Use and Risk of Acoustic Neuroma," Helle Collatz Christensen, Joachim Schüz, Michael Kosteljanetz, Hans Skovgaard Poulsen, Jens Thomsen, and Christoffer Johansen; *American Journal of Epidemiology* 2004; 159:277–283) *Observational study*

11. Experiment 12. Observational study 15. Observational study 16. Observational study 17. Observational study

Applying the Concepts

19. **Future Government Club** The Future Government Club wants to sponsor a panel discussion on the upcoming national election. The club wants four of its members to lead the panel discussion. Obtain a simple random sample of size 4 from the below. Write a short description of the process you used to generate your sample.

20. **Worker Morale** The owner of a private food store is concerned about employee morale at the store level. She decides to survey the employees to see if she can learn about work environment and job satisfaction. Obtain a simple random sample of size 5 from the names below. Write a short description of the process you used to generate your sample.

Blouin	Fallenbuchel	Niemeyer	Rice
Bolden	Grajewski	Nolan	Salihar
Bolt	Haydra	Ochs	Tate
Carter	Keating	Opacian	Thompson
Cooper	Khouri	Pawlak	Trudeau
Debold	Lukens	Pechtold	Washington
De Young	May	Ramirez	Wright
Engler	Motola	Redmond	Zenkel

Archer	Foushi	Kemp	Oliver
Bolcerek	Gow	Lathus	Orsini
Bryant	Grove	Lindsey	Salazar
Carlisle	Hall	Massie	Ullrich
Cole	Hills	McGuffin	Vaneck
Dimas	Houston	Musa	Weber
Ellison	Kats	Nickas	Zavodny
Everhart			

21. Obtaining a Simple Random Sample The following table lists the 50 states.

NW (a) Obtain a simple random sample of size 10 using Table I in Appendix A, a graphing calculator, or computer software.

(b) Obtain a second simple random sample of size 10 using Table I in Appendix A, a graphing calculator, or computer software.

1. Alabama	11. Hawaii	21. Massachusetts	31. New Mexico	41. South Dakota
2. Alaska	12. Idaho	22. Michigan	32. New York	42. Tennessee
3. Arizona	13. Illinois	23. Minnesota	33. North Carolina	43. Texas
4. Arkansas	14. Indiana	24. Mississippi	34. North Dakota	44. Utah
5. California	15. Iowa	25. Missouri	35. Ohio	45. Vermont
6. Colorado	16. Kansas	26. Montana	36. Oklahoma	46. Virginia
7. Connecticut	17. Kentucky	27. Nebraska	37. Oregon	47. Washington
8. Delaware	18. Louisiana	28. Nevada	38. Pennsylvania	48. West Virginia
9. Florida	19. Maine	29. New Hampshire	39. Rhode Island	49. Wisconsin
10. Georgia	20. Maryland	30. New Jersey	40. South Carolina	50. Wyoming

22. Obtaining a Simple Random Sample The following table lists the 43 presidents of the United States.

(a) Obtain a simple random sample of size 8 using Table I in Appendix A, a graphing calculator, or computer software.

(b) Obtain a second simple random sample of size 8 using Table I in Appendix A, a graphing calculator, or computer software.

1. Washington	10. Tyler	19. Hayes	28. Wilson	37. Nixon
2. J. Adams	11. Polk	20. Garfield	29. Harding	38. Ford
3. Jefferson	12. Taylor	21. Arthur	30. Coolidge	39. Carter
4. Madison	13. Fillmore	22. Cleveland	31. Hoover	40. Reagan
5. Monroe	14. Pierce	23. B. Harrison	32. F.D. Roosevelt	41. George H. Bush
6. J.Q. Adams	15. Buchanan	24. Cleveland	33. Truman	42. Clinton
7. Jackson	16. Lincoln	25. McKinley	34. Eisenhower	43. George W. Bush
8. Van Buren	17. A. Johnson	26. T. Roosevelt	35. Kennedy	
9. W.H. Harrison	18. Grant	27. Taft	36. L.B. Johnson	

23. Sampling the Faculty A small community college employs 87 full-time faculty members. To gain the faculty's opinions about an upcoming building project, the college president wishes to obtain a simple random sample that will consist of 9 faculty members. He numbers the faculty from 1 to 87.

(a) Using Table I from Appendix A, the president closes his eyes and drops his ink pen on the table. It points to the digit in row 5, column 22. Using this position as the starting point and proceeding downward, determine the numbers for the 9 faculty members who will be included in the sample.

(b) If the president uses the randInt(feature of a graphing calculator with a seed value of 47, determine the numbers for the 9 faculty members who will be included in the sample.

24. Sampling the Students The same community college from Problem 23 has 7,656 students currently enrolled in classes. To gain the student's opinions about an upcoming building project, the college president wishes to obtain a simple random sample of 20 students. He numbers the students from 1 to 7,656.

(a) Using Table I from Appendix A, the president closes his eyes and drops his ink pen on the table. It points to the digit in row 11, column 32. Using this position as the starting point and proceeding downward, determine the numbers for the 20 students who will be included in the sample.

(b) If the president uses the randInt(feature of a graphing calculator with a seed value of 142, determine the numbers for the 20 students who will be included in the sample.

25. Obtaining a Simple Random Sample Suppose you are the president of the student government. You wish to conduct a survey to determine the student body's opinion regarding student services. The administration provides you with a list of the names and phone numbers of the 19,935 registered students.

(a) Discuss the procedure you would follow to obtain a simple random sample of 25 students.

(b) Obtain this sample.

26. Obtaining a Simple Random Sample Suppose the mayor of Justice, Illinois, asks you to poll the residents of the village. The mayor provides you with a list of the names and phone numbers of the 5,832 residents of the village.

(a) Discuss the procedure you would follow to obtain a simple random sample of 20 residents.

(b) Obtain this sample.

Obtaining a Simple Random Sample

TI-83/84 Plus

Step 1: Enter any nonzero number (the seed) on the HOME screen.

Step 2: Press the STO ⇒ button.

Step 3: Press the MATH button.

Step 4: Highlight the PRB menu and select 1: rand.

Step 5: From the HOME screen press ENTER.

Step 6: Press the MATH button. Highlight PRB menu and select 5: randInt(.

Step 7: With randInt(on the HOME screen enter 1, N, where N is the population size. For example, if N = 500, enter the following:

$$randInt(1,500)$$

Press ENTER to obtain the first individual in the sample. Continue pressing ENTER until the desired sample size is obtained.

MINITAB

Step 1: Select the **Calc** menu and highlight **Set Base** ....

Step 2: Enter any seed number you desire. Note that it is not necessary to set the seed, because MINITAB uses the time of day in seconds to set the seed.

Step 3: Select the **Calc** menu and highlight **Random Data** and select **Integer** ....

Step 4: Fill in the following window with the appropriate values. To obtain a simple random sample for the situation in Example 2, we would enter the following:

The reason we generate 10 rows of data (instead of 5) is in case any of the random numbers repeat. Select OK, and the random numbers will appear in column 1 (C1) in the spreadsheet.

Excel

Step 1: Be sure the Data Analysis Tool Pak is activated. This is done by selecting the **Tools** menu and highlighting **Add – Ins** .... Check the box for the Analysis ToolPak and select OK.

Step 2: Select **Tools** and highlight **Data Analysis** .... Highlight **Random Number Generation** and select OK.

Step 3: Fill in the window with the appropriate values. To obtain a simple random sample for the situation in Example 2, we would fill in the following:

The reason we generate 10 rows of data (instead of 5) is in case any of the random numbers repeat. Notice also that the parameter is between 1 and 31, so any value less than or equal to 31 is possible. In the unlikely event that 31 appears, simply ignore it. Select OK and the random numbers will appear in column 1 (A1) in the spreadsheet. Ignore any values to the right of the decimal place.

1.3 Other Effective Sampling Methods

Objectives

1 **Obtain a stratified sample**

2 **Obtain a systematic sample**

3 **Obtain a cluster sample**

One goal of sampling is to obtain as much information as possible about the population at the least cost. Remember, we are using the word *cost* in a general sense. Cost includes monetary outlays, time, and other resources. With this goal in mind, we may find it advantageous to use sampling techniques other than simple random sampling.

1 **Obtain a Stratified Sample**

Under certain circumstances, *stratified sampling* provides more information about the population for less cost than simple random sampling.

Definition

A **stratified sample** is obtained by separating the population into nonoverlapping groups called *strata* and then obtaining a simple random sample from each stratum. The individuals within each stratum should be homogeneous (or similar) in some way.

For example, suppose Congress was considering a bill that abolishes estate taxes. In an effort to determine the opinion of her constituency, a senator asks a

In Other Words

Stratum is singular, while strata is plural. The word strata means divisions. So a stratified sample is a simple random sample of different divisions of the population.

pollster to conduct a survey within her district. The pollster may divide the population of registered voters within the district into three strata: Republican, Democrat, and Independent. This is because the members within the three party affiliations may have the same opinion regarding estate taxes. The main criterion in performing a stratified sample is that each group (stratum) must have a common attribute that results in the individuals being similar within the stratum.

An advantage of stratified sampling over simple random sampling is that it may allow fewer individuals to be surveyed while obtaining the same or more information. This result occurs because individuals within each subgroup have similar characteristics, so opinions within the group do not vary much from one individual to the next. In addition, a stratified sample guarantees that each stratum is represented in the sample.

EXAMPLE 1   Obtaining a Stratified Sample

Problem: The president of DePaul University wants to conduct a survey to determine the community's opinion regarding campus safety. The president thinks the DePaul community can be divided into three groups: resident students, nonresident (commuting) students, and staff (including faculty). So he will obtain a stratified sample. Suppose there are 6,204 resident students, 13,304 nonresident students, and 2,401 staff, for a total of 21,909 individuals in the population. The president wants to obtain a sample of size 100, with the number of individuals selected from each stratum weighted by the population size. So resident students make up 6,204/21,909 = 28% of the sample, nonresident students account for 61% of the sample, and staff constitute 11% of the sample. To obtain a sample of size 100, the president will obtain a stratified sample of 0.28(100) = 28 resident students, 0.61(100) = 61 nonresident students, and 0.11(100) = 11 staff.

Approach: To obtain the stratified sample, conduct a simple random sample within each group. That is, obtain a simple random sample of 28 resident students (from the 6,204 resident students), a simple random sample of 61 nonresident students, and a simple random sample of 11 staff.

Solution: Using MINITAB, with the seed set to 4032 and the values shown in Figure 3, we obtain the following sample of staff:

240, 630, 847, 190, 2096, 705, 2320, 323, 701, 471, 744

Figure 3

!**CAUTION**

Do not use the same seed for all the groups in a stratified sample, because we want the simple random samples within each stratum to be independent of each other.

Repeat this procedure for the resident and nonresident students.

An advantage of stratified sampling over simple random sampling is that the researcher is able to determine characteristics within each stratum. This allows an analysis to be performed on each subgroup to see if any significant differences between the groups exist. For example, we could analyze the data obtained in Example 1 to see if there is a difference in the opinions of students versus faculty.

Now Work Problem 23.

② Obtain a Systematic Sample

In both simple random sampling and stratified sampling, it is necessary that a list of the individuals in the population being studied (the frame) exists. Therefore, these sampling techniques require some preliminary work before the sample is obtained. A sampling technique that does not require a frame is *systematic sampling.*

Definition

> A **systematic sample** is obtained by selecting every kth individual from the population. The first individual selected corresponds to a random number between 1 and k.

In Other Words

Systematic sampling is like selecting every fifth person out of a line.

Because systematic sampling does not require a frame, it is a useful technique when you can't obtain a list of the individuals in the population that you wish to study.

The idea behind obtaining a systematic sample is relatively simple: Select a number k, randomly select a number between 1 and k and survey that individual, then survey every kth individual thereafter. For example, we might decide to survey every $k = 8$th individual. We randomly select a number between 1 and 8 such as 5. This means we survey the 5th, $5 + 8 = 13$th, $13 + 8 = 21$st, $21 + 8 = 29$th, and so on, individuals until we reach the desired sample size.

EXAMPLE 2

Obtaining a Systematic Sample without a Frame

Problem: The manager of Kroger Food Stores wants to measure the satisfaction of the store's customers. Design a sampling technique that can be used to obtain a sample of 40 customers.

Approach: A frame of Kroger customers would be difficult, if not impossible, to obtain. Therefore, it is reasonable to use systematic sampling by surveying every kth customer who leaves the store.

Solution: The manager decides to obtain a systematic sample by surveying every 7th customer. He randomly determines a number between 1 and 7, say 5. He then surveys the 5th customer exiting the store and every 7th customer thereafter, until a sample of 40 customers is reached. The survey will include customers $5, 12, 19, \ldots, 278.*$

But how do we select the value of k? If the size of the population is unknown, there is no mathematical way to determine k. It must be chosen by determining a value of k that is not so large that we are unable to achieve our desired sample size, but not so small that we do not obtain a sample size that is representative of the population.

To clarify this point, let's revisit Example 2. Suppose we chose a value of k that was too large, say 30. This means that we will survey every 30th shopper, starting with the 5th. To obtain a sample of size 40 would require that 1,175 shoppers visit Kroger on that day. If Kroger does not have 1,175 shoppers, the desired sample size will not be achieved. On the other hand, if k is too small, say 4, we would survey the 5th, 9th, $\ldots$, 161st shopper. It may be that the 161st shopper exits the store at 3 P.M., which means our survey did not include any of the evening shoppers. Certainly, this sample is not representative of *all* Kroger patrons! An estimate of the size of the population would certainly help determine an appropriate value for k.

---

*Because we are surveying 40 customers, the first individual surveyed is the 5th, the second is the $5 + 7 = 12$th, the third is the $5 + (2)7 = 19$th, and so on, until we reach the 40th, which is the $5 + (39)7 = 278$th shopper.

To determine the value of k when the size of the population, N, is known is relatively straightforward. Suppose we wish to survey a population whose size is known to be $N = 20{,}325$ and we desire a sample of size $n = 100$. To guarantee that individuals are selected evenly from the beginning as well as the end of the population (such as early and late shoppers), we compute N/n and round down to the nearest integer. For example, $20{,}325/100 = 203.25$, so $k = 203$. Then we randomly select a number between 1 and 203 and select every 203rd individual thereafter. So, if we randomly selected 90 as our starting point, we would survey the 90th, 293rd, 496th, ..., 20,187th individuals.

We summarize the procedure as follows:

Steps in Systematic Sampling

Step 1: If possible, approximate the population size, N.

Step 2: Determine the sample size desired, n.

Step 3: Compute $\frac{N}{n}$ and round down to the nearest integer. This value is k.

Step 4: Randomly select a number between 1 and k. Call this number p.

Step 5: The sample will consist of the following individuals:

$$p, p + k, p + 2k, \ldots, p + (n - 1)k$$

Because systematic sampling does not require that the size of the population be known, it typically provides more information for a given cost than does simple random sampling. In addition, systematic sampling is easier to employ, so there is less possibility of interviewer error occurring, such as selecting the wrong individual to be surveyed.

Now Work Problem 27.

③ Obtain a Cluster Sample

A fourth sampling method is called *cluster sampling*. The previous three sampling methods discussed have benefits under certain circumstances. So does cluster sampling.

Definition

A **cluster sample** is obtained by selecting all individuals within a randomly selected collection or group of individuals.

Consider a quality-control engineer who wants to verify that a certain machine is filling bottles with 16 ounces of liquid detergent. To obtain a sample of bottles from the machine, the engineer could use systematic sampling by sampling every kth bottle from the machine. However, it would be time consuming waiting next to the filling machine for the bottles to come off the line. Suppose that, as the bottles come off the line, they are placed into cartons of 12 bottles each. An alternative sampling method would be to randomly select a few cartons and measure the contents of all 12 bottles from each carton selected. This would be an example of cluster sampling. It is a good sampling method in this situation because it would speed up the data-collection process.

In Other Words
Imagine a mall parking lot. Each subsection of the lot could be a cluster (Section F-4 for example).

EXAMPLE 3 | **Obtaining a Cluster Sample**

Problem: A sociologist wants to gather data regarding the household income within the city of Boston. Obtain a sample using cluster sampling.

Approach: The city of Boston can be set up so that each city block is a cluster. Once the city blocks have been identified, we obtain a simple random sample of the city blocks and survey all households on the blocks selected.

Solution: Suppose there are 10,493 city blocks in Boston. First, we must number the blocks from 1 to 10,493. Suppose the sociologist has enough time and money to survey 20 clusters (city blocks). Therefore, the sociologist should obtain a simple random sample of 20 numbers between 1 and 10,493 and survey all households from the clusters selected. Cluster sampling is a good choice in this example because it reduces the travel time to households that is likely to occur with both simple random sampling and stratified sampling. In addition, there is no need to obtain a detailed frame with cluster sampling. The only frame needed is one that provides information regarding city blocks.

Recall that in systematic sampling we had to determine an appropriate value for k, the number of individuals to skip between individuals selected to be in the sample. We have a similar problem in cluster sampling. The following are a few of the questions that arise:

- How do I cluster the population?
- How many clusters do I sample?
- How many individuals should be in each cluster?

First, it must be determined whether the individuals within the proposed cluster are homogeneous (similar individuals) or heterogeneous (dissimilar individuals). Consider the results of Example 3. City blocks tend to have similar households. Surveying one house on a city block is likely to result in similar responses from another house on the same block. This results in duplicate information. We conclude the following: If the clusters have homogeneous individuals, it is better to have more clusters with fewer individuals in each cluster.

What if the cluster is heterogeneous? Under this circumstance, the heterogeneity of the cluster likely resembles the heterogeneity of the population. In other words, each cluster is a scaled-down representation of the overall population. For example, a quality-control manager might use shipping boxes that contain 100 light bulbs as a cluster, since the rate of defects within the cluster would closely mimic the rate of defects in the population assuming the bulbs are randomly placed in the box. Thus, when each of the clusters is heterogeneous, fewer clusters with more individuals in each cluster are appropriate.

> **Now Work Problem 11.**

The four sampling techniques just presented are sampling techniques in which the individuals are selected randomly. Often, however, sampling methods are used in which the individuals are not randomly selected, such as *convenience sampling*.

Convenience Sampling

Convenience sampling is probably the easiest sampling method.

Definition

> A **convenience sample** is a sample in which the individuals are easily obtained.

There are many types of convenience samples, but probably the most popular are those in which the individuals in the sample are **self-selected** (the individuals themselves decide to participate in a survey). These are also called **voluntary response** samples. Examples of self-selected sampling include phone-in polling; a radio personality will ask his or her listeners to phone the station to submit their opinions. Another example is the use of the Internet to conduct surveys. For example, *Dateline* will present a story regarding a certain topic and ask its viewers to "tell us what you think" by completing a questionnaire online or phoning in an opinion. Both of these samples are poor designs because the

Note to Instructor

In-Class Activity: Compare and contrast each sampling technique. Discuss a research situation in which each type of sampling would be the most appropriate.

! CAUTION

Stratified and cluster samples are different. In a stratified sample, we divide the population into two or more homogeneous groups. Then we obtain a simple random sample from each group. In a cluster sample, we divide the population into groups, obtain a simple random sample of some of the groups, and survey *all* individuals in the selected groups.

! CAUTION

Studies that use convenience sampling generally have results that are suspect. The results should be looked on with extreme skepticism.

Note to Instructor

Discuss some problems encountered in obtaining a sample from a nearby mall.

individuals who decide to be in the sample generally have strong opinions about the topic. A more typical individual in the population will not bother phoning or logging on to a computer to complete a survey. Any inference made regarding the population from this type of sample should be made with extreme caution.

Multistage Sampling

In practice, most large-scale surveys obtain samples using a combination of the techniques just presented.

As an example of multistage sampling, consider Nielsen Media Research. Nielsen randomly selects households and monitors the television programs these households are watching through a People Meter. The meter is an electronic box placed on each TV within the household. The People Meter measures what program is being watched and who is watching it. Nielsen selects the households with the use of a two-stage sampling process.

Stage 1: Using U.S. Census data, Nielsen divides the country into geographic areas (strata). The strata are typically city blocks in urban areas and geographic regions in rural areas. About 6,000 strata are randomly selected.

Stage 2: Nielsen sends representatives to the selected strata and lists the households within the strata. The households are then randomly selected through a simple random sample.

Nielsen sells this information to television stations and companies. Their results are used to help determine prices for commercials.

As another example of multistage sampling, consider the sample used by the Census Bureau for the Current Population Survey. This survey requires five stages of sampling:

Stage 1: Stratified sample

Stage 2: Cluster sample

Stage 3: Stratified sample

Stage 4: Cluster sample

Stage 5: Systematic sample

This survey is very important because it is used to obtain demographic estimates of the United States in noncensus years. A detailed presentation of the sampling method used by the Census Bureau can be found in *The Current Population Survey: Design and Methodology*, Technical Paper No. 40.

Sample Size Considerations

Throughout the discussion of sampling, we did not mention how to determine the sample size. Determining the sample size is key in the overall statistical process. In other words, the researcher must ask this question: "How many individuals must I survey to draw conclusions about the population within some predetermined margin of error?" The researcher must find the correct balance between the reliability of the results and the cost of obtaining these results. The bottom line is that time and money determine the level of confidence a researcher will place on the conclusions drawn from the sample data. The more time and money the researcher has available, the more accurate will be the results of the statistical inference.

Nonetheless, techniques do exist for determining the sample size required to estimate characteristics regarding the population within some margin of error. We will consider some of these techniques in Sections 9.1 and 9.3. (For a detailed discussion of sample size considerations, consult a text on sampling techniques such as *Elements of Sampling Theory and Methods* by Z. Govindarajulu, Prentice Hall, 1999.)

Summary

Figure 4 provides a summary of the four sampling techniques presented.

Figure 4

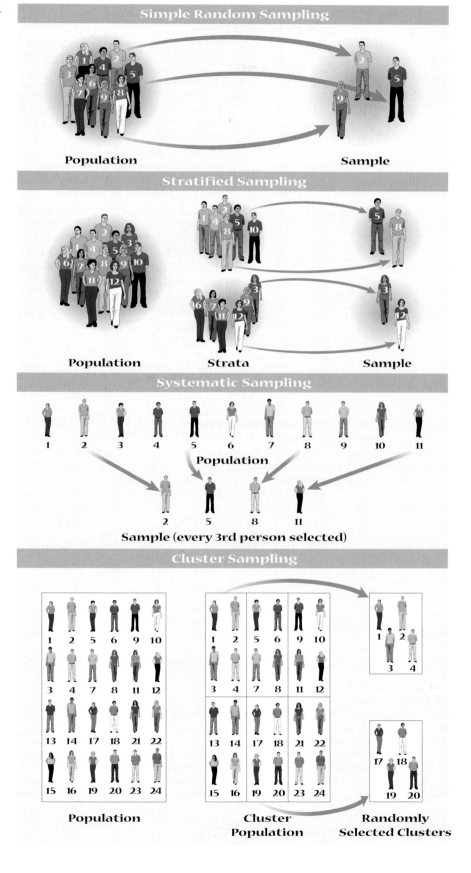

Simple Random Sampling

Population

Sample

Stratified Sampling

Population

Strata

Sample

Systematic Sampling

Population

Sample (every 3rd person selected)

Cluster Sampling

Population

Cluster Population

Randomly Selected Clusters

Note to the Instructor
This activity should take about 20 to 30 minutes.

In-Class Activity: Different Sampling Methods

The following question was recently asked by the Gallup Organization.
"In general, are you satisfied or dissatisfied with the way things are going in the country?"

(a) Number the students in the class from 1 to N, where N is the number of students. Obtain a simple random sample and have them answer this question. Record the number of satisfied responses and the number of dissatisfied responses.

(b) Divide the students in the class by gender. Treat each gender as a stratum. Obtain a simple random sample from each stratum and have them answer this question. Record the number of satisfied responses and the number of dissatisfied responses.

(c) Treat each row of desks as a cluster. Obtain a simple random sample of clusters and have each student in the selected clusters answer this question. Record the number of satisfied responses and the number of dissatisfied responses.

(d) Number the students in the class from 1 to N, where N is the number of students. Obtain a systematic sample and have the selected students answer this question. Record the number of satisfied responses and the number of dissatisfied responses.

(e) Were there any differences in the results of the survey? State some reasons for any differences.

1.3 ASSESS YOUR UNDERSTANDING

Concepts and Vocabulary

1. Describe a circumstance in which stratified sampling would be an appropriate sampling method.
2. Which sampling method does not require a frame?
3. Why are convenience samples ill advised?
4. A _____ is obtained by dividing the population into groups and selecting all individuals from within a random sample of the groups. cluster sample
5. A _____ is obtained by dividing the population into homogeneous groups and randomly selecting individuals from each group. stratified sample

6. *True or False*: When taking a systematic random sample of size n, every group of size n from the population has the same chance of being selected. False
7. *True or False*: A simple random sample is always preferred because it obtains the same information as other sampling plans but requires a smaller sample size. False
8. *True or False*: When conducting a cluster sample, it is better to have fewer clusters with more individuals when the clusters are heterogeneous. True

Skill Building

In Problems 9–22, identify the type of sampling used.

9. To estimate the percentage of defects in a recent manufacturing batch, a quality-control manager at Intel selects every 8th chip that comes off the assembly line starting with the 3rd until she obtains a sample of 140 chips. Systematic

10. To determine the average IQ of ninth-grade students, a school psychologist obtains a list of all high schools in the local public school system. She randomly selects five of these schools and administers an IQ test to all ninth-grade students at the selected schools. Cluster

11. To determine customer opinion of their boarding policy, NW Southwest Airlines randomly selects 60 flights during a certain week and surveys all passengers on the flights. Cluster

12. A member of Congress wishes to determine her constituency's opinion regarding estate taxes. She divides her constituency into three income classes: low-income households, middle-income households, and upper-income households. She then takes a simple random sample of households from each income class. Stratified

13. In an effort to identify if an advertising campaign has been effective, a marketing firm conducts a nationwide poll by randomly selecting individuals from a list of known users of the product. Simple random

14. A radio station asks its listeners to call in their opinion regarding the use of U.S. forces in peacekeeping missions. Convenience

15. A farmer divides his orchard into 50 subsections, randomly selects 4, and samples all the trees within the 4 subsections to approximate the yield of his orchard. Cluster

16. A school official divides the student population into five classes: freshman, sophomore, junior, senior, and graduate student. The official takes a simple random sample from each class and asks the members' opinions regarding student services. Stratified

17. A survey regarding download time on a certain Web site is administered on the Internet by a market research firm to anyone who would like to take it. Convenience

18. A group of lobbyists has a list of the 100 senators of the United States. To determine the Senate's position regarding farm subsidies, they decide to talk with every seventh senator on the list, starting with the third. Systematic

19. A small-town newspaper reporter wants to get local reaction to a controversial new film. She waits outside the theater during an afternoon show and, starting with the second, asks every fifth patron leaving how much they liked the movie. Systematic

20. To determine his DSL Internet connection speed, Shawn divides up the day into four parts: morning, midday, evening, and late night. He then measures his Internet connection speed at 5 randomly selected times during each part of the day. Stratified

21. A statistics instructor with a large number of students attempts to reduce time spent grading by only grading a portion of assigned homework problems. He randomly selects one of the first four problems in the assignment and then grades that problem and every fourth problem thereafter. Systematic

22. 24 Hour Fitness wants to administer a satisfaction survey to its current members. Using its membership roster, the club randomly selects 40 club members and asks them about their level of satisfaction with the club. Simple random

---

Applying the Concepts

23. Stratified Sampling The Future Government Club wants to sponsor a panel discussion on the upcoming national election. The club wants to have four of its members lead the panel discussion. To be fair, however, the panel should consist of two Democrats and two Republicans. Below is a list of the Democrats and Republicans who are currently members of the club. Obtain a stratified sample of two Democrats and two Republicans to serve on the panel.

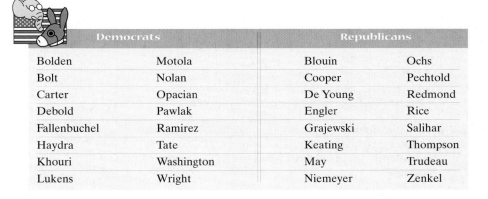

Democrats		Republicans	
Bolden	Motola	Blouin	Ochs
Bolt	Nolan	Cooper	Pechtold
Carter	Opacian	De Young	Redmond
Debold	Pawlak	Engler	Rice
Fallenbuchel	Ramirez	Grajewski	Salihar
Haydra	Tate	Keating	Thompson
Khouri	Washington	May	Trudeau
Lukens	Wright	Niemeyer	Zenkel

24. Stratified Sampling The owner of a private food store is concerned about employee morale. She decides to survey the managers and hourly employees to see if she can learn about work environment and job satisfaction. Below is a list of the managers and hourly workers at the store. Obtain a stratified sample of two managers and four hourly employees to survey.

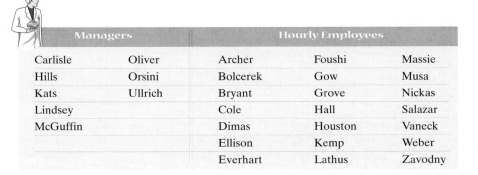

Managers		Hourly Employees		
Carlisle	Oliver	Archer	Foushi	Massie
Hills	Orsini	Bolcerek	Gow	Musa
Kats	Ullrich	Bryant	Grove	Nickas
Lindsey		Cole	Hall	Salazar
McGuffin		Dimas	Houston	Vaneck
		Ellison	Kemp	Weber
		Everhart	Lathus	Zavodny

25. **Stratified Sample** A local youth advisory committee is comprised of 18 youths and 6 adults. The committee needs to select 2 adults and 4 youths to meet with the mayor to discuss renovations of the community center. The members of the committee are:

    Youth: Patrick, Elise, Shawn, Lauren, Sandra, Josh, Erin, Payton, Michelle, Chad, Katy, Amanda, Jim, Janice, Jeff, Amy, Logan, Steve
    Adults: Harold, Kerry, Keith, Greg, Marcie, Debbie

    Obtain a stratified random sample of 4 youths and 2 adults.

26. **Systematic Sample** In 2005 the University of Illinois men's basketball team set a school record for wins in a row (29). Their points per game (listed in chronological order horizontally) for regular and postseason games are given next. Obtain a systematic random sample of size 6 from the scores. Be sure to describe your sampling method.

87	91	85	89	91	72	78	74	83	93
70	105	69	67	84	68	90	78	73	75
89	81	60	57	70	83	75	84	84	64

27. **Systematic Sample** The human resource department at a NW certain company wants to conduct a survey regarding worker morale. The department has an alphabetical list of all 4,502 employees at the company and wants to conduct a systematic sample.
    (a) Determine k if the sample size is 50.
    (b) Determine the individuals who will be administered the survey. More than one answer is possible.

28. **Systematic Sample** To predict the outcome of a county election, a newspaper obtains a list of all 945,035 registered voters in the county and wants to conduct a systematic sample.
    (a) Determine k if the sample size is 130.
    (b) Determine the individuals who will be administered the survey. More than one answer is possible.

29. **Sample Design** The city of Naperville is considering the construction of a new commuter rail station. The city wishes to survey the residents of the city to obtain their opinion regarding the use of tax dollars for this purpose. Design a sampling method to obtain the individuals in the sample. Be sure to support your choice.

30. **Sample Design** A school board at a local community college is considering raising the student services fees. The board wants to obtain the opinion of the student body before proceeding. Design a sampling method to obtain the individuals in the sample. Be sure to support your choice.

31. **Sample Design** Target wants to open a new store in the village of Lockport. Before construction Target's marketers want to obtain some demographic information regarding the area under consideration. Design a sampling method to obtain the individuals in the sample. Be sure to support your choice.

32. **Sample Design** The county sheriff wishes to determine if a certain highway has a high proportion of speeders traveling on it. Design a sampling method to obtain the individuals in the sample. Be sure to support your choice.

33. **Sample Design** A pharmaceutical company wants to conduct a survey of 30 individuals who have high cholesterol. The company has obtained a list from doctors throughout the country of 6,600 individuals who are known to have high cholesterol. Design a sampling method to obtain the individuals in the sample. Be sure to support your choice.

34. **Sample Design** A marketing executive for Coca-Cola Inc., wants to identify television shows that people in the Boston area who typically drink Coke are watching. The executive has a list of all households in the Boston area. Design a sampling method to obtain the individuals in the sample. Be sure to support your choice.

35. Suppose an individual who is selected to be in a survey is not at home when a researcher calls, but the individual's spouse offers to answer the questions. What should the researcher do and why?

36. Research the origins of the Gallup Poll and the current sampling method the organization uses. Report your findings to the class.

37. Research the sampling methods used by a market research firm in your neighborhood. Report your findings to the class. The report should include the types of sampling methods used, number of stages, and sample size.

1.4 Sources of Errors in Sampling

Objective ① **Understand how error can be introduced during sampling**

① ## Understand How Error Can Be Introduced During Sampling

Understand How Error Can Be Introduced During Sampling

Understand how error can be introduced during sampling

Note to Instructor
If you are pressed for time, this section can be given to the students as a reading assignment with light in-class coverage.

Thus far, we have discussed *how* to obtain samples, but have neglected to look at any of the pitfalls that inevitably arise in sampling. In this section we look at problems that can occur in sampling. Some of these problems can be remedied. Some, however, have no solution. Collectively, these errors are called *nonsampling errors*.

Definition

In Other Words
We use the word *bias* in nonsampling error to mean that preference is given to selecting some individuals over others.

Nonsampling errors are errors that result from the survey process. They are due to the nonresponse of individuals selected to be in the survey, to inaccurate responses, to poorly worded questions, to bias in the selection of individuals to be given the survey, and so on.

We contrast nonsampling errors with *sampling errors*.

Definition

In Other Words
We can think of sampling error as error that results from using a subset of the population to describe characteristics of the population. Nonsampling error is error that results from obtaining and recording information collected.

Sampling error is the error that results from using sampling to estimate information regarding a population. This type of error occurs because a sample gives incomplete information about the population.

By incomplete information, we mean that the individuals in the sample cannot reveal all the information about the population. Consider the following: Suppose that we wanted to determine the average age of the students enrolled in an introductory statistics course. To do this, we obtain a simple random sample of 4 students and ask them to write their age on a sheet of paper and turn it in. The average age of these 4 students is found to be 23.25 years. Assume that no students lied about their age, nobody misunderstood the question, and the sampling was done appropriately. If the actual average age of all 30 students in the class (the population) is 22.91 years, then the sampling error is $23.25 - 22.91 = 0.34$ year. Now suppose that the same survey is conducted, but this time one individual lies about his age. Then the results of the survey will have nonsampling error.

A well-designed survey and sampling technique can minimize nonsampling error. However, sampling error is more difficult to control.

Note to Instructor
Make a reference to the Truman–Dewey election of 1948. This election demonstrates the importance of obtaining a representative sample.

When a sampling design is done poorly, the descriptive statistics computed from the data obtained in the sample may not be close to the values that would be obtained if the entire population were surveyed. For example, the *Literary Digest* predicted that Alfred M. Landon would defeat Franklin D. Roosevelt in the 1936 presidential election. The *Literary Digest* conducted a poll by mailing questionnaires based on a list of its subscribers, telephone directories, and automobile owners. On the basis of the results, the *Literary Digest* predicted that Landon would win with 57% of the popular vote. However, Roosevelt won the election with about 62% of the popular vote. The incorrect prediction by the *Literary Digest* was the result of a poor sample design. In 1936, most subscribers to the magazine, households with telephones, and automobile owners were Republican, the party of Landon. Therefore, the choice of the frame used to conduct the survey led to an incorrect prediction. This is an example of nonsampling error.

We now list some sources of nonsampling error. These sources include the frame, nonresponse, data-entry error, and poorly worded questions.

The Frame

Recall that the frame is the list of all individuals in the population under study. For example, in a study regarding voter preference in an upcoming election, the frame would be a list of all registered voters. Sometimes, obtaining the frame would seem to be a relatively easy task, such as obtaining the list of all registered voters. Even under this circumstance, however, the frame may be incomplete. People who recently registered to vote may not be on the published list of registered voters.

Often, it is difficult to gain access to a *complete* list of individuals in a population. For example, in public opinion polls, random telephone surveys are frequently conducted, which implies that the frame is all households with telephones. This method of sampling will exclude any household that does not have a telephone as well as all homeless people. In such a situation, certain segments of the population are *underrepresented*. A part of the population is **underrepresented** when the sampling method used tends to exclude this segment.

In designing any sample, the hope is that the frame used is as complete as possible so that any results inferred regarding the population have as little error as possible.

Nonresponse

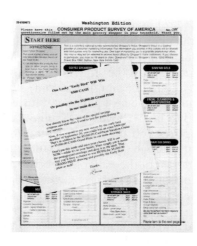

Nonresponse means that an individual selected for the sample does not respond to the survey. Nonresponse can occur because individuals selected for the sample do not wish to respond or because the interviewer was unable to contact them.

This type of error can be controlled using callbacks. The type of callback employed typically depends on the type of survey initially used. For example, if nonresponse occurs because a mailed questionnaire was not returned, a callback might mean phoning the individual to conduct the survey. If nonresponse occurs because an individual was not at home, a callback might mean returning to the home at other times in the day or other days of the week.

Another method to improve nonresponse is using rewards and incentives. Rewards may include cash payments for completing a questionnaire, made only upon receipt of the completed questionnaire. Incentives might also include a cover letter that states that the responses to the questionnaire will dictate future policy. For example, a village may send out questionnaires to households and state in a cover letter that the responses to the questionnaire will be used to decide pending issues within the village.

Interviewer Error

A trained interviewer is essential to obtain accurate information from a survey. A good interviewer will have the skill necessary to elicit responses from individuals within a sample and be able to make the interviewee feel comfortable enough to give truthful responses. For example, a good interviewer should be able to obtain truthful answers to questions as sensitive as "Have you ever cheated on your taxes?" Do not be quick to trust surveys that are conducted by poorly trained interviewers. Do not trust survey results if the sponsor has a vested interest in the results of the survey. Would you trust a survey conducted by a car dealer that reports 90% of customers say they would buy another car from the dealer?

Misrepresented Answers

Watch out for answers to survey questions that misrepresent facts or are flat-out lies. In fact, respondents may lie on surveys even when there is no problem with the interviewer. For example, a survey of recent college graduates may find that self-reported salaries are somewhat inflated.

Data Checks

Once data are collected, the results typically must be entered into a computer. Data entry inevitably results in input errors. It is imperative that data be checked for accuracy at every stage of the statistical analysis. In this text, we present some methodology that can be used to check for data-entry errors.

Questionnaire Design

Appropriate questionnaire design is critical in minimizing the amount of non-sampling error. We will concentrate on the main aspects in the design of a good questionnaire.

Note to Instructor
Have students discuss some problems in coding responses to open questions.

One of the first considerations in designing a question is determining whether the question should be open or closed.

An **open question** is one for which the respondent is free to choose his or her response. For example:

What is the most important problem facing America's youth today?

A **closed question** is one for which the respondent must choose from a list of predetermined responses.

What is the most important problem facing America's youth today?
(a) Drugs
(b) Violence
(c) Single-parent homes
(d) Promiscuity
(e) Peer pressure

When designing an open question, be sure to phrase the question so that the responses are similar. (You don't want a wide variety of responses.) This allows for easy analysis of the responses. The benefit of closed questions is that they limit the number of respondent choices and, therefore, the results are much easier to analyze. However, this limits the choices and does not always allow the respondent to respond the way he or she might want to respond. If the desired answer is not provided as a choice, the respondent will be forced to choose a secondary answer or skip the question.

Survey designers recommend conducting pretest surveys with open questions and then using the most popular answers as the choices on closed question surveys. Another issue to consider in the closed question design is the number of responses the respondent may choose from. It is recommended that the option "no opinion" be omitted, because this option does not allow for meaningful analysis. The bottom line is to try to limit the number of choices in a closed-question format without forcing respondents to choose an option they otherwise would not.

! CAUTION
The wording of questions can significantly affect the responses and, therefore, the validity of a study.

Wording of Questions

The wording of a survey question is vital in obtaining data that are not misrepresentative. Questions must always be asked in balanced form. For example, the "yes/no" question

Do you oppose the reduction of estate taxes?

should be written

Do you favor or oppose the reduction of estate taxes?

The second question is balanced. Do you see the difference? Consider the following report based on studies from Schuman and Presser (*Questions and Answers in Attitude Surveys*, 1981, p. 277), who asked the following two questions:

(A) Do you think the United States should forbid public speeches against democracy?

(B) Do you think the United States should allow public speeches against democracy?

For those respondents presented with question A, 21.4% gave yes responses, while for those given question B, 47.8% gave no responses. The conclusion you may arrive at is that most people are not necessarily willing to forbid something, but more people are willing not to allow something. These results imply that the wording of the question can alter the outcome of a survey.

Another consideration in wording a question is not to be vague. For example, the question "How much do you study?" is too vague. Does the researcher mean how much do I study for all my classes or just for statistics? Does the researcher mean per day or per week? The question should be written "How many hours do you study statistics each week?"

The Order of the Questions, Words, and Responses

Many surveys will rearrange the order of the questions within a questionnaire so that responses are not affected by prior questions. Consider the following example from Schuman and Presser in which the following two questions were asked:

(A) Do you think the United States should let Communist newspaper reporters from other countries come in here and send back to their papers the news as they see it?

(B) Do you think a Communist country such as Russia should let American newspaper reporters come in and send back to America the news as they see it?

For surveys conducted in 1980 in which the questions appeared in the order (A, B), 54.7% of respondents answered yes to A and 63.7% answered yes to B. If the questions were ordered (B, A), then 74.6% answered yes to A and 81.9% answered yes to B. When Americans are first asked if U.S. reporters should be allowed to report Communist news, they are more likely to agree that Communists should be allowed to report American news. Questions should be rearranged as much as possible to help reduce the effects of this type.

Pollsters will also rearrange words within a question. For example, the Gallup Organization asked the following question of 1,017 adults aged 18 years or older:

Do you consider the first six months of the Bush administration to be a [rotated: **success** (or a) **failure**]?

Notice how the words *success* and *failure* were rotated. The purpose of this is to remove the effect that may occur by writing the word *success* first in the question.

Not only should the order of the questions or certain words within the question be rearranged, but, in closed questions, the possible responses should also be rearranged. The reason is that respondents are likely to choose early choices in a list rather than later choices.

MAKING AN INFORMED DECISION

What Movie Should I Go To?
One of the most difficult tasks of surveying is phrasing questions so that they are not misunderstood. In addition, they must be phrased so that the researcher obtains answers that allow for meaningful analysis. We wish to create a questionnaire that can be used to make an informed decision about whether to attend a certain movie. Select a movie that you wish to see. If the movie is still in theaters, make sure that it has been released for at least a couple of weeks so it is likely that a number of people have seen it. Design a questionnaire to be filled out by individuals who have seen the movie. You may wish to include questions regarding the demographics of the respondents first (such as age, gender, level of education, and so on). Ask as many questions as you feel are necessary to obtain an opinion regarding the movie. The questions can be open or closed. Administer the survey to at least 20 randomly selected people who have seen the movie. While administering the survey, keep track of those individuals who have not seen the movie. In particular, keep track of their demographic information. After administering the survey, summarize your findings. On the basis of the survey results, do you think that you will enjoy the movie? Why? Now see the movie. Did you like it? Did the survey accurately predict whether you would enjoy the movie? Now answer the following questions:

(a) What sampling method did you use? Why? Did you have a frame for the population?
(b) Did you have any problems with respondents misinterpreting your questions? How could this issue have been resolved?
(c) What role did the demographics of the respondent have in forming your opinion? Why?
(d) Did the demographics of individuals who did not see the movie play a role while you were forming your opinion regarding the movie?
(e) Look up a review of the movie by a professional movie critic. Did the movie critic's opinion agree with yours? What might account for the similarities or differences in your opinions?
(f) Describe the problems that you had in administering the survey. If you had to do this survey over again, would you change anything? Why?

1.4 ASSESS YOUR UNDERSTANDING

Concepts and Vocabulary

1. Why is it rare for frames to be completely accurate?
2. What are some solutions to nonresponse?
3. What is a closed question? What is an open question? Discuss the advantages and disadvantages of each type of question.
4. What does it mean when a part of the population is underrepresented?
5. Discuss the benefits of having trained interviewers.
6. What are the advantages of having a pretest when constructing a questionnaire that has closed questions?
7. Discuss the pros and cons of telephone interviews that take place during dinner time in the early evening.
8. Why is a high response rate desired? How would a low response rate affect survey results?
9. Discuss why the order of questions or choices within a questionnaire are important in sample surveys.
10. Suppose a survey asks, "Do you own any CDs?" Explain how this could be interpreted in more than one way. Suggest a way in which the question could be improved.

Skill Building

In Problems 11–22, the survey design is flawed. (a) Determine whether the flaw is due to the sampling method or the survey itself. For flawed surveys, identify the cause of the error (wording of question, nonresponse, and so forth). (b) Suggest a remedy to the problem.

11. A college vice-president wants to conduct a study regarding the achievement of undergraduate students. He selects the first 50 students who enter the building on a given day and administers his survey. *Flawed sampling method*

12. The town of Oak Lawn wishes to conduct a study regarding the income level of households within the town. The town manager selects 10 homes in the southwest corner of the town and sends an interviewer to the homes to determine household income. *Flawed sampling method*

13. An antigun advocate wants to estimate the percentage of people who favor stricter gun laws. He conducts a nationwide survey of 1,203 randomly selected adults 18 years old and older. The interviewer asks the respondents, "Do you favor harsher penalties for individuals who sell guns illegally?" *Poorly worded question*

14. Suppose you are conducting a survey regarding students' study habits. From a list of full-time registered students, you obtain a simple random sample of 90 students. One survey question is "How many hours do you study?" *Poorly worded question*

15. A polling organization conducts a study to estimate the percentage of households that speak a foreign language as the primary language. It mails a questionnaire to 1,023 randomly selected households throughout the United States and asks the head of household if a foreign language is the primary language spoken in the home. Of the 1,023 households selected, 12 responded. *Nonresponse*

16. Cold Stone Creamery is considering opening a new store in O'Fallon. Before opening the store, the company would like to know the percentage of households in O'Fallon that regularly visit an ice cream shop. The market researcher obtains a list of households in O'Fallon and randomly selects 150 of them. He mails a questionnaire to the 150 households that asks about ice cream eating habits and flavor preferences. Of the 150 questionnaires mailed, 4 are returned. *Nonresponse*

17. The owner of a shopping mall wishes to expand the number of shops available in the food court. She has a market researcher survey mall customers during weekday mornings to determine what types of food the shoppers would like to see added to the food court. *Flawed sampling method*

18. A produce buyer for a supermarket is inspecting crates of peaches. He examines 20 peaches from the top of each crate and purchases the crate if all 20 peaches are satisfactory. *Flawed sampling method*

19. A magazine is conducting a study on the effects of infidelity in a marriage. The editors randomly select 400 women whose husbands were unfaithful and ask, "Do you believe a marriage can survive when the husband destroys the trust that must exist between husband and wife?" *Poorly worded question*

20. A textbook publisher wants to determine what percentage of college professors either require or recommend that their students purchase textbook packages with supplemental materials, such as study guides, digital media, and online tools. The publisher sends out surveys by e-mail to a random sample of 320 faculty members who have registered with the publisher's Web site and have agreed to receive solicitations. The publisher reports that 80% of college professors require or recommend that their students purchase some type of textbook package. *Flawed sampling method*

21. Suppose you are conducting a survey regarding illicit drug use among teenagers in the Baltimore School District. You obtain a cluster sample of 12 schools within the district and sample all sophomore students in the randomly selected schools. The survey is administered by the teachers. *Interviewer error*

22. To determine public opinion of the police department, the police chief obtains a cluster sample of 15 census tracts within his jurisdiction and samples all households in the randomly selected tracts. Uniformed police officers go door to door to conduct the survey. *Interviewer error*

Applying the Concepts

23. **Order of the Questions** Consider the following two questions.
    A. *Suppose that a rape is committed in which the woman becomes pregnant. Do you think the criminal should or should not face additional charges if the woman becomes pregnant?*
    B. *Do you think abortions should be legal under any circumstances, legal under certain circumstances, or illegal in all circumstances?*

    Do you think the order in which the questions are asked will affect the survey results? If so, what can the pollster do to alleviate this response bias?

24. **Order of the Questions** Consider the following two questions:
    A. *Do you believe that the government should or should not be allowed to prohibit individuals from expressing their religious beliefs at their place of employment?*
    B. *Do you believe that the government should or should not be allowed to prohibit teachers from expressing their religious beliefs in public school classrooms?*

    Do you think the order in which the questions are asked will affect the survey results? If so, what can the pollster do to alleviate this response bias? Discuss the choice of the word *prohibit* in the survey questions.

25. **Rotating Choices** Consider this question from a recent Gallup poll:

    Thinking about how the abortion issue might affect your vote for major offices, would you vote only for a candidate who shares your views on abortion or consider a candidate's position on abortion as just one of many important factors? [rotated]

    Why is it important to rotate the two choices presented in the question?

26. **Overcoverage** An article in the *Los Angeles Times* (Sept. 21, 1996) reported that results of Bosnian elections the previous week were suspect because an independent watchdog agency, International Crisis Group, estimated that voter turnout was 106.7%. Why do you think the agency was concerned?

27. Exit Polls Read the following article from the January 20, 2005 *USA Today*. What types of nonsampling errors led to incorrect exit polls?

FIRMS REPORT FLAWS THAT THREW OFF EXIT POLLS
Kerry backers' willingness, pollsters' inexperience cited

By Mark Memmott, *USA Today*

The exit polls of voters on Election Day so overstated Sen. John Kerry's support that, going back to 1988, they rank as the most inaccurate in a presidential election, the firms that did the work concede.

One reason the surveys were skewed, they say, was because Kerry's supporters were more willing to participate than Bush's. Also, the people they hired to quiz voters were on average too young and too inexperienced and needed more training.

The exit polls, which are supposed to help the TV networks shape their coverage on election night, were sharply criticized. Leaks of preliminary data showed up on the Internet in the early afternoon of Election Day, fueling talk that Kerry was beating President Bush. After the election, some political scientists, pollsters and journalists questioned their value.

In a report to the six media companies that paid them to conduct the voter surveys, pollsters Warren Mitofsky and Joseph Lenski, said Wednesday that "on average, the results from each precinct overstated the Kerry-Bush difference by 6.5 (percentage) points. This is the largest (overstatement) we have observed . . . in the last five presidential elections."

Lenski said Wednesday that issuing the report was like "hanging out your dirty underwear. You hope it's cleaner than people expected."

Among the findings:

- They hired too many relatively young adults to conduct the interviews. Half of the 1,400 interviewers were younger than 35. That may explain in part why Kerry voters were more inclined to participate, since he drew more of the youth vote than did Bush. But Mitofsky and Lenski also found younger interviewers were more likely to make mistakes.

- Early results were skewed by a "programming error" that led to including too many female voters. Kerry outpolled Bush among women.

- Some local officials prevented interviewers from getting close to voters.

For future exit polls, Lenski and Mitofsky recommended hiring more experienced polltakers and giving them better training, and working with election officials to ensure access to polling places.

Lenski and Mitofsky noted that none of the media outlets they worked for—ABC, CBS, CNN, Fox News, NBC and the Associated Press—made any wrong "calls" on election night. Representatives of those six are reviewing the report.

Many other news media, including *USA Today*, also paid to get some of the data.

28. Increasing Response Rates Offering rewards or incentives is one way of attempting to increase response rates. Discuss a possible disadvantage of such a practice.

29. Wording Survey Questions Write a survey question that contains strong wording and a survey question that contains tempered wording. Present the strongly worded question to 10 randomly selected people and the tempered question to 10 different randomly selected people. How does the wording affect the response?

30. Order in Survey Questions Write two questions that could have different responses, depending on the order in which the questions are presented. Randomly select 20 people and present the questions in one order to 10 of the people and in the opposite order to the other 10 people. Did the results differ?

31. Research a survey method used by a company or government branch. Determine the sampling method used, the sample size, the method of collection, and the frame used.

32. Out-of-Class Activity People often respond to survey questions without any knowledge of the subject matter. A common example of this is the discussion on banning dihydrogen monoxide. The Centers for Disease Control (CDC) reports that there were 1,493 deaths due to asbestos in 2002, but over 3,200 deaths were attributed to dihydrogen monoxide in 2000. Articles and Web sites, such as www.dhmo.org tell how this substance is widely used despite the dangers associated with it. Many people have joined the cause to ban this substance without realizing that dihydrogen monoxide is simply water (H_2O). Their eagerness to protect the environment or their fear of seeming uninformed may be part of the problem. Put together a survey that asks individuals whether dihydrogen monoxide should or should not be banned. Give the survey to 20 randomly selected students around campus and report your results to the class. An example survey might look like the following:

Dihydrogen monoxide is colorless, odorless, and kills thousands of people every year. Most of these deaths are caused by accidental inhalation, but the dangers of dihydrogen monoxide do not stop there. Prolonged exposure to its solid form can severely damage skin tissue. Symptoms of ingestion can include excessive sweating and urination and possibly a bloated feeling, nausea, vomiting, and body electrolyte imbalance. Dihydrogen monoxide is a major component of acid rain and can cause corrosion after coming in contact with certain metals.

Do you believe that the government should or should not ban the use of dihydrogen monoxide?

1.5 The Design of Experiments

Objectives **①** **Define designed experiment**
② **Understand the steps in designing an experiment**
③ **Understand the completely randomized design**
④ **Understand the matched-pairs design**

A major theme of this chapter has been data collection. Sections 1.2 through 1.4 discussed techniques for obtaining data through surveys. Data obtained from surveys lead to observational studies. Obtaining data through an experiment is discussed in this section.

① ## Define Designed Experiment

When people hear the word *experiment*, they typically think of a laboratory with a controlled environment. Researchers also have control when they perform experiments.

Definition

> A **designed experiment** is a controlled study conducted to determine the effect that varying one or more **explanatory variables** has on a response variable. The explanatory variables are often called **factors**. The **response variable** represents the variable of interest. Control, manipulation, randomization, and replication are the key ingredients of a well-designed experiment.

Historical Note

Sir Ronald Fisher, often called the Father of Modern Statistics, was born in England on February 17, 1890. He received a BA in astronomy from Cambridge University in 1912. In 1914, he took a position teaching mathematics and physics at a high school. He did this to help serve his country during World War I. (He was rejected by the army because of his poor eyesight.) In 1919, Fisher took a job as a statistician at Rothamsted Experimental Station, where he was involved in agricultural research. In 1933, Fisher became Galton Professor of Eugenics at Cambridge University, where he studied Rh blood groups. In 1943 he was appointed to the Balfour Chair of Genetics at Cambridge. He was knighted by Queen Elizabeth in 1952. Fisher retired in 1957 and died in Adelaide, Australia, on July 29, 1962. One of his famous quotations is, "To call in the statistician after the experiment is done may be no more than asking him to perform a postmortem examination: he may be able to say what the experiment died of."

A **treatment** is any combination of the values of each factor. The **experimental unit** is a person, object, or some other well-defined item to which a treatment is applied. We often refer to the experimental unit as a **subject** when he or she is a person. The subject (experimental unit) is analogous to the individual in a survey.

The goal in an experiment is to determine the effect various treatments have on the response variable. For example, suppose a researcher wants to measure the effect of sleep deprivation on a person's fine-motor skills. The researcher might take a group of 100 individuals and randomly divide them into four groups. The first group might sleep 8 hours a night for four nights, the second will sleep 6 hours per night for four nights, the third group will sleep 4 hours per night for four nights, while the fourth group will sleep 2 hours per night for four nights. The experimental unit or subject is the person, the explanatory variable or factor is the amount of sleep, the treatments are 8, 6, 4, and 2 hours, and the response variable might be the reaction time of the person to some stimulus.

Many designed experiments are **double-blind**. This means that neither the experimental unit nor the experimenter knows what treatment is being administered to the experimental unit. For example, in clinical studies of the cholesterol-lowering drug Lipitor, researchers administered either 10, 20, 40, or 80 mg of Lipitor or a **placebo**, an innocuous medication such as a sugar tablet, to patients (the subjects) with high cholesterol. Because the experiment was double-blind, neither the patients nor the researchers knew which medication was being administered. It is important that double-blind methods be used in this case so that the patients and researchers do not behave in such a way as to affect the results. For example, the researcher might not give as much time to a patient receiving the placebo.

2 **Understand the Steps in Designing an Experiment**

The process of conducting an experiment requires a series of steps.

Step 1: *Identify the Problem to Be Solved.* The statement of the problem should be as explicit as possible. The statement should provide the experimenter with direction. In addition, the statement must identify the response variable and the population to be studied. Often, the statement is referred to as the **claim**.

Step 2: *Determine the Factors That Affect the Response Variable.* The factors are usually identified by an expert in the field of study. In identifying the factors, we must ask, "What things affect the value of the response variable?" Once the factors are identified, it must be determined which factors will be fixed at some predetermined level, which will be manipulated, and which will be uncontrolled.

Step 3: *Determine the Number of Experimental Units.* As a general rule, choose as many experimental units as time and money will allow. Techniques do exist for determining sample size, provided certain information is available. Some of these techniques are discussed later in the text.

Step 4: *Determine the Level of Each Factor.* There are three ways to deal with the factors:

1. **Control:** Fix their level at one predetermined value throughout the experiment. These are variables whose effect on the response variable is not of interest.

2. **Manipulate:** Set them at predetermined levels. These are the variables whose effect on the response variable interests us. The combinations of the levels of these variables constitute the treatments in the experiment.

3. **Randomize:** Randomize the experimental units to various treatment groups so that the effect of factors whose levels cannot be controlled is minimized. The idea is that randomization averages out the effects of uncontrolled factors (explanatory variables).

It is difficult, if not impossible, to identify all factors in an experiment. This is why randomization is so important. It mutes the effect of variation attributable to factors not controlled or manipulated.

Step 5: *Conduct the Experiment.*

(a) The experimental units are randomly assigned to the treatments. **Replication** occurs when each treatment is applied to more than one experimental unit. By using more than one experimental unit for each treatment, we can be assured that the effect of a treatment is not due to some characteristic of a single experimental unit. It is a good idea to assign an equal number of experimental units to each treatment.

(b) Collect and process the data. Measure the value of the response variable for each replication. Then organize the results. The idea is that the value of the response variable for each treatment group is the same before the experiment because of randomization. Then any difference in the value of the response variable among the different treatment groups can be attributed to differences in the level of the treatment.

Step 6: *Test the Claim.* This is the subject of inferential statistics. **Inferential statistics** is a process in which generalizations about a population are made on the basis of results obtained from a sample. In addition, a statement regarding our level of confidence in our generalization is provided. We study methods of inferential statistics in Chapters 9 through 12.

③ **Understand the Completely Randomized Design**

The steps just given apply to any type of designed experiment. We now concentrate on the simplest type of experiment.

Definition   A **completely randomized design** is one in which each experimental unit is randomly assigned to a treatment.

An example will help clarify the process of experimental design.

EXAMPLE 1   **A Completely Randomized Design**

Problem: A farmer wishes to determine the optimal level of a new fertilizer on his soybean crop. Design an experiment that will assist him.

Approach: We follow the steps for designing an experiment.

Solution

Step 1: The farmer wants to identify the optimal level of fertilizer for growing soybeans. We define *optimal* as the level that maximizes yield. So the response variable will be crop yield.

Step 2: Some factors that affect crop yield are fertilizer, precipitation, sunlight, method of tilling the soil, type of soil, plant, and temperature.

Step 3: In this experiment, we will plant 60 soybean plants (experimental units).

Step 4: We list the factors and their levels.

- **Fertilizer.** We manipulate the level of this factor. We wish to measure the effect of varying the level of this variable on the response variable, yield. We will set the level of fertilizer (the treatments) as follows:

    Treatment A: 20 soybean plants receive no fertilizer.

    Treatment B: 20 soybean plants receive 2 teaspoons of fertilizer per gallon of water every 2 weeks.

    Treatment C: 20 soybean plants receive 4 teaspoons of fertilizer per gallon of water every 2 weeks.

See Figure 5.

Figure 5

Treatment A   Treatment B   Treatment C

- **Precipitation.** Although we cannot control the amount of rainfall, we can control the amount of watering we do. This factor will be controlled so that each plant receives the same amount of precipitation.
- **Sunlight.** This is an uncontrollable factor, but it will be roughly the same for each plant.
- **Method of tilling.** We can control this factor. We agree to use the round-up ready method of tilling for each plant.
- **Type of soil.** We can control certain aspects of the soil such as level of acidity. In addition, each plant will be planted within a 1 acre area, so it is reasonable to assume that the soil conditions for each plant are equivalent.

- **Plant.** There may be variation from plant to plant. To account for this, we randomly assign the plants to a treatment.
- **Temperature.** This factor is not within our control, but will be the same for each plant.

Step 5:

(a) We need to assign each plant to a treatment group. To do this, we will number the plants from 1 to 60. To determine which plants get treatment A, we randomly generate 20 numbers. The plants corresponding to these numbers get treatment A. Now number the remaining plants 1 to 40 and randomly generate 20 numbers. The plants corresponding to these numbers get treatment B. The remaining plants get treatment C. Now till the soil, plant the soybean plants, and fertilize according to the schedule prescribed.

(b) At the end of the growing season, determine the crop yield for each plant.

Step 6: Determine whether any differences in yield exist among the three treatment groups.

Figure 6 illustrates the experimental design.

Figure 6

Note to Instructor
It is useful to diagram the experimental designs in this chapter to target the understanding of the visual learner.

Example 1 is a completely randomized design because the experimental units (the plants) were randomly assigned the treatment. It is the most popular experimental design because of its simplicity, but it is not always the best. We discuss inferential procedures for the completely randomized design in which there are two treatments in Section 11.2 and in which there are three or more treatments in Section C.4 on the CD.

Now Work Problem 5.

4 Understand the Matched-Pairs Design

Another type of experimental design is called a *matched-pairs design*.

Definition

> A **matched-pairs design** is an experimental design in which the experimental units are paired up. The pairs are matched up so that they are somehow related (that is, the same person before and after a treatment, twins, husband and wife, same geographical location, and so on). There are only two levels of treatment in a matched-pairs design.

In matched-pairs design, one matched individual will receive one treatment and the other matched individual receives a different treatment. The assignment of the matched pair to the treatment is done randomly using a coin flip or a random-number generator. We then look at the difference in the results of each matched pair. One common type of matched-pairs design is to measure a response variable on an experimental unit before a treatment is applied, and then measure the response variable on the same experimental unit after the treatment is applied. In this way, the individual is matched against itself. These experiments are sometimes called before–after or pretest–posttest experiments.

EXAMPLE 2

A Matched-Pairs Design

Problem: An educational psychologist wanted to determine whether listening to music has an impact on a student's ability to learn. Design an experiment to help the psychologist answer the question.

Approach: We will use a matched-pairs design by matching students according to IQ and gender (just in case gender plays a role in learning with music).

Solution: We match students according to IQ and gender. For example, a female with an IQ in the 110 to 115 range will be matched with a second female with an IQ in the 110 to 115 range.

For each pair of students, we will flip a coin to determine whether the first student in the pair is assigned the treatment of a quiet room or a room with music playing in the background.

Each student will be given a statistics textbook and asked to study Section 1.1. After 2 hours, the students will enter a testing center and take a short quiz on the material in the section. We compute the difference in the scores of each matched pair. Any differences in scores will be attributed to the treatment. Figure 7 illustrates the design.

Figure 7

Match students according to gender and IQ. → Randomly assign a student from each pair to a treatment. → Administer treatment and exam to each matched pair. → For each matched pair, compute the difference in scores on the exam.

We discuss the statistical inference of matched-pairs design in Section 11.1.

Now Work Problem 7.

One note about the relation between a designed experiment and simple random sampling: It is often the case that the experimental units selected to participate in a study are not randomly selected. This is because we often need the experimental units to have some common trait, such as high blood pressure. For this reason, participants in experiments are recruited or volunteer to be in a study. However, once we have the experimental units, we use simple random sampling to assign them to treatment groups. With random assignment we assume that the participants are similar at the start of the experiment. Because the treatment is the only difference between the groups, we can say the treatment *caused* the difference observed in the response variable.

Note to Instructor

The purpose of this activity is to allow students to see why good experiments need control, manipulation, randomization, and replication. In addition, students recognize how much thought needs to go into an experiment, even for something simple like paper frogs.

There are several possible designs for the frogs (e.g., http://www.frogsonice.com/ froggy/origami/index.s/html or http://origami.ousaan.com/studio/ traditional/frog/01/index.html), so choose whichever seems easiest for students to use.

Before the jumping, there should be discussions about why the order of the jumps is randomized, why the same person should jump all the frogs, and so on. In addition, students should decide how they will measure the distance jumped before they start jumping.

A good design might be a matched-pairs design so that the variability due to who is making the frog is eliminated and type of paper is not confounded with person making the frog.

This activity may be done with paper airplanes as well.

In-Class Activity: Experimental Design (Hippity-Hop)

You are commissioned by the board of directors of Paper Toys, Inc. to design a new paper frog for their Christmas catalog. The design for the construction of the frog has already been completed and will be provided to you. However, the material with which to make the frogs has not yet been determined. The Materials Department has narrowed the choices down to either newspaper or brown paper (such as that used in grocery bags). You have decided to test both types of paper. Management decided to build the frogs from sheets of paper 9 inches square.

The goal of the experiment is to determine the material that results in frogs that jump farther.

(a) As a class, design an experiment that will answer the research question.

(b) Make the frogs.

(c) Conduct the experiment.

(d) As a class, discuss the strengths and weaknesses of the design. Would you change anything?

1.5 ASSESS YOUR UNDERSTANDING

Concepts and Vocabulary

1. Define the following:
   (a) Experimental unit
   (b) Treatment
   (c) Response variable
   (d) Factor
   (e) Double-blind
   (f) Placebo

3. *completely randomized; matched pair*

2. What is replication in an experiment?
3. A(n) _____ _____ design is one in which each experimental unit is randomly assigned to a treatment. A(n) _____ _____ design is one in which the experimental units are paired up.
4. Discuss how random sampling is used in the design of an experiment.

(5a) Score on achievement test   (5b) Method of teaching, grade level, intelligence, school district, teacher; Controlled: grade level, school district, teacher; Manipulated: method of teaching   (5e) Completely randomized design   (6a) Proportion of subjects with cold

Applying the Concepts

5. **School Psychology** A school psychologist wants to test the effectiveness of a new method for teaching reading. She recruits 500 first-grade students in District 203 and randomly divides them into two groups. Group 1 is taught by means of the new method, while group 2 is taught via traditional methods. The same teacher is assigned to teach both groups. At the end of the year, an achievement test is administered and the results of the two groups compared.
   (a) What is the response variable in this experiment?
   (b) Think of some of the factors in the study. Which are controlled? Which factor is manipulated?
   (c) What are the treatments? How many treatments are there? *New teaching method and traditional method; 2*
   (d) How are the factors that are not controlled or manipulated dealt with? *Random assignment*
   (e) What type of experimental design is this?
   (f) Identify the subjects. *500 students*
   (g) Draw a diagram similar to Figure 6 or 7 to illustrate the design.

6. **Pharmacy** A pharmaceutical company has developed an experimental drug meant to relieve symptoms associated with the common cold. The company identifies 300 adult males 25 to 29 years old who have a common cold and randomly divides them into two groups. Group 1 is given the experimental drug, while group 2 is given a placebo. After 1 week of treatment, the proportions of each group that still have cold symptoms are compared.
   (a) What is the response variable in this experiment?
   (b) Think of some of the factors in the study. Which are controlled? Which factor is manipulated?
   (c) What are the treatments? How many treatments are there? *Experimental drug and placebo; 2*
   (d) How are the factors that are not controlled or manipulated dealt with? *Random assignment*
   (e) What type of experimental design is this?
   (f) Identify the subjects. *300 males*
   (g) Draw a diagram similar to Figure 6 or 7 to illustrate the design.   *(6e) Completely randomized design*

7. **Assessment** To help assess student learning in her developmental math courses, a mathematics professor at a community college implemented pre- and posttests for her developmental math students. A knowledge-gained score was obtained by taking the difference of the two test scores.   *(7a) Matched pair   (7b) Difference in test scores*
   (a) What type of experimental design is this?
   (b) What is the response variable in this experiment?
   (c) What is the treatment? *Math course*

(8b) Whiteness level

8. **Whiter Teeth** An ad for Crest Whitestrips Premium™ claims that the strips will whiten teeth in 7 days and the results will last for 12 months. A researcher who wishes to test this claim studies 20 sets of identical twins. Within each set of twins, one is randomly selected to use Crest Whitestrips Premium in addition to regular brushing and flossing, while the other just brushes and flosses. Whiteness of teeth is measured at the beginning of the study, after 7 days, and every month thereafter for 12 months.
   (a) What type of experimental design is this? *Matched pair*
   (b) What is the response variable in this experiment?
   (c) What factor is manipulated? What are the treatments?
   (d) What is another factor (controlled or uncontrolled) that could affect the response variable?
   (e) What might be an advantage of using identical twins as subjects in this experiment?

9. **Insomnia** Researchers Jack D. Edinger and associates wanted to test the effectiveness of a new cognitive behavioral therapy (CBT) compared with both an older behavioral treatment and a placebo therapy for treating insomnia. They identified 75 adults with chronic insomnia. Patients were randomly assigned to one of three treatment groups. Twenty-five patients were randomly assigned to receive CBT (sleep education, stimulus control, and time-in-bed restrictions), another 25 received muscle relaxation training (RT), and the final 25 received a placebo treatment. Treatment lasted 6 weeks, with follow-up conducted at 6 months. To measure the effectiveness of the treatment, researchers used wake time after sleep onset (WASO). Cognitive behavioral therapy produced larger improvements than did RT or placebo treatment. For example, the CBT-treated patients achieved an average 54% reduction in their WASO whereas RT-treated and placebo-treated patients, respectively, achieved only 16% and 12% reductions in this measure. Results suggest that CBT treatment leads to significant sleep improvements within 6 weeks, and these improvements appear to endure through 6 months of follow-up.
   (*Source*: "Cognitive Behavioral Therapy for Treatment of Chronic Primary Insomnia," Jack D. Edinger, PhD; William K. Wohlgemuth, PhD; Rodney A. Radtke, MD; Gail R. Marsh, PhD; Ruth E. Quillian, PhD; *Journal of the American Medical Association* 285(2001): 1856–1864)

(6b) Gender, age, geographic location, overall health, drug intervention; Controlled: gender, age, location; Manipulated: drug intervention

(8c) Whitening method; Crest Whitestrips Premium with brushing and brushing alone

(a) What type of experimental design is this?
(b) What is the population being studied?
(c) What is the response variable in this study?
(d) What is the factor? What are the treatments?
(e) Identify the experimental units. 75 adults with insomnia
(f) Draw a diagram similar to Figure 6 or 7 to illustrate the design.

10. **Depression** Researchers wanted to compare the effectiveness and safety of an extract of St. John's wort with placebo in outpatients with major depression. To do this, they recruited 200 adult outpatients diagnosed as having major depression and having a baseline Hamilton Rating Scale for Depression (HAM-D) score of at least 20. Participants were randomly assigned to receive either St. John's wort extract (900 mg/d for 4 weeks, increased to 1200 mg/d in the absence of an adequate response thereafter) or a placebo for 8 weeks. The response variable was the change on the HAM-D over the treatment period. After analysis of the data, it was concluded that St. John's wort was not effective for treatment of major depression.

    (*Source*: "Effectiveness of St. John's Wort in Major Depression," Richard C. Shelton, MD; et. al.; *Journal of the American Medical Association* 285 (2001):1978–1986)

    (a) What type of experimental design is this?
    (b) What is the population that is being studied?
    (c) What is the response variable in this study?
    (d) What is the factor? What are the treatments?
    (e) Identify the experimental units. 200 outpatients
    (f) Draw a diagram similar to Figure 6 or 7 to illustrate the design.

11. **The Memory Drug?** Researchers wanted to evaluate whether ginkgo, an over-the-counter herb marketed as enhancing memory, improves memory in elderly adults as measured by objective tests. To do this, they recruited 98 men and 132 women older than 60 years and in good health. Participants were randomly assigned to receive ginkgo, 40 mg 3 times per day, or a matching placebo. The measure of memory improvement was determined by a standardized test of learning and memory. After 6 weeks of treatment, the data indicated that ginkgo did not increase performance on standard tests of learning, memory, attention, and concentration. These data suggest that, when taken following the manufacturer's instructions, ginkgo provides no measurable increase in memory or related cognitive function to adults with healthy cognitive function.

    (*Source*: "Ginkgo for Memory Enhancement," Paul R. Solomon, et. al.; *Journal of the American Medical Association* 288(2002): 835–840)

    (a) What type of experimental design is this?
    (b) What is the population being studied?
    (c) What is the response variable in this study?
    (d) What is the factor? What are the treatments?

(e) Identify the experimental units.
(f) Draw a diagram similar to Figure 6 or 7 to illustrate the design.

12. **Treating Depression** Researchers wanted to test whether a new drug therapy results in a more rapid response in patients with major depression. To do this, they recruited 63 inpatients with a diagnosis of major depression. Patients were randomly assigned to two treatment groups receiving either placebo (31 patients) or the new drug therapy (32 patients). The response variable was the Hamilton Rating Scale for Depression score. After collecting and analyzing the data, it was concluded that the new drug therapy is effective in the treatment of major depression.

    (*Source*: "Metyrapone as Additive Treatment in Major Depression," Holger Jahn, MD; et. al.; *Archives of General Psychiatry*, 61(2004): 1235–1244)

    (a) What type of experimental design is this?
    (b) What is the population that is being studied?
    (c) What is the response variable in this study?
    (d) What is the factor? What are the treatments?
    (e) Identify the experimental units.
    (f) Draw a diagram similar to Figure 6 or 7 to illustrate the design.

13. **Dominant Hand** Professor Andy Neill wanted to determine if the reaction time in people differs in their dominant hand versus their nondominant hand. To do this, he recruited 15 students. Each student was asked to hold a yardstick between the index finger and thumb. The student was asked to open the hand, release the yardstick, and then asked to catch the yardstick between the index finger and thumb. The distance that the yardstick fell served as a measure of reaction time. A coin flip was used to determine whether the student would use their dominant hand first or the nondominant hand. Results indicated that the reaction time in the dominant hand exceeded that of the nondominant hand.

    (a) What type of experimental design is this?
    (b) What is the response variable in this study?
    (c) What is the factor? What is the treatment?
    (d) Identify the experimental units. 15 students
    (e) Why did Professor Neill use a coin flip to determine whether the student should begin with the dominant hand or the nondominant hand?
    (f) Draw a diagram similar to Figure 6 or 7 to illustrate the design.

14. **Golf Anyone?** A local golf pro wanted to compare two styles of golf club. One golf club had a graphite shaft and the other had the latest style of steel shaft. It is a common belief that graphite shafts allow a player to hit the ball farther, but the manufacturer of the new steel shaft said the ball travels just as far with its new technology. To test this claim, the pro recruited 10 golfers from the driving range. Each player was asked to hit one ball with the graphite-shafted club and one ball with the new steel-shafted club.

(9d) Type of intervention: CBT, progressive muscle relaxation training, placebo   (10a) Completely randomized design   (10b) Adult outpatients diagnosed as having major depression and having a baseline Hamilton Rating Scale for Depression (HAM-D) score of at least 20   (10c) Change in the HAM-D over the treatment period   (11a) Completely randomized design   (11b) Adults older than 60 years in good health   (11c) Standardized test of learning and memory   (11d) Drug; 40 mg of ginkgo 3 times per day or a matching placebo   (11e) 98 men and 132 women older than 60 in good health   (12a) Completely randomized design   (12b) Inpatients with a diagnosis of major depression   (12c) Hamilton Rating Scale for Depression score   (12e) 63 inpatients with a diagnosis of major depression

(13a) Matched-pairs design   (13b) Distance yardstick falls   (13c) Hand dominance; dominant versus nondominant hand

The distance that the ball traveled was determined using a range finder. A coin flip was used to determine whether the player hit with the graphite club or the steel club first. Results indicated that the distance the ball was hit with the graphite club was no different than the distance when using the steel club.

(a) What type of experimental design is this?
(b) What is the response variable in this study?
(c) What is the factor? What is the treatment?
(d) Identify the experimental units. *10 golfers*
(e) Why did the golf pro use a coin flip to determine whether the golfer should hit with the graphite first or the steel first?
(f) Draw a diagram similar to Figure 6 or 7 to illustrate the design.

15. **Drug Effectiveness** A pharmaceutical company wants to test the effectiveness of an experimental drug meant to reduce high cholesterol. The researcher at the pharmaceutical company has decided to test the effectiveness of the drug through a completely randomized design. She has obtained 20 volunteers with high cholesterol: Ann, John, Michael, Kevin, Marissa, Christina, Eddie, Shannon, Julia, Randy, Sue, Tom, Wanda, Roger, Laurie, Rick, Kim, Joe, Colleen, and Bill. Number the volunteers from 1 to 20. Use a random-number generator to randomly assign 10 of the volunteers to the experimental group. The remaining volunteers will go into the control group. List the individuals in each group.

16. **Effects of Alcohol** A researcher has recruited 20 volunteers to participate in a study. The researcher wishes to measure the effect of alcohol on an individual's reaction time. The 20 volunteers are randomly divided into two groups. Group 1 will serve as a control group in which participants drink four 1-ounce cups of a liquid that looks, smells, and tastes like alcohol in 15-minute increments. Group 2 will serve as an experimental group in which participants drink four 1-ounce cups of 80-proof alcohol in 15-minute increments. After drinking the last 1-ounce cup, the participants sit for 20 minutes. After the 20-minute resting period, the reaction time to a stimulus is measured.

(a) What type of experimental design is this?
(b) Use Table I in Appendix A or a random-number generator to divide the 20 volunteers into groups 1 and 2 by assigning the volunteers a number between 1 and 20. Then randomly select 10 numbers between 1 and 20. The individuals corresponding to these numbers will go into group 1.

17. **Tomatoes** An oncologist wants to perform a long-term study on the benefits of eating tomatoes. In particular, she wishes to determine whether there is a significant difference in the rate of prostate cancer among adult males after eating one serving of tomatoes per week for 5 years, eating three servings of tomatoes per week for 5 years, and after eating five servings of tomatoes per week for 5 years. Help the oncologist design the experiment. Include a diagram to illustrate your design. *Completely randomized design*

18.   Completely randomized design
18. **Batteries** An engineer wants to determine the effect of temperature on battery voltage. In particular, he is interested in determining if there is a significant difference in the voltage of the batteries when exposed to temperatures of 90°F, 70°F, and 50°F. Help the engineer design the experiment. Include a diagram to illustrate your design.

19. **The Better Paint** Suppose you are interested in comparing Benjamin Moore's MoorLife Latex house paint with Sherwin Williams' LowTemp 35 Exterior Latex paint. Design an experiment that will answer this question: Which paint is better for painting the exterior of a house? Include a diagram to illustrate your design. *Matched-pair design (match by type of exterior finish)*

20. **Tire Design** An engineer has just developed a new tire design. However, before going into production, the tire company wants to determine if the new tire reduces braking distance on a car traveling 60 miles per hour compared with radial tires. Design an experiment to help the engineer determine if the new tire reduces braking distance. *Matched-pair design (match by car model)*

21. **Designing an Experiment** Researchers wish to know if there is a link between hypertension (high blood pressure) and consumption of salt. Past studies have indicated that the consumption of fruits and vegetables offsets the negative impact of salt consumption. It is also known that there is quite a bit of person-to-person variability as far as the ability of the body to process and eliminate salt. However, no method exists for identifying individuals who have a higher ability to process salt. The U.S. Department of Agriculture recommends that daily intake of salt should not exceed 2400 mg. The researchers want to keep the design simple, so they choose to conduct their study using a completely randomized design.

(a) What is the response variable in the study?
(b) Name three factors that have been identified.
(c) For each factor identified, determine whether the variable can be controlled, cannot be controlled, or should be manipulated. If a factor cannot be controlled, what should be done to reduce variability in the response variable?
(d) How many treatments would you recommend? Why?

22. Search a newspaper, magazine, or other periodical that describes an experiment. Identify the population, experimental unit, response variable, treatment, factors, and their levels.

23. Research the *placebo effect* and the *Hawthorne effect*. Write a paragraph that describes how each affects the outcome of an experiment.

(13e) To eliminate bias due to starting on dominant or nondominant first for each trial   (14a) Matched-pairs design   (14b) Distance ball hit   (14c) Shaft type; graphite shaft versus steel shaft   (14e) To eliminate bias due to type of shaft hit first   (16a) Completely randomized   (21a) Blood pressure   (21b) Daily consumption of salt, daily consumption of fruits and vegetables, body's ability to process salt   (21c) Salt: manipulate; Fruits and vegetables: control; Body: cannot be controlled. To deal with variability in body types, randomize experimental units to each treatment group.   (21d) Answers will vary, but three might be a good choice: one below RDA, one equal to RDA, one above RDA.

Consumer Reports®   Emotional "Aspirin"

Americans have a long history of altering their moods with chemicals ranging from alcohol and illicit drugs to prescription medications, such as diazepam (Valium) for anxiety and fluoxetine (Prozac) for depression. Today, there's a new trend: the over-the-counter availability of apparently effective mood modifiers in the form of herbs and other dietary supplements.

One problem is that many people who are treating themselves with these remedies may be sufficiently anxious or depressed to require professional care and monitoring. Self-treatment can be dangerous, particularly with depression, which causes some 20,000 reported suicides a year in the United States. Another major pitfall is that dietary supplements are largely unregulated by the government, so consumers have almost no protection against substandard preparations.

To help consumers and doctors, *Consumer Reports* tested the amounts of key ingredients in representative brands of several major mood-changing pills. To avoid potential bias, we tested samples from different lots of the pills using a randomized statistical design. The following table contains a subset of the data from this study.

Each of these pills has a label claim of 200 mg of SAM-E. The column labeled Random Code contains a set of 3-digit random codes that were used so that the laboratory did not know which manufacturer was being tested. The column labeled Mg SAM-E contains the amount of SAM-E measured by the laboratory.

(a) Why is it important to label the pills with random codes?

(b) Why is it important to randomize the order in which the pills are tested instead of testing all of brand A first, followed by all of brand B, and so on?

(c) Sort the data by brand. Does it appear that each brand is meeting its label claims?

(d) Design a study that follows the steps presented to answer the following research question: "Is there a difference in the amount of SAM-E contained in brands A, B, C, and D?"

Note to Readers: In many cases, our test protocol and analytical methods are more complicated than described in this example. The data and discussion have been modified to make the material more appropriate for the audience.

Run Order	Brand	Random Code	Mg SAM-E
1	B	461	238.9
2	D	992	219.2
3	C	962	227.1
4	A	305	231.2
5	B	835	263.7
6	D	717	251.1
7	A	206	232.9
8	D	649	192.8
9	C	132	213.4
10	B	923	224.6
11	A	823	261.1
12	C	515	207.8

© 2002 by Consumers Union of U.S., Inc., Yonkers, NY 10703-1057, a nonprofit organization. Reprinted with permission from the Dec. 2002 issue of CONSUMER REORTS® for educational purposes only. No commercial use or photocopying permitted. To learn more about Consumers Union, log onto www.ConsumersReports.org.

CHAPTER 1   # Review

Summary

We defined statistics as a science in which data are collected, organized, summarized, and analyzed to infer characteristics regarding a population. Descriptive statistics consists of organizing and summarizing information, while inferential statistics consists of drawing conclusions about a population based on results obtained from a sample. The population is a collection of individuals on which the study is made, and the sample is a subset of the population.

Data are the observations of a variable. Data can be either qualitative or quantitative. Quantitative data are either discrete or continuous.

Data can be obtained from four sources: a census, existing sources, survey sampling, or a designed experiment. A census will list all the individuals in the population, along with certain characteristics. Due to the cost of obtaining a census, most researchers opt for obtaining a sample. In observational studies, the variable of interest has already been established. For this reason, they are often referred to as *ex post facto* studies. Designed experiments are used when control of the individuals in the study is desired to isolate the effect of a certain treatment on a response variable.

We introduced five sampling methods: simple random sampling, stratified sampling, systematic sampling, cluster sampling, and convenience sampling. All the sampling methods, except for convenience sampling, allow for unbiased statistical inference to be made. Convenience sampling typically leads to an unrepresentative sample and biased results.

Vocabulary

Be sure you can define the following:

Statistics (p. 3)
Data (pp. 3, 8)
Population (p. 4)
Individual (p. 4)
Sample (p. 4)
Descriptive statistics (p. 4)
Inferential statistics (p. 5, 41)
Placebo (pp. 5, 40)
Double-blind (pp. 5, 40)
Variable (p. 6)
Qualitative variable (p. 6)
Quantitative variable (p. 6)
Discrete variable (p. 7)
Continuous variable (p. 7)
Qualitative data (p. 8)

Quantitative data (p. 8)
Discrete data (p. 8)
Continuous data (p. 8)
Census (p. 13)
Observational study (p. 14)
Designed experiment (pp. 14, 40)
Lurking variable (p. 15)
Simple random sampling (p. 16)
Frame (p. 16)
Sampling without replacement (p. 17)
Seed (p. 18)
Stratified sample (p. 23)
Systematic sample (p. 25)
Cluster sample (p. 26)
Convenience sample (p. 27)

Nonsampling error (p. 33)
Sampling error (p. 33)
Underrepresented (p. 34)
Nonresponse (p. 34)
Open question (p. 35)
Closed question (p. 35)
Response variable (p. 40)
Experimental unit (p. 40)
Explanatory variable (p. 40)
Factors (p. 40)
Treatment (p. 40)
Claim (p. 41)
Replication (p. 41)
Completely randomized design (p. 42)
Matched-pairs design (p. 43)

Objectives

Section	You should be able to . . .	Example	Review Exercises
1.1	1 Define statistics and statistical thinking (p. 3)		1
	2 Understand the process of statistics (p. 4)	1	6
	3 Distinguish between qualitative and quantitative variables (p. 6)	2	9–14
	4 Distinguish between discrete and continuous variables (p. 7)	3, 4	9–14
1.2	1 Distinguish between an observational study and an experiment (p. 14)	1	15–18
	2 Obtain a simple random sample (p. 16)	2, 3	25, 29, 34
1.3	1 Obtain a stratified sample (p. 23)	1	26
	2 Obtain a systematic sample (p. 25)	2	27
	3 Obtain a cluster sample (p. 26)	3	28
1.4	1 Understand how error can be introduced during sampling (p. 33)		7
1.5	1 Define designed experiment (p. 40)		5
	2 Understand the steps in designing an experiment (p. 41)		8, 36, 38
	3 Understand the completely randomized design (p. 42)	1	31, 32, 36, 37, 38(a)
	4 Understand the matched-pairs design (p. 43)	2	30, 33, 37, 38(b)

Review Exercises

In Problems 1–5, provide a definition using your own words.

1. Statistics

2. Population

3. Sample

4. Observational study

5. Designed experiment

6. What is meant by the *process of statistics*?

7. State some sources of error in sampling. Provide some methods for correcting these errors. Distinguish sampling and nonsampling error.

8. Describe the components in an appropriately designed experiment.

In Problems 9–14, classify the variable as qualitative or quantitative. If the variable is quantitative, state whether it is discrete or continuous.

9. Number of new automobiles sold at a dealership on a given day Quantitative; discrete

10. Species of tree sold at a nursery Qualitative

11. Weight in carats of an uncut diamond Quantitative; continuous

12. Wait time for an amusement park ride Quantitative; continuous

13. Brand name of a pair of running shoes Qualitative

14. Frequency with which the variable x is used in an algebra book Quantitative; discrete

In Problems 15–18, determine whether the study depicts an observational study or a designed experiment.

15. A parent group examines 25 recently released PG-13 movies and records the number of sexual innuendos and curse words that occur in each. Observational study

16. A sample of 10 new Dodge Neons is randomly divided into two groups. In the first group of cars, SAE 5W-30 motor oil is used. In the second group, SAE 10W-30 motor oil is used. All other variables are controlled to be the same in both groups. Each car is driven 3,000 highway miles. The gasoline mileages of the two groups are then compared. Experiment

17. A sample of 504 patients in early stages of Alzheimer's disease is divided into two groups. One group receives an experimental drug regimen while the other receives a placebo regimen. The advance of the disease in the patients from the two groups is tracked at 1-month intervals over the next year. Experiment

18. A sample of 1,000 registered voters is asked, "Do you approve or disapprove of the president's overall job performance?" Observational study

In Problems 19–22, determine the type of sampling used.

19. On election day, a pollster for Fox News positions herself outside a polling place near her home. She then asks the first 50 voters leaving the facility to complete a survey. Convenience sample

20. The manager of an automobile repair shop compiles a list of all its customers from the past 6 months. He assigns a unique number to each customer and uses a random-number generator to create a list of 40 unique numbers. The customers corresponding to the number are surveyed about satisfaction with services. Simple random sample

21. An Internet service provider randomly selects 15 residential blocks from a large city. It then surveys every household in those 15 blocks to determine the number that would use a high-speed Internet service if it were made available. Cluster sample

22. A hotel manager divides the population of customers into four groups depending on length of stay: 1 night, 2 to 3 nights, 4 to 5 nights, and 6 or more nights. She obtains a simple random sample from each group and conducts a survey on satisfaction. Stratified sample

(23a) Undercoverage or nonrepresentative sample due to a poor sampling frame   (24b) Nonresponse

23. For each error given, identify the nonsampling error.
    (a) A politician sends a survey about tax issues to a random sample of subscribers to a literary magazine.
    (b) An interviewer with little foreign language knowledge is sent to an area where her language is not commonly spoken. Interviewer error
    (c) A data-entry clerk mistypes survey results into his computer. Data checks

24. For each error given, identify the nonsampling error.
    (a) A questionnaire given to students asks, "Are you doing well in school?" Question wording
    (b) Of 1,240 surveys sent out in a study, 17 are returned.
    (c) A warehouse manager doing a study on job commitment surveys the first 20 workers who come in on Monday morning. Undercoverage or nonrepresentative sample due to convenience sampling

25. **Obtaining a Simple Random Sample** The mayor of a small town wants to conduct personal interviews with small business owners to determine if there is anything the mayor could to do to help improve business conditions. The following list gives the names of the companies in the town. Obtain a simple random sample of size 5 from the companies in the town.

Allied Tube and Conduit	Lighthouse Financial	Senese's Winery
Bechstien Construction Co.	Mill Creek Animal Clinic	Skyline Laboratory
Cizer Trucking Co.	Nancy's Flowers	Solus, Maria, DDS
D & M Welding	Norm's Jewelry	Trust Lock and Key
Grace Cleaning Service	Papoose Children's Center	Ultimate Carpet
Jiffy Lube	Plaza Inn Motel	Waterfront Tavern
Levin, Thomas, MD	Risky Business Security	WPA Pharmacy

26. Obtaining a Stratified Sample A congresswoman wants to survey her constituency regarding public policy. She asks one of her staff members to obtain a sample of residents of the district. The frame she has available lists 9,012 Democrats, 8,302 Republicans, and 3,012 Independents. Obtain a stratified random sample of 8 Democrats, 7 Republicans, and 3 Independents.

27. Obtaining a Systematic Sample A quality-control engineer wants to be sure that bolts coming off an assembly line are within prescribed tolerances. He wants to conduct a systematic sample by selecting every 9th bolt to come off the assembly line. The machine produces 30,000 bolts per day, and the engineer wants a sample of 32 bolts. Which bolts will be sampled?

28. Obtaining a Cluster Sample A farmer has a 500-acre orchard in Florida. Each acre is subdivided into blocks of 5. Altogether, there are 2500 blocks of trees on the farm. After a frost, he wants to get an idea of the extent of the damage. Obtain a sample of 10 blocks of trees using a cluster sample.

29. Obtaining a Simple Random Sample Based on the Military Standard 105E (ANS1/ASQC Z1.4, ISO 2859) Tables, a lot of 91 to 150 items with an acceptable quality level (AQL) of 1% and normal inspection plan would require a sample of size 13 to be inspected for defects. If the sample contains no defects, the entire lot is accepted. Otherwise, the entire lot is rejected. A shipment of 100 night vision goggles is received and must be inspected. Discuss the procedure you would follow to obtain a simple random sample of 13 goggles to inspect.

30. Effects of Music To study the effects of music on sleep quality in adults 18 years of age or older, 100 adults were given the Pittsburgh Sleep Quality Index (PSQI) and then asked to listen to 30 minutes of music at bedtime for 3 weeks. At the end of the 3-week study, subjects were given the PSQI again and the change in score was reported.
(a) What type of experimental design in this?
(b) What is the population that is being studied?
(c) What is the factor? *Music at bedtime*
(d) What is the response variable? *Change in PSQI score*
(e) Identify the experimental units. *100 adults*
(f) Draw a diagram similar to Figure 6 or 7 to illustrate the design.

31. Integrated Circuits An integrated circuit manufacturer has developed a new circuit for solid-state ballasts that will be more energy efficient than the circuit they currently manufacture. The company manufactures 100 of each type of circuit and randomly assigns them to be used in the ballast of fluorescent bulbs. The amount of energy required to light each bulb is measured.
(a) What type of experimental design is this?
(b) What is the response variable in this experiment?

(c) What is the factor? What are the treatments?
(d) Identify the experimental units. *200 fluorescent bulbs*
(e) Draw a diagram similar to Figure 6 or 7 to illustrate the design.

32. Atopic Dermatitis In September 2004, Nucryst Pharmaceuticals, Inc., announced the results of its first human trial of NPI 32101, a topical form of its skin ointment. A total of 224 patients diagnosed with skin irritations were randomly divided into three groups as part of a double-blind, placebo-controlled study to test the effectiveness of the new topical cream. The first group received a 0.5% cream, the second group received a 1.0% cream, and the third group received a placebo. Groups were treated twice daily for a 6-week period.
(*Source*: www.nucryst.com)
(a) What type of experimental design is this?
(b) What is the response variable in this experiment?
(c) What is the factor? What are the treatments?
(d) Why do you think it is important for this study to be double-blind?
(e) Identify the experimental units.
(f) Draw a diagram similar to Figure 6 or 7 to illustrate the design.

33. Skylab The four members of Skylab had their lymphocyte count per cubic millimeter measured 1 day before liftoff and measured again on their return to Earth.
(a) What is the response variable in this experiment?
(b) What is the treatment? *Space flight*
(c) What type of experimental design is this? *Matched pair*
(d) Identify the experimental units. *4 members*
(e) Draw a diagram similar to Figure 6 or 7 to illustrate the design.

34. Multiple Choice A common tip for taking multiple-choice tests is to always pick (b) or (c) if you are unsure. The idea is that instructors tend to feel the answer is more hidden if it is surrounded by distractor answers. An astute statistics instructor is aware of this and decides to use a table of random digits to select which choice will be the correct answer. If each question has five choices, use Table I in Appendix A or a random-number generator to determine the correct answers for a 20-question multiple-choice exam.

35. Do Not Call What percent of students on campus have their phone number (either home or cell) on the National Do Not Call Registry? Polling every student on campus is inefficient and would require a fair amount of resources. Design and implement a sampling method to obtain a sample of 50 students at your college or university. Ask each student whether he or she has registered a phone with the National Do Not Call Registry, and use the results to estimate the percentage of all students on campus who have registered with the Registry. Compare your sampling method and results with others in the class. Discuss similarities and differences, as well as the advantages or disadvantages, of your approach.

(30a) Matched pair   (30b) Adults 18 years of age or older   (31a) Completely randomized design   (31b) Energy required to light the bulb   (31c) Integrated circuit; old circuit and new circuit   (32a) Completely randomized design   (32b) Condition (dermatitis) improvement   (32c) Topical cream; 0.5%, 1.0%, 0%   (32e) 224 patients with skin irritations

36. Humor in Advertising A marketing research firm wants to know whether information presented in a commercial is better recalled when presented using humor or serious commentary in adults between 18 and 35 years of age. They will use an exam that asks questions about information presented in the ad. The response variable will be percentage of information recalled. Create a completely randomized design to answer the question. Be sure to include a diagram to illustrate your design.

37. Describe what is meant by a matched-pairs design. Contrast this experimental design with a completely randomized design.

38. The Better Design? Suppose a psychologist wants to determine whether Monster Energy Drinks improve performance in sports. One measure of performance might be reaction time.
(a) Design a completely randomized design to help the psychologist answer her question.
(b) Design a matched-pairs design to help the psychologist answer her question.
(c) In your opinion, which design is better? Justify your opinion.

THE CHAPTER 1 CASE STUDY IS LOCATED ON THE CD THAT ACCOMPANIES THIS TEXT.

PART 2

Descriptive Statistics

CHAPTER 2
Organizing and Summarizing Data

CHAPTER 3
Numerically Summarizing Data

CHAPTER 4
Describing the Relation between Two Variables

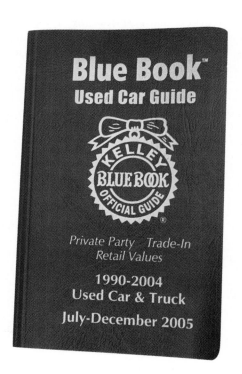

Remember, statistics is a process. The first chapter (Part 1) dealt with the first two steps in the statistical process: (1) Identify the research objective and (2) collect the in-formation needed to answer the questions in the research objective. The next three chapters deal (Part 2) with organizing, summarizing, and presenting the data collected. This step in the process is called descriptive statistics.

Organizing and Summarizing Data

CHAPTER 2

Outline

2.1   Organizing Qualitative Data

2.2   Organizing Quantitative Data

2.3   Graphical Misrepresentations of Data

■ Chapter Review

■ Case Study: The Day the Sky Roared (On CD)

DECISIONS

Suppose that you work for the school newspaper. Your editor approaches you with a special reporting assignment. Your task is to write an article that describes the "typical" student at your school, complete with supporting information. How are you going to do this assignment? See the Decisions project on page 95.

●●● Putting It All Together

Chapter 1 discussed how to collect data. We learned that data can be obtained from four sources: (1) a census, (2) existing data sets, (3) surveys, or (4) designed experiments. When data are obtained, they are referred to as **raw data**. Raw data must be organized into a meaningful form.

Methods for organizing raw data include the creation of tables or graphs, which allow for a quick overview of the information collected. Describing data is the third step in the statistical process. The procedures used in this step depend on whether the data are qualitative, discrete, or continuous.

2.1 Organizing Qualitative Data

Preparing for This Section   Before getting started, review the following:

- Qualitative data (Section 1.1, p. 6)

Objectives

1. **Organize qualitative data in tables**
2. **Construct bar graphs**
3. **Construct pie charts**

In this section we will concentrate on tabular and graphical summaries of qualitative data. In Section 2.2 we discuss methods for summarizing quantitative data.

1. ## Organize Qualitative Data in Tables

Recall that qualitative data provide measures that categorize or classify an individual. When qualitative data are collected, we are often interested in determining the number of individuals observed within each category.

In Other Words
Qualitative data is also known as categorical data.

Definition

> A **frequency distribution** lists each category of data and the number of occurrences for each category of data.

EXAMPLE 1   **Organizing Qualitative Data into a Frequency Distribution**

Problem: A physical therapist wants to get a sense of the types of rehabilitation required by her patients. To do so, she obtains a simple random sample of 30 of her patients and records the body part requiring rehabilitation. See Table 1. Construct a frequency distribution of location of injury.

Table 1					
Back	Back	Hand	Neck	Knee	Knee
Wrist	Back	Groin	Shoulder	Shoulder	Back
Elbow	Back	Back	Back	Back	Back
Back	Shoulder	Shoulder	Knee	Knee	Back
Hip	Knee	Hip	Hand	Back	Wrist

Source: Krystal Catton, student at Joliet Junior College

Approach: To construct a frequency distribution, we create a list of the body parts (categories) and tally each occurrence. Finally, we add up the number of tallies to determine the frequency.

Solution: See Table 2. From the table, we can see that the back is the most common body part requiring rehabilitation, with a total of 12.

Table 2		
Body Part	**Tally**	**Frequency**
Back	++++ ++++ ||	12
Wrist	||	2
Elbow			1
Hip	||	2
Shoulder	||||	4
Knee	++++	5
Hand	||	2
Groin			1
Neck			1

! CAUTION
The data in Table 2 are still qualitative. The frequency represents the count of each category.

With frequency distributions, it is a good idea to add up the frequency column to make sure that it sums to the number of observations. In the case of the data in Example 1, the frequency column adds up to 30, as it should.

Often, rather than being concerned with the frequency with which categories of data occur, we want to know the *relative frequency* of the categories.

Definition

In Other Words

A frequency distribution shows how many observations are in each category. A relative frequency distribution shows the proportion of observations that belong in each category.

The **relative frequency** is the proportion (or percent) of observations within a category and is found using the formula

$$\text{Relative frequency} = \frac{\text{frequency}}{\text{sum of all frequencies}} \qquad (1)$$

A **relative frequency distribution** lists each category of data together with the relative frequency.

EXAMPLE 2 **Constructing a Relative Frequency Distribution of Qualitative Data**

Problem: Using the data in Table 2, construct a relative frequency distribution.

Approach: Add all the frequencies, and then use Formula (1) to compute the relative frequency of each category of data.

Solution: We add the values in the frequency column in Table 2:

Sum of all frequencies = 12 + 2 + 1 + 2 + 4 + 5 + 2 + 1 + 1 = 30

We now compute the relative frequency of each category. For example, the relative frequency of the category "Back" is

$$\frac{12}{30} = 0.4$$

After computing the relative frequency for the remaining categories, we obtain the relative frequency distribution shown in Table 3.

Table 3		
Body Part	**Frequency**	**Relative Frequency**
Back	12	$\frac{12}{30} = 0.4$
Wrist	2	$\frac{2}{30} \approx 0.0667$
Elbow	1	0.0333
Hip	2	0.0667
Shoulder	4	0.1333
Knee	5	0.1667
Hand	2	0.0667
Groin	1	0.0333
Neck	1	0.0333

From the table, we can see that the most common body part for rehabilitation is the back.

USING TECHNOLOGY

Some statistical spreadsheets such as MINITAB have a `Tally` command. This command will construct a frequency and relative frequency distribution of raw qualitative data.

It is a good idea to add up the entries in the relative frequency column to be sure they sum to 1. Sometimes the sum of the relative frequencies will differ slightly from 1 due to rounding. Nonetheless, the check is a good idea.

Now Work Problems 27(a) and (b).

② Construct Bar Graphs

Once raw data are organized in a table, we can create graphs. Creating graphs allows us to see the data and get a sense of what the data are saying about the individuals in the study. In general, pictures of data result in a more powerful message than tables. Try the following exercise for yourself: Open a newspaper and look at a table and graph. Study each. Now put the paper away and close your eyes. What do you see in your mind's eye? Can you recall information obtained from the table or the graph? In general, people are more likely to recall information obtained from a graph than they are from a table.

One of the most common devices for graphically representing qualitative data is a bar graph.

Definition   A **bar graph** is constructed by labeling each category of data on a horizontal axis and the frequency or relative frequency of the category on the vertical axis. Rectangles of equal width are drawn for each category. The height of each rectangle is the category's frequency or relative frequency.

EXAMPLE 3   **Constructing a Frequency and Relative Frequency Bar Graph**

Problem: Use the data summarized in Table 3 to construct the following:

(a) Frequency bar graph
(b) Relative frequency bar graph

Approach: A horizontal axis is used to indicate the categories of the data (body parts, in this case), and a vertical axis is used to represent the frequency or relative frequency. Draw rectangles of equal width to the height that is the frequency or relative frequency for each category. The bars do not touch each other.

Solution

(a) Figure 1(a) shows the frequency bar graph.
(b) Figure 1(b) shows the relative frequency bar graph.

!CAUTION
Watch out for graphs that start the scale at some value other than 0, have bars with unequal widths, or have bars with different colors, because they can misrepresent the data.

Figure 1

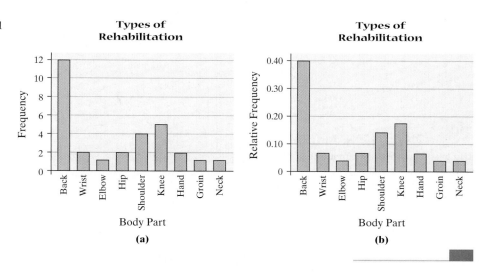

EXAMPLE 4   **Constructing a Frequency or Relative Frequency Bar Graph Using Technology**

Problem: Use a statistical spreadsheet to construct a frequency and relative frequency bar graph.

Approach: We will use Excel to construct the frequency and relative frequency bar graph. The steps for constructing the graphs using MINITAB or Excel are given in the Technology Step by Step on page 70. **Note:** The TI-83 and TI-84 Plus graphing calculators cannot draw frequency or relative frequency bar graphs.

Result: Figure 2(a) shows the frequency bar graph and Figure 2(b) shows the relative frequency bar graph obtained from Excel.

Figure 2

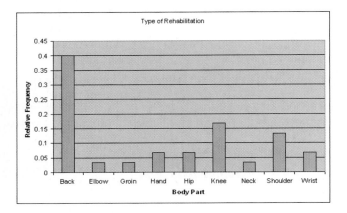

Now Work Problems 27(c) and (d).

Some statisticians prefer to create bar graphs with the categories arranged in decreasing order of frequency.

Definition

A **Pareto chart** is a bar graph whose bars are drawn in decreasing order of frequency or relative frequency.

Figure 3 illustrates a relative frequency Pareto chart for the data in Table 3.

Figure 3

USING TECHNOLOGY

The graphs obtained from a different statistical package should not differ from those in Figure 2. Some packages use the word *count* in place of *frequency* or *percent* in place of *relative frequency*, however.

Side-by-Side Bar Graphs

Graphics provide insight when you are comparing two sets of data. For example, suppose we wanted to know if more people are finishing college today than in 1990. We could draw a **side-by-side bar graph** to compare the two data sets. Data sets should be compared by using relative frequencies, because different sample or population sizes make comparisons using frequencies difficult or misleading.

EXAMPLE 5 **Comparing Two Data Sets**

Problem: The data in Table 4 represent the educational attainment in 1990 and 2003 of adults 25 years and older who are residents of the United States. The data are in thousands. So 16,502 represents 16,502,000.

(a) Draw a side-by-side relative frequency bar graph of the data.
(b) Is a greater proportion of Americans earning bachelor's degrees?

Table 4		
Educational Attainment	**1990**	**2003**
Less than 9th grade	16,502	12,276
9th to 12th grade, no diploma	22,842	16,323
High school diploma	47,643	59,292
Some college, no degree	29,780	31,762
Associate's degree	9,792	15,147
Bachelor's degree	20,833	33,213
Graduate or professional degree	11,478	17,169
Totals	**158,870**	**185,182**

Source: U.S. Census Bureau

Approach: First, we determine the relative frequencies of each category for each year. To construct the side-by-side bar graphs, we draw two bars for each category of data. One of the bars will represent 1990 and the other will represent 2003.

Solution: Table 5 shows the relative frequency for each category.

(a) The side-by-side bar graph is shown in Figure 4.

Table 5		
Educational Attainment	**1990**	**2003**
Less than 9th grade	0.1039	0.0663
9th to 12th grade, no diploma	0.1438	0.0881
High school diploma	0.2999	0.3202
Some college, no degree	0.1874	0.1715
Associate's degree	0.0616	0.0818
Bachelor's degree	0.1311	0.1794
Graduate or professional degree	0.0722	0.0927

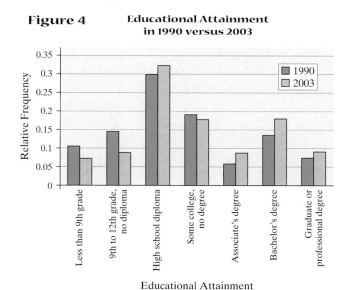

Figure 4 Educational Attainment in 1990 versus 2003

(b) From the graph, we can see that the proportion of Americans 25 years and older who earned a bachelor's degree is higher in 2003. This information is not clear from the frequency table, because the sizes of the populations are different. Increases in the number of Americans who earned a bachelor's degree are due partly to the increases in the size of each population.

Now Work Problem 23.

③ Construct Pie Charts

Pie charts are typically used to present the relative frequency of qualitative data.

Definition

> A **pie chart** is a circle divided into sectors. Each sector represents a category of data. The area of each sector is proportional to the frequency of the category.

EXAMPLE 6

Constructing a Pie Chart

Note to Instructor
Ask students to compare and contrast the similarities and differences of pie charts and bar graphs.

Note to Instructor
The step-by-step approach to constructing pie charts by hand is given so that students will have a conceptual understanding of the process. Encourage students to construct pie charts using technology.

Problem: The data presented in Table 6 represent the educational attainment of residents of the United States 25 years or older in 2003, based on data obtained from the U.S. Census Bureau. The data are in thousands. Construct a pie chart of the data.

Table 6	
Educational Attainment	**2003**
Less than 9th grade	12,276
9th to 12th grade, no diploma	16,323
High school diploma	59,292
Some college, no degree	31,762
Associate's degree	15,147
Bachelor's degree	33,213
Graduate or professional degree	17,169
Totals	**185,182**

Approach: The pie chart will have seven parts, or sectors, corresponding to the seven categories of data. The area of each sector is proportional to the frequency of each category. For example,

$$\frac{12,276}{185,182} = 0.0663$$

of all U.S. residents 25 years or older have less than a 9th-grade education. The category "less than 9th grade" will make up 6.63% of the area of the pie chart. Since a circle has 360 degrees, the degree measure of the sector for the category "less than 9th grade" will be $(0.0663)360° \approx 24°$. Use a protractor to measure each angle.

Solution: We follow the approach presented for the remaining categories of data to obtain Table 7.

Table 7			
Education	**Frequency**	**Relative Frequency**	**Degree Measure of Each Sector**
Less than 9th grade	12,276	0.0663	24
9th to 12th grade, no diploma	16,323	0.0881	32
High school diploma	59,292	0.3202	115
Some college, no degree	31,762	0.1715	62
Associate's degree	15,147	0.0818	29
Bachelor's degree	33,213	0.1794	65
Graduate or professional degree	17,169	0.0927	33

USING TECHNOLOGY
Most statistical spreadsheets are capable of drawing pie charts. See the Technology Step by Step on page 70 for instructions on how to obtain a pie chart using MINITAB or Excel. The TI-83 and TI-84 Plus graphing calculators do not draw pie charts.

Figure 5

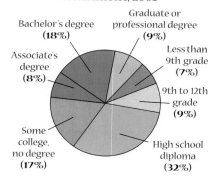

Educational Attainment, 2003

To construct a pie chart by hand, we use a protractor to approximate the angles for each sector. See Figure 5.

Pie charts can be created only if all the categories of the variable under consideration are represented. For example, from the data given in Example 1, we could create a bar graph that lists the proportion of patients requiring rehabilitation on their back, shoulder or knee only, but it would not make sense to construct a pie chart for this situation. Do you see why? Only 70% of the data would be represented.

When should a bar graph be used to display information? When should a pie chart be used? Pie charts are useful for showing the division of all possible values of a qualitative variable into its parts. However, because angles are often hard to judge in pie charts, they are not as useful in comparing two specific values of the qualitative variable. Instead the emphasis is on comparing the part to the whole. Bar graphs are useful when we want to compare the different parts, not the parts to the whole. For example, if we wanted to get the "big picture" regarding educational attainment in 2003, then a pie chart is a good visual summary. However, if we want to compare bachelor's degrees to high school diplomas, then a bar graph is a good visual summary.

Now Work Problem 27(e).

2.1 ASSESS YOUR UNDERSTANDING

Concepts and Vocabulary

1. Define *raw data* in your own words.
2. A frequency distribution lists the _____ of occurrences of each category of data, while a relative frequency distribution lists the _____ of occurrences of each category of data. number, proportion (or percent)
3. When constructing a frequency distribution, why is it a good idea to add up the frequencies?
4. In a relative frequency distribution, what should the relative frequencies add up to? One
5. What is a Pareto chart?
6. Why should relative frequencies be used when comparing two data sets?
7. Explain why Pareto charts might be preferred over bar graphs.

8. Consider the information in the "Why we can't lose weight" chart shown below which is in the *USA Today* style of graph. Could the information provided be organized into a pie chart? Why or why not? No

Why we can't lose weight

63% Metabolism too slow
59% Don't exercise
50% Don't have self-discipline
49% Splurge on favorite foods

Skill Building

9. **Golf Driving Range** The pie chart at the right, the type we see in *USA Today*, depicts the bucket size golfers choose while at the driving range.
   (a) What is the most popular size? What percentage of golfers choose this size? Large; 52%
   (b) What is the least popular size? What percentage of golfers choose this size? X-Large; 9%
   (c) What percentage of golfers choose a medium-sized bucket? 23%

Most popular bucket sizes at the range

Large (52%)
X-Large (9%)
Small (16%)
Medium (23%)

10. Cosmetic Surgery The *USA Today*-type chart shows the most frequent cosmetic surgeries for women in 2003.

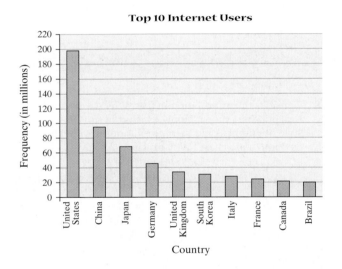

(a) If women had 1,548,000 cosmetic surgeries in 2003, how many of them were for liposuction? *325,080*
(b) How many were for eyelid surgery? *216,720*
(c) What percentage of surgeries are not accounted for in the graph? *38%*

11. Internet Users The following Pareto chart represents the top 10 countries in Internet users as of February 2005. (*Source*: www.internetworldstats.com)

Top 10 Internet Users

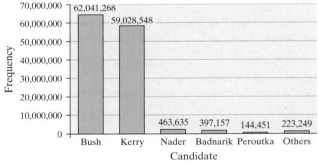

(a) Which country has the most Internet users?
(b) Approximately what is the Internet usage in Canada?
(c) Approximately how many more users are in China than in Japan? *26 million*

12. Poverty Every year the U.S. Census Bureau counts the number of people living in poverty. The bureau uses money income thresholds as its definition of poverty, so noncash benefits such as Medicaid and food stamps do not count toward poverty thresholds. For example, in 2002 the poverty threshold for a family of three was $14,348. The

bar chart represents the number of people living in poverty in the United States in 2003, by ethnicity, based on March 2004 estimates.

Number in Poverty

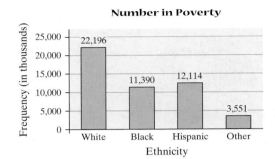

Source: The Henry Kaiser Family Foundation (www.statehealthfacts.kff.org)

(a) How many whites were living in poverty in 2003?
(b) Of the impoverished, what percent were Hispanic?
(c) How might this graph be misleading?

13. 2004 Presidential Election The following bar chart represents the number of voters who cast votes for the top five vote-getters and all others in the 2004 presidential election. The number of votes received for each candidate appears above the bar.

Votes in 2004 Presidential Election

Source: David Leip's Atlas of U.S. Presidential Elections (www.uselectionatlas.org)

(a) How many votes were cast for Michael Badnarik of the Libertarian Party? *397,157*
(b) What percent of the votes was cast for George W. Bush? For John F. Kerry? *50.7%; 48.3%*
(c) To receive presidential election funds, a candidate must garner at least 5% of the vote. Will Ralph Nader receive presidential election funds? *No*

14. Identity Fraud In a study conducted by the Better Business Bureau and Javelin Strategy and Research, victims of identity fraud were asked, "Who was the person who misused your personal information?" The following Pareto chart represents the results for cases in the year 2004 for which the perpetrator's identity was known.

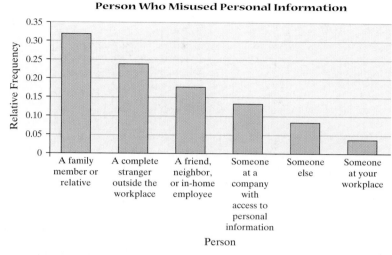

Person Who Misused Personal Information

Source: Javelin Strategy & Research, 2004.

(a) Approximately what percentage of identity-fraud victims were victimized by a family member or relative? 32%

(b) If there were 9.3 million identity-fraud victims in 2004, how many were victimized by someone at a company with access to personal information? 1.2 million

15. Home Heating Fuel The following side-by-side bar graph represents the proportion of households using four common sources of home heating fuel for the years 1978, 1987, 1997, and 2001.

Home Heating Fuel

Source: Energy Information Administration, Residential Energy Consumption Survey, 1978, 1987, 1997, 2001

(a) What proportion of households used electricity as their main source of home heating fuel in 1978? in 1997? 0.16; 0.30

(b) What was the most popular source of home heating fuel in 1987? Natural gas

(c) If there were 107 million U.S. households in 2001, how many used LPG as their main source of home heating fuel? About 5.35 million households

(d) What might account for the rise in homes that use electricity as the main source of home heating fuel?

(e) Which source of home heating fuel has been decreasing rather steadily? What might account for this? Fuel oil or kerosene

(f) Which sources of home heating fuel have remained rather steady? Both natural gas and LPG

16. **Doctorate Recipients** The following side-by-side bar graph represents the number of doctorate recipients from U.S. universities within broad fields of study for the years 1983, 1993, and 2003.

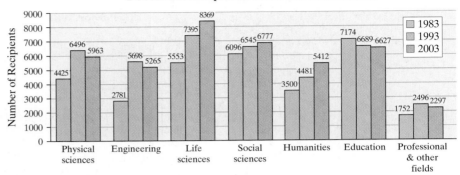

Doctorate Recipients from U.S. Universities

Source: NSF/NIH/USED/NEH/USDA/NASA, 2003 Survey of Earned Doctorates

(16a) 433

(a) How many more engineering doctorates were awarded in 1993 than in 2003?
(b) In 2003, what percentage of doctoral recipients received degrees in physical science? in education? About 14.6%; about 16.3%
(c) What field of study consistently decreased in the number of doctoral degree recipients in the 3 years of the survey? Education
(d) Which field of study had the largest increase in the number of doctoral degree recipients in the 3 years of the survey? Life sciences

Applying the Concepts

17. **Government Income** For fiscal year 2003 (October 1, 2002 to September 30, 2003), the federal government's income was $1,782.3 billion. The various sources of income are broken down in the following table.

Source of Income	Amount (in billions of dollars)
Individual income taxes	793.7
Corporate income taxes	131.8
Social insurance taxes	713.0
Excise, estate and gift taxes, customs, and miscellaneous receipts	143.8

Source: Congressional Budget Office

(a) Construct a relative frequency distribution of the data shown.
(b) What percentage of total income is attributable to individual income taxes? 44.53%
(c) Construct a frequency bar graph of the data.
(d) Construct a relative frequency bar graph of the data.
(e) Construct a pie chart of the data.

(f) In your opinion, which graph appears to place more emphasis on social insurance taxes as a source of income, the relative frequency bar graph or the pie chart? Why?

18. **Government Expenditures** For fiscal year 2003 (October 1, 2002 to September 30, 2003), the federal government spent $2,158 billion. The breakdown of expenditures is given in the following table.

Category	Expenditure (in billions of dollars)
National defense and foreign affairs	474.8
Social programs	1251.6
Physical, human, and community development	215.8
Net interest on the debt	151.1
Law enforcement and general government	64.7

Source: Budget of the United States and Internal Revenue Service

(a) Construct a relative frequency distribution of the data shown.

(b) What percentage of total expenditures is attributable to net interest on debt? 7%
(c) Construct a frequency bar graph of government expenditures.
(d) Construct a relative frequency bar graph of government expenditures.
(e) Construct a pie chart of government expenditures.
(f) In your opinion, which graph appears to place more emphasis on net interest on the debt, the relative frequency bar graph or the pie chart? Why?

19. **College Survey** In a national survey conducted by the Centers for Disease Control to determine health-risk behaviors among college students, college students were asked, "How often do you wear a seat belt when riding in a car driven by someone else?" The frequencies were as follows:

Response	Frequency
Never	125
Rarely	324
Sometimes	552
Most of the time	1257
Always	2518

(a) Construct a relative frequency distribution.
(b) What percentage of respondents answered "Always"? 52.7%
(c) What percentage of respondents answered "Never" or "Rarely"? 9.4%
(d) Construct a frequency bar graph.
(e) Construct a relative frequency bar graph.
(f) Construct a pie chart.
(g) Suppose that a representative from the Centers for Disease Control says, "52.7% of all college students never wear a seat belt." Is this a descriptive or inferential statement? Inferential

20. **College Survey** In a national survey conducted by the Centers for Disease Control to determine health-risk behaviors among college students, college students were asked, "How often do you wear a seat belt when driving a car?" The frequencies were as follows:

Response	Frequency
I do not drive a car	249
Never	118
Rarely	249
Sometimes	345
Most of the time	716
Always	3093

(a) Construct a relative frequency distribution.

(b) What percentage of respondents answered "Always"? 64.8%
(c) What percentage of respondents answered "Never" or "Rarely"? 7.7%
(d) Construct a frequency bar graph.
(e) Construct a relative frequency bar graph.
(f) Construct a pie chart.
(g) Compute the relative frequencies of "Never," "Rarely," "Sometimes," "Most of the time," and "Always," excluding those that do not drive. Compare with those in Problem 19. What might you conclude?
(h) Suppose that a representative from the Centers for Disease Control says, "2.5% of the college students in this survey responded that they never wear a seat belt." Is this a descriptive or inferential statement?

21. **Foreign-Born Population** The following data represent the region of birth of foreign-born residents of the United States in 2003.

Region	Number (thousands)
Caribbean	3,384
Central America	12,362
South America	2,111
Asia	8,375
Europe	4,590
Other Regions	2,680

Source: U.S. Census Bureau

(a) Construct a relative frequency distribution.
(b) What percentage of foreign-born residents was born in Asia? 25.00%
(c) Construct a frequency bar graph.
(d) Construct a relative frequency bar graph.
(e) Construct a pie chart.

22. **Robbery** The following data represent the number of offenses for various robberies in 2003.

Type of Robbery	Number (thousands)
Street or highway	131
Commercial	61
Gas station	10
Convenience store	26
Residence	41
Bank	7

Source: U.S. Federal Bureau of Investigation

(a) Construct a relative frequency distribution.
(b) What percentage of robberies was of gas stations?
(c) Construct a frequency bar graph.
(d) Construct a relative frequency bar graph.
(e) Construct a pie chart.   (22b) 3.62%

23. Educational Attainment On the basis of the 2003 Current
NW Population Survey, there were 88.7 million males and 96.6
million females 25 years old or older in the United States.
The educational attainment of the males and females was
as follows:

Educational Attainment	Males (in millions)	Females (in millions)
Not a high school graduate	14.1	14.5
High school graduate	27.4	31.9
Some college, but no degree	15.2	16.6
Associate's degree	6.4	8.8
Bachelor's degree	16.4	16.9
Advanced degree	9.2	7.9

Source: U.S. Census Bureau

(a) Construct a relative frequency distribution for males.
(b) Construct a relative frequency distribution for females.
(c) Construct a side-by-side relative frequency bar graph.
(d) Compare each gender's educational attainment. Make
a conjecture about the reasons for the differences.

24. Internet Access The following data represent the number of
people who had Internet access in the years 2000 and 2003 by
level of education. Data are in thousands of U.S. residents.

Educational Attainment	2000	2003
No college	24,662	65,862
Some college	31,462	50,931
Graduated college	34,379	49,106

Source: U.S. Statistical Abstract, 2004

(a) Construct a relative frequency distribution for 2000.
(b) Construct a relative frequency distribution for 2003.
(c) Construct a side-by-side relative frequency bar graph.
(d) Compare each year's Internet access. Make some
conjectures about the reasons for any differences or
similarities.

25. Murder Victims A criminologist wanted to know if there
was any relation between age and gender of murder victims. The following data represent the number of male and
female murder victims by age in 2002.

Age	Number of Males	Number of Females
Less than 17	650	444
17–24	3435	704
25–34	2990	704
35–54	2859	1085
55 or older	787	442

Source: U.S. Federal Bureau of Investigation

(a) Construct a relative frequency distribution for males.
(b) Construct a relative frequency distribution for females.
(c) Construct a side-by-side relative frequency bar graph.
(d) Compare each gender's age percentages. Make a conjecture about the reasons for the differences or similarities.

26. Car Color DuPont Automotive is a major supplier of
paint to the automotive industry. It conducted a survey
of 100 randomly selected autos in the luxury car segment and 100 randomly selected autos in the sports car
segment that were recently purchased and obtained the
following colors.

Color	Number of Luxury Cars	Number of Sports Cars
Silver	32	27
White	30	11
Blue	9	13
Black	9	16
Gray	7	7
Red	6	18
Gold	3	4
Green	2	0
Brown	2	4

(a) Construct a relative frequency distribution for each
car type.
(b) Draw a side-by-side relative frequency bar graph.
(c) Compare the colors for the two types of car. Make a
conjecture about the reasons for the differences.

27. 2004 Presidential Election An exit poll was conducted
NW in Los Alamos County, New Mexico, in which a random
sample of 40 voters revealed whom they voted for in
the presidential election. The results of the survey are
as follows:

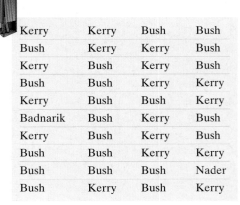

Kerry	Kerry	Bush	Bush
Bush	Kerry	Kerry	Bush
Kerry	Bush	Kerry	Bush
Bush	Bush	Kerry	Kerry
Kerry	Bush	Bush	Kerry
Badnarik	Bush	Kerry	Bush
Kerry	Bush	Kerry	Bush
Bush	Bush	Kerry	Kerry
Bush	Bush	Bush	Nader
Bush	Kerry	Bush	Kerry

(a) Construct a frequency distribution.

(b) Construct a relative frequency distribution.

(c) Construct a frequency bar graph.

(d) Construct a relative frequency bar graph.

(e) Construct a pie chart.

(f) On the basis of the data, make a conjecture about which candidate will win Los Alamos County. Would your conjecture be descriptive statistics or inferential statistics? If George W. Bush wins Los Alamos County, what conclusions might be drawn, assuming that the sample was conducted appropriately? Would you be confident in making this prediction with a sample of 40? If the sample consisted of 100 voters, would your confidence increase? Why? Inferential

28. Hospital Admissions The following data represent the diagnoses of a random sample of 20 patients admitted to a hospital.

Cancer	Motor vehicle accident	Congestive heart failure
Gunshot wound	Fall	Gunshot wound
Gunshot wound	Motor vehicle accident	Gunshot wound
Assault	Motor vehicle accident	Gunshot wound
Motor vehicle accident	Motor vehicle accident	Gunshot wound
Motor vehicle accident	Gunshot wound	Motor vehicle accident
Fall	Gunshot wound	

Source: Tamela Ohm, student at Joliet Junior College

(a) Construct a frequency distribution.

(b) Construct a relative frequency distribution.

(c) Which diagnosis had the most admissions? Gunshot wound

(d) What percentage of diagnoses was motor vehicle accidents? 35%

(e) Construct a frequency bar graph.

(f) Construct a relative frequency bar graph.

(g) Construct a pie chart.

(h) Suppose that an admission specialist at the hospital stated that 40% of all admissions were gunshot wounds. Would this statement be descriptive or inferential? Why? Inferential

29. Which Position in Baseball Pays the Most? You are a prospective baseball agent and are in search of clients. You would like to recruit the highest-paid players as clients, so you perform a study in which you identify the 24 top-paid players for the 2004 season and their positions. The table shows the results of your study.

Player	Position
Manny Ramirez	Left field
Alex Rodriquez	Third base
Carlos Delgado	First base
Derek Jeter	Shortstop
Barry Bonds	Left field
Pedro Martinez	Pitcher
Mo Vaughn	First base
Shawn Green	Right field
Mike Piazza	Catcher
Jeff Bagwell	First base
Randy Johnson	Pitcher
Mike Mussina	Pitcher
Sammy Sosa	Right field
Kevin Brown	Pitcher
Chipper Jones	Third base
Mike Hampton	Pitcher
Magglio Ordonez	Right field
Chan Ho Park	Pitcher
Gary Sheffield	Right field
Larry Walker	Right field
Ken Griffey, Jr.	Center field
Richard Hidalgo	Right field
Andruw Jones	Center field
Matt Morris	Pitcher

Source: usatoday.com

(a) Construct a frequency distribution of position.

(b) Construct a relative frequency distribution of position.

(c) Which position appears to be the most lucrative? For which position would you recruit? Pitcher

(d) Are there any positions that you would avoid recruiting? Why? Second base

(e) Draw a frequency bar graph.

(f) Draw a relative frequency bar graph.

(g) Draw a pie chart.

30. Blood Type A phlebotomist draws the blood of a random sample of 50 patients and determines their blood types as shown:

O	O	A	A	O
B	O	B	A	O
AB	B	A	B	AB
O	O	A	A	O
AB	O	A	B	A
O	A	A	O	A
O	A	O	AB	A
O	B	A	A	O
O	O	O	A	O
O	A	O	A	O

(a) Construct a frequency distribution.
(b) Construct a relative frequency distribution.
(c) According to the data, which blood type is most common? O
(d) According to the data, which blood type is least common? AB
(e) Use the results of the sample to conjecture the percentage of the population that has type O blood. Is this an example of descriptive or inferential statistics?
   44%; inferential

(f) Contact a local hospital and ask them the percentage of the population that is blood type O. Why might the results differ?
(g) Draw a frequency bar graph.
(h) Draw a relative frequency bar graph.
(i) Draw a pie chart.

31. Foreign Language According to the Modern Language Association, the number of college students studying foreign language is increasing. The following data represent the foreign language being studied based on a simple random sample of 30 students learning a foreign language.

Spanish	Chinese	Spanish	Spanish	Spanish
Chinese	German	Spanish	Spanish	French
Spanish	Spanish	Japanese	Latin	Spanish
German	German	Spanish	Italian	Spanish
Italian	Japanese	Chinese	Spanish	French
Spanish	Spanish	Russian	Latin	French

Source: Based on data obtained from the Modern Language Association

(a) Construct a frequency distribution.
(b) Construct a relative frequency distribution.
(c) Construct a frequency bar graph.
(d) Construct a relative frequency bar graph.
(e) Construct a pie chart.

32. The following table lists the presidents of the United States and their state of birth.

Birthplace of U.S. President

President	State of Birth	President	State of Birth	President	State of Birth
Washington	Virginia	Lincoln	Kentucky	Coolidge	Vermont
J. Adams	Massachusetts	A. Johnson	North Carolina	Hoover	Iowa
Jefferson	Virginia	Grant	Ohio	F. D. Roosevelt	New York
Madison	Virginia	Hayes	Ohio	Truman	Missouri
Monroe	Virginia	Garfield	Ohio	Eisenhower	Texas
J. Q. Adams	Massachusetts	Arthur	Vermont	Kennedy	Massachusetts
Jackson	South Carolina	Cleveland	New Jersey	L. B. Johnson	Texas
Van Buren	New York	B. Harrison	Ohio	Nixon	California
W. H. Harrison	Virginia	Cleveland	New Jersey	Ford	Nebraska
Tyler	Virginia	McKinley	Ohio	Carter	Georgia
Polk	North Carolina	T. Roosevelt	New York	Reagan	Illinois
Taylor	Virginia	Taft	Ohio	George H. Bush	Massachusetts
Fillmore	New York	Wilson	Virginia	Clinton	Arkansas
Pierce	New Hampshire	Harding	Ohio	George W. Bush	Connecticut
Buchanan	Pennsylvania				

(a) Construct a frequency bar graph for state of birth.
(b) Which state has yielded the most presidents? Virginia
(c) Explain why the answer obtained in part (b) may be considered to be misleading.

Consumer Reports®   Consumer Reports Rates Treadmills

A study that compared exercisers who worked out equally hard for the same time on several different types of machines found that they generally burned the most calories on treadmills. Our own research has shown that treadmills are less likely than other machines to sit unused. So it should come as no surprise that treadmills are the best-selling home exercise machine in the United States.

In a study by *Consumer Reports* (January 2005), we tested 22 treadmills from best-selling brands, ranging in price from $400 to $3,500. The treadmills were rated on ease of use, ergonomics, quality, and exercise range. Ease of use was based on how easy it was to adjust exercise difficulty, read the display, and operate the monitor functions. Ergonomics was based on how well the machine accommodates users of different sizes, including belt sizes, foot rail, and hand grip design. Quality was based on our judgment of construction quality, sample defects, and the results of our durability test. Exercise range was based on how well each machine pro-

vides an effective workout to users of various fitness levels, from beginners to very fit individuals.

In order to help compare the treadmills, these attribute scores were combined into an overall score. The figure below is a ratings chart for the 22 treadmills based on our test results. In addition to the performance ratings, other useful information, such as the models' price and belt size, is included.

(a) What type of graph is illustrated to display overall score in the figure?

(b) Which model has the highest construction score? Which models have the lowest ease of use score?

(c) For ease of use, how many treadmills rated excellent? very good? good? fair? poor?

(d) Draw a frequency bar graph for each rating category. In other words, draw a bar graph for ease of use, ergonomics, and so on.

- ◉ Excellent
- ◕ Very good
- ○ Good
- ◑ Fair
- ● Poor

Ratings Chart for Threadmills

In performance order. Blue key numbers indicate Quick Picks; see box at left.

Key number	Brand & model (Similar models, in small type, comparable to tested model.)	Price	Belt (L x W, in.)	Overall score	Ease of use	Ergonomics	Quality	Exercise range	Chest-strap heart monitor	Folding deck
1	**Landice** L7 Series Pro Sports Trainer	$2,900	58x20		◉	◉	◉	◉		
2	**True** 400HRC	2,300	54x20		◕	◕	◉	◉	•	
3	**Precor** M 9.33	3,500	58x20		○	◉	◉	◉	•	
4	**Precor** M 9.31	3,000	58x20		○	◉	◉	◉		
5	**Horizon Fitness** Elite 5.1T	1,800	58x20		◕	◉	◕	◕	•	•
6	**Life Fitness** T3i T3	2,800	52x20		◉	◕	◕	◉	•	
7	**Vision Fitness** T9200	1,300	52x20		◕	◕	◕	◉		
8	**Nautilus** NTR 500	2,300	56x20		◕	◕	○	◕		
9	**Life Fitness** Essential FT6	1,400	52x20		◕	◉	◕	◕	•	•
10	**Star Trac** TR901	3,200	50x19		◕	◕	◕	◉	•	
11	**Trimline** T350HR	1,600	56x20		◕	◕	○	◕		
12	**Spirit** Inspire IN400	1,650	55x20		◕	◉	◕	○	•	•
13	**Schwinn** 820p	1,300	51x20		◕	◕	◕	◕		
14	**NordicTrack** E2500	1,200	53x20		◕	◕	○	◉	•	
15	**Pro-Form** 5 Star Interactive Trainer 1200	1,300	54x19		◕	◉	◕	◕	•	
16	**Horizon Fitness** CST4 T53	1,000	53x20		○	◕	○	◉		
17	**Pro-Form** CS11e	800	53x19		◕	◕	○	◕	•	
18	**EVO** by Smooth Fitness EVO2	2,500	60x20		◕	◉	○	○	•	
19	**Keys Fitness** Ironman M4	1,300	54x20		◕	○	◑	○	•	•
20	**Image** 15.0	400	50x18		◉	○	◑	○	•	
21	**Weslo** Cadence C62	500	50x18		◕	○	◑	◑	•	

Not recommended because samples required repeated belt-tension adjustment, a big inconvenience.

| 22 | **NordicTrack** 7600 R | 2,100 | 59x20 | | ◉ | ◕ | ● | ◕ | • | • |

Source: Used with the permission of *Consumer Reports*

Drawing Bar Graphs and Pie Charts

TI-83/84 Plus The TI-83 or TI-84 Plus does not have the ability to draw bar graphs or pie charts.

MINITAB **Frequency or Relative Frequency Distributions from Raw Data**

Step 1: Enter the raw data in C1.

Step 2: Select **Stat** and highlight **Tables** and select Tally Individual Variables . . .

Step 3: Fill in the window with appropriate values. In the "Variables" box, enter C1. Check "counts" for a frequency distribution and/or "percents" for a relative frequency distribution. Click OK.

Bar Graphs from Summarized Data

Step 1: Enter the categories in C1 and the frequency or relative frequency in C2.

Step 2: Select **Graph** and highlight **Bar Chart**.

Step 3: In the "Bars represent" pull-down menu, select "Values from a table" and highlight "Simple." Press OK.

Step 4: Fill in the window with the appropriate values. In the "Graph variables" box, enter C2. In the "Categorical variable" box, enter C1. By pressing Labels, you can add a title to the graph. Click OK to obtain the bar graph.

Bar Graphs from Raw Data

Step 1: Enter the raw data in C1.

Step 2: Select **Graph** and highlight **Bar Chart.**

Step 3: In the "Bars represent" pull-down menu, select "Counts of unique values" and highlight "Simple." Press OK.

Step 4: Fill in the window with the appropriate values. In the "Categorical variable" box, enter C1. By pressing Labels, you can add a title to the graph. Click OK to obtain the bar graph.

Pie Chart from Raw or Summarized Data

Step 1: If the data are in a summarized table, enter the categories in C1 and the frequency or relative frequency in C2. If the data are raw, enter the data in C1.

Step 2: Select **Graph** and highlight **Pie Chart.**

Step 3: Fill in the window with the appropriate values. If the data are summarized, click the "Chart values from a table" radio button; If the data are raw, click the "Chart raw data" radio button. For summarized data, enter C1 in the "Categorical variable" box and C2 in the "Summary variable" box. If the data are raw, enter C1 in the "Categorical variable" box. By pressing Labels, you can add a title to the graph. Click OK to obtain the pie chart.

Excel **Bar Graphs from Summarized Data**

Step 1: Enter the categories in column A and the frequency or relative frequency in column B.

Step 2: Select the chart wizard icon. Click the "column" chart type. Select the chart type in the upper-left-hand corner and hit "Next."

Step 3: Click inside the data range cell. Use the mouse to highlight the data to be graphed. Click "Next."

Step 4: Click the "Titles" tab to enter *x*-axis, *y*-axis, and chart titles. Click "Finish."

Pie Charts from Summarized Data

Step 1: Enter the categories in column A and the frequencies in column B. Select the chart wizard icon and click the "pie" chart type. Select the pie chart in the upper-left-hand corner.

Step 2: Click inside the data range cell. Use the mouse to highlight the data to be graphed. Click "Next."

Step 3: Click the "Titles" tab to the chart title. Click the "Data Labels" tab and select "Show label and percent." Click "Finish."

2.2 Organizing Quantitative Data

Preparing for This Section   Before getting started, review the following:

- Quantitative data (Section 1.1, p. 6)
- Continuous data (Section 1.1, p. 7)

- Discrete data (Section 1.1, p. 7)

Objectives

1. **Organize discrete data in tables**
2. **Construct histograms of discrete data**
3. **Organize continuous data in tables**
4. **Construct histograms of continuous data**
5. **Draw stem-and-leaf plots**
6. **Draw dot plots**
7. **Identify the shape of a distribution**
8. **Draw time-series graphs**

Note to Instructor

Remind students of the differences between discrete and continuous data.

The first step in summarizing quantitative data is to determine whether the data are discrete or continuous. If the data are discrete and there are relatively few different values of the variable, the categories of data will be the observations (as in qualitative data). If the data are discrete, but there are many different values of the variables or, if the data are continuous, categories of data (called *classes*) must be created using intervals of numbers. We will first present the techniques required to organize discrete quantitative data when there are relatively few different values and then proceed to organizing continuous quantitative data.

1 Organize Discrete Data in Tables

The values of a discrete variable are used to create the categories of data.

EXAMPLE 1   **Constructing Frequency and Relative Frequency Distributions from Discrete Data**

Problem: The manager of a Wendy's fast-food restaurant is interested in studying the typical number of customers who arrive during the lunch hour. The data in Table 8 represent the number of customers who arrive at Wendy's for 40 randomly selected 15-minute intervals of time during lunch. For example, during one 15-minute interval, seven customers arrived. Construct a frequency and relative frequency distribution.

Table 8				
Number of Arrivals at Wendy's				
7	6	6	6	4
5	6	6	11	4
2	7	1	2	4
6	5	5	3	7
2	2	9	7	5
6	2	6	5	7
6	8	2	6	5
4	6	9	8	5

Approach: The number of people arriving could be 0, 1, 2, 3, ... From Table 8, we see that there are 11 categories of data from this study: 1, 2, 3, ..., 11. We tally the number of observations for each category, add up each tally, and create the frequency and relative frequency distributions.

Solution: The frequency and relative frequency distributions are shown in Table 9.

Table 9			
Number of Customers	**Tally**	**Frequency**	**Relative Frequency**
1	|	1	$\frac{1}{40} = 0.025$
2	ﬁﬁﬁ |	6	0.15
3	|	1	0.025
4	||||	4	0.1
5	ﬁﬁﬁ ||	7	0.175
6	ﬁﬁﬁ ﬁﬁﬁ |	11	0.275
7	ﬁﬁﬁ	5	0.125
8	||	2	0.05
9	||	2	0.05
10		0	0.0
11	|	1	0.025

On the basis of the relative frequencies, 27.5% of the 15-minute intervals had six customers arrive at Wendy's during the lunch hour.

Now Work Problems 27(a)–(d).

② Construct Histograms of Discrete Data

As with qualitative data, quantitative data may also be represented graphically. We begin our discussion with a graph called the *histogram*, which is similar to the bar graph drawn for qualitative data.

Definition

A **histogram** is constructed by drawing rectangles for each class of data. The height of each rectangle is the frequency or relative frequency of the class. The width of each rectangle is the same and the rectangles touch each other.

EXAMPLE 2   **Drawing a Histogram for Discrete Data**

Problem: Construct a frequency histogram and a relative frequency histogram using the data summarized in Table 9.

Approach: On the horizontal axis, we place the value of each category of data (number of customers). The vertical axis will be the frequency or relative frequency of each category. Rectangles of equal width are drawn, with the center of each rectangle located at the value of each category. For example, the first rectangle is centered at 1. For the frequency histogram, the height of the rectangle will be the frequency of the category. For the relative frequency histogram, the height of the rectangle will be the relative frequency of the category. Remember, the rectangles touch for histograms.

! CAUTION
The rectangles in histograms touch, while the rectangles in bar graphs do not touch.

Solution: Figure 6(a) shows the frequency histogram. Figure 6(b) shows the relative frequency histogram.

Figure 6

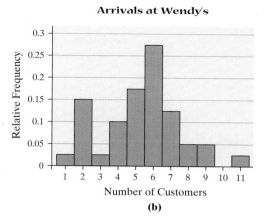

Now Work Problems 27(e) and (f).

③ Organize Continuous Data in Tables

Raw continuous data do not have any predetermined categories that can be used to construct a frequency distribution. Therefore, the categories must be created. Categories of data are created by using intervals of numbers called **classes**.

Table 10 is a typical frequency distribution created from continuous data. The data represent the number of U.S. residents between the ages of 25 and 74 who have earned a bachelor's degree. The data are based on the Current Population Survey conducted in 2003.

In the table, we notice that the data are categorized, or grouped, by intervals of numbers. Each interval represents a class. For example, the first class is 25- to 34-year-old residents of the United States who have a bachelor's degree. We read this interval as follows: "The number of residents of the United States in 2003 who were between 25 and 34 years of age and have a bachelor's degree was 8,849,000." There are five classes in the table, each with a *lower bound* and an *upper bound*. The **lower class limit** of a class is the smallest value within the class, while the **upper class limit** of a class is the largest value within the class. The lower class limit for the first class in Table 10 is 25; the upper class limit is 34. The **class width** is the difference between consecutive lower class limits. The class width for the data in Table 10 is $35 - 25 = 10$.

The classes in Table 10 do not overlap, so there is no confusion as to which class a data value belongs. Notice that the class widths are equal for all classes. One exception to this requirement is in open-ended tables. A table is **open ended** if the first class has no lower class limit or the last class does not have an upper class limit. The data in Table 11 represent the number of persons under sentence of death as of December 31, 2003, in the United States. The last class in the table, "60 and older," is open ended.

Table 10

Age	Number (in thousands)
25–34	8849
35–44	8915
45–54	7576
55–64	4254
65–74	2047

Table 11

Age	Number
20–29	533
30–39	1147
40–49	1090
50–59	493
60 and older	110

Source: U.S. Justice Department

EXAMPLE 3

Organizing Continuous Data into a Frequency and Relative Frequency Distribution

Problem: Suppose you are considering investing in a Roth IRA. You collect the data in Table 12, which represent the three-year rate of return (in percent) for a simple random sample of 40 small-capitalization growth mutual funds. Construct a frequency and relative frequency distribution of the data.

Approach: To construct a frequency distribution, we first create classes of equal width. There are 40 observations in Table 12, and they range from 0.05 to 14.48, so we decide to create the classes such that the lower class limit of the first

In Other Words

For qualitative and many discrete data, the categories of data are formed by using the data. For continuous data, the categories are formed by using an interval of numbers, such as 30–39.

Table 12

Three-Year Rate of Return of Mutual Funds

5.37	4.31	4.13	8.58	5.99	7.90	9.11	6.11
3.06	14.48	12.50	8.33	10.10	8.21	6.83	10.94
2.34	0.97	8.33	8.89	6.07	6.50	5.99	9.38
0.05	13.88	3.71	10.07	9.88	4.93	6.38	10.34
2.27	11.91	11.69	12.06	9.84	7.75	2.86	6.68

Source: Morningstar.com

! CAUTION

Watch out for tables with class widths that overlap, such as a first class of 20–30 and a second class of 30–40.

Historical Note

Florence Nightingale was born in Italy on May 12, 1820. She was named after the city of her birth. Nightingale was educated by her father, who attended Cambridge University. Between 1849 and 1851, she studied nursing throughout Europe. In 1854, she was asked to oversee the introduction of female nurses into the military hospitals in Turkey. While there, she greatly improved the mortality rate of wounded soldiers. She collected data and invented graphs (the polar area diagram), tables, and charts to show that improving sanitary conditions would lead to decreased mortality rates. In 1869, Nightingale founded the Nightingale School Home for Nurses. After a long and eventful life as a reformer of health care and contributor to graphics in statistics, Florence Nightingale died on August 13, 1910.

class is 0 (a little smaller than the smallest data value) and the class width is 2. There is nothing magical about the choice of 2 as a class width. We could have selected a class width of 8 (or any other class width, as well). We choose a class width that we think will nicely summarize the data. If our choice doesn't accomplish this, we can always try another one. The lower class limit of the second class will be $0 + 2 = 2$. Because the classes must not overlap, the upper class limit of the first class is 1.99. Continuing in this fashion, we obtain the following classes:

$$0-1.99$$
$$2-3.99$$
$$\vdots$$
$$14-15.99$$

This gives us eight classes. We tally the number of observations in each class, add up the tallies, and create the frequency distribution. The relative frequency distribution would be created by dividing each class's frequency by 40, the number of observations.

Solution: We tally the data as shown in the second column of Table 13. The third column in the table shows the frequency of each class. From the frequency distribution, we conclude that a three-year rate of return between 8.0% and 9.99% occurs with the most frequency. The fourth column in the table shows the relative frequency of each class. So 22.5% of the small-capitalization growth mutual funds had a three-year rate of return between 8% and 9.99%.

Table 13

Class (3-Year Rate of Return)	Tally	Frequency	Relative Frequency
0–1.99	||	2	2/40 = 0.05
2–3.99	⁙	5	5/40 = 0.125
4–5.99	⁙ |	6	6/40 = 0.15
6–7.99	⁙ |||	8	0.2
8–9.99	⁙ ||||	9	0.225
10–11.99	⁙ |	6	0.15
12–13.99	|||	3	0.075
14–15.99	|	1	0.025

Only one mutual fund had a three-year rate of return between 14% and 15.99%. We might consider this mutual fund worthy of our investment. This type of information would be more difficult to obtain from the raw data.

Notice that the choice of the lower class limit of the first class and the class width was rather arbitrary. While formulas and procedures do exist for creating frequency distributions from raw data, they do not necessarily provide better summaries. It is incorrect to say that a particular frequency distribution is the correct one. Constructing frequency distributions is somewhat of an art form in which the distribution that seems to provide the best overall summary of the data should be used.

Consider the frequency distribution in Table 14, which also summarizes the three-year rate of return data discussed in Example 3. Here, the lower class limit of the first class is 0 and the class width is 4. Do you think Table 13 or Table 14 provides a better summary of the distribution of three-year rates of return? In forming your opinion, consider the following: Too few classes will cause a bunching effect. Too many classes will spread the data out, thereby not revealing any pattern.

In Other Words

Creating the classes for summarizing continuous data is an art form. There is no such thing as the correct frequency distribution. However, there can be less desirable frequency distributions. The larger the class width, the fewer classes a frequency distribution will have.

Table 14		
Class	**Tally**	**Frequency**
0–3.99	‖‖ ‖	7
4–7.99	‖‖ ‖‖ ‖‖	14
8–11.99	‖‖ ‖‖ ‖‖	15
12–15.99	‖‖	4

The goal in constructing a frequency distribution is to reveal interesting features of the data. With that said, when constructing frequency distributions, we typically want the number of classes to be between 5 and 20. When the data set is small, we want fewer classes. When the data set is large, we want more classes. Why do you think this is reasonable?

Now Work Problems 29(a) and (b).

4   Construct Histograms of Continuous Data

We are now ready to draw histograms of continuous data.

EXAMPLE 4   **Drawing a Histogram of Continuous Data**

Problem: Construct a frequency and relative frequency histogram of the three-year rate of return data discussed in Example 3.

Note to Instructor
Have students think about the factors to consider in determining an appropriate class width. Are there any ways that a histogram can be used to distort the data?

Approach: To draw the frequency histogram, we will use the frequency distribution in Table 13. We label the lower class limits of each class on the horizontal axis. Then, for each class, we draw a rectangle whose width is the class width and whose height is the frequency. To construct the relative frequency histogram, we let the height of the rectangle be the relative frequency, instead of the frequency.

Solution: Figure 7(a) represents the frequency histogram, and Figure 7(b) represents the relative frequency histogram.

Figure 7
Frequency and Relative
Frequency Histograms

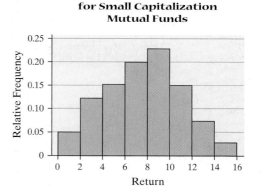

EXAMPLE 5

Drawing a Histogram for Continuous Data Using Technology

Problem: Construct a frequency and relative frequency histogram of the three-year rate of return data discussed in Example 3.

Approach: We will use MINITAB to construct the frequency and relative frequency histograms. The steps for constructing the graphs using the TI-83/84 Plus graphing calculators, MINITAB, and Excel are given in the Technology Step by Step on page 92.

Result: Figure 8(a) shows the frequency histogram and Figure 8(b) shows the relative frequency histogram obtained from MINITAB. Note that MINITAB expresses relative frequencies using percent.

Figure 8

(a)

(b)

Now Work Problems 29(c)–(d).

⑤ Draw Stem-and-Leaf Plots

A **stem-and-leaf plot** is another way to represent quantitative data graphically. Stem-and-leaf plots have some advantages over histograms. Use the following steps to construct a stem-and-leaf plot.

Construction of a Stem-and-Leaf Plot

Step 1: The **stem** of the graph will consist of the digits to the left of the right-most digit. The **leaf** of the graph will be the rightmost digit. Sometimes it is necessary to modify the method of choosing the stem if a different class width is desired.

Step 2: Write the stems in a vertical column in increasing order. Draw a vertical line to the right of the stems.

Step 3: Write each leaf corresponding to the stems to the right of the vertical line.

Step 4: Write the leaves in ascending order.

EXAMPLE 6

Constructing a Stem-and-Leaf Plot

Problem: The data in Table 15 represent the percentage of persons in poverty, by state, in 2002. Draw a stem-and-leaf plot of the data.

Approach:

Step 1: We will treat the integer portion of the number as the stem and the decimal portion as the leaf. For example, the stem of Alabama will be 14 and the leaf will be 6. The stem of 14 will include all data from 14.0 to 14.9.

Step 2: Write the stems vertically in ascending order, and then draw a vertical line to the right of the stems.

Note to Instructor

Spend some time discussing how the stems are chosen.

Table 15

Percentage of Persons Living in Poverty

State	Percent	State	Percent	State	Percent
Alabama	14.6	Kentucky	13.1	North Dakota	11.9
Alaska	8.3	Louisiana	17.0	Ohio	10.1
Arizona	13.3	Maine	11.3	Oklahoma	14.7
Arkansas	18.0	Maryland	7.3	Oregon	11.2
California	12.8	Massachusetts	9.6	Pennsylvania	9.2
Colorado	9.4	Michigan	10.3	Rhode Island	10.3
Connecticut	7.8	Minnesota	6.5	South Carolina	13.5
Delaware	8.1	Mississippi	17.6	South Dakota	10.2
D.C.	16.8	Missouri	9.6	Tennessee	14.2
Florida	12.1	Montana	13.7	Texas	15.3
Georgia	12.1	Nebraska	9.5	Utah	9.3
Hawaii	10.6	Nevada	8.3	Vermont	9.9
Idaho	11.8	New Hampshire	5.6	Virginia	8.7
Illinois	11.2	New Jersey	7.8	Washington	10.8
Indiana	8.7	New Mexico	17.8	West Virginia	16.0
Iowa	8.3	New York	14.0	Wisconsin	8.6
Kansas	9.4	North Carolina	13.1	Wyoming	9.5

Source: *Poverty in the United States, 2002*, Current Population Reports

Step 3: Write the leaves corresponding to the stem.

Step 4: Write the leaves in ascending order.

Solution:

Step 1: The stem from Alabama is 14 and the corresponding leaf is 6. The stem from Alaska is 8 and its leaf is 3, and so on.

Step 2: We write the stems vertically in Figure 9(a), along with a vertical line to the right of the stem.

Step 3: We write the leaves corresponding to each stem. See Figure 9(b).

Step 4: Write the leaves in ascending order. See Figure 9(c).

Figure 9

```
   5 |                    5 | 6                      5 | 6
   6 |                    6 | 5                      6 | 5
   7 |                    7 | 8 3 8                  7 | 3 8 8
   8 |                    8 | 3 1 7 3 3 7 6          8 | 1 3 3 3 6 7 7
   9 |                    9 | 4 4 6 6 5 2 3 9 5      9 | 2 3 4 4 5 5 6 6 9
  10 |                   10 | 6 3 1 3 2 8           10 | 1 2 3 3 6 8
  11 |                   11 | 8 2 3 9 2             11 | 2 2 3 8 9
  12 |                   12 | 8 1 1                 12 | 1 1 8
  13 |                   13 | 3 1 7 1 5             13 | 1 1 3 5 7
  14 |                   14 | 6 0 7 2               14 | 0 2 6 7
  15 |                   15 | 3                     15 | 3
  16 |                   16 | 8 0                   16 | 0 8
  17 |                   17 | 0 6 8                 17 | 0 6 8
  18 |                   18 | 0                     18 | 0

      (a)                   (b)                        (c)
```

EXAMPLE 7   **Constructing a Stem-and-Leaf Plot Using Technology**

Problem: Construct a stem-and-leaf plot of the poverty data discussed in Example 6.

Approach: We will use MINITAB to construct the stem-and-leaf plot. The steps for constructing the graphs using MINITAB are given in the Technology Step by Step on page 92. **Note:** The TI graphing calculators and Excel are not capable of drawing stem-and-leaf plots.

Result: Figure 10 shows the stem-and-leaf plot obtained from MINITAB.

USING TECHNOLOGY

In MINITAB, there is a column of numbers left of the stem. The (6) indicates that there are 6 observations in the class containing the middle value (called the *median*). The values above the (6) represent the number of observations less than or equal to the upper class limit of the class. For example, 12 states have percentage in poverty less than 8.9. The values in the left column below the (6) indicate the number of observations greater than or equal to the lower class limit of the class. For example, 7 states have percentages in poverty greater than 15.0.

Figure 10

```
    1     5     6
    2     6     5
    5     7     388
   12     8     1333677
   21     9     234455669
  (6)    10     123368
   24    11     22389
   19    12     118
   16    13     11357
   11    14     0267
    7    15     3
    6    16     08
    4    17     068
    1    18     0
```

Now Work Problem 35.

If you look at the stem-and-leaf plot carefully, you'll notice that it looks much like a histogram turned on its side. The stem serves as the class. For example, the stem 10 contains all data from 10.0 to 10.9. The leaf serves as the frequency (height of the rectangle). Therefore, it is important to space the leaves equally when drawing a stem-and-leaf plot.

One advantage of the stem-and-leaf plot over frequency distributions and histograms is that the raw data can be retrieved from the stem-and-leaf plot.

Once a frequency distribution or histogram of continuous data is created, the raw data are lost. However, the raw data can be retrieved from the stem-and-leaf plot.

The steps listed for creating stem-and-leaf plots sometimes must be modified to meet the needs of the data. Consider the next example.

In Other Words

The choice of the stem in the construction of a stem-and-leaf diagram is also an art form. It acts just like the class width. For example, the stem of 7 in Figure 10 represents the class 7.0–7.9. The stem of 8 represents the class 8.0–8.9. Notice that the class width is 1.0. The number of leaves is the frequency of each category.

EXAMPLE 8 | **Constructing a Stem-and-Leaf Plot after Modifying the Data**

Problem: Construct a stem-and-leaf plot of the three-year rate of return data listed in Table 12 on page 74.

Approach

Step 1: If we follow the approach from Example 6 and use the integer portion as the stem and the decimals as the leaves, the stems will be 0, 1, 2, ..., 14. This is fine. However, the leaves will be two digits (such as 37, 06, and so on). This is not acceptable. To address this problem, we will round the data to the nearest tenth. Then the stem can be the whole numbers 0, 1, 2, ..., 14, and the leaves will be the decimal portion.

Step 2: Create a vertical column of the whole-number stems in increasing order.

Step 3: Write the leaves corresponding to each stem.

Step 4: Write the leaves in ascending order.

Solution:

Step 1: We round the data to the nearest tenth as shown in Table 16.

Table 16							
5.4	4.3	4.1	8.6	6.0	7.9	9.1	6.1
3.1	14.5	12.5	8.3	10.1	8.2	6.8	10.9
2.3	1.0	8.3	8.9	6.1	6.5	6.0	9.4
0.1	13.9	3.7	10.1	9.9	4.9	6.4	10.3
2.3	11.9	11.7	12.1	9.8	7.8	2.9	6.7

Step 2: Write the stems vertically in ascending order as shown in Figure 11(a).
Step 3: Write the leaves corresponding to each stem as shown in Figure 11(b).
Step 4: Write the leaves in ascending order as shown in Figure 11(c).

Figure 11

```
 0 |              0 | 1                      0 | 1
 1 |              1 |                        1 |
 2 |              2 | 3  3  9                2 | 3  3  9
 3 |              3 | 1  7                   3 | 1  7
 4 |              4 | 3  1  9                4 | 1  3  9
 5 |              5 | 4                      5 | 4
 6 |              6 | 0 1 5 8 0 4 1 7        6 | 0 0 1 1 4 5 7 8
 7 |              7 | 9  8                   7 | 8  9
 8 |              8 | 3 6 3 9 2              8 | 2  3369
 9 |              9 | 9 8 1 4                9 | 1 4 8 9
10 |             10 | 1 1 9 3              10 | 1 1 3 9
11 |             11 | 9  7                 11 | 7  9
12 |             12 | 5  1                 12 | 1  5
13 |             13 | 9                    13 | 9
14 |             14 | 5                    14 | 5

   (a)               (b)                      (c)
```

Note to Instructor
Discuss the similarities and differences of histograms and stem-and-leaf diagrams. Remind students that they can determine the shape of the distribution just as they did with histograms, by turning the stem-and-leaf plot on its side.

Split Stems

In Other Words
Using split stems is like adding more classes to a frequency distribution.

Consider the data shown in Table 17. The data range from 11 to 48. If we drew a stem-and-leaf plot using the tens digit as the stem and the ones digit as the leaf, we would obtain the results shown in Figure 12. The data appear rather "bunched." To resolve this problem, we can use **split stems.** For example, rather than using one stem for the class of data 10–19, we could use two stems, one for the 10–14 interval and the second for the 15–19 interval. We do this in Figure 13.

Table 17				
27	17	11	24	36
13	29	22	18	17
23	30	12	46	17
32	48	11	18	23
18	32	26	24	38
24	15	13	31	22
18	21	27	20	16
15	37	19	19	29

Figure 12

```
1 | 1 1 2 3 3 5 5 6 7 7 7 8 8 8 8 9 9
2 | 0 1 2 2 3 3 4 4 4 6 7 7 9 9
3 | 0 1 2 2 6 7 8
4 | 6 8
```

Figure 13

```
1 | 1 1 2 3 3
1 | 5 5 6 7 7 7 8 8 8 8 9 9
2 | 0 1 2 2 3 3 4 4 4
2 | 6 7 7 9 9
3 | 0 1 2 2
3 | 6 7 8
4 |
4 | 6 8
```

The stem-and-leaf plot shown in Figure 13 reveals the distribution of the data better. As with the construction of class intervals in the creation of frequency histograms, judgment plays a major role. There is no such thing as a correct stem-and-leaf plot. However, a quick comparison of Figures 12 and 13 shows that some are better than others.

One final note: **Stem-and-leaf plots are best used when the data set is small.**

Now Work Problem 41.

Draw Dot Plots

One more graph! A **dot plot** is drawn by placing each observation horizontally in increasing order and placing a dot above the observation each time it is observed.

EXAMPLE 9 ### Drawing a Dot Plot

Problem: Draw a dot plot for the number of arrivals at Wendy's data from Example 1 on page 71.

Approach: The smallest observation in the data set is 1 and the largest is 11. We write the numbers 1 through 11 horizontally. For each observation, we place a dot above the value of the observation.

Solution: Figure 14 shows the dot plot.

Figure 14

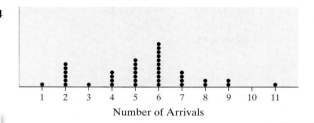

Number of Arrivals

Now Work Problem 45.

Identify the Shape of a Distribution

Note to Instructor
Give examples of data sets that may be uniform, bell shaped, and skewed right. For example, rolling a single die, IQ scores, and income, respectively.

One way that a variable is described is through the shape of its distribution. Distribution shapes are typically classified as symmetric, skewed left, or skewed right. Figure 15 displays various histograms and the shape of the distribution.

Figures 15(a) and (b) display symmetric distributions. These distributions are symmetric because, if we split the histogram down the middle, the right and left sides of the histograms are mirror images. Figure 15(a) is a **uniform distribution**, because the frequency of each value of the variable is evenly spread out across the values of the variable. Figure 15(b) displays a **bell-shaped distribution**, because the highest frequency occurs in the middle and frequencies tail off to the left and right of the middle so the graph looks like the profile of a bell. Figure 15(c) illustrates a distribution that is **skewed right**. Notice that the tail to the right of the peak is longer than the tail to the left of the peak. Finally, Figure 15(d) illustrates a distribution that is **skewed left**, because the tail to the left of the peak is longer than the tail to the right of the peak.

CAUTION
We do not describe qualitative data as skewed left, skewed right, or uniform.

CAUTION
It is important to recognize that data will not always exhibit behavior that perfectly matches any of the shapes given in Figure 15. To identify the shape of a distribution, some flexibility is required. In addition, people may disagree on the shape, since identifying shape is subjective.

Figure 15

(a) Uniform (symmetric)

(b) Bell-shaped (symmetric)

(c) Skewed Right

(d) Skewed Left

EXAMPLE 10

Identifying the Shape of a Distribution

Problem: Figure 16 displays the histogram obtained for the three-year rates of return for small-capitalization stocks. Describe the shape of the distribution.

Approach: We compare the shape of the distribution displayed in Figure 16 with those in Figure 15.

Solution: Since the histogram looks most like Figure 15(b), the distribution is bell shaped.

Figure 16

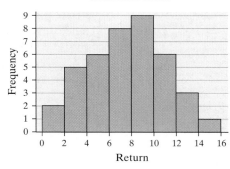

Three-Year Rate of Return for Small Capitalization Mutual Funds

Now Work Problem 29(e).

In-Class Activity: Random-Number Generators, Pulse Rate, and Household Size

1. We saw in Chapter 1 how to use a graphing calculator or statistical spreadsheet to generate random numbers. Using either of these, randomly generate 200 integers between 0 and 20, inclusive. That is, the integer can be any value greater than or equal to 0 or less than or equal to 20. Store these data in your calculator or spreadsheet.

2. Everyone in the class should determine his or her resting pulse rate. Collect these data for the class and store them in your calculator or spreadsheet.

3. Everyone in class should share how many people live in their household. Collect the data for the class, and input them into your calculator or spreadsheet.

   (a) What shape do you expect the distribution of random integers to have? Why?

   (b) What shape do you expect the distribution of pulse rates to have? Why?

   (c) What shape do you expect the distribution of household size to have? Why?

   (d) Draw a histogram of each data set. For the random integer data, use a class width of 2.

   (e) What shape did each have? Are you surprised?

Note to Instructor

This activity should take about 10 to 15 minutes, especially if you are using a calculator or software to draw the histograms.

 Draw Time-Series Graphs

If the value of a variable is measured at different points in time, the data are referred to as **time-series data**. The closing price of Cisco Systems stock each month for the past 12 years is an example of time-series data.

Definition

A **time-series plot** is obtained by plotting the time in which a variable is measured on the horizontal axis and the corresponding value of the variable on the vertical axis. Lines are then drawn connecting the points.

Time-series plots are very useful in identifying trends in the data.

Table 18			
Date	**Closing Price**	**Date**	**Closing Price**
3/03	12.98	3/04	23.57
4/03	15.00	4/04	20.91
5/03	16.41	5/04	22.37
6/03	16.79	6/04	23.70
7/03	19.49	7/04	20.92
8/03	19.14	8/04	18.76
9/03	19.59	9/04	18.10
10/03	20.93	10/04	19.21
11/03	22.70	11/04	18.75
12/03	24.23	12/04	19.32
1/04	25.71	1/05	18.04
2/04	23.16	2/05	17.42

Source: NASDAQ

EXAMPLE 11   Drawing a Time-Series Plot

Problem: The data in Table 18 represent the closing price of Cisco Systems stock at the end of each month from March 2003 through February 2005. Construct a time-series plot of the data.

Approach

Step 1: Plot points for each month, with the date on the horizontal axis and the closing price on the vertical axis.

Step 2: Connect the points with straight lines.

Solution: Figure 17 shows the time-series plot. The trend since June, 2004 does not bode well for investors of Cisco Systems stock.

Figure 17

USING TECHNOLOGY:
Statistical spreadsheets, such as Excel or MINITAB, and certain graphing calculators, such as the TI-83 or TI-84 Plus, have the ability to create time-series graphs.

Now Work Problem 47.

2.2 ASSESS YOUR UNDERSTANDING

Concepts and Vocabulary

1. Discuss circumstances under which it is preferable to use relative frequency distributions, instead of frequency distributions.
2. Why shouldn't classes overlap when one summarizes continuous data?
3. The histogram to the right represents the total rainfall for each time it rained in Chicago during the month of August since 1871. The histogram was taken from the *Chicago Tribune* on August 14, 2001. What is wrong with the histogram?
4. State the advantages and disadvantages of histograms versus stem-and-leaf plots.
5. Contrast the differences between histograms and bar graphs.
6. *True or False*: There is not one particular frequency distribution that is correct, but there are frequency distributions that are less desirable than others. True

Total August Rain Events Since 1871 in Chicago

7. *True or False*: The shape of the distribution shown is best classified as skewed left. False

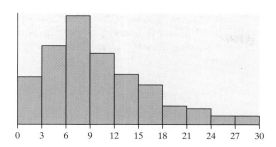

8. *True or False*: The shape of the distribution shown is best classified as uniform. False

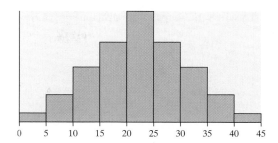

Skill Building

9. Rolling the Dice An experiment was conducted in which two fair dice were thrown 100 times. The sum of the pips showing on the dice was then recorded. The following frequency histogram gives the results.

(a) What was the most frequent outcome of the experiment? 8
(b) What was the least frequent? 2
(c) How many times did we observe a 7? 15
(d) Determine the percentage of time a 7 was observed. 15%
(e) Describe the shape of the distribution. Bell shaped

10. Car Sales A car salesman records the number of cars he sold each week for the past year. The following frequency histogram shows the results.   (10c) 17.3%

(a) What is the most frequent number of cars sold in a week? 4
(b) For how many weeks were two cars sold? 9
(c) Determine the percentage of time two cars were sold.
(d) Describe the shape of the distribution. Skewed right

11. IQ Scores The following frequency histogram represents the IQ scores of a random sample of seventh-grade students. IQs are measured to the nearest whole number. The frequency of each class is labeled above each rectangle.

(a) How many students were sampled? 200
(b) Determine the class width. 10
(c) Identify the classes and their frequencies.
(d) Which class has the highest frequency? 100–109
(e) Which class has the lowest frequency? 150–159

12. Alcohol-Related Traffic Fatalities The following frequency histogram represents the number of alcohol-related traffic fatalities by state in 2003 according to data obtained from the National Highway Traffic Safety Administration.

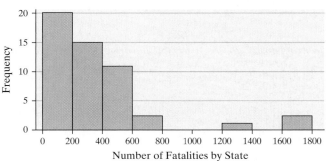

(a) Determine the class width. *200*
(b) Identify the classes.
(c) Which class has the highest frequency? *0–199*
(d) Describe the shape of the distribution. *Skewed right*

(e) A reporter writes the following statement: "According to the data, Texas had 1709 alcohol-related deaths, while Vermont had only 29. So the roads in Vermont are much safer." Explain what is wrong with this statement and how a fair comparison can be made between alcohol-related traffic fatalities in Texas versus Vermont.

Applying the Concepts

13. Predicting School Enrollment To predict future enrollment, a local school district wants to know the number of children under the age of 5. Fifty households within the district were sampled, and the head of household was asked to disclose the number of children under the age of 5 living in the household. The results of the survey are presented in the following table.

Number of Children under 5	Number of Households
0	16
1	18
2	12
3	3
4	1

(a) Construct a relative frequency distribution of the data.
(b) What percentage of households has two children under the age of 5? *24%*
(c) What percentage of households has one or two children under the age of 5? *60%*

14. Free Throws A basketball player habitually makes 70% of her free throws. In an experiment, a researcher asks this basketball player to record the number of free throws she shoots until she misses. The experiment is repeated 50 times. The following table lists the distribution of the number of free throws attempted until a miss is recorded.

Number of Free Throws until a Miss	Frequency
1	16
2	11
3	9
4	7
5	2
6	3
7	0
8	1
9	0
10	1

(a) Construct a relative frequency distribution of the data.
(b) What percentage of the time did she miss on her fourth free throw? *14%*
(c) What percentage of the time did she make nine in a row and then miss the tenth free throw? *2%*
(d) What percentage of the time did she make at least five in a row? *10%*

In Problems 15 and 16, determine the original set of data. The stem represents the tens digit and the leaf represents the ones digit.

15.
```
1 | 0 1 4
2 | 1 4 4 7 9
3 | 3 5 5 5 7 7 8
4 | 0 0 1 2 6 6 8 9 9
5 | 3 3 5 8
6 | 1 2
```

16.
```
4 | 0 4 7
5 | 2 2 3 9 9
6 | 3 4 5 8 8 9
7 | 0 1 1 3 6 6
8 | 2 3 8
```

In Problems 17 and 18, determine the original set of data. The stem represents the ones digits and the leaf represents the tenths digit.

17.
```
1 | 2 4 6
2 | 1 4 7 7 9
3 | 3 3 3 5 7 7 8
4 | 0 1 1 3 6 6 8 8 9
5 | 3 4 5 8
6 | 2 4
```

18.
```
12 | 3 7 9 9
13 | 0 4 5 7 8 9 9
14 | 2 4 4 7 7 8 9
15 | 1 2 2 5 6
16 | 0 3
```

In Problems 19–22, find (a) the number of classes, (b) the class limits, and (c) the class width.

19. Health Insurance The following data represent the number of people aged 25 to 64 covered by health insurance in 2003.

Age	Number (millions)
25–34	28.9
35–44	35.7
45–54	35.1
55–64	24.7

Source: U.S. Census Bureau, Current Population Survey, 2004 Annual Social and Economic Supplement

21. Hospital Admissions The following data represent community hospital admissions for hospitals with 100 to 499 beds during the year 2002.

Number of Beds	Admissions (in thousands)
100–199	6,826
200–299	6,800
300–399	5,607
400–499	3,593

Source: Health, United States, 2004, from the National Center for Health Statistics

20. Earthquakes The following data represent the number of earthquakes worldwide whose magnitude was less than 8.0 in 2004.

Magnitude	Number
0–0.9	3,179
1.0–1.9	1,340
2.0–2.9	6,297
3.0–3.9	7,826
4.0–4.9	10,975
5.0–5.9	1,430
6.0–6.9	139
7.0–7.9	13

Source: U.S. Geological Survey, Earthquake Hazards Program

22. Multiple Births The following data represent the number of live multiple births (three or more babies) in 2002 for women 15 to 44 years old.

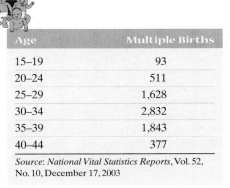

Age	Multiple Births
15–19	93
20–24	511
25–29	1,628
30–34	2,832
35–39	1,843
40–44	377

Source: National Vital Statistics Reports, Vol. 52, No. 10, December 17, 2003

In Problems 23–26, construct (a) a relative frequency distribution, (b) a frequency histogram, and (c) a relative frequency histogram for the given data. Then answer the questions that follow.

23. Using the data in Problem 19, of the people covered by health insurance, what percentage is 25 to 34 years old? Of the people covered by health insurance, what percentage is 44 years or younger? 23.23%; 51.93%

24. Using the data in Problem 20, what percentage of earthquakes registered 4.0 to 4.9? What percentage of earthquakes registered 4.9 or less? 35.18%; 94.93%

25. Using the data in Problem 21, what percentage of the admissions was in hospitals with 300 to 399 beds? What percentage of the admissions was in hospitals with 300 or more beds? 24.56%; 40.30%

26. Using the data in Problem 22, what percentage of multiple births was to women 40 to 44 years old? What percentage of multiple births was to women 24 years or younger? 5.18%; 8.30%

27. Waiting The following data represent the number of customers waiting for a table at 6:00 P.M. for 40 consecutive Saturdays at Bobak's Restaurant:

11	5	11	3
4	5	13	9
13	10	9	6
10	8	7	3
7	9	10	4
6	8	6	7
6	4	14	11
8	10	9	5
8	8	7	8
8	6	11	8

(a) Construct a frequency distribution of the data.
(b) Construct a relative frequency distribution of the data.
(c) What percentage of the Saturdays had 10 or more customers waiting for a table at 6:00 P.M.? 27.5%
(d) What percentage of the Saturdays had five or fewer customers waiting for a table at 6:00 P.M.? 20%
(e) Construct a frequency histogram of the data.
(f) Construct a relative frequency histogram of the data.
(g) Describe the shape of the distribution. Symmetric

28. Highway Repair The following data represent the number of potholes on 50 randomly selected 1-mile stretches of highway in the city of Chicago.

2	7	4	7	2	7
2	2	2	3	4	3
1	2	3	2	1	4
2	2	5	2	3	4
4	1	7	10	3	5
4	3	3	2	2	
1	6	5	7	9	
2	2	2	1	5	
3	5	1	3	5	

(a) Construct a frequency distribution of the data.
(b) Construct a relative frequency distribution of the data.
(c) What percentage of the 1-mile stretches of highway had seven or more potholes? 14%
(d) What percentage of the 1-mile stretches of highway had two or fewer potholes? 42%
(e) Construct a frequency histogram of the data.
(f) Construct a relative frequency histogram of the data.
(g) Describe the shape of the distribution. Skewed right

29. Average Income The following data represent the per NW capita (average) disposable income (income after taxes) for the 50 states and the District of Columbia in 2003.

24,028	30,641	24,293	22,123	29,798	30,507
36,726	28,960	42,345	27,610	26,356	27,837
23,584	30,063	25,929	26,409	27,033	23,567
23,889	25,900	32,637	34,570	27,275	30,397
21,677	26,317	23,528	27,865	28,188	31,251
35,411	23,301	31,527	25,307	26,902	26,684
24,169	26,102	28,557	28,365	23,753	27,149
26,314	26,922	22,581	27,750	29,683	30,288
22,252	27,508	29,600			

Source: U.S. Bureau of Economic Analysis, May 2004

With the first class having a lower class limit of 20,000 and a class width of 2500,

(a) Construct a frequency distribution.
(b) Construct a relative frequency distribution.
(c) Construct a frequency histogram of the data.
(d) Construct a relative frequency histogram of the data.
(e) Describe the shape of the distribution.
(f) Repeat parts (a)–(e) using a class width of 4000. Which frequency distribution seems to provide a better summary of the data?
(g) The highest per capita disposable income exists in the District of Columbia, yet the District of Columbia has one of the highest unemployment rates (7% unemployed). Is this surprising to you? Why?

30. Poverty Every year the federal government adjusts the income level that must be earned to be above the poverty level. The dollar amount depends on the number of people living in the household. In 2002, a family of four needed to earn $18,392 to be above the poverty line. The following data represent the percentage of each state's population (plus the District of Columbia) that is living in poverty.

14.6	8.3	13.3	18.0	12.8	9.4
7.8	8.1	16.8	12.1	12.1	10.6
11.8	11.2	8.7	8.3	9.4	13.1
17.0	11.3	7.3	9.6	10.3	6.5
17.6	9.6	13.7	9.5	8.3	5.6
7.8	17.8	14.0	13.1	11.9	10.1
14.7	11.2	9.2	10.3	13.5	10.2
14.2	15.3	9.3	9.9	8.7	10.8
16.0	8.6	9.5			

Source: U.S. Census Bureau

With the first class having a lower class limit of 5 and a class width of 1,

(a) Construct a frequency distribution.
(b) Construct a relative frequency distribution.
(c) Construct a frequency histogram of the data.
(d) Construct a relative frequency histogram of the data.
(e) Describe the shape of the distribution.
(f) Repeat parts (a)–(e) using a class width of 2. Which frequency distribution seems to provide a better summary of the data?
(g) From Problem 29, we learned that the highest per capita disposable income exists in the District of Columbia. The poverty rate of the District of Columbia is 16.8. Is this surprising to you? Why?

31. **Serum HDL** Dr. Paul Oswiecmiski randomly selects 40 of his 20- to 29-year-old patients and obtains the following data regarding their serum HDL cholesterol:

70	56	48	48	53	52	66	48
36	49	28	35	58	62	45	60
38	73	45	51	56	51	46	39
56	32	44	60	51	44	63	50
46	69	53	70	33	54	55	52

With the first class having a lower class limit of 20 and a class width of 10,
(a) Construct a frequency distribution.
(b) Construct a relative frequency distribution.
(c) Construct a frequency histogram of the data.
(d) Construct a relative frequency histogram of the data.
(e) Describe the shape of the distribution. Bell shaped
(f) Repeat parts (a)–(e) using a class width of 5.
(g) Which frequency distribution seems to provide a better summary of the data?

32. **Volume of Altria Group Stock** The volume of a stock is the number of shares traded on a given day. The following data, in millions, so that 3.78 represents 3,780,000 shares traded, represent the volume of Altria Group stock traded for a random sample of 35 trading days in 2004.

3.78	8.74	4.35	5.02	8.40
6.06	5.75	5.34	6.92	6.23
5.32	3.25	6.57	7.57	6.07
3.04	5.64	5.00	7.16	4.88
10.32	3.38	7.25	6.52	4.43
3.38	5.53	4.74	9.70	3.56
10.96	4.50	7.97	3.01	5.58

Source: Yahoo.finance.com

With the first class having a lower class limit of 3 and a class width of 2,
(a) Construct a frequency distribution.
(b) Construct a relative frequency distribution.
(c) Construct a frequency histogram of the data.
(d) Construct a relative frequency histogram of the data.
(e) Describe the shape of the distribution.
(f) Repeat parts (a)–(e) using a class width of 1.
(g) Which frequency distribution seems to provide a better summary of the data?

33. **Dividend Yield** A dividend is a payment from a publicly traded company to its shareholders. The dividend yield of a stock is determined by dividing the annual dividend of a stock by its price. The following data represent the dividend yields (in percent) of a random sample of 28 publicly traded stocks of companies with a value of at least $5 billion.

1.7	0	1.15	0.62	1.06	2.45	2.38
2.83	2.16	1.05	1.22	1.68	0.89	0
2.59	0	1.7	0.64	0.67	2.07	0.94
2.04	0	0	1.35	0	0	0.41

Source: Yahoo! Finance

With the first class having a lower class limit of 0 and a class width of 0.40,
(a) Construct a frequency distribution.
(b) Construct a relative frequency distribution.
(c) Construct a frequency histogram of the data.
(d) Construct a relative frequency histogram of the data.
(e) Describe the shape of the distribution.
(f) Repeat parts (a)–(e) using a class width of 0.8.
(g) Which frequency distribution seems to provide a better summary of the data?

34. **Violent Crimes** Violent crimes include murder, forcible rape, robbery, and aggravated assault. The following data represent the violent crime rate (crimes per 100,000 population) by state plus the District of Columbia in 2002.

444	563	553	424	593	352
311	599	1,633	770	459	262
255	621	357	286	377	279
662	108	770	484	540	268
343	539	352	314	638	161
375	740	496	470	78	351
503	292	402	285	822	177
717	579	237	107	291	345
234	225	274			

Source: U.S. Federal Bureau of Investigation

With the first class having a lower class limit of 0 and a class width of 150,
(a) Construct a frequency distribution.
(b) Construct a relative frequency distribution.
(c) Construct a frequency histogram of the data.
(d) Construct a relative frequency histogram of the data.
(e) Describe the shape of the distribution.
(f) Repeat parts (a)–(e) using a class width of 300. Which frequency distribution seems to provide a better summary of the data?
(g) Do you believe that the violent crime rate is a good measure of how safe a state is? Why or why not?

In Problems 35–38, construct stem-and-leaf plots.

35. Age at Inauguration The following data represent the ages of the presidents of the United States on inauguration day.

57	61	57	57	58
54	68	51	49	64
65	52	56	46	54
47	55	55	54	42
55	51	54	51	60
55	56	61	52	69
57	50	49	51	62
61	48	50	56	43
64	46	54		

36. Divorce Rate The following data represent the divorce rates (per 1000 population) for most states in the United States in the year 2001. (**Note:** The list includes the District of Columbia but excludes California, Colorado, Indiana, and Louisiana because of failure to report.)

5.3	4.1	4.2	6.6	2.9	4.0	2.3
5.4	3.8	3.8	5.6	3.2	3.2	3.2
5.5	3.9	3.0	2.4	3.9	3.3	5.4
4.3	2.6	3.7	6.8	5.0	3.5	5.1
3.0	4.5	2.7	4.0	3.4	4.9	3.2
3.3	3.5	3.4	5.2	4.1	4.4	4.0
4.3	4.5	5.2	3.2	6.1		

Source: U.S. Census, *Statistical Abstract of the United States*, 2003

37. Grams of Fat in a McDonald's Breakfast The following data represent the number of grams of fat in breakfast meals offered at McDonald's.

12	23	28	2	28	33
31	11	23	40	35	1
23	33	23	16	11	8
8	17	16	15		

Source: McDonald's Corporation, *A Full Serving of Nutrition Facts*, April 2003

38. Miles per Gallon The following data represent the number of miles per gallon achieved on the highway for compact cars for the model year 2005.

30	29	30	21	18	29	27	30	29
34	34	30	28	30	20	32	28	32
34	35	26	26	31	25	35	32	25
19	26	19	24	22	24	19	31	26
34	32	34	25	34	34	32	29	25
31	29	30	30	34	32	29	38	39
46	31	31	30	27	29	26	29	24

Source: U.S. Department of Energy

39. Energy from Coal The following data represent the percentage of energy derived from coal for the 50 states plus the District of Columbia.

54.2	0	75.0	64.3	15.6	8.0	94.2
8.5	32.4	90.4	34.6	87.7	55.7	0
40.6	62.3	23.2	83.1	16.6	0	50.8
48.5	13.3	2.7	60.2	60.4	38.3	8.4
1.3	0.9	59.5	63.1	94.6	42.4	98.1
77.6	46.1	27.4	51.1	90.4	62.1	68.1
10.3	93.7	56.6	23.3	60.8	36.8	95.8
57.7	83.2					

Source: U.S. Energy Information Administration

(a) Round each observation to the nearest percent and draw a stem-and-leaf diagram.

(b) Describe the shape of the distribution.

(c) West Virginia is one of the largest coal mining states. Which number in the data set do you think is West Virginia?

40. Housing Prices The data at the right represent the percentage change in the price of housing from 1998 to 2003 for a random sample of 40 cities.

(a) Round each observation to the nearest percent and draw a stem-and-leaf diagram.

(b) Describe the shape of the distribution.

23.3	20.8	32.5	15.8	47.1	18.9	66.0	22.6
23.4	24.1	16.2	9.1	17.1	22.7	17.0	21.6
29.9	15.6	24.8	52.4	28.3	53.5	17.8	20.6
20.6	37.6	49.4	62.4	11.8	19.2	19.8	59.1
48.1	19.1	35.9	14.7	24.9	25.0	26.1	47.7

Source: Global Insight

41. Dependability Survey J.D. Power and Associates regularly surveys car owners and NW asks them about the reliability of their cars. The following data represent the number of problems per 100 vehicles for the 2000 model year for all makes.

Make	Problems per 100 Vehicles	Make	Problems per 100 Vehicles	Make	Problems per 100 Vehicles
Lexus	163	Subaru	266	Audi	318
Infiniti	174	Nissan	267	Mercedes Benz	318
Buick	179	GMC	269	Jeep	321
Porsche	193	Chevrolet	272	Volvo	330
Acura	196	Saturn	273	Mitsubishi	339
Toyota	201	Oldsmobile	283	Hyundai	342
Cadillac	209	Mazda	288	Isuzu	368
Lincoln	212	Pontiac	293	Volkswagen	391
Honda	218	Chrysler	295	Suzuki	403
Mercury	240	Ford	295	Daewoo	421
Jaguar	247	Plymouth	302	Land Rover	441
Saab	255	Dodge	312	Kia	509
BMW	262				

Source: Associated Press

(a) Round the data to the nearest tens (for example, round 163 as 160).

(b) Draw a stem-and-leaf diagram, treating the hundreds position as the stem and the tens position as the leaf. For example, for the observation 160, 1 is the stem and 6 is the leaf. Do you think this is a good summary of the data? Why?

(c) Redraw the stem-and-leaf diagram using split stems. For example, data between 200 and 240 is one stem and data between 250 and 290 is a second stem. Does this stem-and-leaf diagram better summarize the data? Why?

42. Crime Rates The data on the following page represent the violent crime rates per 100,000 population for the 50 states and the District of Columbia in 2002.

(a) Round the data to the nearest tens (for example, round 563 to 560).

(b) Draw a stem-and-leaf diagram, treating the hundreds position as the stem and the tens position as the leaf. For the observation 560, 5 is the stem and 6 is the leaf. Do you think this is a good summary of the data? Why?

(c) Redraw the stem-and-leaf diagram using split stems. For example, data between 200 and 240 is one stem and data between 250 and 290 is a second stem. Does this stem-and-leaf diagram better summarize the data? Why?

State	Violent Crime Rate	State	Violent Crime Rate	State	Violent Crime Rate
Alabama	444	Kentucky	279	North Dakota	78
Alaska	563	Louisiana	662	Ohio	351
Arizona	553	Maine	108	Oklahoma	503
Arkansas	424	Maryland	770	Oregon	292
California	593	Massachusetts	484	Pennsylvania	402
Colorado	352	Michigan	540	Rhode Island	285
Connecticut	311	Minnesota	268	South Carolina	822
Delaware	599	Mississippi	343	South Dakota	177
District of Columbia	1633	Missouri	539	Tennessee	717
Florida	770	Montana	352	Texas	579
Georgia	459	Nebraska	314	Utah	237
Hawaii	262	Nevada	638	Vermont	107
Idaho	255	New Hampshire	161	Virginia	291
Illinois	621	New Jersey	375	Washington	345
Indiana	357	New Mexico	740	West Virginia	234
Iowa	286	New York	496	Wisconsin	225
Kansas	377	North Carolina	470	Wyoming	274

Source: Federal Bureau of Investigation

In Problems 43 and 44, we compare data sets. A great way to compare two data sets is through back to back stem-and-leaf diagrams. The figure represents the number of grams of fat in 14 sandwiches served at McDonald's and 14 sandwiches served at Burger King. In the figure, 0|1|5 represents 10 for McDonald's and 15 for Burger King.
Source: McDonald's Corporation, *A Full Serving of Nutrition Facts*, April 2003; Burger King Corporation, *Great Taste Table*, January 1999.

```
  McDonald's   Burger King
        7 4 0 | 1 | 5 9
  7 6 6 6 3 1 | 2 | 4 6 7 8
      7 4 2 0 | 3 | 6
            8 | 4 | 0 2 3 3 8
              | 5 | 9
              | 6 | 7
```

43. Academic Award Winners The following data represent the ages of the Academy Award winners for Best Actor and Best Actress in a leading role for the 30 years from 1975 to 2004.

Best Actor Ages					
38	60	30	40	42	37
76	39	52	45	35	61
43	51	32	42	54	52
37	38	31	45	60	45
40	36	47	29	43	37

Best Actress Ages					
41	35	31	41	33	31
74	33	49	38	61	21
41	25	80	42	28	33
35	45	49	39	34	24
25	33	35	35	28	30

(a) Construct a back to back stem-and-leaf display.
(b) Compare the two populations. What can you conclude from the back to back stem-and-leaf display?

44. Home Run Distances In 1998, Mark McGwire of the St. Louis Cardinals set the record for the most home runs hit in a season by hitting 70 home runs. Three years later in 2001, Barry Bonds of the San Francisco Giants broke McGwire's record by hitting 73 home runs. The following data represent the distances of each player's home runs in his record-setting season.

Mark McGwire

360	370	370	430	420	340	460
410	440	410	380	360	350	527
380	550	478	420	390	420	425
370	480	390	430	388	423	410
360	410	450	350	450	430	461
430	470	440	400	390	510	430
450	452	420	380	470	398	409
385	369	460	390	510	500	450
470	430	458	380	430	341	385
410	420	380	400	440	377	370

Barry Bonds

420	417	440	410	390	417	420
410	380	430	370	420	400	360
410	420	391	416	440	410	415
436	430	410	400	390	420	410
420	410	410	450	320	430	380
375	375	347	380	429	320	360
375	370	440	400	405	430	350
396	410	380	430	415	380	375
400	435	420	420	488	361	394
410	411	365	360	440	435	454
442	404	385				

(a) Construct a back to back stem-and-leaf display.
(b) Compare the two populations. What can you conclude from the back to back stem-and-leaf display?

NW **45. Waiting** Draw a dot plot of the waiting data from Problem 27.

46. Highway Repair Draw a dot plot of the highway repair data from Problem 28.

47. Pixar Corporation The data to the right represent the **NW** stock price for Pixar at the end of each month in 2004, adjusted for dividends and splits.
(a) Construct a time-series plot of the data and comment on the trend.
(b) During what month did the stock price increase the most? What might explain this?

Date	Closing Price	Date	Closing Price
1/04	66.39	7/04	68.24
2/04	65.76	8/04	77.72
3/04	64.46	9/04	78.90
4/04	68.27	10/04	80.42
5/04	67.85	11/04	90.67
6/04	69.51	12/04	85.61

48. Google Corporation The following data represent the stock price for Google at the end of each month from August 2004 to February 2005, adjusted for dividends and splits. Construct a time-series plot and comment on any trends. (**Note:** Google stock was traded publicly for the first time in August 2004.)

Google

Date	Closing Price
8/04	102.37
9/04	129.60
10/04	190.64
11/04	181.98
12/04	192.79
1/05	187.99
2/05	185.18

49. College Enrollment The following data represent the percentage of recent high school graduates (graduated within 12 months before the given year-end) who enrolled in college in the fall. Construct a time-series plot of the data.

Year	Percent Enrolled	Year	Percent Enrolled
1988	58.9	1996	65.0
1989	59.6	1997	67.0
1990	60.1	1998	65.6
1991	62.5	1999	62.9
1992	61.9	2000	63.3
1993	62.6	2001	61.7
1994	61.9	2002	65.2
1995	61.9	2003	63.9

Source: U.S. Center for Education Statistics

50. IRS Audits The following data represent the percentage of tax returns audited by the Internal Revenue Service. Construct a time-series plot of the data.

Year	Percent Audited	Year	Percent Audited
1988	1.57	1996	1.67
1989	1.29	1997	1.28
1990	1.04	1998	0.99
1991	1.17	1999	0.90
1992	1.06	2000	0.49
1993	0.92	2001	0.58
1994	1.07	2002	0.57
1995	1.67	2003	0.65

Source: U.S. General Accounting Office

Technology Step by Step

Drawing Histograms and Stem-and-Leaf Plots

TI-83/84 Plus

Histograms

Step 1: Enter the raw data in L1 by pressing STAT and selecting 1: Edit.

Step 2: Press 2ⁿᵈ Y = to access the StatPlot menu. Select 1: Plot1.

Step 3: Place the cursor on "ON" and press ENTER.

Step 4: Place the cursor on the histogram icon (see the figure) and press ENTER. Press 2ⁿᵈ QUIT to exit Plot 1 menu.

Step 5: Press WINDOW. Set Xmin to the lower class limit of the first class. Set Xmax to the lower class limit of the class following the class containing the largest value. For example, if the first class is 0–9, set Xmin to 0. If the class width is 10 and the last class is 90–99, set Xmax to 100. Set Xscl to the class width. Set Ymin to 0. Set Ymax to a value larger than the frequency of the class with the highest frequency.

Step 6: Press GRAPH.

Helpful Hints: To determine each class frequency, press TRACE and use the arrow keys to scroll through each class. If you decrease the value of Ymin to a value such as −5, you can see the values displayed on the screen more easily. The TI graphing calculators do not draw stem-and-leaf plots or dot plots.

MINITAB

Histograms

Step 1: Enter the raw data in C1.

Step 2: Select the **Graph** menu and highlight **Histogram** . . .

Step 3: Highlight the "simple" icon and press OK.

Step 4: Put the cursor in the "Graph variables" box. Highlight C1, and press Select. Click SCALE and select the Y-Scale Type tab. For a frequency histogram, click the frequency radio button. For a relative frequency histogram, click the percent radio button. Click OK twice.

Note: To adjust the class width and to change the labels on the horizontal axis to the lower class limit, double-click inside one of the bars in the histogram. Select the "binning" tab in the window that opens. Click the cutpoint button and the midpoint/cutpoint position radio button. In the midpoint/cutpoint box, enter the lower class limits of each class. Click OK.

Stem-and-Leaf Plots

Step 1: With the raw data entered in C1, select the **Graph** menu and highlight **Stem-and-Leaf**.

Step 2: Select the data in C1 and press OK.

Dot Plots

Step 1: Enter the raw data in C1.

Step 2: Select the **Graph** menu and highlight **Dotplot**.

Step 3: Highlight the "simple" icon and press OK.

Step 4: Put the cursor in the "Graph variables" box. Highlight C1 and press Select. Click OK.

Excel Histograms

Step 1: Enter the raw data in column A.

Step 2: Select **Tools** and **Data Analysis** . . .

Step 3: Select Histogram from the list.

Step 4: With the cursor in the Input Range cell, use the mouse to highlight the raw data. Select the Chart Output box and press OK.

Step 5: Double-click on one of the bars in the histogram. Select the Options tab from the menu that appears. Reduce the gap width to zero.

Excel does not draw stem-and-leaf or dot plots.

2.3 Graphical Misrepresentations of Data

Objective   ① **Describe what can make a graph misleading or deceptive**

① ## Describe What Can Make a Graph Misleading or Deceptive

Often, statistics gets a bad rap for having the ability to manipulate data to support any position desired. One method of distorting the truth is through graphics. Sometimes graphics *mislead*; other times they *deceive*. We will call graphs misleading if they unintentionally create an incorrect impression. We consider graphs deceptive if they purposely create an incorrect impression. We have already discussed the power that graphical representations of data can have, so it is important to be able to recognize misleading and deceptive graphs. The most common graphical misrepresentation of data is accomplished through the manipulation of the scale of the graph.

Note to Instructor
The material in this section can be omitted without loss of continuity. For those instructors who are pressed for time, it can be assigned for reading.

EXAMPLE 1 ## Misrepresentation of Data

Problem: The bar graph illustrated in Figure 18 is a *USA Today* type graph. A survey was conducted by Impulse Research for Quilted Northern Confidential in which individuals were asked how they would flush a toilet when the facilities are not sanitary. What's wrong with the graphic?

Approach: We need to compare the vertical scales of each bar to see if they accurately depict the percentages given.

Solution: First, it is unclear whether the bars include the roll of toilet paper or not. In either case, the roll corresponding to "use shoe" should be 2.4 (= 41/17) times longer than the roll corresponding to "paper towel." If we include the roll of toilet paper, then the bar corresponding to "use shoe" is less than double the length of "paper towel." If we do not include the roll of toilet paper, then the bar corresponding to "use shoe" is almost exactly double the length of the bar corresponding to "paper towel." The vertical scaling is incorrect.

Figure 18

EXAMPLE 2   **Misrepresentation of Data by Manipulating the Vertical Scale**

Problem: The bar graph shown in Figure 19 depicts the average SAT Math scores of college-bound seniors for the years 1991–2004, based on data from the College Board. Determine why this graph might be considered misrepresentative.

Approach: We need to look at the graph for any characteristics that may mislead a reader, such as manipulation of the vertical scale.

Figure 19

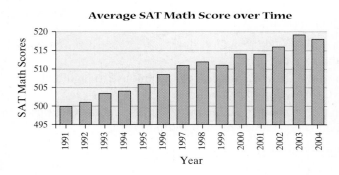

Average SAT Math Score over Time

Solution: The graph in the figure may lead a reader to believe that SAT math scores have increased substantially since 1991. While SAT math scores have been increasing, they have not doubled or tripled, as may be inferred from the graph (since the bar for 1997 is three times as high as the bar for 1991). We notice in the figure that the vertical axis begins its labeling at 495 instead of 0. This type of scaling is common when the smallest observed data value is a rather large number. It is not necessarily done purposely to confuse or mislead the reader. Often, the main purpose in graphs is to discover a trend, rather than the actual differences in the data. The trend is clearer in Figure 19 than in Figure 20, where the vertical axis begins at 0.

Often, instead of beginning the axis of a graph at 0 as in Figure 20, the graph is begun at a value slightly less than the smallest value in the data set. However, special care must be taken to make the reader aware of the vertical-axis scaling. Figure 21 shows the proper construction of the graph of the SAT scores, with the graph beginning at 495. The symbol ⌇ is used to signify that the graph has a gap in it.

Figure 20

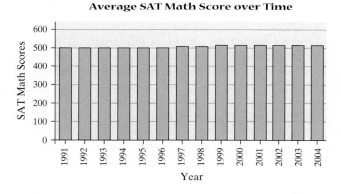

Average SAT Math Score over Time

Figure 21

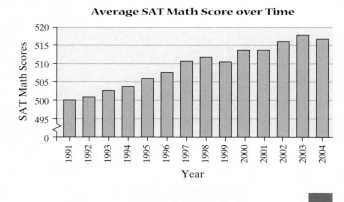

Average SAT Math Score over Time

Now Work Problem 5.

In addition to vertical-axis scaling, figures can be misleading through pictures. Consider the next example.

EXAMPLE 3   **Misleading Graphs**

Problem: The Dow Jones Industrial Average (DJIA) is a collection of 30 stocks from the stock market that are thought to be representative of the U.S. economy. It includes companies such as Intel, Wal-Mart, and General Motors. Had you invested $10,000 in the DJIA on February 20, 1996, it would have been worth $20,000 on February 28, 2005. To illustrate this investment, a brokerage firm might create the graphic shown in Figure 22. Describe how this graph is misleading.

Figure 22

February 20, 1996   February 28, 2005

Approach: Again, we look for characteristics of the graph that seem to manipulate the facts, such as an incorrect depiction of the size of the graphics.

Solution: The graphic on the right of the figure has been doubled in length, width, and height, causing an eightfold increase in the size, thereby misleading the reader into thinking that the size of the investment increased by eight times instead of by two times.

> **Now Work Problem 15.**

Note to Instructor

Class Project: Have students collect graphs from newspapers, magazines, and so on, that are misleading or incorrect. Create a binder of these graphs for the class to refer to and explore.

There are many ways to create graphs that mislead. Two popular texts written about ways that graphs mislead or deceive are *How to Lie with Statistics* (W. W. Norton & Company, Inc., 1982), by Darrell Huff, and *The Visual Display of Quantitative Information* (Graphics Press, 2001), by Edward Tufte.

We conclude this section with some guidelines for constructing good graphics.

Characteristics of Good Graphics

- Label the graphic clearly and provide explanations if needed.
- Avoid distortion. Don't lie about the data.
- Avoid three dimensions. Three-dimensional charts may look nice, but they distract the reader and often result in misinterpretation of the graphic.
- Do not use more than one design in the same graphic. Sometimes graphs use a different design in a portion of the graphic to draw attention to this area. Don't use this technique. Let the numbers speak for themselves.

MAKING AN INFORMED DECISION

Tables or Graphs?

You work for the school newspaper. Your editor approaches you with a special reporting assignment. Your task is to write an article that describes the "typical" student at your school, complete with supporting information. To write this article, you have to survey at least 40 students and ask them to respond to a questionnaire. The editor would like to have at least two qualitative and two quantitative variables that describe the typical student. The results of the survey will be presented in your article, but you are unsure whether you should present tabular or graphical summaries, so you decide to perform the following experiment.

1. Develop a questionnaire that results in obtaining the values of two qualitative and two quantitative variables. Administer the questionnaire to at least 40 students on your campus.
2. Summarize the data in both tabular and graphical form.
3. Select 20 individuals. (They don't have to be students at your school.) Give the tabular summaries

to 10 individuals and the graphical summaries to the other 10. Ask each individual to study the table or graph for 5 seconds. After 1 minute, give a questionnaire that asks various questions regarding the information contained in the table or graph. For example, if you summarized age data, ask the individual which age group had the highest frequency. Record the number of correct answers for each individual. Which summary results in a higher percentage of correct answers, the tables or the graphs? Write a report that discusses your findings.

4. Now use the data collected from the questionnaire to create a couple of misleading graphs. Again, select 20 individuals. Give 10 individuals the misleading graphs and 10 individuals the correct graphs. Ask each individual to study each graph for 5 seconds. After 1 minute has elapsed, give a questionnaire that asks various questions regarding the information contained in the graphs. Record the number of correct answers for each individual. Did the misleading graphs mislead? Write a report that discusses your findings.

2.3 ASSESS YOUR UNDERSTANDING

Applying the Concepts

1. **Inauguration Cost** The following is a *USA Today* type graph. Explain how it is misleading.

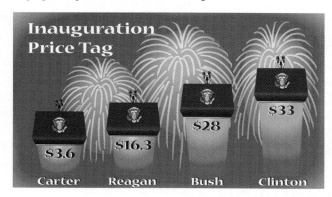

2. **Burning Calories** The following is a *USA Today* type graph.

   (a) Explain how it is misleading.
   (b) What could be done to improve the graphic?

3. **Median Earnings** The following graph shows the median earnings for females from 1998 to 2003.
   (*Source*: U.S. Census Bureau, Income, Poverty, and Health Insurance Coverage in the United States, 2003)

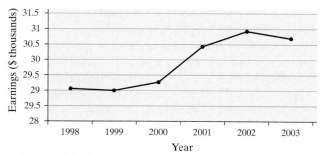

   (a) How is the graph misleading? What does the graph seem to convey?
   (b) Redraw the graph so that it is not misleading. What does the new graph seem to convey?

4. **Union Membership** The following relative frequency histogram represents the proportion of employed people aged 25 to 64 years old who were members of a union. (*Source*: U.S. Bureau of Labor Statistics)

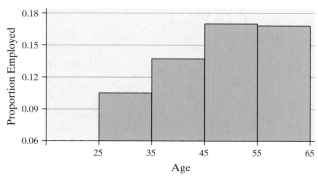

   (a) Describe how this graph is misleading. What might a reader conclude from the graph?
   (b) Redraw the histogram so that it is not misleading.

5. **Health Insurance** The following relative frequency histogram represents the proportion of people aged 25 to 64 years old not covered by any health insurance in 2002. (*Source*: U.S. Census Bureau)

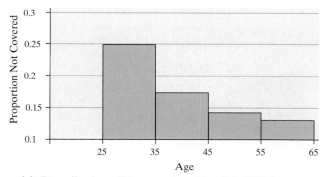

   (a) Describe how this graph is misleading. What might a reader conclude from the graph?
   (b) Redraw the histogram so that it is not misleading.

6. **New Homes** The following time-series plot shows the number of new homes built in the Midwest from 1998 to 2004. (*Source*: U.S. Census Bureau)

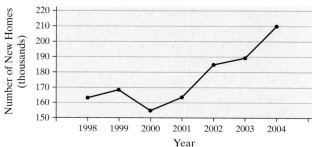

(a) Describe how this graph is misleading.
(b) What is the graph trying to convey?

7. Median Income The following time-series plot shows the median household income for the years 1998 to 2003. (*Source*: U.S. Census Bureau)

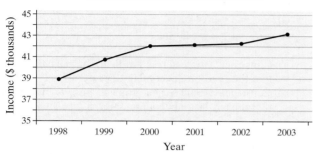

U.S. Median Household Income

(a) Describe how this graph is misleading.
(b) What is the graph trying to convey?

8. National Debt The following graphic is a *USA Today* type graph.

How many times larger should the graphic for 1999 be than the 1900 graphic? 1300 times

9. Cost of Kids The following is a *USA Today* type graph based on data from the Department of Agriculture. It represents the percentage of income a middle-income family will spend on their children.

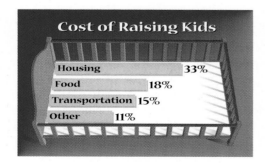

(a) How is the graphic misleading?
(b) What could be done to improve the graphic?

10. SAT Combined Scores The following table gives the average SAT combined scores for the years 1999–2003.

Year	Avg. SAT Combined Score
1999	1016
2000	1019
2001	1020
2002	1020
2003	1026

Note: Beginning with the tests administered in March 2005, the SAT test consists of three parts for a total possible score of 1800.

(a) Construct a misleading time-series plot that indicates the average SAT combined score has risen sharply over the given time period.
(b) Construct a time-series plot that is not misleading.

11. Engineering Degrees The following table gives the number of bachelor's degrees in engineering that were awarded from 1999 to 2003.

Year	Degrees Awarded
1999	62,372
2000	63,731
2001	65,113
2002	67,301
2003	70,949

(a) Construct a misleading graph indicating that the number of degrees awarded has more than doubled since 1999.
(b) Construct a graph that is not misleading.

12. Worker Injury The safety manager at Klutz Enterprises provides the following graph to the plant manager and claims that the rate of worker injuries has been reduced by 67% over a 12-year period. Does the graph support his claim? Explain why or why not.

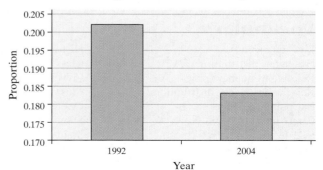

13. **Health Care Expenditures** The following data represent health care expenditures as a percentage of the U.S. gross domestic product (GDP) from 1997 to 2002. Gross domestic product is the total value of all goods and services created during the course of the year.

Year	Health Care as a Percent of GDP
1997	13.1
1998	13.1
1999	13.2
2000	13.3
2001	14.1
2002	14.9

Source: U.S. Health Care Financing Administration

(a) Construct a time-series plot that a politician would create to support the position that health care expenditures, as a percentage of GDP, are increasing and must be slowed.
(b) Construct a time-series plot that the health care industry would create to refute the opinion of the politician.
(c) Construct a time-series plot that is not misleading.

14. **Motor Vehicle Accidents** The following data represent the number of motor vehicle accidents and the traffic death rates (number of deaths per 100,000 licensed drivers) from 1998 to 2002.

Year	Motor Vehicle Deaths (in thousands)	Traffic Death Rate (per 100,000 licensed drivers)
1998	43.5	22.4
1999	43.0	22.3
2000	43.0	22.0
2001	43.7	22.1
2002	44.0	22.0

Source: National Safety Council

(a) Construct a time-series graph to support the belief that the roads are becoming less safe.
(b) Construct a time-series graph to support the belief that the roads are becoming safer.

15. **Distance Learning** Between 1995 and 2001, the enrollment in college-level distance education courses nearly doubled.
(a) Construct a graphic that is not misleading to depict this situation.
(b) Construct a misleading graphic to depict this situation.

16. **Overweight** Between 1980 and 2002, the number of adolescents in the United States who were overweight nearly tripled.
(a) Construct a graphic that is not misleading to depict this situation.
(b) Construct a misleading graphic to depict this situation.

CHAPTER **2** **Review**

Summary

Raw data are first organized into tables. Data are organized by creating classes into which they fall. Qualitative data and discrete data have values that provide clear-cut categories of data. However, with continuous data the categories, called classes, must be created. Typically, the first table created is a frequency distribution, which lists the frequency with which each class of data occurs. Another type of distribution is the relative frequency distribution.

Once data are organized into a table, graphs are created. For data that are qualitative, we can create bar charts and pie charts. For data that are quantitative, we can create histograms or stem-and-leaf plots.

In creating graphs, care must be taken not to draw a graph that misleads or deceives the reader. If a graph's vertical axis does not begin at zero, the symbol ⨎ should be used to indicate the gap that exists in the graph.

Vocabulary

Raw data (p. 54)
Frequency distribution (p. 55)
Relative frequency (p. 56)
Relative frequency distribution (p. 56)
Bar graph (p. 57)
Pareto chart (p. 58)
Side-by-side bar graph (p. 58)
Pie chart (p. 60)

Histogram (p. 72)
Class (p. 73)
Lower and upper class limit (p. 73)
Class width (p. 73)
Open ended (p. 73)
Stem-and-leaf plot (p. 76)
Stem (p. 76)
Leaf (p. 76)

Split stems (p. 79)
Dot plot (p. 80)
Uniform distribution (p. 80)
Bell-shaped distribution (p. 80)
Skewed right (p. 80)
Skewed left (p. 80)
Time-series data (p. 81)
Time series plot (p. 81)

Objectives

Section	You should be able to . . .	Example	Review Exercises
2.1	**1** Organize qualitative data in tables (p. 55)	1, 2	3(a), 4(a), 7(a) and (b), 8(a) and (b)
	2 Construct bar graphs (p. 57)	3 through 5	3(c) and (d), 4(c) and (d), 7(c), 8(c)
	3 Construct pie charts (p. 60)	6	3(e), 4(e), 7(d), 8(d)
2.2	**1** Organize discrete data in tables (p. 71)	1	9(a) and (b), 10(a) and (b)
	2 Construct histograms of discrete data (p. 72)	2	9(c) and (d), 10(c) and (d),
	3 Organize continuous data in tables (p. 73)	3	11(a) and (b), 12(a) and (b), 13 (a) and (b), 14(a) and (b)
	4 Construct histograms of continuous data (p. 75)	4, 5	5(b) and (c), 6(b) and (c), 11(c) and (d); 12(c) and (d), 13(c) and (d), 14(c) and (d)
	5 Draw stem-and-leaf plots (p. 76)	6 through 8	15, 16
	6 Draw dot plots (p. 80)	9	9(g), 10(g)
	7 Identify the shape of a distribution (p. 80)	10	11(c), 12(c), 13(c), 14(c), 15, 16(b)
	8 Draw time-series graphs (p. 81)	11	17, 18
2.3	**1** Describe what can make a graph misleading or deceptive (p. 93)	1 through 3	19, 20, 21

Review Exercises

1. Energy Consumption The following bar chart represents the energy consumption of the United States (in quadrillion Btu) in 2003.

(*Source*: Energy Information Administration)

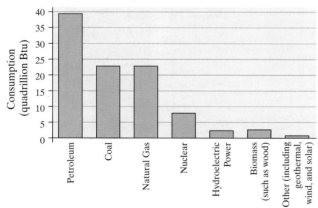

Energy Consumption

(1b) 3 quadrillion Btu

(a) Approximately how much energy did the United States consume from natural gas? 22.5 quadrillion Btu

(b) Approximately how much energy did the United States consume from biomass?

(c) Approximate the total energy consumption of the United States in 2003. 98 quadrillion Btu

(d) Which category has the lowest frequency? Other

(e) Is it appropriate to describe the shape of the distribution as skewed right? Why or why not? No; qualitative data

2. Highway Funds The following frequency histogram represents the dollar amount each state and the District of Columbia gets back in federal highway and mass transit grants for each dollar of gasoline tax that their motorists pay into the federal highway trust fund, based on data obtained from the Federal Highway Administration.

(a) Determine the class width. 0.5

(b) Identify the classes.

(c) Which class has the highest frequency? 1–1.49

Federal Highway and Mass Transit Grants

3. Weapons Used in Homicide The following frequency distribution represents the cause of death in homicides for the year 2002.

Type of Weapon	Frequency
Firearms	9369
Knives or cutting instruments	1767
Blunt objects (clubs, hammers, etc.)	666
Personal weapons (hands, fists, etc.)	933
Strangulation	143
Fire	104
Other weapon or not stated	1176

Source: *Crime in the United States, 2002*, FBI, Uniform Crime Reports

(a) Construct a relative frequency distribution.
(b) What percentage of homicides was committed using a blunt object? 4.7%
(c) Construct a frequency bar graph.
(d) Construct a relative frequency bar graph.
(e) Construct a pie chart.

4. U.S. Greenhouse Emissions The following frequency distribution represents the total greenhouse emissions in millions of metric tons in 2003 in the United States.

Gas	Emissions
Carbon dioxide	1600.8
Methane	164.1
Nitrous oxide	87.3
Hydrofluorocarbons, perfluorocarbons, and sulfur hexafluoride	39.1

Source: Energy Information Administration

(a) Construct a relative frequency distribution.
(b) What percent of emissions was due to carbon dioxide? 84.6%
(c) Construct a frequency bar graph.
(d) Construct a relative frequency bar graph.
(e) Construct a pie chart.

5. Live Births The following frequency distribution represents the number of live births (in thousands) in the United States in 2003 by age of mother.

Age of Mother (years)	Births (thousands)
10–14	7
15–19	415
20–24	1032
25–29	1087
30–34	976
35–39	468
40–44	101

Source: National Center for Health Statistics

(a) Construct a relative frequency distribution.
(b) Construct a frequency histogram. Describe the shape of the distribution.
(c) Construct a relative frequency histogram.
(d) What percentage of live births was to mothers aged 20 to 24? 25.3%
(e) What percentage of live births was to mothers of age 30 or older? 37.8%

6. Suicides The following frequency distribution represents the number of suicides in the United States in 2001, by age.

Age	Suicides	Age	Suicides
5–9	7	55–59	1985
10–14	272	60–64	1332
15–19	1611	65–69	1212
20–24	2360	70–74	1220
25–29	2389	75–79	1219
30–34	2681	80–84	973
35–39	3176	85–89	538
40–44	3459	90–94	200
45–49	3260	95–99	28
50–54	2682		

Source: National Center for Health Statistics

(a) Construct a relative frequency distribution.
(b) Construct a frequency histogram. Describe the shape of the distribution.
(c) Construct a relative frequency histogram.
(d) What percentage of suicides from the distribution was aged 40 to 44? 11.3%
(e) What percentage of suicides from the distribution was 24 years or younger? 13.9%

7. Political Affiliation   A sample of 100 randomly selected registered voters in the city of Naperville was asked their political affiliation: Democrat (D), Republican (R), or Independent (I). The results of the survey are shown in the table below.

(a) Construct a frequency distribution of the data.
(b) Construct a relative frequency distribution of the data.
(c) Construct a relative frequency bar graph of the data.
(d) Construct a pie chart of the data.
(e) What appears to be the most common political affiliation in Naperville?   *Democrat*

D	R	D	R	D	R	D	D	R	D
R	D	D	D	R	R	D	D	D	D
R	R	I	I	D	R	D	R	R	R
I	D	D	R	I	I	R	D	R	R
D	I	R	D	D	D	D	I	I	R
R	I	R	R	I	D	D	D	D	R
D	I	I	D	D	R	R	R	R	D
D	R	R	R	D	D	I	I	D	D
D	D	I	D	R	I	D	D	D	D
R	R	R	R	R	D	R	D	R	D

8. Educational Attainment   The Metra Train Company was interested in knowing the educational background of its customers. The company contracted a marketing firm to conduct a survey with a random sample of 50 commuters at the train station. In the survey, commuters were asked to disclose their educational attainment. The following results were obtained:

(a) Construct a frequency distribution of the data.
(b) Construct a relative frequency distribution of the data.
(c) Construct a relative frequency bar graph of the data.
(d) Construct a pie chart of the data.
(e) What is the most common educational level of a commuter?   *High school graduate*

No high school diploma	Some college	Advanced degree	High school graduate	Advanced degree
High school graduate	High school graduate	High school graduate	High school graduate	No high school diploma
Some college	High school graduate	Bachelor's degree	Associate's degree	High school graduate
No high school diploma	Bachelor's degree	Some college	High school graduate	No high school diploma
Associate's degree	High school graduate	High school graduate	No high school diploma	Some college
Bachelor's degree	Bachelor's degree	Some college	High school graduate	Some college
Bachelor's degree	Advanced degree	No high school diploma	Advanced degree	No high school diploma
High school graduate	Bachelor's degree	No high school diploma	High school graduate	No high school diploma
Associate's degree	Bachelor's degree	High school graduate	Bachelor's degree	Some college
Some college	Associate's degree	High school graduate	Some college	High school graduate

9. Family Size   A random sample of 60 couples married for 7 years were asked to disclose the number of children they have. The results of the survey are shown to the right.

(a) Construct a frequency distribution of the data.
(b) Construct a relative frequency distribution of the data.
(c) Construct a frequency histogram of the data. Describe the shape of the distribution.
(d) Construct a relative frequency histogram of the data.
(e) What percentage of couples married 7 years has two children?   *30%*
(f) What percentage of couples married 7 years has at least two children?   *76.7%*
(g) Draw a dot plot of the data.

0	0	3	1	2	3
3	4	3	3	0	3
1	2	1	3	0	3
4	2	3	2	2	4
2	1	3	4	1	3
0	3	3	3	2	1
2	0	3	1	2	3
4	3	3	5	2	0
4	2	2	2	3	3
2	4	2	2	2	2

10. Waiting in Line The data to the right represent the number of cars that arrived at a McDonald's drive-through between 11:50 A.M. and 12:00 noon each Wednesday for the past 50 weeks.

(a) Construct a frequency distribution of the data.
(b) Construct a relative frequency distribution of the data.
(c) Construct a frequency histogram of the data. Describe the shape of the distribution.
(d) Construct a relative frequency histogram of the data.
(e) What percentage of the time did exactly three cars arrive between 11:50 A.M. and 12:00 noon? 24%
(f) What percentage of the time did three or more cars arrive between 11:50 A.M. and 12:00 noon? 76%
(g) Draw a dot plot of the data.

1	7	3	8	2	3	8	2	6	3
6	5	6	4	3	4	3	8	1	2
5	3	6	3	3	4	3	2	1	2
4	4	9	3	5	2	3	5	5	5
2	5	6	1	7	1	5	3	8	4

11. Crime Rate by State The following data represent the crime rate (per 100,000 population) for each state in 2002.

In (a)–(d), start the first class at a lower class limit of 2000 and maintain a class width of 400.
(a) Construct a frequency distribution.
(b) Construct a relative frequency distribution.

(c) Construct a frequency histogram. Describe the shape of the distribution.
(d) Construct a relative frequency histogram.
(e) Repeat (a)–(d), using a class width of 1000. In your opinion, which class width provides the better summary of the data? Why?

State	Crime Rate	State	Crime Rate	State	Crime Rate
Alabama	4465.2	Kentucky	2902.6	North Dakota	2406.2
Alaska	4309.7	Louisiana	5098.1	Ohio	4107.3
Arizona	6386.3	Maine	2656.0	Oklahoma	4743.2
Arkansas	4157.5	Maryland	4747.4	Oregon	4868.4
California	3943.7	Massachusetts	3094.2	Pennsylvania	2841.0
Colorado	4347.8	Michigan	3874.1	Rhode Island	3589.1
Connecticut	2997.2	Minnesota	3535.1	South Carolina	5297.3
Delaware	3939.0	Mississippi	4159.2	South Dakota	2278.7
District of Columbia	8022.3	Missouri	4602.4	Tennessee	5018.0
Florida	5420.6	Montana	3512.9	Texas	5189.6
Georgia	4507.2	Nebraska	4256.7	Utah	4452.4
Hawaii	6043.7	Nevada	4497.5	Vermont	2530.0
Idaho	3172.5	New Hampshire	2220.0	Virginia	3140.3
Illinois	4016.4	New Jersey	3024.2	Washington	5106.8
Indiana	3750.0	New Mexico	5077.8	West Virginia	2515.2
Iowa	3448.2	New York	2803.7	Wisconsin	3252.7
Kansas	4087.0	North Carolina	4721.4	Wyoming	3580.9

Source: Crime in the United States, 2002. FBI, Uniform Crime Reports.

12. Towing Capacity The data on page 103 represent the towing capacity (in pounds) for selected sport utility vehicles (SUVs).

Start the first class at a lower class limit of 3000 and maintain a class width of 1000:

(a) Construct a frequency distribution.
(b) Construct a relative frequency distribution.
(c) Construct a frequency histogram. Describe the shape of the distribution.
(d) Construct a relative frequency histogram.

SUV	Towing Capacity	SUV	Towing Capacity	SUV	Towing Capacity
Acura MDX	4,500	GMC Yukon	8,700	Land Rover Range Rover	7,700
BMW X5	6,000	GMC Yukon XL	12,000	Lincoln Navigator	8,800
Buick Rendezvous	3,500	Honda Passport	4,500	Mitsubishi Montero	5,000
Chevrolet Blazer	5,600	Hummer	8,300	Nissan Pathfinder	5,000
Chevrolet Suburban	12,000	Infiniti QX4	5,000	Pontiac Aztek	3,500
Chevrolet Tahoe	8,700	Isuzu Axiom	4,500	Suzuki XL-7	3,000
Dodge Durango	7,650	Isuzu Rodeo	4,500	Toyota 4Runner	5,000
Ford Escape	3,500	Jeep Cherokee	5,000	Toyota Highlander	3,500
Ford Excursion	10,000	Jeep Grand Cherokee	6,500	Toyota Land Cruiser	6,500
Ford Expedition	8,100	Jeep Liberty	5,000		
GMC Jimmy	5,900	Land Rover Discovery	7,700		

Source: Manufacturers

13. Diameter of a Cookie   The data to the right represent the diameter (in inches) of a random sample of 34 Keebler Chips Deluxe™ Chocolate Chip Cookies.

In (a)–(d), start the first class at a lower class limit of 2.2000 and maintain a class width of 0.0200:
(a) Construct a frequency distribution.
(b) Construct a relative frequency distribution.
(c) Construct a frequency histogram. Describe the shape of the distribution.
(d) Construct a relative frequency histogram.
(e) Repeat (a)–(d) using a class width of 0.0400. In your opinion, which class width provides the better summary of the data? Why?

2.3414	2.3010	2.2850	2.3015	2.2850	2.3019	2.2400
2.3005	2.2630	2.2853	2.3360	2.3696	2.3300	2.3290
2.2303	2.2600	2.2409	2.2020	2.3223	2.2851	2.2382
2.2438	2.3255	2.2597	2.3020	2.2658	2.2752	2.2256
2.2611	2.3006	2.2011	2.2790	2.2425	2.3003	

Source: Trina S. McNamara, student at Joliet Junior College

14. Home Sales   The data to the right represent the closing price (in U.S. dollars) of homes sold in a midwest city.

Start the first class at a lower class limit of 85,000 and maintain a class width of 10,000.
(a) Construct a frequency distribution.
(b) Construct a relative frequency distribution.
(c) Construct a frequency histogram. Describe the shape of the distribution.
(d) Construct a relative frequency histogram.

138,820	149,143	99,000	115,000	157,216
169,541	140,794	136,924	124,757	149,380
135,512	153,146	136,833	128,429	136,529
147,500	120,936	95,491	115,744	119,900
89,900	102,696	149,634	123,103	126,630
140,269	183,000	133,646	121,225	121,524
146,439	182,000	110,128	109,520	104,640
124,760	134,305	111,220	121,795	170,072
136,550	115,595	155,507	152,600	130,000
152,537	163,165			

Source: Transamerica Intellitech

15. Eat Your Vegetables!   The data to the right represent the number of servings of vegetables per day that a random sample of forty 20- to 39-year-old females consumes. The data are based on a survey conducted by the U.S. Department of Agriculture. Construct a stem-and-leaf diagram of the data, and comment on the shape of the distribution.

1.7	2.7	0.3	3.5	0.7	1.4	5.1	3.9
0.2	2.1	4.1	5.8	3.8	0.4	6.1	0.7
2.4	11.1	3.5	6.7	2.3	4.9	5.9	0.4
3.3	0.8	7.6	10.2	5.8	2.6	0.6	3.2
0.5	2.4	4.9	2.3	8.3	6.0	5.3	3.5

16. Fertility Rate The following data represent the fertility rate (births per 1000 women aged 15 to 44) for each state and the District of Columbia.

(a) Round each observation to the nearest whole number and draw a stem-and-leaf diagram.

(b) Describe the shape of the distribution.

(c) Redraw the stem-and-leaf diagram using split stems. For example, data between 60 and 64 is one stem and data between 65 and 69 is a second stem. Does this stem-and-leaf diagram better summarize the data? Why?

State	Fertility Rate	State	Fertility Rate	State	Fertility Rate
Alabama	61.2	Kentucky	60.5	North Dakota	58.7
Alaska	73.5	Louisiana	65.4	Ohio	61.7
Arizona	77.8	Maine	49.8	Oklahoma	68.8
Arkansas	66.6	Maryland	60.6	Oregon	61.9
California	68.3	Massachusetts	56.7	Pennsylvania	56.4
Colorado	69.3	Michigan	60.7	Rhode Island	54.6
Connecticut	58.8	Minnesota	62.0	South Carolina	60.7
Delaware	62.2	Mississippi	65.7	South Dakota	68.3
District of Columbia	52.9	Missouri	62.1	Tennessee	62.2
Florida	62.5	Montana	60.3	Texas	77.1
Georgia	68.4	Nebraska	69.5	Utah	90.6
Hawaii	68.6	Nevada	72.5	Vermont	48.9
Idaho	73.8	New Hampshire	52.4	Virginia	61.9
Illinois	66.1	New Jersey	63.5	Washington	60.2
Indiana	64.8	New Mexico	70.7	West Virginia	57.0
Iowa	61.7	New York	59.8	Wisconsin	59.0
Kansas	68.7	North Carolina	65.4	Wyoming	63.6

Source: U.S. National Center for Health Statistics

17. Federal Minimum Wage Rates The following data represent the value of the minimum wage for the years 1980 to 2003.

Year	Minimum Wage	Year	Minimum Wage
1980	3.10	1992	4.25
1981	3.35	1993	4.25
1982	3.35	1994	4.25
1983	3.35	1995	4.25
1984	3.35	1996	4.75
1985	3.35	1997	5.15
1986	3.35	1998	5.15
1987	3.35	1999	5.15
1988	3.35	2000	5.15
1989	3.35	2001	5.15
1990	3.80	2002	5.15
1991	4.25	2003	5.15

Source: Economic Policy Institute

(a) Construct a time-series plot of the data.

(b) Comment on the apparent trend.

18. Federal Minimum Wage Rates The following data represent the value of the minimum wage for the years 1980 to 2003 in constant 2003 dollars. Constant dollars are dollars adjusted for inflation.

Year	Minimum Wage	Year	Minimum Wage
1980	6.55	1992	5.46
1981	6.48	1993	5.33
1982	6.11	1994	5.22
1983	5.87	1995	5.09
1984	5.64	1996	5.54
1985	5.46	1997	5.89
1986	5.36	1998	5.80
1987	5.19	1999	5.68
1988	5.01	2000	5.50
1989	4.80	2001	5.35
1990	5.19	2002	5.27
1991	5.60	2003	5.15

Source: Economic Policy Institute

(a) Construct a time-series plot of the data.

(b) Comment on the apparent trend.

(c) Compare this time-series plot with the one in Problem 17. Which graph is misleading? Why?

19. Misleading Graphs   The following graph was found in a magazine advertisement for skin cream. How is this graph misleading?

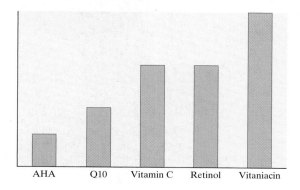

20. Misleading Graphs   The following is a *USA Today* type graph.

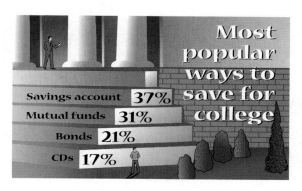

Do you think the graph is misleading? Why? If you think it is misleading, what might be done to improve the graph?

21. Misleading Graphs   In 2002 the average earnings of a high school graduate were $27,280. At $51,194, the average earnings of a recipient of a bachelor's degree were about 88% higher.

(*Source*: U.S. Census Bureau, Current Population Survey, 2003)

(a) Construct a graph that a college recruiter might create to convince high school students that they should attend college.

(b) Construct a graph that does not mislead.

THE CHAPTER 2 CASE STUDY IS LOCATED ON THE CD THAT ACCOMPANIES THIS TEXT.

Numerically Summarizing Data

Outline

3.1   Measures of Central Tendency

3.2   Measures of Dispersion

3.3   Measures of Central Tendency and Dispersion from Grouped Data

3.4   Measures of Position

3.5   The Five-Number Summary and Boxplots

■ Chapter Review

■ Case Study: Who Was "A Mourner"? (On CD)

DECISIONS

Suppose that you are in the market for a used car. To make an informed decision regarding your purchase, you decide to collect as much information as possible. What information is important in helping you make this decision? See the Decisions project on page 164.

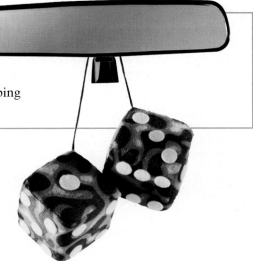

●●● Putting It All Together

When we look at a distribution of data, we should consider three characteristics of the distribution: its shape, its center, and its spread. In the last chapter, we discussed methods for organizing raw data into tables and graphs. These graphs (such as the histogram) allow us to identify the shape of the distribution. Recall that we describe the shape of a distribution as symmetric (in particular, bell shaped or uniform), skewed right, or skewed left.

The center and spread are numerical summaries of the data. The center of a data set is commonly called the average. There are many ways to describe the average value of a distribution. In addition, there are many ways to measure the spread of a distribution. The most appropriate measure of center and spread depends on the shape of the distribution.

Once these three characteristics of the distribution are known, we can analyze the data for interesting features, including unusual data values, called *outliers*.

3.1 Measures of Central Tendency

Preparing for This Section   Before getting started, review the following:

- Quantitative data (Section 1.1, p. 8)
- Qualitative data (Section 1.1, p. 8)
- Population versus sample (Section 1.1, p. 4)
- Simple random sampling (Section 1.2, pp. 16–19)

Objectives

1 **Determine the arithmetic mean of a variable from raw data**

2 **Determine the median of a variable from raw data**

3 **Determine the mode of a variable from raw data**

4 **Use the mean and the median to help identify the shape of a distribution**

Note to Instructor
Discuss with students that we describe distributions using three characteristics: shape, center, and spread. Shape can be determined from pictures of data, like the histogram. Center is discussed in this section, and spread is discussed in Section 3.2.

! **CAUTION**

Whenever you hear the word *average*, be aware that the word may not always be referring to the mean. One average could be used to support one position, while another average could be used to support a different position.

A measure of central tendency numerically describes the average or typical data value of a variable. We hear the word *average* in the news all the time:

- The average miles per gallon of gasoline of the 2006 Chevrolet Camaro in city driving is 19 miles.
- According to the U.S. Census Bureau, the national average commute time to work in 2005 was 24.3 minutes.
- According to the U.S. Census Bureau, the average household income in 2003 was $43,527.
- The average American woman is 5′4″ tall and weighs 142 pounds.

In this chapter, we discuss three measures of central tendency: the *mean*, the *median*, and the *mode*. While other measures of central tendency exist, these three are the most widely used. When the word *average* is used in the media (newspapers, reporters, and so on) it usually refers to the mean. But beware! Some reporters use the term *average* to refer to the median or mode. As we shall see, these three measures of central tendency can give very different results!

Before we discuss measures of central tendency, we must consider whether or not we are computing a measure of central tendency that describes a population or one that describes a sample.

Definitions

A **parameter** is a descriptive measure of a population.

A **statistic** is a descriptive measure of a sample.

In Other Words

To help you remember the difference between a parameter and a statistic, think of the following:

p = parameter = population
s = statistic = sample

For example, if we determine the average test score for *all* the students in a statistics class, our population, the average is a parameter. If we compute the average based on a simple random sample of five students, the average is a statistic.

1 ## Determine the Arithmetic Mean of a Variable from Raw Data

When used in everyday language, the word *average* often stands for the arithmetic mean. To compute the arithmetic mean of a set of data, the data must be quantitative.

Definitions

The **arithmetic mean** of a variable is computed by determining the sum of all the values of the variable in the data set, divided by the number of observations. The **population arithmetic mean**, μ (pronounced "mew"), is computed using all the individuals in a population. The population mean is a parameter. The **sample arithmetic mean**, $\bar{x}$ (pronounced "x-bar"), is computed using sample data. The sample mean is a statistic.

While other types of means exist (see Problems 51 and 52), the arithmetic mean is generally referred to as the **mean**. We will follow this practice for the remainder of the text.

In statistics, Greek letters are used to represent parameters, and Roman letters are used to represent statistics. Statisticians use mathematical expressions to describe the method for computing means.

Definitions

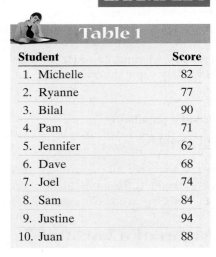

In Other Words
To find the mean of a set of data, add up all the observations and divide by the number of observations.

If $x_1, x_2, \ldots, x_N$ are the N observations of a variable from a population, then the population mean, μ, is

$$\mu = \frac{x_1 + x_2 + \cdots + x_N}{N} = \frac{\sum x_i}{N} \tag{1}$$

If $x_1, x_2, \ldots, x_n$ are n observations of a variable from a sample, then the sample mean, $\bar{x}$, is

$$\bar{x} = \frac{x_1 + x_2 + \cdots + x_n}{n} = \frac{\sum x_i}{n} \tag{2}$$

Note that N represents the size of the population, while n represents the size of the sample. The symbol Σ (the Greek letter capital sigma) tells us the terms are to be added. The subscript i is used to make the various values distinct and does not serve as a mathematical operation. For example, x_1 is the first data value, x_2 is the second, and so on.

Let's look at an example to help distinguish the population mean and sample mean.

EXAMPLE 1 Computing a Population Mean and a Sample Mean

Table 1

Student	Score
1. Michelle	82
2. Ryanne	77
3. Bilal	90
4. Pam	71
5. Jennifer	62
6. Dave	68
7. Joel	74
8. Sam	84
9. Justine	94
10. Juan	88

Problem: The data in Table 1 represent the first exam score of 10 students enrolled in a section of Introductory Statistics.

(a) Compute the population mean.
(b) Find a simple random sample of size $n = 4$ students.
(c) Compute the sample mean of the sample obtained in part (b).

Approach

(a) To compute the population mean, we add up all the data values (test scores) and then divide by the number of individuals in the population.
(b) Recall from Section 1.2 that we can use either Table I in Appendix A, a calculator with a random-number generator, or computer software to obtain simple random samples. We will use a TI-84 Plus graphing calculator.
(c) The sample mean is found by adding the data values that correspond to the individuals selected in the sample and then dividing by $n = 4$, the sample size.

Solution

(a) We compute the population mean by adding the scores of all 10 students:

$$\sum x_i = x_1 + x_2 + x_3 + \cdots + x_{10}$$
$$= 82 + 77 + 90 + 71 + 62 + 68 + 74 + 84 + 94 + 88$$
$$= 790$$

Divide this result by 10, the number of students in the class.

$$\mu = \frac{\sum x_i}{N} = \frac{790}{10} = 79$$

Although it was not necessary in this problem, we will agree to round the mean to one more decimal place than that in the raw data.

(b) To find a simple random sample of size $n = 4$ from a population whose size is $N = 10$, we will use the TI-84 Plus random-number generator with a seed of 54. (Recall that this gives the starting point that the calculator uses to generate the list of random numbers.) Figure 1 shows the students in the sample. Bilal (90), Ryanne (77), Pam (71), and Michelle (82) are in the sample.

(c) We compute the sample mean by first adding the scores of the individuals in the sample.

$$\sum x_i = x_1 + x_2 + x_3 + x_4$$
$$= 90 + 77 + 71 + 82$$
$$= 320$$

Divide this result by 4, the number of individuals in the sample.

$$\bar{x} = \frac{\sum x_i}{n} = \frac{320}{4} = 80$$

Figure 1

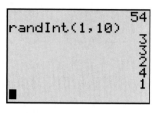

Now Work Problem 25.

Note to Instructor
The purpose of this activity is to introduce the students to sampling variability. The activity should take about 15 to 20 minutes.

In-Class Activity: Population Mean versus Sample Mean

Treat the students in the class as a population. All the students in the class should determine their pulse rates.

(a) Compute the population mean pulse rate.
(b) Obtain a simple random sample of $n = 4$ students and compute the sample mean. Does the sample mean equal the population mean?
(c) Obtain a second simple random sample of $n = 4$ students and compute the sample mean. Does the sample mean equal the population mean?
(d) Are the sample means the same? Why?

It is helpful to think of the mean of a data set as the center of gravity. In other words, the mean is the value such that a histogram of the data is perfectly balanced, with equal weight on each side of the mean. Figure 2 shows a histogram of the data in Table 1 with the mean labeled. The histogram balances at $\mu = 79$.

Figure 2

Scores on First Exam

$\mu = 79$
Score

In-Class Activity: The Mean as the Center of Gravity

Find a yardstick, a fulcrum, and three objects of equal weight (maybe 1-kilogram weights from the physics department). Place the fulcrum at 18 inches so that the yardstick balances like a teeter-totter. Now place one weight on the yardstick at 12 inches, another at 15 inches, and the third at 27 inches. See Figure 3.

Figure 3

Does the yardstick balance? Now compute the mean of the location of the three weights. Compare this result with the location of the fulcrum. Conclude that the mean is the center of gravity of the data set.

② Determine the Median of a Variable from Raw Data

A second measure of central tendency is the median. To compute the median of a set of data, the data must be quantitative.

Definition

In Other Words
To help remember the idea behind the median, think of the median of a highway; it divides the highway in half.

The **median** of a variable is the value that lies in the middle of the data when arranged in ascending order. That is, half the data are below the median and half the data are above the median. We use M to represent the median.

To compute the median of a set of data, we use the following steps:

Steps in Computing the Median of a Data Set

Step 1: Arrange the data in ascending order.
Step 2: Determine the number of observations, n.
Step 3: Determine the observation in the middle of the data set.

- If the number of observations is odd, then the median is the data value that is exactly in the middle of the data set. That is, the median is the observation that lies in the $\left(\dfrac{n+1}{2}\right)$ position.
- If the number of observations is even, then the median is the mean of the two middle observations in the data set. That is, the median is the mean of the data values on either side of the observation that lies in the $\left(\dfrac{n+1}{2}\right)$ position.

Note to Instructor
Suppose that a data set has 10 observations. In class, have students explain in their own words how to find the median. What if the data set has 11 observations?

EXAMPLE 2 **Computing the Median of a Data Set with an Odd Number of Observations**

Problem: The data in Table 2 represent the length (in seconds) of a random sample of songs released in the 1970s. Find the median length of the songs.

Approach: We will follow the steps listed above.

Solution

Step 1: Arrange the data in ascending order:

179, 201, 206, 208, 217, 222, 240, 257, 284

Table 2	
Song Name	**Length**
"Sister Golden Hair"	201
"Black Water"	257
"Free Bird"	284
"The Hustle"	208
"Southern Nights"	179
"Stayin' Alive"	222
"We Are Family"	217
"Heart of Glass"	206
"My Sharona"	240

Step 2: There are $n = 9$ observations.

Step 3: Since there are an odd number of observations, the median will be the observation exactly in the middle of the data set. The median, M, is 217 seconds (the $\frac{n+1}{2} = \frac{9+1}{2} = $ 5th data value). We list the data in ascending order, with the median in blue.

$$179, 201, 206, 208, 217, 222, 240, 257, 284$$

Notice there are four observations to the left and four observations to the right of the median. We conclude that 50% of the songs are less than 217 minutes and 50% of the songs are more than 217 minutes.

EXAMPLE 3   **Computing the Median of a Data Set with an Even Number of Observations**

Problem: Find the median score of the data in Table 1 on page 108.

Approach: We will follow the steps given on page 110.

Solution

Step 1: Arrange the data in ascending order:

$$62, 68, 71, 74, 77, 82, 84, 88, 90, 94$$

Step 2: There are $n = 10$ observations.

Step 3: Because there are $n = 10$ observations, the median will be the mean of the two middle observations. Because $\frac{n+1}{2} = \frac{10+1}{2} = 5.5$, the median is halfway between the fifth and sixth observations. We compute the median, M, by determining the mean of the fifth and sixth observations with the data written in ascending order. So the median is the mean of 77 and 82:

$$M = \frac{77 + 82}{2} = 79.5$$

Notice that there are five observations to the left and five observations to the right of the median, as follows:

$$62, 68, 71, 74, 77, 82, 84, 88, 90, 94$$

$$M = 79.5$$

We conclude that 50% of the students scored less than 79.5 and 50% of the students scored above 79.5.

Now compute the median of the data in Problem 19 by hand.

③ Determine the Mode of a Variable from Raw Data

A third measure of central tendency is the mode. The mode can be computed for either quantitative or qualitative data.

Definition   The **mode** of a variable is the most frequent observation of the variable that occurs in the data set.

To compute the mode, tally the number of observations that occur for each data value. The data value that occurs most often is the mode. A set of data can have no mode, one mode, or more than one mode. If no observation occurs more than once, we say the data have **no mode**.

EXAMPLE 4   Finding the Mode of Quantitative Data

Problem: The following data represent the number of O-ring failures on the shuttle *Columbia* prior to its fatal flight for its seventeen flights:

$$0, 0, 0, 0, 0, 0, 0, 0, 0, 0, 0, 1, 1, 1, 1, 2, 3$$

Find the mode number of O-ring failures.

Approach: We tally the number of times we observe each data value. The data value with the highest frequency is the mode.

Solution: The mode is 0 because it occurs most frequently (eleven times).

EXAMPLE 5   Finding the Mode of Quantitative Data

Problem: Find the mode of the data listed in Table 1 on page 108.

Approach: Tally the number of times we observe each data value. The data value with the highest frequency is the mode. Although not necessary, it is helpful to find the mode of quantitative data by arranging the data in ascending order.

Solution: We arrange the data in ascending order:

$$62, 68, 71, 74, 77, 82, 84, 88, 90, 94$$

Since each data value occurs only once, there is no mode.

> Now compute the mode of the data in Problem 19.

A data set can have more than one mode. For example, suppose the instructor added the scores of Pam and Sam incorrectly and they actually scored 77 and 88, respectively. The data set in Table 1 would now have two modes: 77 and 88. In this case, we say the data are **bimodal**. If a data set has three or more data values that occur with the highest frequency, the data set is **multimodal**. Typically, the mode is not reported for multimodal data because it is not representative of a central tendency or typical value.

We cannot determine the value of the mean or median of data that are qualitative. The only measure of central tendency that can be determined for qualitative data is the mode.

EXAMPLE 6   Determining the Mode of Qualitative Data

Problem: The data in Table 3 represent the location of injuries that required rehabilitation by a physical therapist. Determine the mode area of injury.

Table 3					
Back	Back	Hand	Neck	Knee	Knee
Wrist	Back	Groin	Shoulder	Shoulder	Back
Elbow	Back	Back	Back	Back	Back
Back	Shoulder	Shoulder	Knee	Knee	Back
Hip	Knee	Hip	Hand	Back	Wrist

Source: Krystal Catton, student at Joliet Junior College

Approach: Determine the location of injury that occurs with the highest frequency.

Solution: The mode location of injury is the back, with 12 instances.

Now Work Problem 39.

EXAMPLE 7   Finding the Mean, Median, and Mode Using Technology

Figure 4

	Student Scores
Mean	79
Standard Error	3.272783389
Median	79.5
Mode	#N/A

Problem: Use a statistical spreadsheet or calculator to determine the population mean, median, and mode of the student test score data in Table 1 on page 108.

Approach: We will use Excel to obtain the mean, median, and mode. The steps for calculating measures of central tendency using the TI-83/84 Plus graphing calculator, MINITAB, or Excel are given in the Technology Step by Step on page 123.

Result: Figure 4 shows the output obtained from Excel. The #N/A in the output indicates that the data set has no mode.

 Use the Mean and the Median to Help Identify the Shape of a Distribution

Table 4	
Mean	79
Median	79.5

Often, the mean and the median provide different values. Table 4 shows the mean and median scores on the exam for the data in Table 1 on page 108.

Notice that the median and the mean are close in value. Refer back to Table 1. Suppose Jennifer did not study for the exam and scored 28. The median would not change, but the mean would decrease from 79 to 75.6. We say that the median is **resistant** to extreme values (very large or small), but the mean is not resistant. Therefore, when data sets have unusually large or small values relative to the entire set of data or when the distribution of the data is skewed, the median is the preferred measure of central tendency over the mean because it is more representative of the typical observation.

In fact, the mean and median can be useful in determining the shape of a distribution. It can be shown that, if a distribution is perfectly symmetric and has one mode, then the median will equal the mean (and the mode). So symmetric distributions will have a median and a mean that are close in value. If the mean is substantially larger than the median, the distribution will be skewed right. Do you know why? In distributions that are skewed right, a few data values are substantially larger than the others. These larger data values cause the mean to be inflated while having little, if any, effect on the median. Similarly, distributions that are skewed left will have a mean that is substantially smaller than the median. We summarize these ideas in Table 5 and Figure 5.

Note to Instructor

In-class concept check: Ask students to discuss how extreme observations affect the mean and median. Which measure of center is best when extreme values exist? Which is a better measure of central tendency for the following variables: income, price of housing, height, or IQ?

 CAUTION
Because the mean is not resistant, it should not be reported as a measure of central tendency when the distribution of data is highly skewed.

Note to Instructor

If you collected pulse data, change the largest observation to something extreme, like 120. Ask students to recompute the mean and median. Discuss the results.

Table 5		
Relation between the Mean, Median, and Distribution Shape		
Distribution Shape	**Mean versus Median**	
Skewed left	Mean substantially smaller than median	
Symmetric	Mean roughly equal to median	
Skewed right	Mean substantially larger than median	

Figure 5
Mean/median versus skewness

(a) Skewed Left
Mean < Median

Mean = Median = Mode
(b) Symmetric
Mean = Median

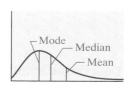
(c) Skewed Right
Mean > Median

EXAMPLE 8   **Describing the Shape of a Distribution**

Problem: In 2004, the New York Yankees had a record $184 million payroll. The data in Table 6 represent the salaries of the players on the opening-day roster in 2004 in thousands of dollars.

Table 6			
Player	**Salary**	**Player**	**Salary**
Brown, Kevin	15,714	Lofton, Kenny	3,100
Cairo, Miguel	900	Matsui, Hideki	7,000
Clark, Tony	750	Mussina, Mike	16,000
Contreras, Jose	8,500	Osborne, Donovan	450
Crosby, Bubba	301	Posada, Jorge	9,000
De Paula, Jorge	303	Quantrill, Paul	3,000
Flaherty, John	775	Rivera, Mariano	10,890
Giambi, Jason	12,429	Rodriguez, Alex	22,000
Gordon, Tom	3,500	Sheffield, Gary	13,000
Heredia, Felix	1,800	Sierra, Ruben	1,000
Hernandez, Orlando	500	Vazquez, Javier	9,000
Jeter, Derek	18,600	White, Gabe	1,925
Karsay, Steve	6,000	Williams, Bernie	12,357
Lee, Travis	2,000	Wilson, Enrique	700
Lieber, Jon	2,700		

Source: usatoday.com

(a) Draw a frequency histogram of the Yankee salaries.
(b) Find the mean and median Yankee salary.
(c) Describe the shape of the distribution of Yankee salaries.
(d) Which measure of central tendency better describes the average salary of a player on the Yankees, the mean or the median?

Approach

(a) We will use MINITAB to draw a histogram of the salaries.
(b) We will use MINITAB to determine the mean and median salary.
(c) We can identify the shape of the distribution by looking at the frequency histogram and comparing the mean to the median. Refer to Table 5 and Figure 5.
(d) If the data are skewed left or skewed right, the median is the better measure of central tendency. If the data are symmetric, the mean is the better measure of central tendency.

Solution

(a) Figure 6 shows a histogram of the data drawn using MINITAB.

Figure 6

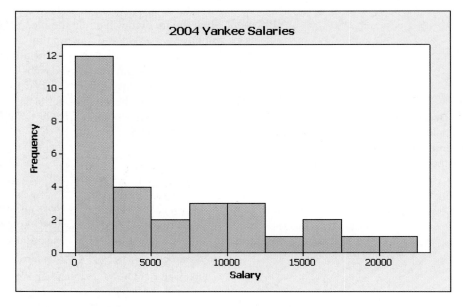

(b) Using MINITAB, we find $\mu = 6352$ and $M = 3100$. See Figure 7.

Figure 7

Descriptive statistics

Variable	N	N*	Mean	SE Mean	StDev	Minimum	Q1	Median	Q3	Maximum
Salaries	29	0	6352	1180	6353	301	838	3100	11624	22000

(c) The shape of the histogram drawn in Figure 6 is skewed right. Notice that the mean is substantially larger than the median because the high salaries (especially Alex Rodriguez and Derek Jeter) push up the value of the mean.

(d) Because the shape of the distribution is skewed right, the median is the better measure of central tendency.

EXAMPLE 9   Describing the Shape of a Distribution

Problem: The data in Table 7 represent the birth weights (in pounds) of 50 randomly sampled babies.

(a) Find the mean and the median.

(b) Describe the shape of the distribution.

(c) Which measure of central tendency better describes the average birth weight?

Approach

(a) Use a TI-84 Plus to compute the mean and the median.

(b) The histogram, along with the mean and the median, is used to identify the shape of the distribution.

(c) If the data are roughly symmetric, the mean is the better measure of central tendency. If the data are skewed, the median is the better measure of central tendency.

Table 7

5.8	7.4	9.2	7.0	8.5	7.6
7.9	7.8	7.9	7.7	9.0	7.1
8.7	7.2	6.1	7.2	7.1	7.2
7.9	5.9	7.0	7.8	7.2	7.5
7.3	6.4	7.4	8.2	9.1	7.3
9.4	6.8	7.0	8.1	8.0	7.5
7.3	6.9	6.9	6.4	7.8	8.7
7.1	7.0	7.0	7.4	8.2	7.2
7.6	6.7				

Solution

(a) Using a TI-84 Plus, we find $\bar{x} = 7.49$ and $M = 7.35$. See Figure 8.

Figure 8

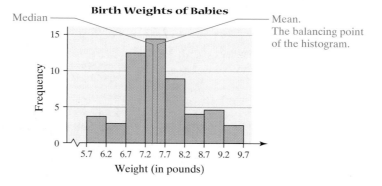

```
1-Var Stats
 x̄=7.488
 Σx=374.4
 Σx²=2835.1
 Sx=.8029638973
 σx=.7948937036
↓n=50
```
(a)

```
1-Var Stats
↑n=50
 minX=5.8
 Q₁=7
 Med=7.35
 Q₃=7.9
 maxX=9.4
```
(b)

(b) See Figure 9 for the frequency histogram with the mean and median labeled. The distribution is bell shaped. We have further evidence of the shape because the mean and median are close to each other.

Figure 9
Birth weights of 50 randomly selected babies

Median ———— **Birth Weights of Babies** ———— Mean.
The balancing point of the histogram.

[Histogram: Frequency (y-axis) vs Weight (in pounds) (x-axis) with axis values 5.7 6.2 6.7 7.2 7.7 8.2 8.7 9.2 9.7]

(c) Because the mean and median are close in value, we use the mean as the measure of central tendency.

Now Work Problem 29.

A question you may be asking yourself is, "Why would I ever compute the mean?" After all, the mean and median are close in value for symmetric data, and the median is the better measure of central tendency for skewed data. The reason we compute the mean is that much of the statistical inference that we perform is based on the mean. We will have more to say about this in Chapter 8.

We conclude this section with the following chart, which addresses the circumstances under which each measure of central tendency should be used.

Measure of Central Tendency	Computation	Interpretation	When to Use
Mean	Population mean: $\mu = \dfrac{\Sigma x_i}{N}$ Sample mean: $\overline{x} = \dfrac{\Sigma x_i}{n}$	Center of gravity	When data are quantitative and the frequency distribution is roughly symmetric
Median	Arrange data in ascending order and divide the data set in half	Divides the bottom 50% of the data from the top 50%	When the data are quantitative and the frequency distribution is skewed left or skewed right
Mode	Tally data to determine most frequent observation	Most frequent observation	When the most frequent observation is the desired measure of central tendency or the data are qualitative

3.1 ASSESS YOUR UNDERSTANDING

Concepts and Vocabulary

1. What does it mean if a statistic is resistant? Why is the median resistant, but the mean is not? Is the mode a resistant measure of center?

2. Describe how the mean and the median can be used to determine the shape of a distribution.

3. In the 2000 census conducted by the U.S. Census Bureau, two average household incomes were reported: $41,349 and $55,263. One of these averages is the mean and the other is the median. Which is the mean? Support your answer.

4. The U.S. Department of Housing and Urban Development (HUD) uses the median to report the average price of a home in the United States. Why do you think HUD uses the median?

5. A histogram of a set of data indicates that the distribution of the data is skewed right. Which measure of central tendency will be larger, the mean or the median? Why?

6. If a data set contains 10,000 values arranged in increasing order, where is the median located?

7. Explain why the mode is used as the measure of central tendency for qualitative data.

8. A(n) _____ is a descriptive measure of a population, and a(n) _____ is a descriptive measure of a sample.

9. *True or False*: A data set will always have exactly one mode. False

10. *True or False*: If the number of observations is odd, the median is $M = \dfrac{n+1}{2}$. False

6. The median is between the 5000th and 5001st ordered values
8. parameter; statistic

Skill Building

In Problems 11–14, find the population mean or sample mean as indicated.

11. Sample: 20, 13, 4, 8, 10 $\overline{x} = 11$

12. Sample: 83, 65, 91, 87, 84 $\overline{x} = 82$

13. Population: 3, 6, 10, 12, 14 $\mu = 9$

14. Population: 1, 19, 25, 15, 12, 16, 28, 13, 6 $\mu = 15$

15. For Super Bowl XXXIX, Fox television sold 59 ad slots for a total revenue of roughly $142 million. What was the mean price per ad slot? $2.4 million

16. The median for the given set of six ordered data values is 26.5. What is the missing value? 7 12 21 _____ 41 50 32

17. **Crash Test Results** The Insurance Institute for Highway Safety crashed the 2001 Honda Civic four times at 5 miles per hour. The costs of repair for each of the four crashes were

$$\$420, \$462, \$409, \$236$$

Compute the mean, median, and mode cost of repair.

18. **Cell Phone Use** The following data represent the monthly cell phone bill for my wife's phone for six randomly selected months.

$$\$35.34, \$42.09, \$39.43, \$38.93 \ \$43.39, \$49.26$$

Compute the mean, median, and mode phone bill.

19. **Concrete Mix** A certain type of concrete mix is designed to
NW withstand 3000 pounds per square inch (psi) of pressure. The strength of concrete is measured by pouring the mix into casting cylinders 6 inches in diameter and 12 inches tall. The

17. $381.75; $414.50; none   18. $41.41; $40.76; none

cylinder is allowed to "set up" for 28 days. The cylinders are then stacked on one another until the cylinders are crushed. The following data represent the strength of nine randomly selected casts (in psi).

3960, 4090, 3200, 3100, 2940, 3830, 4090, 4040, 3780

Compute the mean, median, and mode strength of the concrete (in psi). *3670; 3830; 4090*

20. **Flight Time** The following data represent the flight time (in minutes) of a random sample of seven flights from Las Vegas, Nevada, to Newark, New Jersey, on Continental Airlines.

282, 270, 260, 266, 257, 260, 267

Compute the mean, median, and mode flight time. *266; 266; 260*

21. For each of the three histograms shown, determine whether the mean is greater than, less than, or approximately equal to the median. Justify your answer. (a) $\bar{x} > M$ (b) $\bar{x} = M$ (c) $\bar{x} < M$

(a)

(b)

(c)

22. Match the histograms shown to the summary statistics shown to the right: (a) IV (b) III (c) II (d) I

	Mean	Median
I	42	42
II	31	36
III	31	26
IV	31	32

(a)

(b)

(c)

(d)

Applying the Concepts

23. **ATM Fees** The following data for a random sample of banks in Los Angeles and New York City represent the ATM fee for using another bank's ATM.

Los Angeles: mean = $1.44, median = $1.50, mode = $1.50 and $2.00;
New York City: mean = $1.06, median = $1.13, mode = $1.00

Los Angeles	2.00	1.50	1.50	1.00	1.50	2.00	0.00	2.00
New York City	1.50	1.00	1.00	1.25	1.25	1.50	1.00	0.00

Source: www.bankrate.com

Compute the mean, median, and mode ATM fee for each city. Does there appear to be a difference in the ATM fee charged in Los Angeles versus New York City? Why might this be the case?

24. **Reaction Time** In an experiment conducted online at the University of Mississippi, study participants are asked to react to a stimulus. In one experiment, the participant must press a key upon seeing a blue screen. The time (in seconds) to press the key is measured. The same person is then asked to press a key upon seeing a red screen, again with the time to react measured. The table shows the results for six study participants. Compute the mean, median, and mode reaction time for both blue and red. Does there appear to be a difference in the reaction time? What might account for any difference? How might this information be used?

Blue: 0.551; 0.5315; none; Red: 0.458; 0.432;

Participant Number	Reaction Time to Blue	Reaction Time to Red
1	0.582	0.408
2	0.481	0.407
3	0.841	0.542
4	0.267	0.402
5	0.685	0.456
6	0.45	0.533

Source: PsychExperiments at the University of Mississippi (www.olemiss.edu/psychexps/)

(28d) Method 1: mean = 40.215 km/h; method 2: mean = 40.207 km/h; method 3: mean = 40.207 km/h

25. Pulse Rates The following data represent the pulse rates
NW (beats per minute) of nine students enrolled in a section of
Sullivan's Introductory Statistics course. Treat the nine
students as a population.

Student	Pulse
Perpectual Bempah	76
Megan Brooks	60
Jeff Honeycutt	60
Clarice Jefferson	81
Crystal Kurtenbach	72
Janette Lantka	80
Kevin McCarthy	80
Tammy Ohm	68
Kathy Wojdyla	73

(a) Compute the population mean pulse. 72.2
(b) Determine two simple random samples of size 3 and
compute the sample mean pulse of each sample.
(c) Which samples result in a sample mean that overestimates the population mean? Which samples result in
a sample mean that underestimates the population
mean? Do any samples lead to a sample mean that
equals the population mean?

26. Travel Time The following data represent the travel time
(in minutes) to school for nine students enrolled in Sullivan's College Algebra course. Treat the nine students as a
population.

Student	Travel Time	Student	Travel Time
Amanda	39	Scot	45
Amber	21	Erica	11
Tim	9	Tiffany	12
Mike	32	Glenn	39
Nicole	30		

(a) Compute the population mean for travel time. 26.4
(b) Determine three simple random samples of size 4 and
compute the sample mean for travel time of each
sample.
(c) Which samples result in a sample mean that overestimates the population mean? Which samples result in
a sample mean that underestimates the population
mean? Do any samples lead to a sample mean that
equals the population mean?

27. Soccer Goals Mia Hamm, who retired after the 2004
Olympics, is considered by some to be the most prolific
player in international soccer. The following data represent the number of goals scored over her 18-year career.

0	0	0	4	10	1	10	10	19
9	18	20	13	13	2	7	8	13

Source: www.soccerhall.com

(a) Compute the population mean of the number of goals
she scored. 8.7
(b) Determine two simple random samples of size 3 and
compute the sample mean of the number of goals she
scored. Answers will vary
(c) Which samples result in a sample mean that overestimates the population mean? Which samples result in
a sample mean that underestimates the population
mean? Do any samples lead to a sample mean that
equals the population mean? Answers will vary

28. Tour de Lance Lance Armstrong won the Tour de
France seven consecutive times (1999–2005). The following table gives the winning times, distances, speeds,
and margin of victory.

Year	Winning Time (h)	Distance (km)	Winning Speed (km/h)	Winning Margin (min)
1999	91.538	3687	40.28	7.617
2000	92.552	3662	39.56	6.033
2001	86.291	3453	40.02	6.733
2002	82.087	3278	39.93	7.283
2003	83.687	3427	40.94	1.017
2004	83.601	3391	40.56	6.317
2005	86.251	3593	41.65	44.667

Source: cyclingnews.com

(a) Compute the mean and median of his winning times for
the six races. Mean = 86.572 h; median = 86.251 h
(b) Compute the mean and median of the distances for
the six races. Mean = 3499 km; median = 3453 km
(c) Compute the mean and median of his winning time
margins. Mean = 5.667 min; median = 6.317 min
(d) Compute the mean winning speed by finding the
mean of the data values in the table. Next, compute
the mean winning speed by finding the total of the
six distances and dividing by the total of the six winning times. Finally, compute the mean winning
speed by dividing the mean distance by the mean
winning time. Do the three values agree or are
there differences?

29. Connection Time The following data represent the con-
NW nection time in seconds to an Internet service provider for
30 randomly selected connections.

39.76	36.13	36.61	38.80	39.04	39.09
37.24	35.62	40.07	38.76	39.23	38.38
38.24	36.34	35.89	42.86	36.03	37.03
38.64	41.86	41.22	37.19	40.50	39.81
39.84	39.45	40.91	43.12	40.54	42.02

Source: Nicole Spreitzer, student at Joliet Junior College

A histogram of the data is shown. The mean connection
time is 39.007 seconds and the median connection time
is 39.065 seconds. Use this information to identify the

(32a)Mean = 5.883 million shares; median = 5.58 million shares

shape of the distribution. Which measure of central tendency better describes the "center" of the distribution? Mean

Histogram of Time (seconds)

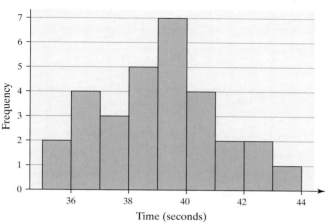

30. **Journal Costs** The following data represent the annual subscription cost (in dollars) for a random sample of 26 biology journals.

1188	778	1970	661	1294	1175
2033	3911	198	8415	796	1840
1141	1050	3643	1407	1092	585
1049	1092	1589	4115	1150	2799
707	2330				

Source: Carol Wesolowski, student at Joliet Junior College

A histogram of the data is shown. The mean subscription cost is $1846 and the median subscription cost is $1182. Use this information to identify the shape of the distribution. Which measure of central tendency better describes the "center" of the distribution? Median

Histogram of Cost

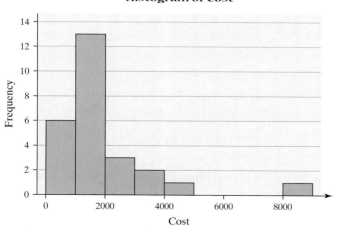

(32b) Skewed right

31. **Serum HDL** Dr. Paul Oswiecmiski randomly selects 40 of his 20- to 29-year-old patients and obtains the following data regarding their serum HDL cholesterol.

(31a) 51.1, 51

70	56	48	48	53	52	66	48
36	49	28	35	58	62	45	60
38	73	45	51	56	51	46	39
56	32	44	60	51	44	63	50
46	69	53	70	33	54	55	52

(a) Compute the mean and the median serum HDL.
(b) Identify the shape of the distribution based on the histogram drawn in Problem 31 in Section 2.2 and the relationship between the mean and the median. Symmetric

32. **Volume of Altria Group Stock** The volume of a stock is the number of shares traded on a given day. The following data represent the volume of Altria Group stock traded for a random sample of 35 trading days in 2004. The data are in millions, so 3.78 represents 3,780,000 shares traded.

3.78	8.74	4.35	5.02	8.40
6.06	5.75	5.34	6.92	6.23
5.32	3.25	6.57	7.57	6.07
3.04	5.64	5.00	7.16	4.88
10.32	3.38	7.25	6.52	4.43
3.38	5.53	4.74	9.70	3.56
10.96	4.50	7.97	3.01	5.58

Source: yahoo.finance.com

(a) Compute the mean and the median number of shares traded.
(b) Identify the shape of the distribution based on the histogram drawn in Problem 32 in Section 2.2 and the relationship between the mean and the median.

33. **M&Ms** The following data represent the weights (in grams) of a simple random sample of 50 M&M plain candies.

0.87	0.88	0.82	0.90	0.90	0.84	0.84
0.91	0.94	0.86	0.86	0.86	0.88	0.87
0.89	0.91	0.86	0.87	0.93	0.88	
0.83	0.94	0.87	0.93	0.91	0.85	
0.91	0.91	0.86	0.89	0.87	0.93	
0.88	0.88	0.89	0.79	0.82	0.83	
0.90	0.88	0.84	0.93	0.76	0.90	
0.88	0.92	0.85	0.79	0.84	0.86	

Source: Michael Sullivan

33. Symmetric; mean

Determine the shape of the distribution of weights of M&Ms by drawing a frequency histogram and computing the mean and median. Which measure of central tendency better describes the weight of a plain M&M?

34. Old Faithful We have all heard of the Old Faithful geyser in Yellowstone National Park. However, there is another, less famous, Old Faithful geyser in Calistoga, California. The following data represent the length of eruption (in seconds) for a random sample of eruptions of the California Old Faithful.

108	108	99	105	103	103	94
102	99	106	90	104	110	110
103	109	109	111	101	101	
110	102	105	110	106	104	
104	100	103	102	120	90	
113	116	95	105	103	101	
100	101	107	110	92	108	

Source: Ladonna Hansen, Park Curator

Determine the shape of the distribution of time between eruptions by drawing a frequency histogram and computing the mean and median. Which measure of central tendency better describes the time between eruptions? Symmetric; mean

35. Hours Working A random sample of 25 college students was asked, "How many hours per week typically do you work outside the home?" Their responses were as follows:

0	0	15	20	30
40	30	20	35	35
28	15	20	25	25
30	5	0	30	24
28	30	35	15	15

Determine the shape of the distribution of hours worked by drawing a frequency histogram and computing the mean and median. Which measure of central tendency better describes hours worked? Skewed left; median

36. A Dealer's Profits The following data represent the profits (in dollars) of a new car dealer for a random sample of 40 sales.

781	1,038	453	1,446	3,082
501	451	1,826	1,348	3,001
1,342	1,889	580	0	2,909
2,883	480	1,664	1,064	2,978
149	1,291	507	261	540
543	87	798	673	2,862
1,692	1,783	2,186	398	526
730	2,324	2,823	1,676	4,148

Source: Ashley Hudson, student at Joliet Junior College

Determine the shape of the distribution of new car profits by drawing a frequency histogram and computing the mean and median. Which measure of central tendency better describes the profit? Skewed right; median

37. Foreign-Born Population The following data represent the region of birth of foreign-born residents of the United States in 2003. Determine the mode region of birth. Central America

Region	Number (thousands)
Caribbean	3,384
Central America	12,362
South America	2,111
Asia	8,375
Europe	4,590
Other Regions	2,680

Source: U.S. Census Bureau

38. Robbery The following data represent the number of offenses for various robberies in 2003. Determine the mode offense. Street or highway

Type of Robbery	Number (thousands)
Street or highway	131
Commercial	61
Gas station	10
Convenience store	26
Residence	41
Bank	7

Source: U.S. Federal Bureau of Investigation

39. 2004 Presidential Election An exit poll was conducted in Los Alamos County, New Mexico, in which a random sample of 40 voters revealed whom they voted for in the presidential election. The results of the survey are shown below. Determine the mode candidate. Bush

Kerry	Kerry	Bush	Bush	Bush
Bush	Kerry	Kerry	Bush	Bush
Kerry	Bush	Kerry	Bush	Kerry
Bush	Bush	Kerry	Kerry	Nader
Kerry	Bush	Bush	Kerry	Kerry
Badnarik	Kerry	Bush	Bush	Bush
Bush	Bush	Bush	Bush	Kerry
Kerry	Kerry	Kerry	Bush	Bush

40. Hospital Admissions The following data represent the diagnosis of a random sample of 20 patients admitted to a hospital. Determine the mode diagnosis. *Gunshot wound*

Cancer	Motor vehicle accident	Congestive heart failure
Gunshot wound	fall	Gunshot wound
Gunshot wound	Motor vehicle accident	Gunshot wound
Assault	Motor vehicle accident	Gunshot wound
Motor vehicle accident	Motor vehicle accident	Gunshot wound
Motor vehicle accident	Gunshot wound	Motor vehicle accident
Fall	Gunshot wound	

Source: Tamela Ohm, student at Joliet Junior College

41. Resistance and Sample Size Each of the following three data sets represents the IQ scores of a random sample of adults. IQ scores are known to have a mean and median of 100. For each data set, compute the mean and median. For each data set recalculate the mean and median, assuming that the individual whose IQ is 106 is accidentally recorded as 160. For each sample size, state what happens to the mean and the median. Comment on the role the number of observations plays in resistance.

Sample of Size 5				
106	92	98	103	100

Sample of Size 12					
106	92	98	103	100	102
98	124	83	70	108	121

Sample of Size 30					
106	92	98	103	100	102
98	124	83	70	108	121
102	87	121	107	97	114
140	93	130	72	81	90
103	97	89	98	88	103

42. Super Bowl XXXIX Champion New England Patriots The following table gives roster information for the offense of the Super Bowl XXXIX Champion New England Patriots.

No.	Name	Position	Age (yr)	Weight (lb)	Years of Experience	College
12	Tom Brady	QB	27	225	5	Michigan
6	Rohan Davey	QB	26	245	3	LSU
13	Jim Miller	QB	33	225	10	Michigan State
27	Ribih Abdullah	RB	29	220	7	Lehigh
34	Cedric Cobbs	RB	24	225	Rookie	Arkansas
28	Corey Dillon	RB	30	225	8	Washington
33	Kevin Faulk	RB	28	202	6	LSU
35	Patrick Pass	RB	27	217	5	Georgia
83	Deion Branch	WR	25	193	3	Louisville
87	David Givens	WR	24	215	3	Notre Dame
81	Bethel Johnson	WR	25	200	2	Texas A&M
10	Kevin Kasper	WR	27	197	4	Iowa
86	David Patten	WR	30	190	8	Western Carolina
88	Christian Fauria	TE	33	250	10	Colorado
82	Daniel Graham	TE	26	257	3	Colorado
85	Jed Weaver	TE	28	258	6	Oregon
67	Daniel Koppen	C	25	296	2	Boston College
66	Lonie Paxton	C	26	260	5	Sacramento State
76	Brandon Gorin	OT	26	308	3	Purdue
72	Matt Light	OT	26	305	4	Purdue
63	Joe Andruzzi	OG	29	312	8	Southern Connecticut State
71	Russ Hochstein	OG	27	305	4	Nebraska
64	Gene Mruczkowski	OG	24	305	2	Purdue
61	Stephen Neal	OG	28	305	3	California State Bakersfield
74	Billy Yates	OG	24	305	2	Texas A&M

Source: ESPN.com

(a) Find the mean, median, and mode age.

(b) Find the mean, median, and mode weight.

(c) Find the mean, median, and mode years of experience. (**Note:** Rookie = 0 years.)

(d) Find the mode college attended. Purdue

(e) Obtain a simple random sample of six members of New England's offense. Compute sample mean age, weight, and years of experience. How do the sample means compare to the population means? Answers will vary

(f) Compute the mean, median, and mode weights of the five offensive guards (OG). Compute the mean, median, and mode weights of the five running backs (RB). Does there appear to be a difference in the weights? What might account for any differences?

(g) Does it make sense to compute the mean player number? Why? No, the data are categorical (qualitative)

43. **Super Bowl XXXIX Champion New England Patriots Revisited** Using the data presented in Problem 42, answer the following.

(a) The skilled positions in football are quarterback (QB), running back (RB), wide receiver (WR), and tight end (TE). Obtain a stratified sample by using skilled positions as one stratum and the remaining positions as a second stratum. Randomly select four skilled players and two "nonskilled" players. Compute the mean weight of the sample data. Compare to the result in part (e) in Problem 42.

(b) Cluster the players by position. Obtain a cluster sample by randomly selecting two clusters. Compute the mean weight of the sample data. Compare to the result in to part (e) in Problem 42. Can you think of any problems with obtaining a cluster sample?

44. You are negotiating a contract for the Players' Association of the NBA. Which measure of central tendency will you use to support your claim that the average player's salary needs to be increased? Why? As the chief negotiator for the owners, which measure would you use to refute the claim made by the Players' Association?

45. In January 2005, the mean amount of money lost per visitor to a local riverboat casino was $135. Do you think the median was more than, less than, or equal to this amount? Why?

46. **Missing Exam Grade** A professor has recorded exam grades for 20 students in his class, but one of the grades is no longer readable. If the mean score on the exam was 82 and the mean of the 19 readable scores is 84, what is the value of the unreadable score? 44

47. Suppose that the mean of a set of six data values is 34. What is the sum of the six data values? 204

48. For each of the following situations, determine which measure of central tendency is most appropriate and justify your reasoning.

(a) Average price of a home sold in Pittsburgh, Pennsylvania, in 2002 Median

(b) Most popular major for students enrolled in a statistics course Mode

(c) Average test score when the scores are distributed symmetrically Mean

(d) Average test score when the scores are skewed right Median

(e) Average income of a player in the National Football League Median

(f) Most requested song at a radio station Mode

49. **Linear Transformations** Benjamin owns a small Internet business. Besides himself, he employs nine other people. The salaries earned by the employees are given below in thousands of dollars (Benjamin's salary is the largest, of course):

$$30, 30, 45, 50, 50, 50, 55, 55, 60, 75$$

(49a) Mean = 50; median = 50; mode = 50

(a) Determine the mean, median, and mode for salary.

(b) Business has been good! As a result, Benjamin has a total of $25,000 in bonus pay to distribute to his employees. One option for distributing bonuses is to give each employee (including himself) $2500. Add the bonuses under this plan to the original salaries to create a new data set. Recalculate the mean, median, and mode. How do they compare to the originals?

(c) As a second option, Benjamin can give each employee a bonus of 5% of his or her original salary. Add the bonuses under this second plan to the original salaries to create a new data set. Recalculate the mean, median, and mode. How do they compare to the originals?

(d) As a third option, Benjamin decides not to give his employees a bonus at all. Instead, he keeps the $25,000 for himself. Use this plan to create a new data set. Recalculate the mean, median, and mode. How do they compare to the originals?

50. **Linear Transformations** Use the five test scores of 65, 70, 71, 75, and 95 to answer the following questions:

(a) Find the sample mean. 75.2

(b) Find the median. 71

(c) Which measure of central tendency best describes the typical test score? Median

(d) Suppose the professor decides to curve the exam by adding 4 points to each test score. Compute the sample mean based on the adjusted scores. 79.2

(e) Compare the unadjusted test score mean with the curved test score mean. What effect did adding 4 to each score have on the mean?

51. **Trimmed Mean** Another measure of central tendency is the trimmed mean. It is computed by determining the mean of a data set after deleting the smallest and largest observed values. Compute the trimmed mean for the data in Problem 33. Is the trimmed mean resistant? Explain. 0.875

(42a) Mean = 27.1 ys; median = 27 ys; mode = 26 yr

(42b) Mean = 249.8 lb; median = 245 lb; mode = 305 lb.

(42c) Mean = 4.6 yr; median = 4 yr; mode = 3 yr

(42f) OG: mean = 306.5 lb, median = 305 lb, mode = 305 lb;
RB: mean = 217.8 lb, median = 220 lb, mode = 225 lb

(49b) mean = 52.5; median = 52.5; mode = 52.5

(49c) mean = 52.5; median = 52.5; mode = 52.5

(49d) mean = 52.5; median = 50; mode = 50

52. Midrange The midrange is also a measure of central tendency. It is computed by adding the smallest and largest observed values of a data set and dividing the result by 2; that is,

$$\text{Midrange} = \frac{\text{largest data value } + \text{ smallest data value}}{2}$$

Compute the midrange for the data in Problem 33. Is the midrange resistant? Explain. *0.85; no*

Technology Step by Step	**Determining the Mean and Median**

TI-83/84 Plus   *Step 1:* Enter the raw data in L1 by pressing STAT and selecting 1:Edit.

Step 2: Press STAT, highlight the CALC menu, and select 1:1-Var Stats.

Step 3: With 1-Var Stats appearing on the HOME screen, press 2nd 1 to insert L1 on the HOME screen. Press ENTER.

MINITAB   *Step 1:* Enter the data in C1.

Step 2: Select the **Stat** menu, highlight **Basic Statistics**, and then highlight **Display Descriptive Statistics**.

Step 3: In the **Variables** window, enter C1. Click OK.

Excel   *Step 1:* Enter the data in column A.

Step 2: Select the **Tools** menu and highlight **Data Analysis . . .**

Step 3: In the Data Analysis window, highlight **Descriptive Statistics** and click OK.

Step 4: With the cursor in the **Input Range** window, use the mouse to highlight the data in column A.

Step 5: Select the **Summary statistics** option and click OK.

3.2   Measures of Dispersion

Objectives

1. **Compute the range of a variable from raw data**
2. **Compute the variance of a variable from raw data**
3. **Compute the standard deviation of a variable from raw data**
4. **Use the Empirical Rule to describe data that are bell shaped**
5. **Use Chebyshev's Inequality to describe any set of data**

In Section 3.1, we discussed measures of central tendency. The purpose of these measures is to describe the typical value of a variable. In addition to measuring the central tendency of a variable, we would also like to know the amount of dispersion in the variable. By *dispersion*, we mean the degree to which the data are "spread out." An example should help to explain why measures of central tendency are not sufficient in describing a distribution.

EXAMPLE 1   **Comparing Two Sets of Data**

Problem: The data in Table 8 represent the IQ scores of a random sample of 100 students from two different universities. For each university, compute the mean IQ score and draw a histogram, using a lower class limit of 55 for the first class and a class width of 15. Comment on the results.

Table 8

University A										University B									
73	103	91	93	136	108	92	104	90	78	86	91	107	94	105	107	89	96	102	96
108	93	91	78	81	130	82	86	111	93	92	109	103	106	98	95	97	95	109	109
102	111	125	107	80	90	122	101	82	115	93	91	92	91	117	108	89	95	103	109
103	110	84	115	85	83	131	90	103	106	110	88	97	119	90	99	96	104	98	95
71	69	97	130	91	62	85	94	110	85	87	105	111	87	103	92	103	107	106	97
102	109	105	97	104	94	92	83	94	114	107	108	89	96	107	107	96	95	117	97
107	94	112	113	115	106	97	106	85	99	98	89	104	99	99	87	91	105	109	108
102	109	76	94	103	112	107	101	91	107	116	107	90	98	98	92	119	96	118	98
107	110	106	103	93	110	125	101	91	119	97	106	114	87	107	96	93	99	89	94
118	85	127	141	129	60	115	80	111	79	104	88	99	97	106	107	112	97	94	107

Approach: We will use MINITAB to compute the mean and draw a histogram for each university.

Solution: We enter the data into MINITAB and determine that the mean IQ score of both universities is 100.0. Figure 10 shows the histograms.

Figure 10

Note to Instructor

You may want to share with students that the smallest data value is also referred to as the minimum and the largest data value is referred to as the maximum, as a preview of the five-number summary.

We notice that both universities have the same mean IQ, but the histograms indicate the IQs from University A are more spread out, that is, more dispersed. While an IQ of 100.0 is typical for both universities, it appears to be a more reliable description of the typical student from University B than from University A. That is, a higher proportion of students have IQ scores within, say, 15 points of the mean of 100.0 from University B than from University A. _____

Our goal in this section is to discuss numerical measures of dispersion so that we can quantify the spread of data. In this section, we discuss three numerical measures for describing the dispersion or spread of data: the range, variance, and standard deviation. In Section 3.4, we will discuss another measure of dispersion, the *interquartile range* (IQR).

 ## Compute the Range of a Variable from Raw Data

The simplest measure of dispersion is the range. To compute the range, the data must be quantitative.

Definition

The **range, R**, of a variable is the difference between the largest data value and the smallest data value. That is,

$$\text{Range} = R = \text{largest data value} - \text{smallest data value}$$

| EXAMPLE 2 | **Computing the Range of a Set of Data** |

Table 9

Student	Score
1. Michelle	82
2. Ryanne	77
3. Bilal	90
4. Pam	71
5. Jennifer	62
6. Dave	68
7. Joel	74
8. Sam	84
9. Justine	94
10. Juan	88

Problem: The data in Table 9 represent the scores on the first exam of 10 students enrolled in a section of Introductory Statistics. Compute the range.

Approach: The range is found by computing the difference between the largest and smallest data values.

Solution: The highest test score is 94 and the lowest test score is 62. The range, R, is

$$94 - 62 = 32$$

All the students in the class scored between 62 and 94 on the exam. The difference between the best score and the worst score is 32 points. ■

Now compute the range of the data in Problem 19.

Notice that the range is affected by extreme values in the data set, so the range is not resistant. If Jennifer did not study and scored 28, the range becomes $R = 94 - 28 = 66$. In addition, the range is computed using only two values in the data set (the largest and smallest). The *variance* and the *standard deviation*, on the other hand, use all the data values in the computations.

In Other Words
The range is not resistant.

2 ## Compute the Variance of a Variable from Raw Data

Note to Instructor
Stress to the students that the variance is difficult to interpret because it is measured in units squared (such as points squared). However, the variance plays an important role in statistical inference.

Just as there is a population mean and sample mean, we also have a population variance and a sample variance. Measures of dispersion are meant to describe how spread out data are. Another way to think about this is to describe how far, on average, each observation is from the mean. Variance is based on the **deviation about the mean**. For a population, the deviation about the mean for the ith observation is $x_i - \mu$. For a sample, the deviation about the mean for the ith observation is $x_i - \bar{x}$. The further an observation is from the mean, the larger the absolute deviation.

The sum of all deviations about the mean must equal zero. That is,

$$\sum (x_i - \mu) = 0 \qquad \text{and} \qquad \sum (x_i - \bar{x}) = 0$$

In other words, observations larger than the mean are offset by observations smaller than the mean. Because the sum of deviations about the mean is zero, we cannot use the average deviation about the mean as a measure of spread. However, squaring a nonzero number always results in a positive number, so we could find the average squared deviation.

Definition

The **population variance** of a variable is the sum of the squared deviations about the population mean divided by the number of observations in the population, N. That is, it is the mean of the squared deviations about the population mean. The population variance is symbolically represented by σ^2 (lowercase Greek sigma squared).

$$\sigma^2 = \frac{(x_1 - \mu)^2 + (x_2 - \mu)^2 + \cdots + (x_N - \mu)^2}{N} = \frac{\sum (x_i - \mu)^2}{N} \quad \text{(1)}$$

where $x_1, x_2, \ldots, x_N$ are the N observations in the population and μ is the population mean.

Note: In using Formula (1), do not round until the last computation. Use as many decimal places as allowed by your calculator to avoid round-off errors. ◀

A formula that is equivalent to Formula (1), called the **computational formula**, for determining the population variance is

$$\sigma^2 = \frac{\sum x_i^2 - \dfrac{\left(\sum x_i\right)^2}{N}}{N}$$

Note to Instructor
Be sure to let students know which, if any, formula you prefer.

where $\sum x_i^2$ means to square each observation and then sum these squared values, and $\left(\sum x_i\right)^2$ means to add up all the observations and then square the sum.

We illustrate how to use both formulas for computing the variance in the next example.

EXAMPLE 3   Computing a Population Variance

Problem: Compute the population variance of the test scores presented in Table 9.

Approach Using Formula (1)

Step 1: Create a table with four columns. Enter the population data in the first column. In the second column, enter the population mean.

Step 2: Compute the deviation about the mean for each data value. That is, compute $x_i - \mu$ for each data value. Enter these values in column 3.

Step 3: Square the values in column 3, and enter the results in column 4.

Step 4: Sum the squared deviations in column 4, and divide this result by the size of the population, N.

Solution

Step 1: See Table 10. Column 1 lists the observations in the data set, and column 2 contains the population mean.

Approach Using the Computational Formula

Step 1: Create a table with two columns. Enter the population data in the first column. Square each value in the first column and enter the result in the second column.

Step 2: Sum the entries in the first column. This is, find $\sum x_i$. Sum the entries in the second column. That is, find $\sum x_i^2$.

Step 3: Substitute the values found in Step 2 into the computational formula and simplify.

Solution

Step 1: See Table 11. Column 1 lists the observations in the data set, and column 2 contains the values in column 1 squared.

Table 10

Score, x_i	Population Mean, μ	Deviation about the Mean, $x_i - \mu$	Squared Deviations about the Mean, $(x_i - \mu)^2$
82	79	$82 - 79 = 3$	$3^2 = 9$
77	79	$77 - 79 = -2$	$(-2)^2 = 4$
90	79	11	121
71	79	-8	64
62	79	-17	289
68	79	-11	121
74	79	-5	25
84	79	5	25
94	79	15	225
88	79	9	81
		$\sum(x_i - \mu) = 0$	$\sum(x_i - \mu)^2 = 964$

Table 11

Score, x_i	Score Squared, x_i^2
82	$82^2 = 6724$
77	$77^2 = 5929$
90	8100
71	5041
62	3844
68	4624
74	5476
84	7056
94	8836
88	7744
$\sum x_i = 790$	$\sum x_i^2 = 63{,}374$

Step 2: Compute the deviations about the mean for each observation, as shown in column 3. For example, the deviation

Step 2: The last row of columns 1 and 2 shows that $\sum x_i = 790$ and $\sum x_i^2 = 63{,}374$.

about the mean for Michelle is $82 - 79 = 3$. It is a good idea to add up the entries in this column to make sure they sum to 0.

Step 3: Column 4 shows the squared deviations about the mean.

Step 4: We sum the entries in column 4 to obtain the numerator of Formula (1). We compute the population variance by dividing the sum of the entries in column 4 by the number of students, 10:

$$\sigma^2 = \frac{\sum(x_i - \mu)^2}{N} = \frac{964}{10} = 96.4$$

Step 3: We substitute 790 for Σx_i, 63,374 for Σx_i^2, and 10 for N into the computational formula.

$$\sigma^2 = \frac{\sum x_i^2 - \dfrac{\left(\sum x_i\right)^2}{N}}{N} = \frac{63,374 - \dfrac{(790)^2}{10}}{10}$$

$$= \frac{964}{10}$$

$$= 96.4$$

The unit of measure of the variance in Example 3 is points squared. This unit of measure results from squaring the deviations about the mean. Because points squared does not have any obvious meaning, the interpretation of variance is limited.

The sample variance is computed using sample data.

Definition

The **sample variance, s^2,** is computed by determining the sum of the squared deviations about the sample mean and dividing this result by $n - 1$. The formula for the sample variance from a sample of size n is

$$s^2 = \frac{\sum(x_i - \overline{x})^2}{n - 1} = \frac{(x_1 - \overline{x})^2 + (x_2 - \overline{x})^2 + \cdots + (x_n - \overline{x})^2}{n - 1} \quad \text{(2)}$$

where $x_1, x_2, \ldots, x_n$ are the n observations in the sample and $\overline{x}$ is the sample mean.

CAUTION
When using Formula (2), be sure to use $\overline{x}$ with as many decimal places as possible to avoid round-off error.

A computational formula that is equivalent to Formula (2) for computing the sample variance is

$$s^2 = \frac{\sum x_i^2 - \dfrac{\left(\sum x_i\right)^2}{n}}{n - 1}$$

where Σx_i^2 means to square each observation and then sum these squared values, whereas $(\Sigma x_i)^2$ means to add up all the observations and then square the sum.

Notice that the sample variance is obtained by dividing by $n - 1$. If we divided by n, as we might expect, the sample variance would consistently underestimate the population variance. Whenever a statistic consistently overestimates or underestimates a parameter, it is called **biased**. To obtain an unbiased estimate of the population variance, we divide the sum of the squared deviations about the mean by $n - 1$.

CAUTION
When computing the sample variance, be sure to divide by $n - 1$, not n.

To help understand the idea of a biased estimator, consider the following situation: Suppose you work for a carnival in which you must guess a person's age. After 20 people come to your booth, you notice that you have a tendency to underestimate people's age. (You guess too low.) What would you do about this? In all likelihood, you would adjust your guesses higher so that you don't underestimate anymore. In other words, before the adjustment, your guesses were biased. To remove the bias, you increase your guess. That is what dividing by $n - 1$ in the sample variance formula accomplishes. Dividing by n results in an underestimate, so we divide by a smaller number to increase our "guess."

Note to Instructor
To help students understand the concept of bias, draw a dartboard on the board. Have all the darts congregate in the upper-left quadrant of the board.

Although a proof that establishes why we divide by $n - 1$ is beyond the scope of the text, we can provide an explanation that has intuitive appeal. We already know that the sum of the deviations about the mean, $\Sigma(x_i - \overline{x})$, must equal zero. Therefore, if the sample mean is known and the first $n - 1$

Note to Instructor
In-class concept check: Ask students to explain why $\Sigma(x_i - \overline{x}) = 0$. When calculating $\Sigma(x_i - \overline{x})$, why does it sometimes turn out to be a little more or less than 0?

Note to Instructor

Here is a visual to help students under-
stand degrees of freedom: Suppose there
are six doughnuts in a box. The first five
people to pick a doughnut have a choice
(or freedom) about which doughnut to
pick. The sixth person does not.

In Other Words

We have $n - 1$ degrees of freedom in
the computation of s^2 because an
unknown parameter, μ, is estimated with
$\bar{x}$. For each parameter estimated, we lose
1 degree of freedom.

observations are known, then the nth observation must be the value that causes the sum of the deviations to equal zero. For example, suppose $\bar{x} = 4$ based on a sample of size 3. In addition, if $x_1 = 2$ and $x_2 = 3$, then we can determine x_3.

$$\frac{x_1 + x_2 + x_3}{3} = \bar{x}$$

$$\frac{2 + 3 + x_3}{3} = 4 \qquad x_1 = 2, x_2 = 3, \bar{x} = 4$$

$$5 + x_3 = 12$$

$$x_3 = 7$$

We call $n - 1$ **degrees of freedom** because the first $n - 1$ observations have freedom to be whatever value they wish, but the nth value has no freedom. It must be whatever value forces the sum of the deviations about the mean to equal zero.

Again, you should notice that Greek letters are used for parameters, while Roman letters are used for statistics. Do not use rounded values of the sample mean in Formula (2).

EXAMPLE 4 **Computing a Sample Variance**

Problem: Compute the sample variance of the sample obtained in Example 1(b) on page 108 from Section 3.1.

Approach: We follow the same approach that we used to compute the population variance, but this time using the sample data. In looking back at Example 1(b) from Section 3.1, we see that Bilal (90), Ryanne (77), Pam (71), and Michelle (82) are in the sample.

Solution Using Formula (2)

Step 1: Create a table with four columns. Enter the sample data in the first column. In the second column, enter the sample mean. See Table 12.

Table 12

Score, x_i	Sample Mean, $\bar{x}$	Deviation about the Mean, $x_i - \bar{x}$	Squared Deviations about the Mean, $(x_i - \bar{x})^2$
90	80	$90 - 80 = 10$	$10^2 = 100$
77	80	-3	9
71	80	-9	81
82	80	2	4
		$\sum(x_i - \bar{x}) = 0$	$\sum(x_i - \bar{x})^2 = 194$

Step 2: Compute the deviations about the mean for each observation, as shown in column 3. For example, the deviation about the mean for Bilal is $90 - 80 = 10$. It is a good idea to add up the entries in this column to make sure they sum to 0.

Step 3: Column 4 shows the squared deviations about the mean.

Step 4: We sum the entries in column 4 to obtain the numerator of Formula (1). We compute the population variance by dividing the sum of the entries in column 4 by one fewer than the number of students, $4 - 1$:

$$s^2 = \frac{\sum(x_i - \bar{x})^2}{n - 1} = \frac{194}{4 - 1} = 64.7$$

Solution Using the Computational Formula

Step 1: See Table 13. Column 1 lists the observations in the data set, and column 2 contains the values in column 1 squared.

Table 13

Score, x_i	Score Squared, x_i^2
90	$90^2 = 8{,}100$
77	$77^2 = 5{,}929$
71	5,041
82	6,724
$\sum x_i = 320$	$\sum x_i^2 = 25{,}794$

Step 2: The last rows of columns 1 and 2 show that $\sum x_i = 320$ and $\sum x_i^2 = 25{,}794$.

Step 3: We substitute 320 for $\sum x_i$, 25,794 for $\sum x_i^2$, and 4 for n into the computational formula.

$$s^2 = \frac{\sum x_i^2 - \dfrac{(\sum x_i)^2}{n}}{n - 1} = \frac{25{,}794 - \dfrac{(320)^2}{4}}{4 - 1}$$

$$= \frac{194}{3}$$

$$= 64.7$$

Notice that the sample variance obtained for this sample is an underestimate of the population variance we found in Example 3. This discrepancy does not violate our definition of an unbiased estimator, however. A biased estimator is one that *consistently* under- or overestimates.

3 Compute the Standard Deviation of a Variable from Raw Data

The standard deviation and the mean are the most popular methods for numerically describing the distribution of a variable. This is because these two measures are used for most types of statistical inference.

Definitions

The **population standard deviation**, σ, is obtained by taking the square root of the population variance. That is,

$$\sigma = \sqrt{\sigma^2}$$

The **sample standard deviation**, s, is obtained by taking the square root of the sample variance. That is,

$$s = \sqrt{s^2}$$

EXAMPLE 5   **Obtaining the Standard Deviation for a Population and a Sample**

Problem: Use the results obtained in Examples 3 and 4 to compute the population and sample standard deviation score on the statistics exam.

Approach: The population standard deviation is the square root of the population variance. The sample standard deviation is the square root of the sample variance.

Solution: The population standard deviation is

$$\sigma = \sqrt{\sigma^2} = \sqrt{\frac{\sum(x_i - \mu)^2}{N}} = \sqrt{\frac{964}{10}} = 9.8 \text{ points}$$

The sample standard deviation for the sample obtained in Example 1 from Section 3.1 is

$$s = \sqrt{s^2} = \sqrt{\frac{\sum(x_i - \bar{x})^2}{n-1}} = \sqrt{\frac{194}{4-1}} = 8.0 \text{ points}$$

! CAUTION
Never use the rounded variance to compute the standard deviation.

To avoid round-off error, never use the rounded value of the variance to compute the standard deviation.

Now Work Problem 25.

In-Class Activity: The Sample Standard Deviation

Using the pulse data from the activity in Section 3.1, page 109, do the following:

(a) Obtain a simple random sample of $n = 4$ students and compute the sample standard deviation.

(b) Obtain a second simple random sample of $n = 4$ students and compute the sample standard deviation.

(c) Are the sample standard deviations the same? Why?

EXAMPLE 6 **Determining the Variance and Standard Deviation Using Technology**

Problem: Use a statistical spreadsheet or calculator to determine the population standard deviation of the data listed in Table 9. Also determine the sample standard deviation of the sample data from Example 4.

Approach: We will use a TI-84 Plus graphing calculator to obtain the population standard deviation and sample standard deviation score on the statistics exam. The steps for determining the standard deviation using the TI-83 or TI-84 Plus graphing calculator, MINITAB, or Excel are given in the Technology Step by Step on page 142.

Solution: Figure 11(a) shows the population standard deviation, and Figure 11(b) shows the sample standard deviation. Notice the TI graphing calculators provide both a population and sample standard deviation as output. This is because the calculator does not know whether the data entered are population data or sample data. It is up to the user of the calculator to choose the correct standard deviation. The results agree with those obtained in Example 5. To get the variance, we need to square the standard deviation. For example, the population variance is $9.818350167^2 = 96.4$ points^2.

Figure 11

Population standard deviation

Sample standard deviation

(a)   (b)

Interpretations of the Standard Deviation

The standard deviation is used in conjunction with the mean to numerically describe distributions that are bell shaped and symmetric. The mean measures the center of the distribution, while the standard deviation measures the spread of the distribution. So how does the value of the standard deviation relate to the dispersion of the distribution? If we are comparing two populations, then **the larger the standard deviation, the more dispersion the distribution has**. This rule is true provided that the variable of interest from the two populations has the same unit of measure. The units of measure must be the same so that we are comparing apples with apples. For example, a standard deviation of $100 is not the same as 100 Japanese yen, because $1 is equivalent to about 109 yen. This means a standard deviation of $100 is substantially higher than a standard deviation of 100 yen.

Note to Instructor
To help students understand the concept of standard deviation, ask them whether they would prefer to invest in a stock with a standard deviation rate of return of 8% or one of 20%. Why? Or would they rather stand in a line with a standard deviation wait time of 3 minutes or 6 minutes? Why? In both examples, assume the means are equal.

EXAMPLE 7 **Comparing the Standard Deviation of Two Data Sets**

Problem: Refer to the data in Example 1. Use the standard deviation to determine whether University A or University B has more dispersion in the IQ scores of its students.

Approach: We will use MINITAB to compute the standard deviation of IQ for each university. The university with the higher standard deviation will be the university with more dispersion in IQ scores. Recall that, on the basis of the histograms, it was apparent that University A had more dispersion. Therefore, we would expect University A to have a higher sample standard deviation.

Solution: We enter the data into MINITAB and compute the descriptive statistics. See Figure 12.

Figure 12
Descriptive Statistics

Descriptive statistics

Variable	N	N*	Mean	SE Mean	StDev	Minimum
Univ A	100	0	100.00	1.61	16.08	60
Univ B	100	0	100.00	0.83	8.35	86

Variable	Q1	Median	Q3	Maximum
Univ A	90	102	110	141
Univ B	94	98	107	119

The sample standard deviation is larger for University A (16.1) than for University B (8.4). Don't forget that we agreed to round the mean and standard deviation to one more decimal place than the original data. Therefore, University A has IQ scores that are more dispersed.

4 Use the Empirical Rule to Describe Data That Are Bell Shaped

If data have a distribution that is bell shaped, the following rule can be used to determine the percentage of data that will lie within k standard deviations of the mean.

The Empirical Rule

If a distribution is roughly bell shaped, then

- Approximately 68% of the data will lie within 1 standard deviation of the mean. That is, approximately 68% of the data lie between $\mu - 1\sigma$ and $\mu + 1\sigma$.
- Approximately 95% of the data will lie within 2 standard deviations of the mean. That is, approximately 95% of the data lie between $\mu - 2\sigma$ and $\mu + 2\sigma$.
- Approximately 99.7% of the data will lie within 3 standard deviations of the mean. That is, approximately 99.7% of the data lie between $\mu - 3\sigma$ and $\mu + 3\sigma$.

Note: We can also use the Empirical Rule based on sample data with $\bar{x}$ used in place of μ and s used in place of σ.   ◄

Figure 13 illustrates the Empirical Rule.

Figure 13

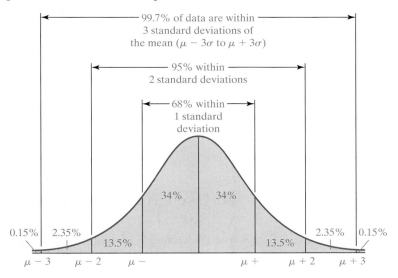

Let's revisit the data from University A in Table 8.

EXAMPLE 8 **Using the Empirical Rule**

Problem: Use the data from University A in Table 8.

(a) Determine the percentage of students who have IQ scores within 3 standard deviations of the mean according to the Empirical Rule.

(b) Determine the percentage of students who have IQ scores between 67.8 and 132.2 according to the Empirical Rule.

(c) Determine the actual percentage of students who have IQ scores between 67.8 and 132.2.

(d) According to the Empirical Rule, what percentage of students will have IQ scores above 132.2?

Figure 14

University A IQ Scores

Approach: To use the Empirical Rule, a histogram of the data must be roughly bell shaped. Figure 14 shows the histogram of the data from University A.

Solution: The histogram of the data drawn in Figure 14 is roughly bell shaped. From Example 7 we know that the mean IQ score of the students enrolled in University A is 100 and the standard deviation is 16.1. To help organize our thoughts and make the analysis easier, we draw a bell-shaped curve like the one in Figure 13, with $\bar{x} = 100$ and $s = 16.1$. See Figure 15.

Figure 15

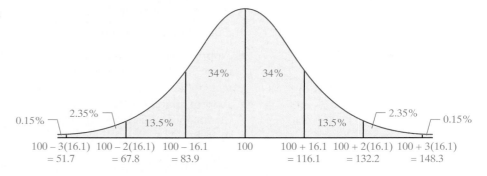

(a) According to the Empirical Rule, approximately 99.7% of the IQ scores will be within 3 standard deviations of the mean. That is, approximately 99.7% of the data will be greater than or equal to $100 - 3(16.1) = 51.7$ and less than or equal to $100 + 3(16.1) = 148.3$.

(b) Since 67.8 is exactly 2 standard deviations below the mean $[100 - 2(16.1) = 67.8]$ and 132.2 is exactly 2 standard deviations above the mean $[100 + 2(16.1) = 132.2]$, we use the Empirical Rule to determine that approximately 95% of all IQ scores lies between 67.8 and 132.2.

(c) Of the 100 IQ scores listed in Table 8, 96, or 96%, are between 67.8 and 132.2. This is very close to the approximation given by the Empirical Rule.

(d) Based on Figure 15, approximately 2.35% + 0.15% = 2.5% of students at University A will have IQ scores above 132.2. ▪

Now Work Problem 39.

5 **Use Chebyshev's Inequality to Describe Any Set of Data**

Note to Instructor

If you are pressed for time, Chebyshev's Inequality can be skipped without loss of continuity.

Chebyshev's Inequality was developed by the Russian mathematician Pafnuty Chebyshev (1821–1894). The inequality is used to determine a lower bound on the percentage of observations that lie within k standard deviations of the mean, where $k > 1$. What's amazing about this result is that these bounds are arrived at regardless of the basic shape of the distribution (skewed left, skewed right, or symmetric).

Chebyshev's Inequality

For any data set, regardless of the shape of the distribution, at least $\left(1 - \frac{1}{k^2}\right)100\%$ of the observations will lie within k standard deviations of the mean, where k is any number greater than 1. That is, at least $\left(1 - \frac{1}{k^2}\right)100\%$ of the data will lie between $\mu - k\sigma$ and $\mu + k\sigma$ for $k > 1$.

Note: We can also use Chebyshev's Inequality based on sample data.

Caution

The Empirical Rule holds only if the distribution is bell shaped. Chebyshev's Inequality holds regardless of the shape of the distribution.

For example, at least $\left(1 - \frac{1}{2^2}\right)100\% = 75\%$ of all observations will lie within $k = 2$ standard deviations of the mean and at least $\left(1 - \frac{1}{3^2}\right)100\% = 88.9\%$ of all observations will lie within $k = 3$ standard deviations of the mean.

Notice the result does not state that exactly 75% of all observations lie within 2 standard deviations of the mean, but instead states that 75% or more of the observations will lie within 2 standard deviations of the mean.

EXAMPLE 9   **Using Chebyshev's Inequality**

Problem: Using the data from University A in Table 8,
(a) Determine the minimum percentage of students who have IQ scores within 3 standard deviations of the mean according to Chebyshev's Inequality.
(b) Determine the minimum percentage of students who have IQ scores between 67.8 and 132.2, according to Chebyshev's Inequality.
(c) Determine the actual percentage of students who have IQ scores between 67.8 and 132.2.

Approach
(a) We use Chebyshev's Inequality with $k = 3$.
(b) We have to determine the number of standard deviations 67.8 and 132.2 are from the mean of 100.0. We then substitute this value of k into Chebyshev's Inequality.
(c) We refer to Table 8 and count the number of observations between 67.8 and 132.2. We divide this result by 100, the number of observations in the data set.

Solution
(a) We use Chebyshev's Inequality with $k = 3$ and determine that at least $\left(1 - \frac{1}{3^2}\right)100\% = 88.9\%$ of all students have IQ scores within 3 standard deviations of the mean. Since the mean of the data set is 100.0 and the standard deviation is 16.1, at least 88.9% of the students have IQ scores between $\bar{x} - ks = 100.0 - 3(16.1) = 51.7$ and $\bar{x} + ks = 100 + 3(16.1) = 148.3$.
(b) Since 67.8 is exactly 2 standard deviations below the mean $[100 - 2(16.1) = 67.8]$ and 132.2 is exactly 2 standard deviations above the mean $[100 + 2(16.1) = 132.2]$, we use Chebyshev's Inequality with $k = 2$

Historical Notes

Pafnuty Chebyshev was born on May 16, 1821, in Okatovo, Russia. In 1847, he began teaching mathematics at the University of St. Petersburg. Some of his more famous work was done on prime numbers. In particular, he discovered a way to determine the number of prime numbers less than or equal to a given number. Chebyshev also studied mechanics, including rotary motion. Chebyshev was elected a Fellow of the Royal Society in 1877. He died on November 26, 1894, in St. Petersburg.

to determine that at least $\left(1 - \dfrac{1}{2^2}\right)100\% = 75\%$ of all IQ scores lie between 67.8 and 132.2.

(c) Of the 100 IQ scores listed, 96 or 96% is between 67.8 and 132.2. Notice that Chebyshev's Inequality provides a rather conservative result. _____ ■

Now Work Problem 43.

Because the Empirical Rule requires that the distribution be bell shaped, while Chebyshev's Inequality applies to all distributions, the Empirical Rule provides results that are more precise.

3.2 ASSESS YOUR UNDERSTANDING

Concepts and Vocabulary

1. Would it be appropriate to say that a distribution with a standard deviation of 10 centimeters is more dispersed than a distribution with a standard deviation of 5 inches? Support your position.
2. What is meant by the phrase *degrees of freedom* as it pertains to the computation of the sample variance?
3. Is the standard deviation resistant?
4. The sum of the deviations about the mean always equals _____. zero
5. What does it mean when a statistic is biased?
6. The simplest measure of dispersion is the _____. range
7. Discuss the relationship between variance and standard deviation.

8. The standard deviation is used in conjunction with the _____ to numerically describe distributions that are bell shaped. The _____ measures the center of the distribution, while the standard deviation measures the _____ of the distribution. mean; mean; spread
9. *True or False*: When comparing two populations, the larger the standard deviation, the more dispersion the distribution has, provided that the variable of interest from the two populations has the same unit of measure. True
10. *True or False*: Chebyshev's Inequality applies to all distributions regardless of shape, but the Empirical Rule holds only for distributions that are bell shaped. True

Skill Building

In Problems 11–16, find the population variance and standard deviation or the sample variance and standard deviation as indicated.

11. Sample: 20, 13, 4, 8, 10 $s^2 = 36$; $s = 6$
12. Sample: 83, 65, 91, 87, 84 $s^2 = 100$; $s = 10$
13. Population: 3, 6, 10, 12, 14 $\sigma^2 = 16$; $\sigma = 4$
14. Population: 1, 19, 25, 15, 12, 16, 28, 13, 6 $\sigma^2 = 64$; $\sigma = 8$
15. Sample: 6, 52, 13, 49, 35, 25, 31, 29, 31, 29 $s^2 = 196$; $s = 14$
16. Population: 4, 10, 12, 12, 13, 21 $\sigma^2 = 25$; $\sigma = 5$
17. **Crash Test Results** The Insurance Institute for Highway Safety crashed the 2001 Honda Civic four times at 5 miles per hour. The cost of repair for each of the four crashes is as follows:

    $420, $462, $409, $236

    Compute the range, sample variance, and sample standard deviation cost of repair. $226; 9962.9; $99.8
18. **Cell Phone Use** The following data represent the monthly cell phone bill for my wife's phone for six randomly selected months:

    $35.34, $42.09, $39.43, $38.93, $43.39, $49.26

    Compute the range, sample variance, and sample standard deviation phone bill. $13.92; 22.584; $4.75

19. **Concrete Mix** A certain type of concrete mix is designed to withstand 3000 pounds per square inch (psi) of pressure. The strength of concrete is measured by pouring the mix into casting cylinders 6 inches in diameter and 12 inches tall. The cylinder is allowed to set up for 28 days. The cylinders are then stacked on one another until the cylinders are crushed. The following data represent the strength of nine randomly selected casts:

    3960, 4090, 3200, 3100, 2940, 3830, 4090, 4040, 3780

    Compute the range, sample variance, and sample standard deviation for strength of the concrete (in psi). 1150; 211; 275; 459.6
20. **Flight Time** The following data represent the flight time (in minutes) of a random sample of seven flights from Las Vegas, Nevada, to Newark, New Jersey, on Continental Airlines.

    282, 270, 260, 266, 257, 260, 267

    Compute the range, sample variance, and sample standard deviation of flight time. 25; 71; 8.4

21. Which histogram depicts a higher standard deviation? Justify your answer. (*b*)

(a)   (b)

22. Match the histograms to the summary statistics given. (*a*) III (*b*) I (*c*) IV (*d*) II

	Mean	Median	Standard Deviation
I	53	53	1.3
II	60	60	11
III	53	53	9
IV	53	53	0.12

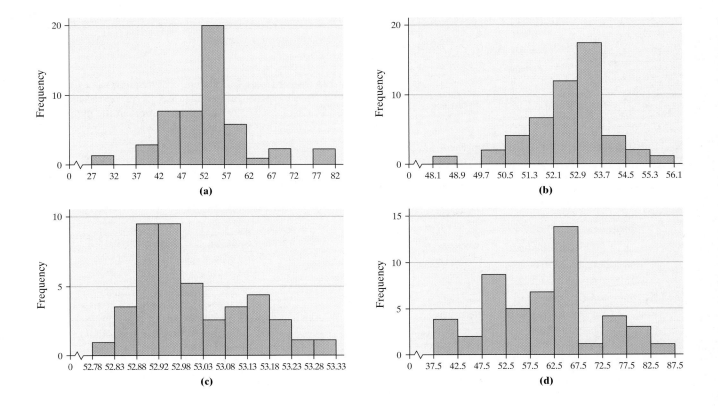

(a)   (b)

(c)   (d)

Applying the Concepts

23. **ATM Fees** The following data for a random sample of banks in Los Angeles and New York City represent the ATM fees for using another bank's ATM.

23. LA: R = $2.00, s = $0.68
    NY: R = $1.50, s = $0.48

Los Angeles	2.00	1.50	1.50	1.00	1.50	2.00	0.00	2.00
New York City	1.50	1.00	1.00	1.25	1.25	1.50	1.00	0.00

Source: www.bankrate.com

Compute the range and sample standard deviation for ATM fees for each city. Which city has more dispersion based on the range? Which city has more dispersion based on the standard deviation?

24. **Reaction Time** In an experiment conducted online at the University of Mississippi, study participants are asked to react to a stimulus. In one experiment, the participant must press a key upon seeing a blue screen. The time (in seconds) to press the key is measured. The same person is then asked to press a key upon seeing a red screen, again with the time to react measured. The results for six study participants are listed in the table. Compute the range and sample standard deviation for reaction time for both blue and red. Does there appear to be a difference in the variability of reaction time? What might account for any difference? Blue: 0.574, 0.1994; red: 0.14, 0.0647

Participant Number	Reaction Time to Blue	Reaction Time to Red
1	0.582	0.408
2	0.481	0.407
3	0.841	0.542
4	0.267	0.402
5	0.685	0.456
6	0.450	0.533

Source: PsychExperiments at the University of Mississippi (www.olemiss.edu/psychexps/)

25. **Pulse Rates** The following data represent the pulse rates (beats per minute) of nine students enrolled in a section of Sullivan's course in Introductory Statistics. Treat the nine students as a population.

Student	Pulse
Perpectual Bempah	76
Megan Brooks	60
Jeff Honeycutt	60
Clarice Jefferson	81
Crystal Kurtenbach	72
Janette Lantka	80
Kevin McCarthy	80
Tammy Ohm	68
Kathy Wojdyla	73

(a) Compute the population variance and population standard deviation. 58.8; 7.7
(b) Determine two simple random samples of size 3, and compute the sample variance and sample standard deviation of each sample.
(c) Which samples underestimate the population standard deviation? Which overestimate the population standard deviation?

26. **Travel Time** The following data represent the travel time (in minutes) to school for nine students enrolled in Sullivan's College Algebra course. Treat the nine students as a population.

Student	Travel Time	Student	Travel Time
Amanda	39	Scot	45
Amber	21	Erica	11
Tim	9	Tiffany	12
Mike	32	Glenn	39
Nicole	30		

(a) Compute the population variance and population standard deviation. 164.9; 12.8
(b) Determine three simple random samples of size 4, and compute the sample variance and sample standard deviation of each sample.
(c) Which samples underestimate the population standard deviation? Which overestimate the population standard deviation?

27. **Soccer Goals** Mia Hamm, considered by some to be the most prolific player in international soccer, retired after the 2004 Olympics. The following data represent the number of goals scored over her 18-year career.

0	0	0	4	10	1	10	10	19
9	18	20	13	13	2	7	8	13

Source: www.soccerhall.org

(a) Compute the population variance and population standard deviation for number of goals scored.
(b) Determine three simple random samples of size 3, and compute the sample variance and sample standard deviation of each sample.
(c) Which samples underestimate the population standard deviation? Which overestimate the population standard deviation?

(27a) $\sigma^2 \approx 41.0$ goals²; $\sigma \approx 6.4$ goals

28. Tour de Lance Lance Armstrong won the Tour de France seven consecutive times (1999–2005). The table gives the winning times, distances, speeds, and margin of victory.

(28c) $R = 6.600$ min; $\sigma^2 \approx 4.390$ min^2; $\sigma \approx 2.095$ min

Year	Winning Time (h)	Distance (km)	Winning Speed (km/h)	Winning Margin (min)
1999	91.538	3687	40.28	7.617
2000	92.552	3662	39.56	6.033
2001	86.291	3453	40.02	6.733
2002	82.087	3278	39.93	7.283
2003	83.687	3427	40.94	1.017
2004	83.601	3391	40.56	6.317
2005	86.251	3593	41.65	4.667

Source: www.cyclinynews.com

(a) Compute the range, population variance, and population standard deviation for winning times for the six races. $R = 10.465$ h; $\sigma^2 \approx 13.981$ h^2; $\sigma \approx 3.739$ h
(b) Compute the range, population variance, and population standard deviation for distances for the six races. $R = 409$ km; $\sigma^2 \approx 19{,}793.3$ km^2; $\sigma \approx 140.7$ km
(c) Compute the range, population variance, and population standard deviation for winning time margins.
(d) Compute the range, population variance, and population standard deviation for winning speeds.

(28d) $R = 2.09$ km/h; $\sigma^2 \approx 0.423$ (km/h)^2; $\sigma \approx 0.651$ km/h

29. A Fish Story Ethan and Drew went on a 10-day fishing trip. The number of smallmouth bass caught and released by the two boys each day was as follows:

Ethan:	9	24	8	9	5	8	9	10	8	10
Drew:	15	2	3	18	20	1	17	2	19	3

(a) Find the population mean and the range for the number of smallmouth bass caught per day by each fisherman. Do these values indicate any differences between the two fishermen's catches per day? Explain. $\mu = 10$; range $= 19$
(b) Find the population standard deviation for the number of smallmouth bass caught per day by each fisherman. Do these values present a different story about the two fishermen's catches per day? Which fisherman has the more consistent record? Explain. Ethan: $\sigma = 4.9$; Drew: $\sigma = 7.9$
(c) Discuss limitations of the range as a measure of dispersion.

30. 2004 NFC Champion Philadelphia Eagles The following data represent the weights (in pounds) of the 33 offensive players and the 24 defensive players for the 2004 NFC Champion Philadelphia Eagles.

Offense			Defense		
195	218	215	281	272	265
240	222	212	265	264	298
210	250	200	303	293	306
243	205	180	294	240	241
210	195	226	254	245	262
180	245	258	200	196	194
255	244	312	210	177	210
300	305	310	211	206	202
340	330	327			
349	330	310			
320	330	325			

Source: ESPN.com

(a) Compute the population mean, the range, and the population standard deviation for the Philadelphia offense. $\mu = 260.3$; $\sigma = 53.8$; $R = 169$
(b) Compute the population mean, the range, and the population standard deviation for the Philadelphia defense. $\mu = 245.4$; $\sigma = 39.2$; $R = 129$
(c) Which player type has more dispersion? Explain how you know. Offense

In Problems 31 and 32, compute the range, sample variance, and sample standard deviation.

31. Serum HDL Dr. Paul Oswiecmiski randomly selects 40 of his 20- to 29-year-old patients and obtains the following data regarding their serum HDL cholesterol: 45; 118.0; 10.9

70	56	48	48	53	52	66	48
36	49	28	35	58	62	45	60
38	73	45	51	56	51	46	39
56	32	44	60	51	44	63	50
46	69	53	70	33	54	55	52

32. Volume of Altria Group Stock The volume of a stock is the number of shares traded on a given day. The following data, given in millions so that 3.78 represents 3,780,000 shares traded, represent the volume of Altria Group stock traded for a random sample 35 trading days in 2004.

32. 7.95; 4.239 million shares^2; 2.059 million shares

3.78	8.74	4.35	5.02	8.40
6.06	5.75	5.34	6.92	6.23
5.32	3.25	6.57	7.57	6.07
3.04	5.64	5.00	7.16	4.88
10.32	3.38	7.25	6.52	4.43
3.38	5.53	4.74	9.70	3.56
10.96	4.50	7.97	3.01	5.58

Source: Yahoo.finance.com

33. The Empirical Rule The following data represent the weights (in grams) of a random sample of 50 M&M plain candies.

0.87	0.88	0.82	0.90	0.90	0.84	0.84
0.91	0.94	0.86	0.86	0.86	0.88	0.87
0.89	0.91	0.86	0.87	0.93	0.88	
0.83	0.94	0.87	0.93	0.91	0.85	
0.91	0.91	0.86	0.89	0.87	0.93	
0.88	0.88	0.89	0.79	0.82	0.83	
0.90	0.88	0.84	0.93	0.76	0.90	
0.88	0.92	0.85	0.79	0.84	0.86	

Source: Michael Sullivan

(a) Determine the sample standard deviation weight. Express your answer rounded to two decimal places. *0.04 gram*
(b) On the basis of the histogram drawn in Section 3.1, Problem 33, comment on the appropriateness of using the Empirical Rule to make any general statements about the weights of M&Ms. *Okay to use*
(c) Use the Empirical Rule to determine the percentage of M&Ms with weights between 0.79 and 0.95 gram. *Hint*: $\bar{x} = 0.87$. *95%*
(d) Determine the actual percentage of M&Ms that weigh between 0.79 and 0.95 gram, inclusive. *98%*
(e) Use the Empirical Rule to determine the percentage of M&Ms with weights more than 0.91 gram. *16%*
(f) Determine the actual percentage of M&Ms that weigh more than 0.91 gram. *14%*

34. The Empirical Rule The following data represent the length of eruption for a random sample of eruptions at the Old Faithful geyser in Calistoga, California.

108	108	99	105	103	103	94
102	99	106	90	104	110	110
103	109	109	111	101	101	
110	102	105	110	106	104	
104	100	103	102	120	90	
113	116	95	105	103	101	
100	101	107	110	92	108	

Source: Ladonna Hansen, Park Curator

(a) Determine the sample standard deviation length of eruption. Express your answer rounded to the nearest whole number. *6 seconds*
(b) On the basis of the histogram drawn in Section 3.1, Problem 34, comment on the appropriateness of using the Empirical Rule to make any general statements about the length of eruptions. *Okay to use*
(c) Use the Empirical Rule to determine the percentage of eruptions that last between 92 and 116 seconds. *Hint*: $\bar{x} = 104$. *95%*

(d) Determine the actual percentage of eruptions that last between 92 and 116 seconds, inclusive. *93%*
(e) Use the Empirical Rule to determine the percentage of eruptions that last less than 98 seconds. *16%*
(f) Determine the actual percentage of eruptions that last less than 98 seconds. *11%*

35. Which Car Would You Buy? Suppose that you are in the market to purchase a car. With gas prices on the rise, you have narrowed it down to two choices and will let gas mileage be the deciding factor. You decide to conduct a little experiment in which you put 10 gallons of gas in the car and drive it on a closed track until it runs out gas. You conduct this experiment 15 times on each car and record the number of miles driven.

Car 1				
228	223	178	220	220
233	233	271	219	223
217	214	189	236	248

Car 2				
277	164	326	215	259
217	321	263	160	257
239	230	183	217	230

Describe each data set. That is, determine the shape, center, and spread. Which car would you buy and why?

36. Which Investment Is Better? You have received a year-end bonus of $5000. You decide to invest the money in the stock market and have narrowed your investment options down to two mutual funds. The following data represent the historical quarterly rates of return of each mutual fund for the past 20 quarters (5 years).

Mutual Fund A				
1.3	−0.3	0.6	6.8	5.0
5.2	4.8	2.4	3.0	1.8
7.3	8.6	3.4	3.8	−1.3
6.4	1.9	−0.5	−2.3	3.1

Mutual Fund B				
−5.4	6.7	11.9	4.3	4.3
3.5	10.5	2.9	3.8	5.9
−6.7	1.4	8.9	0.3	−2.4
−4.7	−1.1	3.4	7.7	12.9

Describe each data set. That is, determine the shape, center, and spread. Which mutual fund would you invest in and why?

37. Rates of Return of Stocks Stocks may be categorized by industry. The following data represent the 5-year rates of return for a simple random sample of financial stocks and energy stocks ending March 4, 2004.

Financial: $\bar{x} = 15.716$, $M = 16.09$;
Energy: $\bar{x} = 22.481$, $M = 19.585$; Energy
Financial: $s = 7.378$;
Energy: $s = 12.751$; Energy

Financial Stocks							
17.10	16.26	22.10	9.96	7.94	10.95	16.34	20.43
7.54	26.84	28.02	15.92	10.80	11.27	20.68	11.09
9.84	11.82	6.28	3.27	21.97	13.74	33.63	25.53
11.01	18.15	17.36	19.14	17.80	26.33	5.35	8.44

Energy Stocks							
11.43	14.52	22.14	7.03	42.31	15.15	9.43	7.39
30.88	19.50	21.17	16.03	53.61	15.38	42.74	26.34
7.51	45.62	19.67	15.17	8.39	43.50	29.97	6.11
23.84	26.18	38.79	15.35	18.42	16.67	20.93	28.23

Source: Morningstar.com

(a) Compute the mean and the median rate of return for each industry. Which sector has the higher mean rate of return? Which sector has the higher median rate of return?
(b) Compute the standard deviation for each industry. In finance, the standard deviation rate of return is called **risk**. Which sector is riskier?

38. American League versus National League The following data represent the earned-run average of a random sample of pitchers in both the American League and the National League during the 2004 season. **Note**: Earned-run average (ERA) is the mean number of runs given up per nine innings pitched. A higher ERA is indicative of a worse pitcher.

American League				
2.22	2.70	2.90	3.13	3.25
3.27	3.30	3.40	3.57	3.60
3.77	3.78	3.78	3.87	3.91
4.02	4.09	4.14	4.14	4.18
4.21	4.29	4.34	4.43	4.47
4.49	4.51	4.51	4.57	4.59
4.61	4.63	4.67	4.68	4.85
5.15	5.20	5.56	5.73	5.75

National League				
2.34	2.39	2.43	2.84	2.95
3.09	3.11	3.20	3.20	3.24
3.28	3.54	3.59	3.62	3.68
3.76	3.78	3.81	3.82	3.84
3.87	3.96	3.99	4.00	4.01
4.08	4.11	4.11	4.13	4.13
4.16	4.19	4.23	4.30	4.43
4.45	4.52	4.52	4.59	4.64

(a) Compute the mean and the median earned-run average for each league. Which league has the higher mean earned-run average? Which league has the higher median earned-run average? AL: $\bar{x} = 4.157$, $M = 4.195$; NL: $\bar{x} = 3.748$, $M = 3.855$; AL, AL
(b) Compute the standard deviation for each league. Which league has more dispersion? AL: $s = 0.787$; NL: $s = 0.610$; AL

39. The Empirical Rule One measure of intelligence is the NW Stanford–Binet Intelligence Quotient (IQ). IQ scores have a bell-shaped distribution with a mean of 100 and a standard deviation of 15.
(a) What percentage of people has an IQ score between 70 and 130? 95%
(b) What percentage of people has an IQ score less than 70 or greater than 130? 5%
(c) What percentage of people has an IQ score greater than 130? 2.5%

40. The Empirical Rule SAT Math scores have a bell-shaped distribution with a mean of 518 and a standard deviation of 114. (*Source*: College Board, 2004)
(a) What percentage of SAT scores is between 404 and 632? 68%
(b) What percentage of SAT scores is less than 404 or greater than 632? 32%
(c) What percentage of SAT scores is greater than 746? 2.5%

41. The Empirical Rule The weight, in grams, of the pair of kidneys in adult males between the ages of 40 and 49 has a bell-shaped distribution with a mean of 325 grams and a standard deviation of 30 grams.

(a) About 95% of kidneys will be between what weights?

(b) What percentage of kidneys weighs between 235 grams and 415 grams? 99.7%

(c) What percentage of kidneys weighs less than 235 grams or more than 415 grams? 0.3%

(d) What percentage of kidneys weighs between 295 grams and 385 grams? 81.5%

42. The Empirical Rule The distribution of the length of bolts has a bell shape with a mean of 4 inches and a standard deviation of 0.007 inch.

(a) About 68% of bolts manufactured will be between what lengths? 3.993 and 4.007 inches

(b) What percentage of bolts will be between 3.986 inches and 4.014 inches? 95%

(c) If the company discards any bolts less than 3.986 inches or greater than 4.014 inches, what percentage of bolts manufactured will be discarded? 5%

(d) What percentage of bolts manufactured will be between 4.007 inches and 4.021 inches? 15.85%

43. Chebyshev's Inequality In December 2004, the average NW price of regular unleaded gasoline excluding taxes in the United States was $1.37 per gallon according to the Energy Information Administration. Assume that the standard deviation price per gallon is $0.05 per gallon to answer the following.

(a) What percentage of gasoline stations had prices within 3 standard deviations of the mean? 88.9%

(b) What percentage of gasoline stations had prices within 2.5 standard deviations of the mean? What are the gasoline prices that are within 2.5 standard deviations of the mean? 84%, $1.245 to $1.495

(c) What is the minimum percentage of gasoline stations that had prices between $1.27 and $1.47? 75%

44. Chebyshev's Inequality According to the U.S. Census Bureau, the mean of the commute time to work for a resident of Boston, Massachusetts, is 27.3 minutes. Assume that the standard deviation of the commute time is 8.1 minutes to answer the following:

(a) What percentage of commuters in Boston has a commute time within 2 standard deviations of the mean? 75%

(b) What percentage of commuters in Boston has a commute time within 1.5 standard deviations of the mean? What are the commute times within 1.5 standard deviations of the mean? 55.6%, 15.15 minutes to 39.45 minutes

(c) What is the minimum percentage of commuters who have commute times between 3 minutes and 51.6 minutes? 89%

45. Comparing Standard Deviations The standard deviation of batting averages of all teams in the American League is 0.008. The standard deviation of all players in the American League is 0.02154. Why is there less variability in team batting averages?

46. Linear Transformations Benjamin owns a small Internet business. Besides himself, he employs nine other people. The salaries earned by the employees are given next in thousands of dollars (Benjamin's salary is the largest, of course):

$$30, 30, 45, 50, 50, 50, 55, 55, 60, 75$$

(a) Determine the range, population variance, and population standard deviation for the data.

(b) Business has been good! As a result, Benjamin has a total of $25,000 in bonus pay to distribute to his employees. One option for distributing bonuses is to give each employee (including himself) $2500. Add the bonuses under this plan to the original salaries to create a new data set. Recalculate the range, population variance, and population standard deviation. How do they compare to the originals? $R = 45, \sigma^2 = 160, \sigma = 12.6$

(c) As a second option, Benjamin can give each employee a bonus of 5% of his or her original salary. Add the bonuses under this second plan to the original salaries to create a new data set. Recalculate the range, population variance, and population standard deviation. How do they compare to the originals?

(d) As a third option, Benjamin decides not to give his employees a bonus at all. Instead, he keeps the $25,000 for himself. Use this plan to create a new data set. Recalculate the range, population variance, and population standard deviation. How do they compare to the originals? $R = 70, \sigma^2 = 341.3, \sigma = 18.5$

47. Resistance and Sample Size Each of the following three data sets represents the IQ scores of a random sample of adults. IQ scores are known to have a mean and median of 100. For each data set, determine the sample standard deviation. Then recompute the sample standard deviation assuming that the individual whose IQ is 106 is accidentally recorded as 160. For each sample size, state what happens to the standard deviation. Comment on the role that the number of observations plays in resistance.

Sample of Size 5				
106	92	98	103	100

Sample of Size 12					
106	92	98	103	100	102
98	124	83	70	108	121

Sample of Size 30					
106	92	98	103	100	102
98	124	83	70	108	121
102	87	121	107	97	114
140	93	130	72	81	90
103	97	89	98	88	103

48. Compute the sample standard deviation of the following test scores: 78, 78, 78, 78. What can be said about a data set in which all the values are identical? $s = 0$

(41a) 265 and 385 grams   (46a) $R = 45, \sigma^2 = 160, \sigma = 12.6$

(46c) $R = 47.25, \sigma^2 = 176.4, \sigma = 13.3$

Consumer Reports® — BASEMENT WATERPROOFING COATINGS

A waterproofing coating can be an inexpensive and easy way to deal with leaking basements. But how effective are they? In a study, *Consumer Reports* tested nine water-proofers to rate their effectiveness in controlling water seepage though concrete foundations.

To compare the products' ability to control water seepage, we applied two coats of each product to slabs cut from concrete block. For statistical validity, this process was repeated at least six times. In each test run, four blocks (each coated with a different product) were simultaneously placed in a rectangular aluminum chamber. See the picture.

The chamber was sealed and filled with water and the blocks were subjected to progressively increasing hydrostatic pressures. Water that leaked out during each period was channeled to the bottom of the chamber opening, collected, and weighed.

The table contains a subset of the data collected for two of the products tested. Using these data,

(a) Calculate the mean, median, and mode weight of water collected for product A.

(b) Calculate the standard deviation of the weight of water collected for product A.

(c) Calculate the mean, median, and mode weight of water collected for product B.

(d) Calculate the standard deviation of the weight of water collected for product B.

(e) Construct a back-to-back stem-and-leaf diagram for these data.

Product	Replicate	Weight of Collected Water (in grams)
A	1	91.2
A	2	91.2
A	3	90.9
A	4	91.3
A	5	90.8
A	6	90.8
B	1	87.1
B	2	87.2
B	3	86.8
B	4	87.0
B	5	87.2
B	6	87.0

Does there appear to be a difference in these two products' ability to mitigate water seepage? Why?

Note to Readers: *In many cases, our test protocol and analytical methods are more complicated than described in these examples. The data and discussions have been modified to make the material more appropriate for the audience.*

Basement Waterproofer Test Chamber

© 2002 by Consumers Union of U.S., Inc., Yonkers, NY 10703-1057, a nonprofit organization. Reprinted with permission from the June 2002 issue of CONSUMER REPORTS® for educational purposes only. No commercial use or photocopying permitted. To learn more about Consumers Union, log onto www.ConsumerReports.org <www.ConsumerReports.org>.

Technology Step by Step
Determining the Range, Variance, and Standard Deviation
The same steps followed to obtain the measures of central tendency from raw data can be used to obtain the measures of dispersion.

3.3 Measures of Central Tendency and Dispersion from Grouped Data

Preparing for This Section   Before getting started, review the following:

- Organizing discrete data in tables (Section 2.2, pp. 71–72)
- Organizing continuous data in tables (Section 2.2, pp. 73–75)

Objectives

1 **Approximate the mean of a variable from grouped data**

2 **Compute the weighted mean**

3 **Approximate the variance and standard deviation of a variable from grouped data**

Note to Instructor
This section can be skipped without loss of continuity. If covered, the formulas used in this section can be overwhelming. Therefore, it is recommended that "by hand" computations be de-emphasized.

We have discussed how to compute descriptive statistics from raw data, but many times the data that we have access to have already been summarized in frequency distributions (grouped data). While we cannot obtain exact values of the mean or standard deviation without raw data, these measures can be approximated using the techniques discussed in this section.

1 Approximate the Mean of a Variable from Grouped Data

Since raw data cannot be retrieved from a frequency table, we assume that, within each class, the mean of the data values is equal to the *class midpoint*. The **class midpoint** is found by adding consecutive lower class limits and dividing the result by 2. We then multiply the class midpoint by the frequency. This product is expected to be close to the sum of the data that lie within the class. We repeat the process for each class and sum the results. This sum approximates the sum of all the data.

Definition

Approximate Mean of a Variable from a Frequency Distribution

Population Mean

$$\mu = \frac{\sum x_i f_i}{\sum f_i} = \frac{x_1 f_1 + x_2 f_2 + \cdots + x_N f_N}{f_1 + f_2 + \cdots + f_N} \tag{1a}$$

Sample Mean

$$\bar{x} = \frac{\sum x_i f_i}{\sum f_i} = \frac{x_1 f_1 + x_2 f_2 + \cdots + x_n f_n}{f_1 + f_2 + \cdots + f_n} \tag{1b}$$

where x_i is the midpoint or value of the *i*th class

f_i is the frequency of the *i*th class

n is the number of classes

In Formula (1), $x_1 f_1$ approximates the sum of all the data values in the first class, $x_2 f_2$ approximates the sum of all the data values in the second class, and so on. Notice that the formulas for the population mean and sample mean are essentially identical, just as they were for computing the mean from raw data.

EXAMPLE 1

Approximating the Mean for Continuous Quantitative Data from the Frequency Distribution

Problem: The frequency distribution in Table 14 represents the three-year rate of return of a random sample of 40 small-capitalization growth mutual funds. Approximate the mean three-year rate of return.

Approach: We perform the following steps to approximate the mean.

Step 1: Determine the class midpoint of each class. The class midpoint is found by adding consecutive lower class limits and dividing the result by 2.

Step 2: Compute the sum of the frequencies, Σf_i.

Step 3: Multiply the class midpoint by the frequency to obtain $x_i f_i$ for each class.

Step 4: Compute $\Sigma x_i f_i$.

Step 5: Substitute into Formula (1b) to obtain the mean from grouped data.

Solution

Step 1: The lower class limit of the first class is 0. The lower class limit of the second class is 2. Therefore, the class midpoint of the first class is $\dfrac{0 + 2}{2} = 1$, so $x_1 = 1$. The remaining class midpoints are listed in column 2 of Table 15.

Step 2: We add the frequencies in column 3 to obtain $\Sigma f_i = 2 + 5 + \cdots + 1 = 40$.

Step 3: Compute the values of $x_i f_i$ by multiplying each class midpoint by the corresponding frequency and obtain the results shown in column 4 of Table 15.

Step 4: We add the values in column 4 of Table 15 to obtain $\Sigma x_i f_i = 304$.

Table 14

Class (3-year rate of return)	Frequency
0–1.99	2
2–3.99	5
4–5.99	6
6–7.99	8
8–9.99	9
10–11.99	6
12–13.99	3
14–15.99	1

Table 15

Class (3-year rate of return)	Class Midpoint, x_i	Frequency, f_i	$x_i f_i$
0–1.99	$\dfrac{0 + 2}{2} = 1$	2	$(1)(2) = 2$
2–3.99	3	5	$(3)(5) = 15$
4–5.99	5	6	30
6–7.99	7	8	56
8–9.99	9	9	81
10–11.99	11	6	66
12–13.99	13	3	39
14–15.99	15	1	15
		$\Sigma f_i = 40$	$\Sigma x_i f_i = 304$

Step 5: Substituting into Formula (1b), we obtain

$$\overline{x} = \frac{\Sigma x_i f_i}{\Sigma f_i} = \frac{304}{40} = 7.6$$

The approximate mean three-year rate of return is 7.6%.

The mean three-year rate of return from the raw data listed in Example 3 on page 74 from Section 2.2 is 7.5%. The approximate mean from grouped data is pretty close to the actual mean.

Note: To compute the mean from a frequency distribution where the data are discrete, treat each category of data as the class midpoint. For discrete data, the mean from grouped data will equal the mean from raw data. ◄

! CAUTION

We computed the mean from grouped data in Example 1 even though the raw data are available. The reason for doing this was to illustrate how close the two values can be. In practice, use raw data whenever possible.

Now compute the mean of the frequency distribution in Problem 3.

② Compute the Weighted Mean

Sometimes, certain data values have a higher importance or weight associated with them. In this case, we compute the *weighted mean*. For example, your grade-point average is a weighted mean, with the weights equal to the number of credit hours in each course. The value of the variable is equal to the grade converted to a point value.

Definition

Note to Instructor
Idea for a classroom discussion: Suppose a company owns three cars that get 20 miles per gallon, two cars that get 22 miles per gallon, and one car that gets 24 miles per gallon. Would the mean miles per gallon be 22? Discuss.

The **weighted mean**, $\bar{x}_w$, of a variable is found by multiplying each value of the variable by its corresponding weight, summing these products, and dividing the result by the sum of the weights. It can be expressed using the formula

$$\bar{x}_w = \frac{\sum w_i x_i}{\sum w_i} = \frac{w_1 x_1 + w_2 x_2 + \cdots + w_n x_n}{w_1 + w_2 + \cdots + w_n} \qquad (2)$$

where w_i is the weight of the ith observation

x_i is the value of the ith observation

EXAMPLE 2

Computing the Weighted Mean

Problem: Marissa just completed her first semester in college. She earned an A in her 4-hour statistics course, a B in her 3-hour sociology course, an A in her 3-hour psychology course, a C in her 5-hour computer programming course, and an A in her 1-hour drama course. Determine Marissa's grade-point average.

Approach: We must assign point values to each grade. Let an A equal 4 points, a B equal 3 points, and a C equal 2 points. The number of credit hours for each course determines its weight. So a 5-hour course gets a weight of 5, a 4-hour course gets a weight of 4, and so on. We multiply the weight of each course by the points earned in the course, sum these products, and divide the sum by the number of credit hours.

Solution

$$\text{GPA} = \bar{x}_w = \frac{\sum w_i x_i}{\sum w_i} = \frac{4(4) + 3(3) + 3(4) + 5(2) + 1(4)}{4 + 3 + 3 + 5 + 1} = \frac{51}{16} = 3.19$$

Marissa's grade-point average for her first semester is 3.19. ▪

> **Now Work Problem 15.**

③ Approximate the Variance and Standard Deviation of a Variable from Grouped Data

The procedure for approximating the variance and standard deviation from grouped data is similar to that of finding the mean from grouped data. Again, because we do not have access to the original data, the variance is approximate.

Definition

Approximate Variance of a Variable from a Frequency Distribution

Population Variance	Sample Variance
$\sigma^2 = \dfrac{\sum (x_i - \mu)^2 f_i}{\sum f_i}$	$s^2 = \dfrac{\sum (x_i - \bar{x})^2 f_i}{\left(\sum f_i\right) - 1}$

$$(3)$$

where x_i is the midpoint or value of the ith class

f_i is the frequency of the ith class

An algebraically equivalent formula for the population variance is $\dfrac{\sum x_i^2 f_i - \dfrac{\left(\sum x_i f_i\right)^2}{\sum f_i}}{\sum f_i}$.

We approximate the standard deviation by taking the square root of the variance.

EXAMPLE 3   ## Approximating the Variance and Standard Deviation from a Frequency Distribution

Problem: The data in Table 14 on page 143 represent the three-year rate of return of a random sample of 40 small-capitalization growth mutual funds. Approximate the variance and standard deviation of the three-year rate of return.

Approach: We will use the sample variance Formula (3).

Step 1: Create a table with the class in the first column, the class midpoint in the second column, the frequency in the third column, and the unrounded mean in the fourth column.

Step 2: Compute the deviation about the mean, $x_i - \bar{x}$, for each class, where x_i is the class midpoint of the ith class and $\bar{x}$ is the sample mean. Enter the results in column 5.

Step 3: Square the deviation about the mean and multiply this result by the frequency to obtain $(x_i - \bar{x})^2 f_i$. Enter the results in column 6.

Step 4: Add the entries in columns 3 and 6 to obtain Σf_i and $\Sigma(x_i - \bar{x})^2 f_i$.

Step 5: Substitute the values obtained in Step 4 into Formula (3) to obtain an approximate value for the sample variance.

Solution

Step 1: We create Table 16. Column 1 contains the classes. Column 2 contains the class midpoint of each class. Column 3 contains the frequency of each class. Column 4 contains the unrounded sample mean obtained in Example 1.

Step 2: Column 5 of Table 16 contains the deviation about the mean, $x_i - \bar{x}$, for each class.

Step 3: Column 6 contains the values of the squared deviation about the mean multiplied by the frequency, $(x_i - \bar{x})^2 f_i$.

Step 4: We add the entries in columns 3 and 6 and obtain $\Sigma f_i = 40$ and $\Sigma(x_i - \bar{x})^2 f_i = 465.6$.

Table 16

Class (3-year rate of return)	Class Midpoint, x_i	Frequency, f_i	$\bar{x}$	$x_i - \bar{x}$	$(x_i - \bar{x})^2 f_i$
0–1.99	1	2	7.6	−6.6	87.12
2–3.99	3	5	7.6	−4.6	105.8
4–5.99	5	6	7.6	−2.6	40.56
6–7.99	7	8	7.6	−0.6	2.88
8–9.99	9	9	7.6	1.4	17.64
10–11.99	11	6	7.6	3.4	69.36
12–13.99	13	3	7.6	5.4	87.48
14–15.99	15	1	7.6	7.4	54.76
		$\Sigma f_i = 40$			$\Sigma(x_i - \bar{x})^2 f_i = 465.6$

Step 5: Substitute these values into Formula (3) to obtain an approximate value for the sample variance.

$$s^2 = \frac{\Sigma(x_i - \bar{x})^2 f_i}{\left(\Sigma f_i\right) - 1} = \frac{465.6}{39} \approx 11.94$$

Take the square root of the unrounded estimate of the sample variance to obtain an approximation of the sample standard deviation.

$$s = \sqrt{s^2} = \sqrt{\frac{465.6}{39}} \approx 3.46\%$$

We approximate the sample standard deviation three-year rate of return to be 3.46%.

EXAMPLE 4 ### Approximating the Mean and Standard Deviation Using Technology

Problem: Approximate the mean and standard deviation of the three-year rate of return data in Table 14 using a TI-83/84 Plus graphing calculator.

Approach: The steps for approximating the mean and standard deviation of grouped data using the TI-83 or TI-84 Plus graphing calculator are given in the Technology Step by Step on page 149.

Result: Figure 16 shows the result from the TI-84 Plus.

Figure 16

Approximate mean

Approximate sample standard deviation

From the output, we can see that the approximate mean is 7.6% and the approximate standard deviation is 3.46%. The results agree with our by-hand solutions.

From the raw data listed in Example 3, we find that the sample standard deviation is 3.46%. The approximate sample standard deviation from grouped data equals the sample standard deviation from the raw data!

> **Now compute the standard deviation from the frequency distribution in Problem 3.**

3.3 ASSESS YOUR UNDERSTANDING

Concepts and Vocabulary

1. Explain the role of the class midpoint in the formulas to approximate the mean and the standard deviation.

2. In Section 3.1, the mean is given by $\bar{x} = \dfrac{\Sigma x_i}{n}$. Explain how this is a special case of the weighted mean, $\bar{x}_w$.

Applying the Concepts

3. **Cell Phones** A sample of college students was asked how much they spent monthly on a cell phone plan (to the nearest NW dollar). Approximate the mean and standard deviation for the cost. $\bar{x} \approx \$32.83$; $s \approx \$11.06$

Monthly Cell Phone Plan Cost ($)	Number of Students
10–19	8
20–29	16
30–39	21
40–49	11
50–59	4

4. Bowl Games The following data represent the difference in scores between the winning and losing teams in the 2004–2005 college football bowl games. Approximate the mean and standard deviation for the point difference. $\mu \approx 14.6$ points; $\sigma \approx 11.0$ points

Point Difference	Number of Bowl Games
1–5	11
6–10	0
11–15	5
16–20	6
21–25	1
26–30	2
31–35	1
36–40	2

Source: espn.com

5. 100-Degree Days The following data represent the annual number of days over 100°F for Dallas–Fort Worth from 1905 to 2004. Approximate the mean and standard deviation annual number of days over 100°F.

$\mu \approx 17.3$ days; $\sigma \approx 12.6$ days

Number of 100° + Days	Number of Years
0–9	31
10–19	39
20–29	17
30–39	6
40–49	4
50–59	2
60–69	1

Source: National Weather Service

6. Working Students The following data represent the number of hours (on average) worked each week for a sample of community college students. Approximate the mean and standard deviation of the number of hours. $\bar{x} \approx 21.6$ hours per week; $s \approx 11.7$ hours per week

Hours Worked (per week)	Number of Students
0–9	24
10–19	14
20–29	39
30–39	18
40–49	5

7. Health Insurance The following data represent the number of people aged 25 to 64 years covered by health insurance (private or government) in 2003. Approximate the mean and standard deviation for age. $\mu = 44.5$ years; $\sigma = 10.5$ years

Age	Number (millions)
25–34	28.9
35–44	35.7
45–54	35.1
55–64	24.7

Source: U.S. Census Bureau

8. Earthquakes The following data represent the magnitude of earthquakes in the United States in 2004. Approximate the mean and standard deviation of the magnitude. $\mu \approx 2.8$; $\sigma \approx 1.1$

Magnitude	Number
0–0.9	539
1.0–1.9	1
2.0–2.9	1336
3.0–3.9	1363
4.0–4.9	289
5.0–5.9	21
6.0–6.9	2

Source: U.S. Geological Survey

9. Meteorology The following data represent the high-temperature distribution for the month of August in Chicago since 1872.

Temperature (°F)	Days
50–59	1
60–69	308
70–79	1519
80–89	1626
90–99	503
100–109	11

Source: National Oceanic and Atmospheric Administration

(a) Approximate the mean and standard deviation for temperature. $\mu = 80.9$; $\sigma = 8.2$
(b) Draw a frequency histogram of the data to verify that the distribution is bell shaped.
(c) According to the Empirical Rule, 95% of days in the month of August will be between what two temperatures? 64.5 and 97.3

10. Rainfall The following data represent the annual rainfall distribution for St. Louis, Missouri, from 1870 to 2004.

Rainfall (inches)	Number of Years
20–24	4
25–29	15
30–34	27
35–39	40
40–44	28
45–49	15
50–54	4
55–59	2

Source: National Oceanic and Atmospheric Administration

(10c) 23.4 inches and 52.2 inches

(a) Approximate the mean and standard deviation for rainfall. $\mu = 37.8$ inches; $\sigma = 7.2$ inches
(b) Draw a frequency histogram of the data to verify that the distribution is bell shaped.
(c) According to the Empirical Rule, 95% of annual rainfalls in St. Louis will be between what two amounts?

11. **Multiple Births** The following data represent the number of live multiple-delivery births (three or more babies) in 2002 for women 15 to 44 years old.

Age	Number of Multiple Births
15–19	93
20–24	511
25–29	1628
30–34	2832
35–39	1843
40–44	377

Source: National Vital Statistics Reports, Vol. 52, No. 10, December 17, 2003

(11c) 21.9 years and 42.7 years

(a) Approximate the mean and standard deviation for age. $\mu = 32.3$ years; $\sigma \approx 5.2$ years
(b) Draw a frequency histogram of the data to verify that the distribution is bell shaped.
(c) According to the Empirical Rule, 95% of mothers of multiple births will be between what two ages?

12. **SAT Scores** The following data represent SAT Verbal scores for ISACS college-bound seniors in 2003.

SAT Verbal Score	Number
400–449	281
450–499	577
500–549	840
550–599	1120
600–649	1166
650–699	900
700–749	518
750–800	394

Source: www.isacs.org

(a) Approximate the mean and standard deviation of the score. $\mu = 603.1$; $\sigma \approx 92.0$
(b) Draw a frequency histogram of the data to verify that the distribution is bell shaped.
(c) According to the Empirical Rule, 95% of these ISACS college-bound seniors will have SAT Verbal scores between what two values? 419.1 and 787.1

13. **Serum HDL** Use the frequency distribution whose class width is 10 obtained in Problem 31 in Section 2.2 to approximate the mean and standard deviation for serum HDL. Compare these results to the values obtained in Problem 31 in Sections 3.1 and 3.2. $\bar{x} = 51.8$; $s = 12.1$

14. **Volume of Altria Group Stock** Use the frequency distribution whose class width is 2 obtained in Problem 32 in

14. $\bar{x} \approx 6$ million shares; $s \approx 1.879$ million shares

Section 2.2 to approximate the mean and standard deviation of the number of shares traded. Compare these results to the values obtained in Problem 32 in Sections 3.1 and 3.2.

15. **Grade-Point Average** Marissa has just completed her second semester in college. She earned a B in her 5-hour calculus course, an A in her 3-hour social work course, an A in her 4-hour biology course, and a C in her 3-hour American literature course. Assuming that an A equals 4 points, a B equals 3 points, and a C equals 2 points, determine Marissa's grade-point average for the semester. 3.27

16. **Computing Class Average** In Marissa's calculus course, attendance counts for 5% of the grade, quizzes count for 10% of the grade, exams count for 60% of the grade, and the final exam counts for 25% of the grade. Marissa had a 100% average for attendance, 93% for quizzes, 86% for exams, and 85% on the final. Determine Marissa's course average. 87.15%

17. **Mixed Chocolates** Michael and Kevin want to buy chocolates. They can't agree on whether they want chocolate-covered almonds, chocolate-covered peanuts, or chocolate-covered raisins. They agree to create a mix. They bought 4 pounds of chocolate-covered almonds at $3.50 per pound, 3 pounds of chocolate-covered peanuts for $2.75 per pound, and 2 pounds of chocolate-covered raisins for $2.25 per pound. Determine the cost per pound of the mix. $2.97

18. **Nut Mix** Michael and Kevin return to the candy store, but this time they want to purchase nuts. They can't decide among peanuts, cashews, or almonds. They again agree to create a mix. They bought 2.5 pounds of peanuts for $1.30 per pound, 4 pounds of cashews for $4.50 per pound, and 2 pounds of almonds for $3.75 per pound. Determine the price per pound of the mix. $3.38

19. **Population** The following data represent the male and female population by age of the United States for residents under 100 years old in July 2003.

Age	Male Resident Pop. (in thousands)	Female Resident Pop. (in thousands)
0–9	20,225	19,319
10–19	21,375	20,295
20–29	20,437	19,459
30–39	21,176	20,936
40–49	22,138	22,586
50–59	16,974	17,864
60–69	10,289	11,563
70–79	6,923	9,121
80–89	3,053	5,367
90–99	436	1,215

Source: U.S. Census Bureau

(a) Approximate the population mean and standard deviation of age for males. $\mu \approx 35.6$ years; $\sigma \approx 21.7$ years
(b) Approximate the population mean and standard deviation of age for females.
(c) Which gender has the higher mean age? women
(d) Which gender has more dispersion in age? women

19. $\mu \approx 38.1$ years; $\sigma \approx 23.0$ years

20. Age of Mother The following data represent the age of the mother at childbirth for 1980 and 2002.

Age of Mother	Number of Births 1980 (thousands)	Number of Births 2002 (thousands)
10–14	1.1	0.7
15–19	53.0	43.0
20–24	115.1	103.6
25–29	112.9	113.6
30–34	61.9	91.5
35–39	19.8	41.4
40–44	3.9	8.3
45–49	0.2	0.5

Source: National Vital Statistics Reports, Vol. 52, No. 10

(a) Approximate the population mean and standard deviation of age for mothers in 1980.
(b) Approximate the population mean and standard deviation of age for mothers in 2002.
(c) Which year has the higher mean age? *2002*
(d) Which year has more dispersion in age? *2002*

(20a) $\mu \approx 26.0$ years; $\sigma \approx 5.7$ years

(20b) $\mu \approx 27.6$ years; $\sigma \approx 6.2$ years

Technology Step by Step

Determining the Mean and Standard Deviation from Grouped Data

TI-83/84 Plus

Step 1: Enter the class midpoint in L1 and the frequency or relative frequency in L2 by pressing STAT and selecting 1:Edit.

Step 2: Press STAT, highlight the CALC menu and select 1:1-Var Stats

Step 3: With 1-Var Stats appearing on the HOME screen, press 2nd 1 to insert L1 on the HOME screen. Then press the comma and press 2nd 2 to insert L2 on the HOME screen. So, the HOME screen should have the following:

$$\text{1-Var Stats L1, L2}$$

Press ENTER to obtain the mean and standard deviation.

3.4 Measures of Position

Objectives

1. **Determine and interpret z-scores**
2. **Determine and interpret percentiles**
3. **Determine and interpret quartiles**
4. **Check a set of data for outliers**

In Section 3.1, we were able to find measures of central tendency. Measures of central tendency are meant to describe the "typical" data value. Section 3.2 discussed measures of dispersion, which describe the amount of spread in a set of data. In this section, we discuss measures of position; that is, we wish to describe the relative position of a certain data value within the entire set of data.

1   Determine and Interpret z-Scores

At the end of the 2004 season, the Boston Red Sox led the American League with 949 runs scored, while the St. Louis Cardinals led the National League with 855 runs scored. A quick comparison might lead one to believe that the Red Sox are the better run-producing team. However, this comparison is unfair because the two teams play in different leagues. The Red Sox play in the American League, where the designated hitter bats for the pitcher, whereas the Cardinals play in the National League, where the pitcher must bat (pitchers are typically poor hitters). To compare the two teams' scoring of runs, we need to determine their relative standings in their respective leagues. This can be accomplished using a *z-score*.

Definition

The **z-score** represents the distance that a data value is from the mean in terms of the number of standard deviations. It is obtained by subtracting the mean from the data value and dividing this result by the standard deviation. There is both a population z-score and a sample z-score; their formulas follow:

$$\text{Population } z\text{-Score} \qquad \text{Sample } z\text{-Score}$$

$$z = \frac{x - \mu}{\sigma} \qquad\qquad z = \frac{x - \bar{x}}{s} \tag{1}$$

The z-score is unitless. It has mean 0 and standard deviation 1.

In Other Words

Z-scores provide a way to compare apples to oranges by converting variables with different centers and/or spreads to variables with the same center (0) and spread (1).

If a data value is larger than the mean, the z-score will be positive. If a data value is smaller than the mean, the z-score will be negative. If the data value equals the mean, the z-score will be zero. Z-scores measure the number of standard deviations an observation is above or below the mean. For example, a z-score of 1.24 is interpreted as "the data value is 1.24 standard deviations above the mean." A z-score of -2.31 is interpreted as "the data value is 2.31 standard deviations below the mean."

We are now prepared to determine whether the Red Sox or Cardinals had a better year in run production.

EXAMPLE 1 **Comparing z-Scores**

Problem: Determine whether the Boston Red Sox or the St. Louis Cardinals had a relatively better run-producing season. The Red Sox scored 949 runs and play in the American League, where the mean number of runs scored was $\mu = 811.3$. The standard deviation was $\sigma = 73.7$. The Cardinals scored 855 runs and play in the National League, where the mean number of runs scored was $\mu = 751.1$. The standard deviation was $\sigma = 78.6$.

Approach: To determine which team had the relatively better run-producing season, we compute each team's z-score. The team with the higher z-score had the better season. Because we know the values of the population parameters, we will compute the population z-score.

Solution: First, we compute the z-score for the Red Sox. Z-scores are typically rounded to two decimal places.

$$z\text{-score} = \frac{x - \mu}{\sigma} = \frac{949 - 811.3}{73.7} = 1.87$$

Next, we compute the z-score for the Cardinals.

$$z\text{-score} = \frac{x - \mu}{\sigma} = \frac{855 - 751.1}{78.6} = 1.32$$

So, the Red Sox had run production 1.87 standard deviations above the mean, while the Cardinals had run production 1.32 standard deviations above the mean. Therefore, the Red Sox had a relatively better year at scoring runs.

Now Work Problem 7.

2 Determine and Interpret Percentiles

Recall that the median divides the lower 50% of a set of data from the upper 50%. In general, the **kth percentile**, denoted P_k, of a set of data divides the lower $k\%$ of a data set from the upper $(100 - k)\%$. Percentiles divide a data set that is written in ascending order into 100 parts, so 99 possible percentiles can be computed. For example, P_1 divides the bottom 1% of the data from the top 99% while P_{99} divides the lower 99% of the data from the top 1%. Figure 17 displays the 99 possible percentiles.

Figure 17

If a data value lies at the 40th percentile, then approximately 40% of the data are less than this value and approximately 60% are higher than this value.

Percentiles are often used to give the relative standing of a data value. Many standardized exams, such as the SAT college entrance exam, use percentiles to provide students with an understanding of how they scored on the exam in relation to all other students who took the exam. For example, in 2004, an SAT verbal score of 580 was at the 73rd percentile. This means approximately 73% of the scores are below 580 and 27% are above 580. Pediatricians use percentiles to describe the progress of a newborn baby's weight gain relative to other newborn babies. A 3- to 5-month-old male child who weighs 14.3 pounds would be at the 15th percentile.

The following steps can be used to compute the kth percentile:

Note to Instructor

Try to cover two examples for finding percentiles: one that uses Step 3(a) and one that uses Step 3(b).

! CAUTION
Don't forget to write the data in ascending order before finding the percentile.

Determining the kth Percentile, P_k

Step 1: Arrange the data in ascending order.
Step 2: Compute an index i using the formula

$$i = \left(\frac{k}{100}\right)(n + 1) \tag{2}$$

where k is the percentile of the data value and n is the number of individuals in the data set.

Step 3:
(a) If i is an integer, the kth percentile, P_k, is the ith data value.
(b) If i is not an integer, find the mean of the observations on either side of i. This number represents the kth percentile, P_k.

An example should clarify the procedure.

EXAMPLE 2   **Determining the Percentile of a Data Value, Index an Integer**

Problem: The data in Table 17 represent the violent crime rate (violent crimes per 100,000 population) for the 50 states and the District of Columbia in 2003. Find the state that corresponds to the 75th percentile.

Table 17

State	Crime Rate	State	Crime Rate	State	Crime Rate
1. North Dakota	77.8	18. Oregon	295.5	35. Missouri	472.8
2. Maine	108.9	19. Connecticut	308.2	36. Oklahoma	505.7
3. Vermont	110.2	20. Mississippi	325.5	37. Michigan	511.2
4. New Hampshire	148.8	21. Ohio	333.2	38. Arizona	513.2
5. South Dakota	173.4	22. Colorado	345.1	39. Texas	552.5
6. Wisconsin	221.0	23. Washington	347.0	40. Illinois	556.8
7. Idaho	242.7	24. Indiana	352.8	41. California	579.3
8. Utah	248.6	25. Montana	365.2	42. Alaska	593.4
9. West Virginia	257.5	26. New Jersey	365.8	43. Nevada	614.2
10. Kentucky	261.7	27. Kansas	395.5	44. Louisiana	646.3
11. Wyoming	262.1	28. Pennsylvania	398.0	45. Delaware	658.0
12. Minnesota	262.6	29. Alabama	429.5	46. New Mexico	665.2
13. Hawaii	270.4	30. Georgia	453.9	47. Tennessee	687.8
14. Iowa	272.4	31. North Carolina	454.9	48. Maryland	703.9
15. Virginia	275.8	32. Arkansas	456.1	49. Florida	730.2
16. Rhode Island	285.6	33. New York	465.2	50. South Carolina	793.5
17. Nebraska	289.0	34. Massachusetts	469.4	51. District of Columbia	1608.1

Source: Federal Bureau of Investigation, Uniform Crime Reports, 2003

Approach: We will follow the steps given on page 151.

Solution

Step 1: The data provided in Table 17 are already listed in ascending order.

Step 2: To find the 75th percentile, P_{75}, we compute the index i with $k = 75$ and $n = 51$.

$$i = \left(\frac{75}{100}\right)(51 + 1) = 39$$

Step 3: The 75th percentile is the 39th observation of the data set written in ascending order. The 39th observation, which corresponds to the state of Texas, is 552.5. Approximately 75% of the states have a violent crime rate less than 552.5 crimes per 100,000 population, and approximately 25% of the states have a violent crime rate above 552.5 crimes per 100,000 population.

EXAMPLE 3   **Determining the Percentile of a Data Value, Index Not an Integer**

Problem: Find the crime rate that corresponds to the 90th percentile for the data in Table 17.

Approach: We will follow the steps given on page 151.

Solution

Step 1: The data provided in Table 17 are listed in ascending order.

Step 2: To find the 90th percentile, P_{90}, we compute the index i with $k = 90$ and $n = 51$.

$$i = \left(\frac{90}{100}\right)(51 + 1) = 46.8$$

Step 3: Because the index, $i = 46.8$, is not an integer, the 90th percentile is the mean of the 46th and 47th data value.

$$P_{90} = \frac{665.2 + 687.8}{2} = 676.5$$

Approximately 90% of the states have violent crime rates below 676.5 crimes per 100,000 population. Approximately 10% of the states have violent crime rates above 676.5 crimes per 100,000 population. ▬

Now Work Problem 13(a).

Often we are interested in knowing the percentile to which a specific data value corresponds. The kth percentile of a data value, x, from a data set that contains n values is computed by using the following steps:

Finding the Percentile That Corresponds to a Data Value

Step 1: Arrange the data in ascending order.
Step 2: Use the following formula to determine the percentile of the score, x.

$$\text{Percentile of } x = \frac{\text{number of data values less than } x}{n} \times 100 \qquad \textbf{(3)}$$

Round this number to the nearest integer.

EXAMPLE 4 **Finding the Percentile of a Specific Data Value**

Problem: Find the percentile rank for the state of Kentucky using the data provided in Table 17.

Approach: We will follow the steps given above.

Solution

Step 1: The data provided in Table 17 are in ascending order.
Step 2: Nine states have a violent crime rate that is less than Kentucky's violent crime rate. So

$$\text{Percentile rank of Kentucky} = \frac{9}{51} \cdot 100 \approx 17.6$$

We round 17.6 to 18. Kentucky's violent crime rate is at the 18th percentile. Approximately 18% of the states have violent crime rates that are less than that of Kentucky, and approximately 82% of the states have violent crime rates that are larger than that of Kentucky. ▬

Now Work Problem 13(d).

③ Determine and Interpret Quartiles

The most common percentiles are quartiles. **Quartiles** divide data sets into fourths, or four equal parts. The first quartile, denoted Q_1, divides the bottom 25% of the data from the top 75%. Therefore, the first quartile is equivalent to the 25th percentile. The second quartile divides the bottom 50% of the data from the top 50%, so the second quartile is equivalent to the 50th percentile, which is equivalent to the median. Finally, the third quartile divides the bottom

75% of the data from the top 25%, so that the third quartile is equivalent to the 75th percentile. Figure 18 illustrates the concept of quartiles.

In Other Words

The first quartile, Q_1, is equivalent to the 25th percentile, P_{25}. The 2nd quartile, Q_2, is equivalent to the 50th percentile, P_{50}, which is equivalent to the median, M. Finally, the third quartile, Q_3, is equivalent to the 75th percentile, P_{75}.

Figure 18

Smallest Data Value		Median		Largest Data Value
	Q_1	Q_2	Q_3	
25% of the data	25% of the data	25% of the data	25% of the data	

EXAMPLE 5 **Finding the Quartiles of a Data Set**

Problem: Find the first, second, and third quartiles for the violent crime rates listed in Table 17.

Approach

Step 1: The first quartile, Q_1, is the 25th percentile, P_{25}. We let $k = 25$ in Formula (2) to obtain the index, i.

Step 2: The second quartile, Q_2, is the 50th percentile, P_{50}. We let $k = 50$ in Formula (2) to obtain the index, i.

Step 3: The third quartile, Q_3, is the 75th percentile, P_{75}. We let $k = 75$ in Formula (2) to obtain the index, i.

Note to Instructor

Ask your students whether quartiles are resistant. What about the interquartile range?

In Other Words

To find Q_2, determine the median of the data set. To find Q_1, determine the median of the "lower half" of the data set. To find Q_3, determine the median of the "upper half" of the data set.

Solution

Step 1: The index for the first quartile, Q_1, is

$$i = \left(\frac{25}{100}\right)(51 + 1) = 13$$

The 13th observation will be the first quartile. So $Q_1 = P_{25} = 270.4$.

Step 2: The index for the second quartile, Q_2, is

$$i = \left(\frac{50}{100}\right)(51 + 1) = 26$$

The 26th observation will be the second quartile. So $Q_2 = P_{50} = M = 365.8$

Step 3: The third quartile, Q_3, is the 75th percentile, which we found in Example 2. The third quartile is $Q_3 = P_{75} = 552.5$.

EXAMPLE 6 **Finding Quartiles Using Technology**

Problem: Find the quartiles of the violent crimes data in Table 17.

USING TECHNOLOGY

Statistical packages may use different formulas for obtaining the quartiles, so results may differ slightly if the index is not an integer.

Approach: We will use a TI-84 Plus graphing calculator to obtain the quartiles. The steps for obtaining quartiles using a TI-83/84 Plus graphing calculator, MINITAB, or Excel are given in the Technology Step by Step on page 159.

Result: Figure 19 shows the results obtained from a TI-84 Plus graphing calculator. Notice that the calculator uses Med for the second quartile.

Figure 19

```
1-Var Stats
↑n=51
  minX=77.8
  Q₁=270.4
  Med=365.8
  Q₃=552.5
  maxX=1608.1
```

Now Work Problem 15(b).

4 Check a Set of Data for Outliers

Whenever performing any type of data analysis, we should always check for extreme observations in the data set. Extreme observations are referred to as **outliers**. Whenever outliers are encountered, their origin must be investigated. They can occur by chance, because of error in the measurement of a variable, during data entry, or from errors in sampling. For example, in the 2000 presidential election, a precinct in New Mexico accidentally recorded 610 absentee ballots for Al Gore as 110. Workers in the Gore camp discovered the data-entry error through an analysis of vote totals.

Sometimes extreme observations are common within a population. For example, suppose we wanted to estimate the mean price of a European car. We might take a random sample of size 5 from the population of all European automobiles. If our sample included a Ferrari 360 Spider (approximately $170,000), it probably would be an outlier, because this car costs much more than the typical European automobile. The value of this car would be considered *unusual* because it is not a typical value from the data set.

We can use the following steps to check for outliers using quartiles.

> **CAUTION**
>
> Outliers distort both the mean and the standard deviation, because neither is resistant. Because these measures often form the basis for most statistical inference, any conclusions drawn from a set of data that contains outliers can be flawed.

Note to Instructor

A second method for identifying outliers uses z-scores. If the z-score is less than −2 or greater than 2, the data value can be considered to be an outlier. However, this method should be used only if the distribution is symmetric, because it is based on the Empirical Rule.

Checking for Outliers by Using Quartiles

Step 1: Determine the first and third quartiles of the data.

Step 2: Compute the interquartile range. The **interquartile range** or **IQR** is the difference between the third and first quartile. That is,

$$IQR = Q_3 - Q_1$$

Step 3: Determine the fences. **Fences** serve as cutoff points for determining outliers.

$$Lower\ fence = Q_1 - 1.5(IQR)$$
$$Upper\ fence = Q_3 + 1.5(IQR)$$

Step 4: If a data value is less than the lower fence or greater than the upper fence, it is considered an outlier.

EXAMPLE 7   **Checking for Outliers**

Problem: Check the data that represent the violent crime rates of the 50 states and the District of Columbia for outliers.

Approach: We follow the preceding steps. Any data value that is less than the lower fence or greater than the upper fence will be considered an outlier.

Solution

Step 1: The quartiles were found in Examples 5 and 6. So $Q_1 = 270.4$ and $Q_3 = 552.5$.

Step 2: The interquartile range, IQR, is

$$IQR = Q_3 - Q_1$$
$$= 552.5 - 270.4$$
$$= 282.1$$

Step 3: The lower fence, LF, is

$$LF = Q_1 - 1.5(IQR)$$
$$= 270.4 - 1.5(282.1)$$
$$= -152.75$$

The upper fence, UF, is

$$UF = Q_3 + 1.5(IQR)$$
$$= 552.5 + 1.5(282.1)$$
$$= 975.65$$

Note to Instructor

In-class discussion: Why is it inappropriate to use only histograms to identify outliers?

Step 4: There are no outliers below the lower fence. However, we do have an outlier above the upper fence corresponding to the District of Columbia (1608.1 violent crimes per 100,000 population).

Figure 20 shows a histogram of the data. We can easily identify the outlier corresponding to the District of Columbia.

Figure 20

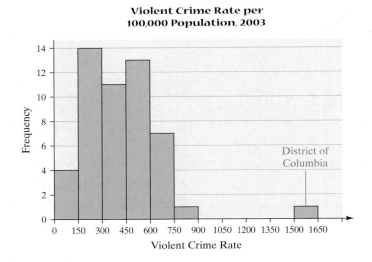

Now Work Problems 15(c) and (d).

3.4 ASSESS YOUR UNDERSTANDING

Concepts and Vocabulary

1. Write a paragraph that explains the meaning of percentiles.
2. Suppose you received the highest score on an exam. Your friend scored the second-highest score, yet you both were in the 99th percentile. How can this be?
3. Morningstar is a mutual fund rating agency. It ranks a fund's performance by using one to five stars. A one-star mutual fund is in the bottom 20% of its investment class; a five-star mutual fund is in the top 20% of its investment class. Interpret the meaning of a four-star mutual fund.

4. When outliers are discovered, should they always be removed from the data set before further analysis?
5. Mensa is an organization designed for people of high intelligence. One qualifies for Mensa if one's intelligence is measured at or above the 98th percentile. Explain what this means.
6. Explain the advantage of using z-scores to compare observations from two different data sets.

Applying the Concepts

7. **Birth Weights** In 2003, babies born after a gestation period of 32 to 35 weeks had a mean weight of 2600 grams and a standard deviation of 670 grams. In the same year, babies born after a gestation period of 40 weeks had a mean weight of 3500 grams and a standard deviation of 475 grams. Suppose a 34-week gestation period baby weighs 2400 grams and a 40-week gestation period baby weighs 3300 grams. Which baby weighs less relative to the gestation period? *40-week gestation*

8. **Birth Weights** In 2003, babies born after a gestation period of 32 to 35 weeks had a mean weight of 2600 grams and a standard deviation of 670 grams. In the same year, babies born after a gestation period of 40 weeks had a mean weight of 3500 grams and a standard deviation of 475 grams. Suppose a 34-week gestation period baby weighs 3000 grams and a 40-week gestation period baby weighs 3900 grams. Which baby weighs less relative to the gestation period? *34-week gestation*

9. **Men versus Women** The average 20- to 29-year-old man is 69.6 inches tall, with a standard deviation of 2.7 inches, while the average 20- to 29-year-old woman is 64.1 inches tall, with a standard deviation of 2.6 inches. Who is relatively taller, a 75-inch man or a 70-inch woman?

   (*Source*: *Vital and Health Statistics*, Advance Data, Number 347, October 27, 2004) Woman

10. **Men versus Women** The average 20- to 29-year-old man is 69.6 inches tall, with a standard deviation of 2.7 inches, while the average 20- to 29-year-old woman is 64.1 inches tall, with a standard deviation of 2.6 inches. Who is relatively taller, a 68-inch man or a 62-inch woman?

    (*Source*: *Vital and Health Statistics*, Advance Data, Oct. 2004) Man

11. **ERA Champions** In 2004, Jake Peavy of the San Diego Padres had the lowest ERA (earned-run average, mean number of runs yielded per nine innings pitched) of any pitcher in the National League, with an ERA of 2.27. Also in 2004, Johann Santana of the Minnesota Twins had the lowest ERA of any pitcher in the American League with an ERA of 2.61. In the National League, the mean ERA in 2004 was 4.198 and the standard deviation was 0.772. In the American League, the mean ERA in 2004 was 4.338 and the standard deviation was 0.785. Which player had the better year relative to his peers, Peavy or Santana? Why? Peavy

12. **Batting Champions** The highest batting average ever recorded was by Ted Williams in 1941 when he hit 0.406. That year, the mean and standard deviation for batting average were 0.28062 and 0.03281. In 2004, Ichiro Suzuki was the American League batting champion, with a batting average of 0.372. In 2004, the mean and standard deviation for batting average were 0.26992 and 0.02154. Who had the better year relative to their peers, Williams or Suzuki? Why? Suzuki

13. **Violent Crime Rates** Use the data in Table 17 regarding
    NW the violent crime rates in 2003 to answer the following:
    (a) Find and interpret the 40th percentile. 329.35
    (b) Find and interpret the 95th percentile. 761.85
    (c) Find and interpret the 10th percentile. 197.2
    (d) What is the percentile rank of the state of Florida? 94th percentile
    (e) What is the percentile rank of the state of California? 78th percentile

14. **Violent Crime Rates** Use the data in Table 17 regarding the violent crime rates in 2003 to answer the following.
    (a) Find and interpret the 30th percentile. 280.7
    (b) Find and interpret the 85th percentile. 652.15
    (c) Find and interpret the 5th percentile. 109.55
    (d) What is the percentile rank of the state of New Mexico? 88th percentile
    (e) What is the percentile rank of the state of Rhode Island? 29th percentile

15. **April Showers** The following data represent the number
    NW of inches of rain in Chicago, Illinois, during the month of
    April for 20 randomly selected years.

0.97	2.78	4.00	5.50
1.14	3.41	4.02	5.79
1.85	3.48	4.11	6.14
2.34	3.94	4.77	6.28
2.47	3.97	5.22	7.69

Source: NOAA, Climate Diagnostics Center

(a) Compute the *z*-score corresponding to the rainfall in 1971 of 0.97 inch. Interpret this result. −1.70
(b) Determine the quartiles. 2.625 in; 3.985 in; 5.36 in
(c) Compute the interquartile range, IQR. 2.735
(d) Determine the lower and upper fences. Are there any outliers, according to this criterion? −1.478 in; 9.463 in

16. **Hemoglobin in Cats** The following data represent the hemoglobin (in g/dL) for 20 randomly selected cats.

5.7	8.9	9.6	10.6	11.7
7.7	9.4	9.9	10.7	12.9
7.8	9.5	10.0	11.0	13.0
8.7	9.6	10.3	11.2	13.4

Source: Joliet Junior College Veterinarian Technology Program

(a) Compute the *z*-score corresponding to the hemoglobin of Blackie, 7.8 g/dL. Interpret this result. −1.21
(b) Determine the quartiles. 9.15 g/dL; 9.95 g/dL; 11.1 g/dL
(c) Compute the interquartile range, IQR. 1.95
(d) Determine the lower and upper fences. Are there any outliers, according to this criterion? 6.225 g/dL; 14.025 g/dL

17. **Concentration of Dissolved Organic Carbon** The following data represent the concentration of organic carbon (mg/L) collected from organic soil.

22.74	29.8	27.1	16.51	6.51
8.81	5.29	20.46	14.9	33.67
30.91	14.86	15.91	15.35	9.72
19.8	14.86	8.09	17.9	18.3
5.2	11.9	14	7.4	17.5
10.3	11.4	5.3	15.72	20.46
16.87	15.42	22.49		

Source: Lisa Emili, Ph.D. candidate, University of Waterloo, Ontario

(17b)   10.01 mg/L; 15.42 mg/L; 20.13 mg/L

(a) Compute the *z*-score corresponding to 20.46. Interpret this result. 0.61
(b) Determine the quartiles.
(c) Compute the interquartile range, IQR. 10.12
(d) Determine the lower and upper fences. Are there any outliers, according to this criterion? −5.17, 35.31; no

18. **Concentration of Dissolved Organic Carbon** The following data represent the concentration of organic carbon (mg/L) collected from mineral soil.

8.5	3.91	9.29	21	10.89
10.3	11.56	7	3.99	3.79
5.5	4.71	7.66	11.72	11.8
8.05	10.72	21.82	22.62	10.74
3.02	7.45	11.33	7.11	9.6
12.57	12.89	9.81	17.99	21.4
8.37	7.92	17.9	7.31	16.92
4.6	8.5	4.8	4.9	9.1
7.9	11.72	4.85	11.97	7.85
9.11	8.79			

Source: Lisa Emili, Ph.D. candidate, University of Waterloo, Ontario

(18b) 7.11 mg/L; 9.1 mg/L; 20.13 mg/L

(a) Compute the z-score corresponding to 17.99. Interpret this result. 1.60
(b) Determine the quartiles.
(c) Compute the interquartile range, IQR. 4.61
(d) Determine the lower and upper fences. Are there any outliers, according to this criterion? 0.195, 18.635; yes

19. **Fraud Detection** As part of its "Customers First" program, a cellular phone company monitors monthly phone usage. The goal of the program is to identify unusual use and alert the customer that their phone may have been used by an unscrupulous individual. The following data represent the monthly phone use in minutes of a customer enrolled in this program for the past 20 months.

346	345	489	358	471
442	466	505	466	372
442	461	515	549	437
480	490	429	470	516

The phone company decides to use the upper fence as the cutoff point for the number of minutes at which the customer should be contacted. What is the cutoff point? 574 minutes

20. **Stolen Credit Card** A credit card company decides to enact a fraud-detection service. The goal of the credit card company is to determine if there is any unusual activity on the credit card. The company maintains a database of daily charges on a customer's credit card. Any day when the card was inactive is excluded from the database. If a day's worth of charges appears unusual, the customer is contacted to make sure that the credit card has not been compromised. Use the following daily charges (rounded to the nearest dollar) to determine the amount the daily charges must exceed before the customer is contacted. $219

143	166	113	188	133
90	89	98	95	112
111	79	46	20	112
70	174	68	101	212

21. **Student Survey of Income** A survey of 50 randomly selected full-time Joliet Junior College students was conducted during the Fall 2005 semester. In the survey, the students were asked to disclose their weekly income from employment. If the student did not work, $0 was entered.

0	262	0	635	0
244	521	476	100	650
12,777	567	310	527	0
83	159	0	547	188
719	0	367	316	0
479	0	82	579	289
375	347	331	281	628
0	203	149	0	403
0	454	67	389	0
671	95	736	300	181

(a) Check the data set for outliers.
(b) Draw a histogram of the data and label the outliers on the histogram.
(c) Provide an explanation for the outliers.

22. **Student Survey of Entertainment Spending** A survey of 40 randomly selected full-time Joliet Junior College students was conducted in the Fall 2005 semester. In the survey, the students were asked to disclose their weekly spending on entertainment. The results of the survey are as follows:

21	54	64	33	65
22	39	67	54	22
115	7	80	59	20
36	10	12	101	1000
28	28	75	50	27
32	51	33	26	35
21	26	13	38	9
16	14	36	8	48

(a) Check the data set for outliers.
(b) Draw a histogram of the data and label the outliers on the histogram.
(c) Provide an explanation for the outliers.

23. **Pulse Rate** Use the results of Problem 25 in Sections 3.1 and 3.2 to compute the z-scores for all the students. Compute the mean and standard deviation of these z-scores.

24. **Travel Time** Use the results of Problem 26 in Sections 3.1 and 3.2 to compute the z-scores for all the students. Compute the mean and standard deviation of these z-scores.

Technology Step by Step	**Determining Percentiles**

TI-83/84 Plus To compute the quartiles, follow the same steps given to compute the mean and median from raw data.

MINITAB MINITAB computes only the quartiles. Follow the same steps given to compute the mean and median from raw data.

Excel *Step 1:* Enter the raw data into column A.

Step 2: With the data analysis Tool Pak enabled, select the **Tools** menu and highlight **Data Analysis. . . .**

Step 3: Select **Rank and Percentile** from the Data Analysis window.

Step 4: With the cursor in the **Input Range** cell, highlight the data. Press OK.

3.5 The Five-Number Summary and Boxplots

Objectives **Compute the five-number summary**

 Draw and interpret boxplots

Some aspects of statistical analysis attempt to verify a conjecture by means of observational studies or designed experiments. In other words, a theory is conjectured, and then data are collected to test the theory. For example, a dietitian might conjecture that exercise will lower an individual's cholesterol. The dietitian would carefully design an experiment that randomly divides study participants into two groups: the control group and the experimental group. She would impose a treatment (exercise or no exercise) on the two groups and then measure the effect on the response variable, cholesterol levels.

Another aspect of statistics looks at data to spot any interesting results that might be concluded from the data. In other words, rather than develop a theory and use data to support or disprove the theory, a researcher starts with data and looks for a theory. This area of statistics is referred to as **exploratory data analysis (EDA)**. The idea behind exploratory data analysis is to draw graphs of data and obtain measures of central tendency and spread to form some conjectures regarding the data.

Many of the methods of exploratory data analysis were developed by John Tukey (1915–2000). A complete presentation of the materials found in this chapter can be found in his text *Exploratory Data Analysis* (Addison-Wesley, 1977).

 ### Compute the Five-Number Summary

Remember that the median is a measure of central tendency that divides the lower 50% from the upper 50% of the data. This particular measure of central tendency is resistant to extreme values and is the preferred measure of central tendency when data are skewed right or left.

The three measures of dispersion presented in Section 3.2 (range, variance, and standard deviation) are not resistant to extreme values. However, the interquartile range, Q_3-Q_1, is resistant. It measures the spread of the data by determining the difference between the 25th and 75th percentiles. It is interpreted as the range of the middle 50% of the data. However, the median, Q_1, and Q_3 do not provide information about the tails of the distribution of the data. To get this information, we need to know the smallest and largest values in the data set.

The **five-number summary** of a set of data consists of the smallest data value, Q_1, the median, Q_3, and the largest data value. Symbolically, the five-number summary is presented as follows:

> **Five-Number Summary**
>
> MINIMUM Q_1 M Q_3 MAXIMUM

EXAMPLE 1   Obtaining the Five-Number Summary

Problem: The data shown in Table 18 show the finishing times (in minutes) of the men in the 60- to 64-year-old age group in a 5-kilometer race.

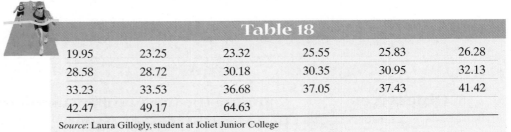

Table 18					
19.95	23.25	23.32	25.55	25.83	26.28
28.58	28.72	30.18	30.35	30.95	32.13
33.23	33.53	36.68	37.05	37.43	41.42
42.47	49.17	64.63			

Source: Laura Gillogly, student at Joliet Junior College

Approach: The five-number summary requires that we determine the minimum data value, Q_1 (the 25th percentile), M (the median), Q_3 (the 75th percentile), and the maximum data value. We need to arrange the data in ascending order and then use the procedures introduced in Section 3.4 to obtain Q_1, M, and Q_3.

Solution: The data in ascending order are as follows:

19.95, 23.25, 23.32, 25.55, 25.83, 26.28, 28.58, 28.72, 30.18, 30.35, 30.95, 32.13, 33.23, 33.53, 36.68, 37.05, 37.43, 41.42, 42.47, 49.17, 64.63

The smallest number (the fastest time) in the data set is 19.95. The largest number in the data set is 64.63. The first quartile, Q_1, is 26.06. The median, M, is 30.95. The third quartile, Q_3, is 37.24. The five-number summary is

19.95   26.06   30.95   37.24   64.63

EXAMPLE 2   Obtaining the Five-Number Summary Using Technology

Problem: Using a statistical spreadsheet or graphing calculator, determine the five-number summary of the data presented in Table 18.

Approach: We will use MINITAB to obtain the five-number summary. The steps for obtaining the five-number summary using a TI-83 or TI-84 Plus graphing calculator, MINITAB, or Excel are given in the Technology Step by Step on page 168.

Result: Figure 21 shows the output supplied by MINITAB. The five-number summary is highlighted.

Figure 21

Descriptive statistics: Times

Variable	N	N*	Mean	SE Mean	StDev	Minimum	Q1	Median	Q3	Maximum
Times	21	0	33.37	2.20	10.10	19.95	26.06	30.95	37.24	64.63

② Draw and Intepret Boxplots

The five-number summary can be used to create another graph, called the **boxplot**.

Drawing a Boxplot

Step 1: Determine the lower and upper fences:

$$\text{Lower fence} = Q_1 - 1.5(\text{IQR})$$

$$\text{Upper fence} = Q_3 + 1.5(\text{IQR})$$

Remember, $\text{IQR} = Q_3 - Q_1$.

Step 2: Draw vertical lines at Q_1, M, and Q_3. Enclose these vertical lines in a box.

Step 3: Label the lower and upper fences.

Step 4: Draw a line from Q_1 to the smallest data value that is larger than the lower fence. Draw a line from Q_3 to the largest data value that is smaller than the upper fence.

Step 5: Any data values less than the lower fence or greater than the upper fence are outliers and are marked with an asterisk (*).

Note to Instructor

Some call the boxplot described here the modified boxplot.

EXAMPLE 3   **Constructing a Boxplot**

Problem: Use the results from Example 1 to a construct a boxplot of the finishing times of the men in the 60- to 64-year-old age group.

Approach: Follow the steps presented above.

Solution: From the results of Example 1, we know that $Q_1 = 26.06$, $M = 30.95$, and $Q_3 = 37.24$. Therefore, the interquartile range = IQR = $Q_3 - Q_1 = 37.24 - 26.06 = 11.18$. The difference between the 75th percentile and 25th percentile is a time of 11.18 minutes.

Step 1: We compute the lower and upper fences:

$$\text{Lower fence} = Q_1 - 1.5(\text{IQR}) = 26.06 - 1.5(11.18) = 9.29$$

$$\text{Upper fence} = Q_3 + 1.5(\text{IQR}) = 37.24 + 1.5(11.18) = 54.01$$

Step 2: Draw a horizontal number line with a scale that will accommodate our graph. Draw vertical lines at $Q_1 = 26.06$, $M = 30.95$, and $Q_3 = 37.24$. Enclose these lines in a box. See Figure 22(a).

Figure 22(a)

Step 3: Temporarily mark the location of the lower and upper fence with brackets ([and]). See Figure 22(b).

Figure 22(b)

Step 4: The smallest data value that is larger than 9.29 (the lower fence) is 19.95. The largest data value that is smaller than 54.01 (the upper fence) is 49.17. We draw horizontal lines from Q_1 to 19.95 and from Q_3 to 49.17. See Figure 22(c).

Figure 22(c)

Step 5: Plot any values less than 9.29 (the lower fence) or greater than 54.01 (the upper fence) using an asterisk (*). These values are outliers. So, 64.63 is an outlier. Remove the brackets from the graph. See Figure 22(d).

Figure 22(d)

Note to Instructor

Emphasize how to read a boxplot. Box-plots will be used throughout the text.

We can describe the shape of the distribution using the boxplot.

Distribution Shape Based on the Boxplot

1. If the median is near the center of the box and each horizontal line is of approximately equal length, the distribution is roughly symmetric.
2. If the median is to the left of the center of the box or the right line is substantially longer than the left line, the distribution is skewed right.
3. If the median is to the right of the center of the box or the left line is substantially longer than the right line, the distribution is skewed left.

CAUTION

Identifying the shape of a distribution from a boxplot (or from a histogram, for that matter) is subjective. When identifying the shape of a distribution from a graph, be sure to support your opinion.

Figure 23 on page 163 provides examples of boxplots that are (a) symmetric, (b) skewed right, and (c) skewed left, along with the corresponding histograms.

The boxplot in Figure 22(d) suggests that the distribution is skewed right, since the right line is longer than the left and the median is to the left of the center of the box.

We stated that the interquartile range (IQR) is a measure of dispersion (just like the standard deviation is a measure of dispersion). If the median is used as the measure of central tendency, then the IQR should be used as the measure of spread. Remember, we use the median to describe the "center" of a set of data when the shape of the distribution is skewed because the mean is distorted by skewness. Likewise, the standard deviation is distorted by skewness. Because the IQR is resistant to extreme values, it is a better measure of spread for skewed distributions.

Summary: Which Measures to Report		
Shape of Distribution	**Measure of Central Tendency**	**Measure of Dispersion**
Symmetric	Mean	Standard deviation
Skewed left or skewed right	Median	Interquartile range

Figure 23

(a) Symmetric

Historical Note

John Tukey was born on July 16, 1915, in New Bedford, Massachusetts. His parents graduated numbers 1 and 2 from Bates College and were elected "the couple most likely to give birth to a genius." In 1936, Tukey graduated from Brown University with an undergraduate degree in chemistry. He went on to earn a master's degree in chemistry at Brown. In 1939, Tukey earned his doctorate in mathematics from Princeton. He remained at Princeton and in 1965 became the founding chair of the Department of Statistics. Among his many accomplishments, Tukey is credited with coining the terms *software* and *bit*. In the early 1970s, he discussed the negative effects of aerosol cans on the ozone layer. Tukey recommended that the 1990 Census be adjusted by means of statistical formulas. John Tukey died in New Brunswick, New Jersey, on July 26, 2000.

(b) Skewed right

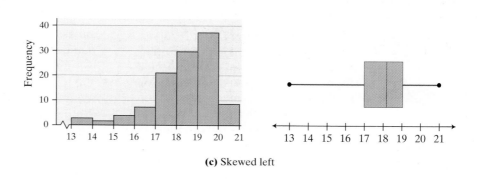

(c) Skewed left

Now Work Problem 11.

EXAMPLE 4 | **Comparing Two Distributions by Using Boxplots**

Problem: In the Spacelab Life Sciences 2, 14 male rats were sent to space. Upon their return, the red blood cell mass (in milliliters) of the rats was determined. A control group of 14 male rats was held under the same conditions (except for spaceflight) as the space rats, and their red blood cell mass was also determined when the space rats returned. The project was led by Paul X. Callahan. The data in Table 19 were obtained. Construct boxplots for red blood cell mass for the flight group and control group. Does it appear that the flight to space affected the red blood cell mass of the rats?

Approach: When comparing two data sets, we draw the boxplots on the same horizontal number line to make the comparison easy. Graphing calculators with advanced statistical features, as well as statistical spreadsheets such as MINITAB and Excel, have the ability to draw boxplots. We will use MINITAB to draw the boxplots. The steps for drawing boxplots using a TI-83 or TI-84 Plus graphing calculator, MINITAB, or Excel are given in the Technology Step by Step on page 168.

Table 19

Flight		Control	
8.59	8.64	8.65	6.99
6.87	7.89	7.62	7.44
7.00	8.80	7.33	8.58
6.39	7.54	7.14	9.14
7.43	7.21	8.40	9.66
9.79	6.85	8.55	8.70
9.30	8.03	9.88	9.94

Source: NASA Life Sciences Data Archive

Solution: Figure 24 shows the boxplots drawn in MINITAB. From the boxplots, it appears that the spaceflight has reduced the red blood cell mass of the rats.

Figure 24

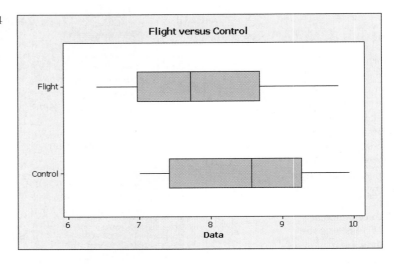

Now Work Problem 15.

MAKING AN INFORMED DECISION

What Car Should I Buy?

Suppose you are in the market to purchase a used car. To make an informed decision regarding your purchase, you would like to collect as much information as possible. Among the information you might consider are the typical price of the car, the typical number of miles the car should have and its crash test results, insurance costs, and expected repair costs.

1. Make a list of at least three cars that you would consider purchasing. To be fair, the cars should be in the same class (such as compact, midsize, and so on). They should also be of the same age.

2. Collect information regarding the three cars in your list by finding at least eight cars of each type that are for sale. Obtain such information as the asking price and the number of miles the car has. Sources of data include your local newspaper, classified ads, and car Web sites (such as www.cars.com and www.vehix.com). Compute summary statistics for asking price, number of

miles, and other variables of interest. Using the same scale, draw side-by-side boxplots of each variable considered.

3. Go to the Insurance Institute for Highway Safety Web site (www.hwysafety.org). Select the Vehicle Ratings link. Choose the make and model for each car you are considering. Obtain information regarding crash testing for each car under consideration. Compare cars in the same class. How does each car compare? Is one car you are considering substantially safer than the others? What about repair costs? Compute summary statistics for crash tests and repair costs.

4. Obtain information about insurance costs. Contact various insurance companies to determine the cost of insuring the cars you are considering. Compute summary statistics for insurance costs and draw boxplots.

5. Write a report supporting your conclusion regarding which car you would purchase.

3.5 ASSESS YOUR UNDERSTANDING

Concepts and Vocabulary

1. Explain the circumstances under which the median and interquartile range would be better measures of central tendency and dispersion than the mean and standard deviation.

2. In a boxplot, if the median is to the left of the center of the box or the right line is substantially longer than the left line, the distribution is skewed _____. *right*

Skill Building

In Problems 3 and 4, (a) identify the shape of the distribution, and (b) determine the five-number summary. Assume that each number in the five-number summary is an integer.

3.

4.

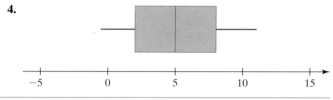

Applying the Concepts

In Problems 5–10, find the five-number summary, and construct a boxplot for the data in the indicated problem. Comment on the shape of the distribution.

5. **Age at Inauguration** The following data represent the age of U.S. presidents on their respective inauguration days. *42, 51, 55, 58, 69*

57	61	57	57	58	57	61
54	68	51	49	64	50	48
65	52	56	46	54	49	50
47	55	55	54	42	51	56
55	51	54	51	60	62	43
55	56	61	52	69	64	46
54						

6. **Grams of Fat in a McDonald's Breakfast** The following data represent the number of grams of fat in breakfast meals offered at McDonald's. *1, 11, 20, 28, 40*

12	23	28	2	28	33
31	11	23	40	35	1
23	33	23	16	11	8
8	17	16	15		

Source: McDonald's Corporation, *A Full Serving of Nutrition Facts*, April 2003

7. **Super Bowl Point Spreads** The following data represent the number of points by which the winning team won Super Bowls I to XXXIX. *1, 7, 15, 22, 45*

25	19	9	16	3	21	7	17
10	4	18	17	4	12	17	5
10	29	22	36	19	32	4	45
1	13	35	17	23	10	14	7
15	7	27	3	27	3	3	

Source: superbowl.com

8. **Miles per Gallon** The following data represent the number of miles per gallon achieved on the highway for compact cars for model year 2005. *18, 26, 30, 32, 46*

30	29	30	21	18	29	27	30	29
34	34	30	28	30	20	32	28	32
34	35	26	26	31	25	35	32	25
19	26	19	24	22	24	19	31	26
34	32	34	25	34	34	32	29	25
31	29	30	30	34	32	29	38	39
46	31	31	30	27	29	26	29	24

Source: U.S. Department of Energy, Office of Energy Efficiency and Renewable Energy, U.S. Environmental Protection Agency, *Fuel Economy Guide, Model Year 2005* (www.fueleconomy.gov)

9. **Got a Headache?** Drugs are made of active ingredients and inactive ingredients. The Food and Drug Administration states that a drug should have the same amount of active ingredient in over-the-counter drugs. The following data represent the weight (in grams) of a random sample of 25 Tylenol tablets. What do you think is the source of the variability in weight? *0.598, 0.604, 0.608, 0.610, 0.612*

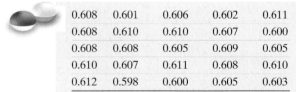

0.608	0.601	0.606	0.602	0.611
0.608	0.610	0.610	0.607	0.600
0.608	0.608	0.605	0.609	0.605
0.610	0.607	0.611	0.608	0.610
0.612	0.598	0.600	0.605	0.603

Source: Kelly Roe, student at Joliet Junior College

10. **Gasoline Expenditures** The following data represent the mean gasoline expenditures per person for each state and the District of Columbia. Wyoming has the highest mean expenditure. What might explain this? New York has the lowest mean expenditure. What might explain this? *421, 653, 684, 713, 971*

971	787	713	704	688	675	660	643	581
830	741	711	698	684	672	654	618	480
821	740	711	698	683	669	653	616	421
816	737	707	692	682	667	649	611	
802	726	707	692	679	666	646	598	
791	715	706	688	678	664	645	583	

Source: Energy Information Administration

11. Serum HDL Dr. Paul Oswiecmiski randomly selects 40 of
NW his 20- to 29-year-old patients and obtains the following
data regarding their serum HDL cholesterol.

70	56	48	48	53	52	66	48
36	49	28	35	58	62	45	60
38	73	45	51	56	51	46	39
56	32	44	60	51	44	63	50
46	69	53	70	33	54	55	52

(a) Compute the five-number summary.
(b) Draw a boxplot of the data.
(c) Determine the shape of the distribution from the box-plot. Refer to the histogram drawn in Problem 31 in Section 2.2 to test your answer. Symmetric
(d) Which measures of central tendency and dispersion should be reported for these data? Mean and s.d.

12. Volume of Altria Group Stock The volume of a stock is
the number of shares traded on a given day. The following
data, given in millions so that 3.78 represents 3,780,000
shares traded, represent the volume of Altria Group stock
traded for a random sample 35 trading days in 2004.

3.78	8.74	4.35	5.02	8.40
6.06	5.75	5.34	6.92	6.23
5.32	3.25	6.57	7.57	6.07
3.04	5.64	5.00	7.16	4.88
10.32	3.38	7.25	6.52	4.43
3.38	5.53	4.74	9.70	3.56
10.96	4.50	7.97	3.01	5.58

Source: yahoo.finance.com

(a) Compute the five-number summary.
(b) Draw a boxplot of the data.
(c) Determine the shape of the distribution from the box-plot. Refer to the histogram drawn in Problem 32 in Section 2.2 to test your answer. Skewed right
(d) Which measures of central tendency and dispersion should be reported for these data? Median and IQR

13. Dividend Yield A dividend is a payment from a publicly
traded company to its shareholders. The dividend yield of
a stock is determined by dividing the annual dividend of a
stock by its price. The following data represent the dividend yields (in percent) of a random sample of 28 publicly
traded stocks with a value of at least $5 billion.

1.7	0	1.15	0.62	1.06	2.45	2.38
2.83	2.16	1.05	1.22	1.68	0.89	0
2.59	0	1.7	0.64	0.67	2.07	0.94
2.04	0	0	1.35	0	0	0.41

Source: Yahoo! Finance

(a) Compute the five-number summary.
(b) Draw a boxplot of the data.
(c) Determine the shape of the distribution from the box-plot. Refer to the histogram drawn in Problem 33 in Section 2.2 to test your answer. Skewed right
(d) Which measures of central tendency and dispersion should be reported for these data? Median an IQR

14. Violent Crimes Violent crimes include murder, forcible
rape, robbery, and aggravated assault. The following data
represent the violent crime rate (per 100,000 population)
by state and the District of Columbia in 2002.

444	563	553	424	593	352
311	599	1633	770	459	262
255	621	357	286	377	279
662	108	770	484	540	268
343	539	352	314	638	161
375	740	496	470	78	351
503	292	402	285	822	177
717	579	237	107	291	345
234	225	274			

Source: U.S. Federal Bureau of Investigation

(a) Compute the five-number summary.
(b) Draw a boxplot of the data.
(c) Determine the shape of the distribution from the box-plot. Refer to the histogram drawn in Problem 34 in Section 2.2 to test your answer. Skewed right
(d) Which measures of central tendency and dispersion should be reported for this data? Median and IQR

(11a) 28, 45, 51, 57.5, 73

(12a) 3.01, 4.43, 5.58, 7.16, 10.96

(13a) 0, 0.205, 1.055, 1.87, 2.83

(14a) 78, 279, 375, 563, 1633

In Problems 15–17, compare the data sets by determining the five-number summary and constructing boxplots on the same scale.

15. Chips per Cookie The data to the right represent the number of chips per cookie in a random sample of Keebler Chips Deluxe Chocolate Chip Cookies and the number of chips per cookie in a store brand's chocolate chip cookies. Does there appear to be a difference in the number of chips per cookie? Does one brand have a more consistent number of chips per cookie?

Keebler			Store Brand		
32	23	28	21	23	24
28	28	29	24	25	27
25	20	25	26	26	21
22	21	24	18	16	24
21	24	21	21	30	17
26	28	24	23	28	31
33	20	31	27	33	29

Source: Trina McNamara, student at Joliet Junior College

16. Tornadoes The following data give the number of tornadoes in Oklahoma, Kansas, and Nebraska for the years 1990 to 2004. Which state appears to have the highest number of tornadoes per year?

Year:	1990	1991	1992	1993	1994	1995	1996	1997	1998	1999	2000	2001	2002	2003	2004
Oklahoma:	30	73	64	64	40	79	47	55	83	145	44	61	18	78	62
Kansas:	88	116	92	113	42	73	68	62	71	64	59	101	95	91	122
Nebraska:	88	63	74	69	55	26	60	30	65	102	61	62	28	81	110

Source: National Oceanic and Atmospheric Administration

17. Home-Run Distances During the 1998 major league baseball season, Mark McGwire of the St. Louis Cardinals and Sammy Sosa of the Chicago Cubs thrilled fans across the country in a race to set the record for the most home runs hit in a season. Sosa ended the season with 66 home runs, and McGwire set the record with 70 home runs. Only 3 years later in 2001, Barry Bonds of the San Francisco Giants broke McGwire's record by hitting 73 home runs. The following data represent the distances of each player's home runs in his record-setting season. Which player appears to have the longest distances? Which player appears to have the most consistent distances?

Mark McGwire						
360	370	370	430	420	340	460
410	440	410	380	360	350	527
380	550	478	420	390	420	425
370	480	390	430	388	423	410
360	410	450	350	450	430	461
430	470	440	400	390	510	430
450	452	420	380	470	398	409
385	369	460	390	510	500	450
470	430	458	380	430	341	385
410	420	380	400	440	377	370

Sammy Sosa						
371	350	430	420	430	434	370
420	440	410	420	460	400	430
410	370	370	410	380	340	350
420	410	415	430	380	380	366
500	380	390	400	364	430	450
440	365	420	350	420	400	380
380	400	370	420	360	368	430
433	388	440	414	482	364	370
400	405	433	390	480	480	434
344	410	420				

Barry Bonds						
420	417	440	410	390	417	420
410	380	430	370	420	400	360
410	420	391	416	440	410	415
436	430	410	400	390	420	410
420	410	410	450	320	430	380
375	375	347	380	429	320	360
375	370	440	400	405	430	350
396	410	380	430	415	380	375
400	435	420	420	488	361	394
410	411	365	360	440	435	454
442	404	385				

| Technology Step by Step | **Drawing Boxplots Using Technology** |

TI-83/84 Plus **Step 1:** Enter the raw data into L1.

Step 2: Press 2nd Y= and select 1:Plot 1.

Step 3: Turn the plots ON. Use the cursor to highlight the modified boxplot icon. Your screen should look as follows:

Step 4: Press ZOOM and select 9: ZoomStat.

MINITAB **Step 1:** Enter the raw data into column C1.

Step 2: Select the **Graph** menu and highlight **Boxplot...**

Step 3: For a single boxplot, select One Y, simple. For two or more boxplots, select Multiple Y's, simple.

Step 4: Select the data to be graphed. If you want the boxplot to be horizontal rather than vertical, select the Scale button, then transpose value and category scales. Click OK.

Excel **Step 1:** Start the PHStat Add-in.

Step 2: Enter the raw data into column A.

Step 3: Select the **PHStat** menu and highlight **Box-and-Whisker Plot** ... With the cursor in the Data Variable Cell Range cell, highlight the data in column A.

Step 4: Click OK.

CHAPTER 3 **Review**

Summary

This chapter concentrated on describing distributions numerically. Measures of central tendency are used to indicate the typical value in a distribution. Three measures of central tendency were discussed. The mean measures the center of gravity of the distribution. The median separates the bottom 50% of the data from the top 50%. Both measures require that the data be quantitative. The mode measures the most frequent observation. The data can be either quantitative or qualitative to compute the mode. The median is resistant to extreme values, while the mean is not. A comparison between the median and mean can help determine the shape of the distribution.

Measures of dispersion describe the spread in the data. The range is the difference between the highest and lowest data value. The variance measures the average squared deviation about the mean. The standard deviation is the square root of the variance. The mean and standard deviation are used in many types of statistical inference.

The mean, median, and mode can be approximated from grouped data. The variance and standard deviation can also be approximated from grouped data.

We can determine the relative position of an observation in a data set using z-scores and percentiles. z-scores denote how many standard deviations an observation is from the mean. Percentiles determine the percent of observations that lie above and below an observation. The upper and lower fences can be used to identify potential outliers. Any potential outlier must be investigated to determine whether it was the result of a data entry error, of some other error in the data collection process, or of an unusual value in the data set.

The interquartile range is also a measure of dispersion. The five-number summary provides an idea about the center and spread of a data set, through the median and the interquartile range. The length of the tails in the distribution can be determined from the smallest and largest data values. The five-number summary is used to construct boxplots. Boxplots can be used to describe the shape of the distribution.

Formulas

Population Mean

$$\mu = \frac{\sum x_i}{N}$$

Sample Mean

$$\bar{x} = \frac{\sum x_i}{n}$$

Population Variance

$$\sigma^2 = \frac{\sum (x_i - \mu)^2}{N} = \frac{\sum x_i^2 - \frac{\left(\sum x_i\right)^2}{N}}{N}$$

Sample Variance

$$s^2 = \frac{\sum (x_i - \bar{x})^2}{n-1} = \frac{\sum x_i^2 - \frac{\left(\sum x_i\right)^2}{n}}{n-1}$$

Population Standard Deviation

$$\sigma = \sqrt{\sigma^2}$$

Sample Standard Deviation

$$s = \sqrt{s^2}$$

Range = Largest Data Value − Smallest Data Value

Weighted Mean

$$\bar{x}_w = \frac{\sum w_i x_i}{\sum w_i}$$

Population Mean from Grouped Data

$$\mu = \frac{\sum x_i f_i}{\sum f_i}$$

Sample Mean from Grouped Data

$$\bar{x} = \frac{\sum x_i f_i}{\sum f_i}$$

Population Variance from Grouped Data

$$\sigma^2 = \frac{\sum (x_i - \mu)^2 f_i}{\sum f_i}$$

Sample Variance from Grouped Data

$$s^2 = \frac{\sum (x_i - \bar{x})^2 f_i}{\left(\sum f_i\right) - 1}$$

Population z-Score

$$z = \frac{x - \mu}{\sigma}$$

Sample z-Score

$$z = \frac{x - \bar{x}}{s}$$

Percentile of x = $\dfrac{\text{number of data values less than } x}{n} \cdot 100$

Interquartile Range

$$\text{IQR} = Q_3 - Q_1$$

Lower and Upper Fences

Lower Fence = $Q_1 - 1.5(\text{IQR})$

Upper Fence = $Q_3 + 1.5(\text{IQR})$

Vocabulary

Parameter (p. 107)
Statistic (p. 107)
Arithmetic mean (p. 107)
Median (p. 110)
Mode (p. 111)
Bimodal (p. 112)
Multimodal (p. 112)
Resistant (p. 113)
Range (p. 124)

Deviation about the mean (p. 125)
Population variance (p. 125)
Sample variance (p. 127)
Biased (p. 127)
Degrees of freedom (p. 128)
Population standard deviation (p. 129)
Sample standard deviation (p. 129)
Class midpoint (p. 142)
Weighted mean (p. 144)

Z-score (p. 150)
kth percentile (p. 151)
Quartiles (p. 153)
Outlier (p. 155)
Interquartile range (p. 155)
Fences (p. 155)
Exploratory data analysis (p. 159)
Five-number summary (p. 160)
Boxplot (p. 161)

Objectives

Section	You should be able to . . .	Example	Review Exercises
3.1	1 Determine the arithmetic mean of a variable from raw data (p. 107)	1, 7	1(a)–8(a) 1(a)–8(a)
	2 Determine the median of a variable from raw data (p. 110)	2, 3, 7	5(a) and 6(a)
	3 Determine the mode of a variable from raw data (p. 111)	4, 5, 6, 7	15(a)–(c), 16(a)–(c)
	4 Use the mean and median to help identify the shape of a distribution (p. 113)	8, 9	
3.2	1 Compute the range of a variable from raw data (p. 124)	2	1(b)–8(b)
	2 Compute the variance of a variable from raw data (p. 125)	3, 4, 6	1(b)–8(b), 15(d), 16(d)
	3 Compute the standard deviation of a variable from raw data (p. 129)	5, 6, 7	1(b)–8(b), 15(d), 16(d)
	4 Use the Empirical Rule to describe data that are bell shaped (p. 131)	8	9(a)–(d), 10 (a)–(c)
	5 Use Chebyshev's Inequality to describe any set of data (p. 132)	9	9(e) and (f), 10(d) and (e)
3.3	1 Approximate the mean of a variable from grouped data (p. 142)	1, 4	11(a), 12(a)
	2 Compute the weighted mean (p. 144)	2	13, 14
	3 Approximate the variance and standard deviation of a variable from grouped data (p. 144)	3, 4	11(b), 12(b)
3.4	1 Determine and interpret z-scores (p. 149)	1	19
	2 Determine and interpret percentiles (p. 151)	2–4	17, 18
	3 Determine and interpret quartiles (p. 153)	5, 6	15(e), 16(e)
	4 Check a set of data for outliers (p. 155)	7	15(h), 16(h)
3.5	1 Compute the five-number summary (p. 159)	1, 2	15(e), 16(e)
	2 Draw and interpret boxplots (p. 161)	3, 4	15(f), 16(f), 20

3 Review Exercises

1. Muzzle Velocity The following data represent the muzzle velocity (in meters per second) of rounds fired from a 155 mm gun.

793.8	793.1	792.4	794.0	791.4
792.4	791.7	792.3	789.6	794.4

Source: Christenson, Ronald, and Blackwood, Larry; "Tests for Precision and Accuracy of Multiple Measuring Devices." *Technometrics*, Nov. 93, Vol. 35, Issue 4, pp. 411–421.

(a) Compute the sample mean and median muzzle velocity. $\bar{x} = 792.51$, $M = 792.40$
(b) Compute the range, sample variance, and sample standard deviation. Range $= 4.8$, $s^2 = 2.03$, $s = 1.42$

2. Pulse Rates The following data represent the pulse rate of eight randomly selected females after stepping up and down on a 6-inch platform for 3 minutes. Pulse is measured in beats per minute. (a) $\bar{x} = 126.8$, $M = 128.5$

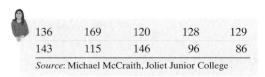

136	169	120	128	129
143	115	146	96	86

Source: Michael McCraith, Joliet Junior College

(a) Compute the sample mean and median pulse.
(b) Compute the range, sample variance, and sample standard deviation.
Range $= 83$, $s^2 = 589.07$, $s = 24.3$

3. Price of Chevy Cavaliers The following data represent the sales price in dollars for nine two-year-old Chevrolet Cavaliers in the Los Angeles area.

14,050	13,999	12,999	10,995	9,980
8,998	7,889	7,200	5,500	

Source: cars.com

(3a) $\bar{x} = \$10,178.89$, $M = \$9,980$ (3b) Range = $\$8,550$, $s = \$3074.92$
(3c) $\bar{x} = \$13,178.89$, $M = \$9,980$, range = $\$35,550$, $s = \$10,797.46$; the median is resistant.

(a) Compute the sample mean and median price.
(b) Compute the range and sample standard deviation.
(c) Redo (a) and (b) if the data value 14,050 was incorrectly entered as 41,050. How does this change affect the mean? The median? The range? The standard deviation? Which of these values is resistant?

4. Home Sales The following data represent the closing prices (in U.S. dollars) of 15 randomly selected homes sold in Joliet, Illinois, in December 2004.

138,820	140,794	136,833	157,216
169,541	153,146	115,000	149,380
135,512	99,000	124,757	136,529
149,143	136,924	128,429	

Source: Transamerica Intellitech

(4a) $\bar{x} = \$138,068.3$; $\$136,924$
(4b) Range = $\$70,541$, $s = \$17,286.3$

(a) Compute the sample mean and median sale price.
(b) Compute the range and sample standard deviation.

5. Chief Justices The following data represent the ages of chief justices of the U.S. Supreme Court when they were appointed.

Justice	Age
John Jay	44
John Rutledge	56
Oliver Ellsworth	51
John Marshall	46
Roger B. Taney	59
Salmon P. Chase	56
Morrison R. Waite	58
Melville W. Fuller	55
Edward D. White	65
William H. Taft	64
Charles E. Hughes	68
Harlan F. Stone	69
Frederick M. Vinson	56
Earl Warren	62
Warren E. Burger	62
William H. Rehnquist	62
John G. Roberts	50

Source: *Information Please Almanac*

(a) Compute the population mean, median, and mode ages. $\mu = 57.8$, $M = 58$, bimodal: 56 and 62
(b) Compute the range and population standard deviation ages. Range = 25, $\sigma = 7.0$
(c) Obtain two simple random samples of size 4, and compute the sample mean and sample standard deviation ages.

6. National League Home Runs The following data represent the number of home runs hit by all teams in the National League in 2004.

Team	Home Runs	Team	Home Runs
1. St. Louis Cardinals	214	9. Los Angeles Dodgers	203
2. San Francisco Giants	183	10. Cincinnati Reds	194
3. Philadelphia Phillies	215	11. Florida Marlins	148
4. Colorado Rockies	202	12. New York Mets	185
5. Atlanta Braves	178	13. Pittsburgh Pirates	142
6. Houston Astros	187	14. Montreal Expos	151
7. Chicago Cubs	235	15. Milwaukee Brewers	135
8. San Diego Padres	139	16. Arizona Diamondbacks	135

Source: Major League Baseball

(6b) $\mu = 177.9$, $M = 184$, mode = 135   (6b) Range = 100, $\sigma = 31.3$

(a) Compute the population mean, median, and mode for number of home runs.
(b) Compute the range and population standard deviation for number of home runs.
(c) Obtain two simple random samples of size 3, and compute the sample mean and sample standard deviation for number of home runs.
(d) If a sports reporter stated that the average number of home runs hit by teams in the National League in 2004 was 135, is he lying? Is he being deceptive? No; yes

7. Family Size A random sample of 36 married couples who had been married 7 years were asked the number of children they had. The results of the survey follow:

0	0	3	1	2	3
3	4	3	3	0	3
1	2	1	3	0	3
4	2	3	2	2	4
2	1	3	4	1	3
0	3	3	3	2	1

(a) Compute the sample mean and the median number of children. $\bar{x} = 2.2$, $M = 2.5$
(b) Compute the range and the sample standard deviation number of children. Range $= 4$, $s = 1.3$

8. Waiting in Line The following data represent the number of cars that arrived at a McDonald's drive-through between 11:50 A.M. and 12:00 noon each Wednesday for the past 30 weeks:

1	3	2	8	6
6	6	3	3	1
5	6	3	3	1
4	9	5	3	5
2	6	7	5	8
7	8	3	2	3

(a) Compute the sample mean and the median number of cars. $\bar{x} = 4.5$, $M = 4.5$
(b) Compute the range and the sample standard deviation number of cars. Range $= 8$, $s = 2.3$

9. Chebyshev's Inequality and the Empirical Rule Suppose that a random sample of 200 lightbulbs has a mean life of 600 hours and a standard deviation of 53 hours.
(a) A histogram of the data indicates the sample data follow a bell-shaped distribution. According to the Empirical Rule, 99.7% of lightbulbs have lifetimes between _____ and _____ hours. 441; 759
(b) Assuming the data are bell shaped, determine the percentage of lightbulbs that will have a life between 494 and 706 hours. 95%
(c) Assuming the data are bell shaped, what percentage of lightbulbs will last between 547 and 706 hours? 81.5%
(d) If the company that manufactures the lightbulb guarantees to replace any bulb that does not last at least 441 hours, what percentage of lightbulbs can the firm expect to have to replace, according to the Empirical Rule? 0.15%
(e) Use Chebyshev's Inequality to determine the minimum percentage of lightbulbs with a life within 2.5 standard deviations of the mean. 84%
(f) Use Chebyshev's Inequality to determine the minimum percentage of lightbulbs with a life between 494 and 706 hours. 75%

10. Chebyshev's Inequality and the Empirical Rule In a random sample of 250 toner cartridges, the mean number of pages a toner cartridge can print is 4302 and the standard deviation is 340.

(a) Suppose a histogram of the data indicates that the sample data follow a bell-shaped distribution. According to the Empirical Rule, 99.7% of toner cartridges will print between _____ and _____ pages. 3282; 5322
(b) Assuming that the distribution of the data is bell shaped, determine the percentage of toner cartridges whose print total is between 3622 and 4982 pages. 95%
(c) If the company that manufactures the toner cartridges guarantees to replace any cartridge that does not print at least 3622 pages, what percent of cartridges can the firm expect to be responsible for replacing, according to the Empirical Rule? 2.5%
(d) Use Chebyshev's Inequality to determine the minimum percentage of toner cartridges with a page count within 1.5 standard deviations of the mean. 55.6%
(e) Use Chebyshev's Inequality to determine the minimum percentage of toner cartridges that print between 3282 and 5322 pages. 88.9%

11. Vehicle Fatalities The frequency distribution listed in the table represents the number of drivers in fatal crashes in 2003, by age, for males 20 to 84 years old.

Age	Number of Drivers	Age	Number of Drivers
20–24	6035	55–59	2355
25–29	4352	60–64	1664
30–34	4083	65–69	1173
35–39	3933	70–74	1025
40–44	4194	75–79	895
45–49	3716	80–84	744
50–54	3005		

Source: NHTSA

(a) Approximate the mean age of a male involved in a traffic fatality. $\mu = 42.28$ years
(b) Approximate the standard deviation age of a male involved in a traffic fatality. $\sigma = 15.93$ years

12. Vehicle Fatalities The frequency distribution listed in the table represents the number of drivers in fatal crashes in 2003, by age, for females 20 to 84 years old.

Age	Number of Drivers	Age	Number of Drivers
20–24	1903	55–59	784
25–29	1415	60–64	599
30–34	1364	65–69	415
35–39	1430	70–74	482
40–44	1409	75–79	456
45–49	1242	80–84	372
50–54	1008		

Source: NHTSA

(a) Approximate the mean age of a female involved in a traffic fatality. $\mu = 43.71$ years
(b) Approximate the standard deviation age of a female involved in a traffic fatality. $\sigma = 16.78$ years
(c) Compare the results to those obtained in Problem 11. How do you think an insurance company might use this information?

13. Weighted Mean   Michael has just completed his first semester in college. He earned an A in his 5-hour calculus course, a B in his 4-hour chemistry course, an A in his 3-hour speech course, and a C in his 3-hour psychology course. Assuming an A equals 4 points, a B equals 3 points, and a C equals 2 points, determine Michael's grade-point average if grades are weighted by class hours. 3.33

14. Weighted Mean   Yolanda wishes to develop a new type of meat loaf to sell at her restaurant. She decides to combine 2 pounds of ground sirloin (cost $2.70 per pound), 1 pound of ground turkey (cost $1.30 per pound), and $\frac{1}{2}$ pound of ground pork (cost $1.80 per pound). What is the cost per pound of the meat loaf? $2.17

15. Mets versus Yankees   The following data represent the 2004 salaries (in dollars) of the players on the rosters of the New York Mets and the New York Yankees.

Yankees		Mets	
Player	**Salary**	**Player**	**Salary**
Bubba Crosby	301,400	Tyler Yates	300,000
Jorge De Paula	302,550	Eric Valent	302,500
Donovan Osborne	450,000	Jose Reyes	307,500
Orlando Hernandez	500,000	Dan Wheeler	311,500
Enrique Wilson	700,000	Ty Wigginton	316,000
Tony Clark	750,000	Orber Moreno	317,500
John Flaherty	775,000	Jason Phillips	318,000
Miguel Cairo	900,000	Grant Roberts	319,500
Ruben Sierra	1,000,000	Joe McEwing	500,000
Felix Heredia	1,800,000	Shane Spencer	537,500
Gabe White	1,925,000	Scott Strickland	650,000
Travis Lee	2,000,000	Scott Erickson	700,000
Jon Lieber	2,700,000	Vance Wilson	715,000
Paul Quintrill	3,000,000	Karim Garcia	800,000
Kenny Lofton	3,100,000	John Franco	1,000,000
Tom Gordon	3,500,000	Todd Zeile	1,000,000
Steve Karsay	6,000,000	Braden Looper	2,000,000
Hideki Matsui	7,000,000	Mike Stanton	3,000,000
Jose Contreras	8,500,000	David Weathers	3,933,333
Jorge Posada	9,000,000	Ricky Gutierrez	4,166,667
Javier Vazquez	9,000,000	Mike Cameron	4,333,333
Mariano Rivera	10,890,000	Steve Trachsel	5,000,000
Bernie Williams	12,357,143	Kazuo Matsui	5,033,333
Jason Giambi	12,428,571	Cliff Floyd	6,500,000
Gary Sheffield	13,000,000	Al Leiter	10,295,600
Kevin Brown	15,714,286	Tom Glavine	10,765,608
Mike Mussina	16,000,000	Mike Piazza	16,071,429
Derek Jeter	18,600,000	Mo Vaughn	17,166,667
Alex Rodriguez	22,000,000		

Source: USATODAY.com

(15a) $\mu_y = \$6,351,516$; $\mu_m = \$3,452,177$

(a) Compute the population mean salary for each team.
(b) Compute the median salary for each team. $M_y = \$3,100,000$; $M_m = \$900,000$
(c) Given the results of (a) and (b), decide whether the distributions are symmetric, skewed right, or skewed left. Skewed right
(d) Compute the population standard deviation salary for each team. Which team has more dispersion in its salaries? $\sigma_y = \$6,242,767.0$; $\sigma_m = \$4,643,606.1$
(e) Compute the five-number summary for each team.
(f) On the same graph, draw boxplots for the two teams. Annotate the graph with some general remarks comparing the team salaries.
(g) Describe the shape of the distribution of each team, as illustrated by the boxplots. Does this confirm the result obtained in (c)? Skewed right
(h) Which measure of central tendency is the better measure of central tendency? Why? Median

16. **Bearing Failures** An engineer is studying bearing failures for two different materials in aircraft gas turbine engines. The following data are failure times (in millions of cycles) for samples of the two material types.

| Material A | 3.17 | 4.31 | 4.52 | 4.66 | 5.69 | 5.88 | 6.91 | 8.01 | 8.97 | 11.92 |
| Material B | 5.78 | 6.71 | 6.84 | 7.23 | 8.20 | 9.65 | 13.44 | 14.71 | 16.39 | 24.37 |

(16a) $\bar{x}_A$ = 6.40 million cycles; $\bar{x}_B$ = 11.33 million cycles

(16b) M_A = 5.79 million cycles; M_B = 8.92 million cycles

(a) Compute the sample mean of the failure time for each material.
(b) Compute the median failure time for each material.
(c) Given the results of parts (a) and (b), decide whether the distributions are symmetric, skewed right, or skewed left. Skewed right
(d) Compute the sample standard deviation of the failure time for each material. Which material has its failure times more dispersed? s_A = 2.63 million cycles; s_B = 5.90 million cycles
(e) Compute the five-number summary for each material.
(f) On the same graph, draw boxplots for the two materials. Annotate the graph with some general remarks comparing the failure times.
(g) Describe the shape of the distribution of each material, as illustrated by the boxplots. Does this confirm the result obtained in part (c)? Skewed right

17. **NASCAR Earnings** The following data represent the total earnings (in dollars) of drivers in the 2004 Nextel Cup Series.

1.	$51,505	23.	227,779	45.	1,095,040	67.	3,695,070
2.	53,465	24.	236,315	46.	1,133,620	68.	3,717,100
3.	53,765	25.	251,813	47.	1,217,520	69.	3,745,240
4.	56,565	26.	252,440	48.	1,259,210	70.	3,872,410
5.	57,450	27.	284,405	49.	1,275,530	71.	3,892,570
6.	57,590	28.	293,704	50.	1,333,520	72.	3,948,500
7.	58,925	29.	303,159	51.	1,349,620	73.	4,025,550
8.	65,175	30.	330,385	52.	1,410,570	74.	4,117,750
9.	70,550	31.	338,332	53.	1,461,640	75.	4,200,330
10.	101,260	32.	341,878	54.	1,985,120	76.	4,245,690
11.	107,090	33.	342,337	55.	2,337,420	77.	4,447,300
12.	111,250	34.	364,460	56.	2,471,940	78.	4,539,330
13.	116,150	35.	366,155	57.	2,666,590	79.	4,570,540
14.	116,359	36.	371,479	58.	2,780,130	80.	4,739,010
15.	116,369	37.	394,489	59.	2,929,400	81.	4,759,020
16.	124,312	38.	399,093	60.	3,044,900	82.	5,152,670
17.	139,614	39.	403,674	61.	3,250,320	83.	5,158,360
18.	144,040	40.	426,994	62.	3,443,350	84.	5,692,620
19.	154,100	41.	567,900	63.	3,483,440	85.	6,221,710
20.	160,261	42.	624,850	64.	3,583,440	86.	6,223,890
21.	171,475	43.	752,386	65.	3,675,880	87.	6,437,660
22.	186,610	44.	945,549	66.	3,676,310	88.	7,201,380

(a) Find and interpret the 40th percentile. $368,817
(b) Find and interpret the 95th percentile. $5,957,165
(c) Find and interpret the 10th percentile. $67,862.50
(d) What is the percentile rank of $4,117,750? 83rd percentile
(e) What is the percentile rank of $116,359? 15th percentile

18. NASCAR Earnings Use the data in Problem 17 to answer the following: (18d) 56th percentile
(a) Find and interpret the 30th percentile. $268,422.50
(b) Find and interpret the 90th percentile. $4,749,015
(c) Find and interpret the 5th percentile. $57,007.50
(d) What is the percentile rank of $1,333,520?
(e) What is the percentile rank of $139,614? 18th percentile

19. Weights of Males versus Females According to the National Center for Health Statistics, the mean weight of a 20- to 29-year-old female is 156.5 pounds, with a standard deviation of 51.2 pounds. The mean weight of a 20- to 29-year-old male is 183.4 pounds, with a standard deviation of 40.0 pounds. Who is relatively heavier: a 20- to 29-year-old female who weights 160 pounds or a 20- to 29-year-old male who weighs 185 pounds? Female

20. Crime Rate Answer the accompanying questions regarding the boxplot, which illustrates crime-rate data per 100,000 population for the 50 states and the District of Columbia in 2002. (*Source*: www.infoplease.com)

Crime Rate

(a) Approximately what is the median crime rate in the United States? 4000
(b) Approximately what is the 25th percentile crime rate in the United States? 3100
(c) Are there any outliers? If so, identify their value(s).
(d) What is the lowest crime rate? 2200
(20c) 8000

THE CHAPTER 3 CASE STUDY IS LOCATED ON THE CD THAT ACCOMPANIES THIS TEXT.

Describing the Relation between Two Variables

CHAPTER 4

Outline

4.1 Scatter Diagrams and Correlation

4.2 Least-Squares Regression

4.3 The Coefficient of Determination

■ Chapter Review

■ Case Study: Thomas Malthus, Population, and
Subsistence (On CD)

DECISIONS

You are still in the market to buy a car. Because cars lose value over time at different rates, you want to look at the depreciation rates of cars you are considering. After all, the higher the depreciation rate, the more value the car loses each year. See the Decision Project on page 204.

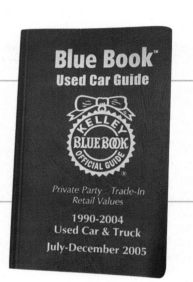

●●● **Putting It All Together**

In Chapters 2 and 3 we examined data in which a single variable was measured for each individual in the study (**univariate data**), such as the three-year rate of return (the variable) for various mutual funds (the individuals). We obtained descriptive measures for the variable that were both graphical and numerical.

However, much research is designed to describe the relation that may exist between two variables. For example, a researcher may be interested in the relationship between the club-head speed of a golf club and the distance the golf ball travels. Here, each swing represents an individual, and the two variables are club-head speed and distance. This type of data is referred to as *bivariate data*. **Bivariate data** are data in which two variables are measured on an individual. To describe the relation between the two quantitative variables, we first graphically represent the data and then obtain some numerical descriptions of the data, just as we did when analyzing univariate data.

4.1 Scatter Diagrams and Correlation

Preparing for This Section   Before getting started, review the following:

- Mean (Section 3.1, pp. 121–124)
- Standard deviation (Section 3.2, pp. 143–144)
- *z*-Scores (Section 3.4, pp. 165–166)

Objectives

1. **Draw and interpret scatter diagrams**
2. **Understand the properties of the linear correlation coefficient**
3. **Compute and interpret the linear correlation coefficient**
4. **Determine whether there is a linear relation between two variables**

Note to Instructor
Remind students that we looked at relations between two qualitative variables when we drew side-by-side bar graphs (Section 2.1). Problems 43 and 44 in Section 2.2 discussed how we can graphically represent relations between a quantitative and qualitative variable. We now look at a graph of two quantitative variables. Also note that Objective 4 is optional and can be skipped without loss of continuity.

Before we can graphically represent bivariate data, a fundamental question must be asked. Am I interested in using the value of one variable to predict the value of the other variable? For example, it seems reasonable to think that as the speed at which a golf club is swung increases, the distance the golf ball travels also increases. Therefore, we might use club-head speed to predict distance. We call distance the *response* (or *dependent*) *variable* and club-head speed the *explanatory* (or *predictor* or *independent*) *variable*.

Definition

> The **response variable** is the variable whose value can be explained by the value of the **explanatory** or **predictor variable**.

In Other Words
We use the term *explanatory variable* because it helps to explain variability in the response variable.

It is important to recognize that, if the data used in the study are observational, we cannot conclude that there is a causal relationship between the two variables. We cannot say that changes in the level of the explanatory variable *cause* changes in the level of the response variable. In fact, it may be that the two are related through some *lurking variable*.

CAUTION
If bivariate data are observational, then we cannot conclude that any relation between the explanatory and response variables is due to cause and effect.

Recall that a **lurking variable** is a variable that may affect the response variable but is excluded from the analysis. For example, air-conditioning bills can be used to predict lemonade sales. As air-conditioning bills rise, the sales of lemonade rise. This relation does not mean that high air-conditioning bills cause high lemonade sales, because both high air-conditioning bills and high lemonade sales are associated with high summer temperatures. Therefore, air temperature is a lurking variable.

1 Draw and Interpret Scatter Diagrams

The first step in identifying the type of relation that might exist between two variables is to draw a picture. Bivariate data can be represented graphically through a *scatter diagram*.

Definition

> A **scatter diagram** is a graph that shows the relationship between two quantitative variables measured on the same individual. Each individual in the data set is represented by a point in the scatter diagram. The explanatory variable is plotted on the horizontal axis and the response variable is plotted on the vertical axis. Do not connect the points when drawing a scatter diagram.

EXAMPLE 1   **Drawing a Scatter Diagram**

Problem: A golf pro wanted to learn the relation between the club-head speed of a golf club (measured in miles per hour) and the distance (in yards) that the ball will travel. He realized that there are other variables besides club-head speed

Table 1	
Club-Head Speed (mph)	**Distance (yards)**
100	257
102	264
103	274
101	266
105	277
100	263
99	258
105	275

Source: Paul Stephenson, student at Joliet Junior College

that determine the distance a ball will travel (such as club type, ball type, golfer, and weather conditions). To eliminate the variability due to these variables, the pro used a single model of club and ball. One golfer was chosen to swing the club on a clear, 70-degree day with no wind. The pro recorded the club-head speed and measured the distance that the ball traveled and collected the data in Table 1. Draw a scatter diagram of the data.

Approach: Because the pro wants to use club-head speed to predict the distance the ball travels, club-head speed is the explanatory variable (horizontal axis) and distance is the response variable (vertical axis). We plot the ordered pairs (100, 257), (102, 264), and so on, in a rectangular coordinate system.

Solution: The scatter diagram is shown in Figure 1.

Figure 1

It would appear from the graph that as club-head speed increases, the distance that the ball travels increases as well.

Note to Instructor
You might want to discuss how a qualitative variable can be included in a scatter diagram. See Problems 31 and 32.

It is not always clear which variable should be considered the response variable and which should be considered the explanatory variable. For example, does high school GPA predict a student's SAT score or can the SAT score be used to predict GPA? The researcher must determine which variable plays the role of explanatory variable based on the questions he or she wants answered. For example, if the researcher is interested in predicting SAT scores on the basis of high school GPA, then high school GPA will play the role of explanatory variable.

Now Work Problems 23(a) and 23(b).

Scatter diagrams show the type of relation that exists between two variables. Our goal in interpreting scatter diagrams will be to distinguish scatter diagrams that imply a linear relation from those that imply a nonlinear relation or those that imply no relation. Figure 2 displays various scatter diagrams and the type of relation implied.

As we compare Figure 2(a) with Figure 2(b), we notice a distinct difference. In Figure 2(a), the data follow a linear pattern that slants upward to the right; the data in Figure 2(b) follow a linear pattern that slants downward to the right. Figures 2(c) and 2(d) show scatter diagrams of nonlinear relations. Figure 2(e) shows a scatter diagram in which there is no relation between the explanatory and response variables.

Figure 2

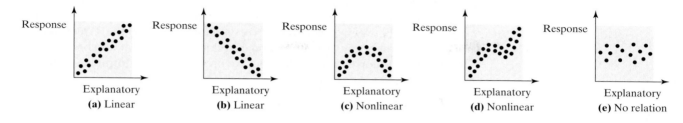

(a) Linear (b) Linear (c) Nonlinear (d) Nonlinear (e) No relation

Definitions

![pen icon] *In Other Words*

If two variables that are linearly related are positively associated, then as one goes up the other also tends to go up. If two variables that are linearly related are negatively associated, then as one goes up the other tends to go down.

Two variables that are linearly related are said to be **positively associated** when above-average values of one variable are associated with above-average values of the other variable. That is, two variables are positively associated if, whenever the value of one variable increases, the value of the other variable also increases.

Two variables that are linearly related are said to be **negatively associated** when above-average values of one variable are associated with below-average values of the other variable. That is, two variables are negatively associated if, whenever the value of one variable increases, the value of the other variable decreases.

So the scatter diagram from Figure 1 implies that club-head speed is positively associated with the distance a golf ball travels.

Now Work Problem 11.

② Understand the Properties of the Linear Correlation Coefficient

It is dangerous to use only a scatter diagram to decide whether two variables follow a linear relation. Suppose we redraw the scatter diagram in Figure 1 using a different scale as shown in Figure 3.

Figure 3

! **CAUTION**
The horizontal or vertical scale of a scatter diagram should be set so that the scatter diagram does not mislead a reader.

From Figure 3, we might conclude that club-head speed and distance are not related. The moral of the story is this: Just as we can manipulate the scale of graphs of univariate data, we can also manipulate the scale of the graphs of bivariate data, thereby encouraging incorrect conclusions. Therefore, numerical summaries of bivariate data should be used in conjunction with graphs to determine the type of relation, if any, that exists between two variables.

Definition

The **linear correlation coefficient** or **Pearson product moment correlation coefficient** is a measure of the strength of linear relation between two quantitative variables. We use the Greek letter ρ (rho) to represent the population correlation coefficient and r to represent the sample correlation coefficient. We present only the formula for the sample correlation coefficient.

Sample Correlation Coefficient*

$$r = \frac{\sum \left(\dfrac{x_i - \overline{x}}{s_x} \right) \left(\dfrac{y_i - \overline{y}}{s_y} \right)}{n - 1} \qquad (1)$$

Historical Note

Karl Pearson was born March 27, 1857. Pearson's proficiency as a statistician was recognized early in his life. It is said that his mother told him not to suck his thumb because otherwise his thumb would wither away. Pearson analyzed the size of each thumb and said to himself, "They look alike to me. I can't see that the thumb I suck is any smaller than the other. I wonder if she could be lying to me."

Karl Pearson graduated from Cambridge University in 1879. From 1893 to 1911, he wrote 18 papers on genetics and heredity. Through this work, he developed ideas regarding correlation and the chi-square test. (See Chapter 12.) In addition, Pearson came up with the term *standard deviation*.

Pearson and Ronald Fisher didn't get along. The dispute between the two was bad enough to have Fisher turn down the post of chief statistician at the Galton Laboratory in 1919 on the grounds that it would have meant working under Pearson. Pearson died on April 27, 1936.

where $\overline{x}$ is the sample mean of the explanatory variable

s_x is the sample standard deviation of the explanatory variable

$\overline{y}$ is the sample mean of the response variable

s_y is the sample standard deviation of the response variable

n is the number of individuals in the sample

The Pearson linear correlation coefficient is named in honor of Karl Pearson (1857–1936).

Properties of the Linear Correlation Coefficient

1. The linear correlation coefficient is always between -1 and 1, inclusive. That is, $-1 \le r \le 1$.
2. If $r = +1$, there is a perfect positive linear relation between the two variables. See Figure 4(a).
3. If $r = -1$, there is a perfect negative linear relation between the two variables. See Figure 4(d).
4. The closer r is to $+1$, the stronger is the evidence of positive association between the two variables. See Figures 4(b) and 4(c).
5. The closer r is to -1, the stronger is the evidence of negative association between the two variables. See Figures 4(e) and 4(f).
6. If r is close to 0, there is little or no evidence of a *linear* relation between the two variables. Because the linear correlation coefficient is a measure of the strength of the linear relation, **r close to 0 does not imply no relation, just no linear relation.** See Figures 4(g) and 4(h).
7. The linear correlation coefficient is a unitless measure of association. So the unit of measure for x and y plays no role in the interpretation of r.

CAUTION

A linear correlation coefficient close to 0 does not imply that there is no relation, just no linear relation. For example, although the scatter diagram drawn in Figure 4(h) indicates that the two variables are related, the linear correlation coefficient of these data is close to 0.

In looking carefully at Formula (1), we should notice that the numerator of the formula is the product of z-scores for the explanatory (x) and response (y) variables. A positive linear correlation coefficient means that the sum of the product of the z-scores for x and y must be positive. Under what circumstances

*An equivalent formula for the linear correlation coefficient is

$$r = \frac{\sum x_i y_i - \dfrac{\sum x_i \sum y_i}{n}}{\sqrt{\left(\sum x_i^2 - \dfrac{(\sum x_i)^2}{n} \right)} \sqrt{\left(\sum y_i^2 - \dfrac{(\sum y_i)^2}{n} \right)}} = \frac{S_{xy}}{\sqrt{S_{xx}} \sqrt{S_{yy}}}$$

Figure 4

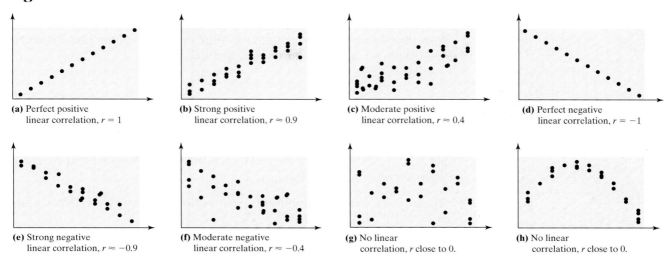

(a) Perfect positive
    linear correlation, $r = 1$

(b) Strong positive
    linear correlation, $r \approx 0.9$

(c) Moderate positive
    linear correlation, $r \approx 0.4$

(d) Perfect negative
    linear correlation, $r = -1$

(e) Strong negative
    linear correlation, $r \approx -0.9$

(f) Moderate negative
    linear correlation, $r \approx -0.4$

(g) No linear
    correlation, r close to 0.

(h) No linear
    correlation, r close to 0.

does this occur? Figure 5 shows a scatter diagram that implies a positive association between x and y. The vertical dashed line represents the value of $\bar{x}$, and the horizontal dashed line represents the value of $\bar{y}$. These two dashed lines divide our scatter diagram into four quadrants, labeled I, II, III, and IV.

Figure 5

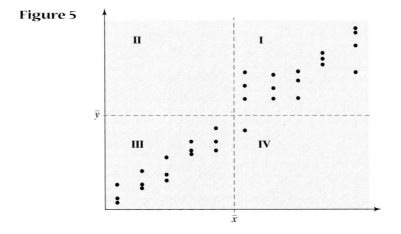

Consider the data in quadrants I and III. If a certain x-value is above its mean, $\bar{x}$, then the corresponding y-value will be above its mean, $\bar{y}$. If a certain x-value is below its mean, $\bar{x}$, then the corresponding y-value will be below its mean, $\bar{y}$. Therefore, for data in quadrant I, we have $\dfrac{x_i - \bar{x}}{s_x}$ positive and $\dfrac{y_i - \bar{y}}{s_y}$ positive, so their product is positive. For data in quadrant III, we have $\dfrac{x_i - \bar{x}}{s_x}$ negative and $\dfrac{y_i - \bar{y}}{s_y}$ negative, so their product is positive. The sum of these products is positive, and therefore we have a positive linear correlation coefficient. A similar argument can be made for negative correlation.

Now suppose the data are equally dispersed in the four quadrants. Then the negative products (resulting from data in quadrants II and IV) will offset the positive products (resulting from data in quadrants I and III). The result is a linear correlation coefficient close to 0.

In Other Words
The correlation coefficient describes the strength and the direction of the linear relationship between two variables.

Now Work Problem 15.

3 ## Compute and Interpret the Linear Correlation Coefficient

Now that we have an understanding of the properties of the linear correlation coefficient, we are ready to compute its value.

EXAMPLE 2 ### Computing and Interpreting the Correlation Coefficient

Problem: In Table 2, columns 1 and 2 represent the club-head speed (in miles per hour) and the distance the ball travels (in yards). Compute and interpret the linear correlation coefficient.

Approach: We treat club-head speed as the explanatory variable, x, and distance as the response variable, y.

Step 1: Compute $\bar{x}$, s_x, $\bar{y}$, and s_y.

Step 2: Determine $\dfrac{x_i - \bar{x}}{s_x}$ and $\dfrac{y_i - \bar{y}}{s_y}$ for each observation.

Step 3: Compute $\left(\dfrac{x_i - \bar{x}}{s_x}\right)\left(\dfrac{y_i - \bar{y}}{s_y}\right)$ for each observation.

Step 4: Determine $\sum\left(\dfrac{x_i - \bar{x}}{s_x}\right)\left(\dfrac{y_i - \bar{y}}{s_y}\right)$ and substitute this value into Formula (1).

Solution

Step 1: We compute $\bar{x}$, s_x, $\bar{y}$, and s_y:

$$\bar{x} = 101.875, \quad s_x = 2.29518, \quad \bar{y} = 266.75, \quad s_y = 7.74135$$

To avoid round-off error when using Formula (1), do not round the statistics.

Step 2: We determine $\dfrac{x_i - \bar{x}}{s_x}$ and $\dfrac{y_i - \bar{y}}{s_y}$ in columns 3 and 4 in Table 2.

Step 3: We multiply the entries in columns 3 and 4 to obtain the entries in column 5.

Table 2				
Club-Head Speed, x_i	**Distance, y_i**	$\dfrac{x_i - \bar{x}}{s_x}$	$\dfrac{y_i - \bar{y}}{s_y}$	$\left(\dfrac{x_i - \bar{x}}{s_x}\right)\left(\dfrac{y_i - \bar{y}}{s_y}\right)$
100	257	-0.816929	-1.259470	1.028898
102	264	0.054462	-0.355235	-0.0193347
103	274	0.490158	0.936529	0.459047
101	266	-0.381234	-0.096882	0.036935
105	277	1.361549	1.324058	1.802770
100	263	-0.816929	-0.484412	0.395726
99	258	-1.252625	-1.130294	1.415835
105	275	1.361549	1.065706	1.451011

$$\sum\left(\frac{x_i - \bar{x}}{s_x}\right)\left(\frac{y_i - \bar{y}}{s_y}\right)$$
$$= 6.570887$$

Step 4: We add the entries in column 5 to obtain

$$\sum \left(\frac{x_i - \overline{x}}{s_x} \right) \left(\frac{y_i - \overline{y}}{s_y} \right) = 6.570887$$

Substitute this value into Formula (1) to obtain the correlation coefficient.

$$r = \frac{\sum \left(\dfrac{x_i - \overline{x}}{s_x} \right) \left(\dfrac{y_i - \overline{y}}{s_y} \right)}{n - 1} = \frac{6.570887}{8 - 1} = 0.9387$$

The linear correlation between club-head speed and distance is 0.9387, indicating a strong positive association between the two variables. The higher the club-head speed, the farther the golf ball tends to travel.

Notice in Example 2 that we carry many decimal places in the computation of the correlation coefficient to avoid rounding error. Also, compare the signs of the entries in columns 3 and 4. Notice that negative values in column 3 correspond to negative values in column 4 and that positive values in column 3 correspond to positive values in column 4 (except for the second trial of the experiment). This means that above-average values of x are associated with above-average values of y, and below-average values of x are associated with below-average values of y. This is why the linear correlation coefficient is positive.

EXAMPLE 3   Drawing a Scatter Diagram and Determining the Linear Correlation Coefficient Using Technology

Problem: Use a statistical spreadsheet or a graphing calculator with advanced statistical features to draw a scatter diagram of the data in Table 1. Then determine the linear correlation between club-head speed and distance.

Approach: We will use Excel to draw the scatter diagram and obtain the linear correlation coefficient. The steps for drawing scatter diagrams and obtaining the linear correlation coefficient using MINITAB, Excel, or the TI-83 and TI-84 Plus graphing calculators are given in the Technology Step by Step on page 194.

Result: Figure 6(a) shows the scatter diagram and Figure 6(b) shows the linear correlation coefficient obtained from Excel. Notice that Excel provides a **correlation matrix**, which means that for every pair of columns in the spreadsheet it will compute and display the correlation in the bottom triangle of the matrix.

Figure 6

	Driving Distance	Club-Head Speed
Driving Distance	1	
Club-Head Speed	0.938695838	1

(a)   (b)

Now Work Problem 23(c).

Correlation versus Causation

In Chapter 1 we stated that there are two types of studies: observational and experimental. The data given in Examples 1 through 3 are the result of an experiment. Therefore, we can claim that a higher club-head speed causes the golf ball to travel a longer distance. However, if data result from an observational study, we cannot claim causation. Consider the scatter diagram shown in Figure 7, which shows the relation between the birthrate (births per 1000 women) of teenagers and the homicide rate (homicides per 100,000 inhabitants) for the years 1993 to 2000.

> ! **CAUTION!**
> A linear correlation coefficient that implies a strong positive or negative association that is computed using observational data does not imply causation.

Figure 7

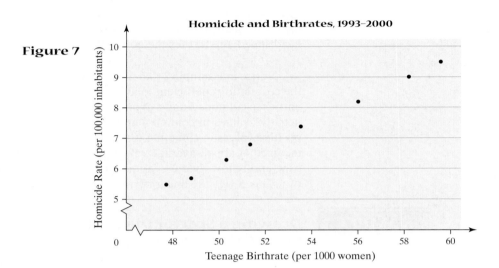

The linear correlation coefficient between these two variables is 0.9987. Does this mean that higher birthrates among teenagers cause a higher homicide rate? Certainly not!

In Chapter 1, we introduced lurking variables. A lurking variable is one that has not been considered in your analysis but is related to both variables in the study. Can you think of any variables that might be related to teen birthrates and homicide rates? Perhaps there is an economic variable, such as poverty rate, proportion of homes with a single parent, or high school dropout rate, that is related to both teenage birthrate and homicide rate.

Note to Instructor

The purpose of this activity, in addition to reviewing correlation, is to have students begin formulating claims about data and then collecting data to check the claim. Time limit: 25 to 30 minutes depending on discussions.

In-Class Activity: Correlation

Randomly select six students from the class and have them determine their at-rest pulse and then discuss the following:

1. When determining the at-rest pulse rate, would it be better to count beats for 30 seconds and multiply by 2 or count beats for 1 full minute? Explain. What are some other ways to find the at-rest pulse rate? Do any of these methods have an advantage?

2. What effect will physical activity have on pulse rate?

3. Do you think the at-rest pulse rate will have any effect on the pulse rate after physical activity? If so, how? If not, why not?

Have the same six students jog in place for 3 minutes and then immediately determine their pulse rate using the same technique as for the at-rest pulse rate.

4. Draw a scatter diagram for the pulse data using the at-rest data as the explanatory variable.

5. Comment on the relationship, if any, between the two variables. Is this consistent with your expectations?

6. Based on the graph, estimate the linear correlation coefficient for the data. Then compute the correlation coefficient using a graphing utility and compare to your estimate.

4 Determine Whether There Is a Linear Relation between Two Variables

Note to Instructor
The material in this objective is optional and can be skipped without loss of continuity.

Other Words
We use two vertical bars to denote absolute value, as in $|5|$ or $|-4|$. Recall, $|5| = 5$, $|-4| = 4$, and $|0| = 0$.

A question you may be asking yourself is, "How do I know the correlation between two variables is strong enough for me to conclude that there is a linear relation between the variables?" While rigorous tests exist that can answer this question, for now we will be content with a simple comparison test that is based on the more rigorous approach.

To test whether the correlation between the explanatory and response variables is strong enough, determine the absolute value of the correlation coefficient. If the absolute value of the correlation coefficient is greater than the critical value in Table VIII in Appendix A for the given sample size, then we say there is a linear relation between the two variables. Otherwise, there is no linear relation.

EXAMPLE 4   **Is There a Linear Relation?**

Problem: Using the data from Example 2, determine whether there is a linear relation between club-head speed and distance. Comment on the type of relation that appears to exist between club-head speed and distance.

Note to Instructor
We are conducting this test using a 0.05 level of significance.

Approach: We compare the absolute value of the linear correlation coefficient to the critical value in Table VIII with $n = 8$. If the absolute value of the linear correlation coefficient is greater than the critical value, we conclude that there is a linear relation between club-head speed and distance.

Solution: The linear correlation coefficient between club-head speed and distance was found to be 0.9387 in Example 2. The absolute value of 0.9387 is 0.9387. We find the critical value for correlation in Table VIII with $n = 8$ to be 0.707. Since 0.9387 is greater than 0.707, we conclude there is a positive linear relation between club-head speed and distance.

Now Work Problem 23(d).

4.1 ASSESS YOUR UNDERSTANDING

Concepts and Vocabulary

1. Describe the difference between univariate and bivariate data.

2. Explain what is meant by a lurking variable. Provide an example.

3. What does it mean to say that two variables are positively associated?

4. What does it mean to say that the linear correlation coefficient between two variables equals 1? What would the scatter diagram look like?

5. What does it mean if $r = 0$?

6. Is the linear correlation coefficient a resistant measure? Support your answer.

7. Explain what is wrong with the following statement: "We have concluded that there is a high correlation between the gender of drivers and rates of automobile accidents."

8. Write a statement that explains the concept of correlation. Include a discussion of the role that $x_i - \bar{x}$ and $y_i - \bar{y}$ play in the computation.

9. Explain what is wrong with the following statement: "A recent study showed that the correlation between the number of acres on a farm and the amount of corn produced was 0.93 bushel."

10. Explain the difference between correlation and causation. When does a linear correlation coefficient that implies a strong positive correlation also imply causation?

Note to Instructor
If you covered the optional Objective 4, then ask students to use the critical values when a problem asks, "Comment on the type of relation that appears to exist between x and y," as in Problems 19(c), 20(c), 21(c), 22(c), 23(d), and so on.

Skill Building

In Problems 11–14, determine whether the scatter diagram indicates that a linear relation may exist between the two variables. If the relation is linear, determine whether it indicates a positive or negative association between the variables.

11. Nonlinear NW

12. Linear, negative

13. Linear, positive

14. Nonlinear

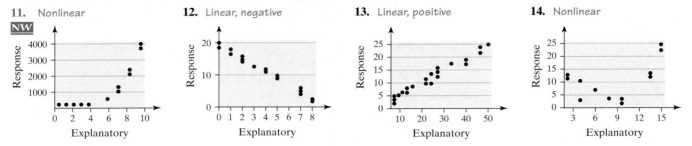

15. Match the linear correlation coefficient to the scatter diagram. The scales on the *x*- and *y*-axes are the same for each scatter NW diagram.

(a) *r* = 0.787 III   (b) *r* = 0.523 IV   (c) *r* = 0.810 II   (d) *r* = 0.946 I

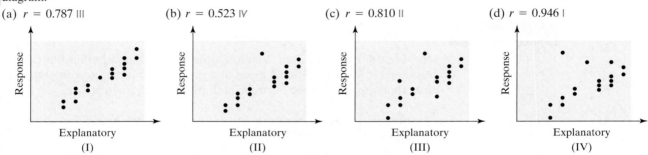

(I)   (II)   (III)   (IV)

16. Match the linear correlation coefficient to the scatter diagram. The scales on the *x*- and *y*-axes are the same for each scatter diagram.

(a) *r* = −0.969 IV   (b) *r* = −0.049 III   (c) *r* = −1 I   (d) *r* = −0.992 II

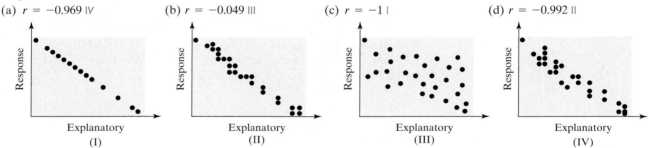

(I)   (II)   (III)   (IV)

17. Does Education Pay? The following scatter diagram drawn in MINITAB shows the relation between the percentage of the population of a state that has at least a bachelor's degree and the median income (in dollars) of the state for 2003.

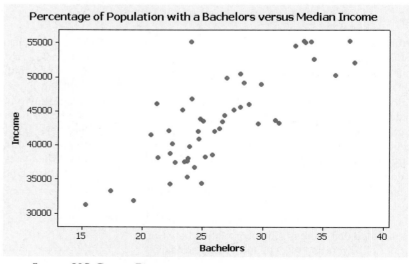

Source: U.S. Census Bureau

(a) Describe the relation that appears to exist between level of education and median income. *Linear, positive association*

(b) One observation appears to stick out from the rest. Which one? This particular observation is for the state of Alaska. Can you think of any reasons why the state of Alaska might have a high median income, given the proportion of the population that has at least a bachelor's degree?

18. **Relation between Income and Birthrate?** The following scatter diagram drawn in Excel shows the relation between median income (in dollars) in a state and birthrate (births per 1000 women 15 to 44 years of age).

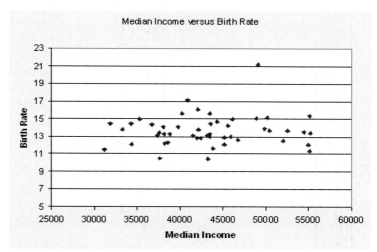

Source: U.S. Census Bureau

No (a) Does there appear to be any relation between median income and birthrate?

(b) One observation sticks out from the rest. Which one? This particular observation is for the state of Utah. Are there any explanations for this result?

In Problems 19–22, (a) draw a scatter diagram of the data, (b) by hand, compute the correlation coefficient, and (c) comment on the type of relation that appears to exist between x and y.

19. *r = 0.9698*

x	2	4	8	8	9
y	1	2	4	5	6

20. *r = −0.8822*

x	2	3	5	6	6
y	10	9	8	3	1

21. *−0.8553*

x	3	5	8	9	12	12
y	18	20	16	10	12	8

22. *r = 0.9506*

x	0	1	1	2	4	7
y	3	5	4	6	8	9

Applying the Concepts

23. **Height versus Head Circumference** A pediatrician wants **NW** to determine the relation that may exist between a child's height and head circumference. She randomly selects 11 three-year-old children from her practice, measures their height and head circumference, and obtains the data shown in the table.

(a) If the pediatrician wants to use height to predict head circumference, determine which variable is the explanatory variable and which is the response variable.

(b) Draw a scatter diagram. *(23a)Explanatory: height*

(c) Compute the linear correlation coefficient between the height and head circumference of a child. *0.911*

(d) Comment on the type of relation that appears to exist between the height and head circumference of a child

on the basis of the scatter diagram and linear correlation coefficient. *Positive association*

Height (inches)	Head Circumference (inches)	Height (inches)	Head Circumference (inches)
27.75	17.5	26.5	17.3
24.5	17.1	27	17.5
25.5	17.1	26.75	17.3
26	17.3	26.75	17.5
25	16.9	27.5	17.5
27.75	17.6		

Source: Denise Slucki, student at Joliet Junior College

24. Gestation Period versus Life Expectancy A researcher wants to know if the gestation period of an animal can be used to predict life expectancy. She collects the following data:

Animal	Gestation (or Incubation) Period (days)	Life Expectancy (years)
Cat	63	11
Chicken	22	7.5
Dog	63	11
Duck	28	10
Goat	151	12
Lion	108	10
Parakeet	18	8
Pig	115	10
Rabbit	31	7
Squirrel	44	9

Source: Time Almanac 2000

(a) Suppose the researcher wants to use the gestation period of an animal to predict its life expectancy. Determine which variable is the explanatory variable and which is the response variable. Explanatory: gestation period
(b) Draw a scatter diagram.
(c) Compute the linear correlation coefficient between gestation period and life expectancy. 0.7257
(d) Comment on the type of relation that appears to exist between gestation period and life expectancy based on the scatter diagram and linear correlation coefficient.
(e) Remove the goat from the data set, and recompute the linear correlation coefficient between the gestation period and life expectancy. What effect did the removal of the data value have on the linear correlation coefficient? Provide a justification for this result. 0.592

26. Bone Length Research performed at NASA and led by Emily R. Morey-Holton measured the lengths of the right humerus and right tibia in 11 rats that were sent to space on Spacelab Life Sciences 2. The following data were collected.

Right Humerus (mm)	Right Tibia (mm)	Right Humerus (mm)	Right Tibia (mm)
24.8	36.05	25.9	37.38
24.59	35.57	26.11	37.96
24.59	35.57	26.63	37.46
24.29	34.58	26.31	37.75
23.81	34.2	26.84	38.5
24.87	34.73		

Source: NASA Life Sciences Data Archive

25. Weight of a Car versus Miles per Gallon An engineer wanted to determine how the weight of a car affects gas mileage. The following data represent the weight of various domestic cars and their gas mileage in the city for the 2005 model year.

Car	Weight (pounds)	Miles per Gallon
Buick LeSabre	3565	20
Cadillac DeVille	3985	18
Chevrolet Corvette	3180	19
Chevrolet Monte Carlo	3340	21
Chrysler PT Cruiser	3100	21
Chrysler Sebring Sedan	3175	22
Dodge Neon	2580	27
Dodge Stratus Sedan	3175	22
Ford Focus	2655	26
Ford Mustang	3300	20
Lincoln LS	3680	20
Mercury Sable	3310	19
Pontiac Bonneville	3590	20
Pontiac Grand Am	3475	20
Pontiac Sunfire	2770	24
Saturn Ion	2690	26

Source: www.roadandtrack.com

(25b) Explanatory: weight

(a) Determine which variable is the likely explanatory variable and which is the likely response variable.
(b) Draw a scatter diagram of the data.
(c) Compute the linear correlation coefficient between the weight of a car and its miles per gallon in the city. −0.892
(d) Comment on the type of relation that appears to exist between the weight of a car and its miles per gallon in the city based on the scatter diagram and the linear correlation coefficient.

(a) Draw a scatter diagram, treating the length of the right humerus as the explanatory variable and the length of the right tibia as the response variable.
(b) Compute the linear correlation coefficient between the length of the right humerus and the length of the right tibia. 0.9513
(c) Comment on the type of relation that appears to exist between the length of the right humerus and the length of the right tibia based on the scatter diagram and the linear correlation coefficient.
(d) Convert the data to inches (1 mm = 0.03937 inch), and recompute the linear correlation coefficient. What effect did the conversion from millimeters to inches have on the linear correlation coefficient? None

27. Attending Class The following data represent the number of days absent and the final grade for a sample of college students in a general education course at a large midwestern state university.

Number of Absences	Final Grade
0	89.2
1	86.4
2	83.5
3	81.1
4	78.2
5	73.9
6	64.3
7	71.8
8	65.5
9	66.2

Source: College Teaching, Winter 2005, Vol. 53, Issue 1

(a) The researcher wants to use the number of days absent to predict the final grade. Determine which variable is the explanatory variable and which is the response variable. Explanatory: absences
(b) Draw a scatter diagram of the data.
(c) Compute the linear correlation coefficient between the number of days absent and the final grade. −0.947
(d) Comment on the type of relation that appears to exist between the number of days absent and the final grade.
(e) Will going to class every day guarantee a passing grade? What other factors might need to be taken into account?

28. Antibiotics A study on antibiotic use among children in Manitoba, Canada, gave the following data for the number of prescriptions per 1000 children x years after 1995.

Year, x	0	1	2	3	4	5	6
Prescriptions (per 1000 children)	1201	1070	944	964	909	949	864

Source: Canadian Medical Association Journal, Vol. 171, Issue 2

(a) Draw a scatter diagram of the data, treating year as the explanatory variable. What type of relation, if any, appears to exist between year and antibiotic prescriptions among children?
(b) Compute the linear correlation coefficient between −0.873 year and antibiotic prescriptions among children.
(c) Comment on the type of relation that appears to exist between year and antibiotic prescriptions among children on the basis of the scatter diagram and the linear correlation coefficient.

29. Age versus HDL Cholesterol A doctor wanted to determine whether there was a relation between a male's age and his HDL (so-called good) cholesterol. He randomly selected 17 of his patients and determined their HDL cholesterol. He obtained the following data.

Age	HDL Cholesterol	Age	HDL Cholesterol
38	57	38	44
42	54	66	62
46	34	30	53
32	56	51	36
55	35	27	45
52	40	52	38
61	42	49	55
61	38	39	28
26	47		

Source: Data based on information obtained from the National Center for Health Statistics

(a) Draw a scatter diagram of the data, treating age as the explanatory variable. What type of relation, if any, appears to exist between age and HDL cholesterol?
(b) Compute the linear correlation coefficient between age and HDL cholesterol. −0.164
(c) Comment on the type of relation that appears to exist between age and HDL cholesterol on the basis of the scatter diagram and the linear correlation coefficient.

30. Intensity of a Lightbulb Cathy is conducting an experiment to measure the relation between a light bulb's intensity and the distance from the light source. She measures a 100-watt lightbulb's intensity 1 meter from the bulb and at 0.1-meter intervals up to 2 meters from the bulb and obtains the following data.

Distance (meters)	Intensity	Distance (meters)	Intensity
1.0	0.29645	1.6	0.11450
1.1	0.25215	1.7	0.10243
1.2	0.20547	1.8	0.09231
1.3	0.17462	1.9	0.08321
1.4	0.15342	2.0	0.07342
1.5	0.13521		

(a) Draw a scatter diagram of the data, treating distance as the explanatory variable.
(b) Do you think that it is appropriate to compute the linear correlation coefficient between distance and intensity? Why?

31. Does Size Matter? Researchers wondered whether the size of a person's brain was related to the individual's mental capacity. They selected a sample of right-handed introductory psychology students who had SAT scores higher than 1350. The subjects took the Wechsler (1981)

Gender	MRI Count	IQ	Gender	MRI Count	IQ
Female	816,932	133	Male	949,395	140
Female	951,545	137	Male	1,001,121	140
Female	991,305	138	Male	1,038,437	139
Female	833,868	132	Male	965,353	133
Female	856,472	140	Male	955,466	133
Female	852,244	132	Male	1,079,549	141
Female	790,619	135	Male	924,059	135
Female	866,662	130	Male	955,003	139
Female	857,782	133	Male	935,494	141
Female	948,066	133	Male	949,589	144

Source: Willerman, L., Schultz, R., Rutledge, J. N., and Bigler, E. (1991). "In Vivo Brain Size and Intelligence," *Intelligence, 15*, 223–228

Adult Intelligence Scale-Revised exam to obtain their IQ scores. Magnetic resonance imaging (MRI) scans were performed at the same facility for the subjects. The scans consisted of 18 horizontal magnetic resonance images. The computer counted all pixels with nonzero gray scale in each of the 18 images, and the total count served as an index for brain size.

(a) Draw a scatter diagram, treating MRI count as the explanatory variable and IQ as the response variable. Comment on what you see.

(b) Compute the linear correlation coefficient between MRI count and IQ. Do you think that MRI count and IQ are linearly related?

(c) A lurking variable in the analysis is gender. Draw a scatter diagram, treating MRI count as the explanatory variable and IQ as the response variable, but use a different plotting symbol for each gender. For example, use a circle for males and a triangle for females. What do you notice?

(d) Compute the linear correlation coefficient between MRI count and IQ for females. Compute the linear correlation coefficient between MRI count and IQ for males. Do you believe that MRI count and IQ are linearly related? What is the moral?

32. Male versus Female Drivers The following data represent the number of licensed drivers in various age groups and the number of accidents within the age group by gender.

Age Group	Number of Male Licensed Drivers (000s)	Number of Crashes Involving a Male (000s)	Number of Female Licensed Drivers (000s)	Number of Crashes Involving a Female (000s)
16	816	244	764	178
17	1,198	233	1,115	175
18	1,342	243	1,212	164
19	1,454	229	1,333	145
20–24	7,866	951	7,394	618
25–29	9,356	899	8,946	595
30–34	10,121	875	9,871	571
35–39	10,521	901	10,439	566
40–44	9,776	692	9,752	455
45–49	8,754	667	8,710	390
50–54	6,840	390	6,763	247
55–59	5,341	290	5,258	165
60–64	4,565	218	4,486	133
65–69	4,234	191	4,231	121
70–74	3,604	167	3,749	104
75–79	2,563	118	2,716	77
80–84	1,400	61	1,516	45
≥85	767	34	767	20

Source: National Highway and Traffic Safety Institute

(a) On the same graph, draw a scatter diagram for both males and females. Be sure to use a different plotting symbol for each group. For example, use a square (□) or an M for males and a plus sign (+) or an F for females. Treat number of licensed drivers as the explanatory variable.

(b) Based on the scatter diagrams, do you think that insurance companies are justified in charging different insurance rates for males and females? Why?

(c) Compute the linear correlation coefficient between number of licensed drivers and number of crashes for males. *0.899*

(d) Compute the linear correlation coefficient between number of licensed drivers and number of crashes for females. *0.863*

(e) Which gender has the stronger linear relation between number of licensed drivers and number of crashes. Why? *Males*

33. **Weight of a Car versus Miles per Gallon** Suppose we add the Ford Taurus to the data in Problem 25. A Ford Taurus weighs 3305 pounds and gets 19 miles per gallon.

(a) Redraw the scatter diagram with the Taurus included.

(b) Recompute the linear correlation coefficient with the Taurus included. *−0.879*

(c) Compare the results of parts (a) and (b) with the results of Problem 25. Why are the results here reasonable?

(d) Now suppose we add the Toyota Prius to the data in Problem 25 (remove the Taurus). A Toyota Prius weighs 2890 pounds and gets 60 miles per gallon. Redraw the scatter diagram with the Prius included. What do you notice?

(e) Recompute the linear correlation coefficient with the Prius included. How did this new value affect your result? *−0.437*

(f) Why does this observation not follow the pattern of the data?

34. **Gestation Period versus Life Expectancy** Suppose we add humans to the data in Problem 24. Humans have a gestation period of 268 days and a life expectancy of 76.5 years.

(a) Redraw the scatter diagram with humans included.

(b) Recompute the linear correlation coefficient with humans included. *0.847*

(c) Compare the results of (a) and (b) with the results of Problem 24. Provide a statement that explains the results.

35. Consider the following four data sets:

Data Set 1		Data Set 2		Data Set 3		Data Set 4	
x	y	x	y	x	y	x	y
10	8.04	10	9.14	10	7.46	8	6.58
8	6.95	8	8.14	8	6.77	8	5.76
13	7.58	13	8.74	13	12.74	8	7.71
9	8.81	9	8.77	9	7.11	8	8.84
11	8.33	11	9.26	11	7.81	8	8.47
14	9.96	14	8.10	14	8.84	8	7.04
6	7.24	6	6.13	6	6.08	8	5.25
4	4.26	4	3.10	4	5.39	8	5.56
12	10.84	12	9.13	12	8.15	8	7.91
7	4.82	7	7.26	7	6.42	8	6.89
5	5.68	5	4.47	5	5.73	19	12.50

Source: Anscombe, Frank, Graphs in statistical analysis, *American Statistician*, 27 (1973):17–21

(a) Compute the linear correlation coefficient for each data set. *r = 0.82 for all four*

(b) Draw a scatter diagram for each data set. Conclude that linear correlation coefficients and scatter diagrams must be used together in any statistical analysis of bivariate data.

36. **The Best Predictor of the Winning Percentage** The ultimate goal in any sport (besides having fun) is to win. One measure of how well a team does is the winning percentage. In baseball, a lot of effort goes into figuring out the variable that best predicts a team's winning percentage. The following data represent the winning percentages of teams in the National League along with potential explanatory variables. Which variable do you think is the best predictor of winning percentage? Why?

Team	Winning Percentage	Runs Scored	Home Runs	Team Batting Average	On-Base Percentage	Batting Average against Team	Team Earned-Run Average
Arizona	0.315	615	135	0.253	0.310	0.266	4.98
Atlanta	0.593	803	178	0.270	0.343	0.265	3.74
Chicago Cubs	0.549	789	235	0.268	0.328	0.247	3.81
Cincinnati	0.469	750	194	0.250	0.331	0.280	5.19
Colorado	0.420	833	202	0.275	0.345	0.290	5.54
Florida	0.512	718	148	0.264	0.329	0.256	4.10
Houston	0.568	803	187	0.267	0.342	0.258	4.05
Los Angeles	0.574	761	203	0.262	0.332	0.254	4.01
Milwaukee	0.416	634	135	0.248	0.321	0.259	4.24
Montreal	0.414	635	151	0.249	0.313	0.266	4.33
New York Mets	0.438	684	185	0.249	0.317	0.261	4.09
Philadelphia	0.531	840	215	0.267	0.345	0.264	4.45
Pittsburgh	0.447	680	142	0.260	0.321	0.267	4.29
San Diego	0.537	768	139	0.273	0.342	0.263	4.03
San Francisco	0.562	850	183	0.270	0.357	0.265	4.29
St. Louis	0.648	855	214	0.278	0.344	0.251	3.75

Source: espn.com

37. **Diversification** One basic theory of investing is diversification. The idea is that you want to have a basket of stocks that do not all "move in the same direction." In other words, if one investment goes down, you don't want a second investment in your portfolio that is also likely to go down. One hallmark of a good portfolio is a low correlation between investments. The following data represent the annual rates of return for various stocks. If you only wish to invest in two of the stocks, which two would you select if your goal is to have low correlation between the two investments? Which two would you select if your goal is to have one stock go up when the other goes down?

	Rate of Return				
Year	Cisco Systems	Walt Disney	General Electric	Exxon Mobil	TECO Energy
1996	0.704	0.204	0.565	0.405	−0.012
1997	0.314	0.448	0.587	0.342	0.223
1998	1.50	−0.080	0.451	0.254	0.050
1999	1.31	−0.015	0.574	0.151	−0.303
2000	−0.286	−0.004	−0.055	0.127	0.849
2001	−0.527	−0.277	−0.151	−0.066	−0.150
2002	−0.277	−0.203	−0.377	−0.089	−0.369
2003	0.850	0.444	0.308	0.206	0.004
2004	−0.203	0.202	0.207	0.281	0.128

Source: Yahoo!Finance

38. Lyme Disease versus Drownings Lyme disease is an inflammatory disease that results in skin rash and flulike symptoms. It is transmitted through the bite of an infected deer tick. The following data represent the number of reported cases of Lyme disease and the number of drowning deaths for a rural county in the United States.

Month	J	F	M	A	M	J	J	A	S	O	N	D
Cases of Lyme Disease	3	2	2	4	5	15	22	13	6	5	4	1
Drowning Deaths		0	1	2	1	2	9	16	5	3	3	1	0

(a) Draw a scatter diagram of the data using cases of Lyme disease as the explanatory variable.
(b) Compute the correlation coefficient for the data. 0.961
(c) Based on your results from parts (a) and (b), what type of relation appears to exist between the number of reported cases of Lyme disease and drowning deaths? Do you believe that an increase in cases of Lyme disease causes an increase in drowning deaths?

39. Television Stations and Life Expectancy Based on data obtained from the *CIA World Factbook,* the linear correlation coefficient between number of television stations in a country and life expectancy of residents of the country is 0.599. What does this correlation imply? Do you believe that the more television stations a country has, the longer its population can expect to live? Why or why not?

40. Caffeine and SIDS A study on the relationship between caffeine consumption during pregnancy and sudden infant death syndrome (SIDS) showed that heavy caffeine consumption during pregnancy was associated with a significant risk of SIDS. The study was later criticized on the claim that parental smoking was not properly assessed. Explain why this might be a concern.

41. Influential Consider the following set of data:

x	2.2	3.7	3.9	4.1	2.6	4.1	2.9	4.7
y	3.9	4.0	1.4	2.8	1.5	3.3	3.6	4.9

(a) Draw a scatter diagram of the data and compute the linear correlation coefficient. 0.2280
(b) Draw a scatter diagram of the data and compute the linear correlation coefficient with the additional data point (10.4, 9.3). Comment on the effect the additional data point has on the linear correlation coefficient. Explain why correlations should always be reported with scatter diagrams. 0.8598

42. Faulty Use of Correlation On the basis of the accompanying scatter diagram, explain what is wrong with the following statement: "Because the linear correlation coefficient between age and median income is 0.012, there is no relation between age and median income."

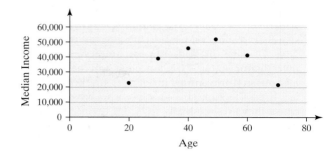

43. Name the Relation, Part I For each of the following statements, explain whether you think the variables will have positive correlation, negative correlation, or no correlation. Support your opinion. (b) Negative
(a) Number of children in the household under the age of 3 and expenditures on diapers Positive
(b) Interest rates on car loans and number of cars sold
(c) Number of hours per week on the treadmill and cholesterol level Negative
(d) Price of a Big Mac and number of McDonald's french fries sold in a week Negative
(e) Shoe size and IQ No correlation

44. Name the Relation, Part II For each of the following statements, explain whether you think the variables will have positive correlation, negative correlation, or no correlation. Support your opinion.
(a) Number of cigarettes smoked by a pregnant woman each week and birth weight of her baby Negative
(b) Annual salary and years of education Positive
(c) Number of doctors on staff at a hospital and number of administrators on staff. Positive
(d) Head circumference and IQ No Correlation
(e) Number of moviegoers and movie ticket price Negative

45. Transformations Consider the following data set:

x	5	6	7	7	8	8	8	8
y	4.2	5	5.2	5.9	6	6.2	6.1	6.9
x	9	9	10	10	11	11	12	12
y	7.2	8	8.3	7.4	8.4	7.8	8.5	9.5

(a) Draw a scatter diagram with the x-axis starting at 0 and ending at 30 and with the y-axis starting at 0 and ending at 20.
(b) Compute the linear correlation coefficient. 0.9518
(c) Now multiply both x and y by 2.
(d) Draw a scatter diagram of the new data with the x-axis starting at 0 and ending at 30 and with the y-axis starting at 0 and ending at 20. Compare the scatter diagrams.
(e) Compute the linear correlation coefficient. 0.9518
(f) Conclude that multiplying each value in the data set does not affect the correlation between the variables. Explain why this is the case.

46. Obesity In a study published in the *Journal of the American Medical Association* (May 16, 2001), researchers found that breast-feeding may help to prevent obesity in kids. In an interview, the head investigator stated, "It's not clear whether breast milk has obesity-preventing properties or the women who are breast-feeding are less likely to have fat kids because they are less likely to be fat themselves and may be more health conscious." Using this researcher's statement, explain what might be wrong with the conclusion that breast-feeding prevents obesity. Identify some lurking variables in the study.

47. How Well Will You Do in College? The College Board is a membership association composed of schools, colleges, universities, and other educational organizations. One of its better-known programs is the administration of the SAT college entrance exam. In a recent study, the College Board wanted to learn what the best predictor of college grade-point average (GPA) was. The following correlations were obtained based on 48,039 students.

Correlation between College GPA and:	Correlation
SAT score combined with high school GPA	0.61
SAT verbal score	0.47
SAT math score	0.48
SAT combined verbal and math score	0.52
High school GPA	0.54

Source: The College Board

(a) Which variable is the best predictor of college GPA?
(b) Which variable is the worst predictor of college GPA?

48. Correlation Applet Load the correlation by eye applet.

(a) In the lower-left corner of the applet, add 10 points that line up with a positive slope so that the linear correlation between the points is about 0.8. Click "show r" to show the correlation.

(b) Add another point in the upper-right corner of the applet that roughly lines up with the 10 points you have in the lower-left corner. Comment on how the linear correlation coefficient changes.

(c) Drag the point in the upper-right corner straight down. Take note of the change in the linear correlation coefficient. Notice how a single point can have a substantial impact on the linear correlation coefficient.

49. Correlation Applet Load the correlation by eye applet. Add about 10 points that form an upside-down U. Certainly, there is a relation between *x* and *y*, but what is the value of the linear correlation coefficient? Conclude that a low linear correlation coefficient does not imply there is no relation between two variables; it means there may be no linear relation between two variables.

50. Correlation Applet Load the correlation by eye applet.

(a) Plot about 10 points that follow a linear trend and have a linear correlation coefficient that is close to 0.8.

(b) Clear the applet. Plot about 6 points vertically on top of each other on the left side of the applet. Add a seventh point to the right of the applet. Move the point until the linear correlation coefficient is close to 0.8.

(c) Clear the applet. Plot about 7 points in a U-shaped curve. Add an eighth point and move it around the applet until the linear correlation coefficient is close to 0.8.

(d) Conclude that a linear correlation coefficient can result from data that have many patterns and so you should always plot your data.

Technology Step by Step

Drawing Scatter Diagrams and Determining the Correlation Coefficient

TI-83/84 Plus   Scatter Diagrams

Step 1: Enter the explanatory variable in L1 and the response variable in L2.

Step 2: Press 2nd Y = to bring up the StatPlot menu. Select 1: Plot1.

Step 3: Turn Plot 1 ON by putting the cursor on the ON button and pressing ENTER.

Step 4: Highlight the scatter diagram icon (see the figure) and press ENTER. Be sure that Xlist is L1 and Ylist is L2.

Step 5: Press ZOOM and select 9: ZoomStat.

Correlation Coefficient

Step 1: Turn the diagnostics on by selecting the catalog (2nd ∅). Scroll down and select `DiagnosticOn`. Hit ENTER to activate diagnostics.

Step 2: With the explanatory variable in `L1` and the response variable in `L2`, press `STAT`, highlight `CALC`, and select `4: LinReg (ax + b)`. With `LinReg` on the HOME screen, press ENTER.

MINITAB   Scatter Diagrams

Step 1: Enter the explanatory variable in C1 and the response variable in C2. You may want to name the variables.

Step 2: Select the **Graph** menu and highlight **Plot** ....

Step 3: With the cursor in the Y column, select the response variable. With the cursor in the X column, select the explanatory variable. Click OK.

Correlation Coefficient

Step 1: With the explanatory variable in C1 and the response variable in C2, select the **Stat** menu and highlight **Basic Statistics**. Highlight **Correlation**.

Step 2: Select the variables whose correlation you wish to determine and click OK.

Excel   Scatter Diagrams

Step 1: Enter the explanatory variable in column A and the response variable in column B. Select the Chart Wizard icon.

Step 2: Follow the instructions in the Chart Wizard.

Correlation Coefficient

Step 1: Be sure the Data Analysis Tool Pak is activated by selecting the **Tools** menu and highlighting **Add-Ins** .... Check the box for the Analysis ToolPak and select OK.

Step 2: Select **Tools** and highlight **Data Analysis** .... Highlight **Correlation** and select OK.

Step 3: With the cursor in the Input Range, highlight the data. Select OK.

4.2 Least-Squares Regression

Preparing for This Section   Before getting started, review the following:

- Lines (Section C.1 on CD, pp. C1–C6)

Objectives

① **Find the least-squares regression line and use the line to make predictions**

② **Interpret the slope and the *y*-intercept of the least-squares regression line**

③ **Compute the sum of squared residuals**

Note to Instructor
Encourage students to review lines on the CD.

Once the scatter diagram and linear correlation coefficient indicate that a linear relation exists between two variables, we proceed to find a linear equation that describes the relation between the two variables. One way to obtain a line that describes the relation is to select two points from the data that appear to provide a good fit and find the equation of the line through these points.

EXAMPLE 1 **Finding an Equation That Describes Linearly Related Data**

Problem: The data in Table 3 represent the club-head speed and the distance a golf ball travels for eight swings of the club. We determined that these data are linearly related in the last section.

! **CAUTION**

The method for obtaining an equation that describes the relation between two variables discussed in Example 1 is *not* the least-squares method. It is used to illustrate a point.

(a) Find a linear equation that relates club-head speed, x (the explanatory variable), and distance, y (the response variable), by selecting two points and finding the equation of the line containing the points.

(b) Graph the line on the scatter diagram.

(c) Use the equation to predict the distance a golf ball will travel if the club-head speed is 104 miles per hour.

Table 3

Club-Head Speed (mph) x	Distance (yards) y	(x, y)
100	257	$(100, 257)$
102	264	$(102, 264)$
103	274	$(103, 274)$
101	266	$(101, 266)$
105	277	$(105, 277)$
100	263	$(100, 263)$
99	258	$(99, 258)$
105	275	$(105, 275)$

Source: Paul Stephenson, student at Joliet Junior College

Approach

(a) To answer part (a), we perform the following steps:

Step 1: Select two points from Table 3 so that a line drawn through the points appears to give a good fit. Call the points (x_1, y_1) and (x_2, y_2). Refer to Figure 1 for the scatter diagram.

In Other Words

A good fit means that the line drawn appears to describe the relation between the two variables well.

Step 2: Find the slope of the line containing these two points using $m = \dfrac{y_2 - y_1}{x_2 - x_1}$.

Step 3: Use the point–slope formula, $y - y_1 = m(x - x_1)$, to find the line through the points selected in Step 1. Express the line in the form $y = mx + b$, where m is the slope and b is the y-intercept.

(b) For part (b), draw a line through the points selected in Step 1 of part (a).

(c) Finally, for part (c), we let $x = 104$ in the equation found in part (a).

Solution

(a) **Step 1:** We will select $(x_1, y_1) = (99, 258)$ and $(x_2, y_2) = (105, 275)$, because a line drawn through these two points seems to give a good fit.

Step 2: $m = \dfrac{y_2 - y_1}{x_2 - x_1} = \dfrac{275 - 258}{105 - 99} = \dfrac{17}{6} = 2.8333$

Step 3: We use the point–slope form of a line to find the equation of the line.

$$y - y_1 = m(x - x_1)$$
$$y - 258 = 2.8333(x - 99) \qquad m = 2.8333, \, x_1 = 99, \, y_1 = 258$$
$$y - 258 = 2.8333x - 280.4967$$
$$y = 2.8333 - 22.4967 \qquad\qquad\qquad \textbf{(1)}$$

The slope of the line is 2.8333 and the y-intercept is -22.4967.

(b) Figure 8 shows the scatter diagram along with the line drawn through the points $(99, 258)$ and $(105, 275)$.

! CAUTION
The line found in Step 3 of Example 1 is not the least-squares regression line.

! CAUTION
Unless otherwise noted, we will round to four decimal places. As always, do not round until the last computation.

Figure 8

(c) We let $x = 104$ in equation (1) to predict the distance a golf ball travels when hit with a club-head speed of 104 miles per hour.

$$y = 2.8333(104) - 22.4967$$
$$= 272.2 \text{ yards}$$

We predict that a golf ball will travel 272.2 yards when it is hit with a club-head speed of 104 miles per hour.

Now Work Problems 11(a), 11(b), and 11(c).

① Find the Least-Squares Regression Line and Use the Line to Make Predictions

Although the line that we found in Example 1 appears to describe the relation between club-head speed and distance well, is there a line that fits the data better? Is there a line that fits the data *best*?

Consider Figure 9. Each y-coordinate on the line corresponds to a predicted distance for a given club-head speed. For example, if club-head speed is 103 miles per hour, the predicted distance is $2.8333(103) - 22.4967 = 269.3$ yards. The observed distance for this club-head is speed is 274 yards. The difference between the observed value of y and the predicted value of y is the error or **residual**. For a swing speed of 103 mph the residual is

$$\text{Residual} = \text{observed } y - \text{predicted } y$$
$$= 274 - 269.3$$
$$= 4.7 \text{ yards}$$

In Other Words
The residual represents how close our prediction comes to actual observation. The smaller the residual, the better the prediction.

The residual for a club-head speed of 103 miles per hour is labeled in Figure 9.

Figure 9

The line that *best* describes the relation between two variables is the one that minimizes the distance between the points and the line. The most popular technique for making the residuals as small as possible is the *method of least squares*, discovered by Adrien Marie Legendre.

Historical Note

Adrien Marie Legendre was born on September 18, 1752, into a wealthy family and was educated in mathematics and physics at the College Mazarin in Paris. From 1775 to 1780, he taught at Ecole Militaire. On March 30, 1783, Legendre was appointed an adjoint in the Académie des Sciences. On May 13, 1791, he became a member of the committee of the Académie des Sciences and was charged with the task of standardizing weights and measures. The committee worked to compute the length of the meter. During the revolution, Legendre lost his small fortune. In 1794, Legendre published *Eléments de géométrie*, which was the leading elementary text in geometry for around 100 years. In 1806, Legendre published a book on orbits, in which he developed the theory of least squares. He died on January 10, 1833.

Definition

Least-Squares Regression Criterion

The **least-squares regression line** is the one that minimizes the sum of the squared errors (or residuals). It is the line that minimizes the square of the vertical distance between the observed values of y and those predicted by the line, $\hat{y}$ (read "y-hat"). We represent this as

$$\text{Minimize } \sum \text{residuals}^2$$

The advantage of the least-squares criterion is that it allows for statistical inference on the predicted value and slope (Chapter 12). Another advantage of the least-squares criterion is explained by Legendre in his text *Nouvelles méthodes pour la détermination des orbites des cométes*, published in 1806.

> Of all the principles that can be proposed for this purpose, I think there is none more general, more exact, or easier to apply, than that which we have used in this work; it consists of making the sum of squares of the errors a *minimum*. By this method, a kind of equilibrium is established among the errors which, since it prevents the extremes from dominating, is appropriate for revealing the state of the system which most nearly approaches the truth.

Equation of the Least-Squares Regression Line

The equation of the least-squares regression line is given by

$$\hat{y} = b_1 x + b_0$$

where

$$b_1 = r \cdot \frac{s_y}{s_x} \text{ is the \textbf{slope} of the least-squares regression line*} \qquad (2)$$

and

$$b_0 = \bar{y} - b_1 \bar{x} \text{ is the \textbf{y-intercept} of the least-squares regression line} \qquad (3)$$

Note: $\bar{x}$ is the sample mean and s_x is the sample standard deviation of the explanatory variable x; $\bar{y}$ is the sample mean and s_y is the sample standard deviation of the response variable y.

*An equivalent formula is

$$b_1 = \frac{S_{xy}}{S_{xx}} = \frac{\sum x_i y_i - \dfrac{\left(\sum x_i\right)\left(\sum y_i\right)}{n}}{\sum x_i^2 - \dfrac{\left(\sum x_i\right)^2}{n}}$$

The notation $\hat{y}$ is used in the least-squares regression line to serve as a reminder that it is a predicted value of y for a given value of x. An interesting property of the least-squares regression line, $\hat{y} = b_1 x + b_0$, is that the line always contains the point $(\overline{x}, \overline{y})$. This property can be useful when drawing the least-squares regression line by hand.

Since s_y and s_x must both be positive, the sign of the linear correlation coefficient and the sign of the slope of the least-squares regression line are the same. For example, if r is positive, then the slope of the least-squares regression line will also be positive.

EXAMPLE 2

Historical Note

Sir Francis Galton was born on February 16, 1822. Galton came from a wealthy and well-known family. Charles Darwin was his first cousin. Galton studied medicine at Cambridge. After receiving a large inheritance, he left the medical field and traveled the world. He explored Africa from 1850 to 1852. In the 1860s, his study of meteorology led him to discover anticyclones. Influenced by Darwin, Galton always had an interest in genetics and heredity. He studied heredity through experiments with sweet peas. He noticed that the weight of the "children" of the "parent" peas reverted or *regressed* to the mean weight of all peas. Hence, the term *regression analysis*. Galton died on January 17, 1911.

Finding the Least-Squares Regression Line

Problem: For the data in Table 3 on page 196,

(a) Find the least-squares regression line.
(b) Predict the distance a golf ball will travel when hit with a club-head speed of 103 miles per hour.
(c) Compute the residual for the prediction made in part (b).
(d) Draw the least-squares regression line on the scatter diagram of the data.

Approach

(a) From Example 2 in Section 4.1, we have the following:

$$r = 0.9387, \overline{x} = 101.875, s_x = 2.2952, \overline{y} = 266.75, \text{ and } s_y = 7.74135$$

We substitute these values into Formula (2) to find the slope of the least-squares regression line. We use Formula (3) to find the intercept of the least-squares regression line.

(b) Substitute $x = 103$ into the least-squares regression line found in part (a) to find $\hat{y}$.

(c) The residual is the difference between the observed y and the predicted y. That is, residual $= y - \hat{y}$.

(d) To draw the least-squares regression line, select two values of x and use the equation to find the predicted values of y. Plot these points on the scatter diagram and draw a line through the points.

Solution

(a) Substituting $r = 0.9387$, $s_x = 2.2952$, and $s_y = 7.74135$ into Formula (2), we obtain

$$b_1 = r \cdot \frac{s_y}{s_x} = 0.9387 \cdot \frac{7.74135}{2.2952} = 3.1661$$

We have that $\overline{x} = 101.875$ and $\overline{y} = 266.75$. Substituting these values into Formula (3), we obtain

$$b_0 = \overline{y} - b_1\overline{x} = 266.75 - 3.1661(101.875) = -55.7964$$

The least-squares regression line is

$$\hat{y} = 3.1661x - 55.7964$$

(b) We let $x = 103$ in the equation $y = 3.1661x - 55.7964$ to predict the distance a golf ball hit with a club-head speed of 103 miles per hour will travel.

$$\hat{y} = 3.1661(103) - 55.7964$$

$$= 270.3 \text{ yards}$$

We predict that the distance the ball will travel is 270.3 yards.

! CAUTION
Throughout the text, we will round the slope and y-intercept values to four decimal places. The predictions should be made to one more decimal place than the response variable.

(c) The actual distance the ball traveled is 274 yards. The residual is

$$\text{Residual} = \text{observed } y - \text{predicted } y$$

$$= y - \hat{y}$$

$$= 274 - 270.3$$

$$= 3.7 \text{ yards}$$

We underestimated the distance by 3.7 yards.

(d) Figure 10 shows the graph of the least-squares regression line drawn on the scatter diagram with the residual labeled.

Figure 10

In Other Words

An underestimate means the residual is positive; an overestimate means the residual is negative. A residual of zero means the prediction is right on!

Notice that an underestimate results in a positive residual, while an overestimate results in a negative residual.

Note to Instructor

The purpose of this activity is to illustrate how regression can be used to forecast or make predictions. The issue of extrapolation should be discussed, particularly when the predicting the thickness of 0 page. Students should consider two ways of estimating the thickness of one page of the text: using the slope or evaluating the model for x = 1. Another good discussion topic is variability in measurement. Students should see that there is variability from student to student, even for the same number of pages. In addition, assuming each page is the same thickness, there is often variability within each student's measurements. (For example, the thickness for 200 pages is not twice the thickness for 100 pages.) Time limit: 45 to 60 minutes, depending on discussions.

In-Class Activity: Paper Thin (Regression)

Each student, or small group of students, should receive a ruler with both inches and centimeters.

1. Use the ruler provided to measure the thickness of only one page of the text. Which unit of measurement did you use and why?
2. Grouping pages (one page is one sheet), complete the following table:

No. of pages	25	50	75	150	200	225
Thickness						

3. Compute the least-squares regression line for your data.
4. Compare your data and your regression line to those around you. Did everyone get the same measurements and model? Explain why or why not.
5. Use your model to estimate the thickness of one page of the text. Is the value you obtained reasonable?
6. Use your model to estimate the thickness of 0 pages of the text. Is the value you obtained reasonable?
7. What, if anything, could you do to improve the model?

We can think of any point on the least-squares regression line as an estimate of the mean value of the response variable for a given value of the explanatory variable. For example, the mean distance a golf ball will travel when hit with a club-head speed of 103 miles per hour is 270.3 yards. In our experiment, when we hit the ball with a club-head speed of 103 miles per hour, the ball traveled 274 yards. So the distance the ball traveled was above the mean. Perhaps we hit the ball in the "sweet spot" of the club face or a breeze kicked up at our back.

② Interpret the Slope and *y*-Intercept of the Least-Squares Regression Line

The definition of the slope of a line is $\dfrac{\text{Rise}}{\text{Run}}$ or $\dfrac{\text{Change in } y}{\text{Change in } x}$. For a line whose slope is $\dfrac{2}{3}$, if *x* increases by 3, *y* will *increase* by 2. Or, if the slope of a line is $-4 = \dfrac{-4}{1}$, if *x* increases by 1, *y* will *decrease* by 4.

The *y*-intercept of any line is the point where the graph intersects the vertical axis. It is found by letting $x = 0$ in an equation and solving for *y*.

We found the regression equation in Example 2 to be $\hat{y} = 3.1661x - 55.7964$. So the slope of the line is 3.1661. We interpret this slope as follows: If the club-head speed increases by 1 mile per hour, the distance the ball travels increases by 3.1661 yards, on average. To interpret the *y*-intercept, we must first ask two questions:

1. Is 0 a reasonable value for the explanatory variable?
2. Do any observations near $x = 0$ exist in the data set?

If the answer to either of these questions is no, we do not give an interpretation to the *y*-intercept. In the regression equation of Example 2, a swing speed of 0 miles per hour does not make sense, so an interpretation of the *y*-intercept is unreasonable. To interpret a *y*-intercept, we would say that it is the value of the response variable when the value of the explanatory variable is 0.

The second condition for interpreting the *y*-intercept is especially important because we should not use the regression model to make predictions **outside the scope of the model**. If this cannot be avoided, be cautious when using the regression model to make predictions for values of the explanatory variable that are much larger or much smaller than those observed, because we cannot be certain of the behavior of data for which we have no observations.

For example, it is inappropriate to use the line we determined in Example 2 to predict distance when club-head speed is 140 miles per hour. The highest observed club-head speed in our data set is 105 miles per hour. We cannot be certain that the linear relation between distance and club-head speed will continue. See Figure 11.

!CAUTION

Be careful when using the least-squares regression line to make predictions for values of the explanatory variable that are much larger or much smaller than those observed.

Figure 11

We have presented the procedure for determining the least-squares regression line by hand. In practice, however, a statistical spreadsheet or calculator with advanced statistical features is used to determine the least-squares regression line.

EXAMPLE 3

Finding the Least-Squares Regression Line Using Technology

Problem: Use a statistical spreadsheet or a graphing calculator with advanced statistical features to find the least-squares regression line of the data in Table 3.

Approach: Because technology plays a major role in obtaining the least-squares regression line, we will use a TI-84 Plus graphing calculator, MINITAB, and Excel to obtain the least-squares regression line. The steps for obtaining these lines are given in the Technology Step by Step on page 208.

Result: Figure 12(a) shows the output obtained from a TI-84 Plus graphing calculator, Figure 12(b) shows the output obtained from MINITAB with the slope and y-intercept highlighted, and Figure 12(c) shows partial output from Excel with the slope and y-intercept highlighted.

Figure 12

Slope

y-intercept

(a) TI-84 Plus output

```
The regression equation is
Distance (yards) = -55.8 + 3.17 Club Head Speed (mph)
Predictor                   Coef     SE Coef         T        P
Constant                  -55.80       48.37     -1.15    0.293
Club Head Speed (mph)      3.1661      0.4747      6.67    0.001

S = 2.88264       R - Sq = 88.1%    R - Sq(adj) = 86.1%
```

(b) MINITAB output

	Coefficients	Standard Error	t Stat	P-value
Intercept	-55.79661017	48.37134953	-1.153505344	0.29257431
Club Head Speed (mph)	3.166101695	0.47470539	6.669613957	0.00054983

(c) Excel output

Note: To get the linear correlation coefficient from MINITAB, use

$r = \sqrt{R - Sq}$. So $r = \sqrt{0.881} = 0.9386$.

Now Work Problems 11(d) and 11(e).

 ## Compute the Sum of Squared Residuals

Note to Instructor

The purpose of this material is to allow students to see that the least-squares regression line has the smallest sum of squared residuals. This section can be skipped without loss of continuity.

Recall that the least-squares regression line is the line that minimizes the sum of the squared residuals. This means that the sum of the squared residuals, Σ residuals^2, for the least-squares line will be smaller than for any other line that may describe the relation between the two variables. In particular, the sum of the squared residuals for the line obtained in Example 2 using the method of least squares will be smaller than the sum of the squared residuals for the line obtained in Example 1. It is worthwhile to verify this result.

EXAMPLE 4   **Comparing the Sum of Squared Residuals**

Problem: Compare the sum of squared residuals for the lines obtained in Examples 1 and 2.

Approach: We compute Σ residuals^2 using the predicted values of y, $\hat{y}$, for the lines obtained in Examples 1 and 2. This is best done by creating a table of values.

Solution: We create Table 4, which contains the value of the explanatory variable in column 1. Column 2 contains the corresponding response variable. Column 3 contains the predicted value using the equation obtained in Example 1, $\hat{y} = 2.8333x - 22.4967$. In column 4, we compute the residuals for each observation: residual = observed y − predicted y. For example, the first residual using the equation found in Example 1 is observed y − predicted $y = 257 - 260.8 = -3.8$. Column 5 contains the squares of the residuals obtained in column 4. Column 6 contains the predicted value using the least-squares regression equation obtained in Example 2: $\hat{y} = 3.1661x - 55.7964$. Column 7 represents the residuals for each observation and column 8 represents the squared residuals.

Table 4

Club-Head Speed (mph)	Distance (yards)	Example 1 ($\hat{y} = 2.8333x - 22.4967$)	Residual	Residual^2	Example 2 ($\hat{y} = 3.1661x - 55.7964$)	Residual	Residual^2
100	257	260.8	−3.8	14.44	260.8	−3.8	14.44
102	264	266.5	−2.5	6.25	267.2	−3.2	10.24
103	274	269.3	4.7	22.09	270.3	3.7	13.69
101	266	263.7	2.3	5.29	264.0	2.0	4.00
105	277	275.0	2.0	4.00	276.6	0.4	0.16
100	263	260.8	2.2	4.84	260.8	2.2	4.84
99	258	258.0	0.0	0.00	257.6	0.4	0.16
105	275	275.0	0.0	0.00	276.6	−1.6	2.56
				Σ residual^2 = 56.91			Σ residual^2 = 50.09

The sum of the squared residuals for the line found in Example 1 is 56.91; the sum of the squared residuals for the least-squares regression line is 50.09. Again, any line that describes the relation between distance and club-head speed will have a sum of squared residuals that is greater than 50.09. ▬

Now Work Problems 11(f), (g) and (h)

We draw the graphs of the two lines obtained in Examples 1 and 2 on the same scatter diagram in Figure 13 to help the reader visualize the difference.

Figure 13

MAKING AN INFORMED DECISION

Blue Book Used Car Guide

1990-2004 Used Car & Truck July-December 2005

What Car Should I Buy?

You are still in the market to buy a car. As we all know, cars lose value over time. Therefore, another item to consider when purchasing a car is its depreciation rate, the more value the car loses each year. Using the same three cars that you used in the Chapter 3 Decision, answer the following questions to determine depreciation rate.

1. Collect information regarding the three cars in your list by finding at least 12 cars of each car model that are for sale. Obtain the asking price and age of the car. Sources of data include your local newspaper classified ads or car Web sites, such as www.cars.com and www.vehix.com.

2. For each car type, draw a scatter diagram, treating age of the car as the explanatory variable and asking price as the response variable. Does the relation between the two variables appear linear?

3. The asking price of a car and the age of a car are related through the exponential equation $y = ab^x$, where y is the price of the car and x is the age of the car. The depreciation rate is $1 - b$. To estimate the values of a and b using least-squares regres-

sion, we need to transform the equation to a linear equation. This is accomplished by computing the logarithm of the asking price. For example, if the asking price of a car is $8000, compute $\log(8000) = 3.9031$. Compute the logarithm of each asking price for each car.

4. For each car model, draw a scatter diagram, treating age of the car as the explanatory variable and the logarithm of asking price as the response variable. Does the relation between age and the logarithm of asking price appear to be linear?

5. For each car model, find the least-squares regression line, treating age of the car as the explanatory variable and the logarithm of asking price as the response variable. The line will be $Y = \log a + (\log b)x = A + Bx$.

6. To determine a and b for the exponential equation $y = ab^x$, let $a = 10^A$ and $b = 10^B$. Graph the exponential equation on each scatter diagram from question 2.

7. For each exponential equation determined in question 6, the depreciation rate is $1 - b$. What are the depreciation rates for the three cars considered? Will this result affect your decision about which car to buy?

4.2 ASSESS YOUR UNDERSTANDING

Concepts and Vocabulary

1. Explain the least-squares regression criterion.
2. What is a residual? What does it mean when a residual is positive?
3. Explain the phrase *outside the scope of the model*. Why is it dangerous to make predictions outside the scope of the model?
4. If the linear correlation between two variables is negative, what can be said about the slope of the regression line?
   Negative

5. In your own words, explain the meaning of Legendre's quote given on page 198.
6. *True or False*: The least-squares regression line always travels through the point $(\bar{x}, \bar{y})$. True
7. In your own words, explain what each point on the least-squares regression line represents.
8. If the linear correlation coefficient is 0, what is the equation of the least-squares regression line? $\hat{y} = \bar{y}$

Skill Building

9. For the data set

x	0	2	3	5	6	6
y	5.8	5.7	5.2	2.8	1.9	2.2

(a) Draw a scatter diagram. Comment on the type of relation that appears to exist between x and y.
(b) Given that $\bar{x} = 3.667$, $s_x = 2.42212$, $\bar{y} = 3.933$, $s_y = 1.8239152$, and $r = -0.9476938$, determine the least-squares regression line. $\hat{y} = -0.7136x + 6.55$
(c) Graph the least-squares regression line on the scatter diagram drawn in part (a).

10. For the data set

x	2	4	8	8	9
y	1.4	1.8	2.1	2.3	2.6

(a) Draw a scatter diagram. Comment on the type of relation that appears to exist between x and y.
(b) Given $\bar{x} = 6.2$, $s_x = 3.03315$, $\bar{y} = 2.04$, that $s_y = 0.461519$, and $r = 0.957241$, determine the least-squares regression line. $\hat{y} = 0.1457x + 1.1370$
(c) Graph the least-squares regression line on the scatter diagram drawn in part (a).

In Problems 11–16,
(a) Draw a scatter diagram treating x as the explanatory variable and y as the response variable.
(b) Select two points from the scatter diagram and find the equation of the line containing the points selected.
(c) Graph the line found in part (b) on the scatter diagram.
(d) Determine the least-squares regression line.
(e) Graph the least-squares regression line on the scatter diagram.
(f) Compute the sum of the squared residuals for the line found in part (b).
(g) Compute the sum of the squared residuals for the least-squares regression line found in part (d).
(h) Comment on the fit of the line found in part (b) versus the least-squares regression line found in part (d).

11.
NW

x	3	4	5	7	8
y	4	6	7	12	14

12.

x	3	5	7	9	11
y	0	2	3	6	9

13.

x	−2	−1	0	1	2
y	−4	0	1	4	5

14.

x	−2	−1	0	1	2
y	7	6	3	2	0

15.

x	20	30	40	50	60
y	100	95	91	83	70

16.

x	5	10	15	20	25
y	2	4	7	11	18

Applying the Concepts

Problems 17–22 use the results from Problems 23–28 in Section 4.1.

17. Height versus Head Circumference (Refer to Problem 23, Section 4.1) A pediatrician wants to determine the relation that exists between a child's height, x, and head circumference, y. She randomly selects 11 children from her practice, measures their height and head circumference, and obtains the following data.

Height, x (inches)	Head Circumference, y (inches)	Height, x (inches)	Head Circumference, y (inches)
27.75	17.5	26.5	17.3
24.5	17.1	27	17.5
25.5	17.1	26.75	17.3
26	17.3	26.75	17.5
25	16.9	27.5	17.5
27.75	17.6		

Source: Denise Slucki, student at Joliet Junior College

(a) Find the least-squares regression line, treating height as the explanatory variable and head circumference as the response variable. $\hat{y} = 0.1827x + 12.4932$
(b) Interpret the slope and intercept, if appropriate.
(c) Use the regression equation to predict the head circumference of a child who is 25 inches tall. 17.06 inches
(d) Compute the residual based on the observed head circumference of the 25-inch-tall child in the table. Is the head circumference of this child above average or below average? −0.16 inches; below
(e) Draw the least-squares regression line on the scatter diagram of the data and label the residual from part (d).
(f) Notice that two children are 26.75 inches tall. One has a head circumference of 17.3 inches; the other has a head circumference of 17.5 inches. How can this be?
(g) Would it be reasonable to use the least-squares regression line to predict the head circumference of a child who was 32 inches tall? Why? No

18. Gestation Period versus Life Expectancy (Refer to Problem 24, Section 4.1) The following data represent the gestation period, x, of various animals along with their life expectancy, y.

Animal	Gestation (or Incubation) Period (days), x	Life Expectancy (years), y
Cat	63	11
Chicken	22	7.5
Dog	63	11
Duck	28	10
Goat	151	12
Lion	108	10
Parakeet	18	8
Pig	115	10
Rabbit	31	7
Squirrel	44	9

Source: Time Almanac 2000

(a) Find the least-squares regression line, treating gestation period as the explanatory variable and life expectancy as the response variable. $\hat{y} = 0.0261x + 7.8738$
(b) Interpret the slope and intercept, if appropriate.
(c) Suppose a new animal species has been discovered. After breeding the species in captivity, it is determined that the gestation period is 95 days. Use the least-squares regression line to predict the life expectancy of the animal. 10.4 years
(d) Use the regression equation to predict the life expectancy of a parakeet. 8.3 years
(e) Use the regression equation to predict the life expectancy of a rabbit. 8.7 years
(f) Compute the residual of the prediction made in part (e). Conclude that the least-squares regression

line sometimes provides accurate predictions (as in the case of the parakeet) and sometimes provides inaccurate predictions (as in the case of the rabbit). Unfortunately, when the value of the response variable is unknown, as in the case of the new animal species from part (c), we don't know the accuracy of the prediction. −1.7 years

19. **Weight of a Car versus Miles per Gallon** (Refer to Problem 25, Section 4.1) An engineer wants to determine how the weight of a car, x, affects gas mileage, y. The following data represent the weight of various domestic cars and their miles per gallon in the city for the 2005 model year.

Car	Weight (pounds), x	Miles per Gallon, y
Buick LeSabre	3565	20
Cadillac DeVille	3985	18
Chevrolet Corvette	3180	19
Chevrolet Monte Carlo	3340	21
Chrysler PT Cruiser	3100	21
Chrysler Sebring Sedan	3175	22
Dodge Neon	2580	27
Dodge Stratus Sedan	3175	22
Ford Focus	2655	26
Ford Mustang	3300	20
Lincoln LS	3680	20
Mercury Sable	3310	19
Pontiac Bonneville	3590	20
Pontiac Grand Am	3475	20
Pontiac Sunfire	2770	24
Saturn Ion	2690	26

Source: www.roadandtrack.com

(a) Find the least-squares regression line treating weight as the explanatory variable and miles per gallon as the response variable. $\hat{y} = -0.0062x + 41.4640$
(b) Interpret the slope and intercept, if appropriate.
(c) Predict the miles per gallon of a Ford Mustang and compute the residual. Is the miles per gallon of a Mustang above average or below average for cars of this weight? 21 mpg; −1; below
(d) Draw the least-squares regression line on the scatter diagram of the data and label the residual.
(e) Would it be reasonable to use the least-squares regression line to predict the miles per gallon of a Toyota Prius, a hybrid gas and electric car? Why or why not? No

20. **Bone Length** (Refer to Problem 26, Section 4.1) Research performed at NASA and led by Emily R. Morey-Holton measured the lengths of the right humerus and right tibia in 11 rats that were sent to space on Spacelab Life Sciences 2. The following data were collected.

Right Humerus (mm)	Right Tibia (mm)	Right Humerus (mm)	Right Tibia (mm)
24.80	36.05	25.90	37.38
24.59	35.57	26.11	37.96
24.59	35.57	26.63	37.46
24.29	34.58	26.31	37.75
23.81	34.20	26.84	38.50
24.87	34.73		

Source: NASA Life Sciences Data Archive

(a) Find the least-squares regression line, treating the length of the right humerus, x, as the explanatory variable and the length of the right tibia, y, as the response variable. $\hat{y} = 1.3902x + 1.1140$
(b) Interpret the slope and intercept, if appropriate.
(c) Determine the residual if the length of the right humerus is 26.11 mm and the actual length of the right tibia is 37.96 mm. Is the length of this tibia above or below average? 0.548 mm; above
(d) Draw the least-squares regression line on the scatter diagram and label the residual from part (c).
(e) Suppose one of the rats sent to space experienced a broken right tibia due to a severe landing. The length of the right humerus is determined to be 25.31 mm. Use the least-squares regression line to estimate the length of the right tibia. 36.30 mm

21. **Attending Class** (Refer to Problem 27, Section 4.1) The following data represent the number of days absent, x, and the final grade, y, for a sample of college students in a general education course at a large state university.

No. of absences, x	0	1	2	3	4	5	6	7	8	9
Final grade, y	89.2	86.4	83.5	81.1	78.2	73.9	64.3	71.8	65.5	66.2

Source: College Teaching, Winter 2005, Vol. 53, Issue 1

(a) Find the least-squares regression line, treating number of absences as the explanatory variable and final grade as the response variable. $\hat{y} = -2.8273x + 88.7327$
(b) Interpret the slope and intercept, if appropriate.
(c) Predict the final grade for a student who misses five class periods and compute the residual. Is the final grade above or below average for this number of absences? 74.60; −0.70; below
(d) Draw the least-squares regression line on the scatter diagram of the data and label the residual.
(e) Would it be reasonable to use the least-squares regression line to predict the final grade for a student who has missed 15 class periods? Why or why not? No

22. Antibiotics (Refer to Problem 28, Section 4.1) A study on antibiotic use among children in Manitoba, Canada, gave the following data for the number of prescriptions per 1000 children x years after 1995.

Year, x	0	1	2	3	4	5	6
Prescriptions (per 1000 children), y	1201	1070	944	964	909	949	864

Source: Canadian Medical Association Journal, Vol. 171, Issue 2

(a) Find the least-squares regression line, treating year as the explanatory variable and prescriptions as the response variable. $\hat{y} = -46x + 1123.8571$
(b) Interpret the slope and intercept, if appropriate.
(c) Predict the number of prescriptions per 1000 children in Manitoba, Canada, in 2002 ($x = 7$). 801.9
(d) Draw the least-squares regression line on the scatter diagram of the data.
(e) Would it be reasonable to use the least-squares regression line to predict the number of prescriptions in Manitoba, Canada, in 2010? Why or why not? No

23. Does Size Matter? Researchers wondered whether the size of a person's brain was related to the individual's mental capacity. They selected a sample of right-handed introductory psychology students who had SAT scores higher than 1350. The subjects were administered the Wechsler (1981) Adult Intelligence Scale-Revised exam to obtain their IQ scores. MRI scans, performed at the same facility, consisted of 18 horizontal magnetic resonance images. The computer counted all pixels with nonzero gray scale in each of the 18 images, and the total count served as an index for brain size. The resulting data are presented in the table.

Gender	MRI Count, x	IQ, y	Gender	MRI Count, x	IQ, y
Female	816,932	133	Male	949,395	140
Female	951,545	137	Male	1,001,121	140
Female	991,305	138	Male	1,038,437	139
Female	833,868	132	Male	965,353	133
Female	856,472	140	Male	955,466	133
Female	852,244	132	Male	1,079,549	141
Female	790,619	135	Male	924,059	135
Female	866,662	130	Male	955,003	139
Female	857,782	133	Male	935,494	141
Female	948,066	133	Male	949,589	144

Source: Willerman, L., Schultz, R., Rutledge, J. N., and Bigler, E. (1991). "In Vivo Brain Size and Intelligence." *Intelligence*, 15, 223–228

(a) Find the least-squares regression line treating MRI count as the explanatory variable and IQ as the response variable. $\hat{y} = 0.00003x + 109.8940$
(b) What do you notice about the value of the slope? Why does this result seem reasonable based on the scatter diagram and linear correlation coefficient obtained in Problem 31 of Section 4.1?
(c) When there is no relation between the explanatory and response variable, we use the mean value of the response variable, $\bar{y}$, to predict. Predict the IQ of an individual whose MRI count is 1,000,000. Predict the IQ of an individual whose MRI count is 830,000. 136; 136

24. Male versus Female Drivers The following data represent the number of licensed drivers in various age groups and the number of accidents within the age group by gender.

Age Group	Number of Male Licensed Drivers (000s)	Number of Crashes Involving a Male (000s)	Number of Female Licensed Drivers (000s)	Number of Crashes Involving a Female (000s)
16	816	244	764	178
17	1,198	233	1,115	175
18	1,342	243	1,212	164
19	1,454	229	1,333	145
20–24	7,866	951	7,394	618
25–29	9,356	899	8,946	595
30–34	10,121	875	9,871	571
35–39	10,521	901	10,439	566
40–44	9,776	692	9,752	455
45–49	8,754	667	8,710	390
50–54	6,840	390	6,763	247
55–59	5,341	290	5,258	165
60–64	4,565	218	4,486	133
65–69	4,234	191	4,231	121
70–74	3,604	167	3,749	104
75–79	2,563	118	2,716	77
80–84	1,400	61	1,516	45
≥85	767	34	767	20

Source: National Highway and Traffic Safety Institute

(a) Find the least-squares regression line for males, treating number of licensed drivers as the explanatory variable, x, and number of crashes, y, as the response variable. Repeat this procedure for females.

(b) Interpret the slope of the least-squares regression line for each gender, if appropriate. How might an insurance company use this information?

(c) Predict the number of accidents for males if there were 8700 thousand licensed drivers. Predict the number of accidents for females if there were 8700 thousand licensed drivers.

25. Mark Twain, in his book *Life on the Mississippi* (1884), makes the following observation:

Therefore, the Mississippi between Cairo and New Orleans was twelve hundred and fifteen miles long one hundred and seventy-six years ago. It was eleven hundred and eighty after the cut-off of 1722. It was one thousand and forty after the American Bend cut-off. It has lost sixty-seven miles since. Consequently its length is only nine hundred and seventy-three miles at present.

Now, if I wanted to be one of those ponderous scientific people, and "let on" to prove what had occurred in the remote past by what had occurred in a given time in the recent past, or what will occur in the far future by what has occurred in late years, what an opportunity is here! Geology never had such a chance, nor such exact data to argue from! Nor "development of species," either! Glacial epochs are great things, but they are vague—vague. Please observe:

In the space of one hundred and seventy-six years the Lower Mississippi has shortened itself two hundred and forty-two miles. That is an average of a trifle over one mile and a third per year. Therefore, any calm person, who is not blind or idiotic, can see that in the Old Oolitic Silurian Period, just a million years ago next November, the Lower Mississippi River was upwards of one million three hundred thousand miles long, and stuck out over the Gulf of Mexico like a fishing-rod. And by the same token any person can see that seven hundred and forty-two years from now the Lower Mississippi will be only a mile and three-quarters long, and Cairo and New Orleans will have joined their streets together, and be plodding comfortably along under a single mayor and a mutual board of aldermen. There is something fascinating about science. One gets such wholesale returns of conjecture out of such a trifling investment of fact.

Discuss how this relates to the material in this section.

26. **Regression Applet** Load the regression by eye applet. Create a scatter diagram with twelve points and a positive linear association. Try to choose the points so that the correlation is about 0.7.

(a) Draw a line that you believe describes the relation between the two variables well.

(b) Now click the Show Least-squares Line on the applet. Compare the sum of squared residuals for the line that you draw to the sum of squared residuals for the least-squares regression line. Repeat parts (a) and (b) as often as you like. Does your eyeballed line ever coincide with the least-squares regression line?

(24a) Males: $\hat{y} = 0.0802x + 8.14$; Females: $\hat{y} = 0.0505x + 15.1611$

(24c) Males: 706 thousand; Females: 455 thousand

Technology Step by Step	**Determining the Least-Squares Regression Line**
TI-83/84 Plus	Use the same steps that were followed to obtain the correlation coefficient.
MINITAB	*Step 1:* With the explanatory variable in C1 and the response variable in C2, select the **Stat** menu and highlight **Regression**. Highlight **Regression** .... *Step 2:* Select the explanatory (predictor) and response variables and click OK.
Excel	*Step 1:* Be sure the Data Analysis Tool Pak is activated by selecting the **Tools** menu and highlighting **Add-Ins** .... Check the box for the Analysis ToolPak and select OK. *Step 2:* Enter the explanatory variable in column A and the response variable in column B. *Step 3:* Select the **Tools** menu and highlight **Data Analysis** .... *Step 4:* Select the **Regression** option. *Step 5:* With the cursor in the Y-range cell, highlight the column that contains the response variable. With the cursor in the X-range cell, highlight the column that contains the explanatory variable. Press OK.

4.3   The Coefficient of Determination

Preparing for This Section   Before getting started, review the following:

• Outliers (Section 3.4, pp. 155–156)

Objectives   **①**   **Compute and interpret the coefficient of determination**

In Section 4.2, we discussed the procedure for obtaining the least-squares regression line. In this section, we discuss another numerical measure of the strength of relation that exists between two quantitative variables.

①   ## Compute and Interpret the Coefficient of Determination

Consider the club-head speed versus distance data introduced in Section 4.1. If we were asked to predict the distance of a randomly selected shot, what would be a good guess? Our best guess might be the average distance of all shots taken. Since we don't know this value, we would use the average distance from the sample data given in Table 1, $\overline{y} = 266.75$ yards.

Now suppose we were told this particular shot resulted from a swing with a club-head speed of 103 mph. We could use the least-squares regression line to adjust our guess to $\hat{y} = 3.1661(103) - 55.7964 = 270.3$ yards. Knowing the linear relation that exists between club-head speed and distance allows us to improve our estimate of the distance of the shot. In statistical terms, we say that some of the variation in distance is explained by the linear relation between club-head speed and distance.

The percentage of variation in distance that is explained by the least-squares regression line is called the *coefficient of determination*.

Definition

> The **coefficient of determination**, R^2, measures the percentage of total variation in the response variable that is explained by the least-squares regression line.

In Other Words

The coefficient of determination is a measure of how well the least-squares regression line describes the relation between the explanatory and response variable. The closer R^2 is to 1, the better the line describes how changes in the explanatory variable affect the value of the response variable.

The coefficient of determination is a number between 0 and 1, inclusive. That is, $0 \leq R^2 \leq 1$. If $R^2 = 0$, the least-squares regression line has no explanatory value. If $R^2 = 1$, the least-squares regression line explains 100% of the variation in the response variable.

Consider Figure 14, where a horizontal line is drawn at $\overline{y} = 266.75$. This value represents the predicted distance of a shot without any knowledge of club-head speed. Armed with the additional information that the club-head speed is 103 miles per hour, we increased our guess to 270.3 yards. The difference between the predicted distance of 266.75 yards and the predicted distance of 270.3 yards is due to the fact that the club-head-speed is 103 miles per hour. In other words, the difference between the prediction of $\hat{y} = 270.3$ and $\overline{y} = 266.75$ is explained by the linear relation between club-head speed and distance. The observed distance when club-head speed is 103 miles per hour is 274 yards (see Table 3 on page 196). The difference between our predicted value, $\hat{y} = 270.3$, and the actual value, $y = 274$, is due to factors (variables) other than the club-head speed and random error. The differences just discussed are called **deviations**.

Figure 14

In Other Words

The word *deviations* comes from *deviate. To deviate* means "to stray".

The deviation between the observed value of the response variable, y, and the mean value of the response variable, $\overline{y}$, is called the **total deviation**, so total deviation $= y - \overline{y}$. The deviation between the predicted value of the response variable, $\hat{y}$, and the mean value of the response variable, $\overline{y}$, is called the **explained deviation**, so explained deviation $= \hat{y} - \overline{y}$. Finally, the deviation between the observed value of the response variable, y, and the predicted value of the response variable, $\hat{y}$, is called the **unexplained deviation**, so unexplained deviation $= y - \hat{y}$. See Figure 15.

Figure 15

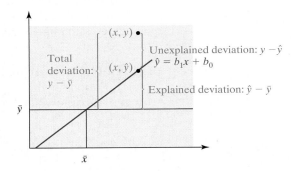

From the figure, it should be clear that

Total deviation = unexplained deviation + explained deviation

or

$$y - \overline{y} = (y - \hat{y}) + (\hat{y} - \overline{y})$$

Although beyond the scope of this text, it can be shown that

$$\sum (y - \overline{y})^2 = \sum (y - \hat{y})^2 + \sum (\hat{y} - \overline{y})^2$$

or

Total variation = unexplained variation + explained variation

Dividing both sides by total variation, we obtain

$$1 = \frac{\text{unexplained variation}}{\text{total variation}} + \frac{\text{explained variation}}{\text{total variation}}$$

Subtracting $\dfrac{\text{unexplained variation}}{\text{total variation}}$ from both sides, we obtain

$$R^2 = \frac{\text{explained variation}}{\text{total variation}} = 1 - \frac{\text{unexplained variation}}{\text{total variation}}$$

Unexplained variation is found by summing the squares of the residuals, $\sum \text{residuals}^2$. So the smaller the sum of squared residuals, the smaller the unexplained variation and, therefore, the larger R^2 will be. Therefore, the closer the observed y's are to the regression line (the predicted y's), the larger R^2 will be.

The coefficient of determination, R^2, is the square of the linear correlation coefficient for the least-squares regression model. Written in symbols, $R^2 = (r)^2$.

EXAMPLE 1

Computing the Coefficient of Determination, R^2

Problem: Compute and interpret the coefficient of determination, R^2, for the club-head speed versus distance data shown in Table 2.

Approach: To compute R^2, we square the linear correlation coefficient, r, found in Example 2 from Section 4.1 on page 182.

Solution: $R^2 = r^2 = 0.9387^2 = 0.8812 = 88.12\%$

Interpretation: 88.12% of the variation in distance is explained by the least-squares regression line, and 11.88% of the variation in distance is explained by other factors.

!**CAUTION**

Squaring the linear correlation coefficient to obtain the coefficient of determination works only for the least-squares linear regression model

$$\hat{y} = b_0 + b_1 x$$

The method does not work in general.

To help reinforce the concept of the coefficient of determination, consider the three data sets in Table 5.

Table 5					
Data Set A		**Data Set B**		**Data Set C**	
x	y	x	y	x	y
3.6	8.9	3.1	8.9	2.8	8.9
8.3	15.0	9.4	15.0	8.1	15.0
0.5	4.8	1.2	4.8	3.0	4.8
1.4	6.0	1.0	6.0	8.3	6.0
8.2	14.9	9.0	14.9	8.2	14.9
5.9	11.9	5.0	11.9	1.4	11.9
4.3	9.8	3.4	9.8	1.0	9.8
8.3	15.0	7.4	15.0	7.9	15.0
0.3	4.7	0.1	4.7	5.9	4.7
6.8	13.0	7.5	13.0	5.0	13.0

Figure 16(a) represents the scatter diagram of data set A, Figure 16(b) represents the scatter diagram of data set B, and Figure 16(c) represents the scatter diagram of data set C.

Figure 16

(a)

$\hat{y} = 1.2992x + 4.2159$

(b)

$\hat{y} = 1.1797x + 4.8438$

(c)

$\hat{y} = 0.4368x + 8.1464$

Notice that the y-values in each of the three data sets are the same. The variance of y is 17.49. If we look at the scatter diagram in Figure 16(a), we notice that almost 100% of the variability in y can be explained by the least-squares regression line, because the data almost lie perfectly on a straight line. In Figure 16(b), a high percentage of the variability in y can be explained by the least-squares regression line because the data have a strong linear relation. Higher x-values are associated with higher y-values. Finally, in Figure 16(c), a low percentage of the variability in y is explained by the least-squares regression line. If x increases, we cannot easily predict the change in y. If we compute the coefficient of determination, R^2, for the three data sets in Table 5, we obtain the following results:

Coefficient of determination for Data Set A: 99.99%

Coefficient of determination for Data Set B: 94.7%

Coefficient of determination for Data Set C: 9.4%

Notice that, as the explanatory ability of the line decreases, so does the coefficient of determination, R^2.

EXAMPLE 2 **Determining the Coefficient of Determination Using Technology**

Figure 17

SUMMARY OUTPUT

Regression Statistics	
Multiple R	0.938695838
R Square	0.881149876
Adjusted R Square	0.861341522
Standard Error	2.882638465
Observations	8

Problem: Determine the coefficient of determination, R^2, for the club-head speed versus distance data found in Example 2 from Section 4.1 using a statistical spreadsheet or graphing calculator with advanced statistical features.

Approach: We will use Excel to determine R^2. The steps for obtaining the coefficient of determination using Excel, MINITAB, and the TI-83/84 Plus graphing calculators are given in the Technology Step by Step on page 215.

Result: Figure 17 shows the results obtained from Excel. The coefficient of determination, R^2, is highlighted.

Now Work Problems 3 and 5.

4.3 ASSESS YOUR UNDERSTANDING

Concepts and Vocabulary

1. Suppose it is determined that $R^2 = 0.75$ when a linear regression is performed. Interpret this result.

2. Explain what is meant by total deviation, explained deviation, and unexplained deviation.

Skill Building

3. Match the coefficient of determination to the scatter diagram. The scales on the horizontal and vertical axis are the same for each scatter diagram.
    (a) $R^2 = 0.58$ III   (b) $R^2 = 0.90$ II   (c) $R^2 = 1$ IV   (d) $R^2 = 0.12$ I

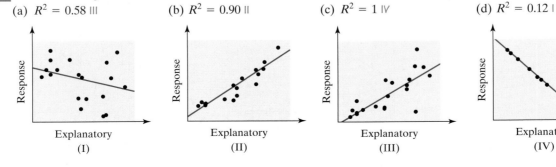

4. Use the linear correlation coefficient given to determine the coefficient of determination, R^2. Interpret each R^2.

    (a) $r = -0.32$ $R^2 = 10.24\%$ (b) $r = 0.13$ $R^2 = 1.69\%$ (c) $r = 0.40$ $R^2 = 16\%$ (d) $r = 0.93$ $R^2 = 86.49\%$

Applying the Concepts

5. The Other Old Faithful Perhaps you are familiar with the
NW famous Old Faithful geyser in Yellowstone National Park.
Another Old Faithful geyser is located in Calistoga in California's Napa Valley. The following data represent the
time between eruptions and the length of eruption for 11
randomly selected eruptions.

Time between Eruptions, x	Length of Eruption, y	Time between Eruptions, x	Length of Eruption, y
12.17	1.88	11.70	1.82
11.63	1.77	12.27	1.93
12.03	1.83	11.60	1.77
12.15	1.83	11.72	1.83
11.30	1.70		

Source: Ladonna Hansen, Park Curator

The coefficient of determination is determined to be
83.0%. Interpret this result.

6. Concrete As concrete cures, it gains strength. The following data represent the 7-day and 28-day strength (in
pounds per square inch) of a certain type of concrete.

7-Day Strength, x	28-Day Strength, y	7-Day Strength, x	28-Day Strength, y
2300	4070	2480	4120
3390	5220	3380	5020
2430	4640	2660	4890
2890	4620	2620	4190
3330	4850	3340	4630

The coefficient of determination, R^2, is determined to be
57.5%. Interpret this result.

7. Calories versus Sugar The following data represent the
number of calories per serving and the number of grams of
sugar per serving for a random sample of high-fiber cereals.

Calories, x	Sugar, y	Calories, x	Sugar, y
200	18	210	23
210	23	210	16
170	17	210	17
190	20	190	12
200	18	190	11
180	19	200	11

Source: Consumer Reports

(a) A scatter diagram with the least-squares regression
line is shown. The least-squares regression equation is
$\hat{y} = 0.0821x + 0.93$. Do you think that calories and
sugar content are linearly related? Why? No

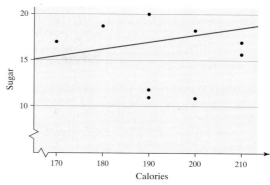

(b) The coefficient of determination, R^2, for these data is
6.8%. Interpret this result. Does this support your conclusion from part (a)? Why or why not?
(c) Suppose that we add Kellogg's All-Bran cereal, which
has 80 calories and 6 grams of sugar per serving, to the
data set. Draw a scatter diagram of the data with this cereal included. The coefficient of determination, R^2, with
Kellogg's All-Bran cereal included, is 42.1%. Interpret
this result. Why do you think that All-Bran cereal has
such a large impact on the value of the coefficient of determination?

8. Consumption versus Income The following data represent the per capita disposable income (income after taxes)
and per capita consumption in constant 2000 dollars in the
United States for 1996–2004.

Year	Per Capita Disposable Income, x	Per Capita Consumption, y
1996	22,546	20,835
1997	23,065	21,365
1998	24,131	22,183
1999	24,564	23,050
2000	25,472	23,862
2001	25,698	24,216
2002	26,229	24,715
2003	26,570	25,270
2004	27,240	25,965

Source: Bureau of Economic Analysis

(a) A scatter diagram is shown. Do you think that per capita
disposable income and per capita consumption are linearly related? Why or why not? Yes

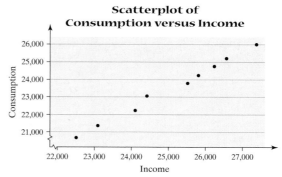

(b) The coefficient of determination, R^2, for the data is
99.4%. Interpret this result. Does this support your conclusion from part (a)? Why?

Problems 9–12 use the results from Problems 23–26 in Section 4.1 and Problems 17–20 in Section 4.2.

9. **Height versus Head Circumference** Use the results from
   NW Problem 23 in Section 4.1 and Problem 17 in Section 4.2 to
   (a) compute the coefficient of determination, R^2. *83.0%*
   (b) interpret the coefficient of determination.

10. **Gestation Period versus Life Expectancy** Use the results from Problem 24 in Section 4.1 and Problem 18 in Section 4.2 to
    (a) compute the coefficient of determination, R^2. *52.7%*
    (b) interpret the coefficient of determination.

11. **Weight of a Car versus Miles per Gallon** Use the results from Problem 25 in Section 4.1 and Problem 19 in Section 4.2 to
    (a) compute the coefficient of determination, R^2. *79.6%*
    (b) interpret the coefficient of determination.

12. **Bone Length** Use the results from Problem 26 in Section 4.1 and Problem 20 in Section 4.2 to

    (a) compute the coefficient of determination, R^2. *90.5%*
    (b) interpret the coefficient of determination.

13. **Weight of a Car versus Miles per Gallon** Suppose we add the Dodge Viper to the data in Problem 19 in Section 4.2. A Dodge Viper weighs 3425 pounds and gets 11 miles per gallon. Compute the coefficient of determination of the expanded data set. What effect does the addition of the Viper to the data set have on R^2?
    52.7%; decreases

14. **Gestation Period versus Life Expectancy** Suppose we add humans to the data in Problem 18 in Section 4.2. Humans have a gestation period of 268 days and a life expectancy of 76.5 years. Compute the coefficient of determination of the expanded data set. What effect does the addition of humans to the data set have on R^2?
    71.7%; increases

Consumer Reports Fit to Drink

The taste, color, and clarity of the water coming out of home faucets have long concerned consumers. Recent reports of lead and parasite contamination have made unappetizing water a health, as well as an esthetic, concern. Water companies are struggling to contain cryptosporidium, a parasite that has caused outbreaks of illness that may be fatal to people with a weakened immune system. Even chlorination, which has rid drinking water of infectious organisms that once killed people by the thousands, is under suspicion as an indirect cause of miscarriages and cancer.

Concerns about water quality and taste have made home filtering increasingly popular. To find out how well they work, technicians at Consumer Reports tested 14 models to determine how well they filtered contaminants and whether they could improve the taste of our cabbage-soup testing mixture.

To test chloroform and lead removal, we added concentrated amounts of both to our water, along with calcium nitrate to increase water hardness. Every few days we analyzed the water to measure chloroform and lead content. The following table contains the lead measurements for one of the models tested.

No. Gallons Processed	% Lead Removed
25	85
26	87
73	86
75	88
123	90
126	87
175	92
177	94

(a) Construct a scatter diagram of the data using % Lead Removed as the response variable.

(b) Does the relationship between No. Gallons Processed and % Lead Removed appear to be linear? If not, describe the relationship.

(c) Calculate the linear correlation coefficient between No. Gallons Processed and % Lead Removed. Based on the scatter diagram constructed in part (a) and your answer to part (b), is this measure useful? What is R^2? Interpret R^2.

(d) Fit a linear regression model to these data.

(e) Using statistical software or a graphing calculator with advanced statistical features, fit a quadratic model to these data. What is R^2? Which model appears to fit the data better?

(f) Given the nature of the variables being measured, describe the type of curve you would expect to show the true relationship between these variables (linear, quadratic, exponential, S-shaped). Support your position.

Note to Readers: In many cases, our test protocol and analytical methods are more complicated than described in these examples. The data and discussions have been modified to make the material more appropriate for the audience.

© by Consumers Union of U.S., Inc., Yonkers, NY 10703-1057, a nonprofit organization. Reprinted with permission.

Technology Step by Step	**Determining R^2**
TI-83/84 Plus	Use the same steps that were followed to obtain the correlation coefficient to obtain R^2. Diagnostics must be on.
MINITAB	This is provided in the standard regression output.
Excel	This is provided in the standard regression output.

CHAPTER 4 # Review

Summary

In this chapter, we introduced techniques that allow us to describe the relation between two quantitative variables. The first step in identifying the type of relation that might exist is to draw a scatter diagram. The explanatory variable is plotted on the horizontal axis and the corresponding response variable on the vertical axis. The scatter diagram can be used to discover whether the relation between the explanatory and the response variables is linear. In addition, for linear relations, we can judge whether the linear relation shows positive or negative association.

A numerical measure for the strength of linear relation between two quantitative variables is the linear correlation coefficient. It is a number between −1 and 1, inclusive. Values of the correlation coefficient near −1 are indicative of a negative linear relation between the two variables. Values of the correlation coefficient near +1 indicate a positive linear relation between the two variables. If the correlation coefficient is near 0, then there is little *linear* relation between the two variables.

Once a linear relation between the two variables has been discovered, we describe the relation by finding the least-squares regression line. This line best describes the linear relation between the explanatory and the response variables. We can use the least-squares regression line to predict a value of the response variable for a given value of the explanatory variable.

The coefficient of determination, R^2, measures the percent of variation in the response variable that is explained by the least-squares regression line. It is a measure between 0 and 1 inclusive. The closer R^2 is to 1, the more explanatory value the line has.

One item worth mentioning again is that a researcher should never claim causation between two variables in a study unless the data are experimental. Observational data allow us to say that two variables might be associated, but we cannot claim causation.

Formulas

Correlation Coefficient

$$r = \frac{\sum\left(\dfrac{x_i - \overline{x}}{s_x}\right)\left(\dfrac{y_i - \overline{y}}{s_y}\right)}{n - 1}$$

Coefficient of Determination, R^2

$$R^2 = \frac{\text{variation explained by explanatory variable}}{\text{total variation}}$$

$$= 1 - \frac{\text{unexplained variation}}{\text{total variation}}$$

$$= r^2 \text{ for the least-squares regression model } \hat{y} = b_1 x + b_0$$

Equation of the Least-Squares Regression Line

The equation of the least-squares regression line is given by

$$\hat{y} = b_1 x + b_0$$

where

$\hat{y}$ is the predicted value of the response variable,

$b_1 = r \cdot \dfrac{s_y}{s_x}$ is the slope of the least-squares regression line, and

$b_0 = \overline{y} - b_1 \overline{x}$ is the intercept of the least-squares regression line.

Vocabulary

Bivariate data (p. 176)
Response variable (p. 177)
Explanatory variable (p. 177)
Predictor variable (p. 177)
Lurking variable (p. 177)
Scatter diagram (p. 177)
Positively associated (p. 179)

Negatively associated (p. 179)
Linear correlation coefficient (p. 180)
Correlation matrix (p. 183)
Residuals (p. 197)
Least-squares regression line (p. 198)
Slope (p. 198)
y-intercept (p. 198)

Outside the scope of the model (p. 201)
Coefficient of determination (p. 209)
Total deviation (p. 210)
Explained deviation (p. 210)
Unexplained deviation (p. 210)

Objectives

Section	You should be able to . . .	Example	Review Exercises
4.1	1 Draw and interpret scatter diagrams (p. 177)	1, 3	1(a)–4(a), 9(a), 10(a), 15(a)
	2 Understand the properties of the linear correlation coefficient (p. 179)		18
	3 Compute and interpret the linear correlation coefficient (p. 182)	2, 3	1(b)–4(b), 15(b)
	4 Determine whether there is a linear relation between two variables (p. 185)	4	1(c)–4(c)
4.2	1 Find the least-squares regression line and use the line to make predictions (p. 197)	2, 3	5(a)–8(a), 9(d)–10(d)
	2 Interpret the slope and y-intercept of the least-squares regression line (p. 201)	Page 218	5(c)–8(c)
	3 Compute the sum of squared residuals (p. 202)	4	9(g), 10(g)
4.3	1 Compute and interpret the coefficient of determination (p. 209)	1, 2	11–14

Review Exercises

1. **Engine Displacement versus Fuel Economy** The following data represent the size of a car's engine (in liters) versus its miles per gallon in the city for various 2005 domestic automobiles.

Car	Engine Displacement (liters), x	City MPG, y	Car	Engine Displacement (liters), x	City MPG, y
Buick Century	3.1	20	Ford Crown Victoria	4.6	18
Buick LeSabre	3.8	20	Ford Focus	2.0	26
Cadillac DeVille	4.6	18	Ford Mustang	3.8	20
Chevrolet Cavalier	2.2	25	Mercury Sable	3.0	19
Chevrolet Impala	3.8	21	Pontiac Grand Am	3.4	20
Chevrolet Malibu	2.2	24	Pontiac Sunfire	2.2	24
Chrysler Sebring Sedan	2.7	22	Saturn Ion	2.2	26
Dodge Magnum	3.5	21			

Source: www.roadandtrack.com

(a) Draw a scatter diagram treating engine displacement as the explanatory variable and miles per gallon as the response variable.

(b) Compute the linear correlation coefficient between engine displacement and miles per gallon. -0.895

(c) Based on the scatter diagram and the linear correlation coefficient, comment on the type of relation that appears to exist between the two variables.

2. **Temperature versus Cricket Chirps** Crickets make a chirping noise by sliding their wings rapidly over each other. Perhaps you have noticed that the number of chirps seems to increase with the temperature. The following data list the temperature (in degrees Fahrenheit) and the number of chirps per second for the striped ground cricket.

(a) Draw a scatter diagram treating temperature as the explanatory variable and chirps per second as the response variable.

(b) Compute the linear correlation coefficient between temperature and chirps per second. 0.835

(c) Based upon the scatter diagram and the linear correlation coefficient, comment on the type of relation that appears to exist between the two variables.

Temperature, x	Chirps per Second, y	Temperature, x	Chirps per Second, y
88.6	20.0	71.6	16.0
93.3	19.8	84.3	18.4
80.6	17.1	75.2	15.5
69.7	14.7	82.0	17.1
69.4	15.4	83.3	16.2
79.6	15.0	82.6	17.2
80.6	16.0	83.5	17.0
76.3	14.4		

Source: Pierce, George W. *The Songs of Insects.* Cambridge, MA: Harvard University Press, 1949, pp. 12–21

3. Apartments The following data represent the square footage and rents for apartments in the Borough of Queens and Nassau County, New York.

Queens (New York City)		Nassau County (Long Island)	
Square Footage, x	Rent Per Month, y	Square Footage, x	Rent Per Month, y
500	650	1100	1875
588	1215	588	1075
1000	2000	1250	1775
688	1655	556	1050
825	1250	825	1300
460	1805	743	1475
1259	2700	660	1315
650	1200	975	1400
560	1250	1429	1900
1073	2350	800	1650
1452	3300	1906	4625
1305	3100	1077	1395

Source: apartments.com

(3b) Queens: 0.909; Nassau: 0.867

(a) On the same graph, draw a scatter diagram for both Queens and Nassau County apartments, treating square footage as the explanatory variable. Use a different plotting symbol for each group.
(b) Compute the linear correlation coefficient between square footage and rent for each location.
(c) Given the scatter diagram and the linear correlation coefficient, comment on the type of relation that appears to exist between the two variables for each group.
(d) Does location appear to be a factor in rent?

4. Boys versus Girls The following data represent the height (in inches) of boys and girls between the ages of 2 and 10 years.

Age	Boy Height, x	Girl Height, y	Age	Boy Height, x	Girl Height, y
2	36.1	39.0	6	49.8	43.7
2	34.2	38.6	7	43.2	50.5
2	31.1	33.6	7	47.9	47.7
3	36.3	41.3	8	51.4	44.0
3	39.5	40.9	8	48.3	62.1
4	41.5	43.2	8	50.9	44.8
4	38.6	39.8	9	52.2	50.9
5	45.6	50.5	9	51.3	55.6
5	44.8	38.3	10	55.6	61.4
5	44.6	43.9	10	59.5	50.8

Source: National Center for Health Statistics

(a) On the same graph, draw a scatter diagram for both boys and girls, treating age as the explanatory variable. Use a different plotting symbol for each gender.
(b) Compute the linear correlation coefficient between age and height for each gender. Boys: 0.9489; girls: 0.8025
(c) Based on the scatter diagram and the linear correlation coefficient, comment on the type of relation that appears to exist between the age and height for each gender.
(d) Does gender appear to be a factor in determining height?

5. Using the data and results from Problem 1, do the following:
(a) Find the least-squares regression line, treating engine displacement as the explanatory variable.
(b) Draw the least-squares regression line on the scatter diagram.
(c) Interpret the slope and y-intercept, if appropriate.
(d) Predict the miles per gallon of a Ford Mustang whose engine displacement is 3.8 liters. 19.8 mpg   (e)0.2 mpg
(e) Compute the residual of the prediction found in part (d).
(f) Is the miles per gallon above or below average for a Ford Mustang? Above
(5a) $\hat{y} = -2.8x + 30.4$

6. Using the data and results from Problem 2, do the following:
(a) Find the least-squares regression line, treating temperature as the explanatory variable and chirps per second as the response variable. $\hat{y} = 0.2119x - 0.309$
(b) Draw the least-squares regression line on the scatter diagram.
(c) Interpret the slope and y-intercept, if appropriate.
(d) Predict the chirps per second if it is 83.3°F. 17.3
(e) Compute the residual of the prediction found in part (d).
(f) Were chirps per second above or below average at 83.3°F? Below   (e) − 1.1

7. Using the Queens data and results from Problem 3, do the following: (7e) −$538
(a) Find the least-squares regression line, treating square footage as the explanatory variable. $\hat{y} = 2.21x - 34$
(b) Draw the least-squares regression line on the scatter diagram.
(c) Interpret the slope and y-intercept, if appropriate.
(d) Predict the rent of an 825-square-foot apartment. $1788
(e) Compute the residual of the prediction found in part (d).
(f) Is this apartment's rent above or below average? Below

8. Using the Boy Height data and results from Problem 4, do the following:
(a) Find the least-squares regression line, treating age as the explanatory variable and height as the response variable. $\hat{y} = 2.6351x + 29.705$
(b) Draw the least-squares regression line on the scatter diagram.   (e) 4.3 inches
(c) Interpret the slope and y-intercept, if appropriate.
(d) Predict the height of a 6-year-old boy. 45.5 inches
(e) Compute the residual of the prediction found in part (d).
(f) Is this boy's height above or below average? Above

In Problems 9 and 10, do the following:
(a) Draw a scatter diagram treating x as the explanatory variable and y as the response variable.
(b) Select two points from the scatter diagram, and find the equation of the line containing the points selected.
(c) Graph the line found in part (b) on the scatter diagram.
(d) Determine the least-squares regression line.
(e) Graph the least-squares regression line on the scatter diagram.
(f) Compute the sum of the squared residuals for the line found in part (b).
(g) Compute the sum of the squared residuals for the least-squares regression line found in part (d).
(h) Comment on the fit of the line found in part (b) versus the least-squares regression line found in part (d).

9.

x	3	4	6	7	9
y	2.1	4.2	7.2	8.1	10.6

10.

x	10	14	17	18	21
y	105	94	82	76	63

11. Use the results from Problems 1 and 5 to compute and interpret R^2. *80.2%*

12. Use the results from Problems 2 and 6 to compute and interpret R^2. *69.7%*

13. Use Queens data and the results from Problems 3 and 7 to compute and interpret R^2. *82.7%*

14. Use the results from Problems 4 and 8 to compute and interpret R^2. *90.0%*

15. 200-Meter Dash The following data represent the gold medal times, in seconds, for men and women in the 200-meter dash at the summer Olympics from 1948 to 2004.

Year	Time (Men)	Time (Women)	Year	Time (Men)	Time (Women)
1948	21.10	24.40	1980	20.19	22.03
1952	20.70	23.70	1984	19.80	21.81
1956	20.60	23.40	1988	19.75	21.34
1960	20.50	24.00	1992	20.01	21.81
1964	20.30	23.00	1996	19.32	22.12
1968	19.80	22.50	2000	20.09	21.84
1972	20.00	22.40	2004	19.79	22.05
1976	20.23	22.37			

Source: www.factmonster.com

(a) Draw a scatter diagram of the data using time for men as the explanatory variable and time for women as the response variable.
(b) Compute the correlation coefficient for the data. *0.835*
(c) Based on your results from parts (a) and (b), what type of relation appears to exist between the gold medal time for men and the gold medal time for women in the 200-meter dash? Do you believe that the gold medal time for men causes the gold medal time for women? *No*

16. Wine and Your Heart The health benefits of moderate wine consumption are well documented. Researchers wanted to determine if alcohol consumption is positively related to heart-rate variability (HRV) in women with coronary heart disease (CHD). The purpose of the study was to shed some doubt on the heart–health benefits of wine. The researchers surveyed female patients who have recently been released from the hospital after successful heart procedures such as bypass surgery or angioplasty. A questionnaire evaluated self-reported consumption of individual alcoholic beverage types: beer, wine, and spirits. Other characteristics, such as age, body mass index, smoking habits, history of diabetes, menopausal status, and educational status, were also assessed. The researchers found that wine intake was associated with increased HRV. Based on this study, can we conclude that increased wine consumption in women with recent heart procedures causes an increase in heart-rate variability? Why?

17. Analyzing a Newspaper Article In a newspaper article written in the *Chicago Tribune* on September 29, 2002, it was claimed that poorer school districts have shorter school days.
(a) The following scatter diagram was drawn using the data supplied in the article. In this scatter diagram, the response variable is length of the school day and the explanatory variable is percent of the population that is low income. The correlation between length and income is −0.461. Do you think that the scatter diagram and correlation coefficient support the position of the article?

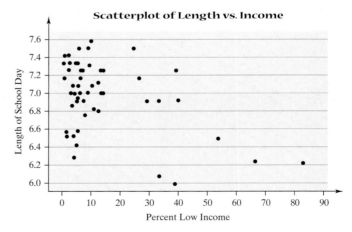

Scatterplot of Length vs. Income

(b) The least-squares regression line between length, y, and income, x, is $\hat{y} = -0.0102x + 7.11$. Interpret the slope of this regression line. Does it make sense to interpret the intercept? If so, interpret the intercept.
(c) Predict the length of the school day for a district in which 20% of the population is low income by letting $x = 20$.

(d) This same article included average Prairie State Achievement Examination (PSAE) scores for each district. The article implied that shorter school days result in lower PSAE scores. The correlation between PSAE score and length of school day is 0.517. A scatter diagram treating PSAE as the response variable is shown below. Do you believe that a longer school day is positively associated with a higher PSAE score?

(e) The correlation between percentage of the population that is low income and PSAE score is −0.720. A scatter diagram treating PSAE score as the response variable is shown below.

Do you believe that percentage of the population that is low income is negatively associated with PSAE score?

(f) Can you think of any lurking variables that are playing a role in this study?

18. List the seven properties of the linear correlation coefficient.

THE CHAPTER 4 CASE STUDY IS LOCATED ON THE CD THAT ACCOMPANIES THIS TEXT.

PART 3 Probability and Probability Distributions

CHAPTER 5
Probability

CHAPTER 6
Discrete Probability Distributions

CHAPTER 7
The Normal Probability Distribution

We now take a break from the statistical process. Why? Recall that we mentioned that inferential statistics uses methods that generalize results obtained from a sample to the population and measures their reliability. But how can we measure their reliability? It turns out that the methods we use to generalize results from a sample to a population are based on *probability* and *probability models*. **Probability** is a measure of the likelihood that something occurs.

Probability

Outline

5.1   Probability Rules

5.2   The Addition Rule and Complements

5.3   Independence and the Multiplication Rule

5.4   Conditional Probability and the General Multiplication Rule

5.5   Counting Techniques

■ Chapter Review

■ Case Study: The Case of the Body in the Bag (On CD)

DECISIONS

Have you ever watched a sporting event on television in which the announcer cites an obscure statistic? Where do these numbers come from? Well, pretend that you are the statistician for your favorite sports team. Your job is to compile strange or obscure probabilities regarding your favorite team and a competing team. See the Decisions project on page 261.

●●● Putting It All Together

In Chapter 1, we learned the methods of collecting data. In Chapters 2 through 4, we learned how to summarize raw data using tables, graphs, and numbers. As far as the statistical process goes, we have discussed the collecting, organizing, and summarizing part of the process.

Before we can proceed with the analysis of data, we introduce probability, which forms the basis of inferential statistics. Why? Well, we can think of the probability of an outcome as the likelihood of observing that outcome. If something has a high likelihood of happening, it has a high probability (close to 1). If something has a small chance of happening, it has a low probability (close to 0). For example, in rolling a single die, it is unlikely that we would roll five straight sixes, so this result has a low probability. In fact, the probability of rolling five straight sixes is 0.0001286. So, if we were playing a game that entailed throwing a single die, and one of the players threw five sixes in a row, we would consider the player to be lucky (or a cheater) because it is such an unusual occurrence. Statisticians use probability in the same way. If something occurs that has a low probability, we investigate to find out "what's up."

5.1 Probability Rules

Preparing for This Section   Before getting started, review the following:

- Relative frequency (Section 2.1, p. 56)

Objectives

  1   **Understand the rules of probabilities**

  2   **Compute and interpret probabilities using the empirical method**

  3   **Compute and interpret probabilities using the classical method**

  4   **Use simulation to obtain data based on probabilities**

  5   **Understand subjective probabilities**

Note to Instructor
Probabilities can be expressed as fractions, decimals, or percents. You may want to give a brief review of how to convert from one form to the other.

In Other Words
Probability describes how likely it is that some event will happen. If we look at the proportion of times an event has occurred over a long period of time (or over a large number of trials), we can be more certain of the likelihood of its occurrence.

Probability is a measure of the likelihood of a random phenomenon or chance behavior. Probability describes the long-term proportion with which a certain **outcome** will occur in situations with short-term uncertainty.

The long-term predictability of chance behavior is best understood through a simple experiment. Flip a coin 100 times and compute the proportion of heads observed after each toss of the coin. Suppose the first flip is tails, so the proportion of heads is $\frac{0}{1}$; the second flip is heads, so the proportion of heads is $\frac{1}{2}$; the third flip is heads, so the proportion of heads is $\frac{2}{3}$; and so on. Plot the proportion of heads versus the number of flips and obtain the graph in Figure 1(a). We repeat this experiment with the results shown in Figure 1(b).

Look at the graphs in Figures 1(a) and (b). Notice that in the short term (fewer flips of the coin) the observed proportion of heads is different and unpredictable for each experiment. As the number of flips of the coin increases, however, both graphs tend toward a proportion of 0.5. This is the basic premise of probability. Probability deals with experiments that yield random short-term results or **outcomes** yet reveal long-term predictability. **The long-term proportion with which a certain outcome is observed is the probability of that outcome**. So, we say that the probability of observing a head is $\frac{1}{2}$ or 50% or 0.5, because as we flip the coin more times, the proportion of heads tends toward $\frac{1}{2}$. This phenomenon is referred to as the *Law of Large Numbers*.

Figure 1

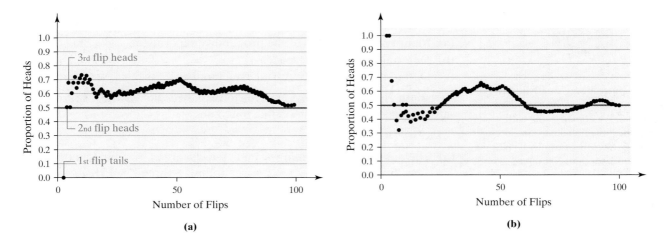

(a)   (b)

The Law of Large Numbers

As the number of repetitions of a probability experiment increases, the proportion with which a certain outcome is observed gets closer to the probability of the outcome.

Note to Instructor

In-class activity: Have each student flip a coin three times. Use the results to find the probability of a head. Repeat this experiment a few times. Use the cumulative results to illustrate the Law of Large Numbers. You may also want to load the probability applet to demonstrate the Law of Large Numbers.

The Law of Large Numbers is illustrated in Figure 1. For a few flips of the coin, the proportion of heads fluctuates wildly around 0.5, but as the number of flips increases, the proportion of heads settles down near 0.5. Jakob Bernoulli (a major contributor to the field of probability) believed that the Law of Large Numbers was common sense. This is evident in the following quote from his text *Ars Conjectandi*: "For even the most stupid of men, by some instinct of nature, by himself and without any instruction, is convinced that the more observations have been made, the less danger there is of wandering from one's goal."

In probability, an **experiment** is any process with uncertain results that can be repeated. The result of any single trial of the experiment is not known ahead of time. However, the results of the experiment over many trials produce regular patterns that enable us to predict with remarkable accuracy. For example, an insurance company cannot know ahead of time whether a particular 16-year-old driver will be involved in an accident over the course of a year. However, based on historical records, the company can be fairly certain that about three out of every ten 16-year-old male drivers will be involved in a traffic accident during the course of a year. Therefore, of the 816,000 16-year-old drivers (816,000 repetitions of the experiment), the insurance company is fairly confident that about 30%, or 244,800, of the drivers will be involved in an accident. This prediction forms the basis for establishing insurance rates for any particular 16-year-old male driver.

We now introduce some terminology that we will need to study probability.

Definitions

In Other Words
An outcome is the result of one trial of a probability experiment. The sample space is a list of all possible results of a probability experiment.

The **sample space**, S, of a probability experiment is the collection of all possible outcomes.

An **event** is any collection of outcomes from a probability experiment. An event may consist of one outcome or more than one outcome. We will denote events with one outcome, sometimes called *simple events*, e_i. In general, events are denoted using capital letters such as E.

We present an example to illustrate the definitions.

EXAMPLE 1

A *fair die* is one in which each possible outcome is equally likely. For example, rolling a 2 is just as likely as rolling a 5. We contrast this with a *loaded* die, in which a certain outcome is more likely. For example, if rolling a 1 is more likely than rolling a 2, 3, 4, 5, or 6, the die is loaded.

Identifying Events and the Sample Space of a Probability Experiment

Problem: A probability experiment consists of rolling a single *fair* die.

(a) Identify the outcomes of the probability experiment.
(b) Determine the sample space.
(c) Define the event E = "roll an even number."

Approach: The outcomes are the possible results of the experiment. The sample space is a list of all possible outcomes.

Solution

(a) The outcomes from rolling a single fair die are e_1 = "rolling a one" = $\{1\}$, e_2 = "rolling a two" = $\{2\}$, e_3 = "rolling a three" = $\{3\}$, e_4 = "rolling a four" = $\{4\}$, e_5 = "rolling a five" = $\{5\}$, and e_6 = "rolling a six" = $\{6\}$.

(b) The set of all possible outcomes forms the sample space, $S = \{1, 2, 3, 4, 5, 6\}$. There are 6 outcomes in the sample space.

(c) The event E = "roll an even number" = $\{2, 4, 6\}$.

1 Understand the Rules of Probabilities

Probabilities have some rules that must be satisfied. In these rules, the notation $P(E)$ means "the probability of event E."

> **Rules of Probabilities**
>
> 1. The probability of any event E, $P(E)$, must be greater than or equal to 0 and less than or equal to 1. If we let E denote any event, then $0 \leq P(E) \leq 1$.
> 2. The sum of the probabilities of all outcomes must equal 1. That is, if the sample space $S = \{e_1, e_2, \ldots, e_n\}$, then
>
> $$P(e_1) + P(e_2) + \cdots + P(e_n) = 1$$

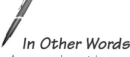

In Other Words

Rule 1 states that probabilities less than 0 or greater than 1 are not possible. Therefore, probabilities such as 1.32 or −0.3 are not possible. Rule 2 states when the probabilities of all outcomes are added, the sum must be 1.

A **probability model** lists the possible outcomes of a probability experiment and each outcome's probability. A probability model must satisfy rules 1 and 2 of the rules of probabilities.

EXAMPLE 2

A Probability Model

In a bag of M&M plain candies, the colors of the candies can be brown, yellow, red, blue, orange, or green. Suppose that a candy is randomly selected from a bag. Table 1 shows each color and the probability of drawing that color.

To verify that this is a probability model, we must show that rules 1 and 2 of the rules of probabilities are satisfied.

Each probability is greater than or equal to 0 and less than or equal to 1, so rule 1 is satisfied.

Because

$$0.13 + 0.14 + 0.13 + 0.24 + 0.20 + 0.16 = 1$$

rule 2 is also satisfied. The table is an example of a probability model. ▪

Table 1

Color	Probability
Brown	0.13
Yellow	0.14
Red	0.13
Blue	0.24
Orange	0.20
Green	0.16

Source: Mars, Incorporated

If an event is **impossible**, the probability of the event is 0. If an event is a **certainty**, the probability of the event is 1.

Now Work Problem 9.

Note to Instructor

The interpretation of probability given here is important to emphasize. We will use this interpretation again when we discuss confidence intervals.

The closer a probability is to 1, the more likely the event will occur. The closer a probability is to 0, the less likely the event will occur. For example, an event with probability 0.8 is more likely to occur than an event with probability 0.75. An event with probability 0.8 will occur about 80 times out of 100 repetitions of the experiment, while an event with probability 0.75 will occur about 75 times out of 100.

Be careful of this interpretation. Just because an event has a probability of 0.75 does not mean that the event *must* occur 75 times out of 100. It means that we *expect* the number of occurrences to be close to 75 in 100 trials of the experiment. The more repetitions of the probability experiment, the closer the proportion with which the event occurs will be to 0.75 (the Law of Large Numbers).

One goal of this course is to learn how probabilities can be used to identify *unusual events*.

In Other Words

An unusual event is an event that is not likely to occur.

Definition

An **unusual event** is an event that has a low probability of occurring.

Note to Instructor

Ask students to cite some real-life examples of unusual events.

Typically, an event with a probability less than 5% is considered unusual, but this *cutoff point* is not set in stone. The researcher and the context of the problem determine the probability that separates unusual events from *not so unusual events*.

For example, suppose the probability of being wrongly convicted of a capital crime punishable by death is 3%. The probability is too high in light of the consequences (death for the wrongly convicted), so the event is not unusual (unlikely) enough. We would want this probability to be as close to zero as possible.

Now suppose that you are planning a picnic on a day for which there is a 3% chance of rain. In this context, you would consider "rain" an unusual (unlikely) event and proceed with the picnic plans.

The point is this: Selecting a probability that separates unusual events from not so unusual events is subjective and depends on the situation. Statisticians typically use cutoff points of 1%, 5%, and 10%. For many circumstances, any event that occurs with probability of 5% or less will be considered unusual.

Next, we introduce three methods for determining the probability of an event: (1) the empirical method, (2) the classical method, and (3) the subjective method.

!CAUTION
A probability of 0.05 should not always be used to separate unusual events from not so unusual events.

② Compute and Interpret Probabilities Using the Empirical Method

Because probabilities deal with the long-term proportion with which a particular outcome is observed, it makes sense that we begin our discussion of determining probabilities using the idea of relative frequency. Probabilities computed in this manner rely on empirical evidence, that is, evidence based on the outcomes of a probability experiment.

Approximating Probabilities Using the Empirical Approach

The probability of an event E is approximately the number of times event E is observed divided by the number of repetitions of the experiment.

$$P(E) \approx \text{ relative frequency of } E = \frac{\text{frequency of } E}{\text{number of trials of experiment}} \quad (1)$$

The probability obtained using the empirical approach is approximate because different *runs* of the probability experiment lead to different outcomes and, therefore, different estimates of $P(E)$. Consider flipping a coin 20 times and recording the number of heads. Use the results of the experiment to estimate the probability of obtaining a head. Now repeat the experiment. Because the results of the second run of the experiment do not necessarily yield the same results, we cannot say the probability *equals* some proportion; rather we say the probability is *approximately* the proportion. As we increase the number of trials of a probability experiment, our estimate becomes more accurate (again, the Law of Large Numbers).

EXAMPLE 3   Using Relative Frequencies to Approximate Probabilities

A pit boss wanted to approximate the probability of rolling a seven using a pair of dice that have been in use for a while. To do this he rolls the dice 100 times and records 15 sevens. The probability of rolling a seven is approximately $\frac{15}{100} = 0.15$.

When we survey a random sample of individuals, the probabilities computed from the survey are approximate. In fact, we can think of a survey as a probability experiment, since the results of a survey are different each time the survey is conducted because different people are included.

EXAMPLE 4 Building a Probability Model from Survey Data

Table 2

Means of Travel	Frequency
Drive alone	153
Carpool	22
Public transportation	10
Walk	5
Other means	3
Work at home	7

Table 3

Means of Travel	Probability
Drive alone	0.765
Carpool	0.11
Public transportation	0.05
Walk	0.025
Other means	0.015
Work at home	0.035

Problem: The data in Table 2 represent the results of a survey in which 200 people were asked their means of travel to work.

(a) Use the survey data to build a probability model for means of travel to work.
(b) Would it be unusual to randomly select an individual who walks to work?

Approach: To build a probability model, we estimate the probability of each outcome by determining the relative frequency of each outcome.

Solution

(a) There are $153 + 22 + \cdots + 7 = 200$ individuals in the survey. The individuals can be thought of as trials of the probability experiment. The relative frequency for "drive alone" is $\dfrac{153}{200} = 0.765$. We compute the relative frequency of the other outcomes similarly and obtain the probability model in Table 3.

(b) The probability that an individual walks to work is 0.025. It is somewhat unusual to randomly choose a person who walks to work. _____

Now Work Problem 37.

③ Compute and Interpret Probabilities Using the Classical Method

When using the empirical method, we obtain an approximate probability of an event by conducting a probability experiment.

The classical method of computing probabilities does not require that a probability experiment actually be performed. Rather, we use counting techniques to determine the probability of an event.

The classical method of computing probabilities requires *equally likely outcomes*. An experiment is said to have **equally likely outcomes** when each outcome has the same probability of occurring. For example, in throwing a fair die once, each of the six outcomes in the sample space, $\{1, 2, 3, 4, 5, 6\}$, has an equal chance of occurring. Contrast this situation with a loaded die in which a five or six is twice as likely to occur as a one, two, three, or four.

Computing Probability Using the Classical Method

If an experiment has n equally likely outcomes and if the number of ways that an event E can occur is m, then the probability of E, $P(E)$, is

$$P(E) = \frac{\text{Number of ways that } E \text{ can occur}}{\text{Number of possible outcomes}} = \frac{m}{n} \qquad (2)$$

So, if S is the sample space of this experiment,

$$P(E) = \frac{N(E)}{N(S)} \qquad (3)$$

where $N(E)$ is the number of outcomes in E, and $N(S)$ is the number of outcomes in the sample space.

EXAMPLE 5 ## Computing Probabilities Using the Classical Method

Problem: A pair of fair dice is rolled.

(a) Compute the probability of rolling a seven.

(b) Compute the probability of rolling "snake eyes"; that is, compute the probability of rolling a two.

(c) Comment on the likelihood of rolling a seven versus rolling a two.

Approach: To compute probabilities using the classical method, we count the number of outcomes in the sample space and count the number of ways the event can occur.

Solution

(a) In rolling a pair of fair dice, there are 36 equally likely outcomes in the sample space, as shown in Figure 2.

Figure 2

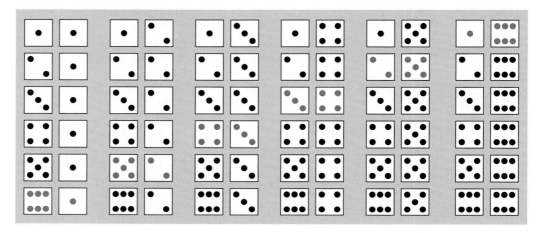

So $N(S) = 36$. The event $E =$ "roll a seven" $= \{(1, 6), (2, 5), (3, 4), (4, 3), (5, 2), (6, 1)\}$ is composed of six outcomes, so $N(E) = 6$. Using Formula (3), the probability of rolling a seven is

$$P(E) = P(\text{roll a seven}) = \frac{N(E)}{N(S)} = \frac{6}{36} = \frac{1}{6}$$

(b) The event $F =$ "roll a two" $= \{(1, 1)\}$ has one outcome, so $N(E) = 1$. Using Formula (3), the probability of rolling a two is

$$P(F) = P(\text{roll a two}) = \frac{N(E)}{N(S)} = \frac{1}{36}$$

(c) Since $P(\text{roll a seven}) = \frac{6}{36}$ and $P(\text{roll a two}) = \frac{1}{36}$, rolling a seven is six times as likely as rolling a two. In other words, in 36 rolls of the dice, we *expect* to observe about 6 sevens and only 1 two. ▬▬▬

If we compare the empirical probability of rolling a seven, 0.15, obtained in Example 3, to the classical probability of rolling a seven, $\frac{1}{6} \approx 0.167$, obtained in Example 5(a), we see that they are not too far apart. In fact, if the dice are fair, we expect the relative frequency of sevens to get closer to 0.167 as we increase the number of rolls of the dice. That is, if the dice are fair, the empirical probability will get closer to the classical probability as the number of trials of the experiment increases. If the two probabilities do not get closer together, we may suspect that the dice are not fair.

In simple random sampling, each individual has the same chance of being selected. Therefore, we can use the classical method to compute the probability of obtaining a specific sample.

EXAMPLE 6

Historical Note

Girolamo Cardano (in English Jerome Cardan) was born in Pavia, Italy, on September 24, 1501. He was an illegitimate child whose father was Fazio Cardano, a lawyer in Milan. Fazio was a part-time mathematician and taught Girolamo. In 1526, Cardano earned his medical degree. Shortly thereafter, his father died. Unable to maintain a medical practice, Cardano spent his inheritance and turned to gambling to help support himself. Cardano developed an understanding of probability that helped him to win. He wrote a booklet on probability, *Liber de Ludo Alaea*, which was not printed until 1663, 87 years after his death. The booklet is a practical guide to gambling, including cards, dice, and cheating. Eventually, Cardano became a lecturer of mathematics at the Piatti Foundation. This position allowed him to practice medicine and develop a favorable reputation as a doctor. In 1545, he published his greatest work, *Ars Magna*.

Now Work Problems 31 and 45.

Computing Probabilities Using Equally Likely Outcomes

Problem: Sophia has three tickets to a concert. Yolanda, Michael, Kevin, and Marissa have all stated they would like to go to the concert with Sophia. To be fair, Sophia decides to randomly select the two people who can go to the concert with her.

(a) Determine the sample space of the experiment. In other words, list all possible simple random samples of size $n = 2$.
(b) Compute the probability of the event "Michael and Kevin attend the concert."
(c) Compute the probability of the event "Marissa attends the concert."

Approach: First, we determine the outcomes in the sample space by making a table. The probability of each event is the number of outcomes in the event divided by the number of outcomes in the sample space.

Solution

(a) The sample space is listed in Table 4.

Table 4		
Yolanda, Michael	Yolanda, Kevin	Yolanda, Marissa
Michael, Marissa	Kevin, Marissa	Michael, Kevin

(b) We have $N(S) = 6$, and there is one way the event "Michael and Kevin attend the concert" can occur. Therefore, the probability that Michael and Kevin attend the concert is $\frac{1}{6}$.

(c) We have $N(S) = 6$, and there are three ways the event "Marissa attends the concert" can occur. The probability that Marissa will attend is $\frac{3}{6} = 0.5 = 50\%$.

EXAMPLE 7

Comparing the Classical Method and Empirical Method

Problem: Suppose a survey is conducted in which 500 families with three children are asked to disclose the gender of their children. Based on the results, it was found that 180 of the families had two boys and one girl.

(a) Estimate the probability of having two boys and one girl in a three-child family using the empirical method.
(b) Compute the probability of having two boys and one girl in a three-child family using the classical method, assuming boys and girls are equally likely.

Approach: To answer part (a), we determine the relative frequency of the event "two boys and one girl." To answer part (b), we must count the number of ways the event "two boys and one girl" can occur and divide this by the number of possible outcomes for this experiment.

Solution

(a) The empirical probability of the event $E =$ "two boys and one girl" is

$$P(E) \approx \text{ relative frequency of } E = \frac{180}{500} = 0.36 = 36\%$$

There is about a 36% probability that a family of three children will have two boys and one girl.

(b) To determine the sample space, we construct a **tree diagram** to list the equally likely outcomes of the experiment. We draw two branches corresponding to the two possible outcomes (boy or girl) for the first repetition of the experiment (the first child). For the second child, we draw four branches: two branches originate from the first boy and two branches originate from the first girl. This is repeated for the third child. See Figure 3, where B stands for boy and G stands for girl.

Figure 3

Historical Note

Pierre de Fermat was born into a wealthy family. His father was a leather merchant and second consul of Beaumont-de-Lomagne. Fermat attended the University of Toulouse. By 1631, Fermat was a lawyer and government official. He rose quickly through the ranks because of all the deaths from the plague. In fact, in 1653, Fermat's death was incorrectly reported. In 1654, Fermat received a correspondence from Blaise Pascal in which Pascal asked Fermat to confirm his ideas on probability. Pascal knew of Fermat through his father, who had died 3 years before. Fermat and Pascal discussed the problem of how to divide the stakes in a game that is interrupted before completion, knowing how many points each player needs to win. Their short correspondence laid the foundation for the theory of probability and, on the basis of it, they are now regarded as joint founders of the subject.

Fermat considered mathematics his passionate hobby and true love. He is most famous for his Last Theorem. This theorem states that the equation $x^n + y^n = z^n$ has no nonzero integer solutions for $n > 2$. The theorem was scribbled in the margin of a book by Diophantus, a Greek mathematician. Fermat stated, "I have discovered a truly marvelous proof of this theorem, which, however, the margin is not large enough to contain." The status of Fermat's Last Theorem baffled mathematicians until Andrew Wiles proved it to be true in 1994.

The sample space S of this experiment is found by following each branch to identify all the possible outcomes of the experiment:

$$S = \{BBB, BBG, BGB, BGG, GBB, GBG, GGB, GGG\}$$

So $N(S) = 8$.

For the event E = "two boys and a girl" = $\{BBG, BGB, GBB\}$, we have $N(E) = 3$. Since the outcomes are equally likely (for example, BBG is just as likely as BGB), the probability of E is

$$P(E) = \frac{N(E)}{N(S)} = \frac{3}{8} = 0.375 = 37.5\%$$

There is a 37.5% probability that a family of three children will have two boys and one girl.

In comparing the results of Examples 7(a) and 7(b), we notice that the two probabilities are slightly different. Empirical probabilities and classical probabilities often differ in value. As the number of repetitions in a probability experiment increases, the empirical probability should get closer to the classical probability. That is, the classical probability is the theoretical relative frequency of an event after a large number of trials of the probability experiment. However, it is also possible that the two probabilities differ because having a boy or girl are not equally likely events. (Maybe the probability of having a boy is 50.5% and the probability of having a girl is 49.5%.) If this is the case, the empirical probability will not get closer to the classical probability.

④ Use Simulation to Obtain Data Based on Probabilities

Suppose we wanted to determine the probability of having a boy. Using classical methods, we assume that having a boy is just as likely as having a girl, so the probability of having a boy is 50%. We could also approximate this probability by looking in the *Statistical Abstract of the United States* under Vital Statistics and determining the number of boys and girls born for the most recent year for which data are available. For example, in 2002, there were 2,058,000 boys born and 1,964,000 girls. Based on empirical evidence, the probability of a boy is approximately $\dfrac{2,058,000}{(2,058,000 + 1,964,000)} = 0.512 = 51.2\%$.

However, instead of obtaining data from existing sources, we could simulate having babies by using a graphing calculator or statistical software to replicate the experiment as many times as we like.

EXAMPLE 8

Simulating Probabilities

Problem

(a) Simulate the experiment of sampling 100 babies.

(b) Simulate the experiment of sampling 1000 babies.

Approach: To simulate probabilities, we use a random-number generator available in statistical software and most calculators. We assume the outcomes "have a boy" and "have a girl" are equally likely.

Solution

(a) We use MINITAB to perform the simulation. Set the seed in MINITAB to any value you wish, say 1204. Use the Integer Distribution* to generate random data that simulate having babies. If we agree to let 0 represent a boy and 1 represent a girl, we can approximate the probability of having a girl by summing the number of 1s (adding up the number of girls) and dividing by the number of repetitions of the experiment, 100. See Figure 4.

Figure 4

*The Integer Distribution involves a mathematical formula that uses a seed number to generate a sequence of equally likely random integers. Consult the technology manuals for setting the seed and generating sequences of integers.

Historical Note

Blaise Pascal was born on June 19, 1623, in Clermont, France. Pascal's father felt that Blaise should not be taught mathematics before age 15. Pascal couldn't resist studying mathematics on his own, and at the age of 12 started to teach himself geometry. In December 1639, the Pascal family moved to Rouen, where Pascal's father had been appointed as a tax collector. Between 1642 and 1645, Pascal worked on developing a calculator to help his father collect taxes. In correspondence with Fermat, he helped develop the theory of probability. This correspondence consisted of five letters written in the summer of 1654. Pascal and Fermat considered the dice problem and the problem of points. The dice problem deals with determining the expected number of times a pair of dice must be thrown before a pair of sixes is observed. The problem of points asks how to divide the stakes if a game of dice is incomplete. They solved the problem of points for a two-player game, but did not solve it for three or more players.

Using MINITAB's Tally command, we can determine the number of 0s and 1s that MINITAB randomly generated. See Figure 5.

Figure 5

Summary Statistics for Discrete Variables

C1	Count	Percent
0	48	48.00
1	52	52.00
N=	100	

Based on Figure 5, we approximate that there is a 48% probability of having a boy and a 52% probability of having a girl.

(b) Again, set the seed to 1204. Figure 6 shows the results of simulating the birth of 1000 babies.

Figure 6

Summary Statistics for Discrete Variables

C1	Count	Percent
0	501	50.10
1	499	49.90
N=	1000	

We approximate that there is a 50.1% probability of having a boy and a 49.9% probability of having a girl. ▬▬▬▬▬▬▬▬▬▬▬▬▬▬

Notice that more repetitions of the experiment (100 babies versus 1000 babies) result in a probability closer to $\frac{1}{2}$ for each gender. Notice also that our simulation results differ from the empirical probability of having a boy, namely 0.512. This serves as evidence against the belief that the probability of having a boy is $\frac{1}{2}$.

> **Now Work Problem 49.**

⑤ Understand Subjective Probabilities

Suppose a sports reporter is asked what he thinks the chances are for the Boston Red Sox to return to the World Series. The sports reporter will likely process information about the Red Sox, their pitching staff, lead-off hitter, and so on, and then come up with an educated guess of the likelihood. The reporter may respond that there is a 20% chance the Red Sox return to the World Series. This forecast is a probability although it is not based on relative frequencies. We cannot, after all, repeat the experiment of playing a season under the same circumstances (same players, schedule, and so on) over and over. Nonetheless, the forecast of 20% does satisfy the criterion that a probability be between 0 and 1, inclusive. This forecast is known as a *subjective probability*.

Definition

> A **subjective probability** of an outcome is a probability obtained on the basis of personal judgment.

It is important to understand that subjective probabilities are perfectly legitimate and are often the only method of assigning likelihood to an outcome. As another example, a financial reporter may ask an economist about the likelihood the economy will fall into recession next year. Again, we cannot conduct an experiment n times to obtain a relative frequency. The economist must use his or her knowledge of the current conditions of the economy and make an educated guess as to the likelihood of recession.

5.1 ASSESS YOUR UNDERSTANDING

Concepts and Vocabulary

1. Describe the difference between classical and empirical probability.
2. What is the probability of an event that is impossible? Suppose a probability is approximated to be zero based on empirical results. Does this mean the event is impossible? *0; no*
3. In computing classical probabilities, all outcomes must be equally likely. Explain what this means.
4. What does it mean for an event to be unusual? Why should the cutoff for identifying unusual events not always be 0.05?

5. *True or False:* In a probability model, the sum of the probabilities of all outcomes must equal 1. *True*
6. *True or False:* Probability is a measure of the likelihood of a random phenomenon or chance behavior. *True*
7. In probability a(n) _____ is any process that can be repeated in which the results are uncertain. *experiment*
8. A(n) _____ is any collection of outcomes from a probability experiment. *event*

Skill Building

9. Verify that the following is a probability model. What do
NW we call the outcome "blue"? *Impossible event*

Color	Probability
red	0.3
green	0.15
blue	0
brown	0.15
yellow	0.2
orange	0.2

10. Verify that the following is a probability model. If the model represents the colors of M&Ms in a bag of plain M&Ms, explain what the model implies. *All the M&Ms are yellow*

Color	Probability
red	0
green	0
blue	0
brown	0
yellow	1
orange	0

11. Why is the following not a probability model? *P(green) < 0*

Color	Probability
red	0.3
green	−0.3
blue	0.2
brown	0.4
yellow	0.2
orange	0.2

12. Why is the following not a probability model?
Sum of probabilities not 1

Color	Probability
red	0.1
green	0.1
blue	0.1
brown	0.4
yellow	0.2
orange	0.3

13. Which of the following numbers could be the probability of an event? *0, 0.01, 0.35, 1*

$$0, 0.01, 0.35, -0.4, 1, 1.4$$

14. Which of the following numbers could be the probability of an event?

$$1.5, \frac{1}{2}, \frac{3}{4}, \frac{2}{3}, 0, -\frac{1}{4} \qquad \frac{1}{2}, \frac{3}{4}, \frac{2}{3}, 0$$

15. In five-card stud poker, a player is dealt five cards. The probability that the player is dealt two cards of the same value and three other cards of different value so that the player has a pair is 0.42. Explain what this probability means. If you play five-card stud 100 times, will you get a pair exactly 42 times? Why or why not? *No*

16. In seven-card stud poker, a player is dealt seven cards. The probability that the player is dealt two cards of the same value and five other cards of different value so that the player has a pair is 0.48. Explain what this probability means. If you play seven-card stud 100 times, will you get a pair exactly 48 times? Why or why not? *No*

17. Suppose you toss a coin 100 times and get 95 heads and 5 tails. Based on these results, what is the probability that the next flip results in a head? *0.95*

18. Suppose you roll a die 100 times and get six 80 times. Based on these results, what is the probability the next roll results in six? *0.8*

19. Bob is asked to construct a probability model for rolling a pair of fair dice. He lists the outcomes as 2, 3, 4, 5, 6, 7, 8, 9,

10, 11, 12. Because there are 11 outcomes, he reasoned, the probability of rolling a two must be $\frac{1}{11}$. What is wrong with Bob's reasoning? *Not equally likely outcomes*

20. Blood Types A person can have one of four blood types: A, B, AB, or O. If people are randomly selected, is the probability they have blood type A equal to $\frac{1}{4}$? Why? *No*

21. If a person rolls a six-sided die and then flips a coin, describe the sample space of possible outcomes using 1, 2, 3, 4, 5, 6 for the die outcomes and H, T for the coin outcomes.

22. If a basketball player shoots three free throws, describe the sample space of possible outcomes using *S* for a made free throw and *F* for a missed free throw.

23. According to the U.S. Department of Education, the proportion of three-year-olds that is enrolled in day care is 0.44. What is the probability a randomly selected three-year-old is enrolled in day care? *0.44*

24. According to the American Veterinary Medical Association, the proportion of households owning a dog is 0.361. What is the probability a randomly selected household owns a dog? *0.361*

For Problems 25–28, let the sample space be $S = \{1, 2, 3, 4, 5, 6, 7, 8, 9, 10\}$. Suppose the outcomes are equally likely.

25. Compute the probability of the event $E = \{1, 2, 3\}$. $\frac{3}{10}$

26. Compute the probability of the event $F = \{3, 5, 9, 10\}$. $\frac{2}{5}$

27. Compute the probability of the event $E =$ "an even number."

28. Compute the probability of the event $F =$ "an odd number." $\frac{1}{2}$ 27. $\frac{1}{2}$

Applying the Concepts

29. Play Sports? In a survey of 500 randomly selected high school students, it was determined that 288 played organized sports. What is the probability that a randomly selected high school student plays organized sports? *0.576*

30. Volunteer? In a survey of 1100 female adults (18 years of age or older), it was determined that 341 volunteered at least once in the past year. What is the probability that a randomly selected adult female volunteered at least once in the past year? *0.31*

31. Planting Tulips A bag of 100 tulip bulbs purchased from a
NW nursery contains 40 red tulip bulbs, 35 yellow tulip bulbs, and 25 purple tulip bulbs.
  (a) What is the probability that a randomly selected tulip bulb is red? *0.4*
  (b) What is the probability that a randomly selected tulip bulb is purple? *0.25*

32. Golf Balls The local golf store sells an "onion bag" that contains 80 "experienced" golf balls. Suppose the bag contains 35 Titleists, 25 Maxflis, and 20 Top-Flites.
  (a) What is the probability that a randomly selected golf ball is a Titleist? *0.4375*
  (b) What is the probability that a randomly selected golf ball is a Top-Flite? *0.25*

33. Roulette In the game of roulette, a wheel consists of 38 slots numbered 0, 00, 1, 2, . . . , 36. (See the photo.) To play the game, a metal ball is spun around the wheel and is allowed to fall into one of the numbered slots.   (33b) $\frac{1}{38}$

  (a) Determine the sample space. $\{0, 00, 1, 2, \ldots, 36\}$
  (b) Determine the probability that the metal ball falls into the slot marked 8. Interpret this probability.
  (c) Determine the probability that the metal ball lands in an odd slot. Interpret this probability. $\frac{9}{19}$

34. Birthdays Exclude leap years from the following calculations and assume each birthday is equally likely:
  (a) Determine the probability that a randomly selected person has a birthday on the 1st day of a month. Interpret this probability. $\frac{12}{365}$
  (b) Determine the probability that a randomly selected person has a birthday on the 31st day of a month. Interpret this probability. $\frac{7}{365}$
  (c) Determine the probability that a randomly selected person was born in December. Interpret this probability. $\frac{31}{365}$
  (d) Determine the probability that a randomly selected person has a birthday on November 8. Interpret this probability. $\frac{1}{365}$
  (e) If you just met somebody and she asked you to guess her birthday, are you likely to be correct? *No*
  (f) Do you think it is appropriate to use the methods of classical probability to compute the probability that a person is born in December?

35. Genetics A gene is composed of two alleles. An allele can be either dominant or recessive. Suppose a husband and wife, who are both carriers of the sickle-cell anemia allele but do not have the disease, decide to have a child. Because both parents are carriers of the disease, each has one dominant normal-cell allele and one recessive sickle-cell allele. Therefore, the genotype of each parent is *Ss*. Each parent contributes one allele to his or her offspring, with each allele being equally likely.   (35a) $\{SS, Ss, sS, ss\}$
  (a) List the possible genotypes of their offspring.
  (b) What is the probability that the offspring will have sickle-cell anemia? In other words, what is the probability the offspring will have genotype *ss*? Interpret this probability. $\frac{1}{4}$

(c) What is the probability that the offspring will not have sickle-cell anemia but will be a carrier? In other words, what is the probability that the offspring will have one dominant normal-cell allele and one recessive sickle-cell allele? Interpret this probability. $\frac{1}{2}$

36. More Genetics In Problem 35, we learned that for some diseases, such as sickle-cell anemia, an individual will get the disease only if he receives both recessive alleles. This is not always the case. For example, Huntington's disease only requires one dominant gene for an individual to contract the disease. Suppose a husband and wife, who both have a dominant Huntington's disease allele (*S*) and a normal recessive allele (*s*), decide to have a child.

(a) List the possible genotypes of their offspring.
(b) What is the probability that the offspring will not have Huntington's disease? In other words, what is the probability the offspring will have genotype *ss*? Interpret this probability. $\frac{1}{4}$
(c) What is the probability that the offspring will have Huntington's disease? $\frac{3}{4}$

37. College Survey In a national survey conducted by the Centers for Disease Control to determine college students' health-risk behaviors, college students were asked, "How often do you wear a seat belt when riding in a car driven by someone else?" The frequencies appear in the following table:

Response	Frequency
Never	125
Rarely	324
Sometimes	552
Most of the time	1257
Always	2518

(a) Construct a probability model for seat-belt use by a passenger.
(b) Would you consider it unusual to find a college student who never wears a seat belt when riding in a car driven by someone else? Why? Yes

38. College Survey In a national survey conducted by the Centers for Disease Control to determine college students' health-risk behaviors, college students were asked, "How often do you wear a seat belt when driving a car?" The frequencies appear in the following table:

Response	Frequency
Never	118
Rarely	249
Sometimes	345
Most of the time	716
Always	3093

(a) Construct a probability model for seat-belt use by a driver.
(b) Is it unusual for a college student never to wear a seat belt when driving a car? Why? Yes

39. Larceny Theft A police officer randomly selected 595 police records of larceny thefts. The following data represent the number of offenses for various types of larceny thefts.

Type of Larceny Theft	Number of Offenses
Pocket picking	5
Purse snatching	5
Shoplifting	118
From motor vehicles	197
Motor vehicle accessories	77
Bicycles	43
From buildings	105
From coin-operated machines	45

Source: U.S. Federal Bureau of Investigation

(a) Construct a probability model for type of larceny theft.
(b) Are purse-snatching larcenies unusual? Yes
(c) Are larcenies from coin-operated machines unusual?

40. Multiple Births The following data represent the number of live multiple-delivery births (three or more babies) in 2002 for women 15 to 44 years old.

Age	Number of Multiple Births
15–19	93
20–24	511
25–29	1628
30–34	2832
35–39	1843
40–44	377

Source: National Vital Statistics Reports, Vol. 52, No. 10, December 17, 2003

(a) Construct a probability model for number of multiple births.
(b) In the sample space of all multiple births, are multiple births for 15- to 19-year-old mothers unusual? Yes
(c) In the sample space of all multiple births, are multiple births for 40- to 44-year-old mothers unusual?

(36a) {SS, Ss, sS, ss}   (40c) Not too unusual

Problems 41–44, use the given table, which lists six possible assignments of probabilities for tossing a coin twice, to answer the following questions.

Assignments	Sample Space			
	HH	**HT**	**TH**	**TT**
A	$\frac{1}{4}$	$\frac{1}{4}$	$\frac{1}{4}$	$\frac{1}{4}$
B	0	0	0	1
C	$\frac{3}{16}$	$\frac{5}{16}$	$\frac{5}{16}$	$\frac{3}{16}$
D	$\frac{1}{2}$	$\frac{1}{2}$	$-\frac{1}{2}$	$\frac{1}{2}$
E	$\frac{1}{4}$	$\frac{1}{4}$	$\frac{1}{4}$	$\frac{1}{8}$
F	$\frac{1}{9}$	$\frac{2}{9}$	$\frac{2}{9}$	$\frac{4}{9}$

41. Which of the assignments of probabilities are consistent with the definition of a probability model? A, B, C, F

42. Which of the assignments of probabilities should be used if the coin is known to be fair? A

43. Which of the assignments of probabilities should be used if the coin is known to always come up tails? B

44. Which of the assignments of probabilities should be used if tails is twice as likely to occur as heads? F

45. Going to Disney World John, Roberto, Clarice, Dominique, and Marco work for a publishing company. The company wants to send two employees to a statistics conference in Orlando. To be fair, the company decides that the two individuals who get to attend will have their names drawn from a hat. This is like obtaining a simple random sample of size 2.
(a) Determine the sample space of the experiment. That is, list all possible simple random samples of size $n = 2$.
(b) What is the probability that Clarice and Dominique attend the conference? 0.1
(c) What is the probability that Clarice attends the conference? 0.4
(d) What is the probability that John stays home? 0.6

46. Six Flags Six Flags over Mid-America in St. Louis has six roller coasters: The Screamin' Eagle, The Boss, River King Mine Train, Batman the Ride, Mr. Freeze, and Ninja. After a long day at the park, Ethan's parents tell him that he can ride two more coasters before leaving (but not the same one twice). Because he likes the rides equally, Ethan decides to randomly select the two coasters by drawing their names from his hat.
(a) Determine the sample space of the experiment. That is, list all possible simple random samples of size $n = 2$.
(b) What is the probability that Ethan will ride Batman and Mr. Freeze? 0.067
(c) What is the probability that Ethan will ride the Screamin' Eagle? 0.333
(d) What is the probability that Ethan will ride neither River King Mine Train nor Ninja? 0.4

47. Barry Bonds On October 5, 2001, Barry Bonds broke Mark McGwire's home-run record for a single season by hitting his 71st and 72nd home runs. Bonds went on to hit one more home run before the season ended, for a total of 73. Of the 73 home runs, 24 went to right field, 26 went to right center field, 11 went to center field, 10 went to left center field, and 2 went to left field. (*Source*: Baseball-almanac.com)
(a) What is the probability that a randomly selected home run was hit to right field? $\frac{24}{73}$
(b) What is the probability that a randomly selected home run was hit to left field? $\frac{2}{73}$
(c) Was it unusual for Barry Bonds to hit a home run to left field? Explain. Yes

48. Rolling a Die
(a) Roll a single die 50 times, recording the result of each roll of the die. Use the results to approximate the probability of rolling a three.
(b) Roll a single die 100 times, recording the result of each roll of the die. Use the results to approximate the probability of rolling a three.
(c) Compare the results of (a) and (b) to the classical probability of rolling a three.

49. Simulation Use a graphing calculator or statistical software to simulate rolling a six-sided die 100 times, using an integer distribution with numbers one through six.
(a) Use the results of the simulation to compute the probability of rolling a one.
(b) Repeat the simulation. Compute the probability of rolling a one.
(c) Simulate rolling a six-sided die 500 times. Compute the probability of rolling a one.
(d) Which simulation resulted in the closest estimate to the probability that would be obtained using the classical method?

50. Classifying Probability Determine whether the following probabilities are computed using classical methods, empirical methods, or subjective methods.
(a) The probability of having eight girls in an eight-child family is 0.390625%. Classical
(b) On the basis of a study of families with eight children, the probability of a family having eight girls is 0.54%.
(c) According to a sports analyst, the probability that the Chicago Bears will win their next game is about 30%.
(d) On the basis of clinical trials, the probability of efficacy of a new drug is 75%. Empirical

51. Checking for Loaded Dice You suspect a pair of dice to be loaded and conduct a probability experiment by rolling each die 200 times. The outcome of the experiment is listed in the following table:
(50b) Empirical (50c) Subjective

Value of Die	Frequency	Value of Die	Frequency
1	105	4	49
2	47	5	51
3	44	6	104

Do you think the dice are loaded? Why?

0

52. Conduct a survey in your school by randomly asking 50 students whether they drive to school. Based on the results of the survey, approximate the probability that a randomly selected student drives to school.

53. In 2004, the median income of families in the United States was $57,500. What is the probability that a randomly selected family has an income greater than $57,500? *0.5*

54. In 2004, 17% of Florida's population was 65 and over (*Source*: U.S. Census Bureau). What is the probability that a randomly selected Floridian is 65 or older? *0.17*

55. The Probability Applet Load the long-run probability applet on your computer.
  (a) Choose the "simulating the probability of a head with a fair coin" applet and simulate flipping a fair coin 10 times. What is the estimated probability of a head based on these 10 trials?

  (b) Reset the applet. Simulate flipping a fair coin 10 times a second time. What is the estimated probability of a head based on these 10 trials? Compare the results to part (a).
  (c) Reset the applet. Simulate flipping a fair coin 1000 times. What is the estimated probability of a head based on these 1000 trials? Compare the results to part (a).
  (d) Reset the applet. Simulate flipping a fair coin 1000 times. What is the estimated probability of a head based on these 1000 trials? Compare the results to part (c).
  (e) Choose the "simulating the probability of head with an unfair coin $(P(H) = 0.2)$" applet and simulate flipping a coin 1000 times. What is the estimated probability of a head based on these 1000 trials? If you did not know that the probability of heads was set to 0.2, what would you conclude about the coin? Why?

Technology Step by Step | **Simulation**

TI-83/84 Plus

Step 1: Set the seed by entering any number on the HOME screen. Press the STO ⇒ button, press the MATH button, highlight the PRB menu, and highlight `1:rand` and hit ENTER. With the cursor on the HOME screen, hit ENTER.

Step 2: Press the MATH button and highlight the PRB menu. Highlight `5:randInt(` and hit ENTER.

Step 3: After the `randInt(` on the HOME screen type `1,n, number of repetitions of experiment)`, where n is the number of equally likely outcomes. For example, to simulate rolling a single die 50 times, we type

```
randInt(1, 6, 50)
```

Step 4: Press the STO ⇒ button and then 2nd 1, and hit ENTER to store the data in L1.

Step 5: Draw a histogram of the data using the outcomes as classes. TRACE to obtain outcomes.

MINITAB

Step 1: Set the seed by selecting the **Calc** menu and highlighting **Set Base** .... Insert any seed you wish into the cell and click OK.

Step 2: Select the **Calc** menu, highlight **Random Data**, and then highlight **Integer**. To simulate rolling a single die 100 times, fill in the window as shown in Figure 4 on page 231.

Step 3: Select the **Stat** menu, highlight **Tables**, and then highlight **Tally** .... Enter C1 into the variables cell. Make sure that the Counts box is checked and click OK.

Excel

Step 1: With cell A1 selected, press the *fx* button.

Step 2: Highlight Math & Trig in the Function category window. Then highlight RANDBETWEEN in the Function Name: window. Click OK.

Step 3: To simulate rolling a die 50 times, enter 1 for the lower limit and 6 for the upper limit. Click OK.

Step 4: Copy the contents of cell A1 into cells A2 through A50.

5.2 The Addition Rule and Complements

Objectives
1. Use the Addition Rule for Disjoint Events
2. Use the General Addition Rule
3. Compute the probability of an event using the Complement Rule

1 Use the Addition Rule for Disjoint Events

Now we introduce some more rules for computing probabilities. However, before we present these rules, we must discuss *disjoint events*.

Definition

Two events are **disjoint** if they have no outcomes in common. Another name for disjoint events is **mutually exclusive** events.

In Other Words
Two events are disjoint if they cannot occur at the same time.

It is often helpful to draw pictures of events. Such pictures, called **Venn diagrams**, represent events as circles enclosed in a rectangle. The rectangle represents the sample space, and each circle represents an event. For example, suppose we randomly select chips from a bag. Each chip is labeled 0, 1, 2, 3, 4, 5, 6, 7, 8, 9. Let E represent the event "choose a number less than or equal to 2," and let F represent the event "choose a number greater than or equal to 8." Because E and F do not have any outcomes in common, they are disjoint. Figure 7 shows a Venn diagram of these disjoint events.

Figure 7

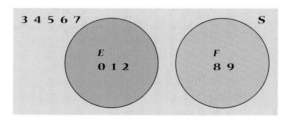

Notice that the outcomes in event E are inside circle E, and the outcomes in event F are inside the circle F. All outcomes that are in the sample space that are not in E nor in F are outside the circles, but inside the rectangle. From this diagram, we know that $P(E) = \dfrac{n(E)}{n(S)} = \dfrac{3}{10} = 0.3$ and $P(F) = \dfrac{n(F)}{n(S)} = \dfrac{2}{10} = 0.2$. In addition, $P(E \text{ or } F) = P(E) + P(F) = 0.3 + 0.2 = 0.5$. This result occurs because of the *Addition Rule for Disjoint Events*.

In Other Words
The Addition Rule for Disjoint Events states that if you have two events that have no outcomes in common, the probability that one or the other occurs is the sum of their probabilities.

Addition Rule for Disjoint Events

If E and F are disjoint (or mutually exclusive) events, then

$$P(E \text{ or } F) = P(E) + P(F)$$

The Addition Rule for Disjoint Events can be extended to more than two disjoint events. In general, if E, F, G, ... each have no outcomes in common (they are pairwise disjoint), then

$$P(E \text{ or } F \text{ or } G \text{ or} \dots) = P(E) + P(F) + P(G) + \cdots$$

Let event G represent "the number is a 5 or 6." The Venn diagram in Figure 8 illustrates the Addition Rule for more than two disjoint events using the chip example. Notice that none of the events has any outcomes in common. So, from the Venn diagram, we can see $P(E) = \dfrac{n(E)}{n(S)} = \dfrac{3}{10} = 0.3$, $P(F) = \dfrac{n(F)}{n(S)} = \dfrac{2}{10} = 0.2$, and $P(G) = \dfrac{n(G)}{n(S)} = \dfrac{2}{10} = 0.2$. In addition, $P(E \text{ or } F \text{ or } G) = P(E) + P(F) + P(G) = 0.3 + 0.2 + 0.2 = 0.7$.

Figure 8

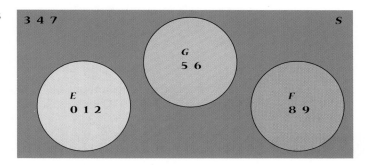

EXAMPLE 1

Benford's Law and the Addition Rule for Disjoint Events

Problem: Our number system consists of the digits 0, 1, 2, 3, 4, 5, 6, 7, 8, and 9. The first significant digit in any number must be 1, 2, 3, 4, 5, 6, 7, 8, or 9, because we do not write numbers such as 12 as 012, for example. Although we may think that each digit appears with equal frequency so that each digit has a $\dfrac{1}{9}$ probability of being the first significant digit, this in fact, is not true. In 1881, Simon Necomb discovered that digits do not occur with equal frequency. This same result was discovered again in 1938 by physicist Frank Benford. After studying lots and lots of data, he was able to assign probabilities of occurrence for each of the first digits, as shown in Table 5.

Table 5								
Digit	1	2	3	4	5	6	7	8	9
Probability	0.301	0.176	0.125	0.097	0.079	0.067	0.058	0.051	0.046

Source: "The First Digit Phenomenon", T. P. Hill, *American Scientist*, July–August, 1998.

The probability model is now known as Benford's Law and plays a major role in identifying fraudulent data on tax returns and accounting books.

(a) Verify that Benford's Law is a probability model.
(b) Use Benford's Law to determine the probability that a randomly selected first digit is 1 or 2.
(c) Use Benford's Law to determine the probability that a randomly selected first digit is at least 6.

Approach: For part (a), we need to verify that each probability is between 0 and 1 and that the sum of all probabilities equals 1. For parts (b) and (c), we use the Addition Rule for Disjoint Events.

Solution

(a) In looking at Table 5, we see that each probability is between 0 and 1. In addition, the sum of all the probabilities is 1.

$$0.301 + 0.176 + 0.125 + \cdots + 0.046 = 1$$

Because rules 1 and 2 are satisfied, Table 5 represents a probability model.

(b)
$$P(1 \text{ or } 2) = P(1) + P(2)$$
$$= 0.301 + 0.176$$
$$= 0.477$$

If we looked at 100 numbers, we would expect about 48 to begin with 1 or 2.

(c)
$$P(\text{at least } 6) = P(6 \text{ or } 7 \text{ or } 8 \text{ or } 9)$$
$$= P(6) + P(7) + P(8) + P(9)$$
$$= 0.067 + 0.058 + 0.051 + 0.046$$
$$= 0.222$$

If we looked at 100 numbers, we would expect about 22 to begin with 6, 7, 8, or 9.

EXAMPLE 2   **A Deck of Cards and the Addition Rule for Disjoint Events**

Problem: Suppose a single card is selected from a standard 52-card deck, such as the one shown in Figure 9.

Figure 9

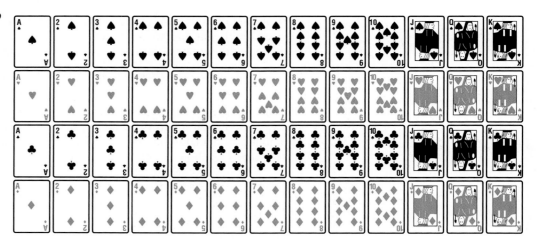

(a) Compute the probability of the event E = "drawing a king."
(b) Compute the probability of the event E = "drawing a king" or F = "drawing a queen."
(c) Compute the probability of the event E = "drawing a king" or F = "drawing a queen" or G = "drawing a jack."

Approach: We will use the classical method for computing the probabilities because the outcomes are equally likely and easy to count. We use the Addition Rule for Disjoint Events to compute the probabilities in parts (b) and (c) because the events are mutually exclusive. For example, you cannot simultaneously draw a king and a queen.

Solution: The sample space consists of the 52 cards in the deck, so $N(S) = 52$.

(a) A standard deck of cards has four kings, so $N(E) = 4$. Therefore,
$$P(\text{king}) = P(E) = \frac{N(E)}{N(S)} = \frac{4}{52} = \frac{1}{13}.$$

(b) A standard deck of cards also has four queens. Because events E and F are mutually exclusive (you cannot draw a king and queen simultaneously), we use the Addition Rule for Disjoint Events. So

$$P(\text{king or queen}) = P(E \text{ or } F)$$

$$= P(E) + P(F)$$

$$= \frac{4}{52} + \frac{4}{52} = \frac{8}{52} = \frac{2}{13}$$

(c) Because events E, F, and G are mutually exclusive, we use the Addition Rule for Disjoint Events extended to two or more disjoint events. So

$$P(\text{king or queen or jack}) = P(E \text{ or } F \text{ or } G)$$

$$= P(E) + P(F) + P(G)$$

$$= \frac{4}{52} + \frac{4}{52} + \frac{4}{52} = \frac{12}{52} = \frac{3}{13}$$

Now Work Problems 25(a)–(c).

② Use the General Addition Rule

A question that you may be asking yourself is this: "What if I need to compute the probability of two events that are not disjoint?" This is a legitimate question and one that has an answer.

Consider the chip example. Suppose we are randomly selecting chips from a bag. Each chip is labeled 0, 1, 2, 3, 4, 5, 6, 7, 8, or 9. Let E represent the event "choose an odd number," and let F represent the event "choose a number less than or equal to 4." Because $E = \{1, 3, 5, 7, 9\}$ and $F = \{0, 1, 2, 3, 4\}$ have the outcomes 1 and 3 in common, the events are not disjoint. Figure 10 shows a Venn diagram of these events.

Note to Instructor
If you wish to design your course so that it minimizes probability coverage, the material in Objective 2 may be skipped without loss of continuity.

Figure 10

The overlapping region is E and F.

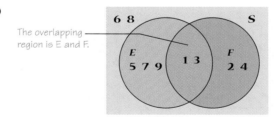

We can compute $P(E \text{ or } F)$ directly by counting because each outcome is equally likely. There are 7 outcomes in E or F and 10 outcomes in the sample space, so

$$P(E \text{ or } F) = \frac{N(E \text{ or } F)}{N(S)}$$

$$= \frac{7}{10}$$

If we attempt to compute $P(E \text{ or } F)$ using the Addition Rule for Disjoint Events, we obtain the following:

$$P(E \text{ or } F) = P(E) + P(F)$$

$$= \frac{5}{10} + \frac{4}{10}$$

$$= \frac{9}{10}$$

This result is incorrect because we counted the outcomes 1 and 3 twice: once for event E and once for event F. To avoid this double counting, we have to subtract the probability corresponding to the overlapping region, E and F. That is, we have to subtract $P(E \text{ and } F) = \dfrac{2}{10}$ from the result and obtain

$$P(E \text{ or } F) = P(E) + P(F) - P(E \text{ and } F)$$

$$= \frac{5}{10} + \frac{4}{10} - \frac{2}{10}$$

$$= \frac{7}{10}$$

which agrees with the result we obtained by counting. These results can be generalized in the following rule:

The General Addition Rule

For any two events E and F,

$$P(E \text{ or } F) = P(E) + P(F) - P(E \text{ and } F)$$

EXAMPLE 3   **Computing Probabilities for Events That Are Not Disjoint**

Problem: Suppose a single card is selected from a standard 52-card deck. Compute the probability of the event E = "drawing a king" or H = "drawing a diamond."

Approach: The events are not disjoint because the outcome "king of diamonds" is in both events, so we use the General Addition Rule.

Solution
$P(\text{king or diamond}) = P(\text{king}) + P(\text{diamond}) - P(\text{king of diamonds})$

$$= \frac{4}{52} + \frac{13}{52} - \frac{1}{52}$$

$$= \frac{16}{52} = \frac{4}{13}$$

Now Work Problem 29.

Consider the data shown in Table 6, which represent the marital status of males and females 18 years old or older in the United States in 2003.

Table 6 is called a **contingency table** or **two-way table**, because it relates two categories of data. The **row variable** is marital status, because each row

Table 6	Males (in millions)	Females (in millions)	Totals (in millions)
Never married	28.6	23.3	51.9
Married	62.1	62.8	124.9
Widowed	2.7	11.3	14.0
Divorced	9.0	12.7	21.7
Totals (in millions)	102.4	110.1	212.5

Source: U.S. Census Bureau, *Current Population Reports*

in the table describes the marital status of each individual. The **column variable** is gender. Each box inside the table is called a **cell**. For example, the cell corresponding to married individuals who are male is in the second row, first column. Each cell contains the frequency of the category: There were 62.1 million married males in the United States in 2003. Put another way, in the United States in 2003, there were 62.1 million individuals who were male *and* married.

EXAMPLE 4   **Using the Addition Rule with Contingency Tables**

Problem: Using the data in Table 6, determine the following:

(a) Determine the probability that a randomly selected U.S. resident 18 years old or older is male.

(b) Determine the probability that a randomly selected U.S. resident 18 years old or older is widowed.

(c) Determine the probability that a randomly selected U.S. resident 18 years old or older is widowed or divorced.

(d) Determine the probability that a randomly selected U.S. resident 18 years old or older is male or widowed.

Approach: We first add up the entries in each row and column so that we get the total number of people in each category. We can then determine the probabilities using either the Addition Rule for Disjoint Events or the General Addition Rule.

Solution: Add the entries in each column. For example, in the "male" column, we find that there are $28.6 + 62.1 + 2.7 + 9.0 = 102.4$ million males 18 years old or older in the United States. Add the entries in each row. For example, in the "never married" row, we find there are $28.6 + 23.3 = 51.9$ million U.S. residents 18 years old or older who have never married. Adding the row totals or column totals, we find there are $102.4 + 110.1 = 51.9 + 124.9 + 14.0 + 21.7 = 212.5$ million U.S. residents 18 years old or older.

(a) There are 102.4 million males 18 years old or older and 212.5 million U.S. residents 18 years old or older. The probability that a randomly selected U.S. resident 18 years old or older is male is $\dfrac{102.4}{212.5} = 0.482$.

(b) There are 14.0 million U.S. residents 18 years old or older who are widowed. The probability that a randomly selected U.S. resident 18 years old or older is widowed is $\dfrac{14.0}{212.5} = 0.066$.

(c) The events widowed and divorced are disjoint. Do you see why? We use the Addition Rule for Disjoint Events.

$$P(\text{widowed or divorced}) = P(\text{widowed}) + P(\text{divorced})$$

$$= \frac{14.0}{212.5} + \frac{21.7}{212.5} = \frac{35.7}{212.5}$$

$$= 0.168$$

(d) The events male and widowed are not mutually exclusive. In fact, there are 2.7 million males who are widowed in the United States. Therefore, we use the General Addition Rule to compute $P(\text{male or widowed})$:

$$P(\text{male or widowed}) = P(\text{male}) + P(\text{widowed}) - P(\text{male and widowed})$$

$$= \frac{102.4}{212.5} + \frac{14.0}{212.5} - \frac{2.7}{212.5} = \frac{113.7}{212.5} = 0.535$$

Now Work Problem 37.

③ Compute the Probability of an Event Using the Complement Rule

Note to Instructor
There are a number of ways to denote the complement of an event, such as $\bar{E}$, E', and $\sim E$. We choose E^c to avoid confusion with the notation for the sample mean.

Suppose the probability of an event E is known and we would like to determine the probability that E does not occur. This can easily be accomplished using the idea of *complements*.

Definition

Complement of an Event

Let S denote the sample space of a probability experiment and let E denote an event. The **complement of** E, denoted E^c, is all outcomes in the sample space S that are not outcomes in the event E.

In Other Words
The Complement Rule is used when you know the probability that some event will occur and you want to know the opposite: the chance it will not occur.

Because E and E^c are mutually exclusive,

$$P(E \text{ or } E^c) = P(E) + P(E^c) = P(S) = 1$$

Subtracting $P(E)$ from both sides, we obtain

$$P(E^c) = 1 - P(E)$$

We have the following result.

Complement Rule

If E represents any event and E^c represents the complement of E, then

$$P(E^c) = 1 - P(E)$$

Figure 11 illustrates the Complement Rule using a Venn diagram.

Figure 11

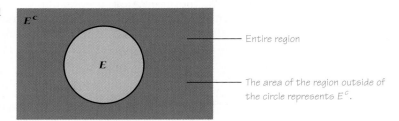

Entire region

The area of the region outside of the circle represents E^c.

EXAMPLE 5 | **Computing Probabilities Using Complements**

Problem: According to the National Gambling Impact Study Commision, 52% of Americans have played state lotteries. What is the probability that a randomly selected American has not played a state lottery?

Approach: Not playing a state lottery is the complement of playing a state lottery. We compute the probability using the Complement Rule.

Solution

$$P(\text{not played state lottery}) = 1 - P(\text{played state lottery}) = 1 - 0.52 = 0.48$$

There is a 48% probability of randomly selecting an American who has not played a state lottery.

EXAMPLE 6 | **Computing Probabilities Using Complements**

Problem: The data in Table 7 represent the income distribution of households in the United States in 2003.

	Table 7			
Annual Income	**Number (in thousands)**	**Annual Income**	**Number (in thousands)**	
Less than $10,000	10,011	$50,000 to $74,999	20,191	
$10,000 to $14,999	7,740	$75,000 to $99,999	12,313	
$15,000 to $24,999	14,649	$100,000 to $149,999	10,719	
$25,000 to $34,999	13,277	$150,000 to $199,999	3,372	
$35,000 to $49,999	16,773	$200,000 or more	2,854	

Source: U.S. Census Bureau

(a) Compute the probability that a randomly selected household earned $200,000 or more in 2003.

(b) Compute the probability that a randomly selected household earned less than $200,000 in 2003.

(c) Compute the probability that a randomly selected household earned at least $10,000 in 2003.

Approach: The probabilities will be determined by finding the relative frequency of each event. We have to find the total number of households in the United States in 2003.

Solution

(a) There was a total of $10,011 + 7740 + \cdots + 2854 = 111,899$ thousand households in the United States in 2003 and 2854 thousand of them earned $200,000 or more. The probability that a randomly selected household in the United States earned $200,000 or more in 2003 is $\dfrac{2854}{111,899} = 0.026$.

(b) We could compute the probability of randomly selecting a household that earned less than $200,000 in 2003 by adding the relative frequencies of each category less than $200,000, but it is easier to use complements. The complement of earning less than $200,000 is earning $200,000 or more. Therefore,

$$P(\text{less than } \$200,000) = 1 - P(\$200,000 \text{ or more})$$

$$= 1 - 0.026 = 0.974$$

There is a 97.4% probability of randomly selecting a household that earned less than $200,000 in 2003.

(c) The phrase *at least* means greater than or equal to. The complement of at least $10,000 is less than $10,000. In 2003, 10,011 thousand households earned less than $10,000. The probability of randomly selecting a household that earned at least $10,000 is

$$P(\text{at least } \$10,000) = 1 - P(\text{less than } \$10,000)$$

$$= 1 - \dfrac{10,011}{111,899} = 0.911$$

There is a 91.1% probability of randomly selecting a household that earned at least $10,000 in 2003.

Now Work Problems 25(d) and 27.

5.2 ASSESS YOUR UNDERSTANDING

Concepts and Vocabulary

1. What does it mean when two events are disjoint?

2. If E and F are disjoint events, then $P(E \text{ or } F) = $ _____.

3. If E and F are not disjoint events, then $P(E \text{ or } F) = $ _____.

4. What does it mean when two events are complements?

Skill Building

In Problems 5–12, a probability experiment is conducted in which the sample space of the experiment is
$S = \{1, 2, 3, 4, 5, 6, 7, 8, 9, 10, 11, 12\}$. *Let event* $E = \{2, 3, 4, 5, 6, 7\}$, *event* $F = \{5, 6, 7, 8, 9\}$, *event* $G = \{9, 10, 11, 12\}$, *and event* $H = \{2, 3, 4\}$. *Assume each outcome is equally likely.*

5. List the outcomes in E and F. Are E and F mutually exclusive? $\{5, 6, 7\}$; No

6. List the outcomes in F and G. Are F and G mutually exclusive? $\{9\}$; No

7. List the outcomes in F or G. Now find $P(F$ or $G)$ by counting the number of outcomes in F or G. Determine $P(F$ or $G)$ using the General Addition Rule. $\frac{2}{3}$

8. List the outcomes in E or H. Now find $P(E$ or $H)$ by counting the number of outcomes in E or H. Determine $P(E$ or $H)$ using the General Addition Rule. $\frac{1}{2}$

9. List the outcomes in E and G. Are E and G mutually exclusive? $\{\quad\}$; Yes

10. List the outcomes in F and H. Are F and H mutually exclusive? $\{\quad\}$; Yes

11. List the outcomes in E^c. Find $P(E^c)$. $\{1, 8, 9, 10, 11, 12\}$; $\frac{1}{2}$

12. List the outcomes in F^c. Find $P(F^c)$.
$\{1, 2, 3, 4, 10, 11, 12\}$; $\frac{7}{12}$

In Problems 13–18, find the probability of the indicated event if $P(E) = 0.25$ *and* $P(F) = 0.45$.

13. Find $P(E$ or $F)$ if $P(E$ and $F) = 0.15$. 0.55

14. Find $P(E$ and $F)$ if $P(E$ or $F) = 0.6$. 0.1

15. Find $P(E$ or $F)$ if E and F are mutually exclusive. 0.7

16. Find $P(E$ and $F)$ if E and F are mutually exclusive. 0

17. Find $P(E^c)$. 0.75

18. Find $P(F^c)$. 0.55

19. If $P(E) = 0.60$, $P(E$ or $F) = 0.85$, and $P(E$ and $F) = 0.05$, find $P(F)$. 0.3

20. If $P(F) = 0.30$, $P(E$ or $F) = 0.65$, and $P(E$ and $F) = 0.15$, find $P(E)$. 0.5

In Problems 21–24, a golf ball is selected at random from a golf bag. If the golf bag contains 9 Titleists, 8 Maxflis, and 3 Top-Flites, find the probability of each event.

21. The golf ball is a Titleist or Maxfli. $\frac{17}{20}$

22. The golf ball is a Maxfli or Top-Flite. $\frac{11}{20}$

23. The golf ball is not a Titleist. $\frac{11}{20}$

24. The golf ball is not a Top-Flite. $\frac{17}{20}$

Applying the Concepts

25. Weapon of Choice The following probability model
NW shows the distribution of murders by type of weapon for murder cases from 1990 to 2002.

Weapon	Probability
Gun	0.668
Knife	0.125
Blunt object	0.048
Personal weapon	0.067
Strangulation	0.017
Fire	0.007
Other	0.068

Source: U.S. Federal Bureau of Investigation

(a) Verify that this is a probability model.
(b) What is the probability a randomly selected murder resulted from a gun or knife? Interpret this probability. 0.793
(c) What is the probability a randomly selected murder resulted from a knife, blunt object, or strangulation? Interpret this probability. 0.19
(d) What is the probability a randomly selected murder resulted from a weapon other than a gun? Interpret this probability. 0.332
(e) Are murders by strangulation unusual? Yes

26. Doctorates Conferred The following probability model shows the distribution of doctoral degrees from U.S. universities in 2003 by area of study.

Area of Study	Probability
Engineering	0.129
Physical sciences	0.101
Life sciences	0.206
Mathematics	0.024
Computer sciences	0.021
Social sciences	0.167
Humanities	0.133
Education	0.163
Professional and other fields	0.056

Source: U.S. National Science Foundation

(a) Verify that this is a probability model.
(b) What is the probability a randomly selected doctoral candidate who earned a degree in 2003 studied physical science or life science? Interpret this probability. 0.307
(c) What is the probability a randomly selected doctoral candidate who earned a degree in 2003 studied physi-

cal science, life science, mathematics, or computer science? Interpret this probability. *0.352*

(d) What is the probability a randomly selected doctoral candidate who earned a degree in 2003 did not study mathematics? Interpret this probability. *0.976*

(e) Are doctoral degrees in mathematics unusual? Does this result surprise you? *Yes*

27. Multiple Births The following data represent the number NW of live multiple-delivery births (three or more babies) in 2002 for women 15 to 54 years old:

Age	Number of Multiple Births
15–19	93
20–24	511
25–29	1628
30–34	2832
35–39	1843
40–44	377
45–54	117

Source: National Vital Statistics Report, Vol. 52, No. 10, December 17, 2003

(a) Determine the probability that a randomly selected multiple birth in 2002 for women 15 to 54 years old involved a mother 30 to 39 years old. Interpret this probability. *0.632*

(b) Determine the probability that a randomly selected multiple birth in 2002 for women 15 to 54 years old involved a mother who was not 30 to 39 years old. Interpret this probability. *0.368*

(c) Determine the probability that a randomly selected multiple birth in 2002 for women 15 to 54 years old involved a mother who was less than 45 years old. Interpret this probability. *0.984*

(d) Determine the probability that a randomly selected multiple birth in 2002 for women 15 to 54 years old involved a mother who was at least 20 years old. Interpret this probability. *0.987*

28. Housing The following probability model shows the distribution for the number of rooms in U.S. housing units.

Rooms	Probability
One	0.004
Two	0.013
Three	0.104
Four	0.209
Five	0.218
Six	0.190
Seven	0.123
Eight or more	0.139

Source: U.S. Census Bureau

(a) Verify that this is a probability model.

(b) What is the probability that a randomly selected housing unit has four or more rooms? Interpret this probability. *0.879*

(c) What is the probability that a randomly selected housing unit has fewer than eight rooms? Interpret this probability. *0.861*

(d) What is the probability that a randomly selected housing unit has from four to six (inclusive) rooms? Interpret this probability. *0.617*

(e) What is the probability that a randomly selected housing unit has at least two rooms? Interpret this probability. *0.996*

29. A Deck of Cards A standard deck of cards contains 52 NW cards as shown in Figure 9. One card is randomly selected from the deck.

(a) Compute the probability of randomly selecting a heart or club from a deck of cards. *1/2*

(b) Compute the probability of randomly selecting a heart or club or diamond from a deck of cards. *3/4*

(c) Compute the probability of randomly selecting an ace or heart from a deck of cards. *4/13*

30. A Deck of Cards A standard deck of cards contains 52 cards as shown in Figure 9. One card is randomly selected from the deck.

(a) Compute the probability of randomly selecting a two or three from a deck of cards. *2/13*

(b) Compute the probability of randomly selecting a two or three or four from a deck of cards. *3/13*

(c) Compute the probability of randomly selecting a two or club from a deck of cards. *4/13*

31. Birthdays Exclude leap years from the following calculations:

(a) Compute the probability that a randomly selected person does not have a birthday on November 8. *364/365*

(b) Compute the probability that a randomly selected person does not have a birthday on the 1st day of a month. *353/365*

(c) Compute the probability that a randomly selected person does not have a birthday on the 31st day of a month. *358/365*

(d) Compute the probability that a randomly selected person was not born in December. *334/365*

32. Roulette In the game of roulette, a wheel consists of 38 slots numbered 0, 00, 1, 2, ... 36. The odd-numbered slots are red, and the even-numbered slots are black. The numbers 0 and 00 are green. To play the game, a metal ball is spun around the wheel and is allowed to fall into one of the numbered slots.

(a) What is the probability that the metal ball lands on green or red? *10/19*

(b) What is the probability that the metal ball does not land on green? *18/19*

33. Health Problems According to the Centers for Disease Control, the probability that a randomly selected citizen of the United States has hearing problems is 0.151. The probability that a randomly selected citizen of the United States has vision problems is 0.093. Can we compute the probability of randomly selecting a citizen of the United States who has hearing problems or vision problems by adding these probabilities? Why or why not? No

34. **Visits to the Doctor** In a National Ambulatory Medical Care Survey administered by the Centers for Disease Control, it was learned that the probability a randomly selected patient visited the doctor for a blood pressure check is 0.601. The probability a randomly selected patient visited the doctor for urinalysis is 0.128. Can we compute the probability of randomly selecting a patient who visited the doctor for a blood pressure check or urinalysis by adding these probabilities? Why or why not? No

35. **Language Spoken at Home** According to the U.S. Census Bureau, the probability a randomly selected household speaks only English at home is 0.82. The probability a randomly selected household speaks only Spanish at home is 0.11.

(a) What is the probability a randomly selected household speaks only English or only Spanish at home? 0.93

(b) What is the probability a randomly selected household speaks a language other than only English or only Spanish at home? 0.07

(c) What is the probability a randomly selected household speaks a language other than only English at home? 0.18

(d) Can the probability that a randomly selected household speaks only Polish at home equal 0.08? Why or why not? No

36. **Getting to Work** According to the U.S. Census Bureau, the probability a randomly selected worker primarily drives a car to work is 0.82. The probability a randomly selected worker primarily takes public transportation to work is 0.053.

(a) What is the probability a randomly selected worker primarily drives a car or takes public transportation to work? 0.873

(b) What is the probability a randomly selected worker neither drives a car nor takes public transportation to work? 0.127

(c) What is the probability a randomly selected worker does not drive a car to work? 0.18

(d) Can the probability a randomly selected worker walks to work equal 0.15? Why or why not? No

37. **Cigar Smoking** The data in the table show the results of a
NW national study of 137,243 U.S. men that investigated the association between cigar smoking and death from cancer. **Note:** Current cigar smoker means cigar smoker at time of death.

	Died from Cancer	Did Not Die from Cancer
Never smoked cigars	782	120,747
Former cigar smoker	91	7,757
Current cigar smoker	141	7,725

Source: Shapiro, Jacobs, and Thun. Cigar Smoking in Men and Risk of Death from Tobacco-Related Cancers, *Journal of the National Cancer Institute*, February 16, 2000.

(a) If an individual is randomly selected from this study, what is the probability that he died from cancer? 0.007

(b) If an individual is randomly selected from this study, what is the probability that he was a current cigar smoker? 0.057

(c) If an individual is randomly selected from this study, what is the probability that he died from cancer and was a current cigar smoker? 0.001

(d) If an individual is randomly selected from this study, what is the probability that he died from cancer or was a current cigar smoker? 0.064

38. **Civilian Labor Force** The following table represents the employment status and gender of the civilian labor force ages 16 to 24 (in millions).

	Male	Female
Employed	11.2	10.3
Unemployed	1.6	1.4

Source: U.S. Bureau of Labor Statistics, August 2002

(a) What is the probability that a randomly selected 16- to 24-year-old individual from the civilian labor force is employed? 0.878

(b) What is the probability that a randomly selected 16- to 24-year-old individual from the civilian labor force is male? 0.522

(c) What is the probability that a randomly selected 16- to 24-year-old individual from the civilian labor force is employed and male? 0.457

(d) What is the probability that a randomly selected 16- to 24-year-old individual from the civilian labor force is employed or male? 0.943

39. **Student Government Satisfaction Survey** The Committee on Student Life at a university conducted a survey of 375 undergraduate students regarding satisfaction with student government. Results of the survey are shown in the table by class rank.

	Freshman	Sophomore	Junior	Senior	Totals
Satisfied	57	49	64	61	**231**
Neutral	23	15	16	11	**65**
Not satisfied	21	18	14	26	**79**
Totals	**101**	**82**	**94**	**98**	**375**

(a) If a survey participant is selected at random, what is the probability that he or she is satisfied with student government? 0.616

(b) If a survey participant is selected at random, what is the probability that he or she is a junior? 0.251

(39c) *0.171*

(c) If a survey participant is selected at random, what is the probability that he or she is satisfied and is a junior?

(d) If a survey participant is selected at random, what is the probability that he or she is satisfied or is a junior? *0.696*

40. The Placebo Effect A company is testing a new medicine for migraine headaches. In the study, 150 women were given the new medicine and an additional 100 women were given a placebo. Each participant was directed to take the medicine when the first symptoms of a migraine occurred and then to record whether the headache went away within 45 minutes or lingered. The results are recorded in the following table:

	Headache Went Away	Headache Lingered	Totals
Given medicine	132	18	**150**
Given placebo	56	44	**100**
Totals	**188**	**62**	**250**

(a) If a study participant is selected at random, what is the probability she was given the placebo? *0.4*

(b) If a study participant is selected at random, what is the probability her headache went away within 45 minutes? *0.752*

(c) If a study participant is selected at random, what is the probability she was given the placebo and her headache went away within 45 minutes? *0.224*

(d) If a study participant is selected at random, what is the probability she was given the placebo or her headache went away within 45 minutes? *0.928*

41. Active Duty The following table represents the number of active-duty military personnel by rank in the four major branches of the military as of February 28, 2005.

	Officers	Enlisted
Army	80,580	409,410
Navy	53,428	310,167
Air Force	73,331	159,164
Marine Corps	18,893	287,328

Source: U.S. Department of Defense

(a) If an active-duty military person is selected at random, what is the probability that the individual is an officer? *0.162*

(b) If an active-duty military person is selected at random, what is the probability that the individual is in the Navy? *0.261*

(c) If an active-duty military person is selected at random, what is the probability that the individual is a Naval officer? *0.038*

(d) If an active-duty military person is selected at random, what is the probability that the individual is an officer or is in the Navy? *0.385*

42. Driver Fatalities The following data represent the number of driver fatalities in the United States in 2002 by age group for male and female drivers:

Age	Male	Female
Under 16	228	108
16–20	5696	2386
21–34	13,553	4148
35–54	14,395	5017
55–69	4937	1708
70 and over	3159	1529

Source: *Traffic Safety Facts 2002*, Federal Highway Administration, 2002

(a) Determine the probability that a randomly selected driver fatality was male. *0.738*

(b) Determine the probability that a randomly selected driver fatality was 16 to 20 years old. *0.142*

(c) Determine the probability that a randomly selected driver fatality was a 16- to 20-year-old male. *0.100*

(d) Determine the probability that a randomly selected driver fatality was male or 16 to 20 years old. *0.780*

5.3 Independence and the Multiplication Rule

Objectives

1 Understand independence

2 Use the Multiplication Rule for independent events

3 Compute at-least probabilities

1 Understand Independence

The Addition Rule for Disjoint Events deals with probabilities involving the word *or*. That is, it is used for computing the probability of observing an outcome in event E or event F. We now describe a probability rule for computing the probability that E and F both occur.

Before we can present this rule, we must discuss the idea of *independent events*.

Definition

> Two events E and F are **independent** if the occurrence of event E in a probability experiment does not affect the probability of event F. Two events are **dependent** if the occurrence of event E in a probability experiment affects the probability of event F.

To help you understand the idea of independence, we again look at a simple situation—flipping a coin. Suppose you flip a coin twice. Does the fact that you obtained a head on the first toss have any impact on the likelihood of obtaining a head on the second toss? Not unless you are a master coin flipper who can manipulate the outcome of a coin flip! For this reason, the outcome from the first flip is independent of the outcome from the second flip. Let's look at other examples.

EXAMPLE 1   **Independent or Not?**

(a) Suppose you flip a coin and roll a die. The events "obtain a head" and "roll a 5" are independent because the results of the coin flip do not impact the results of the die toss.

(b) Are the events "earned a bachelor's degree" and "earn more than $100,000 per year" independent? No, because knowing that an individual has a bachelor's degree affects the likelihood that the individual is earning more than $100,000 per year.

(c) Suppose two 24-year-old male drivers who live in the United States are randomly selected. The events "male 1 gets in a car accident during the year" and "male 2 gets in a car accident during the year" are independent because the males were randomly selected. This means what happens with one of the drivers has nothing to do with what happens to the other driver.

/ **In Other Words**

In determining whether two events are independent, ask yourself whether the probability of one event is affected by the other event. For example, what is the probability that a 29-year-old male has high cholesterol? What is the probability that a 29-year-old male has high cholesterol, given that he eats fast food four times a week? Does the fact that the individual eats fast food four times a week change the likelihood that he has high cholesterol? If yes, the events are not independent.

In Example 1(c), we are able to conclude that the events "male 1 gets in an accident" and "male 2 gets in an accident" are independent because the individuals are randomly selected. By randomly selecting the individuals, it is reasonable to conclude that the individuals are not related in any way (related in the sense that they do not live in the same town, attend the same school, and so on). If the two individuals did have a common link between them (such as they both lived on the same city block), then knowing one of the males had a car accident may affect the likelihood that the other male had a car accident. After all, they could hit each other!

Now Work Problem 7.

Disjoint Events versus Independent Events

It is important that we understand that disjoint events and independent events are different concepts. Recall that two events are disjoint if they have no outcomes in common. In other words, two events are disjoint if knowing that one of the events occurs, we know the other event did not occur. Independence means that one event occurring does not affect the probability of the other event occurring. Therefore, knowing two events are disjoint means that the events are not independent.

! **CAUTION**

Two events that are disjoint are not independent.

Consider the experiment of rolling a single die. Let E represent the event "roll an even number," and let F represent the event "roll an odd number." We can see that E and F are mutually exclusive because they have no outcomes in common. In addition, $P(E) = \dfrac{1}{2}$ and $P(F) = \dfrac{1}{2}$. However, if we are told that the roll of the die is going to be an even number, then what is the probability of event F? Because the outcome will be even, the probability of event F is now 0.

② Use the Multiplication Rule for Independent Events

Suppose that you flip a fair coin twice. What is the probability that you obtain a head on both flips? Put another way, what is the probability that you obtain a head on the first flip *and* you obtain a head on the second flip? We can create a sample space that lists the outcomes of this experiment quite easily. In flipping a coin twice where H represents the outcome heads and T represents the outcome tails, we can obtain

$$S = \{HH, HT, TH, TT\}$$

There is one outcome with both heads. Because each outcome is equally likely, we have

$$P(\text{heads on the 1st toss and heads on the 2nd toss}) = \frac{N(\text{heads on the 1st and heads on the 2nd})}{N(S)}$$

$$= \frac{1}{4}$$

We may have intuitively been able to figure this out by recognizing $P(\text{head}) = \frac{1}{2}$ for each flip. So it seems reasonable that

$$P(\text{heads on the 1st toss and heads on the 2nd toss}) = P(\text{heads on 1st toss}) \cdot P(\text{heads on 2nd toss})$$

$$= \frac{1}{2} \cdot \frac{1}{2}$$

$$= \frac{1}{4}$$

Because both approaches result in the same answer, $\frac{1}{4}$, it seems reasonable to conjecture that $P(E \text{ and } F) = P(E) \cdot P(F)$.

> **Multiplication Rule for Independent Events**
>
> If E and F are independent events, then
>
> $$P(E \text{ and } F) = P(E) \cdot P(F)$$

EXAMPLE 2   **Computing Probabilities of Independent Events**

Problem: In the game of roulette, the wheel has slots numbered 0, 00, and 1 through 36. A metal ball is allowed to roll around a wheel until it falls into one of the numbered slots. You decide to play the game and place a bet on the number 17. What is the probability that the ball will land in the slot numbered 17 two times in a row?

Approach: There are 38 outcomes in the sample space of the experiment. We use the classical method of computing probabilities because the outcomes are equally likely. In addition, we use the Multiplication Rule for Independent Events. The events "17 on first trial" and "17 on second trial" are independent because the ball does not remember it landed on 17 on the first trial, so this cannot affect the probability of landing on 17 on the second trial.

Solution: Because there are 38 possible outcomes to the experiment, the probability of the ball landing on 17 is $\frac{1}{38}$. Because the events "17 on first trial" and "17 on second trial" are independent, we have

P(ball lands in slot 17 in the 1st game and ball lands in slot 17 in the 2nd game)

$= P$(ball lands in slot 17 in the 1st game) $\cdot P$(ball lands in slot 17 in the 2nd game)

$= \dfrac{1}{38} \cdot \dfrac{1}{38} = \dfrac{1}{1,444} \approx 0.0006925$

It is very unlikely that the ball will land on 17 twice in a row. We expect the ball to land on 17 twice in a row about 7 times in 10,000 trials. ▬

We can extend the Multiplication Rule for three or more independent events.

Multiplication Rule for n Independent Events

If events E, F, G, ... are independent, then

$$P(E \text{ and } F \text{ and } G \text{ and } \ldots) = P(E) \cdot P(F) \cdot P(G) \ldots$$

EXAMPLE 3 **Life Expectancy**

Problem: The probability that a randomly selected male 24 years old will survive the year is 0.9985 according to the *National Vital Statistics Report*, Vol. 47, No. 28. What is the probability that three randomly selected 24-year-old males will survive the year? What is the probability that 20 randomly selected 24-year-old males will survive the year?

Approach: We can safely assume that the outcomes of the probability experiment are independent, because there is no indication that the survival of one male affects the survival of the others. For example, if two of the males lived in the same house, a house fire could kill both males and we lose independence. (Knowledge that one male died in a house fire certainly affects the probability that the other died.) By randomly selecting the males, we minimize the chances that they are related in any way.

Solution

P(all three males survive) $= P$(1st survives and 2nd survives and 3rd survives)

$= P$(1st survives) $\cdot P$(2nd survives) $\cdot P$(3rd survives)

$= (0.9985)(0.9985)(0.9985)$

$= 0.9955$

There is a 99.55% probability that all three males survive the year.

P(all 20 males survive) $= P$(1st survives and 2nd survives and ... and 20th survives)

$= P$(1st survives) $\cdot P$(2nd survives) $\cdots P$(20th survives)

Independent events ↑   Multiply 0.9985 by itself 20 times.

$= \overbrace{(0.9985) \cdot (0.9985) \cdots (0.9985)}$

$= (0.9985)^{20}$

$= 0.9704$

Now Work Problems 17(a) and (b).   There is a 97.04% probability that all 20 males survive the year. ▬

Compute At-Least Probabilities

We now present an example in which we compute at-least probabilities. These probabilities use the Complement Rule. The phrase *at least* means "greater than or equal to." For example, a person must be at least 17 years old to see an R-rated movie.

| **EXAMPLE 4** | **Computing At-Least Probabilities** |

Note to Instructor

Remind students that at-least means "greater than or equal to." At-least probabilities are usually determined using the Complement Rule. For Example 4, ask students to distinguish between $1 - 0.9985^{1000}$ and $(1 - 0.9985)^{1000}$.

Problem: Compute the probability that at least 1 male out of 1000 aged 24 years will die during the course of the year if the probability that a randomly selected 24-year-old male survives the year is 0.9985.

Approach: The phrase *at least* means "greater than or equal to," so we wish to know the probability that 1 or 2 or 3 or ... or 1000 males will die during the year. These events are mutually exclusive, so

$$P(1 \text{ or } 2 \text{ or } 3 \text{ or} \dots \text{ or } 1000 \text{ die}) = P(1 \text{ dies}) + P(2 \text{ die}) + P(3 \text{ die}) + \dots + P(1000 \text{ die})$$

Computing these probabilities is very time consuming. However, we notice that the complement of "at least one dying" is "none die." We use the Complement Rule to compute the probability.

Solution

$P(\text{at least one dies}) = 1 - P(\text{none die})$

$\qquad = 1 - P(\text{1st survives and 2nd survives and} \dots \text{and 1000th survives})$

$\qquad = 1 - P(\text{1st survives}) \cdot P(\text{2nd survives}) \cdot \dots \cdot P(\text{1000th survives})$ Independent events

$\qquad = 1 - (0.9985)^{1000}$

$\qquad = 1 - 0.2229$

$\qquad = 0.7771$

$\qquad = 77.71\%$

There is a 77.71% probability that at least one 24-year-old male out of 1000 will die during the course of the year.

> **Now Work Problem 17(c).**

Summary: Rules of Probability

1. The probability of any event must be between 0 and 1, inclusive. If we let E denote any event, then $0 \le P(E) \le 1$.
2. The sum of the probabilities of all outcomes must equal 1. That is, if the sample space $S = \{e_1, e_2, \dots, e_n\}$, then $P(e_1) + P(e_2) + \dots + P(e_n) = 1$.
3. If E and F are disjoint events, then $P(E \text{ or } F) = P(E) + P(F)$. If E and F are not disjoint events, then $P(E \text{ or } F) = P(E) + P(F) - P(E \text{ and } F)$.
4. If E represents any event and E^c represents the complement of E, then $P(E^c) = 1 - P(E)$.
5. If E and F are independent events, then $P(E \text{ and } F) = P(E) \cdot P(F)$.

Notice that *or* probabilities use the Addition Rule, whereas *and* probabilities use the Multiplication Rule. Accordingly, *or* probabilities imply addition, while *and* probabilities imply multiplication.

5.3 ASSESS YOUR UNDERSTANDING

Concepts and Vocabulary

1. Two events E and F are _____ if the occurrence of event E in a probability experiment does not affect the probability of event F. Independent
2. The word *and* in probability implies that we use the _____ Rule. Multiplication
3. The word *or* in probability implies that we use the _____ Rule. Addition
4. *True or False:* When two events are disjoint, they are also independent. False
5. If two events E and F are independent, $P(E \text{ and } F) =$ _____. $P(E) \cdot P(F)$
6. Suppose events E and F are disjoint. What is $P(E \text{ and } F)$? 0

Skill Building

7. Determine whether the events E and F are independent or dependent. Justify your answer.
    (a) E: It rains on June 30.
        F: It is cloudy on June 30. Dependent
    (b) E: Your car has a flat tire. Independent
        F: The price of gasoline increases overnight.
    (c) E: You live at least 80 years.
        F: You smoke a pack of cigarettes every day of your life. Dependent
9. Suppose events E and F are independent, $P(E) = 0.3$, and $P(F) = 0.6$. What is the $P(E \text{ and } F)$? 0.18

8. Determine whether the events A and B are independent or dependent. Justify your answer.
    (a) A: You earn an A on an exam.
        B: You study for an exam. Dependent
    (b) A: You are late for work.
        B: Your car has a flat tire. Dependent
    (c) A: You earn more than $50,000 per year.
        B: You are born in the month of July. Independent
10. Suppose events E and F are independent, $P(E) = 0.7$, and $P(F) = 0.9$. What is the $P(E \text{ and } F)$? 0.63

Applying the Concepts

11. **Flipping a Coin** What is the probability of obtaining five heads in a row when flipping a coin? Interpret this probability. 0.03125

12. **Rolling a Die** What is the probability of obtaining 4 ones in a row when rolling a fair, six-sided die? Interpret this probability. 0.0008

13. **Southpaws** About 13% of the population is left-handed. If two people are randomly selected, what is the probability both are left-handed? What is the probability at least one is right-handed? 0.0169; 0.9831

14. **Investing** Suppose your financial advisor has recommended two stocks, each of which has a 0.6 probability of increasing in value over the next year. Assuming the performance of one stock is independent of the other, what is the probability both stocks will rise over the next year? What is the probability at least one stock will not increase in value? 0.36; 0.64

15. **False Positives** The ELISA is a test to determine whether the HIV antibody is present. The test is 99.5% effective. This means that the test will accurately come back negative if the HIV antibody is not present. The probability of a test coming back positive when the antibody is not present (a false positive) is 0.005. Suppose the ELISA is given to five randomly selected people who do not have the HIV antibody.
    (a) What is the probability that the ELISA comes back negative for all five people? 0.9752
    (b) What is the probability that the ELISA comes back positive for at least one of the five people? 0.0248

16. **Christmas Lights** Christmas lights are often designed with a series circuit. This means that when one light burns out, the entire string of lights goes black. Suppose the lights are designed so that the probability a bulb will last 2 years is 0.995. The success or failure of a bulb is independent of the success or failure of other bulbs.
    (a) What is the probability that in a string of 100 lights all 100 will last 2 years? 0.6058
    (b) What is the probability at least one bulb will burn out in 2 years? 0.3942

17. **Life Expectancy** The probability that a randomly selected 40-year-old male will live to be 41 years old is 0.99718 according to the *National Vital Statistics Report*, Vol. 48, No. 18.
    (a) What is the probability that two randomly selected 40-year-old males will live to be 41 years old? 0.99437
    (b) What is the probability that five randomly selected 40-year-old males will live to be 41 years old? 0.98598
    (c) What is the probability that at least one of five randomly selected 40-year-old males will not live to be 41 years old? Would it be unusual that at least one of five randomly selected 40-year-old males will not live to be 41 years old? 0.014; yes

18. **Life Expectancy** The probability that a randomly selected 40-year-old female will live to be 41 years old is 0.99856 according to the *National Vital Statistics Report*, Vol. 48, No. 18.
    (a) What is the probability that two randomly selected 40-year-old females will live to be 41 years old? 0.99712
    (b) What is the probability that five randomly selected 40-year-old females will live to be 41 years old? 0.99282
    (c) What is the probability that at least one of five randomly selected 40-year-old females will not live to be 41 years old? Would it be unusual that at least one of five randomly selected 40-year-old females will not live to be 41 years old? 0.00718; Yes

19. Blood Types Blood types can be classified as either Rh$^+$ or Rh$^-$. According to the *Information Please Almanac*, 99% of the Chinese population has Rh$^+$ blood.

(a) What is the probability that two randomly selected Chinese people have Rh$^+$ blood? *0.9801*

(b) What is the probability that six randomly selected Chinese people have Rh$^+$ blood? *0.9415*

(c) What is the probability that at least one of six randomly selected Chinese people has Rh$^-$ blood? Would it be unusual that at least one of six randomly selected Chinese people has Rh$^-$ blood? *0.0585*

20. Quality Control Suppose a company selects two people who work independently inspecting two-by-four timbers. Their job is to identify low-quality timbers. Suppose the probability that an inspector does not identify a low-quality timber is 0.20.

(a) What is the probability that both inspectors do not identify a low-quality timber? *0.04*

(b) How many inspectors should be hired to keep the probability of not identifying a low-quality timber below 1%? *3*

21. Cold Streaks Players in sports are said to have "hot streaks" and "cold streaks." For example, a batter in baseball might be considered to be in a slump or cold streak if he has made 10 outs in 10 consecutive at-bats. Suppose a hitter successfully reaches base 30% of the time he comes to the plate.

(a) Find the probability that the hitter makes 10 outs in 10 consecutive at-bats, assuming that at-bats are independent events. *Hint:* The hitter makes an out 70% of the time. *0.028*

(b) Are cold streaks unusual? *Yes*

22. Hot Streaks In a recent basketball game, a player who makes 65% of his free throws made eight consecutive free throws. Assuming free-throw shots are independent, determine whether this feat was unusual. *Unusual*

23. Defense System Suppose a satellite defense system is established in which four satellites acting independently have a 0.9 probability of detecting an incoming ballistic missile. What is the probability at least one of the four satellites detects an incoming ballistic missile? Would you feel safe with such a system? *0.9999*

24. E.P.T. Pregnancy Tests The packaging of an E.P.T. Pregnancy Test states that the test is "99% accurate at detecting typical pregnancy hormone levels." Assume the probability that a test will correctly identify a pregnancy is

0.99. Suppose 12 randomly selected pregnant women with typical hormone levels are each given the test.

(a) What is the probability that all 12 tests will be positive? *0.8864*

(b) What is the probability that at least one test will not be positive? *0.1136*

25. Bowling Suppose Ralph gets a strike when bowling 30% of the time.

(a) What is the probability that Ralph gets two strikes in a row? *0.09*

(b) What is the probability that Ralph gets a turkey (three strikes in a row)? *0.027*

(c) When events are independent, their complements are independent as well. Use this result to determine the probability that Ralph gets a strike and then does not get a strike. *0.21*

26. NASCAR Fans Among Americans who consider themselves auto racing fans, 59% identify NASCAR stock cars as their favorite type of racing. Suppose four auto racing fans are randomly selected. *Source*: ESPN / TNS Sports, reported in *USA Today*

(a) What is the probability that all four will identify NASCAR stock cars as their favorite type of racing? *0.1212*

(b) What is the probability that at least one will not identify NASCAR stock cars as his or her favorite type of racing? *0.8788*

(c) What is the probability that none will identify NASCAR stock cars as his or her favorite type of racing? *0.0283*

(d) What is the probability that at least one will identify NASCAR stock cars as his or her favorite type of racing? *0.9717*

27. Driving under the Influence Among 21- to 25-year-olds, 29% say they have driven while under the influence of alcohol. Suppose three 21- to 25-year-olds are selected at random. *Source*: U.S. Department of Health and Human Services, reported in *USA Today*

(a) What is the probability that all three have driven while under the influence of alcohol? *0.0244*

(b) What is the probability that at least one has not driven while under the influence of alcohol? *0.9756*

(c) What is the probability that none of the three has driven while under the influence of alcohol? *0.3579*

(d) What is the probability that at least one has driven while under the influence of alcohol? *0.6421*

5.4　Conditional Probability and the General Multiplication Rule

Objectives

1　Compute conditional probabilities

2　Compute probabilities using the General Multiplication Rule

1　Compute Conditional Probabilities

In the last section, we learned that when two events are independent the occurence of one event has no impact on the probability of the second event. For

Probability

Note to Instructor

For instructors looking to offer a course that minimizes the coverage of probability, this section can be skipped without loss of continuity.

example, according to data from the Centers for Disease Control, 17% of adults have high cholesterol. So the probability that a randomly selected adult has high cholesterol is 0.17. Now, if we were told that this particular individual eats three Big Macs from McDonald's each week, do you think that the probability he has high cholesterol will change? Of course it will. The probability that the individual has high cholesterol will increase, given the fact that the individual eats three Big Macs each week. This is called a *conditional probability*.

Definition

Conditional Probability

The notation $P(F|E)$ is read "the probability of event F given event E." It is the probability that an event F occurs, given that the event E has occurred.

Let's look at an example.

EXAMPLE 1 **An Introduction to Conditional Probability**

Problem: Suppose that a single die is rolled. What is the probability that the die comes up 3? Now suppose the die is rolled a second time, but we are told the outcome will be an odd number. What is the probability that the die comes up 3?

Approach: We assume that the die is fair and compute the probabilities using equally likely outcomes.

Solution: In the first instance, there are six possibilities in the sample space, $S = \{1, 2, 3, 4, 5, 6\}$, so $P(3) = \dfrac{1}{6}$. In the second instance, there are three possibilities in the sample space, because the only possible outcomes are odd, so $S = \{1, 3, 5\}$. We express this probability symbolically as $P(3|\text{outcome is odd}) = \dfrac{1}{3}$, which is read "the probability of rolling a 3 given that the outcome is odd is one-third."

So conditional probabilities reduce the size of the sample space under consideration. Let's look at another example. The data in Table 8 represent the marital status of males and females 18 years old or older in the United States in 2003.

Table 8	Males (in millions)	Females (in millions)	Totals (in millions)
Never married	28.6	23.3	**51.9**
Married	62.1	62.8	**124.9**
Widowed	2.7	11.3	**14.0**
Divorced	9.0	12.7	**21.7**
Totals (in millions)	102.4	110.1	212.5

Source: U.S. Census Bureau, Current Population Reports

Suppose we want to know the probability that a randomly selected individual is widowed. This probability is found by dividing the number of widowed individuals by the total number of individuals who are 18 years old or older.

$$P(\text{widowed}) = \frac{14.0}{212.5}$$

$$= 0.066$$

Now suppose that we know the individual is female. Does this change the probability that she is widowed? Because the sample space now consists only of females, we can determine the probability that the individual is widowed, given that the individual is female, as follows:

$$P(\text{widowed}|\text{female}) = \frac{N(\text{widowed females})}{N(\text{females})}$$

$$= \frac{11.3}{110.1} = 0.103$$

So knowing that the individual is female increases the likelihood that the individual is widowed. The previous discussion leads to the following.

Conditional Probability Rule

If E and F are any two events, then

$$P(F|E) = \frac{P(E \text{ and } F)}{P(E)} = \frac{N(E \text{ and } F)}{N(E)} \tag{1}$$

The probability of event F occurring, given the occurrence of event E, is found by dividing the probability of E and F by the probability of E. Or the probability of event F occurring, given the occurrence of event E, is found by dividing the number of outcomes in E and F by the number of outcomes in E.

We used the second method for computing conditional probabilities in the widow example.

EXAMPLE 2 **Conditional Probabilities on Marital Status and Gender**

Problem: The data in Table 8 represent the marital status and gender of the residents of the United States aged 18 years old or older in 2003.

(a) Compute the probability that a randomly selected male has never married.
(b) Compute the probability that a randomly selected individual who has never married is male.

Approach

(a) We are given that the randomly selected person is male, so we concentrate on the male column. There are 102.4 million males and 28.6 million people who are male and never married, so $N(\text{male}) = 102.4$ million and $N(\text{male and never married}) = 28.6$ million. Compute the probability using the Conditional Probability Rule.

(b) We are given that the randomly selected person has never married, so we concentrate on the never married row. There are 51.9 million people who have never married and 28.6 million people who are male and have never married, so $N(\text{never married}) = 51.9$ million and $N(\text{male and never married}) = 28.6$ million. Compute the probability using the Conditional Probability Rule.

Solution

(a) Substituting into Formula (1), we obtain

$$P(\text{never married}|\text{male}) = \frac{N(\text{never married and male})}{N(\text{male})} = \frac{28.6}{102.4} \approx 0.279$$

There is a 27.9% probability that the randomly selected individual has never married, given that he is male.

(b) Substituting into Formula (1), we obtain

$$P(\text{male}|\text{never married}) = \frac{N(\text{male and never married})}{N(\text{never married})} = \frac{28.6}{51.9} \approx 0.551$$

There is a 55.1% probability the randomly selected individual is male, given that he or she has never married. ▬

What is the difference between the results of Examples 2(a) and (b)? In Example 2(a), we found that 27.9% of males have never married, whereas in Example 2(b) we found that 55.1% of individuals who have never married are male. Do you see the difference?

> Now Work Problem 17.

EXAMPLE 3   Birth Weights of Preterm Babies

Problem: In 2002, 11.96% of all births were preterm. (The gestation period of the pregnancy was less than 37 weeks.) Also in 2002, 0.23% of all births resulted in a preterm baby that weighed 8 pounds, 13 ounces or more. What is the probability that a randomly selected baby weighs 8 pounds, 13 ounces or more, given that the baby was preterm?

Approach: We want to know the probability that the baby weighs 8 pounds, 13 ounces or more, given that the baby was preterm. We know that 0.23% of all babies weighed 8 pounds, 13 ounces or more and were preterm, so $P(\text{weighs 8 pounds, 13 ounces or more and preterm}) = 0.23\%$. We also know that 11.96% of all births were preterm, so $P(\text{preterm}) = 11.96\%$. We compute the probability by dividing the probability that a baby will weigh 8 pounds, 13 ounces or more *and* be preterm by the probability that a baby will be preterm.

Solution: $P(\text{weighs 8 pounds, 13 ounces or more}|\text{preterm})$

$$= \frac{P(\text{weighs 8 pounds, 13 ounces or more and preterm})}{P(\text{preterm})}$$

$$= \frac{0.23\%}{11.96\%} = \frac{0.0023}{0.1196} \approx 0.0192 = 1.92\%$$

There is a 1.92% probability that a randomly selected baby will weigh 8 pounds, 13 ounces or more, given that the baby is preterm. It is unusual for preterm babies to weigh 8 pounds, 13 ounces or more. ▬

> Now Work Problem 13.

② Compute Probabilities Using the General Multiplication Rule

If we solve the Conditional Probability Rule for $P(E \text{ and } F)$, we obtain the General Multiplication Rule.

> **General Multiplication Rule**
>
> The probability that two events E and F both occur is
>
> $$P(E \text{ and } F) = P(E) \cdot P(F|E)$$
>
> In words, the probability of E and F is the probability of event E occurring times the probability of event F occurring, given the occurrence of event E.

EXAMPLE 4   ## Using the General Multiplication Rule

Problem: The probability that a driver who is speeding gets pulled over is 0.8. The probability that a driver gets a ticket given he/she is pulled over is 0.9. What is the probability that a randomly selected driver who is speeding gets pulled over and gets a ticket?

Approach: Let E represent the event "driver who is speeding gets pulled over," and let F represent the event "driver gets a ticket." We use the General Multiplication Rule to compute $P(E \text{ and } F)$.

Solution: $P(\text{driver who is speeding gets pulled over and gets a ticket}) = P(E \text{ and } F) = P(E) \cdot P(F|E) = 0.8(0.9) = 0.72 = 72\%$. There is a 72% probability that a driver who is speeding gets pulled over and gets a ticket.

> Now Work Problem 29.

EXAMPLE 5   ## Acceptance Sampling

Problem: Suppose that a box of 100 circuits is sent to a manufacturing plant. Of the 100 circuits shipped, 5 are defective. The plant manager receiving the chips randomly selects 2 and tests them. If both chips work, she will accept the shipment. Otherwise, the shipment is rejected. What is the probability the plant manager discovers at least 1 defective circuit and rejects the shipment?

Approach: We wish to determine the probability that at least one of the tested circuits is defective. There are four possibilities in this probability experiment. None of the circuits are defective, the first is defective while the second is not, the first is not defective while the second is defective, or both circuits are defective. We cannot compute the probability that at least 1 is defective using the fact that there are four outcomes and three result in at least 1 defective, because the outcomes are not equally likely. We need a different approach. We could determine the probability of at least 1 defective by computing the probability the first is defective while the second is not, computing the probability the first is not defective while the second is defective, and computing the probability they both are defective and then adding these three probabilities together (because they are disjoint), but this is rather time consuming. It is easier to compute the probability that both circuits are not defective and use the Complement Rule to determine the probability of at least 1 defective.

Solution: We have 100 circuits and 5 of them are defective, so 95 circuits are not defective.

$$P(\text{at least 1 defective}) = 1 - P(\text{none defective})$$
$$= 1 - P(\text{1st not defective}) \cdot P(\text{2nd not defective} \mid \text{1st not defective})$$
$$= 1 - \left(\frac{95}{100}\right) \cdot \left(\frac{94}{99}\right)$$
$$= 1 - 0.9020$$
$$= 0.098$$

There is a 9.8% probability that the shipment will be not be accepted.

> Now Work Problem 21.

Whenever a small random sample is taken from a large population, it is reasonable to compute probabilities of events assuming independence. Consider the following example.

| EXAMPLE 6 | **Sickle-Cell Anemia** |

Problem: In a survey of 10,000 African Americans, it was determined that 27 had sickle-cell anemia.

(a) Suppose we randomly select 1 of the 10,000 African Americans surveyed. What is the probability that he or she will have sickle-cell anemia?

(b) If two individuals from this group are randomly selected, what is the probability that both have sickle-cell anemia?

(c) Compute the probability of randomly selecting two individuals from this group who have sickle-cell anemia, assuming independence.

Approach: We let the event E = "sickle-cell anemia," so $P(E)$ = number of African Americans who have sickle-cell anemia divided by the number in the survey. To answer part (b), we let E_1 = "first person has sickle-cell anemia" and E_2 = "second person has sickle-cell anemia," and then we compute $P(E_1 \text{ and } E_2) = P(E_1) \cdot P(E_2|E_1)$. To answer part (c), we use the Multiplication Rule for Independent Events.

Solution

(a) If one individual is selected, $P(E) = \dfrac{27}{10,000} = 0.0027$.

(b) Using the Multiplication Rule, we have

$$P(E_1 \text{ and } E_2) = P(E_1) \cdot P(E_2|E_1) = \frac{27}{10000} \cdot \frac{26}{9999} \approx 0.00000702$$

Notice $P(E_2|E_1) = \dfrac{26}{9999}$ because we are sampling without replacement, so after event E_1 occurs there is one less person with sickle-cell anemia and one less person in the sample space.

(c) The assumption of independence means that the outcome of the first trial of the experiment does not affect the probability of the second trial. (It is like sampling with replacement.) Therefore, we assume $P(E_1) = P(E_2) = \dfrac{27}{10,000}$. Then

$$P(E_1 \text{ and } E_2) = P(E_1) \cdot P(E_2) = \frac{27}{10,000} \cdot \frac{27}{10,000} \approx 0.00000729$$

The probabilities in Examples 6(b) and 6(c) are extremely close in value. Based on these results, we infer the following principle:

> If small random samples are taken from large populations without replacement, it is reasonable to assume independence of the events. As a rule of thumb, if the sample size is less than 5% of the population size, we treat the events as independent.

For example, in Example 6, we can compute the probability of randomly selecting two African Americans who have sickle-cell anemia using independence because the sample size is less than 5% of the population size $\left(\dfrac{2}{10,000} = 0.0002 = 0.02\% \right)$.

Now Work Problem 35.

Historical Note

Andrei Nikolaevich Kolmogorov was born on April 25, 1903, in Tambov, Russia. His parents were not married. His mother's sister, Vera Yakovlena, raised Kolmogorov. In 1920, Kolmogorov enrolled in Moscow State University. He graduated from the university in 1925. That year he published eight papers, including his first on probability. In 1929, Kolmogorov received his doctorate. By this time he already had 18 publications. He became a professor at Moscow State University in 1931. Kolmogrov is quoted as saying, "The theory of probability as a mathematical discipline can and should be developed from axioms in exactly the same way as Geometry and Algebra." In addition to conducting research, Kolmogorov was interested in helping to educate gifted children. It did not bother him if the students did not become mathematicians; he simply wanted them to be happy. Andrei Kolmogorov died on October 20, 1987.

Conditional Probability and Independence

Two events are independent if the occurrence of event E in a probability experiment does not affect the probability of event F. We can now express independence using conditional probabilities.

Definition   Two events E and F are independent if $P(E|F) = P(E)$.

So, the Multiplication Rule for Independent Events is a special case of the General Multiplication Rule.

Look back at pp. 256–257. Because $P(\text{widowed}) \neq P(\text{widowed} \mid \text{female})$, the events "widowed" and "female" are not independent.

MAKING AN INFORMED DECISION

Sports Probabilities

Have you ever watched a sporting event on television in which the announcer cites an obscure statistic? Where do these numbers come from? Well, pretend that you are the statistician for your favorite sports team. Your job is to compile strange probabilities regarding your favorite team and a competing team. For example, during the 2001 baseball season, the Boston Red Sox won 36% of the games they played on Wednesdays. As statisticians, we represent this as a conditional probability as follows: $P(\text{win}|\text{Wednesday}) = 0.36$. Suppose that Boston was playing the Seattle Mariners on a Wednesday and that Seattle won 68% of the games that it played on Wednesdays. From these statistics, we predict that Seattle will win the game. Other ideas for conditional probabilities include home versus road games, day versus night games, weather, and so on. For basketball, consider conditional probabilities such as

the probability of winning if the team's leading scorer scores fewer than 12 points.

Use the statistics and probabilities that you compile to make a prediction about which team will win. Write an article that presents your predictions along with the supporting numerical facts. Maybe the article could include such "keys to the game" as "Our crack statistician has found that our football team wins 80% of its games when it holds opposing teams to less than 10 points." Repeat this exercise for at least five games. Following each game, determine whether the team you chose has won or lost. Compute your winning percentage for the games you predicted. Did you predict the winner in more than 50% of the games?

A great source for these obscure facts can be found at www.espn.com. For baseball, a great site is www.mlb.com. For basketball, go to www.nba.com. For football, go to www.nfl.com. For hockey, go to www.nhl.com.

5.4 ASSESS YOUR UNDERSTANDING

Concepts and Vocabulary

1. The notation $P(F|E)$ means the probability of event _____ given event _____. *F; E*

2. If $P(E) = 0.6$ and $P(E|F) = 0.34$, are events E and F independent? *No*

Skill Building   4. 0.525   5. 0.568   6. 0.411

3. Suppose that E and F are two events and that $P(E \text{ and } F) = 0.6$ and $P(E) = 0.8$. What is $P(F|E)$? *0.75*

4. Suppose that E and F are two events and that $P(E \text{ and } F) = 0.21$ and $P(E) = 0.4$. What is $P(F|E)$?

5. Suppose that E and F are two events and that $N(E \text{ and } F) = 420$ and $N(E) = 740$. What is $P(F|E)$?

6. Suppose that E and F are two events and that $N(E \text{ and } F) = 380$ and $N(E) = 925$. What is $P(F|E)$?

7. Suppose that E and F are two events and that $P(E) = 0.8$ and $P(F|E) = 0.4$. What is $P(E \text{ and } F)$? *0.32*

8. Suppose that E and F are two events and that $P(E) = 0.4$ and $P(F|E) = 0.6$. What is $P(E \text{ and } F)$? *0.24*

9. According to the U.S. Census Bureau, the probability a randomly selected individual in the United States earns more than $75,000 per year is 18.4%. The probability a randomly selected individual in the United States earns more than $75,000 per year, given that the individual has earned a bachelor's degree, is 35.0%. Are the events "earn more than $75,000 per year" and "earned a bachelor's degree" independent? *No*

10. The probability that a randomly selected individual in the United States 25 years and older has at least a bachelor's degree is 0.272. The probability that an individual in the United States 25 years and older has at least a bachelor's degree, given that the individual is Hispanic, is 0.114. Are the events "bachelor's degree" and "Hispanic" independent? (*Source*: *Educational Attainment in the United States, 2003*. U.S. Census Bureau, June 2004) *No*

Applying the Concepts $\frac{1}{4} \cdot \frac{1}{2}$

11. **Drawing a Card** Suppose a single card is selected from a standard 52-card deck. What is the probability that the card drawn is a club? Now suppose a single card is drawn from a standard 52-card deck, but we are told that the card is black. What is the probability that the card drawn is a club?

12. **Drawing a Card** Suppose a single card is selected from a standard 52-card deck. What is the probability that the card drawn is a king? Now suppose a single card is drawn from a standard 52-card deck, but we are told that the card is a heart. What is the probability that the card drawn is a king? Did the knowledge that the card is a heart change the probability that the card was a king? What is the term used to describe this result? 1/13; 1/13; no; independence

13. **Rainy Days** For the month of June in the city of Chicago, NW 37% of the days are cloudy. Also in the month of June in the city of Chicago, 21% of the days are cloudy and rainy. What is the probability that a randomly selected day in June will be rainy if it is cloudy? 0.568

14. **Cause of Death** According to the U.S. National Center for Health Statistics, in 2002, 0.2% of deaths in the United States were 25- to 34-year-olds whose cause of death was cancer. In addition, 1.97% of all those who died were 25 to 34 years old. What is the probability that a randomly selected death is the result of cancer if the individual is known to have been 25 to 34 years old? 0.102

15. **High School Dropout** According to the U.S. Census Bureau, 9.1% of high school dropouts are 16- to 17-year-olds. In addition, 5.8% of white high school dropouts are 16- to 17-year-olds. What is the probability that a randomly selected dropout is white, given that he or she is 16 to 17 years old? 0.637

16. **Income by Region** According to the U.S Census Bureau, 19.1% of U.S. households are in the Northeast. In addition, 4.4% of U.S. households earn $75,000 per year or more and are located in the Northeast. Determine the probability that a randomly selected U.S. household earns more than $75,000 per year, given that the household is located in the Northeast. 0.230

17. **Health Insurance Coverage** The following data represent, NW in thousands, the type of health insurance coverage of people by age in the year 2002.

	\<18	18–44	45–64	\>64
Private	49,473	76,294	52,520	20,685
Government	19,662	11,922	9,227	32,813
None	8,531	25,678	9,106	258

Source: U.S. Census Bureau

(a) What is the probability that a randomly selected individual who is less than 18 years old has no health insurance? 0.110

(b) What is the probability that a randomly selected individual who has no health insurance is less than 18 years old? 0.196

18. **Cigar Smoking.** The data in the following table show the results of a national study of 137,243 U.S. men that investigated the association between cigar smoking and death from cancer. **Note:** *Current cigar smoker* means "cigar smoker at time of death."

	Died from Cancer	Did Not Die from Cancer
Never smoked cigars	782	120,747
Former cigar smoker	91	7,757
Current cigar smoker	141	7,725

Source: Shapiro, Jacobs, and Thun. "Cigar Smoking in Men and Risk of Death from Tobacco-Related Cancers," *Journal of the National Cancer Institute*, February 16, 2000

(a) What is the probability that a randomly selected individual from the study who died from cancer was a former cigar smoker? 0.090

(b) What is the probability that a randomly selected individual from the study who was a former cigar smoker died from cancer? 0.012

19. **Driver Fatalities** The following data represent the number of driver fatalities in the United States in 2002 by age for male and female drivers:

Age	Male	Female
Under 16	228	108
16–20	5696	2386
21–34	13,553	4148
35–54	14,395	5017
55–69	4937	1708
70 and over	3159	1529

Source: *Traffic Safety Facts 2002*. Federal Highway Administration, 2002

(a) What is the probability that a randomly selected driver fatality who was male was 16 to 20 years old? 0.136

(b) What is the probability that a randomly selected driver fatality who was 16 to 20 was male? 0.705

(c) Suppose you are a police officer called to the scene of a traffic accident with a fatality. The dispatcher states that the victim is 16 to 20 years old, but the gender is not known. Is the victim more likely to be male or female? Why? Male

20. Marital Status The following data, in thousands, represent the marital status of Americans 25 years old or older and their level of education in 2003.

	Did Not Graduate from High School	High School Graduate	Some College	College Graduate
Never married	4,333	8,637	7,403	8,321
Married, spouse present	14,787	35,376	28,892	34,693
Married, spouse absent	2,032	2,534	1,633	1,190
Separated	1,134	1,596	1,103	614
Widowed	4,582	5,155	2,487	1,746
Divorced	2,887	7,612	6,393	4,490

Source: Educational Attainment in the United States: 2003. U.S. Census Bureau, June 2004

(a) What is the probability that a randomly selected individual who has never married is a high school graduate? 0.301

(b) What is the probability that a randomly selected individual who is a high school graduate has never married? 0.142

21. Acceptance Sampling Suppose you just received a shipment of six televisions. Two of the televisions are defective. If two televisions are randomly selected, compute the probability that both televisions work. What is the probability at least one does not work? 0.4; 0.6

22. Committee A committee consists of four women and three men. The committee will randomly select two people to attend a conference in Hawaii. Find the probability that both are women. $\frac{2}{7}$

23. Board Work This past semester, I had a small business calculus section. The students in the class were Mike, Neta, Jinita, Kristin, and Dave. Suppose I randomly select two people to go to the board to work problems. What is the probability that Dave is the first person chosen to go to the board and Neta is the second? $\frac{1}{20}$

24. Party My wife has organized a monthly neighborhood party. Five people are involved in the group: Yolanda (my wife), Lorrie, Laura, Kim, and Anne Marie. They decide to randomly select the first and second home that will host the party. What is the probability that my wife hosts the first party and Lorrie hosts the second? **Note:** Once a home has hosted, it cannot host again until all other homes have hosted. $\frac{1}{20}$

25. Playing a CD on the Random Setting Suppose a compact disk (CD) you just purchased has 13 tracks. After listening to the CD, you decide that you like 5 of the songs. With the random feature on your CD player, each of the 13 songs is played once in random order. Find the probability that among the first two songs played
(a) You like both of them. Would this be unusual?
(b) You like neither of them.
(c) You like exactly one of them.
(d) Redo (a)–(c) if a song can be replayed before all 13 songs are played (if, for example, track 2 can play twice in a row). $\frac{25}{169}, \frac{64}{169}, \frac{80}{169}$

26. Packaging Error Due to a manufacturing error, three cans of regular soda were accidentally filled with diet soda and placed into a 12-pack. Suppose that two cans are randomly selected from the case.

(25a) $\frac{5}{39}$ (25b) $\frac{14}{39}$ (25c) $\frac{20}{39}$

(a) Determine the probability that both contain diet soda. $\frac{1}{22}$

(b) Determine the probability that both contain regular soda. Would this be unusual? $\frac{6}{11}$; no

(c) Determine the probability that exactly one is diet and one is regular. $\frac{9}{22}$

27. Planting Tulips A bag of 30 tulip bulbs purchased from a nursery contains 12 red tulip bulbs, 10 yellow tulip bulbs, and 8 purple tulip bulbs.
(a) What is the probability that two randomly selected tulip bulbs are both red? 0.152
(b) What is the probability that the first bulb selected is red and the second yellow? 0.138
(c) What is the probability that the first bulb selected is yellow and the second is red? 0.138
(d) What is the probability that one bulb is red and the other yellow? 0.276

28. Golf Balls The local golf store sells an "onion bag" that contains 35 "experienced" golf balls. Suppose the bag contains 20 Titleists, 8 Maxflis, and 7 Top-Flites.
(a) What is the probability that two randomly selected golf balls are both Titleists? 0.319
(b) What is the probability that the first ball selected is a Titleist and the second is a Maxfli? 0.134
(c) What is the probability that the first ball selected is a Maxfli and the second is a Titleist? 0.134
(d) What is the probability that one golf ball is a Titleist and the other is a Maxfli? 0.269

29. Smokers According to the National Center for Health Statistics, there is a 23.4% probability that a randomly selected resident of the United States aged 25 years or older is a smoker. In addition, there is a 21.7% probability that a randomly selected resident of the United States aged 25 years or older is female, given that he or she smokes. What is the probability that a randomly selected resident of the United States aged 25 years or older is female and smokes? Would it be unusual to randomly select a resident of the United States aged 25 years or older who is female and smokes? 0.051; no

30. **Multiple Jobs** According to the U.S. Bureau of Labor Statistics, there is a 5.84% probability that a randomly selected employed individual has more than one job (a multiple-job holder). Also, there is a 52.6% probability that a randomly selected employed individual is male, given that he has more than one job. What is the probability that a randomly selected employed individual is a multiple-job holder and male? Would it be unusual to randomly select an employed individual who is a multiple-job holder and male? *0.031; yes*

31. **The Birthday Problem** Determine the probability that at least 2 people in a room of 10 people share the same birthday, ignoring leap years and assuming each birthday is equally likely by answering the following questions:

    (a) Compute the probability that 10 people have different birthdays. (*Hint*: The first person's birthday can occur 365 ways; the second person's birthday can occur 364 ways, because he or she cannot have the same birthday as the first person; the third person's birthday can occur 363 ways, because he or she cannot have the same birthday as the first or second person; and so on.) *0.883*

    (b) The complement of "10 people have different birthdays" is "at least 2 share a birthday." Use this information to compute the probability that at least 2 people out of 10 share the same birthday. *0.117*

32. **The Birthday Problem** Using the procedure given in Problem 31, compute the probability that at least 2 people in a room of 23 people share the same birthday. *0.507*

33. **A Flush** A flush in the card game of poker occurs if a player gets five cards that are all the same suit (clubs, diamonds, hearts, or spades). Answer the following questions to obtain the probability of being dealt a flush in five cards.

    (a) We initially concentrate on one suit, say clubs. There are 13 clubs in a deck. Compute $P(\text{five clubs}) = P(\text{first card is clubs and second card is clubs and third card is clubs and fourth card is clubs and fifth card is clubs})$. *0.000495*

    (b) A flush can occur if we get five clubs or five diamonds or five hearts or five spades. Compute $P(\text{five clubs or five diamonds or five hearts or five spades})$. Note the events are mutually exclusive. *0.002*

34. **A Royal Flush** A royal flush in the game of poker occurs if the player gets the cards Ten, Jack, Queen, King, and Ace all in the same suit. Use the results of Problem 33 to compute the probability of being dealt a royal flush.

35. **Independence in Small Samples from Large Populations**
NW Suppose a computer chip company has just shipped 10,000 computer chips to a computer company. Unfortunately, 50 of the chips are defective.

    (a) Compute the probability that two randomly selected chips are defective using conditional probability.

    (b) There are 50 defective chips out of 10,000 shipped. The probability that the first chip randomly selected is defective is $\frac{50}{10,000} = 0.005 = 0.5\%$. Compute the probability that two randomly selected chips are defective under the assumption of independent events. Compare your results to part (a). Conclude that, when small samples are taken from large populations without replacement, the assumption of independence does not significantly affect the probability. *0.000025*

36. **Independence in Small Samples from Large Populations** Suppose a poll is being conducted in the village of Lemont. The pollster identifies her target population as all residents of Lemont 18 years old or older. This population has 6494 people.

    (a) Compute the probability that the first resident selected to participate in the poll is Roger Cummings and the second is Rick Whittingham. *0.0000000237*

    (b) The probability that any particular resident of Lemont is the first person picked is $\frac{1}{6494}$. Compute the probability that Roger is selected first and Rick is selected second, assuming independence. Compare your results to part (a). Conclude that, when small samples are taken from large populations without replacement, the assumption of independence does not significantly affect the probability. *0.0000000237*

37. **Independent?** Refer to the contingency table in Problem 17 that relates age and health insurance coverage. Determine $P(<18 \text{ years old})$ and $P(<18 \text{ years old}|\text{no health insurance})$. Are the events "<18 years old" and "no health insurance" independent? *No*

38. **Independent?** Refer to the contingency table in Problem 18 that relates cigar smoking and deaths from cancer. Determine $P(\text{died from cancer})$ and $P(\text{died from cancer}|\text{current cigar smoker})$. Are the events "died from cancer" and "current cigar smoker" independent? *No*

39. **Independent?** Refer to the contingency table in Problem 19 that relates age of driving fatality to gender. Determine $P(\text{female})$ and $P(\text{female}|16–20)$. Are the events "female" and "16–20" independent? *No*

40. **Independent?** Refer to the contingency table in Problem 20 that relates marital status and level of education. Determine $P(\text{divorced})$ and $P(\text{divorced}|\text{college graduate})$. Are the events "divorced" and "college graduate" independent? *No*

(34a) *0.00000154* (35a) *0.0000245*

Consumer Reports His 'N' Hers Razor?

With so many men's and women's versions of different products, you might wonder how different they really are. To help answer this question, technicians at Consumers Union compared a new triple-edge razor for women with a leading double-edge razor for women and a leading triple-edge razor for men. The technicians asked 30 women panelists to shave with the razors over a 4-week period, following a random statistical design.

After each shave, the panelists were asked to answer a series of questions related to the performance of the razor. One question involved rating the razor on a 5-point scale, with 1 being Poor and 5 being Excellent. The following table contains a summary of the results for this question.

Using the information in the table, answer the following questions:

(a) Calculate the probability that a randomly selected razor scored Very Good to Excellent.

(b) Calculate the probability that a randomly selected razor scored Poor.

(c) Calculate the probability of randomly selecting Razor B, given that the score was Fair to Good.

(d) Calculate the probability of receiving an Excellent rating, given that Razor C was selected.

(e) Do you think that razor type and rating are independent?

(f) Which razor would you choose based on the information given? Support your decision.

Note to Readers: *In many cases, our test protocol and analytical methods are more complicated than described in these examples. The data and discussions have been modified to make the material more appropriate for the audience.*

© by Consumers Union of U.S., Inc., Yonkers, NY 10703-1057, a nonprofit organization. Reprinted with permission.

Survey Results for Razor Study

Razor	Poor	Fair to Good	Very Good to Excellent
A	1	8	21
B	0	11	19
C	6	11	13

(column header: Rating spans Poor, Fair to Good, Very Good to Excellent)

5.5 Counting Techniques

Objectives

1. Solve counting problems using the Multiplication Rule
2. Solve counting problems using permutations
3. Solve counting problems using combinations
4. Solve counting problems involving permutations with nondistinct items
5. Compute probabilities involving permutations and combinations

Note to Instructor

Most of the material in this section is optional. However, the material on combinations should be covered if you intend to present binomial probabilities (Section 6.2).

1 Solve Counting Problems Using the Multiplication Rule

Counting plays a major role in many diverse areas, including probability. In this section, we look at special types of counting problems and develop general formulas for solving them.

We begin with an example that demonstrates a general counting principle.

EXAMPLE 1 Counting the Number of Possible Meals

Problem: The fixed-price dinner at Mabenka Restaurant provides the following choices:

Appetizer:	soup or salad
Entrée:	baked chicken, broiled beef patty, baby beef liver, or roast beef au jus
Dessert:	ice cream or cheesecake

How many different meals can be ordered?

Approach: Ordering such a meal requires three separate decisions:

Choose an Appetizer	Choose an Entrée	Choose a Dessert
2 choices	4 choices	2 choices

We will draw a tree diagram that lists the possible meals that can be ordered.

Figure 12

Solution: Look at the tree diagram in Figure 12. We see that, for each choice of appetizer, there are 4 choices of entrée, and that for each of these $2 \cdot 4 = 8$ choices, there are 2 choices for dessert. A total of

$$2 \cdot 4 \cdot 2 = 16$$

different meals can be ordered.

Example 1 illustrates a general counting principle.

> **Multiplication Rule of Counting**
>
> If a task consists of a sequence of choices in which there are p selections for the first choice, q selections for the second choice, r selections for the third choice, and so on, then the task of making these selections can be done in
>
> $$p \cdot q \cdot r \cdot \ldots$$
>
> different ways.

EXAMPLE 2   **Counting Airport Codes**

Problem: The International Airline Transportation Association (IATA) assigns three-letter codes to represent airport locations. For example, the airport code for Fort Lauderdale is FLL. How many different airport codes are possible?

Approach: We are choosing 3 letters from 26 letters and arranging them in order. We notice that repetition of letters is allowed. We use the Multiplication Rule of Counting, recognizing that there are 26 ways to choose the first letter, 26 ways to choose the second letter, and 26 ways to choose the third letter.

Solution: By the Multiplication Rule,

$$26 \cdot 26 \cdot 26 = 17{,}576$$

different airport codes are possible.

In Example 2, we were allowed to repeat a letter. For example, a valid airport code is FLL (Ft. Lauderdale International Airport), in which the letter L appears twice. In the next example, repetition is not allowed.

EXAMPLE 3   **Counting without Repetition**

Problem: Three members from a 14-member committee are to be randomly selected to serve as chair, vice-chair, and secretary. The first person selected is the chair; the second person selected, the vice-chair; and the third, the secretary. How many different committee structures are possible?

Approach: The task consists of making three selections. The first selection requires choosing from 14 members. Because a member cannot serve in more than one capacity, the second selection requires choosing from 13 members. The third selection requires choosing from 12 members. (Do you see why?) We use the Multiplication Rule to determine the number of possible committees.

Solution: By the Multiplication Rule,

$$14 \cdot 13 \cdot 12 = 2184$$

different committee structures are possible.

Now Work Problem 31.

The Factorial Symbol

We now introduce a special symbol that can assist us in representing certain types of counting problems.

Definition

If $n \geq 0$ is an integer, the **factorial symbol** $n!$ is defined as follows:

$$0! = 1 \qquad 1! = 1$$
$$n! = n(n-1) \cdot \cdots \cdot 3 \cdot 2 \cdot 1$$

USING TECHNOLOGY:
Your calculator has a factorial key. Use it to see how fast factorials increase in value. Find the value of 69!. What happens when you try to find 70!? In fact, 70! is larger than 10^{100} (a *googol*), the largest number most calculators can display.

For example, $2! = 2 \cdot 1 = 2$, $3! = 3 \cdot 2 \cdot 1 = 6$, $4! = 4 \cdot 3 \cdot 2 \cdot 1 = 24$, and so on. Table 9 lists the values of $n!$ for $0 \leq n \leq 6$.

Table 9						
n	0	1	2	3	4	5	6
$n!$	1	1	2	6	24	120	720

EXAMPLE 4   **The Traveling Salesperson**

Problem: You have just been hired as a book representative for Prentice Hall. On your first day, you must travel to seven schools to introduce yourself. How many different routes are possible?

Approach: The seven schools are different. Let's call the schools A, B, C, D, E, F, and G. School A can be visited first, second, third, fourth, fifth, sixth or seventh. So we have seven choices for school A. We would then have six choices for school B, five choices for school C, and so on. We can use the Multiplication Rule and the factorial to find our solution.

Solution: $7 \cdot 6 \cdot 5 \cdot 4 \cdot 3 \cdot 2 \cdot 1 = 7! = 5040$ different routes are possible.

Now Work Problems 5 and 33.

② **Solve Counting Problems Using Permutations**

Examples 3 and 4 illustrate a type of counting problem referred to as a *permutation*.

Definition

A **permutation** is an ordered arrangement in which r objects are chosen from n distinct (different) objects and repetition is not allowed. The symbol $_nP_r$ represents the number of permutations of r objects selected from n objects.

So we could represent the solution to the question posed in Example 3 as

$$_nP_r = {}_{14}P_3 = 14 \cdot 13 \cdot 12 = 2184$$

and the solution to Example 4 could be represented as

$$_7P_7 = 7 \cdot 6 \cdot 5 \cdot 4 \cdot 3 \cdot 2 \cdot 1 = 5040$$

To arrive at a formula for $_nP_r$, we note that there are n choices for the first selection, $n-1$ choices for the second selection, $n-2$ choices for the third selection, ..., and $n - (r - 1)$ choices for the rth selection. By the Multiplication Rule, we have

$$
\begin{array}{ccccccc}
& \text{1st} & \text{2nd} & \text{3rd} & & \text{rth} \\
{}_nP_r = & n \cdot & (n-1) \cdot & (n-2) & \cdots\cdots & [n-(r-1)] \\
= & n \cdot & (n-1) \cdot & (n-2) & \cdots\cdots & (n-r+1)
\end{array}
$$

This formula for ${}_nP_r$ can be written in **factorial notation**:

$$
{}_nP_r = n\cdot(n-1)\cdot(n-2)\cdots\cdots(n-r+1)
$$

$$
= n\cdot(n-1)\cdot(n-2)\cdots\cdots(n-r+1)\cdot\frac{(n-r)\cdots\cdots 3\cdot 2\cdot 1}{(n-r)\cdots\cdots 3\cdot 2\cdot 1}
$$

$$
= \frac{n!}{(n-r)!}
$$

We have the following result.

> **Number of Permutations of n Distinct Objects Taken r at a Time**
>
> The number of arrangements of r objects chosen from n objects, in which
>
> 1. the n objects are distinct,
> 2. once an object is used it cannot be repeated, and
> 3. order is important,
>
> is given by the formula
>
> $$ {}_nP_r = \frac{n!}{(n-r)!} \qquad (1) $$

EXAMPLE 5   **Computing Permutations**

Problem: Evaluate: **(a)** ${}_7P_5$ **(b)** ${}_8P_2$ **(c)** ${}_5P_5$

Approach: To answer (a), we use Formula (1) with $n = 7$ and $r = 5$. To answer (b), we use Formula (1) with $n = 8$ and $r = 2$. To answer (c), we use Formula (1) with $n = 5$ and $r = 5$.

Solution

(a) ${}_7P_5 = \dfrac{7!}{(7-5)!} = \dfrac{7!}{2!} = \dfrac{7\cdot6\cdot5\cdot4\cdot3\cdot2!}{2!} = \underbrace{7\cdot6\cdot5\cdot4\cdot3}_{5\ factors} = 2520$

(b) ${}_8P_2 = \dfrac{8!}{(8-2)!} = \dfrac{8!}{6!} = \dfrac{8\cdot7\cdot6!}{6!} = \underbrace{8\cdot7}_{2\ factors} = 56$

(c) ${}_5P_5 = \dfrac{5!}{(5-5)!} = \dfrac{5!}{0!} = 5! = 5\cdot4\cdot3\cdot2\cdot1 = 120$

EXAMPLE 6   **Computing Permutations Using Technology**

Problem: Evaluate ${}_7P_5$ using a statistical spreadsheet or graphing calculator with advanced features.

Approach: We will use both Excel and a TI-84 Plus graphing calculator to evaluate ${}_7P_5$. The steps for computing permutations using Excel and the TI-83 or TI-84 Plus graphing calculators can be found in the Technology Step by Step on page 278.

Result: Figure 13(a) shows the result in Excel, and Figure 13(b) shows the result on a TI-84 Plus graphing calculator.

Figure 13

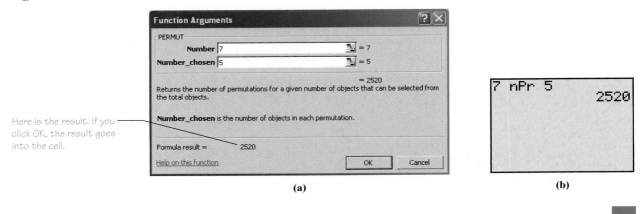

Here is the result. If you click OK, the result goes into the cell.

(a)

(b)

Now Work Problem 11.

Example 5(c) illustrates a general result.

$$_rP_r = r!$$

We need to recognize that, in a permutation, order matters. That is, if we wanted to permute the letters *ABC* by selecting them 3 at a time, the following arrangements are all different:

$$ABC, ACB, BAC, BCA, CAB, CBA$$

EXAMPLE 7 **Betting on the Trifecta**

Problem: In how many ways can horses in a 10-horse race finish first, second, and third?

Approach: The 10 horses are distinct. Once a horse crosses the finish line, that horse will not cross the finish line again, and, in a race, order is important. We have a permutation of 10 objects taken 3 at a time.

Solution: The top three horses can finish a 10-horse race in

$$_{10}P_3 = \frac{10!}{(10-3)!} = \frac{10!}{7!} = \frac{10 \cdot 9 \cdot 8 \cdot 7!}{7!} = \underbrace{\frac{10 \cdot 9 \cdot 8}{3\ factors}} = 720 \text{ ways}$$

Now Work Problem 45.

③ Solve Counting Problems Using Combinations

In a permutation, order is important. For example, the arrangements *ABC, ACB, BAC, BCA, CAB,* and *CBA* are considered different arrangements of the letters *A, B,* and *C*. If order is unimportant, the six arrangements of the letters *A, B,* and *C* given above are not different. That is, we do not distinguish *ABC* from *BAC*. In the card game of poker, the order in which the cards are received does not matter. The *combination* of the cards is what matters.

Definition

A **combination** is a collection, without regard to order, of *n* distinct objects without repetition. The symbol $_nC_r$ represents the number of combinations of *n* distinct objects taken *r* at a time.

EXAMPLE 8   **Listing Combinations**

Problem: Roger, Rick, Randy, and Jay are going to play golf. They will randomly select teams of two players each. List all possible team combinations. That is, list all the combinations of the four people Roger, Rick, Randy, and Jay taken two at a time. What is $_4C_2$?

Approach: We list the possible teams. We note that order is unimportant, so {Roger, Rick} is the same as {Rick, Roger}.

Solution: The list of all such teams (combinations) is

   Roger, Rick; Roger, Randy; Roger, Jay; Rick, Randy; Rick, Jay; Randy, Jay

So

$$_4C_2 = 6$$

There are six ways of forming teams of two from a group of four players.

We can find a formula for $_nC_r$ by noting that the only difference between a permutation and a combination is that we disregard order in combinations. To determine $_nC_r$, we eliminate from the formula for $_nP_r$ the number of permutations that were rearrangements of a given set of *r* objects. In Example 8, for example, selecting {Roger, Rick} was the same as selecting {Rick, Roger}, so there were 2! = 2 rearrangements of the two objects. This can be determined from the formula for $_nP_r$ by calculating $_rP_r = r!$. So, if we divide $_nP_r$ by *r*!, we will have the desired formula for $_nC_r$:

$$_nC_r = \frac{_nP_r}{r!} = \frac{n!}{r!(n-r)!}$$

We have the following result.

Number of Combinations of *n* Distinct Objects Taken *r* at a Time

The number of different arrangements of *n* objects using $r \le n$ of them, in which

  1. the *n* objects are distinct,
  2. once an object is used, it cannot be repeated, and
  3. order is not important

is given by the formula

$$_nC_r = \frac{n!}{r!(n-r)!} \tag{2}$$

Using Formula (2) to solve the problem presented in Example 8, we obtain

$$_4C_2 = \frac{4!}{2!(4-2)!} = \frac{4!}{2!2!} = \frac{4 \cdot 3 \cdot 2!}{2 \cdot 1 \cdot 2!} = \frac{12}{2} = 6$$

EXAMPLE 9 **Using Formula (2)**

Problem: Use Formula (2) to find the value of each expression.

(a) $_4C_1$ **(b)** $_6C_4$ **(c)** $_6C_2$

Approach: We use Formula (2): $_nC_r = \dfrac{n!}{r!(n-r)!}$

Solution

(a) $_4C_1 = \dfrac{4!}{1!(4-1)!} = \dfrac{4!}{1!\cdot 3!} = \dfrac{4\cdot 3!}{1\cdot 3!} = 4$ $n = 4, r = 1$

(b) $_6C_4 = \dfrac{6!}{4!(6-4)!} = \dfrac{6!}{4!\cdot 2!} = \dfrac{6\cdot 5\cdot 4!}{4!\cdot 2\cdot 1} = \dfrac{30}{2} = 15$ $n = 6, r = 4$

(c) $_6C_2 = \dfrac{6!}{2!(6-2)!} = \dfrac{6!}{2!4!} = \dfrac{6\cdot 5\cdot 4!}{2\cdot 1\cdot 4!} = \dfrac{30}{2} = 15$ $n = 6, r = 2$

EXAMPLE 10 **Computing Combinations Using Technology**

Problem: Evaluate $_6C_4$ using a statistical spreadsheet or graphing calculator with advanced features.

Approach: We will use both Excel and a TI-84 Plus graphing calculator to evaluate $_6C_4$. The steps for computing combinations using Excel and the TI-83 or TI-84 Plus graphing calculators can be found in the Technology Step by Step on page 278.

Result: Figure 14(a) shows the result in Excel, and Figure 14(b) shows the result on a TI-84 Plus graphing calculator.

Figure 14

Here is the result. If you click OK, the result goes into the cell.

(a)

(b)

Now Work Problem 19.

Notice in Example 9 that $_6C_4 = {}_6C_2$. This result can be generalized.

$$_nC_r = {}_nC_{n-r}$$

EXAMPLE 11 **Simple Random Samples**

Problem: How many different simple random samples of size 4 can be obtained from a population whose size is 20?

Approach: The 20 individuals in the population are distinct. In addition, the order in which an individual is selected to be in the sample is unimportant. Thus, the number of simple random samples of size 4 from a population of size 20 is a combination of 20 objects taken 4 at a time.

Solution: Use Formula (2) with $n = 20$ and $r = 4$:

$$_{20}C_4 = \frac{20!}{4!(20-4)!} = \frac{20!}{4!16!} = \frac{20 \cdot 19 \cdot 18 \cdot 17 \cdot 16!}{4 \cdot 3 \cdot 2 \cdot 1 \cdot 16!} = \frac{116{,}280}{24} = 4845$$

There are 4845 different simple random samples of size 4 from a population whose size is 20.

Now Work Problem 51.

4   Solve Counting Problems Involving Permutations with Nondistinct Items

Sometimes we wish to arrange objects in order, but some of the objects are not distinguishable.

EXAMPLE 12   **Forming Different Words**

Problem: How many distinguishable strings of letters can be formed by using all the letters in the word REARRANGE?

Approach: Each string formed will have nine letters: three R's, two A's, two E's, one N, and one G. To construct each word, we need to fill in nine positions with the nine letters:

$$\overline{1} \ \ \overline{2} \ \ \overline{3} \ \ \overline{4} \ \ \overline{5} \ \ \overline{6} \ \ \overline{7} \ \ \overline{8} \ \ \overline{9}$$

The process of forming a word consists of five tasks:

Step 1: Choose the positions for the three R's.
Step 2: Choose the positions for the two A's.
Step 3: Choose the positions for the two E's.
Step 4: Choose the position for the one N.
Step 5: Choose the position for the one G.

Task 1 can be done in $_9C_3$ ways. There then remain six positions to be filled, so Task 2 can be done in $_6C_2$ ways. There remain four positions to be filled, so Task 3 can be done in $_4C_2$ ways. There remain two positions to be filled, so Task 4 can be done in $_2C_1$ ways. The last position can be filled in $_1C_1$ way.

Solution: By the Multiplication Rule, the number of possible words that can be formed is

$$_9C_3 \cdot {}_6C_2 \cdot {}_4C_2 \cdot {}_2C_1 \cdot {}_1C_1 = \frac{9!}{6! \cdot 3!} \cdot \frac{6!}{4!2!} \cdot \frac{4!}{2! \cdot 2!} \cdot \frac{2!}{\not{1}! \cdot 1!} \cdot \frac{\not{1}!}{1!0!}$$

$$= \frac{9!}{3! \cdot 2! \cdot 2! \cdot 1! \cdot 1!}$$

$$= 15{,}120$$

The form of the answer to Example 12 is suggestive of a general result. Had the letters in REARRANGE each been different, there would have been $_9P_9 = 9!$ possible words formed. This is the numerator of the answer.

The presence of three R's, two A's, and two E's reduces the number of different words, as the entries in the denominator illustrate. We are led to the following result:

Permutations with Nondistinct Items

The number of permutations of n objects of which n_1 are of one kind, n_2 are of a second kind, ..., and n_k are of a kth kind is given by

$$\frac{n!}{n_1! \cdot n_2! \cdots n_k!} \qquad \textbf{(3)}$$

where $n = n_1 + n_2 + \cdots + n_k$.

EXAMPLE 13

Arranging Flags

Problem: How many different vertical arrangements are there of 10 flags if 5 are white, 3 are blue, and 2 are red?

Approach: We seek the number of permutations of 10 objects, of which 5 are of one kind (white), 3 are of a second kind (blue), and 2 are of a third kind (red).

Solution: Using Formula (3), we find that there are

$$\frac{10!}{5! \cdot 3! \cdot 2!} = \frac{10 \cdot 9 \cdot 8 \cdot 7 \cdot 6 \cdot 5!}{5! \cdot 3! \cdot 2!} = 2520 \text{ different arrangements}$$

> Now Work Problem 55.

Summary

To summarize the differences between combinations and the various types of permutations, we present Table 10.

Table 10		
	Description	**Formula**
Combination	The selection of r objects from a set of n different objects when the order in which the objects is selected does not matter (so AB is the same as BA) and an object cannot be selected more than once (repetition is not allowed)	$_nC_r = \dfrac{n!}{r!(n-r)!}$
Permutation of Distinct Items with Replacement	The selection of r objects from a set of n different objects when the order in which the objects are selected matters (so AB is different from BA) and an object may be selected more than once (repetition is allowed)	n^r
Permutation of Distinct Items without Replacement	The selection of r objects from a set of n different objects when the order in which the objects are selected matters (so AB is different from BA) and an object cannot be selected more than once (repetition is not allowed)	$_nP_r = \dfrac{n!}{(n-r)!}$
Permutation of Nondistinct Items without Replacement	The number of ways n objects can be arranged (order matters) in which there are n_1 of one kind, n_2 of a second kind, ..., and n_k of a kth kind, where $n = n_1 + n_2 + \cdots + n_k$	$\dfrac{n!}{n_1! n_2! \cdots n_k!}$

 Compute Probabilities Involving Permutations and Combinations

The counting techniques presented in this section can be used to determine probabilities of certain events by using the classical method of computing probabilities. Recall that this method stated the probability of an event E is the number of ways event E can occur divided by the number of different possible outcomes of the experiment.

| EXAMPLE 14 | **Winning the Lottery** |

Problem: In the Illinois Lottery, an urn contains balls numbered 1 to 54. From this urn, 6 balls are randomly chosen without replacement. For a $1 bet, a player chooses two sets of 6 numbers. To win, all six numbers must match those chosen from the urn. The order in which the balls are selected does not matter. What is the probability of winning the lottery?

Approach: The probability of winning is given by the number of ways a ticket could win divided by the size of the sample space. Each ticket has two sets of six numbers, so there are two chances (for the two sets of numbers) of winning for each ticket. The size of the sample space S is the number of ways that 6 objects can be selected from 54 objects without replacement and without regard to order, so $N(S) = {}_{54}C_6$.

Solution: The size of the sample space is

$$N(S) = {}_{54}C_6 = \frac{54!}{6! \cdot (54-6)!} = \frac{54 \cdot 53 \cdot 52 \cdot 51 \cdot 50 \cdot 49 \cdot 48!}{6! \cdot 48!} = 25{,}827{,}165$$

Each ticket has two sets of 6 numbers, so a player has two chances of winning for each $1. If E is the event "winning ticket," then $N(E) = 2$. The probability of E is

$$P(E) = \frac{2}{25{,}827{,}165} \approx 0.000000077$$

There is about a 1 in 13,000,000 chance of winning the Illinois Lottery!

| EXAMPLE 15 | **Probabilities Involving Combinations** |

Problem: A shipment of 120 fasteners that contains 4 defective fasteners was sent to a manufacturing plant. The quality-control manager at the manufacturing plant randomly selects 5 fasteners and inspects them. What is the probability that exactly 1 fastener is defective?

Approach: The probability that exactly 1 fastener is defective is found by calculating the number of ways of selecting exactly 1 defective fastener in 5 fasteners and dividing this result by the number of ways of selecting 5 fasteners from 120 fasteners. To choose exactly 1 defective in the 5 requires choosing 1 defective from the 4 defectives and 4 nondefectives from the 116 nondefectives. The order in which the fasteners are selected does not matter, so we use combinations.

Solution: The number of ways of choosing 1 defective fastener from 4 defective fasteners is ${}_4C_1$. The number of ways of choosing 4 nondefective fasteners from 116 nondefectives is ${}_{116}C_4$. Using the Multiplication Rule, we find that the number of ways of choosing 1 defective and 4 nondefective fasteners is

$$({}_4C_1) \cdot ({}_{116}C_4) = 4 \cdot 7{,}160{,}245 = 28{,}640{,}980$$

The number of ways of selecting 5 fasteners from 120 fasteners is ${}_{120}C_5 = 190{,}578{,}024$. The probability of selecting exactly 1 defective fastener is

$$P(1 \text{ defective fastener}) = \frac{({}_4C_1)({}_{116}C_4)}{{}_{120}C_5} = \frac{28{,}640{,}980}{190{,}578{,}024} \approx 0.1503 = 15.03\%$$

There is a 15.03% probability of randomly selecting exactly 1 defective fastener.

Now Work Problem 61.

5.5 ASSESS YOUR UNDERSTANDING

Concepts and Vocabulary

1. A _____ is an ordered arrangement of r objects chosen from n distinct objects without repetition. permutation

2. A _____ is an arrangement of r objects chosen from n distinct objects without repetition and without regard to order. combination

3. *True or False:* In a combination problem, order is not important. True

4. Explain the difference between a combination and a permutation.

Skill Building

In Problems 5–10, find the value of each factorial.

5. $5!$ 120

6. $7!$ 5040

7. $10!$ 3,628,800

8. $12!$ 479,001,600

9. $0!$ 1

10. $1!$ 1

In Problems 11–18, find the value of each permutation.

11. $_6P_2$ 30

12. $_7P_2$ 42

13. $_4P_4$ 24

14. $_7P_7$ 5040

15. $_5P_0$ 1

16. $_4P_0$ 1

17. $_8P_3$ 336

18. $_9P_4$ 3024

In Problems 19–26, find the value of each combination.

19. $_8C_3$ 56

20. $_9C_2$ 36

21. $_{10}C_2$ 45

22. $_{12}C_3$ 220

23. $_{52}C_1$ 52

24. $_{40}C_{40}$ 1

25. $_{48}C_3$ 17,296

26. $_{30}C_4$ 27,405

27. List all the permutations of five objects a, b, c, d, and e taken two at a time without repetition. What is $_5P_2$? 20

28. List all the permutations of four objects a, b, c, and d taken two at a time without repetition. What is $_4P_2$? 12

29. List all the combinations of five objects a, b, c, d, and e taken two at a time. What is $_5C_2$? 10

30. List all the combinations of four objects a, b, c, and d taken two at a time. What is $_4C_2$? 6

Applying the Concepts

31. **Clothing Options** A man has six shirts and four ties. Assuming that they all match, how many different shirt-and-tie combinations can he wear? 24

32. **Clothing Options** A woman has five blouses and three skirts. Assuming that they all match, how many different outfits can she wear? 15

33. **Arranging Songs on a CD** Suppose Dan is going to burn a compact disk (CD) that will contain 12 songs. In how many ways can Dan arrange the 12 songs on the CD? 12!

34. **Arranging Students** In how many ways can 15 students be lined up? 15!

35. **Traveling Salesperson** A salesperson must travel to eight cities to promote a new marketing campaign. How many different trips are possible if any route between cities is possible? 8!

36. **Randomly Playing Songs** A certain compact disk player randomly plays each of 10 songs on a CD. Once a song is played, it is not repeated until all the songs on the CD have been played. In how many different ways can the CD player play the 10 songs? 10!

37. **Stocks on the NYSE** Companies whose stocks are listed on the New York Stock Exchange (NYSE) have their company name represented by either one, two, or three letters (repetition of letters is allowed). What is the maximum number of companies that can be listed on the New York Stock Exchange? 18,278

38. **Stocks on the NASDAQ** Companies whose stocks are listed on the NASDAQ stock exchange have their company name represented by either four or five letters (repetition of letters is allowed). What is the maximum number of companies that can be listed on the NASDAQ? 12,338,352

39. **Garage Door Code** Outside a home, there is a keypad that can be used to open the garage if the correct four-digit code is entered.

(a) How many codes are possible? 10,000

(b) What is the probability of entering the correct code on the first try, assuming that the owner doesn't remember the code? 0.0001

40. **Social Security Numbers** A Social Security number is used to identify each resident of the United States uniquely. The number is of the form xxx–xx–xxxx, where each x is a digit from 0 to 9.

(a) How many Social Security numbers can be formed? 10^9

(b) What is the probability of correctly guessing the Social Security number of the President of the United States? $1/10^9$

41. **User Names** Suppose a local area network requires eight letters for user names. Lower- and uppercase letters are considered the same. How many user names are possible for the local area network? 26^8

42. **Passwords** Suppose a local area network requires eight characters for a password. The first character must be a letter, but the remaining seven characters can be either a letter or a digit (0 through 9). Lower- and uppercase letters are considered the same. How many passwords are possible for the local area network? $26 \cdot 36^7$

43. **Combination Locks** A combination lock has 50 numbers on it. To open it, you turn counterclockwise to a number, then rotate clockwise to a second number, and then counterclockwise to the third number.

(a) How many different lock combinations are there? 50^3

(b) What is the probability of guessing a lock combination on the first try? $1/50^3$

44. **Forming License Plate Numbers** How many different license plate numbers can be made by using one letter followed by five digits selected from the digits 0 through 9?

45. **INDY 500** Suppose 40 cars start at the Indianapolis 500. In how many ways can the top 3 cars finish the race? 59,280

46. **Betting on the Perfecta** In how many ways can the top 2 horses finish in a 10-horse race? 90

47. **Forming a Committee** Four members from a 20-person committee are to be selected randomly to serve as chairperson, vice-chairperson, secretary, and treasurer. The first person selected is the chairperson; the second, the vice-chairperson; the third, the secretary; and the fourth, the treasurer. How many different leadership structures are possible? 116,280

48. **Forming a Committee** Four members from a 50-person committee are to be selected randomly to serve as chairperson, vice-chairperson, secretary, and treasurer. The first person selected is the chairperson; the second, the vice-chairperson; the third, the secretary; and the fourth, the treasurer. How many different leadership structures are possible? $_{50}P_4$

49. **Lottery** Suppose a lottery exists where balls numbered 1 to 25 are placed in an urn. To win, you must match the four balls chosen in the correct order. How many possible outcomes are there for this game? 303,600

50. **Forming a Committee** In the U.S. Senate, there are 21 members on the Committee on Banking, Housing, and

Urban Affairs. Nine of these 21 members are selected to be on the Subcommittee on Economic Policy. How many different committee structures are possible for this subcommittee? 293,930

51. **Simple Random Sample** How many different simple random samples of size 5 can be obtained from a population whose size is 50? 2,118,760

52. **Simple Random Sample** How many different simple random samples of size 7 can be obtained from a population whose size is 100? $_{100}C_7$

53. **Children** A family has six children. If this family has exactly two boys, how many different birth and gender orders are possible? 15

54. **Children** A family has eight children. If this family has exactly three boys, how many different birth and gender orders are possible? 56

55. **Forming Words** How many different 10-letter words (real or imaginary) can be formed from the letters in the word STATISTICS? 50,400

56. **Forming Words** How many different nine-letter words (real or imaginary) can be formed from the letters in the word ECONOMICS? 90,720

57. **Landscape Design** A golf-course architect has four linden trees, five white birch trees, and two bald cypress trees to plant in a row along a fairway. In how many ways can the landscaper plant the trees in a row, assuming that the trees are evenly spaced? 6930

58. **Starting Lineup** A baseball team consists of three outfielders, four infielders, a pitcher, and a catcher. Assuming that the outfielders and infielders are indistinguishable, how many batting orders are possible? 2520

59. **Little Lotto** In the Illinois Lottery game Little Lotto, an urn contains balls numbered 1 to 30. From this urn, 5 balls are chosen randomly, without replacement. For a $1 bet, a player chooses one set of five numbers. To win, all five numbers must match those chosen from the urn. The order in which the balls are selected does not matter. What is the probability of winning Little Lotto with one ticket? 1/142,506

60. **The Big Game** In the Big Game, an urn contains balls numbered 1 to 50, and a second urn contains balls numbered 1 to 36. From the first urn, 5 balls are chosen randomly, without replacement. From the second urn, 1 ball is chosen randomly. For a $1 bet, a player chooses one set of five numbers to match the balls selected from the first urn and one number to match the ball selected from the second urn. To win, all six numbers must match; that is, the player must match the first 5 balls selected from the first urn *and* the single ball selected from the second urn. What is the probability of winning the Big Game with a single ticket? $1/(_{50}C_5 \cdot 36)$

61. **Selecting a Jury** The grade appeal process at a university requires that a jury be structured by selecting five individuals randomly from a pool of eight students and ten faculty.

(a) What is the probability of selecting a jury of all students? 56/8568

44. 2,600,000

(b) What is the probability of selecting a jury of all faculty? *252/8568*

(c) What is the probability of selecting a jury of two students and three faculty? *3360/8568*

62. **Selecting a Committee** Suppose there are 55 Democrats and 45 Republicans in the U.S. Senate. A committee of seven senators is to be formed by selecting members of the Senate randomly.

(a) What is the probability that the committee is composed of all Democrats? *0.0127*

(b) What is the probability that the committee is composed of all Republicans? *0.0028*

(c) What is the probability that the committee is composed of three Democrats and four Republicans? *0.2442*

63. **Acceptance Sampling** Suppose a shipment of 120 electronic components contains 4 defective components. To determine whether the shipment should be accepted, a quality-control engineer randomly selects 4 of the components and tests them. If 1 or more of the components is defective, the shipment is rejected. What is the probability the shipment is rejected? *0.1283*

64. **In the Dark** A box containing twelve 40-watt light bulbs and eighteen 60-watt light bulbs is stored in your basement. Unfortunately, the box is stored in the dark and you need two 60-watt bulbs. What is the probability of randomly selecting two 60-watt bulbs from the box? *0.3517*

65. **Randomly Playing Songs** Suppose a compact disk (CD) you just purchased has 13 tracks. After listening to the CD, you decide that you like 5 of the songs. The random feature on your CD player will play each of the 13 songs once in a random order. Find the probability that among the first 4 songs played

(a) you like 2 of them; *0.3916*

(b) you like 3 of them; *0.1119*

(c) you like all 4 of them. *0.0070*

66. **Packaging Error** Through a manufacturing error, three cans marked "regular soda" were accidentally filled with diet soda and placed into a 12-pack. Suppose that three cans are randomly selected from the 12-pack.

(a) Determine the probability that exactly two contain diet soda. *0.1227*

(b) Determine the probability that exactly one contains diet soda. *0.4909*

(c) Determine the probability that all three contain diet soda. *0.0045*

67. **Three of a Kind** Suppose you are dealt 5 cards from a standard 52-card deck. Determine the probability of being dealt three of a kind (such as three aces or three kings) by answering the following questions:

(a) How many ways can 5 cards be selected from a 52-card deck? *2,598,960*

(b) Each deck contains 4 twos, 4 threes, and so on. How many ways can three of the same card be selected from the deck? *52*

(c) The remaining 2 cards must be different from the 3 chosen and different from each other. For example, if we drew three kings, the 4th card cannot be a king. After selecting the three of a kind, there are 12 different ranks of card remaining in the deck that can be chosen. If we have three kings, then we can choose twos, threes, and so on. Of the 12 ranks remaining, we choose 2 of them and then select one of the 4 cards in each of the two chosen ranks. How many ways can we select the remaining 2 cards? $66 \cdot 4 \cdot 4 = 1056$

(d) Use the General Multiplication Rule to compute the probability of obtaining three of a kind. That is, what is the probability of selecting three of a kind and two cards that are not like? *0.0211*

68. **Two of a Kind** Follow the outline presented in Problem 67 to determine the probability of being dealt exactly one pair. *0.4226*

69. **Acceptance Sampling** Suppose you have just received a shipment of 20 modems. Although you don't know this, 3 of the modems are defective. To determine whether you will accept the shipment, you randomly select 4 modems and test them. If all 4 modems work, you accept the shipment. Otherwise, the shipment is rejected. What is the probability of accepting the shipment? *0.4912*

70. **Acceptance Sampling** Suppose you have just received a shipment of 100 televisions. Although you don't know this, 6 are defective. To determine whether you will accept the shipment, you randomly select 5 televisions and test them. If all 5 televisions work, you accept the shipment; otherwise, the shipment is rejected. What is the probability of accepting the shipment? *0.7291*

Technology Step by Step | **Factorials, Permutations, and Combinations**

TI-83/84 Plus | Factorials

Step 1: To compute 7!, type 7 on the HOME screen.

Step 2: Press MATH, then highlight PRB, and then highlight 4 : ! Press ENTER. With 4! on the HOME screen, press ENTER again.

Permutations and Combinations

Step 1: To compute $_7P_3$, type 7 on the HOME screen.

Step 2: Press MATH, then highlight PRB, and then highlight 2 : $_nP_r$ and press ENTER.

Step 3: Type 3 on the HOME screen, and press ENTER.

Note: To compute $_7C_3$, select 3 : $_nC_r$ instead of 2 : $_nP_r$.

Excel Factorials or Combinations

Step 1: Select Insert and then Function . . . .

Step 2: Highlight Math & Trig in the Function category. For combinations, select COMBIN in the function name and fill in the appropriate cells. For a factorial, select FACT in the function name and fill in the appropriate cells.

Permutations

Step 1: Select Insert and then Function . . . .

Step 2: Highlight Statistical in the Function category. For permutations, select PERMUT in the function name and fill in the appropriate cells.

 CHAPTER 5 **Review**

Summary

In this chapter, we introduced the concept of probability. Probability is a measure of the likelihood of a random phenomenon or chance behavior. Because we are measuring a random phenomenon, there is short-term uncertainty. However, this short-term uncertainty gives rise to long-term predictability.

Probabilities are numbers between zero and one, inclusive. The closer a probability is to one, the more likely the event is to occur. If an event has probability zero, it is said to be impossible. Events with probability one are said to be certain.

We introduced three methods for computing probabilities: (1) the empirical (or relative frequency) method, (2) the classical method, and (3) subjective probabilities. Empirical probabilities rely on the relative frequency with which an event happens. Classical probabilities require the events in the experiment to be equally likely. We count the number of ways an event can happen and divide this by the number of possible outcomes of the experiment. Empirical probabilities actually require that an experiment be performed, whereas classical probability does not. Subjective probabilities are probabilities based on the opinion of the individual providing the probability. They are educated guesses about the likelihood of an event occurring, but still represent a legitimate way of assigning probabilities.

We are also interested in probabilities of multiple outcomes. For example, we might be interested in the probability that either event E or event F happens. The Addition Rule is used to compute the probability of E *or* F; the Multiplication Rule is used to compute the probability that both E *and* F occur. Two events are mutually exclusive (or disjoint) if they do not have any outcomes in common. Two events E and F are independent if the probability of event E occurring does not affect the probability of event F. The complement of an event E, denoted E^c, is all the outcomes in the sample space that are not in E.

Finally, we introduced counting methods. The Multiplication Rule is used to count the number of ways a sequence of events can occur. Permutations are used to count the number of ways r distinct items can be arranged from a set of n items. Combinations are used to count the number of ways r distinct items can be selected from a set of n items without replacement and without regard to order. These counting techniques can be used to calculate probabilities using the classical method.

Formulas

Empirical Probability

$$P(E) \approx \frac{\text{frequency of } E}{\text{number of trials of experiment}}$$

Classical Probability

$$P(E) = \frac{\text{number of ways that } E \text{ can occur}}{\text{number of possible outcomes}} = \frac{N(E)}{N(S)}$$

Addition Rule for Disjoint Events

$$P(E \text{ or } F) = P(E) + P(F)$$

General Addition Rule

$$P(E \text{ or } F) = P(E) + P(F) - P(E \text{ and } F)$$

Probabilities of Complements

$$P(E^c) = 1 - P(E)$$

Multiplication Rule for Independent Events

$$P(E \text{ and } F) = P(E) \cdot P(F)$$

Multiplication Rule for n Independent Events

$$P(E \text{ and } F \text{ and } G \ldots) = P(E) \cdot P(F) \cdot P(G) \cdot \cdots$$

Conditional Probability Rule

$$P(F|E) = \frac{P(E \text{ and } F)}{P(E)} = \frac{N(E \text{ and } F)}{N(E)}$$

General Multiplication Rule

$$P(E \text{ and } F) = P(E) \cdot P(F|E)$$

Factorial Notation

$$n! = n \cdot (n-1) \cdot (n-2) \cdots 3 \cdot 2 \cdot 1$$

Combination

$${}_nC_r = \frac{n!}{r!(n-r)!}$$

Permutation

$${}_nP_r = \frac{n!}{(n-r)!}$$

Permutations with Nondistinct Items

$$\frac{n!}{n_1! \cdot n_2! \cdots \cdot n_k!}$$

Vocabulary

Probability (p. 223)	Equally likely outcomes (p. 227)	Cell (p. 243)
Outcome (p. 223)	Tree diagram (p. 230)	Complement (p. 244)
Experiment (p. 224)	Subjective probability (p. 232)	Independent (p. 250)
Sample space (p. 224)	Disjoint (p. 238)	Dependent (p. 250)
Event (p. 224)	Mutually exclusive (p. 238)	Conditional probability (p. 256)
Probability model (p. 225)	Venn diagram (p. 238)	Factorial (p. 268)
Impossible event (p. 225)	Contingency table (p. 242)	Permutation (p. 268)
Certainty (p. 225)	Row variable (p. 242)	Combination (p. 271)
Unusual event (p. 225)	Column variable (p. 243)	

Objectives

Section	You should be able to . . .	Examples	Review Exercises
5.1	1 Understand the rules of probabilities (p. 225)	2	1, 13(d), 18(a)
	2 Compute and interpret probabilities using the empirical method (p. 226)	3, 4, 7(a)	15(a), 16–18, 19(a) and (b), 20(a) and (b), 21(a), 39
	3 Compute and interpret probabilities using the classical method (p. 227)	5, 6, 7(b)	2–4, 13(a), 14
	4 Use simulation to obtain data based on probabilities (p. 231)	8	36
	5 Understand subjective probabilities (p. 232)		37
5.2	1 Use the Addition Rule for disjoint events (p. 238)	1 and 2	3, 4, 7, 13(b) and (c), 18(b) and (c)
	2 Use the General Addition Rule (p. 241)	3 and 4	6, 19(d), 20(d)
	3 Compute the probability of an event using the Complement Rule (p. 244)	5 and 6	5, 15(b) and (e), 16, 18(d), 21(b), 21(d), 21(f), 22(c)
5.3	1 Understand independence (p. 249)	1	9
	2 Use the Multiplication Rule for independent events (p. 251)	2 and 3	8, 21(c) and (e), 22(a) and (b), 23, 24
	3 Compute at-least probabilities (p. 253)	4	21(d), 21(f), 22(c)
5.4	1 Compute conditional probabilities (p. 255)	1 through 3	11, 19(f), 20(e) and (f)
	2 Compute probabilities using the General Multiplication Rule (p. 258)	4 through 6	10, 15(c) and (d), 25, 26, 38
5.5	1 Solve counting problems using the Multiplication Rule (p. 265)	1 through 4	27, 29
	2 Solve counting problems using permutations (p. 268)	5 through 7	12(e) and (f), 28
	3 Solve counting problems using combinations (p. 270)	8 through 11	12(c) and (d), 31, 32
	4 Solve counting problems involving permutations with nondistinct items (p. 273)	12 and 13	30
	5 Compute probabilities involving permutations and combinations (p. 274)	14 and 15	33–35

Review Exercises

1. (a) Which among the following numbers could be the probability of an event?

$$0, -0.01, 0.75, 0.41, 1.34 \quad {\scriptstyle 0,0.75,0.41}$$

(b) Which among the following numbers could be the probability of an event?

$$\frac{2}{5}, \frac{1}{3}, -\frac{4}{7}, \frac{4}{3}, \frac{6}{7} \quad {\scriptstyle \frac{2}{5}, \frac{1}{3}, \frac{6}{7}}$$

For Problems 2–5, let the sample space be $S = \{$red, green, blue, orange, yellow$\}$. *Suppose the outcomes are equally likely.*

2. Compute the probability of the event $E = \{$yellow$\}$. $\frac{1}{5}$

3. Compute the probability of the event $F = \{$green or orange$\}$. $\frac{2}{5}$

4. Compute the probability of the event $E = \{$red or blue or yellow$\}$. $\frac{3}{5}$

5. Suppose that $E = \{$yellow$\}$. Compute the probability of E^c. $\frac{4}{5}$

6. Suppose that $P(E) = 0.76$, $P(F) = 0.45$, and $P(E \text{ and } F) = 0.32$. What is $P(E \text{ or } F)$? 0.89

7. Suppose that $P(E) = 0.36$, $P(F) = 0.12$, and E and F are mutually exclusive. What is $P(E \text{ or } F)$? 0.48

8. Suppose that events E and F are independent. In addition, $P(E) = 0.45$ and $P(F) = 0.2$. What is $P(E \text{ and } F)$? 0.09

9. Suppose that $P(E) = 0.8$, $P(F) = 0.5$, and $P(E \text{ and } F) = 0.24$. Are events E and F independent? Why? No

10. Suppose that $P(E) = 0.59$ and $P(F|E) = 0.45$. What is $P(E \text{ and } F)$? 0.2655

11. Suppose that $P(E \text{ and } F) = 0.35$ and $P(F) = 0.7$. What is $P(E|F)$? 0.5

12. Determine the value of each of the following:
    (a) $7!$ 5040
    (b) $0!$ 1
    (c) $_9C_4$ 126
    (d) $_{10}C_3$ 120
    (e) $_9P_2$ 72
    (f) $_{12}P_4$ 11,880

13. **Roulette** In the game of roulette, a wheel consists of 38 slots, numbered 0, 00, 1, 2, ..., 36. (See the photo in Problem 33 from Section 5.1.) To play the game, a metal ball is spun around the wheel and is allowed to fall into one of the numbered slots. The slots numbered 0 and 00 are green, the odd numbers are red, and the even numbers are black.
    (a) Determine the probability that the metal ball falls into a green slot. Interpret this probability. $\frac{1}{19}$
    (b) Determine the probability that the metal ball falls into a green or a red slot. Interpret this probability. $\frac{10}{19}$
    (c) Determine the probability that the metal ball falls into 00 or a red slot. Interpret this probability. $\frac{19}{38} = \frac{1}{2}$
    (d) Determine the probability that the metal ball falls into the number 31 and a black slot simultaneously. What term is used to describe this event? 0; Impossible

14. **Craps** Craps is a dice game in which two fair dice are cast. If the roller shoots a 7 or 11 on the first roll, he or she wins. If the roller shoots a 2, 3, or 12 on the first roll, he or she loses.
    (a) Compute the probability that the shooter wins on the first roll. Interpret this probability. $\frac{2}{9}$
    (b) Compute the probability that the shooter loses on the first roll. Interpret this probability. $\frac{1}{9}$

15. **New Year's Holiday** Between 6:00 P.M. December 29, 2001, and 5:59 A.M. January 2, 2002, there were 575 traffic fatalities in the United States. Of these, 301 were alcohol related.
    (a) What is the probability that a randomly selected traffic fatality that happened between 6:00 P.M. December 29,

2001, and 5:59 A.M. January 2, 2002, was alcohol related? 0.523
    (b) What is the probability that a randomly selected traffic fatality that happened between 6:00 P.M. December 29, 2001, and 5:59 A.M. January 2, 2002, was not alcohol related? 0.477
    (c) What is the probability that two randomly selected traffic fatalities that happened between 6:00 P.M. December 29, 2001, and 5:59 A.M. January 2, 2002, were both alcohol related? 0.274
    (d) What is the probability that neither of two randomly selected traffic fatalities that happened between 6:00 P.M. December 29, 2001, and 5:59 A.M. January 2, 2002, were alcohol related? 0.227
    (e) What is the probability that of two randomly selected traffic fatalities that happened between 6:00 P.M. December 29, 2001, and 5:59 A.M. January 2, 2002, at least one was alcohol related? 0.773

16. **Cyclones** According to the National Hurricane Center, about 11% of tropical cyclones occur in the North Atlantic Ocean. What is the probability that a randomly selected cyclone occurs in the North Atlantic Ocean? What is the probability that a randomly selected cyclone occurs somewhere other than in the North Atlantic Ocean? 0.11; 0.89

17. **Louisville Workers** The following data represent the distribution of class of workers in Louisville, Kentucky, in 2003.

Class of Worker	Number of Workers
Private wage and salary worker	99,402
Government worker	14,580
Self-employed worker	3,790
Unpaid family worker	164

Source: U.S. Census Bureau, *American Community Survey*, 2003

    (a) Construct a probability model for Louisville workers.
    (b) Is it unusual for a Louisville worker to be an unpaid family worker? Yes
    (c) Is it unusual for a Louisville worker to be self-employed? Yes

18. **Girl Scout Cookies** The following probability model shows the distribution of the most popular selling Girl Scout Cookies®.

Cookie Type	Probability
Thin Mints	0.25
Samoas®/Caramel deLites	0.19
Peanut Butter Patties®/Tagalongs™	0.13
Peanut Butter Sandwich/Do-si-dos™	0.11
Shortbread/Trefoils	0.09
Other varieties	0.23

Source: www.girlscouts.org

(a) Verify that this is a probability model.

(b) If a girl scout is selling cookies to people who randomly enter a shopping mall, what is the probability that the next box sold will be Peanut Butter Patties®/Tagalongs® or Peanut Butter Sandwich/Do-si-dos™? *0.24*

(c) If a girl scout is selling cookies to people who randomly enter a shopping mall, what is the probability that the next box sold will be Thin Mints, Samoas®/Caramel deLites™, or Shortbread/Trefoils? *0.53*

(d) What is the probability that the next box sold will not be Thin Mints? *0.75*

19. **Gestation Period versus Weight** The following data represent the birth weights (in grams) of babies born in 2002, along with the period of gestation.

	Period of Gestation		
Birth Weight (in grams)	**Preterm**	**Term**	**Postterm**
Less than 1000	28,247	202	32
1000–1999	78,532	10,325	1,051
2000–2999	228,064	606,046	39,481
3000–3999	135,790	2,293,927	192,566
4000–4999	8,974	315,821	34,319
Over 5000	211	4,524	556

Source: National Vital Statistics Report, Vol. 52, No. 10, December 17, 2003

(a) What is the probability that a randomly selected baby born in 2002 was postterm? *0.0674*

(b) What is the probability that a randomly selected baby born in 2002 weighed 3000 to 3999 grams? *0.6591*

(c) What is the probability that a randomly selected baby born in 2002 weighed 3000 to 3999 grams and was postterm? *0.0484*

(d) What is the probability that a randomly selected baby born in 2002 weighed 3000 to 3999 grams or was postterm? *0.6781*

(e) What is the probability that a randomly selected baby born in 2002 weighed less than 1000 grams and was postterm? Is this event impossible? *0.00000008; no*

(f) What is the probability that a randomly selected baby born in 2002 weighed 3000 to 3999 grams, given the baby was postterm? *0.7185*

(g) Are the events "postterm baby" and "weighs 3000 to 3999 grams" independent? Why? *No*

20. **Olympic Medals** The following data represent the medal tallies of the top eight countries at the 2004 Summer Olympics in Athens.

Country	Gold	Silver	Bronze
United States	35	39	29
China	32	17	14
Russia	27	27	38
Australia	17	16	16
Japan	16	9	12
Germany	14	16	18
France	11	9	13
Italy	10	11	11

(a) If a medal is randomly selected from the top eight countries, what is the probability that it is gold? *0.354*

(b) If a medal is randomly selected from the top eight countries, what is the probability that it was won by China? *0.138*

(c) If a medal is randomly selected from the top eight countries, what is the probability that it is gold and was won by China? *0.070*

(d) If a medal is randomly selected from the top eight countries, what is the probability that it is gold or was won by China? *0.422*

(e) If a bronze medal is randomly selected from the top eight countries, what is the probability that it was won by Japan? *0.079*

(f) If a medal that was won by Japan is randomly selected, what is the probability that it is bronze? *0.324*

21. **Better Business Bureau** The Better Business Bureau reported that approximately 63% of consumer complaints in 2003 were filed online.

(a) If a consumer complaint from 2003 is randomly selected, what is the probability that it was filed online? *0.63*

(b) What is the probability that is was not filed online? *0.37*

(c) If a random sample of five consumer complaints was selected, what is the probability that all five were filed online? *0.099*

(d) If a random sample of five consumer complaints was selected, what is the probability that at least one was not filed online? *0.901*

(e) If a random sample of five consumer complaints was selected, what is the probability that none was filed online? *0.007*

(f) If a random sample of five consumer complaints was selected, what is the probability that at least one was filed online? *0.993*

22. **St. Louis Cardinals** During the 2004 season, the St. Louis Cardinals won 64.8% of their games. Assume that the outcomes of the baseball games are independent and that the percentage of wins this season will be the same as in 2004, and answer the following questions:

(a) What is the probability that the Cardinals will win two games in a row? *0.420*

(b) What is the probability that the Cardinals will win seven games in a row? *0.048*

(c) What is the probability that the Cardinals will lose at least one of their next seven games? *0.952*

23. **PICK 3** For the Illinois Lottery's PICK 3 game, a player must match a sequence of three repeatable numbers, ranging from 0 to 9, in exact order (for example, 3–7–2). With a single ticket, what is the probability of matching the three winning numbers? *0.001*

24. **PICK 4** The Illinois Lottery's PICK 4 game is similar to PICK 3, except a player must match a sequence of four repeatable numbers, ranging from 0 to 9, in exact order (for example, 5–8–5–1). With a single ticket, what is the probability of matching the four winning numbers? *0.0001*

25. **Acceptance Sampling** Suppose you just received a shipment of 10 DVD players. One of the DVD players is defective. You will accept the shipment if two randomly selected DVD players work. What is the probability that you will accept the shipment? *0.8*

26. Drawing Cards Suppose you draw 3 cards without replacement from a standard 52-card deck. What is the probability that all 3 cards are aces? *0.00018*

27. Forming License Plates A license plate is designed so that the first two characters are letters and the last four characters are digits (0 through 9). How many different license plates can be formed assuming letters and numbers can be used more then once? *6,760,000*

28. Choosing a Seat If four students enter a classroom that has 10 vacant seats, in how many ways can they be seated? *5040*

29. Jumble In the game of Jumble, the letters of a word are scrambled. The player must form the correct word. In a recent game in a local newspaper, the Jumble "word" was LINCEY. How many different arrangements are there of the letters in this "word"? *6!*

30. Arranging Flags How many different vertical arrangements are there of 10 flags if 4 are white, 3 are blue, 2 are green, and 1 is red? *12,600*

31. Simple Random Sampling How many different simple random samples of size 8 can be obtained from a population whose size is 55? $_{55}C_8$

32. Forming Committees The U.S. Senate Appropriations Committee has 29 members. Suppose that a subcommittee is to be formed by randomly selecting 5 of the members of the Appropriations Committee. How many different committees could be formed? $_{29}C_5$

33. Arizona's Fantasy 5 In one of Arizona's lotteries, balls are numbered 1 to 35. Five balls are selected randomly, without replacement. The order in which the balls are selected does not matter. To win, your numbers must match the five selected. Determine your probability of winning Arizona's Fantasy 5 with one ticket. *1/324,632*

34. Pennsylvania's Cash 5 In one of Pennsylvania's lotteries, balls are numbered 1 to 39. Five balls are selected randomly, without replacement. The order in which the balls are selected does not matter. To win, your numbers must match the five selected. Determine your probability of winning Pennsylvania's Cash 5 with one ticket. *1/575,757*

35. Packaging Error Because of a mistake in packaging, a case of 12 bottles of red wine contained 5 Merlot and 7 Cabernet, each without labels. All the bottles look alike and have an equal probability of being chosen. Three bottles are randomly selected.

(35b) *0.3182*

(a) What is the probability that all three are Merlot? *0.0455*
(b) What is the probability that exactly two are Merlot?
(c) What is the probability that none is a Merlot? *0.1591*

36. Simulation Use a graphing calculator or statistical software to simulate the playing of the game of roulette, using an integer distribution with numbers 1 through 38. Repeat the simulation 100 times. Let the number 37 represent 0 and the number 38 represent 00. Use the results of the simulation to answer the following questions.

(a) What is the probability that the ball lands in the slot marked 7?
(b) What is the probability that the ball lands either in the slot marked 0 or in the one marked 00?

37. Explain what is meant by a subjective probability. List some examples of subjective probabilities.

38. Playing Five-Card Stud In the game of five-card stud, one card is dealt face down to each player and the remaining four cards are dealt face up. After two cards are dealt (one down and one up), the players bet. Players continue to bet after each additional card is dealt. Suppose three cards have been dealt to each of the five players at the table. You currently have three clubs in your hand, so you will attempt to get a flush (all cards in the same suit). Of the cards dealt, there are two clubs showing in other player's hands.

(a) How many clubs are in a standard 52-card deck? *13*
(b) How many cards remain in the deck or are not known by you? Of this amount, how many are clubs? *41; 8*
(c) What is the probability that you get dealt a club on the next card? *8/41*
(d) What is the probability that you get dealt two clubs in a row? *0.034*
(e) Should you stay in the game?

39. Mark McGwire During the 1998 major league baseball season, Mark McGwire of the St. Louis Cardinals hit 70 home runs. Of the 70 home runs, 34 went to left field, 20 went to left center field, 13 went to center field, 3 went to right center field, and 0 went to right field. (*Source*: Miklasz, B., et al. *Celebrating 70: Mark McGwire's Historic Season*, Sporting News Publishing Co., 1998, pp. 179.)

(a) What is the probability that a randomly selected home run was hit to left field? Interpret this probability. *0.529*
(b) What is the probability that a randomly selected home run was hit to right field? *0*
(c) Is it impossible for Mark McGwire to hit a homer to right field? *No*

THE CHAPTER 5 CASE STUDY IS LOCATED ON THE CD THAT ACCOMPANIES THIS TEXT.

Discrete Probability Distributions

Outline

6.1 Discrete Random Variables

6.2 The Binomial Probability Distribution

■ Chapter Review

■ Case Study: The Voyage of the *St. Andrew* (On CD)

DECISIONS

A woman who was shopping in Los Angeles had her purse stolen by a young, blonde female who was wearing a ponytail. Because there were no eyewitnesses and no real evidence, the prosecution used probability to make its case against the defendant. Your job is to play the role of both the prosecution and defense attorneys to make probabilistic arguments both for and against the defendant. See the Decisions project on page 309.

●●● Putting It All Together

In Chapter 5, we discussed the idea of probability. The probability of an event is the long-term proportion with which the event is observed. That is, if we conduct an experiment 1000 times and observe the outcome 300 times, the probability of the outcome is 0.3. The more times we conduct the experiment, the more accurate the empirical probability. This is the Law of Large Numbers. We learned that we can use counting techniques to obtain theoretical probabilities provided that the outcomes in the experiment are equally likely. This is called classical probability.

We also learned that a probability model lists the possible outcomes to a probability experiment and each outcome's probability. A probability model must satisfy the rules of probability. In particular, all probabilities must be between 0 and 1, inclusive, and the sum of the probabilities must equal 1.

In this chapter, we introduce probability models for *random variables*. A random variable is a numerical measure of the outcome to a probability experiment. So, rather than listing specific outcomes to a probability experiment such as heads or tails, we might list the number of heads obtained in, say, three flips of a coin. In Section 6.1, we discuss random variables and describe the distribution of discrete random variables (shape, center, and spread). Then we discuss a specific discrete probability distribution, *the binomial probability distribution*.

6.1 Discrete Random Variables

Preparing for This Section   Before getting started, review the following:

- Discrete versus continuous variables (Section 1.1, pp. 7–9)
- Relative frequency histograms for discrete data (Section 2.2, pp. 72–73)
- Mean (Section 3.1, pp. 107–110)

- Standard deviation (Section 3.2, pp. 129–130)
- Mean from grouped data (Section 3.3, pp. 142–143)
- Standard deviation from grouped data (Section 3.3, pp. 144–146)

Objectives

1. **Distinguish between discrete and continuous random variables**
2. **Identify discrete probability distributions**
3. **Construct probability histograms**
4. **Compute and interpret the mean of a discrete random variable**
5. **Interpret the mean of a discrete random variable as an expected value**
6. **Compute the variance and standard deviation of a discrete random variable**

1 Distinguish between Discrete and Continuous Random Variables

In Chapter 5, we presented the concept of a probability experiment and its outcomes. Suppose we flip a coin two times. The possible outcomes of the experiment are {HH, HT, TH, TT}. Rather than being interested in the outcome, we might be interested in the number of heads. When experiments are conducted in a way such that the outcome is a numerical result, we say the outcome is a *random variable*.

Definition

A **random variable** is a numerical measure of the outcome of a probability experiment, so its value is determined by chance. Random variables are denoted using letters such as X.

Note to Instructor

To help students understand the difference between a discrete and continuous random variable, tell them that a discrete random variable would have values that are points on a number line, while continuous random variables would have all values on the number line for possible values.

So, in our coin-flipping example, if the random variable X represents the number of heads in two flips of a coin, the possible values of X are 0, 1, or 2.

We will follow the practice of using a capital letter to identify the random variable and a small letter to list the possible values of the random variable, that is, the sample space of the experiment. For example, if an experiment is conducted in which a single die is cast, then X represents the number of pips showing on the die and the possible values of X are $x = 1, 2, 3, 4, 5$, or 6. As another example, suppose an experiment is conducted in which the time between arrivals of cars at a drive-through is measured. The random variable T might describe the time between arrivals, so the sample space of the experiment is $t > 0$.

There are two types of random variables, *discrete* and *continuous*.

Definitions

In Other Words

Discrete random variables typically result from counting, such as 0, 1, 2, 3, and so on. Continuous random variables are variables that result from measurement.

A **discrete random variable** has either a finite or countable number of values. The values of a discrete random variable can be plotted on a number line with space between each point. See Figure 1(a) on the next page.

A **continuous random variable** has infinitely many values. The values of a continuous random variable can be plotted on a line in an uninterrupted fashion. See Figure 1(b) on the next page.

Figure 1

(a) Discrete Random Variable   (b) Continuous Random Variable

---

EXAMPLE 1

Distinguishing between Discrete and Continuous Random Variables

(a) The number of A's earned in a section of statistics with 15 students enrolled is a discrete random variable because the value of the random variable results from counting. If we let the random variable X represent the number of A's, then the possible values of X are $x = 0, 1, 2, \ldots, 15$.

(b) The number of cars that travel through a McDonald's drive-through in the next hour is a discrete random variable because the value of the random variable results from counting. If we let the random variable X represent the number of cars through the drive-through in the next hour, the possible values of X are $x = 0, 1, 2, \ldots$.

(c) The speed of the next car that passes a state trooper is a continuous random variable because speed is measured. If we let the random variable S represent the speed of the next car, the possible values of S are all positive real numbers; that is, $s > 0$.

> **! CAUTION**
> Even though a radar gun may report the speed of a car as 37 miles per hour, it is actually any number greater than or equal to 36.5 mph and less than 37.5 mph. That is, $36.5 \leq s < 37.5$.

Now Work Problem 7.

In this chapter, we will concentrate on probabilities of discrete random variables. Probabilities for certain continuous random variables will be discussed in the next chapter.

② Identify Discrete Probability Distributions

Because the value of a random variable is determined by chance, there are probabilities that correspond to the possible values of the random variable.

Definition

> The **probability distribution** of a discrete random variable X provides the possible values of the random variable and their corresponding probabilities. A probability distribution can be in the form of a table, graph, or mathematical formula.

EXAMPLE 2

A Discrete Probability Distribution

Table 1

x	$P(x)$
0	0.01
1	0.10
2	0.38
3	0.51

Suppose we ask a basketball player to shoot three free throws. Let the random variable X represent the number of shots made, so that $x = 0, 1, 2,$ or 3. Table 1 shows a probability distribution for the random variable X assuming the player historically makes 80% of her free-throw attempts.

From the probability distribution in Table 1, we can see that the probability the player makes all three free-throw attempts is 0.51.

We will denote probabilities using the notation $P(x)$, where x is a specific value of the random variable. We read $P(x)$ as "the probability that the random variable X equals x." For example, $P(3) = 0.51$ is read "the probability that the random variable X equals 3 is 0.51."

Recall from Section 5.1 that probabilities must obey certain rules. We repeat the rules for a discrete probability distribution using the notation just introduced.

The first rule states that the sum of the probabilities must equal 1. The second rule states that each probability must be greater than or equal to 0 and less than or equal to 1.

Rules for a Discrete Probability Distribution

Let $P(x)$ denote the probability that the random variable X equals x; then

1. $\sum P(x) = 1$
2. $0 \le P(x) \le 1$

Table 1 from Example 2 is a probability distribution because the sum of the probabilities equals 1 and each probability is between 0 and 1, inclusive. You are encouraged to verify this.

EXAMPLE 3 **Identifying Discrete Probability Distributions**

Problem: Which of the following is a discrete probability distribution?

(a)

x	$P(x)$
1	0.20
2	0.35
3	0.12
4	0.40
5	−0.07

(b)

x	$P(x)$
1	0.20
2	0.25
3	0.10
4	0.14
5	0.49

(c)

x	$P(x)$
1	0.20
2	0.25
3	0.10
4	0.14
5	0.31

Approach: In a discrete probability distribution, the sum of the probabilities must equal 1, and all probabilities must be greater than or equal to 0 and less than or equal to 1.

Solution

(a) This is not a discrete probability distribution because $P(5) = -0.07$, which is less than 0.

(b) This is not a discrete probability distribution because

$$\sum P(x) = 0.2 + 0.25 + 0.10 + 0.14 + 0.49 = 1.18 \neq 1$$

(c) This is a discrete probability distribution because the sum of the probabilities equals 1, and each probability is greater than or equal to 0 and less than or equal to 1.

Now Work Problem 11.

Table 1 is a discrete probability distribution in table form. Probability distributions can also be represented through graphs or mathematical formulas. We discuss discrete probability distributions using graphs now and discuss probability distributions as mathematical formulas in the next section.

③ Construct Probability Histograms

A graphical depiction of a discrete probability distribution is typically done with a *probability histogram*.

Definition

A **probability histogram** is a histogram in which the horizontal axis corresponds to the value of the random variable and the vertical axis represents the probability of each value of the random variable.

EXAMPLE 4 **Constructing a Probability Histogram**

Problem: Construct a probability histogram of the discrete probability distribution given in Table 1 from Example 2.

Approach: Probability histograms are constructed like relative frequency histograms, except that the vertical axis represents the probability of the random variable, rather than its relative frequency. Each rectangle is centered at the value of the discrete random variable.

Solution: Figure 2 presents the probability histogram.

In Other Words

A probability histogram is constructed the same way as a relative frequency histogram for discrete data. The only difference is that the vertical axis is a probability, rather than a relative frequency.

Figure 2

Number of Successful Free Throws

Note to Instructor

Emphasize the relation between area and probability in probability histograms as a prelude to continuous probabilities.

Notice that the area of each rectangle in the probability histogram equals the probability that the random variable assumes the particular value. For example, the area of the rectangle corresponding to the random variable $X = 2$ is $1 \cdot (0.38) = 0.38$, where 1 represents the width of the rectangle and 0.38 represents its height.

Probability histograms help us to determine the shape of the distribution. Recall that we describe distributions as skewed left, skewed right, or symmetric. For example, the probability histogram presented in Figure 2 is skewed left.

Now Work Problems 19(a) and (b).

 Compute and Interpret the Mean of a Discrete Random Variable

Remember, when we describe the distribution of a variable, we describe its center, spread, and shape. We now introduce methods for identifying the center and spread of a discrete random variable. We will use the mean to describe the center of a random variable. The variance and standard deviation are used to describe the spread of a random variable.

To help see where the formula for computing the mean of a discrete random variable comes from, consider the following. One semester I had a small statistics class of 10 students. I asked them to disclose the number of people living in their household and obtained the following:

$$2, 4, 6, 6, 4, 4, 2, 3, 5, 5$$

What is the mean number of people in the 10 households? Of course, we could find the mean by adding the observations and dividing by 10. But we will take a different approach. Let the random variable X represent the number of people in the household and obtain the probability distribution in Table 2.

Table 2	
x	*P(x)*
2	$\frac{2}{10} = 0.2$
3	$\frac{1}{10} = 0.1$
4	$\frac{3}{10} = 0.3$
5	$\frac{2}{10} = 0.2$
6	$\frac{2}{10} = 0.2$

Now we compute the mean as follows:

$$\mu = \frac{\sum x_i}{N} = \frac{2 + 4 + 6 + 6 + 4 + 4 + 2 + 3 + 5 + 5}{10}$$

$$= \frac{\overbrace{2 + 2}^{2} + \overbrace{3}^{1} + \overbrace{4 + 4 + 4}^{3} + \overbrace{5 + 5}^{2} + \overbrace{6 + 6}^{2}}{10}$$

$$= \frac{2 \cdot 2 + 3 \cdot 1 + 4 \cdot 3 + 5 \cdot 2 + 6 \cdot 2}{10}$$

$$= 2 \cdot \frac{2}{10} + 3 \cdot \frac{1}{10} + 4 \cdot \frac{3}{10} + 5 \cdot \frac{2}{10} + 6 \cdot \frac{2}{10}$$

$$= 2 \cdot P(2) + 3 \cdot P(3) + 4 \cdot P(4) + 5 \cdot P(5) + 6 \cdot P(6)$$

$$= 2(0.2) + 3(0.1) + 4(0.3) + 5(0.2) + 6(0.2)$$

$$= 4.1$$

Note to Instructor
Point out to students that 4.1 is not a possible value of the random variable; it just represents the mean number of people in all households of the population.

Based on the preceding computations, we conclude that the mean of a discrete random variable is found by multiplying each possible value of the random variable by its corresponding probability and adding these products.

In Other Words
To find the mean of a discrete random variable, multiply the value of each random variable by its probability. Then add these products.

The Mean of a Discrete Random Variable

The mean of a discrete random variable is given by the formula

$$\mu_X = \sum [x \cdot P(x)] \qquad (1)$$

where x is the value of the random variable and $P(x)$ is the probability of observing the random variable x.

EXAMPLE 5 **Computing the Mean of a Discrete Random Variable**

Problem: Compute the mean of the discrete random variable given in Table 1 from Example 2.

Approach: The mean of a discrete random variable is found by multiplying each value of the random variable by its probability and adding these products.

Solution: Refer to Table 3. The first two columns represent the discrete probability distribution. The third column represents $x \cdot P(x)$.

We substitute into Formula (1) to find the mean number of free throws made.

$$\mu_X = \sum [x \cdot P(x)] = 0(0.01) + 1(0.10) + 2(0.38) + 3(0.51) = 2.39 \approx 2.4$$

We will follow the practice of rounding the mean, variance, and standard deviation to one more decimal place than the values of the random variable. ▄

Table 3		
x	$P(x)$	$x \cdot P(x)$
0	0.01	$0 \cdot 0.01 = 0$
1	0.10	$1 \cdot 0.1 = 0.1$
2	0.38	0.76
3	0.51	1.53

How to Interpret the Mean of a Discrete Random Variable
The mean of a discrete random variable can be thought of as the mean outcome of the probability experiment if we repeated the experiment many times. Consider the result of Example 5. If we repeated the experiment of shooting three free throws many times and recorded the number of free throws made, we would expect the average number of free throws made to be around 2.4.

In Other Words

We can think of the mean of a discrete random variable as the average outcome if the experiment is repeated many, many times.

Interpretation of the Mean of a Discrete Random Variable

Suppose an experiment is repeated n independent times and the value of the random variable X is recorded. As the number of repetitions of the experiment, n, increases, the mean value of the n trials will approach μ_X, the mean of the random variable X. In other words, let x_1 be the value of the random variable X after the first experiment, x_2 be the value of the random variable X after the second experiment, and so on. Then

$$\overline{X} = \frac{x_1 + x_2 + \cdots + x_n}{n}$$

The difference between $\overline{X}$ and μ_X gets closer to 0 as n increases.

EXAMPLE 6

Illustrating the Interpretation of the Mean of a Discrete Random Variable

Problem: The basketball player from Example 2 is asked to shoot three free throws 100 times. Compute the mean number of free throws made.

Approach: The player shoots three free throws and the number made is recorded. We repeat this experiment 99 more times and then compute the mean number of free throws made.

Solution: The results are presented in Table 4.

Table 4									
3	2	3	3	3	3	1	2	3	2
2	3	3	1	2	2	2	2	2	3
3	3	2	2	3	2	3	2	2	2
3	3	2	3	2	3	3	2	3	1
3	2	2	2	2	0	2	3	1	2
3	3	2	3	2	3	2	1	3	2
2	3	3	3	1	3	3	1	3	3
3	2	2	1	3	2	2	2	3	2
3	2	2	2	3	3	2	2	3	3
2	3	2	1	2	3	3	2	3	3

The first time the experiment was conducted, the player made all three free throws. The second time the experiment was conducted, the player made two out of three free throws. The hundreth time the experiment was conducted, the player made three out of three free throws. The mean number of free throws made was

$$\overline{X} = \frac{3 + 2 + 3 + \cdots + 3}{100} = 2.35$$

This is close to the mean of 2.4 (from Example 5). As the number of repetitions of the experiment increases, we expect $\overline{x}$ to get even closer to 2.4.

Figure 3(a) and Figure 3(b) further demonstrate the interpretation of the mean of a discrete random variable. Figure 3(a) shows the mean number of free throws made versus the number of repetitions of the experiment for the data in Table 4. Figure 3(b) shows the mean number of free throws made versus the number of repetitions of the experiment when the same experiment of shooting three free throws 100 times is conducted a second time. In both plots the player starts off "hot," since the mean number of free throws made is above the theoretical level of 2.4. However, both graphs approach the theoretical mean of 2.4 as the number of repetitions of the experiment increases.

Figure 3

(a) Number of Repetitions   (b) Number of Repetitions

Now Work Problem 19(c).

5 **Interpret the Mean of a Discrete Random Variable as an Expected Value**

Note to Instructor
You might want to mention that expected value plays a role in game theory. For example, a negative-sum game has an expected value that is negative.

Because the mean of a random variable represents what we would expect to happen in the long run, the mean of a random variable is also called the **expected value**. The interpretation of expected value is the same as the interpretation of the mean of a discrete random variable.

EXAMPLE 7   **Finding the Expected Value**

In Other Words
The expected value of a discrete random variable is the mean of the discrete random variable.

Problem: A term life insurance policy will pay a beneficiary a certain sum of money upon the death of the policyholder. These policies have premiums that must be paid annually. Suppose a life insurance company sells a $250,000 one-year term life insurance policy to an 18-year-old male for $350. According to the *National Vital Statistics Report*, Vol. 47, No. 28, the probability that the male will survive the year is 0.998789. Compute the expected value of this policy to the insurance company.

Approach: There are two possible outcomes to the experiment: survival or death. Let the random variable X represent the *payout* (money lost or gained), depending on survival or death of the insured. We assign probabilities to each of these random variables and substitute these values into Formula (1).

Solution

Step 1: We have $P(\text{survives}) = 0.998789$, so $P(\text{dies}) = 0.001211$. From the point of view of the insurance company, if the client survives the year, the insurance company makes $350. Therefore, we let $x = \$350$ if the client survives the year. If the client dies during the year, the insurance company must pay $250,000 to the client's beneficiary. However, the company still keeps the $350, so we let $x = \$350 - \$250,000 = -\$249,650$. The value is negative because it is money paid out by the insurance company. The probability distribution is listed in Table 5.

Table 5

x	$P(x)$
$350 (survives)	0.998789
-$249,650 (dies)	0.001211

Step 2: Substituting into Formula (1), we obtain the expected value (from the point of view of the insurance company) of the policy.

$$E(X) = \mu_X = \sum xP(x) = \$350(0.998789) + (-\$249{,}650)(0.001211) = \$47.25$$

Interpretation: The company expects to make $47.25 for each 18-year-old male client it insures. The $47.25 profit of the insurance company is a long-term result. It does not make $47.25 on each person it insures, but rather the average profit per person insured is $47.25. Because this is a long-term result, the insurance "idea" will not work with only a few insured.

Note to Instructor
By law, insurance companies are supposed to set premiums so the expected value of the policy is zero.

Now Work Problem 29.

Historical Note

Christiaan Huygens was born on April 14, 1629, into an influential Dutch family. He studied Law and Mathematics at the University of Leiden from 1645 to 1647. From 1647 to 1649, he continued to study Law and Mathematics at the College of Orange at Breda. Among his many great accomplishments, Huygens discovered the first moon of Saturn in 1655 and the shape of the rings of Saturn in 1656. While in Paris sharing his discoveries, he learned about probability through the correspondence of Fermat and Pascal. In 1657, Huygens published the first book on probability theory. In that text, Huygens introduced the idea of expected value.

In Other Words

The variance of a discrete random variable is a weighted average of the squared deviations where the weights are the probabilities.

Note to Instructor
The purpose of this activity is to reinforce the concept of expected value. Time limit: 50 to 60 minutes depending on discussions.

In-Class Activity: Expected Value

Consider the following game of chance. A player pays $1 and rolls a pair of fair dice. If the player rolls a 2, 3, 4, 10, 11, or 12, the player loses the $1 bet. If the player rolls 5, 6, 8, or 9, there is a "push" and the player gets his or her dollar back. If the player rolls a 7, the player wins $1.

(a) Construct a probability distribution that describes the game.
(b) Compute the expected value of the game from the player's point of view.
(c) Actually play the game in a small group. Keep track of the results on paper (no money should actually change hands). Compute the mean earnings to the player from the game. Are the results close to what you expected? If not, why?

⑥ Compute the Variance and Standard Deviation of a Discrete Random Variable

We now introduce a method for computing the variance and standard deviation of a discrete random variable.

Variance and Standard Deviation of a Discrete Random Variable

The variance of a discrete random variable is given by

$$\sigma_X^2 = \sum[(x - \mu_X)^2 \cdot P(x)] \qquad \text{(2a)}$$
$$= \sum[x^2 \cdot P(x)] - \mu_X^2 \qquad \text{(2b)}$$

where x is the value of the random variable, μ_X is the mean of the random variable, and $P(x)$ is the probability of observing the random variable x.

To find the standard deviation of the discrete random variable, take the square root of the variance. That is, $\sigma_X = \sqrt{\sigma_X^2}$.

EXAMPLE 8   Computing the Variance and Standard Deviation of a Discrete Random Variable

Problem: Find the variance and standard deviation of the discrete random variable given in Table 1 from Example 2.

Approach: We will use Formula (2a) with the un-rounded mean $\mu_X = 2.39$.

Solution: Refer to Table 6. The first two columns represent the discrete probability distribution. The third column represents $(x - \mu_X)^2 \cdot P(x)$. We sum the entries in the third column to get the variance.

Table 6		
x	$P(x)$	$(x - \mu_X)^2 \cdot P(x)$
0	0.01	$(0 - 2.39)^2 \cdot 0.01 = 0.057121$
1	0.10	$(1 - 2.39)^2 \cdot 0.10 = 0.19321$
2	0.38	$(2 - 2.39)^2 \cdot 0.38 = 0.057798$
3	0.51	$(3 - 2.39)^2 \cdot 0.51 = 0.189771$
		$\sum (x - \mu_X)^2 \cdot P(x) = 0.4979$

The variance of the discrete random variable X is

$$\sigma_X^2 = \sum (x - \mu_X)^2 \cdot P(x) = 0.4979 \approx 0.5$$

Approach: We will use Formula (2b) with the un-rounded mean $\mu_X = 2.39$.

Solution: Refer to Table 7. The first two columns represent the discrete probability distribution. The third column represents $x^2 \cdot P(x)$.

Table 7		
x	$P(x)$	$x^2 \cdot P(x)$
0	0.01	$0^2 \cdot 0.01 = 0$
1	0.10	$1^2 \cdot 0.10 = 0.10$
2	0.38	$2^2 \cdot 0.38 = 1.52$
3	0.51	$3^2 \cdot 0.51 = 4.59$
		$\sum x^2 \cdot P(x) = 6.21$

The variance of the discrete random variable X is

$$\sigma_X^2 = \sum [x^2 \cdot P(x)] - \mu_X^2 = 6.21 - 2.39^2 = 0.4979 \approx 0.5$$

The standard deviation of the discrete random variable is found by taking the square root of the variance.

$$\sigma_X = \sqrt{\sigma_X^2} = \sqrt{0.4979} \approx 0.7$$

Now Work Problems 19(d) and (e).

EXAMPLE 9   **Obtaining the Mean and Standard Deviation of a Discrete Random Variable Using Technology**

Problem: Use a statistical spreadsheet or calculator to determine the mean and the standard deviation of the random variable whose distribution is given in Table 1.

Figure 4

Approach: We will use a TI-84 Plus graphing calculator to obtain the mean and standard deviation. The steps for determining the mean and standard deviation using a TI-83 or TI-84 Plus graphing calculator are given in the Technology Step by Step on page 297.

Result: Figure 4 shows the results from a TI-84 Plus graphing calculator. **Note:** The TI does not find s_X when the sum of L_2 is one.

6.1 ASSESS YOUR UNDERSTANDING

Concepts and Vocabulary

1. What is a random variable?
2. What is the difference between a discrete random variable and a continuous random variable? Provide your own examples of each.
3. What are the two requirements for a discrete probability distribution?
4. In your own words, provide an interpretation of the mean of a discrete random variable.
5. Suppose a baseball player historically hits 0.300. (This means that the player averages three hits in every 10 at-bats.) Suppose the player has zero hits in four at-bats in a game and enters the batter's box for the fifth time, whereupon the announcer declares that the player is "due for a hit." What is the flaw in the announcer's reasoning? If the player had four hits in the last four at-bats, is the player "due to make an out"?
6. A game is called a zero-sum game if the expected value of the game is zero. Explain what a game whose expected value is zero means.

Skill Building   (7c) Discrete, x = 0, 1, 2,...   (7d) Continuous, s ≥ 0   (8a) Continuous, t > 0   (9a) Continuous, r ≥ 0
(10a) Discrete, x = 0, 1, 2,...

In Problems 7–10, determine whether the random variable is discrete or continuous. In each case, state the possible values of the random variable.

7. (a) The number of lightbulbs that burn out in the next week in a room of with 20 bulbs. Discrete; x=0, 1,..., 20
   (b) The time it takes to fly from New York City to Los Angeles. Continuous; t > 0
   (c) The number of hits to a Web site in a day.
   (d) The amount of snow in Toronto during the winter.

8. (a) The time it takes for a lightbulb to burn out.
   (b) The weight of a T-bone steak. Continuous, w > 0
   (c) The number of free-throw attempts before the first shot is made. Discrete, x = 1, 2,...
   (d) In a random sample of 20 people, the number who are blood type A. Discrete, x = 0, 1,..., 20

9. (a) The amount of rain in Seattle during April.
   (b) The number of fish caught during a fishing tournament. Discrete, x = 0, 1, 2,...
   (c) The number of customers arriving at a bank between noon and 1:00 P.M. Discrete, x = 0, 1, 2,...
   (d) The time required to download a file from the Internet. Continuous, t > 0

10. (a) The number of defects in a roll of carpet.
    (b) The distance a baseball travels in the air after being hit. Continuous, d > 0
    (c) The number of points scored during a basketball game. Discrete, x = 0, 1, 2,...
    (d) The square footage of a house. Continuous, a > 0

In Problems 11–16, determine whether the distribution is a discrete probability distribution. If not, state why.

NW 11. Yes

x	P(x)
0	0.2
1	0.2
2	0.2
3	0.2
4	0.2

12. Yes

x	P(x)
0	0.1
1	0.5
2	0.05
3	0.25
4	0.1

13. No, $P(50) < 0$

x	P(x)
10	0.1
20	0.23
30	0.22
40	0.6
50	−0.15

14. Yes

x	P(x)
1	0
2	0
3	0
4	0
5	1

15. No, $\sum P(x) \neq 1$

x	P(x)
100	0.1
200	0.25
300	0.2
400	0.3
500	0.1

16. No, $\sum P(x) \neq 1$

x	P(x)
100	0.25
200	0.25
300	0.25
400	0.25
500	0.25

In Problems 17 and 18, determine the required value of the missing probability to make the distribution a discrete probability distribution.

17. $P(4) = 0.3$

x	P(x)
3	0.4
4	?
5	0.1
6	0.2

18. $P(2) = 0.15$

x	P(x)
0	0.30
1	0.15
2	?
3	0.20
4	0.15
5	0.05

Applying the Concepts

19. Parental Involvement In the following probability distribution, the random vari-
NW able X represents the number of activities a parent of a student in kindergarten
through fifth grade is involved in.

(a) Verify that this is a discrete probability distribution.
(b) Draw a probability histogram.
(c) Compute and interpret the mean of the random variable X. $\mu_X = 2.9$
(d) Compute the variance of the random variable X. $\sigma_X^2 = 1.2$
(e) Compute the standard deviation of the random variable X. $\sigma_X = 1.1$
(f) What is the probability that a randomly selected student has a parent involved
in three activities? 0.320
(g) What is the probability that a randomly selected student has a parent involved
in three or four activities? 0.694

x	P(x)
0	0.035
1	0.074
2	0.197
3	0.320
4	0.374

Source: U.S. National
Center for Education
Statistics

20. Parental Involvement In the following probability distribution, the random variable X represents the number of activities a parent of a student in grades 6 through
8 is involved in.

(a) Verify that this is a discrete probability distribution.
(b) Draw a probability histogram.
(c) Compute and interpret the mean of the random variable X. $\mu_X = 2.5$
(d) Compute the variance of the random variable X. $\sigma_X^2 = 1.4$
(e) Compute the standard deviation of the random variable X. $\sigma_X = 1.2$
(f) What is the probability that a randomly selected student has a parent involved
in three activities? 0.322
(g) What is the probability that a randomly selected student has a parent involved
in three or four activities? 0.552

x	P(x)
0	0.073
1	0.117
2	0.258
3	0.322
4	0.230

Source: U.S. National
Center for Education
Statistics

21. Ichiro's Hit Parade In the 2004 baseball season, Ichiro Suzuki of the Seattle Mariners set the record for most hits in a season with a total of 262 hits. In the following probability distribution, the random variable X represents the number of hits Ichiro obtained in a game.

x	P(x)
0	0.1677
1	0.3354
2	0.2857
3	0.1491
4	0.0373
5	0.0248

Source: Chicago Tribune

(a) Verify that this is a discrete probability distribution.
(b) Draw a probability histogram.
(c) Compute and interpret the mean of the random variable X. $\mu_x = 1.6$
(d) Compute the standard deviation of the random variable X. $\sigma_x = 1.2$
(e) What is the probability that in a randomly selected game Ichiro got 2 hits? 0.2857
(f) What is the probability that in a randomly selected game Ichiro got more than 1 hit? 0.4969

22. Waiting in Line A Wendy's manager performed a study to determine a probability distribution for the number of people waiting in line X during lunch. The results were as follows:

(22d) $\sigma_x^2 = 5.0$

x	P(x)	x	P(x)	x	P(x)
0	0.011	5	0.172	10	0.019
1	0.035	6	0.132	11	0.002
2	0.089	7	0.098	12	0.006
3	0.150	8	0.063	13	0.001
4	0.186	9	0.035	14	0.001

(a) Verify that this is a discrete probability distribution.
(b) Draw a probability histogram.
(c) Compute and interpret the mean of the random variable X. $\mu_x = 4.9$
(d) Compute the variance of the random variable X.
(e) Compute the standard deviation of the random variable X. $\sigma_x = 2.2$
(f) What is the probability that there are eight people waiting in line for lunch? 0.063
(g) What is the probability that there are 10 or more people waiting in line for lunch? Would this be unusual? 0.029

In Problems 23–26, (a) construct a discrete probability distribution for the random variable X [Hint: $P(x_i) = \dfrac{f_i}{N}$.], (b) draw the probability histogram, (c) compute and interpret the mean of the random variable X, and (d) compute the standard deviation of the random variable X.

23. The World Series The following data represent the number of games played in each World Series from 1923 to 2005. (c) $\mu_x = 5.8$ (d) $\sigma_x = 1.2$

x (games played)	Frequency
4	16
5	15
6	18
7	33

Source: Information Please Almanac

24. Number of 5- to 9-Year-Old Girls The following data represent (in thousands) the number of 5- to 9-year-old females in the United States in 2000. (c) $\mu_x = 7.0$ (d) $\sigma_x = 1.4$

x (age)	Frequency
5	1934
6	1961
7	2008
8	2041
9	2081

Source: U.S. Census Bureau

25. Grade School Enrollment The following data represent (in thousands) the enrollment levels in grades 1 to 8 in the United States in 2000. (c) $\mu_x = 4.5$ (d) $\sigma_x = 2.3$

x (grade level)	Frequency
1	3635
2	3633
3	3673
4	3708
5	3701
6	3658
7	3624
8	3532

Source: U.S. National Center for Education Statistics

26. **High School Enrollment** The following data represent (in thousands) the enrollment levels in grades 9 to 12 in the United States in 2000. (c) $\mu_x = 10.4$ (d) $\sigma_x = 1.1$

x (grade level)	Frequency
9	3958
10	3487
11	3080
12	2799

Source: U.S. National Center for Education Statistics

27. **Number of Births** The probability histogram that follows represents the number of live births by a mother 50 to 54 years old who had a live birth in 2002. The data are from the *National Vital Statistics Report*, Vol. 52, No. 10, December 17, 2003.

Number of Live Births, 50–54-Year-Old Mother

Probability histogram with values: 1: 0.241, 2: 0.257, 3: 0.172, 4: 0.119, 5: 0.103, 6: 0.027, 7: 0.031, 8: 0.050 (x-axis: Child)

(a) What is the probability that a randomly selected 50- to 54-year-old mother who had a live birth in 2002 has had her fourth live birth? *0.119*

(b) What is the probability that a randomly selected 50- to 54-year-old mother who had a live birth in 2002 has had her fourth or fifth live birth? *0.222*

(c) What is the probability that a randomly selected 50- to 54-year-old mother who had a live birth in 2002 has had her sixth or more live birth? *0.108*

(d) If a 50- to 54-year-old mother who had a live birth in 2002 is randomly selected, how many live births would you expect the mother to have had? *3.0*

28. **Rental Units** The probability histogram that follows represents the number of rooms in rented housing units in 2003. The data are from the U.S. Department of Housing and Urban Development.

(a) What is the probability that a randomly selected rental unit has five rooms? *0.228*

(b) What is the probability that a randomly selected rental unit has five or six rooms? *0.333*

Number of Rooms in Rental Unit

Probability histogram with values: 1: 0.010, 2: 0.028, 3: 0.229, 4: 0.347, 5: 0.228, 6: 0.105, 7: 0.036, 8: 0.012, 9: 0.005 (x-axis: Number of Rooms)

(c) What is the probability that a randomly selected rental unit has seven or more rooms? *0.053*

(d) If a rental unit is randomly selected, how many rooms would you expect the unit to have? *4.3*

29. **Life Insurance** Suppose a life insurance company sells a 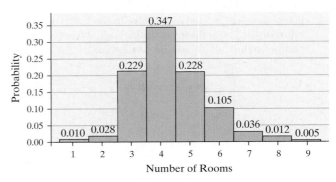$250,000 one-year term life insurance policy to a 20-year-old female for $200. According to the *National Vital Statistics Report*, Vol. 53, No 6, the probability that the female survives the year is 0.999546. Compute and interpret the expected value of this policy to the insurance company. *$86.50*

30. **Life Insurance** Suppose a life insurance company sells a $250,000 one-year term life insurance policy to a 20-year-old male for $350. According to the *National Vital Statistics Report*, Vol. 53, No. 6, the probability that the male survives the year is 0.998611. Compute and interpret the expected value of this policy to the insurance company. *$2.75*

31. **Investment** An investment counselor calls with a hot stock tip. He believes that if the economy remains strong the investment will result in a profit of $50,000. If the economy grows at a moderate pace, the investment will result in a profit of $10,000. However, if the economy goes into recession, the investment will result in a loss of $50,000. You contact an economist who believes there is a 20% probability the economy will remain strong, a 70% probability the economy will grow at a moderate pace, and a 10% probability the economy will slip into recession. What is the expected profit from this investment?

32. **Real Estate Investment** Shawn and Maddie purchase a foreclosed property for $50,000 and spend an additional $27,000 fixing up the property. They feel that they can re-sell the property for $120,000 with probability 0.15, $100,000 with probability 0.45, $80,000 with probability 0.25, and $60,000 with probability 0.15. Compute and interpret the expected profit for reselling the property.

33. **Roulette** In the game of roulette, a player can place a $5 bet on the number 17 and have a $\frac{1}{38}$ probability of winning. If the metal ball lands on 17, the player wins $175. Otherwise, the casino takes the player's $5. What is the expected value of the game to the player? If you played the game 1000 times, how much would you expect to lose?

−$0.26; about $260

34. Connecticut Lottery In the Cash Five Lottery in Connecticut, a player pays $1 for a single ticket with five numbers. Five Ping-Pong balls numbered 1 through 35 are randomly chosen from a bin without replacement. If all five numbers on a player's ticket match the five chosen, the player wins $100,000. The probability of this occurring is $\frac{1}{324,632}$. If four numbers match, the player wins $300. This occurs with probability $\frac{1}{2164}$. If three numbers match, the player wins $10. This occurs with probability $\frac{1}{75}$. Compute and interpret the expected value of the game from the player's point of view. −$0.42

35. Powerball Powerball is a multistate lottery. The following probability distribution represents the cash prizes of Powerball with their corresponding probabilities.

x (cash prize, $)	P(x)
Grand prize	0.00000000684
200,000	0.00000028
10,000	0.000001711
100	0.000153996
7	0.004778961
4	0.007881463
3	0.01450116
0	0.9726824222

Source: www.powerball.com

(a) If the grand prize is $15,000,000, find and interpret the expected cash prize. If a ticket costs $1, what is your expected profit from one ticket? $0.30; −$0.70

(b) To the nearest million, how much should the grand prize be so that you can expect a profit? Assume nobody else wins so that you do not have to share the grand prize. $118,000,000

(c) Does the size of the grand prize affect your chance of winning? Explain. No

36. SAT Test Penalty Some standardized tests, such as the SAT test, incorporate a penalty for wrong answers. For example, a multiple-choice question with five possible answers will have 1 point awarded for a correct answer and $\frac{1}{4}$ deducted point for an incorrect answer. Questions left blank are worth 0 points.

(a) Find the expected number of points received for a multiple-choice question with five possible answers when a student just guesses. 0

(b) Explain why there is a deduction for wrong answers.

37. Simulation Use the probability distribution from Problem 21 and a DISCRETE command for some statistical software to simulate 100 repetitions of the experiment (100 games). The number of hits is recorded. Approximate the mean and standard deviation of the random variable X based on the simulation. Repeat the simulation by performing 500 repetitions of the experiment. Approximate the mean and standard deviation of the random variable. Compare your results to the theoretical mean and standard deviation. What property is being illustrated?

38. Simulation Use the probability distribution from Problem 22 and a DISCRETE command for some statistical software to simulate 100 repetitions of the experiment. Approximate the mean and standard deviation of the random variable X based on the simulation. Repeat the simulation by performing 500 repetitions of the experiment. Approximate the mean and standard deviation of the random variable. Compare your results to the theoretical mean and standard deviation. What property is being illustrated?

Technology Step by Step

Finding the Mean and Standard Deviation of a Discrete Random Variable Using Technology

TI-83/84 Plus

Step 1: Enter the values of the random variable in L1 and their corresponding probabilities in L2.

Step 2: Press STAT, highlight CALC, and select 1: 1-Var Stats.

Step 3: With 1-Var Stats on the HOME screen, type L1 followed by a comma, followed by L2 as follows:

```
1-Var Stats L1, L2
```

Hit ENTER.

6.2 The Binomial Probability Distribution

Preparing for This Section   Before getting started, review the following:

- Independence (Section 5.3, pp. 249–251)
- Combinations (Section 5.5, pp. 270–273)
- Multiplication Rule for Independent Events (Section 5.3, pp. 251–252)
- Addition Rule for Disjoint Events (Section 5.2, pp. 238–241)
- Complement Rule (Section 5.2, pp. 244–245)
- Empirical Rule (Section 3.2, pp. 131–132)

Objectives

1. **Determine whether a probability experiment is a binomial experiment**
2. **Compute probabilities of binomial experiments**
3. **Compute the mean and standard deviation of a binomial random variable**
4. **Construct binomial probability histograms**

---

1. ## Determine Whether a Probability Experiment Is a Binomial Experiment

In Section 6.1, we stated that probability distributions could be presented using tables, graphs, or mathematical formulas. In this section, we introduce a specific type of discrete probability distribution that can be presented using a formula, *the binomial probability distribution*.

The binomial probability distribution is a discrete probability distribution that describes probabilities for experiments in which there are two mutually exclusive (disjoint) outcomes. These two outcomes are generally referred to as *success* and *failure*. For example, a basketball player can either make a free throw (success) or miss (failure). A new surgical procedure can result in either life (success) or death (failure).

Experiments in which there are only two possible outcomes are referred to as *binomial experiments*, provided that certain criteria are met.

In Other Words

The prefix *bi* means "two." This should help remind you that binomial experiments deal with situations in which there are only two outcomes: success and failure.

Criteria for a Binomial Probability Experiment

An experiment is said to be a **binomial experiment** if

1. The experiment is performed a fixed number of times. Each repetition of the experiment is called a **trial**.
2. The trials are independent. This means the outcome of one trial will not affect the outcome of the other trials.
3. For each trial, there are two mutually exclusive (disjoint) outcomes: success or failure.
4. The probability of success is the same for each trial of the experiment.

Let the random variable X be the number of successes in n trials of a binomial experiment. Then X is called a **binomial random variable**. Before introducing the method for computing binomial probabilities, it is worthwhile to introduce some notation.

Notation Used in the Binomial Probability Distribution

- There are n independent trials of the experiment.
- Let p denote the probability of success so that $1 - p$ is the probability of failure.
- Let X denote the number of successes in n independent trials of the experiment. So $0 \le x \le n$.

EXAMPLE 1   ### Identifying Binomial Experiments

Problem: Determine which of the following probability experiments qualify as a binomial experiment. For those that are binomial experiments, identify the number of trials, probability of success, probability of failure, and possible values of the random variable X.

Historical Note

Jacob Bernoulli was born on December 27, 1654, in Basel, Switzerland. He studied philosophy and theology at the urging of his parents. (He resented this.) In 1671, he graduated from the University of Basel with a master's degree in philosophy. In 1676, he received a licentiate in theology. After earning his philosophy degree, Bernoulli traveled to Geneva to tutor. From there, he went to France to study with the great mathematicians of the time. One of Bernoulli's greatest works is *Ars Conjectandi*, published 8 years after his death. In this publication, Bernoulli proved the binomial probability formula. To this day, each observed outcome in a binomial probability experiment is called a *Bernoulli trial*.

(a) An experiment in which a basketball player who historically makes 80% of his free throws is asked to shoot three free throws, and the number of made free throws is recorded.

(b) The number of people with blood type O-negative based on a simple random sample of size 10 is recorded. According to the *Information Please Almanac*, 6% of the human population is blood type O-negative.

(c) A probability experiment in which three cards are drawn from a deck without replacement and the number of aces is recorded.

Approach: We determine whether or not the four conditions for a binomial experiment are satisfied.

1. The experiment is performed a fixed number of times.
2. The trials are independent.
3. There are only two possible outcomes of the experiment.
4. The probability of success for each trial is constant.

Solution

(a) This is a binomial experiment because
  1. There are $n = 3$ trials.
  2. The trials are independent.
  3. There are two possible outcomes: make or miss.
  4. The probability of success (make) is 0.8 and the probability of failure (miss) is 0.2. The probabilities are the same for each trial.

The random variable X is the number of free throws made with $x = 0, 1, 2,$ or 3.

(b) This is a binomial experiment because
  1. There are 10 trials (the 10 randomly selected people).
  2. The trials are independent.*
  3. There are two possible outcomes: finding a person with blood type O-negative or not.
  4. The probability of success is 0.06 and the probability of failure is 0.94.

The random variable X is the number of people with blood type O-negative with $x = 0, 1, 2, 3, \ldots, 10$.

(c) This is not a binomial experiment because the trials are not independent. The probability of an ace on the first trial is $\dfrac{4}{52}$. Because we are sampling without replacement, if an ace is selected on the first trial, the probability of an ace on the second trial is $\dfrac{3}{51}$. If an ace is not selected on the first trial, the probability of an ace on the second trial is $\dfrac{4}{51}$. ━━━━━ ■

Now Work Problem 9.

---

*In sampling from large populations without replacement, the trials are assumed to be independent, provided that the sample size is small in relation to the size of the population. As a rule of thumb, if the sample size is less than 5% of the population size, the trials are assumed to be independent, although they are technically dependent. See Example 6 in Section 5.4.

! **CAUTION**
The probability of success, p, is always associated with the random variable X, the number of successes. So if X represents the number of 18-year-olds involved in an accident, then p represents the probability of an 18-year-old being involved in an accident.

It is worth mentioning that the word *success* does not necessarily imply that something positive has occurred. Success means that an outcome has occurred that corresponds with p, the probability of success. For example, a probability experiment might be to randomly select ten 18-year-old male drivers. We might let X denote the number who have been involved in an accident within the last year. In this case, a success would mean obtaining an 18-year-old male who was involved in an accident. This outcome is certainly not positive, but still represents a success as far as the experiment goes.

② Compute Probabilities of Binomial Experiments

We are now prepared to compute probabilities for a binomial random variable X. We present three methods for obtaining binomial probabilities: (1) the binomial probability distribution formula, (2) a table of binomial probabilities, and (3) technology. We develop the binomial probability formula in Example 2.

EXAMPLE 2 **Constructing a Binomial Probability Distribution**

Problem: According to the *Information Please Almanac*, 6% of the human population is blood type O-negative. A simple random sample of size 4 is obtained, and the number of people X with blood type O-negative is recorded. Construct a probability distribution for the random variable X.

Approach: This is a binomial experiment with $n = 4$ trials. We define a success as selecting an individual with blood type O-negative. The probability of success, p, is 0.06, and X is the random variable representing the number of successes with $x = 0, 1, 2, 3,$ or 4.

Step 1: Construct a tree diagram listing the various outcomes of the experiment by listing each outcome as S (success) or F (failure).

Step 2: Compute the probabilities for each value of the random variable X.

Step 3: Construct the probability distribution.

Solution

Step 1: Figure 5 contains a tree diagram listing the 16 possible outcomes of the experiment.

Step 2: We now compute the probability for each possible value of the random variable X. We start with $P(0)$:

$$P(0) = P(FFFF) = P(F) \cdot P(F) \cdot P(F) \cdot P(F) \quad \text{\small Multiplication Rule for Independent Events}$$
$$= (0.94)(0.94)(0.94)(0.94)$$
$$= (0.94)^4$$
$$= 0.78075$$

$$P(1) = P(SFFF \text{ or } FSFF \text{ or } FFSF \text{ or } FFFS)$$
$$= P(SFFF) + P(FSFF) + P(FFSF) + P(FFFS) \quad \text{\small Addition Rule for Disjoint Events}$$
$$= (0.06)^1(0.94)^3 + (0.06)^1(0.94)^3 + (0.06)^1(0.94)^3 + (0.06)^1(0.94)^3 \quad \text{\small Multiplication Rule for Independent Events}$$
$$= 4(0.06)^1(0.94)^3$$
$$= 0.19934$$

$P(2) = P(SSFF \text{ or } SFSF \text{ or } SFFS \text{ or } FSSF \text{ or } FSFS \text{ or } FFSS)$

$\quad = P(SSFF) + P(SFSF) + P(SFFS) + P(FSSF) + P(FSFS) + P(FFSS)$

$\quad = (0.06)^2(0.94)^2 + (0.06)^2(0.94)^2 + (0.06)^2(0.94)^2 + (0.06)^2(0.94)^2 + (0.06)^2(0.94)^2 + (0.06)^2(0.94)^2$

$\quad = 6(0.06)^2(0.94)^2$

$\quad = 0.01909$

Figure 5

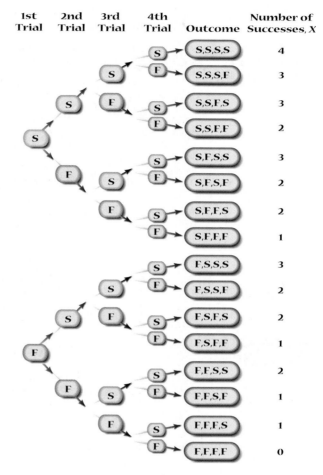

1st Trial	2nd Trial	3rd Trial	4th Trial	Outcome	Number of Successes, X

We compute $P(3)$ and $P(4)$ similarly and obtain $P(3) = 0.00081$ and $P(4) = 0.00001$. You are encouraged to verify these probabilities.

Step 3: We use these results and obtain the probability distribution in Table 8.

Table 8

x	P(x)
0	0.78075
1	0.19934
2	0.01909
3	0.00081
4	0.00001

As we look back at the solution in Example 2, we note some interesting results. Consider the probability of obtaining $X = 1$ success:

$$P(1) = 4(0.06)^1 (0.94)^3$$

"4 is the number of ways we obtain 1 success in 4 trials of the experiment. Here, it is $_4C_1$."

"0.94 is the probability of failure and the exponent 3 is the number of failures."

"0.06 is the probability of success and the exponent 1 is the number of successes."

The coefficient 4 is the number of ways of obtaining one success in four trials. In general, the coefficient will be $_nC_x$, the number of ways of obtaining x successes in n trials. The second factor in the formula, $(0.06)^1$, is the probability of success, p, raised to the number of successes, x. The third factor in the formula, $(0.94)^3$, is the probability of failure, $1 - p$, raised to the number of failures, $n - x$. This formula holds for all binomial experiments, and we have the *binomial probability distribution function (pdf)*.

CAUTION
Before using the binomial probability distribution function, be sure the requirements for a binomial experiment are satisfied.

Binomial Probability Distribution Function

The probability of obtaining x successes in n independent trials of a binomial experiment, where the probability of success is p, is given by

$$P(x) = {_nC_x}\, p^x(1 - p)^{n-x}, \quad x = 0, 1, 2, \ldots, n \tag{1}$$

While reading probability problems, pay special attention to key phrases that translate into mathematical symbols. Table 9 lists various phrases and their corresponding mathematical equivalent.

Table 9

Phrase	Math Symbol
"at least" or "no less than" or "greater than or equal to"	$\geq$
"more than" or "greater than"	$>$
"fewer than" or "less than"	$<$
"no more than" or "at most" or "less than or equal to"	$\leq$
"exactly or "equals" or "is"	$=$

EXAMPLE 3 **Using the Binomial Probability Distribution Function**

Problem: According to the Federal Communications Commission, 70% of all U.S. households have cable television.

(a) In a random sample of 15 households, what is the probability that exactly 10 have cable?

(b) In a random sample of 15 households, what is the probability that at least 13 have cable?

(c) In a random sample of 15 households, what is the probability that fewer than 13 have cable?

(d) In a random sample of 15 households, what is the probability that the number of households with cable is between 10 and 12, inclusive?

Approach: This is a binomial experiment with $n = 15$ independent trials with the probability of success, p, equal to 0.70. The possible values of the random variable X are $x = 0, 1, 2, \ldots, 15$. We use Formula (1) to compute the probabilities.

Solution

(a) $P(10) = {_{15}C_{10}}(0.70)^{10}(1 - 0.70)^{15-10}$ $n = 15, x = 10, p = 0.70$

$$= \frac{15!}{10!(15 - 10)!}(0.70)^{10}(0.30)^5 \quad {_nC_x} = \frac{n!}{x!(n-x)!}$$

$$= 3003(0.02825)(0.00243)$$

$$= 0.2061$$

Note to Instructor
It is important to reinforce the relative frequency approach to interpreting probabilities. This will help students when we get to confidence intervals and P-values.

Interpretation: The probability of getting exactly 10 households out of 15 with cable is 0.2061. In 100 trials of this experiment, we would expect about 21 trials to result in 10 households with cable.

(b) The phrase *at least* means "greater than or equal to." The values of the random variable X greater than or equal to 13 are $x = 13, 14,$ or 15.

$$
\begin{aligned}
P(X \geq 13) &= P(13 \text{ or } 14 \text{ or } 15) \\
&= P(13) + P(14) + P(15) \qquad \text{\small Addition Rule for Disjoint Events} \\
&= {}_{15}C_{13}(0.70)^{13}(1 - 0.70)^{15-13} + {}_{15}C_{14}(0.70)^{14}(1 - 0.70)^{15-14} + {}_{15}C_{15}(0.70)^{15}(1 - 0.70)^{15-15} \\
&= 0.0916 + 0.0305 + 0.0047 \\
&= 0.1268
\end{aligned}
$$

Interpretation: There is a 0.1268 probability that in a random sample of 15 households at least 13 will have cable. In 100 trials of this experiment, we would expect about 13 trials to result in at least 13 households having cable.

(c) The values of the random variable X less than 13 are $x = 0, 1, 2, \ldots, 12$. Rather than compute $P(X \leq 12)$ directly by computing $P(0) + P(1) + \cdots + P(12)$, we can use the Complement Rule.

$$
P(X < 13) = P(X \leq 12) = 1 - P(X \geq 13) = 1 - 0.1268 = 0.8732
$$

Interpretation: There is a 0.8732 probability that in a random sample of 15 households, fewer than 13 will have cable. In 100 trials of this experiment, we expect about 87 trials to result in fewer than 13 households that have cable.

(d) The word *inclusive* means "including," so we want to determine the probability that 10, 11, or 12 households have cable.

$$
\begin{aligned}
P(10 \leq X \leq 12) &= P(10 \text{ or } 11 \text{ or } 12) \\
&= P(10) + P(11) + P(12) \qquad \text{\small Addition Rule for Disjoint Events} \\
&= {}_{15}C_{10}(0.70)^{10}(1 - 0.70)^{15-10} + {}_{15}C_{11}(0.70)^{11}(1 - 0.70)^{15-11} + {}_{15}C_{12}(0.70)^{12}(1 - 0.70)^{15-12} \\
&= 0.2061 + 0.2186 + 0.1700 \\
&= 0.5947
\end{aligned}
$$

Interpretation: The probability that the number of households with cable is between 10 and 12, inclusive, is 0.5947. In 100 trials of this experiment, we expect about 59 trials to result in 10 to 12 households having cable.

Obtaining Binomial Probabilities from Tables

Another method for obtaining probabilities is the binomial probability table. Table II in Appendix A gives probabilities for a binomial random variable X taking on a specific value such as $P(10)$ for select values of n and p. Table III in Appendix A gives cumulative probabilities of a binomial random variable X. This means Table III gives "less than or equal to" binomial probabilities such as $P(X \leq 6)$. We illustrate how to use Tables II and III in Example 4.

EXAMPLE 4   **Computing Binomial Probabilities Using the Binomial Table**

Problem: According to the National Endowment for the Arts, 20% of U.S. women attended a musical play in 2002.

(a) In a random sample of 15 U.S. women, what is the probability that exactly 5 have attended a musical play in 2002?

(b) In a random sample of 15 U.S. women, what is the probability that fewer than 7 attended a musical play in 2002?

(c) In a random sample of 15 U.S. women, what is the probability that 7 or more attended a musical play in 2002?

Approach: We use Tables II and III in Appendix A to obtain the probabilities.

Solution

(a) We have $n = 15$, $p = 0.20$, and $x = 5$. In Table II, we go to the section that contains $n = 15$ and the column that contains $p = 0.20$. Within the $n = 15$ section, we look for the row $x = 5$. The value at which the $x = 5$ row intersects with the $p = 0.20$ column is the probability we seek. See Figure 6. So $P(5) = 0.1032$.

Figure 6

n	x	0.01	0.05	0.10	0.15	0.20	0.25	0.30	0.35	0.40	0.45	0.50	0.55	0.60	0.65	0.70	0.75	0.80	0.85	0.85	0.90	0.95
15	0	0.8601	0.4633	0.2059	0.0874	0.0352	0.0134	0.0047	0.0016	0.0005	0.0001	0.0000	0.0000	0.0000	0.0000	0.0000	0.0000	0.0000	0.0000	0.0000	0.0000	0.0000
	1	0.1303	0.3658	0.3432	0.2312	0.1319	0.0668	0.0305	0.0126	0.0047	0.0015	0.0006	0.0001	0.0000	0.0000	0.0000	0.0000	0.0000	0.0000	0.0000	0.0000	0.0000
	2	0.0092	0.1348	0.2669	0.2856	0.2309	0.1559	0.0916	0.0476	0.0219	0.0090	0.0032	0.0010	0.0003	0.0001	0.0000	0.0000	0.0000	0.0000	0.0000	0.0000	0.0000
	3	0.0004	0.0307	0.1285	0.2184	0.2501	0.2252	0.1700	0.1110	0.0634	0.0318	0.0139	0.0052	0.0016	0.0004	0.0001	0.0000	0.0000	0.0000	0.0000	0.0000	0.0000
	4	0.0000	0.0049	0.0428	0.1156	0.1676	0.2252	0.2186	0.1782	0.1268	0.0780	0.0417	0.0191	0.0074	0.0024	0.0006	0.0001	0.0000	0.0000	0.0000	0.0000	0.0000
	5	0.0000	0.0006	0.0105	0.0449	0.1032	0.1651	0.2061	0.2123	0.1859	0.1404	0.0916	0.0515	0.0245	0.0096	0.0030	0.0007	0.0001	0.0000	0.0000	0.0000	0.0000
	6	0.0000	0.0000	0.0019	0.0132	0.0430	0.0917	0.1472	0.1906	0.2056	0.1914	0.1527	0.1048	0.0612	0.0298	0.0116	0.0034	0.0007	0.0001	0.0000	0.0000	0.0000
	7	1.0000	1.0000	1.0000	0.9994	_(faded)_	_(faded)_	0.0500	_(faded)_	0.7869	_(faded)_	0.500	_(faded)_	_(faded)_	_(faded)_	0.0500	_(faded)_	_(faded)_	_(faded)_	_(faded)_	_(faded)_	_(faded)_

Interpretation: There is a 0.1032 probability that in a random sample of 15 U.S. women, exactly 5 have attended a musical play in 2002. In 100 trials of this experiment, we expect about 10 trials to result in exactly 5 women who have attended a musical play in 2002.

(b) The values of the random variable X that are fewer than 7 are 0, 1, 2, 3, 4, 5, or 6. So $P(X < 7) = P(X \le 6)$. To compute $P(X \le 6)$, we use the cumulative binomial table, Table III in Appendix A. The cumulative binomial table lists binomial probabilities less than or equal to a specified value. We have $n = 15$ and $p = 0.20$. In Table III, we go to the row that contains $n = 15$ and the column that contains $p = 0.20$. Within the $n = 15$ section, we look for the row $x = 6$. This row represents $P(X \le 6)$. The value at which the $x = 6$ row intersects with the $p = 0.20$ column is the probability we seek. See Figure 7. So $P(X \le 6) = 0.9819$.

Figure 7

n	x	0.01	0.05	0.10	0.15	0.20	0.25	0.30	0.35	0.40	0.45	0.50	0.55	0.60	0.65	0.70	0.75	0.80	0.85	0.85	0.90	0.95
15	0	0.8601	0.4633	0.2059	0.0874	0.0352	0.0134	0.0047	0.0016	0.0005	0.0001	0.0000	0.0000	0.0000	0.0000	0.0000	0.0000	0.0000	0.0000	0.0000	0.0000	0.0000
	1	0.9904	0.8290	0.5490	0.3188	0.1671	0.0802	0.0353	0.0142	0.0052	0.0017	0.0005	0.0001	0.0000	0.0000	0.0000	0.0000	0.0000	0.0000	0.0000	0.0000	0.0000
	2	0.9996	0.9538	0.8159	0.6042	0.3980	0.2361	0.1288	0.0617	0.0271	0.0107	0.0037	0.0011	0.0003	0.0001	0.0000	0.0000	0.0000	0.0000	0.0000	0.0000	0.0000
	3	1.0000	0.9945	0.9944	0.8227	0.6482	0.4613	0.2969	0.1727	0.0905	0.0424	0.0178	0.0083	0.0019	0.0005	0.0001	0.0000	0.0000	0.0000	0.0000	0.0000	0.0000
	4	1.0000	0.9994	0.9873	0.9388	0.8358	0.6886	0.5155	0.3519	0.2173	0.1204	0.0592	0.0255	0.0093	0.0028	0.0007	0.0001	0.0000	0.0000	0.0000	0.0000	0.0000
	5	1.0000	0.9999	0.9978	0.9964	0.9389	0.8516	0.7216	0.5843	0.4032	0.2608	0.1509	0.0789	0.0338	0.0124	0.0037	0.0008	0.0001	0.0000	0.0000	0.0000	0.0000
	6	1.0000	1.0000	0.9997	0.9964	0.9819	0.9434	0.9689	0.7546	0.6098	0.4522	0.3036	0.1618	0.0950	0.0422	0.0152	0.0042	0.0008	0.0001	0.0000	0.0000	0.0000
	7	1.0000	1.0000	1.0000	0.9994	0.9958	0.9627	0.9500	0.8868	0.7869	0.6535	0.500	0.3465	0.2131	0.1132	0.0500	0.0173	0.0042	0.0006	0.0000	0.0000	0.0000
	8	1.0000	1.0000	1.0000	0.9994	_(faded)_	_(faded)_	0.0500	_(faded)_	0.7869	_(faded)_	0.500	_(faded)_	_(faded)_	_(faded)_	0.0500	_(faded)_	_(faded)_	_(faded)_	_(faded)_	_(faded)_	_(faded)_

Interpretation: There is a 0.9819 probability that in a random sample of 15 U.S. women, fewer than 7 have attended a musical play in 2002. In 100 trials of this experiment, we would expect about 98 trials to result in fewer than 7 women who have attended a musical play in 2002.

(c) To obtain $P(X \ge 7)$, we use the Complement Rule and the results of part (b) as follows:

$$P(X \ge 7) = 1 - P(X < 7)$$
$$= 1 - P(X \le 6)$$
$$= 1 - 0.9819$$
$$= 0.0181$$

Interpretation: There is a 0.0181 probability that in a random sample of 15 U.S. women, at least 7 have attended a musical play in 2002. In 100 trials of this experiment, we expect about 2 trials to result in at least 7 women who have attended a musical play in 2002. Because this event only happens about 2 out of 100 times, we consider it to be unusual.

Obtaining Binomial Probabilities Using Technology

Statistical software and graphing calculators have the ability to compute binomial probabilities as well. We illustrate this approach to computing probabilities in the next example.

EXAMPLE 5   **Obtaining Binomial Probabilities Using Technology**

Problem: According to the National Endowment for the Arts, 20% of U.S. women attended a musical play in 2002.

(a) In a random sample of 15 U.S. women, what is the probability that exactly 5 have attended a musical play in 2002?

(b) In a random sample of 15 U.S. women, what is the probability that fewer than 7 attended a musical play in 2002?

Approach: Statistical software or graphing calculators with advanced statistical features have the ability to determine binomial probabilities. The steps for determining binomial probabilities using MINITAB, Excel, and the TI-83/84 Plus graphing calculators can be found in the Technology Step by Step on page 313.

Result: We will use Excel to determine the probability for part (a) and a TI-84 Plus to determine the probability for part (b).

(a) Using Excel's formula wizard, we obtain the results in Figure 8(a). So $P(5) = 0.1032$. This agrees with the results of Example 4(a).

(b) To compute probabilities such as $P(X < 7) = P(X \leq 6)$, it is best to use the **cumulative distribution function** (or **cdf**), which computes probabilities less than or equal to a specified value. Using a TI-84 Plus graphing calculator to compute $P(X \leq 6)$ with $n = 15$ and $p = 0.2$, we find $P(X \leq 6) = 0.9819$. See Figure 8(b).

Figure 8

Here is the probability

(a)   (b)

Now Work Problem 35.

3   **Compute the Mean and Standard Deviation of a Binomial Random Variable**

We discussed finding the mean (or expected value) and standard deviation of a discrete random variable in Section 6.1. These formulas can be used to find the mean (or expected value) and standard deviation of a binomial random variable as well. However, there is a faster method.

In Other Words

The mean of a binomial random variable equals the number of trials of the experiment times the probability of success. It can be interpreted as the expected number of successes in n trials of the experiment.

Mean (or Expected Value) and Standard Deviation of a Binomial Random Variable

A binomial experiment with n independent trials and probability of success p has a mean and standard deviation given by the formulas

$$\mu_X = np \qquad \text{and} \qquad \sigma_X = \sqrt{np(1 - p)} \qquad (2)$$

| EXAMPLE 6 | **Finding the Mean and Standard Deviation of a Binomial Random Variable** |

Note to Instructor
You may want to construct a binomial probability distribution with $n = 3$ and $p = 0.4$. Then use the formulas from Section 6.1 to compute μ_X and σ_X. Now use Formula (2) to compute μ_X and σ_X.

Problem: According to the Federal Communications Commission, 70% of all U.S. households had cable television in 2002. In a simple random sample of 300 households, determine the mean and standard deviation number of households that will have cable television.

Approach: This is a binomial experiment with $n = 300$ and $p = 0.70$. We can use Formula (2) to find the mean and standard deviation, respectively.

Solution: $\mu_X = np = 300(0.70) = 210$
and

$$\sigma_X = \sqrt{np(1-p)} = \sqrt{300(0.70)(1-0.70)} = \sqrt{63} = 7.9$$

Interpretation: We expect that in a random sample of 300 households 210 will have cable.

Now Work Problems 29(a), (b), and (c).

 Construct Binomial Probability Histograms

Constructing binomial probability histograms is no different from constructing other probability histograms.

| EXAMPLE 7 | **Constructing Binomial Probability Histograms** |

Problem

(a) Construct a binomial probability histogram with $n = 10$ and $p = 0.2$. Comment on the shape of the distribution.

(b) Construct a binomial probability histogram with $n = 10$ and $p = 0.5$. Comment on the shape of the distribution.

(c) Construct a binomial probability histogram with $n = 10$ and $p = 0.8$. Comment on the shape of the distribution.

Approach: To construct a binomial probability histogram, we will first obtain the probability distribution. We then construct the probability histogram of the probability distribution.

Solution

(a) We obtain the probability distribution with $n = 10$ and $p = 0.2$. See Table 10. Note in Table 10, $P(9) = 0.0000$. The probability is actually 0.000004096 but is written as 0.0000 to four significant digits. The same idea applies to $P(10)$. Figure 9 shows the corresponding probability histogram with the mean $\mu_X = 10(0.2) = 2$ labeled. The distribution is skewed right.

Table 10

x	$P(x)$
0	0.1074
1	0.2684
2	0.3020
3	0.2013
4	0.0881
5	0.0264
6	0.0055
7	0.0008
8	0.0001
9	0.0000
10	0.0000

Figure 9

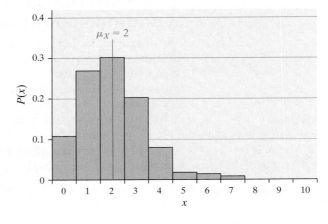

(b) We obtain the probability distribution with $n = 10$ and $p = 0.5$. See Table 11. Figure 10 shows the corresponding probability histogram with the mean $\mu_X = 10(0.5) = 5$ labeled. The distribution is symmetric and approximately bell shaped.

Table 11

x	$P(x)$
0	0.0010
1	0.0098
2	0.0439
3	0.1172
4	0.2051
5	0.2461
6	0.2051
7	0.1172
8	0.0439
9	0.0098
10	0.0010

Figure 10

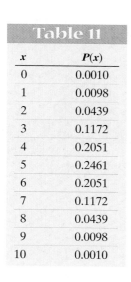

(c) We obtain the probability distribution with $n = 10$ and $p = 0.8$. See Table 12. Figure 11 shows the corresponding probability histogram with the mean $\mu_X = 10(0.8) = 8$ labeled. The distribution is skewed left.

Table 12

x	$P(x)$
0	0.0000
1	0.0000
2	0.0001
3	0.0008
4	0.0055
5	0.0264
6	0.0881
7	0.2013
8	0.3020
9	0.2684
10	0.1074

Figure 11

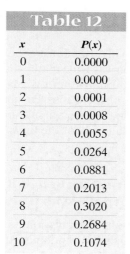

Now Work Problem 29(d).

Based on the results of Example 7, we might conclude that the binomial probability distribution is skewed right if $p < 0.5$, symmetric and approximately bell shaped if $p = 0.5$, and skewed left if $p > 0.5$. Notice that Figure 9 ($p = 0.2$) and Figure 11 ($p = 0.8$) are mirror images.

The binomial probability distribution depends on the parameter p, and n, the number of trials. What role does n play in the shape of the distribution? To answer this question we compare the binomial probability histogram with $n = 10$ and $p = 0.2$. [see Figure 12(a)] to the binomial probability histogram with $n = 30$ and $p = 0.2$. [Figure 12(b)] and the binomial probability histogram with $n = 70$ and $p = 0.2$ [Figure 12(c)].

Figure 12(a) is skewed right. Figure 12(b) is slightly skewed right, and Figure 12(c) appears bell shaped.

Figure 12

We conclude the following:

> As the number of trials n in a binomial experiment increases, the probability distribution of the random variable X becomes bell shaped. As a rule of thumb, if $np(1 - p) \geq 10$,* the probability distribution will be approximately bell shaped.

This result allows us to use the Empirical Rule to identify unusual observations in a binomial experiment. Recall that the Empirical Rule states that in a bell-shaped distribution about 95% of all observations lie within two standard deviations of the mean. That is, about 95% of the observations lie between $\mu - 2\sigma$ and $\mu + 2\sigma$. Any observation that lies outside this interval may be considered unusual because the observation occurs less than 5% of the time.

EXAMPLE 8

Using the Mean, Standard Deviation, and Empirical Rule to Check for Unusual Results in a Binomial Experiment

Problem: According to the Federal Communications Commission, in 2002, 70% of all U.S. households had cable television. In a simple random sample of 300 households, 230 had cable. Is this result unusual?

Approach: Because $np(1 - p) = 300(0.70)(0.30) = 63 \geq 10$, the binomial probability distribution is approximately bell shaped. Therefore, we can use the Empirical Rule to check for unusual observations. If the observation is less than $\mu - 2\sigma$ or greater than $\mu + 2\sigma$, we say it is unusual.

Solution: From Example 6, we have $\mu = 210$ and $\sigma = 7.9$.

$$\mu - 2\sigma = 210 - 2(7.9) = 210 - 15.8 = 194.2$$

and

$$\mu + 2\sigma = 210 + 2(7.9) = 210 + 15.8 = 225.8$$

*Ramsey, P. P., and P. H. Ramsey, Evaluating the Normal Approximation to the Binomial Test, *Journal of Educational Statistics* 13 (1998): 173–182.

Interpretation: Since any value less than 194.2 or greater than 225.8 is unusual, 230 is an unusual result. We should attempt to identify reasons for its value. It may be that the percentage of households that have cable has increased since 2002.

Now Work Problem 43.

MAKING AN INFORMED DECISION

Should We Convict?

A woman who was shopping in Los Angeles had her purse stolen by a young, blonde female who was wearing a ponytail. The blonde female got into a yellow car that was driven by a black male who had a mustache and a beard. The police located a blonde female named Janet Collins who wore her hair in a ponytail and had a friend who was a black male who had a mustache and beard and also drove a yellow car. The police arrested the two subjects.

Because there were no eyewitnesses and no real evidence, the prosecution used probability to make its case against the defendants. The following probabilities were presented by the prosecution for the known characteristics of the thieves.

Characteristic	Probability
Yellow car	$\frac{1}{10}$
Man with a mustache	$\frac{1}{4}$
Woman with a ponytail	$\frac{1}{10}$
Woman with blonde hair	$\frac{1}{3}$
Black man with beard	$\frac{1}{10}$
Interracial couple in car	$\frac{1}{1000}$

(a) Assuming that the characteristics listed are independent of each other, what is the probability that a randomly selected couple would have all these characteristics? That is, what is P ("yellow car" and "man with a mustache" and … and "interracial couple in a car")?

(b) Would you convict the defendants based on this probability? Why or why not?

(c) Now let n represent the number of couples in the Los Angeles area who could have committed the crime. Let p represent the probability a randomly selected couple has all six characteristics listed. Let the random variable X represent the number of couples who have all the characteristics listed in the table. Assuming that the random variable X follows the binomial probability function, we have

$$P(x) = {}_nC_x \cdot p^x (1 - p)^{n-x}, \quad x = 0, 1, 2, \ldots, n$$

Assuming that there were $n = 1{,}000{,}000$ couples in the Los Angeles area, what is the probability that more than one of them have the characteristics listed in the table? Does this result cause you to change your mind regarding the defendants' guilt?

(d) Now let's look at this case from a different point of view. We will compute the probability that more than one couple has the characteristics described, given that at least one couple has the characteristics.

$$P(X > 1 \mid X \geq 1) = \frac{P(X > 1 \text{ and } X \geq 1)}{P(X \geq 1)} \quad \text{Conditional Probability Rule}$$

$$= \frac{P(X > 1)}{P(X \geq 1)}$$

Compute this probability, assuming $n = 1{,}000{,}000$. Compute this probability again, but this time assume that $n = 2{,}000{,}000$. Do you think that the couple should be convicted "beyond all reasonable doubt"? Why or why not?

6.2 ASSESS YOUR UNDERSTANDING

Concepts and Vocabulary

1. State the criteria for a binomial probability experiment.
2. What role does $_nC_x$ play in the binomial probability distribution function?
3. How can the Empirical Rule be used to identify unusual results in a binomial experiment? When can the Empirical Rule be used?
4. Describe how the value of n affects the shape of the binomial probability histogram.
5. Describe how the value of p affects the shape of the binomial probability histogram.
6. Explain what "success" means in a binomial experiment.

Skill Building

In Problems 7–16, determine which of the following probability experiments represents a binomial experiment. If the probability experiment is not a binomial experiment, state why.

7. A random sample of 15 college seniors is obtained, and the individuals selected are asked to state their ages.

8. A random sample of 30 cars in a used car lot is obtained, and their mileage recorded. Not binomial

9. An experimental drug is administered to 100 randomly selected individuals, with the number of individuals responding favorably recorded. Binomial

10. A poll of 1200 registered voters is conducted in which the respondents are asked whether they believe Congress should reform Social Security. Binomial

11. Three cards are selected from a standard 52-card deck without replacement. The number of aces selected is recorded. Not binomial

12. Three cards are selected from a standard 52-card deck with replacement. The number of aces selected is recorded.

13. A basketball player who makes 80% of her free throws is asked to shoot free throws until she misses. The number of free-throw attempts is recorded. Not binomial

14. A baseball player who reaches base safely 30% of the time is allowed to bat until he reaches base safely for the third time. The number of at-bats required is recorded.

15. An investor randomly purchases 10 stocks listed on the New York Stock Exchange. Historically, the probability that a stock listed on the NYSE will increase in value over the course of a year is 48%. The number of stocks that increase in value is recorded. Binomial

16. According to Nielsen Media Research, 70% of all U.S. households have cable television. In a small town of 40 households, a random sample of 10 households is asked whether they have cable television. The number of households with cable television is recorded. Not binomial

In Problems 17–28, a binomial probability experiment is conducted with the given parameters. Compute the probability of x successes in the n independent trials of the experiment.

17. $n = 10$, $p = 0.4$, $x = 3$ 0.2150
18. $n = 15$, $p = 0.85$, $x = 12$ 0.2184
19. $n = 40$, $p = 0.99$, $x = 38$ 0.0532
20. $n = 50$, $p = 0.02$, $x = 3$ 0.0607
21. $n = 8$, $p = 0.35$, $x = 3$ 0.2786
22. $n = 20$, $p = 0.6$, $x = 17$ 0.0123
23. $n = 9$, $p = 0.2$, $x \le 3$ 0.9144
24. $n = 10$, $p = 0.65$, $x < 5$ 0.0949
25. $n = 7$, $p = 0.5$, $x > 3$ 0.5
26. $n = 20$, $p = 0.7$, $x \ge 12$ 0.8867
27. $n = 12$, $p = 0.35$, $x \le 4$ 0.5833
28. $n = 11$, $p = 0.75$, $x \ge 8$ 0.7133

In Problems 29–34, (a) construct a binomial probability distribution with the given parameters; (b) compute the mean and standard deviation of the random variable using the methods of Section 6.1; (c) compute the mean and standard deviation, using the methods of this section; and (d) draw the probability histogram, comment on its shape, and label the mean on the histogram.

29. $n = 6$, $p = 0.3$ $\mu_x = 1.8$, $\sigma_x = 1.1$
30. $n = 8$, $p = 0.5$ $\mu_x = 4$, $\sigma_x = 1.4$
31. $n = 9$, $p = 0.75$ $\mu_x = 6.75$, $\sigma_x = 1.3$
32. $n = 10$, $p = 0.2$ $\mu_x = 2$, $\sigma_x = 1.3$
33. $n = 10$, $p = 0.5$ $\mu_x = 5$, $\sigma_x = 1.6$
34. $n = 9$, $p = 0.8$ $\mu_x = 7.2$, $\sigma_x = 1.2$

7. Not binomial   12. Binomial   14. Not binomial

Applying the Concepts   (35b) 0.3432

35. On-Time Flights According to American Airlines, its
NW flight 215 from Orlando to Los Angeles is on time 90% of
the time. Suppose 15 flights are randomly selected and the
number of on-time flights is recorded.

  (a) Explain why this is a binomial experiment.
  (b) Find the probability that exactly 14 flights are on time.
  (c) Find the probability that at least 14 flights are on time. 0.5490
  (d) Find the probability that fewer than 14 flights are on time. 0.4510
  (e) Find the probability that between 12 and 14 flights, inclusive, are on time. 0.7386

36. Smokers According to the *Information Please Almanac*,
80% of adult smokers started smoking before turning 18
years old. Suppose 10 smokers 18 years old or older are
randomly selected and the number of smokers who started smoking before 18 is recorded.

  (a) Explain why this is a binomial experiment.
  (b) Find the probability that exactly 8 of them started smoking before 18 years of age. 0.3020
  (c) Find the probability that at least 8 of them started smoking before 18 years of age. 0.6778
  (d) Find the probability that fewer than 8 of them started smoking before 18 years of age. 0.3222
  (e) Find the probability that between 7 and 9 of them, inclusive, started smoking before 18 years of age. 0.7718

37. High-Speed Internet According to a report by the Commerce Department in the fall of 2004, 20% of U.S. households had some type of high-speed Internet connection. Suppose 20 U.S. households are selected at random and the number of households with high-speed Internet is recorded.

  (a) Find the probability that exactly 5 households have high-speed Internet. 0.1746
  (b) Find the probability that at least 10 households have high-speed Internet. Would this be unusual?
  (c) Find the probability that fewer than 4 households have high-speed Internet. 0.4114
  (d) Find the probability that between 2 and 5 households, inclusive, have high-speed Internet. 0.7350

38. Allergy Sufferers Clarinex-D is a medication whose purpose is to reduce the symptoms associated with a variety of allergies. In clinical trials of Clarinex-D, 5% of the patients in the study experienced insomnia as a side effect. Suppose a random sample of 20 Clarinex-D users is obtained and the number of patients who experienced insomnia is recorded.

  (a) Find the probability that exactly 3 experienced insomnia as a side effect. 0.0596
  (b) Find the probability that 3 or fewer experienced insomnia as a side effect. 0.9841
  (c) Find the probability that between 1 and 4 patients, inclusive, experienced insomnia as a side effect. 0.6389
  (d) Would it be unusual to find 4 or more patients who experienced insomnia as a side effect? Why? Yes

39. Murder by Firearms According to the *Uniform Crime Report, 2003*, 66.9% of murders are committed with a firearm. Suppose that 25 murders were randomly selected

and the number of murders committed with a firearm is recorded.

  (a) Find the probability that exactly 22 murders were committed using a firearm. 0.0120
  (b) Find the probability that between 14 and 16 murders, inclusive, were committed using a firearm. 0.3648
  (c) Would it be unusual if 22 or more murders were committed using a firearm? Why? Yes

40. Migraine Sufferers Depakote is a medication whose purpose is to reduce the pain associated with migraine headaches. In clinical trials of Depakote, 2% of the patients in the study experienced weight gain as a side effect. Suppose a random sample of 30 Depakote users is obtained and the number of patients who experienced weight gain is recorded.

(*Source*:Abbott Laboratories)

  (a) Find the probability that exactly 3 experienced weight gain as a side effect. 0.0188
  (b) Find the probability that 3 or fewer experienced weight gain as a side effect. 0.9971
  (c) Find the probability that 4 or more patients experienced weight gain as a side effect. 0.0029
  (d) Find the probability that between 1 and 4 patients, inclusive, experienced weight gain as a side effect. 0.4542

41. Airline Satisfaction A CNN/*USA Today*/Gallup poll in April 2005 reported that 75% of adult Americans were satisfied with the job the nation's major airlines were doing. Suppose 10 adult Americans are selected at random and the number who are satisfied is recorded.

  (a) Find the probability that exactly 6 are satisfied with the airlines. 0.1460
  (b) Find the probability that fewer than 7 are satisfied with the airlines. 0.2241
  (c) Find the probability that 5 or more are satisfied with the airlines. 0.9803
  (d) Find the probability that between 5 and 8, inclusive, are satisfied with the airlines. 0.7363

42. College Freshmen According to the Higher Education Research Institute, 55% of college freshmen in 4-year colleges and universities during 2003 were female. Suppose 12 freshmen are randomly selected and the number of females is recorded.

  (a) Find the probability that exactly 7 of them are female. 0.2225
  (b) Find the probability that 5 or more are female. 0.8883
  (c) Find the probability that 8 or fewer are female. 0.8655
  (d) Find the probability that between 7 and 10, inclusive, are female. 0.5186

43. On-Time Flights According to American Airlines, its
NW flight 215 from Orlando to Los Angeles is on time 90% of
the time. Suppose 100 flights are randomly selected.

  (a) Compute the mean and standard deviation of the random variable X, the number of on-time flights in 100 trials of the probability experiment. $\mu_x = 90, \sigma_x = 3$
  (b) Interpret the mean.
  (c) Would it be unusual to observe 80 on-time flights in a random sample of 100 flights from Orlando to Los Angeles? Why? Yes

44. Smokers According to the *Information Please Almanac*, NW 80% of adult smokers started smoking before turning 18 years old.

(a) Compute the mean and standard deviation of the random variable X, the number of smokers who started before 18 in 200 trials of the probability experiment.

(b) Interpret the mean.

(c) Would it be unusual to observe 180 smokers who started smoking before turning 18 years old in a random sample of 200 adult smokers? Why? *Yes*

45. High-Speed Internet According to a report by the Commerce Department in the fall of 2004, 20% of U.S. households had some type of high-speed Internet connection.

(a) Compute the mean and standard deviation of the random variable X, the number of U.S. households with a high-speed Internet connection in 100 households.

(b) Interpret the mean.

(c) Would it be unusual to observe 18 U.S. households that have a high-speed Internet connection in 100 households? Why? *No*

46. Allergy Sufferers Clarinex-D is a medication whose purpose is to reduce the symptoms associated with a variety of allergies. In clinical trials of Clarinex-D, 5% of the patients in the study experienced insomnia as a side effect.

(a) If 240 users of Clarinex-D are randomly selected, how many would we expect to experience insomnia as a side effect? *12*

(b) Would it be unusual to observe 20 patients experiencing insomnia as a side effect in 240 trials of the probability experiment? Why? *Yes*

47. Murder by Firearms According to the *Uniform Crime Report, 2003*, 66.9% of murders are committed with a firearm.

(a) If 100 murders are randomly selected, how many would we expect to be committed with a firearm? *66.9*

(b) Would it be unusual to observe 75 murders by firearm in a random sample of 100 murders? Why? *No*

48. Migraine Sufferers Depakote is a medication whose purpose is to reduce the pain associated with migraine headaches. In clinical trials and extended studies of Depakote, 2% of the patients in the study experienced weight gain as a side effect. Would it be unusual to observe 16 patients who experience weight gain in a random sample of 600 patients who take the medication? Why? *No*

49. Asthma Control Singulair is a medication whose purpose is to control asthma attacks. In clinical trials of Singulair, 18.4% of the patients in the study experienced headaches as a side effect. Would it be unusual to observe 86 patients who experience headaches in a random sample of 400 patients who use this medication? Why? *No*

50. Simulation According to the U.S. National Center for Health Statistics, there is a 98% probability that a 20-year-old male will survive to age 30.

(a) Using statistical software, simulate taking 100 random samples of size 30 from this population.

(b) Using the results of the simulation, compute the probability that exactly 29 of the 30 males survive to age 30.

(c) Compute the probability that exactly 29 of the 30 males survive to age 30, using the binomial probability distribution. Compare the results with part (b). *0.3340*

(d) Using the results of the simulation, compute the probability that at most 27 of the 30 males survive to age 30.

(e) Compute the probability that at most 27 of the 30 males survive to age 30, using the binomial probability distribution. Compare the results with part (d). *0.0217*

(f) Compute the mean number of male survivors in the 100 simulations of the probability experiment. Is it close to the expected value?

(g) Compute the standard deviation of the number of male survivors in the 100 simulations of the probability experiment. Compare the result to the theoretical standard deviation of the probability distribution.

(h) Did the simulation yield any unusual results?

51. Probability Applet Load the binomial applet on your APPLET computer.

(a) Set the probability of success, p, to 0.8 and the number of trials of the binomial experiment, n, to 10. Simulate shooting 10 free throws for $N = 1$. How many were made?

(b) Set the probability of success to 0.8 and the number of trials of the binomial experiment to 10. Simulate shooting 10 free throws $N = 1000$ times. Use the results of the simulation to estimate the probability of making 10 out of 10 free throws.

(c) Use the binomial probability formula to compute the probability of making 10 out of 10 free throws if the probability of success is 0.8. *0.1074*

(d) Use the results of the simulation to estimate the probability of making at least 8 out of 10 free throws.

(e) Use the binomial probability formula to compute the probability of making at least 8 out of 10 free throws. *0.6778*

(f) Determine the mean number of free throws made for the 1000 repetitions of the experiment. Is it close to the expected value?

52. Leisure Activity According to a 2002 survey by the National Endowment for the Arts, 60% of U.S. residents 18 and older attended a movie at least once in the previous year. Suppose you are performing a study and would like at least 12 people in the study to have attended a movie at least once in the past year.

(a) How many residents of the United States 18 years old or older do you expect to have to randomly select? *20*

(b) How many residents of the United States 18 years old or older do you have to randomly select to have a 99% probability that the sample contains at least 12 who have attended a movie in the past year? *30*

53. Educational Attainment According to the U.S. Census Bureau, in 2003 about 27% of residents of the United States 25 years old or older had earned at least a bachelor's degree. Suppose you are performing a study and would like at least 10 people in the study to have earned at least a bachelor's degree.

(a) How many residents of the United States 25 years old or older do you expect to randomly select? *38*

(b) How many residents of the United States 25 years old or older do you have to randomly select to have probability 0.99 that the sample contains at least 10 who have earned at least a bachelor's degree? *65*

(44a) $\mu_x = 160$, $\sigma_x = 5.7$ (45a) $\mu_x = 20$, $\sigma_x = 4$

| Technology Step by Step | **Computing Binomial Probabilities via Technology** |

TI-83/84 Plus   Computing $P(x)$

Step 1: Press 2nd VARS to access the probability distribution menu.

Step 2: Highlight 0: binompdf (for the TI-83 and A: binompdf (for the TI-84 and hit ENTER.

Step 3: With binompdf (on the HOME screen, type the number of trials n, the probability of success, p, and the number of successes, x. For example, with $n = 10$, $p = 0.2$, and $x = 4$, type

$$\texttt{binompdf(10, 0.2, 4)}$$

Then hit ENTER.

Computing $P(X \leq x)$

Step 1: Press 2nd VARS to access the probability distribution menu.

Step 2: Highlight A: binomcdf (for the TI-83 and B: binomcdf (for the TI-84 and hit ENTER.

Step 3: With binomcdf (on the HOME screen, type the number of trials n, the probability of success, p, and the number of successes, x. For example, with $n = 10$, $p = 0.2$, and $x = 4$, type

$$\texttt{binomcdf(10, 0.2, 4)}$$

Then hit ENTER.

MINITAB   Computing $P(x)$

Step 1: Enter the possible values of the random variable x in C1. For example, with $n = 10$, $p = 0.2$, enter $0, 1, 2, \ldots, 10$ into C1.

Step 2: Select the **CALC** menu, highlight **Probability Distributions**, then highlight **Binomial** . . . .

Step 3: Fill in the window as shown to the left. Click OK.

Computing $P(X \leq x)$

Follow the same steps as those for computing $P(x)$. In the window that comes up after selecting Binomial Distribution, select Cumulative probability instead of Probability.

Excel   Computing $P(x)$

Step 1: Select the fx icon. Highlight Statistical in the Function category window. Highlight BINOMDIST in the Function name window.

Step 2: Fill in the window with the appropriate values. For example, if $x = 5$, $n = 10$, and $p = 0.2$, fill in the window as shown in Figure 8(a). Click OK.

Computing $P(X \leq x)$

Follow the same steps as those presented for computing $P(x)$. In the BINOMDIST window, type "TRUE" in the cumulative cell.

Consumer Reports® | Quality Assurance in Customer Relations

The Customer Relations Department at Consumers Union (CU) receives thousands of letters and e-mails from customers each month. Some people write in asking how well a product performed during CU's testing, some people write in sharing their own experiences with their household products, and the remaining people write in for an array of other reasons.

To respond to each letter and e-mail that is received, Customer Relations recently upgraded its customer contact database. Although much of the process has been automated, it still requires employees to manually draft the responses. Given the current size of the department, each Customer Relations representative is required to draft approximately 300 responses each month.

As part of a quality assurance program, the Customer Relations manager would like to develop a plan that allows him to evaluate the performance of his employees. From past experience, he knows that the probability a new employee will write an initial draft of a response that contains errors is approximately 10%. The manager would like to know how many of the 300 responses he should sample to have a cost-effective quality assurance program.

(a) Let X be a discrete random variable that represents the number of the $n = 300$ draft responses that contain errors. Describe the probability distri-bution for X. Be sure to include the name of the probability distribution, possible values for the random variable X, and values of the parameters.

(b) To be effective, suppose the manager would like to have a 95% probability of finding at least one draft document that contains an error. Assuming that the probability that a draft document will have errors is known to be 10%, determine the appropriate sample size to satisfy the manager's requirements. *Hint*: We are required to find the number of draft documents that must be sampled so that the probability of finding at least one document containing an error is 95%. In other words, we have to determine n by solving: $P(X \geq 1) = 0.95$.

(c) Suppose the error rate is really 20%. What sample size will the manager have to review to have a 95% probability of finding one or more documents containing an error?

Note to Readers: In many cases, our test protocol and analytical methods are more complicated than described in these examples. The data and discussions have been modified to make the material more appropriate for the audience.

© by Consumers Union of U.S., Inc., Yonkers, NY 10703-1057, a nonprofit organization. Reprinted with permission.

CHAPTER 6 Review

Summary

In this chapter, we discussed discrete probability distributions. A random variable represents the numerical measurement of the outcome from a probability experiment. Discrete random variables have either a finite or a countable number of outcomes. The term *countable* means that the values result from counting. Probability distributions must satisfy the following two criteria: (1) All probabilities must be between 0 and 1, inclusive, and (2) the sum of all probabilities must equal 1. Discrete probability distributions can be presented in a table, graph, or mathematical formula.

The mean and standard deviation of a random variable describe the center and spread of the distribution. The mean of a random variable is also called its expected value.

We discussed two discrete probability distributions in particular, the binomial and Poisson. A probability experiment is considered a binomial experiment if there are n independent trials of the experiment with only two outcomes. The probability of success, p, is the same for each trial of the experiment. Special formulas exist for computing the mean and standard deviation of a binomial random variable.

Formulas

Mean (or Expected Value) of a Discrete Random Variable
$$\mu_X = E(X) = \sum x P(x)$$

Variance of a Discrete Random Variable
$$\sigma_X^2 = \sum (x - \mu_X)^2 \cdot P(x) = \sum [x^2 P(x)] - \mu_X^2$$

Binomial Probability Distribution Function
$$P(x) = {}_nC_x p^x (1 - p)^{n-x} \quad x = 0, 1, 2, \ldots$$

Mean of a Binomial Random Variable
$$\mu_X = np$$

Standard Deviation of a Binomial Random Variable
$$\sigma_X = \sqrt{np(1 - p)}$$

Vocabulary

Random variable (p. 285)
Discrete random variable (p. 285)
Continuous random variable (p. 285)

Probability distribution (p. 286)
Probability histogram (p. 287)
Expected value (p. 291)

Binomial experiment (p. 298)
Trial (p. 298)
Binomial random variable (p. 298)

Objectives

Section	You should be able to . . .	Examples	Review Exercises
6.1	1 Distinguish between discrete and continuous random variables (p. 285)	1	1, 2
	2 Identify discrete probability distributions (p. 286)	2 and 3	3, 4, 5(a), 6(a)
	3 Construct probability histograms (p. 287)	4	5(b), 6(b)
	4 Compute and interpret the mean of a discrete random variable (p. 288)	5, 6 and 9	5(c), 6(c), 15(b), 16(b)
	5 Interpret the mean of a discrete random variable as an expected value (p. 291)	7	7, 8
	6 Compute the variance and standard deviation of a discrete random variable (p. 292)	8 and 9	5(d), 6(d), 15(b), 16(b)
6.2	1 Determine whether a probability experiment is a binomial experiment (p. 298)	1	9, 10
	2 Compute probabilities of binomial experiments (p. 300)	2 through 5	11(a)–(d), 12(a)–(d), 13(a)–(d), 14(a)–(d)
	3 Compute the mean and standard deviation of a binomial random variable (p. 305)	6	11(e), 12(e), 13(e), 14(e)
	4 Construct binomial probability histograms (p. 306)	7	15(d), 16(d)

Review Exercises

In Problems 1 and 2, determine whether the random variable is discrete or continuous. In each case, state the possible values of the random variable.

1. (a) The number of inches of snow that falls in Buffalo during the winter season. Continuous; $s \geq 0$
   (b) The number of days snow accumulates in Buffalo during the winter season. Discrete; $x = 0, 1, \ldots, 91$
   (c) The number of golf balls hit into the ocean on the famous 18th hole at Pebble Beach on a randomly selected Sunday. Discrete; $x = 0, 1, 2, \ldots$

2. (a) The miles per gallon of gasoline in a 2005 Toyota Sienna. Continuous; $m \geq 0$
   (b) The number of children a randomly selected family has. Discrete, $x = 0, 1, 2, \ldots$
   (c) The number of goals scored by the Edmonton Oilers in a season. Discrete; $x = 0, 1, 2, \ldots$

In Problems 3 and 4, determine whether the distribution is a discrete probability distribution. If not, state why.

3. No, $\sum P(x) \neq 1$

x	$P(x)$
0	0.34
1	0.21
2	0.13
3	0.04
4	0.01

4. Yes

x	$P(x)$
0	0.40
1	0.31
2	0.23
3	0.04
4	0.02

5. **Stanley Cup**   The Stanley Cup is a best-of-seven series to determine the champion of the National Hockey League. The following data represent the number of games played, X, in the Stanley Cup before a champion was determined from 1939 to 2004.

x	Frequency
4	20
5	16
6	17
7	13

Source: Information Please Almanac

   (a) Construct a probability model for the random variable X, the number of games in the Stanley Cup.
   (b) Draw a probability histogram.
   (c) Compute and interpret the mean of the random variable X. $\mu_x = 5.3$
   (d) Compute the standard deviation of the random variable X. $\sigma_x = 1.1$

6. **Property Crime**   In 2003, 77% of crime was property crime, according to the National Crime Victimization Survey. Suppose that four crimes are randomly selected. Let the random variable X represent the number of property crimes.
   (a) Construct a probability model for the random variable X by constructing a tree diagram.
   (b) Draw a probability histogram.

(c) Compute and interpret the mean of the random variable X. $\mu_x = 3.1$
(d) Compute the standard deviation of the random variable X. $\sigma_x = 0.8$

7. Life Insurance Suppose a life insurance company sells a $100,000 one-year term life insurance policy to a 35-year-old male for $200. According to the *National Vital Statistics Report*, Vol. 53, No. 6, the probability that the male survives the year is 0.998592. Compute and interpret the expected value of this policy to the life insurance company. $59.20

8. The Carnival A carnival game is played as follows: You pay $2 to draw a card from an ordinary deck. If you draw an ace, you win $5. You win $3 for a face card and $10 for the seven of spades. If you pick anything else, you lose $2. On average, how much money can the operator expect to make per customer? $0.08

9. Binomial experiment 10. Not binomial

In Problems 9 and 10, determine which of the following probability experiments represents a binomial experiment. If the probability experiment is not a binomial experiment, state why.

9. According to the *Chronicle of Higher Education*, there is a 54% probability that a randomly selected incoming freshman will graduate from college within 6 years. Suppose 10 incoming male freshmen are randomly selected. After 6 years, each student is asked whether he or she graduated.

10. An experiment is conducted in which a single die is cast until a 3 comes up. The number of throws required is recorded.

11. High Cholesterol According to the National Center for Health Statistics, 8% of 20- to 34-year-old females have high serum cholesterol.
(a) In a random sample of 10 females 20 to 34 years old, find the probability that exactly 0 have high serum cholesterol. Interpret this result. 0.4344
(b) In a random sample of 10 females 20 to 34 years old, find the probability that exactly 2 have high serum cholesterol. Interpret this result. 0.1478
(c) In a random sample of 10 females 20 to 34 years old, find the probability that at least 2 have high serum cholesterol. Interpret this result. 0.1879
(d) In a random sample of 10 females 20 to 34 years old, find the probability that exactly 9 will not have high serum cholesterol. Interpret this result. 0.3777
(e) In a random sample of 250 females 20 to 34 years old, what is the expected number with high serum cholesterol? What is the standard deviation? 20; 4.3
(f) If a random sample of 250 females 20 to 34 years old resulted in 12 of them having high serum cholesterol, would this be unusual? Why? No

12. Driving Age According to a Gallup poll conducted December 17 to 19, 2004, 60% of U.S. women 18 years old or older stated that the minimum driving age should be 18 years or older.
(a) In a random sample of 15 U.S. women 18 years old or older, find the probability that exactly 10 believe the minimum driving age should be 18 years or older.
(b) In a random sample of 15 U.S. women 18 years old or older, find the probability that fewer than 5 believe the minimum driving age should be 18 years or older.

(c) In a random sample of 15 U.S. women 18 years old or older, find the probability that at least 5 believe the minimum driving age should be 18 years or older.
(d) In a random sample of 15 U.S. women 18 years old or older, find the probability that exactly 12 *do not* believe the minimum driving age should be 18 years or older.
(e) In a random sample of 200 U.S. women 18 years old or older, what is the expected number who believe the minimum driving age should be 18 years old or older? What is the standard deviation? $\mu_x = 120$; $\sigma_x = 6.928$
(f) If a random sample of 200 U.S. women 18 years old or older resulted in 110 who believed that the minimum driving age should be 18 years or older, would this be unusual? Why? No

13. Nielsen Ratings Nielsen Media Research determines ratings for television programs by placing meters on 5000 televisions throughout the United States. The 2005 NCAA Basketball Championship broadcast resulted in a rating of 15.0, which means 15% of households were tuned into the game.
(a) In a random sample of 20 households, find the probability that exactly 6 were tuned into the 2005 NCAA championship. 0.0454
(b) In a random sample of 20 households, find the probability that fewer than 4 were tuned into the 2005 NCAA championship. 0.6477
(c) In a random sample of 20 households, find the probability that at least 2 were tuned into the 2005 NCAA championship. 0.8244
(d) In a random sample of 20 households, find the probability that exactly 17 were *not* tuned into the 2005 NCAA championship. 0.2428
(e) In a random sample of 500 households, what is the expected number who were tuned into the 2005 NCAA championship? 75
(f) If a random sample of 500 households resulted in 95 that were tuned into the game, would this be unusual? Why? Yes

14. Quit Smoking The drug Zyban is meant to suppress the urge to smoke. In clinical trials, 35% of the study's participants experienced insomnia when taking 300 mg of Zyban per day.
(*Source*: GlaxoSmithKline)
(a) In a random sample of 25 users of Zyban, find the probability that exactly 8 will experience insomnia.
(b) In a random sample of 25 users of Zyban, find the probability that fewer than 4 will experience insomnia.
(c) In a random sample of 25 users of Zyban, find the probability that at least 5 will experience insomnia.
(d) In a random sample of 25 users of Zyban, find the probability that exactly 20 will not experience insomnia.
(e) In a random sample of 1000 users of Zyban, what is the expected number who experience insomnia? What is the standard deviation? 350; 15.1
(f) If a random sample of 1000 users of Zyban results in 330 who experience insomnia, would this be unusual? Why? No

(12a) 0.1859 (12b) 0.0093 (12c) 0.9907 (12d) 0.0016 (14a) 0.1607 (14b) 0.0097 (14c) 0.9680 (14d) 0.0506

In Problems 15 and 16, (a) construct a binomial probability distribution with the given parameters, (b) compute the mean and standard deviation of the random variable by using the methods of Section 6.1, (c) compute the mean and standard deviation by using the methods of Section 6.2, and (d) draw the probability histogram, comment on its shape, and label the mean on the histogram.

15. $n = 5$, $p = 0.2$

16. $n = 8$, $p = 0.75$

17. State the condition required to use the Empirical Rule to check for unusual observations in a binomial experiment.

18. In sampling without replacement, the assumption of independence required for a binomial experiment is violated. Under what circumstances can we sample without replacement and still use the binomial probability formula to approximate probabilities?

THE CHAPTER 6 CASE STUDY IS LOCATED ON THE CD THAT ACCOMPANIES THIS TEXT.

The Normal Probability Distribution

Outline

7.1   Properties of the Normal Distribution

7.2   The Standard Normal Distribution

7.3   Applications of the Normal Distribution

7.4   Assessing Normality

7.5   The Normal Approximation to the Binomial Probability Distribution

■ Chapter Review

■ Case Study: A Tale of Blood, Chemistry, and Health (On CD)

DECISIONS

You are interested in starting your own MENSA-type club. To qualify for the club, the potential member must have intelligence that is in the top 20% of all people. You must decide the baseline score that allows an individual to qualify. See the Decisions project on page 359.

●●● Putting It All Together

In Chapter 6, we introduced discrete probability distributions and, in particular, the binomial probability distribution. We computed probabilities for this discrete distribution using its probability distribution function.

However, we could also determine the probability of any discrete random variable from its probability histogram. For example, the figure shows the probability histogram for the binomial random variable X with $n = 5$ and $p = 0.35$.

From the probability histogram, we can see $P(1) \approx 0.31$. Notice that the width of each rectangle in the probability histogram is 1. Since the area of a rectangle equals height times width, we can think of $P(1)$ as the area of the rectangle corresponding to $X = 1$. Thinking of probability in this fashion makes the transition from computing discrete probabilities to continuous probabilities much easier.

In this chapter, we discuss two continuous distributions, *the uniform distribution* and *the normal distribution*. The greater part of the discussion will focus on the normal distribution, which has many uses and applications.

Binomial Probability Histogram;
n = 5, p = 0.35

7.1 Properties of the Normal Distribution

Preparing for This Section   Before getting started, review the following:

- Continuous variable (Section 1.1, p. 7)
- Rules for a discrete probability distribution (Section 6.1, p. 287)

- *Z*-score (Section 3.4, pp. 149–150)
- The Empirical Rule (Section 3.2, pp. 131–132)

Objectives

1. **Understand the uniform probability distribution**
2. **Graph a normal curve**
3. **State the properties of the normal curve**
4. **Understand the role of area in the normal density function**
5. **Understand the relation between a normal random variable and a standard normal random variable**

1 Understand the Uniform Probability Distribution

We illustrate a uniform distribution using an example. Using the uniform distribution makes it easy to see the relation between area and probability.

EXAMPLE 1   **Illustrating the Uniform Distribution**

Imagine that a friend of yours is always late. Let the random variable X represent the time from when you are supposed to meet your friend until he shows up. Further suppose that your friend could be on time ($x = 0$) or up to 30 minutes late ($x = 30$), with all 1-minute intervals of times between $x = 0$ and $x = 30$ equally likely. That is, your friend is just as likely to be from 3 to 4 minutes late as he is to be 25 to 26 minutes late. The random variable X can be any value in the interval from 0 to 30, that is, $0 \leq x \leq 30$. Because any two intervals of equal length between 0 and 30, inclusive, are equally likely, the random variable X is said to follow a **uniform probability distribution**.

Note to Instructor

To help students understand the relation between area and probability, we begin with a discussion of the uniform distribution. The graph of the uniform distribution is a rectangle. Tell students that we will be using the concept of area in this section and that the area of a rectangle is length times width.

When we compute probabilities for discrete random variables, we usually substitute the value of the random variable into a formula.

Things are not as easy for continuous random variables. Since there are an infinite number of possible outcomes for continuous random variables the probability of observing a particular value of a continuous random variable is zero. For example, the probability that your friend is exactly 12.9438823 minutes late is zero. This result is based on the fact that classical probability is found by dividing the number of ways an event can occur by the total number of possibilities. There is one way to observe 12.9438823, and there are an infinite number of possible values between 0 and 30, so we get a probability that is zero. To resolve this problem, we compute probabilities of continuous random variables over an interval of values. For example, we might compute the probability that your friend is between 10 and 15 minutes late. To find probabilities for continuous random variables, we use *probability density functions*.

Definition

A **probability density function** is an equation used to compute probabilities of continuous random variables that must satisfy the following two properties.

1. The total area under the graph of the equation over all possible values of the random variable must equal 1.

2. The height of the graph of the equation must be greater than or equal to 0 for all possible values of the random variable. That is, the graph of the equation must lie on or above the horizontal axis for all possible values of the random variable.

In Other Words

To find probabilities for continuous random variables, we do not use probability distribution functions (as we did for discrete random variables). Instead, we use probability density functions. The word *density* is used because it refers to the number of individuals per unit of area.

Property 1 is similar to the rule for discrete probability distributions that stated the sum of the probabilities must add up to 1. Property 2 is similar to the rule that stated all probabilities must be greater than or equal to 0.

Figure 1 illustrates the properties for the example about your friend who is always late. Since all possible values of the random variable between 0 and 30 are equally likely, the graph of the probability density function for uniform random variables is a rectangle. Because the random variable is any number between 0 and 30 inclusive, the width of the rectangle is 30. Since the area under the graph of the probability density function must equal 1, and the area of a rectangle equals height times width, the height of the rectangle must be $\frac{1}{30}$.

Figure 1
Uniform Density Function

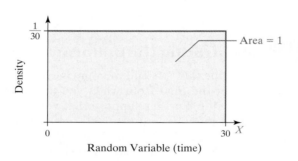

A pressing question remains: How do we use density functions to find probabilities of continuous random variables?

The area under the graph of a density function over some interval represents the probability of observing a value of the random variable in that interval.

The following example illustrates this statement.

EXAMPLE 2 **Area as a Probability**

Problem: Refer to the situation presented in Example 1. What is the probability that your friend will be between 10 and 20 minutes late the next time you meet him?

Approach: Figure 1 presented the graph of the density function. We need to find the area under the graph between 10 and 20 minutes.

Solution: Figure 2 presents the graph of the density function with the area we wish to find shaded in green.

Figure 2

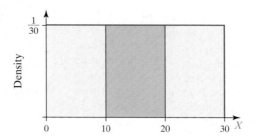

The width of the rectangle is 10 and its height is $\frac{1}{30}$. Therefore, the area between 10 and 20 is $10\left(\frac{1}{30}\right) = \frac{1}{3}$. The probability that your friend is between 10 and 20 minutes late is $\frac{1}{3}$.

Now Work Problem 13.

We introduced the uniform density function so that we could associate probability with area. We are now better prepared to discuss the most popular continuous distribution, the normal distribution.

② Graph a Normal Curve

Many continuous random variables, such as IQ scores, birth weights of babies, or weights of M&Ms, have relative frequency histograms that have a shape similar to Figure 3. Relative frequency histograms that have a shape similar to Figure 3 are said to have the shape of a **normal curve**.

Figure 3

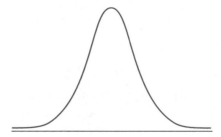

Definition

A continuous random variable is **normally distributed** or has a **normal probability distribution** if its relative frequency histogram of the random variable has the shape of a normal curve.

Figure 4

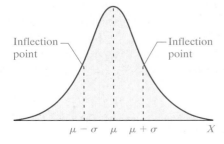

Inflection point — Inflection point

$\mu - \sigma$ μ $\mu + \sigma$ X

Figure 4 shows a normal curve, demonstrating the role μ and σ play in drawing the curve. Look back at Figure 5 on page 113 in Section 3.1. For any distribution, the mode represents the "high point" of the graph of the distribution. The median represents the point where 50% of the area under the distribution is to the left and 50% of the area under the distribution is to the right. The mean represents the balancing point of the graph of the distribution (see Figure 2 on page 109 in Section 3.1). For symmetric distributions, such as the normal distribution, the mean = median = mode. Because of this, the mean, μ, is the "high point" of the graph of the distribution.

The points at $x = \mu - \sigma$ and $x = \mu + \sigma$ are the *inflection points* on the normal curve. The **inflection points** are the points on the curve where the curvature

Historical Note

Karl Pearson coined the phrase *normal curve*. He did not do this to imply that a distribution that is not normal is *abnormal*. Rather, Pearson wanted to avoid giving the name of the distribution a proper name, such as Gaussian (as in Carl Friedrich Gauss).

of the graph changes. To the left of $x = \mu - \sigma$ and to the right of $x = \mu + \sigma$, the curve is drawn upward $\left(\bigcup\text{or}\bigcup\right)$. In between $x = \mu - \sigma$ and $x = \mu + \sigma$, the curve is drawn downward $\left(\bigcap\right)$.*

Figure 5 shows how changes in μ and σ change the position or shape of a normal curve. In Figure 5(a), two normal density curves are drawn with the location of the inflection points labeled. One density curve has $\mu = 0, \sigma = 1$, and the other has $\mu = 3, \sigma = 1$. We can see that increasing the mean from 0 to 3 caused the graph to shift three units to the right. In Figure 5(b), two normal density curves are drawn, again with the inflection points labeled. One density curve has $\mu = 0, \sigma = 1$, and the other has $\mu = 0, \sigma = 2$. We can see that increasing the standard deviation from 1 to 2 causes the graph to become flatter and more spread out.

Figure 5

(a)

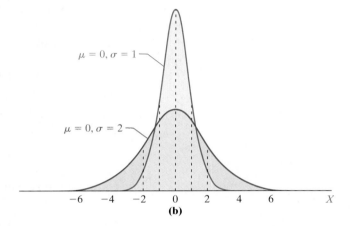

(b)

Now Work Problem 25.

Historical Note

Abraham de Moivre was born in France on May 26, 1667. He is known as a great contributor to the areas of probability and trigonometry. In 1685, he moved to England. De Moivre was elected a fellow of the Royal Society in 1697. He was part of the commission to settle the dispute between Newton and Leibniz regarding who was the discoverer of calculus. He published *The Doctrine of Chance* in 1718. In 1733, he developed the equation that describes the normal curve. Unfortunately, de Moivre had a difficult time being accepted in English society (perhaps due to his accent) and was able to make only a meager living tutoring mathematics. An interesting piece of information regarding de Moivre; he correctly predicted the day of his death, November 27, 1754.

③ **State the Properties of the Normal Curve**

The normal probability density function satisfies all the requirements that are necessary to have a probability distribution. We list the properties of the normal density curve next.

Properties of the Normal Density Curve

1. It is symmetric about its mean, μ.
2. Because mean = median = mode, the highest point occurs at $x = \mu$.
3. It has inflection points at $\mu - \sigma$ and $\mu + \sigma$.
4. The area under the curve is 1.
5. The area under the curve to the right of μ equals the area under the curve to the left of μ, which equals $\frac{1}{2}$.
6. As x increases, without bound (gets larger and larger), the graph approaches, but never reaches, the horizontal axis. As x decreases without bound (gets larger and larger in the negative direction), the graph approaches, but never reaches, the horizontal axis.
7. The Empirical Rule: Approximately 68% of the area under the normal curve is between $x = \mu - \sigma$ and $x = \mu + \sigma$. Approximately 95% of the area under the normal curve is between $x = \mu - 2\sigma$ and $x = \mu + 2\sigma$. Approximately 99.7% of the area under the normal curve is between $x = \mu - 3\sigma$ and $x = \mu + 3\sigma$. See Figure 6.

*The vertical scale on the graph, which indicates density, is purposely omitted. The vertical scale, while important, will not play a role in any of the computations using this curve.

Figure 6

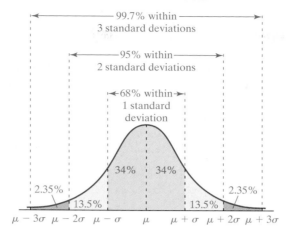

Normal Distribution

4 Understand the Role of Area in the Normal Density Function

Let's look at an example of a normally distributed random variable.

EXAMPLE 3 **A Normal Random Variable**

Problem: The relative frequency distribution given in Table 1 represents the heights of a pediatrician's 200 three-year-old female patients. The raw data indicate that the mean height of the patients is $\mu = 38.72$ inches with standard deviation $\sigma = 3.17$ inches.

(a) Draw a relative frequency histogram of the data. Comment on the shape of the distribution.

(b) Draw a normal curve with $\mu = 38.72$ inches and $\sigma = 3.17$ inches on the relative frequency histogram. Compare the area of the rectangle for heights between 40 and 40.9 inches to the area under the normal curve for heights between 40 and 40.9 inches.

Approach

(a) Draw the relative frequency histogram. If the histogram looks like Figure 4, we say that height is approximately normal. We say "approximately normal," rather than "normal," because the normal curve is an "idealized" description of the data and data rarely follows the curve exactly.

(b) Draw the normal curve on the histogram with the high point at μ and the inflection points at $\mu - \sigma$ and $\mu + \sigma$. Shade the rectangle corresponding to heights between 40 and 40.9 inches, and compare the area of the shaded region to the area under the normal curve between 40 and 40.9.

Solution

(a) Figure 7 shows the relative frequency distribution. The relative frequency histogram is symmetric and bell shaped.

Table 1

Height (Inches)	Relative Frequency
29.0–29.9	0.005
30.0–30.9	0.005
31.0–31.9	0.005
32.0–32.9	0.025
33.0–33.9	0.02
34.0–34.9	0.055
35.0–35.9	0.075
36.0–36.9	0.09
37.0–37.9	0.115
38.0–38.9	0.15
39.0–39.9	0.12
40.0–40.9	0.11
41.0–41.9	0.07
42.0–42.9	0.06
43.0–43.9	0.035
44.0–44.9	0.025
45.0–45.9	0.025
46.0–46.9	0.005
47.0–47.9	0.005

Figure 7 Height of Three-Year-Old Females

! CAUTION

It is rare for a continuous random variable to be exactly normal. Therefore, we usually say that a random variable is approximately normal if its histogram is bell shaped and symmetric.

(b) The normal curve with $\mu = 38.72$ and $\sigma = 3.17$ is superimposed on the relative frequency histogram in Figure 8. The figure demonstrates that the normal curve describes the heights of 3-year-old girls fairly well. We conclude that the heights of 3-year-old girls are approximately normal with $\mu = 38.72$ and $\sigma = 3.17$.

Figure 8 also shows the rectangle corresponding to heights between 40 and 40.9 inches. The area of this rectangle represents the proportion of 3-year-old females between 40 and 40.9 inches. Notice that the area of this shaded region is very close to the area under the normal curve for the same region, so we can use the area under the normal curve to approximate the proportion of 3-year-old females with heights between 40 and 40.9 inches!

Figure 8
Heights of 3-Year-Old Female Patients

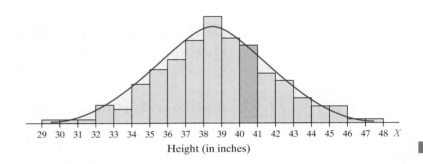

Height (in inches)

In Other Words

Models are not always mathematical. For example, a map can be thought of as a model of a highway system. The model does not show every detail of the highway system (such as traffic lights), but it does serve the purpose of describing how to get from point A to point B. Mathematical models do the same thing: They make assumptions to simplify the mathematics, while still trying to accomplish the goal of accurately describing reality.

The normal curve drawn in Figure 8 is a *model*. In mathematics, a **model** is an equation, table, or graph that is used to describe reality. The normal distribution or normal curve is a model that is used to describe variables that are approximately normally distributed. For example, we saw in Example 3(b) that the normal curve drawn in Figure 8 does a good job of describing the observed distribution of heights of 3-year-old females.

The equation (model) that is used to determine the probability of continuous random variable is called a **probability density function** (or **pdf**). The **normal probability density function** is given by

$$y = \frac{1}{\sigma\sqrt{2\pi}} e^{\frac{-(x-\mu)^2}{2\sigma^2}}$$

where μ is the mean and σ is the standard deviation of the normal random variable. This equation represents the normal distribution. Don't feel threatened by this equation because we will not be using it in this course. Instead, we will use the normal distribution in graphical form by drawing the normal curve, as we did in Figure 5.

We now summarize the role area plays in the normal curve.

The Area under a Normal Curve

Suppose a random variable X is normally distributed with mean μ and standard deviation σ. The area under the normal curve for any interval of values of the random variable X represents either

- the proportion of the population with the characteristic described by the interval of values or
- the probability that a randomly selected individual from the population will have the characteristic described by the interval of values.

In Other Words

The area under a normal curve is a proportion or probability.

EXAMPLE 4

Interpreting the Area under a Normal Curve

Historical Note

The normal probability distribution is often referred to as the Gaussian distribution in honor of Carl Gauss, the individual thought to have discovered the idea. However, it was actually Abraham de Moivre who first wrote down the equation of the normal distribution. Gauss was born in Brunswick, Germany, on April 30, 1777. Mathematical prowess was evident early in Gauss's life. At age 8 he was able to instantly add the first 100 integers. In 1799, Gauss earned his doctorate. The subject of his dissertation was the Fundamental Theorem of Algebra. In 1809, Gauss published a book on the mathematics of planetary orbits. In this book, he further developed the theory of least-squares regression by analyzing the errors. The analysis of these errors led to the discovery that errors follow a normal distribution. Gauss was considered to be "glacially cold" as a person and had troubled relationships with his family. Gauss died on February 23, 1855.

Problem: The serum total cholesterol for males 20 to 29 years old is approximately normally distributed with mean $\mu = 180$ and $\sigma = 36.2$ based on data obtained from the National Health and Nutrition Examination Survey.

(a) Draw a normal curve with the parameters labeled.

(b) An individual with total cholesterol greater than 200 is considered to have high cholesterol. Shade the region under the normal curve to the right of $X = 200$.

(c) Suppose the area under the normal curve to the right of $X = 200$ is 0.2903. (You will learn how to find this area in Section 7.3.) Provide two interpretations of this result.

Approach: **(a)** Draw the normal curve with the mean $\mu = 180$ labeled at the high point and the inflection points at $\mu - \sigma = 180 - 36.2 = 143.8$ and $\mu + \sigma = 180 + 36.2 = 216.2$.

(b) Shade the region under the normal curve to the right of $X = 200$.

(c) The two interpretations of this shaded region are (1) the proportion of 20- to 29-year-old males who have high cholesterol and (2) the probability that a randomly selected 20- to 29-year-old male has high cholesterol.

Solution

(a) Figure 9(a) shows the graph of the normal curve.

Figure 9

(a)

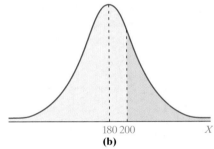

(b)

(b) Figure 9(b) shows the region under the normal curve to the right of $X = 200$ shaded.

(c) The two interpretations for the area of this shaded region are (1) the proportion of 20- to 29-year-old males that have high cholesterol is 0.2903 and (2) the probability that a randomly selected 20- to 29-year-old male has high cholesterol is 0.2903.

> **Now Work Problems 29 and 33.**

5 ## Understand the Relation between a Normal Random Variable and a Standard Normal Random Variable

At this point, we know that a random variable X is approximately normally distributed if its relative frequency histogram has the shape of a normal curve. We use a normal random variable with mean μ and standard deviation σ to model the distribution of X. The area below the normal curve (model of X) represents the proportion of the population with a given characteristic or the probability that a randomly selected individual from the population will have a given characteristic.

In Other Words
The term *normal* refers to the shape of the distribution of a normal random variable.

The question now becomes, "How do I find the area under the normal curve?" Finding the area under a curve requires techniques introduced in calculus, which are beyond the scope of this text. An alternative would be to use a series of tables to find areas. However, this would result in an infinite number of tables being created for each possible mean and standard deviation!

A solution to the problem lies in the Z-score. Recall that the Z-score allows us to transform a random variable X with mean μ and standard deviation σ into a random variable Z with mean 0 and standard deviation 1.

> **Standardizing a Normal Random Variable**
>
> Suppose the random variable X is normally distributed with mean μ and standard deviation σ. Then the random variable
>
> $$Z = \frac{X - \mu}{\sigma}$$
>
> is normally distributed with mean $\mu = 0$ and standard deviation $\sigma = 1$. The random variable Z is said to have the **standard normal distribution**.

In Other Words
To find the area under any normal curve, we first find the Z-score of the normal random variable. Then we use a table to find the area.

This result is powerful! We need only one table of areas corresponding to the standard normal distribution. If a normal random variable has mean different from 0 or standard deviation different from 1, we transform the normal random variable into a standard normal random variable Z, and then we use a table to find the area and, therefore, the probability.

We demonstrate the idea behind standardizing a normal random variable in the next example.

EXAMPLE 5   Relation between a Normal Random Variable and a Standard Normal Random Variable

Problem: The heights of a pediatrician's 200 three-year-old female patients are approximately normal with mean $\mu = 38.72$ inches and $\sigma = 3.17$ inches. We wish to demonstrate that the area under the normal curve between 35 and 38 inches is equal to the area under the standard normal curve between the Z-scores corresponding to heights of 35 and 38 inches.

Approach

Step 1: Draw a normal curve and shade the area representing the proportion of 3-year-old females between 35 and 38 inches tall.

Step 2: Standardize the random variable $X = 35$ and $X = 38$ using

$$Z = \frac{X - \mu}{\sigma}$$

Step 3: Draw the standard normal curve with the standardized versions of $X = 35$ and $X = 38$ labeled. Shade the area that represents the proportion of 3-year-old females between 35 and 38 inches tall. Comment on the relation between the two shaded regions.

Solution

Step 1: Figure 10(a) shows the normal curve with mean $\mu = 38.72$ and $\sigma = 3.17$. The region between $X = 35$ and $X = 38$ is shaded.

Step 2: With $\mu = 38.72$ and $\sigma = 3.17$, the standardized version of $X = 35$ is

$$Z = \frac{X - \mu}{\sigma} = \frac{35 - 38.72}{3.17} = -1.17$$

The standardized version of $X = 38$ is

$$Z = \frac{X - \mu}{\sigma} = \frac{38 - 38.72}{3.17} = -0.23$$

! CAUTION
Recall that we round Z-scores to two decimal places.

Step 3: Figure 10(b) shows the standard normal curve with the region between $Z = -1.17$ and $Z = -0.23$ shaded.

Figure 10

(a) Normal Curve with $\mu = 38.72$ and $\sigma = 3.17$

(b) Standard Normal Curve

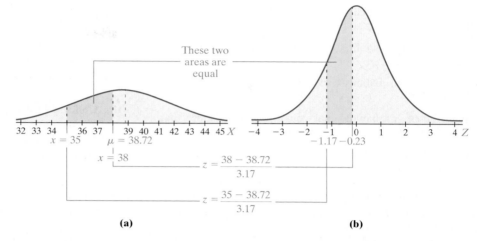

(a)

(b)

The area under the normal curve with $\mu = 38.72$ inches and $\sigma = 3.17$ inches bounded to the left by $X = 35$ and bounded to the right by $X = 38$ is equal to the area under the standard normal curve bounded to the left by $Z = -1.17$ and bounded to the right by $Z = -0.23$.

Now Work Problem 35.

7.1 ASSESS YOUR UNDERSTANDING

Concepts and Vocabulary

1. State the two characteristics of the graph of a probability density function.

2. To find the probabilities for continuous random variables, we do not use probability _____ functions, but instead we use probability _____ functions. *distribution; density*

3. Provide two interpretations of the area under the graph of a probability density function.

4. Why do we standardize normal random variables to find the area under any normal curve?

5. The points at $x =$ _____ and $x =$ _____ are the inflection points on the normal curve. $\mu - \sigma, \mu + \sigma$

6. As σ increases, the normal density curve becomes more spread out. Knowing the area under the density curve must be 1, what effect does increasing σ have on the height of the curve?

For Problems 7–12, determine whether the graph can represent a normal density function. If it cannot, explain why.

7. No

8. Yes

9. No

10. No

11. Yes

12. No

Skill Building

Problems 13–16 use the information presented in Examples 1 and 2.

13. Find the probability that your friend is between 5 and 10
NW minutes late. 1/6

14. Find the probability that your friend is between 15 and 25 minutes late. 1/3

15. Find the probability that your friend is at least 20 minutes late. 1/3

16. Find the probability that your friend is no more than 5 minutes late. 1/6

17. Uniform Distribution The random-number generator on calculators randomly generates a number between 0 and 1. The random variable X, the number generated, follows a uniform probability distribution.

(a) Draw the graph of the uniform density function.
(b) What is the probability of generating a number between 0 and 0.2? 0.2

(c) What is the probability of generating a number between 0.25 and 0.6? 0.35
(d) What is the probability of generating a number greater than 0.95? 0.05
(e) Use your calculator or statistical software to randomly generate 200 numbers between 0 and 1. What proportion of the numbers are between 0 and 0.2? Compare the result with part (b).

18. Uniform Distribution Suppose the reaction time X (in minutes) of a certain chemical process follows a uniform probability distribution with $5 \leq X \leq 10$.

(a) Draw the graph of the density curve.
(b) What is the probability that the reaction time is between 6 and 8 minutes? 2/5
(c) What is the probability that the reaction time is between 5 and 8 minutes? 3/5
(d) What is the probability that the reaction time is less than 6 minutes? 1/5

In Problems 19–22, determine whether or not the histogram indicates that a normal distribution could be used as a model for the variable.

19. Birth Weights The following relative frequency histogram represents the birth weights (in grams) of babies whose term was 36 weeks. Normal

20. Waiting in Line The following relative frequency histogram represents the waiting time in line (in minutes) for the Demon Roller Coaster for 2000 randomly selected people on a Saturday afternoon in the summer. Not normal

21. Length of Phone Calls The following relative frequency histogram represents the length of phone calls on my wife's cell phone during the month of September. Not normal

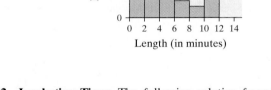

22. Incubation Times The following relative frequency histogram represents the incubation times of a random sample of Rhode Island Red Hens' eggs. Normal

23. One graph in the following figure represents a normal distribution with mean $\mu = 10$ and standard deviation $\sigma = 3$. The other graph represents a normal distribution with mean $\mu = 10$ and standard deviation $\sigma = 2$. Determine which graph is which and explain how you know.

A: $\mu = 10, \sigma = 2$; B: $\mu = 10, \sigma = 3$

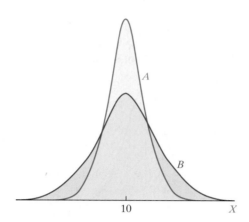

24. One graph in the following figure represents a normal distribution with mean $\mu = 8$ and standard deviation $\sigma = 2$. The other graph represents a normal distribution with mean $\mu = 14$ and standard deviation $\sigma = 2$. Determine which graph is which and explain how you know.

A: $\mu = 8, \sigma = 2$; B: $\mu = 14, \sigma = 2$

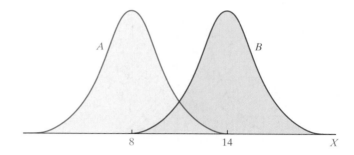

In Problems 25–28, the graph of a normal curve is given. Use the graph to identify the value of μ and σ.

25. $\mu = 2, \sigma = 3$

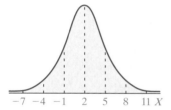

26. $\mu = 5, \sigma = 2$

27. $\mu = 100, \sigma = 15$

28. $\mu = 530, \sigma = 100$

Applying the Concepts

29. Cell Phone Rates Suppose the monthly charge for cell phone plans in the United States is normally distributed with mean $\mu = \$62$ and standard deviation $\sigma = \$18$. (*Source:* Based on information obtained from *Consumer Reports*)

(a) Draw a normal curve with the parameters labeled.

(b) Shade the region that represents the proportion of plans that charge less than \$44.

(c) Suppose the area under the normal curve to the left of $X = \$44$ is 0.1587. Provide two interpretations of this result.

30. Refrigerators Suppose the life of refrigerators is normally distributed with mean $\mu = 14$ years and standard deviation $\sigma = 2.5$ years. (*Source:* Based on information obtained from *Consumer Reports*)

(a) Draw a normal curve with the parameters labeled.

(b) Shade the region that represents the proportion of refrigerators that are kept for more than 17 years.

(c) Suppose the area under the normal curve to the right of $X = 17$ is 0.1151. Provide two interpretations of this result.

31. Birth Weights The birth weights of full-term babies are normally distributed with mean $\mu = 3400$ grams and $\sigma = 505$ grams. (*Source:* Based on data obtained from the *National Vital Statistics Report*, Vol. 48, No. 3)

(a) Draw a normal curve with the parameters labeled.

(b) Shade the region that represents the proportion of full-term babies who weigh more than 4410 grams.

(c) Suppose the area under the normal curve to the right of $X = 4410$ is 0.0228. Provide two interpretations of this result.

32. Height of 10-Year-Old Males The heights of 10-year-old males are normally distributed with mean $\mu = 55.9$ inches and $\sigma = 5.7$ inches.

(a) Draw a normal curve with the parameters labeled.

(b) Shade the region that represents the proportion of 10-year-old males who are less than 46.5 inches tall.

(c) Suppose the area under the normal curve to the left of $X = 46.5$ is 0.0496. Provide two interpretations of this result.

33. Gestation Period The lengths of human pregnancy are NW normally distributed with $\mu = 266$ days and $\sigma = 16$ days.

(a) The following figure represents the normal curve with $\mu = 266$ days and $\sigma = 16$ days. The area to the right of $X = 280$ is 0.1908. Provide two interpretations of this area.

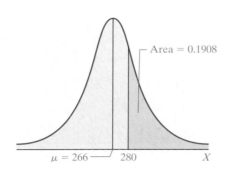

(b) The following figure represents the normal curve with $\mu = 266$ days and $\sigma = 16$ days. The area between $X = 230$ and $X = 260$ is 0.3416. Provide two interpretations of this area.

34. Miles per Gallon Elena conducts an experiment in which she fills up the gas tank on her Toyota Camry 40 times and records the miles per gallon for each fill-up. A histogram of the miles per gallon indicates that a normal distribution with mean of 24.6 miles per gallon and a standard deviation of 3.2 miles per gallon could be used to model the gas mileage for her car.

(a) The following figure represents the normal curve with $\mu = 24.6$ miles per gallon and $\sigma = 3.2$ miles per gallon. The area under the curve to the right of $X = 26$ is 0.3309. Provide two interpretations of this area.

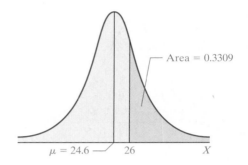

(b) The following figure represents the normal curve with $\mu = 24.6$ miles per gallon and $\sigma = 3.2$ miles per gallon. The area under the curve between $X = 18$ and $X = 21$ is 0.1107. Provide two interpretations of this area.

35. A random variable X is normally distributed with $\mu = 10$ NW and $\sigma = 3$.

(a) Compute $Z_1 = \dfrac{X_1 - \mu}{\sigma}$ for $X_1 = 8$. -0.67

(b) Compute $Z_2 = \dfrac{X_2 - \mu}{\sigma}$ for $X_2 = 12$. 0.67

(c) The area under the normal curve between $X_1 = 8$ and $X_2 = 12$ is 0.495. What is the area between Z_1 and Z_2? 0.495

36. A random variable X is normally distributed with $\mu = 25$ and $\sigma = 6$.

(a) Compute $Z_1 = \dfrac{X_1 - \mu}{\sigma}$ for $X_1 = 18$. -1.17

(b) Compute $Z_2 = \dfrac{X_2 - \mu}{\sigma}$ for $X_2 = 30$. 0.83

(c) The area under the normal curve between $X_1 = 18$ and $X_2 = 30$ is 0.6760. What is the area between Z_1 and Z_2? 0.6760

37. Hitting a Pitching Wedge In the game of golf, distance control is just as important as how far a player hits the ball. Suppose Michael went to the driving range with his range finder and hit 75 golf balls with his pitching wedge and measured the distance each ball traveled (in yards). He obtained the following data:

100	97	101	101	103	100	99	100	100
104	100	101	98	100	99	99	97	101
104	99	101	101	101	100	96	99	99
98	94	98	107	98	100	98	103	100
98	94	104	104	98	101	99	97	103
102	101	101	100	95	104	99	102	95
99	102	103	97	101	102	96	102	99
96	108	103	100	95	101	103	105	100
94	99	95						

(a) Use MINITAB or some other statistical software to construct a relative frequency histogram. Comment on the shape of the distribution.
(b) Use MINITAB or some other statistical software to draw the normal density function on the relative frequency histogram.
(c) Do you think the normal density function accurately describes the distance Michael hits a pitching wedge? Why?

38. Heights of Five-Year-Old Females The following frequency distribution represents the heights (in inches) of eighty randomly selected five-year-old females.

44.5	42.4	42.2	46.2	45.7	44.8	43.3	39.5
45.4	43.0	43.4	44.7	38.6	41.6	50.2	46.9
39.6	44.7	36.5	42.7	40.6	47.5	48.4	37.5
45.5	43.3	41.2	40.5	44.4	42.6	42.0	40.3
42.0	42.2	38.5	43.6	40.6	45.0	40.7	36.3
44.5	37.6	42.2	40.3	48.5	41.6	41.7	38.9
39.5	43.6	41.3	38.8	41.9	40.3	42.1	41.9
42.3	44.6	40.5	37.4	44.5	40.7	38.2	42.6
44.0	35.9	43.7	48.1	38.7	46.0	43.4	44.6
37.7	34.6	42.4	42.7	47.0	42.8	39.9	42.3

(a) Use MINITAB or some other statistical software to construct a relative frequency histogram. Comment on the shape of the distribution.
(b) Use MINITAB or some other statistical software to draw the normal density function on the relative frequency histogram.
(c) Do you think the normal density function accurately describes the heights of 5-year-old females? Why?

7.2 The Standard Normal Distribution

Preparing for This Section   Before getting started, review the following:

• The Complement Rule (Section 5.2, pp. 244–245)

Objectives

1. **Find the area under the standard normal curve**
2. **Find Z-scores for a given area**
3. **Interpret the area under the standard normal curve as a probability**

Note to Instructor

The two main goals of this section are (1) to learn how to find area under the standard normal curve and (2) to learn how to use area to find Z-scores. This can be done through Table IV or technology, whichever you prefer.

In Section 7.1, we introduced the normal distribution. We learned that if X is a normally distributed random variable, we can use the area under the normal density function to obtain the proportion of a population, or the probability that a randomly selected individual from the population has a certain characteristic. To find the area under the normal curve, we first convert the random variable X to a standard normal random variable Z with mean $\mu = 0$ and standard deviation $\sigma = 1$ and find the area under the standard normal curve. This section discusses methods for finding the area under the standard normal curve.

Properties of the Standard Normal Distribution
The standard normal distribution has a mean of 0 and a standard deviation of 1. The standard normal curve, therefore, will have its high point located at 0 and inflection points located at -1 and $+1$. We use the random variable Z to represent a standard normal random variable. The graph of the standard normal curve is presented in Figure 11.

Although we stated the properties of normal curves in Section 7.1, it is worthwhile to restate them here in terms of the standard normal curve.

Figure 11

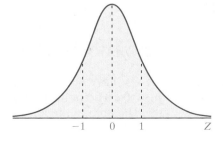

Properties of the Standard Normal Curve

1. It is symmetric about its mean, $\mu = 0$ and has standard deviation $\sigma = 1$.
2. The mean = median = mode = 0. Its highest point occurs at $\mu = 0$.
3. It has inflection points at $\mu - \sigma = 0 - 1 = -1$ and $\mu + \sigma = 0 + 1 = 1$.
4. The area under the curve is 1.
5. The area under the curve to the right of $\mu = 0$ equals the area under the curve to the left of $\mu = 0$, which equals $\frac{1}{2}$.
6. As Z increases, the graph approaches, but never equals, zero. As Z decreases, the graph approaches, but never equals, zero.
7. The Empirical Rule: Approximately $0.68 = 68\%$ of the area under the standard normal curve is between -1 and 1. Approximately $0.95 = 95\%$ of the area under the standard normal curve is between -2 and 2. Approximately $0.997 = 99.7\%$ of the area under the standard normal curve is between -3 and 3. See Figure 12.

Figure 12

We now discuss the procedure for finding area under the standard normal curve.

Figure 13

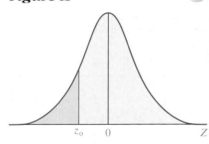

① Find the Area under the Standard Normal Curve

We discuss two methods for finding area under the standard normal curve. The first method uses a table of areas that has been constructed for various values of Z. The second method involves the use of statistical software or a calculator with advanced statistical features.

Table IV, which can be found in the inside back cover of the text or in Appendix A, gives areas under the standard normal curve for values to the left of a specified Z-score, z_0, as shown in Figure 13.

The shaded region represents the area under the standard normal curve to the left of $Z = z_0$. Whenever finding area under a normal curve, you should sketch a normal curve and shade the area you are finding.

EXAMPLE 1 **Finding Area under the Standard Normal Curve to the Left of a Z-Score**

Problem: Find the area under the standard normal curve that lies to the left of $Z = 1.68$.

Approach

Step 1: Draw a standard normal curve with $Z = 1.68$ labeled, and shade the area under the curve to the left of $Z = 1.68$.

Step 2: The rows in Table IV represent the ones and tenths portion of Z, while the columns represent the hundredths portion. To find the area under the curve to the left of $Z = 1.68$, we need to split 1.68 as 1.6 and 0.08. Find the row that represents 1.6 and the column that represents 0.08 in Table IV. Identify where the row and column intersect. This value is the area.

Figure 14

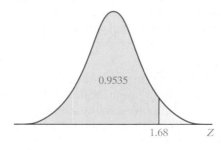

Solution

Step 1: Figure 14 shows the graph of the standard normal curve with $Z = 1.68$ labeled. The area left of $Z = 1.68$ is shaded.

Step 2: A portion of Table IV is presented in Figure 15. We have enclosed the row that represents 1.6 and the column that represents 0.08. The point where the row and column intersect is the area we are seeking. The area to the left of $Z = 1.68$ is 0.9535.

Figure 15

z	.00	.01	.02	.03	.04	.05	.06	.07	.08	.09
0.0	0.5000	0.5040	0.5080	0.5120	0.5160	0.5199	0.5239	0.5279	0.5319	0.5359
0.1	0.5398	0.5438	0.5478	0.5517	0.5557	0.5596	0.5636	0.5675	0.5714	0.5753
0.2	0.5793	0.5832	0.5871	0.5910	0.5948	0.5987	0.6026	0.6064	0.6103	0.6141
0.3	0.6179	0.6217	0.6255	0.6293	0.6331	0.6368	0.6406	0.6443	0.6480	0.6517
1.3	0.9032	0.9049	0.9066	0.9082	0.9099	0.9115	0.9131	0.9147	0.9162	0.9177
1.4	0.9192	0.9207	0.9222	0.9236	0.9251	0.9265	0.9279	0.9292	0.9306	0.9319
1.5	0.9332	0.9345	0.9357	0.9370	0.9382	0.9394	0.9406	0.9418	0.9429	0.9441
1.6	0.9452	0.9463	0.9474	0.9484	0.9495	0.9505	0.9515	0.9525	0.9535	0.9545
1.7	0.9554	0.9564	0.9573	0.9582	0.9591	0.9599	0.9608	0.9616	0.9625	0.9633
1.8	0.9641	0.9649	0.9656	0.9664	0.9671	0.9678	0.9686	0.9693	0.9699	0.9706

The area under a standard normal curve may also be determined using a statistical spreadsheet or graphing calculator with advanced statistical features.

EXAMPLE 2 **Finding the Area under a Standard Normal Curve Using Technology**

Problem: Find the area under the standard normal curve to the left of $Z = 1.68$ using a statistical spreadsheet or graphing calculator with advanced statistical features.

Approach: We will use MINITAB and a TI-84 Plus graphing calculator to find the area. The steps for determining the area under the standard normal curve for MINITAB, Excel, and the TI-83/84 Plus graphing calculators are given in the Technology Step by Step on page 344.

Result: Figure 16(a) shows the results from MINITAB, and Figure 16(b) shows the results from a TI-84 Plus graphing calculator.

Figure 16

Cumulative Distribution Function

Normal with mean = 0 and standard deviation = 1

```
        x        P( X <= x)
      1.68        0.953521
```

```
normalcdf(-1E99,
1.68,0,1)
        .9535213678
```

(a)   (b)

Notice the output for MINITAB is titled Cumulative Distribution Function. Remember, the word *cumulative* means "less than or equal to," so MINITAB is giving the area under the standard normal curve for Z less than or equal to 1.68. The command required by the TI-84 Plus is normalcdf(. The cdf stands for Cumulative Distribution Function. The TI-graphing calculators require a left and right-bound. We use −1E99 for the left bound to obtain areas "less than or equal to" some value.

Now Work Problem 5.

Often, rather than being interested in the area under the standard normal curve to the left of $Z = z_0$, we are interested in obtaining the area under the standard normal curve to the right of $Z = z_0$. The solution to this type of problem uses the fact that the area under the entire standard normal curve is 1 and the Complement Rule. Therefore,

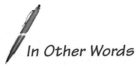

In Other Words

Area right $= 1 -$ area left

$$\left(\begin{array}{c}\text{Area under the normal curve}\\ \text{to the right of } z_0\end{array}\right) = 1 - \left(\begin{array}{c}\text{area to the left}\\ \text{of } z_0.\end{array}\right)$$

$$P(Z > z_0) = 1 - P(Z \le z_0)$$

EXAMPLE 3 **Finding Area under the Standard Normal Curve to the Right of a Z-Score**

Problem: Find the area under the standard normal curve to the right of $Z = -0.46$.

Approach

Step 1: Draw a standard normal curve with $Z = -0.46$ labeled, and shade the area under the curve to the right of $Z = -0.46$.

Step 2: Find the row that represents -0.4 and the column that represents 0.06 in Table IV. Identify where the row and column intersect. This value is the area to the *left* of $Z = -0.46$.

Step 3: The area under the standard normal curve to the right of $Z = -0.46$ is 1 minus the area to the left of $Z = -0.46$.

Figure 17

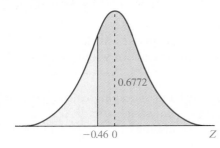

Solution

Step 1: Figure 17 shows the graph of the standard normal curve with $Z = -0.46$ labeled. The area to the right of $Z = -0.46$ is shaded.

Step 2: A portion of Table IV is presented in Figure 18. We have enclosed the row that represents -0.4 and the column that represents 0.06. The point where the row and column intersect is the area to the left of $Z = -0.46$. The area to the left of $Z = -0.46$ is 0.3228.

Figure 18

z	.00	.01	.02	.03	.04	.05	.06	.07	.08	.09
−3.4	0.0003	0.0003	0.0003	0.0003	0.0003	0.0003	0.0003	0.0003	0.0003	0.0002
−3.3	0.0005	0.0005	0.0005	0.0004	0.0004	0.0004	0.0004	0.0004	0.0004	0.0003
−3.2	0.0007	0.0007	0.0006	0.0006	0.0006	0.0006	0.0006	0.0005	0.0005	0.0005
−3.1	0.0010	0.0009	0.0009	0.0009	0.0008	0.0008	0.0008	0.0008	0.0007	0.0007
−3.0	0.0013	0.0013	0.0013	0.0012	0.0012	0.0011	0.0011	0.0011	0.0010	0.0010
−0.5	0.3085	0.3050	0.3015	0.2981	0.2946	0.2912	0.2877	0.2843	0.2810	0.2776
−0.4	0.3446	0.3409	0.3372	0.3336	0.3300	0.3264	0.3228	0.3192	0.3156	0.3121
−0.3	0.3821	0.3783	0.3745	0.3707	0.3669	0.3632	0.3594	0.3557	0.3520	0.3483
−0.2	0.4207	0.4168	0.4129	0.4090	0.4052	0.4013	0.3974	0.3936	0.3897	0.3859
−0.1	0.4602	0.4562	0.4522	0.4483	0.4443	0.4404	0.4364	0.4325	0.4286	0.4247
−0.0	0.5000	0.4960	0.4920	0.4880	0.4840	0.4801	0.4761	0.4721	0.4681	0.4641

Step 3: The area under the standard normal curve to the right of $Z = -0.46$ is 1 minus the area to the left of $Z = -0.46$.

$$\text{Area right of } -0.46 = 1 - (\text{area left of } -0.46)$$
$$= 1 - 0.3228$$
$$= 0.6772$$

USING TECHNOLOGY

Statistical spreadsheets and graphing calculators can also find areas under the standard normal curve to the right of a Z-score.

The area to the right of $Z = -0.46$ is 0.6772.

Now Work Problem 7.

The next example presents a situation in which we are interested in the area between two Z-scores.

EXAMPLE 4 **Find the Area under the Standard Normal Curve between Two Z-Scores**

Problem: Find the area under the standard normal curve between $Z = -1.35$ and $Z = 2.01$.

Approach

Step 1: Draw a standard normal curve with $Z = -1.35$ and $Z = 2.01$ labeled. Shade the area under the curve between $Z = -1.35$ and $Z = 2.01$.

Step 2: Find the area to the left of $Z = -1.35$. Find the area to the left of $Z = 2.01$.

Step 3: The area under the standard normal curve between $Z = -1.35$ and $Z = 2.01$ is the area to the left of $Z = 2.01$ minus the area to the left of $Z = -1.35$.

Solution

Step 1: Figure 19 shows the standard normal curve with the area between $Z = -1.35$ and $Z = 2.01$ shaded.

Step 2: Based upon Table IV, the area to the left of $Z = -1.35$ is 0.0885. The area to the left of $Z = 2.01$ is 0.9778.

Step 3: The area between $Z = -1.35$ and $Z = 2.01$ is

Figure 19

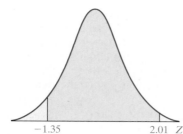

$$(\text{Area between } Z = -1.35 \text{ and } Z = 2.01) = (\text{area left of } Z = 2.01) - (\text{area left of } Z = -1.35)$$

$$= 0.9778 - 0.0885$$
$$= 0.8893$$

The area between $Z = -1.35$ and $Z = 2.01$ is 0.8893.

Now Work Problem 9.

We summarize the methods for obtaining area under the standard normal curve in Table 2 on page 336.

Because the normal curve extends indefinitely in both directions on the Z-axis, there is no Z-value for which the area under the curve to the left of the Z-value is 1. For example, the area to the left of $Z = 10$ is less than 1, even though graphing calculators and statistical software state that the area is 1, because they can compute a limited number of decimal places. We will follow the practice of stating the area to the left of $Z = -3.90$ or to the right of $Z = 3.90$ as <0.0001. The area under the standard normal curve to the left of $Z = 3.90$ or to the right of $Z = -3.90$ will be stated as >0.9999.

! CAUTION

State the area under the standard normal curve to the left of $Z = -3.90$ as <0.0001 (not 0). State the area under the standard normal curve to the left of $Z = 3.90$ as >0.9999 (not 1).

Table 2

Problem	Approach	Solution
Find the area to the left of z_0.	Shade the area to the left of z_0.	Use Table IV to find the row and column that correspond to z_0. The area is the value where the row and column intersect. Or use technology to find the area.
Find the area to the right of z_0.	Shade the area to the right of z_0.	Use Table IV to find the area left of z_0. The area to the right of z_0 is 1 minus the area to the left of z_0. Or use technology to find the area.
Find the area between z_0 and z_1.	Shade the area between z_0 and z_1.	Use Table IV to find the area to the left of z_0 and to the left of z_1. The area between z_0 and z_1 is (area to the left of z_1) − (area to the left of z_0). Or use technology to find the area.

② Find Z-Scores for a Given Area

Up to this point, we have found areas given the value of a Z-score. Often, we are interested in finding a Z-score that corresponds to a given area. The procedure to follow is the reverse of the procedure for finding areas given Z-scores.

EXAMPLE 5 **Finding a Z-Score from a Specified Area to the Left**

Problem: Find the Z-score so that the area to the left of the Z-score is 0.32.

Note to Instructor
If you are using the table to find Z-scores, be sure to emphasize that the area we are looking for may not be in the table. Therefore, the closest area should be used. If the area is equidistant from two areas, the mean Z-score could be used.

Approach

Step 1: Draw a standard normal curve with the area and corresponding unknown Z-score labeled.

Step 2: Look for the area in the table closest to 0.32.

Step 3: Find the Z-score that corresponds to the area closest to 0.32.

Solution

Step 1: Figure 20 shows the graph of the standard normal curve with the area of 0.32 labeled. We know z_0 must be less than 0. Do you know why?

Step 2: We refer to Table IV and look in the body of the table for an area closest to 0.32. The area closest to 0.32 is 0.3192. Figure 21 shows a partial representation of Table IV with 0.3192 labeled.

Figure 20

Area = 0.32
z_0　　　　Z

Figure 21

z	.00	.01	.02	.03	.04	.05	.06	.07	.08	.09
−3.4	0.0003	0.0003	0.0003	0.0003	0.0003	0.0003	0.0003	0.0003	0.0003	0.0002
−3.3	0.0005	0.0005	0.0005	0.0004	0.0004	0.0004	0.0004	0.0004	0.0004	0.0003
−3.2	0.0007	0.0007	0.0006	0.0006	0.0006	0.0006	0.0006	0.0005	0.0005	0.0005
−3.1	0.0010	0.0009	0.0009	0.0009	0.0008	0.0008	0.0008	0.0008	0.0007	0.0007
−3.0	0.0013	0.0013	0.0013	0.0012	0.0012	0.0011	0.0011	0.0011	0.0010	0.0010
−0.7	0.2420	0.2389	0.2358	0.2327	0.2296	0.2266	0.2236	0.2206	0.2177	0.2148
−0.6	0.2743	0.2709	0.2676	0.2643	0.2611	0.2578	0.2546	0.2514	0.2483	0.2451
−0.5	0.3085	0.3050	0.3015	0.2981	0.2946	0.2912	0.2877	0.2843	0.2810	0.2776
−0.4	0.3446	0.3409	0.3372	0.3336	0.3300	0.3264	0.3228	0.3192	0.3156	0.3121
−0.3	0.3821	0.3783	0.3745	0.3707	0.3669	0.3632	0.3594	0.3557	0.3520	0.3483
−0.2	0.4027	0.4168	0.4129	0.4090	0.4052	0.4013	0.3974	0.3936	0.3897	0.3859

Figure 22

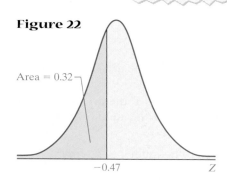

Area = 0.32

−0.47   Z

Step 3: From reading the table in Figure 21, we see that the approximate Z-score that corresponds to an area of 0.32 to its left is −0.47. So $z_0 = -0.47$. See Figure 22.

Statistical spreadsheets or graphing calculators with advanced statistical features can also be used to determine a Z-score corresponding to a specified area.

EXAMPLE 6   **Finding a Z-Score from a Specified Area to the Left Using Technology**

Problem: Find the Z-score such that the area to the left of the Z-score is 0.32 using a statistical spreadsheet or graphing calculator with advanced statistical features.

Approach: We will use Excel and a TI-84 Plus graphing calculator to find the Z-score. The steps for determining the Z-score for MINITAB, Excel, and the TI-83/84 Plus graphing calculators are given in the Technology Step by Step on page 344.

Result: Figure 23(a) shows the results from Excel, and Figure 23(b) shows the results from a TI-84 Plus graphing calculator.

Figure 23

Here is the Z-score

(a)   (b)

The Z-score that corresponds to an area of 0.32 to its left is −0.47, so $z_0 = -0.47$.

Now Work Problem 15.

It is useful to remember that if the area to the left of the Z-score is less than 0.5 the Z-score must be less than 0. If the area to the left of the Z-score is greater than 0.5, the Z-score must be greater than 0.

The next example deals with situations in which the area to the right of some unknown Z-score is given. The solution uses the fact that the area under the normal curve is 1.

EXAMPLE 7 | **Finding a Z-Score from a Specified Area to the Right**

Problem: Find the Z-score so that the area to the right of the Z-score is 0.4332.

Approach

Step 1: Draw a standard normal curve with the area and corresponding unknown Z-score labeled.

Step 2: Determine the area to the left of the unknown Z-score.

Step 3: Look for the area in the table closest to the area determined in Step 2 and record the Z-score that corresponds to the closest area.

! **CAUTION**
To find a Z-score given the area to the right, you must first determine the area to the left if you are using Table IV.

Figure 24

Area = 0.4332

z_0 Z

Solution

Step 1: Figure 24 shows the standard normal curve with the area and unknown Z-score labeled.

Step 2: Since the area under the entire normal curve is 1, the area to the left of the unknown Z-score is 1 minus the area right of the unknown Z-score. Therefore,

$$\text{Area to the left} = 1 - \text{area to the right}$$
$$= 1 - 0.4332$$
$$= 0.5668$$

Step 3: We look in the body of Table IV for an area closest to 0.5668. See Figure 25. The area closest to 0.5668 is 0.5675.

Figure 25

z	.00	.01	.02	.03	.04	.05	.06	.07	.08	.09
0.0	0.5000	0.5040	0.5080	0.5120	0.5160	0.5199	0.5239	0.5279	0.5319	0.5359
0.1	0.5398	0.5438	0.5478	0.5517	0.5557	0.5596	0.5636	0.5675	0.5714	0.5753
0.2	0.5793	0.5832	0.5871	0.5910	0.5948	0.5987	0.6026	0.6064	0.6103	0.6141
0.3	0.6179	0.6217	0.6255	0.6293	0.6331	0.6368	0.6406	0.6443	0.6480	0.6517
0.4	0.6554	0.6591	0.6628	0.6664	0.6700	0.6736	0.6772	0.6808	0.6844	0.6879
0.5	0.6915	0.6950	0.6985	0.7019	0.7054	0.7088	0.7123	0.7157	0.7190	0.7224
0.6	0.7237	0.7291	0.7324	0.7357	0.7389	0.7422	0.7454	0.7486	0.7517	0.7549

The approximate Z-score that corresponds to a right tail area of 0.4332 is 0.17. Therefore, $z_0 = 0.17$. See Figure 26.

Figure 26

Area ≈ 0.5668   Area ≈ 0.4332

0.17 Z

Now Work Problem 19.

In upcoming chapters, we will be interested in finding Z-scores that separate the middle area of the standard normal curve from the area in its tails.

EXAMPLE 8   ## Finding the Z-Score from an Area in the Middle

Problem: Find the Z-score that divides the middle 90% of the area in the standard normal distribution from the area in the tails.

Approach

Step 1: Draw a standard normal curve with the middle $0.9 = 90\%$ of the area separated from the area of $5\% = 0.05$ in each of the two tails. Label the unknown Z-scores z_0 and z_1.

Step 2: Look in the body of Table IV to find the area closest to 0.05.

Step 3: Determine the Z-score in the left tail.

Step 4: The area to the right of z_1 is 0.05. Therefore, the area to the left of z_1 is 0.95. Look in Table IV for an area of 0.95 and find the corresponding Z-value.

Figure 27

Solution

Step 1: Figure 27 shows the standard normal curve with the middle 90% of the area separated from the area in the two tails.

Step 2: We look in the body of Table IV for an area closest to 0.05. See Figure 28.

Figure 28

z	.00	.01	.02	.03	.04	.05	.06	.07	.08	.09
−3.4	0.0003	0.0003	0.0003	0.0003	0.0003	0.0003	0.0003	0.0003	0.0003	0.0002
−3.3	0.0005	0.0005	0.0005	0.0004	0.0004	0.0004	0.0004	0.0004	0.0004	0.0003
−3.2	0.0007	0.0007	0.0006	0.0006	0.0006	0.0006	0.0006	0.0005	0.0005	0.0005
−2.0	0.0228	0.0222	0.0217	0.0212	0.0207	0.0202	0.0197	0.0192	0.0188	0.0183
−1.9	0.0287	0.0281	0.0274	0.0268	0.0262	0.0256	0.0250	0.0244	0.0239	0.0233
−1.8	0.0359	0.0351	0.0344	0.0336	0.0329	0.0322	0.0314	0.0307	0.0301	0.0294
−1.7	0.0446	0.0436	0.0427	0.0418	0.0409	0.0401	0.0392	0.0384	0.0375	0.0367
−1.6	0.0548	0.0537	0.0526	0.0516	0.0505	0.0495	0.0485	0.0475	0.0465	0.0455
−1.5	0.0668	0.0655	0.0643	0.0630	0.0618	0.0606	0.0594	0.0582	0.0571	0.0559

Notice that 0.0495 and 0.0505 are equally close to 0.05. We agree to take the mean of the two Z-scores corresponding to the areas.

Step 3: The Z-score corresponding to an area of 0.0495 is −1.65. The Z-score corresponding to an area of 0.0505 is −1.64. Therefore, the approximate Z-score corresponding to an area of 0.05 to the left is

$$z_0 = \frac{-1.65 + (-1.64)}{2} = -1.645$$

Step 4: The area to the right of z_1 is 0.05. Therefore, the area to the left of $z_1 = 1 - 0.05 = 0.95$. In Table IV, we find an area of 0.9495 corresponding to $z = 1.64$ and an area of 0.9505 corresponding to $z = 1.65$. Consequently, the approximate Z-score corresponding to an area of 0.05 to the right is

$$z_1 = \frac{1.65 + 1.64}{2} = 1.645$$

Figure 29

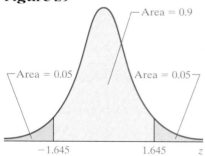

See Figure 29.

Note to Instructor

Remind students that the standard normal curve is symmetric about its mean, O. Therefore, once the left Z-score is known, the right is found by symmetry.

We could also obtain the solution to Example 8 using symmetry. Because the standard normal curve is symmetric about its mean, 0, the Z-score that corresponds to an area to the left of 0.05 will be the additive inverse (i.e., opposite) of the Z-score that corresponds to an area to the right of 0.05. Since the area to the left of $Z = -1.645$ is 0.05, the area to the right of $Z = 1.645$ is also 0.05.

Figure 30

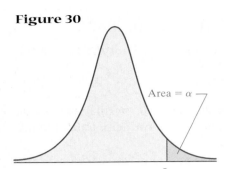

Now Work Problem 23.

We are often interested in finding the Z-score that has a specified area to the right. For this reason, we have special notation to represent this situation.

The notation z_α (pronounced "z sub alpha") is the Z-score such that the area under the standard normal curve to the right of z_α is α. Figure 30 illustrates the notation.

EXAMPLE 9 **Finding the Value of z_α**

Figure 31

Problem: Find the value of $z_{0.10}$.

Approach: We wish to find the Z-value such that the area under the standard normal curve to the right of the Z-value is 0.10.

Solution: The area to the right of the unknown Z-value is 0.10, so the area to the left of the Z-value is $1 - 0.10 = 0.90$. We look in Table IV for the area closest to 0.90. The area closest is 0.8997, which corresponds to a Z-value of 1.28. Therefore, $z_{0.10} = 1.28$. See Figure 31.

Now Work Problem 27.

③ Interpret the Area under the Standard Normal Curve as a Probability

Recall that the area under a normal curve can be interpreted either as a probability or as the proportion of the population with the given characteristic (as represented by an interval of numbers). When interpreting the area under the standard normal curve as a probability, we use the notation introduced in Chapter 6. For example, in Example 8, we found that the area under the standard normal curve to the left of $Z = -1.645$ is 0.05; therefore, the probability of randomly selecting a standard normal random variable that is less than -1.645 is 0.05. We write this statement with the notation $P(Z < -1.645) = 0.05$.

We will use the following notation to denote probabilities of a standard normal random variable, Z.

Notation for the Probability of a Standard Normal Random Variable

$P(a < Z < b)$ represents the probability a standard normal random variable is between a and b.

$P(Z > a)$ represents the probability a standard normal random variable is greater than a.

$P(Z < a)$ represents the probability a standard normal random variable is less than a.

EXAMPLE 10 **Finding Probabilities of Standard Normal Random Variables**

Problem: Evaluate $P(Z < 1.26)$.

Approach

Step 1: Draw a standard normal curve with the area we desire shaded.

Figure 32

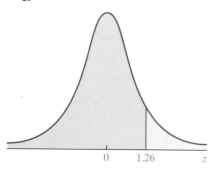

0   1.26   z

Step 2: Use Table IV to find the area of the shaded region. This area represents the probability.

Solution

Step 1: Figure 32 shows the standard normal curve with the area to the left of $Z = 1.26$ shaded.

Step 2: Using Table IV, we find the area under the standard normal curve to the left of $Z = 1.26$ is 0.8962. Therefore, $P(Z < 1.26) = 0.8962$.

Now Work Problem 33.

Note to Instructor

Be sure to emphasize this concept with students.

For any continuous random variable, the probability of observing a specific value of the random variable is 0. For example, for a standard normal random variable, $P(a) = 0$ for any value of a. This is because there is no area under the standard normal curve associated with a single value, so the probability must be 0. So, the following probabilities are equivalent.

$$P(a < Z < b) = P(a \leq Z < b) = P(a < Z \leq b) = P(a \leq Z \leq b)$$

For example, $P(Z < 1.26) = P(Z \leq 1.26) = 0.8962$.

7.2 ASSESS YOUR UNDERSTANDING

Concepts and Vocabulary

1. State the properties of the standard normal curve.
2. If the area under the standard normal curve to the left of $Z = 1.20$ is 0.8849, what is the area under the standard normal curve to the right of $Z = 1.20$? 0.1151

3. *True or False*: The area under the standard normal curve to the left of $Z = 5.30$ is 1. Support your answer.
4. Explain why $P(Z < -1.30) = P(Z \leq -1.30)$.

Skill Building

In Problems 5–12, find the indicated areas. For each problem, be sure to draw a standard normal curve and shade the area that is to be found.

5. Determine the area under the standard normal curve that
NW lies to the left of
  (a) $Z = -2.45$ 0.0071
  (b) $Z = -0.43$ 0.3336
  (c) $Z = 1.35$ 0.9115
  (d) $Z = 3.49$ 0.9998

6. Determine the area under the standard normal curve that lies to the left of
  (a) $Z = -3.49$ 0.0002
  (b) $Z = -1.99$ 0.0233
  (c) $Z = 0.92$ 0.8212
  (d) $Z = 2.90$ 0.9981

7. Determine the area under the standard normal curve that
NW lies to the right of
  (a) $Z = -3.01$ 0.9987
  (b) $Z = -1.59$ 0.9441
  (c) $Z = 1.78$ 0.0375
  (d) $Z = 3.11$ 0.0009

8. Determine the area under the standard normal curve that lies to the right of
  (a) $Z = -3.49$ 0.9998
  (b) $Z = -0.55$ 0.7088
  (c) $Z = 2.23$ 0.0129
  (d) $Z = 3.45$ 0.0003

9. Determine the area under the standard normal curve that
NW lies between
  (a) $Z = -2.04$ and $Z = 2.04$ 0.9586
  (b) $Z = -0.55$ and $Z = 0$ 0.2088
  (c) $Z = -1.04$ and $Z = 2.76$ 0.8479

10. Determine the area under the standard normal curve that lies between
  (a) $Z = -2.55$ and $Z = 2.55$ 0.9892
  (b) $Z = -1.67$ and $Z = 0$ 0.4525
  (c) $Z = -3.03$ and $Z = 1.98$ 0.9749

11. Determine the area under the standard normal curve
   (a) to the left of $Z = -2$ or to the right of $Z = 2$ *0.0456*
   (b) to the left of $Z = -1.56$ or to the right of $Z = 2.56$
   (c) to the left of $Z = -0.24$ or to the right of $Z = 1.20$
(11b) *0.0646* (11c) *0.5203*

12. Determine the area under the standard normal curve
   (a) to the left of $Z = -2.94$ or to the right of $Z = 2.94$
   (b) to the left of $Z = -1.68$ or to the right of $Z = 3.05$
   (c) to the left of $Z = -0.88$ or to the right of $Z = 1.23$
(12a) *0.0032* (12b) *0.0476* (12c) *0.2987*

In Problems 13 and 14, find the area of the shaded region for each standard normal curve.

13. (a) *0.8877*

14. (a) *0.0198*

(b) *0.0500*

(b) *0.3686*

(c) *0.9802*

(c) *0.9500*

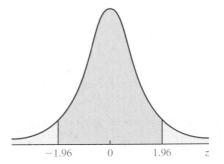

In Problems 15–26, find the indicated Z-score. Be sure to draw a standard normal curve that depicts the solution.

15. Find the Z-score such that the area under the standard normal curve to the left is 0.1. *−1.28*

16. Find the Z-score such that the area under the standard normal curve to the left is 0.2. *−0.84*

17. Find the Z-score such that the area under the standard normal curve to the left is 0.98. *2.05*

18. Find the Z-score such that the area under the standard normal curve to the left is 0.85. *1.04*

19. Find the Z-score such that the area under the standard normal curve to the right is 0.25. *0.67*

20. Find the Z-score such that the area under the standard normal curve to the right is 0.35. *0.39*

21. Find the Z-score such that the area under the standard normal curve to the right is 0.89. *−1.23*

22. Find the Z-score such that the area under the standard normal curve to the right is 0.75. *−0.67*

23. Find the Z-scores that separate the middle 80% of the distribution from the area in the tails of the standard normal distribution. *−1.28; 1.28*

24. Find the Z-scores that separate the middle 70% of the distribution from the area in the tails of the standard normal distribution. *−1.04; 1.04*

25. Find the Z-scores that separate the middle 99% of the distribution from the area in the tails of the standard normal distribution. *−2.575; 2.575*

26. Find the Z-scores that separate the middle 94% of the distribution from the area in the tails of the standard normal distribution. *−1.88; 1.88*

In Problems 27–32, find the value of z_α.

27. $z_{0.05}$ *1.645*

28. $z_{0.35}$ *0.39*

29. $z_{0.01}$ *2.33*

30. $z_{0.02}$ *2.05*

31. $z_{0.20}$ *0.84*

32. $z_{0.15}$ *1.04*

In Problems 33–44, find the indicated probability of the standard normal random variable Z.

33. $P(Z < 1.93)$ *0.9732*

34. $P(Z < -0.61)$ *0.2709*

35. $P(Z > -2.98)$ *0.9986*

36. $P(Z > 0.92)$ *0.1788*

37. $P(-1.20 \le Z < 2.34)$ *0.8753*

38. $P(1.23 < Z \le 1.56)$ *0.0500*

39. $P(Z \ge 1.84)$ *0.0329*

40. $P(Z \ge -0.92)$ *0.8212*

41. $P(Z \le 0.72)$ *0.7642*

42. $P(Z \le -2.69)$ *0.0036*

43. $P(Z < -2.56 \text{ or } Z > 1.39)$ *0.0875*

44. $P(Z < -0.38 \text{ or } Z > 1.93)$ *0.3788*

Applying the Concepts

45. The Empirical Rule The Empirical Rule states that about 68% of the data in a bell-shaped distribution lies within 1 standard deviation of the mean. This means about 68% of the data lie between $Z = -1$ and $Z = 1$. Verify this result. Verify that about 95% of the data lie within 2 standard deviations of the mean. Finally, verify that about 99.7% of the data lie within 3 standard deviations of the mean.

46. According to Table IV, the area under the standard normal curve to the left of $Z = -1.34$ is 0.0901. Without consulting Table IV, determine the area under the standard normal curve to the right of $Z = 1.34$. *0.0901*

47. According to Table IV, the area under the standard normal curve to the left of $Z = -2.55$ is 0.0054. Without consulting Table IV, determine the area under the standard normal curve to the right of $Z = 2.55$. *0.0054*

48. According to Table IV, the area under the standard normal curve between $Z = -1.50$ and $Z = 0$ is 0.4332. Without consulting Table IV, determine the area under the standard normal curve between $Z = 0$ and $Z = 1.50$. *0.4332*

49. According to Table IV, the area under the standard normal curve between $Z = -1.24$ and $Z = -0.53$ is 0.1906. Without consulting Table IV, determine the area under the standard normal curve between $Z = 0.53$ and $Z = 1.24$. *0.1906*

50. (a) Suppose $P(Z < a) = 0.9938$; find a. *2.50*

(b) Suppose $P(Z \ge a) = 0.4404$; find a. *0.15*

(c) Suppose $P(-b < Z < b) = 0.8740$; find b. *1.53*

| Technology Step by Step | **The Standard Normal Distribution** |

TI-83/84 Plus

Finding Areas under the Standard Normal Curve

Step 1: From the HOME screen, press 2nd VARS to access the DISTRibution menu.

Step 2: Select 2:normalcdf(

Step 3: With normalcdf(on the HOME screen, type *lowerbound, upperbound, 0, 1*). For example, to find the area left of $Z = 1.26$ under the standard normal curve, type

Normalcdf(-1E99, 1.26, 0, 1)

and hit ENTER.

Note: When there is no lowerbound, enter $-1E99$. When there is no upperbound, enter 1E99. The E shown is scientific notation; it is $\boxed{2^{nd}}$ $\boxed{,}$ on the keyboard.

Finding Z-Scores Corresponding to an Area

Step 1: From the HOME screen, press 2nd VARS to access the DISTRibution menu.

Step 2: Select 3:invNorm(.

Step 3: With invNorm(on the HOME screen, type *"area left"*, 0, 1). For example, to find the Z-score such that the area under the normal curve left of the Z-score is 0.79, type

InvNorm(0.79, 0 , 1)

and hit ENTER.

MINITAB

Finding Areas under the Standard Normal Curve

Step 1: MINITAB will find an area to the left of a specified Z-score. Select the **Calc** menu, highlight **Probability Distributions**, and highlight **Normal** . . . .

Step 2: Select **Cumulative Probability**. Set the mean to 0 and the standard deviation to 1. Select **Input Constant**, and enter the specified Z-score. Click OK.

Finding Z-Scores Corresponding to an Area

Step 1: MINITAB will find the Z-score for an area to the left of an unknown Z-score. Select the **Calc** menu, highlight **Probability Distributions**, and highlight **Normal** . . . .

Step 2: Select **Inverse Cumulative Probability**. Set the mean to 0 and the standard deviation to 1. Select **Input Constant**, and enter the specified area. Click OK.

Excel

Finding Areas under the Standard Normal Curve

Step 1: Excel will find the area to the left of a specified Z-score. Select the *fx* button from the tool bar. In **Function Category:**, select "Statistical." In **Function Name:**, select "NormsDist." Click OK.

Step 2: Enter the specified Z-score. Click OK.

Finding Z-Scores Corresponding to an Area

Step 1: Excel will find the Z-score for an area to the left of an unknown Z-score. Select the *fx* button from the tool bar. In **Function Category:**, select "Statistical." In **Function Name:**, select "NormsInv." Click OK.

Step 2: Enter the specified area. Click OK.

7.3 Applications of the Normal Distribution

Preparing for This Section Before getting started, review the following:

• Percentiles (Section 3.4, pp. 151–153)

Objectives

① **Find and interpret the area under a normal curve**

② **Find the value of a normal random variable**

① ### Find and Interpret the Area under a Normal Curve

Suppose that a random variable X is normally distributed with mean μ and standard deviation σ. The area below the normal curve represents a proportion or probability.

From the discussions in Section 7.1, we know that finding the area under a normal curve requires that we transform a normal random variable X with mean μ and standard deviation σ into a standard normal random variable Z with mean 0 and standard deviation 1. This is accomplished by letting $Z = \dfrac{X - \mu}{\sigma}$ and using Table IV to find the area under the standard normal curve. This idea is illustrated in Figure 33.

Figure 33

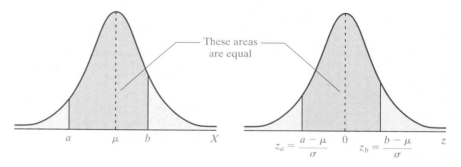

Now that we have the ability to find the area under a standard normal curve, we can find the area under any normal curve. We summarize the procedure next.

Finding the Area under Any Normal Curve

Step 1: Draw a normal curve and shade the desired area.

Step 2: Convert the values of X to Z-scores using $Z = \dfrac{X - \mu}{\sigma}$.

Step 3: Draw a standard normal curve and shade the area desired.

Step 4: Find the area under the standard normal curve. This area is equal to the area under the normal curve drawn in Step 1.

EXAMPLE 1 ### Finding Area under a Normal Curve

Note to Instructor

Be sure to emphasize to students that they draw the nonstandard normal curve and the corresponding standard normal curve to reinforce their conceptual understanding for finding area under any normal curve.

Problem: A pediatrician obtains the heights of her 200 three-year-old female patients. The heights are approximately normally distributed, with mean 38.72 inches and standard deviation 3.17 inches. Use the normal model to determine the proportion of the 3-year-old females that have a height less than 35 inches.

Approach: Follow Steps 1 through 4.

Solution

Step 1: Figure 34 shows the normal curve with the area to the left of 35 shaded.
Step 2: We convert $X = 35$ to a standard normal random variable Z.

$$Z = \frac{X - \mu}{\sigma} = \frac{35 - 38.72}{3.17} = -1.17$$

Step 3: Figure 35 shows the standard normal curve with the area to the left of $Z = -1.17$ shaded. The area to the left of $Z = -1.17$ is equal to the area to the left of $X = 35$.

Figure 34

Figure 35

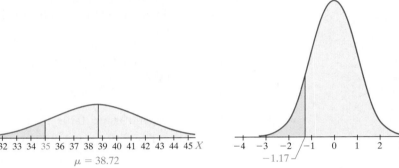

Height (inches)	Relative Frequency
29.0–29.9	0.005
30.0–30.9	0.005
31.0–31.9	0.005
32.0–32.9	0.025
33.0–33.9	0.02
34.0–34.9	0.055
35.0–35.9	0.075
36.0–36.9	0.09
37.0–37.9	0.115
38.0–38.9	0.15
39.0–39.9	0.12
40.0–40.9	0.11
41.0–41.9	0.07
42.0–42.9	0.06
43.0–43.9	0.035
44.0–44.9	0.025
45.0–45.9	0.025
46.0–46.9	0.005
47.0–47.9	0.005

Table 3

Step 4: Using Table IV, we find the area to the left of $Z = -1.17$ is 0.1210. The normal model indicates that the proportion of the pediatrician's 3-year-old females that are less than 35 inches tall is 0.1210.

According to the results of Example 1, the probability that a randomly selected 3-year-old female is shorter than 35 inches is 0.1210. If the normal curve is a good model for determining probabilities (or proportions), then about 12.1% of the 200 three-year-olds in Table 1 should be shorter than 35 inches. For convenience, the information provided in Table 1 is repeated in Table 3.

From the relative frequency distribution in Table 3, we determine that $0.005 + 0.005 + 0.005 + 0.025 + 0.02 + 0.055 = 0.115 = 11.5\%$ of the 3-year-olds are less than 35 inches tall. The results based on the normal curve are in close agreement with the actual results. The normal curve accurately models the heights of 3-year-old females.

Because the area under the normal curve represents a proportion, we can also use the area under the normal curve to find percentile ranks of scores. Recall that the kth percentile divides the lower $k\%$ of a data set from the upper $(100 - k)\%$. In Example 1, 12% of the females have a height less than 35 inches, and 88% of the females have a height greater than 35 inches. Therefore, a child whose height is 35 inches is at the 12th percentile.

Statistical software and graphing calculators with advanced statistical features can also be used to find areas under any normal curve.

EXAMPLE 2   **Finding Area under a Normal Curve Using Technology**

Problem: Find the percentile rank of a 3-year-old female whose height is 43 inches using a statistical spreadsheet or graphing calculator with advanced statistical features. From Example 1, we know the heights are approximately normally distributed with a mean of 38.72 inches and standard deviation of 3.17 inches.

Approach: We will use a TI-84 Plus graphing calculator to find the area. The steps for determining the area under the standard normal curve for MINITAB,

Figure 36

```
normalcdf(-1E99,
43,38.72,3.17)
         .9115172428
```

Excel, and the TI-83/84 Plus graphing calculators are given in the Technology Step by Step on page 353.

Result: Figure 36 shows the results from a TI-84 Plus graphing calculator. The area under the normal curve to the left of 43 is 0.91. Therefore, 91% of the heights are less than 43 inches and 9% of the heights are more than 43 inches. A child whose height is 43 inches is at the 91st percentile.

EXAMPLE 3 Finding the Probability of a Normal Random Variable

Problem: For the pediatrician presented in Example 1, use the normal distribution to compute the probability that a randomly selected 3-year-old female is between 35 and 40 inches tall, inclusive. That is, find $P(35 \leq X \leq 40)$.

Approach: We follow the Steps 1 through 4 on page 345.

Solution

Step 1: Figure 37 shows the normal curve with the area between $X_1 = 35$ and $X_2 = 40$ shaded.

Step 2: Convert the values of $X_1 = 35$ and $X_2 = 40$ to Z-scores.

$$Z_1 = \frac{X_1 - \mu}{\sigma} = \frac{35 - 38.72}{3.17} = -1.17$$

$$Z_2 = \frac{X_2 - \mu}{\sigma} = \frac{40 - 38.72}{3.17} = 0.40$$

Step 3: Figure 38 shows the graph of the standard normal curve with the area between $Z_1 = -1.17$ and $Z_2 = 0.40$ shaded.

Figure 37   **Figure 38**

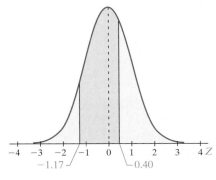

Step 4: Using Table IV, we find that the area to the left of $Z_2 = 0.40$ is 0.6554 and the area to the left of $Z_1 = -1.17$ is 0.1210. Therefore, the area between $Z_1 = -1.17$ and $Z_2 = 0.40$ is $0.6554 - 0.1210 = 0.5344$. We conclude that the probability a randomly selected 3-year-old female is between 35 and 40 inches tall is 0.5344. That is, $P(35 \leq X \leq 40) = P(-1.17 \leq Z \leq 0.40) = 0.5344$. If we randomly selected one 3-year-old female 100 times, we would expect to select a child who is between 35 and 40 inches tall about 53 times.

Now Work Problem 19.

In Other Words

The normal probability density function is used to model random variables that appear to be normal (such as girls' heights). A good model is one that yields results that are close to reality.

According to the relative frequency distribution in Table 3, the proportion of the 200 three-year-old females with heights between 35 inches and 40 inches is $0.075 + 0.09 + 0.115 + 0.15 + 0.12 = 0.55 = 55\%$. This is very close to the probability obtained in Example 3!

Find the Value of a Normal Random Variable

Often, rather than being interested in the proportion or probability of a normal random variable, we are interested in calculating the value of a normal random variable required for the variable to correspond to a certain proportion or probability. For example, we might want to know the height of a 3-year-old girl at the 20th percentile. This means we want to know the height of a 3-year-old girl who is taller than 20% of all 3-year-old girls.

> **Procedure for Finding the Value of a Normal Random Variable Corresponding to a Specified Proportion, Probability, or Percentile**
>
> **Step 1:** Draw a normal curve and shade the area corresponding to the proportion, probability, or percentile.
> **Step 2:** Use Table IV to find the Z-score that corresponds to the shaded area.
> **Step 3:** Obtain the normal value from the fact that $X = \mu + Z\sigma$.*

EXAMPLE 4 | **Finding the Value of a Normal Random Variable**

Problem: The heights of a pediatrician's 200 three-year-old females are approximately normally distributed with mean 38.72 inches and standard deviation 3.17 inches. Find the height of a 3-year-old female at the 20th percentile. That is, find the height of a 3-year-old female that separates the bottom 20% from the top 80%.

Approach: We follow Steps 1 through 3.

Solution

Step 1: Figure 39 shows the normal curve with the unknown value of X separating the bottom 20% of the distribution from the top 80% of the distribution.
Step 2: From Table IV, the area closest to 0.20 is 0.2005. The corresponding Z-score is -0.84.
Step 3: The height of a 3-year-old female that separates the bottom 20% of the data from the top 80% is computed as follows:

$$X = \mu + Z\sigma$$
$$= 38.72 + (-0.84)(3.17)$$
$$= 36.1 \text{ inches}$$

The height of a 3-year-old female at the 20th percentile is 36.1 inches.   ▬

Figure 39

Area = 0.20

Area = 0.80

$X = ?$ X

EXAMPLE 5 | **Finding the Value of a Normal Random Variable Using Technology**

Problem: Use a statistical spreadsheet or a graphing calculator with advanced statistical features to verify the results of Example 4. That is, find the height for a 3-year-old female that is at the 20th percentile, assuming females' heights are approximately normally distributed with a mean of 38.72 inches and a standard deviation of 3.17 inches.

Approach: We will use MINITAB to find the height at the 20th percentile. The steps for determining the area under the standard normal curve for

---

* $Z = \dfrac{X - \mu}{\sigma}$ Formula for standardizing a random variable X
$Z\sigma = X - \mu$ Multiply both sides by σ.
$X = \mu + Z\sigma$ Add μ to both sides.

MINITAB, Excel, and the TI-83/84 Plus graphing calculators are given in the Technology Step by Step on page 353.

Result: Figure 40 shows the results obtained from MINITAB. The height of a three-year-old female at the 20th percentile is 36.1 inches.

Figure 40

Inverse Cumulative Distribution Function

```
Normal with mean = 38.7200 and standard deviation = 3.17000


P ( x <= x)          x
    0.2000      36.0521
```

Now Work Problem 27(a).

EXAMPLE 6 **Finding the Value of a Normal Random Variable**

Problem: The heights of a pediatrician's 200 three-year-old females are approximately normally distributed with mean 38.72 inches and standard deviation 3.17 inches. The pediatrician wishes to determine the heights that separate the middle 98% of the distribution from the bottom 1% and top 1%. In other words, find the 1st and 99th percentiles.

Approach: We follow Steps 1 through 3 given on page 348.

Figure 41

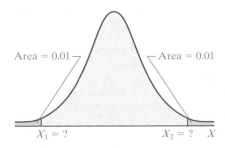

Area = 0.01   Area = 0.01

$X_1 = ?$ $X_2 = ?$ X

Solution

Step 1: Figure 41 shows the normal curve with the unknown values of X separating the bottom and top 1% of the distribution from the middle 98% of the distribution.

Step 2: First, we will find the Z-score that corresponds to the area 0.01 to the left. From Table IV, the area closest to 0.01 is 0.0099. The corresponding Z-score is -2.33. The Z-score that corresponds to the area 0.01 to the right is the Z-score that has the area 0.99 to the left. The area closest to 0.99 is 0.9901. The corresponding Z-score is 2.33.

Step 3: The height of a 3-year-old female that separates the bottom 1% of the distribution from the top 99% is

$$X_1 = \mu + Z\sigma$$
$$= 38.72 + (-2.33)(3.17)$$
$$= 31.3 \text{ inches}$$

The height of a 3-year-old female that separates the top 1% of the distribution from the bottom 99% is

$$X_2 = \mu + Z\sigma$$
$$= 38.72 + (2.33)(3.17)$$
$$= 46.1 \text{ inches}$$

A 3-year-old female whose height is less than 31.3 inches is in the bottom 1% of all 3-year-old females, and a 3-year-old female whose height is more than 46.1 inches is in the top 1% of all 3-year-old females. The pediatrician might use this information to identify those patients who have unusual heights.

Now Work Problem 27(b).

!**CAUTION**

If you are given a value of the random variable and asked to find the probability, proportion, or percentile corresponding to the random variable, convert the random variable to a Z-score using $Z = \dfrac{X - \mu}{\sigma}$ and find the area from the table. If you are using a TI-graphing calculator, use normalcdf.

If you are asked to find the random variable corresponding to a probability, proportion, or percentile, find the area that represents the given probability, proportion, or percentile and use $XZ = \mu + \sigma$ to find the value of the random variable. If you are using a TI-graphing calculator, use invNorm.

7.3 ASSESS YOUR UNDERSTANDING

Concepts and Vocabulary

1. Describe the procedure for finding the area under any normal curve.

2. Describe the procedure for finding the score corresponding to a probability.

Skill Building

In Problems 3–12, assume the random variable X is normally distributed with mean μ = 50 and standard deviation σ = 7. Compute the following probabilities. Be sure to draw a normal curve with the area corresponding to the probability shaded.

3. $P(X > 35)$ 0.9839

4. $P(X > 65)$ 0.0161

5. $P(X \leq 45)$ 0.2375

6. $P(X \leq 58)$ 0.8735

7. $P(40 < X < 65)$ 0.9074

8. $P(56 < X < 68)$ 0.1906

9. $P(55 \leq X \leq 70)$ 0.2354

10. $P(40 \leq X \leq 49)$ 0.3666

11. $P(38 < X \leq 55)$ 0.7192

12. $P(56 \leq X < 66)$ 0.1845

In Problems 13–16, assume the random variable X, is normally distributed with mean μ = 50 and standard deviation σ = 7. Find each indicated percentile for X.

13. The 9th percentile 40.62

14. The 90th percentile 58.96

15. The 81st percentile 56.16

16. The 38th percentile 47.83

Applying the Concepts

17. **Egg Incubation Times** The mean incubation time of fertilized chicken eggs kept at 100.5°F in a still-air incubator is 21 days. Suppose the incubation times are approximately normally distributed with a standard deviation of 1 day. (*Source*: University of Illinois Extension.)

    (a) What is the probability that a randomly selected fertilized chicken egg hatches in less than 20 days? 0.1587
    (b) What is the probability that a randomly selected fertilized chicken egg takes over 22 days to hatch? 0.1587
    (c) What is the probability that a randomly selected fertilized chicken egg hatches between 19 and 21 days? 0.4772
    (d) Would it be unusual for an egg to hatch in less than 18 days? Yes

18. **Medical Residents** In a 2003 study, the Accreditation Council for Graduate Medical Education found that medical residents' mean number of hours worked in a week is 81.7. Suppose the number of hours worked per week by medical residents is approximately normally distributed with a standard deviation of 6.9 hours. (*Source*: www.medrecinst.com)

    (a) What is the probability that a randomly selected medical resident works more than 80 hours in a week? 0.5987
    (b) What is the probability that a randomly selected medical resident works more than 100 hours in a week? 0.0040
    (c) What is the probability that a randomly selected medical resident works less than 60 hours in a week? 0.0008

    (d) Would it be unusual for a medical resident to work less than 70 hours in a week? A little

19. **Chips Ahoy! Cookies** The number of chocolate chips in an 18-ounce bag of Chips Ahoy! chocolate chip cookies is approximately normally distributed with a mean of 1262 chips and standard deviation 118 chips according to a study by cadets of the U.S. Air Force Academy.
    (*Source*: Brad Warner and Jim Rutledge, *Chance*, Vol. 12, No. 1, 1999, pp. 10–14.)

    (a) What is the probability that a randomly selected 18-ounce bag of Chips Ahoy! cookies contains between 1000 and 1400 chocolate chips? 0.8658
    (b) What is the probability that a randomly selected 18-ounce bag of Chips Ahoy! cookies contains fewer than 1000 chocolate chips? 0.0132
    (c) What proportion of 18-ounce bags of Chip Ahoy! cookies contains more than 1200 chocolate chips? 0.7019
    (d) What proportion of 18-ounce bags of Chip Ahoy! cookies contains fewer than 1125 chocolate chips? 0.1230
    (e) What is the percentile rank of an 18-ounce bag of Chip Ahoy! cookies that contains 1475 chocolate chips? 96th percentile
    (f) What is the percentile rank of an 18-ounce bag of Chip Ahoy! cookies that contains 1050 chocolate chips? 4th percentile

20. **Earthquakes** The magnitude of earthquakes since 1900 that measure 0.1 or higher on the Richter scale in California is approximately normally distributed, with μ = 6.2 and σ = 0.5, according to data obtained from the U.S. Geological Survey.

(a) What is the probability that a randomly selected earthquake in California has a magnitude of 6.0 or higher? *0.6554*

(b) What is the probability that a randomly selected earthquake in California has a magnitude less than 6.4? *0.6554*

(c) What is the probability that a randomly selected earthquake in California has a magnitude between 5.8 and 7.1? *0.7522*

(d) The great San Francisco Earthquake of 1906 had a magnitude of 8.25. Is an earthquake of this magnitude unusual in California? *Yes*

(e) What is the percentile rank of a California earthquake that measures 6.8 on the Richter scale?

(f) What is the percentile rank of a California earthquake that measures 5.1 on the Richter scale? *1st percentile*

21. Hybrid Car Introduced in the 2000 model year, the Honda Insight was the first hybrid automobile sold in the United States. The mean gas mileage for the model year 2005 Insight with an automatic transmission is 56 miles per gallon on the highway. Suppose the gasoline mileage of this automobile is approximately normally distributed with a standard deviation of 3.2 miles per gallon. (*Source*: www.fueleconomy.gov)

(a) What proportion of 2005 Honda Insights with automatic transmission gets over 60 miles per gallon on the highway? *0.1056*

(b) What proportion of 2005 Honda Insights with automatic transmission gets 50 miles per gallon or less on the highway? *0.0301*

(c) What proportion of 2005 Honda Insights with automatic transmission gets between 58 and 62 miles per gallon on the highway? *0.2342*

(d) What is the probability that a randomly selected 2005 Honda Insight with an automatic transmission gets less than 45 miles per gallon on the highway? *0.0003*

22. Light Bulbs General Electric manufactures a decorative Crystal Clear 60-watt light bulb that it advertises will last 1500 hours. Suppose the lifetimes of the light bulbs are approximately normally distributed with a mean of 1550 hours and a standard deviation of 57 hours.

(a) What proportion of the light bulbs will last less than the advertised time? *0.1894*

(b) What proportion of the light bulbs will last more than 1650 hours? *0.0401*

(c) What is the probability that a randomly selected GE Crystal Clear 60-watt light bulb lasts between 1625 and 1725 hours? *0.0923*

(d) What is the probability that a randomly selected GE Crystal Clear 60-watt light bulb lasts longer than 1400 hours? *0.9957*

23. Heights of Females As reported by the U.S. National Center for Health Statistics, the mean height of females 20 to 29 years old is $\mu = 64.1$ inches. If height is approximately normally distributed with $\sigma = 2.8$ inches, answer the following questions:

(a) What is the percentile rank of a 20- to 29-year-old female who is 60 inches tall? *7th percentile*

(20e) *88th percentile*

(b) What is the percentile rank of a 20- to 29-year-old female who is 70 inches tall? *98th percentile*

(c) What proportion of 20- to 29-year-old females are between 60 and 70 inches tall? *0.9105*

(d) Would it be unusual for a 20- to 29-year-old female to be taller than 70 inches? *Yes*

24. Gestation Period The length of human pregnancies are approximately normally distributed with mean $\mu = 266$ days and standard deviation $\sigma = 16$ days.

(a) What percent of pregnancies lasts more than 270 days? *40.13%*

(b) What percent of pregnancies lasts less than 250 days? *15.87%*

(c) What percent of pregnancies lasts between 240 and 280 days? *75.90%*

(d) What is the probability that a randomly selected pregnancy lasts more than 280 days? *0.1894*

(e) What is the probability that a randomly selected pregnancy lasts no more than 245 days? *0.0951*

(f) A "very preterm" baby is one whose gestation period is less than 224 days. What proportion of births is "very preterm"? *0.0043*

25. Manufacturing Steel rods are manufactured with a mean length of 25 centimeter (cm). Because of variability in the maufacturing process, the lengths of the rods are approximately normally distributed with a standard deviation of 0.07 cm.

(a) What proportion of rods has a length less than 24.9 cm? *0.0764*

(b) Any rods that are shorter than 24.85 cm or longer than 25.15 cm are discarded. What proportion of rods will be discarded? *0.0324*

(c) Using the results of part (b), if 5000 rods are manufactured in a day, how many should the plant manager expect to discard? *162*

(d) If an order comes in for 10,000 steel rods, how many rods should the plant manager manufacture if the order states that all rods must be between 24.9 cm and 25.1 cm? *11,804*

26. Manufacturing Ball bearings are manufactured with a mean diameter of 5 millimeter (mm). Because of variability in the manufacturing process, the diameters of the ball bearings are approximately normally distributed with a standard deviation of 0.02 mm.

(a) What proportion of ball bearings has a diameter more than 5.03 mm? *0.0668*

(b) Any ball bearings that have a diameter less than 4.95 mm or greater than 5.05 mm are discarded. What proportion of ball bearings will be discarded? *0.0124*

(c) Using the results of part (b), if 30,000 ball bearings are manufactured in a day, how many should the plant manager expect to discard? *372*

(d) If an order comes in for 50,000 ball bearings, how many bearings should the plant manager manufacture if the order states that all ball bearings must be between 4.97 mm and 5.03 mm? *57,711*

27. Egg Incubation Times The mean of the incubation time of fertilized chicken eggs kept at 100.5°F in a still-air incubator is 21 days. Suppose the incubation times are approximately normally distributed with a standard deviation of 1 day.

(*Source*: University of Illinois Extension.)

(a) Determine the 17th percentile for incubation times of fertilized chicken eggs. 20 days

(b) Determine the incubation times that make up the middle 95% of fertilized chicken eggs? 19–23 days

28. **Medical Residents** In a 2003 study, the Accreditation Council for Graduate Medical Education found that medical residents' mean number of hours worked in a week is 81.7. Suppose the number of hours worked per week by medical residents is approximately normally distributed with a standard deviation of 6.9 hours.

(*Source*: www.medrecinst.com)

(a) Determine the 75th percentile for the number of hours worked in a week by medical residents. 86.3 hours

(b) Determine the number of hours worked in a week that makes up the middle 80% of medical residents.

29. **Chips Ahoy! Cookies** The number of chocolate chips in an 18-ounce bag of Chips Ahoy! chocolate chip cookies is approximately normally distributed with a mean of 1262 chips and a standard deviation of 118 chips, according to a study by cadets of the U.S. Air Force Academy.

(*Source*: Brad Warner and Jim Rutledge, *Chance*, Vol. 12, No. 1, 1999, pp. 10–14.)

(a) Determine the 30th percentile for the number of chocolate chips in an 18-ounce bag of Chips Ahoy! cookies. 1201 chips

(b) Determine the number of chocolate chips in a bag of Chips Ahoy! that make up the middle 99% of bags.

30. **Earthquakes** The magnitude of earthquakes since 1900 that measure 0.1 or higher on the Richter scale in California is approximately normally distributed with $\mu = 6.2$ and $\sigma = 0.5$, according to data obtained from the U.S. Geological Survey.

(a) Determine the 40th percentile of the magnitude of earthquakes in California. 6.1

(b) Determine the magnitude of earthquakes that make up the middle 85% of magnitudes. 5.5; 6.9

31. **Hybrid Car** Introduced in the 2000 model year, the Honda Insight was the first hybrid automobile sold in the United States. The mean was mileage for the model year 2005 Insight with an automatic transmission is 56 miles per gallon on the highway. Suppose the gasoline mileages of these automobiles are approximately normally distributed with standard deviation 3.2 miles per gallon.

(*Source*: www.fuel.economy.gov)

(a) Determine the 97th percentile gasoline mileage for the model year 2005 Insight with an automatic transmission. 62.0 mpg

(b) Determine the mileage that makes up the middle 86% gasoline mileage of model year 2005 Insight with an automatic transmission. 51.3–60.7 mpg

32. **Speedy Lube** The time required for Speedy Lube to complete an oil change service on an automobile approximately follows a normal distribution, with a mean of 17 minutes and a standard deviation of 2.5 minutes.

(a) Speedy Lube guarantees customers that the service will take no longer than 20 minutes. If it does take longer, the customer will receive the service for half-price. What percent of customers receives the service for half price? 11.51%

(b) If Speedy Lube does not want to give the discount to more than 3% of its customers, how long should it make the guaranteed time limit? 22 minutes

33. **Multiple Births** The following data represent the relative frequencies of live multiple-delivery births (three or more babies) in 2002 for women 15 to 44 years old.

Age	Relative Frequency
15–19	0.0128
20–24	0.0702
25–29	0.2235
30–34	0.3888
35–39	0.2530
40–44	0.0518

Source: National Vital Statistics Reports, Vol. 52, No. 10, December 17, 2003

Suppose the ages of multiple-birth mothers are approximately normally distributed with $\mu = 31.77$ years and standard deviation $\sigma = 5.19$ years.

(a) Compute the proportion of multiple-birth mothers in each class by finding the area under the normal curve.

(b) Compare the proportion to the actual proportions. Are you convinced that the ages of multiple-birth mothers are approximately normally distributed?

34. **Weather in Chicago** The following frequency distribution represents the daily high temperature in Chicago, November 16 to 30, for the years 1872 to 1999:

(a) Construct a relative frequency distribution.

(b) Draw a relative frequency histogram. Does the distribution of high temperatures appear to be normal?

(c) Compute the mean and standard deviation of high temperature. $\mu = 43.5°, \sigma = 10.5°$

(d) Use the information obtained in part (c) to compute the proportion of high temperatures in each class by finding the area under the normal curve.

(e) Are you convinced that high temperatures are approximately normally distributed?

Temperature	Frequency	Temperature	Frequency
5.0–9.9	1	40.0–44.9	375
10.0–14.9	10	45.0–49.9	281
15.0–19.9	15	50.0–54.9	233
20.0–24.9	40	55.0–59.9	160
25.0–29.9	95	60.0–64.9	101
30.0–34.9	217	65.0–69.9	21
35.0–39.9	325	70.0–74.9	1

Source: *Chicago Tribune*. November 27, 2000

(28b) 72.9–90.5 hours   (29b) 958–1566 chips

Consumer Reports®   Sunscreens

More than 1 million skin cancers are expected to be diagnosed in the United States this year, almost half of all cancers diagnosed. The prevalence of skin cancer is attributable in part to a history of unprotected or underprotected sun exposure. Sunscreens have been shown to prevent certain types of lesions associated with skin cancer. They also protect skin against exposure to light that contributes to premature aging. As a result, sunscreen is now in moisturizers, lip balms, shampoos, hairstyling products, insect repellents, and makeup.

Consumer Reports tested 23 sunscreens and two moisturizers, all with a claimed sun-protection factor (SPF) of 15 or higher. SPF is defined as the degree to which a sunscreen protects the skin from UVB, the ultraviolet rays responsible for sunburn. (Some studies have shown that UVB, along with UVA, can increase the risk of skin cancers.) A person with untreated skin who can stay in the sun for 5 minutes before becoming sunburned should be able to stay in the sun for $15 \times 5 = 75$ minutes using a sunscreen rated at SPF15.

To test whether products met their SPF claims for UVB, we used a solar simulator (basically a sun lamp) to expose people to measured amounts of sunlight. First we determined the exposure time (in minutes) that caused each person's untreated skin to turn pink within 24 hours. Then we applied sunscreen to new areas of skin and made the same determination. To avoid potential sources of bias, samples of the sunscreens were applied to randomly assigned sites on the subjects' skin.

To determine the SPF rating of a sunscreen for a particular individual, the exposure time with sunscreen was divided by the exposure time without sunscreen.

The following table contains the mean and standard deviation of the SPF measurements for two particular sunscreens.

Product	Mean	Std Dev
A	15.5	1.5
B	14.7	1.2

(a) In designing this experiment, why is it important to obtain the exposure time without sunscreen first and then determine the exposure time with sunscreen for each person?

(b) Why is the random assignment of people and application sites to each treatment (the sunscreen) important?

(c) Calculate the percentage of SPF measurements that you expect to be less than 15, the advertised level of protection for each of the products. (Assume that the SPF ratings are approximately normal.)

(d) Calculate the percentage of SPF measurements that you expect to be greater than 17.5 for product A. Repeat this for product B.

(e) Calculate the percentage of SPF measurements that you expect to fall between 14.5 and 15.5 for product A. Repeat this for product B.

(f) Which product appears to be superior, A or B? Support your conclusion.

© by Consumers Union of U.S., Inc., Yonkers, NY 10703-1057, a nonprofit organization. Reprinted with permission.

Note to Readers: In many cases, our test protocol and analytical methods are more complicated than described in these examples. The data and discussions have been modified to make the material more appropriate for the audience.

Technology Step by Step	**The Normal Distribution**

TI-83/84 Plus   Finding Areas under the Normal Curve

Step 1: From the HOME screen, press 2nd VARS to access the DISTRibution menu.

Step 2: Select `2:normalcdf(`

Step 3: With `normalcdf(` on the HOME screen, type *lowerbound, upperbound*, μ, σ). For example, to find the area to the left of $X = 35$ under the normal curve with $\mu = 40$ and $\sigma = 10$, type

```
Normalcdf(-1E99, 35, 40, 10)
```

and hit ENTER.

Note: When there is no lowerbound, enter −1E99. When there is no upperbound, enter 1E99. The E shown is scientific notation; it is ⌈2nd⌉ ⌈ ' ⌉ .

Finding Scores Corresponding to an Area

Step 1: From the HOME screen, press 2nd VARS to access the DISTRibution menu.

Step 2: Select `3:invNorm(`

Step 3: With `invNorm(` on the HOME screen, type *"area left"*, μ, σ). For example, to find the score such that the area under the normal curve to the left of the score is 0.68 with $\mu = 40$ and $\sigma = 10$, type

```
InvNorm(0.68, 40, 10)
```

and hit ENTER.

MINITAB Finding Areas under the Normal Curve

Step 1: Select the **Calc** menu, highlight **Probability Distributions**, and highlight **Normal** ....

Step 2: Select **Cumulative Probability**. Enter the mean, μ, and the standard deviation, σ. Select **Input Constant**, and enter the observation. Click OK.

Finding Scores Corresponding to an Area

Step 1: Select the **Calc** menu, highlight **Probability Distributions**, and highlight **Normal** ....

Step 2: Select **Inverse Cumulative Probability**. Enter the mean, μ, and the standard deviation, σ. Select **Input Constant**, and enter the area left of the unknown score. Click OK.

Excel Finding Areas under the Normal Curve

Step 1: Select the *fx* button from the tool bar. In **Function Category**:. select "Statistical." In **Function Name**:, select NormDist. Click OK.

Step 2: Enter the specified observation, μ, and σ, and set **cumulative** to True. Click OK.

Finding Scores Corresponding to an Area

Step 1: Select the *fx* button from the tool bar. In **Function Category**:, select "Statistical." In **Function Name**:, select NormInv. Click OK.

Step 2: Enter the area left of the unknown score, μ, and σ. Click OK.

7.4 Assessing Normality

Preparing for This Section Before getting started, review the following:

• Shape of a distribution (Section 3.1, pp. 113–116)

Objectives **Draw normal probability plots to assess normality**

Note to Instructor
The goal of this section is for students to be able to understand what a normal probability plot is. In addition, we want students to be able to read the output provided by a computer or calculator. Do not emphasize the by-hand solution. Rather, emphasize how to interpret normal probability plots.

Suppose that we obtain a simple random sample from a population whose distribution is unknown. Many of the statistical tests that we perform on small data sets (sample size less than 30) require that the population from which the sample is drawn be normally distributed.

Up to this point, we have said that a random variable X is normally distributed, or at least approximately normal, provided the histogram of the data is symmetric and bell shaped. This method works well for large data sets, but the shape of a histogram drawn from a small sample of observations does not always accurately represent the shape of the population. For this reason, we need additional methods for assessing the normality of a random variable X when we are looking at a small set of sample data.

In Other Words

Normal probability plots are used to assess normality in small data sets.

① Draw Normal Probability Plots to Assess Normality

A **normal probability plot** is a graph that plots observed data versus *normal scores*. A **normal score** is the expected Z-score of the data value assuming the distribution of the random variable is normal. The expected Z-score of an observed value depends on the number of observations in the data set.

To draw a normal probability plot requires the following steps:

> **Drawing a Normal Probability Plot**
>
> **Step 1:** Arrange the data in ascending order.
>
> **Step 2:** Compute $f_i = \dfrac{i - 0.375}{n + 0.25}$,* where i is the index (the position of the data value in the ordered list) and n is the number of observations. The expected proportion of observations less than or equal to the ith data value is f_i.
>
> **Step 3:** Find the Z-score corresponding to f_i from Table IV.
>
> **Step 4:** Plot the observed values on the horizontal axis and the corresponding expected Z-scores on the vertical axis.

Figure 42

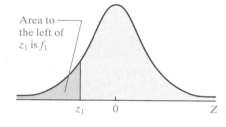

Area to the left of z_1 is f_1

The idea behind finding the expected Z-score is that, if the data come from a population that is normally distributed, we should be able to predict the area to the left of each data value. The value of f_i represents the expected area to the left of the ith observation when the data come from a population that is normally distributed. For example, f_1 is the expected area to the left of the smallest data value, f_2 is the expected area to the left of the second-smallest data value, and so on. Figure 42 illustrates the idea.

Once we determine each f_i, we find the Z-scores corresponding to f_1, f_2, and so on. The smallest observation in the data set will be the smallest expected Z-score, and the largest observation in the data set will be the largest expected Z-score. Also, because of the symmetry of the normal curve, the expected Z-scores are always paired as positive and negative values.

Normal random variables X and their Z-scores are linearly related $(X = \mu + Z\sigma)$, so a plot of observations of normal variables against their expected normal scores will be linear. We conclude the following:

> If sample data are taken from a population that is normally distributed, a normal probability plot of the observed values versus the expected Z-scores will be approximately linear.

Normal probability plots are typically drawn using graphing calculators or statistical software. However, it is worthwhile to go through an example that demonstrates the procedure so that we can better understand the results supplied by software.

EXAMPLE 1   **Constructing a Normal Probability Plot**

Problem: The data in Table 4 represent the finishing time (in seconds) for six randomly selected races of a greyhound named Barbies Bomber in the $\frac{5}{16}$-mile race at Greyhound Park in Dubuque, Iowa. Is there evidence to support the belief that the variable "finishing time" is normally distributed?

Approach: We follow Steps 1 through 4 listed above.

Table 4	
31.35	32.52
32.06	31.26
31.91	32.37

Source: Greyhound Park, Dubuque, IA

*The derivation of this formula is beyond the scope of this text.

Solution

Step 1: The first column in Table 5 represents the index i. The second column represents the observed values in the data set, written in ascending order.

Step 2: The third column in Table 5 represents $f_i = \dfrac{i - 0.375}{n + 0.25}$ for each observation. This is the expected area under the normal curve to the left of the ith observation, assuming normality. For example, $i = 1$ corresponds to the finishing time of 31.26, and

$$f_1 = \frac{1 - 0.375}{6 + 0.25} = 0.1$$

So the area under the normal curve to the left of 31.26 is 0.1 if the sample data come from a population that is normally distributed.

Step 3: We use Table IV to find the Z-scores that correspond to f_i. The expected Z-scores are listed in the fourth column of Table 5. Look in Table IV for the area closest to $f_1 = 0.1$. The expected Z-score is -1.28. Notice that for each negative expected Z-score there is a corresponding positive expected Z-score, as a result of the symmetry of the normal curve.

Note to Instructor

As an alternative to using MINITAB to assess the linearity of the normal probability plot, we can use the linear correlation coefficient, using the following steps.
Step 1: Compute the linear correlation between the data and the expected Z-scores.
Step 2: Compare the linear correlation coefficient to the critical values supplied below ($\alpha = 0.05$). If the linear correlation coefficient is greater than or equal to the critical value, there is evidence to support the belief that the data come from a population that is normally distributed.

Sample Size, n	Critical Value
5	0.880
6	0.888
7	0.898
8	0.906
9	0.912
10	0.918
11	0.923
12	0.928
13	0.932
14	0.935
15	0.939
16	0.941
17	0.944
18	0.946
19	0.949
20	0.951
21	0.952
22	0.954
23	0.956
24	0.957
25	0.959
30	0.960

Source: S. W. Looney and T. R. Gulledge, Jr. Use of the Correlation Coefficient with Normal Probability Plots, *American Statistician*, Vol. 39 (Feb. 1985), pp. 75–79.

\multicolumn{4}{c}{**Table 5**}
Index, i	**Observed Value**	**f_i**	**Expected Z-score**
1	31.26	$\dfrac{1 - 0.375}{6 + 0.25} = 0.1$	-1.28
2	31.35	$\dfrac{2 - 0.375}{6 + 0.25} = 0.26$	-0.64
3	31.91	0.42	-0.20
4	32.06	0.58	0.20
5	32.37	0.74	0.64
6	32.52	0.9	1.28

Step 4: We plot the actual observations on the horizontal axis and the expected Z-scores on the vertical axis. See Figure 43.

Figure 43
Normal Probability Plot

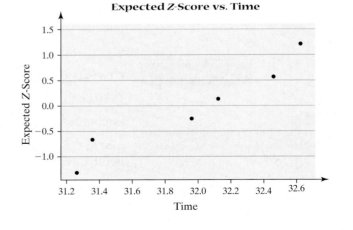

Expected Z-Score vs. Time

Although the normal probability plot in Figure 43 does show some curvature, it is roughly linear.* We conclude that the finishing time of Barbies Bomber in the $\frac{5}{16}$-mile race is approximately normally distributed.

Typically, normal probability plots are drawn using either a graphing calculator with advanced statistical features or statistical software. Certain software, such as MINITAB, provides bounds that the data must lie within to support the belief that the sample data come from a population that is normally distributed.

EXAMPLE 2   ### Assessing Normality Using Technology

Problem: Using MINITAB or some other statistical software, draw a normal probability plot of the data in Table 4 and determine whether there is evidence to support the belief that the sample data come from a population that is normally distributed.

Approach: We will construct a normal probability plot using MINITAB. MINITAB provides curved *bounds* that can be used to assess normality. If the normal probability plot is roughly linear and all the data lie within the bounds provided by the software, we have reason to believe the data come from a population that is approximately normal. The steps for constructing normal probability plots using MINITAB, Excel, or the TI-83/84 Plus graphing calculators can be found on page 361.

Solution: Figure 44 shows the normal probability plot. Notice that MINITAB gives area to the left of the expected Z-score, rather than the Z-score. For example, the area to the left of the expected Z-score of -1.28 is 0.10. MINITAB writes 0.10 as 10 percent.

Figure 44

The normal probability plot is roughly linear, and all the data lie within the bounds provided by MINITAB. We conclude that the sample data could come from a population that is normally distributed.

Throughout the text, we will provide normal probability plots drawn with MINITAB so that assessing normality is straightforward.

---

*In fact, the correlation between the observed value and expected Z-score is 0.970.

EXAMPLE 3 Assessing Normality

Problem: The data in Table 6 represent the time spent waiting in line (in minutes) for the Demon Roller Coaster for 100 randomly selected riders. Is the random variable "waiting time" normally distributed?

Table 6																
7	3	5	107	8	37	16	41	7	25	22	19	1	40	1	29	93
33	76	14	8	9	45	15	81	94	10	115	18	0	18	11	60	34
30	6	21	0	86	6	11	1	1	3	9	79	41	2	9	6	19
4	3	2	7	18	0	93	68	6	94	16	13	24	6	12	121	30
35	39	9	15	53	9	47	5	55	64	51	80	26	24	12	0	
94	18	4	61	38	38	21	61	9	80	18	21	8	14	47	56	

Approach: We will use MINITAB to draw a normal probability plot. If the normal probability plot is roughly linear and the data lie within the bounds provided by MINITAB, conclude that it is reasonable to believe that the sample data come from a population that follows a normal distribution.

Solution: Figure 45 shows a normal probability plot of the data drawn using MINITAB. Since the normal probability plot is not linear, the random variable "wait time" is not normally distributed. Figure 46 shows a histogram of the data in Table 6. The histogram indicates that the data are skewed right.

Figure 45

Figure 46

Now Work Problems 3 and 7.

MAKING AN INFORMED DECISION

Join the Club

Suppose that you are interested in starting your own MENSA-type club. To qualify for the club, the potential member must have intelligence that is in the top 20% of all people. The problem that you face is that you do not have a baseline for measuring what qualifies as a top 20% score. To gather these data, you must obtain a random sample of at least 25 volunteers to take an online intelligence test. There are many online intelligence tests, but you need to make sure that the test will supply scored exams. One suggested site is www.queendom.com/tests/iq/classical_iq_r2_access.html.

Once you have obtained your sample of at least 25 test scores, answer the following questions.

(a) What is the mean test score? What is the standard deviation of the test scores?

(b) Do the sample data come from a population that is normally distributed? How do you know this?

(c) Assuming that the sample data come from a population that is normally distributed, determine the test score that would be required to join your club. That is, determine the test score that serves as a cutoff point for the top 20%. You can use this score to determine which potential members may join!

7.4 ASSESS YOUR UNDERSTANDING

Concepts and Vocabulary

1. Explain why normal probability plots should be linear if the data are normally distributed.

2. What does f_i represent?

Skill Building

In Problems 3–8, determine whether the normal probability plot indicates that the sample data could have come from a population that is normally distributed.

3. Not normal

NW

4. Normal

5. Not normal

6. Not normal

7. Normal

NW

8. Normal

Applying the Concepts

9. Chips per Bag In a 1998 advertising campaign, Nabisco claimed that every 18-ounce bag of Chips Ahoy! cookies contained at least 1000 chocolate chips. Brad Warner and Jim Rutledge (*Chance*, Vol. 12, No. 1, 1999) tried to verify the claim. The following data represent the number of chips in an 18-ounce bag of Chips Ahoy! based on their study.

1087	1098	1103	1121	1132
1185	1191	1199	1200	1213
1239	1244	1247	1258	1269
1307	1325	1345	1356	1363
1135	1137	1143	1154	1166
1214	1215	1219	1219	1228
1270	1279	1293	1294	1295
1377	1402	1419	1440	1514

(a) Use the following normal probability plot to determine if the data could have come from a normal distribution. *Normal*

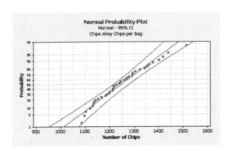

(b) Determine the mean and standard deviation of the sample data. *x̄ = 1247.4, s = 101.0*
(c) Using the sample mean and sample standard deviation obtained in part (b) as estimates for the population mean and population standard deviation, respectively, draw a graph of a normal model for the distribution of chips in a bag a Chips Ahoy!.
(d) Using the normal model from part (c), find the probability that an 18-ounce bag of Chips Ahoy! selected at random contains at least 1000 chips. *0.9929*

(e) Using the normal model from part (c), determine the proportion of 18-ounce bags of Chips Ahoy! that contains between 1200 and 1400 chips. *0.6153*

10. Hours of TV A random sample of college students aged 18 to 24 years was obtained, and the number of hours of television watched last week was recorded.

36.1	30.5	2.9	17.5	21.0
23.5	25.6	16.0	28.9	29.6
7.8	20.4	33.8	36.8	0.0
9.9	25.8	19.5	19.1	18.5
22.9	9.7	39.2	19.0	8.6

(a) Use the following normal probability plot to determine if the data could have come from a normal distribution. *Normal*

(b) Determine the mean and standard deviation of the sample data. *x̄ = 20.90, s = 10.52*
(c) Using the sample mean and sample standard deviation obtained in part (b) as estimates for the population mean and population standard deviation, respectively, draw a graph of a normal model for the distribution of weekly hours of television watched.
(d) Using the normal model from part (c), find the probability that a college student aged 18 to 24 years, selected at random, watches between 20 and 35 hours of television each week. *0.4458*
(e) Using the normal model from part (c), determine the proportion of college students aged 18 to 24 years who watch more than 40 hours of television per week. *0.0344*

In Problems 11–14, use a normal probability plot to assess whether the sample data could have come from a population that is normally distributed.

11. O-Ring Thickness A random sample of O-rings was obtained and the wall thickness (in inches) of each was recorded. Normal

0.276	0.274	0.275	0.274	0.277
0.273	0.276	0.276	0.279	0.274
0.273	0.277	0.275	0.277	0.277
0.276	0.277	0.278	0.275	0.276

13. School Loans A random sample of 20 undergraduate students receiving student loans was obtained, and the amount of their loans for the 2004–2005 school year was recorded. Not normal

2,500	1,000	2,000	14,000	1,800
3,800	10,100	2,200	900	1,600
500	2,200	6,200	9,100	2,800
2,500	1,400	13,200	750	12,000

12. Customer Service A random sample of weekly work logs at an automobile repair station was obtained and the average number of customers per day was recorded. Normal

26	24	22	25	23
24	25	23	25	22
21	26	24	23	24
25	24	25	24	25
26	21	22	24	24

14. Memphis Snowfall A random sample of 25 years between 1890 and 2005 was obtained, and the amount of snowfall, in inches, for Memphis was recorded. Not normal

(*Source:* National Oceanic and Atmospheric Administration)

24.0	7.9	1.5	0.0	0.3
0.4	8.1	4.3	0.0	0.5
3.6	2.9	0.4	2.6	0.1
16.6	1.4	23.8	25.1	1.6
12.2	14.8	0.4	3.7	4.2

Technology Step by Step

Normal Probability Plots

TI-83/84 Plus

Step 1: Enter the raw data into L1.

Step 2: Press 2nd Y = to access STAT PLOTS.

Step 3: Select 1:Plot1.

Step 4: Turn Plot1 ON by highlighting ON and pressing ENTER. Press the down-arrow key. Highlight the *normal probability plot* icon. It is the icon in the lower-right corner under Type:. Press ENTER to select this plot type. The Data List should be set at L1. The data axis should be the *x*-axis.

Step 5: Press ZOOM, and select 9:ZoomStat.

MINITAB

Step 1: Enter the raw data into C1.

Step 2: Select the **Graph** menu. Highlight **Probability Plot** ....

Step 3: In the Variables cell, enter the column that contains the raw data. Make sure Distribution is set to Normal. Click OK.

Excel

Step 1: Load the PHStat Add-in.

Step 2: Enter the raw data into column A.

Step 3: Select the **PHStat** menu. Highlight **Probability Distributions**, then highlight **Normal Probability Plot** ....

Step 4: With the cursor in the "Variable Cell Range:" cell, highlight the raw data. Enter a graph title, if desired. Click OK.

7.5 The Normal Approximation to the Binomial Probability Distribution

Preparing for This Section Before getting started, review the following:

- Binomial probability distribution (Section 6.2, pp. 298–309)

Objective ① Approximate binomial probabilities using the normal distribution

① **Approximate Binomial Probabilities Using the Normal Distribution**

Note to Instructor

The material in this section is optional and can be omitted without loss of continuity.

Note to Instructor

The concepts in this section can be illustrated using the Normal Approximation to the Binomial applet.

In Section 6.2, we discussed the binomial probability distribution. A probability experiment is said to be a binomial experiment if the following conditions are met.

Criteria for a Binomial Probability Experiment

A probability experiment is said to be a binomial experiment if all the following are true:

1. The experiment is performed n independent times. Each repetition of the experiment is called a **trial**. Independence means that the outcome of one trial will not affect the outcome of the other trials.
2. For each trial, there are two mutually exclusive outcomes—success or failure.
3. The probability of success, p, is the same for each trial of the experiment.

The binomial probability formula can be used to compute probabilities of events in a binomial experiment. When there is a large number of trials of a binomial experiment, the binomial probability formula can be difficult to use. For example, suppose there are 500 trials of a binomial experiment and we wish to compute the probability of 400 or more successes. Using the binomial probability formula requires that we compute the following probabilities:

$$P(X \geq 400) = P(400) + P(401) + \cdots + P(500)$$

This would be time consuming to compute by hand! Fortunately, we have other means for approximating binomial probabilities, provided that certain conditions are met.

Recall, as the number of trials, n, in a binomial experiment increases, the probability histogram becomes more nearly symmetric and bell shaped (see page 308). We restate the conclusion here.

As the number of trials n in a binomial experiment increases, the probability distribution of the random variable X becomes more nearly symmetric and bell shaped. As a rule of thumb, if $np(1 - p) \geq 10$, the probability distribution will be approximately symmetric and bell shaped.

Historical Note

The normal approximation to the binomial was discovered by Abraham de Moivre in 1733. With the advance of computing technology, its importance has been diminished.

Because of this result, we might be inclined to think that binomial probabilities can be approximated by the area under the normal curve, provided that $np(1 - p) \geq 10$. This intuition is correct.

The Normal Approximation to the Binomial Probability Distribution

If $np(1 - p) \geq 10$, the binomial random variable X is approximately normally distributed with mean $\mu_X = np$ and standard deviation $\sigma_X = \sqrt{np(1 - p)}$.

Figure 47 shows a probability histogram for the binomial random variable X with $n = 40$ and $p = 0.5$ and a normal curve with $\mu_X = np = 40(0.5) = 20$ and standard deviation $\sigma_X = \sqrt{np(1 - p)} = \sqrt{40(0.5)(0.5)} = \sqrt{10}$. Notice that $np(1 - p) = 40(0.5)(1 - 0.5) = 10$.

Figure 47

We know from Section 6.2 that the area of the rectangle corresponding to $X = 18$ represents $P(18)$. The width of each rectangle is 1, so the rectangle extends from $X = 17.5$ to $X = 18.5$. The area under the normal curve from $X = 17.5$ to $X = 18.5$ is approximately equal to the area of the rectangle corresponding to $X = 18$. Therefore, the area under the normal curve between $X = 17.5$ and $X = 18.5$ is approximately equal to $P(18)$, where X is a binomial random variable with $n = 40$ and $p = 0.5$. We add and subtract 0.5 from $X = 18$ as a **correction for continuity**, because we are using a continuous density function to approximate a discrete probability.

Suppose we want to approximate $P(X \leq 18)$. Figure 48 illustrates the situation.

> ! **CAUTION**
>
> Don't forget about the correction for continuity. It is needed because we are using a continuous density function to approximate the probability of a discrete random variable.

Figure 48

To approximate $P(X \leq 18)$, we compute the area under the normal curve for $X < 18.5$. Do you see why?

If we want to approximate $P(X \geq 18)$, we compute $P(X \geq 17.5)$. Do you see why? Table 7 summarizes how to use the correction for continuity.

Table 7

Exact Probability Using Binomial	Approximate Probability Using Normal	Graphical Depiction
$P(X = a)$	$P(a - 0.5 < X < a + 0.5)$	
$P(X \leq a)$	$P(X < a + 0.5)$	
$P(X \geq a)$	$P(X > a - 0.5)$	
$P(a \leq X \leq b)$	$P(a - 0.5 < X < b + 0.5)$	

A question remains, however. What do we do if the probability is of the form $P(X > a)$, $P(X < a)$, or $P(a < X < b)$? The solution is to rewrite the inequality in a form with $\leq$ or $\geq$. For example, $P(X > 4) = P(X \geq 5)$ and $P(X < 4) = P(X \leq 3)$ for binomial random variables, because the values of the random variables must be whole numbers.

EXAMPLE 1 The Normal Approximation to a Binomial Random Variable

Problem: According to the *Information Please Almanac*, 6% of the human population has blood type O-negative. What is the probability that, in a simple random sample of 500, fewer than 25 have blood type O-negative?

Approach

Step 1: We verify that this is a binomial experiment.

Step 2: Computing the probability by hand would be very tedious. Verify $np(1 - p) \geq 10$. Then we will know that the condition for using the normal distribution to approximate the binomial distribution is met.

Step 3: Approximate $P(X < 25) = P(X \leq 24)$ by using the normal approximation to the binomial distribution.

Solution

Step 1: There are 500 independent trials with each trial having a probability of success equal to 0.06. This is a binomial experiment.

Step 2: We verify $np(1 - p) \geq 10$.

$$np(1 - p) = 500(0.06)(0.94) = 28.2 \geq 10$$

Figure 49

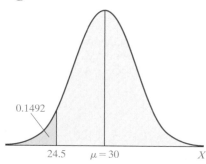

0.1492

24.5 $\mu = 30$ X

We can use the normal distribution to approximate the binomial distribution.

Step 3: We wish to know the probability that fewer than 25 people in the sample have blood type O-negative; that is, we wish to know $P(X < 25) = P(X \leq 24)$. This is approximately equal to the area under the normal curve to the left of $X = 24.5$, with $\mu_X = np = 500(0.06) = 30$ and $\sigma_X = \sqrt{np(1-p)} = \sqrt{500(0.06)(1-0.06)} = \sqrt{28.2} \approx 5.31$. See Figure 49. We convert $X = 24.5$ to a Z-score.

$$Z = \frac{24.5 - 30}{5.31} = -1.04$$

From Table IV, we find the area to the left of $Z = -1.04$ is 0.1492. Therefore, the approximate probability that fewer than 25 people will have blood type O-negative is 0.1492 = 14.92%.

Figure 50

binomcdf(500,.06
,24)
        .1493809338

Using the *binomcdf(* command on a TI-84 Plus graphing calculator, we find that the exact probability is 0.1494. See Figure 50. The approximate result is close indeed!

Now Work Problem 21.

EXAMPLE 2   **A Normal Approximation to the Binomial**

Problem: According to the Federal Communications Commission, 70% of all U.S. households have cable television. Erica conducts a random sample of 1000 households in DuPage County and finds that 734 of them have cable.

(a) Assuming that 70% of households have cable, what is the probability of obtaining a random sample of at least 734 households with cable from a sample of size 1000?

(b) Does the result from part (a) contradict the FCC information? Explain.

Approach: This is a binomial experiment with $n = 1000$ and $p = 0.70$. Erica needs to determine the probability of obtaining a random sample of at least 734 households with cable from a sample of size 1000, assuming 70% of households have cable. Computing this using the binomial probability formula would be difficult, so Erica will compute the probability using the normal approximation to the binomial, since $np(1-p) = 1000(0.70)(0.30) = 210 \geq 10$. We approximate $P(X \geq 734)$ by computing the area under the standard normal curve to the right of $X = 733.5$ with $\mu_X = np = 1000(0.70) = 700$ and $\sigma_X = \sqrt{np(1-p)} = \sqrt{1000(0.70)(1-0.70)} = \sqrt{210} \approx 14.491$.

Solution

Figure 51

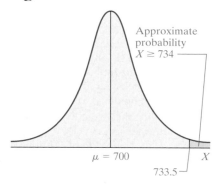

Approximate probability $X \geq 734$

$\mu = 700$ X

733.5

Now Work Problem 27.

(a) Figure 51 shows the area we wish to compute. We convert $X = 733.5$ to a Z-score.

$$Z = \frac{733.5 - 700}{14.491} = 2.31$$

The area under the standard normal curve to the right of $Z = 2.31$ is $1 - 0.9896 = 0.0104$. There is a 1.04% probability of obtaining 734 or more households with cable from a sample of 1000 households, assuming that the percentage of households with cable is 70%.

(b) Yes. The result from part (a) means that about 1 sample in every 100 samples will have 734 or more households with cable if the true proportion is 0.7. Erica is not inclined to believe that her sample is one of the 1 in 100. She would rather believe that the proportion of households in DuPage County with cable is higher than 0.70.

7.5 ASSESS YOUR UNDERSTANDING

Concepts and Vocabulary

1. List the conditions required for a binomial experiment.
2. Under what circumstances can the normal distribution be used to approximate binomial probabilities?
3. Why must we use a correction for continuity when using the normal distribution to approximate binomial probabilities?

4. *True or False*: Suppose X is a binomial random variable. To approximate $P(3 \leq X < 7)$ using the normal probability distribution, we compute $P(3.5 \leq X < 7.5)$. False

Skill Building

In Problems 5–14, a discrete random variable is given. Assume the probability of the random variable will be approximated using the normal distribution. Describe the area under the normal curve that will be computed. For example, if we wish to compute the probability of finding at least five defective items in a shipment, we would approximate the probability by computing the area under the normal curve to the right of $X = 4.5$.

5. The probability that at least 40 households have a gas stove Right of X = 39.5
6. The probability of no more than 20 people who want to see *Roe v. Wade* overturned Left of X = 20.5
7. The probability that exactly eight defective parts are in the shipment Between X = 7.5 and X = 8.5
8. The probability that exactly 12 students pass the course
9. The probability that the number of people with blood type O-negative is between 18 and 24, inclusive
10. The probability that the number of tornadoes that occur in the month of May is between 30 and 40, inclusive

11. The probability that more than 20 people want to see the marriage tax penalty abolished Right of X = 20.5
12. The probability that fewer than 40 households have a pet
13. The probability that more than 500 adult Americans support a bill proposing to extend daylight savings time
14. The probability that fewer than 35 people support the privatization of Social Security Left of 34.5

In Problems 15–20, compute P(x) using the binomial probability formula. Then determine whether the normal distribution can be used as an approximation for the binomial distribution. If so, approximate P(x) and compare the result to the exact probability.

15. $n = 60$, $p = 0.4$, $X = 20$ 0.0616; 0.0618
16. $n = 80$, $p = 0.15$, $X = 18$ 0.0221; 0.0216
17. $n = 40$, $p = 0.25$, $X = 30$ < 0.0001; cannot be used
18. $n = 100$, $p = 0.05$, $X = 50$ < 0.0001; cannot be used
19. $n = 75$, $p = 0.75$, $X = 60$ 0.0677; 0.0630
20. $n = 85$, $p = 0.8$, $X = 70$ 0.0970; 0.0926

Applying the Concepts

21. **On-Time Flights** According to American Airlines, Flight [NW] 215 from Orlando to Los Angeles is on time 90% of the time. Suppose 150 flights are randomly selected. Use the normal approximation to the binomial to
   (a) approximate the probability that exactly 130 flights are on time. 0.0444
   (b) approximate the probability that at least 130 flights are on time. 0.9332
   (c) approximate the probability that fewer than 125 flights are on time. 0.0021
   (d) approximate the probability that between 125 and 135 flights, inclusive, are on time. 0.5536

22. **Smokers** According to *Information Please Almanac*, 80% of adult smokers started smoking before they were 18 years old. Suppose 100 smokers 18 years old or older are randomly selected. Use the normal approximation to the binomial to
   (a) approximate the probability that exactly 80 of them started smoking before they were 18 years old. 0.1034
   (b) approximate the probability that at least 80 of them started smoking before they were 18 years old. 0.5517
   (c) approximate the probability that fewer than 70 of them started smoking before they were 18 years old. 0.0043
   (d) approximate the probability that between 70 and 90 of them, inclusive, started smoking before they were 18 years old. 0.9914

23. **Migraine Sufferers** In clinical trials of a medication whose purpose is to reduce the pain associated with migraine headaches, 2% of the patients in the study experienced weight gain as a side effect. Suppose a random sample of 600 users of this medication is obtained. Use the normal approximation to the binomial to
   (a) approximate the probability that exactly 20 will experience weight gain as a side effect. 0.0077
   (b) approximate the probability that 20 or fewer will experience weight gain as a side effect. 0.9934
   (c) approximate the probability that 22 or more patients will experience weight gain as a side effect. 0.0028
   (d) approximate the probability that between 20 and 30 patients, inclusive, will experience weight gain as a side effect. 0.0143

24. **High-Speed Internet** According to a report by the Commerce Department in the fall of 2004, 20% of U.S. households had some type of high-speed Internet connection. Suppose 80 U.S. households are selected at random. Use the normal approximation to the binomial to
   (a) approximate the probability that exactly 15 households have high-speed Internet access. 0.1071
   (b) approximate the probability that at least 20 households have high-speed Internet access. 0.1635

8. Between X = 11.5 and X = 12.5 9. Between X = 17.5 and X = 24.5 10. Between X = 29.5 and X = 40.5 12. Left of X = 39.5,
13. Right of X = 500.5

(c) approximate the probability that fewer than 10 households have high-speed Internet access. *0.0344*

(d) approximate the probability that between 12 and 18 households, inclusive, have high-speed Internet access. *0.6542*

25. **Allergy Sufferers** Clarinex-D is a medication whose purpose is to reduce the symptoms associated with a variety of allergies. In clinical trials of Clarinex-D, 5% of the patients in the study experienced insomnia as a side effect. Suppose a random sample of 400 Clarinex-D users is obtained. Use the normal approximation to the binomial to

(a) approximate the probability that exactly 20 patients experienced insomnia as a side effect. *0.0876*

(b) approximate the probability that 15 or fewer patients experienced insomnia as a side effect. *0.1515*

(c) approximate the probability that 30 or more patients experienced insomnia as a side effect. *0.0146*

(d) approximate the probability that between 10 and 32 patients, inclusive, experienced insomnia as a side effect. *0.9899*

26. **Murder by Firearms** According to the *Uniform Crime Report, 2003*, 66.9% of murders are committed with a firearm. Suppose that 50 murders are randomly selected. Use the normal approximation to the binomial to

(a) approximate the probability that exactly 40 murders are committed using a firearm. *0.0174*

(b) approximate the probability that at least 35 murders are committed using a firearm. *0.3745*

(c) approximate the probability that fewer than 25 murders are committed using a firearm. *0.0036*

(d) approximate the probability that between 30 and 35 murders, inclusive, are committed using a firearm. *0.6154*

27. **Males Living at Home** According to the *Current Population Survey* (Internet release date: September 15, 2004), 55% of males between the ages of 18 and 24 years lived at home in 2003. (Unmarried college students living in a dorm are counted as living at home.) Suppose that a survey is administered at a community college to 200 randomly selected male students between the ages of 18 and 24 years and that 130 of them respond that they live at home.

(a) Approximate the probability that such a survey will result in at least 130 of the respondents living at home under the assumption that the true percentage is 55%. *0.0028*

(b) Does the result from part (a) contradict the results of the *Current Population Survey*? Explain.

28. **Females Living at Home** According to the *Current Population Survey* (Internet release date: September 15, 2004), 46% of females between the ages of 18 and 24 years lived at home in 2003. (Unmarried college students living in a dorm are counted as living at home.) Suppose that a survey is administered at a community college to 200 randomly selected female students between the ages of 18 and 24 years and that 110 of them respond that they live at home.

(a) Approximate the probability that such a survey will result in at least 110 of the respondents living at home under the assumption that the true percentage is 46%. *0.0066*

(b) Does the result from part (a) contradict the results of the *Current Population Survey*? Explain.

29. **Boys Are Preferred** In a Gallup poll conducted December 2–4, 2000, 42% of survey respondents said that, if they only had one child, they would prefer the child to be a boy. Suppose you conduct a survey of 150 randomly selected students on your campus and find that 80 of them would prefer a boy.

(a) Approximate the probability that, in a random sample of 150 students, at least 80 would prefer a boy, assuming the true percentage is 42%. *0.0032*

(b) Does this result contradict the Gallup poll? Explain. Yes

30. **Liars** According to a *USA Today* "Snapshot," 3% of Americans surveyed lie frequently. Suppose you conduct a survey of 500 college students and find that 20 of them lie frequently.

(a) Compute the probability that, in a random sample of 500 college students, at least 20 lie frequently, assuming the true percentage is 3%. *0.1190*

(b) Does this result contradict the *USA Today* "Snapshot"? Explain. No

 CHAPTER 7 **Review**

Summary

In this chapter, we introduced continuous random variables and the normal probability density function. A continuous random variable is said to be approximately normally distributed if a histogram of its values is symmetric and bell shaped. In addition, we can draw normal probability plots that are based on expected Z-scores. If these normal probability plots are approximately linear, we say the distribution of the random variable is approximately normal. The area under the normal density function can be used to find proportions, probabilities

or percentiles for normal random variables. Also, we can find the value of a normal random variable that corresponds to a specific proportion, probability, or percentile.

If X is a binomial random variable with $np(1 - p) \geq 10$, we can use the area under the normal curve to approximate the probability of a binomial random variable. The parameters of the normal curve are $\mu_X = np$ and $\sigma_X = \sqrt{np(1 - p)}$, where n is the number of trials of the binomial experiment and p is the probability of success.

Formulas

Standardizing a Normal Random Variable

$$Z = \frac{X - \mu}{\sigma}$$

Finding the Score

$$X = \mu + Z\sigma$$

Vocabulary

Uniform probability distribution (p. 319)   Inflection points (p. 321)   Normal score (p. 355)
Probability density function (p. 320)   Normal probability density function (p. 324)   Trial (p. 362)
Normal curve (p. 321)   Standard normal distribution (p. 326)   Correction for continuity (p. 363)
Normal probability distribution (p. 321)   Normal probability plot (p. 355)

Objectives

Section	You should be able to . . .	Examples	Review Exercises
7.1	1 Understand the uniform probability distribution (p. 319)	1 and 2	37
	2 Graph a normal curve (p. 321)	Page 321	19–22
	3 State the properties of the normal curve (p. 322)	Page 322	38
	4 Understand the role of area in the normal density function (p. 323)	3 and 4	1, 2
	5 Understand the relation between a normal random variable and a standard normal random variable (p. 325)	5	3, 4
7.2	1 Find the area under the standard normal curve (p. 332)	1 through 4	5–8
	2 Find the Z-scores for a given area (p. 336)	5 through 9	13–18
	3 Interpret the area under standard normal curve as a probability (p. 340)	10	9–12
7.3	1 Find and interpret the area under a normal curve (p. 345)	1 through 3	19–22, 23(a)–(c), 24(a)–(c), 25(a)–(c), 26(a)–(d), 27, 28
	2 Find the value of a normal random variable (p. 348)	4 through 6	23(d), 24(d), (e), 25(d), (e), 26(e), (f)
7.4	1 Draw normal probability plots to assess normality (p. 355)	1 through 3	31–34
7.5	1 Approximate binomial probabilities using the normal distribution (p. 362)	1 and 2	29, 30

Review Exercises

1. Use the figure to answer the questions that follow:

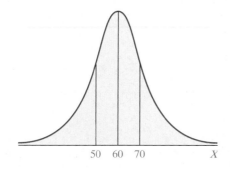

50 60 70   X

(a) What is μ? 60
(b) What is σ? 10
(c) Suppose the area under the normal curve to the right of $X = 75$ is 0.0668. Provide two interpretations for this area.
(d) Suppose the area under the normal curve between $X = 50$ and $X = 75$ is 0.7745. Provide two interpretations for this area.

2. Use the figure to answer the questions that follow:

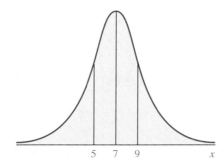

5 7 9   x

(a) What is μ? 7
(b) What is σ? 2
(c) Suppose the area under the normal curve to the left of $X = 10$ is 0.9332. Provide two interpretations for this area.
(d) Suppose the area under the normal curve between $X = 5$ and $X = 8$ is 0.5328. Provide two interpretations for this area.

3. A random variable X is approximately normally distributed with $\mu = 20$ and $\sigma = 4$.

(a) Compute $Z_1 = \dfrac{X_1 - \mu}{\sigma}$ for $X_1 = 18$. -0.5

(b) Compute $Z_2 = \dfrac{X_2 - \mu}{\sigma}$ for $X_2 = 21$. 0.25

(c) The area under the normal curve between $X_1 = 18$ and $X_2 = 21$ is 0.2912. What is the area between Z_1 and Z_2? 0.2912

4. A random variable X is approximately normally distributed with $\mu = 50$ and $\sigma = 8$.

(a) Compute $Z_1 = \dfrac{X_1 - \mu}{\sigma}$ for $X_1 = 48$. -0.25

(b) Compute $Z_2 = \dfrac{X_2 - \mu}{\sigma}$ for $X_2 = 60$. 1.25

(c) The area under the normal curve between $X_1 = 48$ and $X_2 = 60$ is 0.4931. What is the area between Z_1 and Z_2? 0.4931

In Problems 5–8, draw a standard normal curve and shade the area indicated. Then use Table IV to find the area under the normal curve.

5. The area to the left of $Z = -1.04$ 0.1492

6. The area to the right of $Z = 2.04$ 0.0207

7. The area between $Z = -0.34$ and $Z = 1.03$ 0.4816

8. The area between $Z = 1.93$ and $Z = 3.93$ 0.0268

In Problems 9–12, find the indicated probability of the standard normal random variable Z.

9. $P(Z < 1.19)$ 0.8830

10. $P(Z \geq 1.61)$ 0.0537

11. $P(-1.21 < Z \leq 2.28)$ 0.8756

12. $P(0.21 < Z < 1.69)$ 0.3713

13. Find the Z-score such that the area to the left of the Z-score is 0.84. 0.99

14. Find the Z-score such that the area to the right of the Z-score is 0.483. 0.04

15. Find the Z-scores that separate the middle 92% of the data from the area in the tails of the standard normal distribution. $-1.75, 1.75$

16. Find the Z-scores that separate the middle 88% of the data from the area in the tails of the standard normal distribution. $-1.555, 1.555$

17. Find the value of $z_{0.20}$. 0.84

18. Find the value of $z_{0.04}$. 1.75

In Problems 19–22, draw the normal curve with the parameters indicated. Then find the probability of the random variable X. Shade the area that represents the probability.

19. $\mu = 50, \sigma = 6, P(X > 55)$ 0.2023

20. $\mu = 30, \sigma = 5, P(X \leq 23)$ 0.0808

21. $\mu = 70, \sigma = 10, P(65 < X < 85)$ 0.6247

22. $\mu = 20, \sigma = 3, P(22 \leq X \leq 27)$ 0.2415

23. Tire Wear Suppose Dunlop Tire manufactures tires having the property that the mileage the tire lasts approximately follows a normal distribution with mean 70,000 miles and standard deviation 4400 miles.

(a) What percent of the tires will last at least 75,000 miles? 12.71%

(b) Suppose Dunlop warrants the tires for 60,000 miles. What percent of the tires will last 60,000 miles or less? 1.16%

(c) What is the probability that a randomly selected Dunlop tire lasts between 65,000 and 80,000 miles? 0.8613

(d) Suppose that Dunlop wants to warrant no more than 2% of its tires. What mileage should the company advertise as its warranty mileage? $60,980$ miles

24. Talk Time on a Cell Phone Suppose the talk time in digital mode on a Motorola Timeport P8160 is approximately normally distributed with mean 324 minutes and standard deviation 24 minutes. (24d) 344 minutes

(a) What proportion of the time will a fully charged battery last at least 300 minutes? 0.8413

(b) What proportion of the time will a fully charged battery last less than 340 minutes? 0.7486

(c) Suppose you charge the battery fully. What is the probability it will last between 310 and 350 minutes? 0.5789

(d) Determine the talk time that is in the top 20%.

(e) Determine the talk time that makes up the middle 90% of talk time. 285–363 minutes

25. Serum Cholesterol As reported by the U.S. National Center for Health Statistics, the mean serum cholesterol of females 16 to 19 years old is $\mu = 171$. If serum cholesterol is approximately normally distributed with $\sigma = 39.8$, answer the following: (25e) Yes, 144

(a) Determine the proportion of 16- to 19-year-old females with a serum cholesterol above 180. 0.4090

(b) Determine the proportion of 16- to 19-year-old females with a serum cholesterol between 150 and 200. 0.4692

(c) Suppose a 16- to 19-year-old female is randomly selected. Determine the probability her serum cholesterol is below 140. 0.2177

(d) Determine the serum cholesterol that divides the bottom 10% from the top 90% of all serum cholesterol levels of 16- to 19-year-old females. 120

(e) According to the National Center for Health Statistics, the 25th percentile of serum cholesterol for 16- to 19-year-old females is 145. Is the 25th percentile on the normal curve close to the reported value of 145?

26. Wechsler Intelligence Scale The Wechsler Intelligence Scale for Children is approximately normally distributed with mean 100 and standard deviation 15.

(a) What proportion of test takers will score above 125?

(b) What proportion of test takers will score below 90?

(c) What proportion of test takers will score between 110 and 140? 0.2476

(d) If a child is randomly selected, what is the probability that she scores above 150? 0.0004

(e) What intelligence score will place a child in the top 5% of all children? 125

(f) If normal intelligence is defined as scoring in the middle 95% of all test takers, figure out the scores that differentiate normal intelligence from abnormal intelligence. 71–129 (26a) 0.0475 (26b) 0.2514

27. Major League Baseballs According to major league baseball rules, the ball must weigh between 5 and 5.25 ounces. (*Source*: www.baseball-almanac.com) Suppose a factory produces baseballs whose weights are approximately normally distributed with mean 5.11 ounces and standard deviation 0.062 ounces.

(a) What proportion of the baseballs produced by this factory are too heavy for use by major league baseball? 0.0119

(b) What proportion of the baseballs produced by this factory are too light for use by the major league baseball? 0.0384

(c) What proportion of the baseballs produced by this factory can be used by major league baseball? 0.9497

28. Halogen Light Bulbs Feit Electric manufactures a Crystal Clear Halogen 60-watt light bulb that has an average life of 3000 hours. Suppose the lifetimes of the light bulbs are approximately normally distributed with standard deviation 183 hours.

(a) What proportion of the light bulbs will last more than 3300 hours? 0.0505

(b) What proportion of the light bulbs will last less than 2500 hours? 0.0032

(c) What is the probability that a randomly selected Feit Crystal Clear Halogen 60-watt light bulb lasts between 2900 and 3100 hours? 0.4176

(d) What is the probability that a randomly selected Feit Crystal Clear Halogen 60-watt light bulb lasts less than 2600 hours? 0.0143

(e) What is the percentile rank of a Feit Crystal Clear Halogen 60-watt light bulb that lasts 3350 hours? 97th

(f) Would it be unusual for a Feit Crystal Clear Halogen 60-watt light bulb to last longer than 3400 hours? yes

29. High Cholesterol According to the National Center for Health Statistics, 8% of 20- to 34-year-old females have high serum cholesterol. Suppose you conduct a random sample of two hundred 20- to 34-year-old females.

(a) Verify that the conditions for using the normal distribution to approximate the binomial distribution are met.

(b) Approximate the probability that exactly 15 have high serum cholesterol. Interpret this result. 0.1000

(c) Approximate the probability that more than 20 have high serum cholesterol. Interpret this result. 0.1210

(d) Approximate the probability that at least 15 have high serum cholesterol. Interpret this result. 0.6517

(e) Approximate the probability that fewer than 25 have high serum cholesterol. Interpret this result. 0.9864

(f) Approximate the probability that between 15 and 25, inclusive, have high serum cholesterol. Interpret this result. 0.6451

30. America Reads According to a Gallup poll conducted September 10–14, 1999, 56% of Americans 18 years old or older stated they had read at least six books (fiction and nonfiction) within the past year. Suppose you conduct a random sample of 250 Americans 18 years old or older.

(a) Verify that the conditions for using the normal distribution to approximate the binomial distribution are met.

(b) Approximate the probability that exactly 125 read at least six books within the past year. Interpret this result. 0.0078

(c) Approximate the probability that fewer than 120 read at least six books within the past year. Interpret this result. 0.0045

(d) Approximate the probability that at least 140 read at least six books within the past year. Interpret this result. 0.5239

(e) Approximate the probability that between 100 and 120, inclusive, read at least six books within the past year. Interpret this result. 0.0066

In Problems 31 and 32, a normal probability plot of a simple random sample of data from a population whose distribution is unknown was obtained. Given the normal probability plot, is there reason to believe the population is normally distributed?

31. Normal

32. Not normal

In Problems 33 and 34, assess the normality of the sample data.

33. Density of Earth   In 1798, Henry Cavendish obtained 27 measurements of the density of Earth, using a torsion balance. The following data represent his estimates, represented as a multiple of the density of water.   Not normal

5.50	5.57	5.42	5.61	5.53
5.47	4.88	5.62	5.63	4.07
5.29	5.34	5.26	5.44	5.55
5.34	5.30	5.36	5.79	5.29
5.10	5.86	5.58	5.27	5.85
5.65	5.39			

Source: Stigler, S. M. "Do robust estimators work with real data?" *Annals of Statistics*, 5 (1977), 1055–1078.

34. Life Expectancy   The following data represent the life expectancy at birth in 2005 in a random sample of 20 countries. Normal

75.84	77.11	75.91	80.39	78.92
74.23	78.62	77.79	77.83	80.10
79.95	76.58	76.84	79.14	77.62
79.21	78.35	77.71	80.40	78.19

Source: U.S. Census Bureau, International Database

35. Hybrid SUV   As the first hybrid sports utility vehicle (SUV) with gasoline mileage certified by the Environmental Protection Agency, the Ford Escape HEV is the most fuel-efficient SUV in 2005. The mean mileage for the automatic four-wheel-drive Ford Escape HEV is 29 miles per gallon on the highway. (*Source*: www.fueleconomy.gov) Suppose the gasoline mileages of these SUVs are normally distributed with standard deviation 2 miles per gallon.
(a) What proportion of automatic four-wheel-drive 2005 Ford Escape HEVs gets over 25 miles per gallon on the highway? 0.9772

(b) What proportion of automatic four-wheel-drive 2005 Ford Escape HEVs gets less than 30 miles per gallon on the highway? 0.6915
(c) What is the probability that a randomly selected automatic four-wheel-drive 2005 Ford Escape HEV gets between 26 and 34 miles per gallon on the highway? 0.9270
(d) What is the probability that a randomly selected automatic four-wheel-drive 2005 Ford Escape HEV gets over 35 miles per gallon on the highway? 0.0013
(e) What is the percentile rank of an automatic four-wheel-drive 2005 Ford Escape HEV that gets 32 miles per gallon? 93rd
(f) What is the percentile rank of an automatic four-wheel-drive 2005 Ford Escape HEV that gets 25 miles per gallon? 2nd

36. Creative Thinking   According to a *USA Today* "Snapshot," 20% of adults surveyed do their most creative thinking while driving. Suppose you conduct a survey of 250 adults and find that 30 do their most creative thinking while driving.
(a) Compute the probability that, in a random sample of 250 adults, 30 or fewer do their most creative thinking while driving. 0.0010
(b) Does this result contradict the *USA Today* "Snapshot"? Explain. Yes

37. Suppose a continuous random variable X is uniformly distributed with $0 \le X \le 20$.
(a) Draw a graph of the uniform density function.
(b) What is $P(0 \le X \le 5)$? 0.25
(c) What is $P(10 \le X \le 18)$? 0.4

38. List the properties of the standard normal curve.

39. Explain how to use a normal probability plot to assess normality.

THE CHAPTER 7 CASE STUDY IS LOCATED ON THE CD THAT ACCOMPANIES THIS TEXT.

PART 4

Inference: From Samples to Population

CHAPTER 8
Sampling Distributions

CHAPTER 9
Estimating the Value of a Parameter Using Confidence Intervals

CHAPTER 10
Hypothesis Tests Regarding a Parameter

CHAPTER 11
Inference on Two Samples

CHAPTER 12
Additional Inferential Procedures

In Chapter 1, we presented the following process of statistics:

Step 1: Identify a research objective.

Step 2: Collect the information needed to answer the questions posed in Step 1.

Step 3: Organize and summarize the information.

Step 4: Draw conclusions from the information.

The methods for conducting Steps 1 and 2 were discussed in Chapter 1. The methods for conducting Step 3 were discussed in Chapters 2 through 4. We took a break from the statistical process in Chapters 5 through 7 so that we could develop skills that allow us to tackle Step 4.

If the information (data) collected is from a population, we can use the summaries obtained in Step 3 to draw conclusions about the population being studied and the statistical process is over.

However, it is often difficult or impossible to gain access to populations, so the information obtained in Step 2 is often sample data. The sample data are used to make inferences about the population. For example, we might compute a sample mean from the information collected in Step 2 and use this information to draw conclusions regarding the population mean. The last part of this text discusses how sample data are used to draw conclusions about populations.

Sampling Distributions

Outline

8.1   Distribution of the Sample Mean

8.2   Distribution of the Sample Proportion

■ Chapter Review

■ Case Study: Sampling Distribution of the Median (On CD)

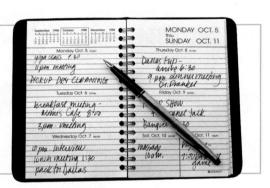

DECISIONS

The American Time Use Survey is a survey of adult Americans conducted by the Bureau of Labor Statistics. The purpose of the survey is to learn how Americans allocate their time during a day. As a reporter for the school newspaper, you wish to file a report that compares the typical student at your school to the rest of Americans. See the Decisions project on page 388.

●●● Putting It All Together

In Chapters 6 and 7, we learned about random variables and their probability distributions. A random variable is a numerical measure of the outcome to a probability experiment. A probability distribution provides a way to assign probabilities to the random variable. For discrete random variables, we discussed the binomial probability distribution. We assigned probabilities using a formula. For continuous random variables, we discussed the normal probability distribution. To compute probabilities for a normal random variable, we found the area under a normal density curve.

In this chapter, we continue our discussion of probability distributions where statistics, such as $\bar{x}$, will be the random variable. Statistics are random variables because the value of a statistic varies from sample to sample. For this reason, statistics have probability distributions associated with them. For example, there is a probability distribution for the sample mean, sample variance, and so on. We use probability distributions to make probability statements regarding the statistic. So this chapter discusses the shape, center, and spread of statistics such as $\bar{x}$.

8.1 Distribution of the Sample Mean

Preparing for This Section   Before getting started, review the following:

- Simple random sampling (Section 1.2, pp. 16–19)
- The mean (Section 3.1, pp. 107–110)
- The standard deviation (Section 3.2, pp. 129–130)
- Applications of the normal distribution (Section 7.3, pp. 345–349)

Objectives

1. **Understand the concept of a sampling distribution**
2. **Describe the distribution of the sample mean for samples obtained from normal populations**
3. **Describe the distribution of the sample mean for samples obtained from a population that is not normal**

Note to Instructor

Remind students that the federal government surveys each household in the union every 10 years with the decennial census.

Suppose that the government wanted to estimate the mean income of all U.S. households. One approach the government could take is to literally survey each household in the United States to determine the population mean, μ. This would be a very expensive and time-consuming survey!

A second approach that the government could (and does) take is to survey a random sample of U.S. households and use the results of the survey to estimate the mean household income. This is done through the American Community Survey. The survey is administered to approximately 250,000 randomly selected households each month. Among the many questions on the survey, respondents are asked to report the income of each individual in the household. From this information, the federal government obtains a sample mean household income for U.S. households. For example, in 2003 the mean annual household income in the United States was estimated to be $\bar{x} = \$58,036$. The government might infer from this result that the mean annual household income of *all* U.S. households in 2003 was $\mu = \$58,036$. This type of statement is an example of **statistical inference**—using information from a sample to draw conclusions about a population.

The households that were administered the American Community Survey were determined by chance (random sampling). A second random sample of households would likely lead to a different sample mean such as $\bar{x} = \$58,132$, and a third random sample of households would likely lead to a third distinct sample mean such as $\bar{x} = \$58,095$. Because the households are selected by chance, the sample mean of household income is also determined by chance. We conclude from this that there is variability in our estimates. This variability leads to uncertainty as to whether our estimates are correct. Therefore, we need a way to assess the reliability of inferences made about a population based on sample data.

Note to Instructor

Remind students about the interpretation of probability as a relative frequency. For example, a probability of 0.8 means that we expect to observe the outcome about 8 times in 10 trials.

The measure of reliability is actually a statement of probability. Probability describes how likely an outcome is to occur. The goal of this chapter is to learn the distribution of statistics such as the sample mean so that our estimates are accompanied by statements that indicate the likelihood that our methods are accurate.

1 ## Understand the Concept of a Sampling Distribution

In general, the sampling distribution of a statistic is a probability distribution for all possible values of the statistic computed from a sample of size n. The **sampling distribution of the sample mean** is the probability distribution of all possible values of the random variable $\bar{x}$ computed from a sample of size n from a population with mean μ and standard deviation σ.

In Other Words

If the number of individuals in a population is a positive integer, we say the population is finite. Otherwise, the population is infinite.

The idea behind obtaining the sampling distribution of the mean is as follows:

Step 1: Obtain a simple random sample of size n.

Step 2: Compute the sample mean.

Step 3: Assuming that we are sampling from a finite population, repeat Steps 1 and 2 until all simple random samples of size n have been obtained. **Note:** Once a particular sample is obtained, it cannot be obtained a second time.

We present an example to illustrate the idea behind a sampling distribution.

EXAMPLE 1   A Sampling Distribution

Problem: One semester, Professor Goehl had a small statistics class of seven students. He asked them the ages of their cars and obtained the following data:

$$2, 4, 6, 8, 4, 3, 7$$

Construct a sampling distribution of the mean for samples of size $n = 2$. What is the probability of obtaining a sample mean between 4 and 6 years, inclusive; that is, what is $P(4 \leq \bar{x} \leq 6)$?

Approach: We follow Steps 1 to 3 listed above to construct the probability distribution.

Solution: There are seven individuals in the population. We are selecting them two at a time without replacement. Therefore, there are $_7C_2 = 21$ samples of size $n = 2$. We list these 21 samples along with the sample means in Table 1.

Table 1

Sample	Sample Mean	Sample	Sample Mean	Sample	Sample Mean
2, 4	3	4, 8	6	6, 7	6.5
2, 6	4	4, 4	4	8, 4	6
2, 8	5	4, 3	3.5	8, 3	5.5
2, 4	3	4, 7	5.5	8, 7	7.5
2, 3	2.5	6, 8	7	4, 3	3.5
2, 7	4.5	6, 4	5	4, 7	5.5
4, 6	5	6, 3	4.5	3, 7	5

Table 2 displays the sampling distribution of the sample mean, $\bar{x}$.

Table 2

Sample Mean	Frequency	Probability	Sample Mean	Frequency	Probability
2.5	1	$\frac{1}{21}$	5.5	3	$\frac{3}{21}$
3	2	$\frac{2}{21}$	6	2	$\frac{2}{21}$
3.5	2	$\frac{2}{21}$	6.5	1	$\frac{1}{21}$
4	2	$\frac{2}{21}$	7	1	$\frac{1}{21}$
4.5	2	$\frac{2}{21}$	7.5	1	$\frac{1}{21}$
5	4	$\frac{4}{21}$			

From Table 2 we can compute

$$P(4 \le \overline{x} \le 6) = \frac{2}{21} + \frac{2}{21} + \frac{4}{21} + \frac{3}{21} + \frac{2}{21} = \frac{13}{21} = 0.619$$

If we took 10 simple random samples of size 2 from this population, about 6 of them would result in sample means between 4 and 6 years, inclusive.   ▆

    The sample mean with the highest probability is $\overline{x} = 5$. This should not be surprising since the population mean of the data in Example 1 is $\mu = 4.9$, rounded to one decimal place. Figure 1 is a probability histogram of the sampling distribution for the sample mean given in Table 2.

Figure 1

Probability Distribution of the Sample Mean

Now Work Problem 31.

In-Class Activity: Sampling Distributions

Note to Instructor

The purpose of this activity is for students to explore sampling distributions of sizes n = 2, 3, 4, and 5. It should take about 30 minutes to complete. Once complete, ask some questions regarding probabilities of the sample mean as is done in Example 1.

Randomly select six students from the class to treat as a population. Choose a quantitative variable (such as pulse rate, age, or number of siblings) to use for this activity, and gather the data appropriately. Compute μ for the population. Divide the class into four groups and have one group list all samples of size $n = 2$, another group list all samples of size $n = 3$, and other groups list all samples of size $n = 4$ and $n = 5$. Each group should do the following:

(a) Compute the sample mean of each sample.
(b) Form the probability distribution for the sample mean.
(c) Draw a probability histogram of the probability distribution.
(d) Verify that $\mu_{\overline{x}} = \mu$.

Compare the spread in each probability distribution based on the probability histogram. What does this result imply about the standard deviation of the sample mean?

---

② **Describe the Distribution of the Sample Mean for Samples Obtained from Normal Populations**

The point of Example 1 is to help you realize that statistics such as $\overline{x}$ are random variables and therefore have probability distributions associated with them. In practice, a single random sample of size n is obtained from a population. The probability distribution of the sample statistic (or sampling distribution) is determined from statistical theory. We will use simulation to help justify the result that statistical theory provides. We consider two possibilities. In the first case, we sample from a population that is known to be normally distributed. In the second case, we sample from a distribution that is not normally distributed.

EXAMPLE 2 ## Sampling Distribution of the Sample Mean: Population Normal

Problem: In Example 3 from Section 7.1, we learned that the height of 3-year-old females is approximately normally distributed with $\mu = 38.72$ inches and $\sigma = 3.17$ inches. Approximate the sampling distribution of $\bar{x}$ by taking 100 simple random samples of size $n = 5$.

Approach: Use MINITAB, Excel, or some other statistical software package to perform the simulation. We will perform the following steps:

Step 1: Obtain 100 simple random samples of size $n = 5$ from the population, using simulation.

Note to Instructor
You may want to use the sampling distribution applet to illustrate this concept.

Step 2: Compute the mean of each sample.

Step 3: Draw a histogram of the sample means.

Step 4: Compute the mean and standard deviation of the sample means.

Solution

Step 1: We obtain 100 simple random samples of size $n = 5$. All the samples of size $n = 5$ are shown in Table 3.

Table 3						
Sample		**Sample of Size** $n = 5$				**Sample Mean**
1	36.48	39.94	42.57	39.53	33.81	38.47
2	43.13	37.97	42.41	39.61	43.30	41.28
3	41.64	39.01	37.77	38.94	41.10	39.69
4	40.37	43.49	37.60	40.14	38.88	40.10
5	38.62	33.43	45.17	42.66	39.98	39.97
6	38.98	41.35	36.80	43.56	39.92	40.12
7	42.48	37.00	35.87	39.62	38.74	38.74
8	39.38	37.02	41.60	40.34	37.62	39.19
9	42.82	45.77	35.16	42.56	39.75	41.21
10	36.19	35.20	37.74	40.46	37.47	37.41
11	36.59	41.62	42.18	39.23	39.26	39.78
12	38.57	42.13	45.39	38.22	46.18	42.10
13	38.40	39.06	43.60	31.46	37.03	37.91
14	34.29	47.73	37.27	41.82	33.33	38.89
15	42.28	43.29	37.69	37.32	40.06	40.13
16	34.31	43.58	40.02	41.13	42.99	40.41
17	38.71	39.03	39.39	42.62	38.41	39.63
18	38.63	39.66	39.47	41.13	38.01	39.38
19	39.09	33.86	37.57	41.65	35.22	37.48
20	40.94	37.50	38.72	41.64	35.48	38.86
21	38.72	35.89	37.82	35.04	37.06	36.91
22	39.64	36.30	35.54	40.40	38.74	38.12
23	38.22	38.49	33.60	40.18	39.07	37.91
24	40.93	40.53	37.55	37.30	37.16	38.69
25	33.27	38.92	37.14	39.90	33.83	36.61
26	39.44	37.28	35.70	41.97	36.80	38.24
27	38.83	41.41	38.87	39.40	37.20	39.14
28	40.10	36.96	35.73	43.00	38.11	38.78

Table 3 (cont'd)

29	41.93	36.57	37.55	35.14	38.75	37.99
30	31.25	38.85	39.25	35.07	39.77	36.84
31	38.47	34.45	30.43	41.76	41.61	37.34
32	37.98	35.56	43.97	44.96	37.81	40.06
33	43.34	40.94	35.17	41.74	37.59	39.76
34	39.80	44.44	37.53	40.52	41.95	40.85
35	41.98	42.02	40.73	40.47	36.81	40.40
36	40.98	35.08	34.61	40.78	37.26	37.74
37	35.75	40.81	40.13	35.99	36.52	37.84
38	36.39	45.97	40.59	37.64	42.42	40.60
39	36.20	35.63	37.43	38.35	34.81	36.48
40	33.58	33.87	41.60	45.10	38.68	38.57
41	31.77	38.34	41.79	37.93	40.83	38.13
42	43.03	33.12	34.98	36.58	37.78	37.10
43	35.76	35.17	42.58	39.10	41.08	38.74
44	38.44	38.45	35.93	35.32	44.60	38.55
45	44.54	41.88	35.84	42.64	42.38	41.46
46	41.89	36.81	41.83	40.24	39.28	40.01
47	38.00	40.08	35.57	34.44	39.51	37.52
48	39.92	38.05	39.96	38.04	32.11	37.62
49	36.37	38.62	32.25	41.35	40.91	37.90
50	34.38	36.65	32.97	39.93	41.34	37.05
51	40.32	39.80	41.00	38.62	38.24	39.60
52	37.95	45.26	38.67	34.96	41.13	39.59
53	36.82	42.63	41.62	39.43	37.48	39.60
54	41.63	37.65	38.58	39.03	37.53	38.88
55	37.91	37.20	38.72	36.87	45.40	39.22
56	41.05	34.01	39.11	38.23	35.74	37.63
57	42.09	45.44	35.52	39.87	37.28	40.04
58	39.31	35.79	37.82	39.15	35.57	37.53
59	41.16	39.98	41.11	39.21	39.98	40.29
60	35.68	45.60	39.34	36.65	43.30	40.11
61	36.07	39.63	42.55	41.72	36.81	39.36
62	38.97	36.83	41.01	38.12	35.27	38.04
63	33.70	39.15	34.81	34.13	39.00	36.16
64	37.19	34.69	36.21	34.34	39.07	36.30
65	33.99	44.87	42.52	40.22	39.26	40.17
66	41.40	27.62	34.57	40.08	34.65	35.66
67	40.14	34.45	38.26	38.09	39.72	38.13
68	33.64	42.62	32.08	34.30	37.34	36.00
69	35.36	39.02	43.98	41.19	32.47	38.40
70	43.26	37.85	35.82	37.11	36.22	38.05
71	36.24	38.07	33.38	38.43	39.88	37.20
72	38.55	43.06	41.07	36.58	37.02	39.26
73	41.26	36.99	36.17	38.98	36.03	37.89
74	37.31	38.41	41.18	39.76	39.64	39.26
75	36.26	41.84	42.50	37.70	41.21	39.90
76	39.27	38.61	44.53	38.08	35.01	39.10

			Table 3 (cont'd)			
77	39.14	40.83	39.83	37.78	36.51	38.82
78	42.53	43.41	41.01	33.71	39.47	40.03
79	45.34	32.61	33.81	39.03	40.32	38.22
80	36.31	35.55	37.12	38.74	40.80	37.70
81	31.40	41.80	40.15	42.53	37.62	38.70
82	41.01	39.02	39.68	36.61	38.44	38.95
83	34.15	36.19	35.98	36.02	36.32	35.73
84	31.50	37.61	43.29	39.82	38.78	38.20
85	43.26	34.01	41.18	40.23	39.28	39.59
86	41.76	41.40	39.02	38.20	39.42	39.96
87	37.06	35.95	39.98	40.00	43.36	39.27
88	41.01	37.56	36.95	39.71	37.97	38.64
89	34.97	38.36	36.30	38.48	34.24	36.47
90	38.38	38.94	40.96	36.13	35.98	38.08
91	39.41	30.78	37.66	37.31	42.04	37.44
92	39.83	35.88	30.20	45.07	40.06	38.21
93	36.25	39.56	34.53	40.69	37.03	37.61
94	45.64	40.66	44.51	40.50	39.43	42.15
95	37.63	44.77	38.31	36.53	38.41	39.13
96	39.78	33.34	43.42	43.63	38.77	39.79
97	41.48	37.39	38.62	43.83	34.26	39.12
98	37.68	40.66	38.93	40.94	37.54	39.15
99	39.72	32.61	32.62	40.35	38.65	36.79
100	39.25	41.06	41.17	38.30	38.24	39.60

Step 2: We compute the sample means for each of the 100 samples as shown in Table 3.

Step 3: We draw a histogram of the 100 sample means. See Figure 2.

Figure 2

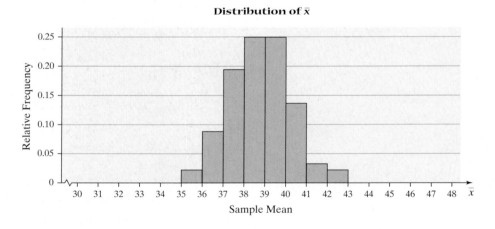

Step 4: The mean of the 100 sample means is 38.72 inches, and the standard deviation is 1.374 inches.

Look back at the histogram of the population data drawn in Figure 7 on page 323 from Section 7.1. Notice the center of the population distribution is the same as the center of the sampling distribution, but the spread of the population distribution is greater than that of the sampling distribution.

In Example 2 we were told that the data are approximately normal, with mean $\mu = 38.72$ inches and $\sigma = 3.17$ inches. The histogram in Figure 2 indicates that the distribution of sample means also appears to be normally distributed. In addition, the mean of the sample means is 38.72 inches, but the standard deviation is only 1.374 inches. We might conclude the following regarding the sampling distribution of $\bar{x}$.

1. *Shape*: It is normally distributed.
2. *Center*: It has mean equal to the mean of the population.
3. *Spread*: It has standard deviation less than the standard deviation of the population.

A question that we might ask is, "What role does n, the sample size, play in the sampling distribution of $\bar{x}$?" Suppose the sample mean is computed for samples of size $n = 1$ through $n = 200$. That is, the sample mean is recomputed each time an additional individual is added to the sample. The sample mean is then plotted against the sample size in Figure 3.

Figure 3

From the graph, we see that, as the sample size n increases, the sample mean gets closer to the population mean. This concept is known as the *Law of Large Numbers*.

The Law of Large Numbers

As additional observations are added to the sample, the difference between the sample mean, $\bar{x}$, and the population mean μ approaches zero.

/ *In Other Words*

As the sample size increases, the sample mean gets closer to the population mean.

So, according to the Law of Large Numbers, the more individuals we sample, the closer the sample mean gets to the population mean. This result implies that there is less variability in the distribution of the sample mean as the sample size increases. We demonstrate this result in the next example.

EXAMPLE 3   **The Impact of Sample Size on Sampling Variability**

Problem: Repeat the problem in Example 2 with a sample of size $n = 15$.

Approach: The approach will be identical to that presented in Example 2, except that we let $n = 15$ instead of $n = 5$.

Solution: Figure 4(a) shows the histogram of the sample means using the same scale as Figure 2. Compare this with the histogram in Figure 2. Notice that the histogram in Figure 4(a) shows less dispersion than the histogram in Figure 2. This implies that there is less variability in the distribution of $\bar{x}$ with $n = 15$.

We redraw the histogram in Figure 4(a) using a different class width in Figure 4(b). The histogram in Figure 4(b) is symmetric and mound shaped. This is an indication that the distribution of the sample mean is approximately normally distributed. The mean of the 100 sample means is 38.72 inches (just as in Example 2); however, the standard deviation is now 0.81 inches.

Figure 4

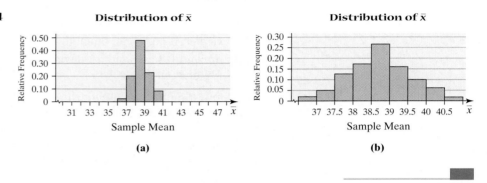

(a)

(b)

From the results of Examples 2 and 3, we conclude that, as the sample size n increases, the standard deviation of the distribution of $\bar{x}$ decreases. Although the proof is beyond the scope of this text, we should be convinced that the following result is reasonable.

The Mean and Standard Deviation of the Sampling Distribution of $\bar{x}$

Suppose that a simple random sample of size n is drawn from a large population* with mean μ and standard deviation σ. The sampling distribution of $\bar{x}$ will have mean $\mu_{\bar{x}} = \mu$ and standard deviation $\sigma_{\bar{x}} = \dfrac{\sigma}{\sqrt{n}}$. The standard deviation of the sampling distribution of $\bar{x}$ is called the **standard error of the mean** and is denoted $\sigma_{\bar{x}}$.

In Other Words

Regardless of the distribution of the population, the sampling distribution of $\bar{x}$ will have a mean equal to the mean of the population and a standard deviation equal to the standard deviation of the population divided by the square root of the sample size!

For the population presented in Example 2, if we draw a simple random sample of size $n = 5$, the sampling distribution $\bar{x}$ will have mean $\mu_{\bar{x}} = 38.72$ inches and standard deviation

$$\sigma_{\bar{x}} = \frac{\sigma}{\sqrt{n}} = \frac{3.17}{\sqrt{5}} \approx 1.418 \text{ inches}$$

Now Work Problem 11.

---

*Technically, we assume that we are drawing a simple random sample from an infinite population. For populations of finite size N, $\sigma_{\bar{x}} = \sqrt{\dfrac{N-n}{N-1}} \cdot \dfrac{\sigma}{\sqrt{n}}$. However, if the sample size is less than 5% of the population size ($n < 0.05N$), the effect of $\sqrt{\dfrac{N-n}{N-1}}$ (the finite population correction factor) can be ignored without affecting the results.

Now that we know how to determine the mean and standard deviation for any sampling distribution of $\bar{x}$, we can concentrate on the shape of the distribution. Refer back to Figures 2 and 4 from Examples 2 and 3. Recall that the population from which the sample was drawn was normal. The shapes of these histograms imply that the sampling distribution of $\bar{x}$ is also normal. This leads us to believe that if the population is normal the distribution of the sample mean is also normal.

Figure 5

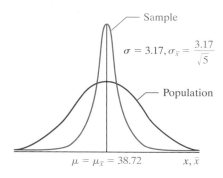

$\sigma = 3.17, \sigma_{\bar{x}} = \dfrac{3.17}{\sqrt{5}}$

Sample

Population

$\mu = \mu_{\bar{x}} = 38.72$ $x, \bar{x}$

> **The Shape of the Sampling Distribution of $\bar{x}$ If X Is Normal**
>
> If a random variable X is normally distributed, the distribution of the sample mean, $\bar{x}$, is normally distributed.

For example, the height of 3-year-old females is modeled by a normal random variable with mean $\mu = 38.72$ inches and standard deviation $\sigma = 3.17$ inches. The distribution of the sample mean, $\bar{x}$, the mean height of a simple random sample of $n = 5$ three-year-old females, is normal with mean $\mu_{\bar{x}} = 38.72$ inches and standard deviation $\sigma_{\bar{x}} = \dfrac{3.17}{\sqrt{5}}$ inches. See Figure 5.

EXAMPLE 4   **Describing the Distribution of the Sample Mean**

Problem: The height, X, of all 3-year-old females is approximately normally distributed with mean $\mu = 38.72$ inches and standard deviation $\sigma = 3.17$ inches. Compute the probability that a simple random sample of size $n = 10$ results in a sample mean greater than 40 inches. That is, compute $P(\bar{x} > 40)$.

Approach: The random variable X is normally distributed, so the sampling distribution of $\bar{x}$ will also be normally distributed. The mean of the sampling distribution is $\mu_{\bar{x}} = \mu$, and its standard deviation is $\sigma_{\bar{x}} = \dfrac{\sigma}{\sqrt{n}}$. We convert the random variable $\bar{x} = 40$ to a Z-score and then find the area under the standard normal curve to the right of this Z-score.

Figure 6

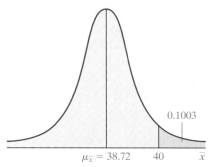

0.1003

$\mu_{\bar{x}} = 38.72$ 40 $\bar{x}$

Solution: The sample mean is normally distributed with mean $\mu_{\bar{x}} = 38.72$ inches and standard deviation $\sigma_{\bar{x}} = \dfrac{\sigma}{\sqrt{n}} = \dfrac{3.17}{\sqrt{10}} = 1.002$ inch.

Figure 6 displays the normal curve with the area we wish to compute shaded. We convert the random variable $\bar{x} = 40$ to a Z-score and obtain

$$Z = \frac{\bar{x} - \mu_{\bar{x}}}{\sigma_{\bar{x}}} = \frac{\bar{x} - \mu_{\bar{x}}}{\dfrac{\sigma}{\sqrt{n}}} = \frac{40 - 38.72}{1.002} = 1.28$$

The area to the right of $Z = 1.28$ is $1 - 0.8997 = 0.1003$.

Interpretation: The probability of obtaining a sample mean greater than 40 inches from a population whose mean is 38.72 inches is 0.1003. That is, $P(\bar{x} \geq 40) = 0.1003$. If we take 1000 simple random samples of $n = 10$ three-year-olds from this population and if the population mean is 38.72 inches, about 100 of the samples will result in a mean height that is 40 inches or more. ▪

Note to Instructor

Emphasize the interpretation of the probabilities as a prelude to P-values.

Now Work Problem 19.

Describe the Distribution of the Sample Mean for Samples Obtained from a Population That Is Not Normal

What if the population from which the sample is drawn is not normal?

EXAMPLE 5   **Sampling from a Population That Is Not Normal**

Figure 7

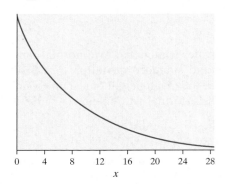

Problem: Figure 7 shows the graph of an *exponential density function* with mean and standard deviation equal to 10. The exponential distribution is used to model lifetimes of electronic components and to model the time required to serve a customer or repair a machine.

Clearly, the distribution of the population is not normal. Approximate the sampling distribution of $\bar{x}$ by obtaining, through simulation, 300 random samples of size (a) $n = 3$, (b) $n = 12$, and (c) $n = 30$ from the probability distribution.

Approach

Step 1: Use MINITAB, Excel, or some other statistical software to obtain 300 random samples for each sample size.

Step 2: Compute the sample mean of each of the 300 random samples.

Step 3: Draw a histogram of the 300 sample means.

Solution

Step 1: Using MINITAB, we obtain 300 random samples of size (a) $n = 3$, (b) $n = 12$, and (c) $n = 30$. For example, in the first random sample of size $n = 30$, we obtained the following results:

Note to Instructor
You may want to use the sampling distribution applet to illustrate this concept.

9.2	20.0	17.0	2.4	2.6	19.9	21.2	5.7	8.1	10.8
1.2	22.3	18.4	4.2	9.9	41.8	4.2	1.2	10.8	2.1
11.3	17.9	28.0	12.1	3.0	0.5	4.5	14.2	5.0	11.4

Step 2: We compute the mean of each of the 300 random samples, using MINITAB. For example, the sample mean of the first sample of size $n = 30$ is 11.36.

Step 3: Figure 8(a) displays the histogram of $\bar{x}$ that results from simulating 300 random samples of size $n = 3$ from an exponential distribution with $\mu = 10$ and $\sigma = 10$. Figure 8(b) displays the histogram of $\bar{x}$ that results from simulating 300 random samples of size $n = 12$, and Figure 8(c) displays the histogram of $\bar{x}$ that results from simulating 300 random samples of size $n = 30$.

Figure 8

(a) $n = 3$

(b) $n = 12$

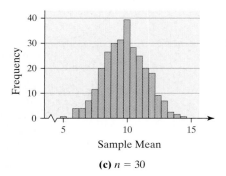

(c) $n = 30$

Notice that, as the sample size increases, the distribution of the sample mean becomes more normal, even though the population clearly is not normal! ■

We formally state the results of Example 5 as the *Central Limit Theorem*.

Note to Instructor

Technically, the parent population must have a finite variance.

In Other Words

For any population, regardless of its shape, as the sample size increases, the shape of the distribution of the sample mean becomes more "normal."

The Central Limit Theorem

Regardless of the shape of the population, the sampling distribution of $\bar{x}$ becomes approximately normal as the sample size n increases.

So, if the random variable X is normally distributed, the sampling distribution of $\bar{x}$ will be normal. If the sample size is large enough, the sampling distribution of $\bar{x}$ will be approximately normal, *regardless of the shape of the distribution of X*. But how large does the sample size need to be before we can say that the sampling distribution of $\bar{x}$ is approximately normal? The answer depends on the shape of the distribution of the population. Distributions that are highly skewed will require a larger sample size for the distribution of $\bar{x}$ to become approximately normal.

For example, from Example 5 we see that this right skewed distribution required a sample size of about 30 before the distribution of the sample mean is approximately normal. However, Figure 9(a) shows a uniform distribution for $0 \leq X \leq 10$. Figure 9(b) shows the distribution of the sample mean for $n = 3$. Figure 9(c) shows the distribution of the sample mean for $n = 12$, and Figure 9(d) shows the distribution of the sample mean for $n = 30$. Notice that even for $n = 3$ the distribution of the sample mean is approximately normal.

Figure 9

(a) Uniform Distribution

(b) Distribution of $\bar{x}$; $n = 3$

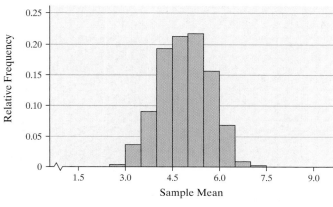

(c) Distribution of $\bar{x}$; $n = 12$

(d) Distribution of $\bar{x}$; $n = 30$

Table 4 shows the distribution of the cumulative number of children for 50- to 54-year-old mothers who had a live birth in 2002.

Table 4	
x (number of children)	P(x)
1	0.241
2	0.257
3	0.172
4	0.119
5	0.103
6	0.027
7	0.031
8	0.050

Source: U.S. Census Bureau

Figure 10(a) shows the probability histogram for this distribution. Figure 10(b) shows the distribution of the sample mean number of children for a random sample of $n = 3$ mothers. Figure 10(c) shows the distribution of the sample mean number of children for a random sample of $n = 12$ mothers, and Figure 10(d) shows the distribution of the sample mean for a random sample of $n = 30$ mothers. In this instance, the distribution of the sample mean is very close to normal for $n = 12$.

Figure 10

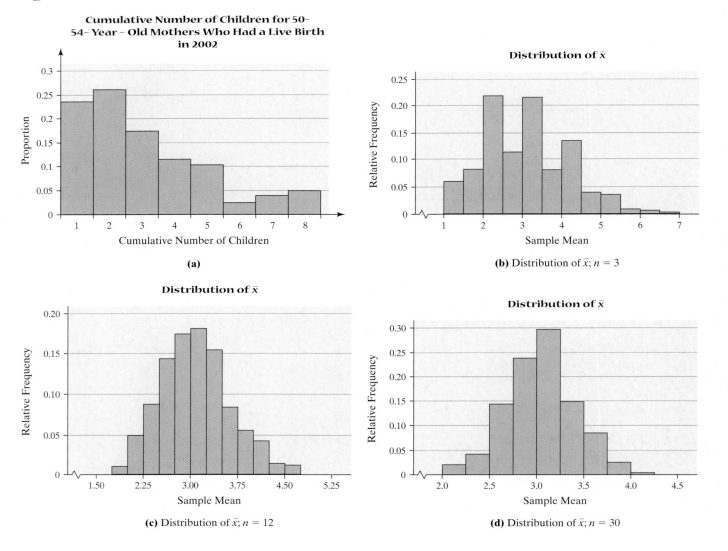

(a)

(b) Distribution of $\bar{x}$; $n = 3$

(c) Distribution of $\bar{x}$; $n = 12$

(d) Distribution of $\bar{x}$; $n = 30$

Note to Instructor

While admittedly conservative, requiring n ≥ 30 for nonnormal distributions provides a clear-cut decision rule for students to use.

The results of Example 5 and Figures 9 and 10 confirm that the shape of the distribution of the population dictates the size of the sample required before the distribution of the sample mean can be called normal. With that said, so that we err on the side of caution, we will say that the distribution of the sample mean is approximately normal provided that the sample size is greater than or equal to 30 if the distribution of the population is unknown or not normal.

EXAMPLE 6   Applying the Central Limit Theorem

!CAUTION

⬤ The Central Limit Theorem has to do only with the *shape* of the distribution of the sample mean, not with its center and spread! The mean of the distribution of $\overline{x}$ is μ and the standard deviation of $\overline{x}$ is $\dfrac{\sigma}{\sqrt{n}}$, regardless of the size of the sample, n.

Problem: According to the U.S. Department of Agriculture, the mean calorie intake of males 20 to 39 years old is $\mu = 2716$, with standard deviation $\sigma = 72.8$. Suppose a nutritionist analyzes a simple random sample of $n = 35$ males between the ages of 20 and 39 years old and obtains a sample mean calorie intake of $\overline{x} = 2750$ calories. What is the probability that a random sample of 35 males between the ages of 20 and 39 years old would result in a sample mean of 2750 calories or higher? Are the results of the survey unusual? Why?

Approach

Step 1: We recognize that we are computing a probability regarding a sample mean, so we need to know the sampling distribution of $\overline{x}$. Because the population from which the sample is drawn is not known to be normal, the sample size must be greater than or equal to 30 to use the results of the Central Limit Theorem.

Step 2: Determine the mean and standard deviation of the sampling distribution of $\overline{x}$.

Step 3: Convert the sample mean to a Z-score.

Step 4: Use Table IV to find the area under the normal curve.

Solution

Step 1: Because the sample size is $n = 35$, the Central Limit Theorem states that the sampling distribution of $\overline{x}$ is approximately normal.

Step 2: The mean of the sampling distribution of $\overline{x}$ will equal the mean of the population, so $\mu_{\overline{x}} = 2716$. The standard deviation of the sampling distribution of $\overline{x}$ will equal the standard deviation of the population divided by the square root of the sample size, so $\sigma_{\overline{x}} = \dfrac{\sigma}{\sqrt{n}} = \dfrac{72.8}{\sqrt{35}} = 12.305$.

Step 3: We convert $\overline{x} = 2750$ to a Z-score.

$$Z = \frac{2750 - 2716}{\dfrac{72.8}{\sqrt{35}}} = 2.76$$

Step 4: We wish to know the probability that a random sample of $n = 35$ from a population whose mean is 2716 results in a sample mean of at least 2750. That is, we wish to know $P(\overline{x} \geq 2750)$. See Figure 11.

Historical Note

Pierre Simon Laplace was born on March 23, 1749 in Normandy, France. At age 16, Laplace attended Caen University, where he studied theology. While there, his mathematical talents were discovered, which led him to Paris, where he got a job as professor of mathematics at the École Militaire. In 1773, Laplace was elected to the Académie des Sciences. Laplace was not humble. It is reported that, in 1780, he stated that he was the best mathematician in Paris. In 1799, Laplace published the first two volumes of *Méchanique céleste*, in which he discusses methods for calculating the motion of the planets. On April 9, 1810, Laplace presented the Central Limit Theorem to the Academy.

Figure 11

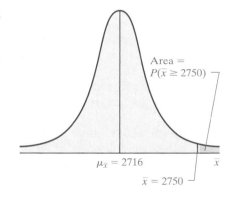

This probability is represented by the area under the standard normal curve to the right of $Z = 2.76$.

$$P(\overline{x} \geq 2750) = P(Z \geq 2.76) = 1 - 0.9971 = 0.0029$$

Interpretation: If the population mean is 2716 calories, the probability that a random sample of 35 males between the ages of 20 and 39 will result in a sample mean calorie intake of 2750 calories or higher is 0.0029. This means that fewer than 1 sample in 100 will result in a sample mean of 2750 calories or higher if the population mean is 2716 calories. We can conclude one of two things based on this result.

1. The mean number of calories for males 20 to 39 years old is 2716, and we just happened to randomly select 35 individuals who, on average, consume more calories.

2. The mean number of calories consumed by 20- to 39-year-old males is higher than 2716 calories.

In statistical inference, we are inclined to accept the second possibility as the more reasonable choice. We recognize there is a possibility that our conclusion is incorrect.

Now Work Problem 25.

Summary: Shape, Center, and Spread of the Distribution of $\overline{x}$

Shape, Center, and Spread of Population	Distribution of the Sample Mean		
	Shape	**Center**	**Spread**
Normal with mean μ and standard deviation σ	Regardless of the sample size n, the shape of the distribution of the sample mean is normal	$\mu_{\overline{x}} = \mu$	$\sigma_{\overline{x}} = \dfrac{\sigma}{\sqrt{n}}$
Population is not normal with mean μ and standard deviation σ	As the sample size n increases, the distribution of the sample mean becomes approximately normal	$\mu_{\overline{x}} = \mu$	$\sigma_{\overline{x}} = \dfrac{\sigma}{\sqrt{n}}$

MAKING AN INFORMED DECISION

How Much Time Do You Spend in a Day . . . ?

The American Time Use Survey is a survey of adult Americans conducted by the Bureau of Labor Statistics. The purpose of the survey is to learn how Americans allocate their time in a day. As a reporter for the school newspaper, you wish to file a report that compares the typical student at your school to the rest of America.

For those Americans who are currently attending school, the mean amount of time spent in class in a day is 5.11 hours, and the mean amount of time spent studying and doing homework is 2.50 hours. The mean amount of time Americans spend watching television each day is 2.57 hours.

Conduct a survey of 35 randomly selected full-time students at your school in which you ask the following questions:

(a) On average, how much time do you spend attending class each day?

(b) On average, how much time do you spend studying and doing homework each day?

(c) On average, how much time do you spend watching television each day? If you do not watch television, write 0 hours.

1. For each question, describe the sampling distribution of the sample mean. Use the national norms as estimates for the population means for each variable. Use the sample standard deviation as an estimate of the population standard deviation.

2. Compute probabilities regarding the values of the statistics obtained from the survey. Are any of the results unusual?

Write an article for your newspaper reporting your findings.

3. Standard error; mean   5. $\mu_{\bar{x}} = \mu; \sigma_{\bar{x}} = \dfrac{\sigma}{\sqrt{n}}$

8.1 ASSESS YOUR UNDERSTANDING

Concepts and Vocabulary

1. Explain what a sampling distribution is.
2. State the Central Limit Theorem.
3. The standard deviation of the sampling distribution of $\bar{x}$, denoted $\sigma_{\bar{x}}$, is called the _____ _____ of the _____.
4. As the sample size increases, the difference between the sample mean, $\bar{x}$, and the population mean, μ, approaches _____. zero
5. What are the mean and standard deviation of the sampling distribution of $\bar{x}$, regardless of the distribution of the population from which the sample was drawn?
6. If a random sample of size $n = 6$ is taken from a population, what is required to say that the sampling distribution of $\bar{x}$ is approximately normal? Population is normal

9. Normal; $\mu_{\bar{x}} = 30$, $\sigma_{\bar{x}} = \dfrac{8}{\sqrt{10}}$ 10. No; $n \geq 30$, $\mu_{\bar{x}} = 50$, $\sigma_{\bar{x}} = \dfrac{4}{\sqrt{40}}$

7. To cut the standard error of the mean in half, the sample size must be increased by a factor of _____. four
8. *True or False*: The mean and standard deviation of the distribution of $\bar{x}$ is $\mu_{\bar{x}} = \mu$ and $\sigma_{\bar{x}} = \dfrac{\sigma}{\sqrt{n}}$, respectively, even if the population is not normal. True
9. Suppose a simple random sample of size $n = 10$ is obtained from a population that is normally distributed with $\mu = 30$ and $\sigma = 8$. What is the sampling distribution of $\bar{x}$?
10. Suppose a simple random sample of size $n = 40$ is obtained from a population with $\mu = 50$ and $\sigma = 4$. Does the population need to be normally distributed for the sampling distribution of $\bar{x}$ to be approximately normally distributed? Why? What is the sampling distribution of $\bar{x}$?

Skill Building

In Problems 11–14, determine $\mu_{\bar{x}}$ and $\sigma_{\bar{x}}$ from the given parameters of the population and the sample size.

11. NW $\mu = 80$, $\sigma = 14$, $n = 49$ $\mu_{\bar{x}} = 80$, $\sigma_{\bar{x}} = 2$

12. $\mu = 64$, $\sigma = 18$, $n = 36$ $\mu_{\bar{x}} = 64$; $\sigma_{\bar{x}} = 3$

13. $\mu = 52$, $\sigma = 10$, $n = 21$ $\mu_{\bar{x}} = 52$; $\sigma_{\bar{x}} = \dfrac{10}{\sqrt{21}} \approx 2.182$

14. $\mu = 27$, $\sigma = 6$, $n = 15$ $\mu_{\bar{x}} = 27$; $\sigma_{\bar{x}} = \dfrac{6}{\sqrt{15}} \approx 1.549$

15. Suppose a simple random sample of size $n = 49$ is obtained from a population with $\mu = 80$ and $\sigma = 14$.
    (a) Describe the sampling distribution of $\bar{x}$.
    (b) What is $P(\bar{x} > 83)$? 0.0668
    (c) What is $P(\bar{x} \leq 75.8)$? 0.0179
    (d) What is $P(78.3 < \bar{x} < 85.1)$? 0.7969
    (15a) Approximately normal with $\mu_{\bar{x}} = 80$ and $\sigma_{\bar{x}} = 2$

16. Suppose a simple random sample of size $n = 36$ is obtained from a population with $\mu = 64$ and $\sigma = 18$.
    (a) Describe the sampling distribution of $\bar{x}$.
    (b) What is $P(\bar{x} < 62.6)$? 0.3192
    (c) What is $P(\bar{x} \geq 68.7)$? 0.0582
    (d) What is $P(59.8 < \bar{x} < 65.9)$? 0.6549
    (16c) Approximately normal with $\mu_{\bar{x}} = 64$ and $\sigma_{\bar{x}} = 3$

(17a) Normal; Normal with $\mu_{\bar{x}} = 64$ and $\sigma_{\bar{x}} = \dfrac{17}{\sqrt{12}}$

17. Suppose a simple random sample of size $n = 12$ is obtained from a population with $\mu = 64$ and $\sigma = 17$.
    (a) What must be true regarding the distribution of the population in order to use the normal model to compute probabilities regarding the sample mean? Assuming this condition is true, describe the sampling distribution of $\bar{x}$.
    (b) Assuming the requirements described in part (a) are satisfied, determine $P(\bar{x} < 67.3)$. 0.7486
    (c) Assuming the requirements described in part (a) are satisfied, determine $P(\bar{x} \geq 65.2)$. 0.4052

18. Suppose a simple random sample of size $n = 20$ is obtained from a population with $\mu = 64$ and $\sigma = 17$.
    (a) What must be true regarding the distribution of the population in order to use the normal model to compute probabilities regarding the sample mean? Assuming this condition is true, describe the sampling distribution of $\bar{x}$.
    (b) Assuming the requirements described in part (a) are satisfied, determine $P(\bar{x} < 67.3)$. 0.8078
    (c) Assuming the requirements described in part (a) are satisfied, determine $P(\bar{x} \geq 65.2)$. 0.3745
    (d) Compare the results obtained in parts (b) and (c) with the results obtained in parts (b) and (c) in Problem 17. What effect does increasing the sample size have on the probabilities? Why do you think this is the case?
    (18a) Normal; Normal with $\mu_{\bar{x}} = 64$ and $\sigma_{\bar{x}} = \dfrac{17}{\sqrt{20}}$

Applying the Concepts

19. **Gestation Period** The length of human pregnancies is approximately normally distributed with mean $\mu = 266$ days and standard deviation $\sigma = 16$ days.
    (a) What is the probability a randomly selected pregnancy lasts less than 260 days? 0.3520
    (b) What is the probability that a random sample of 20 pregnancies has a mean gestation period of 260 days or less? 0.0465

(19c) 0.0040
    (c) What is the probability that a random sample of 50 pregnancies has a mean gestation period of 260 days or less?
    (d) What might you conclude if a random sample of 50 pregnancies resulted in a mean gestation period of 260 days or less?
    (e) What is the probability a random sample of size 15 will have a mean gestation period within 10 days of the mean? 0.9844

(20a) 0.3015

20. **Serum Cholesterol** As reported by the U.S. National Center for Health Statistics, the mean serum high-density-lipoprotein (HDL) cholesterol of females 20 to 29 years old is $\mu = 53$. If serum HDL cholesterol is normally distributed with $\sigma = 13.4$, answer the following questions:

(a) What is the probability that a randomly selected female 20 to 29 years old has a serum cholesterol above 60?

(b) What is the probability that a random sample of 15 female 20- to 29-year-olds has a mean serum cholesterol above 60? 0.0217

(c) What is the probability that a random sample of 20 female 20- to 29-year-olds has a mean serum cholesterol above 60? 0.0096

(d) What effect does increasing the sample size have on the probability? Provide an explanation for this result.

(e) What might you conclude if a random sample of 20 female 20- to 29-year-olds has a mean serum cholesterol above 60?

21. **Old Faithful** The most famous geyser in the world, Old Faithful in Yellowstone National Park, has a mean time between eruptions of 85 minutes. If the interval of time between eruptions is normally distributed with standard deviation 21.25 minutes, answer the following questions: (*Source*: www.unmuseum.org)   (21a) 0.3192

(a) What is the probability that a randomly selected time interval between eruptions is longer than 95 minutes?

(b) What is the probability that a random sample of 20 time intervals between eruptions has a mean longer than 95 minutes? 0.0179

(c) What is the probability that a random sample of 30 time intervals between eruptions has a mean longer than 95 minutes? 0.0049

(d) What effect does increasing the sample size have on the probability? Provide an explanation for this result.

(e) What might you conclude if a random sample of 30 time intervals between eruptions has a mean longer than 95 minutes?

22. **Medical Residents** In a 2003 study, the Accreditation Council for Graduate Medical Education found that medical residents work an average of 81.7 hours per week. Suppose the number of hours worked per week by medical residents is normally distributed with standard deviation 6.9 hours per week. (*Source*: www.medrecinst.com)

(a) What is the probability that a randomly selected medical resident works less than 75 hours per week? 0.1660

(b) What is the probability that the mean number of hours worked per week by a random sample of five medical residents is less than 75 hours? 0.0150

(c) What is the probability that the mean number of hours worked per week by a random sample of eight medical resident is less than 75 hours? 0.0030

(d) What might you conclude if the mean number of hours worked per week by a random sample of eight medical residents is less than 75 hours?

23. **Rates of Return in Stocks** The S&P 500 is a collection of 500 stocks of publicly traded companies. Using data obtained from Yahoo!Finance, the monthly rates of return of the S&P 500 since 1950 are normally distributed. The mean rate of return is 0.007233 (0.7233%), and the standard deviation for rate of return is 0.04135 (4.135%).

(a) What is the probability that a randomly selected month has a positive rate of return? That is, what is $P(x > 0)$? 0.5675

(b) Treating the next 12 months as a simple random sample, what is the probability that the mean monthly rate of return will be positive? That is, with $n = 12$, what is $P(\overline{x} > 0)$? 0.7291

(c) Treating the next 24 months as a simple random sample, what is the probability that the mean monthly rate of return will be positive? 0.8051

(d) Treating the next 36 months as a simple random sample, what is the probability that the mean monthly rate of return will be positive? 0.8531

(e) Use the results of parts (b)–(d) to describe the likelihood of earning a positive rate of return on stocks as the investment time horizon increases.

24. **Gas Mileage** Based on tests of the Chevrolet Cobalt, engineers have found that the miles per gallon in highway driving are normally distributed, with a mean of 32 miles per gallon and a standard deviation 3.5 miles per gallon.

(a) What is the probability that a randomly selected Cobalt gets more than 34 miles per gallon? 0.2843

(b) Suppose that 10 Cobalts are randomly selected and the miles per gallon for each car are recorded. What is the probability that the mean miles per gallon exceed 34 miles per gallon? 0.0351

(c) Suppose that 20 Cobalts are randomly selected and the miles per gallon for each car are recorded. What is the probability that the mean miles per gallon exceed 34 miles per gallon? Would this result be unusual?
                                                                0.0052; yes

25. **Oil Change** The shape of the distribution of the time required to get an oil change at a 10-minute oil-change facility is unknown. However, records indicate that the mean time for an oil change is 11.4 minutes and the standard deviation for oil-change time is 3.2 minutes.

(a) To compute probabilities regarding the sample mean using the normal model, what size sample would be required? At least 30

(b) What is the probability that a random sample of $n = 40$ oil changes results in a sample mean time less than 10 minutes? 0.0028

26. **Time Spent in the Drive-Through** The quality-control manager of a Long John Silver's restaurant wishes to analyze the length of time a car spends at the drive-through window waiting for an order. According to records obtained from the restaurants, it is determined that the mean time spent at the window is 59.3 seconds with a standard deviation of 13.1 seconds. The distribution of time at the window is skewed right (data based on information provided by Danica Williams, student at Joliet Junior College).

(a) To obtain probabilities regarding a sample mean using the normal model, what size sample is required?

(b) The quality-control manager wishes to use a new delivery system designed to get cars through the drive-through system faster. A random sample of 40 cars results in a sample mean time spent at the window of 56.8 seconds. What is the probability of obtaining a sample mean of 56.8 seconds or less assuming the population mean is 59.3 seconds? Do you think that the new system is effective? 0.1131

(26a) At least 30

(27b) $\mu_{\bar{x}} = 3$, $\sigma_{\bar{x}} = \dfrac{\sqrt{3}}{\sqrt{50}}$

27. Insect Fragments The Food and Drug Administration sets Food Defect Action Levels (FDALs) for some of the various foreign substances that inevitably end up in the food we eat and liquids we drink. For example, the FDAL for insect filth in peanut butter is 3 insect fragments (larvae, eggs, body parts, and so on) per 10 grams. A random sample of 50 ten-gram portions of peanut butter is obtained and results in a sample mean of $\bar{x} = 3.6$ insect fragments per ten-gram portion.

(a) Why is the sampling distribution of $\bar{x}$ approximately normal?

(b) What is the mean and standard deviation of the sampling distribution of $\bar{x}$ assuming $\mu = 3$ and $\sigma = \sqrt{3}$.

(c) Suppose a simple random sample of $n = 50$ ten-gram samples of peanut butter results in a sample mean of 3.6 insect fragments. What is the probability a simple random sample of 50 ten-gram portions results in a mean of at least 3.6 insect fragments? Is this result unusual? What might we conclude? *0.0071*

28. Burger King's Drive-Through Suppose cars arrive at Burger King's drive-through at the rate of 20 cars every hour between 12:00 noon and 1:00 P.M. A random sample of 40 one-hour time periods between 12:00 noon and 1:00 P.M. is selected and has 22.1 as the mean number of cars arriving.

(a) Why is the sampling distribution of $\bar{x}$ approximately normal?

(b) What is the mean and standard deviation of the sampling distribution of $\bar{x}$ assuming $\mu = 20$ and $\sigma = \sqrt{20}$.

(c) What is the probability that a simple random sample of 40 one-hour time periods results in a mean of at least 22.1 cars? Is this result unusual? What might we conclude? *0.0015; yes*

29. Blows to the Head In a 2003 study of the long-term effects of concussions in football players, researchers at Virginia Tech concluded that college football players receive a mean of 50 strong blows to the head, each with an average of 40G (40 times the force of gravity). Assume the standard deviation is 16 strong blows to the head. What is the probability that a random sample of 60 college football players results in a mean of 45 or fewer strong blows to the head? Would this be unusual? *0.0078; yes* (*Source:* Neuroscience for Kids, faculty.washington.edu/chudler/nfl.html)

30. Domestic Vacation Costs According to the AAA (American Automobile Association, April 20, 2005), a family of two adults and two children on vacation in the United States will pay an average of $247.02 per day for food and lodging with a standard deviation of $60.41 per day. Suppose a random sample of 50 families of two adults and two children is selected and monitored while on vacation in the United States. What is the probability that the average daily expenses for the sample are over $260.00 per day? Would this be unusual? *0.0643; no*

31. Sampling Distributions The following data represent the ages of the winners of the Academy Award for Best Actor for the years 1999–2004.

2004: Jamie Foxx	37
2003: Sean Penn	43
2002: Adrien Brody	29
2001: Denzel Washington	47
2000: Russell Crowe	36
1999: Kevin Spacey	40

(a) Compute the population mean, μ. $\mu \approx 38.7$ *years*

(b) List all possible samples with size $n = 2$. There should be $_6C_2 = 15$ samples.

(c) Construct a sampling distribution for the mean by listing the sample means and their corresponding probabilities. *(31d)* $\mu_{\bar{x}} \approx 38.7$ *years*

(d) Compute the mean of the sampling distribution.

(e) Compute the probability that the sample mean is within 3 years of the population mean age. *0.6*

(f) Repeat parts (b)–(e) using samples of size $n = 3$. Comment on the effect of increasing the sample size.

32. Sampling Distributions The following data represent the running lengths (in minutes) of the winners of the Academy Award for Best Picture for the years 1999–2004.

2004: *Million Dollar Baby*	132
2003: *The Lord of the Rings: The Return of the King*	201
2002: *Chicago*	112
2001: *A Beautiful Mind*	134
2000: *Gladiator*	155
1999: *American Beauty*	120

(a) Compute the population mean, μ. $\mu \approx 142.3$ *minutes*

(b) List all possible samples with size $n = 2$. There should be $_6C_2 = 15$ samples.

(c) Construct a sampling distribution for the mean by listing the sample means and their corresponding probabilities.

(d) Compute the mean of the sampling distribution.

(e) Compute the probability that the sample mean is within 15 minutes of the population mean running times.

(f) Repeat parts (b)–(e) using samples of size $n = 3$. Comment on the effect of increasing the sample size.

33. Simulation Scores on the Stanford–Binet IQ test are normally distributed with $\mu = 100$ and $\sigma = 16$.

(a) Use MINITAB, Excel, or some other statistical software to obtain 500 random samples of size $n = 20$.

(b) Compute the sample mean for each of the 500 samples.

(c) Draw a histogram of the 500 sample means. Comment on its shape.

(d) What do you expect the mean and standard deviation of the sampling distribution of the mean to be?

(e) Compute the mean and standard deviation of the 500 sample means. Are they close to the expected values?

(f) Compute the probability that a random sample of 20 people results in a sample mean greater than 108.

(g) What proportion of the 500 random samples had a sample mean IQ greater than 108? Is this result close to the theoretical value obtained in part (f)?

(32d) $\mu_{\bar{x}} \approx 142.3$ *minutes*   (32e) *0.467*   (33d) $\mu_{\bar{x}} = 100$, $\sigma_{\bar{x}} = \dfrac{16}{\sqrt{20}}$ (33f) *0.0125*

34. Sampling Distribution Applet Load the sampling distribution applet on your computer. Set the applet so that the population is bell shaped. Take note of the mean and standard deviation.

(a) Obtain 1000 random samples of size $n = 5$. Describe the distribution of the sample mean based on the results of the applet. According to statistical theory, what is the distribution of the sample mean?

(b) Obtain 1000 random samples of size $n = 10$. Describe the distribution of the sample mean based on the results of the applet. According to statistical theory, what is the distribution of the sample mean?

(c) Obtain 1000 random samples of size $n = 30$. Describe the distribution of the sample mean based on the results of the applet. According to statistical theory, what is the distribution of the sample mean?

(d) Compare the results of parts (a)–(c). How are they the same? How are they different?

35. Sampling Distribution Applet Load the sampling distribution applet on your computer. Set the applet so that the population is skewed or draw your own skewed distribution. Take note of the mean and standard deviation.

(a) Obtain 1000 random samples of size $n = 5$. Describe the distribution of the sample mean based on the results of the applet.

(b) Obtain 1000 random samples of size $n = 10$. Describe the distribution of the sample mean based on the results of the applet.

(c) Obtain 1000 random samples of size $n = 50$. Describe the distribution of the sample mean based on the results of the applet. According to statistical theory, what is the distribution of the sample mean?

(d) Compare the results of parts (a)–(c). How are they the same? How are they different? What impact does the sample size have on the shape of the distribution of the sample mean?

8.2 Distribution of the Sample Proportion

Preparing for this Section Before getting started, review the following:

• Applications of the Normal Distribution (Section 7.3, pp. 345–349)

Objectives

1 **Describe the sampling distribution of a sample proportion**

2 **Compute probabilities of a sample proportion**

1 **Describe the Sampling Distribution of a Sample Proportion**

Note to Instructor

Some texts will use π to denote the population proportion to be consistent with the convention of using Greek letters for population parameters. We choose to use p for the population proportion to avoid confusion with the irrational number π.

Suppose we want to determine the proportion of households in a 100-house homeowners association that favor an increase in the annual assessments to pay for neighborhood improvements. One approach that we might take is to survey all households and determine which were in favor of higher assessments. If 65 of the 100 households favor the higher assessment, the population proportion, p, of households in favor of a higher assessment is

$$p = \frac{65}{100}$$

$$= 0.65$$

Of course, it is rare to gain access to all the individuals in a population. For this reason, we usually obtain estimates of population parameters such as p.

Definition

Suppose a random sample of size n is obtained from a population in which each individual either does or does not have a certain characteristic. The **sample proportion**, denoted $\hat{p}$ (read "p-hat"), is given by

$$\hat{p} = \frac{x}{n}$$

where x is the number of individuals in the sample with the specified characteristic.* The sample proportion is a statistic that estimates the population proportion, p.

*For those who studied Section 6.2 on binomial probabilities, x can be thought of as the number of successes in n trials of a binomial experiment.

| EXAMPLE 1 | **Computing a Sample Proportion** |

Problem: Opinion Dynamics Corporation conducted a survey of 1000 adult Americans 18 years of age or older and asked, "Are you currently on some form of a low-carbohydrate diet?" Of the 1000 individuals surveyed, 150 indicated that they were on a low-carb diet. Find the sample proportion of individuals surveyed who are on a low-carb diet.

Approach: The sample proportion of individuals on a low-carb diet is found using the formula $\hat{p} = \dfrac{x}{n}$, where $x = 150$, the number of individuals in the survey with the characteristic "on a low-carb diet," and $n = 1000$.

Solution: Substituting $x = 150$ and $n = 1000$ into the formula $\hat{p} = \dfrac{x}{n}$, we have that $\hat{p} = \dfrac{150}{1000} = 0.15$. Opinion Dynamics Corporation estimates that 0.15 or 15% of adult Americans 18 years of age or older are on some form of low-carbohydrate diet.

If a second survey of 1000 American adults is conducted, it is likely the estimate of the proportion of Americans on a low-carbohydrate diet will be different because there will be different individuals in the sample. Because the value of $\hat{p}$ varies from sample to sample, it is a random variable and has a probability distribution.

To get a sense of the shape, center, and spread of the distribution of $\hat{p}$, we could repeat the exercise of obtaining simple random samples of 1000 adult Americans over and over. This would lead to a list of sample proportions. Each sample proportion would correspond to a simple random sample of 1000. A histogram of the sample proportions will give us a feel for the shape of the distribution of the sample proportion. The mean of the sample proportions will give us an idea of the center of the distribution of the sample proportion. The standard deviation of the sample proportions gives us an idea of the spread of the distribution of the sample proportions.

Rather than literally surveying 1000 adult Americans over and over again, we will use simulation to get an idea of the shape, center, and spread of the sampling distribution of the proportion.

| EXAMPLE 2 | **Using Simulation to Describe the Distribution of the Sample Proportion** |

Problem: According to the Centers for Disease Control, 17% (or 0.17) of Americans have high cholestrol. Simulate obtaining 100 simple random samples of size (a) $n = 10$, (b) $n = 40$, and (c) $n = 80$. Describe the shape, center, and spread of the distribution for each sample size.

Approach: Use MINITAB, Excel, or some other statistical software package to conduct the simulation. We will perform the following steps:

Step 1: Obtain 100 simple random samples of size $n = 10$ from the population.

Step 2: Compute the sample proportion for each of the 100 samples.

Step 3: Draw a histogram of the sample proportions.

Step 4: Compute the mean and standard deviation of the sample proportions.

We then repeat these steps for samples of size $n = 40$ and $n = 80$.

Solution

Step 1: We simulate obtaining 100 simple random samples each of size $n = 10$ using MINITAB.

Step 2: The first sample of size $n = 10$ results in none of the individuals having high cholesterol, so $\hat{p} = \dfrac{0}{10} = 0$. The second sample results in two of the individuals having high cholesterol, so $\hat{p} = \dfrac{2}{10} = 0.2$. Table 5 shows the sample proportions for all 100 simple random samples of size $n = 10$.

Table 5									
0.0	0.1	0.3	0.2	0.0	0.2	0.2	0.0	0.1	0.0
0.2	0.1	0.0	0.3	0.3	0.2	0.1	0.0	0.4	0.3
0.1	0.1	0.0	0.3	0.5	0.3	0.1	0.2	0.3	0.1
0.1	0.2	0.2	0.3	0.3	0.1	0.3	0.5	0.4	0.3
0.0	0.1	0.1	0.2	0.0	0.6	0.3	0.1	0.1	0.2
0.2	0.3	0.2	0.2	0.3	0.1	0.0	0.1	0.2	0.1
0.2	0.1	0.4	0.1	0.2	0.1	0.1	0.2	0.2	0.2
0.1	0.2	0.1	0.0	0.1	0.0	0.3	0.3	0.2	0.2
0.1	0.1	0.4	0.3	0.0	0.3	0.1	0.2	0.1	0.2
0.2	0.0	0.0	0.1	0.3	0.2	0.1	0.1	0.0	0.1

Step 3: Figure 12 shows a histogram of the 100 sample proportions. Notice that the shape of the distribution is skewed right.

Figure 12
Distribution of $\hat{p}$ with $n = 10$

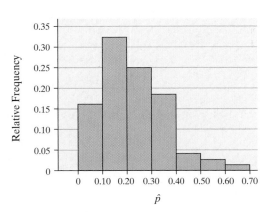

Step 4: The mean of the 100 sample proportions in Table 5 is 0.17. This is the same as the population proportion. The standard deviation of the 100 sample proportions in Table 5 is 0.1262.

We repeat Steps 1 through 4 for samples of size $n = 40$ and $n = 80$. Figure 13 shows the histogram for a sample of size $n = 40$. Notice that the shape of the distribution is skewed right (although not as skewed as the histogram with $n = 10$). The mean of the 100 sample proportions is 0.17 (the same as the population proportion), and the standard deviation is 0.0614 (less than the standard deviation for $n = 10$). Figure 14 shows the histogram for samples of size $n = 80$. Notice that the shape of the distribution is approximately normal. The mean of the 100 sample proportions for samples of size $n = 80$ is 0.17 (the same as the population proportion), and the standard deviation is 0.0408 (less than the standard deviation for $n = 40$).

We notice the following regarding the distribution of the sample proportion:

- *Shape*: As the size of the sample, n, increases, the shape of the distribution of the sample proportion becomes approximately normal.
- *Center*: The mean of the distribution of the sample proportion equals the population proportion, p.
- *Spread*: The standard deviation of the distribution of the sample proportion decreases as the sample size, n, increases.

Figure 13
Distribution of $\hat{p}$ with $n = 40$

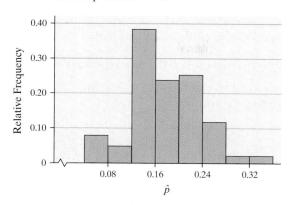

Figure 14
Distribution of $\hat{p}$ with $n = 80$

Although the proof is beyond the scope of this text, we should be convinced that the following result is reasonable.

Sampling Distribution of $\hat{p}$

For a simple random sample of size n such that $n \leq 0.05N$ (that is, the sample size is less than or equal to 5% of the population size)

- The shape of the sampling distribution of $\hat{p}$ is approximately normal provided $np(1 - p) \geq 10$.
- The mean of the sampling distribution of $\hat{p}$ is $\mu_{\hat{p}} = p$.
- The standard deviation of the sampling distribution of $\hat{p}$ is

$$\sigma_{\hat{p}} = \sqrt{\frac{p(1 - p)}{n}}.$$

In Other Words

The reason that the sample size cannot be more than 5% of the population size is because the success or failure of identifying an individual in the population that has the specified characteristic should not be affected by earlier observations. For example, in a population of size 100 where 14 of the individuals have brown hair, the probability that a randomly selected individual has brown hair is 14/100 = 0.14. The probability that a second randomly selected student has brown hair is 13/99 = 0.13. The probability changes because the sampling is done without replacement.

The condition that the sample size is no more than 5% of the population size is needed so that result obtained from one individual in the survey is independent of the result obtained from any other individual in the survey. The condition that $np(1 - p)$ is at least 10 is needed for normality.

Also, regardless of whether $np(1 - p) \geq 10$ or not, the mean of the sampling distribution of $\hat{p}$ is p and the standard deviation of the sampling distribution of $\hat{p}$ is $\sqrt{\dfrac{p(1 - p)}{n}}$.

EXAMPLE 3 | Describing the Distribution of the Sample Proportion

Problem: According to the Centers for Disease Control, 17% (or 0.17) of Americans have high cholesterol. Suppose we obtain a simple random sample of $n = 80$ Americans and determine which have high cholesterol. Describe the shape, center, and spread for the distribution of the sample proportion of Americans who have high cholesterol.

Approach: If the sample size is less than 5% of the population size and $np(1 - p)$ is at least 10, the sampling distribution of $\hat{p}$ is approximately normal, with mean $\mu_{\hat{p}} = p$ and standard deviation $\sigma_{\hat{p}} = \sqrt{\dfrac{p(1 - p)}{n}}$.

Solution: There are about 295 million people in the United States. The sample of $n = 80$ is certainly less than 5% of the population size. Also, $np(1 - p) = 80(0.17)(1 - 0.17) = 11.288 \geq 10$. The distribution of $\hat{p}$ is approximately normal, with mean $\mu_{\hat{p}} = 0.17$ and standard deviation

$$\sigma_{\hat{p}} = \sqrt{\frac{p(1 - p)}{n}} = \sqrt{\frac{0.17(1 - 0.17)}{80}} \approx 0.0420$$

Now Work Problem 7.

2 Compute Probabilities of a Sample Proportion

Now that we can describe the distribution of the sample proportion, we can compute probabilities of obtaining a specific sample proportion.

EXAMPLE 4 **Compute Probabilities of a Sample Proportion**

Problem: According to the National Center for Health Statistics, 15% of all Americans have hearing trouble.

(a) In a random sample of 120 Americans, what is the probability at least 18% have hearing trouble?

(b) Would it be unusual if a random sample of 120 Americans results in 10 having hearing trouble?

Approach: First, we determine whether the distribution of the sampling distribution is approximately normal, with mean $\mu_{\hat{p}} = p$ and standard deviation $\sigma_{\hat{p}} = \sqrt{\dfrac{p(1-p)}{n}}$, by verifying that the sample size is less than 5% of the population size and that $np(1-p) \geq 10$. Then we can use the normal distribution to determine the probabilities.

Solution: There are approximately 295 million people in the United States. The sample size of $n = 120$ is definitely less than 5% of the population size. We are told that $p = 0.15$. Because $np(1-p) = 120(0.15)(1-0.15) = 15.3 \geq 10$, the shape of the distribution of the sample proportion is approximately normal. The mean of the sample proportion $\hat{p}$ is $\mu_{\hat{p}} = 0.15$ and the standard deviation is $\sigma_{\hat{p}} = \sqrt{\dfrac{0.15(1-0.15)}{120}} \approx 0.0326$.

(a) We want to know the probability that a random sample of 120 Americans will result in a sample proportion of at least 18%, or 0.18. That is, we want to know $P(\hat{p} \geq 0.18)$. Figure 15(a) shows the normal curve with the area to the right of 0.18 shaded. To find this area, we convert $\hat{p} = 0.18$ to a standard normal random variable Z by subtracting the mean and dividing by the standard deviation. Don't forget that we round Z to two decimal places.

$$Z = \frac{\hat{p} - \mu_{\hat{p}}}{\sigma_{\hat{p}}} = \frac{0.18 - 0.15}{\sqrt{\dfrac{0.15(1-0.15)}{120}}} = 0.92$$

Figure 15(b) shows a standard normal curve with the area right of 0.92 shaded. Remember, the area to the right of $\hat{p} = 0.18$ is the same as the area to the right of $Z = 0.92$.

Figure 15

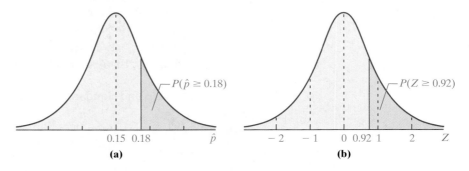

The area to the right of $Z = 0.92$ is 0.1788. Therefore,

$$P(\hat{p} \geq 0.18) = P(Z \geq 0.92) = 0.1788$$

Interpretation: The probability that a random sample of $n = 120$ Americans results in at least 18% having hearing trouble is 0.1788. This means that about 18 out of 100 random samples of size 120 will result in at least 18% having hearing trouble, even though the population proportion of Americans with hearing trouble is 0.15.

(b) A random sample of 120 Americans results in 10 having hearing trouble. The sample proportion of Americans with hearing trouble is $\hat{p} = \dfrac{10}{120} = 0.083$.

To determine whether a sample proportion 0.083 or less is unusual, we compute $P(\hat{p} \leq 0.083)$ because if a sample proportion of 0.083 is unusual, then any sample proportion less than 0.083 is also unusual. Figure 16(a) shows the normal curve with the area to the left of 0.083 shaded. To find this area, we convert $\hat{p} = 0.083$ to a standard normal random variable Z.

$$Z = \frac{\hat{p} - \mu_{\hat{p}}}{\sigma_{\hat{p}}} = \frac{0.083 - 0.15}{\sqrt{\dfrac{0.15(1 - 0.15)}{120}}} = -2.06$$

Figure 16(b) shows a standard normal curve with the area left of -2.06 shaded. The area to the left of $\hat{p} = 0.083$ is the same as the area to the left of $Z = -2.06$.

Figure 16

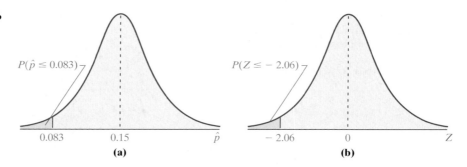

$P(\hat{p} \leq 0.083)$

0.083   0.15 $\hat{p}$

(a)

$P(Z \leq -2.06)$

-2.06 0 Z

(b)

The area to the left of $Z = -2.06$ is 0.0197. Therefore,

$$P(\hat{p} \leq 0.083) = P(Z \leq -2.06) = 0.0197$$

Interpretation: About 2 samples in 100 will result in a sample proportion of 0.083 or less from a population whose proportion is 0.15. We obtained a result that should only happen about 2 times in 100, so the results obtained are indeed unusual.

Now Work Problem 17.

8.2 ASSESS YOUR UNDERSTANDING

Concepts and Vocabulary

1. In a town of 500 households, 220 have a dog. The population proportion of dog owners in this town (expressed as a decimal) is $p = $ _____. 0.44

2. The _____ _____, denoted $\hat{p}$, is given by the formula $\hat{p} = $ _____, where x is the number of individuals with a specified characteristic in a sample of n individuals.

3. *True or False*: The population proportion and sample proportion always have the same value. False

4. *True or False*: The mean of the sampling distribution of $\hat{p}$ is p. True

5. Describe the circumstances under which the shape of the sampling distribution of $\hat{p}$ is approximately normal.

6. What happens to the standard deviation of $\hat{p}$ as the sample size increases? If the sample size is increased by a factor of 4, what happens to the standard deviation of $\hat{p}$? Decreases; halved

2. sample proportion; $\dfrac{x}{n}$

5. $n \leq 0.05N$, $np(1 - p) \geq 10$

Skill Building

(13a) Approximately, normal, $\mu_{\hat{p}} = 0.35, \sigma_{\hat{p}} \approx 0.015$ (14a) Approximately normal, $\mu_{\hat{p}} = 0.42, \sigma_{\hat{p}} \approx 0.013$

In Problems 7–10, describe the sampling distribution of $\hat{p}$. Assume the size of the population is 25,000 for each problem.

7. $n = 500, p = 0.4$
NW Approximately normal, $\mu_{\hat{p}} = 0.4, \sigma_{\hat{p}} \approx 0.022$
8. $n = 300, p = 0.7$
Approximately normal, $\mu_{\hat{p}} = 0.7, \sigma_{\hat{p}} \approx 0.026$
9. $n = 1000, p = 0.103$
Approximately normal, $\mu_{\hat{p}} = 0.103, \sigma_{\hat{p}} \approx 0.010$
10. $n = 1010, p = 0.84$
Approximately normal, $\mu_{\hat{p}} = 0.84, \sigma_{\hat{p}} \approx 0.012$

11. Suppose a simple random sample of size $n = 75$ is obtained from a population whose size is $N = 10,000$ and whose population proportion with a specified characteristic is $p = 0.8$.

(a) Describe the sampling distribution of $\hat{p}$.
(b) What is the probability of obtaining $x = 63$ or more individuals with the characteristic? That is, what is $P(\hat{p} \geq 0.84)$? 0.1922
(c) What is the probability of obtaining $x = 51$ or fewer individuals with the characteristic? That is, what is $P(\hat{p} \leq 0.68)$? 0.0047

(11a) Approximately normal, $\mu_{\hat{p}} = 0.8, \sigma_{\hat{p}} \approx 0.046$

12. Suppose a simple random sample of size $n = 200$ is obtained from a population whose size is $N = 25,000$ and whose population proportion with a specified characteristic is $p = 0.65$.

(12a) Approximately normal, $\mu_{\hat{p}} = 0.65, \sigma_{\hat{p}} \approx 0.034$

(a) Describe the sampling distribution of $\hat{p}$.
(b) What is the probability of obtaining $x = 136$ or more individuals with the characteristic? That is, what is $P(\hat{p} \geq 0.68)$? 0.1867
(c) What is the probability of obtaining $x = 118$ or fewer individuals with the characteristic? That is, what is $P(\hat{p} \leq 0.59)$? 0.0375

13. Suppose a simple random sample of size $n = 1000$ is obtained from a population whose size is $N = 1,000,000$ and whose population proportion with a specified characteristic is $p = 0.35$.

(a) Describe the sampling distribution of $\hat{p}$.
(b) What is the probability of obtaining $x = 390$ or more individuals with the characteristic? 0.0040
(c) What is the probability of obtaining $x = 320$ or fewer individuals with the characteristic? 0.0233

14. Suppose a simple random sample of size $n = 1460$ is obtained from a population whose size is $N = 1,500,000$ and whose population proportion with a specified characteristic is $p = 0.42$.

(a) Describe the sampling distribution of $\hat{p}$.
(b) What is the probability of obtaining $x = 657$ or more individuals with the characteristic? 0.0102
(c) What is the probability of obtaining $x = 584$ or fewer individuals with the characteristic? 0.0606

Applying the Concepts

15. Gardeners Suppose a simple random sample of size $n = 100$ households is obtained from a town with 5000 households. It is known that 30% of the households plant a garden in the spring.

(a) Describe the sampling distribution of $\hat{p}$.
(b) What is the probability that more than 37 households in the sample plant a garden? Is this result unusual?
(c) What is the probability that 18 or fewer households in the sample plant a garden? Is this result unusual?

16. Text Messaging A nationwide study in 2003 indicated that about 60% of college students with cell phones send and receive text messages with their phones. Suppose a simple random sample of $n = 1136$ college students with cell phones is obtained.
(*Source*: promomagazine.com)

(a) Describe the sampling distribution of $\hat{p}$, the sample proportion of college students with cell phones who send or receive text messages with their phones.
(b) What is the probability that 665 or fewer college students in the sample send and receive text messages with their cell phones? Is this result unusual?
0.1562; No

(c) What is the probability that 725 or more college students in the sample send and receive text messages with their cell phone? Is this result unusual?
0.0043; Yes

17. Credit Cards According to a *USA Today* "Snapshot,"
NW 26% of adults do not have any credit cards. Suppose a simple random sample of 500 adults is obtained.

(a) Describe the sampling distribution of $\hat{p}$, the sample proportion of adults who do not have a credit card.
(b) In a random sample of 500 adults, what is the probability that less than 24% have no credit cards? 0.1539
(c) Would it be unusual if a random sample of 500 adults results in 150 or more having no credit cards?

18. Cell Phone Only According to a CNN report, 7% of the population do not have traditional phones and instead rely on only cell phones. Suppose a random sample of 750 telephone users is obtained.

(a) Describe the sampling distribution of $\hat{p}$, the sample proportion that is "cell-phone only."

(15a) Approximately normal, $\mu_{\hat{p}} = 0.3, \sigma_{\hat{p}} = \sqrt{\dfrac{0.3(1-0.3)}{100}}$ (15b) 0.0630; no   (15c) 0.0044; Yes   (16a) Approximately normal,

$\mu_{\hat{p}} = 0.6, \sigma_{\hat{p}} = \sqrt{\dfrac{0.6(1-0.6)}{1136}}$ (17a) Approximately normal, $\mu_{\hat{p}} = 0.26, \sigma_{\hat{p}} = \sqrt{\dfrac{0.26(1-0.26)}{500}}$ (17c) Yes; $P(x \geq 150) = 0.0207$

(18a) Approximately normal, $\mu_{\hat{p}} = 0.07, \sigma_{\hat{p}} = \sqrt{\dfrac{0.07(1-0.07)}{750}}$

(b) In a random sample of 750 telephone users, what is the probability that more than 8% are "cell-phone only"? *0.1423*

(c) Would it be unusual if a random sample of 750 adults results in 40 or fewer being "cell-phone only"?

19. **Phishing** A report released in May 2005 by First Data Corp. indicated that 43% of adults had received a "phishing" contact (a bogus e-mail that replicates an authentic site for the purpose of stealing personal information such as account numbers and passwords). Suppose a random sample of 800 adults is obtained.

(a) In a random sample of 800 adults, what is the probability that no more than 40% have received a phishing contact? *0.0436*

(b) Would it be unusual if a random sample of 800 adults resulted in 45% or more who had received a phishing contact?

20. **Second Homes** According to the National Association of Realtors, 23% of the roughly 8 million homes purchased in 2004 were considered investment properties. Suppose a random sample of 500 homes sold in 2004 is obtained.

(a) In a random sample of 500 homes sold in 2004, what is the probability that at least 125 were purchased as investment properties? *0.1446*

(b) Would it be unusual if a random sample of 500 homes sold in 2004 results in 20% or less being purchased as an investment property? *No; $P(\hat{p} \leq 0.20) = 0.0559$*

21. **Social Security Reform** A researcher studying public opinion of proposed Social Security changes obtains a simple random sample of 50 adult Americans and asks them whether or not they support the proposed changes. To say that the distribution of $\hat{p}$, the sample proportion who respond yes, is approximately normal, how many more adult Americans does the researcher need to sample if

(a) 10% of all adult Americans support the changes?

(b) 20% of all adult Americans support the changes?

22. **ADHD** A researcher studying ADHD among teenagers obtains a simple random sample of 100 teenagers aged 13 to 17 and asks them whether or not they have ever been prescribed medication for ADHD. To say that the distribution of $\hat{p}$, the sample proportion who respond no, is approximately normal, how many more teenagers aged 13 to 17 does the researcher need to sample if

(a) 90% of all teenagers aged 13 to 17 have never been prescribed medication for ADHD? *12 more*

(b) 95% of all teenagers aged 13 to 17 have never been prescribed medication for ADHD? *111 more*

23. **Simulation** The following exercise is meant to illustrate the normality of the distribution of the sample proportion, $\hat{p}$.

(a) Using MINITAB or some other statistical spreadsheet, randomly generate 2000 samples of size 765 from a population with $p = 0.3$. Store the number of successes in a column called *x*.

(b) Determine $\hat{p}$ for each of the 2000 samples by computing $\dfrac{x}{765}$. Store each $\hat{p}$ in a column called *phat*.

(c) Draw a histogram of the 2000 estimates of *p*. Comment on the shape of the distribution.

(d) Compute the mean and standard deviation of the sampling distribution of $\hat{p}$ in the simulation.

(e) Compute the theoretical mean and standard deviation of the sampling distribution of $\hat{p}$. Compare the theoretical results to the results of the simulation. Are they close?

24. **The Sampling Distribution Applet** Load the sampling distribution applet on your computer. Set the applet so that the population is binary with probability of success equal to 0.2.

(a) Obtain 1000 random samples of size $n = 5$. Describe the distribution of the sample proportion based on the results of the applet.

(b) Obtain 1000 random samples of size $n = 30$. Describe the distribution of the sample proportion based on the results of the applet.

(c) Obtain 1000 random samples of size $n = 100$. Describe the distribution of the sample proportion based on the results of the applet.

(d) Compare the results of parts (a)–(c). How are they the same? How are they different?

25. **Finite Population Correction Factor** In this section, we assumed that the sample size was less than 5% of the size of the population. When sampling without replacement from a finite population in which $n > 0.05N$, the standard deviation of the distribution of $\hat{p}$ is given by

$$\sigma_{\hat{p}} = \sqrt{\frac{\hat{p}(1 - \hat{p})}{n - 1} \cdot \left(\frac{N - n}{N} \right)}$$

where N is the size of the population. Suppose a survey is conducted at a college having an enrollment of 6,502 students. The student council wants to estimate the percentage of students in favor of establishing a student union. In a random sample of 500 students, it was determined that 410 were in favor of establishing a student union.

(a) Obtain the sample proportion, $\hat{p}$, of students surveyed who favor establishing a student union.

(b) Calculate the standard deviation of the sampling distribution of $\hat{p}$. *0.017*

(18c) Yes; $P(x \leq 40) = 0.0344$

(19b) No; $P(\hat{p} \geq 0.45) = 0.1271$ (21a) 62 more   (21b) 13 more

Consumer Reports® | Tanning Salons

Medical groups have long warned about the dangers of indoor tanning. The American Medical Association and the American Academy of Dermatology unsuccessfully petitioned the Food and Drug Administration in 1994 to ban cosmetic-tanning equipment. Three years later, the Federal Trade Commission warned the public to beware of advertised claims that "unlike the sun, indoor tanning will not cause skin cancer or skin aging" or that you can "tan indoors with absolutely no harmful side effects."

In February 1999, still under pressure from the medical community, the FDA announced that current recommendations "may allow higher exposures" to UV radiation "than are necessary." The agency proposed reducing recommended exposures and requiring simpler wording on consumer warnings. But it has not yet implemented either of these changes. An FDA spokeswoman told us that "the agency decided to postpone amendment of its standard pending the results of ongoing research and discussions with other organizations."

To make matters worse, only about half the states have any rules for tanning parlors. In some of these states, the regulation is minimal and may not require licensing, inspections, training, record keeping, or parental consent for minors. Despite this, nearly 30 million Americans, including a growing number of teenage girls, are expected to visit a tanning salon in 2005.

In a recent survey of 296 indoor-tanning facilities around the country, to our knowledge the first nationwide survey of its kind, we found evidence of widespread failures to inform customers about the possible risks, including premature wrinkling and skin cancer, and to follow recommended safety procedures, such as wearing eye goggles. Many facilities made questionable claims about indoor tanning: that it's safer than sunlight, for example, and is well controlled.

(a) In designing this survey, why is it important to sample a large number of facilities? And why is it important to sample these facilities in multiple cities?

(b) Given the fact that there are over 150,000 tanning facilities in the United States, is the condition for independence of survey results satisfied? Why?

(c) Sixty-seven of the 296 tanning facilities surveyed stated that "tanning in a salon is the same as tanning in the sun with respect to causing skin cancer." Assuming that the true proportion is 25%, describe the sampling distribution of $\hat{p}$, the sample proportion of tanning facilities that state "tanning in a salon is the same as tanning in the sun with respect to causing skin cancer." Calculate the probability that less than 22.6% of randomly selected tanning salon facilities would state that tanning in a salon is the same as tanning in the sun with respect to causing skin cancer.

(d) Forty-two of the 296 tanning facilities surveyed stated "tanning in a salon does not cause wrinkled skin." Assuming that the true proportion is 18%, describe the sampling distribution of $\hat{p}$, the sample proportion of tanning facilities that state that "tanning in a salon does not cause wrinkled skin." Calculate the probability that at least 14.2% will state that tanning in a salon does not cause wrinkled skin. Would it be unusual for 50 or fewer facilities to state that tanning in a salon does not cause wrinkled skin?

Note to Readers: *In many cases, our test protocol and analytical methods are more complicated than described in this example. The data and discussion have been modified to make the material more appropriate for the audience.*

© by Consumers Union of U.S., Inc., Yonkers, NY 10703-1057, a nonprofit organization. Reprinted with permission.

 CHAPTER 8 **Review**

Summary

This chapter forms the bridge between probability and statistical inference. In Section 8.1, we discussed the distribution of the sample mean. We learned that the mean of the distribution of the sample mean equals the mean of the population ($\mu_{\bar{x}} = \mu$) and that the standard deviation of the distribution of the sample mean is the standard deviation of the population divided by the square root of the sample size $\left(\sigma_{\bar{x}} = \dfrac{\sigma}{\sqrt{n}} \right)$. If the sample is obtained from a population that is known to be normally distributed, the shape of the distribution of the sample mean is also normal. If the sample is obtained from a population that is not normal, the shape of the distribution of the sample mean becomes approximately normal as the sample size increases. This result is known as the Central Limit Theorem.

In Section 8.2, we discussed the distribution of the sample proportion. We learned that the mean of the distribution of the sample proportion is the population proportion ($\mu_{\hat{p}} = p$) and that the standard deviation of the distribution of the sample proportion is $\sigma_{\hat{p}} = \sqrt{\dfrac{p(1 - p)}{n}}$. If $np(1 - p) \geq 10$, then the shape of the distribution of $\hat{p}$ is approximately normal.

Formulas

Mean and Standard Deviation of the Sampling Distribution of $\bar{x}$

$$\mu_{\bar{x}} = \mu \quad \text{and} \quad \sigma_{\bar{x}} = \frac{\sigma}{\sqrt{n}}$$

Sample Proportion

$$\hat{p} = \frac{x}{n}$$

Mean and Standard Deviation of the Sampling Distribution of $\hat{p}$

$$\mu_{\hat{p}} = p \quad \text{and} \quad \sigma_{\hat{p}} = \sqrt{\frac{p(1-p)}{n}}$$

Standardizing a Normal Random Variable

$$Z = \frac{\bar{x} - \mu}{\frac{\sigma}{\sqrt{n}}} \quad \text{or} \quad Z = \frac{\hat{p} - p}{\sqrt{\frac{p(1-p)}{n}}}$$

Vocabulary

Statistical inference (p. 375)
Sampling distribution of the sample mean (p. 375)

Law of Large Numbers (p. 381)
Standard error of the mean (p. 382)
Central Limit Theorem (p. 385)

Sample proportion (p. 392)
Sampling distribution of $\hat{p}$ (p. 395)

Objectives

Section	You should be able to . . .	Example	Review Exercises
8.1	1 Understand the concept of a sampling distribution (p. 375)	1	1
	2 Describe the distribution of the sample mean for samples obtained from normal populations (p. 377)	2 through 4	2, 4, 5, 6
	3 Describe the distribution of the sample mean for samples obtained from a population that is not normal (p. 384)	5 and 6	7, 8, 13, 14
8.2	1 Describe the sampling distribution of a sample proportion (p. 392)	2 and 3	3, 4, 9(a)–12(a)
	2 Compute probabilities of a sample proportion (p. 396)	4	9(b), (c); 10(b), (c); 11(b), (c); 12(b), (c)

Review Exercises

4. $\mu_{\bar{x}} = \mu, \sigma_{\bar{x}} = \dfrac{\sigma}{\sqrt{n}}, \mu_{\hat{p}} = p, \sigma_{\hat{p}} = \sqrt{\dfrac{p(1-p)}{n}}$ (5b) Normal, $\mu_{\bar{x}} = 2600, \sigma_{\bar{x}} = 50/\sqrt{20}$

1. In your own words, explain what a sampling distribution is.

2. Under what conditions is the sampling distribution of $\bar{x}$ normal? *Population normal; large sample size*

3. Under what conditions is the sampling distribution of $\hat{p}$ approximately normal?

4. What is the mean and standard deviation of the sampling distribution of $\bar{x}$? What is the mean and standard deviation of the sampling distribution of $\hat{p}$?

5. **Energy Needs during Pregnancy** Suppose the total energy need during pregnancy is normally distributed with mean $\mu = 2600$ kcal/day and standard deviation $\sigma = 50$ kcal/day.

   (*Source*: American Dietetic Association)

   (a) What is the probability that a randomly selected pregnant woman has an energy need of more than 2625 kcal/day? Is this result unusual? *0.3085; no*

   (b) Describe the sampling distribution of $\bar{x}$, the sample mean daily energy requirement for a random sample of 20 pregnant women.

   (c) What is the probability that a random sample of 20 pregnant women has a mean energy need of more than 2625 kcal/day? Is this result unusual? *0.0125; yes*

6. **Battery Life** The charge life of a certain lithium ion battery for camcorders is normally distributed with mean 90 minutes and standard deviation 35 minutes.

   (a) What is the probability that a battery of this type, randomly selected, lasts more than 100 minutes on a single charge? Is this result unusual? *0.3859; no*

(7a) Approximately normal, $\mu_{\bar{x}} = 0.75, \sigma_{\bar{x}} = 0.004/\sqrt{30}$

   (b) Describe the sampling distribution of $\bar{x}$, the sample mean charge life for a random sample of 10 such batteries. *Normal, $\mu_{\bar{x}} = 90, \sigma_{\bar{x}} = 35/\sqrt{10}$*

   (c) What is the probability that a random sample of 10 such batteries has a mean charge life of more than 100 minutes? Is this result unusual? *0.1841; no*

7. **Copper Tubing** A machine at K&A Tube & Manufacturing Company produces a certain copper tubing component in a refrigeration unit. The tubing components produced by the manufacturer have a mean diameter of 0.75 inch with standard deviation 0.004 inch. The quality-control inspector takes a random sample of 30 components once each week and calculates the mean diameter of these components. If the mean is either less than 0.748 inch or greater than 0.752 inch, the inspector concludes that the machine needs an adjustment.

   (a) Describe the sampling distribution of $\bar{x}$, the sample mean diameter, for a random sample of 30 such components.

   (b) What is the probability that, based on a random sample of 30 such components, the inspector will conclude that the machine needs an adjustment? *0.0062*

8. **Filling Machines** A machine used for filling plastic bottles with a soft drink has a known standard deviation of $\sigma = 0.05$ liter. The target mean fill volume is $\mu = 2.0$ liters.

   (a) Describe the sampling distribution of $\bar{x}$, the sample mean fill volume, for a random sample of 45 such bottles.

   Approximately normal, $\mu_{\bar{x}} = 2.0, \sigma_{\bar{x}} = 0.05/\sqrt{45}$

(b) What is the probability that a random sample of 45 such bottles has a mean fill volume that is less than 1.995 liters? *0.2514*

(c) What is the probability that a random sample of 45 such bottles has a mean fill volume that is more than 2.015 liters? Is this result unusual? What might we conclude? *0.0228; yes*

9. **Entrepreneurship** A Gallup survey in March 2005 indicated that 72% of 18- to 29-year-olds, if given a choice, would prefer to start their own business rather than work for someone else. Suppose a random sample of 600 18- to 29-year-olds is obtained.

(a) Describe the sampling distribution of $\hat{p}$, the sample proportion of 18- to 29-year-olds who would prefer to start their own business.

(b) In a random sample of 600 18- to 29-year-olds, what is the probability that no more than 70% would prefer to start their own business? *0.1379*

(c) Would it be unusual if a random sample of 600 18- to 29-year-olds resulted in 450 or more who would prefer to start their own business? *Yes*

10. **Smokers** According to the National Center for Health Statistics (2004), 22.4% of adults are smokers. Suppose a random sample of 300 adults is obtained.

(a) Describe the sampling distribution of $\hat{p}$, the sample proportion of adults who smoke.

(b) In a random sample of 300 adults, what is the probability that at least 50 are smokers? *0.9913*

(c) Would it be unusual if a random sample of 300 adults results in 18% or less being smokers? *yes*

11. **Advanced Degrees** According to the U.S. Census Bureau, roughly 9% of adults aged 25 years or older have an advanced degree. Suppose a random sample of 200 adults aged 25 years or older is obtained.

(a) Describe the sampling distribution of $\hat{p}$, the sample proportion of adults aged 25 years or older who have an advanced degree.

(b) In a random sample of 200 adults aged 25 years or older, what is the probability that no more than 6% have an advanced degree? *0.0694*

(c) Would it be unusual if a random sample of 200 adults aged 25 years or older results in 25 or more having an advanced degree? *Yes*

12. **Peanut and Tree Nut Allergies** Peanut and tree nut allergies are considered to be the most serious food allergies. According to the National Institute of Allergy and Infectious Diseases, roughly 1% of Americans are allergic to peanuts or tree nuts. Suppose a random sample of 1500 Americans is obtained.

(a) Describe the sampling distribution of $\hat{p}$, the sample proportion of Americans allergic to peanuts or tree nuts.

(b) In a random sample of 1500 Americans, what is the probability that more than 1.5% are allergic to peanuts or tree nuts? *0.0256*

(c) Would it be unusual if a random sample of 1500 Americans results in fewer than 10 with peanut or tree nut allergies? *No*

13. **Principals' Salaries** According to the *National Survey of Salaries and Wages in Public Schools*, the mean salary paid to public high school principals in 2004–2005 was $71,401. Assume the standard deviation is $26,145. What is the probability that a random sample of 100 public high school principals has an average salary under $65,000? *0.0071*

14. **Teaching Supplies** According to the National Education Association, public school teachers spend an average of $443 of their own money each year to meet the needs of their students. Assume the standard deviation is $175. What is the probability that a random sample of 50 public school teachers spends an average of more than $400 each year to meet the needs of their students? *0.9591*

THE CHAPTER 8 CASE STUDY IS LOCATED ON THE CD THAT ACCOMPANIES THIS TEXT.

Estimating the Value of a Parameter Using Confidence Intervals

CHAPTER 9

Outline

9.1 The Logic in Constructing Confidence Intervals about a Population Mean where the Population Standard Deviation Is Known

9.2 Confidence Intervals about a Population Mean in Practice where the Population Standard Deviation Is Unknown

9.3 Confidence Intervals about a Population Proportion

9.4 Putting It All Together: Which Procedure Do I Use?

■ Chapter Review

■ Case Study: The Search for a Fire-Safe Cigarette (On CD)

DECISIONS

One of the most difficult decisions that a person has to make while attending college is choosing a major field of study. This decision plays a large role in the career path of an individual. What major should you choose? The Decisions project on page 440 may help you decide.

●●● Putting It All Together

Chapters 1 through 7 laid the groundwork for the remainder of the course. These chapters dealt with data collection (Chapter 1), descriptive statistics (Chapters 2 through 4), and probability (Chapters 5 through 7). Chapter 8 formed a bridge between probability and statistical inference by giving us a model we can use to make probability statements about the sample mean and sample proportion.

We know from Section 8.1 that $\bar{x}$ is a random variable and so has a distribution associated with it. This distribution is called the sampling distribution of $\bar{x}$, the sample mean. The mean of this distribution is equal to the mean of the population, μ. The standard deviation of $\bar{x}$ is $\frac{\sigma}{\sqrt{n}}$. The shape of the distribution of $\bar{x}$ is normal if either the population is normal or approximately normal if the sample size is large. We learned in Section 8.2 that $\hat{p}$ is also a random variable whose mean is p and standard deviation is $\sqrt{\frac{p(1-p)}{n}}$. If $np(1-p) \geq 10$, the distribution of the random variable $\hat{p}$ is approximately normal.

403

We now discuss inferential statistics—the process of using information obtained in a sample and generalizing it to a population. We will study two areas of inferential statistics: (1) estimation—sample data are used to estimate the value of unknown parameters such as μ or p, and (2) hypothesis testing—statements regarding a characteristic of one or more populations are made and sample data are used to test the statements. In this chapter, we discuss estimation of an unknown parameter, and in the next chapter, we will discuss hypothesis testing.

Because the information collected from a sample does not contain all the information in the population, we will assign probabilities to our estimates. These probabilities serve as a way of measuring what will happen if we estimate the value of the parameter many times and provide a measure of confidence in our results.

9.1 The Logic in Constructing Confidence Intervals about a Population Mean where the Population Standard Deviation Is Known

Preparing for This Section Before getting started, review the following:

- Simple random sampling (Section 1.2, pp. 16–19)
- Sampling error (Section 1.4, p. 33)
- Parameter versus statistic (Section 3.1, p. 107)
- z_α notation (Section 7.2, p. 340)

- Normal probability plots (Section 7.4, pp. 354–358)
- Distribution of the sample mean (Section 8.1, pp. 375–388)

Objectives

1. **Compute a point estimate of the population mean**
2. **Construct and interpret a confidence interval about the population mean (assuming the population standard deviation is known)**
3. **Understand the role of margin of error in constructing a confidence interval**
4. **Determine the sample size necessary for estimating the population mean within a specified margin of error**

Our goal in this section is to estimate the value of an unknown population mean. We begin by introducing the approach to estimation by making the simplifying assumption that the population standard deviation σ is known. Granted, it is unlikely that we know the population standard deviation while not knowing the population mean. However, the assumption is made because it allows us to use the normal distribution to develop the technique of estimation. Put simply, this assumption liberates us from details and allows us to focus on the concept of estimation. We will drop this assumption in the next section.

1 Compute a Point Estimate of the Population Mean

The goal of statistical inference is to use information obtained from a sample and generalize the results to the population that is being studied. The first step in estimating the value of an unknown parameter such as μ, is to obtain a random sample and use the data from the sample to obtain a *point estimate* of the parameter.

Definition

A **point estimate** is the value of a statistic that estimates the value of a parameter.

For example, the sample mean, $\bar{x}$, is a point estimate of the population mean, μ.

EXAMPLE 1   **Computing a Point Estimate**

Problem: A two-lane highway with a posted speed limit of 45 miles per hour is located just outside a small 40-home subdivision. The residents of the neighborhood are concerned that the speed of cars on the highway is excessive and want an estimate of the population mean speed of the cars on the highway.

Approach: The Department of Transportation estimates that 2400 cars travel on the highway between the hours of 9 A.M. and 3 P.M. (nonpeak time). The residents decide to measure the speed of 12 cars on the highway during nonpeak times (congestion slows everyone down). The 12 cars will be selected using a simple random sample by determining 12 numbers between 1 and 2400 and measuring the speed of the cars using a radar gun. The sample mean speed of the 12 cars will be the point estimate of the population mean.

Solution: Table 1 shows the speeds (in miles per hour) of the 12 randomly selected cars. The sample mean is

Table 1			
57.4	56.1	70.3	65.6
44.2	58.6	66.1	57.3
62.2	60.4	64.5	52.7

$$\overline{x} = \frac{57.4 + 44.2 + \cdots + 52.7}{12} = \frac{715.4}{12} = 59.62 \text{ miles per hour}$$

The point estimate of μ is 59.62 miles per hour.

> **Now Work Problem 25(a).**

② Construct and Interpret a Confidence Interval about the Population Mean

What do we do with the point estimate obtained in Example 1? Would you go to the police department and ask for the speed on this highway to be monitored based on these 12 observations? Or would you like to have more evidence? After all, we know that statistics such as $\overline{x}$ vary from sample to sample. So a different random sample of 12 cars might result in a different point estimate of the population mean, such as $\overline{x} = 54.39$ miles per hour. Assuming the method used in selecting the cars was done appropriately, both point estimates would be reasonably good guesses of the population mean. Because of this variability in the random variable $\overline{x}$, we are better off reporting a range (or interval) of values, with some measure of the likelihood that the interval includes the unknown population mean.

To help understand the idea of this interval, consider the following situation. Suppose you were asked to guess the mean age of the students in your statistics class. If the sample mean age based on a survey of five of the students is $\overline{x} = 24$, you might guess the mean age of *all* the students is 24 years. This would be a point estimate of μ, the mean age of all students in the class. As you look around the room, you realize that you did not survey everyone, so your guess is probably off. To account for this error, you might express your guess by producing a range of ages, such as 24 years old give or take 2 years (the *margin of error*). Mathematically, we write this as 24 ± 2. If asked how confident you are that the mean age is between 22 and 26, you might respond, "I am 80% confident that the mean age of all students in my statistics class is between 22 and 26 years." If asked to give a range for which your confidence increases to say, 90%, what do you think will happen to the interval? To have 90% confidence, you may need to increase your interval, say, to 20 to 28 years.

✎ *In Other Words*

The symbol $\pm$ is read "plus or minus." It means "to add and subtract the quantity following the $\pm$ symbol."

Note to Instructor
The U.S. Federal Highway Administration reported that the mean number of miles driven annually was 12,200 miles in 2003 (the last year available at the time of printing). The purpose of this activity is to help students get an intuitive understanding of confidence intervals before getting too deep into computations. Time Limit: 25–35 minutes.

In-Class Activity—Intro to Confidence Intervals

When asked a question whose answer we are unsure of, we typically give an estimate along with some tolerance. For example, one might say the distance to a certain location is 5.5 miles, give or take (or within) a half-mile. The give-or-take amount reflects the certainty of our estimate.

What is the mean number of miles driven annually by individuals in the United States?

(a) Make a reasonable guess to answer the question. On what did you base your answer?

(b) Give an interval centered about your guess in which you are 99% confident the true value lies.

(c) Give an interval centered about your guess in which you are 90% confident the true value lies.

(d) Give an interval centered about your guess in which you are 50% confident the true value lies.

(e) How did the width of your interval change as you changed your level of confidence?

(f) Determine the mean number of miles driven annually by individuals in your class.

(g) Obtain the actual mean miles driven annually from your instructor and compare it to your answer in parts (a) and (f). Discuss the difficulties associated with estimating unknown quantities.

In statistics, we construct intervals for a population mean centered around a guess as well. The *guess* is the sample mean, the point estimate of the population mean.

Definition

In Other Words

A confidence interval is a range of numbers, such as 22–30. The level of confidence is the proportion of times that the interval contains the population mean if repeated samples were obtained.

- A **confidence interval** for an unknown parameter consists of an interval of numbers.
- The **level of confidence** represents the expected proportion of intervals that will contain the parameter if a large number of different samples is obtained. The level of confidence is denoted $(1 - \alpha) \cdot 100\%$.

For example, a 95% level of confidence ($\alpha = 0.05$) implies that if 100 different confidence intervals are constructed, each based on a different sample from the same population, we will expect 95 of the intervals to include the parameter and 5 to not include the parameter.

Confidence interval estimates for the population mean are of the form

Point estimate ± margin of error

The margin of error of a confidence interval estimate of a parameter depends on three factors:

1. *Level of confidence*: As the level of confidence increases, the margin of error also increases. This should feel logical, based on the discussion presented earlier regarding estimating the mean age of students in your statistics class.

2. *Sample size*: As the size of the random sample increases, the margin of error decreases. This is a consequence of the Law of Large Numbers, which states that as the sample size increases, the difference between the statistic and parameter decreases.

3. *Standard deviation of the population*: The more spread there is in the population, the wider our interval will be for a given level of confidence.

Now the question is, "How do we construct an interval?" From Chapter 8, we know the following information about the distribution of the sample mean.

- The shape of the distribution of all possible sample means will be normal provided the population is normal or approximately normal if the sample size is large ($n \geq 30$).
- The mean of the distribution of sample means equals the mean of the population. That is, $\mu_{\bar{x}} = \mu$.
- The standard deviation of the sample mean equals the standard deviation of the population divided by the square root of the sample size. That is, $\sigma_{\bar{x}} = \dfrac{\sigma}{\sqrt{n}}$.

The sample mean is a point estimate of the population mean, so we expect the value of $\bar{x}$ to be close to μ, but we do not know how close. Because $\bar{x}$ is normally distributed, we know 95% of all sample means lie within 1.96 standard deviations of the population mean, μ, and 2.5% of the sample means lie in each tail. See Figure 1. That is, 95% of all sample means are in the interval

$$\mu - 1.96 \cdot \frac{\sigma}{\sqrt{n}} < \bar{x} < \mu + 1.96 \cdot \frac{\sigma}{\sqrt{n}}$$

With a little algebraic manipulation, we can rewrite this inequality with μ in the middle and obtain

$$\bar{x} - 1.96 \cdot \frac{\sigma}{\sqrt{n}} < \mu < \bar{x} + 1.96 \cdot \frac{\sigma}{\sqrt{n}} \qquad \textbf{(1)}$$

This inequality states that 95% of *all* sample means will result in confidence interval estimates that contain the population mean. It is common to write the 95% confidence interval as

$$\bar{x} \pm 1.96 \cdot \frac{\sigma}{\sqrt{n}}$$

so that it is of the form

$$\text{Point estimate} \pm \text{margin of error}$$

Perhaps you are wondering where the 1.96 comes from in a 95% confidence interval. For a 95% confidence interval, we are capturing the middle 95% of all sample means, so 2.5% of the sample means are in each tail. Remember that the notation z_α is used to represent the Z-value such that the area in the right tail is α. So $z_{0.025} = 1.96$ is the Z-value such that 2.5% of the area under the standard normal curve is to the right.

Let's look at an example that illustrates the concept of a 95% confidence interval.

Figure 1
Sampling distribution of $\bar{x}$

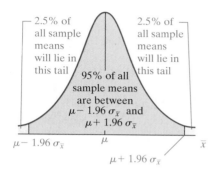

2.5% of all sample means will lie in this tail

2.5% of all sample means will lie in this tail

95% of all sample means are between $\mu - 1.96\, \sigma_{\bar{x}}$ and $\mu + 1.96\, \sigma_{\bar{x}}$

$\mu - 1.96\, \sigma_{\bar{x}}$ μ $\bar{x}$

$\mu + 1.96\, \sigma_{\bar{x}}$

Note to Instructor
To visually illustrate the ideas on this page, draw a normal curve of the sampling distribution of $\bar{x}$. Now create a "slider" that shows $\bar{x} - 1.96\, \sigma_{\bar{x}}$ and $\bar{x} + 1.96\, \sigma_{\bar{x}}$. As $\bar{x}$ goes into nonshaded region, the interval does not capture μ.

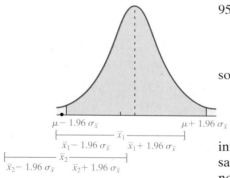

$\mu - 1.96\, \sigma_{\bar{x}}$ $\mu + 1.96\, \sigma_{\bar{x}}$

$\bar{x}_1$
$\bar{x}_1 - 1.96\, \sigma_{\bar{x}}$ $\bar{x}_1 + 1.96\, \sigma_{\bar{x}}$
$\bar{x}_2$
$\bar{x}_2 - 1.96\, \sigma_{\bar{x}}$ $\bar{x}_2 + 1.96\, \sigma_{\bar{x}}$

EXAMPLE 2   **Constructing 95% Confidence Intervals Based on Twenty Samples**

Problem: It is known that scores on the Stanford–Binet IQ test are normally distributed with $\mu = 100$ and $\sigma = 16$. Use MINITAB, Excel, or some other statistical software to simulate obtaining 20 simple random samples of size $n = 15$. Use these 20 different samples to construct 95% confidence intervals for the population mean, μ.

Approach

Step 1: We will use MINITAB to obtain 20 simple random samples of size 15 from a population that is normally distributed with mean $\mu = 100$ and $\sigma = 16$. We then compute the sample means of each of the 20 samples.

Step 2: Construct 95% confidence intervals by computing

$$\bar{x} - 1.96 \cdot \frac{\sigma}{\sqrt{n}} = \bar{x} - 1.96 \cdot \frac{16}{\sqrt{15}}$$

and

$$\bar{x} + 1.96 \cdot \frac{\sigma}{\sqrt{n}} = \bar{x} + 1.96 \cdot \frac{16}{\sqrt{15}}$$

for each sample mean.

Solution

Step 1: Table 2 shows the 20 sample means obtained from MINITAB.

Table 2				
104.32	93.97	108.73	104.11	100.67
96.87	99.74	100.25	101.32	94.24
102.23	94.32	97.66	101.44	98.19
107.15	100.38	95.89	104.43	102.28

Step 2: We construct the 20 confidence intervals for each of the 20 sample means and present the results in Table 3.

		Margin of Error	Lower Bound	Upper Bound
Sample	Sample Mean	$1.96 \cdot \dfrac{\sigma}{\sqrt{n}} = 1.96 \cdot \dfrac{16}{\sqrt{15}} = 8.10$	$\bar{x} - 1.96 \cdot \dfrac{\sigma}{\sqrt{n}}$	$\bar{x} + 1.96 \cdot \dfrac{\sigma}{\sqrt{n}}$
1	104.32	8.10	$104.32 - 8.10 = 96.22$	$104.32 + 8.10 = 112.42$
2	93.97	8.10	$93.97 - 8.10 = 85.87$	$93.97 + 8.10 = 102.07$
3	108.73	8.10	$108.73 - 8.10 = 100.63$	$108.73 + 8.10 = 116.83$
4	104.11	8.10	96.01	112.21
5	100.67	8.10	92.57	108.77
6	96.87	8.10	88.77	104.97
7	99.74	8.10	91.64	107.84
8	100.25	8.10	92.15	108.35
9	101.32	8.10	93.22	109.42
10	94.24	8.10	86.14	102.34
11	102.23	8.10	94.13	110.33
12	94.32	8.10	86.22	102.42
13	97.66	8.10	89.56	105.76
14	101.44	8.10	93.34	109.54
15	98.19	8.10	90.09	106.29
16	107.15	8.10	99.05	115.25
17	100.38	8.10	92.28	108.48
18	95.89	8.10	87.79	103.99
19	104.43	8.10	96.33	112.53
20	102.28	8.10	94.18	110.38

From these results, we can see that 19 of the 20 (or 95%) samples resulted in 95% confidence intervals that contain the value of the population mean, 100. Notice that sample 3 results in a confidence interval that does not contain

Figure 2

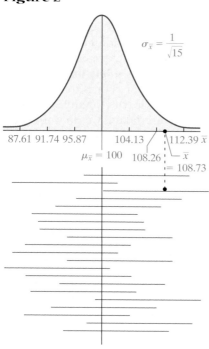

the population mean 100 (lower bound $= 108.73 - 8.10 = 100.63$; upper bound $= 108.73 + 8.10 = 116.83$).

Figure 2 presents the sampling distribution of $\bar{x}$ along with the 20 confidence intervals. Notice that the interval corresponding to $\bar{x} = 108.73$ does not intersect $\mu = 100$.

Note: In this particular simulation, exactly 95% of the samples resulted in intervals that contained μ. It is not always the case that exactly 95% of the sample means results in intervals that contain μ when performing this simulation. It could easily have been the case that all the intervals contained μ or that 17 out of the 20 contained μ. The interpretation of the confidence interval remains: In a 95% confidence interval, as the number of random samples of size n increases, the proportion of intervals that capture the population mean will approach 0.95. When we obtain only 20 samples, the exact proportion may be slightly different from 0.95.   ◀

We are not only interested in constructing 95% confidence intervals, so, we need a method for constructing any $(1 - \alpha) \cdot 100\%$ confidence interval. (Notice that if $\alpha = 0.05$ we are constructing a $(1 - 0.05) \cdot 100\% = 95\%$ confidence interval.)

We generalize Formula (1) given on page 407 by noting that $(1 - \alpha) \cdot 100\%$ of all sample means are in the interval

$$\mu - z_{\frac{\alpha}{2}} \cdot \frac{\sigma}{\sqrt{n}} < \bar{x} < \mu + z_{\frac{\alpha}{2}} \cdot \frac{\sigma}{\sqrt{n}}$$

as shown in Figure 3.

We rewrite this inequality with μ in the middle and obtain

$$\bar{x} - z_{\frac{\alpha}{2}} \cdot \frac{\sigma}{\sqrt{n}} < \mu < \bar{x} + z_{\frac{\alpha}{2}} \cdot \frac{\sigma}{\sqrt{n}}$$

Note to Instructor

Emphasize the relation between the area in the tail of the standard normal distribution and the critical value. Less area means a larger critical value, which implies a wider interval.

So $(1 - \alpha) \cdot 100\%$ of the sample means will result in confidence intervals that contain the population mean. The sample means that are in the tails of the distribution in Figure 3 will not have confidence intervals that include the population mean.

Figure 3

$\frac{\alpha}{2} \cdot 100\%$ of all sample means lie in this tail. These sample means will result in confidence intervals that do not capture the population mean.

$\frac{\alpha}{2} \cdot 100\%$ of all sample means lie in this tail. These sample means will result in confidence intervals that do not capture the population mean.

$(1 - \alpha) \cdot 100\%$ of all sample means lie in this region. These sample means result in confidence intervals that capture the population mean.

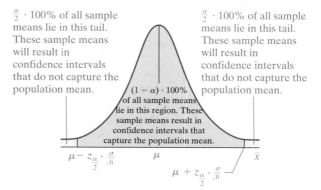

$\mu - z_{\frac{\alpha}{2}} \cdot \frac{\sigma}{\sqrt{n}}$ μ $\bar{x}$

$\mu + z_{\frac{\alpha}{2}} \cdot \frac{\sigma}{\sqrt{n}}$

The value $z_{\frac{\alpha}{2}}$ is called the **critical value** of the distribution. Table 4 shows some of the more common critical values used in the construction of confidence intervals.

Table 4		
Level of Confidence, $(1 - \alpha) \cdot 100\%$	**Area in Each Tail, $\frac{\alpha}{2}$**	**Critical Value, $z_{\frac{\alpha}{2}}$**
90%	0.05	1.645
95%	0.025	1.96
99%	0.005	2.575

! CAUTION

A 95% confidence interval does *not* mean that there is a 95% probability that the interval contains μ. Remember, probability describes the likelihood of undetermined events. Therefore, it does not make sense to talk about the probability that the interval contains μ, since the population mean is a fixed value. Think of it this way: Suppose I flip a coin and obtain a head. If I ask you to determine the probability that the flip resulted in a head, it would not be 0.5, because the outcome has already been determined. Instead, the probability is 0 or 1. Confidence intervals work the same way. Because μ is already determined, we do not say that there is a 95% probability that the interval contains μ.

In Other Words

The interpretation of a confidence interval is this: We are (*insert level of confidence*) confident that the population mean is between (*lower bound*) and (*upper bound*). This is an abbreviated way of saying the method is correct $(1 - \alpha) \cdot 100\%$ of the time.

In a 95% confidence interval, $\alpha = 0.05$, so $z_{\frac{\alpha}{2}} = z_{\frac{0.05}{2}} = z_{0.025} = 1.96$. Any sample mean that lies within 1.96 standard deviations of the population mean will result in a confidence interval that contains μ, and any sample mean that is more than 1.96 standard deviations from the population mean will result in a confidence interval that does not contain μ. This is an extremely important point. **Whether a confidence interval contains μ depends solely on the sample mean, $\bar{x}$.** The population mean is a fixed value that either is or is not in the interval, and we do not know which of these possibilities is true for any computed confidence interval. However, if repeated samples are taken, we do know that $(1 - \alpha) \cdot 100\%$ of the confidence intervals will contain the population mean. This result leads to the following interpretation of a confidence interval.

Interpretation of a Confidence Interval

A $(1 - \alpha) \cdot 100\%$ confidence interval indicates that, if we obtained many simple random samples of size n from the population whose mean, μ, is unknown, then approximately $(1 - \alpha) \cdot 100\%$ of the intervals will contain μ.

For example, if we constructed a 90% confidence interval with a lower bound of 12 and an upper bound of 18, we would interpret the interval as follows: "We are 90% confident that the population mean, μ, is between 12 and 18."

Be sure that you understand that the level of confidence refers to the confidence in the method, not the specific interval. So a 90% confidence interval tells us that the method will result in an interval that contains the population mean, μ, 90% of the time. It does not tell us that there is a 90% probability that μ lies between 12 and 18.

We are now prepared to present a method for constructing a $(1 - \alpha) \cdot 100\%$ confidence interval about μ.

Constructing a $(1 - \alpha) \cdot 100\%$ Confidence Interval about μ, σ Known

Suppose a simple random sample of size n is taken from a population with unknown mean, μ, and known standard deviation σ. A $(1 - \alpha) \cdot 100\%$ confidence interval for μ is given by

$$\text{Lower bound: } \bar{x} - z_{\frac{\alpha}{2}} \cdot \frac{\sigma}{\sqrt{n}} \qquad \text{Upper bound: } \bar{x} + z_{\frac{\alpha}{2}} \cdot \frac{\sigma}{\sqrt{n}}$$

where $z_{\frac{\alpha}{2}}$ is the critical Z-value.

Note: The sample size must be large ($n \geq 30$) or the population must be normally distributed. ◀

When constructing a confidence interval about the population mean, we must verify that the sample is big enough ($n \geq 30$) or comes from a population that is normally distributed. Fortunately, the procedures for constructing confidence intervals presented in this section are **robust**, which means that minor departures from normality will not seriously affect the results. Nevertheless, it is important that the requirements for constructing confidence intervals are verified. We shall follow the practice of verifying the normality assumption for small sample sizes by drawing a normal probability plot and checking for outliers by drawing a boxplot.

Because the construction of the confidence interval with σ known uses Z-scores, it is sometimes referred to as constructing a **Z-interval**.

EXAMPLE 3 **Constructing a Confidence Interval**

Problem: Remember, we want to estimate the speed of cars on the highway outside a subdivision. In Example 1, we obtained a point estimate of 59.62 miles per hour for the population mean. We now wish to construct a 90% confidence interval about the population mean, μ. Assume the population standard deviation speed of vehicles on the highway is known to be 8 miles per hour, based on past studies.

Approach

Step 1: We are treating the cars as a simple random sample of size $n = 12$. We have to verify that the data come from a population that is normally distributed with no outliers. We will verify normality by constructing a normal probability plot. We check for outliers by drawing a boxplot.

Step 2: Since we are constructing a 90% confidence interval, $\alpha = 0.10$. Therefore, we have to identify the value of $z_{\frac{\alpha}{2}} = z_{\frac{0.1}{2}} = z_{0.05}$. This is the Z-score such that the area under the standard normal curve to the right of $z_{0.05}$ is 0.05.

Step 3: We compute the lower and upper bounds on the interval with $\bar{x} = 59.62$, $\sigma = 8$, and $n = 12$.

Step 4: We interpret the result by stating, "We are 90% confident that the mean speed of all cars traveling on the highway outside the subdivision is somewhere between *lower bound* and *upper bound*."

Solution

Step 1: A normal probability plot of the data in Table 1 is provided in Figure 4(a). A boxplot is presented in Figure 4(b).

Figure 4

(a)   (b)

All the data values lie within the bounds on the normal probability plot, indicating that the data could come from a population that is normal. The boxplot does not show any outliers. The requirements for constructing a confidence interval are satisfied.

Step 2: We need to determine the value of $z_{0.05}$. We look in Table IV for an area equal to 0.95 (remember, the table gives areas left of the Z-scores). We have $z_{0.05} = 1.645$.

Step 3: Substitute into the formulas for the lower and upper bound of the confidence interval.

Lower bound: $\bar{x} - z_{\frac{\alpha}{2}} \cdot \dfrac{\sigma}{\sqrt{n}} = 59.62 - 1.645 \cdot \dfrac{8}{\sqrt{12}} = 59.62 - 3.80 = 55.82$

Upper bound: $\bar{x} + z_{\frac{\alpha}{2}} \cdot \dfrac{\sigma}{\sqrt{n}} = 59.62 + 1.645 \cdot \dfrac{8}{\sqrt{12}} = 59.62 + 3.80 = 63.42$

Step 4: We are 90% confident that the mean speed of all cars traveling on the highway outside the subdivision is between 55.82 and 63.42 miles per hour.

Since the speed limit on the highway is 45 miles per hour, the residents now have some powerful evidence to bring to the police department. ▬

EXAMPLE 4

Constructing a Confidence Interval Using Technology

Problem: Construct a 90% confidence interval for the mean speed of all cars on the highway outside the subdivision using the sample data from Table 1.

Approach: We will use MINITAB to construct the confidence interval. The steps for constructing confidence intervals using the TI-83/84 graphing calculators, MINITAB, and Excel are given in the Technology Step by Step on page 422.

Result: Figure 5 shows the confidence interval obtained from MINITAB. The interval is reported in the form (*lower bound, upper bound*).

Figure 5

One-Speed Z: Speed

The assumed standard deviation = 8

Variable	N	Mean	StDev	SE Mean	90% CI
Speed	12	59.616	6.9563	2.3094	(55.8180, 63.4153)

Interpretation: We are 90% confident that the mean speed of all cars traveling on the highway outside the subdivision is between 55.82 and 63.42 miles per hour. ▬

USING TECHNOLOGY
Confidence intervals constructed by hand may differ from those using technology because of rounding.

Now Work Problems 25(b)–(e).

---

③ **Understand the Role of the Margin of Error in Constructing a Confidence Interval**

The width of the interval is determined by the *margin of error*.

Definition

The **margin of error**, E, in a $(1 - \alpha) \cdot 100\%$ confidence interval in which σ is known is given by

$$E = z_{\frac{\alpha}{2}} \cdot \frac{\sigma}{\sqrt{n}}$$

where n is the sample size.

Note: We require that the population from which the sample was drawn be normally distributed or the sample size n be greater than or equal to 30.

✎ **In Other Words**
The margin of error can be thought of as the "give or take" portion of the statement "The mean age of the class is 24, give or take 2 years."

As we look at the formula for obtaining the margin of error, we see that its value depends on three quantities:

1. Level of confidence, $1 - \alpha$
2. Standard deviation of the population, σ
3. Sample size, n

We cannot control the standard deviation of the population, but we certainly can control the level of confidence and/or the sample size. Let's see how changing these values affects the margin of error.

EXAMPLE 5

Role of the Level of Confidence in the Margin of Error

Problem: For the problem of estimating the population mean speed of all cars on the highway outside the subdivision presented in Example 3, determine the effect on margin of error of increasing the level of confidence on the confidence interval to 99%.

Approach: With a 99% level of confidence, we have $\alpha = 0.01$. So, to compute the margin of error, E, we determine the value of $z_{\frac{\alpha}{2}} = z_{\frac{0.01}{2}} = z_{0.005}$. We then substitute this value into the formula for the margin of error with $\sigma = 8$ and $n = 12$.

Solution: After consulting Table IV, we determine $z_{0.005} = 2.575$. Substituting into the formula for margin of error, we obtain

$$E = z_{\frac{\alpha}{2}} \cdot \frac{\sigma}{\sqrt{n}} = 2.575 \cdot \frac{8}{\sqrt{12}} = 5.95$$

In Other Words

As the level of confidence increases, the margin of error also increases.

Notice the margin of error has increased from 3.80 to 5.95 when the level of confidence increases from 90% to 99%. If we want to be more confident that the sample results in an interval contains the population mean, we need to increase the width of the interval. ▬

EXAMPLE 6   **Role of Sample Size in the Margin of Error**

Problem: For the problem of estimating the population mean speed of all cars on the highway outside the subdivision presented in Example 3, determine the effect of increasing the sample size to $n = 48$ on the margin of error. Leave the level of confidence at 90%.

Approach: We compute the margin of error with $n = 48$ instead of $n = 12$.

Solution: Substituting $z_{0.05} = 1.645$, $\sigma = 8$, and $n = 48$ into the formula for margin of error, we obtain

$$E = z_{\frac{\alpha}{2}} \cdot \frac{\sigma}{\sqrt{n}} = 1.645 \cdot \frac{8}{\sqrt{48}} = 1.90$$

In Other Words

As the sample size increases, the margin of error decreases.

By increasing the sample size from $n = 12$ to $n = 48$, the margin of error, E, decreases from 3.80 to 1.90. Notice that by quadrupling the sample size, the margin of error is cut in half.

This result should not be very surprising. Remember that the Law of Large Numbers states that as the sample size n increases the sample mean approaches the value of the population mean. The smaller margin of error supports this idea. ▬

Now Work Problem 17.

---

④ **Determine the Sample Size Necessary for Estimating the Population Mean within a Specified Margin of Error**

Suppose we want to know the number of cars that we should sample to estimate the speed of all cars traveling outside the subdivision within 2 miles per hour with 95% confidence. If we solve the formula for the margin of error, E, for n, we obtain a formula for determining sample size:

$$E = z_{\frac{\alpha}{2}} \cdot \frac{\sigma}{\sqrt{n}}$$

$$E\sqrt{n} = z_{\frac{\alpha}{2}} \cdot \sigma \qquad \text{Multiply both sides by } \sqrt{n}.$$

$$\sqrt{n} = \frac{z_{\frac{\alpha}{2}} \cdot \sigma}{E} \qquad \text{Divide both sides by the margin of error, } E.$$

$$n = \left(\frac{z_{\frac{\alpha}{2}} \cdot \sigma}{E} \right)^2 \qquad \text{Square both sides.}$$

CAUTION
Rounding *up* is different from rounding *off*. We round 5.32 *up* to 6 and *off* to 5.

Note to Instructor
Ask students to identify the reason we round up rather than round off.

Determining the Sample Size *n*

The sample size required to estimate the population mean, μ, with a level of confidence $(1 - \alpha) \cdot 100\%$ with a specified margin of error, E, is given by

$$n = \left(\frac{z_{\frac{\alpha}{2}} \cdot \sigma}{E} \right)^2$$

where n is rounded up to the nearest whole number.

When σ is unknown, it is common practice to conduct a preliminary survey to determine s and use it as an estimate of σ or use results from previous studies to obtain an estimate of σ. When using this approach, the size of the sample should be at least 30.

EXAMPLE 7 **Determining Sample Size**

Problem: We once again consider the problem of estimating the population mean speed of all cars traveling on the highway outside the subdivision. How large a sample is required to estimate the mean speed within 2 miles per hour with 90% confidence?

Approach: The sample size required can be obtained using the formula

$$n = \left(\frac{z_{\frac{\alpha}{2}} \cdot \sigma}{E} \right)^2$$

with $z_{\frac{\alpha}{2}} = z_{0.05} = 1.645$, $\sigma = 8$, and $E = 2$.

Solution: We substitute the values of z, σ, and E into the formula for determining sample size and obtain

$$n = \left(\frac{z_{\frac{\alpha}{2}} \cdot \sigma}{E} \right)^2 = \left(\frac{1.645 \cdot 8}{2} \right)^2 = 43.2964$$

CAUTION
Don't forget to round up when determining sample size.

We round 43.2964 up to 44. A sample size of $n = 44$ results in an interval estimate of the population mean with a margin of error equal to 2 miles per hour with 90% confidence. If we obtain 100 samples of size $n = 44$, we expect about 90 intervals to capture the population mean, while 10 would not. ∎

Now Work Problem 37.

Some Final Thoughts
Many requirements must be satisfied while performing any type of statistical inference. It is worthwhile to list the requirements for constructing a confidence interval about μ with σ known in a single location for quick reference.

Requirements for Constructing a Confidence Interval about μ if σ Is Known

1. *The data obtained come from a simple random sample*. In Chapter 1, we introduced other sampling techniques, such as stratified, cluster, and systematic samples. The techniques introduced in this section apply only to samples obtained through simple random sampling. Although methods do exist for constructing confidence intervals when using the other sampling methods, they are beyond the scope of this text. If the

data are obtained from a suspect sampling method, such as voluntary response or convenience sampling, no methods exist for constructing confidence intervals. If data are collected in a flawed manner, any statistical inference performed on the data is useless!

2. *The data are obtained from a population that is normally distributed, or the sample size, n, is greater than or equal to 30.* When the sample size is small, we can use normal probability plots to help us judge whether the requirement of normality is satisfied. Remember, the techniques introduced in this section are robust. This means that minor departures from requirements will not have a severe effect on the results. However, we have to be aware that the sample mean is not resistant. Any outlier(s) in the data will affect the value of the sample mean and therefore affect the confidence interval. If the data contain outliers, we should proceed with caution when using the methods introduced in this section.

3. *The population standard deviation, σ, is assumed to be known.* However, it is unlikely that the population standard deviation is known when the population mean is not. We will drop this assumption in the next section.

9.1 ASSESS YOUR UNDERSTANDING

Concepts and Vocabulary

1. The construction of a confidence interval depends on three factors. What are they?

2. Why does the margin of error increase as the level of confidence increases?

3. Why does the margin of error decrease as the sample size n increases?

4. Suppose a confidence interval has a lower bound of 10 and an upper bound of 18. What is the sample mean of the data used to construct this interval? What is the margin of error? 14, 4

5. A student constructs a 95% confidence interval for the mean age of students at his college. The lower bound is 21.4 years and the upper bound is 28.8 years. He interprets the interval as, "There is a 95% probability that the mean age of a student is between 21.4 years and 28.8 years." What is wrong with this interpretation? What could be done to increase the precision of the interval?

6. Suppose a professor in a class of 20 students wants to estimate the mean pulse rate of students prior to the final exam. Would it make sense for the professor to construct a 90% confidence interval about the population mean? Explain.

Skill Building

In Problems 7–12, a simple random sample of size n < 30 has been obtained. From the normal probability plot and boxplot, judge whether a Z-interval should be constructed.

7. No

9. No

8. No

10. No

(17b) (32.73, 35.67); *decreases* (17c) (31.89, 36.51); *increases* (18b) (58.36, 60.04); *decreases*

11. Yes

12. Yes

In Problems 13–16, determine the critical value $z_{\frac{\alpha}{2}}$ that corresponds to the given level of confidence.

13. 98% 2.33   **14.** 94% 1.88

15. 85% 1.44   **16.** 80% 1.28

17. A simple random sample of size n is drawn from a popula-
NW tion whose population standard deviation, σ, is known to
be 5.3. The sample mean, $\bar{x}$, is determined to be 34.2.

   (a) Compute the 95% confidence interval about μ if the
   sample size, n, is 35. (32.44, 35.96)

   (b) Compute the 95% confidence interval about μ if the
   sample size, n, is 50. How does increasing the sample
   size affect the margin of error, E?

   (c) Compute the 99% confidence interval about μ if the
   sample size, n, is 35. Compare the results to those obtained in part (a). How does increasing the level of
   confidence affect the size of the margin of error, E?

   (d) Can we compute a confidence interval about μ based
   on the information given if the sample size is $n = 15$?
   Why? If the sample size is $n = 15$, what must be true
   regarding the population from which the sample was
   drawn? Only if the population is normal

18. A simple random sample of size n is drawn from a population whose population standard deviation, σ, is known to
be 3.8. The sample mean, $\bar{x}$, is determined to be 59.2.

   (a) Compute the 90% confidence interval about μ if the
   sample size, n, is 45. (58.27, 60.13)

   (b) Compute the 90% confidence interval about μ if the
   sample size, n, is 55. How does increasing the sample
   size affect the margin of error, E?

   (c) Compute the 98% confidence interval about μ if the
   sample size, n, is 45. Compare the results to those obtained in part (a). How does increasing the level of
   confidence affect the size of the margin of error, E?

   (d) Can we compute a confidence interval about μ based
   on the information given if the sample size is $n = 15$?
   Why? If the sample size is $n = 15$, what must be true
   regarding the population from which the sample was
   drawn? Only if the population is normal

19. A simple random sample of size n is drawn from a population that is normally distributed with population standard
deviation, σ, known to be 13. The sample mean, $\bar{x}$, is found
to be 108.

   (a) Compute the 96% confidence interval about μ if the
   sample size, n, is 25. (102.67, 113.33)

   (b) Compute the 96% confidence interval about μ if the
   sample size, n, is 10. How does decreasing the sample
   size affect the margin of error, E?

   (c) Compute the 88% confidence interval about μ if the
   sample size, n, is 25. Compare the results to those obtained in part (a). How does decreasing the level of
   confidence affect the size of the margin of error, E?

   (d) Could we have computed the confidence intervals in
   parts (a)–(c) if the population had not been normally
   distributed? Why? No

   (e) Suppose an analysis of the sample data revealed three
   outliers greater than the mean. How would this affect
   the confidence interval?

20. A simple random sample of size n is drawn from a population that is normally distributed with population standard
deviation, σ, known to be 17. The sample mean, $\bar{x}$, is found
to be 123.

   (a) Compute the 94% confidence interval about μ if the
   sample size, n, is 20. (115.85, 130.15)

   (b) Compute the 94% confidence interval about μ if the
   sample size, n, is 12. How does decreasing the sample
   size affect the margin of error, E?

   (c) Compute the 85% confidence interval about μ if the
   sample size, n, is 20. Compare the results to those obtained in part (a). How does decreasing the level of
   confidence affect the size of the margin of error, E?

   (d) Could we have computed the confidence intervals in
   parts (a)–(c) if the population had not been normally
   distributed? Why? No

   (e) Suppose an analysis of the sample data revealed one
   outlier greater than the mean. How would this affect
   the confidence interval?

(18c) (57.88, 60.52); *increases* (19b)(99.57, 116.43); *increases*
(19c)(103.96, 112.04); *decreases* (20b) (113.77, 132.23); *increases*
(20c) (117.53, 128.47); *decreases*

Applying the Concepts

21. Time for Bed How much do Americans sleep each
night? Based on a random sample of 1120 Americans
15 years of age or older, the mean amount of sleep
per night is 8.17 hours according to the American Time
Use Survey conducted by the Bureau of Labor Statistics. Assuming the population standard deviation for
amount of sleep per night is 1.2 hours, construct and interpret a 95% confidence interval for the mean amount
of sleep per night of Americans 15 years of age or
older. (8.10, 8.24)

22. Hungry or Thirsty? How much time do Americans spend
eating or drinking? Based on a random sample of 1120
Americans 15 years of age or older, the mean amount of
time spent eating or drinking each day is 1.21 hours according to the American Time Use Survey conducted by the Bureau of Labor Statistics. Assuming the population standard
deviation for amount of time spent eating or drinking is 0.65
hour, construct and interpret a 90% confidence interval for
the mean amount of time spent eating or drinking each day
by Americans 15 years of age or older. (1.18, 1.24)

23. Time with Friends A school social worker wishes to estimate the mean amount of time each week that high school students spend with friends. She obtains a random sample of 1175 high school students and finds that the mean weekly time spent with friends is 9.2 hours. Assuming the population standard deviation amount of time spent with friends is 6.7 hours, construct and interpret a 90% confidence interval for the mean time spent with friends each week. *(8.88, 9.52)*

24. Travel Time to Work An urban economist wishes to estimate the mean amount of time people spend traveling to work. He obtains a random sample of 50 individuals who are in the labor force and finds that the mean travel time is 24.2 minutes. Assuming the population standard deviation of travel time is 18.5 minutes, construct and interpret a 95% confidence interval for the mean travel time to work. **Note:** The standard deviation is large because many people work at home (travel time = 0 minutes) and many have commutes in excess of 1 hour. (*Source*: Based on data obtained from the American Community Survey.) *(19.07, 29.33)*

25. Crash Test Results The Insurance Institute for Highway Safety routinely conducts crash tests on vehicles to determine the cost of repairs. In four crashes of a Chevy Cavalier at 5 miles per hour, the institute found the cost of repairs to be $225, $462, $729, and $753. Treat these data as a simple random sample of four crashes, and answer the following questions, assuming that σ = $220.

(a) Use the data to compute a point estimate for the population mean cost of repairs on a Chevy Cavalier. *$542.25*

(b) Because the sample size is small, we must verify that cost is approximately normally distributed and that the sample does not contain any outliers. The normal probability plot and boxplot are shown next. Are the conditions for constructing a Z-interval satisfied? *Yes*

Cost

(c) Construct a 95% confidence interval for the cost of repairs. Interpret this interval. *($326.65, $757.85)*

(d) Construct a 90% confidence interval for the cost of repairs. Interpret this interval. *($361.30, $723.20)*

(e) What happened to the width of the interval when the level of confidence decreased? Explain why this result is reasonable. *Decreased*

26. Flight Time A researcher for the FAA wants to estimate the average flight time (in minutes) from Albuquerque, New Mexico, to Dallas, Texas, for flights with American Airlines. He randomly selects nine flights between the two cities and obtains the data shown. Assume that σ = 8 minutes.

| 117 | 95 | 109 | 103 | 111 |
| 91 | 100 | 99 | 106 | |

Source: United States Department of Transportation

(a) Use the data to compute a point estimate for the population mean flight time between Albuquerque and Dallas on an American Airlines flight. *103.4 minutes*

(b) Because the sample size is small, we must verify that flight time is approximately normally distributed and that the sample does not contain any outliers. The normal probability plot and boxplot are shown next. Are the conditions for constructing a Z-interval satisfied? *Yes*

Flight Time

(c) Construct a 95% confidence interval for the flight time. Interpret this interval. *(98.2, 108.6)*

(d) Construct a 90% confidence interval for the flight time. Interpret this interval. *(99.0, 107.8)*

(e) What happened to the width of the interval when the level of confidence decreased? Explain why this result is reasonable. *Decreased*

27. Investment Home Sales A study by the National Association of Realtors reports that second-home sales accounted for more than a third of residential real estate transactions in 2004. According to the report, the average age of a person buying a second home as an investment rental property was 47 years. A real estate agent wants to estimate the average age of those buying investment property in his area. He randomly selects 15 of his clients who purchased an investment property and obtains the data shown. Assume that σ = 7.9.

27	49	56	45	49
35	55	40	46	48
38	58	51	42	38

(a) Use the data to compute a point estimate for the population mean age of the real estate agent's clients who purchased investment property. *45.1 years*

(b) Because the sample size is small, we must verify that client age is approximately normally distributed and that the sample does not contain any outliers. The normal probability plot and boxplot are shown next. Are the conditions for constructing a Z-interval satisfied? *Yes*

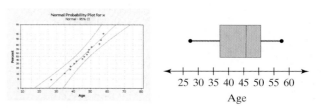

Age

(c) Construct a 95% confidence interval for the mean age for all the real estate agent's clients who purchased investment property. Interpret this interval. *(41.1, 49.1)*

(d) Do the real estate agent's clients appear to differ in age from the general population? Why? *No*

28. Vacation Home Sales A study by the National Association of Realtors reports that second-home sales accounted for more than a third of residential real estate transactions in 2004. According to the report, the average age of a person buying a second home as a vacation home was 55 years. A real estate agent wants to estimate the average age of those buying vacation homes in her area. She randomly selects 15 of her clients who purchased a vacation home and obtains the data shown. Assume that $\sigma = 4.7$.

47	51	51	47	47
35	55	54	37	40
52	49	51	41	48

(a) Use the data to compute a point estimate for the population mean age of the real estate agent's clients who purchased a vacation home. 47 years
(b) Because the sample size is small, we must verify that client age is approximately normally distributed and that the sample does not contain any outliers. The normal probability plot and boxplot are shown next. Are the conditions for constructing a Z-interval satisfied? Yes

Age

(c) Construct a 95% confidence interval for the mean age for all the real estate agent's clients who purchased a vacation home. Interpret this interval. (44.6, 49.4)
(d) Do the real estate agent's clients appear to differ in age from the general population? Why? Yes

29. Serum Cholesterol As reported by the U.S. National Center for Health Statistics, the mean serum high density lipoprotein (HDL) cholesterol of males 20 to 29 years old is $\mu = 47$. Dr. Paul Oswiecmiski wants to estimate the mean serum HDL cholesterol of his 20- to 29-year-old male patients. He randomly selects 15 of his 20- to 29-year-old patients and obtains the data shown. Assume that $\sigma = 12.5$.

48	61	46	35	48
34	64	42	29	71
64	47	56	27	53

Source: Dr. Paul Oswiecmiski

(a) Use the data to compute a point estimate for the population mean serum HDL cholesterol in Dr. Oswiecmiski's patients. 48.3
(b) An analysis of the data indicate that the sample data come from a population that is approximately normally distributed with no outliers. Construct a 95% confidence interval for the mean serum HDL cholesterol for all Dr. Oswiecmiski's 20- to 29-year-old male patients. Interpret this interval. (42.0, 54.6)
(c) Do Dr. Oswiecmiski's patients appear to have a serum HDL different from that of the general population. No
(d) If Dr. Oswiecmiski wanted a more precise confidence interval, what would you recommend?
(29c) Increase sample size or decrease level of confidence

30. Billing Process For a certain billing process, the number of days for customers to pay their bill from the date of invoice is approximately normally distributed, with mean $\mu = 47$ days and $\sigma = 11$ days. A random sample of 10 bills from the billing process during the month of June results in the following data:

55	45	45	42	65
58	35	36	34	60

(a) Use the data to compute a point estimate for the population mean number of days from invoice to pay the bill. 47.5 days
(b) Construct a 95% confidence interval for the mean number of days it takes customers to pay their bill. Interpret this interval. (40.7, 54.3)
(c) Is there evidence that June payments were typical? Yes

31. Volume of Harley-Davidson Stock The volume of a stock is the number of shares traded on a given day. The following data, in millions (so that 3.78 represents 3,780,000 shares traded), represent the volume of Harley-Davidson stock traded for a random sample 40 trading days in 2004. The standard deviation of the number of shares traded in 2004 was $\sigma = 1.00$ million shares.

1.74	0.99	0.96	1.41	2.43	0.77	1.60	1.81	3.42	2.01	1.98
1.15	0.87	1.62	0.97	6.37	1.15	1.43	0.69	3.04	0.97	1.32
2.27	0.51	0.76	3.01	2.44	1.72	1.39	1.72	2.37	1.29	1.70
0.95	1.18	1.05	1.38	1.23	0.68	1.26				

Source: yahoo.finance.com

(a) Use the data to compute a point estimate for the population mean number of shares traded per day in 2004. 1.64 million
(b) Construct a 90% confidence interval for the population mean number of shares traded per day in 2004. Interpret the confidence interval. (1.38, 1.90)

(c) A second random sample of 40 days in 2004 resulted in the following data:

6.23	1.62	3.04	1.13	1.06	1.70	1.37	2.13	1.29	2.02	0.68
1.06	3.79	1.24	1.02	0.85	3.42	1.47	1.60	0.96	1.30	0.96
1.72	1.12	1.33	1.02	0.84	0.80	1.16	1.85	1.79	1.35	3.30
0.82	0.73	1.54	3.98	2.54	2.27	1.60				

Construct a 90% confidence interval for the population mean number of shares traded per day in 2004. Interpret the confidence interval. (1.48, 2.00)

(d) Explain why the confidence intervals obtained in parts (b) and (c) are different.

32. Volume of Google Stock The volume of a stock is the number of shares traded on a given day. The following data, in millions (so that 4.64 represents 4,640,000 shares traded), represent the volume of Google stock traded for a random sample of 35 trading days in 2004. The standard deviation of the number of shares traded in 2004 was $\sigma = 6.17$ million shares.

Google

4.64	7.57	7.49	5.54	21.16	13.36	3.11	3.55	15.27
2.60	22.31	11.19	11.90	4.57	19.83	13.89	6.60	14.85
2.03	2.50	3.91	10.64	6.26	7.57	3.93	9.14	8.77
5.52	7.03	14.41	16.75	8.47	5.84	8.57	5.90	

Source: yahoo.finance.com

(a) Use the data to compute a point estimate for the population mean number of shares traded per day in 2004. 9.05 million

(b) Construct a 90% confidence interval for the population mean of the number of shares traded per day in 2004. Interpret the confidence interval. (7.33, 10.77)

(c) A second random sample of 35 days in 2004 resulted in the following data:

Google

12.41	2.93	13.89	3.93	11.90	14.85	2.60	6.89	5.84
3.61	2.68	5.24	3.79	7.39	10.67	7.06	6.26	4.60
22.31	7.57	3.62	11.06	4.35	7.54	12.37	9.14	7.57
7.67	8.77	15.28	7.63	3.55	4.27	32.76	7.49	

Construct a 90% confidence interval for the population mean for number of shares traded per day in 2004. Interpret the confidence interval. (6.84, 10.27)

(d) Explain why the confidence intervals obtained in parts (b) and (c) are different.

33. Dramas After watching a drama that seemed to last a long time, a student wondered how long the typical drama lasted. She obtained a random sample of 30 dramas and found the mean length of the movies to be 138.3 minutes. Assume the population standard deviation length of a drama is 27.3 minutes.

(a) An analysis of the data indicated that the distribution of lengths of dramas is skewed right. Why does the student have to have a large sample size? Population not normal

(b) Construct and interpret a 99% confidence interval for the mean length of a drama. (125.5, 151.1)

34. What Is Wrong? The mean age of the 42 presidents of the United States on the day of inauguration is 54.8 years with a standard deviation of 6.2 years. A researcher constructed a 95% confidence interval for the mean age of presidents on inauguration day. He wrote that he was 95% confident the mean age of the president on inauguration day is between 53.0 and 56.7 years of age. What is wrong with the researcher's analysis? Census data

35. Miles on a Saturn A researcher is interested in approximating the mean number of miles on 4-year-old Saturn SC1s. She finds a random sample of 33 such Saturn SC1s in the Chicago area and obtains the following results:

45,336	90,574	42,800	84,000	57,506
47,977	10,778	39,176	41,431	86,838
30,114	90,100	26,560	75,312	44,411
29,000	30,447	25,000	49,874	76,576
57,145	38,796	32,004	43,128	51,159
54,000	52,181	51,305	33,867	46,178
51,000	39,174	59,018		

Source: cars.com

(a) Obtain a point estimate of the population mean number of miles on a 4-year-old Saturn SC1. 49,477.7

(b) Construct and interpret a 99% confidence interval for the population mean for number of miles on a 4-year-old Saturn SC1. Assume that $\sigma = 19,700$.

(c) Construct and interpret a 95% confidence interval for the population mean for number of miles on a 4-year-old Saturn SC1. Assume that $\sigma = 19,700$.

(d) What effect does decreasing the level of confidence have on the interval? *Decreases*

(e) Do the confidence intervals computed in parts (b) and (c) represent an estimate for the population mean number of miles on Saturn SC1s in the United States? Why? *No*

(35b) (40,647.2, 58,308.2)   (35c) (42,756, 56,199)

36. Miles on a Cavalier A researcher is interested in approximating the mean number of miles on 3-year-old Chevy Cavaliers. She finds a random sample of 35 Cavaliers in the Orlando, Florida, area and obtains the following results:

37,815	20,000	57,103	46,585	24,822
49,678	30,983	52,969	8,000	39,862
6,000	65,192	34,285	30,906	41,841
39,851	43,000	74,361	52,664	33,587
52,896	45,280	30,000	41,713	76,315
22,442	45,301	52,899	41,526	28,381
55,163	51,812	36,500	31,947	16,529

Source: cars.com

(a) Obtain a point estimate of the population mean number of miles on a 3-year-old Cavalier. *40,520.2 miles*

(b) Construct and interpret a 99% confidence interval for the population mean for number of miles on a 3-year-old Cavalier. Assume that $\sigma = 16,100$. *(33,512.6, 47,527.8)*

(c) Construct and interpret a 95% confidence interval for the population mean for number of miles on a 3-year-old Cavalier. Assume that $\sigma = 16,100$. *(35,186.3, 45,854.1)*

(d) What effect does decreasing the level of confidence have on the width of the interval? *Decreases*

(e) Do the confidence intervals computed in parts (b) and (c) represent an interval estimate for the mean number of miles on Cavaliers in the United States? Why? *No*

37. Sample Size Dr. Paul Oswiecmiski wants to estimate the NW mean serum HDL cholesterol of all 20- to 29-year-old females. How many subjects are needed to estimate the mean serum HDL cholesterol of all 20- to 29-year-old females within 2 points with 99% confidence, assuming that $\sigma = 13.4$? Suppose Dr. Oswiecmiski would be content with 95% confidence. How does the decrease in confidence affect the sample size required? *298; 173; decreases*

38. Sample Size Dr. Paul Oswiecmiski wants to estimate the mean serum HDL cholesterol of all 20- to 29-year-old males. How many subjects are needed to estimate the mean serum HDL cholesterol of all 20- to 29-year-old males within 1.5 points with 90% confidence, assuming that $\sigma = 12.5$? Suppose Dr. Oswiecmiski would prefer 98% confidence. How does the increase in confidence affect the sample size required? *188; 378; increases*

39. How Much Do You Read? A Gallup poll conducted May 20–22, 2005 asked Americans how many books, either hardback or paperback, they read during the previous year. How many subjects are needed to estimate the number of books Americans read the previous year within one book with 95% confidence? Initial survey results indicate that $\sigma = 16.6$ books. *1059*

40. How Long Are You Online? A Gallup poll conducted January 23, 2003–February 10, 2003, asked American teens (aged 13 to 17) how much time they spent each week using the Internet. How many subjects are needed to estimate the time American teens spend on the Internet each week within 0.5 hour with 95% confidence? Initial survey results indicate that $\sigma = 6.6$ hours. *670*

41. Miles on a Saturn A researcher wishes to estimate the mean number of miles on 4-year-old Saturn SC1s.

(a) How many cars should be in a sample to estimate the mean number of miles within 1000 miles with 90% confidence, assuming that $\sigma = 19,700$? *1051*

(b) How many cars should be in a sample to estimate the mean number of miles within 500 miles with 90% confidence, assuming that $\sigma = 19,700$? *4201*

(c) What effect does doubling the required accuracy have on the sample size? Why is this the expected result? *Quadruples*

42. Miles on a Cavalier A researcher wishes to estimate the mean number of miles on 3-year-old Chevy Cavaliers.

(a) How many cars should be in a sample to estimate the mean number of miles within 2000 miles with 98% confidence, assuming that $\sigma = 16,100$? *352*

(b) How many cars should be in a sample to estimate the mean number of miles within 1000 miles with 98% confidence, assuming that $\sigma = 16,100$? *1408*

(c) What effect does doubling the required accuracy have on the sample size? Why is this the expected result? *Quadruples*

43. Simulation IQ scores as measured by the Stanford–Binet IQ test are normally distributed with $\mu = 100$ and $\sigma = 16$.

(a) Simulate obtaining 20 samples of size $n = 15$ from this population.

(b) Construct 95% confidence intervals for each of the 20 samples.

(c) How many of the intervals do you expect to include the population mean? How many actually contain the population mean?

44. Simulation Suppose the arrival of cars at Burger King's drive-through follows a Poisson process with $\mu = 4$ cars every 10 minutes.

(a) Simulate obtaining 30 samples of size $n = 40$ from this population.

(b) Construct 90% confidence intervals for each of the 30 samples. [**Note:** $\sigma = \sqrt{\mu}$ in a Poisson process.]

(c) How many of the intervals do you expect to include the population mean? How many actually contain the population mean?

45. Effect of Nonnormal Data The *exponential probability distribution* is a probability distribution that can be used to model waiting time in line or the lifetime of electronic components. Its density function with $\mu = 5$ is shown in the accompanying figure. We can see that the distribution is skewed right.

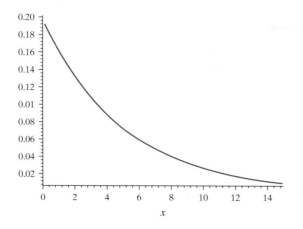

(a) Use MINITAB or some other statistical software to generate 100 random samples of size $n = 6$ from a population that follows the exponential probability distribution with $\mu = 5$. (It turns out that σ also equals 5.)

(b) Use the 100 samples to determine 100 95% confidence intervals with $\sigma = 5$.

(c) How many of the intervals do we expect to contain $\mu = 5$? How many of the 100 intervals contain $\mu = 5$?

(d) What are the consequences of not having data from a normal population when the sample size is small?

46. **Effect of Outliers** Suppose the following small data set represents a simple random sample from a population whose mean is 50 and standard deviation is 10.

43	63	53	50	58	44
53	53	52	41	50	43

(a) A normal probability plot indicates the data come from a population that is normally distributed with no outliers. Compute a 95% confidence interval for this data set, assuming $\sigma = 10$. (44.59, 55.91)

(b) Suppose the observation, 41, is inadvertently entered into the computer as 14. Verify that this observation is an outlier.

(c) Construct a 95% confidence interval on the data set with the outlier. What effect does the outlier have on the confidence interval? (42.34, 53.66)

(d) Consider the following data set, which represents a simple random sample of size 36 from a population whose mean is 50 and standard deviation is 10.

43	63	53	50	58	44
53	53	52	41	50	43
47	65	56	58	41	52
49	56	57	50	38	42
59	54	57	41	63	37
46	54	42	48	53	41

(46e) (46.98, 53.52); *decreases width of interval*
   Verify that the sample mean for the large data set is the same as the sample mean for the small data set.

(e) Compute a 95% confidence interval for the large data set, assuming $\sigma = 10$. Compare the results to part (a). What effect does increasing the sample size have on the confidence interval?

(f) Suppose the last observation, 41, is inadvertently entered as 14. Verify that this observation is an outlier.

(g) Compute a 95% confidence interval for the large data set with the outlier, assuming $\sigma = 10$. Compare the results to part (e). What effect does an outlier have on a confidence interval when the data set is large? (46.23, 52.77)

47. By how many times does the sample size have to be increased to decrease the margin of error by a factor of $\frac{1}{2}$? *4 times*

48. Suppose a certain population, A, has standard deviation $\sigma_A = 5$, and a second population, B, has standard deviation $\sigma_B = 10$. How many times larger than population A's sample size does population B's need to be to estimate μ with the same margin of error? [**Hint:** Compute n_B/n_A.] *4 times*

49. **Resistance and Robustness** The following data sets represent simple random samples from a population whose mean is 100 and standard deviation is 15.

(a) Compute the sample mean of each data set.

Data Set I				
106	122	91	127	88
74	77	108		

Data Set II				
106	122	91	127	88
74	77	108	87	88
111	86	113	115	97
122	99	86	83	102

Data Set III				
106	122	91	127	88
74	77	108	87	88
111	86	113	115	97
122	99	86	83	102
88	111	118	91	102
80	86	106	91	116

(b) For each data set, construct a 95% confidence interval about the population mean with $\sigma = 15$.

(c) What impact does the sample size n have on the width of the interval?
   For parts (d)–(e), suppose the data value 106 was accidentally recorded as 016.

(d) For each data set, construct a 95% confidence interval about the population mean with $\sigma = 15$ using the misentered data.

(e) Which intervals, if any, still capture the population mean, 100? What concept does this illustrate?

50. Confidence Interval Applet: The Role of Level of Confidence Load the confidence interval for a mean (the impact of a confidence level) applet.

(a) Set the shape to normal with mean = 50 and Std. Dev. = 10. Construct at least 1000 confidence intervals with $n = 10$. For 95% confidence, what proportion of the intervals contain the population mean? What proportion did you expect to contain the population mean?

(b) Repeat part (a). Did the same proportion of intervals contain the population mean?

(c) For 99% confidence, what proportion of the intervals contain the population mean? What proportion did you expect to contain the population mean?

51. Confidence Interval Applet: The Role of Sample Size Load the confidence interval for a mean (the impact of a confidence level) applet.

(a) Set the shape to normal with mean = 50 and Std. Dev. = 10. Construct at least 1000 confidence intervals with $n = 10$. What proportion of the 95% confidence intervals contain the population mean? What proportion did you expect to contain the population mean?

(b) Repeat part (a). Did the same proportion of intervals contain the population mean?

(c) Set the shape to normal with mean = 50 and Std. Dev. = 10. Construct at least 1000 confidence intervals with $n = 50$. What proportion of the 95% confidence intervals contain the population mean? What proportion did you expect to contain the population mean? Does sample size have any impact on the proportion of intervals that capture the population mean?

(d) Compare the width of the intervals for the samples of size $n = 50$ obtained in part (c) to the width of the intervals for the samples of size $n = 10$ obtained in part (a). Which are wider? Why?

52. Confidence Interval Applet: The Role of Shape Load the confidence interval applet.

(a) Set the shape to normal with mean = 50 and Std. Dev. = 10. Construct at least 1000 95% confidence intervals with $n = 40$. How many of the intervals contain the population mean? How many did you expect to contain the population mean?

(b) Set the shape to right skewed with mean = 50 and Std. Dev. = 10. Construct at least 1000 95% confidence intervals with $n = 5$. How many of the intervals contain the population mean? How many did you expect to contain the population mean?

(c) What is the impact of nonnormal data on the confidence interval when the sample size is small?

Technology Step by Step   **Confidence Intervals about μ, σ Known**

TI-83/84 Plus   *Step 1*: If necessary, enter raw data in L1.
Step 2: Press STAT, highlight TESTS, and select 7: ZInterval.
Step 3: If the data are raw, highlight DATA. Make sure List1 is set to L1 and Freq to 1. If summary statistics are known, highlight STATS and enter the summary statistics. Following σ:, enter the population standard deviation.
Step 4: Enter the confidence level following C-Level:.
Step 5: Highlight Calculate; press ENTER.

MINITAB   *Step 1*: Enter raw data in column C1.
Step 2: Select the **Stat** menu, highlight **Basic Statistics**, then highlight **1-Sample Z . . . .**
Step 3: Enter C1 in the cell marked "Variables". Select Confidence Interval, and enter a confidence level. In the cell marked "Sigma", enter the value of σ. Click OK.

Excel   *Step 1*: If necessary, enter raw data in column A.
Step 2: Load the PHStat Add-in.
Step 3: Select the **PHStat menu**, highlight **Confidence Intervals . . . ,** then highlight **Estimate for the mean, sigma known . . . .**
Step 4: Enter the value of σ and the confidence level. If the summary statistics are known, click "Sample statistics known" and enter the sample size and sample mean. If summary statistics are unknown, click "Sample statistics unknown". With the cursor in the "Sample cell range" cell, highlight the data in column A. Click OK.

9.2 Confidence Intervals about a Population Mean in Practice where the Population Standard Deviation Is Unknown

Preparing for This Section   Before getting started, review the following:

- Simple random sampling (Section 1.2, pp. 16–19)
- Parameter versus statistic (Section 3.1, p. 107)
- Degrees of freedom (Section 3.2, p. 128)
- Normal probability plots (Section 7.4, pp. 354–358)
- Distribution of the sample mean (Section 8.1, pp. 375–388)

Objectives

① **Know the properties of Student's *t*-distribution**

② **Determine *t*-values**

③ **Construct and interpret a confidence interval about a population mean**

Now we drop the assumption that σ is known and learn how to construct confidence intervals about a population mean in practice. That is, we learn how to construct confidence intervals about a population mean under the more realistic scenario that the population standard deviation is unknown.

① Know the Properties of Student's *t*-Distribution

In Section 9.1, we computed confidence intervals about a population mean, μ, under the assumption that the following conditions were met.

1. σ was known.
2. The population from which the sample was drawn followed a normal distribution or the sample size n was large ($n \geq 30$).
3. The sample was a simple random sample.

A $(1 - \alpha) \cdot 100\%$ confidence interval was then computed as

$$\bar{x} - z_{\frac{\alpha}{2}} \cdot \frac{\sigma}{\sqrt{n}} \quad \text{and} \quad \bar{x} + z_{\frac{\alpha}{2}} \cdot \frac{\sigma}{\sqrt{n}}$$

If σ is unknown, it seems reasonable to replace σ with s and proceed with the analysis. However, while it is true that $Z = \dfrac{\bar{x} - \mu}{\dfrac{\sigma}{\sqrt{n}}}$ is normally distributed with mean 0 and standard deviation 1, we cannot replace σ with s and say that $Z = \dfrac{\bar{x} - \mu}{\dfrac{s}{\sqrt{n}}}$ is normally distributed with mean 0 and standard deviation 1, because s itself is a random variable.

Instead, $\dfrac{\bar{x} - \mu}{\dfrac{s}{\sqrt{n}}}$ follows **Student's *t*-distribution**, developed by William Gosset. Gosset was in charge of conducting experiments at the Guinness brewery to identify the best barley variety. At the time, the only available distribution was the standard normal distribution, but Gosset always performed experiments with small data sets and he did not know σ. This led Gosset to determine the sampling distribution of $\dfrac{\bar{x} - \mu}{\dfrac{s}{\sqrt{n}}}$. He published his findings under the pen name *Student*.

Historical Note

William Sealey Gosset was born on June 13, 1876, in Canterbury, England. Gosset earned a degree in chemistry from New College in Oxford in 1899. He then got a job as a chemist for the Guinness Brewing Company in Dublin. Gosset, along with other chemists, was asked to find a way to make the best beer at the cheapest cost. This allowed him to concentrate on statistics. In 1904, Gosset wrote a paper on the brewing of beer that included a discussion of standard errors. In July 1905, Gosset met with Karl Pearson to learn about the theory of standard errors. Over the next few years, he developed his *t*-distribution. The Guinness Brewery did not allow its employees to publish, so Gosset published his research using the pen name Student. Gosset died on October 16, 1937.

Student's *t*-Distribution

Suppose a simple random sample of size *n* is taken from a population. If the population from which the sample is drawn follows a normal distribution, the distribution of

$$t = \frac{\overline{x} - \mu}{\frac{s}{\sqrt{n}}}$$

follows Student's *t*-distribution with $n - 1$ degrees of freedom,* where $\overline{x}$ is the sample mean and *s* is the sample standard deviation.

Note to Instructor
Consider the following to help students understand degrees of freedom: Suppose you have five pens of different colors and five students want to borrow a pen. The first four students have some freedom as to which pen to select; however, the last person has no freedom, but must choose the last available pen.

The interpretation of the *t*-statistic is the same as that of the *Z*-score. The *t*-statistic represents the number of sample standard errors $\overline{x}$ is from the population mean, μ. It turns out that the shape of the *t*-distribution depends on the sample size, *n*.

To help see how the *t*-distribution differs from the standard normal (or *z*-) distribution and the role the sample size *n* plays, we will go through the following simulation.

EXAMPLE 1 **Comparing the Standard Normal Distribution to the *t*-Distribution Using Simulation**

(a) Using a statistical spreadsheet such as MINITAB, obtain 1000 simple random samples of size $n = 5$ from a normal population with $\mu = 50$ and $\sigma = 10$.

(b) Calculate the sample mean and sample standard deviation for each sample.

(c) Compute $z = \dfrac{\overline{x} - \mu}{\frac{\sigma}{\sqrt{n}}}$ and $t = \dfrac{\overline{x} - \mu}{\frac{s}{\sqrt{n}}}$ for each sample.

(d) Draw histograms for both *z* and *t*.

Results: We use MINITAB to obtain the 1000 simple random samples and compute the 1000 sample means and sample standard deviations. We then compute $z = \dfrac{\overline{x} - \mu}{\frac{\sigma}{\sqrt{n}}} = \dfrac{\overline{x} - 50}{\frac{10}{\sqrt{15}}}$ and $t = \dfrac{\overline{x} - \mu}{\frac{s}{\sqrt{n}}} = \dfrac{\overline{x} - 50}{\frac{s}{\sqrt{5}}}$ for each of the 1000 samples. Figure 6(a) shows the histogram for *z*, and Figure 6(b) shows the histogram for *t*.

Figure 6

Histogram of z

(a)

Histogram of t

(b)

We notice that the histogram in Figure 6(a) is symmetric and bell shaped, with the center of the distribution at 0 and virtually all the rectangles between -3 and 3. In other words, *z* follows a standard normal distribution. The distribution of *t* is also symmetric and bell shaped and has its center at 0, but the

*The reader may wish to review the discussion of degrees of freedom in Section 3.2 on pp. 127–128.

distribution of t has longer tails (that is, t is more dispersed), so it is unlikely that t follows a standard normal distribution. The additional spread in the distribution of t can be attributed to the fact that we divide by $\dfrac{s}{\sqrt{n}}$ to find t instead of by $\dfrac{\sigma}{\sqrt{n}}$. Because the sample standard deviation is itself a random variable (rather than a constant such as σ), we have more dispersion in the distribution of t.

We now introduce the properties of the t-distribution.

Note to Instructor
Ask students to cite some differences between the standard normal and the t-distribution. What are some similarities?

Properties of the t-Distribution

1. The t-distribution is different for different degrees of freedom.
2. The t-distribution is centered at 0 and is symmetric about 0.
3. The area under the curve is 1. The area under the curve to the right of 0 equals the area under the curve to the left of 0 equals $\dfrac{1}{2}$.
4. As t increases without bound, the graph approaches, but never equals, zero. As t decreases without bound, the graph approaches, but never equals, zero.
5. The area in the tails of the t-distribution is a little greater than the area in the tails of the standard normal distribution, because we are using s as an estimate of σ, thereby introducing further variability into the t-statistic.
6. As the sample size n increases, the density curve of t gets closer to the standard normal density curve. This result occurs because, as the sample size n increases, the values of s get closer to the values of σ, by the Law of Large Numbers.

Figure 7

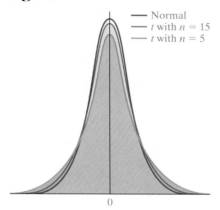

— Normal
— t with $n = 15$
— t with $n = 5$

0

In Figure 7, we show the t-distribution for the sample sizes $n = 5$ and $n = 15$. As a point of reference, we have also drawn the standard normal density curve.

Figure 8

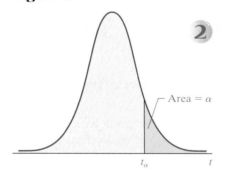

Area $= \alpha$

t_α t

2 Determine t-Values

Recall that the notation z_α is used to represent the z-score whose area under the normal curve to the right of z_α is α. Similarly, we let t_α represent the t-value such that the area under the t-distribution to the right of t_α is α. See Figure 8.

The shape of the t-distribution depends on the sample size, n. Therefore, the value of t_α depends not only on α, but also on the degrees of freedom, $n - 1$. In Table V in Appendix A, the far left column gives the degrees of freedom. The top row represents the area under the t-distribution to the right of some t-value.

EXAMPLE 2   **Finding t-Values**

Problem: Find the t-value such that the area under the t-distribution to the right of the t-value is 0.10, assuming 15 degrees of freedom. That is, find $t_{0.10}$ with 15 degrees of freedom.

Approach: We will perform the following steps.

Step 1: Draw a t-distribution with the unknown t-value labeled. Shade the area under the curve to the right of the t-value, as in Figure 8.

Step 2: Find the row in Table V that corresponds to 15 degrees of freedom and the column that corresponds to an area in the right tail of 0.10. Identify where the row and column intersect. This is the unknown t-value.

Figure 9

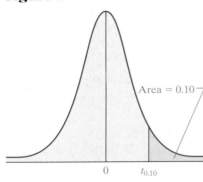

Area = 0.10

0    $t_{0.10}$

Solution

Step 1: Figure 9 shows the graph of the t-distribution with 15 degrees of freedom. The unknown value of t is labeled, and the area under the curve to the right of t is shaded.

Step 2: A portion of Table V is reproduced in Figure 10. We have enclosed the row that represents 15 degrees of freedom and the column that represents the area 0.10 in the right tail. The point where the row and column intersect is the t-value we are seeking. The value of $t_{0.10}$ with 15 degrees of freedom is 1.341; that is, the area under the t-distribution to the right of $t = 1.341$ with 15 degrees of freedom is 0.10.

Figure 10

					Area in Right Tail							
df	0.25	0.20	0.15	0.10	0.05	0.025	0.02	0.01	0.005	0.0025	0.001	0.0005
1	1.000	1.376	1.963	3.078	6.314	12.710	15.890	31.820	63.660	127.300	318.300	636.600
2	0.816	1.061	1.386	1.886	2.920	4.303	4.849	6.965	9.925	14.090	22.330	31.600
3	0.765	0.978	1.250	1.638	2.353	3.182	3.482	4.541	5.841	7.453	10.210	12.920
13	0.694	0.870	1.079	1.350	1.771	2.160	2.282	2.650	3.012	3.372	3.852	4.221
14	0.692	0.868	1.076	1.345	1.761	2.145	2.264	2.624	2.977	3.326	3.787	4.140
15	0.691	0.866	1.074	1.341	1.753	2.131	2.249	2.602	2.947	3.286	3.733	4.073
16	0.690	0.865	1.071	1.337	1.746	2.120	2.235	2.583	2.921	3.252	3.686	4.015

Now Work Problem 7(a).

The critical value of z with an area of to the right of 0.10 is approximately 1.28. Notice that the critical value for t is bigger than the corresponding critical value of z with an area to the right of 0.10.

USING TECHNOLOGY

The TI-84 Plus graphing calculator has an invT feature, which finds the value of t given an area left of the unknown t-value and the degrees of freedom.

If the degrees of fraeedom we desire are not available in Table V, we follow the practice of choosing the closest number of degrees of freedom available in the table. For example, if we have 43 degrees of freedom, we use 40 degrees of freedom from Table V. In addition, the last row of Table V provides the Z-values from the standard normal distribution. We use these values for situations where the degrees of freedom are more than 1000. This is acceptable because the t-distribution starts to behave like the standard normal distribution as n increases.

③ Construct and Interpret a Confidence Interval about a Population Mean

The construction of confidence intervals about μ with σ unknown follows the same logic as the construction of confidence intervals with σ known. The only difference is that we use s in place of σ and t in place of z.

Note to Instructor

Emphasize to students that we construct Z-intervals when σ is known and t-intervals when σ is unknown.

Constructing a $(1 - \alpha) \cdot 100\%$ Confidence Interval about μ, σ Unknown

Suppose a simple random sample of size n is taken from a population with unknown mean μ and unknown standard deviation σ. A $(1 - \alpha) \cdot 100\%$ confidence interval for μ is given by

$$\text{Lower bound: } \bar{x} - t_{\frac{\alpha}{2}} \cdot \frac{s}{\sqrt{n}} \qquad \text{Upper bound: } \bar{x} + t_{\frac{\alpha}{2}} \cdot \frac{s}{\sqrt{n}} \qquad (1)$$

where $t_{\frac{\alpha}{2}}$ is computed with $n - 1$ degrees of freedom.

Note: The interval is exact when the population is normally distributed. It is approximately correct for nonnormal populations, provided that n is large enough.

Notice that a confidence interval about μ with σ unknown can be computed for nonnormal populations even though Student's t-distribution required the population from which the sample was obtained to be normal. This is because the procedure for constructing the confidence interval is robust; that is, the procedure is accurate despite moderate departures from the normality requirement. Notice that we said the procedure is accurate for *moderate* departures from normality. If a small data set has outliers, the results are compromised, because neither the sample mean, $\overline{x}$, nor the sample standard deviation, s, is resistant to outliers. Sample data should always be inspected for serious departures from normality and for outliers. This is easily done with normal probability plots and boxplots.

Because this confidence interval uses the t-distribution, it is often referred to as the ***t*-interval**.

EXAMPLE 3 Constructing a Confidence Interval about a Population Mean

Table 5

64.0	33.4	45.8	56.0
51.5	29.2	63.7	

Source: Matt Gibson, instructor at Joliet West High School

Problem: An arborist is interesting in determining the mean diameter of mature white oak trees. The data in Table 5 represent the diameter (in centimeters) 1 meter above the base for a random sample of seven mature white oak trees in a forest preserve. Treat the sample as a simple random sample of all white oak trees in the forest preserve. Construct a 95% confidence interval for the mean diameter of a mature white oak tree 1 meter from the base. Interpret this interval.

Approach

Step 1: We draw a normal probability plot to verify the data come from a population that is normal and a boxplot to verify that there are no outliers.

Step 2: Compute $\overline{x}$ and s.

Step 3: Find the critical value $t_{\alpha/2}$ with $n - 1$ degrees of freedom.

Step 4: Compute the bounds on a $(1 - \alpha) \cdot 100\%$ confidence interval for μ, using the following:

$$\text{Lower bound:} \quad \overline{x} - t_{\frac{\alpha}{2}} \cdot \frac{s}{\sqrt{n}} \qquad \text{Upper bound:} \quad \overline{x} + t_{\frac{\alpha}{2}} \cdot \frac{s}{\sqrt{n}}$$

Step 5: Interpret the result.

Solution

Step 1: Because the sample size is small, we verify that the sample data come from a population that is normally distributed with no outliers. A normal probability plot of the data in Table 5 is shown in Figure 11(a). A boxplot is shown in Figure 11(b).

Figure 11

(a)

(b)

All the data lie within the bounds on the normal probability plot, indicating that the data could come from a population that is approximately normal. The box-plot does not display any outliers. The requirements for constructing the confidence interval are satisfied.

Step 2: We compute the sample mean and sample standard deviation using a calculator and find that $\bar{x} = 49.09$ centimeters and $s = 13.80$ centimeters.

Step 3: Because we wish to construct a 95% confidence interval, we have $\alpha = 0.05$. The sample size is $n = 7$. So we find $t_{\frac{0.05}{2}} = t_{0.025}$ with $7 - 1 = 6$ degrees of freedom. Referring to Table V, we find that $t_{0.025} = 2.447$.

Step 4: Using Formula (1), we find the lower and upper bounds:

$$\text{Lower bound: } \bar{x} - t_{\frac{\alpha}{2}} \cdot \frac{s}{\sqrt{n}} = 49.09 - 2.447 \cdot \frac{13.80}{\sqrt{7}} = 49.09 - 12.76 = 36.33$$

$$\text{Upper bound: } \bar{x} + t_{\frac{\alpha}{2}} \cdot \frac{s}{\sqrt{n}} = 49.09 + 2.447 \cdot \frac{13.80}{\sqrt{7}} = 49.09 + 12.76 = 61.85$$

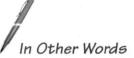

In Other Words

Notice that $t_{0.025} = 2.477$ for $n = 7$, while $z_{0.025} = 1.96$. The use of t results in a larger critical value, so the width of the interval is generally wider.

Step 5: We are 95% confident that the mean diameter 1 meter above the base of mature white oak trees in the forest preserve is between 36.33 and 61.85 centimeters.

EXAMPLE 4   **Constructing a Confidence Interval about a Population Mean Using Technology**

Problem: Construct a 95% confidence interval for the mean diameter of a mature white oak tree 1 meter above the base using the sample data in Table 5. Interpret this interval.

Approach: We will use a TI-84 Plus graphing calculator to construct the confidence interval. The steps for constructing confidence intervals using the TI-83/84 Plus graphing calculators, MINITAB, and Excel are given in the Technology Step by Step on page 434.

Result: Figure 12 shows the results on a TI-84 Plus graphing calculator.

Figure 12

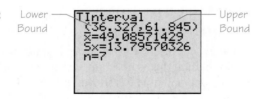

Lower Bound — Upper Bound

```
TInterval
 (36.327,61.845)
 x̄=49.08571429
 Sx=13.79570326
 n=7
```

USING TECHNOLOGY

The results obtained from technology may differ from "by hand" intervals due to rounding error.

Interpretation: We are 95% confident that the mean diameter 1 meter above the base of mature white oak trees in the forest preserve is between 36.33 and 61.85 centimeters.

Now Work Problem 23.

EXAMPLE 5   **The Effect of Outliers**

Problem: The management of Disney World wanted to estimate the mean waiting time at the Dumbo ride. They randomly selected 15 riders and measured the amount of time (in minutes) the riders spent waiting in line. The results are in Table 6. Figure 13 shows a normal probability plot and boxplot for the data in the table. Figure 13(a) demonstrates that the sample data could have come from a population that is normally distributed except for the single outlier. Figure 13(b) shows the outlier as well. Determine a 95% confidence interval for the mean waiting time at the Dumbo ride both with and without the

Table 6

30	29	34	41	37
16	18	30	24	25
29	16	80	19	30

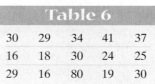

Note to Instructor

Remind students that the mean and standard deviation are not resistant. This is why confidence intervals should not be constructed from sample data that have one or more outliers.

outlier in the data set. Comment on the effect the outlier has on the confidence interval.

Figure 13

(a)

(b)

Approach: We will use MINITAB to construct the confidence intervals.

Solution: Figure 14(a) shows the confidence interval with the outlier included. Figure 14(b) shows the confidence interval with the outlier removed.

Figure 14

One-Sample T: Time(with outlier)

Variable	N	Mean	StDev	SE Mean	95.0% CI
Time	15	30.53	15.59	4.02	(21.90, 39.17)

(a)

One-Sample T: Time(without outlier)

Variable	N	Mean	StDev	SE Mean	95.0% CI
Time	14	27.00	7.75	2.07	(22.53, 31.47)

(b)

The 95% confidence interval with the outlier has a lower bound of 21.90 minutes and an upper bound of 39.17 minutes. The 95% confidence interval with the outlier removed has a lower bound of 22.53 minutes and an upper bound of 31.47. We notice a few things.

- With the outlier included, the sample mean is larger (i.e., drawn toward the outlier), because the sample mean is not resistant.
- With the outlier included, the sample standard deviation is larger, because the sample standard deviation is not resistant.
- With the outlier removed, the width of the interval decreased from $39.17 - 21.90 = 17.27$ minutes to $31.47 - 22.53 = 8.94$ minutes. The confidence interval is nearly twice as wide when the outlier is included.

What should we do if the requirements to compute a *t*-interval are not met? One option is to increase the sample size beyond 30 observations. The other option is to try to use *nonparametric procedures*. **Nonparametric procedures** typically do not require normality and the methods are resistant to outliers.

Now Work Problem 27.

Note to Instructor
This activity allows students to generate their own data that can be used to construct a confidence interval to draw conclusions and reinforce the concept of a confidence interval. Time Limit: 20–30 minutes depending on discussions.

In-Class Activity—Confidence Intervals

What is the mean number of rolls of a die before a 1 is observed? Either as an individual, or in a small group, roll a die until a 1 is observed. Repeat this process 30 times.

(a) Obtain a point estimate of the mean number of rolls of a die before a 1 is observed.

(b) The population standard deviation for the number of rolls before a 1 is observed is $\sqrt{30}$. Use this result to construct a 90% Z-interval for the mean number of rolls required before a 1 is observed.

(c) The population mean number of rolls before a 1 is observed is 6. Does your interval include 6? What proportion of the Z-intervals in the class include 6? How many did you expect to include 6?

(d) Construct a 90% t-interval for the mean number of rolls required before a 1 is observed.

(e) The population mean number of rolls before a 1 is observed is 6. Does your interval include 6? What proportion of the t-intervals in the class include 6? How many did you expect to include 6?

(f) Compare the Z-interval with the t-interval. Which has the smaller margin of error? Why?

9.2 ASSESS YOUR UNDERSTANDING

Concepts and Vocabulary

1. Explain the circumstances under which a Z-interval should be constructed. Under what circumstances should a t-interval be constructed? When can neither a Z- nor a t-interval be constructed?

2. Explain why the t-distribution has less spread as the number of degrees of freedom increases.

3. The procedures for constructing a t-interval are robust. Explain what this means.

4. Consider Figures 6(a) and (b). Why are the t-values more dispersed than the z-values?

5. Discuss the similarities and differences between the standard normal distribution and the t-distribution.

6. Explain what is meant by *degrees of freedom*.

Skill Building

7. (a) Find the t-value such that the area in the right tail is **NW** 0.10 with 25 degrees of freedom. 1.316

(b) Find the t-value such that the area in the right tail is 0.05 with 30 degrees of freedom. 1.697

(c) Find the t-value such that the area left of the t-value is 0.01 with 18 degrees of freedom. [*Hint*: Use symmetry.] −2.552

(d) Find the critical t-value that corresponds to 90% confidence. Assume 20 degrees of freedom. 1.725

8. (a) Find the t-value such that the area in the right tail is 0.02 with 19 degrees of freedom. 2.205

(b) Find the t-value such that the area in the right tail is 0.10 with 32 degrees of freedom. 1.309

(c) Find the t-value such that the area left of the t-value is 0.05 with 6 degrees of freedom. [*Hint*: Use symmetry.] −1.943

(d) Find the critical t-value that corresponds to 95% confidence. Assume 16 degrees of freedom. 2.120

(9b) (100.4, 115.6); increases   (9c) (104.6, 111.4); decreases

9. A simple random sample of size n is drawn from a population that is normally distributed. The sample mean, $\bar{x}$, is found to be 108, and the sample standard deviation, s, is found to be 10.

(a) Construct a 96% confidence interval about μ if the sample size, n, is 25. (103.7, 112.3)

(b) Construct a 96% confidence interval about μ if the sample size, n, is 10. How does decreasing the sample size affect the margin of error, E?

(c) Construct a 90% confidence interval if the sample size, n, is 25. Compare the results to those obtained in

part (a). How does decreasing the level of confidence affect the size of the margin of error, E?

(d) Could we have computed the confidence intervals in parts (a)–(c) if the population had not been normally distributed? Why? No

(10b) (44.6, 55.4); increases   (10c) (46.3, 53.7); decreases

10. A simple random sample of size n is drawn from a population that is normally distributed. The sample mean, $\bar{x}$, is found to be 50, and the sample standard deviation, s, is found to be 8.

(a) Construct a 98% confidence interval about μ if the sample size, n, is 20. (45.5, 54.5)

(b) Construct a 98% confidence interval about μ if the sample size, n, is 15. How does decreasing the sample size affect the margin of error, E?

(c) Construct a 95% confidence interval about μ if the sample size, n, is 20. Compare the results to those obtained in part (a). How does decreasing the level of confidence affect the margin of error, E?

(d) Could we have computed the confidence intervals in parts (a)–(c) if the population had not been normally distributed? Why? No

11. A simple random sample of size n is drawn. The sample mean, $\bar{x}$, is found to be 18.4, and the sample standard deviation, s, is found to be 4.5.

(a) Construct a 95% confidence interval about μ if the sample size, n, is 35. (16.85, 19.95)

(b) Construct a 95% confidence interval about μ if the sample size, n, is 50. How does increasing the sample size affect the margin of error, E? (17.12, 19.68); decreases

(c) Construct a 99% confidence interval about μ if the sample size, n, is 35. Compare the results to those obtained in part (a). How does increasing the level of confidence affect the margin of error, E?

(d) If the sample size is $n = 15$, what conditions must be satisfied to compute the confidence interval?

12. A simple random sample of size n is drawn. The sample mean, $\bar{x}$, is found to be 35.1, and the sample standard deviation, s, is found to be 8.7.

(a) Construct a 90% confidence interval about μ if the sample size, n, is 40. (32.78, 37.42)

(b) Construct a 90% confidence interval about μ if the sample size, n, is 100. How does increasing the sample size affect the margin of error, E?

(c) Construct a 98% confidence interval about μ if the sample size, n, is 40. Compare the results to those obtained in part (a). How does increasing the level of confidence affect the margin of error, E?

(d) If the sample size is $n = 18$, what conditions must be satisfied to compute the confidence interval? Population must be normal

(11c) (16.33, 20.48); increases (11d) Population must be normal (12b) (33.66, 36.55); decreases (12c) (31.76, 38.44); increases

Applying the Concepts

13. How Much Do You Read? A Gallup poll conducted May 20–22, 2005, asked 1006 Americans, "During the past year, about how many books, either hardcover or paperback, did you read either all or part of the way through?" Results of the survey indicated that $\bar{x} = 13.4$ books and $s = 16.6$ books. Construct a 99% confidence interval for the mean number of books Americans read either all or part of during the preceding year. Interpret the interval. (12.05, 14.75)

14. How Much Do You Read? A Gallup poll conducted July 21–August 14, 1978, asked 1006 Americans, "During the past year, about how many books, either hardcover or paperback, did you read either all or part of the way through?" Results of the survey indicated that $\bar{x} = 18.8$ books and $s = 19.8$ books.

(a) Construct a 99% confidence interval for the mean number of books Americans read either all or part of during the preceding year. Interpret the interval. (17.19, 20.41)

(b) Compare these results to those of Problem 13. Were Americans reading more in 1978 than in 2005? Yes

15. The SARS Epidemic Severe acute respiratory syndrome (or SARS) is a viral respiratory illness. It has the distinction of being the first new communicable disease of the 21st century. Researchers wanted to estimate the incubation period of patients with SARS. Based on interviews with 81 SARS patients, they found that the mean incubation period was 4.6 days with a standard deviation of 15.9 days. Based on this information, construct a 95% confidence interval for the mean incubation period of the SARS virus. Interpret the interval. (*Source*: Gabriel M. Leung et al., The Epidemiology of Severe Acute Respiratory Syndrome in the 2003 Hong Kong Epidemic: An Analysis of All 1755 Patients, *Annals of Internal Medicine*, 2004; 141:662–673.) (1.08, 8.12)

16. Tensile Strength Tensile strength is the amount of stress a material can withstand before it breaks. Researchers wanted to determine the tensile strength of a resin cement used in bonding crowns to teeth. The researchers bonded crowns to 72 extracted teeth. Using a tensile resistance test, they found the mean tensile strength of the resin cement to be 242.2 newtons (N) with a standard deviation of 70.6 newtons. Based on this information, construct a 90% confidence interval for the mean tensile strength of the resin cement. (*Source*: Simonides Consani et al., Effect of Cement Types on the Tensile Strength of Metallic Crowns Submitted to Thermocycling, *Brazilian Dental Journal*, Vol. 14, No. 3, 2003.) (228.33, 256.07)

17. How Much TV Do Teenagers Watch? A Gallup poll conducted January 17–February 6, 2005, asked 1028 teenagers aged 13 to 17, "Typically, how many hours per week do you spend watching TV?" Survey results indicate that $\bar{x} = 13.0$ hours and $s = 2.3$ hours. Construct a 95% confidence interval for the number of hours of TV teenagers watch each week. Interpret the interval. (12.86, 13.14)

18. Honda Accord Gas Mileage For a simple random sample of forty 2005 Honda Accords (4 cylinder, 2.4 liter, 5-speed automatic), the mean gas mileage was 23 miles per gallon with a standard deviation of 1.5 miles per gallon. Construct a 95% confidence interval for the mean gas mileage of similar 2005 Honda Accords. Interpret the interval. (22.52, 23.48)

19. Dividend Yields Many companies pay a portion of their profits to shareholders in the form of dividends. A stock's dividend yield is defined as the dividend paid by the company divided by its stock price. A stock analyst wants to estimate the mean dividend yield of financial stocks. He obtains a simple random sample of 11 financial stocks and obtains the following dividend yields. The data are in percent, so 3.17 represents 3.17%. Construct and interpret a 90% confidence interval for the dividend yield of financial stocks. An analysis of the data indicates that it is reasonable to conclude dividend yields follow a normal distribution. The data set has no outliers. (0.763, 1.867)

| 3.19 | 1.86 | 0 | 0.54 | 1.76 | 2.08 |
| 0.67 | 0.67 | 1.57 | 0 | 2.12 | |

Source: Morningstar

20. Tips A server at a restaurant wanted to estimate the mean tip percentage that she earns during dinner. She randomly selects 14 receipts from dinner, records the tip, and computes the tip rate. She obtains the following tip rates. The data are in percent, so 16.5 represents 16.5%. Construct and interpret a 95% confidence interval for the tip percent. An analysis of the data indicates that it is reasonable to conclude tip percent follows a normal distribution. The data has no outliers. (17.12, 20.30)

16.5	20.5	21.4	22.9	21.1
22.6	18.8	18.9	17.7	19.0
17.5	14.2	14.9	15.9	

Source: Pam Pimental, student at Joliet Junior College

21. Concentration of Dissolved Organic Carbon The following data represent the concentration of organic carbon (mg/L) collected from organic soil.

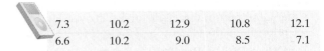

22.74	29.80	27.10	16.51	6.51
8.81	5.29	20.46	14.90	33.67
30.91	14.86	15.91	15.35	9.72
19.80	14.86	8.09	17.90	18.30
5.20	11.90	14.00	7.40	17.50
10.30	11.40	5.30	15.72	20.46
16.87	15.42	22.49		

Source: Lisa Emili, Ph.D. Candidate, University of Waterloo

Construct a 99% confidence interval for the mean concentration of dissolved organic carbon collected from organic soil. Interpret the interval. (**Note:** $\bar{x} = 15.92$ mg/L and $s = 7.38$ mg/L.) (12.40, 19.44)

22. Concentration of Dissolved Organic Carbon The following data represent the concentration of organic carbon (mg/L) collected from mineral soil.

8.50	3.91	9.29	21.00	10.89
10.30	11.56	7.00	3.99	3.79
5.50	4.71	7.66	11.72	11.80
8.05	10.72	21.82	22.62	10.74
3.02	7.45	11.33	7.11	9.60
12.57	12.89	9.81	17.99	21.40
8.37	7.92	17.90	7.31	16.92
4.60	8.50	4.80	4.90	9.10
7.90	11.72	4.85	11.97	7.85
9.11	8.79			

Source: Lisa Emili, Ph.D. candidate, University of Waterloo

(a) Construct a 99% confidence interval for the mean concentration of dissolved organic carbon collected from mineral soil. Interpret the interval. (**Note:** $\bar{x} = 10.03$ mg/L and $s = 4.98$ mg/L.) (8.08, 11.98)
(b) Compare the 99% confidence interval computed for organic soil (Problem 21) to the 99% confidence interval computed for mineral soil. Does there appear to be a difference between the concentration levels for the two soil types? Yes

23. Crawling Babies The following data represent the age (in weeks) at which babies first crawl based on a survey of 12 mothers conducted by Essential Baby.

52	30	44	35
47	37	56	26
39	26	39	28

Source: www.essentialbaby.com

(a) Because the sample size is small, we must verify that the data come from a population that is normally distributed and that the sample size does not contain any

outliers. The normal probability plot and boxplot are shown next. Are the conditions for constructing a confidence interval about the mean satisfied? Yes

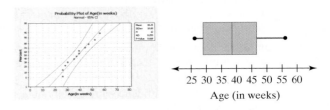

Age (in weeks)

(b) Construct a 95% confidence interval for the mean age at which a baby first crawls. (31.9, 44.7)
(c) What could be done to increase the accuracy of the interval without changing the level of confidence? Increase the sample size

24. Battery Life The following data represent the battery life, in hours, for a random sample of 10 full charges on a fifth-generation iPod™ music player.

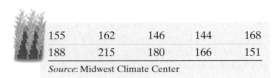

| 7.3 | 10.2 | 12.9 | 10.8 | 12.1 |
| 6.6 | 10.2 | 9.0 | 8.5 | 7.1 |

(a) Because the sample size is small, we must verify that the data come from a population that is normally distributed and that the sample size does not contain any outliers. The normal probability plot and boxplot are shown next. Are the conditions for constructing a confidence interval about the mean satisfied? Yes

Battery Life (hours)

(24b) (8.23, 10.71)

(b) Construct a 90% confidence interval for the mean number hours the battery will last on this player.
(c) Suppose you wanted more accuracy. What can be done to increase the accuracy of the interval without changing the level of confidence? Increase the sample size

25. The Growing Season An agricultural researcher is interested in estimating the mean length of the growing season in the Chicago area. Treating the last 10 years as a simple random sample, he obtains the following data, which represent the number of days of the growing season.

| 155 | 162 | 146 | 144 | 168 |
| 188 | 215 | 180 | 166 | 151 |

Source: Midwest Climate Center

(a) Because the sample size is small, we must verify that the data come from a population that is normally distributed and that the sample size does not contain any outliers. The normal probability plot and boxplot are shown next. Are the conditions for constructing a confidence interval about the mean satisfied? Yes

Length (Days)

(b) Construct a 95% confidence interval for the mean length of the growing season in the Chicago area. (151.9, 183.2)
(c) What could be done to increase the accuracy of the interval without changing the level of confidence?
Increase the sample size

26. Disc Weight When playing disc golf, the weight of the disc has a huge impact on disc performance. Lighter discs tend to fly farther, but are less stable. Heavier discs tend to give better control and are better suited for strong winds. The following data represent the weights, in grams, of a random sample of 12 discs produced by a certain manufacturer.

172	168	170	173	172	173
171	175	173	171	169	173

(a) Because the sample size is small, we must verify that the data come from a population that is normally distributed and that the sample size does not contain any outliers. The normal probability plot and boxplot are shown next. Are the conditions for constructing a confidence interval about the mean satisfied? Yes

Weight (grams)

(b) Construct a 95% confidence interval for disc weight. Interpret the interval. (170.4, 173.0)

27. Effect of Outliers The following data represent the asking price of a simple random sample of homes for sale in Lexington, Kentucky in June 2005.

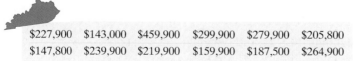

$227,900	$143,000	$459,900	$299,900	$279,900	$205,800
$147,800	$239,900	$219,900	$159,900	$187,500	$264,900

(a) Construct a boxplot to identify the outlier.
(b) Construct a 99% confidence interval with the outlier included. ($158,567.7, $314,148.9)
(c) Construct a 99% confidence interval with the outlier removed. ($165,219.6, $266,853.2)
(d) Comment on the effect the outlier has on the confidence interval.

28. Effect of Outliers The following data represent a simple random sample of ages of West Nile virus victims in the United States during 2004.

64	77	71	54	60	79
76	82	81	77	68	34

(a) Construct a boxplot to identify the outlier.
(b) Construct a 99% confidence interval with the outlier included. (56.1, 81.1)
(c) Construct a 99% confidence interval with the outlier removed. (62.9, 80.5)
(d) Comment on the effect the outlier has on the confidence interval.

29. Simulation IQ scores based on the Wechsler Intelligence Scale for Children (WISC) are known to be normally distributed with $\mu = 100$ and $\sigma = 15$.
(a) Simulate obtaining 20 samples of size $n = 15$ from this population.
(b) Obtain the sample mean and standard deviation for each of the 20 samples.
(c) Construct 95% t-intervals for each of the 20 samples.
(d) How many of the intervals do you expect to include the population mean? How many actually contain the population mean?

30. Simulation Suppose the arrival of cars at Burger King's drive-through follows a Poisson process with $\mu = 4$ cars every 10 minutes.
(a) Simulate obtaining 30 samples of size $n = 35$ from this population.
(b) Obtain the sample mean and standard deviation for each of the 30 samples.
(c) Construct 90% t-intervals for each of the 30 samples.
(d) How many of the intervals do you expect to include the population mean? How many actually contain the population mean?

31. Confidence Interval Applet: The Role of Right Skewed Data Load the confidence intervals for a mean (the impact of not knowing the standard deviation) applet.
(a) Set the shape to normal with mean = 100 and Std. Dev. = 15. Construct at least 1000 confidence intervals with $n = 10$. Compare the proportion of the 95% Z-intervals and t-intervals that contain the population mean. Is this what you would expect?
(b) Set the shape to skewed right with mean = 100 and Std. Dev. = 15. Construct at least 1000 confidence intervals with $n = 10$. Compare the proportion of the 95% Z-intervals and t-intervals that contain the population mean. Is this what you would expect?

Consumer Reports Tests Tires

Consumer Reports' specialized auto-test facility is the largest, most sophisticated consumer-based auto-testing facility in the world. Located on 327 acres in East Haddam, Connecticut, the facility is staffed by a team of experienced engineers and test personnel who buy and test more than 40 new cars, SUVs, minivans, and light trucks each year. For each vehicle, Consumer Reports conducts more than 46 individual tests, ranging from emergency handling, acceleration, and braking to fuel-economy measurements, noise-level evaluations, and bumper-impact tests.

© 2001 by Consumers Union of U.S., Inc., Yonkers, NY 10703-1057, a nonprofit organization. Reprinted with permission from CONSUMER REPORTS® for educational purposes only. No commercial use or photocopying permitted. To learn more about Consumers Union, log onto www.ConsumerReports.org.

As part of testing vehicles, Consumer Reports tests tires. Our tire evaluations include dry and wet braking from 60 mph, braking on ice, snow traction on a flat surface, snow traction on a snow hill, emergency handling, routine cornering, ride comfort, rolling resistance, and noise. All the test data are recorded using an optical instrument that provides precise speed and distance measurements.

The following table contains the dry brake stopping distance (in feet) data for one brand of tires recently tested.

Measurement	Distance
1	131.8
2	123.2
3	132.9
4	139.8
5	140.3
6	128.3
7	129.3
8	129.5
9	134.0

(a) A normal probability plot of the dry brake distance data is shown. What does this plot suggest about the distribution of the brake data?

(b) Does the data set contain any outliers?

(c) Use these data to construct a 95% confidence interval for the mean dry braking distance for this brand of tires. Write a statement that explains the meaning of this confidence interval to the readers of *Consumer Reports* magazine.

Note to Readers: *In many cases, our test protocol and analytical methods are more complicated than described in these examples. The data and discussions have been modified to make the material more appropriate for the audience.*

Technology Step by Step **Confidence Intervals about μ, σ Unknown**

TI-83/84 Plus **Step 1:** If necessary, enter raw data in L1.

Step 2: Press STAT, highlight TESTS, and select 8: TInterval.

Step 3: If the data are raw, highlight DATA. Make sure List1 is set to L1 and Freq to 1. If summary statistics are known, highlight STATS and enter the summary statistics.

Step 4: Enter the confidence level following `C-Level:`.

Step 5: Highlight `Calculate`; press ENTER.

MINITAB   *Step 1:* Enter raw data in column C1.

Step 2: Select the **Stat** menu, highlight **Basic Statistics**, then highlight **1-Sample t . . . .**

Step 3: Enter C1 in the cell marked "Variables". Select Confidence Interval, and enter a confidence level. Click OK.

Excel   *Step 1:* If necessary, enter raw data in column A.

Step 2: Load the PHStat Add-in.

Step 3: Select the **PHStat menu**, highlight **Confidence Intervals . . . ,** then highlight **Estimate for the mean, sigma unknown . . . .**

Step 4: Enter the confidence level. If the summary statistics are known, click "Sample statistics known" and enter the sample size, sample mean, and sample standard deviation. If summary statistics are unknown, click "Sample statistics unknown". With the cursor in the "Sample cell range" cell, highlight the data in column A. Click OK.

9.3 Confidence Intervals about a Population Proportion

Preparing for This Section   Before getting started, review the following:

- Distribution of the sample proportion (Section 8.2, pp. 392–397)

Objectives

1 Obtain a point estimate for the population proportion

2 Construct and interpret a confidence interval for the population proportion

3 Determine the sample size necessary for estimating a population proportion within a specified margin of error

In Other Words

The symbol ± is read "plus or minus." It means "to add and subtract the quantity following the ± symbol."

Probably the most frequently reported confidence interval is one involving the proportion of a population. Researchers are often interested in estimating the proportion of the population that has a certain characteristic. For example, in a poll conducted by the Gallup Organization in early December 2004, a random sample of 1004 Americans resulted in 84% of the respondents stating that they are satisfied with the way things are going in their personal lives. The poll had a margin of error of ±3% with 95% confidence. Based on discussions in Sections 9.1 and 9.2, we know that this means the pollsters are 95% confident the true proportion of Americans who are satisfied with the way things are going in their personal lives is between 81% and 87% (84% ± 3%).

In this section, we will discuss the techniques for estimating the population proportion population, p. In addition, we present methods for determining the sample size required to estimate the population proportion.

1 Obtain a Point Estimate for the Population Proportion

A point estimate is an unbiased estimator of the parameter. The point estimate for the population proportion is $\hat{p} = \dfrac{x}{n}$, where x is the number of individuals in the sample with the specified characteristic and n is the sample size.

EXAMPLE 1 | **Obtaining a Point Estimate of a Population Proportion**

Problem: In a poll conducted March 17–21, 2005, by the Pew Research Center for the People and the Press, a simple random sample of 1505 American adults was asked whether they were in favor of tighter enforcement of government rules on TV content during hours when children are most likely to be watching. Of the 1505 adults, 1129 responded yes. Obtain a point estimate for the population proportion of Americans who are in favor of tighter enforcement of government rules on TV content during hours when children are most likely to be watching.

Approach: The point estimate of the population proportion is $\hat{p} = \dfrac{x}{n}$, where $x = 1129$ and $n = 1505$.

Solution: Substituting into the formula, we obtain $\hat{p} = \dfrac{x}{n} = \dfrac{1129}{1505} = 0.750 = 75.0\%$. We estimate that 75% of adult Americans are in favor of tighter enforcement of government rules on TV content during hours when children are most likely to be watching.

> Now Work Problem 11(a).

② Construct and Interpret a Confidence Interval for the Population Proportion

The point estimate obtained in Example 1 is a single estimate of the unknown parameter, p. We recognize it is unlikely that the sample proportion will equal the population proportion. Therefore, rather than reporting the value of the sample proportion alone, it is preferred to report an interval about the sample proportion.

In Section 8.2, we discussed the sampling distribution of $\hat{p}$. For convenience, we repeat the sampling distribution of $\hat{p}$ here.

Sampling Distribution of $\hat{p}$

For a simple random sample of size n where $n \leq 0.05N$ (that is, the sample size is no more than 5% of the population size), the sampling distribution of $\hat{p}$ is approximately normal with mean $\mu_{\hat{p}} = p$ and standard deviation $\sigma_{\hat{p}} = \sqrt{\dfrac{p(1-p)}{n}}$, provided that $np(1-p) \geq 10$.

We use the sampling distribution of $\hat{p}$ to construct a confidence interval for the population proportion p.

Constructing a $(1 - \alpha) \cdot 100\%$ Confidence Interval for a Population Proportion

Suppose a simple random sample of size n is taken from a population. A $(1 - \alpha) \cdot 100\%$ confidence interval for p is given by the following quantities:

$$\text{Lower bound: } \hat{p} - z_{\alpha/2} \cdot \sqrt{\dfrac{\hat{p}(1-\hat{p})}{n}}$$

$$\text{Upper bound: } \hat{p} + z_{\alpha/2} \cdot \sqrt{\dfrac{\hat{p}(1-\hat{p})}{n}} \qquad \textbf{(1)}$$

Note: It must be the case that $n\hat{p}(1-\hat{p}) \geq 10$ and $n \leq 0.05N$ to construct this interval.

Notice we use $\hat{p}$ in place of p in the standard deviation. This is because p is unknown and $\hat{p}$ is the best point estimate of p.

EXAMPLE 2

Constructing a Confidence Interval for a Population Proportion

Problem: In a poll conducted March 17–21, 2005, by the Pew Research Center for the People and the Press, a simple random sample of 1505 American adults was asked whether they were in favor of tighter enforcement of government rules on TV content during hours when children are most likely to be watching. Of the 1505 adults, 1129 responded yes. Obtain a 95% confidence interval for the proportion of Americans who are in favor of tighter enforcement of government rules on TV content during hours when children are most likely to be watching.

Approach

Step 1: Compute the value of $\hat{p}$.

Step 2: We can compute a 95% confidence interval about p provided that $n\hat{p}(1 - \hat{p}) \geq 10$ and $n \leq 0.05N$.

Step 3: Determine the critical value, $z_{\frac{\alpha}{2}}$.

Step 4: Determine the lower and upper bounds of the confidence interval.

Step 5: Interpret the result.

Solution

Step 1: From Example 1 we have that $\hat{p} = 0.750$.

Step 2: $n\hat{p}(1 - \hat{p}) = 1505(0.750)(1 - 0.750) = 282.1875 \geq 10$. There are over 180 million adult Americans, so our sample size is definitely less than 5% of the population size. We can proceed to construct the confidence interval.

Step 3: Because we want a 95% confidence interval, we have $\alpha = 0.05$, so $z_{\frac{\alpha}{2}} = z_{\frac{0.05}{2}} = z_{0.025} = 1.96$.

Step 4: Substituting into Formula (1) with $n = 1505$, we obtain the lower and upper bounds of the confidence interval:

Lower bound:

$$\hat{p} - z_{\frac{\alpha}{2}} \cdot \sqrt{\frac{\hat{p}(1 - \hat{p})}{n}} = 0.750 - 1.96 \cdot \sqrt{\frac{0.750(1 - 0.750)}{1505}}$$

$$= 0.750 - 0.022 = 0.728$$

Upper bound:

$$\hat{p} + z_{\frac{\alpha}{2}} \cdot \sqrt{\frac{\hat{p}(1 - \hat{p})}{n}} = 0.750 + 1.96 \cdot \sqrt{\frac{0.750(1 - 0.750)}{1505}}$$

$$= 0.750 + 0.022 = 0.772$$

Step 5: We are 95% confident that the proportion of adult Americans who are in favor of tighter enforcement of government rules on TV content during hours when children are most likely to be watching is between 0.728 and 0.772.

EXAMPLE 3

Constructing a Confidence Interval about a Population Proportion Using Technology

Problem: In a poll conducted March 17–21, 2005, by the Pew Research Center for the People and the Press, a simple random sample of 1505 American adults was asked whether they were in favor of tighter enforcement of government rules on TV content during hours when children are most likely to be watching. Of the 1505 adults, 1129 responded yes. Obtain a 95% confidence interval for the proportion of Americans who are in favor of tighter enforcement of government rules on TV content during hours when children are most likely to be watching.

Approach: We will use MINITAB to construct the confidence interval. The steps for constructing confidence intervals using MINITAB, Excel, and the TI-83/84 graphing calculators are given in the Technology Step by Step on page 443.

Result: Figure 15 shows the results obtained from MINITAB.

Figure 15   **Test and CI for One Proportion**
```
Test of p = 0.5 vs p not = 0.5

Sample    X      N    Sample p       95% CI          Z-Value P-Value
1       1129   1505   0.750166 (0.728294,0.772038)    19.41   0.000
```

The lower bound is 0.728 and the upper bound is 0.772.

USING TECHNOLOGY

The results obtained using technology may differ from those obtained by hand due to rounding.

Interpretation: We are 95% confident that the proportion of adult Americans who are in favor of tighter enforcement of government rules on TV content during hours when children are most likely to be watching is between 0.728 and 0.772.

! CAUTION

Beware of surveys that do not report a margin of error. Survey results should also report sample size, sampling technique, and the population that was being studied.

It is important to remember the correct interpretation of a confidence interval. The statement "95% confident" means that, if 100 samples of size 1505 were taken, about 95 of the intervals will contain the parameter p and about 5 will not. Unfortunately, we cannot know whether the interval we computed in Examples 2 and 3 is one of the 95 intervals that contains p or one of the 5 that does not contain p.

Often, polls will report their results by giving the value of $\hat{p}$ obtained from the sample data along with the margin of error, rather than reporting a confidence interval. In Examples 2 and 3, the Pew Research Center might say, "In a survey conducted March 17–21, 2005, 75% of adult Americans are in favor of tighter enforcement of government rules on TV content during hours when children are most likely to be watching. The survey results have a margin of error of 2.2%."

Now Work Problems 11(b), (c), and (d).

Note to Instructor

Remind students not to eat any of the candies until they are counted. It is helpful to state up front how to deal with partial candies. The purpose of this activity is to illustrate the meaning of a confidence interval estimate. Time Limit: 30–40 minutes depending on discussions.

In-Class Activity—Confidence Intervals (M&M's)

Mars, Inc. reports that the proportion of Plain M&M's it produces that are blue is 0.24. However, you suspect that this is not the case.

(a) Each student should obtain a 1-pound bag of plain M&M's candies and bring it to class unopened.
(b) In class, count the number of candies in your bag for each of the six colors (red, brown, blue, yellow, green, and orange).
(c) Obtain a point estimate for the population proportion of blue candies.
(d) Verify the requirements for constructing a confidence interval about $\hat{p}$ are satisfied. If the conditions are not met, what could you do to rectify the situation?
(e) Construct a 95% confidence interval for the proportion of blue candies. Interpret the interval. Does your interval contain 0.24?
(f) Compare your interval with the rest of the class. What proportion of intervals do not contain 0.24? Is this result what you would expect?

③ Determine the Sample Size Necessary for Estimating a Population Proportion within a Specified Margin of Error

In Section 9.1, we introduced a method for determining the sample size n required to estimate the sample mean within a certain margin of error with a specified level of confidence. The formula was obtained by solving the margin of error, $E = z_{\frac{\alpha}{2}} \cdot \dfrac{\sigma}{\sqrt{n}}$, for n. We can follow the same approach to determine sample size when estimating a population proportion. In Formula (1) on page 436, we notice that the margin of error, E, is given by $E = z_{\frac{\alpha}{2}} \cdot \sqrt{\dfrac{\hat{p}(1-\hat{p})}{n}}$. We solve the margin of error for n and obtain $n = \hat{p}(1-\hat{p})\left(\dfrac{z_{\frac{\alpha}{2}}}{E}\right)^2$.

Note to Instructor

You might have students experiment with various values of $\hat{p}$ to see which one maximizes $\hat{p}(1-\hat{p})$.

The problem with this formula is that it depends on $\hat{p}$, and $\hat{p} = \dfrac{x}{n}$ depends on the sample size, n, which is what we are trying to determine in the first place! How do we resolve this issue? There are two possibilities: (1) We could use an estimate of p based on a pilot study or an earlier study, or (2) we could let $\hat{p} = 0.5$. When $\hat{p} = 0.5$ the maximum value of $\hat{p}(1-\hat{p}) = 0.25$ is obtained as illustrated in Figure 16. Using the maximum value gives the largest possible value of n for a given level of confidence and a given margin of error.

Figure 16

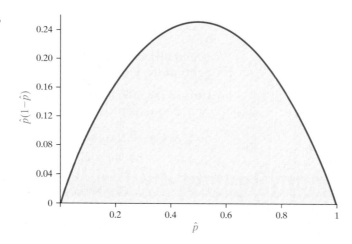

The disadvantage of the second option is that it could lead to a larger sample size than is necessary. Because of the time and expense of sampling, it is desirable to avoid too large a sample.

Sample Size Needed for Estimating the Population Proportion p

The sample size required to obtain a $(1-\alpha) \cdot 100\%$ confidence interval for p with a margin of error E is given by

$$n = \hat{p}(1-\hat{p})\left(\frac{z_{\frac{\alpha}{2}}}{E}\right)^2 \qquad (2)$$

(rounded up to the next integer), where $\hat{p}$ is a prior estimate of p. If a prior estimate of p is unavailable, the sample size required is

$$n = 0.25\left(\frac{z_{\frac{\alpha}{2}}}{E}\right)^2 \qquad (3)$$

rounded up to the next integer. The margin of error should always be expressed as a decimal when using Formulas (2) and (3).

In Other Words

There are two formulas for determining sample size when estimating the population proportion. Formula (2) requires a prior estimate of p; Formula (3) does not.

EXAMPLE 4 **Determining Sample Size**

Problem: A sociologist wishes to estimate the percentage of the U.S. population living in poverty. What size sample should be obtained if she wishes the estimate to be within 2 percentage points with 99% confidence if

(a) she uses the 2003 estimate of 12.7% obtained from the American Community Survey?

(b) she does not use any prior estimates?

Approach: In both cases, we have $E = 0.02$ (2% = 0.02) and $z_{\alpha/2} = z_{0.01/2} = z_{0.005} = 2.575$. To answer part (a), we let $\hat{p} = 0.127$ in Formula (2). To answer part (b), we use Formula (3).

Solution

(a) Substituting $E = 0.02$, $z_{0.005} = 2.575$, and $\hat{p} = 0.127$ into Formula (2), we obtain

$$n = \hat{p}(1 - \hat{p})\left(\frac{z_{\frac{\alpha}{2}}}{E}\right)^2 = 0.127(1 - 0.127)\left(\frac{2.575}{0.02}\right)^2 = 1837.9$$

CAUTION
We always round up when determining sample size.

We round this value up to 1838, so she must survey 1838 randomly selected residents of the United States.

(b) Substituting $E = 0.02$ and $z_{0.005} = 2.575$ into Formula (3), we obtain

$$n = 0.25\left(\frac{z_{\frac{\alpha}{2}}}{E}\right)^2 = 0.25\left(\frac{2.575}{0.02}\right)^2 = 4144.1$$

We round this value up to 4145, so she must survey 4145 randomly selected residents of the United States.

We can see the effect of not having a prior estimate of p: The required sample size more than doubled!

Now Work Problem 17.

MAKING AN INFORMED DECISION

Help Wanted

What's Your Major?
One of the most difficult decisions a person has to make while attending college is choosing a major field of study. This decision plays a huge role in the career path of the individual. The purpose of this project is to help you make a more informed decision regarding your major.

Randomly select people who have recently graduated in the major that you have chosen or randomly select individuals who have recently secured a job in a career that you are considering. If you have not yet chosen a major, select one that interests you. Create a survey that will allow you to make a more informed decision regarding your major field of study. Some suggested sample questions follow:

(a) What was your major in college?

(b) On average, how many hours do you work?

(c) Approximately how many weeks did it take you to find your job?

(d) What was your starting annual salary?

(e) Are you satisfied or dissatisfied with your career?

(f) Do you believe you have job security?

Administer the survey. Be sure to explain the purpose of the survey and make it look professional. Because it is important that we obtain accurate results, make sure that the surveys are completed anonymously. We will treat the randomly selected individuals as a simple random sample.

Estimate the unknown parameter for each question asked. For example, in question (b), estimate the population mean number of hours worked. In question (e), estimate the proportion of individuals satisfied with their career. Then construct 95% confidence intervals about each parameter. Write a report detailing your findings.

9.3 ASSESS YOUR UNDERSTANDING

Concepts and Vocabulary

1. What is the best point estimate of a population proportion?

2. What are the requirements that must be satisfied to construct a confidence interval about a population proportion?

3. When determining the sample size required to obtain an estimate for a population proportion, is the researcher better off using a prior estimate of p or no prior estimate of p? Why? List some pros and cons for each scenario.

4. Explain why you should be wary of surveys that do not report a margin of error.

Skill Building

In Problems 5–10, construct a confidence interval of the population proportion at the given level of confidence.

5. $x = 30, n = 150$, 90% confidence *(0.146, 0.254)*

6. $x = 80, n = 200$, 98% confidence *(0.319, 0.481)*

7. $x = 120, n = 500$, 99% confidence *(0.191, 0.289)*

8. $x = 400, n = 1200$, 95% confidence *(0.306, 0.360)*

9. $x = 860, n = 1100$, 94% confidence *(0.758, 0.805)*

10. $x = 540, n = 900$, 96% confidence *(0.567, 0.633)*

Applying the Concepts

11. **Lipitor** The drug Lipitor™ is meant to lower cholesterol levels. In a clinical trial of 863 patients who received 10-mg doses of Lipitor daily, 47 reported a headache as a side effect.
    (a) Obtain a point estimate for the population proportion of Lipitor users who will experience a headache as a side effect. *0.054*
    (b) Verify that the requirements for constructing a confidence interval about $\hat{p}$ are satisfied.
    (c) Construct a 90% confidence interval for the population proportion of Lipitor users who will report a headache as a side effect. *(0.041, 0.067)*
    (d) Interpret the confidence interval.

12. **Pepcid** A study of 74 patients with ulcers was conducted in which they were prescribed 40 mg of Pepcid™. After 8 weeks, 58 reported ulcer healing.
    (a) Obtain a point estimate for the proportion of patients with ulcers receiving Pepcid who will report ulcer healing. *0.784*
    (b) Verify that the requirements for constructing a confidence interval about $\hat{p}$ are satisfied.
    (c) Construct a 99% confidence interval for the proportion of patients with ulcers receiving Pepcid who will report ulcer healing. *(0.661, 0.907)*
    (d) Interpret the confidence interval.

13. **Defense Spending** In a February 2005 poll conducted by the Gallup Organization, 302 of 1008 randomly selected adults aged 18 or older stated that they believe the United States is spending too little on national defense.
    (a) Obtain a point estimate for the proportion of adults aged 18 or older who feel the United States is spending too little on national defense. *0.300*
    (b) Verify that the requirements for constructing a confidence interval about $\hat{p}$ are satisfied.
    (c) Construct a 98% confidence interval for the proportion of adults aged 18 or older who believe the United States is spending too little on national defense. Interpret the interval. *(0.266, 0.334)*

14. **Partial Birth Abortions** In an October 2003 poll conducted by the Gallup Organization, 684 of 1006 randomly selected adults aged 18 years old or older stated they think the government should make partial birth abortions illegal, except in cases necessary to save the life of the mother.
    (a) Obtain a point estimate for the proportion of adults aged 18 or older who think the government should make partial birth abortions illegal, except in cases necessary to save the life of the mother. *0.680*
    (b) Verify that the requirements for constructing a confidence interval about $\hat{p}$ are satisfied.
    (c) Construct a 98% confidence interval for the proportion of adults aged 18 or older who think the government should make partial birth abortions illegal, except in cases necessary to save the life of the mother. Interpret the interval. *(0.646, 0.714)*

15. **Packer Fans** In a Harris Poll conducted October 20–25, 2004, 381 of 2114 randomly selected adults who follow professional football said the Green Bay Packers were their favorite team.

(a) Verify that the requirements for constructing a confidence interval about $\hat{p}$ are satisfied.

(b) Construct a 90% confidence interval for the proportion of adults who follow professional football who say the Green Bay Packers is their favorite team. Interpret this interval. *(0.166, 0.194)*

(c) Construct a 99% confidence interval for the proportion of adults who follow professional football who say the Green Bay Packers is their favorite team. Interpret this interval. *(0.158, 0.202)*

(d) What is the effect of increasing the level of confidence on the width of the interval? *The interval gets wider*

16. Taking It Easy In a Gallup poll conducted December 11–14, 2003, 455 of 1011 randomly selected adults aged 18 and older said they had too little time for relaxing or doing nothing.

(a) Verify that the requirements for constructing a confidence interval about $\hat{p}$ are satisfied.

(b) Construct a 92% confidence interval for the proportion of adults aged 18 and older who say they have too little time for relaxing or doing nothing. Interpret this interval. *(0.423, 0.477)*

(c) Construct a 96% confidence interval for the proportion of adults aged 18 and older who say they have too little time for relaxing or doing nothing. Interpret this interval. *(0.418, 0.482)*

(d) What is the effect of increasing the level of confidence on the width of the interval? *The interval gets wider*

17. High-Speed Internet Access A researcher wishes to estimate the proportion of adults who have high-speed Internet access. What size sample should be obtained if she wishes the estimate to be within 0.03 with 99% confidence if

(a) she uses a 2004 estimate of 0.44 obtained from a Harris poll? *1816*

(b) she does not use any prior estimates? *1842*

18. Home Ownership An urban economist wishes to estimate the proportion of Americans who own their house. What size sample should be obtained if he wishes the estimate to be within 0.02 with 90% confidence if

(a) he uses an estimate of 0.675 from the fourth quarter of 2000 obtained from the U.S. Census Bureau? *1485*

(b) he does not use any prior estimates? *1692*

19. Affirmative Action A sociologist wishes to conduct a poll to estimate the percentage of Americans who favor affirmative action programs for women and minorities for admission to colleges and universities. What sample size should be obtained if she wishes the estimate to be within 4 percentage points with 90% confidence if

(a) she uses a 2003 estimate of 55% obtained from a Gallup Youth Survey? *419*

(b) she does not use any prior estimates? *423*

20. Affirmative Action A sociologist wishes to conduct a poll to estimate the percentage of Americans who judge that affirmative action programs should require businesses to hire a specific number or quota of minorities and women. What size sample should be obtained if she wishes the estimate to be within 4 percentage points with 90% confidence if

(a) she uses a 1999 estimate of 37% obtained from a *Time*/CNN poll? *395*

(b) she does not use any prior estimates? *423*

21. A Penny for Your Thoughts A researcher for the U.S. Department of the Treasury wishes to estimate the percentage of Americans who support abolishing the penny. What size sample should be obtained if he wishes the estimate to be within 2 percentage points with 98% confidence if

(a) he uses a June 2004 estimate of 23% obtained from a Harris Poll? *2404*

(b) he does not use any prior estimate? *3394*

22. Credit Card Debt A school administrator is concerned about the amount of credit card debt college students have. She wishes to conduct a poll to estimate the percentage of full-time college students who have credit card debt of $2000 or more. What size sample should be obtained if she wishes the estimate to be within 2.5 percentage points with 94% confidence if

(a) a pilot study indicates the percentage is 34%? *1269*

(b) no prior estimates are used? *1414*

23. Death Penalty In a Harris Poll conducted in July 2000, 64% of the people polled answered yes to the following question: "Do you believe in capital punishment, that is, the death penalty, or are you opposed to it?" The margin of error in the poll was $\pm 3\% = \pm 0.03$, and the estimate was made with 95% confidence. How many people were surveyed? *984*

24. Own a Gun? In a Harris Poll conducted in May 2000, 39% of the people polled answered yes to the following question: "Do you happen to have in your home or garage any guns or revolvers?" The margin of error in the poll was $\pm 3\% = \pm 0.03$, and the estimate was made with 95% confidence. How many people were surveyed? *1016*

25. 2004 Presidential Election The Gallup Organization conducted a poll of 2014 likely voters just prior to the 2004 presidential election. The results of the survey indicated that George W. Bush would receive 49% of the popular vote and John Kerry would receive 47% of the popular vote. The margin of error was reported to be 3%. The Gallup Organization reported that the race was too close to call. Use the concept of a confidence interval to explain what this means.

26. Confidence Interval Applet: The Role of Level of Confidence Load the confidence interval for proportions applet.

(a) Construct at least 1000 confidence intervals for the population proportion with $n = 100$, $p = 0.3$. What proportion of the 95% confidence intervals contain the population proportion? What proportion did you expect to contain the population proportion?

(b) Repeat part (a). Did the same proportion of intervals contain the population proportion?

(c) Construct at least 1000 confidence intervals for the population proportion with $n = 100$, $p = 0.3$. What proportion of the 99% confidence intervals contain the population proportion 0.3? What proportion did you expect to contain the population proportion?

27. Confidence Interval Applet: The Role of Sample Size
APPLET Load the confidence interval for proportions applet.

(a) Construct at least 1000 confidence intervals for the population proportion with $n = 10$, $p = 0.3$. What

proportion of the 95% confidence intervals contain the population proportion 0.3?

(b) Construct at least 1000 confidence intervals for the population proportion with $n = 40$, $p = 0.3$. What proportions of the 95% confidence intervals contain the population proportion 0.3?

(c) Construct at least 1000 confidence intervals for the population proportion with $n = 100$, $p = 0.3$. What proportion of the 95% confidence intervals contain the population proportion, 0.3?

(d) What happens to the proportion of intervals that capture the population proportion as the sample size, n, increases?

Technology Step by Step   **Confidence Intervals about p**

TI-83/84 Plus   **Step 1:** Press STAT, highlight TESTS, and select `A:1-PropZInt..`

Step 2: Enter the values of x and n.

Step 3: Enter the confidence level following `C-Level:`

Step 4: Highlight `Calculate`; press ENTER.

MINITAB   **Step 1:** If you have raw data, enter the data in column C1.

Step 2: Select the **Stat** menu, highlight **Basic Statistics**, then highlight **1 Proportion . . . .**

Step 3: Enter C1 in the cell marked "Sample in Columns" if you have raw data. If you have summary statistics, click "Summarized data" and enter the number of trials, n, and the number of successes, x.

Step 4: Click the Options button. Select a Confidence Level. Click "Use test based on a normal distribution" (provided that the assumptions stated are satisfied). Click OK twice.

Excel   **Step 1:** Load the PHStat Add-in.

Step 2: Select the **PHStat menu**, highlight **Confidence Intervals . . . ,** then highlight **Estimate for the proportion . . . .**

Step 3: Enter the confidence level. Enter the sample size, n, and the number of successes, x. Click OK.

 Putting It All Together: Which Procedure Do I Use?

Objective   **1** **Determine the appropriate confidence interval to construct**

1   ### Determine the Appropriate Confidence Interval to Construct

Perhaps the most difficult aspect of constructing a confidence interval is determining which type of confidence interval to construct. To assist in your decision making, we present Figure 17.

Figure 17

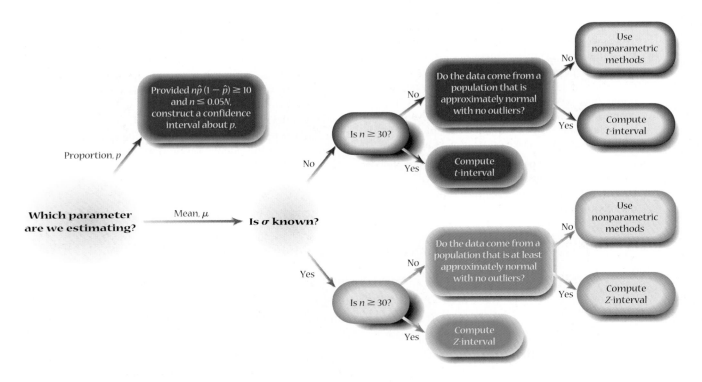

From the flow chart, the first step is to determine which parameter we wish to estimate. If we are estimating a proportion, we verify the requirements to construct the interval and proceed. However, if we are estimating a mean then we have another decision to make: We must determine whether the population standard deviation is known or not. If the population standard deviation is not known, we construct a *t*-interval (provided the sample size is large or the sample data come from a population that is normally distributed with no outliers). If the population standard deviation is known, we construct a *Z*-interval (again provided the sample size is large or the sample data come from a population that is normally distributed with no outliers).

EXAMPLE 1 **Constructing a Confidence Interval:
Which Method Do I Use?**

Problem: Robert wishes to estimate the mean number of miles that his Buick Rendezvous can be driven on a full tank of gas. He fills up his car with regular unleaded gasoline from the same gas station 10 times and records the number of miles that he drives until his low-tank indicator light comes on. He obtains the data shown in Table 7. Construct a 95% confidence interval for the mean number of miles driven on a full tank of gas.

Table 7				
323.9	326.8	370.6	398.8	417.5
450.7	368.8	423.8	382.7	343.1

Approach: We will follow the flow chart given in Figure 17.

Solution: The problem is asking us to construct a confidence interval for the *mean* number of miles driven. So we wish to estimate the population mean, μ. No information is given about the value of the population standard deviation, so σ is unknown. Because the sample size is small, we need to verify that the data come from a population that is normally distributed with no outliers. Figure 18 shows a normal probability plot and boxplot.

Figure 18

The normal probability plot indicates that the sample data come from a population that is normally distributed, and the boxplot indicates that there are no outliers, so we construct a *t*-interval. Based on the sample data from Table 7, we have that $n = 10$, $\bar{x} = 380.67$, and $s = 42.47$. For a 95% confidence interval with $n - 1 = 10 - 1 = 9$ degrees of freedom, we have that $t_{\frac{\alpha}{2}} = t_{\frac{0.05}{2}} = t_{0.025} = 2.262$.

$$\text{Lower bound: } \bar{x} - t_{\frac{\alpha}{2}} \cdot \frac{s}{\sqrt{n}} = 380.67 - 2.262 \cdot \frac{42.47}{\sqrt{10}} = 350.29$$

$$\text{Upper bound: } \bar{x} - t_{\frac{\alpha}{2}} \cdot \frac{s}{\sqrt{n}} = 380.67 + 2.262 \cdot \frac{42.47}{\sqrt{10}} = 411.05$$

Robert is 95% confident that the mean of the miles he can drive on a full tank of gas is between 350.29 miles and 411.05 miles.

9.4 ASSESS YOUR UNDERSTANDING

Concepts and Vocabulary

1. State the circumstances under which we construct a *t*-interval. What are the circumstances under which we construct a *Z*-interval?

2. What are the requirements that must be satisfied before we can construct a confidence interval about a population proportion?

Skill Building

In Problems 3–12, construct the appropriate confidence interval.

3. A simple random sample of size $n = 14$ is drawn from a population that is normally distributed with $\sigma = 20$. The sample mean is found to be $\bar{x} = 60$. Construct a 95% confidence interval about the population mean. (49.52, 70.48)

4. A simple random sample of size $n = 22$ is drawn from a population that is normally distributed with $\sigma = 37$. The sample mean is found to be $\bar{x} = 122.5$. Construct a 90% confidence interval about the population mean. (109.52, 135.48)

5. A simple random sample of size $n = 300$ individuals who are currently employed is asked if they work at home at least once per week. Of the 300 employed individuals surveyed, 35 responded that they did work at home at least once per week. Construct a 99% confidence interval about the population proportion of employed individuals who work at home at least once per week. *(0.069, 0.165)*

6. A simple random sample of size $n = 785$ adults was asked if they follow college football. Of the 785 surveyed, 275 responded that they did follow college football. Construct a 95% confidence interval about the population proportion of adults who follow college football. *(0.317, 0.383)*

7. A simple random sample of size $n = 12$ is drawn from a population that is normally distributed. The sample mean is found to be $\bar{x} = 45$ and the sample standard deviation is found to be $s = 14$. Construct a 90% confidence interval about the population mean. *(37.74, 52.26)*

8. A simple random sample of size $n = 17$ is drawn from a population that is normally distributed. The sample mean is found to be $\bar{x} = 3.25$, and the sample standard deviation is found to be $s = 1.17$. Construct a 95% confidence interval about the population mean. *(2.65, 3.85)*

9. A simple random sample of size $n = 40$ is drawn from a population. The sample mean is found to be $\bar{x} = 120.5$, and the sample standard deviation is found to be $s = 12.9$. Construct a 99% confidence interval about the population mean. *(114.98, 126.02)*

10. A simple random sample of size $n = 210$ is drawn from a population. The sample mean is found to be $\bar{x} = 20.1$, and the sample standard deviation is found to be $s = 3.2$. Construct a 90% confidence interval about the population mean. *(19.73, 20.47)*

Applying the Concepts

11. **Aggravated Assault** In a random sample of 40 felons convicted of aggravated assault, it was determined that the mean length of sentencing was 54 months, with a standard deviation of 8 months. Construct and interpret a 95% confidence interval for the mean length of sentence for an aggravated assault conviction. (*Source*: Based on data obtained from the U.S. Department of Justice.) *(51.4, 56.6)*

12. **Click It** In a February 2005 Harris Poll, 769 of 1010 randomly selected adults said that they always wear their seat belt. Construct and interpret a 98% confidence interval for the proportion of adults who always wear their seat belt. *(0.730, 0.792)*

13. **Estate Tax Returns** In a random sample of 100 estate tax returns that was audited by the Internal Revenue Service, it was determined that the mean amount of additional tax owed was $3137. Assuming the population standard deviation of the additional amount owed is $2694, construct and interpret a 90% confidence interval for the mean additional amount of tax owed for estate tax returns. *($2693.8, $3580.2)*

14. **Muzzle Velocity** Fifty rounds of a new type of ammunition were fired from a test weapon, and the muzzle velocity of the projectile was measured. The sample had a mean muzzle velocity of 863 meters per second, with a standard deviation of 2.7 meters per second. Construct and interpret a 99% confidence interval for the mean muzzle velocity. *(861.98, 864.02)*

15. **Worried about Retirement?** In a survey of 1010 adult Americans, the Gallup Organization asked, "Are you worried or not worried about having enough money for retirement?" Of the 1010 surveyed, 606 stated that they were worried about having enough money for retirement. Construct a 90% confidence interval for the proportion of adult Americans who are worried about having enough money for retirement. *(0.575, 0.625)*

16. **Theme Park Spending** In a random sample of 40 visitors to a certain theme park, it was determined that the mean amount of money spent per person at the park (including ticket price) was $93.43 per day. Assuming the population standard deviation of the amount spent per person is $15, construct and interpret a 95% confidence interval for the mean amount spent daily per person at the theme park. *($88.78, $98.08)*

In Problems 17–22, construct a 95% Z-interval or a 95% t-interval about the population mean. If neither can be constructed, state the reason why. For convenience, a normal probability plot and boxplot are given.

17. **Height of Males** The heights of 20- to 29-year-old males are known to have population standard deviation $\sigma = 2.9$ inches. A simple random sample of $n = 15$ males 20 to 29 years old results in the following data: *Z-Interval; (68.38, 71.32)*

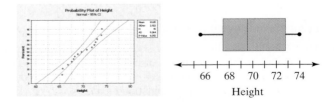

65.5	72.3	68.2	65.6	68.8
66.7	69.6	72.6	72.9	67.5
71.8	73.8	70.7	67.9	73.9

18. **Gestation Period** The gestation period of humans has population standard deviation $\sigma = 16$ days. A simple random sample of $n = 12$ live births results in the following data: *Z-Interval; (258.7, 276.8)*

266	270	277	278	258	275
261	260	270	269	252	277

19. **Officer Friendly** A police officer hides behind a billboard to catch speeders. The following data represent the number of minutes he waits before first observing a car that is exceeding the speed limit by more than 10 miles per hour on 10 randomly selected days: *Neither, outlier*

1.0	5.4	0.8	10.7	0.5
0.9	3.9	0.4	2.5	3.9

20. **M&M's** A quality-control engineer wanted to estimate the mean weight (in grams) of a random sample of 12 plain M&M candies. *t-Interval; (0.852, 0.894)*

0.87	0.88	0.82	0.90
0.84	0.84	0.91	0.94
0.86	0.86	0.88	0.87

21. **Pulse** Fifteen randomly selected women were asked to work on a StairMaster for 3 minutes. After the 3 minutes, their pulses were measured and the following data were obtained: *t-Interval; (101.4, 117.3)*

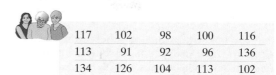

117	102	98	100	116
113	91	92	96	136
134	126	104	113	102

22. **Law Grads' Pay** A random sample of recent graduates of law school was obtained in which the graduates were asked to report their starting salary. The data, based on results reported by the National Association for Law Placement, are as follows: *Neither, nonnormal data*

75,000	49,000	79,000	81,000	38,000
36,500	39,000	41,500	131,000	45,500
92,000	62,500	68,000	37,500	39,500

CHAPTER 9 **Review**

Summary

In this chapter, we discussed estimation methods. We estimated the values of the parameters μ and p. We started by estimating the population mean under the assumption that the population standard deviation was known. This assumption allowed us to construct a confidence interval about μ by utilizing the standard normal distribution. To construct the interval, we required either that the population from which the sample was drawn be normal or that the sample size, n, be greater than or equal to 30. Also, the sampling method had to be simple random sampling. With these requirements satisfied, the $(1 - \alpha) \cdot 100\%$ confidence interval about μ is $\bar{x} \pm z_{\alpha/2} \cdot \dfrac{\sigma}{\sqrt{n}}$. We have $(1 - \alpha) \cdot 100\%$ confidence that the unknown value of μ lies within the interval.

In Section 9.2, we dropped the assumption that the population standard deviation be known. With σ unknown, the sampling distribution of $t = \dfrac{\bar{x} - \mu}{s/\sqrt{n}}$ follows Student's t-distribution with $n - 1$ degrees of freedom. We use the t-distribution to construct the confidence interval about μ. To construct this interval, either the population from which the sample was drawn must be normal or the sample size must be large. Also, the sam-

pling method must be simple random sampling. With these requirements satisfied, the $(1 - \alpha) \cdot 100\%$ confidence interval about μ is $\bar{x} \pm t_{\alpha/2} \cdot \dfrac{s}{\sqrt{n}}$, where $t_{\alpha/2}$ has $n - 1$ degrees of freedom. This means that the procedure results in an interval that contains μ, the population mean, $(1 - \alpha) \cdot 100\%$ of the time.

In Section 9.3, a confidence interval regarding the population proportion, p, was constructed. This confidence interval is constructed about the binomial parameter p. Provided that the sample is obtained by simple random sampling, $n\hat{p}(1 - \hat{p}) \geq 10$, and the sample is less than 5% of the population size the $(1 - \alpha) \cdot 100\%$ confidence interval about p is $\hat{p} \pm z_{\alpha/2} \cdot \sqrt{\dfrac{\hat{p}(1 - \hat{p})}{n}}$. We have $(1 - \alpha) \cdot 100\%$ confidence that the unknown value of p lies within the interval.

Formulas

Confidence Intervals

- A $(1 - \alpha) \cdot 100\%$ confidence interval about μ with σ known is $\bar{x} \pm z_{\frac{\alpha}{2}} \cdot \dfrac{\sigma}{\sqrt{n}}$, provided that the population from which the sample was drawn is normal or that the sample size is large ($n \geq 30$).

- A $(1 - \alpha) \cdot 100\%$ confidence interval about μ with σ unknown is $\bar{x} \pm t_{\frac{\alpha}{2}} \cdot \dfrac{s}{\sqrt{n}}$, where $t_{\frac{\alpha}{2}}$ has $n - 1$ degrees of freedom, provided that the population from which the sample was drawn is normal or that the sample size is large ($n \geq 30$).

- A $(1 - \alpha) \cdot 100\%$ confidence interval about p is $\hat{p} \pm z_{\frac{\alpha}{2}} \cdot \sqrt{\dfrac{\hat{p}(1 - \hat{p})}{n}}$, provided that $n\hat{p}(1 - \hat{p}) \geq 10$ and $n \leq 0.05 N$.

Sample Size

- To estimate the population mean within a margin of error E at a $(1 - \alpha) \cdot 100\%$ level of confidence requires a sample of size $n = \left(\dfrac{z_{\frac{\alpha}{2}} \cdot \sigma}{E}\right)^2$ (rounded up to the next integer).

- To estimate the population proportion within a margin of error E at a $(1 - \alpha) \cdot 100\%$ level of confidence requires a sample of size $n = \hat{p}(1 - \hat{p})\left(\dfrac{z_{\frac{\alpha}{2}}}{E}\right)^2$ (rounded up to the next integer), where $\hat{p}$ is a prior estimate of the population proportion.

- To estimate the population proportion within a margin of error E at a $(1 - \alpha) \cdot 100\%$ level of confidence requires a sample of size $n = 0.25\left(\dfrac{z_{\frac{\alpha}{2}}}{E}\right)^2$ (rounded up to the next integer) when no prior estimate is available.

Vocabulary

Point estimate (p. 404)
Confidence interval (p. 406)
Level of confidence (p. 406)
Critical value (p. 409)

Robust (p. 410)
Z-interval (p. 410)
Margin of error (p. 412)

Student's t-distribution (p. 423)
t-interval (p. 427)

Objectives

Section	You should be able to . . .	Examples	Review Exercises
9.1	1 Compute a point estimate of the population mean (p. 404)	1	8(a), 9(a), 10(b), 13(a), 14(a)
	2 Construct and interpret a confidence interval about a population mean assuming the population standard deviation is known (p. 405)	2–4	3, 5(b), 5(c), 6(b), 6(c), 9(b), 9(c), 10(c), 10(d)
	3 Understand the role of margin of error in constructing a confidence interval (p. 412)	5 and 6	4(c)
	4 Determine the sample size necessary for estimating the population mean within a specified margin of error (p. 413)	7	5(d), 6(d)

9.2	1 Know the properties of Student's *t*-distribution (p. 423)	1	19–21
	2 Determine *t*-values (p. 425)	2	1, 2
	3 Construct and interpret a confidence interval about a population mean (p. 426)	3–5	4, 7(b), 8(b), 11(b), 11(c), 12(b), 12(c), 13(c), 13(d), 14(c), 14(d), 15, 16
9.3	1 Obtain a point estimate for the population proportion (p. 435)	1	17(a), 18(a)
	2 Construct and interpret a confidence interval for the population proportion (p. 436)	2 and 3	17(b), 18(b)
	3 Determine the sample size necessary for estimating a population proportion within a specified margin of error (p. 439)	4	17(c), 17(d), 18(c), 18(d)
9.4	1 Determine the appropriate confidence interval to construct (p. 443)	1	3–18

Review Exercises

In Problems 1 and 2, find the critical t-value for constructing a confidence interval about a population mean at the given level of confidence for the given sample size, n.

1. 99% confidence; $n = 18$ 2.898

2. 90% confidence; $n = 27$ 1.706

3. A simple random sample of size n is drawn from a population that is known to be normally distributed. The sample mean, $\bar{x}$, is determined to be 54.8. (50.94, 58.66)
  (a) Construct the 90% confidence interval about the population mean if the population standard deviation, σ, is known to be 10.5 and the sample size, n, is 20.
  (b) Construct the 90% confidence interval about the population mean if the population standard deviation, σ, is known to be 10.5 and the sample size, n, is 30. How does increasing the sample size affect the width of the interval? (51.65, 57.95); decreases
  (c) Construct the 99% confidence interval about the population mean if the population standard deviation, σ, is known to be 10.5 and the sample size, n, is 20. Compare the results to those obtained in part (a). How does increasing the level of confidence affect the confidence interval? (48.75, 60.85); increases width

4. A simple random sample of size n is drawn from a population that is known to be normally distributed. The sample mean, $\bar{x}$, is determined to be 104.3 and the sample standard deviation, s, is determined to be 15.9.
  (a) Construct the 90% confidence interval about the population mean if the sample size, n, is 15. (97.07, 111.53)
  (b) Construct the 90% confidence interval about the population mean if the sample size, n, is 25. How does increasing the sample size affect the width of the interval? (98.86, 109.74); decreases
  (c) Construct the 95% confidence interval about the population mean if the sample size, n, is 15. Compare the results to those obtained in part (a). How does increasing the level of confidence affect the confidence interval? (95.50, 113.11); increases width

5. Tire Wear Suppose Michelin wishes to estimate the mean mileage for its HydroEdge tire. In a random sample of 40 tires, the sample mean mileage was $\bar{x} = 100{,}294$.

(5b) (99,098, 101,490), (5c) (98,868, 101.720)
  (a) Why can we say that the sampling distribution of $\bar{x}$ is approximately normal? Large sample
  (b) Construct a 90% confidence interval for the mean mileage for all HydroEdge tires, assuming that $\sigma = 4600$ miles. Interpret this interval.
  (c) Construct a 95% confidence interval for the mean mileage for all HydroEdge tires, assuming that $\sigma = 4600$ miles. Interpret this interval.
  (d) How many tires would Michelin require to estimate the mean mileage for all HydroEdge tires within 1500 miles with 99% confidence? 63

6. Talk Time on a Cell Phone Suppose Motorola wishes to estimate the mean talk time for its V505 camera phone before the battery must be recharged. In a random sample of 35 phones, the sample mean talk time was 325 minutes.
  (a) Why can we say that the sampling distribution of $\bar{x}$ is approximately normal? Large sample
  (b) Construct a 94% confidence interval for the mean talk time for all Motorola V505 camera phones, assuming that $\sigma = 31$ minutes. Interpret this interval. (315, 335)
  (c) Construct a 98% confidence interval for the mean talk time for all Motorola V505 camera phones, assuming that $\sigma = 31$ minutes. Interpret this interval. (313, 337)
  (d) How many phones would Motorola need to test to estimate the mean talk time for all V505 camera phones within 5 minutes with 95% confidence? 148

7. Watching TV The Gallup Organization surveyed 1028 teenagers between the ages of 13 and 17 years and asked, "Typically, how many hours per week do you spend watching TV?" The sample mean number of hours was 13, and the sample median number of hours was 7. The sample standard deviation was 5.8 hours.
  (a) Based on the sample mean and median, describe the shape of the distribution of the population. Why is a large sample size necessary to construct a confidence interval about the mean using the methods presented in this chapter?
  (b) Construct and interpret a 90% confidence for the mean number of hours that 13- to 17-year-olds watch television per week.

8. **Bottling Soda** A soft-drink manufacturer is worried that its filling machine is underfilling the bottles that are supposed to have 19.2 ounces. The quality-control manager takes a random sample of 20 bottles from the production line and obtains the following data:

| 18.9 | 18.9 | 19.0 | 18.9 | 19.2 | 19.0 | 19.0 | 19.2 | 19.0 | 19.3 |
| 18.9 | 19.1 | 18.7 | 18.5 | 19.0 | 19.3 | 18.8 | 19.1 | 19.5 | 19.0 |

(a) Use the data to compute a point estimate for the population mean for the content of the bottles produced by the manufacturer. 19.02 ounces
(b) A normal probability plot and boxplot indicate that the sample data come from a population that is normally distributed with no outliers. Construct a 95% confidence interval for the mean content for all bottles produced by the manufacturer. Interpret this interval. (18.92, 19.12)
(c) Does the sample indicate that the bottling process is operating properly? Why? No

9. **Working for a Living** The following data represent the number of hours worked per week for a random sample of 15 adults in the United States and Canada, based on data from an August 2004 study by the Gallup Organization.
(9a) USA: 42.3; Canada: 40.8

USA

70	26	14	41	53
49	40	46	34	33
66	38	46	35	44

Canada

49	22	42	35	36
63	39	40	35	40
31	68	31	53	28

(a) Obtain a point estimate for the population mean number of hours worked each week for each county.
(b) A normal probability plot and boxplot indicate that the U.S. sample data come from a population that is normally distributed with no outliers. Construct a 99% confidence interval for the population mean number of hours worked each week for the United States, assuming that $\sigma = 12.8$ hours. (33.8, 50.8)
(c) A normal probability plot and boxplot indicate that the Canadian sample data come from a population that is normally distributed with no outliers. Construct a 99% confidence interval for the population mean number of hours worked each week for Canada, assuming that $\sigma = 10.8$ hours. (33.6, 48.0)
(d) Does it appear to be the case that Americans work more hours per week than Canadians? Why? No

10. **Math Achievement** The following data represent the mathematics achievement test scores for a random sample of 15 male and 15 female students who had just completed high school in the United States, according to data obtained from the International Association for the Evaluation of Education Achievement study.

Male

488	350	547	488	474
471	443	385	477	452
418	388	441	463	412

Female

433	389	520	479	454
563	411	458	398	337
418	492	442	494	514

(a) Verify that the scores for each gender are normally distributed with no outliers.
(b) Obtain a point estimate for the population mean score of each gender. M: 446.5; F: 453.5
(c) Construct a 95% confidence interval for the population mean achievement score of males, assuming that $\sigma = 64.8$. (413.7, 479.3)
(d) Construct a 95% confidence interval for the population mean achievement score of females, assuming that $\sigma = 56.9$. (424.7, 482.3)
(e) Does there appear to be any difference between the scores of males and those of females? Why? No

11. **Family Size** A random sample of 60 married couples who have been married 7 years was asked the number of children they have. The results of the survey are below. **Note:** $\bar{x} = 2.27$, $s = 1.22$.

0	0	0	3	3	3	1	3	2	2	3	1
3	2	4	0	3	3	3	1	0	2	3	3
1	4	2	3	1	3	3	5	0	2	3	0
4	4	2	2	3	2	2	2	2	3	4	3
2	2	1	4	3	2	4	2	1	2	3	2

(11a) Normal

(a) What is the shape of the distribution of $\bar{x}$? Why?
(b) Compute a 95% confidence interval for the mean number of children of all couples who have been married 7 years. Interpret this interval. (1.95, 2.59)
(c) Compute a 99% confidence interval for the mean number of children of all couples who have been married 7 years. Interpret this interval. (1.85, 2.69)

12. **Waiting in Line** The following data represent the number of cars that arrive at McDonald's drive-through between 11:50 A.M. and 12:00 noon for a random sample of Wednesdays. **Note:** $\bar{x} = 4.08$, $s = 2.12$.

1	7	3	8	2	3	8	2	6	3
6	5	6	4	3	4	3	8	1	2
5	3	6	3	3	4	3	2	1	2
4	4	9	3	5	2	3	5	5	5
2	5	6	1	7	1	5	3	8	4

(a) What is the shape of the distribution of $\bar{x}$? Why? Normal
(b) Compute a 90% confidence interval for the mean number of cars waiting in line between 11:50 A.M. and 12:00 noon on Wednesdays. Interpret this interval. (3.58, 4.58)
(c) Compute a 95% confidence interval for the mean number of cars waiting in line between 11:50 A.M. and 12:00 noon on Wednesdays. Interpret this interval. (3.48, 4.68)

13. **Blood Plasma** In a study of aerobic activity, the blood plasma volume (in liters) of 12 women was measured, and the following data were obtained. (13a) $\bar{x}$ = 3.244, s = 0.487

| 3.15 | 2.99 | 2.77 | 3.12 | 2.45 | 3.85 |
| 2.99 | 3.87 | 4.06 | 2.94 | 3.53 | 3.21 |

Source: Journal of Applied Physiology 65, 6 (December 1988), p. 361, from Hogg and Tanis, *Probability and Statistical Inference*, 6/e

(a) Use the data to compute a point estimate for the population mean and population standard deviation.
(b) Because the sample size is small, we must verify that blood plasma volume is normally distributed and that the data do not contain any outliers. The figures show the normal probability plot and boxplot. Are the conditions for constructing a confidence interval about μ satisfied? Yes

Blood Plasma

(c) Construct a 95% confidence interval for the mean blood plasma volume for all women. Interpret this interval. (2.935, 3.553)
(d) Construct a 99% confidence interval for the mean blood plasma volume for all women. Interpret this interval. (2.807, 3.681)

14. **Water Clarity** The campus at Joliet Junior College has a lake. A Secchi disk is used to measure the water clarity of the lake's water by lowering the dish into the water and measuring the distance below water until the disk is no longer visible. The following measurements (in inches) were taken on the lake at various points in time over the course of a year. (14a) $\bar{x}$ = 57.8, s = 15.4

| 82 | 64 | 62 | 66 | 68 | 43 |
| 38 | 26 | 68 | 56 | 54 | 66 |

Source: Virginia Piekarski, Joliet Junior College

(a) Use the data to compute a point estimate for the population mean and population standard deviation.
(b) Because the sample size is small, we must verify that the data are normally distributed and do not contain any outliers. The figures show the normal probability plot and boxplot. Are the conditions for constructing a confidence interval about μ satisfied? Yes

Depth

(c) Construct a 95% confidence interval for the mean Secchi disk measurement. Interpret this interval.
(d) Construct a 99% confidence interval for the mean Secchi disk measurement. Interpret this interval.

15. **Working for a Living** Redo Problems 9(b)–(d), assuming that σ is unknown. Canada: (31.0, 50.6); U.S.: (31.4, 53.2); no

16. **Math Achievement** Redo Problems 10(c)–(e), assuming that σ is unknown. Male: (418.9, 474.1); female: (421.0, 486.0)

17. **Hypertension** In a random sample of 678 adult males 20 to 34 years of age, it was determined that 58 of them have hypertension (high blood pressure). (*Source:* The Centers for Disease Control.) (17a) 0.086 (17b) (0.064, 0.107)
(a) Obtain a point estimate for the proportion of adult males 20 to 34 years of age who have hypertension.
(b) Construct a 95% confidence interval for the proportion of adult males 20 to 34 years of age who have hypertension. Interpret the confidence interval.
(c) Suppose you wish to conduct your own study to determine the proportion of adult males 20 to 34 years old who have hypertension. What sample size would be needed for the estimate to be within 3 percentage points with 95% confidence if you use the point estimate obtained in part (a)? 336
(d) Suppose you wish to conduct your own study to determine the proportion of adult males 20 to 34 years old who have hypertension. What sample size would be needed for the estimate to be within 3 percentage points with 95% confidence if you don't have a prior estimate? 1068

18. **Carbon Monoxide** From a random sample of 1201 Americans, it was discovered that 1139 of them lived in neighborhoods with acceptable levels of carbon monoxide. (*Source:* The Environmental Protection Agency.)
(a) Obtain a point estimate for the proportion of Americans who live in neighborhoods with acceptable levels of carbon monoxide. 0.948
(b) Construct a 99% confidence interval for the proportion of Americans who live in neighborhoods with acceptable levels of carbon monoxide. (0.932, 0.964)
(c) Suppose you wish to conduct your own study to determine the proportion of Americans who live in neighbor-

hoods with acceptable levels of carbon monoxide. What sample size would be needed for the estimate to be within 1.5 percentage points with 90% confidence if you use the estimate obtained in part (a)? 593

(d) Suppose you wish to conduct your own study to determine the proportion of Americans who live in neighborhoods with acceptable levels of carbon monoxide. What sample size would be needed for the estimate to be within 1.5 percentage points with 90% confidence if you don't have a prior estimate? 3007

19. The area under the t-distribution with 18 degrees of freedom to the right of $t = 1.56$ is 0.0681. What is the area under the t-distribution with 18 degrees of freedom to the left of $t = -1.56$? Why? 0.0681

20. Which is larger, the area under the t-distribution with 10 degrees of freedom to the right of $t = 2.32$ or the area under the standard normal distribution to the right of $z = 2.32$? Why? t-distribution

21. State the properties of Student's t-distribution.

THE CHAPTER 9 CASE STUDY IS LOCATED ON THE CD THAT ACCOMPANIES THIS TEXT.

Hypothesis Tests Regarding a Parameter

Outline

10.1 The Language of Hypothesis Testing

10.2 Hypothesis Tests for a Population Mean Assuming the Population Standard Deviation Is Known

10.3 Hypothesis Tests for a Population Mean in Practice

10.4 Hypothesis Tests for a Population Proportion

10.5 Putting It All Together: Which Method Do I Use?

■ Chapter Review

■ Case Study: How Old Is Stonehenge? (On CD)

DECISIONS

Many of the products we buy have labels that indicate the net weight of the contents. For example, a candy bar wrapper might state that the net weight of the candy bar is 4 ounces. Should we believe the manufacturer? See the Decisions project on page 486.

●●● Putting It All Together

In Chapter 9, we mentioned that there are two areas of inferential statistics: (1) estimation and (2) hypothesis testing. We have already discussed procedures for estimating the population mean and the population proportion.

We now focus our attention on hypothesis testing. Hypothesis testing is used to test statements regarding a characteristic of one or more populations. In this chapter, we will test hypotheses regarding a single population parameter. The hypotheses that we test regard the population mean and the population proportion.

10.1 The Language of Hypothesis Testing

Preparing for This Section Before getting started, review the following:

- Simple random sampling (Section 1.2, pp. 16–19)
- Sampling distribution of $\bar{x}$ (Section 8.1, pp. 374–388)
- Parameter versus statistic (Section 3.1, p. 107)
- Table 9 (Section 6.2, p. 302)

Objectives

① **Determine the null and alternative hypotheses**

② **Understand Type I and Type II errors**

③ **State conclusions to hypothesis tests**

Let's begin with an example that introduces the idea behind hypothesis testing.

EXAMPLE 1 Illustrating Hypothesis Testing

Problem: According to the National Center for Chronic Disease Prevention and Health Promotion, 73.8% of females between the ages of 18 and 29 exercise. Kathleen believes that more women between the ages of 18 and 29 are now exercising, so she obtains a simple random sample of 1000 women and finds that 750 of them are exercising. Is this evidence that the percentage of women between the ages of 18 and 29 who are exercising has increased? What if Kathleen's sample resulted in 920 women exercising?

Approach: Here is the situation Kathleen faces. If 73.8% of 18- to 29-year-old females exercise, she would expect 738 of the 1000 respondents in the sample to exercise. The questions that Kathleen wants to answer are, "How likely is it to obtain a sample of 750 out of 1000 women exercising from a population when the percentage of women who exercise is 73.8%? How likely is a sample that has 920 women exercising?"

Solution: The result of 750 women who exercise is close to what we would expect, so Kathleen is not inclined to believe that the percentage of women exercising has increased. However, the likelihood of obtaining a sample of 920 women who exercise is extremely low if the actual percentage of women who exercise is 73.8%. For the case of obtaining a sample of 920 women who exercise, Kathleen can conclude one of two things: Either the percentage of women who exercise is 73.8% and her sample just happens to include a lot of women who exercise, or the percentage of women who exercise has increased. Provided the sampling was performed in a correct fashion, Kathleen is more inclined to believe that the percentage of women who exercise has increased. ▬▬

Note to Instructor

Before starting this section, divide your students into groups. Each group should define the following (along with the symbolic representation): population mean, sample mean, population standard deviation, sample standard deviation, and standard error of the mean.

① **Determine the Null and Alternative Hypotheses**

Example 1 presents the basic premise behind hypothesis testing: A statement is made regarding the nature of the population, information is collected, and this information is used to test the statement. The steps in conducting a hypothesis test are presented next.

Steps in Hypothesis Testing

1. A statement is made regarding the nature of the population.
2. Evidence (sample data) is collected to test the statement.
3. The data are analyzed to assess the plausibility of the statement.

Note to Instructor

Emphasize that sample data cannot prove that a claim is true.

In this section, we introduce the language of hypothesis testing. Sections 10.2 to 10.4 discuss the formal process of testing a hypothesis.

Definition A **hypothesis** is a statement regarding a characteristic of one or more populations.

In this chapter, we look at hypotheses regarding a single population parameter. Consider the following example.

(A) According to a Gallup poll conducted in 1995, 74% of Americans felt that men were more aggressive than women. A researcher wonders if the percentage of Americans that feel men are more aggressive than women is different today (a statement regarding a population proportion).

(B) The packaging on a lightbulb states that the bulb will last 500 hours under normal use. A consumer advocate would like to know if the mean lifetime of a bulb is less than 500 hours (a statement regarding the population mean).

(C) The standard deviation of the rate of return for a certain class of mutual funds is 0.08. A mutual fund manager believes the standard deviation of the rate of return for his fund is less than 0.08 (a statement regarding the population standard deviation).

We test these types of statements using sample data because it is usually impossible or impractical to gain access to the entire population. The procedure (or process) that we use to test these statements is called *hypothesis testing*.

! CAUTION

If population data are available, there is no need for inferential statistics.

Definition

Hypothesis testing is a procedure, based on sample evidence and probability, used to test statements regarding a characteristic of one or more populations.

Because a statement can either be true or false, hypothesis testing is based on two types of hypotheses.

Definitions

The **null hypothesis**, denoted H_0 (read "H-naught"), is a statement to be tested. The null hypothesis is assumed true until evidence indicates otherwise. In this chapter, it will be a statement regarding the value of a population parameter.

The **alternative hypothesis**, denoted H_1 (read "H-one"), is a statement to be tested. We are trying to find evidence for the alternative hypothesis. In this chapter, it will be a statement regarding the value of a population parameter.

In Other Words

The null hypothesis is a statement of status quo or no difference and always contains a statement of equality. The null hypothesis is assumed to be true until we have evidence to the contrary. We seek evidence for the statement in the alternative hypothesis.

For the lightbulb manufacturer in Situation B, the consumer advocate wishes to know whether the mean lifetime of the bulb is less than 500 hours. Because we are trying to obtain evidence for this statement, it is expressed as the alternative hypothesis using the notation $H_1: \mu < 500$. The statement made by the manufacturer is that the bulb lasts 500 hours on average. We give the manufacturer the benefit of the doubt and assume this statement to be true, so this becomes the statement to be tested. We express the statement to be tested using the notation $H_0: \mu = 500$.

In this chapter, there are three ways to set up the null and alternative hypotheses.

1. Equal hypothesis versus not equal hypothesis **(two-tailed test)**
   H_0: parameter = some value
   H_1: parameter ≠ some value

2. Equal versus less than **(left-tailed test)**
   H_0: parameter = some value
   H_1: parameter < some value

3. Equal versus greater than **(right-tailed test)**
   H_0: parameter = some value
   H_1: parameter > some value

Left- and right-tailed tests are referred to as **one-tailed tests**. Notice that in the left-tailed test the direction of the inequality sign in the alternative hypothesis points to the left ($<$), while in the right-tailed test the direction of the inequality sign in the alternative hypothesis points to the right ($>$). In all three tests the null hypothesis contains a statement of equality. The statement of equality comes from existing information.

Refer to the three claims made on page 455. In Situation A, the null hypothesis is $H_0: p = 0.74$. This is a statement of *status quo* or no difference. The Latin phrase *status quo* means "the existing state or condition." It means that American opinions have not changed from 1995. In Situation B, the null hypothesis is $H_0: \mu = 500$. This is a statement of no difference between the population mean and the lifetime stated on the label. In Situation C, the null hypothesis is $H_0: \sigma = 0.08$. This is a statement of no difference between the population standard deviation rate of return of the manager's mutual fund and all mutual funds.

The statement we are trying to gather evidence for, which is dictated by the researcher before any data are collected, determines the structure of the alternative hypothesis (two-tailed, left-tailed, or right-tailed). For example, the label on a can of soda states that the can contains 12 ounces of liquid. A consumer advocate would be concerned only if the mean contents are less than 12 ounces, so the alternative hypothesis is $H_1: \mu < 12$. However, a quality-control engineer for the soda manufacturer would be concerned if there is too little or too much soda in the can, so the alternative hypothesis would be $H_1: \mu \neq 12$. In both cases, however, the null hypothesis is a statement of no difference between the manufacturer's assertion on the label and the actual mean contents of the can. So the null hypothesis is $H_0: \mu = 12$.

EXAMPLE 2   Forming Hypotheses

In-Class Discussion

Ask students to explain why the alternative in Example 2(a) is $p > 0.02$, rather than $p \neq 0.02$. Ask similar questions for (b) and (c). When would a researcher use a $\neq$ alternative instead of $>$ or $<$? Answer: when he or she is not sure of the direction of the alternative.

Problem: Determine the null and alternative hypotheses. State whether the test is two-tailed, left-tailed, or right-tailed.

(a) The Medco pharmaceutical company has just developed a new antibiotic for children. Among the competing antibiotics, 2% of children who take the drug experience headaches as a side effect. A researcher for the Food and Drug Administration wishes to know if the percentage of children taking the new antibiotic who experience headaches as a side effect is more than 2%.

(b) The *Blue Book* value of a used 3-year-old Chevy Corvette is $37,500. Grant wonders if the mean price of a used 3-year-old Chevy Corvette in the Miami metropolitan area is different from $37,500.

(c) The standard deviation of the contents in a 64-ounce bottle of detergent using an old filling machine was known to be 0.23 ounce. The company purchased a new filling machine and wants to know if the standard deviation for the new filling machine is less than 0.23 ounce.

Approach: In each case, we must determine the parameter to be tested. We then determine the statement of no change or no difference (status quo), and finally we identify the statement we are attempting to gather evidence for.

In Other Words

Structuring the null and alternative hypothesis:

1. Identify the parameter to be tested.
2. Determine the status quo value of the parameter.
3. Determine the statement that reflects what we are trying to gather evidence for.

Solution

(a) The hypothesis deals with a population proportion, p. If the new drug is no different from current drugs on the market, the proportion of individuals taking the new drug who experience a headache will be 0.02, so the null hypothesis is $H_0: p = 0.02$. We are trying to show that the proportion of individuals who experience a headache is "more than" 0.02. Therefore, the alternative hypothesis is $H_1: p > 0.02$. This is a right-tailed test because the alternative hypothesis contains a $>$ symbol.

In Other Words
Look for key phrases when forming the alternative hypothesis. For example, *more than* means $>$; *different from* means $\neq$; *less than* means $<$; and so on. See Table 9 on page 302 for a list of key phrases and the symbols they translate into.

(b) The hypothesis deals with a population mean, μ. If the mean price of a 3-year-old Corvette in Grant's neighborhood is no different from the *Blue Book* price, the population mean in Grant's neighborhood will be $37,500, so the null hypothesis is $H_0: \mu = 37,500$. Grant wishes to determine if the mean price is different from $37,500, so that the alternative hypothesis is $H_1: \mu \neq 37,500$. This is a two-tailed test because the alternative hypothesis contains a $\neq$ symbol.

(c) The hypothesis deals with a population standard deviation, σ. If the new machine is no different from the old machine, the standard deviation of the amount in the bottles filled by the new machine will be 0.23 ounce, so the null hypothesis is $H_0: \sigma = 0.23$. The phrase *less than* is represented symbolically by $<$, so the alternative hypothesis is $H_1: \sigma < 0.23$. This is a left-tailed test because the alternative hypothesis contains a $<$ symbol.

Now Work Problem 17(a).

② **Understand Type I and Type II Errors**

As stated earlier, we use sample data to determine whether to reject or not reject the null hypothesis. Because the decision to reject or not reject the null hypothesis is based on incomplete (sample) information, there is always the possibility of making an incorrect decision. In fact, there are four possible outcomes from hypothesis testing.

In Other Words
When you are testing a hypothesis, there is always the possibility that your conclusion will be wrong. To make matters worse, you won't know whether you are wrong or not! Don't fret, however; we have tools to help manage these incorrect conclusions.

Four Outcomes from Hypothesis Testing

1. We reject the null hypothesis when the alternative hypothesis is true. This decision would be correct.
2. We do not reject the null hypothesis when the null hypothesis is true. This decision would be correct.
3. We reject the null hypothesis when the null hypothesis is true. This decision would be incorrect. This type of error is called a **Type I error**.
4. We do not reject the null hypothesis when the alternative hypothesis is true. This decision would be incorrect. This type of error is called a **Type II error**.

Note to Instructor
Discuss the idea of a false positive or false negative as it relates to Type I and Type II errors.

Figure 1 illustrates the two types of errors that can be made in hypothesis testing.

Figure 1

Conclusion		Reality	
		H_0 Is True	H_1 Is True
	Do Not Reject H_0	Correct Conclusion	Type II Error
	Reject H_0	Type I Error	Correct Conclusion

We illustrate the idea of Type I and Type II errors by looking at hypothesis testing from the point of view of a criminal trial. In any trial, the defendant is assumed to be innocent. (We give the defendant the benefit of the doubt.) The district attorney must present evidence proving that the defendant is guilty.

/ *In Other Words*
A Type I error is like putting an innocent person in jail. A Type II error is like letting a guilty person go free.

Because we are seeking evidence for guilt, it becomes the alternative hypothesis. Innocence is assumed, so this is the null hypothesis. The hypotheses for a trial are written

$$H_0\text{: the defendant is innocent}$$
$$H_1\text{: the defendant is guilty}$$

The trial is the process whereby information (sample data) is obtained. The jury then deliberates about the evidence (the data analysis). Finally, the jury either convicts the defendant (rejects the null hypothesis) or declares the defendant not guilty (fails to reject the null hypothesis).

Note that the defendant is never declared innocent. That is, we never conclude that the null hypothesis is true. Using this analogy, the two correct decisions are to conclude that an innocent person is not guilty or conclude that a guilty person is guilty. The two incorrect decisions are to convict an innocent person (a Type I error) or to let a guilty person go free (a Type II error). It is helpful to think in this way when trying to remember the difference between a Type I and a Type II error.

EXAMPLE 3 **Type I and Type II Errors**

Problem: The Medco pharmaceutical company has just developed a new antibiotic. Among the competing antibiotics, 2% of children who take the drug experience headaches as a side effect. A researcher for the Food and Drug Administration wishes to know if the percentage of children taking the new antibiotic who experience a headache as a side effect is more than 2%. The researcher conducts a hypothesis test with $H_0\text{: } p = 0.02$ and $H_1\text{: } p > 0.02$. Provide statements explaining what it would mean to make (a) a Type I error and (b) a Type II error.

Approach: A Type I error occurs if the null hypothesis is rejected when, in fact, the null hypothesis is true. A Type II error occurs if the null hypothesis is not rejected when, in fact, the alternative hypothesis is true.

Solution

(a) A Type I error is made if the sample evidence leads the researcher to believe that $p > 0.02$ (that is, we reject the null hypothesis) when, in fact, the proportion of children who experience a headache is not greater than 0.02.

(b) A Type II error is made if the researcher does not reject the null hypothesis that the proportion of children experiencing a headache is equal to 0.02 when, in fact, the proportion of children who experience a headache is more than 0.02. For example, the sample evidence led the researcher to believe $p = 0.02$ when in fact the true proportion is some value larger than 0.2.

Now Work Problems 17(b) and 17(c).

Understand the Probability of Making a Type I or Type II Error

Recall that we never know whether a confidence interval contains the unknown parameter. We only know the likelihood that a confidence interval captures the parameter. Similarly, we never know whether or not the outcome of a hypothesis test results in an error. However, just as we place a level of confidence in the construction of a confidence interval, we can determine the probability of making errors. The following notation is commonplace:

$$\alpha = P(\text{Type I error}) = P(\text{rejecting } H_0 \text{ when } H_0 \text{ is true})$$

$$\beta = P(\text{Type II error}) = P(\text{not rejecting } H_0 \text{ when } H_1 \text{ is true})$$

The symbol β is the Greek letter beta (pronounced "BAY tah"). The probability of making a Type I error, α, is chosen by the researcher *before* the sample data are collected. This probability is referred to as the *level of significance*.

Definition

The **level of significance**, α, is the probability of making a Type I error.

The choice of the level of significance depends on the consequences of making a Type I error. If the consequences are severe, the level of significance should be small (say, $\alpha = 0.01$). However, if the consequences of making a Type I error are not severe, a higher level of significance can be chosen (say $\alpha = 0.05$ or $\alpha = 0.10$).

Why is the level of significance not always set at $\alpha = 0.01$? By reducing the probability of making a Type I error, you increase the probability of making a Type II error, β. Using our court analogy, a jury is instructed that the prosecution must provide proof of guilt "beyond all reasonable doubt." This implies that we are choosing to make α small so the probability that we will send an innocent person to jail is very small. The consequence of the small α, however, is a large β, which means many guilty defendants will go free. For now, we are content to recognize the inverse relation between α and β (as one goes up the other goes down).

In Other Words

As the probability of a Type I error increases, the probability of a Type II error decreases, and vice versa.

! CAUTION

We never *accept* the null hypothesis, because, without having access to the entire population, we don't know the exact value of the parameter stated in the null. Rather, we say that we do not reject the null hypothesis. This is just like the court system. We never declare a defendant innocent, but rather say the defendant is not guilty.

③ State Conclusions to Hypothesis Tests

Once the decision to reject or not reject the null hypothesis is made, the researcher must state his or her conclusion. It is important to recognize that we never *accept* the null hypothesis. Again, the court system analogy helps to illustrate the idea. The null hypothesis is H_0: innocent. When the evidence presented to the jury is not enough to convict beyond all reasonable doubt, the jury comes back with a verdict of "not guilty."

Notice that the verdict does not state that the null hypothesis of innocence is true; it states that there is not enough evidence to conclude guilt. This is a huge difference. Being told that you are not guilty is very different from being told that you are innocent!

So sample evidence can never prove the null hypothesis to be true. When we do not reject the null hypothesis, we are saying that the evidence indicates that the null hypothesis *could* be true.

EXAMPLE 4   **Stating the Conclusion**

Problem: The Medco pharmaceutical company has just developed a new antibiotic. Among the competing antibiotics, 2% of children who take the drug experience a headache as a side effect. A researcher for the Food and Drug Administration believes that the proportion of children taking the new antibiotic who experience a headache as a side effect is more than 0.02. From Example 2(a), we know the null hypothesis is H_0: $p = 0.02$ and the alternative hypothesis is H_1: $p > 0.02$.

(a) Suppose the sample evidence indicates that the null hypothesis is rejected. State the conclusion.

(b) Suppose the sample evidence indicates that the null hypothesis is not rejected. State the conclusion.

Approach: When the null hypothesis is rejected, we say that there is sufficient evidence to support the statement in the alternative hypothesis. When the null hypothesis is not rejected, we say that there is not sufficient evidence to support the statement in the alternative hypothesis. We never say that the null hypothesis is true!

Solution

(a) The statement in the alternative hypothesis is that the proportion of children taking the new antibiotic who experience a headache as a side effect is more than 0.02. Because the null hypothesis ($p = 0.02$) is rejected, we conclude there is sufficient evidence to conclude that the proportion of children who experience a headache as a side effect is more than 0.02.

(b) Because the null hypothesis is not rejected, we conclude that there is not sufficient evidence to say that the proportion of children who experience a headache as a side effect is more than 0.02.

Now Work Problem 25.

10.1 ASSESS YOUR UNDERSTANDING

Concepts and Vocabulary

1. Explain what it means to make a Type I error. Explain what it means to make a Type II error.

2. Suppose the consequences of making a Type I error are severe. Would you choose the level of significance, α, to equal 0.01, 0.05, or 0.10? Why?

3. What happens to the probability of making a Type II error, β, as the level of significance, α, decreases? Why is this result intuitive?

4. If a hypothesis is tested at the $\alpha = 0.05$ level of significance, what is the probability of making a Type I error?

5. The following is a quotation from Sir Ronald A. Fisher, a famous statistician.

   "For the logical fallacy of believing that a hypothesis has been proved true, merely because it is not contradicted by the available facts, has no more right to insinuate itself in

statistics than in other kinds of scientific reasoning .... It would, therefore, add greatly to the clarity with which the tests of significance are regarded if it were generally understood that tests of significance, when used accurately, are capable of rejecting or invalidating hypotheses, in so far as they are contradicted by the data: but that they are never capable of establishing them as certainly true ...."

In your own words, explain what this quotation means.

6. In your own words, explain the difference between "beyond all reasonable doubt" and "beyond all doubt."

7. *True or False:* Sample evidence can prove that a null hypothesis is true. False

8. *True or False:* Type I and Type II errors are independent events. False

Skill Building

In Problems 9–14, a null and alternative hypothesis is given. Determine whether the hypothesis test is left-tailed, right-tailed, or two-tailed. What parameter is being tested?

9. $H_0: \mu = 5$ Right-tailed, μ
   $H_1: \mu > 5$

10. $H_0: p = 0.2$ Left-tailed, p
    $H_1: p < 0.2$

11. $H_0: \sigma = 4.2$ Two-tailed, σ
    $H_1: \sigma \neq 4.2$

12. $H_0: p = 0.76$ Right-tailed, p
    $H_1: p > 0.76$

13. $H_0: \mu = 120$ Left-tailed, μ
    $H_1: \mu < 120$

14. $H_0: \sigma = 7.8$ Two-tailed, σ
    $H_1: \sigma \neq 7.8$

In Problems 15–22, (a) determine the null and alternative hypotheses, (b) explain what it would mean to make a Type I error, and (c) explain what it would mean to make a Type II error.

15. **Teenage Mothers** According to the U.S. Census Bureau, 11.8% of registered births in the United States in 2000 were to teenage mothers. A sociologist believes that this percentage has decreased since then.

16. **Charitable Contributions** According to *Giving and Volunteering in the United States, 2001 Edition*, the mean charitable contribution per household in the United States in 2000 was $1623. A researcher believes that the level of giving has changed since then.

17. **Single-Family Home Price** According to the Federal
    [NW] Housing Finance Board, the mean price of a single-family

home in October 2003 was $243,756. A real estate broker believes that because of changes in interest rates, as well as other economic factors, the mean price has increased since then. $H_0: \mu = \$243{,}756; H_1: \mu > \$243{,}756$

18. **Fair Packaging and Labeling** Federal law requires that a jar of peanut butter that is labeled as containing 32 ounces must contain at least 32 ounces. A consumer advocate feels that a certain peanut butter manufacturer is shorting customers by underfilling the jars so that the mean content is less than the 32 ounces stated on the label. $H_0: \mu = 32$ ounces; $H_1: \mu < 32$ ounces

15. $H_0: p = 0.118; H_1: p < 0.118$ 16. $H_0: \mu = \$1623; H_1: \mu \neq \$1623$

19. **Valve Pressure** The standard deviation in the pressure required to open a certain sprung-type valve is known to be $\sigma = 0.7$ psi. Due to changes in the manufacturing process, the quality-control manager feels that the pressure variability has been reduced. $H_0: \sigma = 0.7$ psi; $H_1: \sigma < 0.7$ psi

20. **Overweight** According to the Centers for Disease Control and Prevention, 16% of children aged 6 to 11 years are overweight. A school nurse thinks that the percentage of 6- to 11-year-olds who are overweight is higher in her school district. $H_0: p = 0.16$; $H_1: p > 0.16$

21. **Cell Phone Service** According to the *Statistical Abstract of the United States*, the mean monthly cell phone bill was $49.91 in 2003. A researcher suspects that the mean monthly cell phone bill is different today.

22. **SAT Math Scores** In 2004, the standard deviation SAT math score for all students taking the exam was 114. A teacher believes that, due to changes to the SAT Reasoning Test in 2005, the standard deviation of SAT math scores will increase. $H_0: \sigma = 114$; $H_1: \sigma > 114$

In Problems 23–34, state the conclusion based on the results of the test.

23. For the hypotheses in Problem 15, suppose the null hypothesis is rejected.

24. For the hypotheses in Problem 16, suppose the null hypothesis is not rejected.

25. For the hypotheses in Problem 17, suppose the null hypothesis is not rejected.
NW

26. For the hypotheses in Problem 18, suppose the null hypothesis is rejected.

27. For the hypotheses in Problem 19, suppose the null hypothesis is not rejected.

28. For the hypotheses in Problem 20, suppose the null hypothesis is not rejected.

29. For the hypotheses in Problem 21, suppose the null hypothesis is rejected.

30. For the hypotheses in Problem 22, suppose the null hypothesis is not rejected.

31. For the hypotheses in Problem 15, suppose the null hypothesis is not rejected.

32. For the hypotheses in Problem 16, suppose the null hypothesis is rejected.

33. For the hypotheses in Problem 17, suppose the null hypothesis is rejected.

34. For the hypotheses in Problem 18, suppose the null hypothesis is not rejected.

21. $H_0: \mu = \$49.91$; $H_1: \mu \neq \$49.91$

Applying the Concepts

35. **Fruits and Vegetables** According to the *Statistical Abstract of the United States*, the mean consumption of fruits in 2003 was 98.4 pounds. A dietician believes more people are becoming health conscious and that fruit consumption has risen since then.
    (a) Determine the null and alternative hypotheses.
    (b) Suppose sample data indicate that the null hypothesis should be rejected. State the conclusion of the researcher.
    (c) Suppose, in fact, the mean consumption of fruits is 98.4 pounds. Was a Type I or Type II error committed? If we tested this hypothesis at the $\alpha = 0.05$ level of significance, what is the probability of committing a Type I error? Type I error; 0.05

36. **Test Preparation** The mean score on the SAT Math Reasoning exam is 518. A test preparation company claims that the mean scores of students who take its course are higher than the mean of 518.
    (a) Determine the null and alternative hypotheses.
    (b) Suppose sample data indicate that the null hypothesis should not be rejected. State the conclusion of the company.
    (c) Suppose, in fact, the mean score of students taking the preparatory course is 522. Was a Type I or Type II error committed? If we tested this hypothesis at the $\alpha = 0.01$ level, what is the probability of committing a Type I error? Type II; 0.01

(35a) $H_0: \mu = 98.4$ pounds; $H_1: \mu > 98.4$ pounds

(36a) $H_0: \mu = 518$; $H_1: \mu > 518$

(d) If we wanted to decrease the probability of making a Type II error, would we need to increase or decrease the level of significance? Increase

37. **Marijuana Use** According to the Centers for Disease Control and Prevention, in 2001, 10.2% of high school students had tried marijuana for the first time before the age of 13. The Drug Abuse and Resistance Education (DARE) program underwent several major changes to keep up with technology and issues facing students in the 21st century. After the changes, a school resource officer (SRO) thinks that the proportion of high school students who have tried marijuana for the first time before the age of 13 has decreased from the 2001 level.
    (a) Determine the null and alternative hypotheses.
    (b) Suppose sample data indicate that the null hypothesis should not be rejected. State the conclusion of the SRO.
    (c) Suppose, in fact, the proportion of high school students who have tried marijuana for the first time before the age of 13 was 9.5%. Was a Type I or Type II error committed? Type II

38. **Internet Use** According to the *Statistical Abstract of the United States*, in 2000, 10.7% of Americans over 65 years of age used the Internet. A researcher believes the proportion of Americans over 65 years of age who use the Internet is higher than 10.7% today.
    (a) Determine the null and alternative hypotheses.

(37a) $H_0: p = 0.102$; $H_1: p < 0.102$

(38a) $H_0: p = 0.107$; $H_1: p > 0.107$

(b) Suppose sample data indicate that the null hypothesis should be rejected. State the conclusion of the researcher.

(c) Suppose, in fact, the percentage of Americans over 65 years of age who use the Internet is still 10.7%. Was a Type I or Type II error committed? Type I error

39. Consumer Reports The following is an excerpt from a *Consumer Reports* article from February 2001.

> The Platinum Gasaver makes some impressive claims. The device, $188 for two, is guaranteed to increase gas mileage by 22% says the manufacturer, National Fuelsaver. Also, the company quotes "the government" as concluding, "Independent testing shows greater fuel savings with *Gasaver* than the 22 percent claimed by the developer." Readers have told us they want to know more about it.
>
> The Environmental Protection Agency (EPA), after its lab tests of the Platinum Gasaver, concluded in 1991, "Users of the device would not be expected to realize either an emission or fuel economy benefit." The Federal Trade Commission says, "No government agency endorses gas-saving products for cars."

Determine the null and alternative hypotheses that the EPA used in to draw the conclusion stated in the second paragraph.

40. Prolong Engine Treatment The manufacturer of Prolong Engine Treatment claims that if you add one 12-ounce bottle of its $20 product, your engine will be protected from excessive wear. An infomercial claims that a woman drove 4 hours without oil, thanks to Prolong. *Consumer Reports* magazine tested engines in which they added Prolong to the motor oil, ran the engines, drained the oil, and then determined the time until the engines seized.

(a) Determine the null and alternative hypotheses *Consumer Reports* will test. $H_0: \mu = 4$; $H_1: \mu < 4$

(b) Both engines took exactly 13 minutes to seize. What conclusion might *Consumer Reports* draw based on this evidence? Prolong's claim not true

41. Refer to the claim made in Problem 18. Researchers must choose the level of significance based on the consequences of making a Type I error. In your opinion, is a Type I error or Type II error more serious? Why? On the basis of your answer, decide on a level of significance, α. Be sure to support your opinion.

39. $H_0: \mu = 0$; $H_1: \mu > 0$ where μ is the mean increase

---

10.2 Hypothesis Tests for a Population Mean Assuming the Population Standard Deviation Is Known

Preparing for This Section Before getting started, review the following:

- Using probabilities to identify unusual events (Section 5.1, p. 225)
- z_α Notation (Section 7.2, p. 340)
- Sampling distribution of $\bar{x}$ (Section 8.1, pp. 374–388)
- Computing normal probabilities (Section 7.3, pp. 345–349)

Objectives

1. Understand the logic of hypothesis testing
2. Test hypotheses about a population mean with σ known using the classical approach
3. Test hypotheses about a population mean with σ known using *P*-values
4. Test hypotheses about a population mean with σ known using confidence intervals
5. Understand the difference between statistical significance and practical significance

1 Understand the Logic of Hypothesis Testing

Now that we know the language of hypothesis testing, we are ready to present the methods for conducting a hypothesis test.

In this section, we present three approaches to testing hypotheses about a population mean, μ. As we did for confidence intervals, we begin by assuming that we know the value of the population standard deviation, σ. The assumption is made because it allows us to use the normal model to test hypotheses regarding the population mean. Because the normal model is fairly easy to use, we can concentrate on the techniques of hypothesis testing without getting bogged down with other details. The assumption that σ is known will be dropped in the next section.

Historical Note

Jerzy Neyman was born on April 16, 1894, in Bendery, Russia. In 1921, he moved to Poland. He received his Ph.D. from the University of Warsaw in 1924. He read some of Karl Pearson's works and became interested in statistics; however, Neyman was not impressed with Pearson's mathematical abilities. In 1927, he met Pearson's son, Egon Pearson, who was working on a formal approach to hypothesis testing. It was Neyman who provided the mathematical rigor to their work. Together, they developed the phrases *null hypothesis* and *alternative hypothesis*. In 1938, Neyman joined the faculty at the University of California at Berkeley. He died on August 5, 1981.

To test hypotheses regarding the population mean assuming the population standard deviation is known, two requirements must be satisfied.

- A simple random sample is obtained.
- The population from which the sample is drawn is normally distributed or the sample size is large ($n \geq 30$).

If these requirements are met, then the distribution of $\overline{x}$ is normal with mean μ and standard deviation $\dfrac{\sigma}{\sqrt{n}}$.

The first method that we use in testing hypotheses regarding a population mean is referred to as the classical (traditional) approach, the second method is the *P*-value approach, and the third method uses confidence intervals. Your instructor may choose to cover one, two, or all three approaches to hypothesis testing. Do not be alarmed if one or two of the approaches are not covered by your instructor.

Let's lay out a scenario that will be used to help understand both the classical approach to hypothesis testing and the *P*-value approach. Suppose a consumer advocate is concerned that a manufacturer of potato chips is underfilling its bags. The bag states the contents weigh 12.5 ounces. In hypothesis testing we assume that the manufacturer is "not guilty," which means we assume that the population mean contents of the bags of chips is $\mu = 12.5$ ounces. We are looking for evidence that shows the manufacturer is underfilling the bags. We have the following hypotheses:

$$H_0: \mu = 12.5 \quad \text{versus} \quad H_1: \mu < 12.5$$

Suppose the consumer advocate gathers evidence by obtaining a simple random sample of $n = 36$ bags of chips, weighing the contents, and obtaining a sample mean of 12.45 ounces. Does this sample suggest that the manufacturer is underfilling its bags? What is convincing or *statistically significant* evidence?

Definition　When observed results are unlikely under the assumption that the null hypothesis is true, we say the result is **statistically significant**. When results are found to be statistically significant, we reject the null hypothesis.

Figure 2

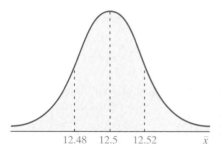

Before we can test the hypothesis, we need to know the distribution of the sample mean, since a different sample of 36 bags of chips will likely result in a different sample mean. Since the sample size is large, the Central Limit Theorem says that the shape of the distribution of the sample mean is approximately normal. Regardless of the size of the sample, the mean of the distribution of the sample mean is $\mu_{\overline{x}} = \mu = 12.5$ ounces because we assume the statement in the null hypothesis to be true, until we have evidence to the contrary. Suppose the population standard deviation is known to be 0.12 ounce; then the standard deviation of the distribution of the sample mean is $\sigma_{\overline{x}} = \dfrac{\sigma}{\sqrt{n}} = \dfrac{0.12}{\sqrt{36}} = 0.02$ ounce.

Figure 2 shows the sampling distribution of the sample mean for our potato chip example.

Now that we have a model that describes the distribution of the sample mean, we can look at two approaches to testing if the potato chip company is underfilling the bags.

Note to Instructor
If you only wish to cover the P-value approach, skip to page 464.

The Logic of the Classical Approach

One criterion we may use for sufficient evidence is to reject the null hypothesis if the sample mean is too many standard deviations below the hypothesized (or status quo) population mean of 12.5 ounces. For example, our criterion might be to reject the null hypothesis if the sample mean is more than 2 standard deviations below the assumed mean of 12.5 ounces.

! CAUTION
● We always test hypotheses assuming that the null hypothesis is true.

Recall that $Z = \dfrac{\bar{x} - \mu}{\sigma/\sqrt{n}}$ represents the number of standard deviations that $\bar{x}$ is from the population mean, μ. Our simple random sample of 36 bags results in sample mean weight of $\bar{x} = 12.45$ ounces, so under the assumption that the null hypothesis is true, we have

$$Z = \frac{\bar{x} - \mu}{\sigma/\sqrt{n}} = \frac{12.45 - 12.5}{0.12/\sqrt{36}} = -2.5$$

The sample mean is 2.5 standard deviations below the hypothesized mean. Because the sample mean is more than 2 standard deviations (that is, "too far") below the hypothesized population mean, we will reject the null hypothesis and conclude that there is sufficient evidence to support the belief that the bag has less than 12.5 ounces of potato chips. This conclusion will lead the consumer advocate to wage a "truth in advertising campaign" against the potato chip manufacturer.

Why does it make sense to reject the null hypothesis if the sample mean is more than two standard deviations away from the hypothesized mean? The area under the standard normal curve to the left of $Z = -2$ is 0.0228, as shown in Figure 3.

Figure 4 shows that, if the null hypothesis is true, 97.72% of all sample means will be 12.46 ounces or more and only 2.28% of the sample means will be less than 12.46 ounces, as indicated by the region shaded blue in Figure 4. Remember, the 12.46 comes from the fact that 12.46 is 2 standard deviations below the hypothesized mean of 12.5 ounces. If a sample mean lies in the green region we are inclined to believe that it came from a population whose mean is less than 12.5, rather than believe that the population mean equals 12.5 and our sample just happened to result in an unusual outcome (a bunch of underfilled bags).

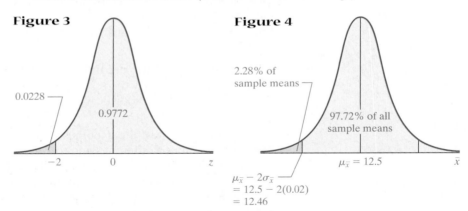

Figure 3

0.0228

0.9772

-2 0 z

Figure 4

2.28% of sample means

97.72% of all sample means

$\mu_{\bar{x}} = 12.5$ $\bar{x}$

$\mu_{\bar{x}} - 2\sigma_{\bar{x}}$
$= 12.5 - 2(0.02)$
$= 12.46$

Notice that our criterion for rejecting the null hypothesis will lead to making a Type I error (rejecting a true null hypothesis) 2.28% of the time. That is, the probability of making a Type I error is 2.28%.

The previous discussion leads to the following premise of hypothesis testing using the classical approach:

> If the sample mean is too many standard deviations from the mean stated in the null hypothesis, we reject the null hypothesis.

The Logic of the *P*-Value Approach

A second criterion we may use for sufficient evidence to support the belief that the manufacturer is underfilling the bags is to compute how likely it is to obtain a sample mean of 12.45 ounces or less from a population whose mean is assumed to be 12.5 ounces. If a sample mean of 12.45 or less is unlikely (or unusual), we have evidence against the null hypothesis. If the sample mean of 12.45 is not unlikely (not unusual), we do not have sufficient evidence against the null hypothesis.

We can compute the probability of obtaining a sample mean of 12.45 or less from a population whose mean is 12.5 using the normal model. Figure 5 shows the area that represents $P(\overline{x} < 12.45)$.

Figure 5

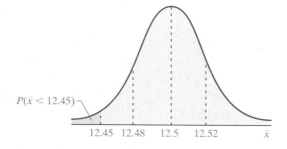

$P(\overline{x} < 12.45)$

12.45 12.48   12.5   12.52 $\overline{x}$

Because

$$Z = \frac{\overline{x} - \mu}{\sigma_{\overline{x}}} = \frac{12.45 - 12.5}{0.02} = -2.5$$

we compute

$$P(\overline{x} \le 12.45) = P(Z \le -2.5) = 0.0062$$

The probability of obtaining a sample mean of 12.45 ounces or less from a population whose mean is 12.5 ounces is 0.0062. This means that less than 1 sample in 100 will give a mean as low or lower than the one we obtained *if* the population mean really is 12.5 ounces. Because these results are so unusual, we take this as evidence against the statement in the null hypothesis.

This discussion leads to the following premise of testing a hypothesis using the *P*-value approach:

> Assuming H_o is true, if the probability of getting a sample mean as extreme or more extreme than the one obtained is small, we reject the null hypothesis.

Figure 6 further illustrates the situation for both the classical and *P*-value approach. The distribution in red shows the distribution of the sample mean assuming the statement in the null hypothesis is true. The sample mean of 12.45 is too far from the assumed population mean of 12.5. Therefore, we reject the null hypothesis that $\mu = 12.5$ and conclude that the sample came from a population with some population mean less than 12.5 ounces, as indicated by the distribution in blue. We don't know what the population mean weight of the bags is, but we have evidence that it is less than 12.5 ounces.

Figure 6

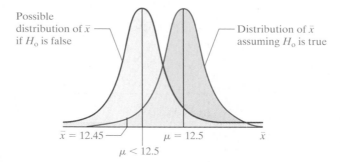

Possible distribution of $\overline{x}$ if H_o is false

Distribution of $\overline{x}$ assuming H_o is true

$\overline{x} = 12.45$ $\mu = 12.5$ $\overline{x}$

$\mu < 12.5$

2

Test Hypotheses about a Population Mean with σ Known Using the Classical Approach

Note to Instructor

If you only wish to teach the *P*-value approach to hypothesis testing, this objective can be skipped without loss of continuity.

We now formalize the procedure for testing hypotheses regarding the population mean when the population standard deviation, σ, is known using the classical approach.

Testing Hypotheses Regarding the Population Mean with σ Known Using the Classical Approach

To test hypotheses regarding the population mean with σ known, we can use the steps that follow provided the following two requirements are satisfied.

1. The sample is obtained using simple random sampling.
2. The sample has no outliers and the population from which the sample is drawn is normally distributed or the sample size, n, is large ($n \geq 30$).

Step 1: Determine the null and alternative hypotheses. The hypotheses can be structured in one of three ways:

Two-Tailed	Left-Tailed	Right-Tailed
$H_0: \mu = \mu_0$	$H_0: \mu = \mu_0$	$H_0: \mu = \mu_0$
$H_1: \mu \neq \mu_0$	$H_1: \mu < \mu_0$	$H_1: \mu > \mu_0$

Note: μ_0 is the assumed or status quo value of the population mean.

Step 2: Select a level of significance α based on the seriousness of making a Type I error.

Step 3: Provided the population from which the sample is drawn is normal or the sample size is large ($n \geq 30$) and the population standard deviation σ, is known, the distribution of the sample mean, $\overline{x}$, is normal with mean μ_0 and standard deviation $\dfrac{\sigma}{\sqrt{n}}$. Therefore,

$$z_0 = \frac{\overline{x} - \mu_0}{\sigma/\sqrt{n}}$$

represents the number of standard deviations the sample mean is from the assumed mean, μ_0. This value is called the **test statistic**.

Step 4: The level of significance is used to determine the *critical value.* The **critical value** represents the maximum number of standard deviations the sample mean can be from μ_0 before the null hypothesis is rejected. For example, the critical value in the left-tailed test is $-z_\alpha$. The shaded region(s) represents the *critical (or rejection) region(s).* The **critical region** or **rejection region** is the set of all values such that the null hypothesis is rejected.

Step 5: Compare the critical value with the test statistic:

Two-Tailed	Left-Tailed	Right-Tailed
If $z_0 < -z_{\frac{\alpha}{2}}$ or $z_0 > z_{\frac{\alpha}{2}}$, reject the null hypothesis.	If $z_0 < -z_\alpha$, reject the null hypothesis.	If $z_0 > z_\alpha$, reject the null hypothesis.

The comparison of the test statistic and critical value is called the **decision rule**.

Step 6: State the conclusion.

The procedure is **robust**, which means that minor departures from normality will not adversely affect the results of the test. However, for small samples, if the data have outliers, the procedure should not be used.

For small samples, we will verify that the data come from a population that is normal by constructing normal probability plots (to assess normality) and boxplots (to determine whether there are outliers). If the normal probability plot indicates that the data do not come from a population that is normally distributed or the boxplot reveals outliers, nonparametric tests, which are not discussed in this text, should be performed.

Note to Instructor

Emphasize to students that they should always draw graphs like Figure 7 so that they can "see" the criteria for rejecting H_0.

EXAMPLE 1

The Classical Approach of Hypothesis Testing: Right-Tailed, Large Sample

Problem: According to the U.S. Federal Highway Administration, the mean number of miles driven annually is 12,200. Patricia believes that residents of the state of Montana drive more than the national average. She obtains a simple random sample of 35 drivers from a list of registered drivers in the state of Montana. The mean number of miles driven for the 35 drivers is 12,895.9. Assuming $\sigma = 3800$ miles, does the sample provide sufficient evidence that residents of the state of Montana drive more than the national average at the $\alpha = 0.1$ level of significance?

Approach: Because the sample size is large, we can proceed to Steps 1 through 6.

Solution

Step 1: Patricia wants to know if people in Montana are driving more than 12,200 miles annually. This can be written $\mu > 12{,}200$. This is a right-tailed test and we have

$$H_0: \mu = 12{,}200 \quad \text{versus} \quad H_1: \mu > 12{,}200$$

Step 2: The level of significance is $\alpha = 0.1$.

Step 3: Patricia found the sample mean, $\bar{x}$, to be 12,895.9 miles. The test statistic is

$$z_0 = \frac{\bar{x} - \mu_0}{\sigma/\sqrt{n}} = \frac{12{,}895.9 - 12{,}200}{3800/\sqrt{35}} = 1.08$$

The sample mean of 12,895.9 miles is 1.08 standard deviations above the mean of 12,200.

Step 4: Because Patricia is performing a right-tailed test, we determine the critical value at the $\alpha = 0.1$ level of significance to be $z_{0.1} = 1.28$. The critical region is displayed in Figure 7.

Figure 7

Critical Region

$z_0 = 1.08$ $z_{0.1} = 1.28$

! CAUTION

In Example 1, we see that the sample mean, $\bar{x} = 12{,}895.9$, is not far enough from the status quo value of the population mean, 12,200 miles. Therefore, we do not have enough evidence to reject the null hypothesis. However, this does not mean we are accepting the null hypothesis that the mean number of miles driven in Montana is 12,200. We are saying that we don't have enough evidence to say it is greater than 12,200 miles. Be sure you understand the difference between these two comments.

Step 5: Because the test statistic $z_0 = 1.08$ is less than the critical value $z_{0.1} = 1.28$, we do not reject the null hypothesis. That is, the value of the test statistic does not fall within the critical region, so we do not reject H_0. We label the test statistic in Figure 7.

Step 6: There is not sufficient evidence at the $\alpha = 0.1$ level of significance to support the belief that residents of the state of Montana drive more than the national average of 12,200 miles.

Now Work Problem 19 Using the Classical Approach.

Now let's look at a two-tailed hypothesis test.

EXAMPLE 2   The Classical Approach of Hypothesis Testing: Two-Tailed, Small Sample

Table 1

94.25	38.94	79.15	56.78
70.07	115.59	77.56	37.01
55.00	76.05	27.29	52.48

Problem: According to CTIA–The Wireless Association, the mean monthly cell phone bill in 2004 was $50.64. A market researcher believes that the mean monthly cell phone bill is different today, but is not sure whether bills have declined because of technological advances or increased due to additional use. The researcher phones a simple random sample of 12 cell phone subscribers and obtains the monthly bills shown in Table 1.

Assuming $\sigma = \$18.49$, use these data to determine whether the mean monthly cell phone bills is different from $50.64 at the $\alpha = 0.05$ level of significance.

Approach: Because the sample size, n, is less than 30, we must verify that the data come from a population that is approximately normal with no outliers. We will construct a normal probability plot and boxplot to verify these requirements. We then proceed to follow Steps 1 through 6.

Solution: Figure 8 displays the normal probability plot and boxplot.

Figure 8

The normal probability plot indicates that the data could come from a population that is normal. The boxplot does not show any outliers.

Step 1: The market researcher wants to know if the mean cell phone bill is different from $50.64, which can be written $\mu \neq 50.64$. This is a two-tailed test and we have

$$H_0: \mu = 50.64 \quad \text{versus} \quad H_1: \mu \neq 50.64$$

Step 2: The level of significance is $\alpha = 0.05$.

Step 3: From the data in Table 1, the sample mean is computed to be $65.014. The test statistic is

$$z_0 = \frac{\bar{x} - \mu_0}{\sigma/\sqrt{n}} = \frac{65.014 - 50.64}{18.49/\sqrt{12}} = 2.69$$

The sample mean of $65.014 is 2.69 standard deviations above the assumed population mean of $50.64.

Step 4: Because this is a two-tailed test, we determine the critical values at the $\alpha = 0.05$ level of significance to be $-z_{0.05/2} = -1.96$ and $z_{0.05/2} = 1.96$. The critical regions are displayed in Figure 9.

Figure 9

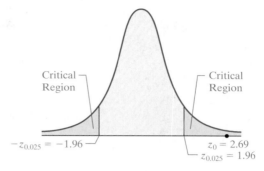

Critical Region

Critical Region

$-z_{0.025} = -1.96$

$z_0 = 2.69$
$z_{0.025} = 1.96$

Note to Instructor
Ask students how this problem will change if the researcher thinks that cell phone bills have risen since 2004.

Step 5: Because the test statistic, $z_0 = 2.69$, is greater than the critical value $z_{0.025} = 1.96$, we reject the null hypothesis. That is, the value of the test statistic falls within the critical region, so we reject H_0. We label this point in Figure 9.

Step 6: There is sufficient evidence at the $\alpha = 0.05$ level of significance to conclude that the mean monthly cell phone bill is different from the mean amount in 2004, $50.64.

> Now Work Problem 25 Using the Classical Approach.

Note to Instructor
Stress to students that the interpretation of P-values is the same for all hypothesis tests. Therefore, P-values are more versatile.

③ **Test Hypotheses about a Population Mean with σ Known Using P-Values**

Now let's look at testing hypotheses with P-values.

Definition

> A **P-value** is the probability of observing a sample statistic as extreme or more extreme than the one observed under the assumption that the null hypothesis is true. Put another way, the P-value is the likelihood or probability that a sample will result in a sample mean such as the one obtained if the null hypothesis is true.

In Other Words

The smaller the P-value, the greater the evidence against the null hypothesis.

A small P-value implies that the sample mean is unlikely if the null hypothesis is true and would be considered as evidence against the null hypothesis.

The following procedures can be used to compute P-values when testing a hypothesis about a population mean with σ known.

Testing Hypotheses Regarding the Population Mean Using P-Values

To test hypotheses regarding the population mean with σ known, we can use the steps that follow to compute the P-value, provided that the following two requirements are satisfied.

1. The sample is obtained using simple random sampling.
2. The sample has no outliers and the population from which the sample is drawn is normally distributed or the sample size, n, is large ($n \geq 30$).

Step 1: Determine the null and alternative hypotheses. The hypotheses can be structured in one of three ways:

Two-Tailed	Left-Tailed	Right-Tailed
$H_0: \mu = \mu_0$	$H_0: \mu = \mu_0$	$H_0: \mu = \mu_0$
$H_1: \mu \neq \mu_0$	$H_1: \mu < \mu_0$	$H_1: \mu > \mu_0$

Note: μ_0 is the assumed value of the population mean.

***Step 2*:** Decide on a level of significance, α, depending on the seriousness of making a Type I error.

***Step 3*:** Compute the test statistic, $z_0 = \dfrac{\overline{x} - \mu_0}{\sigma/\sqrt{n}}$.

***Step 4*:** Determine the P-value.

Two-Tailed	Left-Tailed	Right-Tailed
P-value $= P(Z < -|z_0| \text{ or } Z > |z_0|)$ $= 2P(Z > |z_0|)$	P-value $= P(Z < z_0)$	P-value $= P(Z > z_0)$

The sum of the area in the tails is the P-value

The sum of the area in the tails is the P-value

$-|z_0|$ $|z_0|$

The area left of z_0 is the P-value

z_0

The area right of z_0 is the P-value

z_0

Interpretation: The P-value is the probability of obtaining a sample mean that is more than $|z_0|$ standard deviations from the mean stated in the null hypothesis, μ_0.

Interpretation: The P-value is the probability of obtaining a sample mean of $\overline{x}$ or smaller under the assumption that H_0 is true. In other words, it is the probability of obtaining a sample mean that is more than z_0 standard deviations to the left of μ_0.

Interpretation: The P-value is the probability of obtaining a sample mean of $\overline{x}$ or larger under the assumption that H_0 is true. In other words, it is the probability of obtaining a sample mean that is more than z_0 standard deviations to the right of μ_0.

***Step 5*:** Reject the null hypothesis if the P-value is less than the level of significance, α. The comparison of the P-value and the level of significance is called the **decision rule**.

***Step 6*:** State the conclusion.

EXAMPLE 3
The *P*-Value Approach of Hypothesis Testing: Right-Tailed, Large Sample

Problem: According to the U.S. Federal Highway Administration, the mean number of miles driven annually is 12,200. Patricia believes that residents of the state of Montana drive more than the national average. She obtains a simple random sample of 35 drivers from a list of registered drivers in the state of Montana. The mean number of miles driven for the 35 drivers is 12,895.9. Assuming $\sigma = 3800$ miles, does the sample provide sufficient evidence that residents of the state of Montana drive more than the national average at the $\alpha = 0.1$ level of significance? Use the P-value approach.

Approach: Because the sample size is large, we can proceed to Steps 1 through 6.

Solution

***Step 1*:** Patricia wants to know if people in Montana are driving more than 12,200 miles annually. This can be written $\mu > 12,200$. This is a right-tailed test and we have

$$H_0\text{: } \mu = 12,200 \quad \text{versus} \quad H_1\text{: } \mu > 12,200$$

***Step 2*:** The level of significance is $\alpha = 0.1$.

***Step 3*:** Patricia found the sample mean, $\overline{x}$, to be 12,895.9 miles. The test statistic is

$$z_0 = \frac{\overline{x} - \mu_0}{\sigma/\sqrt{n}} = \frac{12,895.9 - 12,200}{3800/\sqrt{35}} = 1.08$$

The sample mean of 12,895.9 miles is 1.08 standard deviations above the mean of 12,200.

In Other Words

For this test (a right-tailed test), the *P*-value is the area under the standard normal curve to the right of $z_0 = 1.08$.

Note to Instructor

Emphasize to students that they should always draw graphs like Figure 10 so that they can see the area that represents the *P*-value.

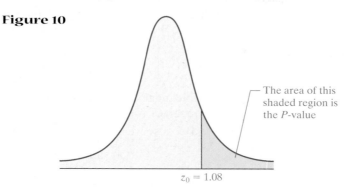

Step 4: Because Patricia is performing a right-tailed test,

$$P\text{-value} = P(Z > z_0) = P(Z > 1.08)$$

We have to determine the area under the standard normal curve to the right of $z_0 = 1.08$, as shown in Figure 10.

Figure 10

The area of this shaded region is the *P*-value

$z_0 = 1.08$

Using Table IV, we have

$$P\text{-value} = P(Z > 1.08) = 1 - P(Z \le 1.08) = 1 - 0.8599 = 0.1401$$

The probability of obtaining a sample mean of 12,895.9 miles or higher from a population whose mean is 12,200 miles is 0.1401. A *P*-value of 0.1401 means we would expect to obtain the results we obtained about 14 times out of 100 if the statement in the null hypothesis were true. The results we obtained are not unusual assuming $\mu = 12,200$.

In Other Words

If *P*-value $< \alpha$, then reject the null hypothesis. Put another way—If the *P*-value is low, the null must go!

Step 5: The decision rule that we use is this: If the *P*-value is less than the level of significance, α, we reject the null hypothesis. Because $0.1401 > 0.1$, we do not reject the null hypothesis.

Step 6: There is not sufficient evidence at the $\alpha = 0.1$ level of significance to support Patricia's belief that residents of the state of Montana drive more than the national average of 12,200 miles. ▮

> **Now Work Problem 19 Using the *P*-Value Approach.**

Table 2			
94.25	38.94	79.15	56.78
70.07	115.59	77.56	37.01
55.00	76.05	27.29	52.48

EXAMPLE 4

The *P*-Value Approach of Hypothesis Testing: Two-Tailed, Small Sample

Problem: According to CTIA—The Wireless Association, the mean monthly cell phone bill in 2004 was $50.64. A market researcher believes that the mean monthly cell phone bill is different today, but is not sure whether bills have declined because of technological advances or increased due to additional use. The researcher phones a simple random sample of 12 cell phone subscribers and obtains the monthly bills shown in Table 2.

Assuming $\sigma = \$18.49$, use these data to determine whether the mean monthly cell phone bill is different from $50.64 at the $\alpha = 0.05$ level of significance. Use the *P*-value approach.

Approach: Because the sample size, n, is less than 30, we must verify that the data come from a population that is approximately normal with no outliers. We will construct a normal probability plot and boxplot to verify these requirements. We then proceed to follow Steps 1 through 6.

Solution: Figure 8 on page 468 displays the normal probability plot and boxplot. The normal probability plot indicates that the data could come from a population that is normal. The boxplot does not show any outliers.

Step 1: The market researcher wants to know if the mean cell phone bill is different from \$50.64, which can be written $\mu \neq 50.64$. This is a two-tailed test and we have

$$H_0\text{: } \mu = 50.64 \quad \text{versus} \quad H_1\text{: } \mu \neq 50.64$$

Step 2: The level of significance is $\alpha = 0.05$.

Step 3: From the data in Table 2, the sample mean is computed to be \$65.014 and $n = 12$. We assume $\sigma = \$18.49$. The test statistic is

$$z_0 = \frac{\overline{x} - \mu_0}{\sigma/\sqrt{n}} = \frac{65.014 - 50.64}{18.49/\sqrt{12}} = 2.69$$

The sample mean of \$65.014 is 2.69 standard deviations above the assumed population mean of \$50.64.

Step 4: Because we are performing a two-tailed test,

$$P\text{-value} = P(Z < -2.69 \text{ or } Z > 2.69)$$

We need to determine the area under the standard normal curve to the right of $Z = 2.69$ and to the left of $Z = -2.69$, as shown in Figure 11.

Figure 11

The area of the shaded region is the P-value

$-z_0 = -2.69$ $z_0 = 2.69$

In Other Words

To find the P-value for a two-tailed test, first determine whether the test statistic, z_0, is positive or negative. If z_0 is negative, determine the area under the standard normal curve to the left of z_0 and then multiply this area by 2. If z_0 is positive, find the area under the standard normal curve to the right of z_0 and double this value.

Using Table IV, we have

$$
\begin{aligned}
P\text{-value} = P(Z < -2.69 \text{ or } Z > 2.69) &= P(Z < -2.69) + P(Z > 2.69) \\
&= 2P(Z > 2.69) \\
&= 2[1 - P(Z \leq 2.69)] \\
&= 2(1 - 0.9964) \\
&= 2(0.0036) \\
&= 0.0072
\end{aligned}
$$

The probability of obtaining a sample mean that is more than 2.69 standard deviations from the status quo population mean of \$50.64 is 0.0072. This means less than 1 sample in 100 will result in a sample mean such as the one we obtained if the statement in the null hypothesis is true.

Step 5: Because the P-value is less than the level of significance $(0.0072 < 0.05)$, we reject the null hypothesis.

Step 6: There is sufficient evidence at the $\alpha = 0.05$ level of significance to conclude that the mean monthly cell phone bill is different from \$50.64, the mean amount in 2004.

Now Work Problem 25 Using the P-Value Approach.

EXAMPLE 5

Testing Hypotheses about a Population Mean Using Technology

Problem: According to CTIA—The Wireless Association, the mean monthly cell phone bill in 2004 was $50.64. A market researcher believes that the mean monthly cell phone bill is different today, but is not sure whether bills have declined because of technological advances, or increased due to additional use. The researcher phones a simple random sample of 12 cell phone subscribers and obtains the data in Table 2.

Assuming $\sigma = \$18.49$, use these data to determine whether the mean monthly cell phone bill is different from $50.64 at the $\alpha = 0.05$ level of significance.

Approach: We will use MINITAB to test the hypothesis. The steps for testing hypotheses about a population mean with σ known using the TI-83/84 Plus graphing calculators, MINITAB, and Excel are given in the Technology Step by Step on page 479.

Result: Figure 12 shows the results obtained from MINITAB.

Figure 12

One-Sample Z: Cell Phone Bill

```
Test of mu = 50.64 vs not = 50.64
The assumed standard deviation = 18.49

Variable        N     Mean   StDev  SE Mean      95% CI          Z      P
Cell Phone
Bill            12  65.0142  25.4587  5.3376  (54.5527, 75.4757)  2.69  0.007
```

USING TECHNOLOGY

The *P*-value obtained from technology may be slightly different from the *P*-value obtained by hand because of rounding.

Note to Instructor

Throughout the text, we will give the level of significance, α, to use in the decision rule. You may want to point out to students that journal articles rarely (if ever) provide levels of significance; they leave it to the reader to judge significance.

The *P*-value is highlighted. From MINITAB, we have *P*-value = 0.007.

Interpretation: Because the *P*-value is less than the level of significance $(0.007 < 0.05)$, we reject the null hypothesis. There is sufficient evidence at the $\alpha = 0.05$ level of significance to conclude that the mean monthly cell phone bill is different from $50.64, the mean amount in 2004.

One advantage of using *P*-values over the classical approach in hypothesis testing is that *P*-values provide information regarding the strength of the evidence. In Example 2, we rejected the null hypothesis, but did not learn anything about the strength of the evidence against the null hypothesis. Example 4 tested the same claim using *P*-values. The *P*-value was 0.0072. This result not only led us to reject the null hypothesis, but also indicates the strength of the evidence against the null hypothesis: Less than 1 sample in 100 would give us the sample mean that we got if the null hypothesis, $H_0: \mu = 50.64$, were true.

Another advantage of *P*-values is that they are interpreted the same way, regardless of the type of hypothesis test being performed. If *P*-value $< \alpha$, reject the null hypothesis.

④ Test Hypotheses about a Population Mean with σ Known Using Confidence Intervals

Recall that the level of confidence in a confidence interval is a probability that represents the percentage of intervals that will contain μ if repeated samples are obtained. The level of confidence is denoted $(1 - \alpha) \cdot 100\%$. We can use confidence intervals to test $H_0: \mu = \mu_0$ versus $H_1: \mu \neq \mu_0$ using the following criterion.

> When testing $H_0: \mu = \mu_0$ versus $H_1: \mu \neq \mu_0$, if a $(1 - \alpha) \cdot 100\%$ confidence interval contains μ_0, we do not reject the null hypothesis. However, if the confidence interval does not contain μ_0, we have sufficient evidence that supports the statement in the alternative hypothesis and conclude $\mu \neq \mu_0$ at the level of significance, α.

EXAMPLE 6

Testing Hypotheses about a Population Mean Using a Confidence Interval

Problem: Test the hypotheses presented in Examples 2 and 4 at the $\alpha = 0.05$ level of significance by constructing a 95% confidence interval about μ, the population mean monthly cell phone bill.

Approach: We construct the 95% confidence interval using the data in Table 1 or Table 2. If the interval contains the status quo mean of $50.64, we do not reject the null hypothesis.

Solution: We use the formula on page 410 to find the lower and upper bounds with $\bar{x} = \$65.014$, $\sigma = \$18.49$, and $n = 12$.

Lower bound: $\bar{x} - z_{\alpha/2} \cdot \dfrac{\sigma}{\sqrt{n}} = \$65.014 - 1.96 \cdot \dfrac{18.49}{\sqrt{12}} = \$65.014 - \$10.462 = \$54.552$

Upper bound: $\bar{x} + z_{\alpha/2} \cdot \dfrac{\sigma}{\sqrt{n}} = \$65.014 + 1.96 \cdot \dfrac{18.49}{\sqrt{12}} = \$65.014 + \$10.462 = \$75.476$

We are 95% confident the mean monthly cell phone bill is between $54.552 and $75.476. Because the mean stated in the null hypothesis, $H_0: \mu = 50.64$, is not included in this interval, we reject the null hypothesis. There is sufficient evidence at the $\alpha = 0.05$ level of significance to conclude that the mean monthly cell phone bill is different from $50.64.

Now Work Problem 35.

In Other Words
Results are statistically significant if the difference between the observed result and the statement made in the null hypothesis is unlikely to occur due to chance alone.

⑤ Understand the Difference between Statistical Significance and Practical Significance

When a large sample size is used in a hypothesis test, the results could be statistically significant even though the difference between the sample statistic and mean stated in the null hypothesis may have no *practical significance*.

Definition

Practical significance refers to the idea that small differences between the statistic and parameter stated in the null hypothesis are statistically significant, while the difference is not large enough to cause concern or be considered important.

EXAMPLE 7

Statistical versus Practical Significance

Problem: According to the American Community Survey, the mean travel time to work in Dallas, Texas, in 2003 was 23.6 minutes. Suppose the Department of Transportation in Dallas just reprogrammed all the traffic lights in an attempt to reduce travel time. To determine if there is evidence that indicates that travel times in Dallas have decreased as a result of the reprogramming, the Department of Transportation obtains a sample of 2500 commuters, records their travel time to work, and obtains a sample mean of 23.3 minutes. Assuming that the population standard deviation travel time to work is known to be 8.4 minutes, determine whether travel times in Dallas have decreased as a result of the reprogramming at the $\alpha = 0.05$ level of significance.

Approach: We will use both the classical approach and *P*-value approach to test the hypotheses.

Solution

Step 1: The Department of Transportation wants to know if the mean travel time to work has decreased from 23.6 minutes, which can be written $\mu < 23.6$. This is a left-tailed test and we have

$$H_0: \mu = 23.6 \quad \text{versus} \quad H_1: \mu < 23.6$$

Step 2: The level of significance is $\alpha = 0.05$.

Step 3: The test statistic is

$$z_0 = \frac{\overline{x} - \mu_0}{\frac{\sigma}{\sqrt{n}}} = \frac{23.3 - 23.6}{\frac{8.4}{\sqrt{2500}}} = -1.79$$

Classical Approach

Step 4: This is a left-tailed test. With $\alpha = 0.05$, the critical value is $-z_{0.05} = -1.645$.

Step 5: Because the test statistic is less than the critical value (the critical value falls in the critical region), we reject the null hypothesis.

P-Value Approach

Step 4: Because this is a left-tailed test, the P-value is $P\text{-value} = P(Z < z_0) = P(Z < -1.79) = 0.0367$.

Step 5: Because the P-value is less than the level of significance $(0.0367 < 0.05)$, we reject the null hypothesis.

Step 6: There is sufficient evidence at the $\alpha = 0.05$ level of significance to conclude that the mean travel time to work has decreased.

While the difference between 23.3 and 23.6 is statistically significant, it really has no practical meaning. After all, is 0.3 minutes (18 seconds) really going to make anyone feel better about his or her commute to work?

The reason that the results from Example 7 were statistically significant had to do with the large sample size. The moral of the story is this:

! CAUTION

Beware of studies with large sample sizes that claim statistical significance because the differences may not have any practical meaning.

Large sample sizes can lead to results that are statistically significant, while the difference between the statistic and parameter in the null hypothesis is not enough to be considered practically significant.

10.2 ASSESS YOUR UNDERSTANDING

Concepts and Vocabulary

1. State the requirements that must be satisfied to test a hypothesis regarding a population mean with σ known.
2. Determine the critical value for a right-tailed test regarding a population mean with σ known at the $\alpha = 0.01$ level of significance. $Z = 2.33$
3. Determine the critical value for a two-tailed test regarding a population mean with σ known at the $\alpha = 0.05$ level of significance. $Z = \pm 1.96$
4. The procedures for testing a hypothesis regarding a population mean with σ known are robust. What does this mean?
5. Explain what a P-value is. What is the criterion for rejecting the null hypothesis using the P-value approach?
6. Suppose that we are testing the hypotheses $H_0: \mu = \mu_0$ versus $H_1: \mu < \mu_0$ and we find the P-value to be 0.23.

Explain what this means. Would you reject H_0? Why?

7. Suppose that we are testing the hypotheses $H_0: \mu = \mu_0$ versus $H_1: \mu \neq \mu_0$ and we find the P-value to be 0.02. Explain what this means. Would you reject H_0? Why?
8. Discuss the advantages and disadvantages of using the classical approach to hypothesis testing. Discuss the advantages and disadvantages of using the P-value approach to hypothesis testing.
9. In your own words, explain the difference between *statistical significance* and *practical significance*.
10. *True or False:* To test $H_0: \mu = \mu_0$ versus $H_1: \mu \neq \mu_0$ using a 5% level of significance, we could construct a 95% confidence interval. True

Skill Building

11. To test H_0: $\mu = 50$ versus H_1: $\mu < 50$, a random sample of size $n = 24$ is obtained from a population that is known to be normally distributed with $\sigma = 12$.

   (a) If the sample mean is determined to be $\bar{x} = 47.1$, compute the test statistic. $Z = -1.18$

   (b) If the researcher decides to test this hypothesis at the $\alpha = 0.05$ level of significance, determine the critical value. $-Z_{0.05} = -1.645$

   (c) Draw a normal curve that depicts the critical region.

   (d) Will the researcher reject the null hypothesis? Why? No

12. To test H_0: $\mu = 40$ versus H_1: $\mu > 40$, a random sample of size $n = 25$ is obtained from a population that is known to be normally distributed with $\sigma = 6$.

   (a) If the sample mean is determined to be $\bar{x} = 42.3$, compute the test statistic. $Z = 1.92$

   (b) If the researcher decides to test this hypothesis at the $\alpha = 0.1$ level of significance, determine the critical value. $Z_{0.1} = 1.28$

   (c) Draw a normal curve that depicts the critical region.

   (d) Will the researcher reject the null hypothesis? Why? Yes

13. To test H_0: $\mu = 100$ versus H_1: $\mu \neq 100$, a random sample of size $n = 23$ is obtained from a population that is known to be normally distributed with $\sigma = 7$.

   (a) If the sample mean is determined to be $\bar{x} = 104.8$, compute the test statistic. $Z = 3.29$

   (b) If the researcher decides to test this hypothesis at the $\alpha = 0.01$ level of significance, determine the critical values. $-Z_{0.005} = -2.575$; $Z_{0.005} = 2.575$

   (c) Draw a normal curve that depicts the critical regions.

   (d) Will the researcher reject the null hypothesis? Why? Yes

14. To test H_0: $\mu = 80$ versus H_1: $\mu < 80$, a random sample of size $n = 22$ is obtained from a population that is known to be normally distributed with $\sigma = 11$.

   (a) If the sample mean is determined to be $\bar{x} = 76.9$, compute the test statistic. $Z = -1.32$

   (b) If the researcher decides to test this hypothesis at the $\alpha = 0.02$ level of significance, determine the critical value. $-Z_{0.02} = -2.05$

   (c) Draw a normal curve that depicts the critical region.

   (d) Will the researcher reject the null hypothesis? Why? No

15. To test H_0: $\mu = 20$ versus H_1: $\mu < 20$, a random sample of size $n = 18$ is obtained from a population that is known to be normally distributed with $\sigma = 3$.

   (a) If the sample mean is determined to be $\bar{x} = 18.3$, compute and interpret the P-value. 0.0082

   (b) If the researcher decides to test this hypothesis at the $\alpha = 0.05$ level of significance, will the researcher reject the null hypothesis? Why? Yes

16. To test H_0: $\mu = 4.5$ versus H_1: $\mu > 4.5$, a random sample of size $n = 13$ is obtained from a population that is known to be normally distributed with $\sigma = 1.2$.

   (a) If the sample mean is determined to be $\bar{x} = 4.9$, compute and interpret the P-value. 0.1151

   (b) If the researcher decides to test this hypothesis at the $\alpha = 0.1$ level of significance, will the researcher reject the null hypothesis? Why? No

17. To test H_0: $\mu = 105$ versus H_1: $\mu \neq 105$, a random sample of size $n = 35$ is obtained from a population whose standard deviation is known to be $\sigma = 12$.

   (a) Does the population need to be normally distributed to compute the P-value? No

   (b) If the sample mean is determined to be $\bar{x} = 101.2$, compute and interpret the P-value. 0.0614

   (c) If the researcher decides to test this hypothesis at the $\alpha = 0.02$ level of significance, will the researcher reject the null hypothesis? Why? No

18. To test H_0: $\mu = 45$ versus H_1: $\mu \neq 45$, a random sample of size $n = 40$ is obtained from a population whose standard deviation is known to be $\sigma = 8$.

   (a) Does the population need to be normally distributed to compute the P-value? No

   (b) If the sample mean is determined to be $\bar{x} = 48.3$, compute and interpret the P-value. 0.009

   (c) If the researcher decides to test this hypothesis at the $\alpha = 0.05$ level of significance, will the researcher reject the null hypothesis? Why? Yes

Applying the Concepts

19. Are Women Getting Taller? A researcher believes that the NW average height of a woman aged 20 years or older is greater than the 1994 mean height of 63.7 inches, on the basis of data obtained from the Centers for Disease Control and Prevention's *Advance Data Report*, No. 347. She obtains a simple random sample of 45 women and finds the sample mean height to be 63.9 inches. Assume that the population standard deviation is 3.5 inches. Test the researcher's belief using either the classical approach or the P-value approach at the $\alpha = 0.05$ level of significance. Do not reject H_0

20. ATM Withdrawals The manufacturer of a certain type of ATM machine reports that the mean ATM withdrawal is $60. The manager of a convenience store with an ATM machine claims that mean withdrawal from his machine is less than this amount. He obtains a simple random sample of 35 withdrawals over the past year and finds the sample mean to be $52. Assume that the population standard deviation is $13. Test the manager's claim using either the classical approach or the P-value approach at the $\alpha = 0.05$ level of significance. Reject H_0

21. SAT Exam Scores A school administrator believes that students whose first language learned is not English score worse on the verbal portion of the SAT exam than students whose first language is English. The mean SAT verbal score of students whose first language is English is 515, on the basis of data obtained from the College Board. Suppose a simple random sample of 20 students whose first language learned was not English results in a sample mean SAT verbal score of 458. SAT verbal scores are normally distributed with a population standard deviation of 112.

   (a) Why is it necessary for SAT verbal scores to be normally distributed to test the hypotheses using the methods of this section? Small sample

   (b) Use the classical approach or the P-value approach at the $\alpha = 0.10$ level of significance to determine if there is evidence to support the administrator's belief. Reject H_0

22. SAT Exam Scores A school administrator wonders if students whose first language learned is not English score differently on the math portion of the SAT exam than students whose first language is English. The mean SAT math score of students whose first language is English is 516, on the basis of data obtained from the College Board. Suppose a simple random sample of 20 students whose first language learned was not English results in a sample mean SAT math score of 522. SAT math scores are normally distributed with a population standard deviation of 114.

(a) Why is it necessary for SAT math scores to be normally distributed to test the hypotheses using the methods of this section? Small sample

(b) Determine whether students whose first language learned is not English score differently on the math portion of the SAT exam using the classical approach or the *P*-value approach at the $\alpha = 0.10$ level of significance. Do not reject H_0

23. Acid Rain In 1990, the mean pH level of the rain in Pierce County, Washington, was 5.03. A biologist wonders if the acidity of rain has increased. (This would mean that the pH level of the rain has decreased.) From a random sample of 19 rain dates in 2004, she obtains the following data:

5.08	4.66	4.70	4.87
4.78	5.00	4.50	4.73
4.79	4.65	4.91	5.07
5.03	4.78	4.77	4.60
4.73	5.05	4.70	

Source: National Atmospheric Deposition Program

(a) Because the sample size is small, she must verify that pH level is normally distributed and the sample does not contain any outliers. The normal probability plot and boxplot are shown next. Are the conditions for testing the hypothesis satisfied? Yes

(b) Test whether the acidity of rain has increased, assuming that $\sigma = 0.2$ at the $\alpha = 0.01$ level of significance. Reject H_0

24. Fuel Catalyst To improve fuel efficiency and reduce pollution, the owner of a trucking fleet decides to install a new fuel catalyst in all his semitrucks. He feels that the catalyst will help to increase the number of miles per gallon. Before the installation, his trucks had a mean gas mileage of 5.6 miles per gallon. A random sample of 12 trucks after the installation gave the following gas mileages:

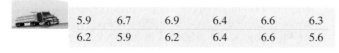

5.9	6.7	6.9	6.4	6.6	6.3
6.2	5.9	6.2	6.4	6.6	5.6

(a) Because the sample size is small, he must verify that mileage is normally distributed and the sample does

not contain any outliers. The normal probability plot and boxplot are shown. Are the conditions for testing the hypothesis satisfied? Yes

(b) Test if the fuel catalyst is effective, assuming that $\sigma = 0.5$ mile per gallon at the 0.05 level of significance. Reject H_0

25. Filling Bottles A certain brand of apple juice is supposed to have 64 ounces of juice. Because the punishment for underfilling bottles is severe, the target mean amount of juice is 64.05 ounces. However, the filling machine is not precise, and the exact amount of juice varies from bottle to bottle. The quality-control manager wishes to verify that the mean amount of juice in each bottle is 64.05 ounces so that she can be sure that the machine is not over- or underfilling. She randomly samples 22 bottles of juice and measures the content and obtains the following data:

64.05	64.05	64.03	63.97	63.95	64.02
64.01	63.99	64.00	64.01	64.06	63.94
63.98	64.05	63.95	64.01	64.08	64.01
63.95	63.97	64.10	63.98		

(a) Because the sample size is small, she must verify that the amount of juice is normally distributed and the sample does not contain any outliers. The normal probability plot and boxplot are shown. Are the conditions for testing the hypothesis satisfied? Yes

(b) Should the assembly line be shut down so that the machine can be recalibrated? Assume $\sigma = 0.06$ ounces and use a 0.01 level of significance. Yes

(c) Explain why a level of significance of $\alpha = 0.01$ might be more reasonable than $\alpha = 0.1$. [**Hint:** Consider the consequences of incorrectly rejecting the null hypothesis.]

26. SAT Reasoning Test In 2005, in response to various criticisms, the College Board implemented changes to the SAT test. In particular, the math portion eliminated quantitative comparisons and expanded the topics covered. A school administrator believes that the new math portion is more difficult and will result in lower scores on the math portion compared to the 2004 average score of

516. A random sample of 25 students taking the new SAT test resulted in the following scores on the math portion.

410	720	480	560	590	340	430	400	440
540	480	450	720	570	380	740	640	600
320	580	450	480	550	470	380		

(a) Because the sample size is small, he must verify that the new math scores are normally distributed and the sample does not contain any outliers. The normal probability plot and boxplot are shown. Are the conditions for testing the hypothesis satisfied? Yes

(b) Do you believe students scored lower? Assume $\sigma = 114$. No

27. **Are Cars Younger?** Suppose you have just been hired by Ford Motor Company. Management wants to know if cars are younger today versus 1995. According to the Nationwide Personal Transportation Survey conducted by the U.S. Department of Transportation, the mean age of a car in 1995 was 8.33 years. Based on a random sample of 18 automobile owners, you obtain the ages as shown in the following table:

8	12	1	2	13	3
5	9	12	6	5	6
10	7	10	11	6	10

(a) Because the sample size is small, you must verify that the data are normally distributed with no outliers by drawing a normal probability plot and boxplot. Based on the following graphs, can you perform a hypothesis test? Yes

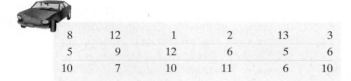

(b) Are cars younger today? Assume that $\sigma = 3.8$ years. Use the $\alpha = 0.1$ level of significance. No

28. **It's a Hot One!** Recently, a friend of mine claimed that the summer of 2000 in Houston, Texas, was hotter than usual. To test his claim, I went to AccuWeather.com and randomly selected 12 days in the summer of 2000. I then recorded the departure from normal, with positive values

indicating above-normal temperatures and negative values indicating below-normal temperatures, as shown in the following:

+4	−1	0	+2
+2	+4	−1	−3
−1	0	+2	+4

Source: AccuWeather.com

(a) Because the sample size is small, I must verify that the temperature departure is normally distributed and the sample does not contain any outliers. The normal probability plot and boxplot are shown. Are the conditions for testing the hypothesis satisfied? Yes

(b) Was the summer of 2000 hotter than normal in Houston? Assume that $\sigma = 1.8$. Use the $\alpha = 0.05$ level of significance. Yes

29. **Farm Size** In 1990, the average farm size in Kansas was 694 acres, according to data obtained from the U.S. Department of Agriculture. A researcher claims that farm sizes are larger now due to consolidation of farms. She obtains a random sample of 40 farms and determines the mean size to be 731 acres. Assume that $\sigma = 212$ acres. Test the researcher's claim at the $\alpha = 0.05$ level of significance. Do not reject H_0

30. **Oil Output** An energy official claims that the oil output per well in the United States has declined from the 1998 level of 11.1 barrels per day. He randomly samples 50 wells throughout the United States and determines the mean output to be 10.7 barrels per day. Assume that $\sigma = 1.3$ barrels. Test the researcher's claim at the $\alpha = 0.05$ level of significance. Reject H_0

31. **Volume of Dell Computer Stock** The average daily volume of Dell Computer stock in 2000 was $\mu = 31.8$ million shares, with a standard deviation of $\sigma = 14.8$ million shares, according to Yahoo!Finance. Based on a random sample of 35 trading days in 2004, the sample mean number of shares traded is found to be 23.5 million. Is the volume of Dell stock different in 2004? Use the $\alpha = 0.05$ level of significance. Reject H_0

32. **Volume of Motorola Stock** The average daily volume of Motorola stock in 2000 was $\mu = 11.4$ million shares, with a standard deviation of $\sigma = 8.3$ million shares, according to Yahoo!Finance. Based on a random sample of 35 trading days in 2004, the sample mean number of shares traded is found to be 13.3 million shares. Is the volume of Motorola stock different in 2004? Use the $\alpha = 0.05$ level of significance. Do not reject H_0

33. Using Confidence Intervals to Test Hypotheses Test the hypotheses in Problem 25 by constructing a 99% confidence interval. (63.974, 64.04); reject H_0

34. Using Confidence Intervals to Test Hypotheses Test the hypotheses in Problem 26 by constructing a 95% confidence interval. (464.1, 553.5); do not reject H_0

35. Using Confidence Intervals to Test Hypotheses Test the NW hypotheses in Problem 31 by constructing a 95% confidence interval. (18.60, 28.40); reject H_0

36. Using Confidence Intervals to Test Hypotheses Test the hypotheses in Problem 32 by constructing a 95% confidence interval. (10.55, 16.05); do not reject H_0

37. Statistical Significance versus Practical Significance A math teacher claims that she has developed a review course that increases the scores of students on the math portion of the SAT exam. Based on data from the College Board, SAT scores are normally distributed with $\mu = 514$ and $\sigma = 113$. The teacher obtains a random sample of 1800 students, puts them through the review class, and finds that the mean SAT math score of the 1800 students is 518.

(a) State the null and alternative hypotheses.

(b) Test the hypothesis at the $\alpha = 0.10$ level of significance. Is a mean SAT math score of 518 significantly higher than 514? Reject H_0

(c) Do you think that a mean SAT math score of 518 versus 514 will affect the decision of a school admissions administrator? In other words, does the increase in the score have any practical significance?

(d) Test the hypothesis at the $\alpha = 0.10$ level of significance with $n = 400$ students. Assume that the sample mean is still 518. Is a sample mean of 518 significantly more than 514? Conclude that large sample sizes cause P-values to shrink substantially, all other things being the same. Do not reject H_0

38. Statistical Significance versus Practical Significance The manufacturer of a daily dietary supplement claims that its product will help people lose weight. The company obtains a random sample of 950 adult males aged 20 to 74 who take the supplement and finds their mean weight loss after eight weeks to be 0.9 pounds. Assume the population standard deviation weight loss is $\sigma = 7.2$ pounds.

(a) State the null and alternative hypotheses.

(b) Test the hypothesis at the $\alpha = 0.1$ level of significance. Is a mean weight loss of 0.9 pound significant?

(c) Do you think that a mean weight loss of 0.9 pounds is worth the expense and commitment of a daily dietary

supplement? In other words, does the weight loss have any practical significance?

(d) Test the hypothesis at the $\alpha = 0.1$ level of significance with $n = 40$ subjects. Assume that the sample mean weight loss is still 0.9 pounds. Is a sample mean weight loss of 0.9 pounds significantly more than 0 pounds? Conclude that large sample sizes cause P-values to shrink substantially, all other things being the same. $Z = 0.79$; P-value $= 0.2148$; do not reject H_0

39. Simulation Simulate drawing 50 simple random samples of size $n = 20$ from a population that is normally distributed with mean 80 and standard deviation 7.

(a) Test the null hypothesis $H_0: \mu = 80$ versus the alternative hypothesis $H_1: \mu \neq 80$.

(b) Suppose we were testing this hypothesis at the $\alpha = 0.1$ level of significance. How many of the 50 samples would you expect to result in a Type I error?

(c) Count the number of samples that lead to a rejection of the null hypothesis. Is it close to the expected value determined in part (b)?

(d) Describe how we know that a rejection of the null hypothesis results in making a Type I error in this situation.

40. Simulation Simulate drawing 40 simple random samples of size $n = 35$ from a population that is exponentially distributed with mean 8 and standard deviation $\sqrt{8}$.

(a) Test the null hypothesis $H_0: \mu = 8$ versus the alternative hypothesis $H_1: \mu \neq 8$.

(b) Suppose we were testing this hypothesis at the $\alpha = 0.05$ level of significance. How many of the 40 samples would you expect to result in a Type I error?

(c) Count the number of samples that lead to a rejection of the null hypothesis. Is it close to the expected value determined in part (b)?

(d) Describe how we know that a rejection of the null hypothesis results in making a Type I error in this situation.

41. Suppose a chemical company has developed a catalyst that is meant to reduce reaction time in a chemical process. For a certain chemical process, reaction time is known to be 150 seconds. The researchers conducted an experiment with the catalyst 40 times and measured reaction time. The researchers reported that the catalysts reduced reaction time with a P-value of 0.02.

(a) Identify the null and alternative hypotheses.

(b) Explain what this result means. Do you believe that the catalyst is effective?

(38a) $H_0: \mu = 0$ pound; $H_1: \mu > 0$ pound   (38b) $Z = 3.85$; P-value $= 0.0001$; reject H_0 (41a) $H_0: \mu = 150$ vs. $H_1: \mu < 150$

Technology Step by Step	**Hypothesis Tests Regarding μ, σ Known**
TI-83/84 Plus	**Step 1:** If necessary, enter raw data in L1.
	Step 2: Press STAT, highlight TESTS, and select `1:Z-Test`.
	Step 3: If the data are raw, highlight DATA; make sure that List1 is set to L1 and Freq is set to 1. If summary statistics are known, highlight STATS and enter

the summary statistics. Following σ, enter the population standard deviation. For the value of μ_0, enter the value of the mean stated in the null hypothesis.

Step 4: Select the direction of the alternative hypothesis.

Step 5: Highlight Calculate and press ENTER. The TI-83/84 Plus gives the *P*-value.

MINITAB   ***Step 1:*** Enter raw data in column C1.

Step 2: Select the **Stat** menu, highlight **Basic Statistics**, and then highlight **1-Sample Z** . . . .

Step 3: Click Options. In the cell marked "Alternative," select the appropriate direction for the alternative hypothesis. Click OK.

Step 4: Enter C1 in the cell marked "Variables." In the cell labeled "Test Mean", enter the value of the mean stated in the null hypothesis. In the cell labeled "standard deviation," enter the value of σ. Click OK.

Excel   ***Step 1:*** If necessary, enter raw data in column A.

Step 2: Load the PHStat Add-in.

Step 3: Select the **PHStat** menu, highlight **One Sample Tests . . .**, and then highlight **Z Test for the mean, sigma known . . . .**

Step 4: Enter the value of the null hypothesis, the level of significance, α, and the value of σ. If the summary statistics are known, click "Sample statistics known" and enter the sample size and sample mean. If summary statistics are unknown, click "Sample statistics unknown." With the cursor in the "Sample cell range" cell, highlight the data in column A. Click the option corresponding to the desired test [two-tail, upper (right) tail, or lower (left) tail]. Click OK.

10.3   Hypothesis Tests for a Population Mean in Practice

Preparing for This Section   Before getting started, review the following:

- Sampling distribution of $\bar{x}$ (Section 8.1, pp. 374–388)
- The *t*-distribution (Section 9.2, pp. 423–426)
- Using probabilities to identify unusual events (Section 5.1, p. 225)

Objective   ①   **Test hypotheses about a population mean with σ unknown**

In Section 10.2, we assumed that the population standard deviation, σ, was known when testing hypotheses regarding the population mean. We now introduce procedures for testing hypotheses regarding a population mean when σ is not known. The only difference from the situation where σ is known is that we must use the *t*-distribution rather than the *z*-distribution.

We do not replace σ with s and say that $z = \dfrac{\bar{x} - \mu}{s/\sqrt{n}}$ is normally distributed

In Other Words

When σ is known, use *z*; when σ is unknown, use *t*.

with mean 0 and standard deviation 1. Instead, $t = \dfrac{\bar{x} - \mu}{s/\sqrt{n}}$ follows **Student's *t*-distribution** with $n - 1$ degrees of freedom. Let's review the properties of the *t*-distribution.

Properties of the *t*-Distribution

1. The *t*-distribution is different for different degrees of freedom.
2. The *t*-distribution is centered at 0 and is symmetric about 0.

3. The area under the curve is 1. Because of the symmetry, the area under the curve to the right of 0 equals the area under the curve to the left of 0, which equals $\frac{1}{2}$.

4. As t increases without bound, the graph approaches, but never equals, zero. As t decreases without bound, the graph approaches, but never equals, zero.

5. The area in the tails of the t-distribution is a little greater than the area in the tails of the standard normal distribution, because using s as an estimate of σ introduces more variability to the t-statistic.

6. As the sample size n increases, the density curve of t gets closer to the standard normal density curve. This result occurs because, as the sample size n increases, the values of s get closer to the values of σ, by the Law of Large Numbers.

① Test Hypotheses about a Population Mean with σ Unknown

From Section 10.2, we know that there are two approaches (besides using confidence intervals) we can use to test claims regarding a population mean, the classical approach and the P-value approach. We will present both methods here. Your instructor may choose one or both approaches.

Both the classical approach and the P-value approach to testing hypotheses about μ with σ unknown follow the exact same logic as testing hypotheses about μ with σ known. The only difference is that we use Student's t-distribution, rather than the normal distribution.

Testing Hypotheses Regarding a Population Mean with σ Unknown

To test hypotheses regarding the population mean with σ unknown, we use the following steps, provided that

1. The sample is obtained using simple random sampling.
2. The sample has no outliers and the population from which the sample is drawn is normally distributed or the sample size, n, is large ($n \geq 30$).

Step 1: Determine the null and alternative hypotheses. The hypotheses can be structured in one of three ways:

Two-Tailed	Left-Tailed	Right-Tailed
$H_0: \mu = \mu_0$	$H_0: \mu = \mu_0$	$H_0: \mu = \mu_0$
$H_1: \mu \neq \mu_0$	$H_1: \mu < \mu_0$	$H_1: \mu > \mu_0$

Note: μ_0 is the assumed value of the population mean.

Step 2: Select a level of significance α, depending on the seriousness of making a Type I error.

Step 3: Compute the test statistic

$$t_0 = \frac{\bar{x} - \mu_0}{s/\sqrt{n}}$$

which follows Student's t-distribution with $n - 1$ degrees of freedom.

Classical Approach

Step 4: Use Table V to determine the critical value using $n - 1$ degrees of freedom.

	Two-Tailed	**Left-Tailed**	**Right-Tailed**
Critical value(s)	$-t_{\frac{\alpha}{2}}$ and $t_{\frac{\alpha}{2}}$	$-t_\alpha$	t_α
Critical region			

Step 5: Compare the critical value with the test statistic.

Two-Tailed	**Left-Tailed**	**Right-Tailed**
If $t_0 < -t_{\frac{\alpha}{2}}$ or $t_0 > t_{\frac{\alpha}{2}}$, reject the null hypothesis	If $t_0 < -t_\alpha$, reject the null hypothesis	If $t_0 > t_\alpha$, reject the null hypothesis

P-Value Approach

Step 4: Use Table V to estimate the *P*-value using $n - 1$ degrees of freedom.

Step 5: If the *P*-value $<\alpha$, reject the null hypothesis.

Step 6: State the conclusion.

Notice that the procedure just presented requires either that the population from which the sample was drawn be normal or that the sample size be large ($n \geq 30$). The procedure is robust, so minor departures from normality will not adversely affect the results of the test. However, if the data include outliers, the procedure should not be used.

Just as we did for hypothesis tests with σ known, we will verify these assumptions by constructing normal probability plots (to assess normality) and boxplots (to discover whether there are outliers). If the normal probability plot indicates that the data do not come from a normal population or if the boxplot reveals outliers, nonparametric tests should be performed, which are not discussed in this text.

Before we look at a couple of examples, it is important to understand that we cannot find exact *P*-values using the *t*-distribution table (Table V) because the table provides *t*-values only for certain areas. However, we can use the table to calculate lower and upper bounds on the *P*-value. To find exact *P*-values, we use statistical software or a graphing calculator with advanced statistical features.

In Other Words

When σ is unknown, exact *P*-values can be found using technology.

EXAMPLE 1

Testing a Hypothesis about a Population Mean, Large Sample

Problem: According to the Centers for Disease Control, the mean number of cigarettes smoked per day by individuals who are daily smokers is 18.1. A researcher wonders if retired adults smoke less than the general population of daily smokers, so she obtains a random sample of 40 retired adults who are current smokers and records the number of cigarettes smoked on a randomly selected day. The data result in a sample mean of 16.8 cigarettes and a standard deviation of 4.7 cigarettes. Is there sufficient evidence at the $\alpha = 0.1$ level of significance to conclude that retired adults who are daily smokers smoke less than the general population of daily smokers?

Approach: Because the sample size is large, we can follow the steps to testing a claim about a population mean given on pages 481–482.

Solution

Step 1: The researcher wants to know if retired adults smoke less than the general population. The mean number of cigarettes smoked per day by individuals who are daily smokers is 18.1, so we have

$$H_0: \mu = 18.1 \quad \text{versus} \quad H_1: \mu < 18.1$$

This is a left-tailed test.

Step 2: The level of significance is $\alpha = 0.1$.

Step 3: The sample mean is $\bar{x} = 16.8$, and the sample standard deviation is $s = 4.7$. The test statistic is

$$t_0 = \frac{\bar{x} - \mu_0}{s/\sqrt{n}} = \frac{16.8 - 18.1}{4.7/\sqrt{40}} = -1.749$$

Classical Approach

Step 4: Because this is a left-tailed test, we determine the critical t-value at the $\alpha = 0.1$ level of significance with $n - 1 = 40 - 1 = 39$ degrees of freedom to be $-t_{0.1} = -1.304$. The critical region is displayed in Figure 13.

Figure 13

Critical Region

-1.749 $-t_{0.1} = -1.304$

Step 5: Because the test statistic $t_0 = -1.749$ is less than the critical value $-t_{0.1} = -1.304$, the researcher rejects the null hypothesis. We label this point in Figure 13.

P-Value Approach

Step 4: Because this is a left-tailed test, the P-value is the area under the t-distribution with $40 - 1 = 39$ degrees of freedom to the left of the test statistic, $t_0 = -1.749$, as shown in Figure 14(a). That is, P-value $= P(t < t_0) = P(t < -1.749)$, with 39 degrees of freedom.

Because of the symmetry of the t-distribution, the area under the distribution to the left of -1.749 equals the area under the distribution to the right of 1.749. So P-value $= P(t < -1.749) = P(t > 1.749)$. See Figure 14(b).

Using Table V, we find the row that corresponds to 39 degrees of freedom. The value 1.749 lies between 1.685 and 2.023. The value of 1.685 has an area under the t-distribution of 0.05 to the right, with 39 degrees of freedom. The area under the t-distribution with 39 degrees of freedom to the right of 2.023 is 0.025. See Figure 15 on page 484.

Because 1.749 is between 1.685 and 2.023, the P-value is between 0.025 and 0.05. So

$$0.025 < P\text{-value} < 0.05$$

Figure 14

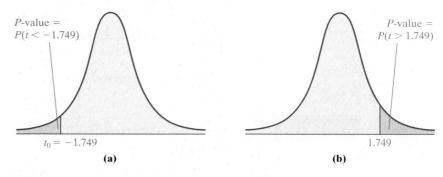

P-value $= P(t < -1.749)$

$t_0 = -1.749$

(a)

P-value $= P(t > 1.749)$

1.749

(b)

Step 5: Because the P-value is less than the level of significance $\alpha = 0.1$, we reject the null hypothesis.

Step 6: There is sufficient evidence to conclude that retired adults smoke less than the general population of daily smokers at the $\alpha = 0.1$ level of significance.

Figure 15

					Area in Right Tail							
df	0.25	0.20	0.15	0.10	0.05	0.025	0.02	0.01	0.005	0.0025	0.001	0.0005
1	1.000	1.376	1.963	3.078	6.314	12.706	15.894	31.821	63.657	127.321	318.289	636.558
2	0.816	1.061	1.386	1.886	2.920	4.303	4.849	6.965	9.925	14.089	22.328	31.600
3	0.765	0.978	1.250	1.638	2.353	3.182	3.482	4.541	5.841	7.453	10.214	12.924
37	0.681	0.851	1.051	1.305	1.687	2.026	2.129	2.431	2.715	2.985	3.326	3.574
38	0.681	0.851	1.051	1.304	1.686	2.024	2.127	2.429	2.712	2.980	3.339	3.566
39	0.641	0.851	1.050	1.304	1.685	2.023	2.125	2.426	2.704	2.976	3.313	3.558
40	0.681	0.851	1.050	1.303	1.654	2.021	2.123	2.423	2.706	2.971	3.307	3.551
50	0.679	0.849	1.047	1.299	1.676	2.009	2.109	2.403	2.678	2.937	3.261	3.496
60	0.679	0.848	1.045	1.296	1.671	2.000	2.099	2.390	2.660	2.915	3.232	3.460

Obtaining the approximate P-value in Example 1 was somewhat challenging. With the aid of technology, we can find the exact P-value quite painlessly.

EXAMPLE 2   Testing a Hypothesis about a Population Mean Using Technology

Problem: Obtain an exact P-value for the problem in Example 1 using statistical software or a graphing calculator with advanced statistical features.

Approach: We will use a TI-84 Plus graphing calculator to obtain the P-value. The steps for testing hypotheses about a population mean with σ unknown using the TI-83/84 Plus graphing calculators, MINITAB, and Excel are given in the Technology Step by Step on page 493.

Result: Figure 16(a) shows the results from the TI-84 Plus using the Calculate option. Figure 16(b) shows the results using the Draw option. The P-value is 0.044. Notice the P-value is between 0.025 and 0.05. This agrees with the results in Example 1.

Figure 16

P-value ⎯

(a) Calculate Option   (b) Draw Option

Now Work Problem 13.

EXAMPLE 3   Testing a Hypothesis about a Population Mean, Small Sample

Table 3

19.68	20.66	19.56
19.98	20.65	19.61
20.55	20.36	21.02
21.5	19.74	

Source: Michael Carlisle, student at Joliet Junior College

Problem: The "fun size" of a Snickers bar is supposed to weigh 20 grams. Because the punishment for selling candy bars that weigh less than 20 grams is so severe, the manufacturer calibrates the machine so that the mean weight is 20.1 grams. The quality-control engineer at M&M–Mars, the company that manufactures Snickers bars, is concerned that the machine that manufacturers the candy is miscalibrated. She obtains a random sample of 11 candy bars, weighs them, and obtains the data in Table 3. Should the machine be shut down and calibrated? Because shutting down the plant is very expensive, she decides to conduct the test at the $\alpha = 0.01$ level of significance.

Approach: Before we can perform the hypothesis test, we must verify that the data come from a population that is normally distributed with no outliers. We will construct a normal probability plot and boxplot to verify these requirements. We then proceed to follow Steps 1 through 6.

Solution: Figure 17 displays the normal probability plot and boxplot.

Figure 17

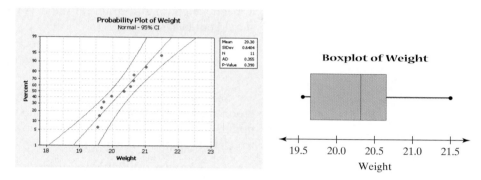

The normal probability plot indicates that the data come from a population that is approximately normal. The boxplot does not show any outliers. We can proceed with the hypothesis test.

Step 1: The quality-control engineer wishes to determine whether the Snickers have a mean weight of 20.1 grams or not. The hypotheses can be written

$$H_0: \mu = 20.1 \quad \text{versus} \quad H_1: \mu \neq 20.1$$

This is a two-tailed test.

Step 2: The level of significance is $\alpha = 0.01$.

Step 3: From the data in Table 3, the sample mean is $\bar{x} = 20.3$, and the sample standard deviation is $s = 0.64$. The test statistic is

$$t_0 = \frac{\bar{x} - \mu_0}{s/\sqrt{n}} = \frac{20.3 - 20.1}{0.64/\sqrt{11}} = 1.036$$

Classical Approach

Step 4: Because this is a two-tailed test, we determine the critical t-values at the $\alpha = 0.01$ level of significance with $n - 1 = 11 - 1 = 10$ degrees of freedom to be $-t_{0.01/2} = -t_{0.005} = -3.169$ and $t_{0.01/2} = t_{0.005} = 3.169$. The critical regions are displayed in Figure 18.

Figure 18

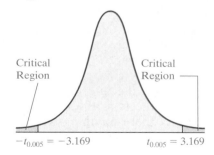

Critical Region   Critical Region

$-t_{0.005} = -3.169$ $t_{0.005} = 3.169$

P-Value Approach

Step 4: Because this is a two-tailed test, the P-value is the area under the t-distribution with $11 - 1 = 10$ degrees of freedom to the left of the test statistic $-t_0 = -1.036$ and to the right of $t_0 = 1.036$, as shown in Figure 19. That is, $P\text{-value} = P(t < -1.036) + P(t > 1.036) = 2P(t > 1.036)$, with 10 degrees of freedom.

Using Table V, we find the row that corresponds to 10 degrees of freedom. The value 1.036 lies between 0.879 and 1.093. The value of 0.879 has an area under the t-distribution of 0.20 to the right. The area under the t-distribution with 10 degrees of freedom to the right of 1.093 is 0.015.

Because 1.036 is between 0.879 and 1.093, the P-value is between $2(0.20)$ and $2(0.15)$. So

$$0.30 < P\text{-value} < 0.40$$

Using MINITAB, we find that the exact P-value is 0.323.

Figure 19

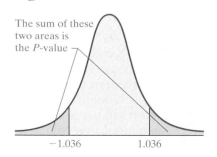

The sum of these two areas is the P-value

-1.036 1.036

Step 5: The test statistic is $t_0 = -1.036$. Because the test statistic is between the critical values $-t_{0.005} = -3.169$ and $t_{0.005} = 3.169$, the quality-control engineer does not reject the null hypothesis.

Step 5: Because the P-value is greater than the level of significance $(0.323 > 0.01)$, the quality-control engineer does not reject the null hypothesis.

Step 6: There is not sufficient evidence to conclude that the Snickers do not have a mean weight of 20.1 grams at the $\alpha = 0.01$ level of significance. The machine should not be shut down.

> Now Work Problem 21.

Sections 10.2 and 10.3 discussed performing hypothesis tests about a population mean. The main criterion for choosing which test to use is whether the population standard deviation, σ, is known. Provided that the population from which the sample is drawn is normal or that the sample size is large,

- if σ is known, use the z-test procedures from Section 10.2;
- if σ is unknown, use the t-test procedures from Section 10.3.

In Section 10.4, we will discuss testing hypotheses about a population proportion.

In-Class Activity: Stringing Them Along (Part I)

Note to Instructor

The purpose of this activity is to give students a hands-on experience with hypothesis testing involving means. Be sure to tell students the length of the rope only after all estimates have been reported. The activity could be elaborated to include discussion of bias and reliability in measurements or other related topics. Time limit: 30–45 minutes depending on discussion.

How skilled are people at estimating the length of a piece of rope? Do you think that they will tend to overestimate its length? Underestimate? Or are you not sure?

1. Look at the piece of rope that your instructor is holding and estimate the length of the rope in inches.
2. Using the null hypothesis $H_0: \mu = \mu_0$, where μ_0 represents the actual length of the rope, select an appropriate alternative hypothesis and a level of significance based on your responses to the questions posed at the beginning of the activity.
3. Obtain the actual length of the rope from your instructor. Combine the data for the entire class and test the hypothesis formed in part (b). What did you conclude?

[**Note:** Save the class data for use in another activity.]

MAKING AN INFORMED DECISION

What Does It Really Weigh?

Many consumer products that we purchase have labels that describe the net weight of the contents. For example, the net weight of a candy bar might be listed as 4 ounces. Choose any consumer product that reports the net weight of the contents on the packaging.

(a) Obtain a random sample of size 8 or more of the consumer product. We will treat the random purchases as a simple random sample. Weigh the contents without the packaging.
(b) If your sample size is less than 30, verify that the population from which the sample was drawn is

normal and that the sample does not contain any outliers.
(c) As the consumer, you are concerned only with situations in which you are getting ripped off. Determine the null and alternative hypotheses from the point of view of the consumer.
(d) Test whether the consumer is getting ripped off at the $\alpha = 0.05$ level of significance. Are you getting ripped off? What makes you say so?
(e) Suppose you are the quality-control manager. How would you structure the alternative hypothesis? Test this hypothesis at the $\alpha = 0.05$ level of significance. Is there anything wrong with the manufacturing process? What makes you say so?

10.3 ASSESS YOUR UNDERSTANDING

Concepts and Vocabulary

1. State the requirements that must be satisfied to test hypotheses about a population mean with σ unknown.
2. Determine the critical value for a right-tailed test of a population mean with σ unknown at the $\alpha = 0.01$ level of significance with 15 degrees of freedom.
3. Determine the critical value for a two-tailed test of a population mean with σ unknown at the $\alpha = 0.05$ level of significance with 12 degrees of freedom.
4. Determine the critical value for a left-tailed test of a population mean with σ unknown at the $\alpha = 0.05$ level of significance with 19 degrees of freedom.

Skill Building

5. To test H_0: $\mu = 50$ versus H_1: $\mu < 50$, a simple random sample of size $n = 24$ is obtained from a population that is known to be normally distributed. (5a) $t = -1.379$
   (a) If $\bar{x} = 47.1$ and $s = 10.3$, compute the test statistic.
   (b) If the researcher decides to test this hypothesis at the $\alpha = 0.05$ level of significance, determine the critical value. $-t_{0.05} = -1.714$
   (c) Draw a t-distribution that depicts the critical region.
   (d) Will the researcher reject the null hypothesis? Why? No

6. To test H_0: $\mu = 40$ versus H_1: $\mu > 40$, a simple random sample of size $n = 25$ is obtained from a population that is known to be normally distributed.(6a) $t = 2.674$
   (a) If $\bar{x} = 42.3$ and $s = 4.3$, compute the test statistic.
   (b) If the researcher decides to test this hypothesis at the $\alpha = 0.1$ level of significance, determine the critical value. $t_{0.1} = 1.318$
   (c) Draw a t-distribution that depicts the critical region.
   (d) Will the researcher reject the null hypothesis? Why? Yes

7. To test H_0: $\mu = 100$ versus H_1: $\mu \neq 100$, a simple random sample of size $n = 23$ is obtained from a population that is known to be normally distributed. (7a) $t = 2.502$
   (a) If $\bar{x} = 104.8$ and $s = 9.2$, compute the test statistic.
   (b) If the researcher decides to test this hypothesis at the $\alpha = 0.01$ level of significance, determine the critical values. $-t_{0.005} = -2.819$; $t_{0.005} = 2.819$
   (c) Draw a t-distribution that depicts the critical region.
   (d) Will the researcher reject the null hypothesis? Why? No

8. To test H_0: $\mu = 80$ versus H_1: $\mu < 80$, a simple random sample of size $n = 22$ is obtained from a population that is known to be normally distributed. (8a) $t = -1.711$
   (a) If $\bar{x} = 76.9$ and $s = 8.5$, compute the test statistic.
   (b) If the researcher decides to test this hypothesis at the $\alpha = 0.02$ level of significance, determine the critical value. $-t_{0.02} = -2.189$
   (c) Draw a t-distribution that depicts the critical region.
   (d) Will the researcher reject the null hypothesis? Why? No

9. To test H_0: $\mu = 20$ versus H_1: $\mu < 20$, a simple random sample of size $n = 18$ is obtained from a population that is known to be normally distributed. (9a) $t = -1.677$

   (a) If $\bar{x} = 18.3$ and $s = 4.3$, compute the test statistic.
   (b) Draw a t-distribution with the area that represents the P-value shaded.
   (c) Approximate and interpret the P-value. 0.0559
   (d) If the researcher decides to test this hypothesis at the $\alpha = 0.05$ level of significance, will the researcher reject the null hypothesis? Why? No

10. To test H_0: $\mu = 4.5$ versus H_1: $\mu > 4.5$, a simple random sample of size $n = 13$ is obtained from a population that is known to be normally distributed. (10a) $t = 1.109$
    (a) If $\bar{x} = 4.9$ and $s = 1.3$, compute the test statistic.
    (b) Draw a t-distribution with the area that represents the P-value shaded.
    (c) Approximate and interpret the P-value. 0.1445
    (d) If the researcher decides to test this hypothesis at the $\alpha = 0.1$ level of significance, will the researcher reject the null hypothesis? Why? No

11. To test H_0: $\mu = 105$ versus H_1: $\mu \neq 105$, a simple random sample of size $n = 35$ is obtained.
    (a) Does the population have to be normally distributed to test this hypothesis by using the methods presented in this section? No
    (b) If $\bar{x} = 101.9$ and $s = 5.9$, compute the test statistic.
    (c) Draw a t-distribution with the area that represents the P-value shaded.
    (d) Determine and interpret the P-value.
    (e) If the researcher decides to test this hypothesis at the $\alpha = 0.01$ level of significance, will the researcher reject the null hypothesis? Why? Yes

12. To test H_0: $\mu = 45$ versus H_1: $\mu \neq 45$, a simple random sample of size $n = 40$ is obtained.
    (a) Does the population have to be normally distributed to test this hypothesis by using the methods presented in this section?
    (b) If $\bar{x} = 48.3$ and $s = 8.5$, compute the test statistic.
    (c) Draw a t-distribution with the area that represents the P-value shaded.
    (d) Determine and interpret the P-value.
    (e) If the researcher decides to test this hypothesis at the $\alpha = 0.01$ level of significance, will the researcher reject the null hypothesis? Why? No

Applying the Concepts

13. Effects of Alcohol on the Brain In a study published in NW the *American Journal of Psychiatry* (157:737–744, May 2000), researchers wanted to measure the effect of alcohol on the development of the hippocampal region in adolescents. The hippocampus is the portion of the brain responsible for long-term memory storage. The researchers randomly selected 12 adolescents with alcohol use disorders. They wanted to determine whether the hippocampal volumes in the alcoholic adolescents were less than the normal volume of 9.02 cm^3. An analysis of the sample data revealed that the hippocampal volume is approximately normal with $\bar{x} = 8.10$ and $s = 0.7$. Conduct the appropriate test at the $\alpha = 0.01$ level of significance. Reject H_0

14. Effects of Plastic Resin Para-nonylphenol is found in polyvinyl chloride (PVC) used in the food processing and packaging industries. Researchers wanted to determine the effect this substance had on the organ weight of first-generation mice when both parents were exposed to 50 μg/L of para-nonylphenol in drinking water for 4 weeks. After 4 weeks, the mice were bred. After 100 days, the offspring of the exposed parents were sacrificed and the kidney weights were determined. The mean weight of the 12 offspring was found to be 396.9 mg with a standard deviation of 45.4 mg. Is there significant evidence to conclude that the kidney weight of the offspring whose parents were exposed to 50 μg/L of para-nonylphenol in drinking water for 4 weeks is greater than 355.7 mg, the mean weight of kidneys in normal 100-day old mice at the $\alpha = 0.05$ level of significance? Reject H_0
(*Source*: Vendula Kyselova et al., Effects of *p*-nonylphenol and resveratrol on body and organ weight and in vivo fertility of outbred CD-1 mice, *Reproductive Biology and Endocrinology*, 2003)

15. No, do not reject H_0

15. Got Milk? The U.S. Food and Drug Administration recommends that individuals consume 1000 mg of calcium daily. The International Dairy Foods Association (IDFA) sponsors an advertising campaign aimed at male teenagers. After the campaign, the IDFA obtained a random sample of 50 male teenagers and found that the mean amount of calcium consumed was 1081 mg, with a standard deviation of 426 mg. Conduct a test to determine if the campaign was effective. Use the $\alpha = 0.05$ level of significance.

16. Too Much Salt? A nutritionist believes that children under the age of 10 years are consuming more than the U.S. Food and Drug Administration's recommended daily allowance of sodium, which is 2400 mg. She obtains a random sample of 75 children under the age of 10 and measures their daily consumption of sodium. The mean amount of sodium consumed was determined to be 2993 mg, with a standard deviation of 1489 mg. Is there significant evidence to conclude that children under the age of 10 years are consuming too much sodium? Use the $\alpha = 0.05$ level of significance. Yes, reject H_0

17. Normal Temperature Carl Reinhold August Wunderlich said that the mean temperature of humans is 98.6°F. Researchers Philip Mackowiak, Steven Wasserman, and Myron Levine [*JAMA*, Sept. 23–30 1992; 268(12):1578–80] thought that the mean temperature of humans is less than 98.6°F. They measured the temperature of 148 healthy adults 1 to 4 times daily for 3 days, obtaining 700 measurements. The sample data resulted in a sample mean of 98.2°F and a sample standard deviation of 0.7°F.
(a) Test whether the mean temperature of humans is less than 98.6°F at the $\alpha = 0.01$ level of significance using the classical approach. $t = -15.119$; reject H_0
(b) Determine and interpret the *P*-value.

18. Normal Temperature Carl Reinhold August Wunderlich said that the mean temperature of humans is 98.6°F. Researchers Philip Mackowiak, Steven Wasserman, and Myron Levine [*JAMA*, Sept. 23–30 1992; 268(12):1578–80] measured the temperatures of 26 females 1 to 4 times daily for 3 days to get a total of 123 measurements. The sample data yielded a sample mean of 98.4°F and a sample standard deviation of 0.7°F.
(a) Using the classical approach, judge whether the normal temperature of women is less than 98.6°F at the $\alpha = 0.01$ level of significance. $t = -3.169$; reject H_0
(b) Determine and interpret the *P*-value. P-value = 0.001

19. Age of Death-Row Inmates In 2002, the mean age of an inmate on death row was 40.7 years, according to data obtained from the U.S. Department of Justice. A sociologist wondered whether the mean age of a death-row inmate has changed since then. She randomly selects 32 death-row inmates and finds that their mean age is 38.9, with a standard deviation of 9.6.
(a) Do you believe the mean age has changed? Use the $\alpha = 0.05$ level of significance. Do not reject H_0
(b) Construct a 95% confidence interval about the mean age. What does the interval imply?

20. Energy Consumption In 2001, the mean household expenditure for energy was $1493, according to data obtained from the U.S. Energy Information Administration. An economist wanted to know whether this amount has changed significantly from its 2001 level. In a random sample of 35 households, he found the mean expenditure (in 2001 dollars) for energy during the most recent year to be $1618, with standard deviation $321.
(a) Do you believe that the mean expenditure has changed significantly from the 2001 level at the $\alpha = 0.05$ level of significance? Reject H_0
(b) Construct a 95% confidence interval about the mean energy expenditure. What does the interval imply?

21. Conforming Golf Balls The United States Golf Associa-NW tion requires that golf balls have a diameter that is 1.68 inches. To determine if Maxfli XS golf balls conform to USGA standards, a random sample of Maxfli XS golf balls was selected. Their diameters are shown in the table.

1.683	1.677	1.681
1.685	1.678	1.686
1.684	1.684	1.673
1.685	1.682	1.674

Source: Michael McCraith, Joliet Junior College

(a) Because the sample size is small, the engineer must verify that the diameter is normally distributed and the sample does not contain any outliers. The normal probability plot and boxplot are shown. Are the conditions for testing the hypothesis satisfied? Yes

(21b) $t = 0.770$; do not reject

(b) Do the golf balls conform to USGA standards? Use the $\alpha = 0.05$ level of significance.

(c) Determine and interpret the P-value. P-value = 0.45

22. **Conforming Golf Balls** The USGA requires that golf balls have a weight that is less than 1.62 ounces. An engineer for the USGA wants to verify that Maxfli XS balls conform to USGA standards. He obtains a random sample of 12 Maxfli XS golf balls and obtains the weights in the table.

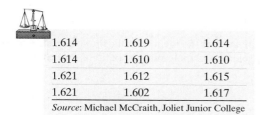

1.614	1.619	1.614
1.614	1.610	1.610
1.621	1.612	1.615
1.621	1.602	1.617

Source: Michael McCraith, Joliet Junior College

(a) Because the sample size is small, the engineer must verify that weight is normally distributed and that the sample does not contain any outliers. The normal probability plot and boxplot are shown. Are the conditions for testing the hypothesis satisfied? Yes

(b) Decide whether the golf balls meet Maxfli's standard at the $\alpha = 0.1$ level of significance. Reject H_0

23. **Waiting in Line** The mean waiting time at the drive-through of a fast-food restaurant from the time an order is placed to the time the order is received is 84.3 seconds. A manager devises a new drive-through system that he

believes will decrease wait time. He initiates the new system at his restaurant and measures the wait time for 10 randomly selected orders. The wait times are provided in the table.

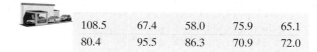

| 108.5 | 67.4 | 58.0 | 75.9 | 65.1 |
| 80.4 | 95.5 | 86.3 | 70.9 | 72.0 |

(a) Because the sample size is small, the manager must verify that wait time is normally distributed and the sample does not contain any outliers. The normal probability plot and boxplot are shown. Are the conditions for testing the hypothesis satisfied? Yes

(b) Is the new system effective? Use the $\alpha = 0.1$ level of significance.

24. **Calibrating a pH Meter** An engineer wants to measure the bias in a pH meter. She uses the meter to measure the pH in 14 neutral substances (pH = 7.0) and obtains the data shown in the table.

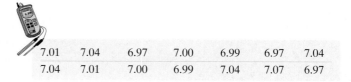

| 7.01 | 7.04 | 6.97 | 7.00 | 6.99 | 6.97 | 7.04 |
| 7.04 | 7.01 | 7.00 | 6.99 | 7.04 | 7.07 | 6.97 |

(a) Because the sample size is small, the engineer must verify that pH is normally distributed and the sample does not contain any outliers. The normal probability plot and boxplot are shown. Are the conditions for testing the hypothesis satisfied? Yes

(b) Is the pH meter correctly calibrated? Use the $\alpha = 0.05$ level of significance. Do not reject H_0

25. **P/E Ratio** A stock analyst believes that the price-to-earnings (P/E) ratio of companies listed on the Standard and Poor's 500 (S&P 500) Index is less than its December 1, 2000, level of 22.0, in response to economic uncertainty. The P/E ratio is the price an investor is willing to pay for $1 of earnings. For example, a P/E of 23 means the investor pays $23 for each $1 of earnings. A higher P/E is an indication of investor optimism. Lower P/Es are generally assigned to companies with lower earnings growth. To test his claim, he randomly samples 14 companies listed on the S&P 500 and calculates their P/E ratios. He obtains the following data:

Company	P/E Ratio	Company	P/E Ratio
Boeing	25.1	Dow Chemical	13.7
General Motors	7.8	Citigroup	18.5
Halliburton	35.8	Merck and Co.	26.8
Norfolk Southern	25.6	Sara Lee	11.9
Agilent Technologies	22.5	Harley-Davidson	37.4
Old Kent Financial	20.0	Circuit City	14.5
Cendent	15.0	Minnesota Mining and Manufacturing	23.8

Source: Checkfree Corporation

(a) Verify that P/E ratios are normally distributed and check for outliers by drawing a normal probability plot and boxplot.
(b) Determine whether the P/E ratio is below the December 1, 2000 level. Use the $\alpha = 0.05$ level of significance. *t = −0.30; do not reject H_0*
(c) Determine and interpret the *P*-value. *P-value = 0.39*

26. **Enlarged Prostate** Benign prostatic hyperplasia is a common cause of urinary outflow obstruction in aging males. The efficacy of Cardura (doxazosin mesylate) was measured in clinical trials of 173 patients with benign prostatic hyperplasia. Researchers wanted to discover whether Cardura significantly increased the urinary flow rate. It was found that an average increase of 0.8 mL/sec was obtained. This was said to be significant, with a *P*-value less than 0.01. State the null and alternative hypotheses of the researchers and interpret the *P*-value.

27. **Systolic Blood Pressure of Surgical Patients** A nursing student maintained that the mean systolic blood pressure of her male patients on the surgical floor was less than 130 mm Hg. She randomly selected 19 male surgical patients and collected the systolic blood pressures shown in the table.

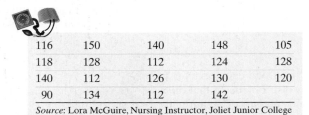

116	150	140	148	105
118	128	112	124	128
140	112	126	130	120
90	134	112	142	

Source: Lora McGuire, Nursing Instructor, Joliet Junior College

(a) Because the sample size is small, the student must verify that the systolic blood pressure is normally distributed and the sample does not contain any outliers. The normal probability plot and boxplot are shown. Are the conditions for testing the hypothesis satisfied? *Yes*

Systolic

(b) The student enters the data into MINITAB and obtains the following results:

T-Test of the Mean

```
Test of mu = 130.00 vs mu < 130.00

Variable        N     Mean   StDev   SE Mean      T       P
Systolic       19      125   15.47      3.55   -1.41   0.088
```

What are the null and alternative hypotheses? Identify the *P*-value. Will the nursing student reject the null hypothesis at the $\alpha = 0.05$ level of significance? State her conclusion. *H_0: $\mu = 130$ vs. H_1: $\mu < 130$; do not reject H_0*

28. **Temperature of Surgical Patients** A nursing student suspects that the mean temperature of surgical patients is above the normal temperature, 98.2°F. (See Problem 17.) She randomly selects 32 surgical patients and obtains the temperatures shown in the table.

97.4	98.6	98.2	98.2	98.4	98.6
99.8	97.7	97.8	98.9	97.8	97.8
96.7	97.8	98.3	98.5	98.0	98.7
96.8	98.6	98.4	97.4	99.1	98.7
97.8	99.2	99.2	98.1	98.6	98.4
99.2	98.6				

Source: Lora McGuire, Nursing Instructor, Joliet Junior College

(a) What are the null and alternative hypotheses of the student? $H_0: \mu = 98.2$ vs. $H_1: \mu > 98.2$
(b) The student enters the data into MINITAB and obtains the following results:

T-Test of the Mean

Test of mu = 98.200 vs mu > 98.200

Variable	N	Mean	StDev	SE Mean	T	P
Temperat	32	98.291	0.689	0.122	0.74	0.23

What is the *P*-value of the test? State the nursing student's conclusion. *P*-value = 0.23; do not reject H_0

---

29. **Soybean Yield** The mean yield per acre of soybeans on farms in the United States in 2003 was 33.5 bushels, according to data obtained from the U.S. Department of Agriculture. A farmer in Iowa claimed the yield was higher than the reported mean. He randomly sampled 35 acres on his farm and determined the mean yield to be 37.1 bushels, with a standard deviation of 2.5 bushels. He computed the *P*-value to be less than 0.0001 and concluded that the U.S. Department of Agriculture was wrong. Why should his conclusions be looked on with skepticism?

30. **Significance Test Applet** Load the hypothesis tests for a mean applet.
   (a) Set the shape to normal, the mean to 100, and the standard deviation to 15. These parameters describe the distribution of IQ scores. Obtain 1000 simple random samples of size $n = 10$ from this population, and test whether the mean is different from 100. How many samples led to a rejection of the null hypothesis if $\alpha = 0.05$? How many would we expect to lead to a rejection of the null hypothesis? For this level of significance, what is the probability of a Type I error?
   (b) Set the shape to normal, the mean to 100, and the standard deviation to 15. These parameters describe the distribution of IQ scores. Obtain 1000 simple random samples of size $n = 30$ from this population, and test whether the mean is different from 100. How many samples led to a rejection of the null hypothesis if $\alpha = 0.05$? How many would we expect to lead to a rejection of the null hypothesis? For this level of significance, what is the probability of a Type I error?
   (c) Compare the results of parts (a) and (b). Did the sample size have any impact on the number of samples that incorrectly rejected the null hypothesis?

31. **Significance Test Applet: Violating Assumptions** Load the hypothesis tests for a mean applet.
   (a) Set the shape to right skewed, the mean to 50, and the standard deviation to 10. Obtain 1000 simple random samples of size $n = 8$ from this population, and test whether the mean is different from 50. How many of the samples led to a rejection of the null hypothesis if $\alpha = 0.05$? How many would we expect to lead to a rejection of the null hypothesis if $\alpha = 0.05$? What might account for any discrepancies?
   (b) Set the shape to right skewed, the mean to 50, and the standard deviation to 10. Obtain 1000 simple random samples of size $n = 40$ from this population, and test whether the mean is different from 50. How many of the samples led to a rejection of the null hypothesis if $\alpha = 0.05$? How many would we expect to lead to a rejection of the null hypothesis if $\alpha = 0.05$?

32. **Simulation** Simulate drawing 40 simple random samples of size $n = 20$ from a population that is normally distribution with mean 50 and standard deviation 10.
   (a) Test the null hypothesis $H_0: \mu = 50$ versus the alternative hypothesis $H_1: \mu \neq 50$ for each of the 40 samples using a *t*-test.
   (b) Suppose we were testing this hypothesis at the $\alpha = 0.05$ level of significance. How many of the 40 samples would you expect to result in a Type I error?
   (c) Count the number of samples that lead to a rejection of the null hypothesis. Is it close to the expected value determined in part (b)?
   (d) Describe why we know a rejection of the null hypothesis results in making a Type I error in this situation.

Consumer Reports® **Eyeglass Lenses**

Eyeglasses are part medical device and part fashion statement, a marriage that has always made them a tough buy. Aside from the thousands of different frames the consumer has to choose from, various lens materials and coatings can add to the durability, and the cost, of a pair of eyeglasses. One manufacturer even goes so far as to claim that its lenses are "the most scratch-resistant plastic lenses ever made." With a claim like that, we had to test the lenses.

One test involved tumbling the lenses in a drum containing scrub pads of grit of varying size and hardness. Afterward, readings of the lenses' haze were taken on a spectrometer to determine how scratched they had become. To evaluate their scratch resistance, we measured the difference between the haze reading before and after tumbling.

The photo illustrates the difference between an uncoated lens (on the left) and the manufacturer's "scratch-resistant" lens (on the right).

The following table contains the haze measurements both before and after the scratch resistance test for this manufacturer. Haze difference is measured by subtracting the before score from the after score. In other words, haze difference is computed as After–Before.

(a) Suppose it is known that the closest competitor to the manufacturer's lens has a mean haze difference of 1.0. Do the data support the manufacturer's scratch resistance claim?

(b) Write the null and alternative hypotheses, letting μ_{hdiff} represent the mean haze difference for the manufacturer's lens.

(c) We used MINITAB to perform a one-sample t-test. The results are shown below.

Using the MINITAB output, answer the following questions:

1. What is the value of the test statistic?

2. What is the P-value of the test?

3. What is the conclusion of this test? Write a paragraph for the readers of *Consumer Reports* magazine that explains your findings.

Note to Readers: In many cases, our test protocol and analytical methods are more complicated than described in these examples. The data and discussions have been modified to make the material more appropriate for the audience.

Before	After	Difference
0.18	0.72	0.54
0.16	0.85	0.69
0.20	0.71	0.51
0.17	0.42	0.25
0.21	0.76	0.55
0.21	0.51	0.30

One-Sample T: Difference

```
Test of mu = 1 vs mu < 1

Variable          N              Mean      StDev    SE Mean
Difference        6           0.4733     0.1665     0.0680

Variable     95.0%     Upper Bound          T          P
Difference             0.6103             -7.75      0.000
```

© 2001 by Consumers Union of U.S., Inc., Yonkers, NY 10703-1057, a nonprofit organization. Reprinted with permission from the June 2001 issue of CONSUMER REPORTS® for educational purposes only. No commercial use or photocopying permitted. To learn more about Consumers Union, log onto www.ConsumerReports.org.

Technology Step by Step

Hypothesis Tests Regarding μ, σ Unknown

TI-83/84 Plus

Step 1: If necessary, enter raw data in L1.

Step 2: Press STAT, highlight TESTS, and select `2:T-Test`.

Step 3: If the data are raw, highlight DATA; make sure that List1 is set to L1 and Freq is set to 1. If summary statistics are known, highlight STATS and enter the summary statistics. For the value of μ_0, enter the value of the mean stated in the null hypothesis.

Step 4: Select the direction of the alternative hypothesis.

Step 5: Highlight **Calculate** and press ENTER. The TI-83/84 gives the *P*-value.

MINITAB

Step 1: Enter raw data in column C1.

Step 2: Select the **Stat** menu, highlight **Basic Statistics**, then highlight **1-Sample t . . . .**

Step 3: Enter C1 in the cell marked "samples in columns." Enter the value of the mean stated in the null hypothesis in the cell marked "Test mean:" Click Options. In the cell marked "Alternative," select the direction of the alternative hypothesis. Click OK twice.

Excel

Step 1: If necessary, enter raw data in column A.

Step 2: Load the PHStat Add-in.

Step 3: Select the **PHStat menu**, highlight **One Sample Tests . . .,** and then highlight **t Test for the mean, sigma known . . . .**

Step 4: Enter the value of the null hypothesis and the level of significance, α. If the summary statistics are known, click "Sample statistics known" and enter the sample size, sample mean, and sample standard deviation. If summary statistics are unknown, click "Sample statistics unknown." With the cursor in the "Sample cell range" cell, highlight the data in column A. Click the option corresponding to the desired test [two-tail, upper (right) tail, or lower (left) tail]. Click OK.

10.4 Hypothesis Tests for a Population Proportion

Preparing for This Section   Before getting started, review the following:

- Binomial probability distribution (Section 6.2, pp. 298–309)
- Distribution of the sample proportion (Section 8.2, pp. 392–397)

Objective   ① **Test hypotheses about a population proportion**

① ## Test Hypotheses about a Population Proportion

Note to Instructor

You may want to use the hypothesis tests for a proportion applet to illustrate some of the concepts of this section.

Recall that the best point estimate of p, the proportion of the population with a certain characteristic, is given by

$$\hat{p} = \frac{x}{n}$$

where x is the number of individuals in the sample with the specified characteristic and n is the sample size. Recall from Section 8.2 that the sampling distribution of $\hat{p}$ is approximately normal, with mean $\mu_{\hat{p}} = p$ and standard deviation $\sigma_{\hat{p}} = \sqrt{\frac{p(1 - p)}{n}}$, provided that the following requirements are satisfied.

- The sample is a simple random sample.
- $np(1 - p) \geq 10$.
- $n \leq 0.05N$ (that is, the sample size is no more than 5% of the population size).

Testing hypotheses about the population proportion, p, follows the same logic as the testing of hypotheses about a population mean with σ known. The only difference is that the **test statistic** is

$$z_0 = \frac{\hat{p} - p_0}{\sqrt{\dfrac{p_0(1 - p_0)}{n}}}$$

> **! CAUTION**
> When determining the standard error for the sampling distribution of $\hat{p}$ for hypothesis testing, use the assumed value of the population proportion, p_0.

where p_0 is the value of the population proportion stated in the null hypothesis.

Notice that we are using p_0 in computing the standard error rather than $\hat{p}$ (as we did in computing confidence intervals about p). This is because, when we test a hypothesis, the null hypothesis is always assumed true. Therefore, we are assuming that the population proportion is p_0.

Testing Hypotheses Regarding a Population Proportion, p

To test hypotheses regarding the population proportion, we can use the following steps, provided that

1. The sample is obtained by simple random sampling.
2. $np_0(1 - p_0) \geq 10$ with $n \leq 0.05N$ (the sample size, n, is no more than 5% of the population size, N).

Step 1: Determine the null and alternative hypotheses. The hypotheses can be structured in one of three ways:

Two-Tailed	Left-Tailed	Right-Tailed
$H_0: p = p_0$	$H_0: p = p_0$	$H_0: p = p_0$
$H_1: p \neq p_0$	$H_1: p < p_0$	$H_1: p > p_0$

Note: p_0 is the assumed value of the population proportion.

Step 2: Select a level of significance α, depending on the seriousness of making a Type I error.

Step 3: Compute the test statistic

$$z_0 = \frac{\hat{p} - p_0}{\sqrt{\dfrac{p_0(1 - p_0)}{n}}}$$

Classical Approach

Step 4: Use Table IV to determine the critical value.

P-Value Approach

Step 4: Use Table IV to determine the P-value.

Step 5: Compare the critical value with the test statistic.

Two-Tailed	Left-Tailed	Right-Tailed
If $z_0 < -z_{\frac{\alpha}{2}}$ or $z_0 > z_{\frac{\alpha}{2}}$, reject the null hypothesis.	If $z_0 < -z_\alpha$, reject the null hypothesis.	If $z_0 > z_\alpha$, reject the null hypothesis.

Step 5: If P-value $< \alpha$, reject the null hypothesis.

Step 6: State the conclusion.

EXAMPLE 1

Testing Hypotheses about a Population Proportion: Right-Tailed Test

Problem: In 2004, 65% of adult Americans thought that the death penalty was morally acceptable. In a poll conducted by the Gallup Organization May 2–5, 2005, a simple random sample of 1005 adult Americans resulted in 704 respondents stating that they believe the death penalty was morally acceptable when asked, "Do you believe the death penalty is morally acceptable or morally wrong?" The choices "morally acceptable" and "morally wrong" were randomly interchanged for each interview. Is there significant evidence to indicate that the proportion of adult Americans who believe that the death penalty is morally acceptable has increased from the level reported in 2004 at the $\alpha = 0.05$ level of significance?

! **CAUTION**
Always verify the requirements before conducting a hypothesis test.

Approach: We must verify the requirements to perform the hypothesis test; that is, the sample must be a simple random sample, $np_0(1 - p_0) \geq 10$, and the sample size cannot be more than 5% of the population size. Then we follow Steps 1 through 6.

Solution: We want to know if the proportion of adult Americans who believe that the death penalty is morally acceptable has increased since 2004. This can be written $p > 0.65$. The sample is a simple random sample. Also, $np_0(1 - p_0) = (1005)(0.65)(1 - 0.65) = 228.6 > 10$. There are over 200 million adult Americans, so the sample size is less than 5% of the population size. The requirements are satisfied, so we now proceed to Steps 1 through 6.

Step 1: We want to show that $p > 0.65$. We assume "no difference" between opinions in 2004 and 2005, so we have

$$H_0: p = 0.65 \quad \text{versus} \quad H_1: p > 0.65$$

Step 2: The level of significance is $\alpha = 0.05$.

Step 3: The assumed value of the population proportion is $p_0 = 0.65$. The point estimate of the population proportion is $\hat{p} = \dfrac{x}{n} = \dfrac{704}{1005} = 0.70$. The test statistic is

$$z_0 = \frac{\hat{p} - p_0}{\sqrt{\dfrac{p_0(1 - p_0)}{n}}} = \frac{0.70 - 0.65}{\sqrt{\dfrac{0.65(1 - 0.65)}{1005}}} = 3.32$$

Classical Approach

Step 4: Because this is a right-tailed test, we determine the critical value at the $\alpha = 0.05$ level of significance to be $z_{0.05} = 1.645$. The critical region is displayed in Figure 20.

Figure 20

$z_{0.05} = 1.645$ — Critical Region — 3.32

Step 5: The test statistic is $z_0 = 3.32$. We label this point in Figure 20. Because the test statistic is greater than the critical value ($3.32 > 1.645$), we reject the null hypothesis.

P-Value Approach

Step 4: Because this is a right-tailed test, the *P*-value is the area under the standard normal distribution to the right of the test statistic, $z_0 = 3.32$, as shown in Figure 21. That is, *P*-value $= P(Z > z_0) = P(Z > 3.32) = 1 - P(Z \le 3.32) = 1 - 0.9995 = 0.0005$.

Figure 21

P-value $= P(Z > 3.32)$

3.32

Step 5: Because the *P*-value is less than the level of significance $\alpha = 0.05$ ($0.0005 < 0.05$), we reject the null hypothesis.

Step 6: There is sufficient evidence to conclude that the proportion of adult Americans who believe that the death penalty is morally acceptable has increased since 2004 at the $\alpha = 0.05$ level of significance.

Now Work Problem 9.

EXAMPLE 2

Testing Hypotheses about a Population Proportion, Two-Tailed Test

Problem: The drug Prevnar is a vaccine meant to prevent meningitis. (It also helps control ear infections.) It is typically administered to infants. In clinical trials, the vaccine was administered to 710 randomly sampled infants between 12 and 15 months of age. Of the 710 infants, 121 experienced a loss of appetite. Is there significant evidence to conclude that the proportion of infants who receive Prevnar and experience loss of appetite is different from 0.135, the proportion of children who experience a loss of appetite with competing medications at the $\alpha = 0.01$ level of significance?

Approach: We must verify the requirements to perform the hypothesis test: the sample must be a simple random sample with $np_0(1 - p_0) \ge 10$, and the sample size cannot be more than 5% of the population. Then we follow Steps 1 through 6, as laid out previously.

Solution: We want to know if the proportion of infants who experience a loss of appetite is different from 0.135; that is, $p \ne 0.135$. The sample is a simple random sample. Also, $np_0(1 - p_0) = (710)(0.135)(1 - 0.135) = 82.9 > 10$. Because there are about 1 million babies between 12 and 15 months of age, the sample size is less than 5% of the population size. The requirements are satisfied, so we now proceed to follow Steps 1 through 6.

Step 1: We want to know whether $p \neq 0.135$. This is a two-tailed test.

$$H_0: p = 0.135 \quad \text{versus} \quad H_1: p \neq 0.135$$

Step 2: The level of significance is $\alpha = 0.01$.

Step 3: The assumed value of the population proportion is $p_0 = 0.135$. The point estimate of the population proportion is $\hat{p} = \dfrac{x}{n} = \dfrac{121}{710} = 0.170$. The test statistic is

$$z_0 = \frac{\hat{p} - p_0}{\sqrt{\dfrac{p_0(1 - p_0)}{n}}} = \frac{0.170 - 0.135}{\sqrt{\dfrac{0.135(1 - 0.135)}{710}}} = 2.73$$

Classical Approach

Step 4: Because this is a two-tailed test, we determine the critical values at the $\alpha = 0.01$ level of significance to be $-z_{0.01/2} = -z_{0.005} = -2.575$ and $z_{0.01/2} = z_{0.005} = 2.575$. The critical regions are displayed in Figure 22.

Figure 22

Step 5: The test statistic is $z_0 = 2.73$. We label this point in Figure 22. Because the test statistic is greater than the critical value ($2.73 > 2.575$), we reject the null hypothesis.

P-Value Approach

Step 4: Because this is a two-tailed test, the P-value is the area under the standard normal distribution to the left of $-z_0 = -2.73$ and to the right of $z_0 = 2.73$ as shown in Figure 23. That is, $P\text{-value} = P(Z < -|z_0|) + P(Z > |z_0|) = 2P(Z < -2.73) = 2(0.0032) = 0.0064$.

Figure 23

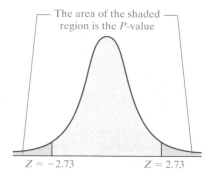

Step 5: Because the P-value is less than the level of significance $\alpha = 0.01$ ($0.0064 < 0.01$), we reject the null hypothesis.

Step 6: There is sufficient evidence to conclude that the proportion of infants who experienced a loss of appetite when receiving Prevnar is different from 0.135 at the $\alpha = 0.01$ level of significance.

Now Work Problem 15.

EXAMPLE 3

Testing Hypotheses Regarding a Population Proportion Using Technology

Problem: Test the hypotheses presented in Example 2 by obtaining the P-value using statistical software or a graphing calculator with advanced statistical features.

Approach: We will use MINITAB to obtain the P-value. The steps for testing hypotheses regarding a population proportion using the TI-83/84 Plus graphing calculator, MINITAB, and Excel are given in the Technology Step by Step on page 500.

Result: Figure 24 shows the results using MINITAB. The P-value is 0.006. Because the P-value is less than the level of significance ($0.006 < 0.01$), we reject the null hypothesis.

Figure 24 **Test and CI for One Proportion**

```
Test of p = 0.135 vs p not = 0.135

Sample    X     N    Sample      99% CI           Z-Value   P-Value
1        121   710   0.170423   (0.134075, 0.206770)   2.76    0.006
```

In-Class Activity (Hypothesis Testing): Taste the Rainbow

The advertising campaign for Skittles® Brand candy in 2005 said to "Taste the Rainbow!". While the original candies do not have all the colors of the rainbow, they do come in red, orange, yellow, green, and purple (violet). But are the proportions of each color the same? If so, the proportion of each color would be $p = 0.2$.

Note to Instructor

The purpose of this activity is to give students a hands-on experience with hypothesis testing involving proportions. The activity could be elaborated to include review questions involving binomial or normal probabilities. Time limit: 40 to 45 minutes.

(a) Obtain a 1-pound bag of Skittles (original flavor).
(b) Select one of the five original colors. Count the total number of candies in your bag, as well as the number for the color you selected.
(c) Is $np_0(1 - p_0) \geq 10$? If not, what could you do?
(d) Determine whether the proportion of the selected color is different from $p = 0.2$.
(e) Compare your results to others in the class. Did everyone arrive at the same conclusion for the same color? What about different colors?

10.4 ASSESS YOUR UNDERSTANDING

Concepts and Vocabulary

1. State the assumptions required to test a hypothesis about a population proportion.
2. A poll conducted by CNN, *USA Today*, and Gallup reported the following results: "According to the most recent CNN/*USA Today*/Gallup poll, conducted June 28–July 1, a majority of Americans (52%) approve of the job Bush is doing as president." The poll results were obtained by conducting simple random sample of 1014 adults aged 18 years old or older, with a margin of error of ±3 percentage points. State what is wrong with the conclusions presented by the pollsters.

Skill Building

In Problems 3–8, test the hypothesis, using (a) the classical approach and then (b) the P-value approach. Be sure to verify the requirements of the test.

3. H_0: $p = 0.3$ versus H_1: $p > 0.3$
   $n = 200$; $x = 75$; $\alpha = 0.05$ P-value = 0.0104

4. H_0: $p = 0.6$ versus H_1: $p < 0.6$
   $n = 250$; $x = 124$; $\alpha = 0.01$ P-value = 0.0004

5. H_0: $p = 0.55$ versus H_1: $p < 0.55$
   $n = 150$; $x = 78$; $\alpha = 0.1$ P-value = 0.2296

6. H_0: $p = 0.25$ versus H_1: $p < 0.25$
   $n = 400$; $x = 96$; $\alpha = 0.1$ P-value = 0.3228

7. H_0: $p = 0.9$ versus H_1: $p \neq 0.9$
   $n = 500$; $x = 440$; $\alpha = 0.05$ P-value = 0.1362

8. H_0: $p = 0.4$ versus H_1: $p \neq 0.4$
   $n = 1000$; $x = 420$; $\alpha = 0.01$ P-value = 0.1970

Applying the Concepts

10. Do not reject H_0; H_0: $p = 0.94$, H_1: $p > 0.94$, $Z = 0.69$, P-value = 0.2451
11. No; H_0: $p = 0.81$, H_1: $p \neq 0.81$, $Z = 1.62$, P-value = 0.1052
12. Yes; H_0: $p = 0.33$, H_1: $p > 0.33$, $Z = 2.63$, P-value = 0.0043

9. Lipitor The drug Lipitor is meant to reduce total choles-
NW terol and LDL cholesterol. In clinical trials, 19 out of 863
patients taking 10 mg of Lipitor daily complained of flulike
symptoms. Suppose that it is known that 1.9% of patients
taking competing drugs complain of flulike symptoms. Is
there sufficient evidence to conclude that more than 1.9%
of Lipitor users experience flulike symptoms as a side effect at the $\alpha = 0.01$ level of signifi-cance?

H_0: $p = 0.019$ vs. H_1: $p > 0.019$; P-value = 0.2578

10. Nexium Nexium is a drug that can be used to reduce the
acid produced by the body and heal damage to the esoph-
agus due to acid reflux. Suppose the manufacturer of Nexium claims that more than 94% of patients taking Nexium
are healed within 8 weeks. In clinical trials, 213 of 224 patients suffering from acid reflux disease were healed after
8 weeks. Test the manufacturer's claim at the $\alpha = 0.01$
level of significance.

11. Americans Reading In a Gallup Poll conducted May
20–22, 2005, 835 of 1006 adults aged 18 or older said
they had read at least one book during the previous
year. In December 1990, 81% of adults aged 18 or older
had read at least one book during the previous year. Is
there sufficient evidence to conclude that the percent of
adults who have read at least one book in the last year
is different from 1990 at the $\alpha = 0.05$ significance level?

12. Haunted Houses In September 1996, 33% of adult Americans believed in haunted houses. In a Gallup Poll conducted June 6–8, 2005, 370 of 1002 adult Americans aged
18 or older believed in haunted houses. Is there sufficient
evidence to conclude that the proportion of adult Americans who believe in haunted houses has increased at the
$\alpha = 0.05$ level of significance?

13. Eating Together In December 2001, 38% of adults with
children under the age of 18 reported that their family ate
dinner together 7 nights a week. Suppose in a recent poll,
337 of 1122 adults with children under the age of 18 reported that their family ate dinner together 7 nights a
week. Has the proportion of families with children under
the age of 18 who eat dinner together 7 nights a week decreased? Use the $\alpha = 0.05$ significance level.

14. Critical Job Skills In August 2003, 56% of employed
adults in the United States reported that basic mathematical skills were critical or very important to their job.
The supervisor of the job placement office at a 4-year
college thinks this percentage has increased due to increased use of technology in the workplace. He takes a
random sample of 480 employed adults and finds that
297 of them feel that basic mathematical skills are critical or very important to their job. Has the percentage of
employed adults who feel basic mathematical skills are
critical or very important to their job increased? Use the
$\alpha = 0.05$ level of significance.

15. Chance for Promotion In October 1998, 30% of em-
NW ployed adults were satisfied with their chances for promotion. A human resource manager wants to determine
if this percentage has changed significantly since then.
She randomly selects 280 employed adults and finds that
112 of them are completely satisfied with their chances
for promotion. Is the percentage of employed adults satisfied with their chances for promotion changed significantly from the percentage in 1998, at the $\alpha = 0.1$ level
of significance?

16. Living Alone? In 2000, 58% of females aged 15 years of
age and older lived alone, according to the U.S. Census
Bureau. A sociologist wants to know if this percentage is
different today, so she obtains a random sample of 500
females aged 15 years of age and older and finds that 285
are living alone. Is there evidence at the $\alpha = 0.1$ level of
significance to support belief in a change?

17. Confidence in Schools In 1995, 40% of adults aged 18 years
or older reported that they had "a great deal" of confidence
in the public schools. On June 1, 2005, the Gallup Organization released results of a poll in which 372 of 1004 adults
aged 18 years or older stated that they had "a great deal" of
confidence in the public schools. Does the evidence suggest
at the $\alpha = 0.05$ level of significance that the proportion of
adults aged 18 years or older having "a great deal" of confidence in the public schools is significantly lower in 2005 than
the 1995 proportion?

18. Pathological Gamblers Pathological gambling is an im-
pulse-control disorder. The American Psychiatric Association lists 10 characteristics that indicate the disorder in
its DSM-IV manual. The National Gambling Impact
Study Commission randomly selected 2417 adults and
found that 35 were pathological gamblers. Is there evidence to conclude that more than 1% of the adult population are pathological gamblers at the $\alpha = 0.05$ level of
significance?

19. Talk to the Animals In a survey conducted by the
American Animal Hospital Association, 37% of respondents stated that they talk to their pets on the answering machine or telephone. A veterinarian found
this result hard to believe, so he randomly selected 150
pet owners and discovered that 54 of them spoke to
their pet on the answering machine or telephone. Does
the veterinarian have a right to be skeptical? Use the
$\alpha = 0.05$ level of significance.

20. Eating Salad According to a survey conducted by the Association for Dressings and Sauces (this is an actual association!), 85% of American adults eat salad at least once a
week. A nutritionist suspects that the percentage is higher
than this. She conducts a survey of 200 American adults
and finds that 171 of them eat salad at least once a week.
Conduct the appropriate test that addresses the nutritionist's suspicions. Use the $\alpha = 0.10$ level of significance.

13. Yes; H_0: $p = 0.38$, H_1: $p < 0.38$, $Z = -5.52$, P-value < 0.0001 14. Yes; H_0: $p = 0.56$, H_1: $p > 0.56$, $Z = 2.60$,
P-value = 0.0047   15. Yes; H_0: $p = 0.3$, H_1: $p \neq 0.3$, $Z = 3.65$, P-value < 0.0002
16. No; H_0: $p = 0.58$, H_1: $p \neq 0.58$, $Z = -0.45$, P-value = 0.6528 17. Yes; H_0: $p = 0.4$, H_1: $p < 0.4$, $Z = -1.88$,
P-value = 0.0301   18. Yes; H_0: $p = 0.01$ vs. H_1: $p > 0.01$; P-value = 0.0132
19. H_0: $p = 0.37$ vs. H_1: $p < 0.37$; P-value = 0.4013 20. H_0: $p = 0.85$ vs. H_1: $p > 0.85$; P-value = 0.4207

21. Statistics in the Media One of the more popular statistics reported in the media is the president's job approval rating. The approval rating is reported as the proportion of Americans who approve of the job that the sitting president is doing and is typically based on a random sample of about 1100 Americans.

(a) This proportion tends to fluctuate from week to week. Name some reasons for the fluctuation in the statistic.

(b) A recent article had the headline "Bush Ratings Show Decline." This headline was written because an April poll showed President Bush's approval rating to be 0.48 (48%). A poll in June based on 1100 randomly selected Americans showed that 506 approved of the job Bush was doing. Do the results of the June poll indicate that the proportion of Americans who approve of the job Bush is doing is significantly less than April's level? Explain.

22. Statistics in the Media In May 2002, 71% (0.71) of Americans favored the death penalty for a person convicted of murder. In May 2005, 1005 adult Americans were asked by the Gallup Organization, "Are you in favor of the death penalty for a person convicted of murder?" Of the 1005 adults surveyed, 744 responded yes. The headline in the article reporting the survey results stated, "Americans' Views of Death Penalty More Positive This Year." Use a test of significance to support or refute this headline.

Technology Step by Step	**Hypothesis Tests Regarding a Population Proportion**

TI-83/84 Plus

Step 1: Press STAT, highlight TESTS, and select `5:1-PropZTest`.

Step 2: For the value of p_0, enter the "status quo" value of the population proportion.

Step 3: Enter the number of successes, x, and the sample size, n.

Step 4: Select the direction of the alternative hypothesis.

Step 5: Highlight Calculate or Draw, and press ENTER. The TI-83 or TI-84 gives the *P*-value.

MINITAB

Step 1: Select the **Stat** menu, highlight **Basic Statistics**, then highlight **1-Proportion**.

Step 2: Select "Summarized data."

Step 3: Enter the number of trials, n, and the number of successes, x.

Step 4: Click Options. Enter the "status quo" value of the population proportion in the cell "Test proportion." Enter the direction of the alternative hypothesis. If $np_0(1 - p_0) \geq 10$, check the box marked "Use test and interval based on normal distribution." Click OK twice.

Excel

Step 1: Load the PHStat Add-in.

Step 2: Select the **PHStat menu**, highlight **One Sample Tests . . .**, then highlight **Z Test for proportion**.

Step 3: Enter the value of the null hypothesis, the level of significance, α, the number of successes, x, and the number of trials, n. Click the option corresponding to the desired test [two-tail, upper (right) tail, or lower (left) tail]. Click OK.

10.5 Putting It All Together: Which Method Do I Use?

Objective ① Determine the appropriate hypothesis test to perform

① Determine the Appropriate Hypothesis Test to Perform

Perhaps the most difficult aspect of testing hypotheses is determining which hypothesis test to conduct. To assist in the decision making, we present Figure 25 which shows which approach to take in testing hypotheses for the three parameters discussed in this chapter.

Figure 25

Provided $np_0(1 - p_0) \geq 10$ and the sample size is no more than 5% of the population size, use the normal distribution with

$$z_0 = \frac{\dot{p} - p_0}{\sqrt{\frac{p_0(1 - p_0)}{n}}} \text{ where } \dot{p} = \frac{x}{n}$$

Section 10.4

Proportion, p

What parameter is addressed in the claim?   Mean,   **Is σ known?**

Yes

Provided the sample size is greater than 30 or the data came from a population that is normally distributed, use the normal distribution with

$$z_0 = \frac{\bar{x} - \mu_0}{\frac{\sigma}{\sqrt{n}}}$$

Section 10.2

No

Provided the sample size is greater than 30 or the data came from a population that is normally distributed, use Student's t-distribution with $n - 1$ degrees of freedom with

$$t = \frac{\bar{x} - \mu_0}{\frac{s}{\sqrt{n}}}$$

Section 10.3

3. $H_0: \mu = 70$, $H_1: \mu < 70$, $Z = -1.87$, *P*-value = 0.0307; reject H_0
4. *P*-value = 0.0214; do not reject H_0
5. $H_0: p = 0.5$, $H_1: p > 0.5$, $Z = 2.12$, *P*-value = 0.0169; reject H_0
6. $H_0: \mu = 600$, $H_1: \mu \neq 600$, $t = -1.186$, *P*-value = 0.2401; do not reject H_0
7. $H_0: \mu = 25$, $H_1: \mu \neq 25$, $t = -0.738$, *P*-value = 0.4729; do not reject H_0

10.5 ASSESS YOUR UNDERSTANDING

Concepts and Vocabulary

1. What are the requirements that must be satisfied to test a hypothesis about a population mean? When do we use the normal model to test a hypothesis about a population mean? When do we use Student's *t*-distribution to test a hypothesis about a population mean?

2. What are the requirements that must be satisfied before we can test a hypothesis about a population proportion?

Skill Building

8. $H_0: \mu = 55$, $H_1: \mu \neq 55$, $Z = -1.29$, *P*-value = 0.1970; do not reject H_0
9. $H_0: \mu = 100$, $H_1: \mu > 100$, $t = 3.003$, *P*-value = 0.0023; reject H_0
10. $H_0: p = 0.25$, $H_1: p < 0.25$, $Z = -2.85$, *P*-value = 0.0022; reject H_0

In Problems 3–12, conduct the appropriate test.

3. A simple random sample of size $n = 14$ is drawn from a population that is normally distributed with $\sigma = 20$. The sample mean is found to be $\bar{x} = 60$. Test whether the population mean is less than 70 at the $\alpha = 0.1$ level of significance.

4. A simple random sample of size $n = 19$ is drawn from a population that is normally distributed. The sample mean is found to be 0.8, and the sample standard deviation is found to be 0.4. Test whether the population mean is less than 1.0 at the $\alpha = 0.01$ level of significance.

5. A simple random sample of size $n = 200$ individuals with a valid driver's license is asked if they drive an American-made automobile. Of the 200 individuals surveyed, 115 responded that they drive an American-made automobile. Test whether more than half of those with a valid driver's license drive an American-made automobile at the $\alpha = 0.05$ level of significance.

6. A simple random sample of size $n = 65$ is drawn from a population. The sample mean is found to be 583.1, and the sample standard deviation is found to be 114.9. Test the claim that the population mean is different from 600 at the $\alpha = 0.1$ level of significance.

7. A simple random sample of size $n = 15$ is drawn from a population that is normally distributed. The sample mean is found to be 23.8, and the sample standard deviation is found to be 6.3. Test whether the population mean is different from 25 at the $\alpha = 0.01$ level of significance.

8. A simple random sample of size $n = 25$ is drawn from a population that is normally distributed with $\sigma = 7$. The sample mean is found to be 53.2. Test whether the population mean is different from 55 at the $\alpha = 0.05$ level of significance.

9. A simple random sample of size $n = 40$ is drawn from a population. The sample mean is found to be 108.5, and the sample standard deviation is found to be 17.9. Test the claim that the population mean is greater than 100 at the $\alpha = 0.05$ level of significance.

10. A simple random sample of size $n = 320$ adults was asked their favorite ice cream flavor. Of the 320 individuals surveyed, 58 responded that they preferred mint chocolate chip. Test the claim that less than 25% of adults prefer mint chocolate chip ice cream at the $\alpha = 0.01$ level of significance.

(11a) H_0: $\mu = 2.5$, H_1: $\mu \neq 2.5$, $Z = 2.64$, P-value = 0.0082; reject H_0
12. H_0: $p = 0.66$, H_1: $p > 0.66$, $Z = 7.59$, P-value < 0.0001; reject H_0

Applying the Concepts

11. **Family Size** In 1985, the mean for the ideal number of children for a family was considered to be 2.5. A Gallup Poll of 1006 adults aged 18 or older conducted February 16–17, 2004, reported the mean for the ideal number of children to be 2.6.

    (a) Assuming $\sigma = 1.2$, is there sufficient evidence to conclude that the mean ideal number of children has changed since 1985 at the $\alpha = 0.1$ level of significance?

    (b) Do the results have any practical significance? Explain.

12. **Americans Online** In spring 2002, 66% of adults in the United States aged 18 years or older had Internet access. A Harris Interactive poll in February and May of 2005 found that 1496 of 2022 adults surveyed had Internet access. Is this enough evidence to conclude that more adults had Internet access in 2005 than in 2002 at the $\alpha = 0.05$ level of significance?

13. **Tattoos** In 2001, 23% of American university undergraduate students had at least one tattoo. A health practitioner obtains a random sample of 1026 university undergraduates and finds that 254 have at least one tattoo. Has the proportion of American university undergraduate students with at least one tattoo changed since 2001? Use the $\alpha = 0.1$ level of significance.

14. **Mortgage Rates** In 2001, the mean contract interest rate for a conventional 30-year first loan for the purchase of a single-family home was 6.3 percent, according to the U.S. Federal Housing Board. A real estate agent believes that interest rates are lower today and obtains a random sample of 41 recent 30-year conventional loans. The mean interest rate was found to be 6.05 percent, with a standard deviation of 1.75 percent. Is this enough evidence to conclude that interest rates are lower at the $\alpha = 0.05$ level of significance?

15. **Auto Insurance** According to the Insurance Information Institute, the mean expenditure for auto insurance in the United States was $774 for 2002. An insurance salesman obtains a random sample of 35 auto insurance policies and determines the mean expenditure to be $735 with a standard deviation of $48.31. Is there enough evidence to conclude that the mean expenditure for auto insurance is different from the 2002 amount at the $\alpha = 0.01$ level of significance?

16. **Toner Cartridge** The manufacturer of a toner cartridge claims the mean number of printouts is 10,000 for each cartridge. A consumer advocate believes the actual mean number of printouts is lower. He selects a random sample of 14 such cartridges and obtains the following number of printouts:

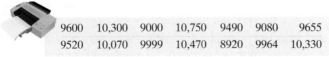

| 9600 | 10,300 | 9000 | 10,750 | 9490 | 9080 | 9655 |
| 9520 | 10,070 | 9999 | 10,470 | 8920 | 9964 | 10,330 |

(a) Because the sample size is small, he must verify that the number of printouts is normally distributed and the sample does not contain any outliers. The normal probability plot and boxplot are shown. Are the conditions for testing the hypothesis satisfied? Yes

(b) Are the consumer advocate's beliefs justified? Use the $\alpha = 0.05$ level of significance.

17. **Vehicle Emission Inspection** A certain vehicle emission inspection station states that the mean wait time for customers is less than 8 minutes. A local resident is skeptical and collects a random sample of 49 wait times for customers at the testing station. He finds that the sample mean is 7.34 minutes, with a standard deviation of 3.2 minutes. Is the resident's skepticism justified? Use the $\alpha = 0.01$ level of significance.

18. **Lights Out** With a previous contractor, the mean time to replace a streetlight was 3.2 days. A city councilwoman thinks that the new contractor is not getting the streetlights replaced as quickly. She selects a random sample of 12 streetlight service calls and obtains the following times to replacement (in days).

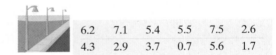

| 6.2 | 7.1 | 5.4 | 5.5 | 7.5 | 2.6 |
| 4.3 | 2.9 | 3.7 | 0.7 | 5.6 | 1.7 |

(a) Because the sample size is small, she must verify that replacement time is normally distributed and the sample does not contain any outliers. The normal probability plot and boxplot are shown below. Are the conditions for testing the hypothesis satisfied? Yes

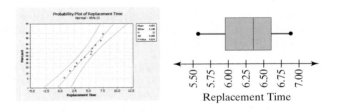

(b) Is there evidence that the contractor is not performing? Use the $\alpha = 0.05$ level of significance.
H_0: $\mu = 3.2$, H_1: $\mu > 3.2$, $t = 1.989$, P-value = 0.0361; reject H_0

13. H_0: $p = 0.23$, H_1: $p \neq 0.23$, $Z = 1.37$, P-value = 0.1706; do not reject H_0
14. H_0: $\mu = 6.3$, H_1: $\mu < 6.3$, $t_0 = -0.91$, P-value = 0.1829; do not reject H_0 15. H_0: $\mu = \$774$, H_1: $\mu \neq \$774$, $t = -4.776$, P-value < 0.0001; reject H_0 (16b) H_0: $\mu = 10,000$, H_1: $\mu < 10,000$, $t = -1.343$, P-value = 0.1012, do not reject H_0
17. H_0: $\mu = 8$ min, H_1: $\mu < 8$ min, $t = -1.444$, P-value = 0.0777; do not reject H_0

CHAPTER **10** **Review**

Summary

In this chapter, we discussed hypothesis testing. A statement is made regarding a population parameter, which leads to a null and an alternative hypothesis. The null hypothesis is assumed true. Given sample data, we either reject or do not reject the null hypothesis. In performing a hypothesis test, there is always the possibility of making a Type I error (rejecting the null hypothesis when it is true) or of making a Type II error (not rejecting the null hypothesis when it is false). The probability of making a Type I error is equal to the level of significance, α, of the test.

We discussed three types of hypothesis tests in this chapter. First, we performed tests about a population mean with σ known. Second, we performed tests about a population mean with σ unknown.

In both of these cases, we required that either the sample size be large ($n \geq 30$) or the population be approximately normal with no outliers. For small sample sizes, we verified the normality of the data by using normal probability plots. Boxplots were used to check for outliers.

The third test we performed regarded hypothesis tests about a population proportion. To perform these tests, we required a large sample size so that $np(1 - p) \geq 10$, yet the sample size could be no more than 5% of the population size.

All three hypothesis tests were performed by using classical methods and the P-value approach. The P-value approach to testing hypotheses has appeal, because the rejection rule is always to reject the null hypothesis if the P-value is less than the level of significance, α.

Formulas

Test Statistics

- $z_0 = \dfrac{\bar{x} - \mu_0}{\sigma/\sqrt{n}}$ follows the standard normal distribution if the population from which the sample was drawn is normal or if the sample size is large ($n \geq 30$).

- $t_0 = \dfrac{\bar{x} - \mu_0}{s/\sqrt{n}}$ follows Student's t-distribution with $n - 1$ degrees of freedom if the population from which the sample was drawn is normal or if the sample size is large ($n \geq 30$).

- $z_0 = \dfrac{\hat{p} - p_0}{\sqrt{\dfrac{p_0(1 - p_0)}{n}}}$ follows the standard normal distribution if $np_0(1 - p_0) \geq 10$ and $n \leq 0.05N$.

Type I and Type II Errors

- $\alpha = P(\text{Type I error}) = P(\text{rejecting } H_0 \text{ when } H_0 \text{ is true})$
- $\beta = P(\text{Type II error}) = P(\text{not rejecting } H_0 \text{ when } H_1 \text{ is true})$

Vocabulary

Hypothesis (p. 454)
Hypothesis testing (p. 455)
Null hypothesis (p. 455)
Alternative hypothesis (p. 455)
Two-tailed test (p. 455)
Left-tailed test (p. 455)

Right-tailed test (p. 455)
Type I error (p. 457)
Type II error (p. 457)
Level of significance (p. 459)
Statistically significant (p. 463)
Test statistic (p. 466)

Critical value (p. 466)
Critical region (p. 466)
Decision rule (p. 466, 470)
Robust (p. 467)
P-value (p. 469)
Practical significance (p. 474)

Objectives

Section	You should be able to . . .	Examples	Review Exercises
10.1	1 Determine the null and alternative hypothesis (p. 454)	2	1, 2
	2 Understand Type I and Type II errors (p. 457)	3	1–4, 15(b), (c); 16(b), (c)
	3 State conclusions to hypothesis tests (p. 459)	4	1–2
10.2	1 Understand the logic of hypothesis testing (p. 462)	pp. 462–465	22
	2 Test hypotheses about a population mean with σ known using the classical approach (p. 465)	1 and 2	5, 6, 13, 14, 17, 18
	3 Test hypotheses about a population mean with σ known using P-values (p. 469)	3–5	5, 6, 13, 14, 17, 18

	4 Test hypotheses about a population mean with σ known using confidence intervals (p. 473)	6	6(f)
	5 Understand the difference between statistical significance and practical significance (p. 474)	7	19
10.3	1 Test hypotheses about a population mean with σ unknown (p. 480)	1–3	7, 8, 11, 12, 15(a), 16(a), 21
10.4	1 Test hypotheses about a population proportion (p. 493)	1–3	9, 10, 19, 20
10.5	1 Determine the appropriate hypothesis test to perform (p. 500)	pp. 500–501	5–21

Review Exercises

1. $H_0: \mu = \$4277$, $H_1: \mu > \$4277$

In Problems 1 and 2, (a) determine the null and alternative hypotheses, (b) explain what it would mean to make a Type I error, (c) explain what it would mean to make a Type II error, (d) state the conclusion that would be drawn if the null hypothesis is not rejected, and (e) state the conclusion that would be reached if the null hypothesis is rejected.

1. **Credit Card Debt** According to the *Statistical Abstract of the United States*, the mean outstanding credit card debt per cardholder was $4277 in 2000. A consumer credit counselor believes that the mean outstanding credit card debt per cardholder is now more than the 2000 amount.

2. **Downloading Music** According to a study by Ipsos-Reid, 61% of Internet users aged 18 to 24 years old had downloaded music from the Internet by the end of 2000. A researcher believes that the percentage is now higher than 61%. $H_0: p = 0.61$, $H_1: p > 0.61$

3. Suppose a test is conducted at the $\alpha = 0.05$ level of significance. What is the probability of a Type I error? 0.05

4. Suppose β is computed to be 0.113. What is the probability of a Type II error? 0.113

5. To test $H_0: \mu = 30$ versus $H_1: \mu < 30$, a simple random sample of size $n = 12$ is obtained from a population that is known to be normally distributed with $\sigma = 4.5$.
   (a) If the sample mean is determined to be $\bar{x} = 28.6$, compute the test statistic. $Z = -1.08$
   (b) If the researcher decides to test this hypothesis at the $\alpha = 0.05$ level of significance, determine the critical value. $-Z_{0.05} = -1.645$
   (c) Draw a normal curve that depicts the rejection region.
   (d) Will the researcher reject the null hypothesis? Why? No
   (e) What is the P-value? 0.1401

6. To test $H_0: \mu = 65$ versus $H_1: \mu \neq 65$, a simple random sample of size $n = 23$ is obtained from a population that is known to be normally distributed with $\sigma = 12.3$.
   (a) If the sample mean is determined to be $\bar{x} = 70.6$, compute the test statistic. $Z = 2.18$
   (b) If the researcher decides to test this hypothesis at the $\alpha = 0.1$ level of significance, determine the critical values. $-Z_{0.05} = -1.645$, $Z_{0.05} = 1.645$
   (c) Draw a normal curve that depicts the rejection region.
   (d) Will the researcher reject the null hypothesis? Why? Yes
   (e) What is the P-value? 0.0290
   (f) Conduct the test by constructing a 90% confidence interval. (66.38, 74.82)

7. To test $H_0: \mu = 8$ versus $H_1: \mu \neq 8$, a simple random sample of size $n = 15$ is obtained from a population that is known to be normally distributed.
   (a) If $\bar{x} = 7.3$ and $s = 1.8$, compute the test statistic.
   (b) If the researcher decides to test this hypothesis at the $\alpha = 0.02$ level of significance, determine the critical values. $-t_{0.01} = -2.624$, $t_{0.01} = 2.624$
   (c) Draw a t-distribution that depicts the rejection region.
   (d) Will the researcher reject the null hypothesis? Why? No
   (e) Determine the P-value. 0.1543

8. To test $H_0: \mu = 3.9$ versus $H_1: \mu < 3.9$, a simple random sample of size $n = 25$ is obtained from a population that is known to be normally distributed.
   (a) If $\bar{x} = 3.5$ and $s = 0.9$, compute the test statistic.
   (b) If the researcher decides to test this hypothesis at the $\alpha = 0.05$ level of significance, determine the critical value. $-t_{0.05} = -1.711$
   (c) Draw a t-distribution that depicts the rejection region.
   (d) Will the researcher reject the null hypothesis? Why? Yes
   (e) Determine the P-value. 0.0180

In Problems 9 and 10, test the hypothesis at the $\alpha = 0.05$ level of significance, using (a) the classical approach and (b) the P-value approach. Be sure to verify the requirements of the test.

9. $H_0: p = 0.6$ versus $H_1: p > 0.6$
   $n = 250$; $x = 165$; $\alpha = 0.05$

10. $H_0: p = 0.35$ versus $H_1: p \neq 0.35$
    $n = 420$; $x = 138$; $\alpha = 0.01$

11. **Linear Rotary Bearing** A linear rotary bearing is designed so that the distance between the retaining rings is 0.875 inch. The quality-control manager suspects that the manufacturing process needs to be recalibrated and that the mean distance between the retaining rings is longer than 0.875 inch. In a random sample of 36 bearings, he finds the sample mean distance between the retaining rings to be 0.876 inch with standard deviation 0.005 inch. Should the machine be recalibrated? Use the $\alpha = 0.05$ level of significance. Do not reject H_0

12. **Education Pays** The U.S. Census Bureau reported that the mean annual salary in 2002 was $51,194 for an individual whose highest degree was a bachelor's. A government economist believes that the mean annual salary for individuals whose highest degree is a bachelor's is different today. She obtains a random sample of 300 employed adults whose highest degree is a bachelor's and determines

(7a) $t = -1.506$ (8a) $t = -2.222$

the mean annual salary to be $55,988 with a standard deviation of $26,855 (both in 2002 dollars). Does the evidence indicate salaries are different today? Use the $\alpha = 0.05$ level of significance.

13. SAT Math Scores A mathematics instructor wanted to know whether or not use of a calculator improves SAT math scores. In 2000, the SAT math scores of students who used a calculator once or twice weekly were normally distributed, with a mean of 474 and a standard deviation 103. In a random sample of 50 students who use a calculator every day, the mean score was 539. Does the evidence support the instructor's assertion that students who use a calculator "frequently" score better on the SAT math portion than those who use a calculator "infrequently"? Use the $\alpha = 0.01$ level of significance. Reject H_0

14. Birth Weight An obstetrician wants to determine whether a new diet significantly increases the birth weight of babies. In 2002, birth weights of full-term babies (gestation period of 37 to 41 weeks) were normally distributed, with mean 7.53 pounds and standard deviation 1.15 pounds, according to the *National Vital Statistics Report*, Vol. 48, No. 3. The obstetrician randomly selects 50 recently pregnant mothers and persuades them to partake of this new diet. The obstetrician then records the birth weights of the babies and obtains a mean of 7.79 pounds. Does the new diet increase the birth weights of newborns? Use the $\alpha = 0.01$ level of significance. Do not reject H_0

15. High Cholesterol A nutritionist maintains that 20- to 39-year-old males consume too much cholesterol. The USDA-recommended daily allowance of cholesterol is 300 mg. In a survey conducted by the U.S. Department of Agriculture of 404 20- to 39-year-old males, it was determined the mean daily cholesterol intake was 326 milligrams, with standard deviation 342 milligrams.
(a) Is there evidence to support the nutritionist's belief at the $\alpha = 0.05$ level of significance? No
(b) What would it mean for the nutritionist to make a Type I error? A Type II error?
(c) What is the probability the nutritionist will make a Type I error? 0.05

16. Sodium A nutritionist thinks that 20- to 39-year-old females consume too much sodium. The USDA-recommended daily allowance of sodium is 2400 mg. In a survey conducted by the U.S. Department of Agriculture of 257 20- to 39-year-old females, it was determined the mean daily sodium intake was 2919 milligrams and the standard deviation was 906 milligrams.
(a) Is there evidence to support the nutritionist's belief at the $\alpha = 0.10$ level of significance? Yes
(b) What would it mean for the nutritionist to make a Type I error? A Type II error?
(c) What is the probability the nutritionist will make a Type I error? 0.1

17. Acid Rain In 1990, the mean pH level of the rain in Barnstable County, Massachussetts, was 4.61. A biologist fears that the acidity of rain has increased (in other words

that the pH level of the rain has decreased). She draws a random sample of 25 rain dates in 2004 and obtains the following data:

4.80	4.27	4.09	4.55	5.08	4.34	4.08
4.36	4.82	4.70	4.40	4.73	4.62	4.48
4.28	4.28	4.48	4.72	4.12	4.00	4.93
4.91	4.32	4.63	4.03			

Source: National Atmospheric Deposition Program

(a) Because the sample size is small, she must verify that the pH level is normally distributed and the sample does not contain any outliers. The normal probability plot and boxplot are shown. Are the conditions for testing the hypothesis satisfied? Yes

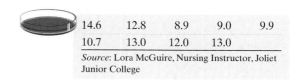

(b) Assuming $\sigma = 0.26$, judge whether the biologist's fears are justified. Use the $\alpha = 0.01$ level of significance. Reject H_0

18. Hemoglobin A medical researcher maintains that the mean hemoglobin reading of surgical patients is different from 14.0 grams per deciliter. She randomly selects nine surgical patients and obtains the following data:

14.6	12.8	8.9	9.0	9.9
10.7	13.0	12.0	13.0	

Source: Lora McGuire, Nursing Instructor, Joliet Junior College

(a) Because the sample size is small, the student must verify that hemoglobin is normally distributed and the sample does not contain any outliers. The normal probability plot and boxplot are shown. Are the conditions for testing the hypothesis satisfied? Yes

(b) Assuming that $\sigma = 2.001$, test the researcher's beliefs. Use the $\alpha = 0.01$ level of significance. Reject H_0

19. **Tuberculosis** According to the Centers for Disease Control, 56% of all tuberculosis cases in 1999 were of foreign-born residents of the United States. A researcher believes that this proportion has increased from its 1999 level. She obtains a simple random sample of 300 tuberculosis cases in the United States and determines that 170 of them are foreign-born. Is there sufficient evidence to support the belief that the percentage of cases of tuberculosis of foreign-born residents has increased at the $\alpha = 0.01$ level of significance? *No, do not reject H_0*

20. **Phone Purchases** In 1997, 39.4% of females had ordered merchandise or services by phone in the last 12 months. A market research analyst feels that the percentage of females ordering merchandise or services by phone has declined from the 1997 level because of Internet purchases. She obtains a random sample of 500 females and determines that 191 of them have ordered merchandise or services by phone in the last 12 months. Does the evidence suggest that the percentage of females ordering merchandise or services by phone has decreased from its 1997 proportion at the $\alpha = 0.10$ level of significance? *Do not reject H_0*

21. **A New Teaching Method** A large university has a college algebra enrollment of 5000 students each semester. Because of space limitations, the university decides to offer its college algebra courses in a self-study format in which students learn independently but have access to tutors and other help in a lab setting. Historically, students in traditional college algebra scored 73.2 points on the final exam and the coordinator of this course is concerned that test scores are going to decrease in the new format. At the end of the first semester using the new delivery system, 3851 students took the final exams and had a mean score of 72.8 and a standard deviation of 12.3. Treating these students as a simple random sample of all students, determine whether or not the scores decreased significantly at the $\alpha = 0.05$ level of significance. Do you think that the decrease in scores has any practical significance?

22. In your own words, explain the procedure for conducting a test regarding a population mean when assuming the population standard deviation is known.

THE CHAPTER 10 CASE STUDY IS LOCATED ON THE CD THAT ACCOMPANIES THIS TEXT.

Inferences on Two Samples

Outline

11.1 Inference about Two Means: Dependent Samples

11.2 Inference about Two Means: Independent Samples

11.3 Inference about Two Population Proportions

■ Chapter Review

■ Case Study: Control in the Design of an Experiment (On CD)

DECISIONS

Suppose that you have just received an inheritance of $10,000 and decide that the money should be invested, rather than blown on frivolous items. You have decided that you will invest the money in one of two types of mutual funds. But which type? See the Decisions project on page 528.

●●● Putting It All Together

In Chapters 9 and 10 we discussed inferences regarding a single population parameter. The inferential methods presented in these chapters will be modified slightly in this chapter so that we can compare two population parameters.

The first two sections of this chapter deal with testing for the difference of two population means. The methods presented in this chapter can be used to determine whether a certain treatment results in significantly different sample statistics. From a design-of-experiments point of view, the methods presented in Section 11.1 are used to handle matched-pairs designs (Section 1.5, pages 43–44) with a quantitative-response variable. For example, we might want to know whether married couples have similar IQs. To test this theory, we could randomly select 20 married couples and determine the difference in their IQs.

Section 11.2 presents inferential methods used to handle completely randomized designs when there is a single treatment that has two levels and the response variable is quantitative. (Section 1.5, pages 42–43) For example, we might randomly divide 100 volunteers who have the common cold into two groups, a control group and an experimental group. The control group would receive a placebo and the experimental group would receive a predetermined amount of some experimental drug. The response variable might be the time until the cold symptoms go away.

Section 11.3 discusses the difference between two population proportions. Again, we can use a completely randomized design to compare two population proportions. However, rather than having a quantitative response variable, we would have a binomial response variable; that is, either the experimental unit has a characteristic or it does not.

11.1 Inference about Two Means: Dependent Samples

Preparing for This Section   Before getting started, review the following:

- Matched-pairs design (Section 1.5, pp. 43–44)
- Confidence intervals about μ, σ unknown (Section 9.2, pp. 423–430)

- Hypothesis tests about μ, σ unknown (Section 10.3, pp. 480–486)
- Type I and Type II errors (Section 10.1, pp. 457–458)

Objectives

1. **Distinguish between independent and dependent sampling**
2. **Test hypotheses regarding matched-pairs data**
3. **Construct and interpret confidence intervals about the population mean difference of matched-pairs data**

1 Distinguish between Independent and Dependent Sampling

In Other Words

If the individuals in two samples are somehow related (husband–wife, siblings, similar characteristics, or even the same person), the sampling is dependent.

To perform inference on the difference of two population means, we must first determine whether the data come from an *independent* or *dependent* sample. A sampling method is **independent** when the individuals selected for one sample do not dictate which individuals are to be in a second sample. A sampling method is **dependent** when the individuals selected to be in one sample are used to determine the individuals to be in the second sample. For example, if we are conducting a study that compares the IQs of husbands and wives, once a husband is selected to be in the study, his wife is automatically matched with him, so this is dependent sampling. Dependent samples are often referred to as **matched-pairs** samples.

EXAMPLE 1   **Distinguishing between Independent and Dependent Sampling**

Problem: For each of the following experiments, determine whether the sampling method is independent or dependent.

(a) Researcher Steven J. Sperber, MD, and his associates wanted to determine the effectiveness of a new medication* in the treatment of discomfort associated with the common cold. They randomly divided 430 subjects into two groups: Group 1 received the new medication and Group 2 received a placebo. The goal of the study was to determine whether the mean of the symptom assessment scores of the individuals receiving the new medication (Group 1) was less than that of the placebo group (Group 2).

(b) In an experiment conducted in a biology class, Professor Andy Neill measured the time required for 12 students to catch a falling meter stick using their dominant hand and nondominant hand. The goal of the study was to determine whether the reaction time in an individual's dominant hand is different from the reaction time in the nondominant hand.

Approach: We must determine whether the individuals in one group were used to determine the individuals in the other group. If so, the sampling method is dependent. If not, the sampling method is independent.

Solution

(a) The sampling method is independent because the individuals in Group 1 were not used to determine which individuals are in Group 2.

*The medication was a combination of pseudoephedrine and acetaminophen. The study is published in the *Archives of Family Medicine* 9(2000): 979–985.

(b) The sampling method is dependent because the individuals are related. The measurements for the dominant and nondominant hand are on the same individual.

> **Now Work Problem 5.**

In this section, we will discuss inference on the difference of two means for dependent sampling. Section 11.2 addresses inference when the sampling is independent.

② Test Hypotheses Regarding Matched-Pairs Data

Inference on matched-pairs data is very similar to inference regarding a population mean when the population standard deviation is unknown. Recall that if the population from which the sample was drawn is normally distributed or the sample size is large ($n \geq 30$), we said that

$$t = \frac{\bar{x} - \mu}{\frac{s}{\sqrt{n}}}$$

follows Student's t-distribution with $n - 1$ degrees of freedom.

When analyzing matched-pairs data, we compute the difference in each matched pair and then perform inference on the differenced data using the methods of Section 9.2 or 10.3.

In Other Words
Statistical inference methods on matched-pairs data use the same methods as inference on a single population mean with σ unknown, except that the *differences* are analyzed.

> ### Testing Hypotheses Regarding the Difference of Two Means Using a Matched-Pairs Design
>
> To test hypotheses regarding the mean difference of matched-pairs data, we can use the following steps, provided that
>
> 1. the sample is obtained using simple random sampling;
> 2. the sample data are matched pairs;
> 3. the differences are normally distributed with no outliers or the sample size, n, is large ($n \geq 30$).
>
> *Step 1*: Determine the null and alternative hypotheses. The hypotheses can be structured in one of three ways, where μ_d is the population mean difference of the matched-pairs data.
>
> | Two-Tailed | Left-Tailed | Right-Tailed |
> |---|---|---|
> | $H_0: \mu_d = 0$ | $H_0: \mu_d = 0$ | $H_0: \mu_d = 0$ |
> | $H_1: \mu_d \neq 0$ | $H_1: \mu_d < 0$ | $H_1: \mu_d > 0$ |
>
> *Step 2*: Select a level of significance α, depending on the seriousness of making a Type I error.
>
> *Step 3*: Compute the test statistic
>
> $$t_0 = \frac{\bar{d} - 0}{\frac{s_d}{\sqrt{n}}} = \frac{\bar{d}}{\frac{s_d}{\sqrt{n}}}$$
>
> which approximately follows Student's t-distribution with $n - 1$ degrees of freedom. The values of $\bar{d}$ and s_d are the mean and standard deviation of the differenced data.

Note to Instructor
Ask students to explain the differences in the meaning of the following symbols:

$$\mu, \mu_{\bar{x}}, \mu_0, \mu_d$$

Classical Approach

Step 4: Use Table V to determine the critical value using $n - 1$ degrees of freedom.

	Two-Tailed	Left-Tailed	Right-Tailed
Critical value(s)	$-t_{\frac{\alpha}{2}}$ and $t_{\frac{\alpha}{2}}$	$-t_\alpha$	t_α
Critical region(s)	Critical Region, Critical Region at $-t_{\alpha/2}$, $t_{\alpha/2}$	Critical Region at $-t_\alpha$	Critical Region at t_α

Step 5: Compare the critical value with the test statistic.

Two-Tailed	Left-Tailed	Right-Tailed
If $t_0 < -t_{\frac{\alpha}{2}}$ or $t_0 > t_{\frac{\alpha}{2}}$, reject the null hypothesis.	If $t_0 < -t_\alpha$, reject the null hypothesis.	If $t_0 > t_\alpha$, reject the null hypothesis.

P-Value Approach

Step 4: Use Table V to estimate the P-value using $n - 1$ degrees of freedom.

Two-Tailed	Left-Tailed	Right-Tailed
The sum of the area in the tails is the P-value, $-|t_0|$, $|t_0|$	The area left of t_0 is the P-value, t_0	The area right of t_0 is the P-value, t_0

Step 5: If P-value $< \alpha$, reject the null hypothesis.

Step 6: State the conclusion.

The procedures just presented are **robust**, which means that minor departures from normality will not adversely affect the results of the test. If the data have outliers, however, the procedure should not be used.

We will verify the assumption that the differenced data come from a population that is normally distributed by constructing normal probability plots. We use boxplots to determine whether there are outliers. If the normal probability plot indicates that the differenced data are not normally distributed or the boxplot reveals outliers, nonparametric tests should be performed, which are not discussed in this text.

EXAMPLE 2 **Testing Hypotheses Regarding Matched-Pairs Data**

Problem: Professor Andy Neill measured the time (in seconds) required to catch a falling meter stick for 12 randomly selected students' dominant hand and nondominant hand. Professor Neill wants to know if the reaction time in an individual's dominant hand is less than the reaction time in his or her nondominant hand. Conduct the test at the $\alpha = 0.05$ level of significance. The data obtained are presented in Table 1.

Table 1		
Student	Dominant Hand, X_i	Nondominant Hand, Y_i
1	0.177	0.179
2	0.210	0.202
3	0.186	0.208
4	0.189	0.184
5	0.198	0.215
6	0.194	0.193
7	0.160	0.194
8	0.163	0.160
9	0.166	0.209
10	0.152	0.164
11	0.190	0.210
12	0.172	0.197

Source: Professor Andy Neill, Joliet Junior College

Note to Instructor

Remind students of some of the key phrases in the claim that help them to determine whether the test is right-tailed, left-tailed, or two-tailed. Examples include "less than," "more than," "no different."

Approach: This is a matched-pairs design because the variable is measured on the same subject for both the dominant and nondominant hand, the treatment in this experiment. We compute the difference between the dominant time and the nondominant time. So, for the first student we compute $X_1 - Y_1$, for the second student we compute $X_2 - Y_2$, and so on. If the reaction time in the dominant hand is less than the reaction time in the nondominant hand, we would expect the values of $X_i - Y_i$ to be negative. Before we perform the hypothesis test, we must verify that the differences come from a population that is approximately normally distributed with no outliers because the sample size is small. We will construct a normal probability plot and boxplot of the differenced data to verify these requirements. We then proceed to follow Steps 1 through 6.

Note to Instructor

Emphasize the importance of constructing a table such as Table 2. Not only does it help organize the information, but it can be used to help students see the direction of the alternative hypothesis.

Solution: We compute the differences as $d_i = X_i - Y_i$ = time of dominant hand for ith student minus time of nondominant hand for ith student. We expect these differences to be negative, so we wish to determine if $\mu_d < 0$. Table 2 displays the differences.

Table 2

Student	Dominant Hand, X_i	Nondominant Hand, Y_i	Difference, d_i
1	0.177	0.179	$0.177 - 0.179 = -0.002$
2	0.210	0.202	$0.210 - 0.202 = 0.008$
3	0.186	0.208	-0.022
4	0.189	0.184	0.005
5	0.198	0.215	-0.017
6	0.194	0.193	0.001
7	0.160	0.194	-0.034
8	0.163	0.160	0.003
9	0.166	0.209	-0.043
10	0.152	0.164	-0.012
11	0.190	0.210	-0.020
12	0.172	0.197	-0.025

$$\sum d_i = -0.158$$

! **CAUTION**

The way that we define the difference determines the direction of the alternative hypothesis in one-tailed tests. In Example 1, we expect $X_i < Y_i$, so the difference $X_i - Y_i$ is expected to be negative. Therefore, the alternative hypothesis is $H_1: \mu_d < 0$, and we have a left-tailed test. However, if we computed the differences as $Y_i - X_i$, we'd expect the differences to be positive, and we have a right-tailed test!

We compute the mean and standard deviation of the differences and obtain $\bar{d} = -0.0132$ rounded to four decimal places and $s_d = 0.0164$ rounded to four decimal places. We must verify that the data come from a population that is approximately normal with no outliers. Figure 1 shows the normal probability plot and boxplot of the differenced data.

Figure 1

The normal probability plot is roughly linear and the boxplot does not show any outliers. We can proceed with the hypothesis test.

Step 1: Professor Neill wants to know if the reaction time in the dominant hand is less than the reaction time in the nondominant hand. We express this claim as $\mu_d < 0$. We have

$$H_0: \mu_d = 0 \quad \text{versus} \quad H_1: \mu_d < 0$$

This test is left-tailed.

Step 2: The level of significance is $\alpha = 0.05$.

Step 3: The sample mean is $\overline{d} = -0.0132$ second, and the sample deviation is $s_d = 0.0164$ second. The test statistic is

$$t_0 = \frac{\overline{d}}{\frac{s_d}{\sqrt{n}}} = \frac{-0.0132}{\frac{0.0164}{\sqrt{12}}} = -2.788$$

Classical Approach

Step 4: Because this is a left-tailed test, we determine the critical t-value at the $\alpha = 0.05$ level of significance with $n - 1 = 12 - 1 = 11$ degrees of freedom to be $-t_{0.05} = -1.796$. The critical region is displayed in Figure 2.

Figure 2

Critical Region

$-t_{0.05} = -1.796$

Test Statistic $t = -2.788$

Step 5: Because the test statistic $t_0 = -2.788$ is less than the critical value $-t_{0.05} = -1.796$, Professor Neill rejects the null hypothesis. We show the test statistic in Figure 2.

P-Value Approach

Step 4: Because this is a left-tailed test, the P-value is the area under the t-distribution with $12 - 1 = 11$ degrees of freedom to the left of the test statistic, $t_0 = -2.788$, as shown in Figure 3(a). That is, P-value $= P(t < t_0) = P(t < -2.788)$, with 11 degrees of freedom.

Because of the symmetry of the t-distribution, the area under the t-distribution to the left of -2.788 equals the area under the t-distribution to the right of 2.788. So P-value $= P(t_0 < -2.788) = P(t_0 > 2.788)$. See Figure 3(b).

Using Table V, we find the row that corresponds to 11 degrees of freedom. The value 2.788 lies between 2.718 and 3.106. The value of 2.718 has an area under the t-distribution with 11 degrees of freedom of 0.01 to the right. The value of 3.106 has an area under the t-distribution with 11 degrees of freedom of 0.005 to the right.

Because 2.788 is between 2.718 and 3.106, the P-value is between 0.005 and 0.01. So $0.005 < P$-value < 0.01.

Figure 3

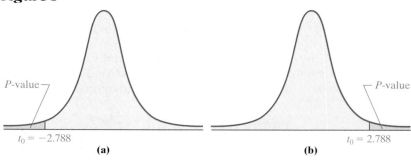

P-value

$t_0 = -2.788$

(a)

P-value

$t_0 = 2.788$

(b)

Step 5: Because the P-value is less than the level of significance $\alpha = 0.05$, Professor Neill rejects the null hypothesis.

Step 6: There is sufficient evidence to conclude that the reaction time in the dominant hand is less than the reaction time in the nondominant hand at the $\alpha = 0.05$ level of significance.

EXAMPLE 3   **Testing Hypotheses Regarding Matched-Pairs Data Using Technology**

Problem: Obtain an exact *P*-value for the problem in Example 2 using statistical software or a graphing calculator with advanced statistical features.

Approach: We will use MINITAB to obtain the *P*-value. The steps for testing hypotheses regarding matched-pairs data using the TI-83/84 Plus graphing calculators, MINITAB, and Excel are given in the Technology Step by Step on page 520.

Result: Figure 4 shows the results obtained from MINITAB. Notice that the *P*-value is 0.009.

Figure 4   **Paired T-Test and Confidence Interval**

Paired T for Dominant − Non-dominant

	N	Mean	StDev	SE Mean
Dominant	12	0.17975	0.01752	0.00506
Non-domi	12	0.19292	0.01799	0.00519
Difference	12	−0.01317	0.01643	0.00474

95% CI for mean difference: (−0.02361, −0.00273)
T-Test of mean difference = 0 (vs < 0): T-Value = −2.78 P-Value = 0.009

Interpretation: There is a 0.009 probability of obtaining a sample mean difference of −0.0132 or less from a population whose mean difference is 0. The results that we obtained would happen about 1 time in 100 samples if the null hypothesis were true. At the $\alpha = 0.05$ level of significance, there is sufficient evidence to reject the null hypothesis because the *P*-value is less than α. We conclude that the reaction time in the dominant hand is faster than the reaction time in the nondominant hand.

Note to Instructor
The purpose of this activity is to have students experience a situation where a matched comparison is necessary. In addition to discussing why matching is important, it is also useful to discuss how earlier concepts, such as randomization, come into play. Time limit: 45–50 minutes.

In-Class Activity: Measuring Reaction Time

We just saw that reaction time is different between your dominant and nondominant hand. Let's try to reproduce the results of Example 2. We can estimate reaction time by measuring reaction distance, in particular, the distance a ruler will fall before being caught. Would you expect this distance to be larger or smaller with your dominant hand?

(a) Pair up with another student in the class.
(b) Randomly determine whether to work first with the dominant hand or the nondominant hand. Why is this step important?
(c) One student holds his or her hand perpendicular to the floor so that the thumb and forefinger are roughly 5 cm apart. The other student holds a ruler (with centimeters) vertically just above the first student's fingers. The second student releases the ruler so that it falls vertically between the first student's thumb and forefinger (without hinting at when it will be dropped), and the first student grasps the ruler as quickly as possible. Record the number of centimeters required to grasp the ruler.
(d) Repeat the process for the other hand of the first student and then for both hands of the second student.
(e) Compute the difference in each student's reaction distance by computing $\text{distance}_{\text{dom. hand}} - \text{distance}_{\text{nondom. hand}}$
(f) Combine your team data with the rest of the class and conduct an appropriate hypothesis test to answer the question.

Now Work Problem 13(a) and (b).

③ ## Construct and Interpret Confidence Intervals about the Population Mean Difference of Matched-Pairs Data

We can also create a confidence interval for the population mean difference, μ_d, using the sample mean difference, $\overline{d}$, the sample standard deviation difference, s_d, the sample size, and $t_{\frac{\alpha}{2}}$. Remember, a confidence interval about a population mean is given in the following form:

$$\text{Point estimate} \pm \text{margin of error}$$

Based on the preceding formula, we compute the confidence interval about μ_d as follows:

> **Confidence Interval for Matched-Pairs Data**
>
> A $(1 - \alpha) \cdot 100\%$ confidence interval for μ_d is given by
>
> Lower bound: $\overline{d} - t_{\frac{\alpha}{2}} \cdot \dfrac{s_d}{\sqrt{n}}$ Upper bound: $\overline{d} + t_{\frac{\alpha}{2}} \cdot \dfrac{s_d}{\sqrt{n}}$ **(1)**
>
> The critical value $t_{\alpha/2}$ is determined using $n - 1$ degrees of freedom. The values of $\overline{d}$ and s_d are the mean and standard deviation of the differenced data.
> **Note:** The interval is exact when the population is normally distributed and approximately correct for nonnormal populations, provided that n is large.

EXAMPLE 4 ## Constructing a Confidence Interval for Matched-Pairs Data

Problem: Using the data from Table 2, construct a 95% confidence interval estimate of the mean difference, μ_d.

Approach

Step 1: Compute the differenced data. Because the sample size is small, we must verify that the differenced data come from a population that is approximately normal with no outliers.

Step 2: Compute the sample mean difference, $\overline{d}$, and the sample standard deviation difference, s_d.

Step 3: Determine the critical value, $t_{\frac{\alpha}{2}}$, with $\alpha = 0.05$ and $n - 1$ degrees of freedom.

Step 4: Use Formula (1) to determine the lower and upper bounds.

Step 5: Interpret the results.

Solution

Step 1: We computed the differenced data and verified that they come from a population that is approximately normally distributed with no outliers in Example 2.

Step 2: We computed the sample mean difference, $\overline{d}$, to be -0.0132 and the sample standard deviation of the difference, s_d, to be 0.0164 in Example 2.

Step 3: Using Table V with $\alpha = 0.05$ and $12 - 1 = 11$ degrees of freedom, we find $t_{\frac{\alpha}{2}} = t_{0.025} = 2.201$.

Step 4: Substituting into Formula (1), we find

$$\text{Lower bound: } \overline{d} - t_{\frac{\alpha}{2}} \cdot \frac{s_d}{\sqrt{n}} = -0.0132 - 2.201 \cdot \frac{0.0164}{\sqrt{12}} = -0.0236$$

$$\text{Upper bound: } \overline{d} + t_{\frac{\alpha}{2}} \cdot \frac{s_d}{\sqrt{n}} = -0.0132 + 2.201 \cdot \frac{0.0164}{\sqrt{12}} = -0.0028$$

Step 5: We are 95% confident that the mean difference between the dominant hand's reaction time and the nondominant hand's reaction time is between

-0.0236 and -0.0028 second. In other words, we are 95% confident that the dominant hand has a mean reaction time that is somewhere between 0.0028 second and 0.0236 second faster than the nondominant hand. Notice that the confidence interval does not contain zero. This evidence supports the belief that the reaction time of a person's dominant hand is different from the reaction time of the nondominant hand.

Now Work Problem 13(c).

We can see that the results of Example 4 agree with the 95% confidence interval determined by MINITAB in Figure 4.

11.1 ASSESS YOUR UNDERSTANDING

Concepts and Vocabulary

1. A sampling method is _____ when the individuals selected for one sample do not dictate which individuals are in a second sample. *independent*

2. A sampling method is _____ when the individuals selected to be in one sample are used to determine the individuals in the second sample. *dependent*

3. Suppose a researcher believes the mean from population 1 is less than the mean from population 2 in matched-pairs data. How would you define μ_d? How would you determine d_i?

4. What are the requirements to test hypotheses regarding the difference of two means with dependent sampling?

Skill Building

In Problems 5–10, determine whether the sampling is dependent or independent.

5. A sociologist wishes to compare the annual salaries of married couples. She obtains a random sample of 50 married couples in which both spouses work and determines each spouse's annual salary. *Dependent*

6. A researcher wishes to determine the effects of alcohol on people's reaction times to a stimulus. She randomly divides 100 people aged 21 or older into two groups. Group 1 is asked to drink 3 ounces of alcohol, while group 2 drinks a placebo. Both drinks taste the same, so the individuals in the study do not know which group they belong to. Thirty minutes after consuming the drink, the subjects in each group perform a series of tests meant to measure reaction time. *Independent*

7. An educator wants to determine whether a new curriculum significantly improves standardized test scores for third-grade students. She randomly divides 80 third-graders into two groups. Group 1 is taught using the new curriculum, while group 2 is taught using the traditional curriculum. At the end of the school year, both groups are given the standardized test and the mean scores are compared. *Independent*

8. A psychologist wants to know whether subjects respond faster to a go/no go stimulus or a choice stimulus. With the

go/no go stimulus, subjects must respond to a particular stimulus by pressing a button and disregard other stimuli. In the choice stimulus, the subjects respond differently depending on the stimulus. The psychologist randomly selects 20 subjects and each subject is presented a series of go/no go stimuli and choice stimuli. The mean reaction time to each stimulus is compared. *Dependent*

9. A study was conducted by researchers designed "to determine the genetic and nongenetic factors to structural brain abnormalities on schizophrenia." The researchers examined the brains of 29 patients diagnosed with schizophrenia and compared them with 29 healthy patients. The whole-brain volumes of the two groups were compared. *Independent*

 (*Source:* William F. C. Baare et al., Volumes of Brain Structures in Twins Discordant for Schizophrenia, *Archives of General Psychiatry* **58**: (2000) 33–40.)

10. An agricultural researcher wanted to determine whether there were any significant differences in the plowing method used on crop yield. He divided a parcel of land that had uniform soil quality into 30 subplots. He then randomly selected 15 of the plots to be chisel plowed and 15 plots to be fall plowed. He recorded the crop yield at the end of the growing season to determine whether there was a significant difference in the mean crop yield. *Independent*

In Problems 11 and 12, assume that the paired data came from a population that is normally distributed.

11.

(11c) Reject H_0

Observation	1	2	3	4	5	6	7
X_i	7.6	7.6	7.4	5.7	8.3	6.6	5.6
Y_i	8.1	6.6	10.7	9.4	7.8	9.0	8.5

 (a) Determine $d_i = X_i - Y_i$ for each pair of data.
 (b) Compute $\overline{d}$ and s_d. $\overline{d} = -1.614$, $s_d = 1.915$
 (c) Test if $\mu_d < 0$ at the $\alpha = 0.05$ level of significance.
 (d) Compute a 95% confidence interval about the population mean difference $\mu_d = \mu_x - \mu_y$. $(-3.39, 0.16)$

12.

(12c) Do not reject H_0

Observation	1	2	3	4	5	6	7	8
X_i	19.4	18.3	22.1	20.7	19.2	11.8	20.1	18.6
Y_i	19.8	16.8	21.1	22.0	21.5	18.7	15.0	23.9

(a) Determine $d_i = X_i - Y_i$ for each pair of data.
(b) Compute $\bar{d}$ and s_d. $\bar{d} = -1.075$, $s_d = 3.833$
(c) Test if $\mu_d \neq 0$ at the $\alpha = 0.01$ level of significance.
(d) Compute a 99% confidence interval about the population mean difference $\mu_d = \mu_x - \mu_y$. $(-5.82, 3.67)$

Applying the Concepts

13. Muzzle Velocity The following data represent the muzzle velocity (in feet per second) of rounds fired from a 155-mm gun. For each round, two measurements of the velocity were recorded using two different measuring devices, with the following data obtained:

Observation	1	2	3	4	5	6	7	8	9	10	11	12
A	793.8	793.1	792.4	794.0	791.4	792.4	791.7	792.3	789.6	794.4	790.9	793.5
B	793.2	793.3	792.6	793.8	791.6	791.6	791.6	792.4	788.5	794.7	791.3	793.5

Source: Christenson, Ronald and Blackwood, Larry. Tests for Precision and Accuracy of Multiple Measuring Devices. *Technometrics*, Nov. 93, Vol 35, Issue 4, pp. 411–421.

(a) Why are these matched-pairs data?
(b) Is there a difference in the measurement of the muzzle velocity between device A and device B at the $\alpha = 0.01$ level of significance? **Note:** A normal probability plot and boxplot of the data indicate that the differences are approximately normally distributed with no outliers. Do not reject H_0
(c) Construct a 99% confidence interval about the population mean difference. Interpret your results. $(-0.309, 0.542)$
(d) Draw a boxplot of the differenced data. Does this visual evidence support the results obtained in part (b)?

14. Reaction Time In an experiment conducted online at the University of Mississippi, study participants are asked to react to a stimulus. In one experiment, the participant must press a key on seeing a blue screen. Reaction time (in seconds) to press the key is measured. The same person is then asked to press a key on seeing a red screen, again with reaction time measured. The results for six randomly sampled study participants are as follows:

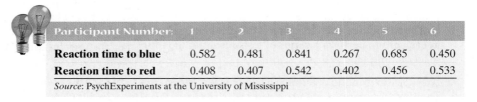

Participant Number:	1	2	3	4	5	6
Reaction time to blue	0.582	0.481	0.841	0.267	0.685	0.450
Reaction time to red	0.408	0.407	0.542	0.402	0.456	0.533

Source: PsychExperiments at the University of Mississippi

(a) Why are these matched-pairs data?
(b) Is the reaction time to the blue stimulus different from the reaction time to the red stimulus at the $\alpha = 0.01$ level of significance? **Note:** A normal probability plot and boxplot of the data indicate that the differences are approximately normally distributed with no outliers. Do not reject H_0
(c) Construct a 98% confidence interval about the population mean difference. Interpret your results. $(-0.3316, 0.1456)$
(d) Draw a boxplot of the differenced data. Does this visual evidence support the results obtained in part (b)?

15. Secchi Disk A Secchi disk is an 8-inch-diameter weighted disk that is painted black and white and attached to a rope. The disk is lowered into water and the depth (in inches) at which it is no longer visible is recorded. The measurement is an indication of water clarity. An environmental biologist is interested in determining whether the water clarity of the lake at Joliet Junior College is improving. She takes measurements at the same location on the same dates during the course of a year and repeats the measurements on the same dates 5 years later. She obtains the following results:

Observation:	1	2	3	4	5	6	7	8
Date:	5/11	6/7	6/24	7/8	7/27	8/31	9/30	10/12
Initial depth	38	58	65	74	56	36	56	52
Depth 5 years later	52	60	72	72	54	48	58	60

Source: Virginia Piekarski, Joliet Junior College

(a) Why is it important to take the measurements on the same date?

(b) Does the evidence suggest that the clarity of the lake is improving at the $\alpha = 0.05$ level of significance? **Note:** A normal probability plot and boxplot of the data indicate that the differences are approximately normally distributed with no outliers. *Reject H_0*

(c) Construct a 95% confidence interval about the population mean difference. Interpret your results. *(0.04, 10.21)*

(d) Draw a boxplot of the differenced data. Does this visual evidence support the results obtained in part (b)?

16. **Rat's Hemoglobin.** Hemoglobin helps the red blood cells transport oxygen and remove carbon dioxide. Researchers at NASA wanted to determine the effects of space flight on a rat's hemoglobin. The following data represent the hemoglobin (in grams per deciliter) at lift-off minus 3 days (H-L3) and immediately upon the return (H-R0) for 12 randomly selected rats sent to space on the Spacelab Sciences 1 flight.

Rat No.:	1	2	3	4	5	6	7	8	9	10	11	12
H-L3	15.2	16.1	15.3	16.4	15.7	14.7	14.3	14.5	15.2	16.1	15.1	15.8
H-R0	15.8	16.5	16.7	15.7	16.9	13.1	16.4	16.5	16.0	16.8	17.6	16.9

Source: NASA Life Sciences Data Archive

(a) Does the evidence suggest that the hemoglobin levels at lift-off minus 3 days are less than the hemoglobin levels upon return at the $\alpha = 0.05$ level of significance? **Note:** A normal probability plot and boxplot of the data indicate that the differences are approximately normally distributed with no outliers. *Reject H_0*

(b) Construct a 90% confidence interval about the population mean difference. Interpret your results. *(0.27, 1.48)*

(c) Draw a boxplot of the differenced data. Does this visual evidence support the results obtained in part (a)?

17. **Getting Taller?** To test the belief that sons are taller than their fathers, a researcher randomly selects 13 fathers who have adult male children. She records the height of both the father and son in inches and obtains the following data.

	1	2	3	4	5	6	7	8	9	10	11	12	13
Height of father	70.3	67.1	70.9	66.8	72.8	70.4	71.8	70.1	69.9	70.8	70.2	70.4	72.4
Height of son	74.1	69.2	66.9	69.2	68.9	70.2	70.4	69.3	75.8	72.3	69.2	68.6	73.9

Source: Anna Behounek, student at Joliet Junior College

Are sons taller than their fathers? Use the $\alpha = 0.1$ level of significance. **Note:** A normal probability plot and boxplot of the data indicate that the differences are approximately normally distributed with no outliers. *Do not reject H_0*

18. **Waiting in Line** A quality-control manager at an amusement park feels that the amount of time that people spend waiting in line for the American Eagle roller coaster is too long. To determine if a new loading/unloading procedure is effective in reducing wait time in line, he measures the amount of time (in minutes) people are waiting in line on 7 days. After implementing the new procedure, he again measures the amount of time (in minutes) people are waiting in line on 7 days and obtains the data shown on page 518. To make a reasonable comparison, he chooses days when the weather conditions are similar.

Day:	Mon (2 P.M.)	Tues (2 P.M.)	Wed (2 P.M.)	Thurs (2 P.M.)	Fri (2 P.M.)	Sat (11 A.M.)	Sat (4 P.M.)	Sun (12 noon)	Sun (4 P.M.)
Wait time before new procedure	11.6	25.9	20.0	38.2	57.3	32.1	81.8	57.1	62.8
Wait time after new procedure	10.7	28.3	19.2	35.9	59.2	31.8	75.3	54.9	62.0

Conduct the appropriate test to determine whether the new loading/unloading procedure is effective in reducing wait time at the $\alpha = 0.05$ level of significance. **Note:** A normal probability plot and boxplot of the data indicate that the differences are approximately normally distributed with no outliers. *Do not reject H_0*

19. **Hardness Testing** The manufacturer of hardness testing equipment uses steel-ball indenters to penetrate metal that is being tested. However, the manufacturer thinks it would be better to use a diamond indenter so that all types of metal can be tested. Because of the differences between the two types of indenters, it is suspected that the two methods will produce different hardness readings. The metal specimens to be tested are large enough so that two indentions can be made. Therefore, the manufacturer wants to use both indenters on each specimen and compare the readings.

					Specimen				
	1	2	3	4	5	6	7	8	9
Steel ball	50	57	61	71	68	54	65	51	53
Diamond	52	56	61	74	69	55	68	51	56

Do the two indenters result in different measurements at the $\alpha = 0.05$ level of significance? **Note:** A normal probability plot and boxplot of the data indicate that the differences are approximately normally distributed with no outliers. *Reject H_0*

20. **Car Rentals** The following data represent the daily rental for a compact automobile charged by two car rental companies, Thrifty and Hertz, in 10 locations.

City	Thrifty	Hertz	City	Thrifty	Hertz
Chicago	21.81	18.99	Seattle	21.96	22.99
Los Angeles	29.89	48.99	Pittsburgh	20.90	19.99
Houston	17.90	19.99	Phoenix	47.75	36.99
Orlando	27.98	35.99	New Orleans	33.81	26.99
Boston	24.61	25.60	Minneapolis	33.49	20.99

Source: Yahoo!Travel

Test whether Thrifty is less expensive than Hertz at the $\alpha = 0.1$ level of significance. **Note:** A normal probability plot and boxplot of the data indicate that the differences are approximately normally distributed with no outliers. *Do not reject H_0*

21. **DUI Simulator** To illustrate the effects of driving under the influence (DUI) of alcohol, a police officer brought a DUI simulator to a local high school. Student reaction time in an emergency was measured with unimpaired vision and also while wearing a pair of special goggles to simulate the effects of alcohol on vision. For a random sample of nine teenagers, the time (in seconds) required to bring the vehicle to a stop from a speed of 60 miles per hour was recorded. Construct and interpret a 95% confidence interval for the mean difference in braking time with impaired vision and normal vision where the differences are computed as "impaired minus normal." **Note:** A normal probability plot and boxplot of the data indicate that the differences are approximately normally distributed with no outliers. *(0.69, 1.24)*

	Subject								
	1	**2**	**3**	**4**	**5**	**6**	**7**	**8**	**9**
Normal	4.47	4.24	4.58	4.65	4.31	4.80	4.55	5.00	4.79
Impaired	5.77	5.67	5.51	5.32	5.83	5.49	5.23	5.61	5.63

22. **Braking Distance** An automotive researcher wanted to estimate the difference in distance required to come to a complete stop while traveling 40 miles per hour on wet versus dry pavement. Because car type plays a role, the researcher used eight different cars with the same driver and tires. The braking distance (in feet) on both wet and dry pavement is shown in the following table. Construct a 95% confidence interval for the mean difference in braking distance on wet versus dry pavement where the differences are computed as "wet minus dry." Interpret the interval. **Note:** A normal probability plot and boxplot of the data indicate that the differences are approximately normally distributed with no outliers. (29.55, 35.85)

Car No.:	1	2	3	4	5	6	7	8
Wet braking distance	106.9	100.9	108.8	111.8	105.0	105.6	110.6	107.9
Dry braking distance	71.8	68.8	74.1	73.4	75.9	75.2	75.7	81.0

23. **Does Octane Matter?** Octane is a measure of how much the fuel can be compressed before it spontaneously ignites. Some people believe that higher-octane fuels result in better gas mileage for their car. To test this claim, a researcher randomly selected 11 individuals (and their cars) to participate in the study. Each participant received 10 gallons of gas and drove his or her car on a closed course that simulated both city and highway driving. The number of miles driven until the car ran out of gas was recorded. A coin flip was used to determine whether the car was filled up with 87-octane or 92-octane fuel first, and the driver did not know which type of fuel was in the tank. The results are in the following table:

Driver:	1	2	3	4	5	6	7	8	9	10	11
Miles on 87 octane	234	257	243	215	114	287	315	229	192	204	547
Miles on 92 octane	237	238	229	224	119	297	351	241	186	209	562

(a) Why is it important that the matching be done by driver and car?

(b) Why is it important to conduct the study on a closed track?

(c) The normal probability plots for miles on 87 octane and miles on 92 octane are shown. Are either of these variables normally distributed? No

(d) The differences are computed as 92 octane minus 87 octane. The normal probability plot of the differences is shown. Is there reason to believe that the differences are normally distributed? Conclude that the differences can be normally distributed even though the original data are not. *Yes*

(e) The researchers used MINITAB to determine whether the mileage from 92 octane is greater than the mileage from 87 octane. The results are as follows:

Paired T-Test and CI: 92 Octane, 87 Octane
Paired T for 92 Octane - 87 Octane

	N	Mean	StDev	SE Mean
92 Octane	11	263.000	115.041	34.686
87 Octane	11	257.909	109.138	32.906
Difference	11	5.09091	14.87585	4.48524

95% lower bound for mean difference: -3.03841
T-Test of mean difference = 0 (vs > 0): T-Value = 1.14 P-Value = 0.141

What do you conclude? Why? *Fail to reject H_0; P-value is too large.*

Technology Step by Step	**Two-Sample *t*-Tests, Dependent Sampling**
TI-83/84 Plus	**Hypothesis Tests**

Step 1: If necessary, enter raw data in L1 and L2. Let L3 = L1 − L2 (or L2 − L1), depending on how the alternative hypothesis was defined.

Step 2: Press STAT, highlight TESTS, and select 2: T-Test.

Step 3: If the data are raw, highlight Data, making sure that List is set to L3 with frequency set to 1. If summary statistics are known, highlight Stats and enter the summary statistics.

Step 4: Highlight the appropriate relation between μ_1 and μ_2 in the alternative hypothesis.

Step 5: Highlight Calculate or Draw and press ENTER. Calculate gives the test statistic and *P*-value. Draw will draw the *t*-distribution with the *P*-value shaded.

Confidence Intervals

Follow the same steps as those given for hypothesis tests, except select 8: TInterval. Also, select a confidence level (such as 95% = 0.95). |
| **MINITAB** | **Step 1:** Enter raw data in columns C1 and C2.

Step 2: Select the **Stat** menu, highlight **Basic Statistics**, and then highlight **Paired-t** . . . .

Step 3: Enter C1 in the cell marked "First Sample" and enter C2 in the cell marked "Second Sample." Under OPTIONS, select the direction of the alternative hypothesis and select a confidence level. Click OK. |
| **Excel** | **Step 1:** Enter raw data in columns A and B.

Step 2: Select the **Tools** menu and highlight **Data Analysis** . . . .

Step 3: Select "*t*-test: Paired Two-Sample for Means." With the cursor in the "Variable 1 Range" cell, highlight the data in column A. With the cursor in the "Variable 2 Range" cell, highlight the data in column B. Enter the hypothesized difference in the means (usually 0) and a value for alpha. Click OK. |

11.2 Inference about Two Means: Independent Samples

Preparing for This Section   Before getting started, review the following:

- The Completely Randomized Design (Section 1.5, pp. 42–43)
- Confidence intervals about μ, σ unknown (Section 9.2, pp. 423–430)
- Hypothesis tests about μ, σ unknown (Section 10.3, pp. 480–486)
- Type I and Type II errors (Section 10.1, pp. 457–458)

Objectives

1. **Test hypotheses regarding the difference of two independent means**
2. **Construct and interpret confidence intervals regarding the difference of two independent means**

Note to Instructor

Students should be familiar with the concept of hypothesis testing by now, so the pace can be a little quicker.

We now turn our attention to inferential methods for comparing means from two independent samples. For example, suppose we wish to know whether a new experimental drug relieves symptoms attributable to the common cold. The response variable might be the time until the cold symptoms go away. If the drug is effective, the mean time until the cold symptoms go away should be less for individuals taking the drug than for those not taking the drug. If we let μ_1 represent the mean time until cold symptoms go away for the individuals taking the drug, and μ_2 represent the mean time until cold symptoms go away for individuals taking a placebo, the null and alternative hypotheses will be

$$H_0\!: \mu_1 = \mu_2 \quad \text{versus} \quad H_1\!: \mu_1 < \mu_2$$

or, equivalently,

$$H_0\!: \mu_1 - \mu_2 = 0 \quad \text{versus} \quad H_1\!: \mu_1 - \mu_2 < 0$$

Note to Instructor

The computations in this section become quite tedious. You might consider emphasizing technology solutions. This allows for more emphasis on the P-value approach to hypothesis testing.

To conduct this test, we might randomly divide 500 volunteers who have a common cold into two groups: an experimental group (Group 1) and a control group (Group 2). The control group will receive a placebo and the experimental group will receive a predetermined amount of the experimental drug. Next, determine the time until the cold symptoms go away. Compute $\bar{x}_1$, the sample mean time until cold symptoms go away in the experimental group, and $\bar{x}_2$, the sample mean time until cold symptoms go away in the control group. Now, we determine whether the difference in the sample means, $\bar{x}_1 - \bar{x}_2$, is significantly different from 0, the assumed difference stated in the null hypothesis. To do this, we need to know the sampling distribution of $\bar{x}_1 - \bar{x}_2$.

It is unreasonable to expect to know information regarding σ_1 and σ_2 without knowing information regarding the population means. Therefore, we must develop a sampling distribution for the difference of two means when the population standard deviations are unknown.

The comparison of two means with unequal (and unknown) population variances is called the Behrens–Fisher problem. While an exact method for performing inference on the equality of two means with unequal population standard deviations does not exist, an approximate solution is available. The approach that we use is known as **Welch's approximate *t***, in honor of English statistician Bernard Lewis Welch (1911–1989).

Note to Instructor

We do not discuss comparing the difference between two means using the pooled standard deviation. This is because the test requires that the two independent samples come from populations with the same variance. Verifying the requirement of equal population variances suggests using the F-test comparing two variances. Numerous statisticians (including George Box) have shown that the F-test is extremely sensitive to departures from normality. In fact, Box states, "it would seem logical to replace the usual analysis . . . which uses a pooled estimate . . . by the alternative criterion proposed by Welch."

(Source: Nonnormality and Tests on Variances, Biometrica, 40, 318–340.)

Sampling Distribution of the Difference of Two Means: Independent Samples with Population Standard Deviations Unknown (Welch's *t*)

Suppose a simple random sample of size n_1 is taken from a population with unknown mean μ_1 and unknown standard deviation σ_1. In addition, a simple random sample of size n_2 is taken from a second population with unknown mean μ_2 and unknown standard deviation σ_2. If the two populations are normally distributed or the sample sizes are sufficiently large ($n_1 \geq 30$ and $n_2 \geq 30$), then

$$t = \frac{(\bar{x}_1 - \bar{x}_2) - (\mu_1 - \mu_2)}{\sqrt{\dfrac{s_1^2}{n_1} + \dfrac{s_2^2}{n_2}}} \tag{1}$$

approximately follows Student's *t*-distribution with the smaller of $n_1 - 1$ or $n_2 - 1$ degrees of freedom, where $\bar{x}_1$ is the sample mean and s_1 is the sample standard deviation from population 1, and $\bar{x}_2$ is the sample mean and s_2 is the sample standard deviation from population 2.

 ## Test Hypotheses Regarding the Difference of Two Independent Means

Now that we know the approximate sampling distribution of $\bar{x}_1 - \bar{x}_2$, we can introduce a procedure that can be used to test hypotheses regarding two population means.

Testing Hypotheses Regarding the Difference of Two Means

To test hypotheses regarding two population means, μ_1 and μ_2, with unknown population standard deviations, we can use the following steps, provided that

1. the samples are obtained using simple random sampling;
2. the samples are independent;
3. the populations from which the samples are drawn are normally distributed or the sample sizes are large ($n_1 \geq 30$ and $n_2 \geq 30$).

Step 1: Determine the null and alternative hypotheses. The hypotheses are structured in one of three ways:

Two-Tailed	Left-Tailed	Right-Tailed
$H_0: \mu_1 = \mu_2$	$H_0: \mu_1 = \mu_2$	$H_0: \mu_1 = \mu_2$
$H_1: \mu_1 \neq \mu_2$	$H_1: \mu_1 < \mu_2$	$H_1: \mu_1 > \mu_2$

Note: μ_1 is the population mean for population 1, and μ_2 is the population mean for population 2.

Step 2: Select a level of significance α, depending on the seriousness of making a Type I error.

Step 3: Compute the test statistic

$$t_0 = \frac{(\bar{x}_1 - \bar{x}_2) - (\mu_1 - \mu_2)}{\sqrt{\dfrac{s_1^2}{n_1} + \dfrac{s_2^2}{n_2}}}$$

which approximately follows Student's *t*-distribution.

Classical Approach

Step 4: Use Table V to determine the critical value using the smaller of $n_1 - 1$ or $n_2 - 1$ degrees of freedom.

	Two-Tailed	Left-Tailed	Right-Tailed
Critical Value	$-t_{\frac{\alpha}{2}}$ and $t_{\frac{\alpha}{2}}$	$-t_\alpha$	t_α
Critical Region	Critical Region $\quad$ Critical Region $-t_{\alpha/2} \quad t_{\alpha/2}$	Critical Region $-t_\alpha$	Critical Region t_α

Step 5: Compare the critical value with the test statistic.

Two-Tailed	Left-Tailed	Right-Tailed
If $t_0 < -t_{\frac{\alpha}{2}}$ or $t_0 > t_{\frac{\alpha}{2}}$ reject the null hypothesis.	If $t_0 < -t_\alpha$, reject the null hypothesis.	If $t_0 > t_\alpha$ reject the null hypothesis.

P-Value Approach

Step 4: Use Table V to estimate the *P*-value using the the smaller of $n_1 - 1$ or $n_2 - 1$ degrees of freedom.

Two-Tailed	Left-Tailed	Right-Tailed
The sum of the area in the tails is the *P*-value $-|t_0| \quad |t_0|$	The area left of t_0 is the *P*-value t_0	The area right of t_0 is the *P*-value t_0

Step 5: If *P*-value $< \alpha$, reject the null hypothesis.

Step 6: State the conclusion.

The procedure just presented is robust, so minor departures from normality will not adversely affect the results of the test. If the data have outliers, however, the procedure should not be used.

We will verify these requirements by constructing normal probability plots (to assess normality) and boxplots (to determine whether there are outliers). If the normal probability plot indicates that the data came from a population that is not normally distributed or the boxplot reveals outliers, nonparametric tests should be performed, which are not discussed in this text.

EXAMPLE 1   **Testing Hypotheses Regarding Two Means**

Problem: In the Spacelab Life Sciences 2 payload, 14 male rats were sent to space. Upon their return, the red blood cell mass (in milliliters) of the rats was determined. A control group of 14 male rats was held under the same conditions (except for space flight) as the space rats, and their red blood cell mass was also determined when the space rats returned. The project, led by Dr. Paul X. Callahan, resulted in the data listed in Table 3. Does the evidence suggest that the flight animals have a different red blood cell mass from the control animals at the $\alpha = 0.05$ level of significance?

Note that this experiment is a completely randomized design with two levels of treatment: flight and control.

Table 3							
Flight				**Control**			
8.59	8.64	7.43	7.21	8.65	6.99	8.40	9.66
6.87	7.89	9.79	6.85	7.62	7.44	8.55	8.70
7.00	8.80	9.30	8.03	7.33	8.58	9.88	9.94
6.39	7.54			7.14	9.14		

Source: NASA Life Sciences Data Archive

Approach: We verify that each sample comes from a population that is approximately normal with no outliers by drawing normal probability plots and boxplots. The boxplots will be drawn on the same graph so that we can visually compare the two samples. We then follow Steps 1 through 6, listed on page 522.

Solution: Figure 5 shows normal probability plots of the data, which indicate that the data could come from populations that are normal. On the basis of the boxplots, it seems that there is not much difference in the red blood cell mass of the two samples, although the flight group might have a slightly lower red blood cell mass. We have to determine if this difference is significant or due to chance.

Figure 5

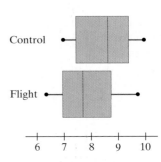

Red Blood Cell Mass

Step 1: We want to know whether the flight animals have a different red blood cell mass from the control animals. Let μ_1 represent the mean red blood cell mass of the flight animals and μ_2 represent the mean red blood cell mass of the control animals. Then the claim can be expressed as $\mu_1 \neq \mu_2$, and we have the hypotheses

$$H_0: \mu_1 = \mu_2 \qquad\qquad H_0: \mu_1 - \mu_2 = 0$$
$$\text{versus} \qquad \text{or} \qquad \text{versus}$$
$$H_1: \mu_1 \neq \mu_2 \qquad\qquad H_1: \mu_1 - \mu_2 \neq 0$$

Step 2: The level of significance $\alpha = 0.05$.

Step 3: The sample statistics for the data in Table 3 are shown in Table 4. The test statistic is

Table 4		
	Flight Animals	**Control Animals**
Sample size	$n_1 = 14$	$n_2 = 14$
Sample mean	$\bar{x}_1 = 7.881$	$\bar{x}_2 = 8.430$
Sample standard deviation	$s_1 = 1.017$	$s_2 = 1.005$

$$t_0 = \frac{(\bar{x}_1 - \bar{x}_2) - (\mu_1 - \mu_2)}{\sqrt{\dfrac{s_1^2}{n_1} + \dfrac{s_2^2}{n_2}}} = \frac{(7.881 - 8.430) - 0}{\sqrt{\dfrac{1.017^2}{14} + \dfrac{1.005^2}{14}}}$$

$$= \frac{-0.549}{0.3821288115} = -1.437$$

Classical Approach

Step 4: This is a two-tailed test with $\alpha = 0.05$. Since the sample sizes of the experimental group and control group are both 14, we have $n_1 - 1 = 14 - 1 = 13$ degrees of freedom. The critical values are $t_{\frac{\alpha}{2}} = t_{\frac{0.05}{2}} = t_{0.025} = 2.160$ and $-t_{0.025} = -2.160$.

The critical region is displayed in Figure 6.

Figure 6

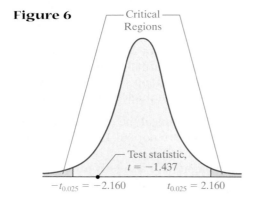

Critical Regions

Test statistic, $t = -1.437$

$-t_{0.025} = -2.160$ $t_{0.025} = 2.160$

Step 5: Because the test statistic does not lie within a critical region, we do not reject the null hypothesis.

P-Value Approach

Step 4: Because this is a two-tailed test, the P-value is the area under the t-distribution to the left of $t_0 = -1.437$ plus the area under the t-distribution to the right of $t_0 = 1.437$. See Figure 7.

Figure 7

The sum of the area in the tails is the P-value

The sum of the area in the tails is the P-value

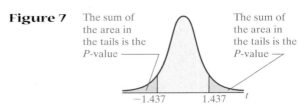

-1.437 1.437 t

Since the sample size of the experimental group and control group are both 14, we have $n_1 - 1 = 14 - 1 = 13$ degrees of freedom. Because of symmetry, we use Table V to estimate the area under the t-distribution to the right of $t_0 = 1.437$ and double it.

$$\text{P-value} = P(t_0 < -1.437 \text{ or } t_0 > 1.437)$$
$$= 2P(t_0 > 1.437)$$

Using Table V, we find the row that corresponds to 13 degrees of freedom. The value 1.437 lies between 1.350 and 1.771. The area under the t-distribution with 13 degrees of freedom to the right of 1.350 is 0.10. The area under the t-distribution with 13 degrees of freedom to the right of 1.771 is 0.05. After doubling these values, we have

$$0.10 < \text{P-value} < 0.20$$

Step 5: Because P-value $> \alpha$, we do not reject the null hypothesis.

Step 6: There is not sufficient evidence to conclude that the flight animals have a different red blood cell mass from the control animals at the $\alpha = 0.05$ level of significance.

The degrees of freedom used to determine the critical value(s) presented in Example 1 are conservative. Results that are more accurate can be obtained by using the following degrees of freedom:

$$\mathrm{df} = \frac{\left(\dfrac{s_1^2}{n_1} + \dfrac{s_2^2}{n_2}\right)^2}{\dfrac{\left(\dfrac{s_1^2}{n_1}\right)^2}{n_1 - 1} + \dfrac{\left(\dfrac{s_2^2}{n_2}\right)^2}{n_2 - 1}} \qquad (2)$$

When using Formula (2) to compute degrees of freedom, round down to the nearest integer to use Table V. For hand inference, it is recommended that we use the smaller of $n_1 - 1$ or $n_2 - 1$ as the degrees of freedom to ease computation. However, computer software will use Formula (2) when computing the degrees of freedom for increased precision in determining the P-value.

EXAMPLE 2   **Testing Hypotheses Regarding Two Means Using Technology**

Problem: Obtain an exact P-value for the problem in Example 1 using statistical software or a graphing calculator with advanced statistical features.

Approach: We will use Excel to obtain the P-value. The steps for testing hypotheses regarding two means using the TI-83/84 Plus graphing calculator, MINITAB, and Excel are given in the Technology Step by Step on page 534.

Result: Figure 8 shows the results obtained from Excel. The P-value is 0.1627.

Figure 8

t-Test: *Two-Sample Assuming Unequal Variances*

	Flight	Control
Mean	7.880714286	8.43
Variance	1.035207143	1.010969231
Observations	14	14
Hypothesized Mean Difference	0	
df	26	
t Stat	−1.436781704	
P(T<=t) one-tail	0.081352709	
t Critical one-tail	1.705616341	
P(T<=t) two-tail	0.162705419	
t Critical two-tail	2.055530786	

Interpretation: There is a 0.1627 probability of obtaining a sample mean difference that is $|t_0| = |-1.437| = 1.44$ sample standard deviations from the hypothesized mean difference of 0. The results obtained assuming the statement in the null hypothesis is true are not unusual. There is not sufficient evidence to conclude that the flight animals have a different red blood cell mass from the control animals at the $\alpha = 0.05$ level of significance.

Notice that the degrees of freedom in the technology solution are 26* versus 13 in the conservative solution done by hand in Example 1. With the lower degrees of freedom, the critical t is larger (2.160 with 13 degrees of freedom versus 2.056 with approximately 26 degrees of freedom). The larger critical value increases the number of standard deviations the difference in the sample means must be from the hypothesized mean difference before the null hypothesis is rejected. Therefore, in using the smaller of $n_1 - 1$ or $n_2 - 1$ degrees of

! **CAUTION**

⬤ The degrees of freedom in by-hand solutions will not equal the degrees of freedom in technology solutions unless you use Formula (2) to compute degrees of freedom.

*Actually, the degrees of freedom are 25.996, but Excel rounded to 26.

freedom, we need more substantial evidence to reject the null hypothesis. This requirement decreases the probability of a Type I error (rejecting the null when the null is true) below the actual level of α chosen by the researcher. This is what we mean when we say that the method of using the lesser of $n_1 - 1$ and $n_2 - 1$ as a proxy for degrees of freedom is conservative compared with using Formula (2).

Now Work Problems 13(a) and (b).

Note to Instructor
The purpose of this activity is to give students a hands on experience with hypothesis testing involving two means. The second rope should be clearly longer or shorter than the first rope. Be sure to tell students the lengths of the ropes only after all estimates have been reported. Time limit: 30–45 minutes depending on discussion.

In-Class Activity: Stringing Them Along (Part II)

Do you believe that it is easier to estimate the length of a longer rope or shorter rope by sight?

(a) Look at the piece of rope that your instructor is holding and estimate the length of the rope in inches. Now look at the second piece of rope and estimate its length in inches. After your instructor tells you the actual length of each piece of rope, compute the absolute value of the difference between your estimate and the actual length. That is, compute |actual−estimate| for each student's estimate.

(b) Assuming that estimates are just as good for long ropes as they are for short ropes, the null hypothesis is H_0: $\mu_{longer} = \mu_{shorter}$ (where μ represents the population mean absolute difference between the actual length and the estimated length). Based on your answer to the question posed at the beginning of the activity, select an appropriate alternative hypothesis and a level of significance.

(c) Combine your data with the rest of the class. Conduct the hypothesis test you outlined in part (b). What did you conclude?

(d) Could this test have been done as a matched-pairs test? Explain.

Note: Save the class data for use in another activity.

② **Construct and Interpret Confidence Intervals Regarding the Difference of Two Independent Means**

Constructing a confidence interval about the difference of two means is an extension of the results presented in Section 9.2.

Constructing a $(1 - \alpha) \cdot 100\%$ Confidence Interval about the Difference of Two Means

Suppose a simple random sample of size n_1 is taken from a population with unknown mean μ_1 and unknown standard deviation σ_1. Also, a simple random sample of size n_2 is taken from a population with unknown mean μ_2 and unknown standard deviation σ_2. If the two populations are normally distributed or the sample sizes are sufficiently large ($n_1 \geq 30$ and $n_2 \geq 30$), a $(1 - \alpha) \cdot 100\%$ confidence interval about $\mu_1 - \mu_2$ is given by

$$\text{Lower bound:} \quad (\bar{x}_1 - \bar{x}_2) - t_{\frac{\alpha}{2}} \cdot \sqrt{\frac{s_1^2}{n_1} + \frac{s_2^2}{n_2}}$$

and

$$\text{Upper bound:} \quad (\bar{x}_1 - \bar{x}_2) + t_{\frac{\alpha}{2}} \cdot \sqrt{\frac{s_1^2}{n_1} + \frac{s_2^2}{n_2}} \quad \text{(3)}$$

where $t_{\frac{\alpha}{2}}$ is computed using the smaller of $n_1 - 1$ or $n_2 - 1$ degrees of freedom or Formula (2).

EXAMPLE 3

Constructing a Confidence Interval about the Difference of Two Means

Problem: Construct a 95% confidence interval about $\mu_1 - \mu_2$ using the data presented in Table 3.

Approach: A normal probability plot and boxplot (Figure 5) indicate that the data are approximately normal with no outliers. We compute the confidence interval with $\alpha = 0.05$ using Formula (3).

Solution: We have already found the sample statistics in Example 1. In addition, we found $t_{\alpha/2} = t_{0.025}$ with 13 degrees of freedom to be 2.160. Substituting into Formula (3), we obtain the following results:

Lower bound:

$$(\bar{x}_1 - \bar{x}_2) - t_{\frac{\alpha}{2}} \cdot \sqrt{\frac{s_1^2}{n_1} + \frac{s_2^2}{n_2}} = (7.881 - 8.430) - 2.160 \cdot \sqrt{\frac{1.017^2}{14} + \frac{1.005^2}{14}}$$

$$= -0.549 - 0.825 = -1.374$$

Upper bound:

$$(\bar{x}_1 - \bar{x}_2) + t_{\frac{\alpha}{2}} \cdot \sqrt{\frac{s_1^2}{n_1} + \frac{s_2^2}{n_2}} = (7.881 - 8.430) + 2.160 \cdot \sqrt{\frac{1.017^2}{14} + \frac{1.005^2}{14}}$$

$$= -0.549 + 0.825 = 0.276$$

Interpretation: We are 95% confident that the mean difference between the red blood cell mass of the flight animals and control animals is between -1.374 and 0.276 mL. Because the confidence interval contains zero, there is not sufficient evidence to conclude that there is a difference in the red blood cell mass of the flight group and the control group.

Now Work Problems 13(c) and (d).

! CAUTION

We would use the pooled two-sample t-test when the two samples come from populations that have the same variance. *Pooling* refers to finding a weighted average of the two sample variances from the independent samples. It is difficult to verify that two sample variances might be equal, so we will always use Welch's t when comparing two means.

What about the Pooled Two-Sample t-tests?

Perhaps you noticed that statistical software and graphing calculators with advanced statistical features provide an option for two types of two-sample t-tests: one that assumes equal population variances (pooling) and one that does not assume equal population variances. Welch's t-statistic does not assume that the population variances are equal and can be used whether the population variances are equal or not. The test that assumes equal population variances is referred to as the *pooled t-statistic*.

The **pooled t-statistic** is computed by finding a weighted average of the sample variances and uses this average in the computation of the test statistic. The advantage of this test statistic is that it exactly follows Student's t-distribution with $n_1 + n_2 - 2$ degrees of freedom.

The disadvantage of the test statistic is that it requires that the population variances be equal. How is this requirement to be verified? While a test for determining the equality of variances does exist (F-test, not covered in this text), the test *requires* that each population be normally distributed. However, the F-test is not robust. Any minor departures from normality will make the results of the F-test unreliable. It has been recommended by many statisticians[*] that a preliminary F-test to check the requirement of equality of variance not be performed. In fact, George Box once said, "To make preliminary tests on variances is rather like putting to sea in a rowing boat to find out whether conditions are sufficiently calm for an ocean liner to leave port!"

[*]Moser and Stevens, Homogeneity of Variance in the Two-Sample Means Test, *American Statistician*, Vol. 46, No. 1.

Because the formal F-test for testing the equality of variances is so volatile, we are content to use Welch's t. This test is more conservative than the pooled t. The price that must be paid for the conservative approach is that the probability of a Type II error is higher in Welch's t than in the pooled t when the population variances are equal. However, the two tests typically provide the same conclusion, even if the assumption of equal population standard deviations seems reasonable.

MAKING AN INFORMED DECISION

Where Should I Invest?

Suppose that you have just received an inheritance of $10,000 and you decide that you should invest the money rather than blow it on frivolous items. You have decided that you will invest the money in one of two types of mutual funds. The first type you are considering follows a *large-value* approach to investing. This means that the mutual fund invests only in large, established companies that are considered to be a good bargain. The second type of mutual fund you are considering follows a *large-growth* approach to investing. This means that the mutual fund invests in large companies that are experiencing solid revenue growth.

To make an informed decision, you decide to research the rate of return of the past 3 years for each of these types of mutual funds. The mutual fund must have a Morningstar rating of four or five stars.

The Morningstar mutual-fund rating system ranks mutual funds, using one to five stars. The stars divide the mutual-fund performance into quintiles; that is, a mutual fund with a one-star rating is in the bottom 20% of mutu-

al funds in its category, a mutual fund with a two-star rating has an investment performance between the 21st and 40th percentile, and so on. These data can be found at www.morningstar.com or screen.yahoo.com/funds.html.

(a) Obtain a simple random sample of at least 15 mutual funds for each investment category. Determine the 3-year rate of return for each fund.

(b) Verify that the 3-year rates of return come from a population that is normally distributed. Also, verify that the data have no outliers. If the data do not come from a population that is normally distributed, you will have to increase the sample size so that the Central Limit Theorem can be used.

(c) Construct a boxplot for the rate of return of each fund category using the same scale. Which investment category, if any, seems superior?

(d) Obtain a 95% confidence interval for the difference between the mean rates of return. Interpret the interval.

(e) Write a report that details which investment category seems to be superior.

11.2 ASSESS YOUR UNDERSTANDING

Concepts and Vocabulary

1. What are the requirements that need to be satisfied to test a hypothesis regarding the difference of two means with σ unknown?

2. Explain why using the smaller of $n_1 - 1$ or $n_2 - 1$ degrees of freedom to determine the critical t instead of Formula (2) is conservative.

Skill Building*

In Problems 3–8, assume that the populations are normally distributed.

3. (a) Test whether $\mu_1 \neq \mu_2$ at the $\alpha = 0.05$ level of significance for the given sample data. Do not reject H_0
   (b) Construct a 95% confidence interval about $\mu_1 - \mu_2$.

4. (a) Test whether $\mu_1 \neq \mu_2$ at the $\alpha = 0.05$ level of significance for the given sample data. Reject H_0
   (b) Construct a 95% confidence interval about $\mu_1 - \mu_2$.

	Population 1	Population 2
n	15	15
$\overline{x}$	15.3	14.2
s	3.2	3.5

	Population 1	Population 2
n	20	20
$\overline{x}$	111	104
s	8.6	9.2

(3b) $(-1.53, 3.73)$

(4b) (1.1, 12.9)

*The confidence intervals in the back of the text were computed using the smaller of $n_1 - 1$ or $n_2 - 1$ degrees of freedom. These intervals will be wider than those obtained using technology.

5. (a) Test whether $\mu_1 > \mu_2$ at the $\alpha = 0.1$ level of significance for the given sample data. Reject H_0
(b) Construct a 90% confidence interval about $\mu_1 - \mu_2$.

	Population 1	Population 2
n	25	18
$\overline{x}$	50.2	42.0
s	6.4	9.9

6. (a) Test whether $\mu_1 < \mu_2$ at the $\alpha = 0.05$ level of significance for the given sample data. Reject H_0
(b) Construct a 95% confidence interval about $\mu_1 - \mu_2$.

	Population 1	Population 2
n	40	32
$\overline{x}$	94.2	115.2
s	15.9	23.0

7. (a) Test whether $\mu_1 < \mu_2$ at the $\alpha = 0.02$ level of significance for the given sample data. Reject H_0
(b) Construct a 90% confidence interval about $\mu_1 - \mu_2$.

	Population 1	Population 2
n	32	25
$\overline{x}$	103.4	114.2
s	12.3	13.2

8. (a) Test whether $\mu_1 > \mu_2$ at the $\alpha = 0.05$ level of significance for the given sample data. Do not reject H_0
(b) Construct a 95% confidence interval about $\mu_1 - \mu_2$.

	Population 1	Population 2
n	23	13
$\overline{x}$	43.1	41.0
s	4.5	5.1

(5b) $(3.57, 12.83)$ (6b) $(-30.75, -11.25)$ (7b) $(-16.65, -4.95)$ (8b) $(-1.60, 5.80)$

Applying the Concepts

9. Treating Bipolar Mania In a study published in the *Archives of General Psychiatry* entitled "Efficacy of Olanzapine in Acute Bipolar Mania" (Vol. 57, No. 9, pp. 841–849), researchers conducted a randomized, double-blind study to measure the effects of the drug olanzapine on patients diagnosed with bipolar disorder. A total of 115 patients with a DSM-IV diagnosis of bipolar disorder were randomly divided into two groups. Group 1 ($n_1 = 55$) received 5 to 20 mg per day of olanzapine, while Group 2 ($n_2 = 60$) received a placebo. The effectiveness of the drug was measured using the Young–Mania Rating Scale total score with the net improvement in the score recorded. The results are presented in the table.

	Experimental Group	Control Group
n	55	60
Mean improvement	14.8	8.1
Sample standard deviation	12.5	12.7

(a) Does the evidence suggest that the experimental group experienced a larger mean improvement than the control group at the $\alpha = 0.01$ level of significance? Reject H_0
(b) Construct a 95% confidence interval about $\mu_1 - \mu_2$ and interpret the results. $(1.98, 11.42)$

10. Hormone Replacement Therapy Coronary heart disease is the leading cause of death among older women. In observational studies, the number of deaths due to coronary heart disease has been reduced in postmenopausal women who take hormone replacement therapy. Low levels of serum high-density lipoprotein (HDL) cholesterol are considered to be one of the risk factors predictive of death from coronary heart disease.

Researchers at the Washington University School of Medicine claimed that serum HDL increases when patients participate in hormone replacement therapy. The researchers randomly divided 59 sedentary women 75 years of age or older into two groups. The 30 patients in Group 1 (the experimental group) took hormone replacement pills for 9 months. The 29 patients in Group 2 (the control group) took a placebo for 9 months. At the conclusion of the treatment, the patient's serum HDL was recorded. The experiment was double-blind. The following results were obtained, where the means and standard deviations are in milligrams per deciliter (mg/dL).

	Experimental Group	Control Group
Sample size	$n_1 = 30$	$n_2 = 29$
Mean increase in HDL	$\overline{x}_1 = 8.1$	$\overline{x}_2 = 2.4$
Sample standard deviation	$s_1 = 10.5$	$s_2 = 4.3$

Source: Ellen F. Binder et al., Effects of Hormone Replacement Therapy on Serum Lipids in Elderly Women, *Annals of Internal Medicine* 134 (May 2001), pp. 754–760.

(a) What type of experimental design is this? What are the treatments? How many levels does the factor have?
(b) Does the evidence suggest that the experimental group had a larger mean increase in serum HDL levels than the control group at the $\alpha = 0.01$ level of significance (serum HDL is normally distributed)? Reject H_0
(c) Construct a 95% confidence interval about $\mu_1 - \mu_2$ and interpret the results. $(1.45, 9.95)$

11. **Walking in the Airport, Part I** Do people walk faster in the airport when they are departing (getting on a plane) or when they are arriving (getting off a plane)? Researcher Seth B. Young measured the walking speed of travelers in San Francisco International Airport and Cleveland Hopkins International Airport. His findings are summarized in the table.

Direction of Travels	Departure	Arrival
Mean speed (feet per minute)	260	269
Standard deviation (feet per minute)	53	34
Sample size	35	35

Source: Young, Seth B., Evaluation of Pedestrian Walking Speeds in Airport Terminals, *Transportation Research Record*, Paper 99-0824.

(a) Is this an observational study or a designed experiment? Why? *Observational*

(b) Explain why it is reasonable to use Welch's *t*-test.

(c) Do individuals walk at different speeds depending on whether they are departing or arriving at the $\alpha = 0.05$ level of significance? *Do not reject H_0*

(d) Construct a 95% confidence interval about $\mu_{arrival} - \mu_{departure}$. Interpret the interval.

12. **Walking in the Airport, Part II** Do business travelers walk at a different pace than leisure travelers? Researcher Seth B. Young measured the walking speed of business and leisure travelers in San Francisco International Airport and Cleveland Hopkins International Airport. His findings are summarized in the table.

Type of Traveler	Business	Leisure
Mean speed (feet per minute)	272	261
Standard deviation (feet per minute)	43	47
Sample size	20	20

Source: Young, Seth B., Evaluation of Pedestrian Walking Speeds in Airport Terminals, *Transportation Research Record*, Paper 99-0824.

(a) Is this an observational study or a designed experiment? Why? *Observational*

(b) What must be true regarding the populations to use Welch's *t*-test to compare the means?

(c) Assuming the requirements listed in part (b) are satisfied, determine whether business travelers walk at a different speed from leisure travelers at the $\alpha = 0.05$ level of significance. *Do not reject H_0*

(d) Construct a 95% confidence interval about $\mu_{business} - \mu_{leisure}$. Interpret the interval.

13. **Concrete Strength** An engineer wanted to know whether NW the strength of two different concrete mix designs differed significantly. He randomly selected 9 cylinders, measuring 6 inches in diameter and 12 inches in height, into which mixture 67-0-301 was poured. After 28 days, he measured the strength (in pounds per square inch) of the cylinder. He also randomly selected 10 cylinders of mixture 67-0-400 and performed the same test. The results are as follows:

Mixture 67-0-301			Mixture 67-0-400			
3960	4090	3100	4070	4890	5020	4330
3830	3200	3780	4640	5220	4190	3730
4080	4040	2940	4120	4620		

(a) Is it reasonable to use Welch's *t*-test? Why? *Yes*

   Note: Normal probability plots indicate that the data are approximately normal and boxplots indicate that there are no outliers.

(b) Determine whether mixture 67-0-400 is stronger than mixture 67-0-301 at the $\alpha = 0.05$ level of significance.

(c) Construct a 90% confidence interval about $\mu_{400} - \mu_{301}$ and interpret the results. *(416, 1212)*

(d) Draw boxplots of each data set using the same scale. Does this visual evidence support the results obtained in part (b)?

14. **Measuring Reaction Time** Researchers at the University of Mississippi wanted to determine whether the reaction time (in seconds) of males differed from that of females to a go/no go stimulus. The researchers randomly selected 20 females and 15 males to participate in the study. The go/no go stimulus required the student to respond to a particular stimulus and not to respond to other stimuli. The results are as follows:

Female Students				Male Students		
0.588	0.652	0.442	0.293	0.375	0.256	0.427
0.340	0.636	0.391	0.367	0.654	0.563	0.405
0.377	0.646	0.403	0.377	0.374	0.465	0.402
0.380	0.403	0.617	0.434	0.373	0.488	0.337
0.443	0.481	0.613	0.274	0.224	0.477	0.655

Source: PsychExperiments at the University of Mississippi

(a) Is it reasonable to use Welch's *t*-test? Why? *Yes*

   Note: Normal probability plots indicate that the data are approximately normal and boxplots indicate that there are no outliers.

(b) Test whether there is no difference in the reaction time of males and females at the $\alpha = 0.05$ level of significance. *Do not reject H_0*

(c) Construct a 90% confidence interval about $\mu_f - \mu_m$ and interpret the results. *(−0.0482, 0.1006)*

(d) Draw boxplots of each data set using the same scale. Does this visual evidence support the results obtained in part (b)?

(11d) $(-12.63, 30.63)$ (12b) *Normally distributed*   (12d) $(-18.81, 40.81)$

15. Bacteria in Hospital Carpeting Researchers wanted to determine if carpeted rooms contained more bacteria than uncarpeted rooms. To determine the amount of bacteria in a room, researchers pumped the air from the room over a Petri dish at the rate of 1 cubic foot per minute for eight carpeted rooms and eight uncarpeted rooms. Colonies of bacteria were allowed to form in the 16 Petri dishes. The results are presented in the table. A normal probability plot and boxplot indicate that the data are approximately normally distributed with no outliers.

Carpeted Rooms (bacteria/cubic foot)		Uncarpeted Rooms (bacteria/cubic foot)	
11.8	10.8	12.1	12.0
8.2	10.1	8.3	11.1
7.1	14.6	3.8	10.1
13.0	14.0	7.2	13.7

Source: Walter, William G., and Stober, Angie, Microbial Air Sampling in a Carpeted Hospital. *Journal of Environmental Health*, 30 (1968), p. 405.

(a) Do carpeted rooms have more bacteria than uncarpeted rooms at the $\alpha = 0.05$ level of significance?

(b) Construct a 95% confidence interval about $\mu_{carpet} - \mu_{uncarpeted}$ and interpret the results.

16. Visual versus Textual Learners Researchers wanted to know whether there was a difference in comprehension among students learning a computer program based on the style of the text. They randomly divided 36 students into two groups of 18 each. The researchers verified that the 36 students were similar in terms of educational level, age, and so on. Group 1 individuals learned the software using a visual manual (*multimodal instruction*), while Group 2 individuals learned the software using a textual manual (*unimodal instruction*). The following data represent scores the students received on an exam given to them after they studied from the manuals.

Visual Manual		Textual Manual	
51.08	60.35	64.55	56.54
57.03	76.60	57.60	39.91
44.85	70.77	68.59	65.31
75.21	70.15	50.75	51.95
56.87	47.60	49.63	49.07
75.28	46.59	43.58	48.83
57.07	81.23	57.40	72.40
80.30	67.30	49.48	42.01
52.20	60.82	49.57	61.16

Source: Mark Gellevij et al., Multimodal Versus Unimodal Instruction in a Complex Learning Context, *Journal of Experimental Education*, 2002, 70(3), pp. 215–239.

(a) What type of experimental design is this?

(b) What are the treatments?

(c) A normal probability plot and boxplot indicate it is reasonable to use Welch's *t*-test. Is there a difference in test scores at the $\alpha = 0.05$ level of significance?

(d) Construct a 95% confidence interval about $\mu_{visual} - \mu_{textual}$ and interpret the results.

17. Does the Designated Hitter Help? In baseball, the American League allows a designated hitter (DH) to bat for the pitcher, who is typically a weak hitter. In the National League, the pitcher must bat. The common belief is that this results in American League teams scoring more runs. In interleague play, when American League teams visit National League teams, the American League pitcher must bat. So, if the DH does result in more runs, we would expect that American League teams will score fewer runs when visiting National League parks. To test this claim, a random sample of runs scored by American League teams with and without their DH is given in the following table. Does the designated hitter results in more runs scored at the $\alpha = 0.05$ level of significance? **Note:** $\overline{x}_{NL} = 4.3$, $s_{NL} = 2.6$, $\overline{x}_{AL} = 6.0$, $s_{AL} = 3.5$. Reject H_0

National League Park (without DH)					American League Park (with DH)				
1	5	5	4	7	6	2	3	6	8
2	6	2	9	2	1	3	7	6	4
8	8	2	10	4	4	12	5	6	13
4	3	4	1	9	6	9	5	6	7
3	5	1	3	3	4	3	2	5	5
3	5	2	7	2	6	14	14	7	0

Source: espn.com

18. **Rhythm & Blues versus Alternative** A music industry producer wondered whether there is a difference in lengths (in seconds) of rhythm & blues songs versus alternative songs. He obtained a random sample of each music category and documented song lengths. The results are in the following table. Test whether the length of rhythm & blues songs is different from the length of alternative songs at the $\alpha = 0.1$ level of significance. **Note:** $\overline{x}_{RB} = 242.7$, $s_{RB} = 26.9$, $\overline{x}_{ALT} = 238.3$, $s_{ALT} = 28.9$.
Do not reject H_0

Rhythm & Blues (in seconds)					Alternative (in seconds)				
267	244	233	293	231	246	279	226	255	249
224	271	246	258	255	225	216	197	216	232
281	256	236	231	224	256	307	237	216	187
203	258	237	228	217	258	253	223	264	255
205	217	227	211	235	227	274	192	213	272
241	211	257	321	264	226	251	202	278	216

Source: www.yahoo.com/music

19. **Kids and Leisure** Young children require a lot of time. This time commitment cuts into a parent's leisure time. A sociologist wanted to estimate the difference in the amount of daily leisure time (in hours) of adults who do not have children under the age of 18 years and the amount of daily leisure time (in hours) of adults who have children under the age of 18 years. A random sample of 40 adults with no children under the age of 18 years results in a mean daily leisure time of 5.62 hours with a standard deviation of 2.43 hours. A random sample of 40 adults with children under the age of 18 years results in a mean daily leisure time of 4.10 hours with a standard deviation of 1.82 hours. Construct and interpret a 90% confidence interval for the mean difference in leisure time between adults with no children and adults with children. (*Source*: American Time Use Survey) (0.71, 2.33)

20. **Aluminum Bottles** The aluminum bottle, first introduced in 1991 by CCL Container for mainly personal and household items such as lotions, has become popular with beverage manufacturers. Besides being lightweight and requiring less packaging, the aluminum bottle is reported to cool faster and stay cold longer than typical glass bottles. A small brewery tests this claim and obtains the following information regarding the time (in minutes) required to chill a bottle of beer from room temperature (75°F) to serving temperature (45°F). Construct and interpret a 90% confidence interval for the mean difference in cooling time for clear glass versus aluminum. (38.08, 44.72)

	Clear Glass	Aluminum
Sample size	$n_1 = 42$	$n_2 = 35$
Mean time to chill	133.8	92.4
Sample standard deviation	9.9	7.3

21. **Comparing Step Pulses** A physical therapist wanted to know whether the mean step pulse of men was less than the mean step pulse of women. She randomly selected 51 men and 70 women to participate in the study. Each subject was required to step up and down onto a 6-inch platform for 3 minutes. The pulse of each subject (in beats per minute) was then recorded. After the data were entered into MINITAB, the following results were obtained.

Two Sample T-Test and Confidence Interval
Two sample T for Men vs Women

	N	Mean	StDev	SE Mean
Men	51	112.3	11.3	1.6
Women	70	118.3	14.2	1.7

95% CI for mu Men − mu Women: (−10.7, −1.5)
T-Test mu Men = mu Women (vs <):T = −2.61 P = 0.0051 DF = 118

(a) State the null and alternative hypotheses.
(b) Identify the *P*-value and state the researcher's conclusion if the level of significance was $\alpha = 0.01$. 0.0051
(c) What is the 95% confidence interval for the mean difference in pulse rates of men versus women? Interpret this interval. (−10.7, −1.5)

22. **Comparing Flexibility** A physical therapist believes that women are more flexible than men. She measures the flexibility of 31 randomly selected women and 45 randomly selected men by determining the number of inches subjects could reach while sitting on the floor with their legs straight out and back perpendicular to the ground. The more flexible an individual is, the higher the measured flexibility will be. After entering the data into MINITAB, she obtained the following results:

Two Sample T-Test and Confidence Interval
Two sample T for Men vs Women

	N	Mean	StDev	SE Mean
Men	45	18.64	3.29	0.49
Women	31	20.99	2.07	0.37

95% CI for mu Men − mu Women: (−3.58, −1.12)
T-Test mu Men = mu Women (vs <):T = −3.82 P = 0.0001 DF = 73

(a) State the null and alternative hypotheses.
(b) Identify the *P*-value and state the researcher's conclusion if the level of significance was $\alpha = 0.01$. 0.0001
(c) What is the 95% confidence interval for the mean difference in flexibility of men versus women? Interpret this interval. (−3.58, −1.12)

(21a) $H_0: \mu_{men} = \mu_{women}$ vs $H_1: \mu_{men} < \mu_{women}$ (22a) $H_0: \mu_{men} = \mu_{women}$ vs $H_1: \mu_{men} < \mu_{women}$

Consumer Reports   **The High Cost of Convenience**

Consumer Reports was interested in comparing a name-brand paper towel with a new version packaged in a box. The towels in the box, which cost nearly twice as much as the traditional roll, are marketed for their convenience. Given the difference in cost, one might wonder if the boxed version performs better than the traditional roll. To help answer this question, technicians at Consumers Union subjected both types of towels to five physical tests: absorption time in water, absorption time in oil, absorption capacity in water, absorption capacity in oil, and wet strength. For brevity, we will discuss only the results of the absorption time in water test.

The absorption time in water was defined as the amount of time necessary for a single sheet to absorb a predetermined amount of water. To compare the absorption times of the two types of towels, we tested six randomly selected sheets of paper towels. To avoid potential sources of bias, the individual sheets were taken from different samples of the products, and the tests were conducted in a randomly chosen order.

Consumer Reports wanted to determine whether the water absorption time for the boxed version is less than the water absorption time for the traditional roll.

(a) Write the null and alternative hypotheses, letting μ_{box} represent the mean absorption time for the boxed version and μ_{roll} represent the mean absorption time for the roll version.

(b) Normal probability plots of the water absorption times for the two products are shown next. Based on the normal probability plots, is it reasonable to conduct a two-sample hypothesis test?

(c) A boxplot of the water absorption times for the two products follows:

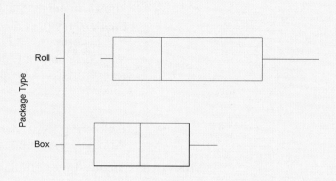

Does the data set have any outliers? Based on the boxplots, do you think that the absorption times for the boxed version are lower than the absorption times for the roll?

(d) To conduct the test, we used MINITAB to perform a two-sample *t*-test. The results are as shown. Using the MINITAB output, determine the value of the test statistic. What is the *P*-value of the test? Although they are not discussed here, the other physical tests provided similar results. Write an article that states your conclusion and any recommendations that you would make regarding the purchase of the two products.

Two-Sample T-Test and CI: Absorption Time In Water, CU

```
Two-sample  T  for Absorption     Time   In Water

CU-Text     N            Mean    StDev   SE Mean
Box         6          0.9717   0.0538     0.022
Roll        6          1.0200   0.0942     0.038

Difference = mu (Box) - mu (Roll)
Estimate for difference: -0.0483
95% upper bound for difference: 0.0320
T-Test of difference = 0 (vs <):
T-Value = -1.09 P-Value = 0.150 DF = 10
Both use Pooled StDev = 0.0767
```

© by Consumers Union of U.S., Inc., Yonkers, NY 10703-1057, a nonprofit organization. Reprinted with permission.

Note to Readers: *In many cases, our test protocol and analytical methods are more complicated than described in these examples. The data and discussions have been modified to make the material more appropriate for the audience.*

| Technology Step by Step | **Two-Sample *t*-Tests, Independent Sampling** |

TI-83/84 Plus   Hypothesis Tests

Step 1: If necessary, enter raw data in L1 and L2.

Step 2: Press STAT, highlight TESTS, and select `4:2-SampTTest ....`

Step 3: If the data are raw, highlight Data, making sure that List1 is set to L1 and List2 is set to L2, with frequencies set to 1. If summary statistics are known, highlight Stats and enter the summary statistics.

Step 4: Highlight the appropriate relation between μ_1 and μ_2 in the alternative hypothesis. Set Pooled to NO.

Step 5: Highlight Calculate or Draw and press ENTER. Calculate gives the test statistic and *P*-value. Draw will draw the *t*-distribution with the *P*-value shaded.

Confidence Intervals

Follow the steps given for hypothesis tests, except select `0::2-SampTInt`. Also, select a confidence level (such as 95% = 0.95).

MINITAB   *Step 1:* Enter raw data in columns C1 and C2.

Step 2: Select the **Stat** menu, highlight **Basic Statistics**, then highlight **2-Sample t . . . .**

Step 3: Select "Samples in different columns." Enter C1 in the cell marked "First" and enter C2 in the cell marked "Second." Select the direction of the alternative hypothesis and select a confidence level. Click OK.

Excel   *Step 1:* Enter raw data in columns A and B.

Step 2: Select the **Tools menu** and highlight **Data Analysis . . . .**

Step 3: Select "*t*-test: Two-Sample Assuming Unequal Variances." With the cursor in the "Variable 1 Range" cell, highlight the data in column A. Enter the hypothesized difference in the means (usually 0) and a value for alpha. Click OK.

11.3 Inference about Two Population Proportions

Preparing for This Section   Before getting started, review the following:

- Confidence intervals about a population proportion (Section 9.3, pp. 435–438)
- Hypothesis tests about a population proportion (Section 10.4, pp. 493–498)

Objectives

① **Test hypotheses regarding two population proportions**

② **Construct and interpret confidence intervals for the difference between two population proportions**

③ **Determine the sample size necessary for estimating the difference between two population proportions within a specified margin of error**

In Sections 9.3 and 10.4, we discussed inference regarding a single population proportion. We will now discuss inferential methods for comparing two population proportions. For example, in clinical trials of the drug Nasonex, a drug that is meant to relieve allergy symptoms, 26% of patients receiving 200 micrograms (μg) of Nasonex reported a headache as a side effect while 22% of patients receiving a placebo reported a headache as a side effect. Researchers want to determine whether the proportion of patients receiving the treatment and complaining of headaches is significantly higher than the proportion of patients receiving the placebo and complaining of headaches.

To conduct inference about two population proportions, we must first determine the sampling distribution of the difference of two proportions. Recall that the point estimate of a population proportion, p, is given by $\hat{p} = \dfrac{x}{n}$, where x is the number of the n individuals in the sample that have a specific characteristic. In addition, we recall that the sampling distribution of $\hat{p}$ is approximately normal with mean $\mu_{\hat{p}} = p$ and standard deviation $\sigma_{\hat{p}} = \sqrt{\dfrac{p(1-p)}{n}}$, provided that $np(1-p) \geq 10$, so

$$z = \frac{\hat{p} - p}{\sqrt{\dfrac{p(1-p)}{n}}}$$

is approximately normal with mean 0 and standard deviation 1. Using this information along with the idea of independent sampling from two populations, we obtain the sampling distribution of the difference between two proportions:

Sampling Distribution of the Difference between Two Proportions

Suppose a simple random sample of size n_1 is taken from a population where x_1 of the individuals have a specified characteristic, and a simple random sample of size n_2 is independently taken from a different population where x_2 of the individuals have a specified characteristic. The sampling distribution of $\hat{p}_1 - \hat{p}_2$, where $\hat{p}_1 = \dfrac{x_1}{n_1}$ and $\hat{p}_2 = \dfrac{x_2}{n_2}$, is approximately normal, with mean $\mu_{\hat{p}_1 - \hat{p}_2} = p_1 - p_2$, and standard deviation

$$\sigma_{\hat{p}_1 - \hat{p}_2} = \sqrt{\frac{p_1(1 - p_1)}{n_1} + \frac{p_2(1 - p_2)}{n_2}},$$ provided that $n_1\hat{p}_1(1 - \hat{p}_1) \geq 10$ and $n_2\hat{p}_2(1 - \hat{p}_2) \geq 10$. The standardized version of $\hat{p}_1 - \hat{p}_2$ is then written as

$$z = \frac{(\hat{p}_1 - \hat{p}_2) - (p_1 - p_2)}{\sqrt{\dfrac{p_1(1 - p_1)}{n_1} + \dfrac{p_2(1 - p_2)}{n_2}}}$$

which has an approximate standard normal distribution.

---

① Test Hypotheses Regarding Two Population Proportions

Now that we know the approximate sampling distribution of $\hat{p}_1 - \hat{p}_2$, we can introduce a procedure that can be used to test hypotheses regarding two population proportions. We first consider the test statistic. Following the discussion for comparing two means, it seems reasonable that the test statistic for the difference of two population proportions would be

$$z = \frac{(\hat{p}_1 - \hat{p}_2) - (p_1 - p_2)}{\sqrt{\dfrac{p_1(1 - p_1)}{n_1} + \dfrac{p_2(1 - p_2)}{n_2}}} \tag{1}$$

When comparing two population proportions, the null hypothesis will always be $H_0\colon p_1 = p_2$. Because the null hypothesis is assumed to be true, the test assumes $p_1 = p_2$, so $p_1 - p_2 = 0$. We also assume both p_1 and p_2 equal p, where p is the common population proportion. If we substitute this value of p into Equation (1), we obtain

$$z = \frac{(\hat{p}_1 - \hat{p}_2) - (p_1 - p_2)}{\sqrt{\dfrac{p_1(1 - p_1)}{n_1} + \dfrac{p_2(1 - p_2)}{n_2}}} = \frac{\hat{p}_1 - \hat{p}_2 - 0}{\sqrt{\dfrac{p(1 - p)}{n_1} + \dfrac{p(1 - p)}{n_2}}} = \frac{\hat{p}_1 - \hat{p}_2}{\sqrt{p(1 - p)}\sqrt{\dfrac{1}{n_1} + \dfrac{1}{n_2}}} \quad \textbf{(2)}$$

In Other Words

The pooled estimate of p is obtained by summing the individuals in the sample that have a certain characteristic and dividing this result by the sum of the two sample sizes.

We need a point estimate of p because it is unknown. The best point estimate of p is called the **pooled estimate of p**, denoted $\hat{p}$, where

$$\hat{p} = \frac{x_1 + x_2}{n_1 + n_2}$$

Substituting the pooled estimate of p into Equation (2), we obtain

$$z = \frac{\hat{p}_1 - \hat{p}_2}{\sigma_{\hat{p}_1 - \hat{p}_2}} = \frac{\hat{p}_1 - \hat{p}_2}{\sqrt{\hat{p}(1 - \hat{p})}\sqrt{\dfrac{1}{n_1} + \dfrac{1}{n_2}}}$$

This test statistic will be used to test claims regarding two population proportions.

Hypothesis Test Regarding the Difference between Two Population Proportions

To test hypotheses regarding the two population proportions, p_1 and p_2, we can use the steps that follow, provided that

1. the samples are independently obtained using simple random sampling,
2. $n_1 \hat{p}_1 (1 - \hat{p}_1) \geq 10$ and $n_2 \hat{p}_2 (1 - \hat{p}_2) \geq 10$,
3. $n_1 \leq 0.05 N_1$ and $n_2 \leq 0.05 N_2$ (the sample size is no more than 5% of the population size); this requirement ensures the independence necessary for a binomial experiment.

Step 1: Determine the null and alternative hypotheses. The hypotheses can be structured in one of three ways:

Two-Tailed	Left-Tailed	Right-Tailed
H_0: $p_1 = p_2$	H_0: $p_1 = p_2$	H_0: $p_1 = p_2$
H_1: $p_1 \neq p_2$	H_1: $p_1 < p_2$	H_1: $p_1 > p_2$

Note: p_1 is the population proportion for population 1, and p_2 is the population proportion for population 2.

Step 2: Select a level of significance α, depending on the seriousness of making a Type I error.

Step 3: Compute the test statistic

$$z_0 = \frac{(\hat{p}_1 - \hat{p}_2)}{\sqrt{\hat{p}(1 - \hat{p})}\sqrt{\dfrac{1}{n_1} + \dfrac{1}{n_2}}}$$

where $\hat{p} = \dfrac{x_1 + x_2}{n_1 + n_2}$.

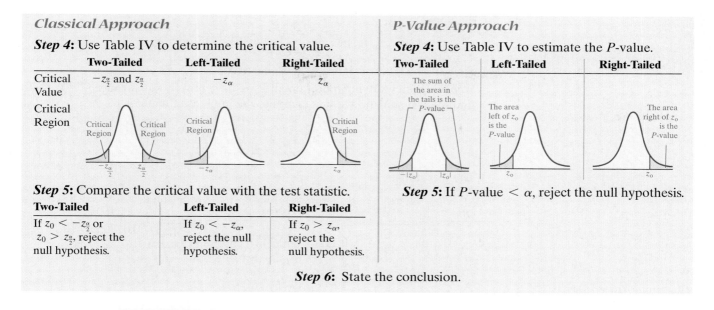

Classical Approach

Step 4: Use Table IV to determine the critical value.

	Two-Tailed	Left-Tailed	Right-Tailed
Critical Value	$-z_{\frac{\alpha}{2}}$ and $z_{\frac{\alpha}{2}}$	$-z_{\alpha}$	z_{α}

Step 5: Compare the critical value with the test statistic.

Two-Tailed	Left-Tailed	Right-Tailed
If $z_0 < -z_{\frac{\alpha}{2}}$ or $z_0 > z_{\frac{\alpha}{2}}$, reject the null hypothesis.	If $z_0 < -z_{\alpha}$, reject the null hypothesis.	If $z_0 > z_{\alpha}$, reject the null hypothesis.

P-Value Approach

Step 4: Use Table IV to estimate the P-value.

Two-Tailed	Left-Tailed	Right-Tailed
The sum of the area in the tails is the P-value	The area left of z_0 is the P-value	The area right of z_0 is the P-value

Step 5: If P-value $< \alpha$, reject the null hypothesis.

Step 6: State the conclusion.

EXAMPLE 1

Testing a Hypothesis Regarding Two Population Proportions

Problem: In clinical trials of Nasonex, 3774 adult and adolescent allergy patients (patients 12 years and older) were randomly divided into two groups. The patients in Group 1 (experimental group) received 200 mcg of Nasonex, while the patients in Group 2 (control group) received a placebo. Of the 2103 patients in the experimental group, 547 reported headaches as a side effect. Of the 1671 patients in the control group, 368 reported headaches as a side effect. Is there significant evidence to conclude that the proportion of Nasonex users that experienced headaches as a side effect is greater than the proportion in the control group at the $\alpha = 0.05$ level of significance?

Approach: We must verify the requirements to perform the hypothesis test. That is, the sample must be a simple random sample and $n_1\hat{p}_1(1 - \hat{p}_1) \geq 10$ and $n_2\hat{p}_2(1 - \hat{p}_2) \geq 10$. In addition, the sample size cannot be more than 5% of the population size. Then we follow the preceding Steps 1 through 6.

Solution: First we verify that the requirements are satisfied.

1. The samples are independently obtained using simple random sampling.
2. We have $x_1 = 547$, $n_1 = 2103$, $x_2 = 368$, and $n_2 = 1671$, so

$$\hat{p}_1 = \frac{x_1}{n_1} = \frac{547}{2103} = 0.26 \text{ and } \hat{p}_2 = \frac{x_2}{n_2} = \frac{368}{1671} = 0.22. \text{ Therefore,}$$

$$n_1\hat{p}_1(1 - \hat{p}_1) = 2103(0.26)(1 - 0.26) = 404.6172 \geq 10$$
$$n_2\hat{p}_2(1 - \hat{p}_2) = 1671(0.22)(1 - 0.22) = 286.7436 \geq 10$$

3. More than 10 million Americans 12 years old or older are allergy sufferers, so the sample sizes are less than 5% of the population size.

   All three requirements are satisfied, so we now proceed to follow Steps 1 through 6.

Step 1: We want to know whether the proportion of patients taking Nasonex who experience a headache is greater than the proportion of patients taking the placebo who experience a headache. Letting p_1 represent the population proportion of patients taking Nasonex who experience a headache and p_2 represent the population proportion of patients taking the placebo who experience a headache, we want to know if $p_1 > p_2$. This is a right-tailed hypothesis with

$$H_0: p_1 = p_2 \quad \text{versus} \quad H_1: p_1 > p_2$$

or, equivalently,

$$H_0: p_1 - p_2 = 0 \quad \text{versus} \quad H_1: p_1 - p_2 > 0$$

Step 2: The level of significance is $\alpha = 0.05$.

Step 3: From verifying requirement 2, we have that $\hat{p}_1 = 0.26$ and $\hat{p}_2 = 0.22$. To find the test statistic, we first compute the pooled estimate of $\hat{p}$:

$$\hat{p} = \frac{x_1 + x_2}{n_1 + n_2} = \frac{547 + 368}{2103 + 1671} = 0.242$$

The test statistic is

$$z_0 = \frac{(\hat{p}_1 - \hat{p}_2)}{\sqrt{\hat{p}(1 - \hat{p})}\sqrt{\frac{1}{n_1} + \frac{1}{n_2}}} = \frac{0.26 - 0.22}{\sqrt{0.242(1 - 0.242)}\sqrt{\frac{1}{2103} + \frac{1}{1671}}} = \frac{0.04}{0.0140357419} = 2.85$$

Classical Approach

Step 4: This is a right-tailed test with $\alpha = 0.05$. The critical value is $z_{0.05} = 1.645$. The critical region is displayed in Figure 9.

Figure 9

Critical Region

Test statistic, $z_0 = 2.85$

$z_{0.05} = 1.645$

Step 5: Because $z_0 > z_{0.05}$ (the test statistic lies within critical region), we reject the null hypothesis.

P-Value Approach

Step 4: Because this is a right-tailed test, the P-value is the area under the standard normal distribution to the right of $z_0 = 2.85$. See Figure 10.

Figure 10

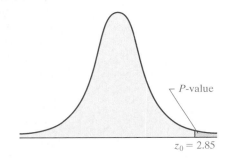

P-value

$z_0 = 2.85$

P-value $= P(z_0 > 2.85)$
$= 0.0022$

Step 5: Because P-value $< \alpha$ (0.0022 < 0.05), we reject the null hypothesis.

Step 6: There is sufficient evidence at the $\alpha = 0.05$ level of significance to conclude that the proportion of individuals 12 years and older taking 200 mcg of Nasonex who experience headaches is greater than the proportion of individuals 12 years and older taking a placebo who experience headaches.

! CAUTION

In any statistical study, be sure to consider practical significance. Many statistically significant results can be produced simply by increasing the sample size.

In looking back at the results of Example 1, we notice that the proportion of individuals taking 200 mcg of Nasonex who experience headaches is *statistically significantly* greater than the proportion of individuals 12 years and older taking a placebo who experience headaches. However, we need to ask ourselves a pressing question. Would you not take an allergy medication because 26% of patients experienced a headache taking the medication versus 22% who experienced a headache taking a placebo? Most people would be willing to accept the additional risk of a headache to relieve their allergy symptoms. While the difference of 4% is statistically significant, it does not have any *practical significance*.

EXAMPLE 2

Testing Hypotheses Regarding the Difference of Two Population Proportions Using Technology

Problem: Obtain the exact *P*-value for the problem in Example 1 using statistical software or a graphing calculator with advanced statistical features.

Approach: We will use MINITAB to obtain the *P*-value. The steps for testing hypotheses regarding two proportions using the TI-83/84 Plus graphing calculator, MINITAB, and Excel are given in the Technology Step by Step on page 545.

Result: Figure 11 shows the results obtained from MINITAB. The *P*-value is 0.002.

USING TECHNOLOGY

The *P*-value obtained from MINITAB differs from the *P*-value obtained by hand in Example 1 because of rounding.

Figure 11

Test and CI for Two Proportions

```
Sample       X      N   Sample p
1          547   2103   0.260105
2          368   1671   0.220227


Difference = p (1) − p (2)
Estimate for difference: 0.0398772
90% CI for difference: (0.0169504, 0.0628040)
Test for difference = 0 (vs > 0): Z = 2.86 P−Value = 0.002
```

Interpretation: There is a 0.002 probability of obtaining a difference in sample proportions that is 2.68 standard deviations from the hypothesized mean difference of 0. There is sufficient evidence to conclude that the proportion of individuals 12 years and older taking 200 mcg of Nasonex who experience headaches is greater than the proportion of individuals 12 years and older taking a placebo who experience headaches at the $\alpha = 0.05$ level of significance.

Now Work Problem 13(a).

② Construct and Interpret Confidence Intervals for the Difference between Two Population Proportions

The sampling distribution of the difference of two proportions, $\hat{p}_1 - \hat{p}_2$, can also be used to construct confidence intervals for the difference of two proportions.

> ### Constructing a $(1 - \alpha) \cdot 100\%$ Confidence Interval for the Difference between Two Population Proportions
>
> To construct a $(1 - \alpha) \cdot 100\%$ confidence interval for the difference between two population proportions, the following requirements must be satisfied:
>
> 1. the samples are obtained independently, using simple random sampling,
> 2. $n_1 \hat{p}_1 (1 - \hat{p}_1) \geq 10$ and $n_2 \hat{p}_2 (1 - \hat{p}_2) \geq 10$,
> 3. $n_1 \leq 0.05 N_1$ and $n_2 \leq 0.05 N_2$ (the sample size is no more than 5% of the population size); this ensures the independence necessary for a binomial experiment.
>
> Provided that these requirements are met, a $(1 - \alpha) \cdot 100\%$ confidence interval for $p_1 - p_2$ is given by
>
> $$\text{Lower bound:} \quad (\hat{p}_1 - \hat{p}_2) - z_{\frac{\alpha}{2}} \cdot \sqrt{\frac{\hat{p}_1(1 - \hat{p}_1)}{n_1} + \frac{\hat{p}_2(1 - \hat{p}_2)}{n_2}}$$
>
> $$\text{Upper bound:} \quad (\hat{p}_1 - \hat{p}_2) + z_{\frac{\alpha}{2}} \cdot \sqrt{\frac{\hat{p}_1(1 - \hat{p}_1)}{n_1} + \frac{\hat{p}_2(1 - \hat{p}_2)}{n_2}} \qquad (3)$$

Notice that we do not pool the sample proportions. This is because we are not making any assumptions regarding their equality, as we did in hypothesis testing.

EXAMPLE 3 **Constructing a Confidence Interval for the Difference between Two Population Proportions**

Problem: In clinical trials of Nasonex, 750 randomly selected pediatric patients (ages 3 to 11 years old) were randomly divided into two groups. The patients in Group 1 (experimental group) received 100 mcg of Nasonex, while the patients in Group 2 (control group) received a placebo. Of the 374 patients in the experimental group, 64 reported headaches as a side effect. Of the 376 patients in the control group, 68 reported headaches as a side effect. Construct a 90% confidence interval for the difference between the two population proportions, $p_1 - p_2$.

Approach: We can compute a 90% confidence interval about $p_1 - p_2$, provided that the requirements stated above are satisfied. We then construct the interval by using Formula (3).

Solution

Step 1: We have to verify the requirements for constructing a confidence interval about the difference between two population proportions. (1) The samples were randomly divided into two groups. (2) For the experimental group (Group 1), we have $n_1 = 374$ and $x_1 = 64$, so $\hat{p}_1 = \dfrac{x_1}{n_1} = \dfrac{64}{374} = 0.171$. For the control group (Group 2), we have $n_2 = 376$ and $x_2 = 68$, so that $\hat{p}_2 = \dfrac{x_2}{n_2} = \dfrac{68}{376} = 0.181$. Therefore,

$$n_1 \hat{p}_1 (1 - \hat{p}_1) = 374(0.171)(1 - 0.171) = 53.02 \geq 10$$

$$n_2 \hat{p}_2 (1 - \hat{p}_2) = 376(0.181)(1 - 0.181) = 55.74 \geq 10$$

USING TECHNOLOGY

Graphing calculators with advanced statistical features and statistical spreadsheets can be used to construct confidence intervals about the difference between two population proportions. Figure 12 shows the results using a TI-84 Plus graphing calculator.

Figure 12

```
2-PropZInt
(-.0555,.03601)
p̂1=.1711229947
p̂2=.1808510638
n1=374
n2=376
```

(3) The samples were independently obtained and the sample sizes are less than 5% of the population size. (There are over 20 million children between the ages of 3 and 11 in the United States.)

Step 2: Because we want a 90% confidence interval, we have $\alpha = 0.10$, so $z_{\frac{\alpha}{2}} = z_{0.05} = 1.645$.

Step 3: Substituting into Formula (3) with $\hat{p}_1 = 0.171$, $n_1 = 374$, $\hat{p}_2 = 0.181$, and $n_2 = 376$, we obtain the lower and upper bounds on the confidence interval:

Lower bound:

$$(\hat{p}_1 - \hat{p}_2) - z_{\frac{\alpha}{2}} \cdot \sqrt{\frac{\hat{p}_1(1 - \hat{p}_1)}{n_1} + \frac{\hat{p}_2(1 - \hat{p}_2)}{n_2}}$$

$$= (0.171 - 0.181) - 1.645 \cdot \sqrt{\frac{0.171(1 - 0.171)}{374} + \frac{0.181(1 - 0.181)}{376}}$$

$$= -0.010 - 0.046 = -0.056$$

Upper bound:

$$(\hat{p}_1 - \hat{p}_2) + z_{\frac{\alpha}{2}} \cdot \sqrt{\frac{\hat{p}_1(1 - \hat{p}_1)}{n_1} + \frac{\hat{p}_2(1 - \hat{p}_2)}{n_2}}$$

$$= \left(0.171 - 0.181\right) + 1.645 \cdot \sqrt{\frac{0.171(1 - 0.171)}{374} + \frac{0.181(1 - 0.181)}{376}}$$

$$= -0.010 + 0.046 = 0.036$$

Based on the results of the study, we are 90% confident that the difference between the proportion of headaches in the experimental group and the control group is between -0.056 and 0.036. Because the confidence interval contains 0, there is no evidence to conclude that the proportion of 3- to 11-year old patients complaining of headaches who receive Nasonex is different from those who do not receive Nasonex at the $\alpha = 0.1$ level of significance. ▪

Now Work Problem 13(b).

3 Determine the Sample Size Necessary for Estimating the Difference between Two Population Proportions within a Specified Margin of Error

In Section 9.3, we introduced a method for determining the sample size n required to estimate a single population proportion within a specified margin of error, E, with a specified level of confidence. This formula was obtained by solving the margin of error, $E = z_{\frac{\alpha}{2}} \cdot \sqrt{\dfrac{\hat{p}(1 - \hat{p})}{n}}$, for n. We can follow the same approach to determine the sample size when we want to estimate two population proportions. Notice that the margin of error, E, in Formula (3) is given by $E = z_{\frac{\alpha}{2}} \cdot \sqrt{\dfrac{\hat{p}_1(1 - \hat{p}_1)}{n_1} + \dfrac{\hat{p}_2(1 - \hat{p}_2)}{n_2}}$. Assuming that $n_1 = n_2 = n$, we can solve this expression for $n = n_1 = n_2$ and obtain the following result:

CAUTION
When doing sample size calculations, always round up.

> **Sample Size for Estimating $p_1 - p_2$**
>
> The sample size required to obtain a $(1 - \alpha) \cdot 100\%$ confidence interval with a margin of error, E, is given by
>
> $$n = n_1 = n_2 = [\hat{p}_1(1 - \hat{p}_1) + \hat{p}_2(1 - \hat{p}_2)]\left(\frac{z_{\frac{\alpha}{2}}}{E}\right)^2 \qquad (4)$$
>
> rounded up to the next integer, if prior estimates of P_1 and P_2, $\hat{p}_1$ and $\hat{p}_2$, are available. If prior estimates of P_1 and P_2 are unavailable, the sample size is
>
> $$n = n_1 = n_2 = 0.5\left(\frac{z_{\frac{\alpha}{2}}}{E}\right)^2 \qquad (5)$$
>
> rounded up to the next integer. The margin of error should always be expressed as a decimal when using Formulas (4) and (5).

Note to Instructor
The 0.5 in Formula (5) results from adding 0.25 and 0.25, the maximum value of $p(1 - p)$.

EXAMPLE 4   **Determining Sample Size**

Problem: A nutritionist wishes to estimate the difference between the proportion of males and females who consume the USDA's recommended daily intake of calcium. What sample size should be obtained if she wishes the estimate to be within 3 percentage points with 95% confidence, assuming that

(a) she uses the results of the USDA's 1994–1996 Diet and Health Knowledge Survey, according to which 51.1% of males and 75.2% of females consume the USDA's recommended daily intake of calcium,

(b) she does not use any prior estimates?

Approach: We have $E = 0.03$ and $z_{\frac{\alpha}{2}} = z_{\frac{0.05}{2}} = z_{0.025} = 1.96$. To answer part (a), we let $\hat{p}_1 = 0.511$ (for males) and $\hat{p}_2 = 0.752$ (for females) in Formula (4). To answer part (b), we use Formula (5).

Solution

(a) Substituting $E = 0.03$, $z_{0.025} = 1.96$, $\hat{p}_1 = 0.511$, and $\hat{p}_2 = 0.752$ into Formula (4), we obtain

$$n_1 = n_2 = [\hat{p}_1(1 - \hat{p}_1) + \hat{p}_2(1 - \hat{p}_2)]\left(\frac{z_{\frac{\alpha}{2}}}{E}\right)^2 = [0.511(1 - 0.511) + 0.752(1 - 0.752)]\left(\frac{1.96}{0.03}\right)^2$$

$$= 1862.6$$

We round this value up to 1863. The nutritionist must survey 1863 randomly selected males and 1863 randomly selected females.

(b) Substituting $E = 0.03$ and $z_{0.025} = 1.96$ into Formula (5), we obtain

$$n_1 = n_2 = 0.5\left(\frac{z_{\frac{\alpha}{2}}}{E}\right)^2 = 0.5\left(\frac{1.96}{0.03}\right)^2 = 2134.2$$

We round this value up to 2135. The nutritionist must survey 2135 randomly selected males and 2135 randomly selected females.

Now Work Problem 23.

In Other Words

If possible, obtain a prior estimate of $\hat{p}$ when doing sample size computations.

We can see that having prior estimates of the population proportions reduces the number of individuals that need to be surveyed.

11.3 ASSESS YOUR UNDERSTANDING

Concepts and Vocabulary

1. Explain why we determine a pooled estimate of the population proportion when testing hypotheses regarding the difference of two proportions, but do not pool when constructing confidence intervals about the difference of two proportions.

2. State the requirements that must be satisfied to test hypotheses regarding two population proportions.

Skill Building

In Problems 3–6, conduct each test at the $\alpha = 0.05$ level of significance by determining (a) the null and alternative hypotheses, (b) the test statistic, (c) the critical value, and (d) the P-value. Assume the samples were obtained independently using simple random sampling.

3. Test whether $p_1 > p_2$. Sample data:
$x_1 = 368, n_1 = 541, x_2 = 351, n_2 = 593$

4. Test whether $p_1 < p_2$. Sample data:
$x_1 = 109, n_1 = 475, x_2 = 78, n_2 = 325$

5. Test whether $p_1 \neq p_2$. Sample data:
$x_1 = 28, n_1 = 254, x_2 = 36, n_2 = 301$

6. Test whether $p_1 \neq p_2$. Sample data:
$x_1 = 804, n_1 = 874, x_2 = 902, n_2 = 954$

In Problems 7–10, construct a confidence interval for $p_1 - p_2$ at the given level of confidence.

7. $x_1 = 368, n_1 = 541, x_2 = 421, n_2 = 593, 90\%$ confidence

8. $x_1 = 109, n_1 = 475, x_2 = 78, n_2 = 325, 99\%$ confidence

9. $x_1 = 28, n_1 = 254, x_2 = 36, n_2 = 301, 95\%$ confidence

10. $x_1 = 804, n_1 = 874, x_2 = 892, n_2 = 954, 95\%$ confidence

7. $(-0.075, 0.015)$ 8. $(-0.090, 0.068)$
9. $(-0.063, 0.043)$ 10. $(-0.039, 0.009)$

Applying the Concepts

11. Prevnar The drug Prevnar is a vaccine meant to prevent certain types of bacterial meningitis. It is typically administered to infants starting around 2 months of age. In randomized, double-blind clinical trials of Prevnar, infants were randomly divided into two groups. Subjects in Group 1 received Prevnar while subjects in Group 2 received a control vaccine. After the first dose, 107 of 710 subjects in the experimental group (Group 1) experienced fever as a side effect. After the first dose, 67 of 611 of the subjects in the control group (Group 2) experienced fever as a side effect.

(a) Does the evidence suggest that a higher proportion of subjects in Group 1 experienced fever as a side effect than subjects in Group 2 at the $\alpha = 0.05$ level of significance?

(b) Construct a 90% confidence interval for the difference between the two population proportions, $p_1 - p_2$. (0.011, 0.071)

12. Prevnar The drug Prevnar is a vaccine meant to prevent certain types of bacterial meningitis. It is typically administered to infants starting around 2 months of age. In randomized, double-blind clinical trials of Prevnar, infants were randomly divided into two groups. Subjects in Group 1 received Prevnar, while subjects in Group 2 received a control vaccine. After the second dose, 137 of 452 subjects in the experimental group (Group 1) experienced drowsiness as a side effect. After the second dose, 31 of 99 subjects in the control group (Group 2) experienced drowsiness as a side effect.

(a) Does the evidence suggest that a different proportion of subjects in Group 1 experienced drowsiness as a side effect than subjects in Group 2 at the $\alpha = 0.05$ level of significance? Do not reject H_0; P-value = 0.8414

(b) Construct a 99% confidence interval for the difference between the two population proportions, $p_1 - p_2$. (−0.142, 0.122)

13. Too Much Cholesterol in Your Diet? A nutritionist claims that the proportion of individuals who have at most an eighth-grade education and consume more than the USDA's recommended daily allowance of 300 mg of cholesterol is higher than the proportion of individuals who have at least some college and consume too much cholesterol. In interviews with 320 individuals who have at most an eighth-grade education, she determined that 114 of them consumed too much cholesterol. In interviews with 350 individuals with at least some college, she determined that 112 of them consumed too much cholesterol per day. (Based on data obtained from the USDA's Diet and Health Knowledge Survey.)

(a) Determine whether the proportion of individuals with at most an eighth-grade education who consume too much cholesterol is higher than the proportion of individuals who have at least some college and

consume too much cholesterol at the $\alpha = 0.1$ level of significance. Do not reject H_0

(b) Construct a 95% confidence interval for the difference between the two population proportions, $p_8 - p_c$.

14. Consumption of Saturated Fat A nutritionist claims that the proportion of females who consume too much saturated fat is lower than the proportion of males who consume too much saturated fat. In interviews with 513 randomly selected females, she determines that 300 consume too much saturated fat. In interviews with 564 randomly selected males, she determines that 391 consume too much saturated fat, based on data obtained from the USDA's Diet and Health Knowledge Survey.

(a) Determine whether a lower proportion of females than males consume too much saturated fat at the $\alpha = 0.05$ level of significance. Reject H_0

(b) Construct a 95% confidence interval for the difference between the two population proportions, $p_f - p_m$. (−0.165, −0.051)

15. Percentage of Americans Who Smoke on the Decline? On November 3–5, 2005, the Gallup Organization surveyed 1028 adults and found that 226 of them had smoked at least one cigarette in the past week. In 1990, Gallup also asked 1028 adults the same question and determined that 278 adults had smoked at least one cigarette in the past week. Can we say the proportion of American adults who smoked at least one cigarette in the past week has decreased since 1990? Why? Yes

16. Life on Mars? On March 19–21, 1999, the Gallup Organization surveyed 535 adults aged 18 years old or older and asked, "Do you think there is life of some form on other planets in the universe or not?" Of the 535 individuals surveyed, 326 responded yes. When the same question was asked on September 3–5, 1996, 385 of the 535 individuals surveyed responded yes.

(a) Can we conclude that the proportion of adults who believe that there is life on other planets has decreased since September, 1996, at the $\alpha = 0.10$ level of significance? Reject H_0

(b) Construct a 90% confidence interval for the difference between the two population proportions, $p_{1996} - p_{1999}$. (−0.158, −0.064)

17. Salk Vaccine On April 12, 1955, Dr. Jonas Salk released the results of clinical trials for his vaccine to prevent polio. In these clinical trials, 400,000 children were randomly divided in two groups. The subjects in Group 1 (the experimental group) were given the vaccine, while the subjects in Group 2 (the control group) were given a placebo. Of the 200,000 children in the experimental group, 33 developed polio. Of the 200,000 children in the control group, 115 developed polio.

(11a) Reject H_0: P-value = 0.0139   (13b) (−0.036, 0.108)   (17a) Reject H_0

(a) Test whether the proportion of subjects in the experimental group who contracted polio is less than the proportion of subjects in the control group who contracted polio at the $\alpha = 0.01$ level of significance.

(b) Construct a 90% confidence interval for the difference between the two population proportions, $p_1 - p_2$. $(-0.0005, -0.0003)$

18. **Mind over Body** On May 10–14, 2001, the Gallup Organization surveyed 1002 adult Americans and asked them if they believed in psychic or spiritual healing. Of the 1002 individuals surveyed, 551 said yes. When the same question was asked on June 6–8, 2005, 541 of the 1002 individuals surveyed responded yes.

(a) Test whether the proportion of adult Americans who believe in psychic or spiritual healing has changed since May 2001 at the $\alpha = 0.05$ level of significance. Do not reject H_0

(b) Construct a 90% confidence interval for the difference between the two population proportions, $p_{2001} - p_{2005}$. $(-0.027, 0.047)$

19. **Side Effects** In clinical trials of the allergy medicine Clarinex (5 mg), it was reported that 50 out of 1655 individuals in the Clarinex group and 31 out of 1652 individuals in the placebo group experienced dry mouth as a side effect of their respective treatments.
(*Source*: www.clarinex.com).

(a) Is the proportion of individuals experiencing dry mouth is greater for those taking Clarinex than for those taking a placebo at the $\alpha = 0.05$ level of significance? Reject H_0

(b) Is the difference between the groups practically significant? No

20. **Practical versus Statistical Significance** Suppose that in clinical trials for treatment of a skin disorder 642 of 2105 patients receiving the current standard treatment were cured of the disorder and 697 of 2115 patients receiving a new proposed treatment were cured of the disorder.

(a) Does the new procedure cure a higher percentage of patients at the $\alpha = 0.05$ level of significance? Reject H_0

(b) Do you think that the difference in success rates is practically significant? What factors might influence your decision?

21. **Own a Gun?** In October 2004, the Gallup Organization surveyed 1134 American adults and found that 431 owned a gun. In February 1999, the Gallup Organization had surveyed 1134 American adults and found that 408 owned a gun. Suppose that a newspaper article has a headline that reads, "Percentage of American Gun Owners on the Rise." Is this an accurate headline? Why? No

22. **Accupril** Accupril, a medication supplied by Pfizer Pharmaceuticals, is meant to control hypertension. In

clinical trials of Accupril, 2142 subjects were divided into two groups. The 1563 subjects in Group 1 (the experimental group) received Accupril. The 579 subjects in Group 2 (the control group) received a placebo. Of the 1563 subjects in the experimental group, 61 experienced dizziness as a side effect. Of the 579 subjects in the control group, 15 experienced dizziness as a side effect. To test whether the proportion experiencing dizziness in the experimental group is greater than that in the control group, the researchers entered the data into MINITAB statistical software and obtained the following results:

Test and Confidence Interval for Two Proportions

```
Sample     X      N   Sample p
1         61   1563   0.039028
2         15    579   0.025907

Estimate for p(1) - p(2): 0.0131208
95% CI for p(1) - p(2): (-0.00299150, 0.0292330)
Test for p(1) - p(2) = 0 (vs > 0): Z = 1.46 P-Value = 0.072
```

What conclusion can be drawn from the clinical trials?

23. **Determining Sample Size** A physical therapist wants to determine the difference in the proportion of men and women who participate in regular sustained physical activity. What sample size should be obtained if she wishes the estimate to be within three percentage points with 95% confidence, assuming that

(a) she uses the 1998 estimates of 21.9% male and 19.7% female from the U.S. National Center for Chronic Disease Prevention and Health Promotion? 1406

(b) she does not use any prior estimates? 2135

24. **Determining Sample Size** An educator wants to determine the difference between the proportion of males and females who have completed 4 or more years of college. What sample size should be obtained if she wishes the estimate to be within two percentage points with 90% confidence, assuming that

(a) she uses the 1999 estimates of 27.5% male and 23.1% female from the U.S. Census Bureau? 2551

(b) she does not use any prior estimates? 3383

25. **Treating Apnea** An individual with apnea stops breathing while sleeping. In a randomized, double-blind study to evaluate the efficacy of caffeine citrate for treatment of apnea, infants with six or more episodes of apnea were administered 10 mg/kg of caffeine citrate intravenously, followed by 2.5 mg/kg/day for up to 10 days or a placebo. (Caffeine Citrate for the Treatment of Apnea in Prematurity: A Double-blind, Placebo Controlled Study, *Pharmacotherapy*, June 2000, 20(6):644–652.) With successful treatment defined as a 50% or more reduction in apnea episodes, it was determined that caffeine citrate was significantly more effective than the placebo in reducing apnea episodes by at least 50% in 6 days with the *P*-value less than 0.05. State any conclusions, using this *P*-value.

| Technology Step by Step | **Inference about Two Population Proportions** |

TI-83/84 Plus

Hypothesis Tests

Step 1: Press STAT, highlight TESTS, and select `6:2-PropZTest` ....

Step 2: Enter the values of x_1, n_1, x_2, and n_2.

Step 3: Highlight the appropriate relation between p_1 and p_2 in the alternative hypothesis.

Step 4: Highlight Calculate or Draw and press ENTER. Calculate gives the test statistic and *P*-value. Draw will draw the *Z*-distribution with the *P*-value shaded.

Confidence Intervals

Follow the same steps given for hypothesis tests, except select `B:2-PropZInt`.... Also, select a confidence level (such as 95% = 0.95).

MINITAB
Step 1: Select the **Stat** menu, highlight **Basic Statistics**, then highlight **2 Proportions** . . . .

Step 2: Click the Summarized Data Option. Enter the number of trials, n_1, and the number of successes, x_1. Enter the number of trials, n_2, and the number of successes, x_2.

Step 3: Under OPTIONS, select the direction of the alternative hypothesis and select a confidence level. Click OK.

Excel
Step 1: Load the PHStat add-in.

Step 2: Select the **PHStat** menu. Highlight **Two-Sample Tests** and then highlight **Z Test for Differences in Two Proportions** . . . .

Step 3: Enter the hypothesized difference in the two proportions (ususally 0). Enter the level of significance, α. For the Population 1 sample, enter the number of successes, x_1, and the sample size, n_1. For the Population 2 sample, enter the number of successes, x_2, and the sample size, n_2. Select the appropriate test option. Click OK.

CHAPTER 11 **Review**

Summary

This chapter discussed performing statistical inference by comparing two population parameters. We began with a discussion regarding the comparison of two population means. To determine the method to use, we must know whether the sampling was dependent or independent. A sampling method is independent when the choice of individuals for one sample does not dictate which individuals will be in a second sample. A sampling method is dependent when the individuals selected for one sample are used to determine the individuals in the second sample. For dependent sampling, we use the paired *t*-test to perform statistical inference. For independent sampling, we use Welch's two-sample *t*. For both tests, the population must be normally distributed or the sample sizes must be large.

Section 11.3 dealt with statistical inference for comparing two population proportions. To perform these tests, $n\hat{p}(1 - \hat{p})$ must be greater than or equal to 10 for each population; and each sample size should be no more than 5% of the population size. If these requirements are satisfied the distribution of $\hat{p}_1 - \hat{p}_2$ is approximately normal, with mean $p_1 - p_2$ and standard deviation

$$\sqrt{\frac{p_1(1 - p_1)}{n_1} + \frac{p_2(1 - p_2)}{n_2}}.$$

To help determine which test to use, we include the flow chart in Figure 13.

Figure 13

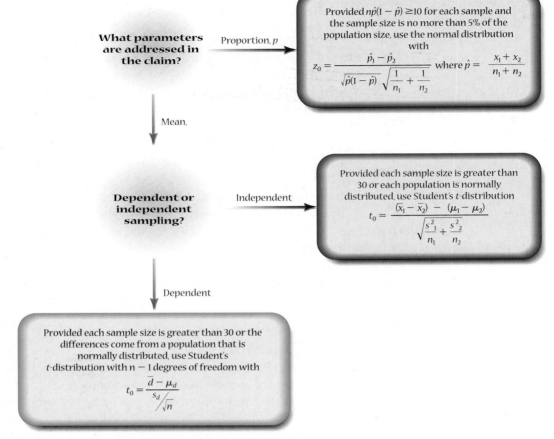

Formulas

- Test statistic for matched-pairs data:

$$t_0 = \frac{\overline{d}}{\frac{s_d}{\sqrt{n}}}$$

where $\overline{d}$ is the mean and s_d is the standard deviation of the differenced data

- Confidence interval for matched-pairs data:

Lower bound: $\overline{d} - t_{\alpha/2} \cdot \dfrac{s_d}{\sqrt{n}}$

Upper bound: $\overline{d} + t_{\alpha/2} \cdot \dfrac{s_d}{\sqrt{n}}$

- Test statistic comparing two means (independent sampling):

$$t_0 = \frac{(\overline{x}_1 - \overline{x}_2) - (\mu_1 - \mu_2)}{\sqrt{\dfrac{s_1^2}{n_1} + \dfrac{s_2^2}{n_2}}}$$

- Confidence interval for the difference of two means (independent samples):

Lower bound: $(\overline{x}_1 - \overline{x}_2) - t_{\alpha/2} \cdot \sqrt{\dfrac{s_1^2}{n_1} + \dfrac{s_2^2}{n_2}}$

Upper bound: $(\overline{x}_1 - \overline{x}_2) + t_{\alpha/2} \cdot \sqrt{\dfrac{s_1^2}{n_1} + \dfrac{s_2^2}{n_2}}$

- Test statistic comparing two population proportions:

$$z_0 = \frac{\hat{p}_1 - \hat{p}_2}{\sqrt{\hat{p}(1 - \hat{p})}\sqrt{\dfrac{1}{n_1} + \dfrac{1}{n_2}}}$$

where $\hat{p} = \dfrac{x_1 + x_2}{n_1 + n_2}$

- Confidence interval for the difference of two proportions:

Lower bound: $(\hat{p}_1 - \hat{p}_2) - z_{\alpha/2} \cdot \sqrt{\dfrac{\hat{p}_1(1 - \hat{p}_1)}{n_1} + \dfrac{\hat{p}_2(1 - \hat{p}_2)}{n_2}}$

Upper bound: $(\hat{p}_1 - \hat{p}_2) + z_{\alpha/2} \cdot \sqrt{\dfrac{\hat{p}_1(1 - \hat{p}_1)}{n_1} + \dfrac{\hat{p}_2(1 - \hat{p}_2)}{n_2}}$

- Sample size for estimating $p_1 - p_2$:

$$n = n_1 = n_2 = [\hat{p}_1(1 - \hat{p}_1) + \hat{p}_2(1 - \hat{p}_2)]\left(\frac{z_{\alpha/2}}{E}\right)^2$$

$$n = n_1 = n_2 = 0.5\left(\frac{z_{\alpha/2}}{E}\right)^2$$

Vocabulary

Dependent sampling (p. 508)

Independent sampling (p. 508)

Matched pairs (p. 508)

Robust (p. 510)

Welch's approximate t (p. 521)

Pooled t-statistic (p. 527)

Pooled estimate of p (p. 536)

Objectives

Section	You should be able to . . .	Examples	Review Exercises
11.1	**1** Distinguish between independent and dependent sampling (p. 508)	1	1–4, 13(a), 14(a), 15(a), 16(a)
	2 Test hypotheses regarding matched-pairs data (p. 509)	2 and 3	5(c), 6(c), 13(b), 16(b)
	3 Construct and interpret confidence intervals about the population mean difference of matched-pairs data (p. 514)	4	16(c), 21
11.2	**1** Test hypotheses regarding the difference of two independent means (p. 522)	1 and 2	7(a), 8(a), 9(a), 10(a), 14(b), 15(b)
	2 Construct and interpret confidence intervals regarding the difference of two independent means (p. 526)	3	22, 23
11.3	**1** Test hypotheses regarding two population proportions (p. 535)	1 and 2	11, 12, 17(a), 18(a)
	2 Construct and interpret confidence intervals for the difference between two population proportions (p. 539)	3	17(b), 18(b)
	3 Determine the sample size necessary for estimating the difference between two population proportions within a specified margin of error (p. 541)	4	19, 20

Review Exercises

In Problems 1–4, determine if the sampling is dependent or independent.

1. A researcher wants to know if the mean length of stay in for-profit hospitals is different from the mean length of stay in not-for-profit hospitals. He randomly selected 20 individuals in the for-profit hospital and matched them with 20 individuals in the not-for-profit hospital by diagnosis. Dependent

2. An urban economist believes that commute times to work in the South are less than commute times to work in the Midwest. He randomly selects 40 employed individuals in the South and 45 employed individuals in the Midwest and determines their commute times. Independent

3. A stock analyst wants to know if there is a difference between the mean rate of return from energy stocks and that from financial stocks. He randomly selects 13 energy stocks and computes the rate of return for the past year. He randomly selects 13 financial stocks and computes the rate of return for the past year. Independent

4. A prison warden wants to know if men receive longer sentences for crimes than women. He randomly samples 30 men and matches them with 30 women by type of crime committed and records their lengths of sentence. Dependent

---

In Problems 5 and 6, assume that the paired data came from a population that is normally distributed. (6a) 2.06 (7b) (0.73, 7.67)
(8b) ($-12.44, 0.04$)

5.

Observation	1	2	3	4	5	6
X_i	34.2	32.1	39.5	41.8	45.1	38.4
Y_i	34.9	31.5	39.5	41.9	45.5	38.8

(a) Compute $d_i = X_i - Y_i$ for each pair of data.
(b) Compute $\bar{d}$ and s_d. $\bar{d} = -0.167$, $s_d = 0.450$
(c) Test whether $\mu_d < 0$ at the $\alpha = 0.05$ level of significance. Do not reject H_0
(d) Compute a 98% confidence interval about the population mean difference μ_d. ($-0.79, 0.45$)

6.

Observation	1	2	3	4	5	6	7
X_i	18.5	21.8	19.4	22.9	18.3	20.2	23.1
Y_i	18.3	22.3	19.2	22.3	18.9	20.7	23.9

(a) Compute $d_i = X_i - Y_i$ for each pair of data.
(b) Compute $\bar{d}$ and s_d. $\bar{d} = -0.2$, $s_d = 0.526$
(c) Test whether $\mu_d \neq 0$ at the $\alpha = 0.01$ level of significance. Do not reject H_0
(d) Compute a 95% confidence interval about the population mean difference μ_d. ($-0.69, 0.29$)

In Problems 7–10, assume that the populations are normally distributed and that independent sampling occurred.

7.

	Population 1	Population 2
n	13	8
$\bar{x}$	32.4	28.2
s	4.5	3.8

(a) Test whether $\mu_1 \neq \mu_2$ at the $\alpha = 0.1$ level of significance for the given sample data. Reject H_0
(b) Construct a 90% confidence interval about $\mu_1 - \mu_2$.

8.

	Population 1	Population 2
n	24	27
$\bar{x}$	104.2	110.4
s	12.3	8.7

(a) Test whether $\mu_1 \neq \mu_2$ at the $\alpha = 0.1$ level of significance for the given sample data. Reject H_0
(b) Construct a 95% confidence interval about $\mu_1 - \mu_2$.

9.

	Population 1	Population 2
n	45	41
$\overline{x}$	48.2	45.2
s	8.4	10.3

(a) Test whether $\mu_1 > \mu_2$ at the $\alpha = 0.01$ level of signifi-cance for the given sample data. *Do not reject H_0*

(b) Construct a 90% confidence interval about $\mu_1 - \mu_2$.
(9) (−0.43, 6.43)

10.

	Population 1	Population 2
n	13	8
$\overline{x}$	96.6	98.3
s	3.2	2.5

(a) Test whether $\mu_1 < \mu_2$ at the $\alpha = 0.05$ level of signifi-cance for the given sample data. *Do not reject H_0*

(b) Construct a 99% confidence interval about $\mu_1 - \mu_2$.
(10b) (−6.08, 2.68)

In Problems 11 and 12, test the hypotheses at the $\alpha = 0.05$ level of significance by (a) determining the null and alternative hypotheses, (b) computing the test statistic, (c) computing the critical value, and (d) computing the P-value. Assume that the samples were obtained independently by simple random sampling.

11. Test whether $p_1 \ne p_2$. Sample data: $x_1 = 451$, $n_1 = 555$, $x_2 = 510$, $n_2 = 600$

12. Test whether $p_1 < p_2$. Sample data: $x_1 = 156$, $n_1 = 650$, $x_2 = 138$, $n_2 = 540$

13. Height versus Arm Span   A statistics student believes that an individual's arm span is equal to the individual's height. The student used a random sample of 10 students and obtained the following data.

Student:	1	2	3	4	5
Height (inches)	59.5	69	77	59.5	74.5
Arm span (inches)	62	65.5	76	63	74

Student:	6	7	8	9	10
Height (inches)	63	61.5	67.5	73	69
Arm span (inches)	66	61	69	70	71

Source: John Climent, Cecil Community College

(a) Is the sampling method dependent or independent? Why?

(b) Does the evidence suggest that an individual's height and arm span are different at the $\alpha = 0.05$ level of sig-nificance? **Note:** A normal probability plot indicates that the data and differenced data are normally dis-tributed. A boxplot indicates that the data and differ-enced data have no outliers.

14. Acid Rain   A researcher wants to know whether the acidi-ty of rain (pH) near Houston, Texas, is significantly different from that near Chicago, Illinois. He randomly selects 12 rain dates in Texas and 14 rain dates in Illinois and obtains the following data:

Texas

4.69	5.10	5.22	4.46	4.93	4.65
5.22	4.76	4.25	5.14	4.11	4.71

Illinois

4.40	4.69	4.22	4.64	4.54	4.35	4.69
4.40	4.75	4.63	4.45	4.49	4.36	4.52

Source: National Atmospheric Deposition Program

(a) Is the sampling method dependent or independent? Why? *Independent*

(b) Test whether there is no significant difference between the acidity of rain in Houston and that in Chicago at the $\alpha = 0.05$ level of significance. *Reject H_0*

(c) Draw boxplots of each data set, using the same scale. Does this visual evidence support the results obtained in part (b)?

15. McDonald's versus Wendy's   A student wanted to deter-mine whether the wait time in the drive-through at Mc-Donald's differed from that at Wendy's. She used a random sample of 30 cars at McDonald's and 27 cars at Wendy's and obtained these results:

Wait Time at McDonald's Drive-Through (in seconds)

151.09	227.38	111.84	131.21	128.75
191.60	126.91	137.90	195.44	246.59
141.78	127.35	121.21	101.03	95.09
122.06	122.62	100.04	71.37	153.34
140.44	126.62	116.72	131.69	100.94
115.66	147.28	81.43	86.31	156.34

Wait Time at Wendy's Drive-Through (in seconds)

281.90	71.02	204.29	128.59	133.56
187.53	199.86	190.91	110.55	110.64
196.84	233.65	171.01	182.54	183.79
284.48	363.34	270.82	390.50	471.62
123.66	174.43	385.90	386.71	155.53
203.62	119.61			

Source: Catherine M. Simmons, student at Joliet Junior College

Note to Instructor: In Problem 15, the sample size for Wendy's is less than 30. However, the data does not contain any outliers, so the Central Limit Theorem can be used.

(a) Is the sampling method dependent or independent?

(b) Is there a difference in wait times at each restaurant's drive-through at the $\alpha = 0.1$ level of significance?

(c) Draw boxplots of each data set using the same scale. Does this visual evidence support the results obtained in part (b)?

16. Gastric Bypass Laparoscopic gastric bypass surgery reduces the size of the stomach and can cause malabsorption of nutrients, leading to weight loss, according to webmd.com. Researchers wanted to test the claim that body mass index (BMI) reduction is significantly greater than zero 10 months after surgery. They selected 110 patients who elected to have laparoscopic gastric bypass surgery and measured their BMI prior to surgery. After 10 months they again measured the BMI. The mean reduction in BMI in the 110 patients was found to be 12.61 with a standard deviation of 4.90 (*Source*: St. Peter, Shawn D., et al., Impact of Advanced Age on Weight Loss and Health Benefits after Laparoscopic Gastric Bypass, *Archives of Surgery*, Vol. 140, No. 2, February, 2005)

(a) What type of experimental design is this? Why?

(b) Is the BMI reduction greater than zero at the $\alpha = 0.05$ level of significance? Reject H_0

(c) Construct a 90% confidence interval for the mean BMI reduction after laparoscopic gastric bypass surgery. Write a sentence that a doctor could read to his or her patients that explains the expected BMI reduction 10 months after the surgery. (11.834, 13.386)

17. Treatment for Osteoporosis Osteoporosis is a condition in which people experience decreased bone mass and an increase in the risk of bone fracture. Actonel is a drug that helps combat osteoporosis in postmenopausal women. In clinical trials, 1374 postmenopausal women were randomly divided into experimental and control groups. The subjects in the experimental group were administered 5 mg of Actonel, while the subjects in the control group were administered a placebo. The number of women who experienced a bone fracture over the course of 1 year was recorded. Of the 696 women in the experimental group, 27 experienced a fracture during the course of the year. Of the 678 women in the control group, 49 experienced a fracture during the course of the year.

(a) Do a lower proportion of women in the experimental group experience a bone fracture than the women in the control group at the $\alpha = 0.01$ level of significance?

(b) Construct a 95% confidence interval for the difference between the two population proportions, $p_{exp} - p_{control}$.

(c) What type of experimental design is this? What is the treatment? How many levels does it have?

18. Zoloft Zoloft is a drug that is used to treat obsessive–compulsive disorder (OCD). In randomized, double-blind clinical trials, 926 patients diagnosed with OCD were randomly divided into two groups. Subjects in Group 1 (experimental group) received 200 mg per day of Zoloft, while subjects in Group 2 (control group) received a placebo. Of the 553 subjects in the experimental group, 77 experienced dry mouth as a side effect. Of the 373 subjects in the control group, 34 experienced dry mouth as a side effect.

(a) Do a higher proportion of the subjects in the experimental group experience dry mouth than the subjects in the control group at the $\alpha = 0.05$ level of significance?

(b) Construct a 90% confidence interval for the difference between the two population proportions, $p_1 - p_2$.

19. Determining Sample Size A nutritionist wants to estimate the difference between the percentage of men and women who have high cholesterol. What sample size should be obtained if she wishes the estimate to be within 2 percentage points with 90% confidence, assuming

(a) that she uses the 1994 estimates of 18.8% male and 20.5% female from the National Center for Health Statistics? 2136

(b) that she does not use any prior estimates? 3383

20. Determining Sample Size A researcher wants to estimate the difference between the percentage of individuals without a high school diploma who smoke and the percentage of individuals with bachelor's degrees who smoke. What sample size should be obtained if she wishes the estimate to be within 4 percentage points with 95% confidence, assuming

(a) that she uses the 1999 estimates of 32.2% of those without a high school diploma and 11.1% of those with a bachelor's degree, from the National Center for Health Statistics? 762

(b) that she does not use any prior estimates? 1201

21. Height versus Arm Span Construct and interpret a 95% confidence interval about the population mean difference between height and arm span using the data from Problem 13. What does the interval lead us to conclude regarding any differences between height and arm span?

22. Acid Rain Construct and interpret a 90% confidence interval about $\mu_T - \mu_I$ using the data from Problem 14.

23. McDonald's versus Wendy's Construct and interpret a 95% confidence interval about $\mu_M - \mu_W$ using the data from Problem 15. How might a marketing executive with McDonald's use this information?

24. Explain when the matched-pairs t should be used instead of Welch's t in comparing two population means. What are some advantages in designing a matched-pairs experiment versus using Welch's t?

(15b) Reject H_0 (16a) Matched pairs   (17a)Reject H_0 (17b) $(-0.057, -0.009)$ (17c) Completely randomized design: drug: 2
(18a) Reject H_0 (18b) (0.01, 0.08)

Additional Inferential Procedures

CHAPTER 12

Outline

12.1 Goodness-of-Fit Test

12.2 Tests for Independence and the Homogeneity of Proportions

12.3 Testing the Significance of the Least-Squares Regression Model

12.4 Confidence and Prediction Intervals

■ Chapter Review

■ Case Study: Feeling Lucky? Well, Are You? (On CD)

DECISIONS

Are there benefits to attending college? If so, what are they? See the Decisions project on page 573.

●●● Putting It All Together

In Chapters 9–11, we introduced statistical methods that can be used to test hypotheses regarding a population parameter such as μ or p. This chapter can be considered in two parts. The first part, Sections 12.1 and 12.2, introduces inference using the chi-square distribution.

Often, rather than being interested in testing a hypotheses regarding a parameter of a probability distribution, we are interested in testing hypotheses regarding the entire probability distribution. For example, we might wish to test whether the distribution of colors in a bag of plain M&M candies is 13% brown, 14% yellow, 13% red, 20% orange, 24% blue, and 16% green. We introduce methods for testing hypotheses such as this in Section 12.1.

In Section 12.2, we discuss a method that can be used to determine whether two variables are independent based on a sample. We conclude Section 12.2 by introducing tests

for homogeneity. This procedure is used to compare proportions from two or more populations. It is an extension of the two-sample z-test for proportions discussed in Section 11.3.

The second part of this chapter extends inferential statistics to the least-squares regression line. Sections 12.3 and 12.4 introduce inferential methods that can be used on the least-squares regression line.

In Chapter 4, we presented methods for describing the relation between two variables-bivariate data.

In Section 12.3, we use the methods of hypothesis testing first presented in Chapter 10 to test whether a linear relation exists between two quantitative variables.

In Section 12.4, we construct confidence intervals about the predicted value of the least-squares regression line.

12.1 Goodness-of-Fit Test

Preparing for This Section Before getting started, review the following:

- Expected value (Section 6.1, pp. 291–292)
- Mean of a binomial random variable (Section 6.2, pp. 305–306)
- Mutually exclusive (Section 5.2, pp. 238–241)

Objective

 1 Perform a goodness-of-fit test

1 Perform a Goodness-of-Fit Test

In this section, we present a procedure that can be used to test hypotheses regarding a probability distribution. For example, we might want to test whether the distribution of plain M&M candies in a bag is 13% brown, 14% yellow, 13% red, 20% orange, 24% blue, and 16% green. Or we might want to test if the number of hits a player gets in his next four at-bats follows a binomial distribution with $n = 4$ and $p = 0.298$.

We use the symbol χ^2, chi-square, pronounced "kigh-square" (to rhyme with "sky-square"), to represent values of the chi-square distribution. We can find critical values of the chi-square distribution in Table VI in Appendix A of the text. Before discussing how to read Table VI, we introduce characteristics of the chi-square distribution.

Figure 1

Characteristics of the Chi-Square Distribution

1. It is not symmetric.
2. The shape of the chi-square distribution depends on the degrees of freedom, just like Student's t-distribution.
3. As the number of degrees of freedom increases, the chi-square distribution becomes more symmetric, as illustrated in Figure 1.
4. The values of χ^2 are nonnegative. That is, the values of χ^2 are greater than or equal to 0.

Table VI is structured similarly to Table V for the t-distribution. The left column represents the degrees of freedom, and the top row represents the area under the chi-square distribution to the right of the critical value. We use the notation χ^2_α to denote the critical χ^2-value such that the area under the chi-square distribution to the right of χ^2_α is α.

EXAMPLE 1 Finding Critical Values for the Chi-Square Distribution

Figure 2

Problem: Find the critical values that separate the middle 90% of the chi-square distribution from the 5% area in each tail, assuming 15 degrees of freedom.

Approach: We perform the following steps to obtain the critical values.

Step 1: Draw a chi-square distribution with the critical values and areas labeled.
Step 2: Use Table VI to find the critical values.

Solution

Step 1: Figure 2 shows the chi-square distribution with 15 degrees of freedom and the unknown critical values labeled. The area to the right of $\chi^2_{0.05}$ is 0.05. The area to the right of $\chi^2_{0.95}$ is 0.95.

Step 2: Figure 3 shows a partial representation of Table VI. The row containing 15 degrees of freedom is boxed. The columns corresponding to an area to the right of 0.95 and 0.05 are also boxed. The critical values are $\chi^2_{0.95} = 7.261$ and $\chi^2_{0.05} = 24.996$.

Figure 3

Degrees of Freedom	Area to the Right of the Critical Value									
	0.995	0.99	0.975	0.95	0.90	0.10	0.05	0.025	0.01	0.005
1	—	—	0.001	0.004	0.016	2.706	3.841	5.024	6.635	7.879
2	0.010	0.020	0.051	0.103	0.211	4.605	5.991	7.378	9.210	10.597
3	0.072	0.115	0.216	0.352	0.584	6.251	7.815	9.348	11.345	12.838
12	3.074	3.571	4.404	5.226	6.304	18.549	21.026	23.337	26.217	28.299
13	3.565	4.107	5.009	5.892	7.042	19.812	22.362	24.736	27.688	29.819
14	4.075	4.660	5.629	6.571	7.790	21.064	23.685	26.119	29.141	31.319
15	4.601	5.229	6.262	7.261	8.547	22.307	24.996	27.488	30.578	32.801
16	5.142	5.812	6.908	7.962	9.312	23.542	26.296	28.845	32.000	34.267
17	5.697	6.408	7.564	8.672	10.085	24.769	27.587	30.191	33.409	35.718
18	6.365	7.215	8.231	9.288	10.265	25.200	28.601	31.595	24.265	36.456

In studying Table VI, we notice that the degrees of freedom are numbered 1 to 30 inclusive, then 40, 50, 60, . . . , 100. If the number of degrees of freedom is not found in the table, we follow the practice of choosing the degrees of freedom closest to that desired. If the degrees of freedom are exactly between two values, find the mean of the values. For example, to find the critical value corresponding to 75 degrees of freedom, compute the mean of the critical values corresponding to 70 and 80 degrees of freedom.

Now that we understand the chi-square distribution, we can discuss the *goodness-of-fit* test .

Definition

A **goodness-of-fit test** is an inferential procedure used to determine whether a frequency distribution follows a claimed distribution.

As an example, we might want to test whether a die is fair. This would mean the probability of each outcome is $\frac{1}{6}$ when a die is cast. We express this as

$$H_0: p_1 = p_2 = p_3 = p_4 = p_5 = p_6 = \frac{1}{6}$$

Here's another example: According to the U.S. Census Bureau, in 2000, 19.0% of the population of the United States resided in the Northeast; 22.9% resided in the Midwest, 35.6% resided in the South, and 22.5% resided in the West. We might want to test whether the distribution of U.S. residents is the same today as it was in 2000. Remember, the null hypothesis is a statement of "no change," so for this test, the null hypothesis is

H_0: The distribution of residents in the United States is the same today as it was in 2000.

The idea behind testing these types of hypotheses is to compare the actual number of observations for each category of data with the number of observations we would expect if the null hypothesis were true. If a significant difference between the observed counts and expected counts exists, we have evidence against the null hypothesis.

The method for obtaining the expected counts is an extension of the expected value of a binomial random variable. Recall that the mean (and therefore expected value) of a binomial random variable with n independent trials and probability of success, p, is given by $E = \mu = np$.

In Other Words

The expected count for each category is the number of trials of the experiment times the probability of success in the category.

Expected Counts

Suppose there are n independent trials of an experiment with $k \geq 3$ mutually exclusive possible outcomes. Let p_1 represent the probability of observing the first outcome and E_1 represent the expected count of the first outcome, p_2 represent the probability of observing the second outcome and E_2 represent the expected count of the second outcome, and so on. The expected counts for each possible outcome are given by

$$E_i = \mu_i = np_i \quad \text{for} \quad i = 1, 2, \ldots, k$$

EXAMPLE 2   **Finding Expected Counts**

Problem: An urban economist wishes to determine whether the distribution of residents in the United States is the same today as it was in 2000. That year, 19.0% of the population of the United States resided in the Northeast, 22.9% resided in the Midwest, 35.6% resided in the South, and 22.5% resided in the West (based on data obtained from the Census Bureau). If the economist randomly selects 1500 households in the United States, compute the expected number of households in each region, assuming that the distribution of households did not change from 2000.

Approach

Step 1: Determine the probabilities for each outcome.

Step 2: There are $n = 1500$ trials (the 1500 households surveyed) of the experiment. We expect $np_{\text{northeast}}$ of the households surveyed to reside in the Northeast, np_{midwest} of the households to reside in the Midwest, and so on.

Solution

Step 1: The probabilities are the relative frequencies from the 2000 distribution: $p_{\text{northeast}} = 0.190$, $p_{\text{midwest}} = 0.229$, $p_{\text{south}} = 0.356$, and $p_{\text{west}} = 0.225$.

Step 2: The expected counts for each location within the United States are as follows:

Expected count of Northeast: $np_{\text{northeast}} = 1500(0.190) = 285$

Expected count of Midwest: $np_{\text{midwest}} = 1500(0.229) = 343.5$

Expected count of South: $np_{\text{south}} = 1500(0.356) = 534$

Expected count of West: $np_{\text{west}} = 1500(0.225) = 337.5$

Of the 1500 households surveyed, the economist expects 285 households in the Northeast, 343.5 households in the Midwest, 534 households in the South, and 337.5 households in the West if the distribution of residents of the United States is the same today as it was in 2000.

Now Work Problem 5.

Historical Note

The goodness-of-fit test was invented by Karl Pearson (the Pearson of correlation coefficient fame). Pearson believed that statistics should be done by determining the distribution of a random variable. Such a determination could be made only by looking at large numbers of data. This philosophy caused Pearson to "butt heads" with Ronald Fisher, because Fisher believed in analyzing small samples.

To test a hypothesis, we compare the observed counts with the expected counts. If the observed counts are significantly different from the expected counts, we have evidence against the null hypothesis. To perform this test, we need a test statistic and sampling distribution.

! CAUTION
Goodness-of-fit tests are used to test hypotheses regarding the distribution of a variable based on a single population. If you wish to compare two or more populations, you must use the tests for homogeneity presented in Section 12.2.

Test Statistic for Goodness-of-Fit Tests

Let O_i represent the observed counts of category i, E_i represent the expected counts of category i, k represent the number of categories, and n represent the number of independent trials of an experiment. Then the formula

$$\chi^2 = \sum \frac{(O_i - E_i)^2}{E_i} \qquad i = 1, 2, \dots, k$$

approximately follows the chi-square distribution with $k - 1$ degrees of freedom, provided that

1. all expected frequencies are greater than or equal to 1 (all $E_i \geq 1$) and
2. no more than 20% of the expected frequencies are less than 5.

Note: $E_i = np_i$ for $i = 1, 2, \dots, k$.

From Example 2, there were $k = 4$ categories (Northeast, Midwest, South, and West). For the Northeast, the expected frequency, E, is 295.

Now that we know the distribution of goodness-of-fit tests, we can present a method for testing hypotheses regarding the distribution of a random variable.

The Goodness-of-Fit Test

To test hypotheses regarding a distribution, we can use the steps that follow.

Step 1: Determine the null and alternative hypotheses:

  H_0: The random variable follows a certain distribution.

  H_1: The random variable does not follow a certain distribution.

Step 2: Decide on a level of significance, α, depending on the seriousness of making a Type I error.

Step 3:

(a) Calculate the expected counts for each of the k categories. The expected counts are $E_i = np_i$ for $i = 1, 2, \dots, k$, where n is the number of trials and p_i is the probability of the ith category, assuming that the null hypothesis is true.

! CAUTION
If requirements in Step 3(b) are not satisfied, one option is to combine two of the low-frequency categories into a single category.

(b) Verify that the requirements for the goodness-of-fit test are satisfied.

  1. All expected counts are greater than or equal to 1 (all $E_i \geq 1$).
  2. No more than 20% of the expected counts are less than 5.

(c) Compute the **test statistic:**

$$\chi_0^2 = \sum \frac{(O_i - E_i)^2}{E_i}$$

Note: O_i is the observed count for the ith category.

Classical Approach

Step 4: Determine the critical value. All goodness-of-fit tests are right-tailed tests, so the critical value is χ_α^2 with $k - 1$ degrees of freedom. See Figure 4.

P-Value Approach

Step 4: Use Table VI to obtain an approximate P-value by determining the area under the chi-square distribution with $k - 1$ degrees of freedom to the right of the test statistic. See Figure 5.

(continued on next page)

Classical Approach

Figure 4

Critical Region
Area = α

χ_α^2
(critical value)

Step 5: Compare the critical value to the test statistic. If $\chi_0^2 > \chi_\alpha^2$, reject the null hypothesis.

P-value Approach

Figure 5

P-value

χ_0^2

Step 5: If P-value < α, reject the null hypothesis.

Step 6: State the conclusion.

| EXAMPLE 3 | **Testing Hypotheses Using the Goodness-of-Fit Test** |

Problem: An urban economist wonders if the distribution of residents in the United States is different today than it was in 2000. In 2000, 19.0% of the population of the United States resided in the Northeast, 22.9% resided in the Midwest, 35.6% resided in the South, and 22.5% resided in the West (based on data obtained from the Census Bureau). The economist randomly selects 1500 households in the United States and obtains the frequency distribution shown in Table 1.

Conduct the appropriate test to determine if the distribution of residents in the United States is different today from the distribution in 2000 using the $\alpha = 0.05$ level of significance.

Table 1

Region	Frequency
Northeast	274
Midwest	303
South	564
West	359

Approach: We follow Steps 1 through 6 just presented.

Solution

Step 1: We want to know if the distribution of residents is different today from the distribution in 2000. The null hypothesis is a statement of "no difference", which means the distribution today is "no different" from the distribution in 2000.

H_0: The distribution of residents of the United States is the same today as it was in 2000.

H_1: The distribution of residents of the United States is different today from what it was in 2000.

Step 2: The level of significance is $\alpha = 0.05$.

Step 3:

(a) The expected counts were computed in Example 2. For convenience, we show the observed and expected counts in Table 2.

> **In Other Words**
> Remember, the null hypothesis is always a statement of "no change." Therefore, the null hypothesis is that there is no change in the distribution from 2000.

Table 2

Region	Observed Counts	Expected Counts
Northeast	274	285
Midwest	303	343.5
South	564	534
West	359	337.5

(b) Since all expected counts are greater than or equal to 5, the requirements for the goodness-of-fit test are satisfied.

(c) The test statistic is

$$\chi_0^2 = \sum \frac{(O_i - E_i)^2}{E_i} = \frac{(274 - 285)^2}{285} + \frac{(303 - 343.5)^2}{343.5} + \frac{(564 - 534)^2}{534} + \frac{(359 - 337.5)^2}{337.5}$$
$$= 8.255$$

Classical Approach

Step 4: There are $k = 4$ categories, so we find the critical value using $4 - 1 = 3$ degrees of freedom. The critical value is $\chi_{0.05}^2 = 7.815$. See Figure 6.

Figure 6

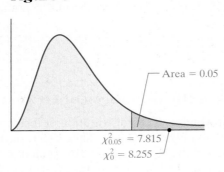

Area = 0.05

$\chi_{0.05}^2 = 7.815$
$\chi_0^2 = 8.255$

Step 5: Because the test statistic, 8.255, is greater than the critical value, 7.815, we reject the null hypothesis.

P-Value Approach

Step 4: There are $k = 4$ categories. The P-value is the area under the chi-square distribution with $4 - 1 = 3$ degrees of freedom to the right of $\chi_0^2 = 8.255$, as shown in Figure 7.

Figure 7

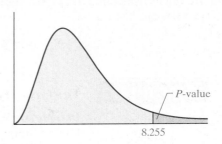

P-value

8.255

Using Table VI, we find the row that corresponds to 3 degrees of freedom. The value of 8.255 lies between 7.815 and 9.348. The value of 7.815 has an area under the chi-square distribution of 0.05 to the right. The area under the chi-square distribution with 3 degrees of freedom to the right of 9.348 is 0.025. Because 8.255 is between 7.815 and 9.348, the P-value is between 0.025 and 0.05. So $0.025 < P\text{-value} < 0.05$.

Step 5: Because the P-value is less than the level of significance, $\alpha = 0.05$, we reject the null hypothesis.

Step 6: There is sufficient evidence at the $\alpha = 0.05$ level of significance to conclude that the distribution of U. S. residents today is different from the distribution in 2000.

If we compare the observed and expected counts, we notice that the Northeast and Midwest regions of the United States have observed counts lower than expected, while the South and West regions of the United States have observed counts higher than expected. So we might conclude that residents of the United States are moving to southern and western locations.

Now Work Problem 11.

In the next example, each of the k categories is equally likely.

EXAMPLE 4 **Testing Hypotheses Using a Goodness-of-Fit Test**

Problem: An obstetrician wants to know whether or not the proportion of children born each day of the week is the same. She randomly selects 500 birth records and obtains the data shown in Table 3 (based on data obtained from *Vital Statistics of the United States*, 2002).

Is there reason to believe that the day on which a child is born occurs with equal frequency at the $\alpha = 0.01$ level of significance?

Table 3

Day of Week	Frequency
Sunday	57
Monday	78
Tuesday	74
Wednesday	76
Thursday	71
Friday	81
Saturday	63

Approach: We follow Steps 1 through 6 presented on pages 554–555.

Solution:

Step 1: A statement of "no difference" would mean that the day on which a child is born occurs with equal frequency. If we let 1 represent Sunday, 2 represent Monday, and so on, we can express this statement as $p_1 = p_2 = \cdots = p_7 = \frac{1}{7}$. Because the null hypothesis is a statement of "no difference", we have

$$H_0: p_1 = p_2 = p_3 = p_4 = p_5 = p_6 = p_7 = \frac{1}{7}$$

H_1: At least one of the proportions is different from the others.

Step 2: The level of significance is $\alpha = 0.01$.

Step 3:

(a) The expected counts for each category (day of the week), assuming the null hypothesis is true, are shown in Table 4.

Table 4

Day of the Week	Frequency (observed count)	Assumed Probability	Expected Count
Sunday	57	$\frac{1}{7}$	$500\left(\frac{1}{7}\right) = \frac{500}{7}$
Monday	78	$\frac{1}{7}$	$\frac{500}{7}$
Tuesday	74	$\frac{1}{7}$	$\frac{500}{7}$
Wednesday	76	$\frac{1}{7}$	$\frac{500}{7}$
Thursday	71	$\frac{1}{7}$	$\frac{500}{7}$
Friday	81	$\frac{1}{7}$	$\frac{500}{7}$
Saturday	63	$\frac{1}{7}$	$\frac{500}{7}$

(b) Since all expected counts are greater than or equal to $5\left(\frac{500}{7} \approx 71.4\right)$, the requirements for the goodness-of-fit test are satisfied.

(c) The test statistic is

$$\chi_0^2 = \frac{(57 - 500/7)^2}{500/7} + \frac{(78 - 500/7)^2}{500/7} + \frac{(74 - 500/7)^2}{500/7} + \frac{(76 - 500/7)^2}{500/7} +$$

$$\frac{(71 - 500/7)^2}{500/7} + \frac{(81 - 500/7)^2}{500/7} + \frac{(63 - 500/7)^2}{500/7} = 6.184$$

Classical Approach

Step 4: There are $k = 7$ categories, so we find the critical value using $7 - 1 = 6$ degrees of freedom. The critical value is $\chi_{0.01}^2 = 16.812$. See Figure 8.

P-Value Approach

Step 4: There are $k = 7$ categories. The P-value is the area under the chi-square distribution with $7 - 1 = 6$ degrees of freedom to the right of $\chi_0^2 = 6.184$, as shown in Figure 9.

(*continued on next page*)

Classical Approach

Figure 8

Step 5: Because the test statistic, 6.184, is less than the critical value, 16.812, we do not reject the null hypothesis.

P-Value Approach

Figure 9

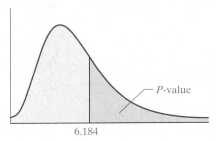

Using Table VI, we find the row that corresponds to 6 degrees of freedom. The value of 6.184 is less than 10.546, which has an area under the chi-square distribution of 0.10 to the right. The *P*-value is greater than 0.10. So the *P*-value > 0.10.

Step 5: Because the *P*-value is greater than the level of significance, $\alpha = 0.01$, we do not reject the null hypothesis.

Step 6: There is no reason to believe that the day of the week on which a child is born does not occur with equal frequency.

Now Work Problem 17.

! **CAUTION**
⬤ To find exact *P*-values, we must use technology.

Finding Exact *P*-Values Using Technology
We cannot obtain exact *P*-values by hand, but statistical software will provide *P*-values based on the value of the test statistic and the degrees of freedom. We can compute the *P*-value of the hypothesis tested in Example 4, using the χ^2-cdf command on a TI-84 Plus graphing calculator. See Figure 10. The area under the chi-square distribution to the right of 6.184 with 6 degrees of freedom is 0.403, so the *P*-value is 0.403. See Figure 11. The *P*-value is greater than the level of significance, so we do not reject the null hypothesis.

Figure 10

```
X²cdf(6.184,1E99
,6)
        .402897295
```

Figure 11

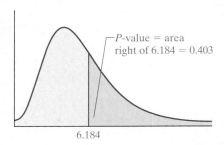

Note to Instructor
This activity is intended to give students a hands-on experience with a goodness-of-fit test and to illustrate how researchers compare new data to known distributions to draw conclusions. As an incentive, you may want to hand out small fun-size packs to those who correctly identify the candy type. Time limit: 30 to 40 minutes

In-Class Activity: If the Candy Fits—Eat It

Researchers can sometimes help determine or refute the authorship of a text or play by looking at the distribution of word lengths and sentence lengths. On a tastier note, could you determine the type of M&Ms candy just by knowing how many of each color are in the bag?

(continued on next page)

In-Class Activity (*continued*)

(a) Each student should purchase a bag of M&Ms that are either plain, peanut, almond, crispy, or peanut butter. Make sure the bag is large enough to obtain a decent sample size.

(b) At home, record the number of each color from your bag (ignore broken candies).

(c) In class, give your data to your partner and see if he or she can determine which type of candy you purchased while you try to determine his or her type.
[**Note:** Actual distributions can be found at us.mms.com/us/about/products.]

(d) Compare your results to the rest of the class. Were there types that were harder to distinguish? Why do you think this was the case?

12.1 ASSESS YOUR UNDERSTANDING

Concepts and Vocabulary

1. Why is goodness of fit a good choice for the title of the procedures used in this section?

2. Explain why chi-square goodness-of-fit tests are always right tailed.

3. State the requirements to perform a goodness-of-fit test.

4. Suppose the expected count of a category is less than 1. What can be done to the categories so that a goodness-of-fit test can be performed?

Skill Building

In Problems 5 and 6, determine the expected counts for each outcome.

5.

$n = 500$				
p_i	0.2	0.1	0.45	0.25
Expected counts	100	50	225	125

6.

$n = 700$				
p_i	0.15	0.3	0.35	0.20
Expected counts	105	210	245	140

In Problems 7–10, determine (a) the χ^2 test statistic, (b) the degrees of freedom, and (c) the critical value using $\alpha = 0.05$, and (d) test the hypothesis at the $\alpha = 0.05$ level of significance.

7. $H_0: p_A = p_B = p_C = p_D = \dfrac{1}{4}$

H_1: At least one of the proportions is different from the others. Do not reject H_0

Outcome	A	B	C	D
Observed	30	20	28	22
Expected	25	25	25	25

8. $H_0: p_A = p_B = p_C = p_D = p_E = \dfrac{1}{5}$

H_1: At least one of the proportions is different from the others. Do not reject H_0

Outcome	A	B	C	D	E
Observed	38	45	41	33	43
Expected	40	40	40	40	40

9. H_0: The random variable X is binomial with $n = 4$, $p = 0.8$.

H_1: The random variable X is not binomial with $n = 4$, $p = 0.8$. Reject H_0

X	0	1	2	3	4
Observed	1	38	132	440	389
Expected	1.6	25.6	153.6	409.6	409.6

10. H_0: The random variable X is binomial with $n = 4$, $p = 0.3$.

H_1: The random variable X is not binomial with $n = 4$, $p = 0.8$. Reject H_0

X	0	1	2	3	4
Observed	260	400	280	50	10
Expected	240.1	411.6	264.6	75.6	8.1

Applying the Concepts

11. Plain M&Ms According to the manufacturer of M&Ms,
NW 13% of the plain M&Ms in a bag should be brown, 14% yellow, 13% red, 20% orange, 24% blue, and 16% green. A student wanted to determine whether a randomly selected bag of plain M&Ms had contents that followed this distribution. He counted the number of M&Ms that were each color and obtained the results shown in the table. Test whether plain M&Ms follow the distribution stated by M&M/Mars at the $\alpha = 0.05$ level of significance. Reject H_0

Color	Frequency
Brown	61
Yellow	64
Red	54
Blue	61
Orange	96
Green	64

12. Peanut M&Ms According to the manufacturer of M&Ms, 12% of the peanut M&Ms in a bag should be brown, 15% yellow, 12% red, 23% blue, 23% orange, and 15% green. A student wanted to determine whether a randomly selected bag of peanut M&Ms had contents that followed this distribution. He counted the number of M&Ms that were each color and obtained the results shown in the table. Test whether peanut M&Ms follow the distribution stated by M&M/Mars at the $\alpha = 0.05$ level of significance. Do not reject H_0

Color	Frequency	Color	Frequency
Brown	53	Blue	96
Yellow	66	Orange	88
Red	38	Green	59

---

13. Benford's Law, Part I Our number system consists of the digits 0, 1, 2, 3, 4, 5, 6, 7, 8, and 9. The first significant digit in any number must be 1, 2, 3, 4, 5, 6, 7, 8, or 9 because we do not write numbers such as 12 as 012. Although we may think that each digit appears with equal frequency so that each digit has a $\frac{1}{9}$ probability of being the first significant digit, this is not true. In 1881, Simon Newcomb discovered that digits do not occur with equal frequency. This same result was discovered again in 1938 by physicist Frank Benford. After studying much data, he was able to assign probabilities of occurrence to the first digit in a number as shown.

Digit	1	2	3	4	5	6	7	8	9
Probability	0.301	0.176	0.125	0.097	0.079	0.067	0.058	0.051	0.046

Source: T. P. Hill, The First Digit Phenomenon, *American Scientist*, July–August, 1998.

The probability distribution is now known as Benford's Law and plays a major role in identifying fraudulent data on tax returns and accounting books. For example, the distribution below represents the first digits in 200 allegedly fraudulent checks written to a bogus company by an employee attempting to embezzle funds from his employer.

First digit of fraudulent check	1	2	3	4	5	6	7	8	9
Frequency	36	32	28	26	23	17	15	16	7

Source: State of Arizona v. Wayne James Nelson

(a) Because these data are meant to prove that someone is guilty of fraud, what would be an appropriate level of significance when performing a goodness-of-fit test?

(b) Using the level of significance chosen in part (a), test whether the first digits in the allegedly fraudulent checks obey Benford's Law.

(c) Based on the results of part (b), do you think that the employee is guilty of embezzlement?

14. Benford's Law, Part II Refer to Problem 13. The following distribution lists the first digit of the surface area (in square miles) of 335 rivers. Is there evidence at the $\alpha = 0.05$ level of significance that the distribution follows Benford's Law?

First digit of surface area	1	2	3	4	5	6	7	8	9
Frequency	104	55	36	38	24	29	18	14	17

Source: Eric W. Weisstein, Benford's Law, From *MathWorld*—A Wolfram Web Resource.

15. Are Some Months Busier than Others? A researcher wants to know whether the distribution of birth month is uniform. The following data, based on results obtained from *Vital Statistics of the United States*, 2005, Volume 1, represent the distribution of months in which 500 randomly selected children were born:

Month	Frequency	Month	Frequency
Jan.	40	July	45
Feb.	38	Aug.	44
March	41	Sep.	44
April	40	Oct.	43
May	42	Nov.	39
June	41	Dec.	43

Is there reason to believe that each birth month occurs with equal frequency at the $\alpha = 0.05$ level of significance? Do not reject H_0

16. Bicycle Deaths A researcher wanted to determine whether bicycle deaths were uniformly distributed over the days of the week. She randomly selected 200 deaths that involved a bicycle, recorded the day of the week on which the death occurred, and obtained the following results (the data are based on information obtained from the Insurance Institute for Highway Safety).

Day of the Week	Frequency	Day of the Week	Frequency
Sunday	16	Thursday	34
Monday	35	Friday	41
Tuesday	16	Saturday	30
Wednesday	28		

Is there reason to believe that bicycle fatalities occur with equal frequency with respect to day of the week at the $\alpha = 0.05$ level of significance? Reject H_0

17. Pedestrian Deaths A researcher wanted to determine whether pedestrian deaths were uniformly distributed over the days of the week. She randomly selected 300 pedestrian deaths, recorded the day of the week on which the death occurred, and obtained the following results (the data are based on information obtained from the Insurance Institute for Highway Safety).

Day of the Week	Frequency	Day of the Week	Frequency
Sunday	39	Thursday	41
Monday	40	Friday	49
Tuesday	30	Saturday	61
Wednesday	40		

Test the belief that the day of the week on which a fatality happens involving a pedestrian occurs with equal frequency at the $\alpha = 0.05$ level of significance. Reject H_0

18. Is the Die Loaded? A player in a craps game suspects that one of the dice being used in the game is loaded. A loaded die is one in which all the possibilities (1, 2, 3, 4, 5, and 6) are not equally likely. The player throws the die 400 times, records the outcome after each throw, and obtains the following results: (a) Do not reject H_0

Outcome	Frequency	Outcome	Frequency
1	62	4	62
2	76	5	57
3	76	6	67

(a) Test whether the die is loaded. Use the $\alpha = 0.01$ level of significance.
(b) Why do you think the player might use the $\alpha = 0.01$ level of significance rather than, say, the $\alpha = 0.1$ level of significance?

19. Home Schooling A school social worker wants to determine if the grade distribution of home-schooled children is different in her district than nationally. The U.S. National Center for Education Statistics provided her with the following data, which represent the relative frequency of home-schooled children by grade level.

Grade	Relative Frequency
K	0.076
1–3	0.253
4–5	0.158
6–8	0.235
9–12	0.278

She obtains a sample of 25 home-schooled children within her district that yields the following data:

Grade	Frequency
K	6
1–3	9
4–5	3
6–8	4
9–12	3

(a) Because of the low cell counts, combine cells into three categories K–3, 4–8, and 9–12.
(b) Test whether the grade distribution of home-schooled children is different in her district than nationally at the $\alpha = 0.05$ level of significance. Reject H_0

20. Golden Benford The Fibonacci sequence is a famous sequence of numbers whose elements commonly occur in nature. The terms in the Fibonacci sequence are 1, 1, 2, 3, 5, 8, 13, 21, ... The ratio of consecutive terms approaches the "golden ratio," $\Phi = \dfrac{1 + \sqrt{5}}{2}$. If we examine the first digit of the first 85 terms in the Fibonacci sequence, the distribution of digits is as shown in the following table:

Digit	1	2	3	4	5	6	7	8	9
Frequency	25	16	11	7	7	5	4	6	4

Is there evidence to support the belief that the first digit of the Fibonacci numbers follow the Benford distribution (shown in Problem 13) at the $\alpha = 0.05$ level of significance? *Do not reject* H_0

21. **Testing the Random-Number Generator** Statistical spreadsheets and graphing calculators with advanced statistical features have random-number generators that create random numbers conforming to a specified distribution.
    (a) Use a random-number generator to create a list of 500 randomly selected integers numbered 1 to 5.
    (b) What proportion of the numbers generated should be 1? 2? 3? 4? 5?
    (c) Test if the random-number generator is generating random integers between 1 and 5 with equal likelihood by performing a chi-square goodness-of-fit test at the $\alpha = 0.01$ level of significance.

22. **Testing the Random-Number Generator** Statistical spreadsheets and graphing calculators with advanced statistical features have random-number generators that create random numbers conforming to a specified distribution.
    (a) Use a random-number generator to create a list of 500 trials of a binomial experiment with $n = 5$ and $p = 0.2$.
    (b) What proportion of the numbers generated should be 0? 1? 2? 3? 4? 5?
    (c) Test if the random-number generator is generating random outcomes of a binomial experiment with $n = 5$ and $p = 0.2$ by performing a chi-square goodness-of-fit test at the $\alpha = 0.01$ level of significance.

In Section 10.4, we tested hypotheses regarding a population proportion using a z-test. However, we can also use the chi-square goodness-of-fit test to test hypotheses with $k = 2$ possible outcomes. In Problems 23 and 24, we test hypotheses with the use of both methods.

23. **Low Birth Weight** According to the U.S. Census Bureau, 7.1% of all babies born to nonsmoking mothers are of low birth weight (<5 lb, 8 oz). An obstetrician wanted to know whether mothers between the ages of 35 and 39 years give birth to a higher percentage of low-birth-weight babies. She randomly selected 160 births for which the mother was 35 to 39 years old and found 15 low-birth-weight babies.
    (a) If the proportion of low-birth-weight babies for mothers in this age group is 0.071, compute the expected number of low-birth-weight births to 35- to 39-year old mothers. What is the expected number of births to mothers 35 to 39 years old that are not low birth weight? *11.36; 148.64*
    (b) Test the hypothesis at the $\alpha = 0.05$ level of significance using the chi-square goodness-of-fit test.
    (c) Test the hypothesis using the approach presented in Section 10.4. *Do not reject* H_0

24. **Living Alone?** In 2000, 25.8% of Americans 15 years of age or older lived alone, according to the Census Bureau. A sociologist who believes that this percentage is greater today, conducts a random sample of 400 Americans 15 years of age or older, and finds that 164 are living alone.
    (a) If the proportion of Americans aged 15 years or older living alone is 0.258, compute the following expected numbers: Americans 15 years of age or older who live alone; Americans 15 years of age or older who do not live alone. *103.2; 296.8*
    (b) Test the hypothesis at the $\alpha = 0.05$ level of significance using the goodness-of-fit test. *Reject* H_0
    (c) Test the hypothesis using the approach presented in Section 10.4. *Reject* H_0

25. Using the results of Problem 7, compute $\Sigma(O - E)$. Explain why this result is reasonable. *0*

(23b) *Do not reject* H_0

 ## 12.2 Tests for Independence and the Homogeneity of Proportions

Preparing for This Section Before getting started, review the following:

• The language of hypothesis tests (Section 10.1, pp. 454–460)

• Independent events (Section 5.3, pp. 249–252)

• Mean of a binomial random variable (Section 6.2, pp. 305–306)

• Testing a hypothesis about two population proportions (Section 11.3, pp. 534–539)

Objectives

1. **Perform a test for independence**
2. **Perform a test for homogeneity of proportions**

As we saw in Section 12.1, data, whether qualitative or quantitative, can be organized into categories. For example, a person might be categorized as a male or as a female. A person might also be categorized as a 20- to 29-year-old.

Consider the data (measured in thousands) in Table 5, which represent the employment status and level of education of all U.S. residents 25 years old or older in 2004. By definition, an individual is unemployed if he or she is actively seeking work but is unable to find work. An individual is considered not to be in the labor force if he or she is not employed and is not actively seeking employment.

Table 5

Employment Status	Level of Education			
	Did Not Finish High School	High School Graduate	Some College	Four or More Years of College
Employed	11,537	35,857	32,977	39,293
Unemployed	1,109	2,069	1,461	1,098
Not in the labor force	15,509	21,554	13,117	11,462

Source: Bureau of Labor Statistics

Table 5 is referred to as a **contingency table** or a **two-way table**, because it relates two categories of data. The data presented in Table 5 describe two characteristics regarding the population of U.S. residents who are 25 years or older: their employment status and their level of education. The **row variable** is employment status, because each row in the table describes the employment status of a group. The **column variable** is level of education. Each box inside the table is referred to as a **cell**. For example, the cell corresponding to employed individuals who are high school graduates is in the first row, second column. Each cell contains the frequency of the category: There were 11,537 thousand employed individuals who did not finish high school in 2004.

① Perform a Test for Independence

In this section, we develop methods for performing statistical inference on two categorical variables to determine whether there is any association between two variables. We call the method the *chi-square test for independence.*

Definition

> The **chi-square test for independence** is used to determine whether there is an association between a row variable and column variable in a contingency table constructed from sample data. The null hypothesis is that the variables are not associated; in other words, they are independent. The alternative hypothesis is that the variables are associated, or dependent.

In Other Words

In a chi-square independence test, the null hypothesis is always

H_0: The variables are independent.

The alternative hypothesis is always

H_1: The variables are not independent.

The idea behind testing these types of hypotheses is to compare actual counts to the counts we would expect if the null hypothesis were true (if the variables are independent). If a significant difference between the actual counts and expected counts exists, we take this as evidence against the null hypothesis.

The method for obtaining the expected counts requires that we compute the number of observations expected within each cell under the assumption of independence. Recall, if two events E and F are independent, then $P(E \text{ and } F) = P(E) \cdot P(F)$. We can use the Multiplication Rule for Independent Events to obtain the expected proportion of observations within each cell under the assumption of independence. We then multiply this result by n, the sample size, to obtain the expected count within each cell.* We present an example to introduce the method for obtaining expected counts.

---

*Recall that the expected value of a binomial random variable for n independent trials of a binomial experiment with probability of success p is given by $E = \mu = np$.

EXAMPLE 1   **Determining the Expected Counts in a Test for Independence**

Problem: Blood type is classified as A, B, AB, or O. In addition, blood can be classified as Rh$^+$ or Rh$^-$. In a survey of 500 randomly selected individuals, a phlebotomist obtained the results shown in Table 6.

Table 6

| | Blood Type | | | |
Rh-status	A	B	AB	O
Rh$^+$	176	28	22	198
Rh$^-$	30	12	4	30

Compute the expected counts within each cell, assuming that Rh-status and blood type are independent.

Approach

Step 1: Compute the row and column totals.

Step 2: Compute the relative marginal frequencies for the row variable and column variable.

Step 3: Use the Multiplication Rule for Independent Events to compute the proportion of observations within each cell under the assumption of independence.

Step 4: Multiply the proportions by 500, the sample size, to obtain the expected counts within each cell.

Solution

Step 1: The row totals (blue) and column totals (red) are presented in Table 7.

Table 7

	A	B	AB	O	Row Totals
Rh$^+$	176	28	22	198	424
Rh$^-$	30	12	4	30	76
Column totals	206	40	26	228	500

Step 2: The relative frequencies for the row variable (Rh-status) and column variable (blood type) are presented in Table 8.

Table 8

	A	B	AB	O	Relative Frequency
Rh$^+$	176	28	22	198	$\frac{424}{500} = 0.848$
Rh$^-$	30	12	4	30	$\frac{76}{500} = 0.152$
Relative frequency	$\frac{206}{500} = 0.412$	$\frac{40}{500} = 0.08$	$\frac{26}{500} = 0.052$	$\frac{228}{500} = 0.456$	1

Step 3: Assuming blood type and Rh-status are independent, we use the Multiplication Rule for Independent Events to compute the proportion of observations we would expect in each cell. For example, the proportion of individuals who are Rh^+ and of blood type A would be

$$\left(\begin{array}{c}\text{Proportion } Rh^+ \text{ and}\\ \text{blood type A}\end{array}\right) = (\text{proportion } Rh^+)\cdot(\text{proportion blood type A})$$

$$= (0.848)(0.412)$$

$$= 0.349376$$

Table 9 contains the expected proportion in each cell, under the assumption of independence.

Table 9	A	B	AB	O
Rh^+	0.349376	0.06784	0.044096	0.386688
Rh^-	0.062624	0.01216	0.007904	0.069312

Step 4: We multiply the expected proportions in Table 9 by 500, the sample size, to obtain the expected counts under the assumption of independence. The results are presented in Table 10.

Table 10	A	B	AB	O
Rh^+	500(0.349376) = 174.688	500(0.06784) = 33.92	500(0.044096) = 22.048	193.344
Rh^-	31.312	6.08	3.952	34.656

If blood type and Rh-status are independent, we would expect a random sample of 500 individuals to contain about 175 who are of blood type A and are Rh^+.

The technique used in Example 1 to find the expected counts might seem rather tedious. It certainly would be more pleasant if we could determine a shortcut formula that could be used to obtain the expected counts. Let's consider the expected count for blood type A and Rh^+. This expected count was obtained by multiplying the proportion of individuals who were of blood type A, the proportion of individuals who were Rh^+, and the number of individuals in the sample. That is,

$$\text{Expected count} = (\text{proportion } Rh^+)(\text{proportion blood type A})(\text{sample size})$$

$$= \frac{424}{500}\cdot\frac{206}{500}\cdot 500$$

$$= \frac{424\cdot 206}{500} \qquad \textit{Cancel the 500s.}$$

$$= \frac{(\text{row total for } Rh^+)(\text{column total for blood type A})}{\text{table total}}$$

This leads to the following general result:

Expected Frequencies in a Chi-Square Test for Independence

To find the expected frequencies in a cell when performing a chi-square independence test, multiply the row total of the row containing the cell by the column total of the column containing the cell and divide this result by the table total. That is,

$$\text{Expected frequency} = \frac{(\text{row total})(\text{column total})}{\text{table total}} \qquad \textbf{(1)}$$

For example, to calculate the expected frequency for Rh^{+}, blood type A, we compute

$$\text{Expected frequency} = \frac{(\text{row total})(\text{column total})}{\text{table total}} = \frac{(424)(206)}{500} = 174.688$$

Now Work Problem 7(a).

This result agrees with the result obtained in Table 10.

To perform a test regarding the independence of two variables, we compare the actual (observed) counts to those expected. If the observed counts are significantly different from the expected counts, we take this as evidence against the null hypothesis. We need a test statistic and sampling distribution to conduct the test.

Test Statistic for the Test of Independence

Let O_i represent the observed number of counts in the ith cell and E_i represent the expected number of counts in the ith cell. Then

$$\chi^2 = \sum \frac{(O_i - E_i)^2}{E_i}$$

approximately follows the chi-square distribution with $(r - 1)(c - 1)$ degrees of freedom, where r is the number of rows and c is the number of columns in the contingency table, provided that (1) all expected frequencies are greater than or equal to 1 and (2) no more than 20% of the expected frequencies are less than 5.

From Example 1, there were $r = 2$ rows and $c = 4$ columns.

We now present a method for testing hypotheses regarding the association between two variables in a contingency table.

Chi-Square Test for Independence

To test the association between (or independence of) two variables in a contingency table, we use the steps that follow.

Step 1: Write the null and alternative hypotheses.

   H_0: The row variable and column variable are independent.

   H_1: The row variable and column variable are dependent.

Step 2: Choose a level of significance, α, depending on the seriousness of making a Type I error.

Step 3:

(a) Calculate the expected frequencies (counts) for each cell in the contingency table using Formula (1).

(b) Verify that the requirements for the goodness-of-fit test are satisfied:

   1. All expected frequencies are greater than or equal to 1 (all $E_i \geq 1$).

   2. No more than 20% of the expected frequencies are less than 5.

(c) Compute the **test statistic**:

$$\chi_0^2 = \sum \frac{(O_i - E_i)^2}{E_i}$$

Note: O_i is the observed frequency for the ith category.

Classical Approach

Step 4: Determine the critical value. All chi-square tests for independence are right-tailed tests, so the critical value is χ_α^2 with $(r - 1)(c - 1)$ degrees of freedom, where r is the number of rows and c is the number of columns in the contingency table. See Figure 12.

Figure 12

Classical Approach Step 5 follows:

Step 5: Compare the critical value to the test statistic. If $\chi_0^2 > \chi_\alpha^2$, reject the null hypothesis.

P-Value Approach

Step 4: Use Table VI to determine an approximate P-value by determining the area under the chi-square distribution with $(r - 1)(c - 1)$ degrees of freedom, where r is the number of rows and c is the number of columns in the contingency table to the right of the test statistic. See Figure 13.

Figure 13

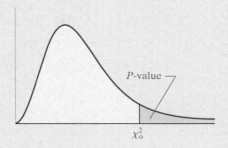

Step 5: If P-value $< \alpha$, reject the null hypothesis.

Step 6: State the conclusion.

EXAMPLE 2   **Performing a Chi-Square Test for Independence**

Problem: Blood is classified as A, B, AB, or O. In addition, blood can be classified as Rh$^+$ or Rh$^-$. In a survey of 500 randomly selected individuals, a phlebotomist obtained the results shown in Table 11.

Table 11				
	Blood Type			
Rh-status	**A**	**B**	**AB**	**O**
Rh$^+$	176	28	22	198
Rh$^-$	30	12	4	30

Test whether blood type and Rh-status are independent at the $\alpha = 0.05$ level of significance.

Approach: We follow Steps 1 through 6 just given.

Solution

Step 1: We wish to determine whether blood type and Rh-status are independent. We state the hypotheses as follows:

   H_0: Blood type and Rh-status are independent (or not related).
   H_1: Blood type and Rh-status are dependent (or somehow related).

Step 2: The level of significance is $\alpha = 0.05$.

Step 3:

(a) The expected frequencies were computed in Example 1. Table 12 shows the observed frequencies, with the expected frequencies in parentheses.

Table 12			
Observed (and Expected) Frequencies			
	A	**B**	**AB**	**O**
---	---	---	---	---
Rh$^+$	176	28	22	198
	(174.688)	(33.92)	(22.048)	(193.344)
Rh$^-$	30	12	4	30
	(31.312)	(6.08)	(3.952)	(34.656)

(b) Since all expected frequencies are greater than or equal to 1 and only 1 out of 8 (=12.5%) expected frequencies is less than 5, the requirements for the goodness-of-fit test are satisfied.

(c) The test statistic is

$$\chi_0^2 = \frac{(176 - 174.688)^2}{174.688} + \frac{(28 - 33.92)^2}{33.92} + \frac{(22 - 22.048)^2}{22.048} + \frac{(198 - 193.344)^2}{193.344}$$

$$+ \frac{(30 - 31.312)^2}{31.312} + \frac{(12 - 6.08)^2}{6.08} + \frac{(4 - 3.952)^2}{3.952} + \frac{(30 - 34.656)^2}{34.656}$$

$$= 7.601$$

Classical Approach

Step 4: There are $r = 2$ rows and $c = 4$ columns, so we find the critical value using $(r - 1)(c - 1) = (2 - 1)(4 - 1) = 3$ degrees of freedom. The critical value is $\chi_{0.05}^2 = 7.815$. See Figure 14.

Figure 14

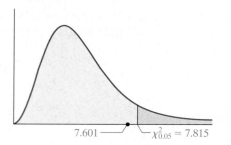

Step 5: Because the test statistic, 7.601, is less than the critical value, 7.815, we do not reject the null hypothesis.

P-Value Approach

Step 4: There are $r = 2$ rows and $c = 4$ columns, so we find the *P*-value using $(r - 1)(c - 1) = (2 - 1)(4 - 1) = 3$ degrees of freedom. The *P*-value is the area under the chi-square distribution with 3 degrees of freedom to the right of $\chi_0^2 = 7.601$, as shown in Figure 15.

Figure 15

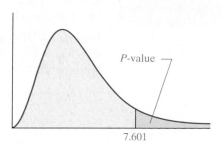

Using Table VI, we find the row that corresponds to 3 degrees of freedom. The value of 7.601 lies between 6.251 and 7.815. The area under the chi-square distribution with 3 degrees of freedom to the right of 6.251 is 0.10. The value of 7.815 has an area under the chi-square distribution of 0.05 to the right. Because 7.601 is between 6.251 and 7.815, the *P*-value is between 0.05 and 0.10. So $0.05 < P\text{-value} < 0.10$.

The exact *P*-value using the TI-84 Plus's χ_{cdf}^2 command is 0.055.

Step 5: Because the *P*-value is greater than the level of significance, $\alpha = 0.05$, we do not reject the null hypothesis.

Step 6: There is not enough evidence, at the $\alpha = 0.05$ level of significance, to support the belief that Rh-status and blood type are dependent.

Now Work Problems 7(b), (c), (d), and (e).

While we did not reject the null hypothesis that Rh-status and blood type are independent at the $\alpha = 0.05$ level of significance, the closeness of the test statistic to the critical value could be considered suggestive of some type of relation between Rh-status and blood type.

To see the relation between Rh-status and blood type, we draw bar graphs of the *conditional distributions* of Rh-status by blood type. A **conditional distribution** lists the relative frequency of each category of a variable, given a specific value of the other variable in a contingency table. For example, we can calculate the relative frequency of Rh^+ and Rh^-, given that an individual is of blood type A. We repeat this for the remaining blood types.

EXAMPLE 3   **Constructing a Conditional Distribution and Bar Graph**

Problem: Find the conditional distribution of the variable Rh-status by blood type for the data in Table 11. Then draw a bar graph that represents the conditional distribution of Rh-status by blood type.

Approach: First, compute the relative frequency for Rh-status, given that the individual is blood type A. Then compute the relative frequency for Rh-status, given that the individual is blood type B, and so on. We will draw two bars, side by side, for each blood type. The horizontal axis represents the blood type, and the vertical axis represents the relative frequency of Rh-status for each blood type.

Solution: We start with the individuals who are blood type A. From Table 11, we see there are $176 + 30 = 206$ individuals who are blood type A. The relative frequency with which we observe an individual who is Rh^+, given that the individual is blood type A, is $\frac{176}{206} = 0.854$. Therefore, 85.4% of individuals who are blood type A are Rh^+. The relative frequency with which we observe an individual who is Rh^-, given that the individual is blood type A, is $\frac{30}{206} = 0.146$. We now proceed to compute the relative frequency for each Rh-status, given that the individual is blood type B. The relative frequency with which we observe an individual who is Rh^+, given that the individual is blood type B, is $\frac{28}{40} = 0.70$.

Repeat this process for individuals who are Rh^-. Finally, compute the relative frequency for each Rh-status, given that the individual is blood type AB and then given that the individual is blood type O. We obtain Table 13.

Table 13			
	A	**B**	**AB**	**O**
Rh^+	$\frac{176}{206} = 0.854$	$\frac{28}{40} = 0.700$	$\frac{22}{26} = 0.846$	$\frac{198}{228} = 0.868$
Rh^-	$\frac{30}{206} = 0.146$	$\frac{12}{40} = 0.300$	$\frac{4}{26} = 0.154$	$\frac{30}{228} = 0.132$

From the conditional distributions by blood type, the association between blood type and Rh-status should be apparent. The proportion of individuals who are Rh^+ is less for those individuals who are of blood type B than for the other blood types. However, the difference is not significant at the $\alpha = 0.05$ level of significance.

Figure 16 contains the bar graph of the conditional distribution. The blue bars represent the proportion of individuals with Rh^+ blood for each blood type, and the green bars represent the proportion of individuals with Rh^- blood for each blood type.

Figure 16

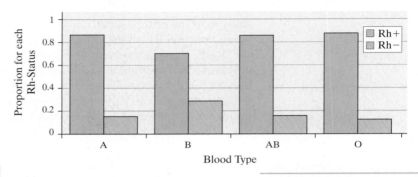

Proportion of Each Rh-Status by Blood Type

Now Work Problem 7(f).

EXAMPLE 4 Chi-square Test for Independence Using Technology

Problem: Using the data presented in Table 11 from Example 2, test whether blood type and Rh-status are independent at the $\alpha = 0.05$ level of significance using statistical software or a graphing calculator with advanced statistical features.

Approach: We will use MINITAB to test the claim. The steps for testing for independence using the TI-83/84 Plus graphing calculators, MINITAB, and Excel are given in the Technology Step by Step on page 580.

Result: Figure 17 shows the results obtained from MINITAB.

Figure 17

Chi-Square Test
```
Expected counts are printed below observed counts

            C1        C2        C3        C4      Total
    1      176        28        22       198       424
        174.69     33.92     22.05    193.34

    2       30        12         4        30        76
         31.31      6.08      3.95     34.66

Total      206        40        26       228       500

Chi-Sq =  0.010 + 1.033 + 0.000 + 0.112 +
          0.055 + 5.764 + 0.001 + 0.626 = 7.601

DF = 3, P-Value = 0.055
1 cells with expected counts less than 5.0
```

Interpretation: Because the *P*-value is greater than the level of significance ($0.055 > 0.05$), we do not reject the null hypothesis. There is not sufficient evidence at the $\alpha = 0.05$ level of significance to conclude that blood type and Rh-status are dependent.

② **Perform a Test for Homogeneity of Proportions**

The chi-square test for independence is a test regarding a sample from a *single* population. Each individual in this single population is classified in two ways (such as blood type and Rh-status). We now discuss a second type of chi-square test, which can be used to compare the population proportions from *different* populations. The test is an extension of the two-sample *z*-test introduced in Section 11.3, where we compared two population proportions.

In Other Words
The chi-square test for homogeneity of proportions is used to compare proportions from two or more populations.

Definition

In a **chi-square test for homogeneity of proportions**, we test whether different populations have the same proportion of individuals with some characteristic.

For example, we might look at the proportion of individuals who experience headaches as a side effect for a placebo group (Group 1), for an experimental group that receives 50 mg per day of a medication (Group 2), and for an experimental group that receives 100 mg per day of the medication (Group 3). Under this circumstance, we might want to know if the proportion of individuals in each group who experience a headache as a side effect is different. Because the null hypothesis is always a statement of "no difference" our null hypothesis would be

$$H_0: p_1 = p_2 = p_3$$

versus the alternative,

H_1: At least one of the population proportions is different from the others.

The procedures for performing a test of homogeneity are identical to those for a test of independence.

EXAMPLE 5   A Test for Homogeneity of Proportions

Problem: Zocor is a drug manufactured by Merck and Co. that is meant to reduce the level of LDL (bad) cholesterol while increasing the level of HDL (good) cholesterol. In clinical trials of the drug, patients were randomly divided into three groups. Group 1 received Zocor, Group 2 received a placebo, and Group 3 received cholestyramine, a cholesterol-lowering drug currently available. Table 14 contains the number of patients in each group who did and did not experience abdominal pain as a side effect.

Table 14

	Group 1 (Zocor)	Group 2 (Placebo)	Group 3 (Cholestyramine)
Number of people who experienced abdominal pain	51	5	16
Number of people who did not experience abdominal pain	1532	152	163

Source: Merck and Co.

Is there evidence to indicate that the proportion of subjects in each group who experienced abdominal pain is different at the $\alpha = 0.01$ level of significance?

Approach: We will follow Steps 1 through 6 on pages 566–567.

Solution

Step 1: Because the null hypothesis is a statement of "no difference", the null hypothesis will be that the proportion of subjects in each group who experienced abdominal pain are equal. So,

$H_0: p_1 = p_2 = p_3$
H_1: At least one of the proportions is different from the others.

Here, p_1, p_2, and p_3 are the proportions in groups 1, 2, and 3, respectively.
Step 2: The level of significance is $\alpha = 0.01$.
Step 3:

(a) The expected frequency of subjects who experienced abdominal pain in Group 1 is computed by multiplying the row total of individuals who experienced abdominal pain by the column total of number of individuals in Group 1 and dividing this result by the total number of subjects in the study. There were $51 + 5 + 16 = 72$ subjects who experienced abdominal pain and $51 + 1532 = 1583$ subjects in Group 1. There were a total of $51 + 5 + 16 + 1532 + 152 + 163 = 1919$ subjects in the study. The number of subjects expected to experience abdominal pain in Group 1 is

$$E = \frac{72 \cdot 1583}{1919} = 59.393$$

Table 15 contains the row and column totals along with the observed frequencies. The expected frequencies are in parentheses.

Table 15			
Observed (and Expected) Frequencies			
	Group 1 (Zocor)	**Group 2 (Placebo)**	**Group 3 (Cholestyramine)**	**Row Totals**
Number of people who experienced abdominal pain	51 (59.393)	5 (5.891)	16 (6.716)	72
Number of people who did not experience abdominal pain	1532 (1523.607)	152 (151.109)	163 (172.284)	1847
Column totals	1583	157	179	1919

(b) All the expected frequencies are greater than 5. The requirements have been satisfied.

(c) The test statistic is

$$\chi_0^2 = \frac{(51 - 59.393)^2}{59.393} + \frac{(5 - 5.891)^2}{5.891} + \frac{(16 - 6.716)^2}{6.716} +$$

$$\frac{(1532 - 1523.607)^2}{1523.607} + \frac{(152 - 151.109)^2}{151.109} + \frac{(163 - 172.284)^2}{172.284} = 14.707$$

Classical Approach

Step 4: There are $r = 2$ rows and $c = 3$ columns, so we find the critical value using $(2 - 1)(3 - 1) = 2$ degrees of freedom. The critical value is $\chi_{0.01}^2 = 9.210$. See Figure 18.

Figure 18

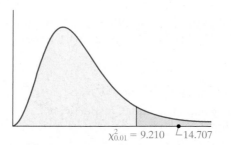

$\chi_{0.01}^2 = 9.210$ 14.707

Step 5: Because the test statistic, 14.707, is greater than the critical value, 9.210, we reject the null hypothesis.

P-Value Approach

Step 4: There are $r = 2$ rows and $c = 3$ columns, so we find the P-value using $(2 - 1)(3 - 1) = 2$ degrees of freedom. The P-value is the area under the chi-square distribution with 2 degrees of freedom to the right of $\chi_0^2 = 14.707$, as shown in Figure 19.

Figure 19

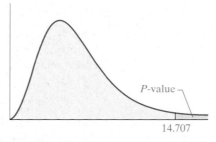

P-value

14.707

Using Table VI, we find the row that corresponds to 2 degrees of freedom. The area under the chi-square distribution with 2 degrees of freedom to the right of 10.597 is 0.005. Because 14.707 is to the right of 10.597, the P-value is less than 0.005. So P-value < 0.005.

Step 5: Because the P-value is less than the level of significance, $\alpha = 0.01$, we reject the null hypothesis.

CAUTION

If we reject the null hypothesis in a chi-square test for homogeneity, we are saying that there is sufficient evidence for us to believe that at least one proportion is different from the others. However, rejecting the null hypothesis does not tell us which proportions differ.

Step 6: There is sufficient evidence at the $\alpha = 0.01$ level of significance to conclude that the proportion of subjects in each group who experience abdominal pain are different. At least one of the three groups experiences abdominal pain at a rate different from the other two groups.

Figure 20 shows the bar graph, with the blue bars representing the proportion of individuals who experienced abdominal pain for each group and the green bars representing the proportion of individuals who did not experience abdominal pain for each group.

Figure 20

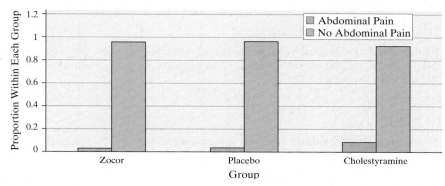

Patients Reporting Abdominal Pain by Treatment

Now Work Problem 15.

From the graph, it is apparent that a higher proportion of patients taking cholestyramine experience abdominal pain as a side effect.

Recall that the requirements for performing a chi-square test are that all expected frequencies are greater than 1 and that at most 20% of the expected frequencies can be less than 5. If these requirements are not satisfied, the researcher has one of two options: (1) combine two columns (or rows) to increase the expected frequencies or (2) increase the sample size.

MAKING AN INFORMED DECISION

Benefits of College

Are there benefits to attending college? If so, what are they? In this project, we will identify some of the perks that a college education provides. Obtain a random sample of at least 50 people aged 21 years or older and administer the following survey:

Please answer the following questions:

1. What is the highest level of education you have attained?
_____ Have not completed high school
_____ High school graduate
_____ College graduate

2. What is your employment status?
_____ Employed
_____ Unemployed, but actively seeking work
_____ Unemployed, but not actively seeking work

3. If you are employed, what is your annual income?
_____ Less than $20,000
_____ $20,000–$39,999
_____ $40,000–$60,000
_____ More than $60,000

4. If you are employed, which statement best describes the level of satisfaction you have with your career? Answer this question only if you are employed.
_____ Satisfied—I enjoy my job and am happy with my career.
_____ Somewhat satisfied—Work is work, but I am not unhappy with my career.
_____ Somewhat dissatisfied—I do not enjoy my work, but I also have no intention of leaving.
_____ Dissatisfied—Going to work is painful. I would quit tomorrow if I could.

(continued on next page)

MAKING AN INFORMED DECISION

(a) Use the results of the survey to create a contingency table for each of the following categories:
- Level of education/employment status
- Level of education/annual income
- Level of education/job satisfaction
- Annual income/job satisfaction

(b) Perform a chi-square test for independence on each contingency table from part (a).

(c) Draw bar graphs for each contingency table from part (a).

(d) Write a report that details your findings.

12.2 ASSESS YOUR UNDERSTANDING

Concepts and Vocabulary

1. Explain the differences between the chi-square test for independence and the chi-square test for homogeneity. What are the similarities?

2. Why does the test for homogeneity follow the same procedures as the test for independence?

Skill Building

3. The following table contains observed values and expected values in parentheses for two categorical variables, X and Y, where variable X has three categories and variable Y has two categories:

	X_1	X_2	X_3
Y_1	34 (36.26)	43 (44.63)	52 (48.11)
Y_2	18 (15.74)	21 (19.37)	17 (20.89)

(a) Compute the value of the chi-square test statistic. 1.701
(b) Test whether X and Y are independent at the $\alpha = 0.05$ level of significance. Do not reject H_0
(c) What is the P-value? 0.427

4. The following table contains observed values and expected values in parentheses for two categorical variables, X and Y, where variable X has three categories and variable Y has two categories: (4a) 13.053

	X_1	X_2	X_3
Y_1	87 (75.12)	74 (80.43)	34 (39.46)
Y_2	12 (23.88)	32 (25.57)	18 (12.54)

(a) Compute the value of the chi-square test statistic.
(b) Test whether X and Y are independent at the $\alpha = 0.05$ level of significance. Reject H_0
(c) What is the P-value? 0.001

5. The following table contains the number of successes and failures for three categories of a variable.

	Category 1	Category 2	Category 3
Success	76	84	69
Failure	44	41	49

Test whether the proportions are equal for each category at the $\alpha = 0.01$ level of significance. What is the P-value? Do not reject H_0; P-value = 0.370

6. The following table contains the number of successes and failures for three categories of a variable.

	Category 1	Category 2	Category 3
Success	204	199	214
Failure	96	121	98

Test whether the proportions are equal for each category at the $\alpha = 0.01$ level of significance. What is the P-value? Do not reject H_0; P-value = 0.171

Applying the Concepts

7. Family Structure and Sexual Activity A sociologist wants [NW] to discover whether the sexual activity of females between the ages of 15 and 19 years of age and family structure are associated. She randomly selects 380 females between 15 and 19 years of age and asks each to disclose her family structure at age 14 and whether she has had sexual intercourse. The results are shown in the table on the next page. Data are based on information obtained from the U.S. National Center for Health Statistics.

(a) Compute the expected values of each cell under the assumption of independence.

(b) Verify that the requirements for performing a chi-square test of independence are satisfied.

(c) Compute the chi-square test statistic. 10.357

(d) Test whether family structure and sexual activity of 15- to 19-year-old females are independent at the $\alpha = 0.05$ level of significance. Reject H_0

Sexual Activity	Family Structure			
	Both Biological Adoptive Parents	Single Parent	Parent and Stepparent	Nonparental Guardian
Had sexual intercourse	64	59	44	32
Did not have sexual intercourse	86	41	36	18

(e) Compare the observed frequencies with the expected frequencies. Which cell contributed most to the test statistic? Was the expected frequency greater than or less than the observed frequency? What does this information tell you?

(f) Construct a conditional distribution by family structure and draw a bar graph. Does this evidence support your conclusion in part (d)?

(g) Compute the P-value for this test by finding the area under the chi-square distribution to the right of the test statistic. 0.016

8. **Prenatal Care** An obstetrician wants to learn whether the amount of prenatal care and the wantedness of the pregnancy are associated. He randomly selects 939 women who had recently given birth and asks them to disclose whether their pregnancy was intended, unintended, or mistimed. In addition, they were to disclose when they started receiving prenatal care, if ever. The results of the survey are as follows:

Wantedness of Pregnancy	Months Pregnant before Prenatal Care Began		
	Less Than 3 Months	3 to 5 Months	More Than 5 Months (or never)
Intended	593	26	33
Unintended	64	8	11
Mistimed	169	19	16

(a) Compute the expected values of each cell under the assumption of independence.

(b) Verify that the requirements for performing a chi-square test of independence are satisfied.

(c) Compute the chi-square test statistic. 21.357

(d) Test whether prenatal care and the wantedness of pregnancy are independent at the $\alpha = 0.05$ level of significance. Reject H_0

(e) Compare the observed frequencies with the expected frequencies. Which cell contributed most to the test statistic? Was the expected frequency greater than or less than the observed frequency? What does this information tell you?

(f) Construct a conditional distribution by wantedness of the pregnancy and draw a bar graph. Does this evidence support your conclusion in part (d)?

9. **Education versus Area of Country** An urban economist wants to determine whether the region of the United States a resident lives in is associated with level of education. He randomly selects 1804 residents of the United States and asks them to disclose the region of the United States in which they reside and their level of education. He obtains the data in the following table:

Area of Country	Level of Education			
	Not a High School Graduate	High School Graduate	Some College	Bachelor's Degree or Higher
Northeast	52	123	70	94
Midwest	123	146	102	96
South	119	204	148	144
West	62	106	111	104

Source: U.S. Census Bureau

(a) Does the sample evidence suggest that level of education and region of the United States are independent at the $\alpha = 0.05$ level of significance? Reject H_0

(b) Compare the observed frequencies with the expected frequencies. Which cell contributed most to the test statistic? Was the expected frequency greater than or less than the observed frequency? What does this information tell you?

(c) Construct a conditional distribution by level of education and draw a bar graph. Does this evidence support your conclusion in part (a)?

10. **Profile of Smokers** The following data represent the smoking status by level of education for residents of the United States 18 years old or older from a random sample of 1054 residents. *(10a) Do not reject H_0*

	Smoking Status		
Number of Years of Education	**Current**	**Former**	**Never**
<12	178	88	208
12	137	69	143
13–15	44	25	44
16 or more	34	33	51

Source: National Health Interview Survey

(a) Does the evidence suggest that smoking status and level of education are independent at the $\alpha = 0.05$ level of significance?

(b) Compute the *P*-value for this test by finding the area under the chi-square distribution to the right of the test statistic. *0.253*

(c) Construct a conditional distribution by number of years of education, and draw a bar graph. Does this evidence support your conclusion in part (a)?

11. **Legalization of Marijuana** On May 14, 2001, the Supreme Court, by a vote of 8 to 0, struck down state laws that legalized marijuana for medicinal purposes. The Gallup Organization later conducted surveys of randomly selected Americans 18 years old or older and asked whether they support the limited use of marijuana when prescribed by physicians to relieve pain and suffering. The results of the survey, by age group, are as follows:

	Age		
Opinion	**18–29 Years Old**	**30–49 Years Old**	**50 Years or Older**
For	172	313	258
Against	52	103	119

(a) Test whether age and opinion regarding the legalization of marijuana are independent at the $\alpha = 0.05$ level of significance. *Reject H_0*

(b) Compute the *P*-value for this test by finding the area under the chi-square distribution to the right of the test statistic. *0.035*

(c) Construct a conditional distribution by age and draw a bar graph. Does this evidence support your conclusion in part (a)?

12. **Pro Life or Pro Choice** A recent Gallup Organization Poll asked male and female Americans whether they were pro life or pro choice when it comes to abortion issues. The results of the survey are as follows:

	Pro Life/Pro Choice	
Gender	**Pro Life**	**Pro Choice**
Men	196	199
Women	239	249

(a) Test whether an individual's opinion regarding abortion is independent of gender at the $\alpha = 0.1$ level of significance. *Do not reject H_0*

(b) Compute the *P*-value for this test by finding the area under the chi-square distribution to the right of the test statistic. *0.849*

(c) Construct a conditional distribution by gender and draw a bar graph. Does this evidence support your conclusion in part (a)?

13. **Delinquencies** A delinquency offense is an act committed by a juvenile for which an adult could be prosecuted in a criminal court. The following data represent the number of various types of delinquencies by gender in a random sample of 750 delinquencies. *(13a) Do not reject H_0*

	Delinquency			
Gender	**Person**	**Property**	**Drugs**	**Public Order**
Female	24	85	7	28
Male	97	367	39	103

(a) Does the evidence suggest that gender is independent of type of delinquency at the $\alpha = 0.05$ level of significance?

(b) Compute the *P*-value for this test by finding the area under the chi-square distribution to the right of the test statistic. *0.814*

(c) Construct a conditional distribution by type of delinquency and draw a bar graph. Does this evidence support your conclusion in part (a)?

14. **Visits to the Emergency Room** The following data represent the gender and age of 489 randomly selected patients who visited the emergency room with an injury-related emergency.

	Age (years)					
Gender	**Under 15**	**15–24**	**25–44**	**45–64**	**65–74**	**75 and Older**
Male	66	56	92	37	8	11
Female	44	39	66	34	12	24

(a) Does the evidence suggest that the age of an individual visiting the emergency room is independent of gender at the $\alpha = 0.1$ level of significance? *Reject H_0*

(b) Compute the *P*-value for this test by finding the area under the chi-square distribution to the right of the test statistic. *0.031*

(c) Construct a conditional distribution by age and draw a bar graph. Does this evidence support your conclusion in part (a)?

15. **Smoked Lately?** Suppose a researcher wants to investigate whether the proportion of smokers within different age groups is the same. He divides the American population into four age groups: 18 to 29 years old, 30 to 49 years old, 50 to 64 years old, and 65 years or older. Within each age group, he surveys 80 individuals and asks, "Have you smoked at least one cigarette in the past week?" The results of the survey are as follows:

Smoking Status	Age (years)			
	18–29	30–49	50–64	65 or Older
Smoked at least one cigarette in past week	24	21	23	12
Did not smoke at least one cigarette in past week	56	59	57	68

Source: Gallup Organization

(a) Is there evidence to indicate that the proportion of individuals within each age group who have smoked at least one cigarette in the past week is different at the $\alpha = 0.05$ level of significance? *Do not reject*

(b) Compute the P-value for this test by finding the area under the chi-square distribution to the right of the test statistic. *0.112*

(c) Construct a conditional distribution by age and draw a bar graph. Does this evidence support your conclusion in part (a)?

16. Are You Satisfied? Suppose an economist wants to gauge the level of satisfaction of Americans. He randomly samples 150 people 18 years old or older from four geographic regions of the United States: East, South, Midwest, and West. He asks the individuals selected, "Are you satisfied or dissatisfied with the way things are going in the United States at this time?" The following table gives results of the survey.

Satisfaction	Region			
	East	South	Midwest	West
Satisfied	77	84	93	83
Dissatisfied	73	66	57	67

Source: Gallup Organization

(a) Test whether the proportions of Americans who are satisfied with the way things are going in the United States for each region of the country are equal at the $\alpha = 0.1$ level of significance. *Do not reject*

(b) Compute the P-value for this test by finding the area under the chi-square distribution to the right of the test statistic. *0.316*

(c) Construct a conditional distribution by region of the country and draw a bar graph. Does this evidence support your conclusion in part (a)?

17. Celebrex Celebrex is a drug manufactured by Pfizer, Inc., used to relieve symptoms associated with osteoarthritis and rheumatoid arthritis in adults. It is considered to be one of the nonsteroidal anti-inflammatory drugs. These types of drugs are known to be associated with gastrointestinal toxicity, such as bleeding, ulceration, and perforation of the stomach, small intestine, or large intestine. In clinical trials, researchers wanted to learn whether the proportion of subjects taking Celebrex who experienced these side effects differed significantly from that in other treatment groups. The following data were collected (Naproxen is a nonsteroidal anti-inflammatory drug that is also used in the treatment of arthritis).

Side Effect	Placebo	Treatment			
		Celebrex (50 mg per day)	Celebrex (100 mg per day)	Celebrex (200 mg per day)	Naproxen (500 mg per day)
Experienced ulcers	5	8	7	13	34
Did not experience ulcers	212	225	220	208	176

Source: Pfizer, Inc.

(a) Test whether the proportion of subjects within each treatment group is the same at the $\alpha = 0.01$ level of significance. *Reject H_0*

(b) Compute the P-value for this test by finding the area under the chi-square distribution to the right of the test statistic. *<0.001*

(c) Construct a conditional distribution by treatment and draw a bar graph. Does this evidence support your conclusion in part (a)?

18. Celebrex Celebrex is a drug manufactured by Pfizer, Inc., used to relieve symptoms associated with osteoarthritis and rheumatoid arthritis in adults. In clinical trials of the medication, some subjects reported dizziness as a side effect. The researchers wanted to discover whether the proportion of subjects taking Celebrex who reported dizziness as a side effect differed significantly from that for other treatment groups. The following data were collected.

Side Effect	Drug				
	Celebrex	Placebo	Naproxen	Diclofenac	Ibuprofen
Dizziness	83	32	36	5	8
No dizziness	4063	1832	1330	382	337

Source: Pfizer, Inc.

(a) Test whether the proportion of subjects within each treatment group who experienced dizziness are the same at the $\alpha = 0.01$ level of significance. Do not reject
(b) Compute the P-value for this test by finding the area under the chi-square distribution to the right of the test statistic. 0.323

(c) Construct a conditional distribution by treatment and draw a bar graph. Does this evidence support your conclusion in part (a)?

---

19. Dropping a Course A survey of 52 randomly selected students who dropped a course in the current semester was conducted at a community college. The goal of the survey was to learn why students drop courses. The following data were collected: "Personal" drop reasons include financial, transportation, family issues, health issues, and lack of child care. "Course" drop reasons include reducing one's load, being unprepared for the course, the course was not what was expected, dissatisfaction with teaching, and not getting the desired grade. "Work" drop reasons include an increase in hours, a change in shift, and obtaining full-time employment. "Career" drop reasons include not needing the course and a change of plans. The results of the survey are as follows:

Gender	Drop Reason	Gender	Drop Reason
Male	Personal	Male	Work
Female	Personal	Male	Work
Male	Work	Female	Course
Male	Personal	Male	Work
Male	Course	Female	Course
Male	Course	Female	Course
Female	Course	Female	Course
Female	Course	Male	Work
Male	Course	Male	Personal
Female	Course	Male	Course
Male	Personal	Female	Course
Male	Work	Female	Course
Male	Work	Male	Course
Male	Course	Female	Course
Male	Course	Male	Work
Male	Work	Male	Course
Female	Personal	Female	Work
Male	Course	Male	Personal
Female	Work	Male	Work
Male	Work	Female	Course
Male	Work	Male	Course
Female	Course	Male	Personal
Female	Personal	Female	Course
Female	Personal	Female	Work
Female	Personal	Male	Work

(a) Construct a contingency table for the two variables.
(b) Is gender independent of drop reason at the $\alpha = 0.1$ level of significance? Reject H_0
(c) Compute the P-value for this test by finding the area under the chi-square distribution to the right of the test statistic. 0.061
(d) Construct a conditional distribution by drop reason and draw a bar graph. Does this evidence support your conclusion in part (a)?

20. Political Affiliation A political scientist wanted to learn whether there is any association between the education level of a registered voter and his or her political party affiliation. He randomly selected 46 registered voters and obtained the following data:

Education	Political Party	Education	Political Party
Grade school	Democrat	High school	Democrat
College	Republican	College	Republican
High school	Democrat	College	Republican
High school	Republican	Grade school	Democrat
High school	Democrat	High school	Republican
Grade school	Democrat	High school	Democrat
College	Republican	High school	Democrat
Grade school	Democrat	College	Republican
High school	Democrat	High school	Republican
High school	Democrat	Grade school	Democrat
Grade school	Democrat	High school	Democrat
College	Republican	College	Democrat
Grade school	Democrat	College	Republican
College	Democrat	High school	Republican
College	Democrat	College	Democrat
Grade school	Republican	College	Democrat
College	Republican	High school	Democrat
Grade school	Republican	College	Republican
College	Republican	College	Democrat
High school	Democrat	High school	Republican
College	Democrat	College	Republican
College	Republican	High school	Republican
College	Democrat	College	Democrat

(a) Construct a contingency table for the two variables.
(b) Is level of education independent of political affiliation at the $\alpha = 0.1$ level of significance? *Do not reject H_0*
(c) Compute the P-value for this test by finding the area under the chi-square distribution to the right of the test statistic. *0.175*

(d) Construct a conditional distribution by level of education and draw a bar graph. Does this evidence support your conclusion in part (a)?

In Problem 21, we demonstrate that the z-test for comparing two population proportions is equivalent to the chi-square test for homogeneity when there are two possible outcomes.

21. **Percentage of Americans Who Smoke on the Decline?** On November 3, 2004, the Gallup Organization surveyed 1007 adults and found that 222 of them had smoked at least one cigarette in the past week. In 1998, they also asked 1007 adults the same question and determined that 282 adults had smoked at least one cigarette in the past week. The results are presented in the following table:

Smoking Status	Year 1998	Year 2004
Smoked	282	222
Did not smoke	725	785

(a) Compute the expected number of adult Americans who have smoked at least one cigarette in the past week and the expected number who have not, assuming $p_{1990} = p_{2004}$. *252; 755*
(b) Compute the chi-square test statistic. $\chi^2 = 9.53$
(c) Use a chi-square test to determine whether the proportion of adults that smoked at least one cigarette in the past week has changed at the $\alpha = 0.05$ level of significance. *Reject H_0*
(d) Compute the Z-test statistic. Now compute z^2. Compare z^2 with the chi-square test statistic. Conclude that $z^2 = \chi^2$. $z = 3.087$; $z^2 = 9.53$

Consumer Reports   Dirty Birds?

Hungry for a cheap, low-fat alternative to beef, Americans are eating more chicken than ever. Although precise figures are impossible to obtain, the U.S. Centers for Disease Control and Prevention reported that the number of cases of outbreaks of illness caused by chicken rose threefold between 1988 and 1992. *Salmonella* bacteria were the most common cause of these outbreaks.

In a study for *Consumer Reports*, we purchased 1000 fresh, whole broiler chickens at grocery stores in 36 cities across the United States over a 5-week period. Our shoppers packed the birds in coolers and shipped them overnight to the lab. There, tests were conducted to determine the presence of salmonella and campylobacter, another chicken-related bug. The results of the study for salmonella were as follows:

Brand	Salmonella Present	Salmonella Absent	Total
A	8	192	200
B	17	183	200
C	27	173	200
D	14	186	200
E	20	180	200
Total	86	914	1000

Assuming that the chickens represent a random sample from each brand included in the study, use the information presented in the table to answer the following:

(a) Calculate the proportion of incidence for each brand shown in the table.

(b) Compute a 95% confidence interval for the incidence of salmonella for brand C.

(c) Using a chi-square test of homogeneity of proportions, is there evidence that the five brands have the same incidence rate for salmonella?

(d) Brands A and D are major competitors in the same market. The manufacturer of brand A claims to have improved its cleanliness and claims that it is substantially cleaner than brand D. Is there evidence to support this contention?

(e) Write a paragraph for the readers of *Consumer Reports* magazine that explains your conclusions.

Note to Readers: In many cases, our test protocol and analytical methods are more complicated than described in these examples. The data and discussions have been modified to make the material more appropriate for the audience.

© by Consumers Union of U.S., Inc., Yonkers, NY 10703-1057, a nonprofit organization. Reprinted with permission.

Technology Step by Step	**Chi-Square Tests**

TI-83/84 Plus

Step 1: Access the MATRX menu. Highlight the EDIT menu, and select `1: [A]`.

Step 2: Enter the number of rows and columns of the matrix.

Step 3: Enter the cell entries for the observed matrix, and press 2nd QUIT. Repeat Steps 1–3 for the expected values, but enter the expected frequencies in matrix B.

Step 4: Press STAT, highlight the TESTS menu, and select `C: χ²-Test`

Step 5: With the cursor after the **Observed:**, enter matrix [A] by accessing the MATRX menu, highlighting NAMES, and selecting `1: [A]`.

Step 6: With the cursor after the **Expected:**, enter matrix [B] by accessing the MATRX menu, highlighting NAMES, and selecting `2: [B]`.

Step 7: Highlight **Calculate** or **Draw**, and press ENTER.

MINITAB

Step 1: Enter the data into the MINITAB spreadsheet.

Step 2: Select the **Stat** menu, highlight **Tables**, and select

Chi-Square Test . . . .

Step 3: Select the columns that contain the data, and press OK.

Excel

Step 1: Enter the observed frequencies in the spreadsheet.

Step 2: Compute the expected frequencies and enter them in a different location in the spreadsheet.

Step 3: Select *fx* from the tool bar. Select **Statistical** for the function category and highlight **CHITEST** in the function name.

Step 4: With the cursor in the actual cell, highlight the observed data. With the cursor in the expected cell, highlight the expected frequencies. Click OK. The output provided is the *P*-value.

Note: This test can also be performed by using the PHStat add-in. See the *Excel Technology Manual.*

12.3 Testing the Significance of the Least-Squares Regression Model

Preparing for This Section Before getting started, review the following:

- Scatter diagrams; correlation (Section 4.1, pp. 177–185)

- Least-squares regression (Section 4.2, pp. 195–203)

- Sampling distribution of the sample mean $\bar{x}$ (Section 8.1, pp. 375–388)

- Testing a hypothesis about μ, σ unknown (Section 10.3, pp. 480–486)

- Confidence intervals about a mean (Section 9.2, pp. 423–429)

Objectives

1. **Understand the requirements of the least-squares regression model**

2. **Compute the standard error of the estimate**

3. **Verify that the residuals are normally distributed**

4. **Conduct inference on the slope**

5. **Construct a confidence interval about the slope of the least-squares regression model**

Note to Instructor
You might want to continue using the pulse data collected in Chapter 4 for Sections 12.3 and 12.4.

As a quick review of the topics discussed in Chapter 4, we present the following example:

EXAMPLE 1   **Least-Squares Regression**

Problem: A family doctor is interested in examining the relationship between a patient's age and total cholesterol. He randomly selects 14 of his female patients and obtains the data presented in Table 16. The data are based on results obtained from the National Center for Health Statistics. Draw a scatter diagram, compute the correlation coefficient, and find the least-squares regression equation and the coefficient of determination.

Figure 21

Figure 22

Table 16			
Age, x	**Total Cholesterol, y**	**Age, x**	**Total Cholesterol y**
25	180	42	183
25	195	48	204
28	186	51	221
32	180	51	243
32	210	58	208
32	197	62	228
38	239	65	269

Approach: We will use a TI-84 Plus graphing calculator to obtain the information requested.

Solution: Figure 21 displays the scatter diagram. Figure 22 displays the output obtained from the calculator. The linear correlation coefficient is 0.7178. The least-squares regression equation for these data is $\hat{y} = 151.3537 + 1.3991x$, where $\hat{y}$ represents the predicted total cholesterol for a female whose age is x. The coefficient of determination, R^2, is 0.515. So 51.5% of the variation in total cholesterol is explained by the regression line. Figure 23 shows a graph of the least-squares regression equation on the scatter diagram in order to get a feel for the fit. ▬▬▬

Figure 23

The information obtained in Example 1 is descriptive in nature. Notice that the descriptions are both graphical (as in the scatter diagram) and numerical (as in the correlation coefficient and least-squares regression equation).

① Understand the Requirements of the Least-Squares Regression Model

In the least-squares regression equation $\hat{y} = b_0 + b_1x$, the values for the slope, b_1, and intercept, b_0, are statistics, just as the sample mean, $\bar{x}$, and sample standard deviation, s, are statistics. The statistics b_0 and b_1 are estimates for the population intercept, β_0, and the population slope, β_1. The true linear relation between the explanatory variable, x, and the response variable, y, is given by $y = \beta_0 + \beta_1x$.

Because b_0 and b_1 are statistics, their values vary from sample to sample, so there is a sampling distribution associated with each of them. We use this sampling distribution to perform inference on b_0 and b_1. For example, we might believe that β_1 is different from 0. If we have evidence that supports this belief, we conclude that there is a linear relation between the explanatory variable, x, and response variable, y.

To find the sampling distributions of b_0 and b_1, we have some requirements about the population from which the bivariate data (x_i, y_i) were sampled. Just as

In Other Words

Because b_0 and b_1 are statistics, they have sampling distributions.

in Section 8.1 when we discussed the sampling distribution of $\bar{x}$, we start by asking what would happen if we took many samples for a given value of the explanatory variable, x. For example, in looking back at Table 16, we notice that our sample included three women aged 32 years, so x has the same value, 32, for all three women in our sample, but the corresponding values of y for these three women are different: 180, 210, and 197. There is a distribution of total cholesterol levels for $x = 32$ years of age. Suppose we looked at *all* women aged 32 years. From these population data, we could find the population mean total cholesterol for *all* 32-year-old women, denoted $\mu_{y|32}$. The notation $\mu_{y|32}$ is read, "the mean value of the response variable y given that the explanatory variable is 32." We could repeat this process for any other age. In general, different ages will have different population mean total cholesterol. This brings us to our first requirement regarding inference on the least-squares regression model.

Requirement 1 for Inference on the Least-Squares Regression Model

For any particular value of the explanatory variable x (such as 32 in Example 1), the mean of the corresponding responses in the population depends linearly on x. That is,

$$\mu_{y|x} = \beta_0 + \beta_1 x$$

for some numbers β_0 and β_1, where $\mu_{y|x}$ represents the population mean response when the explanatory variable is x.

We also have a requirement regarding the distribution of the response variable for any particular value of the explanatory variable.

Requirement 2 for Inference on the Least-Squares Regression Model

The response variables are normally distributed with mean $\mu_{y|x} = \beta_0 + \beta_1 x$ and standard deviation σ.

In Other Words
When doing inference on the least-squares regression model, we require (1) that for any explanatory variable, x, the mean of the response variable, y, depends on the value of x through a linear equation, and (2) that the response variable, y, is normally distributed with a constant standard deviation, σ. The mean increases/decreases at a constant rate depending on the slope, while the variance remains constant.

This requirement states that the mean of the response variable changes linearly, but the variance remains constant, and the distribution of the response variable for any given value of the explanatory variable is normal. For example, if we obtained a sample of many 32-year-old females and measured their total cholesterol, the distribution would be normal with mean $\mu_{y|32} = \beta_0 + \beta_1(32)$ and standard deviation σ. If we obtained a sample of many 43-year-old females and measured their total cholesterol, the distribution would be normal with mean $\mu_{y|43} = \beta_0 + \beta_1(43)$ and standard deviation σ. See Figure 24.

Figure 24

$\mu_{y|32} = \beta_0 + \beta_1(32)$ $\mu_{y|43} = \beta_0 + \beta_1(43)$

In Other Words
The larger σ is, the more spread out the data are around the regression line.

A large value of σ, the population standard deviation, indicates that the data are widely dispersed about the regression line, and a small σ indicates that the data lie fairly close to the regression line. Figure 25 illustrates the ideas just presented. The regression line represents the mean value of each normal distribution at a specified value of x. The standard deviation of each distribution is σ.

Figure 25

Of course, not all the observed values of the response variable will lie on the true regression line, $\mu_{y|x} = \beta_0 + \beta_1 x$. The difference between the observed and predicted value of the response variable is an error term or residual, ε_i. We now present the least-squares regression model.

Definition

The **least-squares regression model** is given by

$$y_i = \beta_0 + \beta_1 x_i + \varepsilon_i \qquad (1)$$

where

y_i is the value of the response variable for the ith individual,

β_0 and β_1 are the parameters to be estimated based on sample data,

x_i is the value of the explanatory variable for the ith individual,

ε_i is a random error term with mean 0 and variance $\sigma_{\varepsilon_i}^2 = \sigma^2$. The error terms are independent, and

$i = 1, \dots, n$, where n is the sample size (number of ordered pairs in the data set).

Because the expected value or mean of y_i is $\beta_0 + \beta_1 x_i$ and the expression on the left side of Equation (1) equals the expression on the right side, the expected value or mean of the error term, ε_i, is 0.

> Now Work Problem 13(a).

2 Compute the Standard Error of the Estimate

In Section 4.2, we learned how to obtain estimates for β_0 and β_1. We now present the method for obtaining the estimate of σ, the standard deviation of the response variable y for any given value of x. The unbiased estimator of σ is called the *standard error of the estimate.*

Remember the formula for the sample standard deviation presented in Section 3.2?

$$s = \sqrt{\frac{\sum (x_i - \overline{x})^2}{n - 1}}$$

We compute the deviations about the mean, square them, add up the squared deviations, and divide by $n - 1$. We divide by $n - 1$ because we lose 1 degree of freedom since one parameter, $\overline{x}$, is estimated. Exactly the same logic is used to compute the standard error of the estimate.

As we mentioned, the predicted values of y, denoted $\hat{y}_i$, represent the mean value of the response variable for any given value of the explanatory variable, x_i. So $y_i - \hat{y}_i$ represents the difference between the observed value, y_i, and the mean value, $\hat{y}_i$. This calculation is used to get the standard error of the estimate.

Definition

The **standard error of the estimate**, s_e, is found using the formula

$$s_e = \sqrt{\frac{\sum (y_i - \hat{y}_i)^2}{n - 2}} = \sqrt{\frac{\sum \text{residuals}^2}{n - 2}} \qquad (2)$$

Notice that we divide by $n - 2$ because we have estimated two parameters, β_0 and β_1.

EXAMPLE 2 **Computing the Standard Error**

Problem: Compute the standard error for the data in Table 16.

Approach: We use the following steps to compute the standard error.
Step 1: Find the least-squares regression line.
Step 2: Obtain predicted values for each of the observations in the data set.
Step 3: Compute the residuals for each of the observations in the data set.
Step 4: Compute $\Sigma \text{residuals}^2$.
Step 5: Compute the standard error, using Formula (2).

Solution
Step 1: The least-squares regression line was found in Example 1.
Step 2: Column 3 of Table 17 represents the predicted values for each of the $n = 14$ observations.
Step 3: Column 4 of Table 17 represents the residuals for each of the 14 observations.

Table 17				
Age, x	Total Cholesterol, y	$\hat{y} = 1.3991x + 151.3537$	Residuals, $y - \hat{y}$	Residuals^2, $(y - \hat{y})^2$
25	180	186.33	−6.33	40.0689
25	195	186.33	8.67	75.1689
28	186	190.53	−4.53	20.5209
32	180	196.12	−16.12	259.8544
32	210	196.12	13.88	192.6544
32	197	196.12	0.88	0.7744
38	239	204.52	34.48	1188.8704
42	183	210.12	−27.12	735.4944
48	204	218.51	−14.51	210.5401
51	221	222.71	−1.71	2.9241
51	243	222.71	20.29	411.6841
58	208	232.50	−24.50	600.25
62	228	238.10	−10.10	102.01
65	269	242.30	26.70	712.89

$$\Sigma \text{residuals}^2 = 4553.705$$

CAUTION!
Be sure to divide by $n - 2$ when computing the standard error.

Step 4: Column 5 of Table 17 contains the squared residuals. We sum the entries in column 5 to obtain the sum of squared errors. So

$$\Sigma \text{residuals}^2 = 4553.705$$

Step 5: We use Formula (2) to compute the standard error.

$$s_e = \sqrt{\frac{\Sigma \text{residuals}^2}{n - 2}} = \sqrt{\frac{4553.705}{14 - 2}} = 19.48$$

EXAMPLE 3 **Obtaining the Standard Error Using Technology**

Figure 26

Regression Statistics	
Multiple R	0.7178106
R Square	0.5152521
Adjusted R Square	0.4748564
Standard Error	19.480535
Observations	14

Problem: Obtain the standard error for the data in Table 16 using statistical software.

Approach: We will use Excel to obtain the standard error. The steps for obtaining standard errors using the TI-83/84 graphing calculators, MINITAB, and Excel are given in the Technology Step by Step on page 594.

Result: Figure 26 shows the partial output obtained from Excel. Notice that the results agree with the by-hand computation.

Now Work Problem 13(b).

! CAUTION

The residuals must be normally distributed to perform inference on the least-squares regression line.

3 Verify That the Residuals Are Normally Distributed

For the least-squares regression model $y_i = \beta_0 + \beta_1 x_i + \varepsilon_i$, we require that the response variable, y_i, be normally distributed. Because $\beta_0 + \beta_1 x_i$ is constant for any x_i, the requirement that y_i is normal means that the residuals, ε_i must also be normal. To perform statistical inference on the regression, we must verify that the residuals are normally distributed by examining a normal probability plot.

EXAMPLE 4   **Verifying That the Residuals Are Normally Distributed**

Problem: Verify that the residuals obtained in Table 17 from Example 2 are normally distributed.

Approach: We construct a normal probability plot to assess normality. If the normal probability plot is roughly linear, the residuals are said to be normal.

Solution: Figure 27 contains the normal probability plot obtained from MINITAB.

Figure 27

Because the points follow a linear pattern (all the points lie within the bands created by MINITAB), the residuals are normally distributed. We can perform inference on the least-squares regression equation.

Now Work Problem 13(c).

4 Conduct Inference on the Slope

At this point, we know how to estimate the intercept and slope of the least-squares regression model. In addition, we can compute the standard error, s_e, which is an estimate of σ, the standard deviation of the response variable about the true least-squares regression model, and we know how to assess the normality of the residuals. We will now use this information to test whether a linear relation exists between the explanatory and the response variables.

This is the question that we want to answer: Do the sample data provide sufficient evidence to conclude that a linear relation exists between the two variables? If there is no linear relation between the response and explanatory variables, the slope of the true regression line will be zero. Do you know why? A slope of zero means that information about the explanatory variable, x, does not change our "guess" as to the value of the response variable, y.

Using the notation of hypothesis testing, we can perform one of three tests

Two-Tailed	Left-Tailed	Right-Tailed
$H_0: \beta_1 = 0$	$H_0: \beta_1 = 0$	$H_0: \beta_1 = 0$
$H_1: \beta_1 \neq 0$	$H_1: \beta_1 < 0$	$H_1: \beta_1 > 0$

In the two-tailed test, we are testing whether a linear relation exists between two variables without regard to the sign of the slope. In the left-tailed test, we are testing whether the slope of the true regression line is negative. In the right-tailed test, we are testing whether the slope of the true regression line is positive.

To test any one of these hypotheses, we need to know the sampling distribution of b_1. It turns out that

$$t = \frac{b_1 - \beta_1}{\dfrac{s_e}{\sqrt{\Sigma(x_i - \overline{x})^2}}} = \frac{b_1 - \beta_1}{s_{b_1}}$$

follows Student's t-distribution with $n - 2$ degrees of freedom, where n is the number of observations, b_1 is the estimate of the the slope of the regression line β_1, and s_{b_1} is the sample standard error of b_1.

Hypothesis Test Regarding the Slope Coefficient, β_1

To test whether two quantitative variables are linearly related, we use the following steps provided that

1. The sample is obtained using random sampling.
2. The residuals are normally distributed with constant error variance.

Note to Instructor
Remind students that the procedures presented here are similar to those presented for testing hypotheses about μ with σ unknown.

Step 1: Determine the null and alternative hypotheses. The hypotheses can be structured in one of three ways:

Two-Tailed	Left-Tailed	Right-Tailed
$H_0: \beta_1 = 0$	$H_0: \beta_1 = 0$	$H_0: \beta_1 = 0$
$H_1: \beta_1 \neq 0$	$H_1: \beta_1 < 0$	$H_1: \beta_1 > 0$

Step 2: Select a level of significance α, depending on the seriousness of making a Type I error.

Step 3: Compute the test statistic

$$t_0 = \frac{b_1 - \beta_1}{s_{b_1}} = \frac{b_1}{s_{b_1}}$$

which follows Student's t-distribution with $n - 2$ degrees of freedom. Remember, when computing the test statistic, we assume the null hypothesis to be true. So we assume $\beta_1 = 0$.

Classical Approach

Step 4: Use Table V to determine the critical value using $n - 2$ degrees of freedom.

	Two-Tailed	Left-Tailed	Right-Tailed
Critical value	$-t_{\frac{\alpha}{2}}$ and $t_{\frac{\alpha}{2}}$	$-t_\alpha$	t_α
Critical region			

Step 5: Compare the critical value with the test statistic.

Two-Tailed	Left-Tailed	Right-Tailed
If $t_o < -t_{\frac{\alpha}{2}}$ or $t_o > t_{\frac{\alpha}{2}}$, reject the null hypothesis.	If $t_o < -t_\alpha$, reject the null hypothesis.	If $t_o > t_\alpha$, reject the null hypothesis.

P-Value Approach

Step 4: Use Table V to estimate the P-value using $n - 2$ degrees of freedom.

Two-Tailed	Left-Tailed	Right-Tailed

Step 5: If P-value $< \alpha$, reject the null hypothesis.

Step 6: State the conclusion.

The procedures just presented are **robust**, which means that minor departures from normality will not adversely affect the results of the test. In fact, for large samples ($n \geq 30$), inferential procedures regarding b_1 can be used even with significant departures from normality.

EXAMPLE 5   **Testing for a Linear Relation**

Problem: Test whether there is a linear relation between age and total cholesterol at the $\alpha = 0.05$ level of significance, using the data given in Table 16.

Approach: We verify that the requirements to perform the inference are satisfied. We then follow Steps 1–6.

Solution: In Example 1, we were told that the individuals were randomly selected. In Example 3, we confirmed that the residuals were normally distributed by constructing a normal probability plot.

We can now follow Steps 1–6 to conduct the test.

Step 1: We are testing if there is no linear relation between age and total cholesterol. Because we are not concerned about the direction of the relation, we are testing

$$H_0: \beta_1 = 0 \qquad \text{versus} \qquad H_1: \beta_1 \neq 0$$

Step 2: We are using an $\alpha = 0.05$ level of significance.

Step 3: We obtained an estimate of β_1 in Example 1, and we computed the standard error, s_e, in Example 2. To determine the standard deviation of b_1, we need to compute $\sum(x_i - \bar{x})^2$, where the x_i are the values of the explanatory variable, age, and $\bar{x}$ is the sample mean. We compute this value in Table 18.

<table>
<tr><th colspan="4">Table 18</th></tr>
<tr><th>Age, x</th><th>$\bar{x}$</th><th>$x_i - \bar{x}$</th><th>$(x_i - \bar{x})^2$</th></tr>
<tr><td>25</td><td>42.07143</td><td>−17.07143</td><td>291.4337</td></tr>
<tr><td>25</td><td>42.07143</td><td>−17.07143</td><td>291.4337</td></tr>
<tr><td>28</td><td>42.07143</td><td>−14.07143</td><td>198.0051</td></tr>
<tr><td>32</td><td>42.07143</td><td>−10.07143</td><td>101.4337</td></tr>
<tr><td>32</td><td>42.07143</td><td>−10.07143</td><td>101.4337</td></tr>
<tr><td>32</td><td>42.07143</td><td>−10.07143</td><td>101.4337</td></tr>
<tr><td>38</td><td>42.07143</td><td>−4.07143</td><td>16.5765</td></tr>
<tr><td>42</td><td>42.07143</td><td>−0.07143</td><td>0.0051</td></tr>
<tr><td>48</td><td>42.07143</td><td>5.92857</td><td>35.1479</td></tr>
<tr><td>51</td><td>42.07143</td><td>8.92857</td><td>79.7194</td></tr>
<tr><td>51</td><td>42.07143</td><td>8.92857</td><td>79.7194</td></tr>
<tr><td>58</td><td>42.07143</td><td>15.92857</td><td>253.7193</td></tr>
<tr><td>62</td><td>42.07143</td><td>19.92857</td><td>397.1479</td></tr>
<tr><td>65</td><td>42.07143</td><td>22.92857</td><td>525.7193</td></tr>
<tr><td colspan="4" align="right">$\sum(x_i - \bar{x})^2 = 2472.9284$</td></tr>
</table>

> ! **CAUTION**
> Use unrounded values of the sample mean in the computation of $\sum(x_i - \bar{x})^2$ to avoid round-off error.

We have

$$s_{b_1} = \frac{s_e}{\sqrt{\sum(x_i - \bar{x})^2}} = \frac{19.48}{\sqrt{2472.9284}} = 0.3917$$

The test statistic is

$$t_0 = \frac{b_1}{s_{b_1}} = \frac{1.3991}{0.3917} = 3.572$$

Classical Approach

Step 4: Because this is a two-tailed test, we determine the critical *t*-values at the $\alpha = 0.05$ level of significance with $n - 2 = 14 - 2 = 12$ degrees of freedom to be $-t_{0.05/2} = -t_{0.025} = -2.179$ and $t_{0.05/2} = t_{0.025} = 2.179$ The critical regions are displayed in Figure 28.

Figure 28

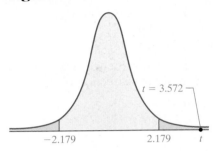

Step 5: The test statistic $t_0 = 3.572$. We label this point in Figure 28. Because the test statistic is greater than the critical value $t_{0.025} = 2.179$, we reject the null hypothesis.

P-Value Approach

Step 4: Because this is a two-tailed test, the *P*-value is the area under the *t*-distribution with $14 - 2 = 12$ degrees of freedom to the left of the test statistic, $-t_0 = -3.572$, and to the right of $t_0 = 3.572$, as shown in Figure 29. That is, *P*-value $= P(t < -3.572) + P(t > 3.572) = 2P(t > 3.572)$, with 12 degrees of freedom.

Figure 29

Using Table V, we find the row that corresponds to 12 degrees of freedom. The value 3.572 lies between 3.428 and 3.930. The value of 3.428 has an area under the *t*-distribution with 12 degrees of freedom of 0.0025 to the right. The area under the *t*-distribution with 12 degrees of freedom to the right of 3.930 is 0.001.

Because 3.572 is between 3.428 and 3.930, the *P*-value is between 2(0.001) and 2(0.0025). So $0.002 < P\text{-value} < 0.005$.

Step 5: Because the *P*-value is less than the level of significance $\alpha = 0.05$, we reject the null hypothesis.

Step 6: There is sufficient evidence at the $\alpha = 0.05$ level of significance to conclude that there is a linear relation between age and total cholesterol. ■

EXAMPLE 6 | **Testing for a Linear Relation Using Technology**

Problem: Test whether a linear relation exists between age and total cholesterol at the $\alpha = 0.05$ level of significance using statistical software.

Approach: We will use MINITAB, Excel, and a TI-84 Plus graphing calculator to test for a linear relation. The steps for testing whether a linear relation exists using the TI-83/84 graphing calculators, MINITAB, and Excel are given in the Technology Step by Step on page 594.

Result: Figure 30(a) shows the results obtained from MINITAB, Figure 30(b) shows the results obtained from Excel, and Figure 30(c) shows the results obtained from a TI-84 Plus graphing calculator.

Figure 30 | **Regression Analysis**

```
The regression equation is
total cholesterol = 151 + 1.40 age

Predictor          Coef       StDev         T          P
Constant         151.35       17.28      8.76      0.000
Age               1.3991      0.3917     3.57      0.004

S = 19.48         R-Sq = 51.5%       R-Sq(adj) = 47.5%

Analysis of Variance

Source            Df         SS          MS         F         P
Regression         1       4840.5      4840.5     12.76     0.004
Residual Error    12       4553.9       379.5
Total             13       9394.4
```

(a) MINITAB output

Figure 30 (cont'd)

SUMMARY OUTPUT

Regression Statistics	
Multiple R	0.7178106
R Square	0.5152521
Adjusted R Square	0.4748564
Standard Error	19.480535
Observations	14

ANOVA

	df	SS	MS	F
Regression	1	4840.462	4840.462	12.75514
Residual	12	4553.895	379.4912	
Total	13	9394.357		

	Coefficients	Standard Error	t Stat	P-value
Intercept	151.35366	17.28376	8.756987	1.47E-06
Age	1.3990642	0.391737	3.571433	0.003842

(b) Excel output

P-value

(c) TI-84 Plus output

> ! **CAUTION**
> If we do not reject H_0, then we use the sample mean of y to predict the value of the response for any value of the explanatory variable.

In all three sets of output, the P-value for the slope is given as 0.004. This means that there is a 0.004 probability of obtaining a slope estimate as extreme as or more extreme than the one obtained if the null hypothesis of no linear relation was true. Because the P-value is less than the level of significance, $\alpha = 0.05$, we reject the null hypothesis of no linear relation. ▬

Now Work Problems 13(d) and 13(e).

5   Construct a Confidence Interval about the Slope of the Least-Squares Regression Model

We can also obtain confidence intervals for the slope of the least-squares regression line. The procedure is identical to that for obtaining confidence intervals about a mean. As was the case with confidence intervals about a population mean, the confidence interval for the slope of the least-squares regression line is of the form

Point estimate $\pm$ margin of error

Confidence Intervals for the Slope of the Regression Line

A $(1 - \alpha) \cdot 100\%$ confidence interval for the slope of the true regression line, β_1, is given by the following formulas:

$$\text{Lower bound:}\quad b_1 - t_{\frac{\alpha}{2}} \cdot \frac{s_e}{\sqrt{\sum(x_i - \bar{x})^2}}$$

$$\text{Upper bound:}\quad b_1 + t_{\frac{\alpha}{2}} \cdot \frac{s_e}{\sqrt{\sum(x_i - \bar{x})^2}} \quad \textbf{(3)}$$

(*continued on next page*)

Here, $t_{\frac{\alpha}{2}}$ is computed with $n-2$ degrees of freedom.

Note: This interval can be computed only if the data are randomly obtained, the residuals are normally distributed, and there is constant error variance.

EXAMPLE 7   **Constructing a Confidence Interval about the Slope of the True Regression Line**

Problem: Compute a 95% confidence interval about the slope of the true regression line for the data presented in Table 17.

Approach

Step 1: Determine the least-squares regression line.

Step 2: Verify that the requirements for inference on the regression line are satisfied.

Step 3: Compute s_e.

Step 4: Determine the critical value $t_{\frac{\alpha}{2}}$ with $n-2$ degrees of freedom.

Step 5: Compute the bounds on the $(1-\alpha)\cdot 100\%$ confidence interval for β_1 using Formula (3).

Step 6: Interpret the result by stating, "We are 95% confident that β_1 is somewhere between *lower bound* and *upper bound*."

Solution

Step 1: The least-squares regression line was determined in Example 1 and is $\hat{y} = 151.3537 + 1.3991x$.

Step 2: The requirements were verified in Examples 1, 3, and 5.

Step 3: We computed s_e in Example 2, obtaining $s_e = 19.48$.

Step 4: Because we wish to determine a 95% confidence interval, we have $\alpha = 0.05$. Therefore, we need to find $t_{0.05/2} = t_{0.025}$ with $14-2 = 12$ degrees of freedom. Referring to Table V, we find that $t_{0.025} = 2.179$.

Step 5: We find the lower and upper bounds.

$$\text{Lower bound:} \quad b_1 - t_{\frac{\alpha}{2}}\cdot\frac{s_e}{\sqrt{\Sigma(x_i - \overline{x})^2}} = 1.3991 - 2.179\cdot\frac{19.48}{\sqrt{2472.9284}}$$

$$= 1.3991 - 0.8536$$

$$= 0.5455$$

$$\text{Upper bound:} \quad b_1 + t_{\frac{\alpha}{2}}\cdot\frac{s_e}{\sqrt{\Sigma(x_i - \overline{x})^2}} = 1.3991 + 2.179\cdot\frac{19.48}{\sqrt{2472.9284}}$$

$$= 1.3991 + 0.8536$$

$$= 2.2527$$

Step 6: We are 95% confident that the mean increase in cholesterol for each additional year of life is somewhere between 0.5455 and 2.2527.

CAUTION
It is best that the explanatory variables be spread out when doing regression analysis.

In looking carefully at the formula for the standard deviation of b_1, we should notice that the larger the value of $\Sigma(x_i - \overline{x})^2$, the smaller the value of s_{b_1}. This result implies that whenever we are finding a least-squares regression line, we should attempt to make the values of the explanatory variable, x, as evenly spread out as possible so that b_1, our estimate of β_1, is as precise as possible.

Now Work Problem 13(f).

12.3 ASSESS YOUR UNDERSTANDING

Concepts and Vocabulary

1. State the requirements to perform inference on the least-squares regression model. How are these requirements verified?

2. Why is it important to perform graphical as well as analytical analysis when analyzing relations between two quantitative variables?

3. What do the y-coordinates on the least-squares regression line represent?

4. Why is it desirable to have the explanatory variables spread out to test a claim regarding β_1 or construct confidence intervals about β_1?

5. If H_0: $\beta_1 = 0$ is not rejected, what is the best guess for the value of the response variable for any value of the explanatory variable?

6. Why don't we conduct inference on the linear correlation coefficient?

Skill Building

In Problems 7–12, use the results of Problems 11–16, respectively, from Section 4.2 to answer the following questions:

(a) *What are the estimates of β_0 and β_1?*

(b) *Compute the standard error, the point estimate for σ.*

(c) *Assuming the residuals are normally distributed, determine s_{b_1}.*

(d) *Assuming the residuals are normally distributed, test H_0: $\beta_1 = 0$ versus H_1: $\beta_1 \neq 0$ at the $\alpha = 0.05$ level of significance.*

7.

x	3	4	5	7	8
y	4	6	7	12	14

8.

x	3	5	7	9	11
y	0	2	3	6	9

9.

x	−2	−1	0	1	2
y	−4	0	1	4	5

10.

x	−2	−1	0	1	2
y	7	6	3	2	0

11.

x	20	30	40	50	60
y	100	95	91	83	70

12.

x	5	10	15	20	25
y	2	4	7	11	18

Applying the Concepts

13. Height versus Head Circumference A pediatrician
NW wants to determine the relation that may exist between a child's height and head circumference. She randomly selects 11 children from her practice, measures their height and head circumference, and obtains the following data:

Height (inches), x	Head Circumference (inches), y
27.75	17.5
24.5	17.1
25.5	17.1
26	17.3
25	16.9
27.75	17.6
26.5	17.3
27	17.5
26.75	17.3
26.75	17.5
27.5	17.5

Source: Denise Slucki, Student at Joliet Junior College

Use the results from Problem 17 in Section 4.2 to answer the following questions:

(a) Treating height as the explanatory variable, x, determine the estimates of β_0 and β_1.

(b) Compute the standard error of the estimate, s_e. 0.0954

(c) Determine whether the residuals are normally distributed.

(d) If the residuals are normally distributed, determine s_{b_1}. 0.0276

(e) If the residuals are normally distributed, test whether a linear relation exists between height and head circumference at the $\alpha = 0.01$ level of significance.

(f) If the residuals are normally distributed, construct a 95% confidence interval about the slope of the true least-squares regression line. (0.1230, 0.2451)

(g) A child comes in for a physical, and the nurse determines his height to be 26.5 inches. However, the child is being rather uncooperative, so the nurse is unable to measure the head circumference of the child. What would be a good guess as to this child's head circumference? Why is this a good guess? 17.34 inches

(13a) $b_0 = 12.4932$; $b_1 = 0.1827$ (13e) Reject H_0

14. Bone Length Research performed at NASA and led by Dr. Emily R. Morey-Holton measured the lengths of the right humerus and right tibia in 11 rats that were sent into space on Spacelab Life Sciences 2. The following data were collected:

Right Humerus (mm), x	Right Tibia (mm), y	Right Humerus (mm), x	Right Tibia (mm), y
24.80	36.05	25.90	37.38
24.59	35.57	26.11	37.96
24.59	35.57	26.63	37.46
24.29	34.58	26.31	37.75
23.81	34.20	26.84	38.50
24.87	34.73		

Source: NASA Life Sciences Data Archive

Use the results from Problem 20 in Section 4.2 to answer the following questions:

(a) Treating the length of the right humerus as the explanatory variable, x, determine the estimates of β_0 and β_1. $b_0 = 1.1140$; $b_1 = 1.3902$

(b) Compute the standard error of the estimate. 0.4944

(c) Determine whether the residuals are normally distributed.

(d) If the residuals are normally distributed, determine s_{b_1}. 0.1501

(e) If the residuals are normally distributed, test whether a linear relation exists between the explanatory variable, x, and response variable, y, at the $\alpha = 0.01$ level of significance. Reject H_0

(f) If the residuals are normally distributed, construct a 99% confidence interval about the slope of the true least-squares regression line. (0.9024, 1.8780)

(g) What is the mean length of the right tibia on a rat whose right humerus is 25.93 mm? 37.162 mm

15. Concrete As concrete cures, it gains strength. The following data represent the 7-day and 28-day strength (in pounds per square inch) of a certain type of concrete:

7-Day Strength, x	28-Day Strength, y	7-Day Strength, x	28-Day Strength, y
2300	4070	2480	4120
3390	5220	3380	5020
2430	4640	2660	4890
2890	4620	2620	4190
3330	4850	3340	4630

(a) Treating the 7-day strength as the explanatory variable, x, determine the estimates of β_0 and β_1.

(b) Compute the standard error of the estimate. 271.04

(c) Determine whether the residuals are normally distributed.

(d) If the residuals are normally distributed, determine s_{b_1}. 0.2055

(e) If the residuals are normally distributed, test whether a linear relation exists between 7-day strength and 28-day strength at the $\alpha = 0.05$ level of significance.

(f) If the residuals are normally distributed, construct a 95% confidence interval about the slope of the true least-squares regression line. (0.2025, 1.1503)

(g) What is the estimated mean 28-day strength of this concrete if the 7-day strength is 3000 psi? 4704.8 psi

16. Tar and Nicotine Every year the Federal Trade Commission (FTC) must report tar and nicotine levels in cigarettes to Congress. The FTC obtains the tar and nicotine levels in over 1200 brands of cigarettes. A random sample from those reported to Congress is given in the following table:

Brand	Tar (mg), x	Nicotine (mg), y
Barclay 100	5	0.4
Benson and Hedges King	16	1.1
Camel Regular	24	1.7
Chesterfield King	24	1.4
Doral	8	0.5
Kent Golden Lights	9	0.8
Kool Menthol	9	0.8
Lucky Strike	24	1.5
Marlboro Gold	15	1.2
Newport Menthol	18	1.3
Salem Menthol	17	1.3
Virginia Slims Ultra Light	5	0.5
Winston Light	10	0.8

Source: Federal Trade Commission

(a) Treating the amount of tar as the explanatory variable, x, determine the estimates of β_0 and β_1.

(b) Compute the standard error of the estimate. 0.1121

(c) Determine whether the residuals are normally distributed.

(d) If the residuals are normally distributed, determine s_{b_1}. 0.0046

(e) If the residuals are normally distributed, test whether a linear relation exists between the explanatory variable, x, and response variable, y, at the $\alpha = 0.1$ level of significance. Reject H_0

(f) If the residuals are normally distributed, construct a 90% confidence interval about the slope of the true least-squares regression line. (0.0492, 0.0658)

(g) What is the mean amount of nicotine in a cigarette that has 12 mg of tar? 0.90 mg

17. United Technologies versus the S&P 500 United Technologies is a conglomerate that includes companies such as Otis Elevators and Carrier Heating and Cooling. The ticker symbol of the company is UTX. The following data represent the rate of return of UTX stock for 11 months, compared with the rate of return of the Standard and Poor's Index of 500 stocks. Both are in percent.

Month	Rate of Return of S&P 500, x	Rate of Return in United Technologies, y
Aug-04	0.23	1.21
Sept-04	0.94	−0.57
Oct-04	1.40	−0.59
Nov-04	3.86	5.89
Dec-04	3.25	5.90
Jan-05	−2.53	−2.58
Feb-05	1.89	0.06
Mar-05	−1.91	1.78
Apr-05	−2.01	0.06
May-05	3.00	5.79
Jun-05	0.90	−1.57

Source: Yahoo! Finance

(a) Treating the rate of return of the S&P 500 as the explanatory variable, x, determine the estimates of β_0 and β_1. $b_0 = 0.5975, b_1 = 0.9764$
(b) Compute the standard error of the estimate. 2.3518
(c) Determine whether the residuals are normally distributed.
(d) If the residuals are normally distributed, determine s_{b_1}. 0.3379
(e) If the residuals are normally distributed, test whether a linear relation exists between the explanatory variable, x, and response variable, y, at the $\alpha = 0.1$ level of significance. Reject H_0
(f) If the residuals are normally distributed, construct a 90% confidence interval about the slope of the true least-squares regression line. $(0.357, 1.596)$
(g) What is the mean rate of return for United Technologies stock if the rate of return of the S&P 500 is 4.2%? 4.698%

18. Fat-Free Mass versus Energy Expenditure In an effort to measure the dependence of energy expenditure on body build, researchers used underwater weighing techniques to determine the fat-free body mass in seven men. In addition, they measured the total 24-hour energy expenditure during inactivity. The results are as follows:

Fat-Free Mass (kg), x	Energy Expenditure (Kcal), y
49.3	1894
59.3	2050
68.3	2353
48.1	1838
57.6	1948
78.1	2528
76.1	2568

Source: Webb, P. Energy expenditure and fat-free mass in men and women. *American Journal of Clinical Nutrition*, 34, 1816–1826.

(18a) $b_0 = 607.7, b_1 = 25.012$ (18b) 64.8477

(a) What are the estimates of β_0 and β_1?
(b) Compute the standard error of the estimate.
(c) Determine whether the residuals are normally distributed.
(d) If the residuals are normally distributed, determine s_{b_1}. 2.189
(e) If the residuals are normally distributed, test whether a linear relation exists between the explanatory variable, x, and response variable, y, at the $\alpha = 0.01$ level of significance. Reject H_0
(f) If the residuals are normally distributed, construct a 99% confidence interval about the slope of the true least-squares regression line. $(16.186, 33.838)$
(g) What is the mean energy expenditure of a man if his fat-free mass is 57.3 kg? 2040.9

19. Calories versus Sugar The following data represent the number of calories per serving and the number of grams of sugar per serving for a random sample of high-fiber cereals.

Calories, x	Sugar, y	Calories, x	Sugar, y
200	18	210	23
210	23	210	16
170	17	210	17
190	20	190	12
200	18	190	11
180	19	200	11

Source: Consumer Reports, October 1999

(a) Draw a scatter diagram of the data, treating calories as the explanatory variable. What type of relation, if any, appears to exist between calories and sugar?
(b) Determine the least-squares regression equation from the sample data. $\hat{y} = 0.93 + 0.0821x$
(c) Compute the standard error of the estimate. 4.1512
(d) Determine whether the residuals are normally distributed.
(e) Determine s_{b_1}. 0.0961
(f) If the residuals are normally distributed, test whether a linear relation exists between calories and sugar content at the $\alpha = 0.01$ level of significance. Do not reject H_0
(g) If the residuals are normally distributed, construct a 95% confidence interval about the slope of the true least-squares regression line. $(-0.1320, 0.2962)$
(h) Suppose a high-fiber cereal is randomly selected. Would you recommend using the least-squares regression line obtained in part (b) to predict the sugar content of the cereal? Why? What would be a good guess as to the sugar content of the cereal? $\bar{y} = 17.1$ grams

20. Influential Observations The following data represent the heights and weights of a random sample of professional baseball players.

Player	Height (inches)	Weight (pounds)
Alex Rodriguez	75	210
Derek Jeter	75	195
Greg Maddux	72	185
Randy Johnson	82	230
David Justice	75	200
Al Leiter	75	220
Barry Bonds	74	210
Ray Lankford	71	200
Jason Isringhausen	75	210

Source: Yahoo! Sports

(a) Draw a scatter diagram of the data, treating height as the explanatory variable and weight as the response variable.

(b) Determine the least-squares regression line. Test whether there is a linear relation between height and weight at the $\alpha = 0.05$ level of significance.

(c) Remove the values listed for Randy Johnson. Test whether there is a linear relation between height and weight. What effect does Randy Johnson have on the hypothesis test?

21. The output shown was obtained from MINITAB.

```
The regression equation is
y = 12.4 + 1.40 x

Predictor    Coef     StDev      T        P
Constant    12.396    1.381     8.97    0.000
x           1.3962   0.1245    11.21    0.000

S = 2.167    R-Sq = 91.3%   R-Sq(adj) = 90.6%
```

(a) The least-squares regression equation is $\hat{y} = 12.396 + 1.3962x$. What is the predicted value of y at $x = 10$? 26.36

(b) What is the mean of y at $x = 10$? 26.36

(c) The standard error, s_e, is 2.167. What is the standard deviation of y at $x = 10$? 2.167

(d) If the requirements for inference on the least-squares regression model are satisfied, what is the distribution of y at $x = 10$?

(20b) $\hat{y} = 3.361x - 45$; reject H_0
(20c) $\hat{y} = 3.611x - 63.472$; do not reject H_0

(21d) Normal; $\mu \approx 26.36$, $\sigma \approx 2.167$

Technology Step by Step	Testing the Least-Squares Regression Model

TI-83/84 Plus
Step 1: Enter the explanatory variable in L1 and the response variable in L2.
Step 2: Press STAT, highlight TESTS, and select E:LinRegTTest....
Step 3: Be sure that Xlist is L1 and Ylist is L2. Make sure that Freq: is set to 1. Select the direction of the alternative hypothesis. Place the cursor on Calculate and press ENTER.

MINITAB
Step 1: With the explanatory variable in C1 and the response variable in C2, select the **Stat** menu and highlight **Regression**. Highlight **Regression** ....
Step 2: Select the explanatory variable (MINITAB calls them predictors) and response variable and click OK.

Excel
Step 1: Make sure the Data Analysis Tool Pak is activated by selecting the **Tools** menu and highlighting **Add-Ins** .... Check the box for the Analysis ToolPak and click OK.
Step 2: Enter the explanatory variable in column A and the response variable in column B.
Step 3: Select the **Tools** menu and highlight **DataAnalysis** ....
Step 4: Select the **Regression** option.
Step 5: With the cursor in the Y-range cell, highlight the column that contains the response variable. With the cursor in the X-range cell, highlight the column that contains the explanatory variable. Click OK.

12.4 Confidence and Prediction Intervals

Preparing for This Section Before getting started, review the following:

• Confidence intervals (Section 9.2, pp. 423–429)

Objectives ① **Construct confidence intervals for a mean response**
② **Construct prediction intervals for an individual response**

Note to Instructor

Given the tedious calculations in this section, it is recommended that hand solutions be de-emphasized. Let software do the work so students can concentrate on the results.

We know how to obtain the least-squares regression equation from data. We also know how to use the least-squares regression equation to obtain a predicted value. For example, the least-squares regression equation for the cholesterol data introduced in Example 1 from Section 12.3 is

$$\hat{y} = 151.3537 + 1.3991x$$

where $\hat{y}$ represents the predicted total cholesterol for a female whose age is x. The predicted value of total cholesterol for a given age x actually has two interpretations:

1. It represents the mean total cholesterol for all females whose age is x.
2. It represents the predicted total cholesterol for a randomly selected female whose age is x.

So, if we let $x = 42$ in the least-squares regression equation $\hat{y} = 151.3537 + 1.3991x$, we obtain $\hat{y} = 151.3537 + 1.3991(42) = 210.1$. We can interpret this result in one of two ways:

1. The mean total cholesterol for all females 42 years old is 210.1.
2. Our best guess as to the total cholesterol for a randomly selected 42-year-old female is 210.1.

Of course, there is a margin of error in making predictions, so we construct intervals about any predicted value to describe the accuracy of the prediction. The type of interval constructed will depend on whether we are predicting a mean total cholesterol for all 42-year-old females or the total cholesterol for an individual 42-year-old female. In other words, the margin of error is going to be different for predicting the mean total cholesterol for all females who are 42 years old versus the total cholesterol for one individual who is 42 years old. Which prediction (the mean or the individual) do you think will be more accurate? It seems logical that the distribution of means should have less variability (and therefore a lower margin of error) than the distribution of individuals. After all, in the distribution of means, high total cholesterols can be offset by low total cholesterols.

Definitions

> **Confidence intervals for a mean response** are intervals constructed about the predicted value of y, at a given level of x, that are used to measure the accuracy of the mean response of all the individuals in the population.
>
> **Prediction intervals for an individual response** are intervals constructed about the predicted value of y that are used to measure the accuracy of a single individual's predicted value.

In Other Words

Confidence intervals are intervals for the mean of the population. Prediction intervals are intervals for an individual from the population.

If we use the least-squares regression equation to predict the mean total cholesterol for all 42-year-old females, we construct a confidence interval for a mean response. If we use the least-squares regression equation to predict the total cholesterol for a single 42-year-old female, we construct a prediction interval for an individual response.

① Construct Confidence Intervals for a Mean Response

The structure of a confidence interval is the same as it was in Section 9.1. The interval is of the form

$$\text{Point estimate} \pm \text{margin of error}$$

The following formula can be used to construct a confidence interval about $\hat{y}$.

Confidence Interval for the Mean Response of y, $\hat{y}$

A $(1 - \alpha) \cdot 100\%$ confidence interval for $\hat{y}$, the mean response of y, is given by

$$\text{Lower bound:} \quad \hat{y} - t_{\frac{\alpha}{2}} \cdot s_e \sqrt{\frac{1}{n} + \frac{(x^* - \bar{x})^2}{\Sigma(x_i - \bar{x})^2}}$$

$$\text{Upper bound:} \quad \hat{y} + t_{\frac{\alpha}{2}} \cdot s_e \sqrt{\frac{1}{n} + \frac{(x^* - \bar{x})^2}{\Sigma(x_i - \bar{x})^2}} \tag{1}$$

where x^* is the given value of the explanatory variable, n is the number of observations, and $t_{\frac{\alpha}{2}}$ is the critical value with $n - 2$ degrees of freedom.

EXAMPLE 1

Constructing a Confidence Interval for a Mean Response

Problem: Construct a 95% confidence interval about the predicted mean total cholesterol of all 42-year-old females, using the data in Table 16.

Approach: We wish to determine the predicted mean total cholesterol at $x^* = 42$ using Formula (1) since our estimate is for the mean cholesterol of all 42-year-old females.

Solution: The least-squares regression equation is $\hat{y} = 151.3537 + 1.3991x$. To find the predicted mean total cholesterol of all 42-year-olds, let $x^* = 42$ in the regression equation and obtain $\hat{y} = 151.3537 + 1.3991(42) = 210.1$. From Example 2 in Section 12.3, we found that $s_e = 19.48$, and from Example 5 in Section 12.3, we found that $\Sigma(x_i - \bar{x})^2 = 2472.9284$ and $\bar{x} = 42.07143$. The critical t value, $t_{\frac{\alpha}{2}} = t_{0.025}$, with $n - 2 = 14 - 2 = 12$ degrees of freedom is 2.179. The 95% confidence interval about the predicted mean total cholesterol for all 42-year-old females is therefore

$$\text{Lower bound:} \quad \hat{y} - t_{\frac{\alpha}{2}} \cdot s_e \cdot \sqrt{\frac{1}{n} + \frac{(x^* - \bar{x})^2}{\Sigma(x_i - \bar{x})^2}} = 210.1 - 2.179 \cdot 19.48 \cdot \sqrt{\frac{1}{14} + \frac{(42 - 42.07143)^2}{2472.9284}}$$

$$= 198.8$$

$$\text{Upper bound:} \quad \hat{y} + t_{\frac{\alpha}{2}} \cdot s_e \cdot \sqrt{\frac{1}{n} + \frac{(x^* - \bar{x})^2}{\Sigma(x_i - \bar{x})^2}} = 210.1 + 2.179 \cdot 19.48 \cdot \sqrt{\frac{1}{14} + \frac{(42 - 42.07143)^2}{2472.9284}}$$

$$= 221.4$$

We are 95% confident that the mean total cholesterol of all 42-year-old females is between 198.8 and 221.4.

Now Work Problems 3(a) and 3(b).

In Other Words

Prediction intervals are wider than confidence intervals because it is tougher to guess the value of an individual than the mean of a population.

② ## Construct Prediction Intervals for an Individual Response

The procedure for obtaining a prediction interval for an individual response is identical to that for finding a confidence interval for a mean response. The only difference is the standard error. More variability is associated with individuals than with means. Therefore, the computation of the interval must account for this increased variability. Again, the form of the interval is

$$\text{Point estimate } \pm \text{ margin of error}$$

The following formula can be used to construct a prediction interval about $\hat{y}$.

Prediction Interval for an Individual Response about $\hat{y}$

A $(1 - \alpha) \cdot 100\%$ prediction interval for $\hat{y}$, the individual response of y, is given by

$$\text{Lower bound:} \quad \hat{y} - t_{\frac{\alpha}{2}} \cdot s_e \sqrt{1 + \frac{1}{n} + \frac{(x^* - \overline{x})^2}{\Sigma(x_i - \overline{x})^2}}$$

$$\text{Upper bound:} \quad \hat{y} + t_{\frac{\alpha}{2}} \cdot s_e \sqrt{1 + \frac{1}{n} + \frac{(x^* - \overline{x})^2}{\Sigma(x_i - \overline{x})^2}} \qquad \textbf{(2)}$$

where x^* is the given value of the explanatory variable, n is the number of observations, and $t_{\frac{\alpha}{2}}$ is the critical value with $n - 2$ degrees of freedom.

Notice the only difference between Formula (1) and Formula (2) is the "1 + " under the radical in Formula (2).

EXAMPLE 2   **Constructing a Prediction Interval for an Individual Response**

Problem: Construct a 95% prediction interval about the predicted total cholesterol of a 42-year-old female.

Approach: We need to determine the predicted total cholesterol at $x^* = 42$ and use Formula (2) since our estimate is for a particular 42-year-old female.

Solution: The least-squares regression equation is $\hat{y} = 151.3537 + 1.3991x$. To find the predicted total cholesterol of a 42-year-old, let $x^* = 42$ in the regression equation and obtain $\hat{y} = 151.3537 + 1.3991(42) = 210.1$. From Example 2 in Section 12.3, we found that $s_e = 19.48$; from Example 5 in Section 12.3, we found that $\Sigma(x_i - \overline{x})^2 = 2472.9284$ and $\overline{x} = 42.07143$. We find $t_{\frac{\alpha}{2}} = t_{0.025}$ with $n - 2 = 14 - 2 = 12$ degrees of freedom to be 2.179.

The 95% confidence interval about the predicted total cholesterol for a 42-year-old female is

$$\text{Lower bound:} \quad \hat{y} - t_{\frac{\alpha}{2}} \cdot s_e \sqrt{1 + \frac{1}{n} + \frac{(x^* - \overline{x})^2}{\Sigma(x_i - \overline{x})^2}} = 210.1 - 2.179 \cdot 19.48 \cdot \sqrt{1 + \frac{1}{14} + \frac{(42 - 42.07143)^2}{2472.9284}} = 166.2$$

$$\text{Upper bound:} \quad \hat{y} + t_{\frac{\alpha}{2}} \cdot s_e \sqrt{1 + \frac{1}{n} + \frac{(x^* - \overline{x})^2}{\Sigma(x_i - \overline{x})^2}} = 210.1 + 2.179 \cdot 19.48 \cdot \sqrt{1 + \frac{1}{14} + \frac{(42 - 42.07143)^2}{2472.9284}} = 254.0$$

We are 95% confident that the total cholesterol of a randomly selected 42-year-old female is between 166.2 and 254.0.

Now Work Problems 3(c) and 3(d).

Notice that the interval about the individual (prediction interval for an individual response) is wider than the interval about the mean (confidence interval for a mean response). The reason for this should be clear: more variability is associated with individuals than with groups of individuals. That is, it is more difficult to predict a single 42-year-old female's total cholesterol than it is to predict the mean total cholesterol for all 42-year-old females.

EXAMPLE 3 Confidence and Prediction Intervals Using Technology

USING TECHNOLOGY

The bounds for confidence and prediction intervals obtained using statistical software may differ from bounds computed by hand due to rounding error.

Problem: Construct a 95% confidence interval about the predicted total cholesterol of all 42-year-old females using statistical software. Construct a 95% prediction interval about the predicted total cholesterol for a 42-year-old female using statistical software.

Approach: We will use MINITAB to obtain the intervals. The steps for obtaining confidence and prediction intervals using MINITAB and Excel are given in the Technology Step by Step on page 599.

Result: Figure 31 shows the results obtained from MINITAB.

Figure 31

```
Predicted Values

       Fit    StDev Fit           95.0% CI             95.0% PI
    210.11         5.21    ( 198.77, 221.46)    ( 166.18, 254.05)
```

12.4 ASSESS YOUR UNDERSTANDING

Concepts and Vocabulary

1. Explain the difference between a confidence interval and prediction interval.

2. Suppose a normal probability plot of residuals indicates that the requirement of normally distributed residuals is violated. Explain the circumstances under which confidence and prediction intervals could still be constructed.

Skill Building

In Problems 3–6, use the results of Problems 11–14 in Section 4.2 and Problems 7–10 in Section 12.3.

3. Using the sample data from Problem 7 in Section 12.3,
   NW (a) Predict the mean value of y if $x = 7$. 11.8
   (b) Construct a 95% confidence interval about the mean value of y if $x = 7$. (10.8, 12.8)
   (c) Predict the value of y if $x = 7$. 11.8
   (d) Construct a 95% prediction interval about the value of y if $x = 7$. (9.9, 13.7)
   (e) Explain the difference between the prediction in parts (a) and (c).

4. Using the sample data from Problem 8 in Section 12.3,
   (a) Predict the mean value of y if $x = 8$. 5.1
   (b) Construct a 95% confidence interval about the mean value of y if $x = 8$. (4.0, 6.2)
   (c) Predict the value of y if $x = 8$. 5.1

   (d) Construct a 95% prediction interval about the value of y if $x = 8$. (2.5, 7.7)
   (e) Explain the difference between the prediction in parts (a) and (c).

5. Using the sample data from Problem 9 in Section 12.3,
   (a) Predict the mean value of y if $x = 1.4$. 4.3
   (b) Construct a 95% confidence interval about the mean value of y if $x = 1.4$. (2.5, 6.1)
   (c) Predict the value of y if $x = 1.4$. 4.3
   (d) Construct a 95% prediction interval about the value of y if $x = 1.4$. (0.9, 7.6)

6. Using the sample data from Problem 10 in Section 12.3,
   (a) Predict the mean value of y if $x = 1.8$. 0.4
   (b) Construct a 90% confidence interval about the mean value of y if $x = 1.8$. (−0.5, 1.3)
   (c) Predict the value of y if $x = 1.8$. 0.4
   (d) Construct a 90% prediction interval about the value of y if $x = 1.8$. (−1.1, 1.9)

Applying the Concepts   (7b) (17.12, 17.28)

7. **Height versus Head Circumference** Use the results of Problem 13 from Section 12.3 to answer the following questions:
   (a) Predict the mean head circumference of children who are 25.75 inches tall. 17.20 inches
   (b) Construct a 95% confidence interval about the mean head circumference of children who are 25.75 inches tall
   (c) Predict the head circumference of a randomly selected child who is 25.75 inches tall. 17.20 inches

   (d) Construct a 95% prediction interval about the head circumference of a child who is 25.75 inches tall. (16.97, 17.43)
   (e) Explain the difference between the prediction in part (a) and the prediction in part (c).

8. **Bone Length** Use the results of Problem 14 in Section 12.3 to answer the following questions:

(a) Predict the mean length of the right tibia of all rats whose right humerus is 25.83 mm. 37.023 mm
(b) Construct a 95% confidence interval about the mean length found in part (a). (36.646, 37.398)
(c) Predict the length of the right tibia of a randomly selected rat whose right humerus is 25.83 mm. 37.023 mm
(d) Construct a 95% prediction interval about the length found in part (c). (35.842, 38.202)
(e) Explain why the predicted lengths found in parts (a) and (c) are the same, yet the intervals constructed in parts (b) and (d) are different.

9. **Concrete** Use the results of Problem 15 from Section 12.3 to answer the following questions:
(a) Predict the mean 28-day strength of concrete whose 7-day strength is 2550 psi. 4400.4 psi
(b) Construct a 95% confidence interval about the mean 28-day strength of concrete whose 7-day strength is 2550 psi. (4147.8, 4653.1)
(c) Predict the 28-day strength of concrete whose 7-day strength is 2550 psi. 4400.4 psi
(d) Construct a 95% prediction interval about the 28-day strength of concrete whose 7-day strength is 2550 psi. (3726.3, 5074.6)
(e) Explain the difference between the prediction in part (a) and the prediction in part (c).

10. **Tar and Nicotine** Use the results of Problem 16 in Section 12.3 to answer the following questions:
(a) Predict the mean nicotine content of all cigarettes whose tar content is 12 mg. 0.90 mg
(b) Construct a 95% confidence interval about the tar content found in part (a). (0.83, 0.97)
(c) Predict the nicotine content of a randomly selected cigarette whose tar content is 12 mg. 0.899 mg
(d) Construct a 95% prediction interval about the nicotine content found in part (c). (0.64, 1.16)
(e) Explain why the predicted nicotine contents found in parts (a) and (c) are the same, yet the intervals constructed in parts (b) and (d) are different.

11. **United Technologies versus the S&P 500** Use the results of Problem 17 in Section 12.3 to answer the following questions:
(a) What is the mean rate of return for United Technologies stock if the rate of return of the S&P 500 is 4.2%? 4.698 percent
(b) Construct a 90% confidence interval about the mean rate of return found in part (a). (2.234, 7.163)
(c) Predict the rate of return on United Technologies stock if the rate of return on the S&P 500 for a randomly selected month is 4.2%. 4.698 percent
(d) Construct a 90% prediction interval about the rate of return found in part (c). (−0.267, 9.664)
(e) Explain why the predicted rates of return found in parts (a) and (c) are the same, yet the intervals constructed in parts (b) and (d) are different.

12. **Fat-Free Mass versus Energy Expenditure** Use the results of Problem 18 in Section 12.3 to answer the following questions:
(a) What is the mean energy expenditure for individuals whose fat-free mass is 57.3 kg? 2040.9 Kcal
(b) Construct a 99% confidence interval about the mean energy expenditure found in part (a). (1932.3, 2149.5)
(c) Predict the energy expenditure of a randomly selected individual whose fat-free mass is 57.3 kg. 2040.9 Kcal
(d) Construct a 99% prediction interval about the energy expenditure found in part (c). (1757.7, 2324.0)
(e) Explain why the predicted energy expenditures found in parts (a) and (c) are the same, yet the intervals constructed in parts (b) and (d) are different.

13. **Calories versus Sugar** Use the results of Problem 19 from Section 12.3 to answer the following:
(a) Explain why it does not make sense to construct confidence or prediction intervals based on the least-squares regression equation.
(b) Construct a 95% confidence interval for the mean sugar content of high-fiber cereals. (14.5, 19.7)

Technology Step by Step | **Confidence and Prediction Intervals**

TI-83/84 Plus The TI-83/84 Plus does not compute confidence or prediction intervals.

MINITAB **Step 1:** With the predictor variable in C1 and the response variable in C2, select the **Stat** menu and highlight **Regression**. Highlight **Regression . . . .**

Step 2: Select the explanatory and response variables.

Step 3: Click the Options . . . button.

Step 4: In the cell marked "Prediction intervals for new observations:", enter the value of x^*. Select a confidence level. Click OK twice.

Excel **Step 1:** Load the PhStat Add-in.

Step 2: Enter the values of the explanatory variable in column A and the corresponding values of the response variable in column B.

Step 3: Select the **PHStat** menu. Highlight **Regression**. Highlight **Simple Linear Regression**.

Step 4: With the cursor in the Y variable cell range, highlight the data in column B. With the cursor in the X variable cell range, highlight the data in column A. Select Confidence & Prediction Interval for X and enter the value of x^*. Choose a level of confidence. Click OK.

Summary

In this chapter, we introduced chi-square methods. The first chi-square method involved tests for goodness of fit. We used the chi-square distribution to test whether a random variable followed a certain distribution. This is done by comparing values expected based on the distribution of the random variable to observed values.

Next, we introduced chi-square methods that allowed us to perform tests for independence and homogeneity. In a test for independence, the researcher obtains random data for two variables and tests whether the variables are associated. The null hypothesis in these tests is always that the variables are not associated (independent). The test statistic compares the values expected if the variables were independent to those observed. If the expected and observed values differ significantly, we reject the null hypothesis and conclude that there is evidence to support the belief that the variables are not independent (they are associated). We draw bar graphs of the marginal distributions to help us see the association, if any.

The last chi-square test was the test for homogeneity of proportions. This test is similar to the test for independence, except we are testing that the proportion of individuals in the study with a certain characteristic are equal ($p_1 = p_2 = \cdots = p_k$). To perform this test, we take random samples of a predetermined size from each group under consideration (a random sample of size n_1 for group 1, a random sample of size n_2 for group 2, and so on).

The last two sections of this chapter dealt with inferential techniques that can be used on the least-squares regression model $y_i = \beta_0 + \beta_1 x_i + \varepsilon_i$.

In Section 12.3, we used sample data to obtain estimates of an intercept and slope. The residuals are required to be normally distributed, with mean 0 and constant variance σ^2. We verified this requirements through a normal probability plot of the residuals. Provided that this requirement is satisfied, we can test hypotheses regarding the slope to determine whether or not the relation between the explanatory and response variables is linear.

In Section 12.4, we learned how to construct confidence and prediction intervals about a predicted value. We construct confidence intervals about a mean response and prediction intervals about an individual response.

Formulas

Expected Counts In a Goodness-of-Fit Test

$$E_i = \mu_i = np_i \quad \text{for} \quad i = 1, 2, \ldots, k$$

Chi-Square Test Statistic

$$\chi^2 = \sum \frac{(O_i - E_i)^2}{E_i} \qquad i = 1, 2, \ldots, k$$

Standard Error of the Estimate

$$s_e = \sqrt{\frac{\sum(y_i - \hat{y}_i)^2}{n-2}} = \sqrt{\frac{\sum\text{residuals}^2}{n-2}}$$

Standard Error of b_1

$$s_{b_1} = \frac{s_e}{\sqrt{\sum(x_i - \bar{x})^2}}$$

Confidence Intervals for the Slope of the Regression Line

A $(1 - \alpha) \cdot 100\%$ confidence interval for the slope of the true regression line, β_1, is given by the following formulas:

Lower bound: $\quad b_1 - t_{\frac{\alpha}{2}} \cdot \dfrac{s_e}{\sqrt{\sum(x_i - \bar{x})^2}} = b_1 - t_{\frac{\alpha}{2}} \cdot s_{b_1}$

Upper bound: $\quad b_1 + t_{\frac{\alpha}{2}} \cdot \dfrac{s_e}{\sqrt{\sum(x_i - \bar{x})^2}} = b_1 + t_{\frac{\alpha}{2}} \cdot s_{b_1}$

Here, $t_{\frac{\alpha}{2}}$ is computed with $n - 2$ degrees of freedom.

Expected Frequencies in a Test for Independence

$$\text{Expected frequency} = \frac{(\text{row total})(\text{column total})}{\text{table total}}$$

Confidence Interval about the Mean Response of $\hat{y}$

A $(1 - \alpha) \cdot 100\%$ confidence interval for the mean response of y, $\hat{y}$, is given by the following formulas:

Lower bound: $\quad \hat{y} - t_{\frac{\alpha}{2}} \cdot s_e \sqrt{\dfrac{1}{n} + \dfrac{(x^* - \bar{x})^2}{\sum(x_i - \bar{x})^2}}$

Upper bound: $\quad \hat{y} + t_{\frac{\alpha}{2}} \cdot s_e \sqrt{\dfrac{1}{n} + \dfrac{(x^* - \bar{x})^2}{\sum(x_i - \bar{x})^2}}$

Here, x^* is the given value of the explanatory variable, and $t_{\frac{\alpha}{2}}$ is the critical value with $n - 2$ degrees of freedom.

Prediction Interval about an Individual Response, $\hat{y}$

A $(1 - \alpha) \cdot 100\%$ prediction interval for the individual response of y, $\hat{y}$, is given by

Lower bound: $\quad \hat{y} - t_{\frac{\alpha}{2}} \cdot s_e \sqrt{1 + \dfrac{1}{n} + \dfrac{(x^* - \bar{x})^2}{\sum(x_i - \bar{x})^2}}$

Upper bound: $\quad \hat{y} + t_{\frac{\alpha}{2}} \cdot s_e \sqrt{1 + \dfrac{1}{n} + \dfrac{(x^* - \bar{x})^2}{\sum(x_i - \bar{x})^2}}$

where x^* is the given value of the explanatory variable and $t_{\frac{\alpha}{2}}$ is the critical value with $n - 2$ degrees of freedom.

Vocabulary

Goodness-of-fit test (p. 552)
Expected counts (p. 553)
Contingency (or two-way) table (p. 563)
Row variable (p. 563)
Column variable (p. 563)
Cell (p. 563)

Chi-square independence test (p. 563)
Conditional distribution (p. 569)
Chi-square test for homogeneity of
    proportions (p. 570)
Least-squares regression model (p. 583)
Standard error of the estimate (p. 583)

Robust (p. 587)
Confidence interval about a mean
    response (p. 595)
Prediction interval about an individual
    response (p. 595)

Objectives

Section	You should be able to . . .	Examples	Review Exercises
12.1	**1** Perform a goodness-of-fit test (p. 551)	3	1–4
12.2	**1** Perform a test for independence (p. 563)	1–4	5–8
	2 Perform a test for homogeneity of proportions (p. 570)	5	9, 10, 11
12.3	**1** Understand the requirements of the least-squares regression model (p. 581)	pp. 581-583	13–17
	2 Compute the standard error of the estimate (p. 583)	2 and 3	13(b), 14(b), 15(b), 16(b), 17(b)
	3 Verify that the residuals are normally distributed (p. 585)	4	13(c), 14(c), 15(c), 16(c), 17(c)
	4 Conduct inference on the slope (p. 585)	5 and 6	13(e), 14(e), 15(e), 16(e), 17(e)
	5 Construct a confidence interval about the slope of the least-squares regression model (p. 589)	7	13(f), 14(f), 15(f), 16(f)
12.4	**1** Construct confidence intervals for a mean response (p. 595)	1 and 3	13(g), 14(g), 15(g), 16(g)
	2 Construct prediction intervals for an individual response (p. 596)	2 and 3	13(i), 14(i), 15(i), 16(i)

Review Exercises

1. Roulette Wheel A pit boss suspects that a roulette wheel is out of balance. A roulette wheel has 18 black slots, 18 red slots, and 2 green slots. The pit boss spins the wheel 500 times and records the following frequencies:

Outcome	Frequency
Black	233
Red	237
Green	30

Test whether that the wheel is out of balance at the $\alpha = 0.05$ level of significance. *Do not reject* H_0

2. Fair Dice? A pit boss is concerned that a pair of dice being used in a craps game is not fair. The distribution of the expected sum of two fair dice is as follows:

Sum of Two Dice	Probability	Sum of Two Dice	Probability
2	$\frac{1}{36}$	8	$\frac{5}{36}$
3	$\frac{2}{36}$	9	$\frac{4}{36}$
4	$\frac{3}{36}$	10	$\frac{3}{36}$
5	$\frac{4}{36}$	11	$\frac{2}{36}$
6	$\frac{5}{36}$	12	$\frac{1}{36}$
7	$\frac{6}{36}$		

The pit boss rolls the dice 400 times and records the sum of the dice. The following are the results:

Sum of Two Dice	Frequency	Sum of Two Dice	Frequency
2	16	8	59
3	23	9	45
4	31	10	34
5	41	11	19
6	62	12	11
7	59		

Does the evidence suggest that the dice are not fair at the $\alpha = 0.01$ level of significance? *Do not reject H_0*

3. **Educational Attainment** A researcher wanted to test whether the distribution of educational attainment of Americans today is different from the distribution in 1994. The distribution of educational attainment in 1994 is as follows:

Education	Relative Frequency
Not a high school graduate	0.191
High school graduate	0.344
Some college	0.174
Associate's degree	0.070
Bachelor's degree	0.147
Advanced degree	0.075

Source: Statistical Abstract of the United States

The researcher randomly selects 500 Americans, learns their levels of education, and obtains the following data:

Education	Frequency
Not a high school graduate	89
High school graduate	152
Some college	83
Associate's degree	39
Bachelor's degree	93
Advanced degree	44

At the $\alpha = 0.1$ level of significance, test whether the distribution of educational attainment today is different from that of educational attainment in 1994. *Reject H_0*

4. **School Violence** A school administrator is concerned that the distribution of school violence has changed and has become more violent. The distribution of school crime for 1992 is as follows:

Crime	Theft	Violent	Serious Violent
Relative frequency	0.619	0.314	0.067

Source: National Center for Educational Statistics

The researcher randomly selects 800 school crimes, finds out whether they were theft, violent, or serious violent crimes, and obtains the following data:

Crime	Theft	Violent	Serious Violent
Frequency	421	311	68

At the $\alpha = 0.05$ level of significance, test whether the distribution of school crime today is different from the 1992 distribution. *Reject H_0*

5. **Evolution or Creation?** The Gallup Organization conducted a poll of 1016 randomly selected Americans aged 18 years old or older in February 2001 and 1017 randomly selected Americans aged 18 years old or older in June 1993 and asked them the following question:

> Which of the following statements comes closest to your views on the origin and development of human beings? (1) Human beings have developed over millions of years from less advanced forms of life, but God guided this process. (2) Human beings have developed over millions of years from less advanced forms of life, and God had no part in this process. (3) God created human beings pretty much in their present form at one time within the last 10,000 years or so.

The results of the survey are as follows:

Date	Humans Developed, with God Guiding	Humans Developed, but God Had No Part in Process	God Created Humans in Present Form	Other/No Opinion
February 2001	376	122	457	61
June 1993	356	112	478	71

(a) Compute the expected values of each cell under the assumption of independence.

(b) Verify that the requirements for performing a chi-square test of independence are satisfied.

(c) Compute the chi-square test statistic. 2.203

(d) Test whether people's opinion regarding human origin is independent of the date the question is asked at the $\alpha = 0.05$ level of significance.

(e) Compare the observed frequencies with the expected frequencies. Which cell contributed most to the test statistic? Was the expected frequency greater than or less than the observed frequency? What does this information tell you?

(f) Construct a conditional distribution by date and draw a bar graph. Does this evidence support your conclusion in part (d)?

(g) Compute the *P*-value for this test by finding the area under the chi-square distribution to the right of the test statistic. 0.531

(5d) Do not reject H_O

---

6. **Caesarean Sections** An obstetrician wanted to discover whether the method of delivering a baby was independent of race. The following data represent the race of the mother and the method of delivery for 365 randomly selected births.

(6d) Do not reject H_O

Race	Delivery Method	
	Vaginal	Caesarean
White	242	63
Black	47	13

Source: *Statistical Abstract of the United States*

(a) Compute the expected values of each cell under the assumption of independence.

(b) Verify that the requirements to perform a chi-square test of independence are satisfied.

(c) Compute the chi-square test statistic. 0.031

(d) Test whether method of delivery is independent of the race of the mother at the $\alpha = 0.05$ level of significance.

(e) Compare the observed frequencies with the expected frequencies. Which cell contributed most to the test statistic? Was the expected frequency greater than or less than the observed frequency? What does this information tell you?

(f) Construct a conditional distribution by method of delivery and draw a bar graph. Does this evidence support your conclusion in part (d)?

(g) Compute the *P*-value for this test by finding the area under the chi-square distribution to the right of the test statistic. 0.860

7. **Race versus Region of the United States** A sociologist wanted to determine whether the locations in which individuals live are independent of their races. He randomly selects 2712 U.S. residents and asks them to disclose their race and the location in which they live. He obtains the following data:

Race	Region			
	Northeast	Midwest	South	West
White	421	520	656	400
Black	56	57	158	28
American Indian, Eskimo, Aleut	2	4	6	9
Asian or Pacific Islander	13	8	11	40
Hispanic	38	17	68	101
Other	17	8	24	50

Source: *Statistical Abstract of the United States*

At the $\alpha = 0.01$ level of significance, test whether race is independent of the location in which a resident lives within the United States. Reject H_O

8. **Marital Status and Gender** A sociologist wanted to determine whether marital status and gender were independent. He randomly sampled 201 residents of the United States who were 18 years old or older and asked them to disclose their gender and marital status. The following data were collected.

Marital Status	Gender	
	Male	Female
Never married	26	22
Married	59	60
Divorced	9	11
Widowed	3	11

At the $\alpha = 0.05$ level of significance, test whether gender is independent of marital status. Do not reject H_O

9. **The Common Cold** A doctor wanted to determine whether the proportion of Americans who have had symptoms associated with the common cold is the same for all four regions of the United States. He randomly sampled 120 individuals from each region of the United States and obtained the following data:

	Region			
Symptoms	Northeast	Midwest	South	West
Symptoms within last year	26	31	23	35
No symptoms within last year	94	89	97	85

Source: National Center for Health Statistics

At the $\alpha = 0.05$ level of significance, test whether the proportion of Americans who have had symptoms associated with the common cold is different for the regions of the United States. Do not reject H_0

10. **Hardworking?** In a *Newsweek* poll conducted on June 24, 1999, 750 randomly selected adults were asked, "Do you believe Americans today are as willing to work hard at their jobs to get ahead as they were in the past, or are not as willing to work hard to get ahead?" The results of the survey are as follows:

	Age (years)		
Response	18–29	30–49	50 and Older
As willing	58	75	70
Not as willing	192	175	180

Source: pollingreport.com

At the $\alpha = 0.05$ level of significance, test whether the proportion of Americans who believe that Americans are willing to work just as hard today is different for the age groups. Do not reject H_0

11. **Night Owls** For a Gallup poll conducted on June 6–8, 2005, 1279 randomly selected adults aged 18 and older were asked, "Are you a morning person or a night person?" The results of the survey are as follows:

	Age			
Response	18–29	30–49	50–64	65 or Older
Morning person	97	177	210	210
Night person	234	147	107	97

At the $\alpha = 0.05$ level of significance, test whether the proportion of adults aged 18 years or older who are "morning people" are different for the age groups. Reject H_0

(12a) Runners in scoring position: Mohr
No runners in scoring position: Mohr

12. **Simpson's Paradox in Baseball** The following tables summarize the number of hits and at-bats for Dustan Mohr (Minnesota Twins/San Francisco Giants) and Darin Erstad (Anaheim Angels) during the 2003 MLB season when accounting for runners in scoring position.
    (*Source*: ite.pubs.informs.org/Vol5No1/KvamSokol/index.php)

Runner(s) in Scoring Position	Mohr	Erstad
Hits	19	9
At-bats	97	50

No Runners in Scoring Position	Mohr	Erstad
Hits	68	56
At-bats	251	208

   (a) Which player had the better batting average (hits divided by at-bats) in each situation?
   (b) Which player had the better batting average overall (total hits divided by total at-bats)? Erstad
   (c) Explain why your answer to part (b) differs from part (a).

13. **Engine Displacement versus Fuel Economy** The following data represent the size of a car's engine (in liters) versus its miles per gallon in the city for various 2005 domestic automobiles.

Car	Engine Displacement (liters), x	City MPG, y
Buick Century	3.1	20
Buick LeSabre	3.8	20
Cadillac DeVille	4.6	18
Chevrolet Cavalier	2.2	25
Chevrolet Impala	3.8	21
Chevrolet Malibu	2.2	24
Chrysler Sebring Sedan	2.7	22
Dodge Magnum	3.5	21
Ford Crown Victoria	4.6	18
Ford Focus	2.0	26
Ford Mustang	3.8	20
Mercury Sable	3.0	19
Pontiac Grand Am	3.4	20
Pontiac Sunfire	2.2	24
Saturn Ion	2.2	26

Source: www.roadandtrack.com

Using the results from Problems 1, 5, and 15 from the Chapter Review of Chapter 4, answer the following questions:

   (a) What are the estimates of β_0 and β_1? What is the mean number of miles per gallon of all cars that have a 3.8-liter engine? $b_0 = 30.3848$, $b_1 = -2.7977$, 19.8 mpg
   (b) Compute the standard error of the estimate, σ. 1.2692

(13c) Yes

(c) Determine whether the residuals are normally distributed.

(d) If the residuals are normally distributed, determine s_{b_1}.

(e) If the residuals are normally distributed, test whether a linear relation exists between the explanatory variable, x, and the response variable, y, at the $\alpha = 0.05$ level of significance.

(f) If the residuals are normally distributed, construct a 95% confidence interval about the slope of the true least-squares regression line.

(g) Construct a 90% confidence interval about the mean miles per gallon of all cars that have a 3.8-liter engine.

(h) Predict the miles per gallon of a particular car with a 3.8-liter engine.

(i) Construct a 90% prediction interval about the miles per gallon of a particular car that has a 3.8-liter engine.

(j) Explain why the predicted miles per gallon found in parts (a) and (h) are the same, yet the intervals are different.

14. **Temperature versus Cricket Chirps** Crickets make a chirping noise by sliding their wings rapidly over each other. Perhaps you have noticed that the number of chirps seems to increase with the temperature. The following table lists the temperature (in degrees Fahrenheit) and the number of chirps per second for the striped ground cricket.

Temperature, x	Chirps per Second, y	Temperature, x	Chirps per Second, y
88.6	20.0	71.6	16.0
93.3	19.8	84.3	18.4
80.6	17.1	75.2	15.5
69.7	14.7	82.0	17.1
69.4	15.4	83.3	16.2
79.6	15.0	82.6	17.2
80.6	16.0	83.5	17.0
76.3	14.4		

Source: *The Songs of Insects*, Pierce, George W., *Cambridge*, MA: Harvard University Press, 1949. pp. 12–21.

Using the results from Problems 2, 6, and 16 from the Chapter Review of Chapter 4, answer the following questions:

(a) What are the estimates of β_0 and β_1? What is the mean number of chirps when the temperature is 80.2°F? $b_0 = -0.309$, $b_1 = 0.2119$; 16.69

(b) Compute the standard error of the estimate, σ. 0.9715

(c) Determine whether the residuals are normally distributed.

(d) If the residuals are normally distributed, determine s_{b_1}. 0.0387

(e) If the residuals are normally distributed, test whether a linear relation exists between the explanatory variable, x, and response variable, y, at the $\alpha = 0.05$ level of significance. Reject H_0

(f) If the residuals are normally distributed, construct a 95% confidence interval about the slope of the true least-squares regression line. (0.1283, 0.2955)

(g) Construct a 90% confidence interval about the mean number of chirps found in part (a). (16.24, 17.13)

(h) Predict the number of chirps on a day when the temperature is 80.2°F. 16.69

(i) Construct a 90% prediction interval about the number of chirps found in part (h). (14.91, 18.46)

(j) Explain why the predicted number of chirps found in parts (a) and (h) are the same, yet the intervals are different.

15. **Apartments** The following data represent the square footage and rents for apartments in Queens, New York.

Square Footage, x	Rent per Month($), y
500	650
588	1215
1000	2000
688	1655
825	1250
1259	2700
650	1200
560	1250
1073	2350
1452	3300
1305	3100

Source: apartments.com

(15a) $b_0 = -399.2$, $b_1 = 2.5315$, $1879

(a) What are the estimates of β_0 and β_1? What is the mean rent of a 900-square-foot apartment in Queens?

(b) Compute the standard error of the estimate, σ. 229.547

(c) Determine whether the residuals are normally distributed. Yes

(d) If the residuals are normally distributed, determine s_{b_1}. 0.2166

(e) If the residuals are normally distributed, test whether a linear relation exists between the explanatory variable, x, and response variable, y, at the $\alpha = 0.05$ level of significance. Reject H_0

(f) If the residuals are normally distributed, construct a 95% confidence interval about the slope of the true least-squares regression line.

(g) Construct a 90% confidence interval about the mean rent of all 900-square-foot apartments in Queens.

(h) Predict the rent of a particular 900-square-foot apartment in Queens.

(i) Construct a 90% prediction interval about the rent of a particular 900-square-foot apartment in Queens.

(j) Explain why the predicted rents found in parts (a) and (h) are the same, yet the intervals are different.

16. Boys' Heights The following data represent the height (in inches) of boys between the ages of 2 and 10 years.

(16a) $b_0 = 29.705$; $b_1 = 2.6351$; 45.2 inches

Age, x	Boy Height, y	Age, x	Boy Height, y	Age, x	Boy Height, y
2	36.1	5	45.6	8	48.3
2	34.2	5	44.8	8	50.9
2	31.1	5	44.6	9	52.2
3	36.3	6	49.8	9	51.3
3	39.5	7	43.2	10	55.6
4	41.5	7	47.9	10	59.5
4	38.6	8	51.4		

Source: National Center for Health Statistics

(a) Treating age as the explanatory variable, determine the estimates of β_0 and β_1. What is the mean height of a 7-year-old boy?

(b) Compute the standard error of the estimate, σ. 2.450

(c) Determine whether the residuals are normally distributed.

(d) If the residuals are normally distributed, determine s_{b_1}. 0.2067

(e) If the residuals are normally distributed, test whether a linear relation exists between the explanatory variable, age, and response variable, height, at the $\alpha = 0.05$ level of significance. Reject H_0

(f) If the residuals are normally distributed, construct a 95% confidence interval about the slope of the true least-squares regression line. (2.2008, 3.0694)

(g) Construct a 90% confidence interval about the mean height found in part (a). (47.11, 49.19)

(h) Predict the height of a 7-year-old boy. 48.15

(i) Construct a 90% prediction interval about the height found in part (h). (43.78, 52.52)

(j) Explain why the predicted heights found in parts (a) and (h) are the same, yet the intervals are different.

17. Grip Strength A researcher believes that as age increases the grip strength (in pounds per square inch) of an individual's dominant hand decreases. From a random sample of 17 females, he obtains the following data:

Age, x	Grip Strength, y	Age, x	Grip Strength, y
15	65	34	45
16	60	37	58
28	58	41	70
61	60	43	73
53	46	49	45
43	66	53	60
16	56	61	56
25	75	68	30
28	46		

Source: Kevin McCarthy, student at Joliet Junior College

(17a) $b_0 = 67.388$; $b_1 = -0.2632$

(a) Treating age as the explanatory variable, determine the estimates of β_0 and β_1.

(b) Compute the standard error of the estimate, σ. 11.19

(c) Determine whether the residuals are normally distributed.

(d) If the residuals are normally distributed, determine s_{b_1}. 0.1680

(e) If the residuals are normally distributed, test whether a linear relation exists between the explanatory variable, age, and response variable, grip strength, at the $\alpha = 0.05$ level of significance. Do not reject H_0

(f) Based on your answers to (d) and (e), what would be a good guess as to the grip strength of a randomly selected 42-year-old female? 57 psi

THE CHAPTER 12 CASE STUDY IS LOCATED ON THE CD THAT ACCOMPANIES THIS TEXT.

APPENDIX A TABLES

				Table I						
				Random Numbers						
				Column Number						
Row Number	**01–05**	**06–10**	**11–15**	**16–20**	**21–25**	**26–30**	**31–35**	**36–40**	**41–45**	**46–50**
01	89392	23212	74483	36590	25956	36544	68518	40805	09980	00467
02	61458	17639	96252	95649	73727	33912	72896	66218	52341	97141
03	11452	74197	81962	48443	90360	26480	73231	37740	26628	44690
04	27575	04429	31308	02241	01698	19191	18948	78871	36030	23980
05	36829	59109	88976	46845	28329	47460	88944	08264	00843	84592
06	81902	93458	42161	26099	09419	89073	82849	09160	61845	40906
07	59761	55212	33360	68751	86737	79743	85262	31887	37879	17525
08	46827	25906	64708	20307	78423	15910	86548	08763	47050	18513
09	24040	66449	32353	83668	13874	86741	81312	54185	78824	00718
10	98144	96372	50277	15571	82261	66628	31457	00377	63423	55141
11	14228	17930	30118	00438	49666	65189	62869	31304	17117	71489
12	55366	51057	90065	14791	62426	02957	85518	28822	30588	32798
13	96101	30646	35526	90389	73634	79304	96635	06626	94683	16696
14	38152	55474	30153	26525	83647	31988	82182	98377	33802	80471
15	85007	18416	24661	95581	45868	15662	28906	36392	07617	50248
16	85544	15890	80011	18160	33468	84106	40603	01315	74664	20553
17	10446	20699	98370	17684	16932	80449	92654	02084	19985	59321
18	67237	45509	17638	65115	29757	80705	82686	48565	72612	61760
19	23026	89817	05403	82209	30573	47501	00135	33955	50250	72592
20	67411	58542	18678	46491	13219	84084	27783	34508	55158	78742

Table II

Binomial Probability Distribution

This table computes the probability of obtaining x successes in n trials of a binomial experiment with probability of success p.

n	x	0.01	0.05	0.10	0.15	0.20	0.25	0.30	0.35	0.40	0.45	0.50	0.55	0.60	0.65	0.70	0.75	0.80	0.85	0.90	0.95
2	0	0.9801	0.9025	0.8100	0.7225	0.6400	0.5625	0.4900	0.4225	0.3600	0.3025	0.2500	0.2025	0.1600	0.1225	0.0900	0.0625	0.0400	0.0225	0.0100	0.0025
	1	0.0198	0.0950	0.1800	0.2550	0.3200	0.3750	0.4200	0.4550	0.4800	0.4950	0.5000	0.4950	0.4800	0.4550	0.4200	0.3750	0.3200	0.2550	0.1800	0.0950
	2	0.0001	0.0025	0.0100	0.0225	0.0400	0.0625	0.0900	0.1225	0.1600	0.2025	0.2500	0.3025	0.3600	0.4225	0.4900	0.5625	0.6400	0.7225	0.8100	0.9025
3	0	0.9703	0.8574	0.7290	0.6141	0.5120	0.4219	0.3430	0.2746	0.2160	0.1664	0.1250	0.0911	0.0640	0.0429	0.0270	0.0156	0.0080	0.0034	0.0010	0.0001
	1	0.0294	0.1354	0.2430	0.3251	0.3840	0.4219	0.4410	0.4436	0.4320	0.4084	0.3750	0.3341	0.2880	0.2389	0.1890	0.1406	0.0960	0.0574	0.0270	0.0071
	2	0.0003	0.0071	0.0270	0.0574	0.0960	0.1406	0.1890	0.2389	0.2880	0.3341	0.3750	0.4084	0.4320	0.4436	0.4410	0.4219	0.3840	0.3251	0.2430	0.1354
	3	0.0000+	0.0001	0.0010	0.0034	0.0080	0.0156	0.0270	0.0429	0.0640	0.0911	0.1250	0.1664	0.2160	0.2746	0.3430	0.4219	0.5120	0.6141	0.7290	0.8574
4	0	0.9606	0.8145	0.6561	0.5220	0.4096	0.3164	0.2401	0.1785	0.1296	0.0915	0.0625	0.0410	0.0256	0.0150	0.0081	0.0039	0.0016	0.0005	0.0001	0.0000+
	1	0.0388	0.1715	0.2916	0.3585	0.4096	0.4219	0.4116	0.3845	0.3456	0.2995	0.2500	0.2005	0.1536	0.1115	0.0756	0.0469	0.0256	0.0116	0.0036	0.0005
	2	0.0006	0.0135	0.0486	0.0975	0.1536	0.2109	0.2646	0.3105	0.3456	0.3675	0.3760	0.3675	0.3456	0.3105	0.2646	0.2109	0.1536	0.0975	0.0486	0.0135
	3	0.0000+	0.0005	0.0036	0.0115	0.0256	0.0469	0.0756	0.1115	0.1536	0.2005	0.2500	0.2995	0.3456	0.3845	0.4116	0.4219	0.4096	0.3685	0.2916	0.1715
	4	0.0000+	0.0000+	0.0001	0.0005	0.0016	0.0039	0.0081	0.0160	0.0256	0.0410	0.0625	0.0915	0.1296	0.1785	0.2401	0.3164	0.4096	0.5220	0.6561	0.8145
5	0	0.9510	0.7738	0.5905	0.4437	0.3277	0.2373	0.1681	0.1160	0.0778	0.0503	0.0313	0.0185	0.0102	0.0053	0.0024	0.0010	0.0003	0.0001	0.0000+	0.0000+
	1	0.0480	0.2036	0.3281	0.3915	0.4096	0.3955	0.3602	0.3124	0.2592	0.2059	0.1563	0.1128	0.0768	0.0488	0.0284	0.0146	0.0064	0.0022	0.0005	0.0000+
	2	0.0010	0.0214	0.0729	0.1382	0.2048	0.2637	0.3087	0.3364	0.3456	0.3369	0.3125	0.2757	0.2304	0.1811	0.1323	0.0879	0.0512	0.0244	0.0081	0.0011
	3	0.0000+	0.0011	0.0081	0.0244	0.0512	0.0879	0.1323	0.1811	0.2304	0.2757	0.3125	0.3369	0.3456	0.3364	0.3087	0.2637	0.2048	0.1382	0.0729	0.0214
	4	0.0000+	0.0000+	0.0005	0.0022	0.0064	0.0146	0.0284	0.0488	0.0768	0.1128	0.1563	0.2059	0.2592	0.3124	0.3602	0.3955	0.4096	0.3915	0.3281	0.2036
	5	0.0000+	0.0000+	0.0000+	0.0001	0.0003	0.0010	0.0024	0.0053	0.0102	0.0185	0.0313	0.0503	0.0778	0.1160	0.1681	0.2373	0.3277	0.4437	0.5905	0.7738
6	0	0.9415	0.7351	0.5314	0.3771	0.2621	0.1780	0.1176	0.0754	0.0467	0.0277	0.0156	0.0083	0.0041	0.0018	0.0007	0.0002	0.0001	0.0000+	0.0000+	0.0000+
	1	0.0571	0.2321	0.3543	0.3993	0.3932	0.3560	0.3025	0.2437	0.1866	0.1359	0.0938	0.0609	0.0369	0.0205	0.0102	0.0044	0.0015	0.0004	0.0001	0.0000+
	2	0.0014	0.0305	0.0984	0.1762	0.2458	0.2966	0.3241	0.3280	0.3110	0.2780	0.2344	0.1861	0.1382	0.0951	0.0595	0.0330	0.0154	0.0056	0.0012	0.0001
	3	0.0000+	0.0021	0.0146	0.0415	0.0819	0.1318	0.1852	0.2355	0.2765	0.3032	0.3125	0.3032	0.2765	0.2355	0.1852	0.1318	0.0819	0.0415	0.0146	0.0021
	4	0.0000+	0.0001	0.0012	0.0056	0.0154	0.0330	0.0595	0.0951	0.1382	0.1861	0.2344	0.2780	0.3110	0.3280	0.3241	0.2966	0.2458	0.1762	0.0984	0.0305
	5	0.0000+	0.0000+	0.0001	0.0004	0.0015	0.0044	0.0102	0.0205	0.0369	0.0609	0.0938	0.1359	0.1866	0.2437	0.3025	0.3560	0.3932	0.3993	0.3543	0.2321
	6	0.0000+	0.0000+	0.0000+	0.0000+	0.0001	0.0002	0.0007	0.0018	0.0041	0.0083	0.0156	0.0277	0.0467	0.0754	0.1176	0.1780	0.2621	0.3771	0.5314	0.7351
7	0	0.9321	0.6983	0.4783	0.3206	0.2097	0.1335	0.0824	0.0490	0.0280	0.0152	0.0078	0.0037	0.0016	0.0006	0.0002	0.0001	0.0000+	0.0000+	0.0000+	0.0000+
	1	0.0659	0.2573	0.3720	0.3960	0.3670	0.3115	0.2471	0.1848	0.1306	0.0872	0.0547	0.0320	0.0172	0.0084	0.0036	0.0013	0.0004	0.0001	0.0000+	0.0000+
	2	0.0020	0.0406	0.1240	0.2097	0.2753	0.3115	0.3177	0.2985	0.2613	0.2140	0.1641	0.1172	0.0774	0.0466	0.0250	0.0115	0.0043	0.0012	0.0002	0.0000+
	3	0.0000+	0.0036	0.0230	0.0617	0.1147	0.1730	0.2269	0.2679	0.2903	0.2918	0.2734	0.2388	0.1935	0.1442	0.0972	0.0577	0.0287	0.0109	0.0026	0.0002
	4	0.0000+	0.0002	0.0026	0.0109	0.0287	0.0577	0.0972	0.1442	0.1935	0.2388	0.2734	0.2918	0.2903	0.2679	0.2269	0.1730	0.1147	0.0617	0.0230	0.0036
	5	0.0000+	0.0000+	0.0002	0.0012	0.0043	0.0115	0.0250	0.0466	0.0774	0.1172	0.1641	0.2140	0.2613	0.2985	0.3177	0.3115	0.2753	0.2097	0.1240	0.0406
	6	0.0000+	0.0000+	0.0000+	0.0001	0.0004	0.0013	0.0036	0.0084	0.0172	0.0320	0.0547	0.0872	0.1306	0.1848	0.2471	0.3115	0.3670	0.3960	0.3720	0.2573
	7	0.0000+	0.0000+	0.0000+	0.0000+	0.0000+	0.0001	0.0002	0.0006	0.0016	0.0037	0.0078	0.0152	0.0280	0.0490	0.0824	0.1335	0.2097	0.3206	0.4783	0.6983

Note: 0.0000+ means the probability is 0.0000 rounded to four decimal places. However, the probability is *not* zero.

Table II (continued)

n	x											p										
		0.01	**0.05**	**0.10**	**0.15**	**0.20**	**0.25**	**0.30**	**0.35**	**0.40**	**0.45**	**0.50**	**0.55**	**0.60**	**0.65**	**0.70**	**0.75**	**0.80**	**0.85**	**0.90**	**0.95**
8	0	0.9227	0.6634	0.4305	0.2725	0.1678	0.1001	0.0576	0.0319	0.0168	0.0084	0.0039	0.0017	0.0007	0.0002	0.0001	0.0000+	0.0000+	0.0000+	0.0000+	0.0000+
	1	0.0746	0.2793	0.3826	0.3847	0.3355	0.2670	0.1977	0.1373	0.0896	0.0548	0.0313	0.0164	0.0079	0.0033	0.0012	0.0004	0.0001	0.0000+	0.0000+	0.0000+
	2	0.0026	0.0515	0.1488	0.2376	0.2936	0.3115	0.2965	0.2587	0.2090	0.1569	0.1094	0.0703	0.0413	0.0217	0.0100	0.0038	0.0011	0.0002	0.0000+	0.0000+
	3	0.0001	0.0054	0.0331	0.0839	0.1468	0.2076	0.2541	0.2786	0.2787	0.2568	0.2188	0.1719	0.1239	0.0808	0.0467	0.0231	0.0092	0.0026	0.0004	0.0000+
	4	0.0000+	0.0004	0.0046	0.0185	0.0459	0.0865	0.1361	0.1875	0.2322	0.2627	0.2734	0.2627	0.2322	0.1675	0.1361	0.0865	0.0459	0.0185	0.0046	0.0004
	5	0.0000+	0.0000+	0.0004	0.0026	0.0092	0.0231	0.0467	0.0808	0.1239	0.1719	0.2188	0.2568	0.2787	0.2786	0.2541	0.2076	0.1468	0.0839	0.0331	0.0054
	6	0.0000+	0.0000+	0.0000+	0.0002	0.0011	0.0038	0.0100	0.0217	0.0413	0.0703	0.1094	0.1569	0.2090	0.2587	0.2965	0.3115	0.2936	0.2376	0.1488	0.0515
	7	0.0000+	0.0000+	0.0000+	0.0000+	0.0001	0.0004	0.0012	0.0033	0.0079	0.0164	0.0313	0.0548	0.0896	0.1373	0.1977	0.2670	0.3355	0.3647	0.3826	0.2793
	8	0.0000+	0.0000+	0.0000+	0.0000+	0.0000+	0.0000+	0.0001	0.0002	0.0007	0.0017	0.0039	0.0084	0.0168	0.0319	0.0576	0.1001	0.1678	0.2725	0.4305	0.6634
9	0	0.9135	0.6302	0.3874	0.2316	0.1342	0.0751	0.0404	0.0207	0.0101	0.0046	0.0020	0.0008	0.0003	0.0001	0.0000+	0.0000+	0.0000+	0.0000+	0.0000+	0.0000+
	1	0.0830	0.2985	0.3874	0.3679	0.3020	0.2253	0.1556	0.1004	0.0605	0.0339	0.0176	0.0083	0.0035	0.0013	0.0004	0.0001	0.0000+	0.0000+	0.0000+	0.0000+
	2	0.0034	0.0629	0.1722	0.2597	0.3020	0.3003	0.2668	0.2162	0.1612	0.1110	0.0703	0.0407	0.0212	0.0098	0.0039	0.0012	0.0003	0.0000+	0.0000+	0.0000+
	3	0.0001	0.0077	0.0446	0.1069	0.1762	0.2336	0.2668	0.2716	0.2508	0.2119	0.1641	0.1160	0.0743	0.0424	0.0210	0.0087	0.0028	0.0006	0.0001	0.0000+
	4	0.0000+	0.0006	0.0074	0.0283	0.0661	0.1168	0.1715	0.2194	0.2508	0.2600	0.2461	0.2128	0.1672	0.1181	0.0735	0.0389	0.0165	0.0050	0.0008	0.0000+
	5	0.0000+	0.0000+	0.0008	0.0050	0.0165	0.0389	0.0735	0.1181	0.1672	0.2128	0.2461	0.2600	0.2508	0.2194	0.1715	0.1168	0.0661	0.0283	0.0074	0.0006
	6	0.0000+	0.0000+	0.0001	0.0006	0.0028	0.0087	0.0210	0.0424	0.0743	0.1160	0.1641	0.2119	0.2508	0.2716	0.2668	0.2336	0.1762	0.1069	0.0446	0.0077
	7	0.0000+	0.0000+	0.0000+	0.0000+	0.0003	0.0012	0.0039	0.0098	0.0212	0.0407	0.0703	0.1110	0.1612	0.2162	0.2668	0.3003	0.3020	0.2597	0.1722	0.0629
	8	0.0000+	0.0000+	0.0000+	0.0000+	0.0000+	0.0001	0.0004	0.0013	0.0035	0.0083	0.0176	0.0339	0.0605	0.1004	0.1556	0.2253	0.3020	0.3679	0.3874	0.2985
	9	0.0000+	0.0000+	0.0000+	0.0000+	0.0000+	0.0000+	0.0000+	0.0001	0.0003	0.0008	0.0020	0.0046	0.0101	0.0207	0.0404	0.0751	0.1342	0.2316	0.3874	0.6302
10	0	0.9044	0.5987	0.3487	0.1969	0.1074	0.0563	0.0282	0.0135	0.0060	0.0025	0.0010	0.0003	0.0001	0.0000+	0.0000+	0.0000+	0.0000+	0.0000+	0.0000+	0.0000+
	1	0.0914	0.3151	0.3874	0.3474	0.2684	0.1877	0.1211	0.0725	0.0403	0.0207	0.0098	0.0042	0.0016	0.0005	0.0001	0.0000+	0.0000+	0.0000+	0.0000+	0.0000+
	2	0.0042	0.0748	0.1937	0.2759	0.3020	0.2816	0.2335	0.1757	0.1209	0.0763	0.0439	0.0229	0.0106	0.0043	0.0014	0.0004	0.0001	0.0000+	0.0000+	0.0000+
	3	0.0001	0.0105	0.0574	0.1298	0.2013	0.2503	0.2668	0.2522	0.2150	0.1665	0.1172	0.0746	0.0425	0.0212	0.0090	0.0031	0.0008	0.0001	0.0000+	0.0000+
	4	0.0000+	0.0010	0.0112	0.0401	0.0881	0.1460	0.2001	0.2377	0.2508	0.2384	0.2051	0.1596	0.1115	0.0689	0.0368	0.0162	0.0055	0.0012	0.0001	0.0000+
	5	0.0000+	0.0001	0.0015	0.0085	0.0264	0.0584	0.1029	0.1536	0.2007	0.2340	0.2461	0.2340	0.2007	0.1536	0.1029	0.0584	0.0264	0.0085	0.0015	0.0001
	6	0.0000+	0.0000+	0.0001	0.0012	0.0055	0.0162	0.0368	0.0689	0.1115	0.1596	0.2051	0.2384	0.2508	0.2377	0.2001	0.1460	0.0881	0.0401	0.0112	0.0010
	7	0.0000+	0.0000+	0.0000+	0.0001	0.0008	0.0031	0.0090	0.0212	0.0425	0.0746	0.1172	0.1665	0.2150	0.2522	0.2668	0.2503	0.2013	0.1298	0.0574	0.0105
	8	0.0000+	0.0000+	0.0000+	0.0000+	0.0001	0.0004	0.0014	0.0043	0.0106	0.0229	0.0439	0.0763	0.1209	0.1757	0.2335	0.2816	0.3020	0.2759	0.1937	0.0746
	9	0.0000+	0.0000+	0.0000+	0.0000+	0.0000+	0.0000+	0.0001	0.0005	0.0016	0.0042	0.0098	0.0207	0.0403	0.0725	0.1211	0.1877	0.2684	0.3474	0.3874	0.3151
	10	0.0000+	0.0000+	0.0000+	0.0000+	0.0000+	0.0000+	0.0000+	0.0000+	0.0001	0.0003	0.0010	0.0025	0.0060	0.0135	0.0282	0.0563	0.1074	0.1969	0.3487	0.5987

Note: 0.0000+ means the probability is 0.0000 rounded to four decimal places. However, the probability is *not* zero.

Table II (continued)

p

n	x	0.01	0.05	0.10	0.15	0.20	0.25	0.30	0.35	0.40	0.45	0.50	0.55	0.60	0.65	0.70	0.75	0.80	0.85	0.90	0.95
11	0	0.8953	0.5688	0.3138	0.1673	0.0859	0.0422	0.0198	0.0088	0.0036	0.0014	0.0005	0.0002	0.0000+	0.0000+	0.0000+	0.0000+	0.0000+	0.0000+	0.0000+	0.0000+
	1	0.0995	0.3293	0.3835	0.3248	0.2362	0.1549	0.0932	0.0518	0.0266	0.0125	0.0054	0.0021	0.0007	0.0002	0.0000+	0.0000+	0.0000+	0.0000+	0.0000+	0.0000+
	2	0.0050	0.0867	0.2131	0.2866	0.2953	0.2581	0.1998	0.1395	0.0887	0.0513	0.0269	0.0126	0.0052	0.0018	0.0005	0.0001	0.0000+	0.0000+	0.0000+	0.0000+
	3	0.0002	0.0137	0.0710	0.1517	0.2215	0.2581	0.2568	0.2254	0.1774	0.1259	0.0806	0.0462	0.0234	0.0102	0.0037	0.0011	0.0002	0.0000+	0.0000+	0.0000+
	4	0.0000+	0.0014	0.0158	0.0536	0.1107	0.1721	0.2201	0.2428	0.2365	0.2060	0.1611	0.1128	0.0701	0.0379	0.0173	0.0064	0.0017	0.0003	0.0000+	0.0000+
	5	0.0000+	0.0001	0.0025	0.0132	0.0388	0.0803	0.1321	0.1830	0.2207	0.2360	0.2256	0.1931	0.1471	0.0985	0.0566	0.0268	0.0097	0.0023	0.0003	0.0000+
	6	0.0000+	0.0000+	0.0003	0.0023	0.0097	0.0268	0.0566	0.0985	0.1471	0.1931	0.2256	0.2360	0.2207	0.1830	0.1321	0.0803	0.0388	0.0132	0.0025	0.0001
	7	0.0000+	0.0000+	0.0000+	0.0003	0.0017	0.0064	0.0173	0.0379	0.0701	0.1128	0.1611	0.2060	0.2365	0.2428	0.2201	0.1721	0.1107	0.0536	0.0158	0.0014
	8	0.0000+	0.0000+	0.0000+	0.0000+	0.0002	0.0011	0.0037	0.0102	0.0234	0.0462	0.0806	0.1259	0.1774	0.2254	0.2568	0.2581	0.2215	0.1517	0.0710	0.0137
	9	0.0000+	0.0000+	0.0000+	0.0000+	0.0000+	0.0001	0.0005	0.0018	0.0052	0.0126	0.0269	0.0513	0.0887	0.1395	0.1998	0.2581	0.2953	0.2866	0.2131	0.0867
	10	0.0000+	0.0000+	0.0000+	0.0000+	0.0000+	0.0000+	0.0000+	0.0002	0.0007	0.0021	0.0054	0.0125	0.0266	0.0518	0.0932	0.1549	0.2362	0.3248	0.3835	0.3293
	11	0.0000+	0.0000+	0.0000+	0.0000+	0.0000+	0.0000+	0.0000+	0.0000+	0.0000+	0.0002	0.0005	0.0014	0.0036	0.0088	0.0198	0.0422	0.0859	0.1673	0.3138	0.5688
12	0	0.8864	0.5404	0.2824	0.1422	0.0687	0.0317	0.0138	0.0057	0.0022	0.0008	0.0002	0.0001	0.0000+	0.0000+	0.0000+	0.0000+	0.0000+	0.0000+	0.0000+	0.0000+
	1	0.1074	0.3413	0.3766	0.3012	0.2062	0.1267	0.0712	0.0368	0.0174	0.0075	0.0029	0.0010	0.0003	0.0001	0.0000+	0.0000+	0.0000+	0.0000+	0.0000+	0.0000+
	2	0.0060	0.0988	0.2301	0.2924	0.2835	0.2323	0.1678	0.1088	0.0639	0.0339	0.0161	0.0068	0.0025	0.0008	0.0002	0.0000+	0.0000+	0.0000+	0.0000+	0.0000+
	3	0.0002	0.0173	0.0852	0.1720	0.2362	0.2581	0.2397	0.1954	0.1419	0.0923	0.0537	0.0277	0.0125	0.0048	0.0015	0.0004	0.0001	0.0000+	0.0000+	0.0000+
	4	0.0000+	0.0021	0.0213	0.0683	0.1329	0.1936	0.2311	0.2367	0.2128	0.1700	0.1208	0.0762	0.0420	0.0199	0.0078	0.0024	0.0005	0.0001	0.0000+	0.0000+
	5	0.0000+	0.0002	0.0038	0.0193	0.0532	0.1032	0.1585	0.2039	0.2270	0.2225	0.1934	0.1489	0.1009	0.0591	0.0291	0.0115	0.0033	0.0006	0.0000+	0.0000+
	6	0.0000+	0.0000+	0.0005	0.0040	0.0155	0.0401	0.0792	0.1281	0.1766	0.2124	0.2256	0.2124	0.1766	0.1281	0.0792	0.0401	0.0155	0.0040	0.0005	0.0000+
	7	0.0000+	0.0000+	0.0000+	0.0006	0.0033	0.0115	0.0291	0.0591	0.1009	0.1489	0.1934	0.2225	0.2270	0.2039	0.1585	0.1032	0.0532	0.0193	0.0038	0.0002
	8	0.0000+	0.0000+	0.0000+	0.0001	0.0005	0.0024	0.0078	0.0199	0.0420	0.0762	0.1208	0.1700	0.2128	0.2367	0.2311	0.1936	0.1329	0.0683	0.0213	0.0021
	9	0.0000+	0.0000+	0.0000+	0.0000+	0.0001	0.0004	0.0015	0.0048	0.0125	0.0277	0.0537	0.0923	0.1419	0.1954	0.2397	0.2581	0.2362	0.1720	0.0852	0.0173
	10	0.0000+	0.0000+	0.0000+	0.0000+	0.0000+	0.0000+	0.0002	0.0008	0.0025	0.0068	0.0161	0.0339	0.0639	0.1088	0.1678	0.2323	0.2835	0.2924	0.2301	0.0988
	11	0.0000+	0.0000+	0.0000+	0.0000+	0.0000+	0.0000+	0.0000+	0.0001	0.0003	0.0010	0.0029	0.0075	0.0174	0.0368	0.0712	0.1267	0.2062	0.3012	0.3766	0.3413
	12	0.0000+	0.0000+	0.0000+	0.0000+	0.0000+	0.0000+	0.0000+	0.0000+	0.0000+	0.0001	0.0002	0.0008	0.0022	0.0057	0.0138	0.0317	0.0687	0.1422	0.2824	0.5404

Note: 0.0000+ means the probability is 0.0000 rounded to four decimal places. However, the probability is *not* zero.

Table II (continued)

n	x	0.01	0.05	0.10	0.15	0.20	0.25	0.30	0.35	0.40	0.45	0.50	0.55	0.60	0.65	0.70	0.75	0.80	0.85	0.90	0.95
15	0	0.8601	0.4633	0.2059	0.0874	0.0352	0.0134	0.0047	0.0016	0.0005	0.0001	0.0000+	0.0000+	0.0000+	0.0000+	0.0000+	0.0000+	0.0000+	0.0000+	0.0000+	0.0000+
	1	0.1303	0.3658	0.3432	0.2312	0.1319	0.0668	0.0305	0.0126	0.0047	0.0016	0.0005	0.0001	0.0000+	0.0000+	0.0000+	0.0000+	0.0000+	0.0000+	0.0000+	0.0000+
	2	0.0092	0.1348	0.2669	0.2856	0.2309	0.1559	0.0916	0.0476	0.0219	0.0090	0.0032	0.0010	0.0003	0.0001	0.0000+	0.0000+	0.0000+	0.0000+	0.0000+	0.0000+
	3	0.0004	0.0307	0.1285	0.2184	0.2501	0.2262	0.1700	0.1110	0.0634	0.0318	0.0139	0.0052	0.0016	0.0004	0.0001	0.0000+	0.0000+	0.0000+	0.0000+	0.0000+
	4	0.0000+	0.0049	0.0428	0.1156	0.1876	0.2252	0.2186	0.1792	0.1268	0.0780	0.0417	0.0191	0.0074	0.0024	0.0006	0.0001	0.0000+	0.0000+	0.0000	0.0000
	5	0.0000+	0.0006	0.0105	0.0449	0.1032	0.1651	0.2081	0.2123	0.1859	0.1404	0.0916	0.0515	0.0245	0.0096	0.0030	0.0007	0.0001	0.0000+	0.0000+	0.0000+
	6	0.0000+	0.0000+	0.0019	0.0132	0.0430	0.0917	0.1472	0.1906	0.2066	0.1914	0.1527	0.1048	0.0612	0.0298	0.0116	0.0034	0.0007	0.0001	0.0000+	0.0000+
	7	0.0000+	0.0000+	0.0003	0.0030	0.0138	0.0393	0.0811	0.1319	0.1771	0.2013	0.1964	0.1647	0.1181	0.0710	0.0348	0.0131	0.0035	0.0005	0.0000+	0.0000+
	8	0.0000+	0.0000+	0.0000+	0.0005	0.0035	0.0131	0.0348	0.0710	0.1181	0.1647	0.1964	0.2013	0.1771	0.1319	0.0811	0.0393	0.0138	0.0030	0.0003	0.0000+
	9	0.0000+	0.0000+	0.0000+	0.0001	0.0007	0.0034	0.0116	0.0298	0.0612	0.1048	0.1527	0.1914	0.2066	0.1906	0.1472	0.0917	0.0430	0.0132	0.0019	0.0006
	10	0.0000+	0.0000+	0.0000+	0.0000+	0.0001	0.0007	0.0030	0.0096	0.0245	0.0515	0.0916	0.1404	0.1859	0.2123	0.2061	0.1651	0.1032	0.0449	0.0105	0.0049
	11	0.0000+	0.0000+	0.0000+	0.0000+	0.0000+	0.0001	0.0006	0.0024	0.0074	0.0191	0.0417	0.0780	0.1268	0.1792	0.2186	0.2252	0.1876	0.1156	0.0428	0.0307
	12	0.0000+	0.0000+	0.0000+	0.0000+	0.0000+	0.0000+	0.0001	0.0004	0.0016	0.0052	0.0139	0.0318	0.0634	0.1110	0.1700	0.2252	0.2501	0.2184	0.1285	0.1348
	13	0.0000+	0.0000+	0.0000+	0.0000+	0.0000+	0.0000+	0.0000+	0.0001	0.0003	0.0010	0.0032	0.0090	0.0219	0.0476	0.0916	0.1559	0.2309	0.2856	0.2669	0.3658
	14	0.0000+	0.0000+	0.0000+	0.0000+	0.0000+	0.0000+	0.0000+	0.0000+	0.0000+	0.0001	0.0005	0.0016	0.0047	0.0126	0.0305	0.0668	0.1319	0.2312	0.3432	0.3658
	15	0.0000+	0.0000+	0.0000+	0.0000+	0.0000+	0.0000+	0.0000+	0.0000+	0.0000+	0.0000+	0.0000+	0.0001	0.0005	0.0016	0.0047	0.0134	0.0352	0.0874	0.2059	0.4633
20	0	0.8179	0.3585	0.1216	0.0388	0.0115	0.0032	0.0008	0.0002	0.0000+	0.0000+	0.0000+	0.0000+	0.0000+	0.0000+	0.0000+	0.0000+	0.0000+	0.0000+	0.0000+	0.0000+
	1	0.1652	0.3774	0.2702	0.1368	0.0576	0.0211	0.0068	0.0020	0.0005	0.0001	0.0000+	0.0000+	0.0000+	0.0000+	0.0000+	0.0000+	0.0000+	0.0000+	0.0000+	0.0000+
	2	0.0159	0.1887	0.2852	0.2293	0.1369	0.0669	0.0278	0.0100	0.0031	0.0008	0.0002	0.0000+	0.0000+	0.0000+	0.0000+	0.0000+	0.0000+	0.0000+	0.0000+	0.0000+
	3	0.0010	0.0596	0.1901	0.2428	0.2054	0.1339	0.0716	0.0323	0.0123	0.0040	0.0011	0.0002	0.0000+	0.0000+	0.0000+	0.0000+	0.0000+	0.0000+	0.0000+	0.0000+
	4	0.0000+	0.0133	0.0898	0.1821	0.2182	0.1897	0.1304	0.0738	0.0350	0.0139	0.0046	0.0013	0.0003	0.0000+	0.0000+	0.0000+	0.0000+	0.0000+	0.0001	0.0000+
	5	0.0000+	0.0022	0.0319	0.1028	0.1746	0.2023	0.1789	0.1272	0.0746	0.0365	0.0148	0.0049	0.0013	0.0003	0.0000+	0.0000+	0.0000+	0.0000+	0.0004	0.0000+
	6	0.0000+	0.0003	0.0089	0.0454	0.1091	0.1686	0.1916	0.1712	0.1244	0.0746	0.0370	0.0150	0.0049	0.0012	0.0002	0.0000+	0.0000+	0.0000+	0.0020	0.0003
	7	0.0000+	0.0000+	0.0020	0.0160	0.0545	0.1124	0.1643	0.1844	0.1659	0.1221	0.0739	0.0366	0.0146	0.0045	0.0010	0.0002	0.0000+	0.0000+	0.0089	0.0022
	8	0.0000+	0.0000+	0.0004	0.0046	0.0222	0.0609	0.1144	0.1614	0.1797	0.1623	0.1201	0.0727	0.0355	0.0136	0.0039	0.0008	0.0001	0.0000+	0.0319	0.0133
	9	0.0000+	0.0000+	0.0001	0.0011	0.0074	0.0271	0.0654	0.1158	0.1597	0.1771	0.1602	0.1185	0.0710	0.0336	0.0120	0.0030	0.0005	0.0000+	0.0898	0.0596
	10	0.0000+	0.0000+	0.0000+	0.0002	0.0020	0.0099	0.0308	0.0686	0.1171	0.1593	0.1762	0.1593	0.1171	0.0686	0.0308	0.0099	0.0020	0.0002	0.1901	0.1887
	11	0.0000+	0.0000+	0.0000+	0.0000+	0.0005	0.0030	0.0120	0.0336	0.0710	0.1185	0.1602	0.1771	0.1597	0.1158	0.0654	0.0271	0.0074	0.0011	0.2702	0.3774
	12	0.0000+	0.0000+	0.0000+	0.0000+	0.0001	0.0008	0.0039	0.0136	0.0355	0.0727	0.1201	0.1623	0.1797	0.1614	0.1144	0.0609	0.0222	0.0046	0.2852	0.1887
	13	0.0000+	0.0000+	0.0000+	0.0000+	0.0000+	0.0002	0.0010	0.0045	0.0146	0.0366	0.0739	0.1221	0.1659	0.1844	0.1643	0.1124	0.0545	0.0160	0.1901	0.0596
	14	0.0000+	0.0000+	0.0000+	0.0000+	0.0000+	0.0000+	0.0002	0.0012	0.0049	0.0150	0.0370	0.0746	0.1244	0.1712	0.1916	0.1686	0.1091	0.0454	0.0898	0.0133
	15	0.0000+	0.0000+	0.0000+	0.0000+	0.0000+	0.0000+	0.0000+	0.0003	0.0013	0.0049	0.0148	0.0365	0.0746	0.1272	0.1789	0.2023	0.1746	0.1028	0.0319	0.0022
	16	0.0000+	0.0000+	0.0000+	0.0000+	0.0000+	0.0000+	0.0000+	0.0000+	0.0003	0.0013	0.0046	0.0139	0.0350	0.0738	0.1304	0.1897	0.2182	0.1821	0.0898	0.0133
	17	0.0000+	0.0000+	0.0000+	0.0000+	0.0000+	0.0000+	0.0000+	0.0000+	0.0000+	0.0002	0.0011	0.0040	0.0123	0.0323	0.0716	0.1339	0.2054	0.2428	0.1901	0.0596
	18	0.0000+	0.0000+	0.0000+	0.0000+	0.0000+	0.0000+	0.0000+	0.0000+	0.0000+	0.0000+	0.0002	0.0008	0.0031	0.0100	0.0278	0.0669	0.1369	0.2293	0.2852	0.1887
	19	0.0000+	0.0000+	0.0000+	0.0000+	0.0000+	0.0000+	0.0000+	0.0000+	0.0000+	0.0000+	0.0000+	0.0001	0.0005	0.0020	0.0068	0.0211	0.0576	0.1368	0.2702	0.3774
	20	0.0000+	0.0000+	0.0000+	0.0000+	0.0000+	0.0000+	0.0000+	0.0000+	0.0000+	0.0000+	0.0000+	0.0000+	0.0000+	0.0002	0.0008	0.0032	0.0115	0.0388	0.1216	0.3585

Note: 0.0000+ means the probability is 0.0000 rounded to four decimal places. However, the probability is *not* zero.

Table III

Cumulative Binomial Probability Distribution

This table computes the cumulative probability of obtaining x successes in n trials of a binomial experiment with probability of success p.

n	x	\(p\) 0.01	0.05	0.10	0.15	0.20	0.25	0.30	0.35	0.40	0.45	0.50	0.55	0.60	0.65	0.70	0.75	0.80	0.85	0.90	0.95
2	0	0.9801	0.9025	0.8100	0.7225	0.6400	0.5625	0.4900	0.4225	0.3600	0.3025	0.2500	0.2025	0.1600	0.1225	0.0900	0.0625	0.0400	0.0225	0.0100	0.0025
	1	0.9999	0.9975	0.9900	0.9775	0.9600	0.9375	0.9100	0.8775	0.8400	0.7975	0.7500	0.6975	0.6400	0.5775	0.5100	0.4375	0.3600	0.2775	0.1900	0.0975
	2	1.0000	1.0000	1.0000	1.0000	1.0000	1.0000	1.0000	1.0000	1.0000	1.0000	1.0000	1.0000	1.0000	1.0000	1.0000	1.0000	1.0000	1.0000	1.0000	1.0000
3	0	0.9703	0.8574	0.7290	0.6141	0.5120	0.4219	0.3430	0.2746	0.2160	0.1664	0.1250	0.0911	0.0640	0.0429	0.0270	0.0156	0.0080	0.0034	0.0010	0.0001
	1	0.9997	0.9928	0.9720	0.9393	0.8960	0.8438	0.7840	0.7183	0.6480	0.5748	0.5000	0.4253	0.3520	0.2818	0.2160	0.1563	0.1040	0.0608	0.0280	0.0073
	2	1.0000−	0.9999	0.9990	0.9966	0.9920	0.9844	0.9730	0.9571	0.9360	0.9089	0.8750	0.8336	0.7840	0.7254	0.6570	0.5781	0.4880	0.3859	0.2710	0.1426
	3	1.0000	1.0000	1.0000	1.0000	1.0000	1.0000	1.0000	1.0000	1.0000	1.0000	1.0000	1.0000	1.0000	1.0000	1.0000	1.0000	1.0000	1.0000	1.0000	1.0000
4	0	0.9606	0.8145	0.6561	0.5220	0.4096	0.3164	0.2401	0.1785	0.1296	0.0915	0.0625	0.0410	0.0256	0.0150	0.0081	0.0039	0.0016	0.0005	0.0001	0.0000+
	1	0.9994	0.9860	0.9477	0.8905	0.8192	0.7383	0.6517	0.5630	0.4752	0.3910	0.3125	0.2415	0.1792	0.1265	0.0837	0.0508	0.0272	0.0120	0.0037	0.0005
	2	1.0000−	0.9995	0.9963	0.9880	0.9728	0.9492	0.9163	0.8735	0.8208	0.7585	0.6875	0.6090	0.5248	0.4370	0.3483	0.2617	0.1808	0.1095	0.0523	0.0140
	3	1.0000−	1.0000−	0.9999	0.9995	0.9984	0.9961	0.9919	0.9850	0.9744	0.9590	0.9375	0.9085	0.8704	0.8215	0.7599	0.6836	0.5904	0.4780	0.3439	0.1855
	4	1.0000	1.0000	1.0000	1.0000	1.0000	1.0000	1.0000	1.0000	1.0000	1.0000	1.0000	1.0000	1.0000	1.0000	1.0000	1.0000	1.0000	1.0000	1.0000	1.0000
5	0	0.9510	0.7738	0.5905	0.4437	0.3277	0.2373	0.1681	0.1160	0.0778	0.0503	0.0313	0.0185	0.0102	0.0053	0.0024	0.0010	0.0003	0.0001	0.0000+	0.0000+
	1	0.9990	0.9774	0.9185	0.8352	0.7373	0.6328	0.5282	0.4284	0.3370	0.2562	0.1875	0.1312	0.0870	0.0540	0.0308	0.0156	0.0067	0.0022	0.0005	0.0000+
	2	1.0000−	0.9988	0.9914	0.9734	0.9421	0.8965	0.8369	0.7648	0.6826	0.5931	0.5000	0.4069	0.3174	0.2352	0.1631	0.1035	0.0579	0.0266	0.0086	0.0012
	3	1.0000−	1.0000−	0.9995	0.9978	0.9933	0.9844	0.9692	0.9460	0.9130	0.8688	0.8125	0.7438	0.6630	0.5716	0.4718	0.3672	0.2627	0.1648	0.0815	0.0226
	4	1.0000−	1.0000−	1.0000−	0.9999	0.9997	0.9990	0.9976	0.9947	0.9898	0.9815	0.9688	0.9497	0.9222	0.8840	0.8319	0.7627	0.6723	0.5563	0.4095	0.2262
	5	1.0000	1.0000	1.0000	1.0000	1.0000	1.0000	1.0000	1.0000	1.0000	1.0000	1.0000	1.0000	1.0000	1.0000	1.0000	1.0000	1.0000	1.0000	1.0000	1.0000
6	0	0.9415	0.7351	0.5314	0.3771	0.2621	0.1780	0.1176	0.0754	0.0467	0.0277	0.0156	0.0083	0.0041	0.0018	0.0007	0.0002	0.0001	0.0000+	0.0000+	0.0000+
	1	0.9985	0.9672	0.8857	0.7765	0.6554	0.5339	0.4202	0.3191	0.2333	0.1636	0.1094	0.0692	0.0410	0.0223	0.0109	0.0046	0.0016	0.0004	0.0001	0.0000+
	2	1.0000−	0.9978	0.9842	0.9527	0.9011	0.8306	0.7443	0.6471	0.5443	0.4415	0.3438	0.2553	0.1792	0.1174	0.0705	0.0376	0.0170	0.0059	0.0013	0.0001
	3	1.0000−	0.9999	0.9987	0.9941	0.9830	0.9624	0.9295	0.8826	0.8208	0.7447	0.6563	0.5585	0.4557	0.3529	0.2557	0.1694	0.0989	0.0473	0.0159	0.0022
	4	1.0000−	1.0000−	0.9999	0.9996	0.9984	0.9954	0.9891	0.9777	0.9590	0.9308	0.8906	0.8364	0.7667	0.6809	0.5798	0.4661	0.3446	0.2235	0.1143	0.0328
	5	1.0000−	1.0000−	1.0000−	1.0000−	0.9999	0.9998	0.9993	0.9982	0.9959	0.9917	0.9844	0.9723	0.9533	0.9246	0.8824	0.8220	0.7379	0.6229	0.4686	0.2649
	6	1.0000	1.0000	1.0000	1.0000	1.0000	1.0000	1.0000	1.0000	1.0000	1.0000	1.0000	1.0000	1.0000	1.0000	1.0000	1.0000	1.0000	1.0000	1.0000	1.0000
7	0	0.9321	0.6983	0.4783	0.3206	0.2097	0.1335	0.0824	0.0490	0.0280	0.0152	0.0078	0.0037	0.0016	0.0006	0.0002	0.0001	0.0000+	0.0000+	0.0000+	0.0000+
	1	0.9980	0.9556	0.8503	0.7166	0.5767	0.4449	0.3294	0.2338	0.1586	0.1024	0.0625	0.0357	0.0188	0.0090	0.0038	0.0013	0.0004	0.0001	0.0000+	0.0000+
	2	1.0000−	0.9962	0.9743	0.9262	0.8520	0.7564	0.6471	0.5323	0.4199	0.3164	0.2266	0.1529	0.0963	0.0556	0.0288	0.0129	0.0047	0.0012	0.0002	0.0000+
	3	1.0000−	0.9998	0.9973	0.9879	0.9667	0.9294	0.8740	0.8002	0.7102	0.6083	0.5000	0.3917	0.2898	0.1998	0.1260	0.0706	0.0333	0.0121	0.0027	0.0002
	4	1.0000−	1.0000−	0.9998	0.9988	0.9953	0.9871	0.9712	0.9444	0.9037	0.8471	0.7734	0.6836	0.5801	0.4677	0.3529	0.2436	0.1480	0.0738	0.0257	0.0038
	5	1.0000−	1.0000−	1.0000−	0.9999	0.9996	0.9987	0.9962	0.9910	0.9812	0.9643	0.9375	0.8976	0.8414	0.7662	0.6706	0.5551	0.4233	0.2834	0.1497	0.0444
	6	1.0000−	1.0000−	1.0000−	1.0000−	1.0000−	0.9999	0.9998	0.9994	0.9984	0.9963	0.9922	0.9848	0.9720	0.9510	0.9176	0.8665	0.7903	0.6794	0.5217	0.3017
	7	1.0000	1.0000	1.0000	1.0000	1.0000	1.0000	1.0000	1.0000	1.0000	1.0000	1.0000	1.0000	1.0000	1.0000	1.0000	1.0000	1.0000	1.0000	1.0000	1.0000

Note: 0.0000+ means the probability is 0.0000 rounded to four decimal places. However, the probability is *not* zero.
1.0000− means the probability is 1.0000 rounded to four decimal places. However, the probability is *not* one.

Table III (continued)

n	x	0.01	0.05	0.10	0.15	0.20	0.25	0.30	0.35	0.40	0.45	0.50	0.55	0.60	0.65	0.70	0.75	0.80	0.85	0.90	0.95
										p											
8	0	0.9227	0.6634	0.4305	0.2725	0.1678	0.1001	0.0576	0.0319	0.0168	0.0084	0.0039	0.0017	0.0007	0.0002	0.0001	0.0000+	0.0000+	0.0000+	0.0000+	0.0000+
	1	0.9973	0.9428	0.8131	0.6572	0.5033	0.3671	0.2553	0.1691	0.1064	0.0632	0.0352	0.0181	0.0085	0.0036	0.0013	0.0004	0.0001	0.0000+	0.0000+	0.0000+
	2	0.9999	0.9942	0.9619	0.8948	0.7969	0.6785	0.5518	0.4278	0.3154	0.2201	0.1445	0.0885	0.0498	0.0253	0.0113	0.0042	0.0012	0.0002	0.0000+	0.0000+
	3	1.0000-	0.9996	0.9950	0.9786	0.9437	0.8862	0.8059	0.7064	0.5941	0.4770	0.3633	0.2604	0.1737	0.1061	0.0580	0.0273	0.0104	0.0029	0.0004	0.0000+
	4	1.0000-	1.0000-	0.9996	0.9971	0.9896	0.9727	0.9420	0.8939	0.8263	0.7396	0.6367	0.5230	0.4059	0.2936	0.1941	0.1138	0.0563	0.0214	0.0050	0.0004
	5	1.0000-	1.0000-	1.0000-	0.9998	0.9988	0.9958	0.9887	0.9747	0.9502	0.9115	0.8556	0.7799	0.6846	0.5722	0.4482	0.3215	0.2031	0.1052	0.0381	0.0058
	6	1.0000-	1.0000-	1.0000-	1.0000-	0.9999	0.9996	0.9987	0.9964	0.9915	0.9819	0.9648	0.9368	0.8936	0.8309	0.7447	0.6329	0.4967	0.3428	0.1869	0.0572
	7	1.0000-	1.0000-	1.0000-	1.0000-	1.0000-	1.0000-	0.9999	0.9998	0.9993	0.9983	0.9961	0.9916	0.9832	0.9681	0.9424	0.8999	0.8322	0.7275	0.5695	0.3366
	8	1.0000-	1.0000-	1.0000-	1.0000-	1.0000-	1.0000-	1.0000-	1.0000-	1.0000-	1.0000-	1.0000-	1.0000-	1.0000-	1.0000-	1.0000-	1.0000-	1.0000-	1.0000-	1.0000-	1.0000-
9	0	0.9135	0.6302	0.3874	0.2316	0.1342	0.0751	0.0404	0.0207	0.0101	0.0046	0.0020	0.0008	0.0003	0.0001	0.0000+	0.0000+	0.0000+	0.0000+	0.0000+	0.0000+
	1	0.9966	0.9288	0.7748	0.5995	0.4362	0.3003	0.1960	0.1211	0.0705	0.0385	0.0195	0.0091	0.0038	0.0014	0.0004	0.0001	0.0000+	0.0000+	0.0000+	0.0000+
	2	0.9999	0.9916	0.9470	0.8591	0.7382	0.6007	0.4628	0.3373	0.2318	0.1495	0.0898	0.0498	0.0250	0.0112	0.0043	0.0013	0.0003	0.0000+	0.0000+	0.0000+
	3	1.0000-	0.9994	0.9917	0.9661	0.9144	0.8343	0.7297	0.6089	0.4826	0.3614	0.2539	0.1658	0.0994	0.0536	0.0253	0.0100	0.0031	0.0006	0.0001	0.0000+
	4	1.0000-	1.0000-	0.9991	0.9944	0.9804	0.9511	0.9012	0.8283	0.7334	0.6214	0.5000	0.3786	0.2666	0.1717	0.0988	0.0489	0.0196	0.0056	0.0009	0.0000+
	5	1.0000-	1.0000-	0.9999	0.9994	0.9969	0.9900	0.9747	0.9464	0.9006	0.8342	0.7461	0.6386	0.5174	0.3911	0.2703	0.1657	0.0856	0.0339	0.0083	0.0006
	6	1.0000-	1.0000-	1.0000-	1.0000-	0.9997	0.9987	0.9957	0.9888	0.9750	0.9502	0.9102	0.8505	0.7682	0.6627	0.5372	0.3993	0.2618	0.1409	0.0530	0.0077
	7	1.0000-	1.0000-	1.0000-	1.0000-	1.0000-	0.9999	0.9996	0.9986	0.9962	0.9909	0.9805	0.9615	0.9295	0.8789	0.8040	0.6997	0.5638	0.4005	0.2252	0.0629
	8	1.0000-	1.0000-	1.0000-	1.0000-	1.0000-	1.0000-	1.0000-	0.9999	0.9997	0.9992	0.9980	0.9954	0.9899	0.9793	0.9596	0.9249	0.8658	0.7684	0.6126	0.2985
	9	1.0000-	1.0000-	1.0000-	1.0000-	1.0000-	1.0000-	1.0000-	1.0000-	1.0000-	1.0000-	1.0000-	1.0000-	1.0000-	1.0000-	1.0000-	1.0000-	1.0000-	1.0000-	1.0000-	1.0000-
10	0	0.9044	0.5987	0.3487	0.1969	0.1074	0.0563	0.0282	0.0135	0.0060	0.0025	0.0010	0.0003	0.0001	0.0000+	0.0000+	0.0000+	0.0000+	0.0000+	0.0000+	0.0000+
	1	0.9957	0.9139	0.7361	0.5443	0.3758	0.2440	0.1493	0.0860	0.0464	0.0233	0.0107	0.0046	0.0017	0.0005	0.0001	0.0000+	0.0000+	0.0000+	0.0000+	0.0000+
	2	0.9999	0.9885	0.9298	0.8202	0.6778	0.5256	0.3828	0.2616	0.1673	0.0996	0.0547	0.0274	0.0123	0.0048	0.0016	0.0004	0.0001	0.0000+	0.0000+	0.0000+
	3	1.0000-	0.9990	0.9872	0.9500	0.8791	0.7759	0.6496	0.5138	0.3823	0.2660	0.1719	0.1020	0.0548	0.0280	0.0106	0.0035	0.0009	0.0001	0.0000+	0.0000+
	4	1.0000-	0.9999	0.9984	0.9901	0.9672	0.9219	0.8497	0.7515	0.6331	0.5044	0.3770	0.2616	0.1662	0.0949	0.0473	0.0197	0.0064	0.0014	0.0001	0.0000+
	5	1.0000-	1.0000-	0.9999	0.9986	0.9936	0.9803	0.9527	0.9051	0.8338	0.7384	0.6230	0.4956	0.3669	0.2485	0.1503	0.0781	0.0328	0.0099	0.0016	0.0001
	6	1.0000-	1.0000-	1.0000-	0.9999	0.9991	0.9965	0.9894	0.9740	0.9452	0.8980	0.8281	0.7340	0.6177	0.4862	0.3504	0.2241	0.1209	0.0500	0.0128	0.0010
	7	1.0000-	1.0000-	1.0000-	1.0000-	0.9999	0.9996	0.9984	0.9952	0.9877	0.9726	0.9453	0.9004	0.8327	0.7384	0.6172	0.4744	0.3222	0.1798	0.0702	0.0115
	8	1.0000-	1.0000-	1.0000-	1.0000-	1.0000-	1.0000-	0.9999	0.9995	0.9983	0.9955	0.9893	0.9767	0.9536	0.9140	0.8507	0.7560	0.6242	0.4557	0.2639	0.0861
	9	1.0000-	1.0000-	1.0000-	1.0000-	1.0000-	1.0000-	1.0000-	1.0000-	0.9999	0.9997	0.9990	0.9975	0.9940	0.9865	0.9718	0.9437	0.8926	0.8031	0.6513	0.4013
	10	1.0000-	1.0000-	1.0000-	1.0000-	1.0000-	1.0000-	1.0000-	1.0000-	1.0000-	1.0000-	1.0000-	1.0000-	1.0000-	1.0000-	1.0000-	1.0000-	1.0000-	1.0000-	1.0000-	1.0000-

Note: 0.0000+ means the probability is 0.0000 rounded to four decimal places. However, the probability is *not zero*. 1.0000− means the probability is 1.0000 rounded to four decimal places. However, the probability is *not one*.

Table III (continued)

n	x	0.01	0.05	0.10	0.15	0.20	0.25	0.30	0.35	0.40	0.45	0.50	0.55	0.60	0.65	0.70	0.75	0.80	0.85	0.90	0.95
11	0	0.8953	0.5688	0.3138	0.1673	0.0859	0.0422	0.0198	0.0088	0.0036	0.0014	0.0005	0.0002	0.0000+	0.0000+	0.0000+	0.0000+	0.0000+	0.0000+	0.0000+	0.0000+
	1	0.9948	0.8981	0.6974	0.4922	0.3221	0.1971	0.1130	0.0606	0.0302	0.0139	0.0059	0.0022	0.0007	0.0002	0.0000+	0.0000+	0.0000+	0.0000+	0.0000+	0.0000+
	2	0.9998	0.9848	0.9104	0.7788	0.6174	0.4552	0.3127	0.2001	0.1189	0.0652	0.0327	0.0148	0.0059	0.0020	0.0006	0.0001	0.0000+	0.0000+	0.0000+	0.0000+
	3	1.0000-	0.9984	0.9815	0.9306	0.8389	0.7133	0.5696	0.4256	0.2963	0.1911	0.1133	0.0610	0.0293	0.0122	0.0043	0.0012	0.0002	0.0000+	0.0000+	0.0000+
	4	1.0000-	0.9999	0.9972	0.9841	0.9496	0.8854	0.7897	0.6683	0.5326	0.3971	0.2744	0.1738	0.0994	0.0501	0.0216	0.0076	0.0020	0.0003	0.0000+	0.0000+
	5	1.0000-	1.0000-	0.9997	0.9973	0.9883	0.9657	0.9218	0.8513	0.7535	0.6331	0.5000	0.3669	0.2465	0.1487	0.0782	0.0343	0.0117	0.0027	0.0003	0.0000+
	6	1.0000-	1.0000-	1.0000-	0.9997	0.9980	0.9924	0.9784	0.9499	0.9006	0.8262	0.7258	0.6029	0.4672	0.3317	0.2103	0.1146	0.0504	0.0159	0.0028	0.0001
	7	1.0000-	1.0000-	1.0000-	1.0000-	0.9998	0.9988	0.9957	0.9878	0.9707	0.9390	0.8867	0.8089	0.7037	0.5744	0.4304	0.2867	0.1611	0.0694	0.0185	0.0016
	8	1.0000-	1.0000-	1.0000-	1.0000-	1.0000-	0.9999	0.9994	0.9980	0.9941	0.9852	0.9673	0.9348	0.8811	0.7999	0.6873	0.5448	0.3826	0.2212	0.0896	0.0152
	9	1.0000-	1.0000-	1.0000-	1.0000-	1.0000-	1.0000-	1.0000-	0.9998	0.9993	0.9976	0.9941	0.9861	0.9698	0.9394	0.8870	0.8029	0.6779	0.5078	0.3028	0.1019
	10	1.0000-	1.0000-	1.0000-	1.0000-	1.0000-	1.0000-	1.0000-	1.0000-	1.0000-	0.9996	0.9995	0.9986	0.9964	0.9912	0.9802	0.9578	0.9141	0.8327	0.6862	0.4312
	11	1.0000-	1.0000-	1.0000-	1.0000-	1.0000-	1.0000-	1.0000-	1.0000-	1.0000-	1.0000-	1.0000-	1.0000-	1.0000-	1.0000-	1.0000-	1.0000-	1.0000-	1.0000-	1.0000-	1.0000-
12	0	0.8864	0.5404	0.2824	0.1422	0.0687	0.0317	0.0138	0.0057	0.0022	0.0008	0.0002	0.0001	0.0000+	0.0000+	0.0000+	0.0000+	0.0000+	0.0000+	0.0000+	0.0000+
	1	0.9938	0.8816	0.6590	0.4435	0.2749	0.1584	0.0850	0.0424	0.0196	0.0083	0.0032	0.0011	0.0003	0.0001	0.0000+	0.0000+	0.0000+	0.0000+	0.0000+	0.0000+
	2	0.9998	0.9804	0.8891	0.7358	0.5583	0.3907	0.2528	0.1513	0.0834	0.0421	0.0193	0.0079	0.0028	0.0008	0.0002	0.0000+	0.0000+	0.0000+	0.0000+	0.0000+
	3	1.0000-	0.9978	0.9744	0.9078	0.7946	0.6488	0.4925	0.3467	0.2253	0.1345	0.0730	0.0356	0.0153	0.0056	0.0017	0.0004	0.0001	0.0000+	0.0000+	0.0000+
	4	1.0000-	0.9998	0.9957	0.9761	0.9274	0.8424	0.7237	0.5833	0.4382	0.3044	0.1938	0.1117	0.0573	0.0255	0.0095	0.0028	0.0006	0.0001	0.0000+	0.0000+
	5	1.0000-	1.0000-	0.9995	0.9954	0.9806	0.9456	0.8822	0.7873	0.6652	0.5269	0.3872	0.2607	0.1582	0.0846	0.0386	0.0143	0.0039	0.0007	0.0001	0.0000+
	6	1.0000-	1.0000-	0.9999	0.9993	0.9961	0.9857	0.9614	0.9154	0.8418	0.7393	0.6128	0.4731	0.3348	0.2127	0.1178	0.0544	0.0194	0.0046	0.0005	0.0000+
	7	1.0000-	1.0000-	1.0000-	0.9999	0.9994	0.9972	0.9905	0.9745	0.9427	0.8883	0.8062	0.6956	0.5618	0.4167	0.2763	0.1576	0.0726	0.0239	0.0043	0.0002
	8	1.0000-	1.0000-	1.0000-	1.0000-	0.9999	0.9996	0.9983	0.9944	0.9847	0.9644	0.9270	0.8655	0.7747	0.6533	0.5075	0.3512	0.2054	0.0922	0.0256	0.0022
	9	1.0000-	1.0000-	1.0000-	1.0000-	1.0000-	1.0000-	0.9998	0.9992	0.9972	0.9921	0.9807	0.9579	0.9166	0.8487	0.7472	0.6093	0.4417	0.2642	0.1109	0.0196
	10	1.0000-	1.0000-	1.0000-	1.0000-	1.0000-	1.0000-	1.0000-	0.9999	0.9997	0.9989	0.9968	0.9917	0.9804	0.9576	0.9150	0.8416	0.7251	0.5565	0.3410	0.1184
	11	1.0000-	1.0000-	1.0000-	1.0000-	1.0000-	1.0000-	1.0000-	1.0000-	1.0000-	0.9999	0.9998	0.9992	0.9978	0.9943	0.9862	0.9683	0.9313	0.8578	0.7176	0.4596
	12	1.0000-	1.0000-	1.0000-	1.0000-	1.0000-	1.0000-	1.0000-	1.0000-	1.0000-	1.0000-	1.0000-	1.0000-	1.0000-	1.0000-	1.0000-	1.0000-	1.0000-	1.0000-	1.0000-	1.0000-

Note: 0.0000+ means the probability is 0.0000 rounded to four decimal places. However, the probability is *not* zero. 1.0000− means the probability is 1.0000 rounded to four decimal places. However, the probability is *not* one.

Table III (continued)

p

n = 15

x	0.01	0.05	0.10	0.15	0.20	0.25	0.30	0.35	0.40	0.45	0.50	0.55	0.60	0.65	0.70	0.75	0.80	0.85	0.90	0.95
0	0.8601	0.4633	0.2059	0.0874	0.0352	0.0134	0.0047	0.0016	0.0005	0.0001	0.0000+	0.0000+	0.0000+	0.0000+	0.0000+	0.0000+	0.0000+	0.0000+	0.0000+	0.0000+
1	0.9904	0.8290	0.5490	0.3186	0.1671	0.0802	0.0353	0.0142	0.0052	0.0017	0.0005	0.0001	0.0000+	0.0000+	0.0000+	0.0000+	0.0000+	0.0000+	0.0000+	0.0000+
2	0.9996	0.9638	0.8159	0.6042	0.3980	0.2361	0.1268	0.0617	0.0271	0.0107	0.0037	0.0011	0.0003	0.0001	0.0000+	0.0000+	0.0000+	0.0000+	0.0000+	0.0000+
3	1.0000−	0.9945	0.9444	0.8227	0.6482	0.4613	0.2969	0.1727	0.0905	0.0424	0.0176	0.0063	0.0019	0.0005	0.0001	0.0000+	0.0000+	0.0000+	0.0000+	0.0000+
4	1.0000−	0.9994	0.9873	0.9383	0.8358	0.6865	0.5155	0.3519	0.2173	0.1204	0.0592	0.0255	0.0093	0.0028	0.0007	0.0001	0.0000+	0.0000+	0.0000+	0.0000+
5	1.0000−	0.9999	0.9978	0.9832	0.9389	0.8516	0.7216	0.5643	0.4032	0.2608	0.1509	0.0769	0.0338	0.0124	0.0037	0.0008	0.0001	0.0000+	0.0000+	0.0000+
6	1.0000−	1.0000−	0.9997	0.9964	0.9819	0.9434	0.8689	0.7548	0.6098	0.4522	0.3036	0.1818	0.0950	0.0422	0.0152	0.0042	0.0008	0.0001	0.0000+	0.0000+
7	1.0000−	1.0000−	1.0000−	0.9994	0.9958	0.9827	0.9500	0.8868	0.7869	0.6535	0.5000	0.3465	0.2131	0.1132	0.0500	0.0173	0.0042	0.0006	0.0000+	0.0000+
8	1.0000−	1.0000−	1.0000−	0.9999	0.9992	0.9958	0.9848	0.9578	0.9050	0.8182	0.6964	0.5478	0.3902	0.2452	0.1311	0.0566	0.0181	0.0036	0.0003	0.0000+
9	1.0000−	1.0000−	1.0000−	1.0000−	0.9999	0.9992	0.9963	0.9876	0.9662	0.9231	0.8491	0.7392	0.5968	0.4357	0.2784	0.1484	0.0611	0.0168	0.0022	0.0001
10	1.0000−	1.0000−	1.0000−	1.0000−	1.0000−	0.9999	0.9993	0.9972	0.9907	0.9745	0.9408	0.8796	0.7827	0.6481	0.4845	0.3135	0.1642	0.0617	0.0127	0.0006
11	1.0000−	1.0000−	1.0000−	1.0000−	1.0000−	1.0000−	0.9999	0.9995	0.9981	0.9937	0.9824	0.9576	0.9095	0.8273	0.7031	0.5387	0.3518	0.1773	0.0556	0.0055
12	1.0000−	1.0000−	1.0000−	1.0000−	1.0000−	1.0000−	1.0000−	0.9999	0.9997	0.9989	0.9963	0.9893	0.9729	0.9383	0.8732	0.7639	0.6020	0.3958	0.1841	0.0362
13	1.0000−	1.0000−	1.0000−	1.0000−	1.0000−	1.0000−	1.0000−	1.0000−	1.0000−	0.9999	0.9995	0.9983	0.9948	0.9858	0.9647	0.9198	0.8329	0.6814	0.4510	0.1710
14	1.0000−	1.0000−	1.0000−	1.0000−	1.0000−	1.0000−	1.0000−	1.0000−	1.0000−	1.0000−	1.0000−	0.9999	0.9995	0.9984	0.9953	0.9866	0.9648	0.9126	0.7941	0.5367
15	1.0000−	1.0000−	1.0000−	1.0000−	1.0000−	1.0000−	1.0000−	1.0000−	1.0000−	1.0000−	1.0000−	1.0000−	1.0000−	1.0000−	1.0000−	1.0000−	1.0000−	1.0000−	1.0000−	1.0000−

n = 20

x	0.01	0.05	0.10	0.15	0.20	0.25	0.30	0.35	0.40	0.45	0.50	0.55	0.60	0.65	0.70	0.75	0.80	0.85	0.90	0.95
0	0.8179	0.3585	0.1216	0.0388	0.0115	0.0032	0.0008	0.0002	0.0000+	0.0000+	0.0000+	0.0000+	0.0000+	0.0000+	0.0000+	0.0000+	0.0000+	0.0000+	0.0000+	0.0000+
1	0.9831	0.7358	0.3917	0.1756	0.0692	0.0243	0.0076	0.0021	0.0005	0.0001	0.0000+	0.0000+	0.0000+	0.0000+	0.0000+	0.0000+	0.0000+	0.0000+	0.0000+	0.0000+
2	0.9990	0.9245	0.6769	0.4049	0.2061	0.0913	0.0355	0.0121	0.0036	0.0009	0.0002	0.0000+	0.0000+	0.0000+	0.0000+	0.0000+	0.0000+	0.0000+	0.0000+	0.0000+
3	1.0000−	0.9841	0.8670	0.6477	0.4114	0.2252	0.1071	0.0444	0.0160	0.0049	0.0013	0.0003	0.0000+	0.0000+	0.0000+	0.0000+	0.0000+	0.0000+	0.0000+	0.0000+
4	1.0000−	0.9974	0.9568	0.8298	0.6296	0.4148	0.2375	0.1182	0.0510	0.0189	0.0059	0.0015	0.0003	0.0000+	0.0000+	0.0000+	0.0000+	0.0000+	0.0000+	0.0000+
5	1.0000−	0.9997	0.9887	0.9327	0.8042	0.6172	0.4164	0.2454	0.1256	0.0553	0.0207	0.0064	0.0016	0.0003	0.0000+	0.0000+	0.0000+	0.0000+	0.0000+	0.0000+
6	1.0000−	1.0000−	0.9976	0.9781	0.9133	0.7858	0.6080	0.4166	0.2500	0.1299	0.0577	0.0214	0.0065	0.0015	0.0003	0.0000+	0.0000+	0.0000+	0.0000+	0.0000+
7	1.0000−	1.0000−	0.9996	0.9941	0.9679	0.8982	0.7723	0.6010	0.4159	0.2520	0.1316	0.0580	0.0210	0.0060	0.0013	0.0002	0.0000+	0.0000+	0.0000+	0.0000+
8	1.0000−	1.0000−	0.9999	0.9987	0.9900	0.9591	0.8867	0.7624	0.5956	0.4143	0.2517	0.1308	0.0565	0.0196	0.0051	0.0009	0.0001	0.0000+	0.0000+	0.0000+
9	1.0000−	1.0000−	1.0000−	0.9998	0.9974	0.9861	0.9520	0.8782	0.7553	0.5914	0.4119	0.2493	0.1275	0.0532	0.0171	0.0039	0.0006	0.0000+	0.0000+	0.0000+
10	1.0000−	1.0000−	1.0000−	1.0000−	0.9994	0.9961	0.9829	0.9468	0.8725	0.7507	0.5881	0.4086	0.2447	0.1218	0.0480	0.0139	0.0026	0.0002	0.0000+	0.0000+
11	1.0000−	1.0000−	1.0000−	1.0000−	0.9999	0.9991	0.9949	0.9804	0.9435	0.8692	0.7483	0.5857	0.4044	0.2376	0.1133	0.0409	0.0100	0.0013	0.0001	0.0000+
12	1.0000−	1.0000−	1.0000−	1.0000−	1.0000−	0.9998	0.9987	0.9940	0.9790	0.9420	0.8684	0.7480	0.5841	0.3990	0.2277	0.1018	0.0321	0.0059	0.0004	0.0000+
13	1.0000−	1.0000−	1.0000−	1.0000−	1.0000−	1.0000−	0.9997	0.9985	0.9935	0.9786	0.9423	0.8701	0.7500	0.5834	0.3920	0.2142	0.0867	0.0219	0.0024	0.0000+
14	1.0000−	1.0000−	1.0000−	1.0000−	1.0000−	1.0000−	1.0000−	0.9997	0.9984	0.9936	0.9793	0.9447	0.8744	0.7546	0.5836	0.3828	0.1958	0.0673	0.0113	0.0003
15	1.0000−	1.0000−	1.0000−	1.0000−	1.0000−	1.0000−	1.0000−	1.0000−	0.9997	0.9985	0.9941	0.9811	0.9490	0.8818	0.7625	0.5852	0.3704	0.1702	0.0432	0.0026
16	1.0000−	1.0000−	1.0000−	1.0000−	1.0000−	1.0000−	1.0000−	1.0000−	1.0000−	0.9997	0.9987	0.9951	0.9840	0.9556	0.8929	0.7748	0.5886	0.3523	0.1330	0.0159
17	1.0000−	1.0000−	1.0000−	1.0000−	1.0000−	1.0000−	1.0000−	1.0000−	1.0000−	1.0000−	0.9998	0.9991	0.9964	0.9879	0.9645	0.9087	0.7939	0.5951	0.3231	0.0755
18	1.0000−	1.0000−	1.0000−	1.0000−	1.0000−	1.0000−	1.0000−	1.0000−	1.0000−	1.0000−	1.0000−	0.9999	0.9995	0.9979	0.9924	0.9757	0.9308	0.8244	0.6083	0.2642
19	1.0000−	1.0000−	1.0000−	1.0000−	1.0000−	1.0000−	1.0000−	1.0000−	1.0000−	1.0000−	1.0000−	1.0000−	1.0000−	0.9998	0.9992	0.9968	0.9885	0.9612	0.8784	0.6415
20	1.0000−	1.0000−	1.0000−	1.0000−	1.0000−	1.0000−	1.0000−	1.0000−	1.0000−	1.0000−	1.0000−	1.0000−	1.0000−	1.0000−	1.0000−	1.0000−	1.0000−	1.0000−	1.0000−	1.0000−

Note: 0.0000+ means the probability is 0.0000 rounded to four decimal places. However, the probability is *not* zero. 1.0000− means the probability is 1.0000 rounded to four decimal places. However, the probability is *not* one.

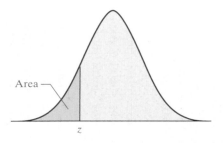

Area

z

Table IV

Standard Normal Distribution

z	.00	.01	.02	.03	.04	.05	.06	.07	.08	.09
−3.4	0.0003	0.0003	0.0003	0.0003	0.0003	0.0003	0.0003	0.0003	0.0003	0.0002
−3.3	0.0005	0.0005	0.0005	0.0004	0.0004	0.0004	0.0004	0.0004	0.0004	0.0003
−3.2	0.0007	0.0007	0.0006	0.0006	0.0006	0.0006	0.0006	0.0005	0.0005	0.0005
−3.1	0.0010	0.0009	0.0009	0.0009	0.0008	0.0008	0.0008	0.0008	0.0007	0.0007
−3.0	0.0013	0.0013	0.0013	0.0012	0.0012	0.0011	0.0011	0.0011	0.0010	0.0010
−2.9	0.0019	0.0018	0.0018	0.0017	0.0016	0.0016	0.0015	0.0015	0.0014	0.0014
−2.8	0.0026	0.0025	0.0024	0.0023	0.0023	0.0022	0.0021	0.0021	0.0020	0.0019
−2.7	0.0035	0.0034	0.0033	0.0032	0.0031	0.0030	0.0029	0.0028	0.0027	0.0026
−2.6	0.0047	0.0045	0.0044	0.0043	0.0041	0.0040	0.0039	0.0038	0.0037	0.0036
−2.5	0.0062	0.0060	0.0059	0.0057	0.0055	0.0054	0.0052	0.0051	0.0049	0.0048
−2.4	0.0082	0.0080	0.0078	0.0075	0.0073	0.0071	0.0069	0.0068	0.0066	0.0064
−2.3	0.0107	0.0104	0.0102	0.0099	0.0096	0.0094	0.0091	0.0089	0.0087	0.0084
−2.2	0.0139	0.0136	0.0132	0.0129	0.0125	0.0122	0.0119	0.0116	0.0113	0.0110
−2.1	0.0179	0.0174	0.0170	0.0166	0.0162	0.0158	0.0154	0.0150	0.0146	0.0143
−2.0	0.0228	0.0222	0.0217	0.0212	0.0207	0.0202	0.0197	0.0192	0.0188	0.0183
−1.9	0.0287	0.0281	0.0274	0.0268	0.0262	0.0256	0.0250	0.0244	0.0239	0.0233
−1.8	0.0359	0.0351	0.0344	0.0336	0.0329	0.0322	0.0314	0.0307	0.0301	0.0294
−1.7	0.0446	0.0436	0.0427	0.0418	0.0409	0.0401	0.0392	0.0384	0.0375	0.0367
−1.6	0.0548	0.0537	0.0526	0.0516	0.0505	0.0495	0.0485	0.0475	0.0465	0.0455
−1.5	0.0668	0.0655	0.0643	0.0630	0.0618	0.0606	0.0594	0.0582	0.0571	0.0559
−1.4	0.0808	0.0793	0.0778	0.0764	0.0749	0.0735	0.0721	0.0708	0.0694	0.0681
−1.3	0.0968	0.0951	0.0934	0.0918	0.0901	0.0885	0.0869	0.0853	0.0838	0.0823
−1.2	0.1151	0.1131	0.1112	0.1093	0.1075	0.1056	0.1038	0.1020	0.1003	0.0985
−1.1	0.1357	0.1335	0.1314	0.1292	0.1271	0.1251	0.1230	0.1210	0.1190	0.1170
−1.0	0.1587	0.1562	0.1539	0.1515	0.1492	0.1469	0.1446	0.1423	0.1401	0.1379
−0.9	0.1841	0.1814	0.1788	0.1762	0.1736	0.1711	0.1685	0.1660	0.1635	0.1611
−0.8	0.2119	0.2090	0.2061	0.2033	0.2005	0.1977	0.1949	0.1922	0.1894	0.1867
−0.7	0.2420	0.2389	0.2358	0.2327	0.2296	0.2266	0.2236	0.2206	0.2177	0.2148
−0.6	0.2743	0.2709	0.2676	0.2643	0.2611	0.2578	0.2546	0.2514	0.2483	0.2451
−0.5	0.3085	0.3050	0.3015	0.2981	0.2946	0.2912	0.2877	0.2843	0.2810	0.2776
−0.4	0.3446	0.3409	0.3372	0.3336	0.3300	0.3264	0.3228	0.3192	0.3156	0.3121
−0.3	0.3821	0.3783	0.3745	0.3707	0.3669	0.3632	0.3594	0.3557	0.3520	0.3483
−0.2	0.4207	0.4168	0.4129	0.4090	0.4052	0.4013	0.3974	0.3936	0.3897	0.3859
−0.1	0.4602	0.4562	0.4522	0.4483	0.4443	0.4404	0.4364	0.4325	0.4286	0.4247
−0.0	0.5000	0.4960	0.4920	0.4880	0.4840	0.4801	0.4761	0.4721	0.4681	0.4641

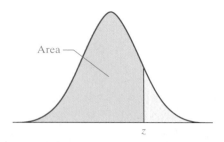

Area

z

Table IV (continued)

Standard Normal Distribution

z	.00	.01	.02	.03	.04	.05	.06	.07	.08	.09
0.0	0.5000	0.5040	0.5080	0.5120	0.5160	0.5199	0.5239	0.5279	0.5319	0.5359
0.1	0.5398	0.5438	0.5478	0.5517	0.5557	0.5596	0.5636	0.5675	0.5714	0.5753
0.2	0.5793	0.5832	0.5871	0.5910	0.5948	0.5987	0.6026	0.6064	0.6103	0.6141
0.3	0.6179	0.6217	0.6255	0.6293	0.6331	0.6368	0.6406	0.6443	0.6480	0.6517
0.4	0.6554	0.6591	0.6628	0.6664	0.6700	0.6736	0.6772	0.6808	0.6844	0.6879
0.5	0.6915	0.6950	0.6985	0.7019	0.7054	0.7088	0.7123	0.7157	0.7190	0.7224
0.6	0.7257	0.7291	0.7324	0.7357	0.7389	0.7422	0.7454	0.7486	0.7517	0.7549
0.7	0.7580	0.7611	0.7642	0.7673	0.7704	0.7734	0.7764	0.7794	0.7823	0.7852
0.8	0.7881	0.7910	0.7939	0.7967	0.7995	0.8023	0.8051	0.8078	0.8106	0.8133
0.9	0.8159	0.8186	0.8212	0.8238	0.8264	0.8289	0.8315	0.8340	0.8365	0.8389
1.0	0.8413	0.8438	0.8461	0.8485	0.8508	0.8531	0.8554	0.8577	0.8599	0.8621
1.1	0.8643	0.8665	0.8686	0.8708	0.8729	0.8749	0.8770	0.8790	0.8810	0.8830
1.2	0.8849	0.8869	0.8888	0.8907	0.8925	0.8944	0.8962	0.8980	0.8997	0.9015
1.3	0.9032	0.9049	0.9066	0.9082	0.9099	0.9115	0.9131	0.9147	0.9162	0.9177
1.4	0.9192	0.9207	0.9222	0.9236	0.9251	0.9265	0.9279	0.9292	0.9306	0.9319
1.5	0.9332	0.9345	0.9357	0.9370	0.9382	0.9394	0.9406	0.9418	0.9429	0.9441
1.6	0.9452	0.9463	0.9474	0.9484	0.9495	0.9505	0.9515	0.9525	0.9535	0.9545
1.7	0.9554	0.9564	0.9573	0.9582	0.9591	0.9599	0.9608	0.9616	0.9625	0.9633
1.8	0.9641	0.9649	0.9656	0.9664	0.9671	0.9678	0.9686	0.9693	0.9699	0.9706
1.9	0.9713	0.9719	0.9726	0.9732	0.9738	0.9744	0.9750	0.9756	0.9761	0.9767
2.0	0.9772	0.9778	0.9783	0.9788	0.9793	0.9798	0.9803	0.9808	0.9812	0.9817
2.1	0.9821	0.9826	0.9830	0.9834	0.9838	0.9842	0.9846	0.9850	0.9854	0.9857
2.2	0.9861	0.9864	0.9868	0.9871	0.9875	0.9878	0.9881	0.9884	0.9887	0.9890
2.3	0.9893	0.9896	0.9898	0.9901	0.9904	0.9906	0.9909	0.9911	0.9913	0.9916
2.4	0.9918	0.9920	0.9922	0.9925	0.9927	0.9929	0.9931	0.9932	0.9934	0.9936
2.5	0.9938	0.9940	0.9941	0.9943	0.9945	0.9946	0.9948	0.9949	0.9951	0.9952
2.6	0.9953	0.9955	0.9956	0.9957	0.9959	0.9960	0.9961	0.9962	0.9963	0.9964
2.7	0.9965	0.9966	0.9967	0.9968	0.9969	0.9970	0.9971	0.9972	0.9973	0.9974
2.8	0.9974	0.9975	0.9976	0.9977	0.9977	0.9978	0.9979	0.9979	0.9980	0.9981
2.9	0.9981	0.9982	0.9982	0.9983	0.9984	0.9984	0.9985	0.9985	0.9986	0.9986
3.0	0.9987	0.9987	0.9987	0.9988	0.9988	0.9989	0.9989	0.9989	0.9990	0.9990
3.1	0.9990	0.9991	0.9991	0.9991	0.9992	0.9992	0.9992	0.9992	0.9993	0.9993
3.2	0.9993	0.9993	0.9994	0.9994	0.9994	0.9994	0.9994	0.9995	0.9995	0.9995
3.3	0.9995	0.9995	0.9995	0.9996	0.9996	0.9996	0.9996	0.9996	0.9996	0.9997
3.4	0.9997	0.9997	0.9997	0.9997	0.9997	0.9997	0.9997	0.9997	0.9997	0.9998

Area in right tail

Table V												

t-Distribution
Area in Right Tail

df	0.25	0.20	0.15	0.10	0.05	0.025	0.02	0.01	0.005	0.0025	0.001	0.0005
1	1.000	1.376	1.963	3.078	6.314	12.706	15.894	31.821	63.657	127.321	318.289	636.558
2	0.816	1.061	1.386	1.886	2.920	4.303	4.849	6.965	9.925	14.089	22.328	31.600
3	0.765	0.978	1.250	1.638	2.353	3.182	3.482	4.541	5.841	7.453	10.214	12.924
4	0.741	0.941	1.190	1.533	2.132	2.776	2.999	3.747	4.604	5.598	7.173	8.610
5	0.727	0.920	1.156	1.476	2.015	2.571	2.757	3.365	4.032	4.773	5.893	6.869
6	0.718	0.906	1.134	1.440	1.943	2.447	2.612	3.143	3.707	4.317	5.208	5.959
7	0.711	0.896	1.119	1.415	1.895	2.365	2.517	2.998	3.499	4.029	4.785	5.408
8	0.706	0.889	1.108	1.397	1.860	2.306	2.449	2.896	3.355	3.833	4.501	5.041
9	0.703	0.883	1.100	1.383	1.833	2.262	2.398	2.821	3.250	3.690	4.297	4.781
10	0.700	0.879	1.093	1.372	1.812	2.228	2.359	2.764	3.169	3.581	4.144	4.587
11	0.697	0.876	1.088	1.363	1.796	2.201	2.328	2.718	3.106	3.497	4.025	4.437
12	0.695	0.873	1.083	1.356	1.782	2.179	2.303	2.681	3.055	3.428	3.930	4.318
13	0.694	0.870	1.079	1.350	1.771	2.160	2.282	2.650	3.012	3.372	3.852	4.221
14	0.692	0.868	1.076	1.345	1.761	2.145	2.264	2.624	2.977	3.326	3.787	4.140
15	0.691	0.866	1.074	1.341	1.753	2.131	2.249	2.602	2.947	3.286	3.733	4.073
16	0.690	0.865	1.071	1.337	1.746	2.120	2.235	2.583	2.921	3.252	3.686	4.015
17	0.689	0.863	1.069	1.333	1.740	2.110	2.224	2.567	2.898	3.222	3.646	3.965
18	0.688	0.862	1.067	1.330	1.734	2.101	2.214	2.552	2.878	3.197	3.611	3.922
19	0.688	0.861	1.066	1.328	1.729	2.093	2.205	2.539	2.861	3.174	3.579	3.883
20	0.687	0.860	1.064	1.325	1.725	2.086	2.197	2.528	2.845	3.153	3.552	3.850
21	0.686	0.859	1.063	1.323	1.721	2.080	2.189	2.518	2.831	3.135	3.527	3.819
22	0.686	0.858	1.061	1.321	1.717	2.074	2.183	2.508	2.819	3.119	3.505	3.792
23	0.685	0.858	1.060	1.319	1.714	2.069	2.177	2.500	2.807	3.104	3.485	3.768
24	0.685	0.857	1.059	1.318	1.711	2.064	2.172	2.492	2.797	3.091	3.467	3.745
25	0.684	0.856	1.058	1.316	1.708	2.060	2.167	2.485	2.787	3.078	3.450	3.725
26	0.684	0.856	1.058	1.315	1.706	2.056	2.162	2.479	2.779	3.067	3.435	3.707
27	0.684	0.855	1.057	1.314	1.703	2.052	2.158	2.473	2.771	3.057	3.421	3.690
28	0.683	0.855	1.056	1.313	1.701	2.048	2.154	2.467	2.763	3.047	3.408	3.674
29	0.683	0.854	1.055	1.311	1.699	2.045	2.150	2.462	2.756	3.038	3.396	3.659
30	0.683	0.854	1.055	1.310	1.697	2.042	2.147	2.457	2.750	3.030	3.385	3.646
31	0.682	0.853	1.054	1.309	1.696	2.040	2.144	2.453	2.744	3.022	3.375	3.633
32	0.682	0.853	1.054	1.309	1.694	2.037	2.141	2.449	2.738	3.015	3.365	3.622
33	0.682	0.853	1.053	1.308	1.692	2.035	2.138	2.445	2.733	3.008	3.356	3.611
34	0.682	0.852	1.052	1.307	1.691	2.032	2.136	2.441	2.728	3.002	3.348	3.601
35	0.682	0.852	1.052	1.306	1.690	2.030	2.133	2.438	2.724	2.996	3.340	3.591
36	0.681	0.852	1.052	1.306	1.688	2.028	2.131	2.435	2.719	2.990	3.333	3.582
37	0.681	0.851	1.051	1.305	1.687	2.026	2.129	2.431	2.715	2.985	3.326	3.574
38	0.681	0.851	1.051	1.304	1.686	2.024	2.127	2.429	2.712	2.980	3.319	3.566
39	0.681	0.851	1.050	1.304	1.685	2.023	2.125	2.426	2.708	2.976	3.313	3.558
40	0.681	0.851	1.050	1.303	1.684	2.021	2.123	2.423	2.704	2.971	3.307	3.551
50	0.679	0.849	1.047	1.299	1.676	2.009	2.109	2.403	2.678	2.937	3.261	3.496
60	0.679	0.848	1.045	1.296	1.671	2.000	2.099	2.390	2.660	2.915	3.232	3.460
70	0.678	0.847	1.044	1.294	1.667	1.994	2.093	2.381	2.648	2.899	3.211	3.435
80	0.678	0.846	1.043	1.292	1.664	1.990	2.088	2.374	2.639	2.887	3.195	3.416
90	0.677	0.846	1.042	1.291	1.662	1.987	2.084	2.368	2.632	2.878	3.183	3.402
100	0.677	0.845	1.042	1.290	1.660	1.984	2.081	2.364	2.626	2.871	3.174	3.390
1000	0.675	0.842	1.037	1.282	1.646	1.962	2.056	2.330	2.581	2.813	3.098	3.300
z	0.674	0.841	1.036	1.282	1.645	1.960	2.054	2.326	2.576	2.807	3.091	3.291

Table VI

Chi-Square (χ^2) Distribution
Area to the Right of Critical Value

Degrees of Freedom	0.995	0.99	0.975	0.95	0.90	0.10	0.05	0.025	0.01	0.005
1	—	—	0.001	0.004	0.016	2.706	3.841	5.024	6.635	7.879
2	0.010	0.020	0.051	0.103	0.211	4.605	5.991	7.378	9.210	10.597
3	0.072	0.115	0.216	0.352	0.584	6.251	7.815	9.348	11.345	12.838
4	0.207	0.297	0.484	0.711	1.064	7.779	9.488	11.143	13.277	14.860
5	0.412	0.554	0.831	1.145	1.610	9.236	11.071	12.833	15.086	16.750
6	0.676	0.872	1.237	1.635	2.204	10.645	12.592	14.449	16.812	18.548
7	0.989	1.239	1.690	2.167	2.833	12.017	14.067	16.013	18.475	20.278
8	1.344	1.646	2.180	2.733	3.490	13.362	15.507	17.535	20.090	21.955
9	1.735	2.088	2.700	3.325	4.168	14.684	16.919	19.023	21.666	23.589
10	2.156	2.558	3.247	3.940	4.865	15.987	18.307	20.483	23.209	25.188
11	2.603	3.053	3.816	4.575	5.578	17.275	19.675	21.920	24.725	26.757
12	3.074	3.571	4.404	5.226	6.304	18.549	21.026	23.337	26.217	28.299
13	3.565	4.107	5.009	5.892	7.042	19.812	22.362	24.736	27.688	29.819
14	4.075	4.660	5.629	6.571	7.790	21.064	23.685	26.119	29.141	31.319
15	4.601	5.229	6.262	7.261	8.547	22.307	24.996	27.488	30.578	32.801
16	5.142	5.812	6.908	7.962	9.312	23.542	26.296	28.845	32.000	34.267
17	5.697	6.408	7.564	8.672	10.085	24.769	27.587	30.191	33.409	35.718
18	6.265	7.015	8.231	9.390	10.865	25.989	28.869	31.526	34.805	37.156
19	6.844	7.633	8.907	10.117	11.651	27.204	30.144	32.852	36.191	38.582
20	7.434	8.260	9.591	10.851	12.443	28.412	31.410	34.170	37.566	39.997
21	8.034	8.897	10.283	11.591	13.240	29.615	32.671	35.479	38.932	41.401
22	8.643	9.542	10.982	12.338	14.042	30.813	33.924	36.781	40.289	42.796
23	9.260	10.196	11.689	13.091	14.848	32.007	35.172	38.076	41.638	44.181
24	9.886	10.856	12.401	13.848	15.659	33.196	36.415	39.364	42.980	45.559
25	10.520	11.524	13.120	14.611	16.473	34.382	37.652	40.646	44.314	46.928
26	11.160	12.198	13.844	15.379	17.292	35.563	38.885	41.923	45.642	48.290
27	11.808	12.879	14.573	16.151	18.114	36.741	40.113	43.194	46.963	49.645
28	12.461	13.565	15.308	16.928	18.939	37.916	41.337	44.461	48.278	50.993
29	13.121	14.257	16.047	17.708	19.768	39.087	42.557	45.722	49.588	52.336
30	13.787	14.954	16.791	18.493	20.599	40.256	43.773	46.979	50.892	53.672
40	20.707	22.164	24.433	26.509	29.051	51.805	55.758	59.342	63.691	66.766
50	27.991	29.707	32.357	34.764	37.689	63.167	67.505	71.420	76.154	79.490
60	35.534	37.485	40.482	43.188	46.459	74.397	79.082	83.298	88.379	91.952
70	43.275	45.442	48.758	51.739	55.329	85.527	90.531	95.023	100.425	104.215
80	51.172	53.540	57.153	60.391	64.278	96.578	101.879	106.629	112.329	116.321
90	59.196	61.754	65.647	69.126	73.291	107.565	113.145	118.136	124.116	128.299
100	67.328	70.065	74.222	77.929	82.358	118.498	124.342	129.561	135.807	140.169

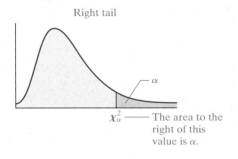

Right tail

χ^2_α —— The area to the right of this value is α.

Left tail
Area = $1-\alpha$

$\chi^2_{1-\alpha}$ —— The area to the right of this value is $1-\alpha$.

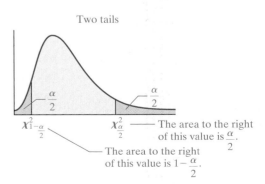

Two tails

$\frac{\alpha}{2}$ $\frac{\alpha}{2}$

$\chi^2_{1-\frac{\alpha}{2}}$ $\chi^2_{\frac{\alpha}{2}}$ —— The area to the right of this value is $\frac{\alpha}{2}$.

The area to the right of this value is $1-\frac{\alpha}{2}$.

Table VIII

Critical Values for Correlation Coefficient

n	
3	0.997
4	0.950
5	0.878
6	0.811
7	0.754
8	0.707
9	0.666
10	0.632
11	0.602
12	0.576
13	0.553
14	0.532
15	0.514
16	0.497
17	0.482
18	0.468
19	0.456
20	0.444
21	0.433
22	0.423
23	0.413
24	0.404
25	0.396
26	0.388
27	0.381
28	0.374
29	0.367
30	0.361

ANSWERS

CHAPTER 1 Data Collection

1.1 Assess Your Understanding (page 9)

1. Statistics is the science of collecting, organizing, summarizing, and analyzing information to answer questions or draw conclusions.

2. The group to be studied as defined by the research objective is the population. Any subset of the population is a sample.

3. information   4. individual   5. Descriptive; inferential

6. Neither the person conducting an experiment nor the experimental unit know which treatment they are receiving.

7. An experimental group usually receives some sort of experimental drug or treatment, while the control group receives a placebo or a treatment whose effect on the response variable is known.

8. Placebos are innocuous tablets such as sugar tablets. They are needed so that the treatments each group receives appear to be the same.

9. Variables

10. Qualitative variables describe characteristics of individuals. Quantitative variables provide numerical measures of individuals.

11. A discrete variable is a quantitative variable that has a finite or countable number of possible values. Continuous variables are also quantitative variables, but there are an infinite number of possible values that are not countable.

12. True

13. The process of statistics is to (1) identify the research objective, (2) collect the information needed to answer the questions posed in (1), (3) organize and summarize the information, and (4) draw conclusions from the information.

14. No. We measure age to as much accuracy as we wish.

15. Qualitative   16. Quantitative   17. Qualitative
18. Qualitative   19. Quantitative   20. Quantitative
21. Quantitative   22. Qualitative   23. Quantitative
24. Quantitative   25. Qualitative   26. Qualitative
27. Discrete   28. Continuous   29. Continuous
30. Discrete   31. Discrete   32. Continuous
33. Continuous   34. Continuous   35. Discrete
36. Discrete   37. Discrete   38. Continuous

39. Population: teenagers 13 to 17 years of age who live in the United States; Sample: 1,028 teenagers 13 to 17 years of age who live in the United States

40. Population: all Coca-Cola bottles filled on October 15; Sample: 50 Coca-Cola bottles filled on October 15

41. Population: entire soybean crop; Sample: 100 plants selected

42. Population: all United States households; Sample: 50,000 households surveyed

43. Population: women 27 to 44 years of age with hypertension; Sample: 7,373 women 27 to 44 years of age with hypertension

44. Population: full-time students enrolled at the community college; Sample: the 128 full-time students surveyed

45. (a) To determine if application of duct tape is effective as cryotherapy in the treatment of common warts
    (b) 51 patients with warts
    (c) 85% of patients in group 1 and 60% of the patients in group 2 had complete resolution of their warts.
    (d) Duct tape is significantly more effective in treating warts than cryotherapy.

46. (a) To determine if pregnant women in first-time labor could receive low-dose epidurals early without raising their chances of a Caesarean section
    (b) 728 women in first-time labor

(c) The C-section rate in the early epidural group was 18%, and the C-section rate in the delayed group was 21%.
(d) Pregnant women in first-time labor can be given low-dose epidural early without raising their chances of a C-section.

47. (a) To determine whether music cognition and cognitions pertaining to abstract operations such as mathematical or spatial reasoning and related
    (b) Thirty-six college students
    (c) The mean test score following the Mozart piece was 119, while the mean test score following silence was 110.
    (d) Subjects perform better on abstract and spatial reasoning tests after listening to Mozart.

48. (a) To determine whom adult Americans believe was the greatest president
    (b) 800 adult Americans
    (c) 20% chose Lincoln, 15% chose Reagan, 12% chose F. Roosevelt, 11% chose Kennedy, 10% chose Clinton, 8% chose G.W. Bush, 6% chose Washington, 18% chose other presidents.
    (d) Lincoln ranks first among all Americans as the greatest president.

49. (a) To determine the number of movies U.S. adults aged 18 years or older have attended in a movie theater in the past 12 months
    (b) 1,003 U.S. adults 18 years or older
    (c) About 65% of the 1,003 adults surveyed stated that they have attended at least one movie in a movie theater in the past 12 months.
    (d) 65% of all adult Americans 18 years old or older have attended at least one movie in a movie theater in the past 12 months.

50. (a) To determine if U.S. adults aged 18 years or older believe that having a baby outside of marriage is morally acceptable or morally wrong
    (b) 1,000 U.S. adults aged 18 years or older
    (c) 49% of the 1,000 adults surveyed said having a baby outside of marriage is morally acceptable.
    (d) 49% of all American adults aged 18 years of age or older believe that having a baby outside of marriage is morally acceptable.

51. Individuals: Sanyo #PDP42H2W, Panasonic #PT-47WX54, Tatung #P50BSAT, RCA #HD50LPW42, RCA #D52W19, JVC #HD52Z575, Sony #KDF-60XS955. Variables: size, screen type price. Data for size: 42, 47, 50, 50, 52, 52, 60; data for screen type: plasma, projection, plasma, projection, projection, projection, projection; data for price: $2994, $1072, $4248, $2696, $1194, $2850, $4100. The variable size is continuous; the variable screen type is qualitative; the variable price is discrete.

52. Individuals: M/Z3 Coupe, M/Z3 Roadster, 3 Series, 5 Series, 7 Series, Z8; Variables: body style, weight, number of seats; Data for body style; coupe, convertible, coupe, sedan, sedan, convertible: Data for weight: 2945, 2690, 2780, 3450, 4255, 3600; data for number of seats: 2, 2, 5, 5, 5, 2. The variable body style is qualitative; the variable weight is continuous; the variable number of seats is discrete.

53. Individuals: Colorado, Missouri, Montana, New York, Texas. Variables: minimum age for driver's license (unrestricted); blood-alcohol concentration limit, mandatory belt-use law seating positions, maximum allowable speed limit in 2003. Data for minimum age for driver's license: 17, 18, 15, 17, 16.5; data for blood-alcohol concentration limit: 0.10, 0.08, 0.08, 0.08, 0.08; data for mandatory belt-use law seating positions: front, front, all, all, front; data for maximum allowable speed limit in 2003: 75, 70, 75, 65, 75. The variable minimum age for driver's license is continuous; the variable blood-alcohol concentration limit is continuous; the variable mandatory belt-use law seating positions is qualitative; the variable maximum allowable speed limit in 2003 is continuous.

54. Individuals: RCA Lyra™ 256 MB MP3 Player. iRiver 256 MB MP3 Player. Samsung 256 MB MP3 Player, Creative® 256 MuVo® Micro N200. iRiver 512 MB MP3 Player, Creative® 512 MuVo® Micro N200, SanDisk 1 GB MP3 Player, Variables; memory size, weight, price. Data for memory size: 256 MB, 256 MB, 256 MB, 256 MB. 512 MB, 512 MB, 512 MB, 1 GB: data for weight: 2 oz, 2.1 oz, 0.85 oz, 0.8 oz, 2.1 oz, 1.4 oz, 0.8 oz, 1.4 oz; data

for price: $149.99, $99.99, $129.99, $99.99, $149.99, $139.99, $139.99, $169.99. The variable memory size is continuous; the variable weight is continuous; the variable price is discrete.

55. (a) Are levels of the substance BNP associated with heart trouble?
(b) All people (presumably adults)
(c) The 1,034 patients who were actually included in the study
(d) People with the highest protein levels were 2.5 times more likely to die than those with the lowest protein levels.
(e) Higher levels of BNP are associated with worse outcomes.

56. (a) Does vigorous physical activity in teens and young adults affect the risk of Parkinson's disease later in life?
(b) All middle-aged men and women
(c) The 48,000 men and 77,000 women who were relatively healthy and middle-aged or older who were included in the study.
(d) Men who participated in vigorous activity at least twice a week in high school, college, and up to age 40 had a 60% reduced risk of getting Parkinson's.

57. (a) Nominal **(b)** Ordinal **(c)** Nominal
(d) Ordinal **(e)** Nominal

1.2 Assess Your Understanding (page 19)

1. An observational study does not attempt to manipulate or influence the individuals in the study. In an experiment, a treatment is applied to individuals in an attempt to isolate the effects of the treatment on the individuals.

2. The frame is necessary because it is the list we use to assign numbers to the individuals in the population.

3. Sampling is used because it is often difficult or impractical to attempt to study each individual in the population under consideration.

4. Once an individual is selected, he or she cannot be selected again.

5. frame **6.** False **7.** True

8. So that the individuals are truly selected at random. The seed represents the starting point for the random-number generator.

9. Observational study **10.** Experiment
11. Experiment **12.** Observational study
13. Observational study **14.** Observational study
15. Experiment **16.** Observational study
17. Observational study **18.** Observational study
19. Answers will vary **20.** Answers will vary
21. (a) Answers will vary. **(b)** Answers will vary.
22. (a) Answers will vary. **(b)** Answers will vary.
23. (a) 83, 67, 84, 38, 22, 24, 36, 58, 34
(b) Answers may vary depending on the type of graphing calculator used. If using a TI-84, the sample will be 4, 20, 52, 5, 24, 87, 67, 86, 39.
24. (a) 2869, 5518, 6635, 2182, 0603, 2654, 2686, 0135, 4080, 6621, 3774, 0826, 0916, 3188, 0876, 5418, 0037, 3130, 2882, 0662
(b) Answers may vary depending on the type of graphing calculator used. If using a TI-84, the sample will be 6658, 4118, 9, 4828, 3905, 454, 2825, 2381, 495, 4445, 4455, 5795, 5397, 7066, 3404, 6667, 5074, 3777, 3206, 5216.
25. (a) First, find a list of all currently enrolled students. This list will serve as the frame. Number the students on the list from 1 through 19,935. Using a random-number generator. Set a seed (or select a starting point if using Table 1). Generate 25 different numbers randomly. The students corresponding to these numbers will be the 25 students in the sample.
(b) Answers will vary.
26. (a) First, find a list of all current residents of the village. This list will serve as the frame. Number the residents on the list from 1 through 5832. Using a random-number generator, set a seed or select a starting point if using Table 1. Generate 20 different numbers randomly. The residents corresponding to these numbers will be the 20 residents in the sample.
(b) Answers will vary.

1.3 Assess Your Understanding (page 30)

1. If the population can be divided into homogeneous, nonoverlapping groups

2. Systematic

3. Convenience samples are not probability samples. Individuals that participate are often self-selected, so the results are not likely representative of the population.

4. cluster sample **5.** stratified sample **6.** False
7. False **8.** True **9.** Systematic
10. Cluster **11.** Cluster **12.** Stratified
13. Simple random **14.** Convenience **15.** Cluster
16. Stratified **17.** Convenience **18.** Systematic
19. Systematic **20.** Stratified **21.** Systematic
22. Simple random **23.** Answers will vary.
24. Answers will vary. **25.** Answers will vary.
26. Answers will vary.
27. (a) 90
(b) Randomly select a number between 1 and 90. Suppose that we randomly select 15; then the individuals in the survey will be 15, 105, 195, 285, ..., 4425.
28. (a) 7269
(b) Randomly select a number between 1 and 7269. Suppose that we randomly select 2000; then the individuals in the survey will be 2000, 9269, 16538, ..., 939, 701.
29. Answers will vary. A good choice might be stratified sampling with the strata being commuters and noncommuters.
30. Answers will vary. One option would be cluster sampling. Each cluster would be a class. Randomly select clusters and then survey all the students in the selected classes.
31. Answers will vary. One option would be cluster sampling. The clusters could be the city blocks. Randomly select clusters and then survey all the households in the selected city blocks.
32. Answers will vary. Probably the most logical choice is systematic sampling.
33. Answers will vary. Simple random sampling will work fine here, especially because a list of 6,600 individuals who meet the needs of our study already exists (the frame).
34. Answers will vary. Simple random sampling will work because a list of all households in the population exists. Just number the households 1 to N and use a random-number generator to obtain the households in the survey.
35. The researcher should make a return visit or phone call. Allowing the spouse to answer is a form of convenience sampling and can lead to incorrect results.
36. Answers will vary.

1.4 Assess Your Understanding (page 37)

1. Frames are printed periodically, whereas populations are constantly changing.
2. Callbacks, incentives
3. A closed question has fixed choices for answers, whereas an open question is a free-response question. Closed questions are easier to analyze, but limit the responses. Open questions allow respondents to state exactly how they feel, but are harder to analyze due to the variety of answers and possible misinterpretation of answers.
4. A certain segment of the population is not represented in the sample proportionally to their size in the population.
5. Better survey results. Trained interviewers can elicit responses to difficult questions.
6. The researcher can learn common answers.
7. Pro: Higher chance of making contact with individuals. Con: People upset being interrupted during dinner.
8. Better survey results. A low response rate may mean that some segments of the population are underrepresented or that only individuals with strong opinions have participated.
9. Changing the order helps prevent bias due to previous question answers or situations where respondents are more likely to pick earlier choices.
10. The question is ambiguous since CD could mean compact disc or certificate of deposit. The question could be improved by not using the abbreviation.
11. (a) Flawed sampling method because only the first 50 students who enter the building have a chance of being surveyed

(b) Could use systematic sampling, but a better choice would be cluster or simple random sampling because the vice-president should have access to a complete frame.

12. **(a)** Flawed sampling method because only homes in the SW corner have a chance to be interviewed
    (b) Cluster sampling should be used here. The clusters would be households with similar income.

13. **(a)** Flawed survey due to the wording of the question
    (b) The survey should begin by stating the current penalty for selling a gun illegally. The question might be rewritten as "Do you approve or disapprove of harsher penalties for individuals that sell guns illegally?" The words *approve* and *disapprove* should be rotated from individual to individual.

14. **(a)** Flawed survey because the wording of the question is ambiguous.
    (b) The question might be rewritten as "How many hours per week do you study for your statistics class?"

15. **(a)** Flawed sampling method because (assuming the survey is written in English), nonEnglish speaking homes will not read the survey. This is why the response rate may be low.
    (b) The survey can be improved through face-to-face or telephone interviews.

16. **(a)** Flawed sampling method because of nonresponse
    (b) The survey could be improved by using face-to-face or telephone interviews.

17. **(a)** Flawed sampling method because the sample of weekday shoppers may not represent the eating habits of all shoppers at the mall. There may be different tastes for those who shop in the evening or on the weekends.
    (b) A stratified sample would be better (e.g., weekday, weeknight, weekends).

18. **(a)** Flawed sampling method because the peaches on top may not be indicative of the peaches on the bottom. The produce supplier may pack low-quality peaches on the bottom to disguise them.
    (b) Cluster sampling by obtaining a simple random sample of some crates and inspecting all the peaches.

19. **(a)** Flawed survey due to the wording of the question
    (b) The question should be reworded so that it doesn't imply the opinion of the editors. One possibility might be 'Do you believe that a marriage can be maintained after an extramarital relation?'

20. **(a)** Flawed sampling method because the frame is not necessarily representative of all college professors
    (b) The company could use cluster sampling and obtain a list of faculty from the human resources department at selected colleges.

21. **(a)** Flawed survey because the students are not likely to respond truthfully when their teachers administer the survey
    (b) The survey should be administered by an impartial party so that the students will be more likely to respond truthfully.

22. **(a)** Flawed survey because people are not likely to respond truthfully to uniformed police officers.
    (b) Send a trained interviewer to the house.

23. The ordering of the questions is likely to affect the survey results. Perhaps question B should be asked first. Another possibility is to rotate the questions randomly.

24. The ordering of the questions is likely to affect the survey results. The questions should be rotated randomly. Prohibit is a strong word. People don't like to be prohibited from doing things. If the word must be used, it should be offset by the word *allow*, and the use of the words *prohibit* and *allow* should be rotated within the question.

25. The choices need to be rotated so that any response bias due to the ordering of the questions is minimized.

26. The agency was concerned because estimates indicated that more people voted than were registered to vote. This indicates a counting error or perhaps ballot stuffing.

27. Poorly trained interviewers; interviewer bias, surveyed too many female voters

28. This has the potential of indicating that responses can be bought. Respondents may attempt to respond the way they think the interviewer wants them to respond as repayment for the reward.

1.5 Assess Your Understanding (page 45)

1. **(a)** A person, object, or some other well-defined item upon which a treatment is applied
   (b) Any combination of the explanatory variables

(c) A quantitative or qualitative variable that represents the variable of interest
(d) The variable whose effect on the response variable is to be assessed by the experimenter
(e) Neither the experimental unit nor the experimenter knows which treatment is being administered to the experimental unit.
(f) An innocuous treatment such as a sugar tablet
(g) The effect of two explanatory variables that cannot be distinguished.

2. This occurs when each treatment is applied to more than one experimental unit.

3. Completely randomized; matched-pair

4. The blocks act like the homogeneous groups in stratified sampling.

5. **(a)** Score on achievement test.
   (b) Method of teaching, grade level, intelligence, school district, teacher; Controlled: grade level, school district, teacher. Manipulated: method of teaching
   (c) New teaching method and traditional method; 2
   (d) Random assignment
   (e) Completely randomized design
   (f) 500 students
   (g)

6. **(a)** Proportion of subjects with a cold
   (b) Gender, age, geographic location, overall health, drug intervention; Controlled: gender, age, location; Manipulated: drug intervention
   (c) Experimental drug and placebo; 2
   (d) Random assignment
   (e) Completely randomized design
   (f) 300 males
   (g)

7. **(a)** Matched pair
   (b) Difference in test scores
   (c) Math course

8. **(a)** Matched pair
   (b) Whiteness level
   (c) Whitening method; Crest Whitestrips Premium with brushing and brushing alone
   (d) Answers will vary. One possible variable would be diet. Certain foods and tobacco products are more likely to stain teeth. This could affect the whiteness level.
   (e) Answers will vary. One possible answer is that using twins helps control for genetic factors (like weak teeth) that may affect the results of the study.

9. **(a)** Completely randomized design
   (b) Adults with insomnia
   (c) Terminal wake time after sleep onset (WASO)
   (d) Type of intervention; CBT, RT, placebo
   (e) 75 adults with insomnia
   (f)

10. **(a)** Completely randomized design
    (b) Adult outpatients diagnosed as having major depression and having a baseline Hamilton Rating Scale for Depression (HAM-D) score of at least 20

(c) Change in the HAM-D over the treatment period
(d) Type of drug: St. John's wort extract or placebo
(e) 200 outpatients diagnosed with depression
(f)

11. (a) Completely randomized design
(b) Adults older than 60 years in good health
(c) Score on a standardized test of learning and memory
(d) Drug; 40 mg of ginkgo 3 times per day or a matching placebo
(e) 98 men and 132 women older than 60 in good health
(f)

12. (a) Completely randomized design
(b) Inpatients with a diagnosis of major depression
(c) Hamilton Rating Scale for Depression score
(d) Drug; placebo or new drug therapy
(e) 63 inpatients with a diagnosis of major depression
(f)

13. (a) Matched-pairs design
(b) Distance yardstick falls
(c) Hand dominance; dominant versus nondominant hand
(d) 15 students
(e) To eliminate bias due to starting on dominant or nondominant first for each trial
(f)

14. (a) Matched-pairs design
(b) Distance ball hit
(c) Shaft type; graphite shaft versus steel shaft
(d) 10 golfers
(e) To eliminate bias due to type of shaft used first
(f)

15. Answers will vary.

16. (a) Completely randomized design
(b) Answers will vary.

17. Answers will vary. Completely randomized design is likely best.

18. Answers will vary. Completely randomized design is likely best.

19. Answers will vary. Matched-pairs design is likely best (match by type of exterior finish).

20. Answers will vary. Matched-pairs design is likely best (match by car model).

21. (a) Blood pressure
(b) Daily consumption of salt, daily consumption of fruits and vegetables, body's ability to process salt

(c) Salt: manipulate; Fruits/vegetables; control; body; cannot be controlled. To deal with variability in body types, randomly assign experimental units to each treatment group.
(d) Answers will vary. Three might be a good choice; one level below RDA, one equal to RDA, one above RDA.

Chapter Review Exercises (page 49)

1. The science of collecting, organizing, summarizing, and analyzing data in order in draw conclusions or answer questions.

2. A group of individuals that a researcher wishes to study

3. A subset of the population

4. Often called ex post factor studies because the value of the response variable has already been determined. In observational studies, there is no attempt to influence the value of the response variable.

5. Applies a treatment to the individuals in the study to isolate the effect of the treatment on the response variable.

6. (1) Identify the research objective. (2) Collect data needed in answer the research questions. (3) Organize and summarize the data. (4) Draw conclusions from the data.

7. There are two main types of errors: nonsampling errors and sampling errors. Nonsampling errors are errors that result from the survey process, such as an incomplete frame, poorly worded questions, inaccurate responses, and so on. Sampling errors are errors that result from using a sample in estimate the characteristics of a population. They result because samples contain incomplete information regarding a population

8. Control, manipulation, randomization, and replication

9. Quantitative; discrete   **10.** Qualitative

11. Quantitative; continuous   **12.** Quantitative; continuous

13. Qualitative   **14.** Quantitative; discrete

15. Observational study, since no treatment has been imposed.

16. Experiment   **17.** Experiment

18. Observational study, since no treatment has been imposed.

19. Convenience sample

20. Simple random sample

21. Cluster sample

22. Stratified sample

23. (a) Undercoverage or nonrepresentative sample due to a poor sampling frame
(b) Interviewer error
(c) Data checks

24. (a) Question wording
(b) Nonresponse
(c) Undercoverage or nonrepresentative sample due to convenience sampling.

25. Answers will vary.   **26.** Answers will vary.

27. Answers will vary.   **28.** Answers will vary.

29. Answers will vary. Label each goggle with pairs of digits from 00 to 99. Using row 12, column 1 of Table 1 and reading across, the selected labels would be 55, 36, 65, 10, 57, 90, 06, 51, 47, 91, 62, 42, 60. The goggles with these labels would be inspected for defects.

30. (a) Matched pair design
(b) Adults
(c) Music at bedtime
(d) Change in PSQI score
(e) 100 adults
(f)

31. (a) Completely randomized design
(b) Energy required to light the bulb
(c) Integrated circuit; old circuit and new circuit
(d) 200 fluorescent bulbs
(e)

32. (a) Completely randomized design
(b) Condition (dermatitis) improvement
(c) NPI 32101 topical crème; 0.5%, 1.0%, 0%
(d) So that the experimenter does not inadvertently indicate which cream the subject received. For example, the experimenter may spend less time with the placebo group because he or she does not expect to see much change.
(e) 224 patients with mild to moderate atopic dermatitis
(f)

33. (a) Lymphocyte count
(b) Space flight
(c) Matched pairs
(d) Four flight members
(e)

34. Answers will vary.   **35.** Answers will vary.
36. Answers will vary.

37. In a completely randomized design the experimental units are randomly assigned to one of the treatments. The value of the response variable is compared for each treatment. In a matched-pairs design, experimental units are matched up on the basis of some common characteristic (such as husband–wife, pre–post, or twins). The difference in the paired experimental units is analyzed.

38. (a) Identify 60 volunteers. Use a random number generator to divide the 60 individuals into groups of 30 each. Let the first 30 individuals drink 1 Monster Energy Drink and the other 30 drink a placebo that looks and tastes like a Monster Energy Drink. After 30 minutes each individual can be given a test to measure reaction time to some stimulus (in seconds). Compare the reaction times of the experimental group to the control group.
(b) Identify 30 volunteers. For each volunteer use a random number generator to assign them either Monster Energy Drink or a placebo that looks and tastes like a Monster Energy Drink. After 30 minutes each individual can be given a test to measure reaction time to some stimulus (in seconds). After one week, have each of the 30 individuals return. This time give the individual the other drink and repeat the experiment. Compare the difference in reaction times.
(c) The matched pair is probably the better design since it accounts for differences in individual performance by blocking by person.

CHAPTER 2

2.1 Assess Your Understanding (page 61)

1. Raw data are data that are not organized.
2. number; proportion (or percent)

3. It is a good idea to make sure that the frequencies add up to the correct number of observations.
4. 1 (although rounding may cause the result to vary slightly)
5. A Pareto chart is a bar graph whose bars are drawn in decreasing order of frequency or relative frequency.
6. Relative frequencies should be used when comparing two data sets, because different sample or population sizes make comparisons using frequencies difficult or misleading.
7. Answers vary.
8. No. A pie chart cannot be used to describe these data because the percentages sum to more than 100%, indicating that some people selected more than one category as a reason for not losing weight.
9. (a) Large; 52%
(b) X-Large; 9%
(c) 23%
10. (a) 325,080 were liposuctions.
(b) 216,720 were eyelid surgery.
(c) 38% of cosmetic surgeries for women are not accounted for in the graph.
11. (a) United States has the most Internet users.
(b) Approximately 20 million Internet users are in Canada.
(c) Approximately 25 million more Internet users are in China than in Japan.
12. (a) 22,196,000 whites were living in poverty.
(b) 24.6% of those living in poverty were Hispanic.
(c) The graph could be misleading because it does not account for the different population sizes of each ethnicity.
13. (a) Badnarik received 397,157 votes.
(b) Bush received 50.7% of the vote; Kerry received 48.3%.
(c) No. He only received 0.4% of the votes.
14. (a) 32% were victimized by a family member or relative.
(b) About 1.2 million people
15. (a) 0.16; 0.30
(b) Natural gas was the most popular source.
(c) About 5.35 million households used LPG.
(d) Answers will vary.
(e) Fuel oil or kerosene; answers will vary.
(f) Both natural gas and LPG have remained steady.
16. (a) 433 more doctorates were awarded.
(b) About 14.6% of doctorates were awarded in physical science; about 16.3% of doctorates were awarded in education.
(c) The number of degree recipients in education has consistently decreased.
(d) Life sciences had the largest increase in the number of degree recipients.

17. (a)

Source of Income	Percentage of Income
Individual income taxes	0.4453
Corporate income taxes	0.0739
Social insurance taxes	0.4000
Excise, estate and gift taxes, customs, and miscellaneous receipts	0.0807

(b) 44.53% of the government's income is attributable to individual income taxes.

(c)

(d)

(e)

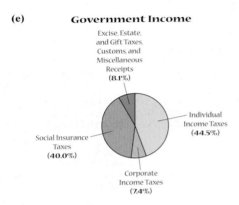

(f) Answers will vary.

18. (a)

Category	Percentage of Expenditures
National defense and foreign affairs	0.220
Social programs	0.580
Physical, human, and community development	0.100
Net interest on the debt	0.070
Law enforcement and general government	0.030

(b) 7% of total expenditures is attributable to net interest on the debt.

(c)

(d)

(e)

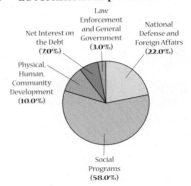

(f) Answers will vary.

19. (a)

Response	Relative Frequency
Never	0.0262
Rarely	0.0678
Sometimes	0.1156
Most of the time	0.2632
Always	0.5272

(b) 52.72%
(c) 9.4%

(d)

(e)

(f)

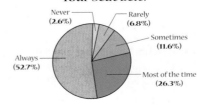

(g) This is inferential statistics because it is taking a result of the sample and generalizing it to the population "all college students."

20. (a)

Response	Frequency
I do not drive a car	0.0522
Never	0.0247
Rarely	0.0522
Sometimes	0.0723
Most of the time	0.1501
Always	0.6484

(b) 64.84%
(c) 7.69%

(d)

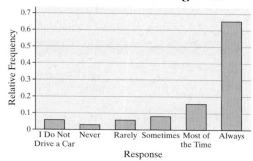

"How Often Do You Wear a Seat Belt when Driving a Car?"

(e)

"How Often Do You Wear a Seat Belt when Driving a Car?"

(f)

"How Often Do You Wear a Seat Belt when Driving a Car?"

(g)

Response	Relative Frequency
Never	0.0261
Rarely	0.0551
Sometimes	0.0763
Most of the time	0.1584
Always	0.6841

(h) This is a descriptive statement since it is a summary of the data collected.

21. (a)

Region	Relative Frequency
Caribbean	0.1010
Central America	0.3690
South America	0.0630
Asia	0.2500
Europe	0.1370
Other Regions	0.0800

(b) 25.00% of foreign-born residents were born in Asia.

(c)

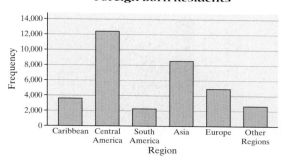

Birthplace of Foreign-Born Residents

(d)

Birthplace of Foreign-Born Residents

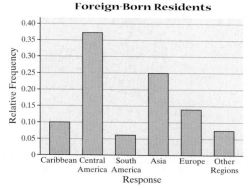

(e)

Birthplace of Foreign-Born Residents

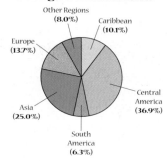

22. (a)

Type of Robbery	Relative Frequency
Street or highway	0.4746
Commercial	0.2210
Gas station	0.0362
Convenience store	0.0942
Residence	0.1486
Bank	0.0254

(b) About 3.62% of robberies were of gas stations.

(c)

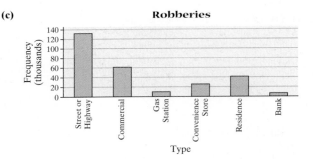

(d) Answers will vary.

24. (a)

Educational attainment	2000
No college	0.2725
Some college	0.3476
Graduated college	0.3799

(b)

Educational attainment	2003
No college	0.3970
Some college	0.3070
Graduated college	0.2960

(d)

(c)

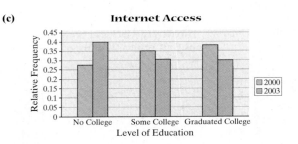

(d) Answers will vary.

(e)

25. (a)

Age of Victim (male)	Relative Frequency
Less than 17	0.0606
17–24	0.3204
24–35	0.2789
35–54	0.2667
55 or older	0.0734

(b)

Age of Victim (female)	Relative Frequency
Less than 17	0.1314
17–24	0.2083
24–35	0.2083
35–54	0.3211
55 or older	0.1308

23. (a)

Educational Attainment	Relative Frequency (males)
Not a high school graduate	0.159
High school graduate	0.309
Some college, but no degree	0.171
Associate's degree	0.072
Bachelor's degree	0.185
Advanced degree	0.104

(b)

Educational Attainment	Relative Frequency (females)
Not a high school graduate	0.150
High school graduate	0.330
Some college, but no degree	0.172
Associate's degree	0.091
Bachelor's degree	0.175
Advanced degree	0.082

(c)

(c)

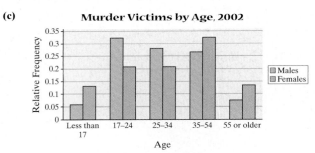

(d) Answers will vary.

26. (a)

Color	Relative Frequency Luxury Cars	Sports Cars
Silver	0.32	0.27
White	0.30	0.11
Blue	0.09	0.13
Black	0.09	0.16
Gray	0.07	0.07
Red	0.06	0.18
Gold	0.03	0.04
Green	0.02	0.00
Brown	0.02	0.04

(b)

Popular Automobile Colors

(c) Answers will vary.

27. (a)

Candidate	Frequency
Bush	21
Kerry	17
Nader	1
Badnarik	1

(b)

Candidate	Relative Frequency
Bush	0.525
Kerry	0.425
Nader	0.025
Badnarik	0.025

(c)

Votes in Los Alamos County, NM

(d)

Votes in Los Alamos County, NM

(e) **Votes in Los Alamos County, NM**

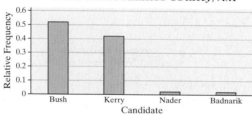

(f) Answers will vary. The conjecture is inferential.

28. (a), (b)

Diagnosis	Frequency	Relative Frequency
Assault	1	0.05
Cancer	1	0.05
Congestive heart failure	1	0.05
Fall	2	0.10
Gunshot wound	8	0.40
Motor vehicle accident	7	0.35

(c) Gunshot wound
(d) 35%

(e)

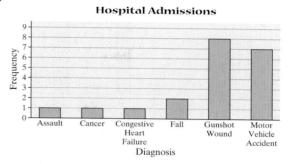

Hospital Admissions

(f)

Hospital Admissions

(g)

Hospital Admissions

Assault (5%) Cancer (5%)
Congestive Heart Failure (5%)
Motor Vehicle Accident (35%)
Fall (10%)
Gunshot Wound (40%)

(h) The admissions specialist's comment is inferential, because the results from a sample were generalized and attributed to the population of hospital admissions.

29. (a)

Position	Frequency
First base	3
Second base	0
Third base	2
Shortstop	1
Pitcher	7
Catcher	1
Right field	6
Center field	2
Left field	2

(b)

Position	Relative Frequency
First base	0.1250
Second base	0.0000
Third base	0.0833
Short stop	0.0417
Pitcher	0.2917
Catcher	0.0417
Right field	0.2500
Center field	0.0833
Left field	0.0833

(c) Pitcher
(d) Second base

(e)

Highest-Paying Positions, 2004

(f)

Highest-Paying Positions, 2004

(g) Highest-Paying Positions, 2004

30. (a), (b)

Blood Type	Frequency	Relative Frequency
A	18	0.36
AB	4	0.08
B	6	0.12
O	22	0.44

(c) O   **(d)** AB   **(e)** 44%; inferential
(f) Answers will vary.

(g)

Blood Types

(h)

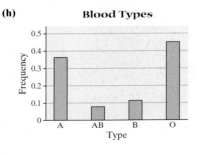

Blood Types

(i) Blood Types

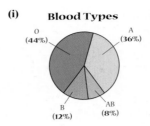

31. (a), (b)

Language Studied	Frequency	Relative Frequency
Chinese	3	0.100
French	3	0.100
German	3	0.100
Italian	2	0.067
Japanese	2	0.067
Latin	2	0.067
Russian	1	0.033
Spanish	14	0.467

(c)

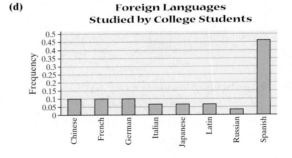

Foreign Languages Studied by College Students

(d)

Foreign Languages Studied by College Students

(e) Foreign Languages Studied by College Students

32. (a)

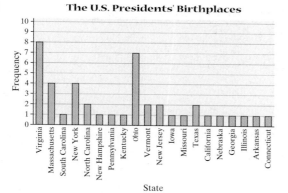

The U.S. Presidents' Birthplaces

(b) More presidents were born in Virginia than in any other state.
(c) Virginia is one of the original 13 colories.

2.2 Assess Your Understanding (page 82)

1. Answers will vary.

2. The classes do not overlap, so there is no confusion as to which class a datum value belongs.

3. The class widths are not all the same, and the rectangles in a histogram should touch.

4. An advantage is that the raw data can be retrieved in a stem-and-leaf plot, but they are lost in a histogram. Advantages of histograms over stem-and-leaf plots include more flexibility in choosing class widths in a histogram and the ability of a histogram to represent large data sets.

5. Answers will vary. Some possibilities follow: Histograms represent quantitative data, while bar graphs represent qualitative data. The bars in a histogram touch, but the bars in a bar graph do not touch. The width of the bars of a histogram has meaning, but the width of the bars in a bar graph is meaningless.

6. True

7. False. The distribution shape shown is skewed right.

8. False. The distribution shape shown is symmetric.

9. (a) 8   **(b)** 2   **(c)** 15
(d) 15%   **(e)** Roughly symmetric

10. (a) 4 cars   **(b)** 9 weeks   **(c)** 17.3%
(d) The distribution is skewed right.

11. (a) 200   **(b)** 10
(c) 60–69, 2; 70–79, 3; 80–89, 13; 90–99, 42; 100–109, 58; 110–119, 40; 120–129, 31; 130–139, 8; 140–149, 2; 150–159, 1
(d) 100–109   **(e)** 150–159

12. (a) Class width is 200.
(b) Classes: 0–199; 200–399; 400–599; 600–799; 800–999; 1000–1199; 1200–1399; 1400–1599; 1600–1799
(c) Highest frequency is in class 0–199.
(d) Distribution is skewed right.
(e) Answers will vary.

13. (a)

Number of Children under 5	Relative Frequency
0	0.32
1	0.36
2	0.24
3	0.06
4	0.02

(b) 24%   **(c)** 60%

14. (a)

Number of Free Throws until a Miss	Relative Frequency
1	0.32
2	0.22
3	0.18
4	0.14
5	0.04
6	0.06
7	0
8	0.02
9	0
10	0.02

(b) 14% of the time she first missed on the fourth try.
(c) 2% of the time she first missed on the tenth try.
(d) 10% of the time she made at least five in a row.

15. 10, 11, 14, 21, 24, 24, 27, 29, 33, 35, 35, 35, 37, 37, 38, 40, 40, 41, 42, 46, 46, 48, 49, 49, 53, 53, 55, 58, 61, 62

16. 40, 44, 47, 52, 52, 53, 59, 59, 63, 64, 65, 68, 68, 69, 70, 71, 71, 73, 76, 76, 82, 83, 88

17. 1.2, 1.4, 1.6, 2.1, 2.4, 2.7, 2.7, 2.9, 3.3, 3.3, 3.3, 3.5, 3.7, 3.7, 3.8, 4.0, 4.1, 4.1, 4.3, 4.6, 4.6, 4.8, 4.8, 4.9, 5.3, 5.4, 5.5, 5.8, 6.2, 6.4

18. 12.3, 12.7, 12.9, 12.9, 13.0, 13.4, 13.5, 13.7, 13.8, 13.9, 13.9, 14.2, 14.4, 14.4, 14.7, 14.7, 14.8, 14.9, 15.1, 15.2, 15.2, 15.5, 15.6, 16.0, 16.3

19. (a) Four classes
(b) Lower class limits: 25, 35, 45, 55; upper class limits: 34, 44, 54, 64
(c) Class width: 10 years

20. (a) Eight classes
(b) Lower class limits: 0, 1.0, 2.0, 3.0, 4.0, 5.0, 6.0, 7.0; upper class limits: 0.9, 1.9, 2.9, 3.9, 4.9, 5.9, 6.9, 7.9
(c) Class width: 1.0

21. (a) Four classes
(b) Lower class limits: 100, 200, 300, 400; upper class limits: 199, 299, 399, 499
(c) Class width: 100 beds

22. (a) Six classes
(b) Lower class limits: 15, 20, 25, 30, 35, 40; upper class limits: 19, 24, 29, 34, 39, 44
(c) Class width: 5 years

23. (a)

Age	Relative Frequency
25–34	0.2323
35–44	0.2870
45–54	0.2822
55–64	0.1986

(b)

Covered by Health Insurance

(c)

Covered by Health Insurance

23.23% is between the ages of 25 and 34; 51.93% is 44 years or younger.

24. (a)

Magnitude	Relative Frequency
0–0.9	0.1019
1.0–1.9	0.0430
2.0–2.9	0.2018
3.0–3.9	0.2508
4.0–4.9	0.3518
5.0–5.9	0.0458
6.0–6.9	0.0045
7.0–7.9	0.0004

(b)

Earthquakes in 2004

(c)

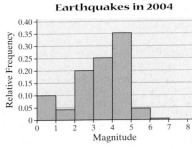

Earthquakes in 2004

35.18% of the earthquakes measured from 4.0 to 4.9 in magnitude; 94.93% of the earthquakes measured 4.9 in magnitude or less.

25. (a)

Number of Beds	Relative Frequency
100–199	0.2990
200–299	0.2979
300–399	0.2456
400–499	0.1574

(b)

Hospital Admissions

(c)

Hospital Admissions

24.56% of admissions was to hospitals with 300 to 399 beds; 40.30% of admissions was to hospitals with more than 300 beds.

26. (a)

Age	Relative Frequency
15–19	0.0128
20–24	0.0702
25–29	0.2235
30–34	0.3888
35–39	0.2530
40–44	0.0518

(b)

Age of Mother of Multiple Births

26. (c)

Age of Mother of Multiple Births

5.18% of multiple births was to women 40 to 44 years of age; 8.30% of multiple births was to women 24 years of age or younger.

27. (a),(b)

Number of Customers	Frequency	Relative Frequency	Number of Customers	Frequency	Relative Frequency
3	2	0.050	9	4	0.1
4	3	0.075	10	4	0.1
5	3	0.075	11	4	0.1
6	5	0.125	12	0	0
7	4	0.1	13	2	0.05
8	8	0.2	14	1	0.025

(c) 27.5%   **(d)** 20%

(e)

Customers Waiting for a Table

(f)

Customers Waiting for a Table

(g) Symmetric

28. (a), (b)

Number of Potholes	Frequency	Relative Frequency	Number of Potholes	Frequency	Relative Frequency
1	6	0.12	6	1	0.02
2	15	0.30	7	5	0.10
3	9	0.18	8	0	0
4	6	0.12	9	1	0.02
5	6	0.12	10	1	0.02

(c) 14%   **(d)** 42%

(e)

Potholes per 1-Mile Stretch of Highway

(f)

Potholes per 1-Mile Stretch of Highway

(g) Skewed right

29. (a), (b)

2003 per Capita Disposable Income		
Class Intervals	Frequency	Relative Frequency
20,000–22,499	3	0.0588
22,500–24,999	10	0.1961
25,000–27,499	14	0.2745
27,500–29,999	12	0.2353
30,000–32,499	7	0.1373
32,500–34,999	2	0.0392
35,000–37,499	2	0.0392
37,500–39,999	0	0
40,000–42,499	1	0.0196

(c)

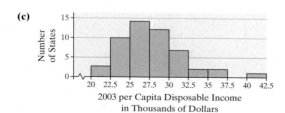

2003 per Capita Disposable Income in Thousands of Dollars

(d)

2003 per Capita Disposable Income (in thousands of dollars) by State

(e) The distribution is skewed right.

(f)

2003 per Capita Disposable Income		
Class Intervals	Frequency	Relative Frequency
20,000–23,999	10	0.1961
24,000–27,999	22	0.4314
28,000–31,999	14	0.2745
32,000–35,999	3	0.0588
36,000–39,999	1	0.0196
40,000–43,999	1	0.0196

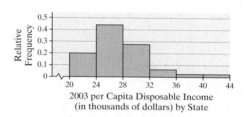

2003 per Capita Disposable Income (in thousands of dollars) by State

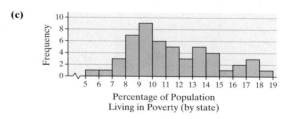

2003 per Capita Disposable Income (in thousands of dollars) by State

30. (a), (b)

Percentage of the Population Living in Poverty		
Class Interval	Frequency	Relative Frequency
5–5.9	1	0.0196
6–6.9	1	0.0196
7–7.9	3	0.0588
8–8.9	7	0.1373
9–9.9	9	0.1765
10–10.9	6	0.1177
11–11.9	5	0.0980
12–12.9	3	0.0588
13–13.9	5	0.0980
14–14.9	4	0.0784
15–15.9	1	0.0196
16–16.9	2	0.0392
17–17.9	3	0.0589
18–18.9	1	0.0196

(c)

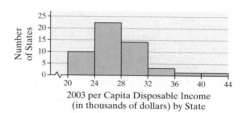

Percentage of Population Living in Poverty (by state)

(d)

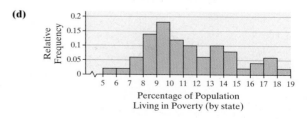

Percentage of Population Living in Poverty (by state)

(e) The distribution is skewed to the right.

(f)

Percentage of Population Living in Poverty		
Class Interval	Frequency	Relative Frequency
5–6.9	2	0.0392
7–8.9	10	0.1961
9–10.9	15	0.2941
11–12.9	8	0.1569
13–14.9	9	0.1765
15–16.9	3	0.0588
17–18.9	4	0.0784

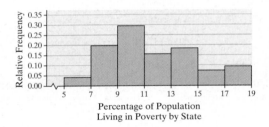

The distribution is skewed right.
(g) Answers will vary.

31. (a), (b)

Class	Frequency	Relative Frequency
20–29	1	$\frac{1}{40} = 0.025$
30–39	6	$\frac{6}{40} = 0.15$
40–49	10	$\frac{10}{40} = 0.25$
50–59	14	0.35
60–69	6	0.15
70–79	3	0.075

(c)

Serum HDL of 20–29-Year-Olds

(d)

Serum HDL of 20–29-Year-Olds

(e) Bell shaped

(f)

Class	Frequency	Relative Frequency
20–24	0	0
25–29	1	0.025
30–34	2	0.05
35–39	4	0.1
40–44	2	0.05
45–49	8	0.2
50–54	9	0.225
55–59	5	0.125
60–64	4	0.1
65–69	2	0.05
70–74	3	0.075
75–79	0	0

Serum HDL of 20–29-Year-Olds

Serum HDL of 20–29-Year-Olds

Bell shaped
(g) Answers will vary.

32. (a), (b)

Class	Frequency	Relative Frequency
3–4.99	12	0.3429
5–6.99	14	0.4000
7–8.99	6	0.1714
9–10.99	3	0.0857

(c)

Altria Stock

(d)

Altria Stock

(e) The distribution is skewed right

(f)

Class	Frequency	Relative Frequency
3–3.99	7	0.2000
4–4.99	5	0.1429
5–5.99	8	0.2286
6–6.99	6	0.1714
7–7.99	4	0.1143
8–8.99	2	0.0571
9–9.99	1	0.0286
10–10.99	2	0.0571

Altria Stock

Altria Stock

The distribution is skewed right.

(g) Answers will vary.

33. (a), (b)

Dividend Yield		
Class Interval	**Number of Stocks**	**Relative Frequency**
0–0.39	7	0.2500
0.40–0.79	4	0.1429
0.80–1.19	5	0.1786
1.20–1.59	2	0.0714
1.60–1.99	3	0.1071
2.0–2.39	4	0.1429
2.4–2.79	2	0.0714
2.8–3.19	1	0.0357

(c)

(d)

(e) The distribution is skewed right.

(f)

Dividend Yield (%)		
Class	**Number of Stocks**	**Relative Frequency**
0.0–0.79	11	0.3929
0.8–1.59	7	0.2500
1.6–2.39	7	0.2500
2.4–3.19	3	0.1071

The distribution is skewed right.

34. (a), (b)

Violent Crimes by State		
Class	**Frequency**	**Relative Frequency**
0–149	3	0.0588
150–299	14	0.2745
300–449	13	0.2549
450–599	12	0.2353
600–749	5	0.0980
750–899	3	0.0588
900–1049	0	0
1050–1199	0	0
1200–1349	0	0
1350–1499	0	0
1500–1649	1	0.0196

(c)

(d)

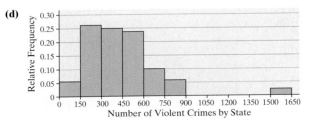

(e) The distribution is skewed to the right.

(f)

Number of Violent Crimes by State		
Class	Number of States	Relative Frequency
0–299	17	0.3333
300–599	25	0.4902
600–899	8	0.1569
900–1199	0	0
1200–1499	0	0
1500–1799	1	0.0196

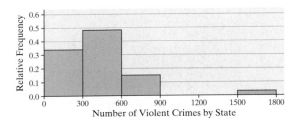

The distribution is skewed right.
(g) Answers will vary.

35.
```
4 | 23
4 | 667899
5 | 0011112244444
5 | 555566677778
6 | 0111244
6 | 589
```

36.
```
2 | 34679
3 | 00222223344557889 9
4 | 000112334559
5 | 012234456
6 | 168
```

37.
```
0 | 1288
1 | 1125667
2 | 333388
3 | 1335
4 | 0
```

38.
```
1 | 8999
2 | 012444555566667788999999999
3 | 00000000111112222224444444445589
4 | 6
```

39. (a) The energy derived from coal rounded to the nearest percent:

54	0	75	64	16	8	94
9	32	90	35	88	56	0
41	62	23	83	17	0	51
49	13	3	60	60	38	8
1	1	60	63	95	42	98
78	46	27	51	90	62	68
10	94	57	23	61	37	96
58	83					

```
0 | 000113889
1 | 0367
2 | 337
3 | 2578
4 | 1269
5 | 114678
6 | 000122348
7 | 58
8 | 338
9 | 0044568
```

(b) Fairly uniform
(c) Answers will vary.

40. (a) Percentage of change in the price of housing rounded to the nearest percent:

23	21	33	16	47	19	66	23
23	24	16	9	17	23	17	22
30	16	25	52	28	54	18	21
21	38	49	62	12	19	20	59
48	19	36	15	25	25	26	48

```
0 | 9
1 | 25666778999
2 | 011123333455568
3 | 0368
4 | 7889
5 | 249
6 | 26
```

(b) The distribution is skewed right.

41. (a)

Make	Problems per 100 Vehicles	Make	Problems per 100 Vehicles	Make	Problems per 100 Vehicles
Lexus	160	Subaru	270	Audi	320
Infiniti	170	Nissan	270	Mercedes Benz	320
Buick	180	GMC	270	Jeep	320
Porsche	190	Chevrolet	270	Volvo	330
Acura	200	Saturn	270	Mitsubishi	340
Toyota	200	Oldsmobile	280	Hyundai	340
Cadillac	210	Mazda	290	Isuzu	370
Lincoln	210	Pontiac	290	Volkswagen	390
Honda	220	Chrysler	300	Suzuki	400
Mercury	240	Ford	300	Daewoo	420
Jaguar	250	Plymouth	300	Land Rover	440
Saab	260	Dodge	310	Kia	510
BMW	260				

(b)
```
1 | 6789
2 | 00112456677777899
3 | 000122234479
4 | 024
5 | 1
```
Answers will vary.

(c)
```
1 | 6789
2 | 001124
2 | 56677777899
3 | 0001222344
3 | 79
4 | 024
4 |
5 | 1
```
Answers will vary.

42. (a)

State	Violent Crime Rate	State	Violent Crime Rate	State	Violent Crime Rate
Alabama	440	Kentucky	280	North Dakota	80
Alaska	560	Louisiana	660	Ohio	350
Arizona	550	Maine	110	Oklahoma	500
Arkansas	420	Maryland	770	Oregon	290
California	590	Massachusetts	480	Pennsylvania	400
Colorado	350	Michigan	540	Rhode Island	290
Connecticut	310	Minnesota	270	South Carolina	820
Delaware	600	Mississippi	340	South Dakota	180
District of Columbia	1630	Missouri	540	Tennessee	720
Florida	770	Montana	350	Texas	580
Georgia	460	Nebraska	310	Utah	240
Hawaii	260	Nevada	640	Vermont	110
Idaho	260	New Hampshire	160	Virginia	290
Illinois	620	New Jersey	380	Washington	350
Indiana	360	New Mexico	740	West Virginia	230
Iowa	290	New York	500	Wisconsin	230
Kansas	380	North Carolina	470	Wyoming	270

Source: Federal Bureau of Investigation

(b)
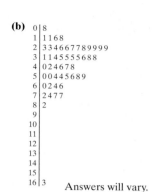
```
 0 | 8
 1 | 1168
 2 | 334667789999
 3 | 1145555688
 4 | 024678
 5 | 00445689
 6 | 0246
 7 | 2477
 8 | 2
 9 |
10 |
11 |
12 |
13 |
14 |
15 |
16 | 3
```
Answers will vary.

(c)

```
 0 |
 0 | 8
 1 | 11
 1 | 68
 2 | 334
 2 | 667789999
 3 | 114
 3 | 5555688
 4 | 024
 4 | 678
 5 | 0044
 5 | 5689
 6 | 024
 6 | 6
 7 | 24
 7 | 77
 8 | 2
 8 |
 9 |
 9 |
10 |
10 |
11 |
11 |
12 |
12 |
13 |
13 |
14 |
14 |
15 |
15 |
16 | 3
16 |
```
Answers will vary.

43. (a)

Best Actor Age		Best Actress Age
9	2	145588
98877765210	3	01133334555589
7555332200	4	1112599
4221	5	
100	6	1
6	7	4
	8	0

(b) Answers will vary.

44. (a)

McGwire		Bonds
	32	00
	33	
10	34	7
00	35	0
9000	36	00015
70000	37	005555
85500000	38	000005
80000	39	00146
900	40	000045
00000	41	0000000000155677
5300000	42	000000009
0000000	43	00000556
000	44	00002
820000	45	04
100	46	
8000	47	
0	48	8
	49	
0	50	
00	51	
7	52	
	53	
	54	
0	55	

(b) Answers will vary.

45.

46.

47. (a)

(b) The stock price increased the most during November. Explanations for why will vary.

48.

49.

50.
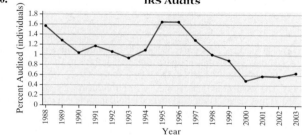

2.3 Assess Your Understanding (page 96)

1. The lengths of the bars are not proportional. For example, the bar representing the cost of Clinton's inauguration should be slightly more than 9 times as long as the one for Carter's cost and twice as long as the bar representing Reagan's cost.

2. (a) Unclear where the start of the graph is - at the base of each figure, or the bottom of the platform. The top of cheeseburger should be 2.2. times as high as the soda.
   (b) Answers will vary.

3. (a) The vertical axis starts at 28 instead of 0. This tends to indicate that the median earnings for females increased at a faster rate than it actually did.

(b) This graph indicates that the median earnings for females has increased slightly over the given 5-year period.

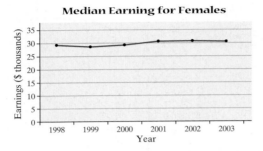

Median Earning for Females

4. (a) The vertical axis starts at 0.06 instead of 0. Answers will vary.

(b)

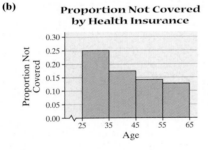

Union Membership

5. (a) The vertical axis starts at 0.1 instead of 0. This might cause the reader to conclude, for example, that the proportion of people aged 25 to 34 who are not covered by health insurance is more than 4 times the proportion for those aged 55 to 64.

(b)

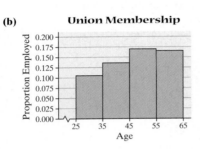

Proportion Not Covered by Health Insurance

6. (a) The vertical axis starts at 150 without indicating a gap.
(b) It may be trying to convey that new homes in the Midwest increased significantly in the past 5 years.

7. (a) The vertical axis starts at 35 without indicating a gap.
(b) It may convey that median household income is increasing more rapidly than it really is.

8. 1300 times

9. (a) The bar for housing should be a little more than twice the length of the bar for transportation, but it is not.
(b) Adjust the graphs so that the lengths of the bars are proportional.

10. (a)

Average SAT Combined Score

(b)

Average SAT Combined Score

11. (a)

Engineering Bachelor's Degrees

(b)

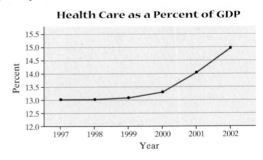

Engineering Bachelor's Degrees

12. The graph does not support the safety manager's claim. The vertical scale starts at 0.17 instead of 0, so the difference between the bars is distorted. While there has been a decrease, it appears that the decrease is roughly 10% of the 1992 rate.

13. (a) The politician's view:

Health Care as a Percent of GDP

(b) The health care industry's view:

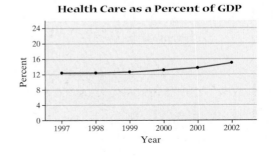

Health Care as a Percent of GDP

(c) An honest view:

Health Care as a Percent of GDP

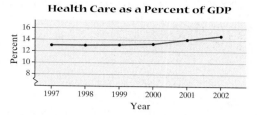

14. (a) Graph that supports belief roads are becoming less safe:

Motor Vehicle Deaths

(b) Graph that supports the belief that roads are becoming safer:

Traffic Death Rate

15. (a) Graphic that is not misleading:

Enrollment in Distance Learning

(b) Graphics will vary.

16. (a) Graphic that is not misleading:

U.S. Adolescents Who Are Overweight

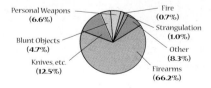

(b) Graphics will vary.

Chapter Review Exercises (page 99)

1. (a) 22.5 quadrillion Btu used was from natural gas.
(b) 3 quadrillion Btu's used was from biomass.
(c) Approximately 98 quadrillion Btu's of energy was consumed.

(d) "Other" has the lowest frequency.
(e) No, these are qualitative data.

2. (a) Class width is 0.5.
(b) The classes are 0.5–0.99, 1–1.49, 1.5–1.99. 2.0–2.49, 2.5–2.99, 3.0–3.49, 3.5–3.99, 4.0–4.49, 4.5–4.99, 5.0–5.49, 5.5–5.99, and 6.0–6.49
(c) 1–1.49 has the highest frequency.

3. (a)

Type of Weapon	Relative Frequency
Firearms	0.6617
Knives or cutting instruments	0.1248
Blunt objects (clubs, hammers, etc.)	0.0470
Personal weapons (hands, fists, etc.)	0.0659
Strangulation	0.0101
Fire	0.0073
Other weapon or not stated	0.0831

(b) 4.7% of homicides was committed with a blunt object.

(c)

Weapons Used in Homocides

(d)

Weapons Used in Homocides

(e)

Weapons Used in Homicides

Personal Weapons (6.6%)
Fire (0.7%)
Strangulation (1.0%)
Blunt Objects (4.7%)
Other (8.3%)
Knives, etc. (12.5%)
Firearms (66.2%)

4. (a)

Gas	Relative Frequency
Carbon dioxide	0.8464
Methane	0.0868
Nitrous oxide	0.0462
Hydrofluorocarbons, perfluorocarbons, and sulfur hexafluoride	0.0207

(b) 84.6% of emissions was due to carbon dioxide.

(c)

U.S. Greenhouse Gas Emissions

(d)

U.S. Greenhouse Gas Emissions

(e) U.S. Greenhouse Gas Emissions

5. (a)

Age of Mother (years)	Relative Frequency
10–14	0.0017
15–19	0.1016
20–24	0.2526
25–29	0.2660
30–34	0.2389
35–39	0.1145
40–44	0.0247

(b)

Live Births in the U.S. by Age of Mother

The distribution is symmetric.

(c)

Live Births in the U.S. by Age of Mother

(d) 25.3% of the births was to mothers aged 20 to 24.

(e) 37.8% of the live births was to mothers age 30 or above.

6. (a)

Age (years)	Relative Frequency
5–9	0.0002
10–14	0.0089
15–19	0.0526
20–24	0.0771
25–29	0.0781
30–34	0.0876
35–39	0.1038
40–44	0.1130
45–49	0.1065
50–54	0.0876
55–59	0.0649
60–64	0.0435
65–69	0.0396
70–74	0.0399
75–79	0.0398
80–84	0.0318
85–89	0.0176
90–94	0.0065
95–99	0.0009

(b)

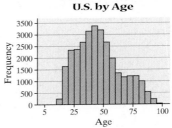

Suicides in the U.S. by Age

The distribution is skewed right.

(c)

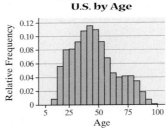

Suicides in the U.S. by Age

(d) 11.3% of the suicides was by persons aged 40 to 44.

(e) 13.9% of U.S. suicides was by persons 24 or younger.

7. (a), (b)

Affiliation	Frequency	Relative Frequency
Democrat	46	0.46
Independent	16	0.16
Republican	38	0.38

(c)

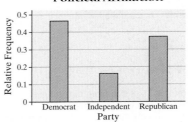

Political Affiliation

(d)

Political Affiliation

(e) Democrat appears to be the most common affiliation.

8. (a), (b)

Educational Attainment	Frequency	Relative Frequency
No high school diploma	9	0.18
High school graduate	16	0.32
Some college	9	0.18
Association's degree	4	0.08
Bachelor's degree	8	0.16
Advanced degree	4	0.08

(c)

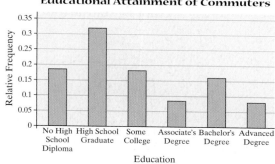

Educational Attainment of Commuters

(d)

Educational Attainment

(e) High school graduate is the most common educational level.

9. (a), (b)

Number of Children	Frequency	Relative Frequency
0	7	0.1167
1	7	0.1167
2	18	0.3000
3	20	0.3333
4	7	0.1167
5	1	0.0167

(c)

Number of Children for Couples Married 7 Years

The distribution is symmetric.

(d)

Number of Children for Couples Married 7 Years

(e) 30% of the couples has two children.

(f) 76.7% of the couples has at least two children.

(g)

10. (a), (b)

Number of Customers	Frequency	Relative Frequency
1	5	0.10
2	7	0.14
3	12	0.24
4	6	0.12
5	8	0.16
6	5	0.10
7	2	0.04
8	4	0.08
9	1	0.02

(c)

Number of Cars Waiting in Line at McDonald's

The distribution is skewed right.

(d)

Number of Cars Waiting in Line at McDonald's

(e) 24% of the time three cars arrived between 11:50 A.M. and noon.

(f) 76% of the time at least three cars arrived between 11:50 A.M. and noon.

(g)

11. (a), (b)

Crime Rate by State		
Class	Frequency	Relative Frequency
2000–2399	2	0.0392
2400–2799	4	0.0784
2800–3199	8	0.1569
3200–3599	6	0.1176
3600–3999	4	0.0784
4000–4399	8	0.1569
4400–4799	8	0.1569
4800–5199	6	0.1176
5200–5599	2	0.0392
5600–5999	0	0.0000
6000–6399	2	0.0392
6400–6799	0	0.0000
6800–7199	0	0.0000
7200–7599	0	0.0000
7600–7999	0	0.0000
8000–8399	1	0.0196

(c)

The distribution is skewed right.

(d)

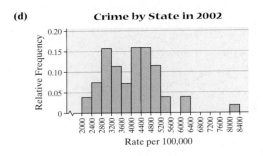

(e) For (a) and(b):

Crime Rate (per 100,000 population)		
Class	Frequency	Relative Frequency
2000–2999	10	0.1961
3000–3999	14	0.2745
4000–4999	17	0.3333
5000–5999	7	0.1373
6000–6999	2	0.0392
7000–7999	0	0.0000
8000–8999	1	0.0196

For (c):

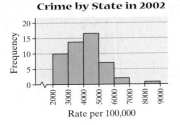

The distribution is skewed right.

For (d):

Crime by State in 2002

For (e): Answers will vary.

12. (a), (b)

Class	Frequency	Relative Frequency	Class	Relative Frequency	Frequency
3000–3999	5	0.1613	8000–8999	5	0.1613
4000–4999	4	0.1290	9000–9999	0	0.0000
5000–5999	8	0.2581	10,000–10,999	1	0.0323
6000–6999	3	0.0968	11,000–11,999	0	0.0000
7000–7999	3	0.0968	12,000–12,999	2	0.0645

(c)

Distribution is skewed right.

(d)

Towing Capacity of SUVs

(e) For (a) and (b):

Diameter of a Chocolate Chip Cookie		
Class	Frequency	Relative Frequency
2.2000–2.2399	5	0.1471
2.2400–2.2799	11	0.3235
2.2800–2.3199	11	0.3235
2.3200–2.3599	6	0.1765
2.3600–2.3999	1	0.0294

13. (a), (b)

Diameter of a Cookie		
Class	Frequency	Relative Frequency
2.2000–2.2199	2	0.0588
2.2200–2.2399	3	0.0882
2.2400–2.2599	5	0.1471
2.2600–2.2799	6	0.1765
2.2800–2.2999	4	0.117
2.3000–2.3199	7	0.2059
2.3200–2.3399	5	0.1471
2.3400–2.3599	1	0.0294
2.3600–2.3799	1	0.0294

For (c):

Diameter of Chocolate Chip Cookies

The distribution is symmetric.

(c)

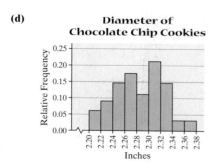

Diameter of Chocolate Chip Cookies

The distribution is symmetric.

For (d):

Diameter of Chocolate Chip Cookies

Answers will vary.

(d)

Diameter of Chocolate Chip Cookies

14. (a), (b)

Class	Frequency	Relative Frequency
85,000–94,999	1	0.0213
95,000–104,999	4	0.0851
105,000–114,999	3	0.0638
115,000–124,999	11	0.2340
125,000–134,999	5	0.1064
135,000–144,999	8	0.1702
145,000–154,999	8	0.1702
155,000–164,999	3	0.0638
165,000–174,999	2	0.0426
175,000–184,999	2	0.0426

(c)

Closing Price of Home Sales

(b)
```
4 | 9
5 | 0 2 3 5 6 6 7 7 9 9 9
6 | 0 0 0 1 1 1 1 1 2 2 2 2 2 2 2 2 2 3 4 4 5 5 5 6 6 7 8 8 8 9 9 9 9
7 | 0 1 3 4 4 7 8
8 |
9 | 1
```

The distribution is skewed right.

(c)
```
4 |
4 | 9
5 | 0 2 3
5 | 5 6 7 7 9 9 9
6 | 0 0 0 1 1 1 1 1 2 2 2 2 2 2 2 2 2 3 4 4
6 | 5 5 5 6 6 7 8 8 8 9 9 9 9
7 | 0 1 3 4 4
7 | 7 8
8 |
8 |
9 | 1
```

Answers will vary.

(d)

Closing Price of Home Sales

15.
```
 0 | 2 3 4 4 5 6 7 7 8
 1 | 4 7
 2 | 1 3 3 4 4 6 7
 3 | 2 3 5 5 5 8 9
 4 | 1 9 9
 5 | 1 3 8 8 9
 6 | 0 1 7
 7 | 6
 8 | 3
 9 |
10 | 2
11 | 1
```

The distribution is skewed right.

17. (a)

Federal Minimum Wage

(b) The trend indicates that the federal minimum wage remains constant.

18. (a)

Federal Minimum Wage (in 2003 dollars)

(b) Trend indicates that the value of the federal minimum wage is decreasing.
(c) Answers will vary.

19. No vertical scale.

20. Answers will vary.

21. (a) Graphs will vary.
(b) An example of a graph that does not mislead:

2002 Average Earnings

16. (a)

State	Fertility Rate	State	Fertility Rate	State	Fertility Rate
Alabama	61	Kentucky	61	North Dakota	59
Alaska	74	Louisiana	65	Ohio	62
Arizona	78	Maine	50	Oklahoma	69
Arkansas	67	Maryland	61	Oregon	62
California	68	Massachusetts	57	Pennsylvania	56
Colorado	69	Michigan	61	Rhode Island	55
Connecticut	59	Minnesota	62	South Carolina	61
Delaware	62	Mississippi	66	South Dakota	68
District of Columbia	53	Missouri	62	Tennessee	62
Florida	63	Montana	60	Texas	77
Georgia	68	Nebraska	70	Utah	91
Hawaii	69	Nevada	73	Vermont	49
Idaho	74	New Hampshire	52	Virginia	62
Illinois	66	New Jersey	64	Washington	60
Indiana	65	New Mexico	71	West Virginia	57
Iowa	62	New York	60	Wisconsin	59
Kansas	69	North Carolina	65	Wyoming	64

CHAPTER 3

3.1 Assess Your Understanding (page 116)

1. A statistic is resistant if it is not sensitive to extreme values. The median is resistant because it is a positional measure of central tendency, and increasing the largest value or decreasing the smallest value does not affect the position of the center. The mean is not resistant because it is a function of the sum of the values of data. Changing the magnitude of one value changes the sum of the values. The mode is a resistant measure of center.

2. The mean and the median are approximately equal when the data are symmetric. If the mean is significantly greater than the median, the data are skewed right. If the mean is significantly less than the median, the data are skewed left.

3. $55,263 is the mean household income.

4. HUD uses the median because the data are skewed. Explanations will vary.

5. The mean will be larger. Explanations will vary.

6. The median is between the 5000th and the 5001st ordered values.

7. The mode is used with qualitative data because they cannot be ordered to find the median or given numerical values needed to calculate the mean.

8. Parameter; statistic

9. False

10. False

11. $\bar{x} = 11$

12. $\bar{x} = 12$

13. $\mu = 9$

14. $\mu = 15$

15. Mean price per ad slot is $2.4 million.

16. The missing value is 32.

17. Mean cost is $381.75; median cost is $414.50; there is no mode cost.

18. Mean cost is $41.41; median cost is $40.76; there is no mode cost.

19. The mean, median, and mode strengths are 3670, 3830, and 4090 pounds per square inch, respectively.

20. The mean, median, and mode flight times are 266, 266, and 260 minutes, respectively.

21. (a) $\mu > M$
    (b) $\mu = M$
    (c) $\mu < M$
    Justification will vary.

22. (a) IV
    (b) III
    (c) II
    (d) I

23. Los Angeles ATM fees: $\bar{x} = \$1.44$, $M = \$1.50$, mode = $1.50 and $2.00. New York ATM fees: $\bar{x} = \$1.06$, $M = \$1.13$, mode = $1.00. It appears that ATM fees in Los Angeles are higher than they are in New York. Explanations will vary.

24. Blue stimulus: $\bar{x} = 0.551$ second, $M = 0.5315$ second, no mode time. Red stimulus: $\bar{x} = 0.458$ second, $M = 0.432$ second, no mode time. Answers will vary.

25. (a) The mean pulse rate is 72.2 beats per minute.
    (b) Samples and sample means will vary.
    (c) Answers will vary.

26. (a) The mean travel time is 26.4 minutes.
    (b) Samples and sample means will vary.
    (c) Answers will vary.

27. (a) The mean number of goals she scored is 8.7 per year.
    (b) Samples and sample means will vary.
    (c) Answers will vary.

28. (a) Mean winning time = 86.251 min
        Median winning time = 86.251 min
    (b) Mean winning distance = 3453 km
        Median winning distance = 3453 km
    (c) Mean winning margin = 5.667 min
        Median winning margin = 6.317 min
    (d) Mean winning speed: $\bar{x} = 40.420$ km/h

    Winning speed: $\dfrac{\sum \text{distance}}{\sum \text{time}} = 40.414$ km/h

    Winning speed: $\dfrac{\text{mean distance}}{\text{mean time}} = 40.414$ km/h

29. The mean is the better measure of central tendency because the distribution is symmetric.

30. The distribution is skewed right since the mean is significantly greater than the median. The median is the better measure of central tendency of the distribution.

31. (a) $\bar{x} = 51.1$; $M = 51$ (b) The distribution is symmetric.

32. (a) $\bar{x} = 5.883$ million shares; $M = 5.58$ million shares
    (b) The distribution is skewed right.

33. $\bar{x} = 0.874$ gram; $M = 0.88$ gram. The distribution is symmetric; so the mean is the better measure of central tendency.

Weight of Plain M&Ms

34. $\bar{x} = 104.1$ seconds; $M = 104$ seconds. The mean is approximately equal to the median, so the distribution is symmetric. The mean is the better measure of central tendency.

Length of Eruptions

35. $\bar{x} = 22$ hours; $M = 25$ hours. The distribution is skewed left, so the median is the better measure of central tendency.

Hours Worked per Week

36. $\bar{x} = \$1392.83$; $M = \$1177.50$. The distribution is skewed right, so the median is the better measure of central tendency.

Car Dealer's Profit

37. The mode region of birth is Central America.

38. The mode offence is street or highway robbery.

39. The mode candidate is Bush.

40. The mode diagnosis of admission is gunshot wound.

41. Sample of size 5: All data recorded correctly: $\bar{x} = 99.8$; $M = 100$; 106 recorded as 160: $\bar{x} = 110.6$; $M = 100$

Sample of size 12: All data recorded correctly: $\bar{x} = 100.4$; $M = 101$; 106 recorded as 160: $\bar{x} = 104.9$; $M = 101$

Sample of size 30: All data recorded correctly: $\bar{x} = 100.6$; $M = 99$; 106 recorded as 160: $\bar{x} = 102.4$; $M = 99$

For each sample size, the mean becomes larger, but the median remains constant. As the sample size increases, the impact of the misrecorded data on the mean decreases.

42. (a) $\mu = 27.1$ years; $M = 27$ years; mode $= 26$ years
(b) $\mu = 249.8$ lbs; $M = 245$ lb; mode $= 305$ lb
(c) $\mu = 4.6$ years; $M = 4$ years; mode $= 3$ years
(d) Mode college: Purdue
(e) Answers will vary.
(f) Offensive guards: $\mu = 306.4$ lb; $M = 305$ lb; mode $= 305$ lb. Running backs: $\mu = 217.8$ lb; $M = 220$ lb; mode $= 225$ lb.
(g) It does not make sense to compute the mean player number. Player number is a qualitative variable.

43. Answers will vary.

44. The salary distribution is skewed right, so the players' negotiator would want to use the median salary; the owner's negotiator would use the mean salary to refute the players' claim.

45. The distribution is skewed right, so the median amount of money lost is less than the mean amount lost.

46. The unreadable score is 44.

47. The sum is 204.

48. (a) Median; the distribution of home prices tends to be skewed.
(b) Mode; the data are qualitative.
(c) Mean; the data are quantitative and symmetric.
(d) Median; the data are quantitative and skewed.
(e) Median; NFL salaries are skewed right.
(f) Mode; the data are qualitative.

49. (a) Mean $=$ $50,000; median $=$ $50,000; mode $=$ $50,000
(b) New data set: 32.5, 32.5, 47.5, 52.5, 52.5, 52.5, 57.5, 57.5, 62.5, 77.5; new mean $=$ $52,500; new median $=$ $52,500; new mode $=$ $52,500. All three measures increased by $2500.
(c) New data set: 31.5, 31.5, 47.25, 52.5, 52.5, 52.5, 57.75, 57.75, 63, 78.75. new mean $=$ $52,500; new median $=$ $52,500; new mode $=$ $52,500. All three measures increased by 5%.
(d) New data set: 30, 30, 45, 50, 50, 50, 55, 55, 60, 100; new mean $=$ $52,500; new median $=$ $50,000; new mode$=$ $50,000; The mean increased by $2500, but the median and the mode remained at $50,000.

50. (a) $\bar{x} = 75.2$
(b) $M = 71$
(c) The median is the better measure of central tendency since the distribution is skewed right.
(d) $\bar{x} = 79.2$
(e) Adding 4 to each score increases the mean by 4.

51. The trimmed mean is 0.875. Explanations will vary.

52. Midrange $= 0.85$. Midrange is not resistant; it only considers extreme data points.

3.2 Assess Your Understanding (page 134)

1. It is not appropriate because the units are different.

2. The degrees of freedom are the number of data values that are free to vary.

3. No, none of the measures of dispersion is resistant.

4. Zero

5. A statistic is biased when it consistently overestimates or underestimates the population parameter.

6. Range

7. The standard deviation is the square root of the variance.

8. Mean; mean; spread

9. True

10. True

11. $s^2 = 36$; $s = 6$

12. $s^2 = 100$; $s = 10$

13. $\sigma^2 = 16$; $\sigma = 4$

14. $\sigma^2 = 64$; $\sigma = 8$

15. $s^2 = 196$; $s = 14$

16. $\sigma^2 = 25$; $\sigma = 5$

17. $R = \$226$; $s^2 = 9962.9$; $s = \$99.81$

18. $R = \$13.92$; $s^2 = 22.584$; $s = \$4.75$

19. $R = 1150$ psi; $s^2 = 211{,}275$ psi^2; $s = 459.6$ psi

20. $R = 25$ minutes; $s^2 = 71$ minutes^2; $s = 8.4$ minutes

21. (b) The data go from 30 to 75, in (a) they are clustered between 40 and 60.

22. (a) III **(b)** I **(c)** IV **(d)** II

23. Los Angeles: $R = \$2.00$, $s = \$0.68$; New York: $R = \$1.50$, $s = \$0.48$; Los Angeles has more dispersion based on both the range and standard deviation.

24. Blue: $R = 0.574$ second, $s = 0.1994$ second; Red: $R = 0.140$ second, $s = 0.0647$ second; The response time to the blue stimulus appears to have more variability. Reasons will vary.

25. (a) $\sigma^2 = 58.8$, $\sigma = 7.7$ beats per minute
(b) (c) Answers will vary.

26. (a) $\sigma^2 = 164.9$, $\sigma = 12.8$ minutes
(b) (c) Answers will vary.

27. (a) $\sigma^2 = 41.0$, $\sigma = 6.4$ goals
(b) (c) Answers will vary.

28. (a) $R = 10.465$ hours, $\sigma^2 = 13.981$, $\sigma = 3.739$ hours
(b) $R = 409$ km, $\sigma^2 = 19{,}793.3$, $\sigma = 140.7$ km
(c) $R = 6.6$ minutes, $\sigma^2 = 4.390$, $\sigma = 2.095$ minutes
(d) $R = 2.09$ km/h, $\sigma^2 = 0.423$, $\sigma = 0.651$ km/h

29. (a) Ethan: $\mu = 10$ fish, $R = 19$ fish; Drew: $\mu = 10$ fish, $R = 19$ fish
(b) Ethan: $\sigma = 4.9$ fish, Drew: $\sigma = 7.9$ fish
(c) Answers will vary.

30. (a) Offense: $\mu = 260.3$ lb, $R = 169$ lb, $\sigma = 53.8$ lb
(b) Defense: $\mu = 245.4$ lb, $R = 129$ lb, $\sigma = 39.2$ lb
(c) The weights of the offense have more dispersion.

31. $R = 45$, $s^2 = 118.0$, $s = 10.9$

32. $R = 7.95$ million shares, $s^2 = 4.238$, $s = 2.059$ million shares

33. (a) $s = 0.04$ gram
(b) The histogram is approximately symmetric, so the Empirical Rule is applicable.
(c) 95% of the M&Ms should weigh between 0.79 and 0.95 gram.
(d) 98% of the M&Ms actually weigh between 0.79 and 0.95 gram.
(e) 16% of the M&Ms should weigh more than 0.91 gram.
(f) 14% of the M&Ms actually weigh more than 0.91 gram.

34. (a) $s = 6$ seconds
(b) The histogram is approximately symmetric, so the Empirical Rule is applicable.
(c) 95% of the eruptions should last between 92 and 116 seconds.
(d) 93% of the observed eruptions actually lasted between 92 and 116 seconds.
(e) 16% of the eruptions should last less than 98 seconds.
(f) 11% of the observed eruptions actually lasted fewer than 98 seconds.

35.

	Car 1	Car 2	
Sample mean	$\bar{x} = 223.5$	$\bar{x} = 237.2$	mi./10 gal
Median	$M = 223$	$M = 230$	mi./10 gal
Mode	none	none	mi./10 gal
Range	$R = 93$	$R = 166$	mi./10 gal
Sample variance	$s^2 = 475.1$	$s^2 = 2406.9$	
Sample standard deviation	$s = 21.8$	$s = 49.1$	mi./10 gal

Answers will vary.

36.

	Rates of Return	
	Fund A	**Fund B**
Sample mean	$\bar{x} = 3.05$	$\bar{x} = 3.41$
Median	$M = 3.05$	$M = 3.65$
Mode	None	4.3
Range	$R = 10.9$	$R = 19.6$
Sample variance	$s^2 = 8.95$	$s^2 = 31.23$
Sample standard deviation	$s = 2.99$	$s = 5.59$

Answers will vary.

37. (a) Financial stocks: $\bar{x} = 15.716$, $M = 16.09$; energy stocks: $\bar{x} = 22.481$, $M = 19.585$. Energy stocks have the higher median rate of return.
(b) Financial stocks: $s = 7.378$; energy stocks: $s = 12.751$. Energy stocks are riskier.

38. (a) American League: $\bar{x} = 4.157$, $M = 4.195$; National League: $\bar{x} = 3.748$, $M = 3.855$. The American League has both higher mean and median earned-run averages.
(b) American League: $s = 0.787$; National League: $s = 0.610$ The American League has more dispersion.

39. (a) 95% of people has an IQ score between 70 and 130.
(b) 5% of people has an IQ score either less than 70 or greater than 130.
(c) 2.5% of people has an IQ score greater than 130.

40. (a) 68% of SAT Math scores is between 404 and 632.
(b) 32% of SAT Math scores is either less than 404 or greater than 632.
(c) 2.5% of SAT Math scores is greater than 746.

41. (a) 95% of pairs of kidneys weighs between 265 and 385 grams.
(b) 99.7% of pairs of kidneys weighs between 235 and 415 grams.
(c) 0.3% of pairs of kidneys weighs either less than 235 or more than 415 grams.
(d) 81.5% of pairs of kidneys weighs between 295 and 385 grams.

42. (a) 68% of bolts is between 3.993 and 4.007 inches long.
(b) 95% of bolts is between 3.986 and 4.014 inches long.
(c) 5% of the bolts is discarded.
(d) 15.85% of bolts is between 4.007 and 4.021 inches long.

43. (a) 88.9% of gas stations has prices within 3 standard deviations of the mean.
(b) 84% of gas stations has prices within 2.5 standard deviations of the mean. Gasoline priced from \$1.245 to \$1.495 is within 2.5 standard deviations of the mean.
(c) At least 75% of gas stations has prices between \$1.27 and \$1.47.

44. (a) 75% of commuters in Boston has a commute time within 2 standard deviations of the mean.
(b) 55.6% of commuters has a commute time within 1.5 standard deviations of the mean. They commute between 15.15 and 39.45 minutes to work.
(c) 88.9% of commuters commutes between 3 and 51.6 minutes to work.

45. There is more variation among individuals than among means.

46. (a) $R = \$45$, $\sigma^2 = 160$, $\sigma = \$12.6$
(b) Option 1: $R = \$45$, $\sigma^2 = 160$, $\sigma = \$12.6$
(c) Option 2: $R = \$47.25$, $\sigma^2 = 176.4$, $\sigma = \$13.3$
(d) Option 3: $R = \$70$, $\sigma^2 = 341.3$, $\sigma = \$18.5$

47. Sample of size 5: correct $s = 5.3$, incorrect $s = 27.9$
Sample of size 12: correct $s = 14.7$, incorrect $s = 22.7$
Sample of size 30: correct $s = 15.9$, incorrect $s = 19.2$
As the sample size increases, the impact of the misrecorded observation on standard deviation decreases.

48. $s = 0$. If all values in a data set are identical, there is zero variance.

49. (a) There is more variability in blood pressure after exercise.
(b) There is more variability in calcium concentration in the group with high blood pressure.

3.3 Assess Your Understanding (page 146)

1. The class midpoint is meant to represent the average value of all the data points in the class.

2. $\bar{x} = \dfrac{\sum x_i}{n}$ is a weighted average for which the value of each weight is 1.

3. $\bar{x} = \$32.83$, $s = \$11.06$

4. $\mu = 14.6$ points, $\sigma = 11.0$ points

5. $\mu = 17.3$ days, $\sigma = 12.6$ days

6. $\bar{x} = 21.6$ hours per week, $s = 11.7$ hours per week

7. $\mu = 44.5$ years, $\sigma = 10.5$ years

8. $\mu = 2.8$, $\sigma = 1.1$

9. (a) $\mu = 80.9°F$, $\sigma = 8.2°F$
(b)

High Temperature in August in Chicago
(c) 95% of the time it is between 64.5° and 97.3°F.

10. (a) $\mu = 37.8$ inches, $\sigma = 7.2$ inches
(b)

Annual Rainfall for St. Louis, MO
(c) 95% of the time annual rainfall is between 23.4 and 52.2 inches.

11. (a) $\mu = 32.3$ years of age, $\sigma = 5.2$ years
(b)
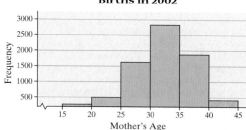
Number of Multiple Births in 2002
(c) 95% of mothers of multiple births is between 21.9 and 42.7 years of age.

12. (a) $\mu = 603.1$, $\sigma = 92.0$
(b)

SAT Verbal Scores, 2003
(c) 95% of the SAT Verbal scores is between 419 and 787.

13. Grouped data: $\bar{x} = 51.8$, $s = 12.1$

14. Grouped data: $\bar{x} = 6$ million shares, $s = 1.879$ million shares

15. GPA $= 3.27$

16. Course average $= 87.15\%$

17. Cost per pound $= \$2.97$

18. Cost per pound $= \$3.38$

19. (a) Males: $\mu = 35.6$ years, $\sigma = 21.7$ years
    (b) Females: $\mu = 38.1$ years, $\sigma = 23.0$ years
    (c) Females have the higher mean age.
    (d) Females have more dispersion in age.

20. (a) 1980: $\mu = 26.0$ years, $\sigma = 5.7$ years
    (b) 2002: $\mu = 27.6$ years, $\sigma = 6.2$ years
    (c) 2002 has the higher mean age of new mothers.
    (d) 2002 has more dispersion in the age of new mothers.

3.4 Assess Your Understanding (page 156)

1. Answers will vary. The kth percentile separates the lower k percent of the data from the upper $(100 - k)$ percent of the data.

2. Since the percentile of a score is rounded to the nearest integer, it is possible for two scores to have the same percentile.

3. A four-star fund is in the 4th quintile of the funds. That is, it is above the bottom 60%, but below the top 20% of the ranked funds.

4. No; when an outlier is discovered, it should be investigated to find its cause.

5. One qualifies for Mensa if one's intelligence is in the top 2% of people.

6. Comparing z-scores allows a unitless comparison of the number of standard deviations an observation is from the mean.

7. The 40-week gestation baby weighs less relative to the gestation period.

8. The 34-week gestation baby weighs less relative to the gestation period.

9. The woman is relatively taller.

10. The man is relatively taller.

11. Peavy had the better year because his ERA was 2.50 standard deviations below the National League's mean ERA, while Santana's ERA was 2.21 standard deviations below the American League's mean ERA.

12. Suzuki had the better year relative to his peers, because his batting average was 4.88 standard deviations above the American League's mean. Williams's batting average was 3.82 standard deviations above the mean 1941 batting average.

13. (a) Approximately 40% of the states has violent crime rates less than 329.35 crimes per 100,000 population.
    (b) Approximately 95% of the states has violent crime rates less than 761,85 crimes per 100,000 population.
    (c) Approximately 10% of the states has violent crime rates less than 187.2 crimes per 100,000 population.
    (d) Florida is at the 94th percentile, meaning that approximately 94% of the states has violent crime rates less than 730.2 crimes per 100,000 population.
    (e) California is at the 78th percentile, meaning that approximately 78% of the states has violent crime rates less than 579.3 violent crimes per 100,000 population.

14. (a) Approximately 30% of the states has violent crime rates less than 280.7 crimes per 100,000 population.
    (b) Approximately 85% of the states has violent crime rates less than 652.15 crimes per 100,000 population.
    (c) Approximately 5% of the states has violent crime rates less than 109.55 crimes per 100,000 population.
    (d) New Mexico is at the 88th percentile, meaning that approximately 88% of the states has violent crime rates less than 665.2 crimes per 100,000 population.
    (e) Rhode Island is at the 29th percentile, meaning that approximately 29% of the states has violent crimes rates less than 285.6 crimes per 100,000 population.

15. (a) $z = -1.70$. The rainfall in 1971 was 1.70 standard deviations below the mean April rainfall.
    (b) $Q_1 = P_{25} = 2.625$ in.; $Q_2 = P_{50} = M = 3.985$ in.; $Q_3 = P_{75} = 5.36$ in.

    (c) $IQR = 2.735$ in.
    (d) Lower fence $= -1.478$ in., upper fence $= 9.463$ in. There are no outliers according to this criterion.

16. (a) $z = -1.21$. Blackie's hemoglobin level is 1.21 standard deviations below the mean.
    (b) $Q_1 = P_{25} = 9.15$ g/dL, $Q_2 = P_{50} = M = 9.95$ g/dL, $Q_3 = P_{75} = 11.1$ g/dL
    (c) $IQR = 1.95$ g/dL
    (d) Lower fence $= 6.225$ g/dL, upper fence $= 14.025$ g/dL. According to this criterion, 5.7 g/dL is an outlier.

17. (a) $z = 0.61$. The organic carbon concentration in this soil sample is 0.61 standard deviations above the mean concentration.
    (b) $Q_1 = P_{25} = 10.01$ mg/L, $Q_2 = P_{50} = M = 15.42$ mg/L, $Q_3 = P_{75} = 20.13$ mg/L
    (c) $IQR = 10.12$ mg/L
    (d) Lower fence $= -5.17$ mg/L, upper fence $= 35.31$ mg/L. There are no outliers according to this criterion.

18. (a) $z = 1.60$. The organic carbon concentration in this soil sample is 1.60 standard deviations above the mean concentration.
    (b) $Q_1 = P_{25} = 7.11$ mg/L, $Q_2 = P_{50} = M = 9.1$ mg/L, $Q_3 = P_{75} = 11.72$ mg/L
    (c) $IQR = 4.61$ mg/L
    (d) Lower fence $= 0.195$ mg/L, upper fence $= 18.635$ mg/L. According to this criterion, 21, 21.4, 21.82, and 22.62 are outliers.

19. The cutoff point is 574 minutes. If more minutes are used, the customer is contacted.

20. Daily charges must exceed $219.00 before the customer is contacted.

21. (a) $12,777 is an outlier.
    (b)

Students' Weekly Income

    (c) Answers will vary.

22. (a) $115 and $1000 are outliers.
    (b)

Students' Weekly Expenditures

    (c) Answers will vary.

23. Mean of the z-scores is 0.0.

    Standard deviation of the z-scores is 1.0.

Student	z-Score
Perpectual Bempah	0.49
Megan Brooks	-1.58
Jeff Honeycutt	-1.58
Clarice Jefferson	1.14
Crystal Kurtenbach	-0.03
Janette Lantka	1.07
Kevin McCarthy	1.01
Tammy Ohm	-0.55
Kathy Wojdyla	0.10

24. Mean of the z-scores is 0.0.
Standard deviation of the z-scores is 1.0.

Student	z-Score	Student	z-Score
Amanda	0.98	Scot	1.45
Amber	−0.42	Erica	−1.20
Tim	−1.36	Tiffany	−1.13
Mike	0.44	Glenn	0.98
Nicole	0.28		

3.5 Assess Your Understanding (page 164)

1. The median and interquartile range are better measures of central tendency and dispersion if the data are skewed or if the data contain outliers.

2. Right

3. (a) The distribution is skewed right.
(b) Five-number summary: 0, 1, 3, 6, 16

4. (a) The distribution is symmetric.
(b) Five-number summary: −1, 2, 5, 8, 11

5. Five-number summary: 42, 51, 55, 58, 69

Age of Presidents at Inauguration

Graph is symmetric.

6. Five-number summary: 1, 11, 20, 28, 40

Grams of Fat

The distribution is symmetric.

7. Five-number summary: 1, 7, 15, 22, 45

Super Bowl Point Spreads

The distribution is skewed right.

8. Five-number summary: 18, 26, 30, 32, 46

Miles per Gallon—2005 Compacts

This distribution is skewed left.

9. Five-number summary: 0.598, 0.604, 0.608, 0.610, 0.612

Weight (in grams) of Tylenol Tablets

The distribution is skewed left.

10. Five-number summary: 421, 653, 684, 713, 971

Gas Expenditures per Person (by state)

The distribution is symmetric.

11. (a) Five-number summary: 28, 45, 51, 57.5, 73

(b) **Serum HDL Cholesterol Level**

(c) The distribution is symmetric.
(d) Mean and standard deviation

12. (a) Five-number summary: 3.01, 4.43, 5.58, 7.16, 10.96

(b) **Daily Volume (in millions) of Altria Group Traded**

(c) The distribution is skewed right.
(d) Median and IQR

13. (a) Five-number summary: 0, 0.205, 1.055, 1.87, 2.83

(b) **Dividend Yield**

(c) The distribution is skewed right.
(d) Median and IQR

14. (a) Five-number summary: 78, 279, 375, 563, 1633

(b) **Violent Crimes by State in 2002**

(c) The distribution is skewed right.
(d) Median and IQR

15. Five-number summaries:

Keebler	20,	21.5,	25,	28,	33
Store brand	16,	21,	24,	27.5,	33

Chips per Cookie

Answers will vary.

16. Five-number summaries:

Oklahoma:	18,	44,	62,	78,	145
Kansas:	42,	64,	88,	101,	122
Nebraska:	26,	55,	63,	81,	110

Number of Tornados, 1990–2004

17. Five-number summaries:

McGwire:	340,	380,	420,	450,	550
Sosa:	340,	370.5,	410,	430,	500
Bonds:	320,	380,	410,	420,	488

Homerun Distances

Chapter Review Exercises (page 170)

1. (a) Mean = 792.51 m/s, median = 792.40 m/s
(b) Range = 4.8 m/s, s^2 = 2.03, s = 1.42 m/s

2. (a) Mean = 126.8 beats/min, median = 128.5 beats/min
(b) Range = 83 beats/min, s^2 = 589.07, s = 24.3 beats/min

3. (a) $\bar{x}$ = \$10,178.89, M = \$9980
(b) Range: R = \$8550, s = \$3074.92
(c) $\bar{x}$ = \$13,178.89, M = \$9980; range: R = \$35,550, s = \$10,797.46
The median is resistant.

4. (a) Mean = \$138,068, median = \$136,924
(b) Range = \$70,541, s = \$17,286.30

5. (a) μ = 57.8 years, median = 58 years, bimodal: 56 and 62 years
(b) Range = 25 years, σ = 7.0 years
(c) Answers will vary.

6. (a) μ = 177.9 home runs, M = 184 home runs, mode = 135 home runs
(b) Range: R = 100 home runs, σ = 31.3 home runs
(c) Answers will vary.
(d) He is not lying because the mode is an average, but he is being deceptive because usually the word average is used for the mean.

7. (a) Mean = 2.2 children, median = 2.5 children
(b) Range = 4 children, s = 1.3 children

8. (a) Mean = 4.5 cars, median = 4.5 cars
(b) Range = 8 cars, s = 2.3 cars

9. (a) 441, 759
(b) 95% of the light bulbs has a life between 494 and 706 hours.
(c) 81.5% of the light bulbs has a life between 547 and 706 hours.
(d) The firm can expect to replace 0.15% of the light bulbs.
(e) At least 84% of the light bulbs has a life within 2.5 standard deviations of the mean.
(f) At least 75% of the light bulbs has a life between 494 and 706 hours.

10. (a) 3282, 5322
(b) 95% of the cartridges prints between 3622 and 4982 pages.
(c) The firm can expect to replace 2.5% of the cartridges.
(d) At least 55.6% of the cartridges has a page count within 1.5 standard deviations of the mean.
(e) At least 88.9% of the cartridges prints between 3282 and 5322 pages.

11. (a) μ = 42.28 years of age
(b) σ = 15.93 years

12. (a) μ = 43.71 years of age
(b) σ = 16.78 years
(c) Answers will vary.

13. Michael's GPA is 3.33.

14. \$2.17

15. (a) Yankees: μ = \$6,351,515.52; Mets: μ = \$3,452,177.50
(b) Yankees: M = \$3,100,000; Mets: M = \$900,000
(c) The distributions are skewed right since their means are larger than their medians.
(d) Yankees: σ = \$6,242,767.24; Mets σ = \$4,643,606.15
(e) Yankees: \$301,400, \$837,500, \$3,100,000, \$11,623,571.50, \$22,000,000
Mets: \$300,000, \$318,375, \$900,000, \$4,666,666.50, \$17,166,667

(f)

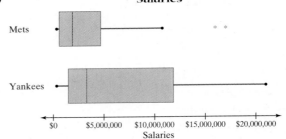

Salaries

The Mets data contain two outliers: \$17,166,667 (Mo Vaughn) and \$16,071,429 (Mike Piazza).

(g) Both distributions are skewed right.
(h) The median is the better measure of central tendency since the distributions are skewed.

16. (a) Material A: $\bar{x}$ = 6.404 million cycles; material B: $\bar{x}$ = 11.332 million cycles
(b) Material A: M = 5.785 million cycles; material B: M = 8.925 million cycles
(c) Both distributions are skewed right since the mean is larger than the median.
(d) Material A: s = 2.626 million cycles; material B: s = 5.900 million cycles
(e) Machine A (in million cycles): 3.17, 4.52, 5.785, 8.01, 11.92
Machine B (in million cycles): 5.78, 6.84, 8.925, 14.71, 24.37

(f)

Bearing Failures

Annotations will vary.
(g) The distributions shown in the boxplots are skewed right.

17. (a) \$368,817; approximately 40% of the drivers earned less than \$368,817.
(b) \$5,957,165; approximately 95% of the drivers earned less than \$5,957,165.
(c) \$67,862.50; approximately 10% of the drivers earned less than \$67,862.50.
(d) \$4,117,750 is at the 83rd percentile.
(e) \$116,359 is at the 15th percentile.

18. (a) $268,422,50; approximately 30% of the drivers earned less than $268,422.50.
  (b) $4,749,015; approximately 90% of the drivers earned less than $4,749,015.
  (c) $57,007.50; approximately 5% of the drivers earned less than $57,007.50.
  (d) $1,333,520 is at the 56th percentile.
  (e) $139,614 is at the 18th percentile.

19. The female is relatively heavier.

20. (a) The median crime rate is approximately 4050.
  (b) The 25th percentile crime rate is approximately 3100.
  (c) There is one outlier, at approximately 8000.
  (d) The lowest crime rate is approximately 2200.

CHAPTER 4

4.1 Assess Your Understanding (page 185)

1. Univariate data measure the value of a single variable for each individual in the study. Bivariate data measure values of two variables for each individual.

2. A lurking variable is a variable that is related either to the response variable or to the explanatory variable, or both, but is not measured in the study. Examples will vary.

3. Two variables that are linearly related are positively associated if increases in the explanatory variable tend to be associated with increases in the response variable.

4. If $r = +1$, then there is a perfect positive linear relation between the variables, and the points of the scatter diagram lie exactly on a straight line with a positive slope.

5. If $r = 0$, then there is no *linear* relationship between the variables.

6. No, the linear correlation coefficient is sensitive to outliers. Supporting examples will vary.

7. The linear correlation coefficient can only be calculated from bivariate *quantitative* data, and the gender of a driver is a qualitative variable.

8. Answers will vary. Answers should include that correlation measures the strength of the linear association between two quantitative variables.

9. Correlation is a unitless measure of association; it cannot be 0.93 *bushel*.

10. Correlation describes a relation between two variables in an observational study. Causation describes a conditional (if–then) relation in an experimental study.

11. Nonlinear

12. Linear, negative

13. Linear, positive

14. Nonlinear

15. (a) III   **(b)** IV
  (c) II   **(d)** I

16. (a) IV   **(b)** III
  (c) I   **(d)** II

17. (a) The relation is linear and positively associated.
  (b) The point (24, 55,000) appears to stick out. Reasons will vary.

18. (a) There is no linear relation between median income and birth rate.
  (b) The point (49,000, 21.25) appears to stick out. Reasons will vary.

19. (a)

  (b) $r = 0.9698$
  (c) Positive association

20. (a)

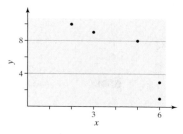

  (b) $r = -0.8822$
  (c) Negative association

21. (a)

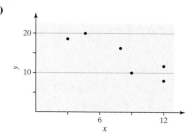

  (b) $r = -0.8553$
  (c) Negative association

22. (a)

  (b) $r = 0.9506$
  (c) Positive association

23. (a) Explanatory height; Response; head circumference
  (b)

  (c) $r = 0.9111$
  (d) Positive association

24. (a) Explanatory gestation period; Response: life expectancy
  (b)

  (c) $r = 0.7257$
  (d) Positive association
  (e) $r = 0.5923$; Correlation coefficient decreased. Correlation coefficients are not resistant to extreme values.

25. (a) Explanatory variable: weight; Response variable: gas mileage

(b)

(c) $r = -0.892$
(d) There is a negative linear relation between the weight of a car and its gas mileage.

26. (a)

(b) $r = 0.9513$
(c) Positive association
(d) Converting units has no effect on the correlation coefficient.

27. (a) Explanatory variable: Number of absences; Response variable: Average final grade

(b)

(c) $r = -0.947$
(d) There appears to be a strong negative linear relation between days absent and final grade.
(e) Going to class every day does not guarantee a passing grade. Factors may vary.

28. (a)

(b) $\rho = -0.873$
(c) There appears to be a strong negative linear relation between years since 1995 and antibiotic prescription rate.

29. (a)

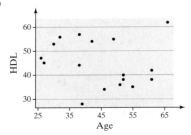

(b) -0.1637
(c) There appears to be little, if any, linear association between age and HDL cholesterol level.

30. (a)

(b) No, the relation between distance and intensity is not linear.

31. (a)

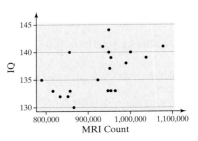

(b) $r = 0.548$. It would appear that there is a positive association between MRI count and IQ.

(c)

The females have lower MRI counts. Looking at each gender's plot, we see that the relation that appeared to exist between brain size and IQ disappears.
(d) Females: $r = 0.359$; Males: $r = 0.236$. There appears to be no linear relation between brain size and IQ.

32. (a)

(b) Answers will vary.
(c) Males: $\rho = 0.899$
(d) Females: $\rho = 0.863$
(e) Males have a slightly stronger linear relation between number of licensed drivers and number of crashes than females. Reasons will vary.

33. (a)

(b) Correlation coefficient (with Taurus included): $r = -0.879$
(c) The results are reasonable because the Taurus follows the overall pattern of the data.

(d)

Weight (in pounds)

(e) Correlation coefficient (with Prius included): $r = -0.437$
(f) The Prius is a hybrid car; the other cars are not hybrids.

34. (a)

Gestation Period

(b) $r = 0.847$
(c) The point corresponding to humans is an extreme point on the scatter diagram. A correlation coefficient is not resistant to extreme values.

35. (a) $r = 0.82$ for all four data sets.
(b)

36. Answers may vary. Compare correlation coefficients and scatter diagrams to choose the one that appears to have the strongest association.

37. To have the lowest correlation between the two stocks, invest in General Electric and TECO Energy. $r = 0.0169$. If your goal is to have one stock go up when the other goes down, invest in Cisco Systems and TECO Energy: $r = -0.235$.

38. (a)

Cases of Lyme Disease

(b) $r = 0.961$

(c) There appears to be a strong, positive, linear relation between cases of Lyme disease and number of drownings. No, Lyme disease is a summer disease. This is an instance of correlation, but not causation.

39. $r = 0.599$ implies that there is a positive linear relation between the number of television stations and life expectancy, but this is correlation, not causation. The more television stations a country has, the more affluent it is. The more affluent, the better the health care.

40. Parental smoking might be a lurking variable.

41. (a)

$r = 0.2280$

(b)

$r = 0.8598$

42. Although there is no linear relation between age and median income, the two variables are related.

43. (a) Positive   **(b)** Negative   **(c)** Negative
    (d) Negative   **(e)** No correlation

44. (a) Negative   **(b)** Positive   **(c)** Positive
    (d) No correlation   **(e)** Negative

45. (a)

(b) 0.9518

x	10	12	14	14	16	16	16	16
y	8.4	10	10.4	11.8	12	12.4	12.2	13.8
x	18	18	20	20	22	22	24	24
y	14.4	16	16.6	14.8	16.8	15.6	17	19.0

(d)

(e) 0.9518
(f) By doubling the values of x and y, we double the standard deviations and deviations about the mean. Therefore, the z-scores for each point are unchanged.

46. Some lurking variables might include parental weight or health consciousness of parents.

47. (a) It appears that the SAT score combined with the high school GPA is the best predictor of college GPA. They have the highest correlation coefficient.

    (b) It appears that SAT Verbal score is the worst predictor of college GPA. They have the lowest correlation coefficient.

4.2 Assess Your Understanding (page 204)

1. The least-squares regression line is the line that minimizes the sum of the squared errors (residuals).

2. A residual is the difference between the observed value of y and the predicted value of y. If the residual is positive, the prediction underestimates the actual value.

3. Values of the explanatory variable that are much larger or much smaller than those observed are considered *outside the scope of the model*. It is dangerous to make such predictions because we do not know the behavior of the data for which we have no observations.

4. If the linear correlation between two variables is negative, the slope of the regression line will also be negative.

5. Answers will vary.

6. True

7. Rounding error can prevent the sum of the residuals from equaling 0.

8. If $r = 0$, then the slope of the linear regression line will also be zero, and the line will be $y = \bar{y}$ since all regression lines contain the point $(\bar{x}, \bar{y})$.

9. (a)

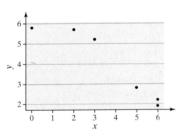

    (b) $\hat{y} = -0.7136x + 6.5498$
    (c)

10. (a)

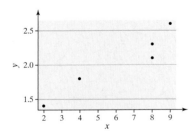

    (b) $\hat{y} = 0.1457x + 1.1370$

(c)

11. (a)

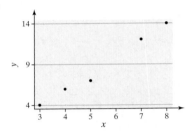

    (b) Using points $(3, 4)$ and $(8, 14)$: $\hat{y} = 2x - 2$
    (c)

    (d) $\hat{y} = 2.0233x - 2.3256$
    (e)

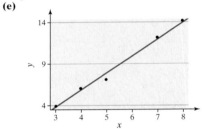

    (f) Sum of squared residuals (computed line): 1
    (g) Sum of squared residuals (least-squares line): 0.7907

12. (a)

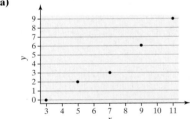

    (b) Using points $(3, 0)$ and $(11, 9)$: $\hat{y} = \dfrac{9}{8}x - \dfrac{27}{8}$
    (c)

    (d) $\hat{y} = 1.1x - 3.7$

(e)

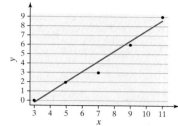

(f) Sum of squared residuals (computed line): 2.875
(g) Sum of squared residuals (least-squares line) : 1.6

13. (a)

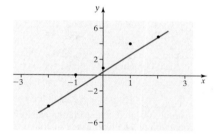

(b) Using points $(-2, -4)$ and $(2, 5)$: $\hat{y} = \dfrac{9}{4}x + \dfrac{1}{2} = 2.25x + 0.5$
(c)

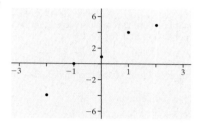

(d) $\hat{y} = 2.2x + 1.2$
(e)

(f) Sum of squares residual from my line: 4.875
(g) Sum of squares residual from least-squares line: 2.4
(h) Answers will vary.

14. (a)

(b) Using the points $(-2, 7)$ and $(2, 0)$: $\hat{y} = -\dfrac{7}{4}x + \dfrac{7}{2}$
  $= -1.75x + 3.5$

(c)

(d) $\hat{y} = -1.8x + 3.6$
(e)

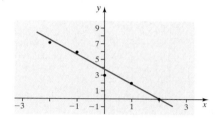

(f) Sum of squares residual from my line: 0.875
(g) Sum of squares residual from least-squares line: 0.8
(h) Answers will vary.

15. (a)

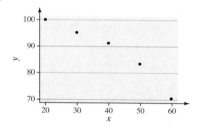

(b) Using points $(30, 95)$ and $(60, 70)$: $\hat{y} = -\dfrac{5}{6}x + 120$
(c)

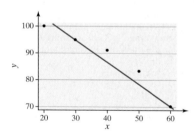

(d) $\hat{y} = -0.72x + 116.6$
(e)

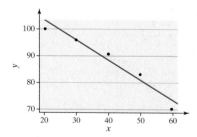

(f) Sum of squares residual from my line: 51.6667
(g) Sum of squares residual from least-squares line: 32.4

16. (a)

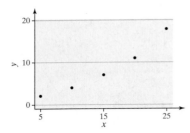

(b) Using points $(10, 4)$ and $(25, 18)$: $\hat{y} = \dfrac{14}{15}x - \dfrac{16}{3}$

(c)

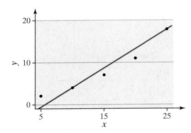

(d) $\hat{y} = 0.78x - 3.3$

(e)

(f) Sum of squares residual from my line: 15.3333
(g) Sum of squares residual from least-squares line: 9.1

17. (a) $\hat{y} = 0.1827x + 12.4932$
(b) If height increases by 1 inch, head circumference increases by about 0.1827 inch, on average. It is not appropriate to interpret the y-intercept. It is outside the scope of the model.
(c) $\hat{y} = 17.06$ inches
(d) Residual $= -0.16$ inch; below
(e)

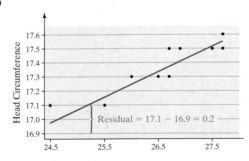

(f) For children who are 26.75 inches tall, head circumference varies.
(g) No, 32 inches is outside the scope of the model.

18. (a) $\hat{y} = 0.0261x + 7.8738$
(b) If the gestation period increases by 1 day, the life expectancy increases by 0.0261 year, on average. It is not appropriate to interpret the intercept, because a gestation period of 0 days does not make sense.
(c) 10.4 years
(d) 8.3 years
(e) 8.7 years
(f) -1.7 years

19. (a) $\hat{y} = -0.0062x + 41.4640$
(b) For every pound added to the weight of the car, gas mileage in the city will decrease by 0.0062 mile per gallon, on average. It is not appropriate to interpret the y-intercept.
(c) $\hat{y} = 21.0$ miles per gallon; residual: -1.0 miles per gallon; Mustang mileage is below average.
(d)

(e) It is not reasonable to use this least-squares regression to predict the miles per gallon of a Toyota Prius.

20. (a) $\hat{y} = 1.3902x + 1.1140$
(b) If the humerus increases by 1 mm, the length of the tibia increases by about 1.3902 mm. It does not make sense to interpret the intercept.
(c) Residual $= 0.548$ mm; above
(d)

(e) $\hat{y} = 36.30$ mm

21. (a) $\hat{y} = -2.8273x + 88.7327$
(b) For every additional absence, the student's final grade drops 2.8273 points, on average. The final grade of students who miss no classes is 88.7327.
(c) $\hat{y} = 74.60$; residual is -0.70; the final grade is below average for 5 absences.
(d)

(e) No, 15 absences is out of the scope of the model.

22. (a) $\hat{y} = -46x + 1123.857$
(b) For every year since 1995, 46 fewer prescriptions are written, on average. The y-intercept, 1123.857, can be interpreted as the number of prescriptions written in 1995.
(c) $\hat{y} = 801.9$, so the predicted number of prescriptions per 1000 children is 801.9.
(d)

(e) No, 2010 $(x = 15)$ is out of the scope of the model.

23. (a) $\hat{y} = 0.00003x + 108.8940$
(b) The slope is close to zero.
(c) MRI count 1,000,000: $\hat{y} = \bar{y} = 136$; MRI count 830,000: $\hat{y} = \bar{y} = 136$

24. (a) Male drivers: $\hat{y} = 0.0802x + 8.1400$; female drivers: $\hat{y} = 0.0505x + 15.1611$

(b) Males: If the number of drivers increases by 1, then the number of crashes increases by about 0.08 thousand, on average. Females: If the number of drivers increases by 1, then the number of crashes increases by 0.05 thousand, on average.

(c) Males: 705.9 thousand; females: 454.5 thousand

25. Answers will vary.

4.3 Assess Your Understanding (page 212)

1. 75% of total variation in the response variable is explained by the regression.

2. The total deviation of a data point is the difference between the observed value of the response variable, $\hat{y}$, and the mean of the response variable, $\bar{y}$. The explained deviation is the difference between the predicted value $\hat{y}$ of the response variable and the mean value, $\bar{y}$. The unexplained deviation is the difference between the observed value y and the predicted value $\hat{y}$.

3. (a) III
(b) II
(c) IV
(d) I

4. (a) $R^2 = 10.24\%$; 10.24% of the variance in the explanatory variable is explained by the least-squares regression line.
(b) $R^2 = 1.69\%$; 1.69% of the variance in the explanatory variable is explained by the least-squares regression line.
(c) $R^2 = 16\%$; 16% of the variance in the explanatory variable is explained by the least-squares regression line.
(d) $R^2 = 86.49\%$; 86.49% of the variance in the explanatory variable is explained by the least-squares regression line.

5. 83.0% of the variation in length of eruption is explained by the least-squares regression equation.

6. 57.5% of the variation in 28-day strength is explained by the least-squares regression equation.

7. (a) No, the data do not appear to follow a linear pattern.
(b) 6.8% of the variation in sugar content is explained by the least-squares regression equation. Yes.
(c)

42.1% of the variability in sugar content is explained by the least-squares regression equation.

8. (a) It appears that the data are linearly related; the data lie in an approximately straight line.
(b) $R^2 = 0.994$ indicates that 99.4% of the variation is explained by the linear model. R^2 close to 1 supports the conclusion that the data are linearly related.

9. (a) $R^2 = 83.0\%$
(b) 83.0% of the variation in head circumference is explained by the least-squares regression equation.

10. (a) $R^2 = 52.7\%$
(b) 52.7% of the variation in life expectancy is explained by the least-squares regression line. The least-squares regression model may not be adequate. The variance of the residuals may be decreasing as the gestation period increases. Further analysis may be warranted.

11. (a) $R^2 = 0.796$
(b) 79.6% of the variance in gas mileage is explained by the linear model. Residuals may be patterned.

12. (a) $R^2 = 90.5\%$
(b) 90.5% of the variation in the length of the right humerus is explained by the least-squares regression line. The least-squares regression model appears to be appropriate.

13. The coefficient of determination with the Viper is $R^2 = 0.527$. Adding the Viper reduces the amount of variance explained by the model by approximately 33%.

14. $R^2 = 71.7\%$

Chapter Review Exercises (page 216)

1. (a)

(b) $r = -0.8954$
(c) There appears to be a strong negative linear relation between engine displacement and fuel economy.

2. (a)

(b) $r = 0.835$
(c) There appears to be a positive relation between temperature and chirps per second.

3. (a)

(b) Queens: $r = 0.9093$; Nassau County: $r = 0.8674$
(c) Both groups appear to have a positive association between square footage and monthly rent.
(d) For small apartments (those less than 1000 square feet in area), there seems to be no difference in rent between Queens and Nassau County. In larger apartments, Queens seems to have higher rents than Nassau County.

4. (a)

(b) Boys: $r = 0.9489$; girls: $r = 0.8025$
(c) There appears to be a positive association between age and height that exists for both boys and girls.
(d) Gender does not appear to be a factor in predicting height.

5. (a) $\hat{y} = -2.7977x + 30.3848$
(b)

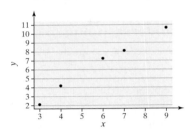

Engine Displacement (liters)

(c) The slope of the least-squares regression line indicates that, for each additional liter of engine displacement, fuel economy is lessened by 2.8 miles per gallon in city driving, on average. It is not appropriate to interpret the y-intercept because it is not possible to have zero displacement.
(d) A Ford Mustang with a 3.8-liter engine is predicted to get 19.8 miles per gallon when driven in the city.
(e) The residual is 0.2 mile per gallon.
(f) The miles per gallon is above average for a Ford Mustang.

6. (a) $\hat{y} = 0.2119x - 0.3091$
(b)

Temperature (°F)

(c) If the temperature increases 1°F, the number of chirps per second increases by 0.2119, on average. Since there are no observations for 0°F, it does not make sense to interpret the intercept.
(d) At 83.3°F, $\hat{y} = 17.3$ chirps per second.
(e) At 83.3°F, $y - \hat{y} = -1.1$ chirps per second.
(f) At 83.3°F, the number of chirps was below normal.

7. (a) $\hat{y} = 2.2092x - 34.3148$
(b)

Rents in Queens, New York

Square Footage

(c) The slope of the least-squares regression line indicates that, for each additional square foot of floor area, the rent increases by $2.21, on average. It is not appropriate to interpret the y-intercept since it is not possible to have an apartment with 0 square footage.
(d) For 825 square feet, $\hat{y} = \$1788.27$.
(e) The residual $y - \hat{y} = -538.233$.
(f) This apartment's rent is below average.

8. (a) $\hat{y} = 2.6351\,(x) + 29.705$
(b)

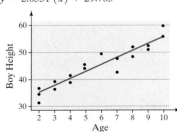

Age

(c) If age increases by 1 year, then height increases by 2.6 inches, on average. Although an age of 0 makes sense (birth), we do not have any observations near $x = 0$, so we will not interpret the intercept.
(d) For a 6-year-old boy, $\hat{y} = 45.515$ inches.
(e) The residual $y - \hat{y} = 4.285$ inches.
(f) This boy's height is above average.

9. (a)

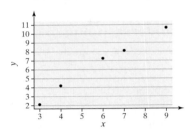

(b) Using points $(3, 2.1)$ and $(7, 8.1)$, $\hat{y} = 1.5x - 2.4$.
(c)

(d) $\hat{y} = 1.3877x - 1.6088$
(e)

(f) Computed line: sum of squared residuals $= 0.97$
(g) Least-squares line: sum of squared residuals $= 0.585$
(h) Answers will vary.

10. (a)

(b) Using points $(10, 105)$ and $(18, 76)$, $\hat{y} = -\dfrac{29}{8}x + \dfrac{565}{4}$.
(c)

(d) $\hat{y} = -3.8429x + 145.4857$
(e)

(f) Computed line: sum of squared residuals = 22.406
(g) Least-squares line: sum of squared residuals = 16.271
(h) Answers will vary.

11. $R^2 = 0.8017$; 80.2% of the variation in gas mileage is explained by the least-squares regression model.

12. $R^2 = 69.7\%$; 69.7% of the variation in chirps per second is explained by the least-squares regression equation.

13. $R^2 = 0.8268$; 82.7% of the variance in rent is explained by the least-squares regression model.

14. $R^2 = 0.9003$; 90.0% of the variation in height is explained by age.

15. (a)

(b) $r = 0.8352$
(c) The relation between men's and women's time seems to be positively associated. Men's times do not cause women's times.

16. No, correlation does not imply causation.

17. (a) Answers will vary.
(b) The slope can be interpreted as "the school day decreases by 0.01 hour for each 1% increase in percent low income, on average. additional thousand dollars of income." The intercept can be interpreted as the length of the school day for those districts with 0% low income.
(c) $\hat{y} = 6.91$ hours

18. The seven properties of a linear correlation coefficient are the following:
1. The linear correlation coefficient is always between -1 and 1, inclusive. That is, $-1 \leq r \leq 1$.
2. If $r = +1$, there is a perfect positive linear relation between the two variables. See Figure 4(a) on page 181.
3. If $r = -1$, there is a perfect negative linear relation between the two variables. See Figure 4(d).
4. The closer r is to $+1$, the stronger is the evidence of positive association between the two variables. See Figures 4(b) and 4(c).
5. The closer r is to -1, the stronger is the evidence of negative association between the two variables. See Figures 4(e) and 4(f).
6. If r is close to 0, there is no evidence of a *linear* relation between the two variables. Because the linear correlation coefficient is a measure of strength of linear relation, r close to 0 does not imply no relation, just no linear relation. See Figures 4(g) and 4(h).
7. The linear correlation coefficient is a unitless measure of association. So the unit of measure for x and y plays no role in the interpretation of r.

CHAPTER 5

5.1 Asess Your Understanding (page 233)

1. Empirical probability is based on the outcome of a probability experiment and is approximately equal to the relative frequency of the event. Classical probability is based on counting techniques and is equal to the ratio of the number of ways an event can occur to the number of possible outcomes in the experiment.

2. The probability of an impossible event is zero. No, when a probability is based on an empirical experiment, a probability of zero does not mean the event cannot occur.

3. Outcomes are equally likely when each outcome has the same probability of occurring.

4. An event is unusual if it has a low probability of occurring. The choice of a cutoff should consider the context of the problem.

5. True   **6.** True

7. Experiment   **8.** Event

9. Rule 1: All probabilities in the model are greater than or equal to zero and less than or equal to one.
Rule 2: The sum of the probabilities in the model is 1.
The outcome blue is called an impossible event.

10. Rule 1: All probabilities in the model are greater than or equal to zero and less than or equal to one.
Rule 2: The sum of the probabilities in the model is 1.
All the M&Ms in the bag are yellow.

11. This is not a probability model because $P(\text{green}) < 0$.

12. This is not a probability model because the sum of the probabilities is greater than 1.

13. The numbers $0, 0.01, 0.35$, and 1 could be probabilities.

14. The numbers $\dfrac{1}{2}, \dfrac{1}{4}, \dfrac{2}{3}$, and 0 could be probabilities.

15. The probability of 0.42 means that approximately 42 out of every 100 hands will contain two cards of the same value and three cards of different value.

16. The probability of 0.48 means that approximately 48 out of every 100 hands will contain two cards of the same value and five cards of different value.

17. The empirical probability is 0.95. **18.** The empirical probability is 0.80.

19. The $P(2) \neq \dfrac{1}{11}$ since the outcomes are not equally likely.

20. No, blood types are not equally likely.

21. $S = \{1H, 1T, 2H, 2T, 3H, 3T, 4H, 4T, 5H, 5T, 6H, 6T\}$

22. $S = \{SSS, SSF, SFS, FSS, SFF, FSF, FFS, FFF\}$

23. 0.44   **24.** 0.361   **25.** $P(E) = \dfrac{3}{10} = 0.3$

26. $P(F) = \dfrac{2}{5} = 0.4$ **27.** $P(E) = \dfrac{1}{2} = 0.5$ **28.** $P(F) = \dfrac{1}{2} = 0.5$

29. $P(\text{plays sports}) = \dfrac{288}{500} = 0.576$

30. $P(\text{female volunteered}) = \dfrac{341}{1100} = 0.31$

31. (a) $P(\text{red}) = \dfrac{40}{100} = 0.4$ **(b)** $P(\text{purple}) = \dfrac{25}{100} = 0.25$

32. (a) $P(\text{Titleist}) = \dfrac{35}{80} = 0.4375$ **(b)** $P(\text{Top-Flite}) = \dfrac{20}{80} = 0.25$

33. (a) $S = \{0, 00, 1, 2, 3, 4, \ldots, 35, 36\}$
(b) $P(8) = \dfrac{1}{38} = 0.0263$; there is a 2.63% probability that the ball will land in slot 8.
(c) $P(\text{odd}) = \dfrac{18}{38} = 0.4737$

34. (a) $P(\text{1st day of month}) = \dfrac{12}{365}$
(b) $P(\text{31st day of the month}) = \dfrac{7}{365}$

(c) $P(\text{born in December}) = \dfrac{31}{365}$

(d) $P(\text{born on November 8}) = \dfrac{1}{365}$

(e) No, it is unlikely that you would guess her birthday.

(f) No, outcomes are not equally likely.

35. (a) $\{SS, Ss, sS, ss\}$

(b) $P(ss) = \dfrac{1}{4}$; there is a 25% probability that a randomly selected offspring will have sickle-cell anemia.

(c) $P(Ss \text{ or } sS) = \dfrac{1}{2} = 0.5$; there is 50% probability that a randomly selected offspring will be a carrier of sickle-cell anemia.

36. (a) $\{SS, Ss, sS, ss\}$

(b) $P(ss) = \dfrac{1}{4} = 0.25$; there is a 25% probability that a randomly selected offspring will not have Huntington's disease.

(c) $P(SS, Ss, sS) = \dfrac{3}{4} = 0.75$; there is a 75% probability that a randomly selected offspring will have Huntington's disease.

37. (a)

Response	Probability
Never	0.026
Rarely	0.068
Sometimes	0.116
Most of the time	0.263
Always	0.527

(b) It would be unusual to randomly find someone who never wears a seatbelt, because $P(\text{never}) < 0.05$.

38. (a)

Response	Probability
Never	0.026
Rarely	0.055
Sometimes	0.076
Most of the time	0.158
Always	0.684

(b) It would be unusual to randomly find someone who never wears a seatbelt, because $P(\text{never}) < 0.05$.

39. (a)

Type of Larceny Theft	Probability
Pocket picking	0.008
Purse snatching	0.008
Shoplifting	0.198
From motor vehicles	0.331
Motor vehicle accessories	0.129
Bicycles	0.072
From buildings	0.176
From coin-operated machines	0.076

(b) Purse snatching larcenies are unusual.

(c) Larcenies from coin-operated machines are not unusual.

40. (a)

Age	Probability
15–19	0.013
20–24	0.070
25–29	0.224
30–34	0.389
35–39	0.253
40–44	0.052

(b) Multiple births are unusual for 15- to 19-year-old mothers.

(c) Multiple births are not unusual for 40- to 44-year-old mothers.

41. A, B, C, and F are consistent with the definition of a probability model.

42. Use model A if the coin is known to be fair.

43. Use model B if the coin is known to always come up tails.

44. If tails are twice as likely to come up, model F should be used.

45. (a) $S = \{\text{JR, JC, JD, JM, RC, RD, RM, CD, CM, DM}\}$

(b) $P(\text{Clarese and Dominique attend}) = \dfrac{1}{10} = 0.10$

(c) $P(\text{Clarice attends}) = \dfrac{4}{10} = 0.40$

(d) $P(\text{John stays home}) = \dfrac{6}{10} = 0.60$

46. (a) $S = \{$Screamin' Eagle—The Boss, Screamin' Eagle—Train, Screamin' Eagle—Batman, Screamin' Eagle—Mr. Freeze, Screamin' Eagle—Ninja, The Boss—Train, The Boss—Batman, The Boss—Mr. Freeze, The Boss—Ninja, Train—Batman, Train—Mr. Freeze, Train—Ninja, Batman—Mr. Freeze, Batman—Ninja, Mr. Freeze—Ninja$\}$

(b) $P(\text{Batman and Mr. Freeze}) = 0.067$

(c) $P(\text{Screamin' Eagle}) = 0.333$

(d) $P(\text{neither Train nor Ninja}) = 0.4$

47. (a) $P(\text{right field}) = \dfrac{24}{73} = 0.329$

(b) $P(\text{left field}) = \dfrac{2}{73} = 0.027$

(c) It is unusual for Barry Bonds to hit a home run to left field; the probability is less than 5%.

48. (a) Answers will vary. **(b)** Answers will vary. **(c)** $P(3) = \dfrac{1}{6}$

49. Answers will vary.

50. (a) Classical method is used. **(b)** Empirical method is used.
(c) Subjective method is used. **(d)** Empirical method is used.

51. Answers will vary.

52. Answers will vary.

53. $P(\text{income is greater than } \$57,500) = \dfrac{1}{2}$

54. $P(\text{Floridian is 65 or older}) = 0.17$

5.2 Assess Your Understanding (page 245)

1. Two events are mutually exclusive (disjoint) if they have no outcomes in common.

2. $P(E) + P(F)$ **3.** $P(E) + P(F) - P(E \text{ and } F)$

4. Two events E and E^c are complementary if they are disjoint, that is, if $P(E \text{ and } E^c) = 0$ and if $P(E \text{ or } E^c) = 1$.

5. E and $F = \{5, 6, 7\}$; E and F are not mutually exclusive.

6. F and $G = \{9\}$; F and G are not mutually exclusive.

7. F or $G = \{5, 6, 7, 8, 9, 10, 11, 12\}$; $P(F \text{ or } G) = \dfrac{2}{3}$

8. E or $H = \{2, 3, 4, 5, 6, 7\}$; $P(E \text{ or } H) = \dfrac{1}{2}$

9. There are no outcomes in event "E and G." Events E and G are mutually exclusive.

10. There are no outcomes in event "F and H." Events F and H are mutually exclusive.

11. $E^c = \{1, 8, 9, 10, 11, 12\}$; $P(E^c) = \dfrac{1}{2}$

12. $F^c = \{1, 2, 3, 4, 10, 11, 12\}$; $P(F^c) = \dfrac{7}{12}$

13. $P(E \text{ or } F) = 0.55$ **14.** $P(E \text{ and } F) = 0.1$

15. $P(E \text{ or } F) = 0.70$ **16.** $P(E \text{ and } F) = 0$

17. $P(E^c) = 0.75$ **18.** $P(F^c) = 0.55$

19. $P(F) = 0.30$ **20.** $P(E) = 0.5$

21. $P(\text{Titleist or Maxfli}) = \dfrac{17}{20} = 0.85$

22. $P(\text{Maxfli or Top-Flite}) = \dfrac{11}{20} = 0.55$

23. $P(\text{not a Titleist}) = \dfrac{11}{20} = 0.55$

24. $P(\text{not a Top-Flite}) = \dfrac{17}{20} = 0.85$

25. (a) All the probabilities are nonnegative; the sum of the probabilities equals 1.
(b) $P(\text{gun or knife}) = 0.793$. There is a 79.3% probability of randomly selecting a murder committed with a gun or a knife.
(c) $P(\text{knife, blunt object, or strangulation}) = 0.190$. There is 19.0% probability of randomly selecting a murder committed with a knife or blunt object or by strangulation.
(d) $P(\text{not a gun}) = 0.332$. There is 33.2% chance of randomly selecting a murder that was not committed with a gun.
(e) A murder by strangulation is unusual.

26. (a) All the probabilities are nonnegative; the sum of the probabilities equals 1.
(b) $P(\text{earning a physical or life sciences degree}) = 0.307$. There is a 30.7% probability that a randomly selected doctoral candidate received a degree in either the physical or life sciences.
(c) $P(\text{doctorate in physical, life, or computer sciences or mathematics}) = 0.352$. There is a 35.2% probability that a randomly selected doctoral candidate receive a degree in the physical, life, or computer sciences or in mathematics.
(d) $P(\text{not in mathematics}) = 0.976$. The probability is 97.6% that a randomly selected doctoral candidate did not receive a degree in mathematics.
(e) It is unusual to earn a doctoral degree in mathematics.

27. (a) $P(\text{30- to 39-year-old mother}) = 0.632$. There is a 63.2% probability that a mother involved in a multiple birth was between 30 and 39 years old.
(b) $P(\text{not 30 to 39 years old}) = 0.368$. There is a 36.8% probability that a mother involved in a multiple birth was either younger than 30 or older than 39.
(c) $P(\text{younger than 45 years}) = 0.984$. The probability that a randomly selected mother involved in a multiple birth was younger than 45 years old is 98.4%.
(d) $P(\text{at least 20 years old}) = 0.987$. There is a 98.7% probability that a randomly selected mother involved with a multiple birth was at least 20 years old.

28. (a) All the probabilities are nonnegative; the sum of the probabilities equals 1.
(b) $P(\text{4 or more rooms}) = 0.879$. The probability is 87.9% that a randomly selected house has 4 or more rooms.
(c) $P(\text{fewer than 8 rooms}) = 0.861$. There is an 86.1% probability that a randomly selected house will have fewer than 8 rooms.
(d) $P(\text{from 4 to 6 rooms}) = 0.617$. There is a 61.7% probability that a randomly selected house will have 4, 5, or 6 rooms.
(e) $P(\text{at least 2 rooms}) = 0.996$. The probability is 99.6% that a randomly selected house will have at least 2 rooms.

29. (a) $P(\text{heart or club}) = \frac{1}{2} = 0.5$
(b) $P(\text{heart or club or diamond}) = \frac{3}{4} = 0.75$
(c) $P(\text{ace or heart}) = \frac{4}{13} = 0.3077$

30. (a) $P(\text{two or three}) = \frac{2}{13} = 0.1538$
(b) $P(\text{two or three or four}) = \frac{3}{13} = 0.2308$
(c) $P(\text{two or club}) = \frac{4}{13} = 0.3077$

31. (a) $P(\text{birthday is not November 8}) = \frac{364}{365} = 0.9973$
(b) $P(\text{birthday is not 1st of month}) = \frac{353}{365} = 0.9671$
(c) $P(\text{birthday is not 31st of month}) = \frac{358}{365} = 0.9808$
(d) $P(\text{birthday is not in December}) = \frac{334}{365} = 0.9151$

32. (a) $P(\text{green or red}) = \frac{10}{19} = 0.5263$
(b) $P(\text{not green}) = \frac{18}{19} = 0.9474$

33. No, some people have both vision and hearing problems, but we don't know the proportion.

34. No, some people visit the doctor for both blood pressure check and urine analysis, but we don't know the proportion.

35. (a) $P(\text{only English or Spanish is spoken}) = 0.93$
(b) $P(\text{language other than only English or only Spanish}) = 0.07$
(c) $P(\text{not only English is spoken}) = 0.18$
(d) No, the sum of the probabilities would be greater than 1, and there would be no probability model.

36. (a) $P(\text{employee drives or takes public transportation}) = 0.873$. There is an 87.3% probability that a randomly selected employee either drives or takes public transportation to work.
(b) $P(\text{employee neither drives nor takes public transportation}) = 0.127$. There is a 12.7% probability that a randomly selected employee neither drives nor takes public transportation to work.
(c) $P(\text{employee does not drive}) = 0.18$. The probability is 18% that a randomly selected employee does not drive to work.
(d) No, the probability a person drives or walks or takes public transportation would be 102.3%, which is greater than 1.

37. (a) $P(\text{died from cancer}) = 0.007$
(b) $P(\text{current smoker}) = 0.057$
(c) $P(\text{died from cancer and smoked cigars until death}) = 0.001$
(d) $P(\text{died from cancer or was current cigar smoker}) = 0.064$

38. (a) $P(\text{employed}) = 0.878$
(b) $P(\text{male}) = 0.522$
(c) $P(\text{employed and male}) = 0.457$
(d) $P(\text{employed or male}) = 0.943$

39. (a) $P(\text{student is satisfied}) = \frac{77}{125} = 0.616$
(b) $P(\text{student is a junior}) = \frac{94}{375} = 0.251$
(c) $P(\text{student is satisfied and is a junior}) = \frac{64}{375} = 0.171$
(d) $P(\text{student is satisfied or is a junior}) = \frac{87}{125} = 0.696$

40. (a) $P(\text{placebo}) = \frac{2}{5} = 0.40$
(b) $P(\text{headache went away}) = \frac{94}{125} = 0.752$
(c) $P(\text{placebo and headache went away}) = \frac{28}{125} = 0.224$
(d) $P(\text{placebo or headache went away}) = \frac{116}{125} = 0.928$

41. (a) $P(\text{officer}) = 0.162$
(b) $P(\text{Navy}) = 0.261$
(c) $P(\text{Navy officer}) = 0.038$
(d) $P(\text{officer or in the Navy}) = 0.385$

42. (a) $P(\text{male}) = 0.738$
(b) $P(\text{16 to 20 years old}) = 0.142$
(c) $P(\text{16 to 20 years old and male}) = 0.100$
(d) $P(\text{male or 16 to 20 years old}) = 0.780$

5.3 Assess Your Understanding (page 254)

1. Independent **2.** Multiplication
3. Addition **4.** False
5. $P(E) \cdot P(F)$ **6.** 0
7. (a) Dependent **(b)** Independent **(c)** Dependent
8. (a) Dependent **(b)** Dependent **(c)** Independent
9. $P(E \text{ and } F) = 0.18$
10. $P(E \text{ and } F) = 0.63$
11. $P(\text{5 heads}) = \frac{1}{32} = 0.03125$; 3.125% of the time tossing a coin 5 times will result in 5 heads.
12. $P(\text{tossing 4 ones in a row}) = \frac{1}{6^4} = 0.0008$. If a die is tossed, 0.08% of the time all 4 tosses will be one.
13. $P(\text{two left-handed people are chosen}) = 0.0169$
$P(\text{at least one person chosen is right-handed}) = 0.9831$
14. $P(\text{both increase}) = 0.36$
$P(\text{at least one does not increase}) = 0.64$

15. $P(\text{all five are negative}) = 0.9752$

$P(\text{at least one is positive}) = 0.0248$

16. $P(\text{all 100 last}) = 0.6058$; $P(\text{at least one burns out}) = 0.3942$

17. (a) $P(\text{two will live to 41 years}) = 0.99437$
(b) $P(\text{five will live to 41 years}) = 0.98598$
(c) $P(\text{at least one of five dies}) = 0.01402$; it would be unusual that at least one of the five dies before age 41.

18. (a) $P(\text{two will live to 41 years}) = 0.99712$
(b) $P(\text{five will live to 41 years}) = 0.99282$
(c) $P(\text{at least one of five dies}) = 0.00718$; it would be unusual that at least one of the five dies before age 41.

19. (a) $P(\text{two will have Rh}^+ \text{ blood}) = 0.9801$
(b) $P(\text{six will have Rh}^+ \text{ blood}) = 0.9415$
(c) $P(\text{at least out of six will have Rh}^- \text{ blood}) = 0.0585$; it would not be unusual to select a person with Rh$^-$ blood.

20. (a) $P(\text{two inspectors do not identify low-quality timber}) = 0.04$
(b) Three inspectors should be hired to keep the probability of failing to identify low-quality timber less than 1%.

21. (a) $P(\text{batter makes 10 consecutive outs}) = 0.02825$
(b) A cold streak would be unusual.

22. $P(\text{makes eight consecutive free shots}) = 0.032$; it is unusual for the player to make eight consecutive free shots.

23. $P(\text{at least 1}) = 0.9999$; answers will vary.

24. (a) $P(\text{twelve positive tests}) = 0.8864$
(b) $P(\text{at least one negative test}) = 0.1136$

25. (a) $P(\text{two strikes in a row}) = 0.09$
(b) $P(\text{turkey}) = 0.027$
(c) $P(\text{strike followed by nonstrike}) = 0.21$

26. (a) $P(\text{four Nascar fans}) = 0.1212$
(b) $P(\text{at least one non-Nascar fan}) = 0.8788$
(c) $P(\text{no one likes Nascar}) = 0.0283$
(d) $P(\text{at least one likes Nascar}) = 0.9717$

27. (a) $P(\text{all three have driven under the influence of alchohol}) = 0.0244$
(b) $P(\text{at least one has not driven under the influence of alcohol}) = 0.9756$
(c) $P(\text{no one has driven under the influence of alcohol}) = 0.3579$
(d) $P(\text{at least one has driven under the influence of alcohol}) = 0.6421$

5.4 Assess Your Understanding (page 261)

1. F occurring; E has occurred

2. No

3. $P(F|E) = 0.75$ **4.** $P(B|A) = 0.525$

5. $P(F|E) = 0.5676$ **6.** $P(B|A) = 0.4108$

7. $P(E \text{ and } F) = 0.32$ **8.** $P(E \text{ and } F) = 0.24$

9. The events are not independent.

10. The events are not independent.

11. $P(\text{club}) = \dfrac{1}{4}$; $P(\text{club}|\text{black card}) = \dfrac{1}{2}$

12. $P(\text{king}) = \dfrac{1}{13}$; $P(\text{king}|\text{heart}) = \dfrac{1}{13}$; no; independent

13. $P(\text{rains}|\text{cloudy}) = \dfrac{0.21}{0.37} = 0.568$

14. $P(\text{cancer}|25\text{-}34 \text{ years old}) = \dfrac{20}{197} = 0.102$

15. $P(\text{W}|\text{DO}) = \dfrac{58}{91} = 0.637$

16. $P(>\$75{,}000|\text{live in Northeast}) = \dfrac{44}{191} = 0.230$

17. (a) $P(\text{no health insurance}|\text{under 18 years of age}) = 0.110$
(b) $P(\text{under 18 years of age}|\text{no health insurance}) = 0.196$

18. (a) $P(\text{former cigar smoker}|\text{died from cancer}) = 0.090$
(b) $P(\text{died from cancer}|\text{former cigar smoker}) = 0.012$

19. (a) $P(16 \text{ to } 20 \text{ years old}|\text{male}) = 0.136$
(b) $P(\text{male}|16 \text{ to } 20 \text{ years old}) = 0.705$
(c) Probably the victim is male.

20. (a) $P(\text{HS graduate}|\text{never married}) = 0.301$
(b) $P(\text{never married}|\text{given a HS graduate}) = 0.142$

21. $P(\text{both TVs work}) = 0.40$; $P(\text{at least one TV doesn't work}) = 0.6$

22. $P(\text{two women are chosen}) = \dfrac{2}{7} = 0.2857$

23. $P(\text{Dave and then Neta is chosen}) = \dfrac{1}{20} = 0.05$

24. $P(\text{Yolanda hosts, then Lorrie hosts the parties}) = \dfrac{1}{20} = 0.05$

25. (a) $P(\text{like both songs}) = \dfrac{5}{39} = 0.128$; it is not unusual to like both songs.
(b) $P(\text{like neither song}) = \dfrac{14}{39} = 0.359$
(c) $P(\text{like exactly one song}) = \dfrac{20}{39} = 0.513$
(d) $P(\text{like both songs}) = \dfrac{25}{169} = 0.148$;
$P(\text{like neither song}) = \dfrac{64}{169} = 0.379$;
$P(\text{like exactly one song}) = \dfrac{80}{169} = 0.473$

26. (a) $P(\text{both are diet soda}) = \dfrac{1}{22} = 0.046$
(b) $P(\text{both are regular soda}) = \dfrac{6}{11} = 0.545$; it is not unusual to choose two regular sodas.
(c) $P(\text{one regular and one diet soda are chosen}) = \dfrac{9}{22} = 0.409$

27. (a) $P(\text{two red bulbs are chosen}) = 0.152$
(b) $P(\text{first a red and then a yellow bulb is chosen}) = 0.138$
(c) $P(\text{first a yellow and then a red bulb is chosen}) = 0.138$
(d) $P(\text{a red bulb and a yellow bulb are chosen}) = 0.276$

28. (a) $P(\text{two Titleists are selected}) = 0.319$
(b) $P(\text{first a Titleist and then a Maxfli is picked}) = 0.134$
(c) $P(\text{first a Maxfli and then a Titleist is selected}) = 0.134$
(d) $P(\text{a Maxfli and a Titleist are selected}) = 0.269$

29. $P(\text{female and smokes}) = 0.051$; it would not be unusual to randomly select a female who smokes.

30. $P(\text{male and holds multiple jobs}) = 0.031$; it would be unusual to randomly select an employed male who has more than one job.

31. (a) $P(10 \text{ people each have a different birthday}) = 0.883$
(b) $P(\text{at least two of the ten people have the same birthday}) = 0.117$

32. $P(\text{at least 2 people in a group of 23 share the same birthday}) = 0.507$

33. (a) $P(\text{being dealt five clubs}) = 0.000495$
(b) $P(\text{being dealt a flush}) = 0.0020$

34. $P(\text{being dealt a royal flush}) = 0.00000154$

35. (a) $P(\text{two defective chips}) = 0.0000245$
(b) Assuming independence: $P(\text{two defective chips}) = 0.000025$

36. (a) $P(\text{first is Roger Cummings; second is Rick Whittingham}) = 2.37 \times 10^{-8}$
(b) $P(\text{first is Rick Whittingham; second is Roger Cummings}) = 2.37 \times 10^{-8}$

37. $P(<18 \text{ years old}) = 0.246$; $P(<18 \text{ years old}|\text{no health insurance}) = 0.196$; the events are not independent.

38. $P(\text{died from cancer}) = 0.007$; $P(\text{died from cancer}|\text{current cigar smoker}) = 0.018$; the events "died from cancer" and "smoking cigars" are not independent.

39. $P(\text{female}) = 0.262$; $P(\text{female}|16 \text{ to } 20 \text{ years old}) = 0.295$; the events are not independent.

40. $P(\text{divorced}) = 0.113$; $P(\text{divorced}|\text{graduated from college}) = 0.088$; the events "being divorced" and having "graduated from college" are not independent.

5.5 Assess Your Understanding (page 276)

1. Permutation   **2.** Combination   **3.** True

4. In a permutation the order in which the objects are chosen matters; in a combination the order in which the objects are chosen is unimportant.

5. $5! = 120$ **6.** $7! = 5040$ **7.** $10! = 3,628,800$

8. $12! = 479,001,600$ **9.** $0! = 1$ **10.** $1! = 1$

11. $_6P_2 = 30$ **12.** $_7P_2 = 42$ **13.** $_4P_4 = 24$

14. $_7P_7 = 5040$ **15.** $_5P_0 = 1$ **16.** $_4P_0 = 1$

17. $_8P_3 = 336$ **18.** $_9P_4 = 3024$ **19.** $_8C_3 = 56$

20. $_9C_2 = 36$ **21.** $_{10}C_2 = 45$ **22.** $_{12}C_3 = 220$

23. $_{52}C_1 = 52$ **24.** $_{40}C_{40} = 1$ **25.** $_{48}C_3 = 17,296$

26. $_{30}C_4 = 27,405$

27. $ab, ac, ad, ae, ba, bc, bd, be, ca, cb, cd, ce, da, db, dc, de, ea, eb, ec, ed$; $_5P_2 = 20$

28. $ab, ac, ad, ba, bc, bd, ca, cb, cd, da, db, dc$; $_4P_2 = 12$

29. $ab, ac, ad, ae, bc, bd, be, cd, ce, de$; $_5C_2 = 10$

30. ab, ac, ad, bc, bd, cd; $_4C_2 = 6$

31. He can wear 24 different shirt-and-tie combinations.

32. She can wear 15 different outfits.

33. Dan can arrange the songs $12! = 479,001,600$ ways.

34. Fifteen students can be lined up $15! \approx 1.30767 \times 10^{12}$ ways.

35. The salesman can take $8! = 40,320$ different routes.

36. The 10 songs can be played $10! = 3,628,800$ ways.

37. At most 18,278 companies can be listed on the NYSE.

38. At most 12,338,352 companies can be listed on the NASDAQ.

39. (a) 10,000 different codes are possible.

   (b) $P(\text{correct code is guessed}) = \dfrac{1}{10,000}$

40. (a) 10^9 different Social Security numbers are possible.

   (b) $P(\text{guessing the President's Social Security number}) = \dfrac{1}{10^9}$

41. 26^8 different user names are possible.

42. $26 \cdot 36^7$ different user names are possible.

43. (a) There are 50^3 lock combinations.

   (b) $P(\text{guessing the correct combination}) = \dfrac{1}{50^3}$

44. There are $26 \cdot 10^5 = 2,600,000$ different license plates.

45. The top three cars can finish in $_{40}P_3 = 59,280$ ways.

46. The top two horses can finish in $_{10}P_2 = 90$ ways.

47. The officers can be chosen in $_{20}P_4 = 116,280$ ways.

48. $_{50}P_4 = 5,527,200$ different leadership structures are possible.

49. There are $_{25}P_4 = 303,600$ possible outcomes.

50. There are $_{21}C_9 = 293,930$ possible committee structures.

51. There are $_{50}C_5 = 2,118,760$ possible random samples of size 5.

52. There are $_{100}C_7 \approx 1.60076 \times 10^{10}$ possible random samples of size 7.

53. $_6C_2 = 15$ different birth and gender orders are possible.

54. $_8C_3 = 56$ different birth and gender orders are possible.

55. $\dfrac{10!}{3! \cdot 3! \cdot 2! \cdot 1! \cdot 1!} = 50,400$ different 10-letter words can be formed.

56. $\dfrac{9!}{2! \cdot 2! \cdot 1! \cdot 1! \cdot 1! \cdot 1! \cdot 1!} = 90,720$ different 9-letter words can be formed.

57. The trees can be planted $\dfrac{11!}{4! \cdot 5! \cdot 2!} = 6930$ ways.

58. $\dfrac{9!}{3! \cdot 4!} = 2520$ different batting orders are possible.

59. $P(\text{winning}) = \dfrac{1}{_{30}C_5} = \dfrac{1}{142,506}$

60. $P(\text{winning}) = \dfrac{1}{_{50}C_5 \cdot 36} = 0.00000001$

61. (a) $P(\text{jury has only students}) = \dfrac{1}{153} = 0.0065$

   (b) $P(\text{jury has only faculty}) = \dfrac{1}{34} = 0.0294$

   (c) $P(\text{jury has two students and three faculty}) = \dfrac{20}{51} = 0.3922$

62. (a) $P(\text{committee is all Democrats}) = 0.0127$
   (b) $P(\text{committee is all Republicans}) = 0.0028$
   (c) $P(\text{committee has three Democrats and four Republicans}) = 0.2442$

63. $P(\text{shipment is rejected}) = 0.1283$

64. $P(\text{selecting two 60-watt light bulbs}) = 0.3517$

65. (a) $P(\text{you like 2 of 4 songs}) = 0.3916$
   (b) $P(\text{you like 3 of 4 songs}) = 0.1119$
   (c) $P(\text{you like all 4 songs}) = 0.0070$

66. (a) $P(\text{2 of the 3 cans are diet soda}) = 0.1227$
   (b) $P(\text{1 of the 3 cans is diet soda}) = 0.4909$
   (c) $P(\text{all 3 cans are diet soda}) = 0.0045$

67. (a) Five cards can be selected from a deck $_{52}C_5 = 2,598,960$ ways.
   (b) Three of the same card can be chosen $13 \cdot {_4C_3} = 52$ ways.
   (c) The remaining two cards can be chosen $_{12}C_2 \cdot {_4C_1} \cdot {_4C_1} = 1056$ ways.
   (d) $P(\text{three of a kind}) = \dfrac{52 \cdot 1056}{2,598,960} = 0.0211$

68. (a) Five cards can be selected from a deck $_{52}C_5 = 2,598,960$ ways.
   (b) Two of the same card can be chosen $13 \cdot {_4C_2} = 78$ ways.
   (c) The remaining three cards can be chosen $_{12}C_3 \cdot {_4C_1} \cdot {_4C_1} \cdot {_4C_1} = 14,080$ ways.
   (d) $P(\text{two of a kind}) = \dfrac{78 \cdot 14,080}{2,598,960} = 0.4226$

69. $P(\text{all 4 modems tested worked}) = 0.4912$

70. $P(\text{all 5 TVs tested worked}) = 0.7291$

Chapter Review Exercises (page 280)

1. (a) Possible probabilities are $0, 0.75, 0.41$.
   (b) Possible probabilities are $\dfrac{2}{5}, \dfrac{1}{3}, \dfrac{6}{7}$.

2. $P(E) = \dfrac{1}{5} = 0.2$ **3.** $P(F) = \dfrac{2}{5} = 0.4$

4. $P(E) = \dfrac{3}{5} = 0.6$ **5.** $P(E^c) = \dfrac{4}{5} = 0.8$

6. $P(E \text{ or } F) = 0.89$ **7.** $P(E \text{ or } F) = 0.48$

8. $P(E \text{ and } F) = 0.09$

9. Events E and F are not independent because $P(E \text{ and } F) \neq P(E) \cdot P(F)$.

10. $P(E \text{ and } F) = 0.2655$ **11.** $P(E|F) = 0.5$

12. (a) $7! = 5040$ **(b)** $0! = 1$ **(c)** $_9C_4 = 126$
   (d) $_{10}C_3 = 120$ **(e)** $_9P_2 = 72$ **(f)** $_{12}P_4 = 11,880$

13. (a) $P(\text{green}) = \dfrac{1}{19} = 0.0526$; there is a 5.26% probability that the next spin will end with the ball in a green slot.

   (b) $P(\text{green or red}) = \dfrac{10}{19} = 0.5263$; there is a 52.63% probability that the next spin will end with the ball in either a green slot or a red slot.

   (c) $P(\text{00 or red}) = \dfrac{19}{38} = 0.5$; there is a 50% probability that the next spin will end with the ball either in 00 or in a red slot.

   (d) $P(\text{31 and black}) = 0$; this is called an impossible event.

14. (a) $P(\text{7 or 11}) = \dfrac{8}{36} = \dfrac{2}{9} = 0.2222$; there is a 22.2% probability that the player will win on the first roll.

   (b) $P(\text{2, 3, or 12}) = \dfrac{4}{36} = \dfrac{1}{9} = 0.1111$; there is a 11.1% probability that the player will lose on the first roll.

15. (a) $P(\text{fatality was alcohol related}) = \dfrac{301}{575} = 0.5235$

(b) $P(\text{fatality was not alcohol related}) = \dfrac{274}{575} = 0.4765$

(c) $P(\text{both fatalities were alcohol related}) = 0.2736$
(d) $P(\text{neither fatality was alcohol related}) = 0.2266$
(e) $P(\text{at least one fatality was alcohol related}) = 0.7734$

16. The probability is 0.11 that a randomly selected cyclone occurs in the North Atlantic Ocean. The probability is 0.89 that a randomly selected cyclone does not occur in the North Atlantic Ocean.

17. (a)

Class of Worker	Probability
Private wage and salary worker	0.8428
Government worker	0.1236
Self-employed worker	0.0321
Unpaid family worker	0.0014

(b) It is unusual for a Louisville worker to be an unpaid family worker.
(c) It is unusual for a Louisville worker to be self-employed.

18. (a) All the probabilities are greater than or equal to zero and less than or equal to one, and the sum of the probabilities is 1.
(b) $P(\text{a peanut butter cookie}) = 0.24$
(c) $P(\text{next box sold is mint, caramel, or shortbread}) = 0.53$
(d) $P(\text{next box sold is not mint}) = 0.75$

19. (a) $P(\text{baby was postterm}) = 0.0674$
(b) $P(\text{baby weighed 3000 to 3999 grams}) = 0.6591$
(c) $P(\text{baby weighed 3000 to 3999 grams and was postterm}) = 0.0484$
(d) $P(\text{baby weighed 3000 to 3999 grams or was postterm}) = 0.6781$
(e) $P(\text{baby weighed} <1000 \text{ grams and was postterm}) = 0.000008$; this event is not impossible.
(f) $P(\text{baby weighed 3000 to 3999 grams}\mid$ baby was postterm$) = 0.7185$
(g) The events "postterm baby" and "weighs 3000 to 3999 grams" are not independent.

20. (a) $P(\text{gold}) = 0.3545$
(b) $P(\text{China won medal}) = 0.1379$
(c) $P(\text{gold and China won}) = 0.0700$
(d) $P(\text{gold or China won}) = 0.4224$
(e) $P(\text{Japan won}\mid\text{medal was bronze}) = 0.0795$
(f) $P(\text{medal was bronze}\mid\text{Japan won}) = 0.3243$

21. (a) $P(\text{complaint filed online}) = 0.63$
(b) $P(\text{complaint not filed online}) = 0.37$
(c) $P(\text{all five complaints filed online}) = 0.099$
(d) $P(\text{at least one of five complaints not filed online}) = 0.901$
(e) $P(\text{none of five complaints filed online}) = 0.007$
(f) $P(\text{at least one of five complaints filed online}) = 0.993$

22. (a) $P(\text{win two in a row}) = 0.420$
(b) $P(\text{win seven in a row}) = 0.048$
(c) $P(\text{lose at least one of next seven}) = 0.952$

23. $P(\text{matching the three winning numbers}) = 0.001$

24. $P(\text{matching the four winning numbers}) = 0.0001$

25. $P(\text{shipment accepted}) = 0.8$

26. $P(\text{drawing three aces}) = 0.00018$

27. $26^2 \cdot 10^4 = 6{,}760{,}000$ different license plates can be formed.

28. The students can be seated in $_{10}P_4 = 5040$ different arrangements.

29. There are $6! = 720$ different arrangements of the letters LINCEY.

30. There are $\dfrac{10!}{4! \cdot 3! \cdot 2!} = 12{,}600$ different vertical arrangements of the flags.

31. There are $_{55}C_8 = 1{,}217{,}566{,}350$ possible random samples.

32. $_{29}C_5 = 118{,}755$ different subcommittees can be formed.

33. $P(\text{winning Fantasy 5}) = \dfrac{1}{_{35}C_5} = 0.000003$

34. $P(\text{winning Cash 5}) = \dfrac{1}{_{39}C_5} = 0.000002$

35. (a) $P(\text{three Merlot}) = 0.0454$ **(b)** $P(\text{two Merlot}) = 0.3182$
(c) $P(\text{no Merlot}) = 0.1591$

36. Answers will vary.   **37.** Answers will vary.

38. (a) There are 13 clubs in the deck.
(b) Thirty-seven cards remain in the deck. Forty-one cards are not known by you. Eight of the unknown cards are clubs.
(c) $P(\text{next card dealt is a club}) = \dfrac{8}{41}$
(d) $P(\text{two clubs in a row}) = 0.0341$
(e) No

39. (a) $P(\text{home run to left field}) = \dfrac{34}{70} = 0.4857$; there is a 48.6% probability that a randomly selected home run by McGwire went to left field.
(b) $P(\text{home run to right field}) = 0$
(c) No, it is not impossible for McGwire to hit a home run to right field.

CHAPTER 6

6.1 Assess Your Understanding (page 293)

1. A random variable is a numerical measure of the outcome of a probability experiment, so its value is determined by chance.

2. A discrete random variable has either a finite number of values or a countable number of values. A continuous random variable has an infinite number of values that can be represented as an interval on a number line.

3. When $P(x)$ denotes the probability that the random variable X equals x, each probability must be between 0 and 1, inclusive, and the sum of the probabilities must equal 1.

4. Answers will vary.

5. A batting average of 0.300 means the probability of the player's getting a hit at any at-bat is 0.3, despite what the results of the previous at-bats.

6. In a zero-sum game, the gains of the winner always equal the losses of the loser. So, in the long run, a player's mean gain (loss) is zero.

7. (a) Discrete, $x = 0, 1, 2, \dots, 20$ **(b)** Continuous, $t > 0$
(c) Discrete, $x = 0, 1, 2, \dots$ **(d)** Continuous, $s \geq 0$

8. (a) Continuous, $t > 0$ **(b)** Continuous, $w > 0$
(c) Discrete, $x = 0, 1, 2, \dots$ **(d)** Discrete, $x = 0, 1, 2, \dots, 20$

9. (a) Continuous, $r \geq 0$ **(b)** Discrete, $x = 0, 1, 2, 3, \dots$
(c) Discrete, $x = 0, 1, 2, 3, \dots$ **(d)** Continuous, $t > 0$

10. (a) Discrete, $x = 0, 1, 2, 3, \dots$ **(b)** Continuous, $d > 0$
(c) Discrete, $x = 0, 1, 2, 3, \dots$ **(d)** Continuous, $a > 0$

11. Yes, it is a probability distribution.

12. Yes, it is a probability distribution.

13. No, $P(50) < 0$.

14. Yes, it is a probability distribution.

15. No, $\Sigma P(x) \neq 1$ **16.** No, $\Sigma P(x) \neq 1$

17. $P(4) = 0.3$ **18.** $P(2) = 0.15$

19. (a) Each probability is between 0 and 1, inclusive, and the sum of the probabilities equals 1.

(b)

**Parental Involvement
in School
(child grades K–5)**

(c) $\mu_X = 2.9$; the mean of the random variable X is the mean outcome if the experiment were repeated many times.
(d) $\sigma_X^2 = 1.2$ **(e)** $\sigma_X = 1.1$
(f) $P(3) = 0.320$ **(g)** $P(3 \text{ or } 4) = 0.694$

20. (a) Each probability is between 0 and 1, inclusive, and the sum of the probabilities equals 1.

(b)

Parental Involvement in School (child grades 6–8)

(c) $\mu_X = 2.5$; the mean of the random variable X is the mean outcome if the experiment were repeated many times.
(d) $\sigma_X^2 = 1.4$ **(e)** $\sigma_X = 1.2$
(f) $P(3) = 0.322$ **(g)** $P(3 \text{ or } 4) = 0.552$

21. (a) Each probability is between 0 and 1, inclusive, and the sum of the probabilities equals 1.

(b)

Ichiro's Hit Parade

(c) $\mu_X = 1.6$ **(d)** $\sigma_X = 1.2$
(e) $P(2) = 0.2857$ **(f)** $P(X > 1) = 0.4969$

22. (a) All probabilities are between 0 and 1, inclusive, and the sum of the probabilities is 1.

(b)

Number of People Waiting in Line

(c) $\mu_X = 4.9$; the mean of the random variable X is the mean outcome that would be obtained if the experiment were repeated many times.
(d) $\sigma_X^2 = 5.0$ **(e)** $\sigma_X = 2.2$
(f) $P(8) = 0.063$ **(g)** $P(X \geq 10) = 0.029$

23. (a)

x (games played)	P(x)
4	0.1852
5	0.1852
6	0.2222
7	0.4074

(b)

Games Played

(c) $\mu_X = 5.8$ **(d)** $\sigma_X = 1.2$

24. (a)

x (age)	P(x)
5	0.1929
6	0.1956
7	0.2003
8	0.2036
9	0.2076

(b)

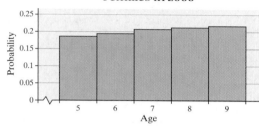

Distribution of 5–9-Year-Old Females in 2000

(c) $\mu_X = 7.0$ **(d)** $\sigma_X = 1.4$

25. (a)

x (grade level)	P(x)
1	0.1246
2	0.1246
3	0.1259
4	0.1271
5	0.1269
6	0.1254
7	0.1243
8	0.1211

(b)

Grade School Enrollment (2000)

(c) $\mu_X = 4.5$; if the experiment were repeated many times, the mean grade would be 4.5.
(d) $\sigma_X = 2.3$

26. (a)

x (grade level)	P(x)
9	0.2971
10	0.2617
11	0.2312
12	0.2101

(b)

High School Enrollment

(c) $\mu_X = 10.4$; if the experiment were repeated many times, the mean high school grade would be 10.4.
(d) $\sigma_X = 1.1$

27. (a) $P(4) = 0.119$ **(b)** $P(4 \text{ or } 5) = 0.222$
(c) $P(X \geq 6) = 0.108$
(d) $\mu_X = 3.0$; you would expect that this is the mother's third live birth.

28. (a) $P(5) = 0.228$ **(b)** $P(5 \text{ or } 6) = 0.333$
(c) $P(7, 8, \text{ or } 9) = 0.053$
(d) $\mu_X = 4.3$; you would expect a rental unit to have 4.3 rooms.

29. $E(X) = \$86.50$; the insurance company expects to make an average profit of $86.50 on every 20-year-old female it insures for 1 year.

30. $E(X) = \$2.75$; the insurance company expects to earn an average profit of $2.75 on every 20-year-old male it insures for 1 year.

31. The expected profit is $12,000.

32. The expected profit is $15,000.

33. $E(X) = -\$0.26$; if you played 1000 times, you would expect to lose about $260.

34. $E(X) = -\$0.42$; if the lottery is played many times, a player expects to lose $0.42 per ticket.

35. (a) The expected cash prize is $0.30. After paying $1.00 to play, your expected profit is $-\$0.70$.
(b) A grand prize of $118,000,000 has an expected profit greater than zero.
(c) The size of the grand prize does not affect the chance of winning provided the probabilities remain constant.

36. (a) A student who guesses will expect zero points.
(b) Answers will vary.

37. Answers will vary. The simulations illustrate the Law of Large Numbers.

38. Answers will vary. The simulations illustrate the Law of Large Numbers.

6.2 Assess Your Understanding (page 310)

1. An experiment is binomial provided:
  1. The experiment consists of a fixed number, n, of trials.
  2. The trials are independent.
  3. Each trial has two possible mutually exclusive outcomes: success and failure.
  4. The probability of success, p, remains constant for each trial of the experiment.

2. $_nC_x$ represents the number of ways x successes can be obtained from n trials.

3. When the binomial distribution is approximately bell shaped, about 95% of the outcomes are in the interval from $\mu - 2\sigma$ to $\mu + 2\sigma$. The Empirical Rule can be used when $np(1 - p) \geq 10$.

4. As n increases, the binomial distribution becomes more bell shaped.

5. When n is small, the shape of the binomial distribution is determined by p. If p is close to zero, the distribution is skewed right, if p is close to 0.5, the distribution is approximately symmetric; and if p is close to 1, the distribution is skewed left.

6. Success means the outcome you are measuring.

7. Not binomial, because the random variable is continuous

8. Not binomial, because the random variable is continuous

9. Binomial   **10.** Binomial

11. Not binomial, because the trials are not independent

12. Binomial

13. Not binomial, because the number of trials is not fixed

14. Not binomial, because the number of trials is not fixed

15. Binomial

16. Not binomial, because the trials are not independent

17. $P(3) = 0.2150$ **18.** $P(12) = 0.2184$ **19.** $P(38) = 0.0532$

20. $P(3) = 0.0607$ **21.** $P(3) = 0.2786$ **22.** $P(17) = 0.0123$

23. $P(X \leq 3) = 0.9144$ **24.** $P(X < 5) = 0.0949$

25. $P(X > 3) = 0.5$ **26.** $P(X \geq 12) = 0.8867$

27. $P(X \leq 4) = 0.5833$ **28.** $P(X \geq 8) = 0.7133$

29. (a)

x	P(x)	x	P(x)
0	0.1176	4	0.0595
1	0.3025	5	0.0102
2	0.3241	6	0.0007
3	0.1852		

(b) $\mu_X = 1.8$; $\sigma_X = 1.1$ **(c)** $\mu_X = 1.8$; $\sigma_X = 1.1$

(d)

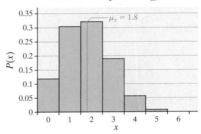

Probability Histogram

The distribution is skewed right.

30. (a)

x	P(x)	x	P(x)
0	0.0039	5	0.2188
1	0.0313	6	0.1094
2	0.1094	7	0.0313
3	0.2188	8	0.0039
4	0.2734		

(b) $\mu_x = 4$; $\sigma_x = 1.4$ **(c)** $\mu_x = 4$; $\sigma_x = 1.4$

(d)

Probability Histogram

The distribution is symmetric.

31. (a)

x	P(x)	x	P(x)
0	0.0000038	5	0.1168
1	0.0001	6	0.2336
2	0.0012	7	0.3003
3	0.0087	8	0.2253
4	0.0389	9	0.0751

(b) $\mu_x = 6.75$; $\sigma_x = 1.3$ **(c)** $\mu_x = 6.75$; $\sigma_x = 1.3$

(d)

Probability Histogram

The distribution is skewed left.

32. (a)

x	P(x)	x	P(x)
0	0.1074	6	0.0055
1	0.2684	7	0.0008
2	0.3020	8	0.0001
3	0.2013	9	0.000004
4	0.0881	10	0.0000001
5	0.0264		

(b) $\mu_x = 2$; $\sigma_x = 1.3$ **(c)** $\mu_x = 2$; $\sigma_x = 1.3$

(d)

Probability Histogram

The distribution is skewed right.

33. (a)

x	P(x)	x	P(x)
0	0.0010	6	0.2051
1	0.0098	7	0.1172
2	0.0439	8	0.0439
3	0.1172	9	0.0098
4	0.2051	10	0.0010
5	0.2461		

(b) $\mu_x = 5$; $\sigma_x = 1.6$ **(c)** $\mu_x = 5$; $\sigma_x = 1.6$

(d)

Probability Histogram

The distribution is symmetric.

34. (a)

x	P(x)	x	P(x)
0	0.0000005	5	0.0660603
1	0.0000184	6	0.1761608
2	0.0002949	7	0.3019899
3	0.0027525	8	0.3019899
4	0.0165151	9	0.1342177

(b) $\mu_x = 7.2$; $\sigma_x = 1.2$ **(c)** $\mu_x = 7.2$; $\sigma_x = 1.2$

(d)

Probability Histogram

The distribution is skewed left.

35. (a) This is a binomial experiment because
  1. It is performed a fixed number, $n = 15$, of times.
  2. The trials are independent.
  3. For each trial, there are two possible mutually exclusive outcomes: on time and not on time.
  4. The probability of "on time" is fixed at $p = 0.90$.
(b) $P(14) = 0.3432$ **(c)** $P(X \geq 14) = 0.5490$
(d) $P(X < 14) = 0.4510$ **(e)** $P(12 \leq X \leq 14) = 0.7386$

36. (a) This is a binomial experiment because
  1. It is performed a fixed number, $n = 10$, of times.
  2. The trials are independent.
  3. For each trial, there are two possible mutually exclusive outcomes: started smoking before turning 18 years old, started smoking at or after 18 years old
  4. The probability "starting to smoke before 18 years of age" is fixed at $p = 0.80$.

(b) $P(8) = 0.3020$ **(c)** $P(X \geq 8) = 0.6778$
(d) $P(X < 8) = 0.3222$ **(e)** $P(7 \leq X \leq 9) = 0.7718$

37. (a) $P(5) = 0.1746$
  (b) $P(X \geq 10) = 0.0026$; it would be unusual for at least 10 of the 20 households selected to have high-speed Internet.
  (c) $P(X < 4) = 0.4114$ **(d)** $P(2 \leq X \leq 5) = 0.7350$

38. (a) $P(3) = 0.0596$ **(b)** $P(X \leq 3) = 0.9841$
  (c) $P(1 \leq X \leq 4) = 0.6389$
  (d) It would be unusual to find 4 or more patients who experienced insomnia as a side effect because $P(X \geq 4) < 0.05$.

39. (a) $P(22) = 0.0120$ **(b)** $P(14 \leq X \leq 16) = 0.3648$
  (c) It would be unusual if 22 or more murders were committed using a firearm because $P(X \geq 22) < 0.05$.

40. (a) $P(3) = 0.0188$ **(b)** $P(X \leq 3) = 0.9971$
  (c) $P(X \geq 4) = 0.0029$ **(d)** $P(1 \leq X \leq 4) = 0.4542$

41. (a) $P(6) = 0.1460$ **(b)** $P(X < 7) = 0.2241$
  (c) $P(X \geq 5) = 0.9803$ **(d)** $P(5 \leq X \leq 8) = 0.7363$

42. (a) $P(7) = 0.2225$ **(b)** $P(X \geq 5) = 0.8883$
  (c) $P(X \leq 8) = 0.8655$ **(d)** $P(7 \leq X \leq 10) = 0.5186$

43. (a) $\mu_X = 90$; $\sigma_X = 3$
  (b) We expect that, in a random sample of 100 flights from Orlando to Los Angeles, 90 of them will be on time.
  (c) It would be unusual to have 80 on-time flights because $P(80) < 0.05$.

44. (a) $\mu_X = 160$; $\sigma_X = 5.7$
  (b) In a random sample of 200 adult smokers, we expect 160 to report that they started smoking before turning 18 years old.
  (c) Yes, because 180 is more than two standard deviations above the mean.

45. (a) $\mu_X = 20$; $\sigma_X = 4$
  (b) In a random sample of 100 U.S. households, we expect 20 of them to have a high-speed Internet connection.
  (c) It would not be unusual to observe 18 U.S. households with a high-speed Internet connection.

46. (a) $E(X) = 12$ patients with insomnia.
  (b) It would be unusual to observe 20 patients with insomnia.

47. (a) $E(X) = 66.9$ murders.
  (b) It would not be unusual to observe 75 murders by firearm.

48. It would not be unusual to observe 16 patients who experience weight gain because 16 is within two standard deviations of the mean.

49. It would not be unusual to observe 86 patients who experience headaches because 86 is within two standard deviations of the mean.

50. (a), (b) Answers will vary.
  (c) $P(29) = 0.3340$ **(d)** Answers will vary.
  (e) $P(X \leq 27) = 0.0217$
  (f) Simulated mean will vary. $E(X) = 29.4$ males.
  (g) Simulated standard deviation will vary. $\sigma_X = 0.8$.
  (h) Answers will vary.

51. (a), (b) Answers will vary.
  (c) $P(10) = 0.1074$ **(d)** Answers will vary.
  (e) $P(X \geq 8) = 0.6778$
  (f) Simulated mean will vary. $E(X) = 8$ free throws

52. (a) You would select 20 residents.
  (b) You would select 30 residents.

53. (a) You would select 38 residents 25 years and older.
  (b) You would select 65 residents 25 years and older.

Chapter Review Exercises (page 315)

1. (a) Continuous, $s \geq 0$
  (b) Discrete, $x = 0, 1, \ldots, 91$ (assuming 91 days of winter)
  (c) Discrete, $x = 0, 1, 2, \ldots$

2. (a) Continuous, $m \geq 0$
  (b) Discrete, $x = 0, 1, 2, \ldots$
  (c) Discrete, $x = 0, 1, 2, \ldots$

3. It is not a probability distribution because $\sum P(x) \neq 1$.

4. This is a probability distribution.

5. (a)

x	P(x)
4	0.3030
5	0.2424
6	0.2576
7	0.1970

(b)

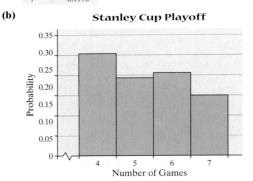

Stanley Cup Playoff

(c) $\mu_X = 5.3$ **(d)** $\sigma_X = 1.1$

6. (a)

Number of Property Crimes

x	P(x)
0	0.0028
1	0.0375
2	0.1882
3	0.4200
4	0.3515

(b)

Victims of Property Crimes

(c) $\mu_X = 3.1$ **(d)** $\sigma_X = 0.8$

7. $E(X) = 59.2$; the insurance company expects to earn an average profit of $59.20 for each 35-year-old male that it insures for 1 year at $100,000.

8. $E(X) = 0.08$ **9.** Binomial experiment

10. Not a binomial experiment, because the number of trials is not fixed.

11. (a) $P(0) = 0.4344$ **(b)** $P(2) = 0.1478$
(c) $P(X \geq 2) = 0.1879$ **(d)** $P(1) = 0.3777$
(e) $\mu_X = 20; \sigma_X = 4.3$
(f) It would not be unusual to have 12 females in a sample of 250 with high cholesterol.

12. (a) $P(10) = 0.1859$ **(b)** $P(X < 5) = 0.0093$
(c) $P(X \geq 5) = 0.9907$ **(d)** $P(3) = 0.0016$
(e) $\mu_X = 120; \sigma_X = 6.9$
(f) It would not be unusual to have 110 females in a sample of 200 believe that the driving age should be 18 years or older.

13. (a) $P(6) = 0.0454$ **(b)** $P(X < 4) = 0.6477$
(c) $P(X \geq 2) = 0.8244$ **(d)** $P(3) = 0.2428$
(e) $\mu_X = 75$
(f) It would be unusual to have 95 households in a sample of 500 tuned into the game.

14. (a) $P(8) = 0.1607$ **(b)** $P(X < 4) = 0.0097$
(c) $P(X \geq 5) = 0.9680$ **(d)** $P(3) = 0.0506$
(e) $\mu_X = 350; \sigma_X = 15.1$
(f) It would not be unusual to have 330 users in a sample of 1000 to suffer from insomnia because 330 is less than two standard deviations from the mean.

15. (a)

x	P(x)	x	P(x)
0	0.3277	3	0.0512
1	0.4096	4	0.0064
2	0.2048	5	0.0003

(b) $\mu_X = 1; \sigma_X = 0.9$ **(c)** $\mu_X = 1; \sigma_X = 0.9$

(d)

Binomial Probability Distribution

The distribution is skewed right.

16. (a)

x	P(x)	x	P(x)
0	0.00002	5	0.20764
1	0.00037	6	0.31146
2	0.00385	7	0.26697
3	0.02307	8	0.10011
4	0.08652		

(b) $\mu_X = 6; \sigma_X = 1.2$ **(c)** $\mu_X = 6; \sigma_X = 1.2$

(d)

Binomial Probability Distribution

The distribution is skewed left.

17. As a rule of thumb, if X is binomially distributed, the Empirical Rule can be used when $np(1 - p) \geq 10$.

18. We can sample without replacement and use the binomial probability distribution to approximate probabilities when the sample size is small in relation to the population size. As a rule of thumb, if the sample size is less than 5% of the population size, the trials can be considered nearly independent.

CHAPTER 7

7.1 Assess Your Understanding (page 327)

1. For a graph to be that of a probability density function, the area under the graph must equal 1, and the graph must be on or above the horizontal axis for all possible values of the random variable.

2. distribution; density

3. The area under the graph of a probability density function can be interpreted either as the proportion of the population with the characteristic described by the interval or as the probability that a randomly selected individual from the population has the characteristic described by the interval.

4. We standardize normal random variables so that one table can be used to find the area under the curve of any normal density function.

5. $\mu - \sigma; \mu + \sigma$

6. Increasing σ decreases the height of the graph of the normal density function.

7. This graph is not symmetric; it cannot represent a normal density function.

8. The graph can represent a normal density function.

9. This graph is not always greater than or equal to zero; it cannot represent a normal density function.

10. This graph increases as X becomes very large; it cannot represent a normal density function.

11. The graph can represent a normal density function.

12. This graph is not bell shaped; it cannot represent a normal density function.

13. $P(5 \leq X \leq 10) = \dfrac{1}{6}$ 14. $P(15 \leq X \leq 25) = \dfrac{1}{3}$

15. $P(X \geq 20) = \dfrac{1}{3}$ 16. $P(X \leq 5) = \dfrac{1}{6}$

17. **(a)**

(b) $P(0 \leq X \leq 0.2) = 0.2$
(c) $P(0.25 \leq X \leq 0.6) = 0.35$
(d) $P(X \geq 0.95) = 0.05$
(e) Answers will vary.

18. **(a)**

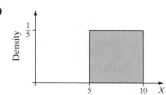

(b) $P(6 \leq X \leq 8) = \dfrac{2}{5} = 0.4$

(c) $P(5 \leq X \leq 8) = \dfrac{3}{5} = 0.6$

(d) $P(X \leq 6) = \dfrac{1}{5} = 0.2$

19. Normal

20. Not normal

21. Not normal

22. Normal

23. Graph A: $\mu = 10, \sigma = 2$; graph B: $\mu = 10, \sigma = 3$. A larger standard deviation makes the graph lower and more spread out.

24. Graph A: $\mu = 8, \sigma = 2$; graph B: $\mu = 14, \sigma = 2$. A larger mean shifts the graph to the right.

25. $\mu = 2, \sigma = 3$ 26. $\mu = 5, \sigma = 2$ 27. $\mu = 100, \sigma = 15$

28. $\mu = 530, \sigma = 100$

29. **(a)**

(b)

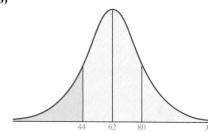

(c) (i) 15.87% of the cell phone plans in the United States are less than $44.00 per month.
(ii) The probability is 0.1587 that a randomly selected cell phone plan in the United States is less than $44.00 per month.

30. **(a)**

(b)

(c) (i) 11.51% of all refrigerators are kept for more than 17 years.
(ii) The probability is 0.1151 that a randomly selected refrigerator is more than 17 years old.

31. **(a)**

(b)

2895   3400   3905   4410   *X*

(c) (i) 2.28% of all full-term babies have a birth weight of at least 4410 grams.
(ii) The probability is 0.0228 that the birth weight of a randomly chosen full-term baby are at least 4410 grams.

32. (a), (b)

46.5 50.2 55.9 61.6   *X*

(c) (i) The proportion of 10-year-old males who are less than 46.5 inches tall is 0.0496.
(ii) The probability that a randomly selected 10-year-old male is less than 46.5 inches tall is 0.0496.

33. (a) (i) The proportion of human pregnancies that last more than 280 days is 0.1908.
(ii) The probability that a randomly selected human pregnancy lasts more than 280 days is 0.1908.
(b) (i) The proportion of human pregnancies that last between 230 and 260 days is 0.3416.
(ii) The probability that a randomly selected human pregnancy lasts between 230 and 260 days is 0.3416.

34. (a) (i) The proportion of times that Elena gets more than 26 miles per gallon is 0.3309.
(ii) The probability that a randomly selected fill-up yields at least 26 miles per gallon is 0.3309.
(b) (i) The proportion of times that Elena gets between 18 and 21 miles per gallon is 0.1107.
(ii) The probability that a randomly selected fill-up yields between 18 and 21 miles per gallon is 0.1107.

35. (a) $Z_1 = -0.67$
(b) $Z_2 = 0.67$
(c) The area between Z_1 and Z_2 is 0.495

36. (a) $Z_1 = -1.17$
(b) $Z_2 = 0.83$
(c) The area between Z_1 and Z_2 is 0.6760

37. (c) Answers will vary.
38. (c) Answers will vary.

7.2 Assess Your Understanding (page 341)

1. A standard normal curve has the following properties:
 1. It is symmetric about its mean, $\mu = 0$.
 2. Its highest point occurs at $\mu = 0$.
 3. It has inflection points at -1 and 1.
 4. The area under the curve is 1.
 5. The area under the curve to the right of $\mu = 0$ equals the area under the curve to the left of $\mu = 0$. Both equal $\frac{1}{2}$.

6. As *z* increases, the graph approaches but never equals zero. When *z* decreases, the graph approaches but never equals zero.
7. The Empirical Rule:
 Approximately 68% of the area under the standard normal curve is between -1 and 1. Approximately 95% of the area under the standard normal curve is between -2 and 2. Approximately 99.7% of the area under the standard normal curve is between -3 and 3.

2. The area under the curve to the right of $Z = 1.20$ is $1 - 0.8849 = 0.1151$

3. False. Explanations will vary.
4. The $P(Z < -1.30) = P(Z \le -1.30)$ because there is no area under the standard normal curve associated with a single value.

5. (a) Area = 0.0071   **(b)** Area = 0.3336
 (c) Area = 0.9115   **(d)** Area = 0.9998
6. (a) Area = 0.0002   **(b)** Area = 0.0233
 (c) Area = 0.8212   **(d)** Area = 0.9981
7. (a) Area = 0.9987   **(b)** Area = 0.9441
 (c) Area = 0.0375   **(d)** Area = 0.0009
8. (a) Area = 0.9998   **(b)** Area = 0.7088
 (c) Area = 0.0129   **(d)** Area = 0.0003
9. (a) Area = 0.9586   **(b)** Area = 0.2088
 (c) Area = 0.8479
10. (a) Area = 0.9892   **(b)** Area = 0.4525
 (c) Area = 0.9749
11. (a) Area = 0.0456   **(b)** Area = 0.0646
 (c) Area = 0.5203
12. (a) Area = 0.0032   **(b)** Area = 0.0476
 (c) Area = 0.2987
13. (a) Area = 0.8877   **(b)** Area = 0.0500
 (c) Area = 0.9802
14. (a) Area = 0.0198   **(b)** Area = 0.3686
 (c) Area = 0.9500

15. $z = -1.28$ **16.** $z = -0.84$ **17.** $z = -2.05$
18. $z = 1.04$ **19.** $z = 0.67$ **20.** $z = 0.39$
21. $z = -1.23$ **22.** $z = -0.67$
23. $z_0 = -1.28; z_1 = 1.28$ **24.** $z_0 = -1.04; z_1 = 1.04$
25. $z_0 = -2.575; z_1 = 2.575$ **26.** $z_0 = -1.88; z_1 = 1.88$
27. $z_{0.05} = 1.645$ **28.** $z_{0.35} = 0.39$ **29.** $z_{0.01} = 2.33$
30. $z_{0.02} = 2.05$ **31.** $z_{0.20} = 0.84$ **32.** $z_{0.15} = 1.04$
33. $P(Z < 1.93) = 0.9732$ **34.** $P(Z < -0.61) = 0.2709$
35. $P(Z < 1.93) = 0.9986$ **36.** $P(Z > 0.92) = 0.1788$
37. $P(-1.20 \le Z < 2.34) = 0.8753$
38. $P(1.23 < Z \le 1.56) = 0.0500$
39. $P(Z \ge 1.84) = 0.0329$ **40.** $P(Z \ge -0.92) = 0.8212$
41. $P(Z \le 0.72) = 0.7642$ **42.** $P(Z \le -2.69) = 0.0036$
43. $P(Z < -2.56 \text{ or } Z > 1.39) = 0.0875$
44. $P(Z < -0.38 \text{ or } Z > 1.93) = 0.3788$
45. $P(z < 1) - P(z < -1) = 0.8413 - 0.1687 = 0.6826$. So about 68% of the data lies within 1 standard deviation of the mean.
 $P(z < 2) - P(z < -2) = 0.9772 - 0.0228 = 0.9544$. So about 95% of the data lies within 2 standard deviations of the mean.
 $P(z < 3) - P(z < -3) = 0.9987 - 0.0013 = 0.9974$. So about 99.7% of the data lies within 3 standard deviations of the mean.
46. Area = 0.0901 because the graph is symmetric around the mean $\mu = 0$.
47. Area = 0.0054 because the graph is symmetric around the mean $\mu = 0$.
48. Area = 0.4332 because the graph is symmetric around the mean $\mu = 0$.
49. Area = 0.1906 because the graph is symmetric around the mean $\mu = 0$.
50. (a) $a = 2.50$ **(b)** $a = 0.15$ **(c)** $b = 1.53$

7.3 Assess Your Understanding (page 350)

1. To find the area under any normal curve:
 1. Draw the curve, label the random variables, and shade the desired area.
 2. Convert the values of *X* to *Z*-scores using the formula
 $$Z = \frac{X - \mu}{\sigma}.$$
 3. Draw a standard normal curve, label the *Z*-scores, and shade the desired area.
 4. Find the area under the standard normal curve.

2. To find the score that corresponds to a given probability:
   1. Draw a normal curve and shade the area corresponding to the probability that is given.
   2. Use Table IV to find the Z-score corresponding to the shaded area.
   3. Obtain the normal value from the formula $X = \mu + Z\sigma$.

3.

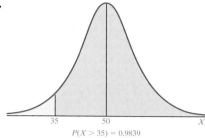

$P(X > 35) = 0.9839$

4.

$P(X > 65) = 0.0161$

5.

$P(X \leq 45) = 0.2375$

6.

$P(X \leq 58) = 0.8735$

7.

$P(40 < X < 65) = 0.9074$

8.

$P(56 < X < 68) = 0.1906$

9.

$P(55 \leq X \leq 70) = 0.2354$

10.

$P(40 \leq X \leq 49) = 0.3666$

11.

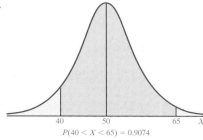

$P(38 < X \leq 55) = 0.7192$

12.

$P(56 \leq X < 66) = 0.1845$

13. $X = 40.62$ is at the 9th percentile.

14. $X = 58.96$ is at the 90th percentile.

15. $X = 56.16$ is at the 81st percentile.

16. $X = 47.83$ is at the 38th percentile.

17. (a) $P(X < 20) = 0.1587$ **(b)** $P(X > 22) = 0.1587$
(c) $P(19 \le X \le 21) = 0.4772$ **(d)** Yes, $P(X < 18) = 0.0013$

18. (a) $P(X > 80) = 0.5987$ **(b)** $P(X > 100) = 0.0040$
(c) $P(X < 60) = 0.0008$ **(d)** Yes, $P(X < 70) = 0.0446$

19. (a) $P(1000 \le X \le 1400) = 0.8658$
(b) $P(X < 1000) = 0.0132$
(c) 0.7019 of the bags has more than 1200 chocolate chips.
(d) 0.1230 of the bags has fewer than 1125 chocolate chips.
(e) A bag that contains 1475 chocolate chips is at the 96th percentile.
(f) A bag that contains 1050 chocolate chips is at the 4th percentile.

20. (a) $P(X \ge 6.0) = 0.6554$
(b) $P(X < 6.4) = 0.6554$
(c) $P(5.8 \le X \le 7.1) = 0.7522$
(d) Yes, $P(X \ge 8.25) = 0.00002$
(e) A California earthquake measuring 6.8 on the Richter scale is at the 88th percentile.
(f) A California earthquake measuring 5.1 on the Richter scale is at the 1st percentile.

21. (a) 0.1056 of the 2005 Honda Insights gets more than 60 miles per gallon.
(b) 0.0301 of the 2005 Honda Insights gets 50 miles per gallon or less.
(c) 0.2342 of the 2005 Honda Insights gets between 58 and 62 miles per gallon.
(d) $P(X < 45) = 0.0003$

22. (a) 0.1894 of the light bulbs lasts less than the advertised time.
(b) 0.0401 of the light bulbs lasts more than 1650 hours.
(c) $P(1625 \le X \le 1725) = 0.0923$
(d) $P(X > 1400) = 0.9957$

23. (a) A 20- to 29-year-old female who is 60 inches tall is at the 7th percentile.
(b) A 20- to 29-year-old female who is 70 inches tall is at the 98th percentile.
(c) 0.9105 of 20- to 29-year-old females are between 60 and 70 inches tall.
(d) Yes, $P(X > 70) = 0.0176$

24. (a) 40.13% of pregnancies lasts more than 270 days.
(b) 15.87% of pregnancies lasts fewer than 250 days.
(c) 75.9% of pregnancies lasts between 240 and 280 days.
(d) $P(X > 280) = 0.1894$
(e) $P(X \le 245) = 0.0951$
(f) 0.0043 of births is very preterm.

25. (a) 0.0764 of the rods has a length of less than 24.9 cm.
(b) 0.0324 of the rods is discarded.
(c) The plant manager expects to discard 162 of the 5000 rods manufactured.
(d) To meet the order, the plant manager should manufacture 11,804 rods.

26. (a) 0.0668 of the ball bearings has a diameter more than 5.03 mm.
(b) 0.0124 of the ball bearings will be discarded.
(c) Each day the plant manager expects to discard 372 of the 30,000 ball bearings manufactured.
(d) To meet the order, the plant manufacturer should manufacture 57,711 bearings.

27. (a) The 17th percentile for incubation time is 20 days.
(b) From 19 to 23 days make up the middle 95% of incubation times of the eggs.

28. (a) The 75th percentile for hours worked is 86.4 hours.
(b) From 72.9 to 90.5 hours make up the middle 80% of the hours worked by medical residents.

29. (a) The 30th percentile for the number of chips in an 18-ounce bag is 1201 chips.
(b) The middle 99% of the bags contains between 958 and 1566 chocolate chips.

30. (a) An earthquake measuring 6.1 on the Richter scale is at the 40th percentile.

(b) The middle 85% of the earthquakes in California measures between 5.5 and 6.9 on the Richter scale.

31. (a) The 97% gasoline mileage for the 2005 Honda Insight is 62 miles per gallon.
(b) From 51.3 to 60.7 miles per gallon is the middle 84% Honda Insight's gas mileage.

32. (a) 11.51% of customers receives the service for half-price.
(b) So that no more than 3% receives the discount, the guaranteed time limit should be 22 minutes.

33. (a)

Age	Proportion
15–19.99	0.0110
20–24.99	0.0852
25–29.99	0.2701
30–34.99	0.3655
35–39.99	0.2117
40–44.99	0.0505

(b) Answers will vary.

34. (a)

Temperature	Relative Frequency	Temperature	Relative Frequency
5.0–9.9	0.0005	40.0–44.9	0.2000
10.0–14.9	0.0053	45.0–49.9	0.1499
15.0–19.9	0.0080	50.0–54.9	0.1243
20.0–24.9	0.0213	55.0–59.9	0.0853
25.0–29.9	0.0507	60.0–64.9	0.0539
30.0–34.9	0.1157	65.0–69.9	0.0112
35.0–39.9	0.1733	70.0–74.9	0.0005

(b)

Chicago Temperatures, Nov. 16–30

(c) $\mu = 43.5°F$, $\sigma = 10.5°F$

(d)

Temperature	Theoretical Proportion
5.0–9.9	0.0006
10.0–14.9	0.0025
15.0–19.9	0.0090
20.0–24.9	0.0256
25.0–29.9	0.0586
30.0–34.9	0.1071
35.0–39.9	0.1567
40.0–44.9	0.1836
45.0–49.9	0.1721
50.0–54.9	0.1291
55.0–59.9	0.0776
60.0–64.9	0.0373
65.0–69.9	0.0143
70.0–74.9	0.0044

(e) Answers will vary.

7.4 Assess Your Understanding (page 359)

1. Explanations will vary.

2. f_i represents the expected proportion of observations less than or equal to the ith observation.

3. The sample data do not come from a normally distributed population.

4. The sample data come from a normally distributed population.

5. The sample data do not come from a normally distributed population.

6. The sample data do not come from a normally distributed population.

7. The sample data come from a normally distributed population.

8. The sample data come from a normally distributed population.

9. (a) The sample data come from a normally distributed population.
(b) $\bar{x} = 1247.4$ chips, $s = 101.0$ chips
(c)

1146.4 1247.4 1348.4 X

(d) $P(X \geq 1000) = 0.9929$
(e) $P(1200 \leq X \leq 1400) = 0.6153$

10. (a) The sample data come from a normally distributed population.
(b) $\bar{x} = 20.90$ hours, $s = 10.52$ hours
(c)

10.38   20.9   31.42 X

(d) $P(20 \leq X \leq 35) = 0.4458$
(e) $P(X > 40) = 0.0344$

11. The normal probability plot is approximately linear. So the sample data come from a normally distributed population.

12. The normal probability plot is approximately linear. So the sample data come from a normally distributed population.

13. The normal probability plot is not approximately linear. So the sample data do not come from a normally distributed population.

14. The normal probability plot is not approximately linear. So the sample data do not come from a normally distributed population.

7.5 Assess Your Understanding (page 366)

1. A probability experiment is binomial if:
 1. The experiment is performed n independent times.
 2. For each trial there are two mutually exclusive outcomes—success and failure.
 3. The probability of success p is the same for each trial of the experiment.

2. The normal distribution can be used to approximate binomial probabilities if $np(1 - p) \geq 10$.

3. We use a correction for continuity because we are using a continuous density function to approximate a discrete probability.

4. False

5. Area under the normal curve to the right of $X = 39.5$

6. Area under the normal curve to the left of $X = 20.5$

7. Area under the normal curve between $X = 7.5$ and $X = 8.5$

8. Area under the normal curve between $X = 11.5$ and $X = 12.5$

9. Area under the normal curve between $X = 17.5$ and $X = 24.5$

10. Area under the normal curve between $X = 29.5$ and $X = 40.5$

11. Area under the normal curve to the right of $X = 20.5$

12. Area under the normal curve to the left of $X = 39.5$

13. Area under the normal curve to the right of $X = 500.5$

14. Area under the normal curve to the left of $X = 34.5$

15. Using the binomial formula, $P(20) = 0.0616$; $np(1 - p) = 14.4$, the normal distribution can be used. Approximate probability is 0.0618.

16. Using the binomial formula, $P(18) = 0.0221$; $np(1 - p) = 10.2$, the normal distribution can be used. Approximate probability is 0.0216

17. Using the binomial formula, $P(30) < 0.0001$; $np(1 - p) = 7.5$, the normal distribution cannot be used.

18. Using the binomial formula, $P(50) < 0.0001$; $np(1 - p) = 4.75$, the normal distribution cannot be used.

19. Using the binomial formula, $P(60) = 0.0677$; $np(1 - p) = 14.1$, the normal distribution can be used. Approximate probability is 0.0630.

20. Using the binomial formula, $P(70) = 0.0970$; $np(1 - p) = 13.6$, the normal distribution can be used. Approximate probability is 0.0926.

21. (a) $P(30) \approx 0.0444$ **(b)** $P(X \geq 130) \approx 0.9332$
(c) $P(X < 125) \approx 0.0021$
(d) $P(125 \leq X \leq 135) \approx 0.5536$

22. (a) $P(80) \approx 0.1034$ **(b)** $P(X \geq 80) \approx 0.5517$
(c) $P(X < 70) \approx 0.0043$
(d) $P(70 \leq X \leq 90) \approx 0.9914$

23. (a) $P(20) \approx 0.0077$ **(b)** $P(X \leq 20) \approx 0.9934$
(c) $P(X \geq 22) \approx 0.0028$
(d) $P(20 \leq X \leq 30) \approx 0.0143$

24. (a) $P(15) \approx 0.1071$ **(b)** $P(X \geq 20) \approx 0.1635$
(c) $P(X < 10) \approx 0.0344$
(d) $P(12 \leq X \leq 18) \approx 0.6542$

25. (a) $P(20) \approx 0.0876$ **(b)** $P(X \leq 15) \approx 0.1515$
(c) $P(X \geq 30) \approx 0.0146$
(d) $P(10 \leq X \leq 32) \approx 0.9899$

26. (a) $P(40) \approx 0.0174$ **(b)** $P(X \geq 35) \approx 0.3745$
(c) $P(X < 25) \approx 0.0036$
(d) $P(30 \leq X \leq 35) \approx 0.6154$

27. (a) $P(X \geq 130) \approx 0.0028$
(b) Result is unusual. Reasons will vary.

28. (a) $P(X \geq 110) \approx 0.0066$
(b) The result is unusual. Reasons will vary.

29. (a) $P(X \geq 80) \approx 0.0032$ **(b)** Yes. Answers will vary.

30. (a) $P(X \geq 20) \approx 0.1190$ **(b)** No. Answers will vary.

Chapter Review Exercises (page 368)

1. (a) $\mu = 60$ **(b)** $\sigma = 10$
(c) (i) The proportion of random variables to the right of $X = 75$ is 0.0668.
(ii) The probability that a randomly selected random variable is greater than $X = 75$ is 0.0668.
(d) (i) The proportion of random variables between $X = 50$ and $X = 75$ is 0.7745.
(ii) The probability that a randomly selected random variable is between $X = 50$ and $X = 75$ is 0.7745.

2. (a) $\mu = 7$
(b) $\sigma = 2$
(c) (i) The proportion of random variables to the left of $X = 10$ is 0.9332.
(ii) The probability that a randomly selected random variable is less than $X = 10$ is 0.9332.
(d) (i) The proportion of random variables between $X = 5$ and $X = 8$ is 0.5328.
(ii) The probability that a randomly selected random variable is between $X = 5$ and $X = 8$ is 0.5328.

3. (a) $Z_1 = -0.5$
(b) $Z_2 = 0.25$
(c) The area between Z_1 and Z_2 is 0.2912.

4. (a) $Z_1 = -0.25$
(b) $Z_2 = 1.25$
(c) The area between Z_1 and Z_2 is 0.4931.

5.

Area under the normal curve to the left of −1.04 is 0.1492.

6.

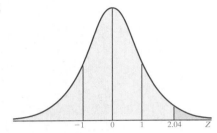

Area under the normal curve to the right of 2.04 is 0.0207.

7.

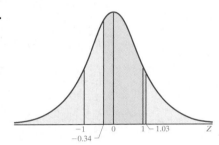

Area under the normal curve between −0.34 and 1.03 is 0.4816.

8.

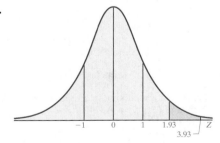

Area under the normal curve between 1.93 and 3.93 is 0.0268.

9. $P(Z < 1.19) = 0.8830$

10. $P(Z \geq 1.61) = 0.0537$

11. $P(-1.21 < Z \leq 2.28) = 0.8756$

12. $P(0.21 < Z < 1.69) = 0.3713$

13. $Z = 0.99$

14. $Z = 0.04$

15. $Z_1 = -1.75$ and $Z_2 = 1.75$

16. $Z_1 = -1.555$ and $Z_2 = 1.555$

17. $z_{0.20} = 0.84$

18. $z_{0.04} = 1.75$

19.

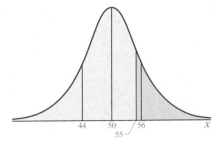

$P(X > 55) = 0.2023$

20.

$P(X \leq 23) = 0.0808$

21.

$P(65 < X < 85) = 0.6247$

22.

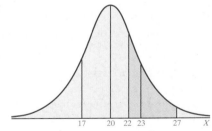

$P(22 \leq X \leq 27) = 0.2415$

23. (a) 12.71% of the tires lasts at least 75,000 miles.
  (b) 1.16% of the tires lasts at most 60,000 miles.
  (c) $P(65,000 \leq X \leq 80,000) = 0.8613$
  (d) The company should advertise 60,980 miles as its mileage warranty.

24. (a) 0.8413 of the time a fully charged battery lasts for at least 300 minutes.
  (b) 0.7486 of the time a fully charged battery lasts less than 340 minutes.
  (c) $P(310 \leq X \leq 350) = 0.5789$
  (d) 344 minutes is in the top 20% of talk time.
  (e) The middle 90% of talk time is between 285 and 363 minutes.

25. (a) 0.4090 of 16- to 19-year-old females has serum cholesterol above 180.
  (b) 0.4692 of 16- to 19-year-old females has serum cholesterol between 150 and 200.
  (c) $P(X < 140) = 0.2177$
  (d) A 16- to 19-year-old female with a cholesterol level of 120 is at the 10th percentile.
  (e) Yes, the 25th percentile on the normal curve is 144.

26. (a) 0.0475 of the test takers scores above 125.
(b) 0.2514 of the test takers scores below 90.
(c) 0.2476 of the test takers scores between 110 and 140.
(d) $P(X > 150) = 0.0004$
(e) A score of 125 places a child in the top 5% of all children.
(f) Normal children score between 71 and 129 points.

27. (a) 0.0119 of the baseballs produced is too heavy for use.
(b) 0.0384 of the baseballs produced is too light to use.
(c) 0.9497 of the baseballs produced is within acceptable weight limits.

28. (a) 0.0505 of the light bulbs lasts more than 3300 hours.
(b) 0.0032 of the light bulbs lasts less than 2500 hours.
(c) $P(2900 \le X \le 3100) = 0.4176$
(d) $P(X < 2600) = 0.0143$
(e) A bulb lasting 3350 hours is at the 97th percentile.
(f) Yes, $P(X > 3400) = 0.0143$

29. (a) $np(1 - p) = 14.7 > 10$, so the normal distribution can be used to approximate the binomial probabilities.
(b) $P(15) \approx 0.1000$ **(c)** $P(X > 20) \approx 0.1210$
(d) $P(X > 15) \approx 0.6517$ **(e)** $P(X < 25) \approx 0.9864$
(f) $P(15 \le X \le 25) \approx 0.6451$

30. (a) $np(1 - p) = 61.6 > 10$, so the normal distribution can be used to approximate the binomial probabilities.
(b) $P(125) \approx 0.0078$ **(c)** $P(X < 120) \approx 0.0045$
(d) $P(X \ge 140) \approx 0.5239$ **(e)** $P(100 \le X \le 120) \approx 0.0066$

31. The sample appears to be from a normally distributed population.
32. The sample is not from a normal population.
33. Not normal

34.

The sample appears to be from a normally distributed population.

35. (a) 0.9772 of the Ford Escape HEVs gets more than 25 miles per gallon.
(b) 0.6915 of the Ford Escape HEVs gets less than 30 miles per gallon.
(c) $P(26 \le X \le 34) = 0.9270$ **(d)** $P(X > 35) = 0.0013$
(e) A Ford Escape HEV that gets 32 miles per gallon is at the 93rd percentile.
(f) A Ford Escape HEV that gets 25 miles per gallon is at the 2nd percentile.

36. (a) $P(X \le 30) = 0.0010$
(b) This contradicts the *USA Today* "Snapshot."

37. (a)

(b) $P(0 \le X \le 5) = 0.25$ **(c)** $P(10 \le X \le 18) = 0.4$

38. A standard normal curve has the following properties:
1. It is symmetric about its mean, $\mu = 0$.
2. Its highest point occurs at $\mu = 0$.
3. It has inflection points at -1 and 1.
4. The area under the curve is 1.
5. The area under the curve to the right of $\mu = 0$ equals the area under the curve to the left of $\mu = 0$. Both equal $\frac{1}{2}$.
6. As z increases, the graph approaches but never equals zero. When z decreases, the graph approaches but never equals zero.
7. The Empirical Rule:
Approximately 68% of the area under the standard normal curve is between -1 and 1.
Approximately 95% of the area under the standard normal curve is between -2 and 2.
Approximately 99.7% of the area under the standard normal curve is between -3 and 3.

39. Answers will vary.

CHAPTER 8

8.1 Assess Your Understanding (page 389)

1. A sampling distribution is a probability distribution for all possible values of a statistic computed from a sample of size n.

2. The Central Limit Theorem states that regardless of the distribution of the population, the sampling distribution of the sample mean, $\bar{x}$, becomes approximately normal as the sample size, n, increases.

3. Standard error; mean   **4.** Zero

5. $\mu_{\bar{x}} = \mu, \sigma_{\bar{x}} = \dfrac{\sigma}{\sqrt{n}}$

6. The population must be normally distributed.

7. Four   **8.** True

9. The sampling distribution of $\bar{x}$ is normal with $\mu_{\bar{x}} = 30$ and $\sigma_{\bar{x}} = \dfrac{8}{\sqrt{10}}$.

10. The population does not need to be normally distributed because the sample is sufficiently large, $n = 40$. The sampling distribution of $\bar{x}$ is approximately normal with $\mu_{\bar{x}} = 50$, $\sigma_{\bar{x}} = \dfrac{4}{\sqrt{40}} = \dfrac{2}{\sqrt{10}}$.

11. $\mu_{\bar{x}} = 80, \sigma_{\bar{x}} = 2$ **12.** $\mu_{\bar{x}} = 64, \sigma_{\bar{x}} = 3$

13. $\mu_{\bar{x}} = 52, \sigma_{\bar{x}} = \dfrac{10}{\sqrt{21}}$ **14.** $\mu_{\bar{x}} = 27, \sigma_{\bar{x}} = \dfrac{6}{\sqrt{15}}$

15. (a) $\bar{x}$ is approximately normal with $\mu_{\bar{x}} = 80, \sigma_{\bar{x}} = 2$.
(b) $P(\bar{x} > 83) = 0.0668$ **(c)** $P(\bar{x} \le 75.8) = 0.0179$
(d) $P(78.3 < \bar{x} < 85.1) = 0.7969$

16. (a) $\bar{x}$ is approximately normal with $\mu_{\bar{x}} = 64, \sigma_{\bar{x}} = 3$.
(b) $P(\bar{x} < 62.6) = 0.3192$ **(c)** $P(\bar{x} \ge 68.7) = 0.0582$
(d) $P(59.8 < \bar{x} < 65.9) = 0.6549$

17. (a) The population must be normally distributed to compute probabilities regarding the sample mean. If the population is normally distributed, then the sampling distribution of $\bar{x}$ is normally distributed with $\mu_{\bar{x}} = 64$ and $\sigma_{\bar{x}} = \dfrac{17}{\sqrt{12}}$.
(b) $P(\bar{x} < 67.3) = 0.7486$ **(c)** $P(\bar{x} \ge 65.2) = 0.4052$

18. (a) The population must be normally distributed to compute probabilities regarding the sample mean. If the population is normally distributed, then the sampling distribution of $\bar{x}$ is normally distributed with $\mu_{\bar{x}} = 64$ and $\sigma_{\bar{x}} = \dfrac{17}{\sqrt{20}}$.

(b) $P(\bar{x} < 67.3) = 0.8078$
(c) $P(\bar{x} \geq 65.2) = 0.3745$
(d) Answers will vary.

19. (a) $P(x < 260) = 0.3520$ (b) $P(\bar{x} < 260) = 0.0465$
(c) $P(\bar{x} < 260) = 0.0040$ (d) Answers will vary.
(e) $P(256 \leq \bar{x} \leq 276) = 0.9844$

20. (a) $P(x > 60) = 0.3015$
(b) $P(\bar{x} > 60) = 0.0217$
(c) $P(\bar{x} > 60) = 0.0096$
(d) Increasing the sample size reduces the probability that $\bar{x} > 60$.
(e) Answers will vary.

21. (a) $P(x > 95) = 0.3192$
(b) $P(\bar{x} > 95) = 0.0179$
(c) $P(\bar{x} > 95) = 0.0049$
(d) Increasing the sample size reduces the probability that $\bar{x} > 95$.
(e) Answers will vary.

22. (a) $P(x < 75) = 0.1660$ (b) $P(\bar{x} < 75) = 0.0150$
(c) $P(\bar{x} < 75) = 0.0030$ (d) Answers will vary.

23. (a) $P(x > 0) = 0.5675$ (b) $P(\bar{x} > 0) = 0.7291$
(c) $P(\bar{x} > 0) = 0.8051$ (d) $P(\bar{x} > 0) = 0.8531$
(e) Answers will vary.

24. (a) $P(x > 34) = 0.2843$
(b) $P(\bar{x} > 34) = 0.0351$
(c) $P(\bar{x} > 34) = 0.0052$. Yes, it would be unusual for a random sample of 20 Chevrolet Cobalts to have a mean gas mileage of more than 34 miles per gallon.

25. (a) A sample size of at least 30 is needed to compute the probabilities.
(b) $P(\bar{x} < 10) = 0.0028$

26. (a) A sample size of at least 30 is needed to compute the probabilities.
(b) $P(\bar{x} \leq 56.8) = 0.1131$; answers will vary.

27. (a) From the Central Limit Theorem, as the sample size increases, the sampling distribution of the mean becomes approximately normal.
(b) $\mu_{\bar{x}} = 3, \sigma_{\bar{x}} = \sqrt{\dfrac{3}{50}}$
(c) $P(\bar{x} \geq 3.6) = 0.0071$. This result is unusual. Conclusions will vary.

28. (a) From the Central Limit Theorem, as the sample size increases, the sampling distribution of the mean becomes approximately normal.
(b) $\mu_{\bar{x}} = 20, \sigma_{\bar{x}} = \sqrt{\dfrac{1}{2}}$
(c) $P(\bar{x} \geq 22.1) = 0.0015$. This result is unusual. Conclusions will vary.

29. $P(\bar{x} \leq 45) = 0.0078$. It would be unusual for the random sample to have a mean number of blows less than or equal to 45.

30. $P(\bar{x} > 260) = 0.0643$. It would not be unusual for the average daily expenses for the sample to be above $260.00.

31. (a) $\mu = 38.7$ years old
(b) Samples: (37, 43); (37, 29); (37, 47); (37, 36); (37, 40); (43, 29); (43, 47); (43, 36); (43, 40); (29, 47); (29, 36); (29, 40); (47, 36); (47, 40); (36, 40)

(c)

$\bar{x}$	Probability	$\bar{x}$	Probability
32.5	$\dfrac{1}{15}$	39.5	$\dfrac{1}{15}$
33	$\dfrac{1}{15}$	40	$\dfrac{1}{15}$
34.5	$\dfrac{1}{15}$	41.5	$\dfrac{2}{15}$
36	$\dfrac{1}{15}$	42	$\dfrac{1}{15}$
36.5	$\dfrac{1}{15}$	43.5	$\dfrac{1}{15}$
38	$\dfrac{2}{15}$	45	$\dfrac{1}{15}$
38.5	$\dfrac{1}{15}$		

(d) $\mu_{\bar{x}} = 38.7$ years old

(e) $P(35.7 \leq \bar{x} \leq 41.7) = \dfrac{9}{15} = 0.6$

37, 47, 40	29, 36, 40
37, 36, 40	47, 36, 40

(f)

Samples	
37, 43, 29	43, 29, 47
37, 43, 47	43, 29, 36
37, 43, 36	43, 29, 40
37, 43, 40	43, 47, 36
37, 29, 47	43, 47, 40
37, 29, 36	43, 36, 40
37, 29, 40	29, 47, 36
37, 47, 36	29, 47, 40

Sampling Distribution	
$\bar{x}$	Probability
32	0.05
35	0.05
35.3	0.05
36	0.05
36.3	0.05
37.3	0.10
37.7	0.10
38.7	0.10
39.7	0.10
40	0.10
41	0.05
41.3	0.05
42	0.05
42.3	0.05
43.3	0.05

$\mu_{\bar{x}} = 38.7$; $P(35.7 \leq \bar{x} \leq 41.7) = 0.7$

32. (a) $\mu = 142.3$ minutes

(b)

Samples	
132, 201	201, 120
132, 112	112, 134
132, 134	112, 155
132, 155	112, 120
132, 120	134, 155
201, 112	134, 120
201, 134	155, 120
201, 155	

(c)

Sampling Distribution	
$\bar{x}$	Probability
116	0.0667
122	0.0667
126	0.0667
127	0.0667
129	0.0667
133	0.0667
133.5	0.0667
137.5	0.0667
143.5	0.0667
144.5	0.0667
156.5	0.0667
160.5	0.0667
166.5	0.0667
167.5	0.0667
178	0.0667

(d) $\mu_{\bar{x}} = 142.3$ minutes
(e) $P(127.3 \leq \bar{x} \leq 157.3) = 0.467$

(f)

Samples	
132, 201, 112	201, 112, 134
132, 201, 134	201, 112, 155
132, 201, 155	201, 112, 120
132, 201, 120	201, 134, 155
132, 112, 134	201, 134, 120
132, 112, 155	201, 155, 120
132, 112, 120	112, 134, 155
132, 134, 155	112, 134, 120
132, 134, 120	112, 155, 120
132, 155, 120	134, 155, 120

Chapter Review Exercises   Answers   AN-57

Sampling Distribution	
$\bar{x}$	Probability
121.3	0.05
122	0.05
126	0.05
128.7	0.05
129	0.05
133	0.05
133.7	0.05
135.7	0.05
136.3	0.05
140.3	0.05
144.3	0.05
148.3	0.05
149	0.05
151	0.05
151.7	0.05
155.7	0.05
156	0.05
158.7	0.05
162.7	0.05
163.7	0.05

$\mu_{\bar{x}} = 142.3$ minutes; $P(127.3 \le \bar{x} \le 157.3) = 0.7$

33. (a) Answers will vary.   (b) Answers will vary.
(c) Answers will vary.
(d) $\mu_{\bar{x}} = 100$, $\sigma_{\bar{x}} = \dfrac{8}{\sqrt{5}}$
(e) Answers will vary.
(f) $P(\bar{x} > 108) = 0.0125$ (g) Answers will vary.

34. (a) Answers will vary.   (b) Answers will vary.
(c) Answers will vary.   (d) Answers will vary.
35. (a) Answers will vary.   (b) Answers will vary.
(c) Answers will vary.   (d) Answers will vary.

8.2 Assess Your Understanding (page 397)

1. 0.44   **2.** Sample proportion; $\dfrac{x}{n}$
3. False   **4.** True
5. The sampling distribution of $\hat{p}$ is approximately normal when $n \le 0.05N$ and $np(1 - p) \ge 10$.
6. The standard deviation of $\hat{p}$ decreases as the sample size increases. If the sample size is increased by a factor of 4, the standard deviation is reduced by a factor of 2.
7. The sampling distribution of $\hat{p}$ is approximately normal with $\mu_{\hat{p}} = 0.4$ and $\sigma_{\hat{p}} = \sqrt{\dfrac{0.4(0.6)}{500}}$.
8. The sampling distribution of $\hat{p}$ is approximately normal with $\mu_{\hat{p}} = 0.7$ and $\sigma_{\hat{p}} = \sqrt{\dfrac{0.7(0.3)}{300}}$.
9. The sampling distribution of $\hat{p}$ is approximately normal with $\mu_{\hat{p}} = 0.103$ and $\sigma_{\hat{p}} = \sqrt{\dfrac{0.103(0.897)}{1000}}$.
10. The sampling distribution of $\hat{p}$ is approximately normal with $\mu_{\hat{p}} = 0.84$ and $\sigma_{\hat{p}} = \sqrt{\dfrac{0.84(0.16)}{1010}}$.
11. (a) The sampling distribution of $\hat{p}$ is approximately normal with $\mu_{\hat{p}} = 0.8$ and $\sigma_{\hat{p}} = \sqrt{\dfrac{0.8(0.2)}{75}}$.
(b) $P(\hat{p} \ge 0.84) = 0.1922$ (c) $P(\hat{p} \le 0.68) = 0.0047$
12. (a) The sampling distribution of $\hat{p}$ is approximately normal with $\mu_{\hat{p}} = 0.65$ and $\sigma_{\hat{p}} = \sqrt{\dfrac{0.65(0.35)}{200}}$.
(b) $P(\hat{p} \ge 0.68) = 0.1867$ (c) $P(\hat{p} \le 0.59) = 0.0375$
13. (a) The sampling distribution of $\hat{p}$ is approximately normal with

$\mu_{\hat{p}} = 0.35$ and $\sigma_{\hat{p}} = \sqrt{\dfrac{0.35(0.65)}{1000}}$.
(b) $P(\hat{p} \ge 0.39) = 0.0040$ (c) $P(\hat{p} \le 0.320) = 0.0233$
14. (a) The sampling distribution of $\hat{p}$ is approximately normal with
$\mu_{\hat{p}} = 0.42$ and $\sigma_{\hat{p}} = \sqrt{\dfrac{0.42(0.58)}{1460}}$.
(b) $P(\hat{p} \ge 0.45) = 0.0102$ (c) $P(\hat{p} \le 0.4) = 0.0606$
15. (a) The sampling distribution of $\hat{p}$ is approximately normal with
$\mu_{\hat{p}} = 0.3$ and $\sigma_{\hat{p}} = \sqrt{\dfrac{0.42(0.58)}{1460}}$.
(b) $P(\hat{p} > 0.37) = 0.0630$; no   (c) $P(\hat{p} \le 0.18) = 0.0043$; yes
16. (a) The sampling distribution of $\hat{p}$ is approximately normal with
$\mu_{\hat{p}} = 0.6$ and $\sigma_{\hat{p}} = \sqrt{\dfrac{0.6(0.4)}{1136}}$.
(b) $P(\hat{p} \le 0.585) = 0.1562$. The result is not unusual.
(c) $P(\hat{p} \ge 0.638) = 0.0043$. This result is rather unusual.
17. (a) The sampling distribution of $\hat{p}$ is approximately normal with
$\mu_{\hat{p}} = 0.26$ and $\sigma_{\hat{p}} = \sqrt{\dfrac{0.26(0.74)}{500}}$.
(b) $P(\hat{p} < 0.24) = 0.1539$
(c) It would be unusual for at least 150 adults to have no credit card. $P(\hat{p} \ge 0.3) = 0.0207$.
18. (a) The sampling distribution of $\hat{p}$ is approximately normal with
$\mu_{\hat{p}} = 0.07$ and $\sigma_{\hat{p}} = \sqrt{\dfrac{0.07(0.93)}{750}}$.
(b) $P(\hat{p} > 0.08) = 0.1423$
(c) Yes; $P(X \le 40) = P(\hat{p} \le 0.053) = 0.0344$
19. (a) $P(\hat{p} \le 0.4) = 0.0436$
(b) It would not be unusual, $P(\hat{p} \ge 0.45) = 0.1271$.
20. (a) $P(X \ge 125) = P(\hat{p} \ge 0.25) = 0.1446$
(b) It would not be unusual; $P(\hat{p} \le 0.20) = 0.0559$.
21. (a) 62 more adult Americans must be polled to make $np(1 - p) \ge 10$.
(b) 13 more adult Americans must be polled if $p = 0.2$.
22. (a) 12 more teenagers must be sampled if $p = 0.9$.
(b) 111 more teenagers must be sampled if $p = 0.95$.
23. (a) Answers will vary.   (b) Answers will vary.
(c) Answers will vary.   (d) Answers will vary.
(e) $\mu_{\hat{p}} = 0.3$, $\sigma_{\hat{p}} = \sqrt{\dfrac{0.3(0.7)}{765}}$.
24. Answers will vary.
25. (a) $\hat{p} = \dfrac{410}{500} = 0.82$
(b) $\sigma_{\hat{p}} = 0.017$

Chapter Review Exercises (page 401)

1. Answers will vary.
2. The sampling distribution of $\bar{x}$ is exactly normal when the underlying population distribution is normal. The sampling distribution of $\bar{x}$ is approximately normal when the sample size is large, usually greater than 30, regardless of how the population is distributed.
3. The sampling distribution of $\hat{p}$ is approximately normal when the sample size, n, is less than or equal to 5% of the population size, N, and $np(1 - p) \ge 10$.
4. $\mu_{\bar{x}} = \mu$, $\sigma_{\bar{x}} = \dfrac{\sigma}{\sqrt{n}}$, and $\mu_{\hat{p}} = p$, $\sigma_{\hat{p}} = \sqrt{\dfrac{p(1 - p)}{n}}$
5. (a) $P(x > 2625) = 0.3085$. It is not unusual to randomly select a pregnant woman who needs more than 2625 kcal per day.
(b) $\bar{x}$ is normally distributed with $\mu_{\bar{x}} = 2600$ kcal and $\sigma_{\bar{x}} = \dfrac{25}{\sqrt{5}}$ kcal.
(c) $P(\bar{x} > 2625) = 0.0125$. It would be unusual to find a random sample of 20 pregnant women with a mean energy need of more than 2625 kcal per day.

6. (a) $P(x > 100) = 0.3859$

(b) $\bar{x}$ is normally distributed with $\mu_{\bar{x}} = 90$ minutes and $\sigma_{\bar{x}} = \dfrac{35}{\sqrt{10}}$ minutes.

(c) $P(\bar{x} > 100) = 0.1841$. It would not be unusual to select a random sample of 10 batteries with a mean charge life of more than 100 minutes.

7. (a) $\bar{x}$ is approximately normally distributed with $\mu_{\bar{x}} = 0.75$ inch and $\sigma_{\bar{x}} = \dfrac{0.004}{\sqrt{30}}$ inch.

(b) There is a 0.0061 probability that the inspector will conclude the machine needs adjustment.

8. (a) $\bar{x}$ is approximately normally distributed with $\mu_{\bar{x}} = 2$ liters and $\sigma_{\bar{x}} = \dfrac{0.05}{\sqrt{45}}$ liter.

(b) $P(\bar{x} < 1.995) = 0.2514$

(c) $P(\bar{x} > 2.015) = 0.0228$. It is unusual that the mean fill volume is more than 2.015 liters. Conclusions may vary.

9. (a) $\hat{p}$ is approximately normally distributed with $\mu_{\hat{p}} = 0.72$ and $\sigma_{\hat{p}} = \sqrt{\dfrac{0.72(0.28)}{600}}$.

(b) $P(\hat{p} \le 0.70) = 0.1379$

(c) It would a little be unusual for 450 of the 600 randomly selected individuals to prefer to start their own business. $P(\hat{p} \ge 0.75) = 0.0505$.

10. (a) $\hat{p}$ is approximately normally distributed with $\mu_{\hat{p}} = 0.224$ and $\sigma_{\hat{p}} = \sqrt{\dfrac{0.224(0.776)}{300}}$.

(b) $P(X \ge 50) = P(\hat{p} \ge 0.167) = 0.9913$

(c) It would be unusual for a random sample of 300 adults to result in 18% smokers. $P(\hat{p} \le 0.18) = 0.0336$.

11. (a) $\hat{p}$ is approximately normally distributed with $\mu_{\hat{p}} = 0.09$ and $\sigma_{\hat{p}} = \sqrt{\dfrac{0.09(0.91)}{200}}$.   **(b)** $P(\hat{p} \le 0.06) = 0.0694$

(c) It would be unusual for 25 people in a random sample of 200 adults to have an advanced degree. $P(\hat{p} \ge 0.125) = 0.0418$.

12. (a) $\hat{p}$ is approximately normally distributed with $\mu_{\hat{p}} = 0.01$ and $\sigma_{\hat{p}} = \sqrt{\dfrac{0.01(0.99)}{1500}}$.   **(b)** $P(\hat{p} > 0.015) = 0.0256$

(c) It would not be unusual for a random sample of 150 Americans to have no one allergic to peanuts or tree nuts. $P(\hat{p} \le 0.0067) = 0.0968$.

13. $P(\bar{x} < -65,000) = 0.0071$ **14.** $P(\bar{x} > 400) = 0.9591$

CHAPTER 9   Estimating the Value of a Parameter

9.1 Assess Your Understanding (page 405)

1. Level of confidence, sample size, and the standard deviation of the population.

2. The margin of error increases because to be more confident that the interval contains the population mean we need to make the interval wider.

3. The margin of error decreases as the sample size increases because the Law of Large Numbers states that as the sample size increases the sample mean approaches the value of the population mean.

4. $\bar{x} = 14$; the margin of error, $E = 4$.

5. The mean age of the population is a fixed value and so is not probabilistic.

6. If the n is rounded down, the confidence interval will be smaller than $(1 - \alpha) \cdot 100\%$.

7. No, the data are not normal and contain an outlier.

8. No, the data are not normal and contain outliers.

9. No, the data are not normal.   **10.** No, the data are not normal.

11. Yes, the data are normal, and there are no outliers.

12. Yes, the data are normal, and there are no outliers.

13. $z_{\alpha/2} = z_{0.01} = 2.33$ **14.** $z_{\alpha/2} = z_{0.03} = 1.88$

15. $z_{\alpha/2} = z_{0.075} = 1.44$ **16.** $z_{\alpha/2} = z_{0.1} = 1.28$

17. (a) Lower bound: 32.44, upper bound: 35.96
(b) Lower bound: 32.73, upper bound: 35.67; increasing the sample size decreases the margin of error.
(c) Lower bound: 31.89, upper bound: 36.51; the interval in (c) is wider than that in (a). Increasing the confidence level increases the margin of error from 1.76 to 2.31.
(d) If the sample size is 15, the data must come from a normally distributed population.

18. (a) Lower bound: 58.27, upper bound: 60.13
(b) Lower bound: 58.36, upper bound: 60.04; increasing the sample size decreases the margin of error.
(c) Lower bound: 57.88, upper bound: 60.52; the interval in (c) is wider than that in (a). Increasing the confidence level increases the margin of error from 0.93 to 1.32.
(d) If the sample size is 15, the data must come from a normally distributed population.

19. (a) Lower bound: 102.67, upper bound: 113.33
(b) Lower bound: 99.57, upper bound: 116.43; decreasing the sample size increases the margin of error.
(c) Lower bound: 103.96, upper bound: 112.04; decreasing the level of confidence reduces the margin of error.
(d) We could not compute the confidence intervals because the samples are too small.
(e) The outliers would have increased the mean, shifting the confidence intervals to the right.

20. (a) Lower bound: 115.85, upper bound: 130.15
(b) Lower bound: 113.77, upper bound: 132.23; decreasing the sample size increases the margin of error.
(c) Lower bound: 117.53, upper bound: 128.47; decreasing the level of confidence reduces the margin of error.
(d) We could not compute the confidence intervals because the samples are too small.
(e) The outlier would have increased the mean, shifting the confidence intervals to the right.

21. Lower bound: 8.10, upper bound: 8.24; we are 95% confident that the population mean time sleeping is between 8.10 and 8.24 hours.

22. Lower bound: 1.18, upper bound: 1.24; we are 90% confident that the population mean time eating is between 1.18 hours and 1.24 hours.

23. Lower bound: 8.88, upper bound: 9.52; she is 90% confident that the population mean time spent with friends is between 8.88 hours and 9.52 hours.

24. Lower bound: 19.07, upper bound: 29.33; the economist is 95% confident that the population mean commute time is between 19.07 minutes and 29.33 minutes.

25. (a) $\bar{x} = 542.25$ dollars
(b) The conditions are met. The data are approximately normal, and there are no outliers.
(c) Lower bound: 326.65, upper bound: 757.85; the institute is 95% confident that the population mean cost of repairs is between $326.65 and $757.85.
(d) Lower bound: 361.30, upper bound: 723.20; the institute is 90% confident that the population mean cost of repairs is between $361.30 and $723.20.
(e) When the level of confidence is decreased, the width of the confidence interval is decreased.

26. (a) $\bar{x} = 103.4$ minutes
(b) The conditions are met. The data are approximately normal, and there are no outliers.
(c) Lower bound: 98.2, upper bound: 108.6; the researcher is 95% confident that the mean flight time is between 98.2 minutes and 108.6 minutes.
(d) Lower bound: 99.0, upper bound: 107.8; the researcher is 90% confident that the mean flight time is between 99.0 minutes and 107.8 minutes.
(e) When the level of confidence is decreased, the width of the confidence interval is decreased.

27. (a) $\bar{x} = 45.1$ years of age
(b) The conditions are met. The data are approximately normal, and there are no outliers.

(c) Lower bound: 41.1, upper bound: 49.1; the realtor is 95% confident that the mean age of a client is between 41.1 and 49.1 years of age.

(d) The realtor's clients do not appear to differ in age from the general population.

28. (a) $\bar{x} = 47$ years old
(b) The conditions are met. The data are approximately normal, and there are no outliers.
(c) Lower bound: 44.6, upper bound: 49.4; the agent is 95% confident that the mean age of her clients is between 44.6 and 49.4 years.
(d) The real estate agent's clients appear to differ in age from the general population.

29. (a) $\bar{x} = 48.3$
(b) Lower bound: 42.0, upper bound: 54.6; Dr. Oswiecmiski is 95% confident that the mean HDL of his male patients is between 42.0 and 54.6.
(c) Dr. Oswiecmiski's patients do not appear to differ from the general population.
(d) To obtain a more precise (smaller) interval, he should either increase the sample size or decrease his confidence level.

30. (a) $\bar{x} = 47.5$
(b) Lower bound: 40.7, upper bound: 54.3; we can be 95% confident that mean number of days it takes customers to pay their bills is between 40.7 and 54.3 days.
(c) Yes, June payments appear to be typical.

31. (a) $\bar{x} = 1.64$ million shares
(b) Lower bound: 1.38, upper bound: 1.90; we are 90% confident that the mean number of shares traded is between 1.38 and 1.90 million.
(c) Lower bound: 1.48, upper bound: 2.00; we are 90% confident that the mean number of shares traded is between 1.48 and 2.00 million.
(d) The intervals are different because the sample mean in (c) is 100 thousand shares larger than the sample mean in (a).

32. (a) $\bar{x} = 9.05$ million shares
(b) Lower bound: 7.33, upper bound: 10.27; we are 90% confident that the mean number of Google shares traded is between 7.33 and 10.27 million shares.
(c) Lower bound: 6.84, upper bound: 10.27; we are 90% confident that the mean number of Google shares traded is between 6.84 and 10.27 million shares.
(d) The intervals are different because the sample mean in (c) is 490 thousand shares less than the sample mean in (a).

33. (a) Since the length of dramas is not normally distributed, the sample must be large.
(b) Lower bound: 125.5; upper bound: 151.1; she is 99% confident that the mean length of a drama is between 125.5 minutes and 151.1 minutes.

34. The mean age of the presidents is known, so it makes no sense to talk about a confidence interval.

35. (a) $\bar{x} = 49,477.7$ miles
(b) Lower bound: 40,607.2, upper bound: 58,308.2; she is 99% confident that the mean number of miles on 4-year-old Saturns in Chicago is between 40,607.2 and 58,308.2.
(c) Lower bound: 42,756.2, upper bound: 56,199.2; she is 95% confident that the mean number of miles on 4-year-old Saturns in Chicago is between 42,756.2 and 56,199.2.
(d) Decreasing the level of confidence decreases the width of the interval.
(e) No, because she only sampled cars in Chicago, the results cannot be generalized to the entire United States.

36. (a) $\bar{x} = 40,520.2$ miles
(b) Lower bound: 33,512.6, upper bound: 47,527.8; she is 99% confident that the mean number of miles on 3-year-old Cavaliers in Orlando is between 33,512.6 and 47,527.8 miles.
(c) Lower bound: 35,186.3, upper bound: 45,854.1; she is 95% confident that the mean number of miles on 3-year-old Cavaliers in Orlando is between 35,186.3 and 45,854.1 miles.
(d) Decreasing the level of confidence decreases the width of the interval.
(e) No, because she only sampled cars in Orlando, the results cannot be generalized to the entire United States.

37. For a 99% confidence level, $n = 298$ patients; for a 95% confidence level, $n = 173$ patients. Decreasing the confidence level decreases the sample size needed.

38. For a 90% confidence level, $n = 188$ subjects are needed; for a 98% confidence level, $n = 378$ subjects are needed. Increasing the confidence level increases the sample size needed.

39. For a 95% confidence level, $n = 1059$ subjects are needed.

40. For a 95% confidence level, $n = 670$ subjects are needed.

41. (a) For a 90% confidence level, $n = 1051$ cars are needed in the sample.
(b) For a 90% confidence level, $n = 4201$ cars are needed in the sample.
(c) Doubling the required accuracy quadruples the required sample size.

42. (a) For a 98% confidence level, $n = 352$ cars are needed in the sample.
(b) For a 90% confidence level, $n = 1408$ cars are needed in the sample.
(c) Doubling the required accuracy quadruples the required sample size.

43. (a)–(b) Answers will vary.
(c) Expect 19 intervals to include the population mean; actual number will vary.

44. (a)–(b) Answers will vary.
(c) Expect 27 intervals to include the population mean; actual number will vary.

45. Answers will vary.

46. (a) Lower bound: 44.59, upper bound: 55.91
(b) 14 is less than the lower fence; $14 < Q_1 - 1.5(\text{IQR}) = 29.25$
(c) Lower bound: 42.34; upper bound: 53.66; the outlier shifts the confidence interval to the left.
(d) $\bar{x} = 50.25$
(e) Lower bound: 46.98; upper bound: 53.52; increasing the sample size reduces the width of the interval.
(f) 14 is less than the lower fence, $14 < Q_1 - 1.5(\text{IQR}) = 23.5$
(g) Lower bound: 46.23; upper bound: 52.77; the outlier shifts the confidence interval only slightly to the left.

47. The sample size must be increased by a factor of 4 to decrease the margin of error by one-half.

48. Sample B must be 4 times larger than sample A.

49. (a) Set I: $\bar{x} = 99.125$; set II: $\bar{x} = 99.1$; set III: $\bar{x} = 99.033$
(b) Set I: Lower bound: 88.73, upper bound: 109.52. Set II: Lower bound: 92.53, upper bound: 105.67. Set III: Lower bound: 93.67, upper bound: 104.40.
(c) As the size of the sample increases, the width of the confidence interval decreases.
(d) Set I: Lower bound: 77.48, upper bound: 98.27. Set II: Lower bound: 88.03, upper bound: 101.17. Set III: Lower bound: 90.67; upper bound: 101.40.
(e) Each interval except set I, which has an outlier, contains the population mean.

50. (a) Answers will vary. Expect 95% of the intervals to contain the population mean.
(b) Answers will vary.
(c) Answers will vary. Expect 99% of the intervals to contain the population mean.

51. (a) Answers will vary. Expect 95% of the intervals to contain the population mean.
(b) Answers will vary.
(c) Answers will vary. Expect 95% of the intervals to contain the population mean.
(d) Confidence intervals with $n = 10$ should be wider than those with $n = 50$.

52. (a) Answers will vary. Expect 95% of the intervals to contain the population mean.
(b) Answers will vary. Expect less than 95% of the intervals to contain the population mean.
(c) The true proportion of intervals that contain the population mean is less than what is expected.

9.2 Assess Your Understanding (page 430)

1. A Z-interval should be constructed if the sample is random, the population from which the sample is drawn is normal or the sample size is large ($n \geq 30$), and if σ is known. A t-interval should be constructed if the sample is random, the population from which the sample is drawn is normal, but if σ is not known. Neither interval can be used if the sample is not random, the population is not normal and the sample size is small, or when there are outliers.

2. t-Distribution has less spread as the degrees of freedom increase because as n increases s becomes closer to σ, by the Law of Large Numbers.

3. Robust means that the procedure is accurate when there are moderate departures from normality in the distribution of the population.

4. t-Values are more dispersed than z-values because the sample standard deviation, s, is a random variable.

5. Similarities: Both are probability density distributions; both have $\mu = 0$; both are symmetric around their means.

Differences: t-distributions vary for different sample sizes, n; there is only one standard normal distribution; t-distributions have longer, thicker tails; the z-distribution has most of its area between -3 and 3.

6. Answers will vary.

7. **(a)** $t_{0.10} = 1.316$
    (b) $t_{0.05} = 1.697$
    (c) $t_{0.99} = -2.552$
    (d) $t_{0.05} = 1.725$

8. **(a)** $t_{0.02} = 2.205$ **(b)** $t_{0.10} = 1.309$
    (c) $t_{0.95} = -1.943$ **(d)** $t_{0.025} = 2.120$

9. **(a)** Lower bound: 103.7, upper bound: 112.3
    (b) Lower bound: 100.4, upper bound: 115.6; decreasing the sample size increases the margin of error.
    (c) Lower bound: 104.6, upper bound: 111.4; decreasing the level of confidence lessens the margin of error.
    (d) No, the sample sizes were too small.

10. **(a)** Lower bound: 45.5, upper bound: 54.5
    (b) Lower bound: 44.6, upper bound: 55.4; decreasing the sample size increases the margin of error.
    (c) Lower bound: 46.3, upper bound: 53.7; decreasing the level of confidence lessens the margin of error.
    (d) No, the sample sizes were too small.

11. **(a)** Lower bound: 16.85, upper bound: 19.95
    (b) Lower bound: 17.12, upper bound: 19.68; increasing the sample size decreases the margin of error.
    (c) Lower bound: 16.32, upper bound: 20.48; increasing the level of confidence increases the margin of error.
    (d) If $n = 18$, the population must be normal.

12. **(a)** Lower bound: 32.78, upper bound: 37.42
    (b) Lower bound: 33.66, upper bound: 36.54; increasing the sample size decreases the margin of error.
    (c) Lower bound: 31.76, upper bound: 38.44; increasing the level of confidence increases the margin of error.
    (d) If $n = 18$, the population must be normal.

13. Lower bound: 12.05, upper bound: 14.75; we can be 99% confident that the mean number of books read by Americans during 2004 was between 12.05 and 14.75.

14. **(a)** Lower bound: 17.19, upper bound: 20.41; we can be 99% confident that the mean number of books read by Americans during 1978 was between 17.19 and 20.41.
    (b) Americans were reading more in 1978 than in 2005.

15. Lower bound: 1.08, upper bound: 8.12; we can be 95% confident that the mean incubation period of patients with SARS is between 1.08 and 8.12 days.

16. Lower bound: 228.33, upper bound: 256.07; researchers are 90% confident that the mean tensile strength of the resin cement is between 228.33 and 256.07 newtons.

17. Lower bound: 12.86, upper bound: 13.14; the pollsters are 95% confident that the mean number of hours teenagers watch TV each week is between 12.86 and 13.14 hours.

18. Lower bound: 22.52, upper bound: 23.48; we can be 95% confident that the mean gas mileage of the 2005 Honda Accord is between 22.52 and 23.48 miles per gallon.

19. Lower bound: 0.763, upper bound: 1.867; the stock analyst is 90% confident that the mean dividend yield of financial stocks is between 0.763% and 1.867%.

20. Lower bound: 17.12, upper bound: 20.30; the server can be 95% confident that the mean tip rate per dinner is between 17.12% and 20.3%.

21. Lower bound: 12.40, upper bound: 19.44; we can be 99% confident that the mean concentration of dissolved organic carbon in organic soil is between 12.40 and 19.44 mg/L.

22. **(a)** Lower bound: 8.08, upper bound: 11.98; we can be 99% confident that the mean concentration of dissolved organic carbon in mineral soil is between 8.08 and 11.98 mg/L.
    (b) Yes, there appears to be a higher concentration of dissolved organic carbon in organic soil than in mineral soil.

23. **(a)** Yes, the data are approximately normal, and there are no outliers.
    (b) Lower bound: 31.9, upper bound: 44.7
    (c) To increase the accuracy of the interval, increase the sample size.

24. **(a)** Yes, the data are approximately normal, and there are no outliers.
    (b) Lower bound: 8.23, upper bound: 10.71
    (c) To increase the accuracy of the interval, increase the sample size.

25. **(a)** Yes, the data are approximately normal, and there are no outliers.
    (b) Lower bound: 151.9, upper bound: 183.2
    (c) To increase the accuracy of the interval, increase the sample size.

26. **(a)** Yes, the data are approximately normal, and there are no outliers.
    (b) Lower bound: 170.4, upper bound: 173.0; we can be 95% confident that the mean weight of the discs produced by the manufacturer is between 170.4 and 173.0 grams.

27. **(a)**

    (b) Lower bound: $158,567.7, upper bound: $314,148.9
    (c) Lower bound: $165,219.6, upper bound: $266,853.2
    (d) Wider interval

28. **(a)**

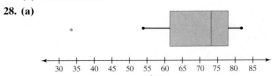

    (b) Lower bound: 56.1, upper bound: 81.1
    (c) Lower bound: 62.9, upper bound: 80.5
    (d) Wider interval

29. Answers will vary. You would expect 19 of the 20 samples to include the population mean.

30. Answers will vary. You would expect 27 of the 30 samples to include the population mean.

31. Answers will vary.

9.3 Assess Your Understanding (page 441)

1. $\hat{p}$ is the best point estimate of a population proportion.

2. To construct a confidence interval about a population proportion, the sample size n must be less than 5% of the population, and $np(1 - p) \geq 10$.

3. Answers will vary.   **4.** Answers will vary.

5. Lower bound: 0.146, upper bound: 0.254

6. Lower bound: 0.319, upper bound: 0.481

7. Lower bound: 0.191, upper bound: 0.289

8. Lower bound: 0.306, upper bound: 0.360

9. Lower bound: 0.758, upper bound: 0.805

10. Lower bound: 0.567, upper bound: 0.633

11. (a) $\hat{p} = 0.054$
 (b) $n\hat{p}(1 - \hat{p}) = 44.44 \geq 10$, and the sample is less than 5% of the population.
 (c) Lower bound: 0.041, upper bound: 0.067
 (d) We are 90% confident that the proportion of Lipitor users who experience a headache as a side effect is between 0.041 and 0.067.

12. (a) $\hat{p} = 0.784$
 (b) $n\hat{p}(1 - \hat{p}) = 12.54 \geq 10$, and the sample is less than 5% of the population.
 (c) Lower bound: 0.661, upper bound: 0.907
 (d) The researchers are 99% confident that the proportion of ulcer patients who receive Pepcid and then have their ulcers heal is between 0.661 and 0.907.

13. (a) $\hat{p} = 0.300$
 (b) $n\hat{p}(1 - \hat{p}) = 211.68 \geq 10$, and the sample is less than 5% of the population.
 (c) Lower bound: 0.266, upper bound: 0.334; the Gallup Organization is 98% confident that the proportion of adults aged 18 or older who believe the United States is spending too little on national defense and military purposes is between 0.266 and 0.334.

14. (a) $\hat{p} = 0.680$
 (b) $n\hat{p}(1 - \hat{p}) = 218.9 \geq 10$, and the sample is less than 5% of the population.
 (c) Lower bound: 0.646, upper bound: 0.714; the Gallup Organization is 98% confident that the proportion of adults aged 18 or older who think the government should make "partial birth" abortions illegal, except in cases necessary to save the life of the mother, is between 0.646 and 0.714.

15. (a) $n\hat{p}(1 - \hat{p}) = 312.03 \geq 10$, and the sample is less than 5% of the population.
 (b) Lower bound: 0.166, upper bound: 0.194; the Harris Organization is 90% confident that the proportion of adults who follow professional football whose favorite team is the Green Bay Packers is between 0.166 and 0.194.
 (c) Lower bound: 0.158, upper bound: 0.202; the Harris Organization is 99% confident that the proportion of adults who follow professional football whose favorite team is the Green Bay Packers is between 0.158 and 0.202.
 (d) Increasing the level of confidence widens the interval.

16. (a) $n\hat{p}(1 - \hat{p}) = 250.2 \geq 10$, and the sample is less than 5% of the population.
 (b) Lower bound: 0.423, upper bound: 0.477; the Gallup Organization is 92% confident that adults aged 18 and older who feel they have too little time for relaxing or doing nothing is between 0.423 and 0.477.
 (c) Lower bound: 0.418, upper bound: 0.482; the Gallup Organization is 96% confident that adults aged 18 and older who feel they have too little time for relaxing or doing nothing is between 0.418 and 0.482.
 (d) Increasing the level of confidence widens the interval.

17. (a) Using $\hat{p} = 0.44$, $n = 1816$.
 (b) Using $\hat{p} = 0.5$, $n = 1842$.

18. (a) Using $\hat{p} = 0.675$, $n = 1485$.
 (b) Using $\hat{p} = 0.5$, $n = 1692$.

19. (a) Using $\hat{p} = 0.55$, $n = 419$.
 (b) Using $\hat{p} = 0.5$, $n = 423$.

20. (a) Using $\hat{p} = 0.37$, $n = 395$.
 (b) Using $\hat{p} = 0.5$, $n = 423$.

21. (a) Using $\hat{p} = 0.23$, $n = 2404$.
 (b) Using $\hat{p} = 0.5$, $n = 3394$.

22. (a) Using $\hat{p} = 0.34$, $n = 1269$.
 (b) Using $\hat{p} = 0.5$, $n = 1414$.

23. $n = 984$ people were surveyed.

24. $n = 1016$ people were surveyed.

25. Answers will vary.

26. (a) Answers will vary. Expect 95% of the intervals to contain the population proportion.
 (b) Answers will vary.
 (c) Answers will vary. Expect 0.99 of the intervals to contain the population proportion.

27. (a) Answers will vary.
 (b) Answers will vary.
 (c) Answers will vary. About 0.95 of the intervals should contain the population proportion.
 (d) As n increases, the proportion of intervals that capture p approaches the level of confidence.

9.4 Assess Your Understanding (page 445)

1. We construct a t-interval when we are estimating the population mean, we do not know the population standard deviation, and the underlying population is normally distributed. If the underlying population is not normally distributed, we can construct a t-interval to estimate the population mean provided the sample size is large ($n \geq 30$).

We construct a Z-interval when we are estimating the population mean, we know the population standard deviation, and the underlying population is normally distributed. If the underlying population is not normally distributed, we can construct a Z-interval to estimate the population mean provided the sample size is large ($n \geq 30$). We also construct a Z-interval when we are estimating the population proportion, provided the sample size is smaller than 5% of the population and $np(1 - p) \geq 10$.

2. To construct a confidence interval about a population proportion, the sample size must be less than 5% of the population size and $n\hat{p}(1 - \hat{p}) \geq 10$.

3. Lower bound: 49.52, upper bound: 70.48

4. Lower bound: 109.52, upper bound: 135.48

5. Lower bound: 0.069, upper bound: 0.165

6. Lower bound: 0.317, upper bound: 0.383

7. Lower bound: 37.74, upper bound: 52.26

8. Lower bound: 2.65, upper bound: 3.85

9. Lower bound: 114.98, upper bound: 126.02

10. Lower bound: 19.73, upper bound: 20.47

11. Lower bound: 51.4, upper bound: 56.6; we can be 95% confident that felons convicted of aggravated assault serve a mean sentence between 51.4 and 56.6 months.

12. Lower bound: 0.730, upper bound: 0.792; the Harris Organization can be 98% confident that the proportion of adults who always wear their seatbelts is between 0.730 and 0.792.

13. Lower bound: 2693.8, upper bound: 3580.2; the Internal Revenue Service can be 90% confident that the mean additional tax owed is between $2693.8 and $3580.2.

14. Lower bound: 861.98, upper bound: 864.02; we can be 99% confident that the mean muzzle velocity of the projectile is between 861.98 and 864.02 meters per second.

15. Lower bound: 0.575, upper bound: 0.625; the Gallup Organization can be 90% confident that the proportion of adult Americans who are worried about having enough money for retirement is between 0.575 and 0.625.

16. Lower bound: 88.78, upper bound: 98.08; we can be 95% confident that the mean amount spent per person in one day at a certain theme park is between $88.78 and $98.08.

17. Use a Z-interval. The distribution is normal, and we know the population standard deviation. Lower bound: 69.38, upper bound: 71.32; we are 95% confident that the mean height of 20- to 29-year-old males is between 69.38 and 71.32 inches.

18. Use a Z-interval. The distribution is normal, and we know the population standard deviation. Lower bound: 258.7, upper bound: 276.8; we are 95% confident that the mean gestation period of humans is between 258.7 and 276.8 days.

19. Use neither interval. There is an outlier in the data set.

20. Use a t-interval. The distribution is normal, but we do not know the population standard deviation. Lower bound: 0.852, upper bound: 0.894; we are 95% confident that the mean weight of plain M&Ms is between 0.852 gram and 0.894 gram.

21. Use a *t*-interval. The distribution is normal, but we do not know the population standard deviation. Lower bound: 101.4, upper bound: 117.3;, we are 95% confident that the mean pulse rate of women after 3 minutes of exercise is between 101.4 and 117.3 beats per minute.

22. Use neither interval. The data are not normally distributed.

Chapter Review Exercises (page 449)

1. $t_{0.005} = 2.898$ **2.** $t_{0.05} = 1.706$

3. (a) Lower bound: 50.94, upper bound: 58.66
(b) Lower bound: 51.65, upper bound: 57.95; increasing the sample size decreases the width of the interval.
(c) Lower bound: 48.75, upper bound: 60.85; increasing the level of confidence increases the width of the interval.

4. (a) Lower bound: 97.07, upper bound: 111.53
(b) Lower bound: 98.86, upper bound: 109.74; increasing the sample size decreases the width of the interval.
(c) Lower bound: 95.49, upper bound: 113.11; increasing the level of confidence increases the width of the interval.

5. (a) According to the Central Limit Theorem, the sampling distribution of $\bar{x}$ is approximately normal because the sample size is large.
(b) Lower bound: 99,097, upper bound: 101,490; Michelin can be 90% confident that the mean mileage for its HydroEdge tire is between 99,098 and 101,490 miles.
(c) Lower bound: 98,868, upper bound: 101,720; Michelin can be 95% confident that the mean mileage for its HydroEdge tire is between 98,868 and 101,720 miles.
(d) Michelin would need to sample 63 tires to be within 1500 miles of the mean mileage with 99% confidence.

6. (a) According to the Central Limit Theorem, the sampling distribution of $\bar{x}$ is approximately normal because the sample size is large.
(b) Lower bound: 315, upper bound: 335; Motorola can be 94% confident that the mean talk time for its V505 camera phone before the battery must be recharged is between 315 and 335 minutes.
(c) Lower bound: 313, upper bound: 337; Motorola can be 98% confident that the mean talk time for its V505 camera phone before the battery must be recharged is between 313 and 337 minutes.
(d) Motorola would need to sample 148 phones to be within 5 minutes of the mean talk time with 95% confidence.

7. (a) Since the mean is larger than the median, the distribution is skewed right. A large sample is necessary to construct a confidence interval about the mean because the population is not normally distributed.
(b) Lower bound: 12.70, upper bound: 13.30; the Gallup Organization is 90% confident that the mean number of hours that 13- to 17-year-olds watch television is between 12.7 and 13.3 hours per week.

8. (a) $\bar{x} = 19.02$ ounces
(b) Lower bound: 18.92, upper bound: 19.12; the soft-drink manufacturer can be 95% confident that the mean content of the bottles of soda produced is between 18.92 and 19.12 ounces.
(c) The sample indicates that the bottling process is not operating properly because the desired mean content 19.2 ounces, is outside the confidence interval.

9. (a) United States: $\bar{x} = 42.3$ hours; Canada: $\bar{x} = 40.8$ hours
(b) Lower bound: 33.82, upper bound: 50.84
(c) Lower bound: 33.6, upper bound: 48.0
(d) It does not appear that Americans work more than Canadians.

10. (a)

Normal Probability Plot for Male

Male Achievement Scores

Normal Probability Plot for Female

Female Achievement Scores

Both normal probability plots and both boxplots indicate that the data come from normally distributed populations.
(b) Males: $\bar{x} = 446.5$; females: $\bar{x} = 453.5$
(c) Lower bound: 413.71, upper bound: 479.29
(d) Lower bound: 424.70, upper bound: 482.30
(e) There does not appear to be a difference between the males' mean score and the females' mean score because the confidence intervals are approximately the same.

11. (a) From the Central Limit Theorem, when the sample is large, $\bar{x}$ is approximately normally distribution with $\mu_{\bar{x}} = 2.27$ and
$$\sigma_{\bar{x}} = \frac{1.22}{\sqrt{60}}.$$
(b) Lower bound: 1.95, upper bound: 2.59; we can be 95% confident that couples who have been married for 7 years have a mean number of children between 1.95 and 2.59.
(c) Lower bound: 1.85, upper bound: 2.69; we can be 99% confident that couples who have been married for 7 years have a mean number of children between 1.85 and 2.69.

12. (a) From the Central Limit Theorem, when the sample is large, $\bar{x}$ is approximately normally distribution with $\mu_{\bar{x}} = 4.08$ and
$$\sigma_{\bar{x}} = \frac{2.12}{\sqrt{50}}.$$
(b) Lower bound: 3.58, upper bound: 4.58; we can be 90% confident that the mean number of cars arriving at McDonald's drive-through between 11:50 A.M. and 12:00 noon on Wednesdays is between 3.58 and 4.58 cars.

(c) Lower bound: 3.48, upper bound: 4.68; we can be 95% confident that the mean number of cars arriving at McDonald's drive-through between 11:50 A.M. and 12:00 noon on Wednesdays is between 3.48 and 4.68 cars.

13. (a) $\bar{x} = 3.244$ liters; $s = 0.487$ liter

(b) Yes

(c) Lower bound: 2.935 liters, upper bound: 3.553 liters; the researchers are 95% confident that the mean blood plasma volume of women is between 2.935 and 3.553 liters.

(d) Lower bound: 2.807 liters, upper bound: 3.681 liters; the researchers are 99% confident that the mean blood plasma volume of women is between 2.807 and 3.681 liters.

14. (a) $\bar{x} = 57.8$ inches; $s = 15.4$ inches

(b) Yes, the conditions are met. The distribution is approximately normal, and there are no outliers.

(c) Lower bound: 48.0, upper bound: 67.6; the researcher is 95% confident that the mean depth of visibility of the Secchi disk is between 48.0 and 67.6 inches.

(d) Lower bound: 44.0, upper bound: 71.6; the researcher is 99% confident that the mean depth of visibility of the Secchi disk is between 44.0 and 71.6 inches.

15. (a) United States: Lower bound: 31.4 hours, upper bound: 53.2 hours

(b) Canada: Lower bound: 31.0 hours, upper bound: 50.6 hours

(c) It does not appear that Americans work more than Canadians.

16. (a) Males: Lower bound: 418.9, upper bound: 474.1

(b) Females: Lower bound: 421.0, upper bound: 486.0

(c) There does not appear to be a difference between the males' mean score and the females' mean score because the confidence intervals are approximately the same.

17. (a) $\hat{p} = 0.086$

(b) Lower bound: 0.065, upper bound: 0.107; the Centers for Disease Control are 95% confident that the proportion of adult males 20–34-years old who have hypertension is between 0.065 and 0.107.

(c) 336 subjects would be needed if we use the point estimate of the proportion found in (a).

(d) 1068 subjects would be needed to obtain 95% confidence to be within 3% points if no prior estimate is available.

18. (a) $\hat{p} = 0.948$

(b) Lower bound: 0.932, upper bound: 0.964; the EPA is 99% confident that the proportion of Americans who live in neighborhoods with acceptable levels of carbon monoxide is between 0.932 and 0.964.

(c) 593 Americans must be sampled if the prior estimate of p is used.

(d) 3007 Americans must be sampled for the estimate to be within 1.5 percentage points with 90% confidence if no prior estimate is available.

19. Area to the left of $t = -1.56$ is 0.0681 because the t-distribution is symmetric about zero.

20. There is more area under the t-distribution to the right of $t = 2.32$ than under the standard normal distribution to the right of $z = 2.32$, because the t-distribution uses s to approximate σ making it more dispersed than the Z-distribution.

21. The properties of the Student's t-distribution:

1. It is symmetric around $t = 0$.

2. It is different for different sample sizes.

3. The area under the curve is 1; half the area is to the right of 0 and half the area is to the left of 0.

4. As t gets extremely large, the graph approaches, but never equals, zero. Similarly, as t gets extremely small (negative), the graph approaches, but never equals, zero.

5. The area in the tails of the t-distribution is greater than the area in the tails of the standard normal distribution.

6. As the sample size n increases, the distribution (and the density curve) of the t-distribution becomes more like those of the standard normal distribution.

CHAPTER 10

10.1 Assess Your Understanding (page 460)

1. To reject the null hypothesis, when it is true. To not reject the null hypothesis when the alternative is true.

2. $\alpha = 0.01$

3. As the level of significance, α, decreases, the probability of making a Type II error, β, increases.

4. If the level of significance is $\alpha = 0.05$, the probability of making a Type I error is 0.05.

5. Answers will vary.   **6.** Answers will vary.

7. False   **8.** False   **9.** Right-tailed, μ

10. Left-tailed, p **11.** Two tailed, σ **12.** Right-tailed, p

13. Left-tailed, μ **14.** Two tailed, σ

15. (a) $H_0: p = 0.118$, $H_1: p < 0.118$

(b) Type I error: The sociologist rejects the hypothesis that the percentage of births to teenaged mothers is 11.8%, when in fact it is the true percentage of births to teenaged mothers.

(c) Type II error: The sociologist fails to reject the hypothesis that the percentage of births to teenaged mothers is 11.8%, when in fact the true percentage of births to teenaged mothers is less than 11.8%.

16. (a) $H_0: \mu = \$1623$, $H_1: \mu \neq \$1623$

(b) Type I error: The researcher rejects the hypothesis that the average charitable contribution per household is $1623, when in fact it is $1623.

(c) Type II error: The researcher fails to reject the hypothesis that the average charitable contribution per household is $1623, when in fact it is different from $1623.

17. (a) $H_0: \mu = \$243,756$, $H_1: \mu > \$243,756$

(b) Type I error: The real estate broker rejects the hypothesis that the mean price of a single-family home is $243,756, when in fact $243,756 is the true mean cost of a single-family home.

(c) Type II error: The real estate broker fails to reject the hypothesis that the mean price of a single-family home is $243,756, when in fact the mean cost of a single-family home is greater than $243,756.

18. (a) $H_0: \mu = 32$ ounces, $H_1: \mu < 32$ ounces

(b) Type I error: The consumer advocate rejects the hypothesis that the mean net weight of the peanut butter is 32 ounces, when in fact the true mean weight of the peanut butter is 32 ounces.

(c) Type II error: The consumer advocate fails to reject the hypothesis that the mean net weight of the peanut butter is 32 ounces, when in fact the mean weight of the peanut butter is less than 32 ounces.

19. (a) $H_0: \sigma = 0.7$ psi, $H_1: \sigma < 0.7$ psi

(b) Type I error: We reject that the variability in the pressure required is 0.7 psi, when the true variability is 0.7 psi.

(c) Type II error: We fail to reject that the variability in the pressure required is 0.7 psi, when the variability is less than 0.7 psi.

20. (a) $H_0: p = 0.16$, $H_1: p > 0.16$

(b) Type I error: The school nurse rejects the Centers for Disease Control and Prevention's claim that the mean percentage of children who is overweight is 16%, when in fact the true mean percentage of overweight children is 16%.

(c) Type II error: The school nurse fails to reject the Centers for Disease Control and Prevention's claim that the mean percentage of children who is overweight is 16%, when the true mean percentage of overweight children is more than 16%.

21. (a) $H_0: \mu = \$49.91$, $H_1: \mu \neq \$49.91$

(b) Type I error: The researcher rejects the claim that the average monthly cell phone bill is $49.91, when in fact the true mean cell phone bill is $49.91.

(c) Type II error: The researcher fails to reject the claim that the average monthly cell phone bill is $49.91, when the true mean cell phone bill is different from $49.91.

22. (a) $H_0: \sigma = 114$, $H_1: \sigma > 114$

(b) Type I error: The teacher rejects the hypothesis that the standard deviation of SAT math scores is 114, when in fact the true standard deviation is 114 points.

(c) Type II error: The teacher fails to reject the hypothesis that the standard deviation of SAT math scores is 114, when the true standard deviation is greater than 114 points.

23. There is sufficient evidence to conclude that the percentage of births to teenage mothers is less than 11.8%.

24. There is not sufficient evidence to conclude that the level of charitable giving per household has changed from $1623.

25. There is not sufficient evidence to conclude that the mean price of a single-family home has increased from $243,756.

26. There is sufficient evidence to conclude that the average content in a jar of peanut butter is less than 32 ounces.

27. There is not sufficient evidence to conclude that the variability in pressure has been reduced.

28. There is not sufficient evidence to conclude that the percentage of 6- to 11-year-olds is greater than 16%.

29. There is sufficient evidence to conclude that the average monthly cell phone bill is different from $49.91.

30. There is not sufficient evidence to conclude that the standard deviation in SAT math scores is more than 114 points.

31. There is not sufficient evidence to conclude that the percentage of births to teenage mothers is less than 11.8%.

32. There is sufficient evidence to conclude that the level of charitable giving per household has changed from $1623.

33. There is sufficient evidence to conclude that the mean price of a single-family home has increased from $243,756.

34. There is not sufficient evidence to conclude that the average content in a jar of peanut butter is less than 32 ounces.

35. (a) $H_0: \mu = 98.4$ pounds, $H_1: \mu > 98.4$ pounds
(b) There is sufficient evidence to conclude that the mean per capita consumption of fruits and vegetables is more than 98.4 pounds.
(c) A Type I error was committed because the dietician rejected a true null hypothesis. When $\alpha = 0.05$, the probability of committing a Type I error is 0.05.

36. (a) $H_0: \mu = 518$, $H_1: \mu > 518$
(b) There is not sufficient evidence to conclude that the mean score on the SAT Math Reasoning exam is higher than 518.
(c) A Type II error was committed because the researcher failed to reject a false null hypothesis. When $\alpha = 0.01$, the probability of committing a Type I error is 0.01.
(d) To decrease the probability of making a Type II error, we need to increase the level of significance.

37. (a) $H_0: p = 0.102$, $H_1: p < 0.102$
(b) There is not sufficient evidence to conclude that the percentage of high school students who tried marijuana for the first time before the age of 13 has decreased from 10.2%.
(c) A Type II error was committed because the SRO failed to reject a false null hypothesis.

38. (a) $H_0: p = 0.107$, $H_1: p > 0.107$
(b) There is sufficient evidence to conclude that more than 10.7% of Americans over 65 years of age use the Internet.
(c) A Type I error was committed because the researcher rejected a true null hypothesis.

39. $H_0: \mu = 0$, $H_1: \mu > 0$, where μ is the mean increase in gas mileage when using the device.

40. (a) $H_0: \mu = 4$ hours, $H_1: \mu < 4$ hours
(b) Consumer Reports might reject the null hypothesis and conclude that a car with Prolong will not run 4 hours without oil.

41. Answers will vary.

10.2 Assess Your Understanding (page 475)

1. We must have a simple random sample drawn from a normal population to test hypotheses regarding a population mean when we know σ. If the distribution of the population is not normal, or if it is unknown, the sample must be large.

2. When $\alpha = 0.01$ and σ is known, the critical value for a right-tailed test is $Z = 2.33$.

3. When $\alpha = 0.05$ and σ is known, the critical values for a two-tailed test regarding a population mean are $Z = 1.96$ and $-Z = -1.96$.

4. Methods for testing hypotheses regarding a population mean if σ is known are robust means that minor departures from normality will not adversely affect the results of the test.

5. P-value is the probability of observing a sample statistic as extreme or more extreme than the one observed under the assumption that the null hypothesis is true. If the P-value is less than the level of significance, α, the null hypothesis is rejected.

6. $P = 0.23$ is the probability of obtaining a sample mean of $\bar{x}$ or smaller assuming that μ_0 is the true mean. Since the outcome is not unusual, we do not reject H_0.

7. $P = 0.02$ is the probability of obtaining a sample mean more than $|Z_0|$ standard deviations from the hypothesized mean μ_0. Decisions and reasons will vary.

8. Answers will vary.

9. Answers will vary.

10. True

11. (a) $Z = -1.18$ **(b)** $-z_{0.05} = -1.645$
(c)

(d) No, the test statistic does not lie in the critical region.

12. (a) $Z = 1.92$ **(b)** $z_{0.1} = 1.28$
(c)

(d) Yes, the test statistic lies in the critical region.

13. (a) $Z = 3.29$
(b) $-z_{0.005} = -2.575$; $z_{0.005} = 2.575$
(c)

(d) Yes, the test statistic lies in the critical region.

14. (a) $Z = -1.32$ **(b)** $-z_{0.02} = -2.05$
(c)

(d) No, the test statistic does not lie in the critical region.

15. (a) P-value $= 0.0082$, assuming $\mu = 20$; $P(\bar{d} \le 18.3) = 0.0082$

(b) Yes, the null hypothesis would be rejected at the $\alpha = 0.05$ level of significance, because the P-value is less than the level of significance ($0.0082 < 0.05$).

16. (a) P-value $= 0.1151$, assuming $\mu = 4.5$; P-value $= 0.1151$

(b) No, the null hypothesis would not be rejected at the $\alpha = 0.1$ level of significance, because the P-value is greater than the level of significance ($0.1151 > 0.1$).

17. (a) The population does not need to be normally distributed to compute the P-value since the sample size is larger than 30.

(b) P-value $= 0.0614$, assuming $\mu = 105$; the probability of obtaining a sample mean that is more than 1.87 standard deviations from the hypothesized mean of 105 is 0.0614.

(c) No, the null hypothesis would not be rejected at the $\alpha = 0.02$ level of significance, because the P-value is greater than the level of significance ($0.0614 > 0.02$).

18. (a) The population does not need to be normally distributed to compute the P-value since the sample size is larger than 30.

(b) P-value $= 0.0090$, assuming $\mu = 45$; the probability of obtaining a sample mean that is more than 2.61 standard deviations from the hypothesized mean of 45 is 0.0090.

(c) Yes, the null hypothesis would be rejected at the $\alpha = 0.05$ level of significance, because the P-value is less than the level of significance ($0.0090 < 0.05$).

19. Hypotheses: H_0: $\mu = 63.7$, H_1: $\mu > 63.7$
Classical approach: $Z_0 = 0.38 < z_{0.05} = 1.645$; do not reject the null hypothesis.
P-value approach: P-value $= 0.3520 > \alpha = 0.05$; do not reject the null hypothesis.
There is not sufficient evidence to support the claim that the average height of a woman aged 20 years or older is greater than it was in 1994.

20. Hypotheses: H_0: $\mu = 60$, H_1: $\mu < 60$
Classical approach: $Z_0 = -3.64 < -z_{0.05} = -1.645$; reject the null hypothesis.
P-value approach: P-value $< 0.0001 < \alpha = 0.05$; reject the null hypothesis.
There is sufficient evidence to conclude that the mean withdrawal from the ATM machine is less than \$60.

21. (a) The scores must be normally distributed to test this hypotheses because the sample size, $n = 20$, is small.

(b) Hypotheses: H_0: $\mu = 515$, H_1: $\mu < 515$
Classical approach: $Z_0 = -2.28 < -z_{0.10} = -1.28$; reject the null hypothesis.
P-value approach: P-value $= 0.0113 < \alpha = 0.10$; reject the null hypothesis.
There is sufficient evidence to conclude that students whose first language learned is not English score worse on the verbal portion of the SAT exam than students whose first language is English.

22. (a) The scores must be normally distributed to test this hypotheses because the sample size, $n = 20$, is small.

(b) Hypotheses: H_0: $\mu = 516$, H_1: $\mu \neq 516$
Classical approach: $Z_0 = 0.24$ is between $-z_{0.05} = -1.645$ and $z_{0.05} = 1.645$; do not reject the null hypothesis.
P-value approach: P-value $= 0.8184 > \alpha = 0.10$; do not reject the null hypothesis.
There is not sufficient evidence to conclude that students whose first language learned is not English score differently on the math portion of the SAT exam than students whose first language is English.

23. (a) Yes, the conditions are satisfied. The data appear to be normally distributed, and there are no outliers.

(b) Hypotheses: H_0: $\mu = 5.03$, H_1: $\mu < 5.03$
Classical approach: $Z_0 = -4.74 < -z_{0.10} = -1.28$; reject the null hypothesis.
P-value approach: P-value $< 0.0001 < \alpha = 0.20$; reject the null hypothesis.
There is sufficient evidence to conclude that the acidity of rain has increased since 1990.

24. (a) Yes, the conditions are satisfied. The data appear to be normally distributed, and there are no outliers.

(b) Hypotheses: H_0: $\mu = 5.6$, H_1: $\mu > 5.6$
Classical approach: $Z_0 = 4.91 > z_{0.05} = 1.645$; reject the null hypothesis.

P-value approach: P-value $< 0.0001 < \alpha = 0.05$; reject the null hypothesis.
There is sufficient evidence to conclude that the fuel catalyst is effective.

25. (a) Yes, the conditions are satisfied. The data appear to be normally distributed, and there are no outliers.

(b) Hypotheses: H_0: $\mu = 64.05$, H_1: $\mu \neq 64.05$
Classical approach: $Z_0 = -3.36 < -z_{0.005} = -1.645$; reject the null hypothesis.
P-value approach: P-value $= 0.0008 < \alpha = 0.01$; reject the null hypothesis.
Yes, the machine should be recalibrated because the consequences for underfilling bottles are severe.

(c) Explanations will vary.

26. (a) Yes, the conditions are satisfied. The data appear to be normally distributed, and there are no outliers.

(b) Hypotheses: H_0: $\mu = 516$, H_1: $\mu < 516$
Classical approach: $Z_0 = -0.32 > -z_{0.10} = -1.28$; do not reject the null hypothesis.
P-value approach: P-value $= 0.3745 > \alpha = 0.10$; do not reject the null hypothesis. No, it does not appear to be the case that students scored lower on the new exam.

27. (a) Yes, the conditions are satisfied. The data appear to be normally distributed, and there are no outliers.

(b) Hypotheses: H_0: $\mu = 8.33$, H_1: $\mu < 8.33$
Classical approach: $Z_0 = -0.86 > -z_{0.10} = -1.28$; do not reject the null hypothesis.
P-value approach: P-value $= 0.1949 > \alpha = 0.10$; do not reject the null hypothesis.
No, it does not appear to be the case that cars are younger today than they were in 1995.

28. (a) Yes, the conditions are satisfied. The data appear to be normally distributed, and there are no outliers.

(b) Hypotheses: H_0: $\mu = 0$, H_1: $\mu > 0$
Classical approach: $Z_0 = 1.92 > z_{0.05} = 1.645$; reject the null hypothesis.
P-value approach: P-value $= 0.0274 < \alpha = 0.05$; reject the null hypothesis.
Yes, the summer of 2000 in Houston appears to be hotter than normal.

29. Hypotheses: H_0: $\mu = 694$, H_1: $\mu > 694$
Classical approach: $Z_0 = 1.10 < z_{0.05} = 1.645$; do not reject the null hypothesis.
P-value approach: P-value $= 0.1357 > \alpha = 0.05$; do not reject the null hypothesis.
There is not sufficient evidence to conclude that the average farm size in Kansas is greater than it was in 1990.

30. Hypotheses: H_0: $\mu = 11.1$, H_1: $\mu < 11.1$
Classical approach: $Z_0 = -2.17 < -z_{0.05} = -1.645$; reject the null hypothesis.
P-value approach: P-value $= 0.015 < \alpha = 0.05$; reject the null hypothesis.
There is sufficient evidence to conclude that the oil output per well in the United States has declined from the 1998 level of 11.1 barrels per day.

31. Hypotheses: H_0: $\mu = 31.8$, H_1: $\mu \neq 31.8$
Classical approach: $Z_0 = -3.32 < -z_{0.025} = -1.96$; reject the null hypothesis.
P-value approach: P-value $= 0.0010 < \alpha = 0.05$; reject the null hypothesis.
There is sufficient evidence to conclude that the volume of Dell stock is different in 2004.

32. Hypotheses: H_0: $\mu = 11.4$, H_1: $\mu \neq 11.4$
Classical approach: $Z_0 = 1.35$ is between $-z_{0.025} = -1.96$ and $z_{0.025} = 1.96$; do not reject the null hypothesis.
P-value approach: P-value $= 0.1770 > \alpha = 0.05$; do not reject the null hypothesis.
There is not sufficient evidence to conclude that the volume of Motorola stock is different in 2004.

33. Lower bound: 63.974, Upper bound: 64.040; since the confidence interval does not include $\mu = 64.05$ ounces, reject the null hypothesis.

34. Lower bound: 464.1, Upper bound: 553.5; since the confidence interval includes the hypothesized $\mu = 516$, do not reject the null hypothesis.

35. Lower bound: 18.60, Upper bound: 28.40; since the confidence interval does not include $\mu = 31.8$ million shares, reject the null hypothesis.

36. Lower bound: 10.55, Upper bound: 16.05; since the confidence interval includes $\mu = 11.4$ million shares, do not reject the null hypothesis.

37. (a) $H_0: \mu = 514$, $H_1: \mu > 514$
 (b) Classical approach: $Z_0 = 1.50 > z_{0.10} = 1.28$; reject the null hypothesis.
  P-value approach: P-value $= 0.0668 < \alpha = 0.10$; reject the null hypothesis.
 (c) Answers will vary.
 (d) With $n = 400$ students:
  Classical approach: $Z_0 = 0.71 < z_{0.10} = 1.28$; do not reject the null hypothesis.
  P-value approach: P-value $= 0.2389 > \alpha = 0.10$; do not reject the null hypothesis.

38. (a) $H_0: \mu = 0$, $H_1: \mu > 0$
 (b) Classical approach: $Z_0 = 3.85 > z_{0.10} = 1.28$; reject the null hypothesis.
  P-value approach: P-value $= 0.0001 < \alpha = 0.10$; reject the null hypothesis.
 (c) Answers will vary.
 (d) With $n = 40$ subjects:
  Classical approach: $Z_0 = 0.79 < z_{0.10} = 1.28$; do not reject the null hypothesis.
  P-value approach: P-value $= 0.2148 > \alpha = 0.10$; do not reject the null hypothesis.

39. (a), (c), (d) Answers will vary.
 (b) Expect 5 samples to result in a Type I error.

40. (a), (c), (d) Answers will vary.
 (b) Expect 2 samples to result in a Type I error.

41. (a) Hypotheses: $H_0: \mu = 150$, $H_1: \mu < 150$
 (b) Answers will vary.

10.3 Assess Your Understanding (page 487)

1. A hypothesis about a population mean with σ unknown can be tested provided that the sample is obtained using simple random sampling, and the population from which the sample is drawn is either normally distributed with no outliers or the sample size n is larger than 30.

2. The critical value for a right-tailed test at an $\alpha = 0.01$ level of significance when σ is unknown and there are 15 degrees of freedom is $t = 2.602$.

3. The critical values for a two-tailed test at an $\alpha = 0.05$ level of significance when σ is unknown and there are 12 degrees of freedom are $t = 2.179$ and $-t = -2.179$.

4. The critical value for a left-tailed test at an $\alpha = 0.05$ level of significance when σ is unknown and there are 19 degrees of freedom is $-t = -1.729$.

5. (a) $t = -1.379$ **(b)** $-t_{0.05} = -1.714$
 (c)

 (d) There is not enough evidence for the researcher to reject the null hypothesis because it is a left-tailed test and the test statistic is greater than the critical value, $(-1.379 > -1.714)$.

6. (a) $t = 2.674$ **(b)** $t_{0.1} = 1.318$
 (c)

 (d) The researcher will reject the null hypothesis because it is a right-tailed test and the test statistic is greater than the critical value $(2.674 > 1.318)$.

7. (a) $t = 2.502$
 (b) $-t_{0.005} = -2.819$; $t_{0.005} = 2.819$
 (c)

 (d) There is not sufficient evidence for the researcher to reject the null hypothesis, since the test statistic is between the critical values $(-2.819 < 2.502 < 2.819)$.

8. (a) $t = -1.711$ **(b)** $-t_{0.02} = -2.189$
 (c)

 (d) The researcher will not reject the null hypothesis because it is a left-tailed test and the test statistic is greater than the critical value $(-1.711 > -2.189)$.

9. (a) $t = -1.677$
 (b)

 (c) $0.10 < P$-value < 0.15; the P-value is the area under the t-distribution with 17 degrees of freedom to the left of the test statistic, $t_0 = -1.677$.
 (d) The researcher will not reject the null hypothesis at the $\alpha = 0.05$ level of significance because the P-value is greater than the level of significance $(0.10 > 0.05)$.

10. (a) $t = 1.109$
 (b)

 (c) $0.10 < P$-value < 0.15; the P-value is the area under the t-distribution with 12 degrees of freedom to the right of the test statistic, $t_0 = 1.109$.
 (d) The researcher will not reject the null hypothesis at the $\alpha = 0.10$ level of significance because the P-value is greater than the level of significance $(P$-value $> 0.10)$.

11. (a) No, $n \geq 30$ **(b)** $t = -3.108$

(c)

The sum of the areas is the P-value

-3.108 3.108

(d) P-value $= 0.0038$; the P-value is the sum of the areas under the t-distribution with 34 degrees of freedom to the left of the test statistic $-t_0 = -3.108$ and to the right of the test statistic $t_0 = 3.108$.

(e) The researcher will reject the null hypothesis at the $\alpha = 0.01$ level of significance because the P-value is less than the level of significance $(0.0038 < 0.01)$.

12. (a) No, $n \geq 30$ **(b)** $t = 2.455$

(c)

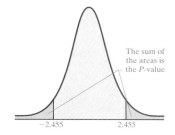

The sum of the areas is the P-value

-2.455 2.455

(d) P-value $= 0.0186$; the P-value is the sum of the areas under the t-distribution with 39 degrees of freedom to the left of the test statistic $-t_0 = -2.455$ and to the right of the test statistic $t_0 = 2.455$.

(e) The researcher will not reject the null hypothesis at the $\alpha = 0.01$ level of significance because the P-value is greater than the level of significance $(0.0186 > 0.01)$.

13. Hypotheses: $H_0: \mu = 9.02$, $H_1: \mu < 9.02$
Classical approach: $t_0 = -4.553 < -t_{0.01} = -2.718$; reject the null hypothesis.
P-value approach: P-value $= 0.0004 < \alpha = 0.01$; reject the null hypothesis.
There is sufficient evidence to conclude that the mean hippocampal volume in alcoholic adolescents is less than the mean normal volume of 9.02 cm^3.

14. Assuming the population of kidney weights is normally distributed, with no outliers,
Hypotheses: $H_0: \mu = 355.7$, $H_1: \mu > 355.7$
Classical approach: $t_0 = 3.14 > t_{0.05} = 1.796$; reject the null hypothesis.
P-value approach: P-value $= 0.0047 < \alpha = 0.05$; reject the null hypothesis.
There is sufficient evidence to conclude that the mean weight of kidneys of first-generation mice whose both parents were exposed to para-nonylphenol is greater than the mean weight of the kidneys of nonexposed mice.

15. Hypotheses: $H_0: \mu = 1000$, $H_1: \mu > 1000$
Classical approach: $t_0 = 1.344 < t_{0.05} = 1.676$; do not reject the null hypothesis.
P-value approach: P-value $= 0.0825 > \alpha = 0.05$; do not reject the null hypothesis.
There is not sufficient evidence to conclude that the campaign was effective.

16. Hypotheses: $H_0: \mu = 2400$ mg, $H_1: \mu > 2400$ mg
Classical approach: $t_0 = 3.449 > t_{0.05} = 1.667$; reject the null hypothesis.
P-value approach: P-value $= 0.0005 < \alpha = 0.05$; reject the null hypothesis.
There is sufficient evidence to conclude that children under the age of 10 years are consuming more than the USFDA's recommended daily allowance of 2400 mg.

17. (a) $t_0 = -15.119 < -t_{0.01} = -1.282$; reject the null hypothesis. There is sufficient evidence to conclude the mean temperatures of humans is less than 98.6°F.

(b) P-value < 0.0001; the probability of obtaining a sample mean of 98.2°F or less from a population whose mean is 98.6°F is less than 0.0001. Less than 1 sample in 10,000 would give us the sample results we obtained if the null hypothesis were true.

18. (a) $t_0 = -3.164 < -t_{0.01} = -2.364$; reject the null hypothesis. There is sufficient evidence to conclude the mean temperature of women is less than 98.6°F.

(b) P-value $= 0.0010$; the probability of obtaining a sample mean of 98.4°F or less from a population whose mean is 98.6°F is 0.001. Less than 1 sample in 1000 would give us the sample results we obtained if the null hypothesis were true.

19. (a) Hypotheses: $H_0: \mu = 40.7$ years, $H_1: \mu \neq 40.7$ years
Classical approach: $t_0 = -1.061$ is between $-t_{0.05} = -1.696$ and $t_{0.05} = 1.696$ with 31 degrees of freedom; do not reject the null hypothesis.
P-value Approach: P-value $= 0.2970 > \alpha = 0.05$; do not reject the null hypothesis.
There is not sufficient evidence to conclude that mean age of a death row inmate is different from 40.7 years of age.

(b) Lower bound: 35.44, upper bound: 42.36; do not reject H_0 since 40.4 years is contained in the 95% confidence interval.

20. (a) Hypotheses: $H_0: \mu = \$1493$, $H_1: \mu \neq \$1493$
Classical approach: $t_0 = 2.304$ is less than $-t_{0.05} = -1.691$ with 34 degrees of freedom; reject the null hypothesis.
P-value approach: P-value $= 0.02707 < \alpha = 0.05$; reject the null hypothesis.
There is sufficient evidence conclude that the mean household expenditure for energy has changed since 2001.

(b) Lower bound: \$1507.7, upper bound: \$1728.3; reject H_0 since \$1493 is not contained in the 95% confidence interval.

21. (a) Yes, the conditions are satisfied. The data appear to be normally distributed, and there are no outliers.

(b) $t_0 = 0.778$ is between $-t_{0.05} = -1.796$ and $t_{0.05} = 1.796$ with 11 degrees of freedom; do not reject the null hypothesis. There is not evidence to suggest that the golf balls do not conform to USGA standards.

(c) P-value $= 0.4529$; the probability of obtaining a sample mean as extreme or more extreme than the one obtained assuming the null hypothesis is true is 0.4529. About 45 samples out of 100 would give us the results we obtained if the null hypothesis were true.

22. (a) Yes, the conditions are satisfied. The data appear to be normally distributed, and there are no outliers.

(b) Classical approach: $t_0 = -3.843 < -t_{0.10} = -1.363$ with 11 degrees of freedom; reject the null hypothesis
P-value approach: P-value $= 0.0014 < \alpha = 0.10$; reject the null hypothesis
There is sufficient evidence to conclude that the Maxfli XS golf balls conform to USGA standards.

23. (a) Yes, the conditions are satisfied. The data appear to be normally distributed, and there are no outliers.

(b) Hypotheses: $H_0: \mu = 84.3$ seconds, $H_1: \mu < 84.3$ seconds
Classical approach: $t_0 = -1.310 > -t_{0.10} = -1.383$ with 9 degrees of freedom; do not reject the null hypothesis.
P-value approach: P-value $= 0.1113 > \alpha = 0.10$; do not reject the null hypothesis.
There is not sufficient evidence to conclude that the new system is effective.

24. (a) Yes, the conditions are satisfied. The data appear to be normally distributed, and there are no outliers.

(b) Hypotheses: $H_0: \mu = 7.0$, $H_1: \mu \neq 7.0$
Classical approach: $t_0 = 1.183$ is between $-t_{0.05} = -1.771$ and $t_{0.05} = 1.771$ with 13 degrees of freedom; do not reject the null hypothesis.
P-value approach: P-value $= 0.2579 > \alpha = 0.05$; do not reject the null hypothesis.
There is not sufficient evidence to conclude that there is bias in the pH meter.

25. (a) The data appear to be normally distributed, and there are no outliers.

Normal Probability Plot for p/e

(b) Hypotheses: H_0: $\mu = 22.0$, H_1: $\mu < 22.0$
Classical approach: $t_0 = -0.2980 > -t_{0.05} = -1.771$ with 13 degrees of freedom; do not reject the null hypothesis.
P-value approach: P-value $= 0.3852 > \alpha = 0.05$; do not reject the null hypothesis.
There is not sufficient evidence to conclude that the mean P/E ratio of the S&P 500 Index has decreased from its December 2000 level.
(c) The area under the t-distribution with 13 degrees of freedom to the left of the test statistic $t_0 = -0.2980$ is 0.3852 (the P-value).

26. H_0: $\mu = 0$, H_1: $\mu > 0$, the area under the t-distribution with 172 degrees of freedom to the right of the test statistic t_0 is less than 0.10 (the P-value).

27. (a) Yes, the conditions are satisfied. The data appear to be normally distributed, and there are no outliers.
(b) Hypotheses: H_0: $\mu = 130$, H_1: $\mu < 130$; the P-value $= 0.088$. The nursing student will not reject the null hypothesis at the $\alpha = 0.05$ level of significance. There is not sufficient evidence to conclude that male surgical patients have a mean systolic blood pressure less than 130.

28. (a) H_0: $\mu = 98.2$, H_1: $\mu > 98.2$
(b) P-value $= 0.23$; the nursing student will not reject the null hypothesis at the $\alpha = 0.05$ level of significance. There is not sufficient evidence to conclude that mean temperature of surgical patients is above 98.2° F.

29. Answers will vary.

30. (a) Answers will vary. Expect 5 samples to result in a rejection of the null hypothesis at $\alpha = 0.05$. The probability of Type I error is 0.05.
(b) Answers will vary. Expect 5 samples to result in a rejection of the null hypothesis at $\alpha = 0.05$. The probability of Type I error is 0.05.
(c) Answers will vary.

31. (a) Answers will vary. Experiment does not meet the requirements for hypothesis testing.
(b) Answers will vary. Expect 5 samples to result in a rejection of the null hypothesis at $\alpha = 0.05$.

32. (a), (c), and (d) answers will vary.
(b) Expect 2 samples to result in a Type I error.

10.4 Assess Your Understanding (page 498)

1. A hypothesis about a population proportion can be tested if the sample is obtained by simple random sampling and $np(1 - p) \geq 10$, with $n \leq 0.05N$ (the sample size is no more than 5% of the population size).

2. Answers will vary.

3. $np_0(1 - p_0) = 42 > 10$
(a) Classical approach: $Z_0 = 2.31 > z_{0.05} = 1.645$; reject the null hypothesis.
(b) P-value approach: P-value $= 0.0104 < \alpha = 0.05$; reject the null hypothesis. There is sufficient evidence at the $\alpha = 0.05$ level of significance to reject the null hypothesis.

4. $np_0(1 - p_0) = 60 > 10$
(a) Classical approach: $Z_0 = -3.36 < -z_{0.01} = -2.33$; reject the null hypothesis.
(b) P-value approach: P-value $= 0.0004 < \alpha = 0.01$; reject the null hypothesis. There is sufficient evidence at the $\alpha = 0.01$ level of significance to reject the null hypothesis.

5. $np_0(1 - p_0) = 37.1 > 10$
(a) Classical approach: $Z_0 = -0.74 > -z_{0.10} = -1.28$; do not reject the null hypothesis.
(b) P-value approach: P-value $= 0.2296 > \alpha = 0.10$; do not reject the null hypothesis. There is not sufficient evidence at the $\alpha = 0.10$ level of significance to reject the null hypothesis.

6. $np_0(1 - p_0) = 75 > 10$
(a) Classical approach: $Z_0 = -0.46 > -z_{0.10} = -1.28$; do not reject the null hypothesis.
(b) P-value approach: P-value $= 0.3228 > \alpha = 0.10$; do not reject the null hypothesis. There is not sufficient evidence at the $\alpha = 0.10$ level of significance to reject the null hypothesis.

7. $np_0(1 - p_0) = 45 > 10$
(a) Classical approach: $Z_0 = -1.49$ is between $-z_{0.025} = -1.96$ and $z_{0.025} = 1.96$; do not reject the null hypothesis.
(b) P-value approach: P-value $= 0.1362 > \alpha = 0.05$; do not reject the null hypothesis. There is not sufficient evidence at the $\alpha = 0.05$ level of significance to reject the null hypothesis.

8. $np_0(1 - p_0) = 240 > 10$
(a) Classical approach: $Z_0 = 1.29$ is between $-z_{0.005} = -2.575$ and $z_{0.005} = 2.575$; do not reject the null hypothesis.
(b) P-value approach: P-value $= 0.1970 > \alpha = 0.01$; do not reject the null hypothesis. There is not sufficient evidence at the $\alpha = 0.01$ level of significance to reject the null hypothesis.

9. $np_0(1 - p_0) = 16.1 > 10$
Hypotheses: H_0: $p = 0.019$, H_1: $p > 0.019$
(a) Classical approach: $Z_0 = 0.65 < z_{0.01} = 2.33$; do not reject the null hypothesis.
(b) P-value approach: P-value $= 0.2578 > \alpha = 0.01$; do not reject the null hypothesis. There is not sufficient evidence at the $\alpha = 0.01$ level of significance to conclude that more than 1.9% of Lipitor users experience flulike symptoms as a side effect.

10. $np_0(1 - p_0) = 12.6 > 10$
Hypotheses: H_0: $p = 0.94$, H_1: $p > 0.94$
(a) Classical approach: $Z_0 = 0.69 < z_{0.01} = 2.33$; do not reject the null hypothesis.
(b) P-value approach: P-value $= 0.2451 > \alpha = 0.01$; do not reject the null hypothesis. There is not sufficient evidence at the $\alpha = 0.01$ level of significance to conclude that more than 94% of patients taking Nexium are healed within 8 weeks.

11. $np_0(1 - p_0) = 154.8 > 10$
Hypotheses: H_0: $p = 0.81$, H_1: $p \neq 0.81$
(a) Classical approach: $Z_0 = 1.62$ is between $-z_{0.025} = -1.96$ and $z_{0.025} = 1.96$; do not reject the null hypothesis.
(b) P-value approach: P-value $= 0.1052 > \alpha = 0.05$; do not reject the null hypothesis. There is not sufficient evidence at the $\alpha = 0.05$ level of significance to conclude that the proportion of adults who have read at least one book in the last year is different from 1990.

12. $np_0(1 - p_0) = 221.1 > 10$
Hypotheses: H_0: $p = 0.33$, H_1: $p > 0.33$
(a) Classical approach: $Z_0 = 2.69 > z_{0.05} = 1.645$; reject the null hypothesis.
(b) P-value approach: P-value $= 0.0036 < \alpha = 0.05$; reject the null hypothesis. There is sufficient evidence at the $\alpha = 0.05$ level of significance to conclude that the proportion of adults who believe in haunted houses has increased since 1996.

13. $np_0(1 - p_0) = 264.3 > 10$
Hypotheses: H_0: $p = 0.38$, H_1: $p < 0.38$
(a) Classical approach: $Z_0 = -5.52 < -z_{0.05} = -1.645$; reject the null hypothesis.
(b) P-value approach: P-value $< 0.0001 < \alpha = 0.05$; reject the null hypothesis. There is sufficient evidence at the $\alpha = 0.05$ level

of significance to conclude that the proportion of families with children under the age of 18 who eat together 7 nights a week has decreased.

14. $np_0(1 - p_0) = 118.272 > 10$
Hypotheses: $H_0: p = 0.56$, $H_1: p > 0.56$
(a) Classical approach: $Z_0 = 2.59 > Z_{0.05} = 1.96$; reject the null hypothesis.
(b) P-value approach: P-value $= 0.0048 < \alpha = 0.05$; reject the null hypothesis There is sufficient evidence at the $\alpha = 0.05$ level of significance to conclude that the proportion of employed adults who feel basic mathematical skills are critical or very important to their job is greater than 0.56.

15. $np_0(1 - p_0) = 58.8 > 10$
Hypotheses: $H_0: p = 0.30$, $H_1: p \neq 0.30$
(a) Classical approach: $Z_0 = 3.65 > z_{0.05} = 1.645$; reject the null hypothesis.
(b) P-value approach: P-value $= 0.0002 < \alpha = 0.10$; reject the null hypothesis. There is sufficient evidence at the $\alpha = 0.10$ level of significance to conclude that the percentage of employed adults who are satisfied with their chances for promotion is significantly different from the percentage in 1998.

16. $np_0(1 - p_0) = 121.8 > 10$
Hypotheses: $H_0: p = 0.58$, $H_1: p \neq 0.58$
(a) Classical approach: $Z_0 = -0.45$ is between $-z_{0.05} = -1.645$ and $z_{0.05} = 1.645$; do not reject the null hypothesis.
(b) P-value approach: P-value $= 0.6528 > \alpha = 0.10$; do not reject the null hypothesis. There is not sufficient evidence at the $\alpha = 0.10$ level of significance to conclude that the percentage of females aged 15 years or older who live alone has changed.

17. $np_0(1 - p_0) = 241 > 10$
Hypotheses: $H_0: p = 0.40$, $H_1: p < 0.40$
(a) Classical approach: $Z_0 = -1.91 < -z_{0.05} = -1.645$; reject the null hypothesis.
(b) P-value approach: P-value $= 0.0281 < \alpha = 0.05$; reject the null hypothesis. There is sufficient evidence at the $\alpha = 0.05$ level of significance to conclude that the proportion of 2005 adults aged 18 years or older having a great deal of confidence in the public schools is significantly lower than the 1995 proportion.

18. $np_0(1 - p_0) = 23.9 > 10$
Hypotheses: $H_0: p = 0.01$, $H_1: p > 0.01$
(a) Classical approach: $Z_0 = 2.22 > z_{0.05} = 1.645$; reject the null hypothesis.
(b) P-value approach: P-value $= 0.0132 < \alpha = 0.05$; reject the null hypothesis. There is sufficient evidence at the $\alpha = 0.05$ level of significance to conclude that more than 1% of the adult population is pathological gamblers.

19. $np_0(1 - p_0) = 35 > 10$
Hypotheses: $H_0: p = 0.37$, $H_1: p < 0.37$
Classical approach: $Z_0 = -0.25 > -z_{0.05} = -1.645$; do not reject the null hypothesis.
P-value approach: P-value $= 0.4013 > \alpha = 0.05$; do not reject the null hypothesis. No, there is not sufficient evidence at the $\alpha = 0.05$ level of significance to conclude that less than 37% of pet owners speaks to their pets on the answering machine or telephone.

20. $np_0(1 - p_0) = 25.5 > 10$
Hypotheses: $H_0: p = 0.85$, $H_1: p > 0.85$
(a) Classical approach: $Z_0 = 0.20 < z_{0.10} = 1.28$; do not reject the null hypothesis.
(b) P-value approach: P-value $= 0.4207 > \alpha = 0.10$; do not reject the null hypothesis. There is not sufficient evidence at the $\alpha = 0.10$ level of significance to conclude that more than 85% of the adult population eats salad at least once a week.

21. (a) Answers will vary.
(b) $H_0: p = 0.48$, $H_1: p < 0.48$; $np_0(1 - p_0) = 274.6 > 10$; $Z_0 = -1.66$; P-value $= 0.0918$. The results of the poll do not indicate that the proportion declined significantly.

22. $H_0: p = 0.71$ vs. $H_1: p > 0.71$; P-value $= 0.0179$. The test of significance supports this headline.

10.5 Assess Your Understanding (page 501)

1. To test a hypotheses about a population mean, we must use a simple random sample that is either drawn from a normally distributed population, or the sample must have at least 30 subjects. Assuming the prerequisites are met, use a normal model to test the hypotheses if the population standard deviation (or variance) is known, and use a Student's t-distribution if the population standard deviation (or variance) is not known.

2. Before testing hypotheses about a population proportion we must be sure that $np(1 - p) \geq 10$ and that the sample size is no larger than 5% of the population ($n \leq 0.05N$).

3. Hypotheses: $H_0: \mu = 70$, $H_1: \mu < 70$
Classical approach: $Z_0 = -1.87 < -z_{0.10} = -1.28$; reject the null hypothesis.
P-value approach: P-value $= 0.0307 < \alpha = 0.10$; reject the null hypothesis.
There is sufficient evidence at the $\alpha = 0.10$ level of significance to conclude that the population mean is less than 70.

4. Hypotheses: $H_0: \mu = 1$, $H_1: \mu < 1$
Classical approach: $t_0 = -2.179 > -t_{0.10} = -2.552$ with 18 degrees of freedom; do not reject the null hypothesis.
P-value approach: P-value $= 0.0214 > \alpha = 0.01$; do not reject the null hypothesis.
There is not sufficient evidence at the $\alpha = 0.01$ level of significance to conclude that the mean is less than 1.

5. Hypotheses: $H_0: p = 0.5$, $H_1: p > 0.5$; $np_0(1 - p_0) = 50 > 10$
Classical approach: $Z_0 = 2.12 > z_{0.05} = 1.645$; reject the null hypothesis.
P-value approach: P-value $= 0.0170 < \alpha = 0.05$; reject the null hypothesis.
There is sufficient evidence at the $\alpha = 0.05$ level of significance to conclude that more than half the individuals with valid drivers licenses drives an American-made automobile.

6. Hypotheses: $H_0: \mu = 600$, $H_1: \mu \neq 600$
Classical approach: $t_0 = -1.186$ is between $-t_{0.05} = -1.671$ and $t_{0.05} = 1.671$ with 60 degrees of freedom; do not reject the null hypothesis.
P-value approach: P-value $= 0.2401 > \alpha = 0.10$; do not reject the null hypothesis.
There is not sufficient evidence at the $\alpha = 0.10$ level of significance to conclude that the population mean is different from 600.

7. Hypotheses: $H_0: \mu = 25$, $H_1: \mu \neq 25$
Classical approach: $t_0 = -0.738$ is between $-t_{0.005} = -2.977$ and $t_{0.005} = 2.977$ with 14 degrees of freedom; do not reject the null hypothesis.
P-value approach: P-value $= 0.4729 > \alpha = 0.01$; do not reject the null hypothesis.
There is not sufficient evidence at the $\alpha = 0.01$ level of significance to conclude that the population mean is different from 25.

8. Hypotheses: $H_0: \mu = 55$, $H_1: \mu \neq 55$
Classical approach: $Z_0 = -1.29$ is between $-z_{0.025} = -1.96$ and $z_{0.025} = 1.96$; do not reject the null hypothesis.
P-value approach: P-value $= 0.1970 > \alpha = 0.05$; do not reject the null hypothesis.
There is not sufficient evidence at the $\alpha = 0.05$ level of significance to conclude that the population mean is different from 55.

9. Hypotheses: $H_0: \mu = 100$, $H_1: \mu > 100$
Classical approach: $t_0 = 3.003 > t_{0.05} = 1.685$; reject the null hypothesis.
P-value approach: P-value $= 0.0023 < \alpha = 0.05$; reject the null hypothesis.
There is sufficient evidence at the $\alpha = 0.05$ level of significance to conclude that the population mean is greater than 100.

10. Hypotheses: $H_0: p = 0.25$, $H_1: p < 0.25$; $np_0(1 - p_0) = 60 > 10$
Classical approach: $Z_0 = -2.80 < -z_{0.01} = -2.33$; reject the null hypothesis.
P-value approach: P-value $= 0.0026 < \alpha = 0.01$; reject the null hypothesis.
There is sufficient evidence at the $\alpha = 0.01$ level of significance to conclude that less than 25% of adults prefers mint chocolate chip ice cream.

11. (a) Hypotheses: H_0: $\mu = 2.5$, H_1: $\mu \neq 2.5$
Classical approach: $Z_0 = 2.64 > z_{0.05} = 1.645$; reject the null hypothesis.
P-value approach: *P*-value $= 0.0082 < \alpha = 0.10$; reject the null hypothesis.
There is sufficient evidence at the $\alpha = 0.10$ level of significance to support the claim that the mean ideal number of children has changed since 1985.
(b) Answers will vary.

12. Hypotheses: H_0: $p = 0.66$, H_1: $p > 0.66$;
$np_0(1 - p_0) = 453.7 > 10$
Classical approach: $Z_0 = 7.58 > z_{0.05} = 1.645$; reject the null hypothesis.
P-value approach: *P*-value $< 0.0001 < \alpha = 0.05$; reject the null hypothesis.
There is sufficient evidence at the $\alpha = 0.05$ level of significance to conclude that more adults were online than in 2002.

13. Hypotheses: H_0: $p = 0.23$, H_1: $p \neq 0.23$;
$np_0(1 - p_0) = 181.7 > 10$
Classical approach: $Z_0 = 1.37$, which is between $-z_{0.05} = -1.645$ and $z_{0.05} = 1.645$; do not reject the null hypothesis.
P-value approach: *P*-value $= 0.1706 > \alpha = 0.10$; do not reject the null hypothesis.
There is not sufficient evidence at the $\alpha = 0.10$ level of significance to conclude that the percentage of university undergraduate students who has at least one tattoo has changed since 2001.

14. Hypotheses: H_0: $\mu = 0.063$, H_1: $\mu < 0.063$
Classical approach: $t_0 = -0.9147 > -t_{0.05} = -1.684$; do not reject the null hypothesis.
P-value approach: *P*-value $= 0.1829 > \alpha = 0.05$; do not reject the null hypothesis.
There is not sufficient evidence at the $\alpha = 0.05$ level of significance to conclude that interest rates are lower than in 2001.

15. Hypotheses: H_0: $\mu = 774$, H_1: $\mu \neq 774$
Classical approach: $t_0 = -4.776 < -t_{0.005} = -2.728$; reject the null hypothesis.
P-value approach: *P*-value $< 0.0001 < \alpha = 0.01$; reject the null hypothesis.
There is sufficient evidence at the $\alpha = 0.01$ level of significance to conclude that the mean expenditure for auto insurance is different from the 2002 amount.

16. (a) Yes, the conditions are satisfied. The data appear to be normally distributed, and there are no outliers.
(b) Hypotheses: H_0: $\mu = 10{,}000$, H_1: $\mu < 10{,}000$
Classical approach: $t_0 = -1.343 > -t_{0.05} = -1.771$; do not reject the null hypothesis.
P-value approach: *P*-value $= 0.1012 > \alpha = 0.05$; do not reject the null hypothesis.
There is not sufficient evidence at the $\alpha = 0.05$ level of significance to conclude that the toner cartridges have a mean of less than 10,000 copies.

17. Hypotheses: H_0: $\mu = 8$, H_1: $\mu < 8$
Classical approach: $t = -1.443$, $t_{0.01} = -2.403$; do not reject the null hypothesis.
P-value approach: *P*-value $= 0.0777$; do not reject the null hypothesis.
There is not sufficient evidence at the $\alpha = 0.01$ level of significance to conclude that the mean wait time is less than eight minutes.

18. (a) Yes, the conditions are satisfied. The data appear to be normally distributed, and there are no outliers.
(b) Hypotheses: H_0: $\mu = 3.2$, H_1: $\mu > 3.2$
Classical approach: $t_0 = 1.989 > t_{0.05} = 1.796$; reject the null hypothesis.
P-value approach: *P*-value $= 0.0361 < \alpha = 0.05$; reject the null hypothesis.
There is sufficient evidence at the $\alpha = 0.05$ level of significance to conclude that the new contractor is not getting the streetlights replaced as quickly.

Chapter Review Exercises (page 504)

1. (a) H_0: $\mu = \$4277$, H_1: $\mu > \$4277$

(b) If the null hypothesis is rejected, but H_0 is true, a Type I error is made.
(c) A Type II error is made when the null hypothesis is not rejected but H_1 is true.
(d) There is not sufficient evidence at the α level of significance to conclude that the mean outstanding credit card debt per cardholder is more than the 2000 amount.
(e) There is sufficient evidence at the α level of significance conclude that the mean outstanding credit card debt per cardholder is more than the 2000 amount.

2. (a) H_0: $\mu = 0.61$, H_1: $\mu > 0.61$
(b) If null hypothesis is rejected, but 0.61 is the true population proportion, a Type I error is made.
(c) A Type II error is made when the null hypothesis is not rejected, but the true population proportion is greater than 0.61.
(d) There is not sufficient evidence at the α level of significance to conclude that the percentage of Internet users who downloaded music is greater than 61%.
(e) There is sufficient evidence at the α level of significance to conclude that the percentage of Internet users who downloaded music is greater than 61%.

3. The probability of a Type I error is 0.05.

4. The probability of a Type II error is 0.113.

5. (a) $Z = -1.08$ **(b)** $-z_{0.05} = -1.645$
(c)

Critical Region

-1.645

(d) No, because $Z > -z_{0.05}$.   **(e)** *P*-value $= 0.1401$
6. (a) $Z_0 = 2.18$
(b) $-z_{0.05} = -1.645$; $z_{0.05} = 1.645$
(c)

Critical Region   Critical Region

-1.645 1.645

(d) Yes, because the test statistic lies in the critical region.
(e) *P*-value $= 0.0292$
(f) Lower bound: 66.38, upper bound: 74.82. We reject the null hypothesis because the confidence interval does not contain $\mu = 65$.

7. (a) $t_0 = -1.506$
(b) $-t_{0.01} = -2.624$; $t_{0.01} = 2.624$
(c)

Critical Region   Critical Region

-2.624 2.624

(d) No, because $-t_{0.01} = -2.624 < t_0 = -1.506 < t_{0.01} = 2.624$.
(e) *P*-value $= 0.1543$

8. (a) $t_0 = -2.222$ **(b)** $-t_{0.05} = -1.711$

(c)

Critical Region

-1.711

(d) Yes, because the test statistic lies in the critical region.

(e) P-value $= 0.0180$

9. Hypotheses: H_0: $p = 0.6$, H_1: $p > 0.6$; $np_0(1 - p_0) = 60 > 10$
Classical approach: $Z_0 = 1.94 > z_{0.05} = 1.645$; reject the null hypothesis.
P-value approach: P-value $= 0.0262 < \alpha = 0.05$; reject the null hypothesis.
There is sufficient evidence at the $\alpha = 0.05$ level of significance to conclude that $p > 0.6$.

10. Hypotheses: H_0: $p = 0.35$, H_1: $p \neq 0.35$; $np_0(1 - p_0) = 95.6 > 10$
Classical approach: $Z_0 = -0.92$ is between $-z_{0.005} = -2.575$ and $z_{0.005} = 2.575$; do not reject the null hypothesis.
P-value approach: P-value $= 0.3576 > \alpha = 0.01$; do not reject the null hypothesis.
There is not sufficient evidence at the $\alpha = 0.01$ level of significance to conclude that $p \neq 0.35$.

11. Hypotheses: H_0: $\mu = 0.875$, H_1: $\mu > 0.875$
Classical approach: $t_0 = 1.2 < t_{0.05} = 1.690$; do not reject the null hypothesis.
P-value approach: P-value $= 0.1191 > \alpha = 0.05$; do not reject the null hypothesis.
There is not sufficient evidence at the $\alpha = 0.05$ level of significance to conclude that the mean distance between retaining rings is longer than 0.875 inch.

12. Hypotheses: H_0: $\mu = \$51{,}194$, H_1: $\mu \neq \$51{,}194$
Classical approach: $t_0 = 3.092 > t_{0.025} = 1.984$; reject the null hypothesis.
P-value approach: P-value $= 0.0021 < \alpha = 0.05$; reject the null hypothesis.
There is sufficient evidence at the $\alpha = 0.05$ level of significance to conclude that the mean annual salary of individuals whose highest degree is a bachelor's is different from $51,194.

13. Hypotheses: H_0: $\mu = 474$, H_1: $\mu > 474$
Classical approach: $Z_0 = 4.46 > z_{0.01} = 2.33$; reject the null hypothesis.
P-value approach: P-value $< 0.0001 < \alpha = 0.01$; reject the null hypothesis.
There is sufficient evidence at the $\alpha = 0.01$ level of significance to conclude that students who use calculators frequently score better on the SAT math portion than those who use a calculator infrequently.

14. Hypotheses: H_0: $\mu = 7.53$, H_1: $\mu > 7.53$
Classical approach: $Z_0 = 1.60 < z_{0.01} = 2.33$, do not reject the null hypothesis.
P-value approach: P-value $= 0.0548 > \alpha = 0.01$; do not reject the null hypothesis.
There is not sufficient evidence at the $\alpha = 0.01$ level of significance to conclude that the new diet increases the birth weights of newborns.

15. (a) Hypotheses: H_0: $\mu = 300$, H_1: $\mu > 300$
Classical approach: $t_0 = 1.528 < t_{0.05} = 1.660$; do not reject the null hypothesis.
P-value approach: P-value $= 0.0636 > \alpha = 0.05$; do not reject the null hypothesis.
There is not sufficient evidence at the $\alpha = 0.05$ level of significance to conclude that mean cholesterol consumption of 20- to 39-year-old males is greater than 300 mg.

(b) Type I error: The nutritionist rejects the null hypothesis that the mean cholesterol consumption is 300 mg when, in fact, the mean consumption is 300 mg. Type II error: The nutritionist does not reject the null hypothesis that the mean cholesterol consumption is 300 mg when, in fact, the mean consumption is more than 300 mg.

(c) The probability of making a Type I error is $\alpha = 0.05$.

16. (a) Hypotheses: H_0: $\mu = 2400$, H_1: $\mu > 2400$
Classical approach: $t_0 = 9.183 > t_{0.10} = 1.290$; reject the null hypothesis.
P-value approach: P-value $< 0.001 < \alpha = 0.10$; reject the null hypothesis.
There is sufficient evidence at the $\alpha = 0.10$ level of significance to conclude that mean daily sodium intake of 20- to 39-year-old females is greater than 2400 mg.

(b) Type I error: The nutritionist rejects the null hypothesis that the mean sodium intake is 2400 mg when, in fact, the mean consumption is 2400 mg. Type II error: The nutritionist does not reject the null hypothesis that the mean sodium intake is 2400 mg when, in fact, the mean sodium intake is more than 2400 mg.

(c) The probability of making a Type I error is $\alpha = 0.10$.

17. (a) Yes, the conditions are satisfied. The data appear to be normally distributed, and there are no outliers.

(b) Hypotheses: H_0: $\mu = 4.61$, H_1: $\mu < 4.61$
Classical approach: $Z_0 = -2.48 < -z_{0.01} = -2.33$; reject the null hypothesis.
P-value approach: P-value $= 0.0066 < \alpha = 0.01$; reject the null hypothesis.
There is sufficient evidence at the $\alpha = 0.01$ level of significance to conclude that the acidity of the rain has increased since 1990.

18. (a) Yes, the conditions are satisfied. The data appear to be normally distributed, and there are no outliers.

(b) Hypotheses: H_0: $\mu = 14.0$, H_1: $\mu \neq 14.0$
Classical approach: $Z_0 = -3.68 < -z_{0.005} = -2.575$; reject the null hypothesis.
P-value approach: P-value $= 0.0002 < \alpha = 0.01$; reject the null hypothesis.
There is sufficient evidence at the $\alpha = 0.01$ level of significance to conclude that the mean hemoglobin reading of surgical patients is different from 14.0 grams per deciliter.

19. Hypotheses: H_0: $p = 0.56$, H_1: $p > 0.56$; $np_0(1 - p_0) = 73.9 > 10$
Classical approach: $Z_0 = 0.24 < z_{0.01} = 2.33$; do not reject the null hypothesis.
P-value approach: P-value $= 0.4052 > \alpha = 0.01$; do not reject the null hypothesis.
There is not sufficient evidence at the $\alpha = 0.01$ level of significance to conclude that the percentage of cases of tuberculosis of foreign-born residents has increased.

20. Hypotheses: H_0: $p = 0.394$, H_1: $p < 0.394$;
$np_0(1 - p_0) = 119.3 > 10$
Classical approach: $Z_0 = -0.55 > -z_{0.10} = -1.28$; do not reject the null hypothesis.
P-value approach: P-value $= 0.2912 > \alpha = 0.10$; do not reject the null hypothesis.
There is not sufficient evidence at the $\alpha = 0.10$ level of significance to conclude that the percentage of females ordering merchandise or services by phone has decreased from its 1997 proportion.

21. Hypotheses: H_0: $\mu = 73.2$, H_1: $\mu < 73.2$
Classical approach: $t_0 = -2.018 < -t_{0.05} = -1.645$; reject the null hypothesis.
P-value approach: P-value $= 0.0218 < \alpha = 0.05$; reject the null hypothesis.
There is sufficient evidence at the $\alpha = 0.05$ level of significance to conclude that the mean test scores decreased.

22. Answers will vary

CHAPTER 11

11.1 Assess Your Understanding (page 515)

1. Independent **2.** Dependent

3. $H_1: u_d < 0$ where $d_j = x_j - y_j$

4. To test a claim regarding the difference of two means with dependent sampling, we need a simple random sample for which the differences are normally distributed with no outliers or for which the sample size n is large.

5. Dependent **6.** Independent **7.** Independent

8. Dependent **9.** Independent **10.** Independent

11. (a)

Observation	1	2	3	4	5	6	7
d_t	−0.5	1	−3.3	−3.7	0.5	−2.4	−2.9

(b) $\bar{d} = -1.614$; $s_d = 1.915$

(c) Classical approach: $t_0 = -2.230 < -t_{0.05} = -1.943$; reject the null hypothesis

P-value approach: P-value $= 0.0336 < \alpha = 0.05$; reject the null hypothesis

There is sufficient evidence at the $\alpha = 0.05$ level of significance to reject the null hypothesis that $\mu_d = 0$.

(d) We can be 95% confident that the mean difference is between −3.385 and 0.157.

12. (a)

Observation	1	2	3	4	5	6	7	8
d_t	−0.4	1.5	1	−1.3	−2.3	−6.9	5.1	−5.3

(b) $\bar{d} = -1.075$; $s_d = 3.833$

(c) Classical approach: $t_0 = -0.7932$; do not reject the null hypothesis since $-t_{0.005} = -3.499 < -0.7932 < t_{0.005} = 3.499$

P-value approach: P-value $= 0.4537 > \alpha = 0.01$; do not reject the null hypothesis

There is not sufficient evidence at the $\alpha = 0.01$ level of significance to support the claim that $\mu_d \neq 0$.

(d) We can be 99% confident that the mean difference is between −5.817 and 3.667.

13. (a) This is matched-pairs data because two measurements (A and B) are taken on the same round.

(b) Hypotheses: $H_0: \mu_d = 0$, $H_1: \mu_d \neq 0$

Classical approach: $t_0 = 0.8517$; do not reject the null hypothesis since $-t_{0.005} = -3.106 < 0.8517 < t_{0.005} = 3.106$

P-value approach: P-value $= 0.4125 > \alpha = 0.01$; do not reject the null hypothesis

There is not sufficient evidence at the $\alpha = 0.01$ level of significance to conclude that there is a difference in the measurements of velocity between device A and device B.

(c) Lower bound: −0.309, upper bound: 0.543. We are 99% confident that the mean difference in measurement is between −0.309 and 0.543 feet per second.

(d)

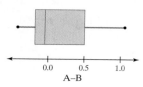

A–B

14. (a) These are matched-pairs data because the same person is being timed for both the red and blue stimuli.

(b) Hypotheses: $H_0: \mu_d = 0$, $H_1: \mu_d \neq 0$

Classical approach: $t_0 = -1.312$; do not reject the null hypothesis since $-t_{0.005} = -4.032 < 1.312 < t_{0.005} = 4.032$

P-value approach: P-value $= 0.2466 > \alpha = 0.01$; do not reject the null hypothesis

There is not sufficient evidence at the $\alpha = 0.01$ level of significance to conclude that there is a difference in the reaction time to the blue and red stimuli.

(c) We can be 98% confident that the mean difference in reaction time between blue and red stimuli is between −0.3316 second and 0.1456 second.

(d)

Red–Blue

15. (a) Answers may vary.

(b) Hypotheses: $H_0: \mu_d = 0$, $H_1: \mu_d > 0$

Classical approach: $t_0 = 2.3837 > t_{0.05} = 1.895$; reject the null hypothesis

P-value approach: P-value $= 0.0243 < \alpha = 0.05$; reject the null hypothesis

No. There is sufficient evidence at the $\alpha = 0.05$ level of significance to conclude that there has been an improvement in the clarity of the water in the 5-year period.

(c) We can be 95% confident that the mean difference in clarity between the initial measurements and those taken 5 years later is between 0.04 inch and 10.213 inches.

(d)

5 Years Minus Initial Measure

16. (a) Hypotheses: $H_0: \mu_d = 0$, $H_1: \mu_d > 0$

Classical approach: $t_0 = 2.615 > t_{0.05} = 1.796$; reject the null hypothesis

P-value approach: P-value $= 0.0120 < \alpha = 0.05$; reject the null hypothesis

No. There is sufficient evidence at the $\alpha = 0.05$ level of significance to conclude that the hemoglobin levels at lift-off minus three days are less than the hemoglobin levels upon return.

(b) Lower bound: 0.27, upper bound: 1.48. We are 90% confident that the mean difference in hemoglobin readings is between 0.27 gram per deciliter and 1.48 grams per deciliter.

(c)

R0–L3

17. Hypotheses: $H_0: \mu_d = 0$, $H_1: \mu_d > 0$

Classical approach: $t_0 = 0.398 < t_{0.10} = 1.356$; do not reject the null hypothesis

P-value approach: P-value $= 0.3508 > \alpha = 0.10$; do not reject the null hypothesis

No. There is not sufficient evidence at the $\alpha = 0.10$ level of significance to conclude that sons are taller than their fathers.

18. Hypotheses: $H_0: \mu_d = 0$, $H_1: \mu_d < 0$

Classical approach: $t_0 = -1.220 > -t_{0.05} = -1.860$; do not reject the null hypothesis

P-value approach: P-value $= 0.1286 > \alpha = 0.05$; do not reject the null hypothesis

There is not sufficient evidence at the $\alpha = 0.05$ level of significance to conclude that the new loading procedure reduces wait time.

19. Hypotheses: $H_0: \mu_d = 0$, $H_1: \mu_d \neq 0$

Classical approach: $t_0 = 2.6 > t_{0.025} = 2.306$; reject the null hypothesis

P-value approach: P-value $= 0.0285 < \alpha = 0.05$; reject the null hypothesis

Yes. There is sufficient evidence at the $\alpha = 0.05$ level of significance to conclude that the two different indenters result in different measurements.

20. Hypotheses: $H_0: \mu_d = 0$, $H_1: \mu_d < 0$

Classical approach: $t_0 = 0.0890 > -t_{0.10} = -1.383$; do not reject the null hypothesis

P-value approach: P-value $= 0.5345 > \alpha = 0.10$; do not reject the null hypothesis

There is not sufficient evidence at the $\alpha = 0.10$ level of significance to conclude that Thrifty is less expensive than Hertz.

21. Lower bound: 0.69, upper bound: 1.24. We can be 95% confident that the mean difference in reaction time when teenagers are driving impaired from when driving normally is between 0.69 second and 1.24 seconds.

22. Lower bound: 29.55, upper bound: 35.85. We can be 95% confident that the mean difference in braking distance on a wet road from a dry road is between 29.55 feet and 35.85 feet when driving at 40 miles per hour.

23. (a), (b) Answers will vary.
    (c) Neither variable is normally distributed.
    (d) The difference in mileage appears to be approximately normally distributed.
    (e) Classical approach: $t_0 = 1.14 < t_{0.05} = 1.812$; do not reject the null hypothesis
    P-value approach: P-value $= 0.141 > \alpha = 0.05$; do not reject the null hypothesis
    There is not sufficient evidence at the $\alpha = 0.05$ level of significance to conclude that cars get better mileage when using 92-octane gasoline than when using 87-octane gasoline.

11.2 Assess Your Understanding (page 528)

1. To test a hypothesis regarding the difference of two means, we need two independent random samples that are drawn from normally distributed populations. If the populations are not normally distributed each sample must be large ($n_1 \geq 30$ and $n_2 \geq 30$).

2. The degrees of freedom obtained from Formula (2) are larger than the smaller of $n_1 - 1$ or $n_2 - 1$, and t_α decreases as the degrees of freedom increase. The larger the critical value is, the harder it is to reject the null hypothesis.

3. (a) $H_0: \mu_1 = \mu_2$, $H_1: \mu_1 \neq \mu_2$
    Classical approach: $t_0 = 0.8783$ is between $-t_{0.025} = -2.145$ and $t_{0.025} = 2.145$; do not reject H_0
    P-value approach: P-value $= 0.3767 > \alpha = 0.05$; do not reject H_0
    There is not sufficient evidence at the $\alpha = 0.05$ level of significance to conclude that the population means are different.
    (b) Lower bound: -1.526, upper bound: 3.726

4. (a) $H_0: \mu_1 = \mu_2$, $H_0: \mu_1 \neq \mu_2$
    Classical approach: $t_0 = 2.4858 > t_{0.025} = 2.093$; reject H_0
    P-value approach: P-value $= 0.0175 < \alpha = 0.05$; reject H_0
    There is sufficient evidence at the $\alpha = 0.05$ level of significance to conclude that the population means are different.
    (b) Lower bound: 1.106, upper bound: 12.894

5. (a) $H_0: \mu_1 = \mu_2$, $H_1: \mu_1 > \mu_2$
    Classical approach: $t_0 = 3.081 > t_{0.10} = 1.333$; reject H_0
    P-value approach: P-value $= 0.0024 < \alpha = 0.10$; reject H_0
    There is sufficient evidence at the $\alpha = 0.10$ level of significance to conclude that $\mu_1 > \mu_2$.
    (b) Lower bound: 3.569, upper bound: 12.831

6. (a) $H_0: \mu_1 = \mu_2$, $H_1: \mu_1 < \mu_2$
    Classical approach: $t_0 = -4.393 < -t_{0.05} = -1.696$; reject H_0
    P-value approach: P-value $< 0.0001 < \alpha = 0.05$; reject H_0
    There is sufficient evidence at the $\alpha = 0.05$ level of significance to conclude that $\mu_1 < \mu_2$.
    (b) Lower bound: -30.75, upper bound: -11.248

7. (a) $H_0: \mu_1 = \mu_2$; $H_1: \mu_1 < \mu_2$
    Classical approach: $t_0 = -3.158 < -t_{0.02} = -2.172$; reject H_0
    P-value approach: P-value $= 0.0013 < \alpha = 0.02$; reject H_0
    There is sufficient evidence at the $\alpha = 0.02$ level of significance to conclude that $\mu_1 < \mu_2$.
    (b) Lower bound: -16.652, upper bound: -4.948

8. (a) $H_0: \mu_1 = \mu_2$, $H_1: \mu_1 > \mu_2$
    Classical approach: $t_0 = 1.237 < t_{0.05} = 1.782$; do not reject H_0
    P-value approach: P-value $= 0.1144 > \alpha = 0.05$; do not reject H_0
    There is not sufficient evidence at the $\alpha = 0.05$ level of significance to conclude that $\mu_1 > \mu_2$.
    (b) Lower bound: -1.599, upper bound: 5.799

9. (a) $H_0: \mu_1 = \mu_2$, $H_1: \mu_1 > \mu_2$
    Classical approach: $t_0 = 2.849 > t_{0.01} = 2.403$; reject H_0
    P-value approach: P-value $= 0.0026 < \alpha = 0.01$; reject H_0

Yes. There is sufficient evidence at the $\alpha = 0.01$ level of significance to conclude that mean improvement in the treatment group was greater than the mean improvement in the control group. The drug appears to be effective in improving the Young–Mania Rating Scale score.
    (b) Lower bound: 1.98, upper bound: 11.42. The researchers are 95% confident that the mean Young–Mania Rating Scale score for the treatment group is between 1.98 and 11.42 points higher than that of the control group.

10. (a) This is a completely randomized design. The treatment is the drug. It has two levels: hormone replacement pills or a placebo.
    (b) $H_0: \mu_1 = \mu_2$, $H_1: \mu_1 > \mu_2$
    Classical approach: $t_0 = 2.745 > t_{0.01} = 2.467$; reject H_0
    P-value approach: P-value $= 0.0046 < \alpha = 0.01$; reject H_0
    Yes. There is sufficient evidence at the $\alpha = 0.01$ level of significance to conclude that the mean increase in serum HDL levels of women in the treatment group is greater than the mean increase in serum HDL levels of women in the control group.
    (c) Lower bound: 1.45, upper bound: 9.95. We are 95% confident that the mean difference in the increase in serum HDL levels in the treatment group over that of the control group is between 1.45 and 9.95 mg/dL.

11. (a) This is an observational study.
    (b) Answers will vary.
    (c) $H_0: \mu_{\text{arrival}} = \mu_{\text{departure}}$, $H_1: \mu_{\text{arrival}} \neq \mu_{\text{departure}}$
    Classical approach: $t_0 = 0.8456$ is between $-t_{0.025} = -2.032$ and $t_{0.025} = 2.032$; do not reject H_0
    P-value approach: P-value $= 0.4013 > \alpha = 0.05$; do not reject H_0
    No. There is not sufficient evidence at the $\alpha = 0.05$ level of significance to conclude that travelers walk at different speeds depending on whether they are arriving or departing an airport.
    (d) Lower bound: -12.63, upper bound: 30.63. We can be 95% confident that the mean difference in speed walked between passengers arriving and departing an airport is between -12.63 and 30.63 feet per minute.

12. (a) This is an observational study.
    (b) Since the samples are small, the populations must be normally distributed.
    (c) $H_0: \mu_{\text{business}} = \mu_{\text{leisure}}$, $H_1: \mu_{\text{business}} \neq \mu_{\text{leisure}}$
    Classical approach: $t_0 = 0.7722$ is between $-t_{0.025} = -2.093$ and $t_{0.025} = 2.093$; do not reject H_0
    P-value approach: P-value $= 0.4448 > \alpha = 0.05$; do not reject H_0
    There is not sufficient evidence at the $\alpha = 0.05$ level of significance to conclude that business travelers walk at different speeds from leisure travelers when walking in an airport.
    (d) Lower bound: -18.81, upper bound: 40.81. We can be 95% confident that the mean difference in speed walked between business travelers and leisure travelers in an airport is between -18.81 and 40.81 feet per minute.

13. (a) Yes, we can treat each sample as a simple random sample of all mixtures of each type. The samples were obtained independently. We are told that a normal probability plot indicates that the data could come from a population that is normal, with no outliers.
    (b) $H_0: \mu_{67-0-301} = \mu_{67-0-400}$, $H_1: \mu_{67-0-301} < \mu_{67-0-400}$
    Classical approach: $t_0 = -3.804 < -t_{0.05} = -1.860$; reject H_0
    P-value approach: P-value $= 0.0007 < \alpha = 0.05$; reject H_0
    There is sufficient evidence at the $\alpha = 0.05$ level of significance to conclude that mixture 67-0-400 is stronger than mixture 67-0-301.
    (c) Lower bound: 416, upper bound: 1212. We are 90% confident that the mean strength of mixture 67-0-400 is between 416 and 1212 psi stronger than the mean strength of mixture 67-0-301.
    (d)

14. (a) Yes, each sample is a simple random sample. The samples were obtained independently. We are told that a normal probability plot indicates that the data could come from a population that is normal, with no outliers.

(b) $H_0: \mu_{\text{female}} = \mu_{\text{male}}$, $H_1: \mu_{\text{female}} \neq \mu_{\text{male}}$
Classical approach: $t_0 = 0.6199$ is between $-t_{0.025} = -2.145$ and $t_{0.025} = 2.145$; do not reject H_0
P-value approach: P-value $= 0.5403 > \alpha = 0.05$; do not reject H_0
There is not sufficient evidence at the $\alpha = 0.05$ level of significance to conclude that there is a difference in reaction time between females and males.

(c) Lower bound: -0.0482, upper bound: 0.1006. We are 90% confident that the mean difference in reaction time for females and males is between -0.0482 and 0.1006 second.

(d)

15. (a) $H_0: \mu_{\text{carpet}} = \mu_{\text{nocarpet}}$, $H_1: \mu_{\text{carpet}} > \mu_{\text{nocarpet}}$
Classical approach: $t_0 = 0.954 < t_{0.05} = 1.895$; do not reject H_0
P-value approach: P-value $= 0.1780 > \alpha = 0.05$; do not reject H_0
No. There is not sufficient evidence at the $\alpha = 0.05$ level of significance to conclude that carpeted rooms have more bacteria than uncarpeted rooms.

(b) Lower bound: -2.09, upper bound: 4.91. We are 95% confident that the mean difference in the number of bacteria per cubic foot in a carpeted room versus that in an uncarpeted room is between -2.08 and 4.91.

16. (a) This is an experiment using a completely randomized design.
(b) The treatments are visual manual style and textual manual style.
(c) $H_0: \mu_{\text{visual}} = \mu_{\text{textual}}$, $H_1: \mu_{\text{visual}} \neq \mu_{\text{textual}}$
Classical approach: $t_0 = 2.378 > t_{0.025} = 2.110$; reject H_0
P-value approach: P-value $= 0.0236 < \alpha = 0.05$; reject H_0
Yes. There is sufficient evidence at the $\alpha = 0.05$ level of significance to conclude that there is a difference in the test scores of people who use the visual manual and those who use the textual manual.
(d) Lower bound: 0.958, upper bound: 16.038. We are 95% confident that the mean difference between the test scores of visual manual users versus those of textual manual users is between 0.958 point and 16.038 points. Because the interval does not contain zero, we take this as evidence against the null hypothesis.

17. $H_0: \mu_{\text{AL}} = \mu_{\text{NL}}$, $H_1: \mu_{\text{AL}} > \mu_{\text{NL}}$
Classical approach: $t_0 = 2.1356 > t_{0.05} = 1.699$; reject H_0
P-value approach: P-value $= 0.0187 < \alpha = 0.05$; reject H_0
Yes. There is sufficient evidence at the $\alpha = 0.05$ level of significance to conclude that games played with a designated hitter result in more runs.

18. $H_0: \mu_{\text{RB}} = \mu_{\text{ALT}}$, $H_1: \mu_{\text{RB}} \neq \mu_{\text{ALT}}$
Classical approach: $t_0 = 0.6104$ is between $-t_{0.05} = -1.699$ and $t_{0.05} = 1.699$; do not reject H_0
P-value approach: P-value $= 0.5440 > \alpha = 0.10$; do not reject H_0
There is not sufficient evidence at the $\alpha = 0.10$ level of significance to conclude that the lengths of rhythm & blues songs differ from the length of alternative songs.

19. Lower bound: 0.71, upper bound: 2.33. We can be 90% confident that the mean difference in daily leisure time between adults without children and those with children is between 0.71 and 2.33 hours. Since the confidence interval does not include zero, we can conclude that there is a significant difference in the leisure time of adults without children and those with children.

20. Lower bound: 38.08, upper bound: 44.72. We can be 90% confident that the mean difference in cooling time between filled clear glass bottle and filled aluminum bottles is between 38.08 and 44.72 minutes. Since the confidence interval does not include zero, we can conclude that there is a significant difference in the cooling times of glass bottles and aluminum bottles.

21. (a) $H_0: \mu_{\text{men}} = \mu_{\text{women}}$ versus $H_1: \mu_{\text{men}} < \mu_{\text{women}}$
(b) P-value $= 0.0051$. Because P-value $< \alpha$, we reject the null hypothesis. There is sufficient evidence at the $\alpha = 0.01$ level of significance to conclude that the mean step pulse of men is lower than the mean step pulse of women.
(c) Lower bound: -10.7, upper bound: -1.5. We are 95% confident that the mean step pulse of men is between 1.5 and 10.7 beats per minute lower than the mean step pulse of women.

22. (a) $H_0: \mu_{\text{men}} = \mu_{\text{women}}$ versus $H_1: \mu_{\text{men}} < \mu_{\text{women}}$
(b) P-value $= 0.0001$. Because P-value $< \alpha$, we reject the null hypothesis. There is sufficient evidence at the $\alpha = 0.01$ level of significance to conclude that the women are more flexible than the men.
(c) Lower bound: -3.58, upper bound: -1.12. We are 95% confident that the mean flexibility of women is between 1.12 and 3.58 inches higher than the mean flexibility of men.

11.3 Assess Your Understanding (page 542)

1. A pooled estimate of p is the best point estimate of the common population proportion p. However, when finding a confidence interval, the sample proportions are not pooled because no assumption about their equality is made.

2. To test a hypotheses regarding two population proportions, we must have two independent random samples, $n_1 \hat{p}_1(1 - \hat{p}_1) \geq 10$ and $n_2 \hat{p}_2(1 - \hat{p}_2) \geq 10$, and $n_1 \leq 0.05 N_1$ and $n_2 \leq 0.05 N_2$.

3. (a) $H_0: p_1 = p_2$ versus $H_1: p_1 > p_2$ **(b)** $z_0 = 3.07$
(c) $z_{0.05} = 1.645$
(d) P-value $= 0.0011$. Because $z_0 > z_{0.05}$ (or P-value $< \alpha$), we reject the null hypothesis. There is sufficient evidence to conclude that $p_1 > p_2$.

4. (a) $H_0: p_1 = p_2$ versus $H_1: p_1 < p_2$ **(b)** $z_0 = -0.36$
(c) $-z_{0.05} = -1.645$
(d) P-value $= 0.3594$. Because $z_0 > -z_{0.05}$ (or P-value $> \alpha$), we do not reject the null hypothesis. There is not sufficient evidence to conclude that $p_1 < p_2$.

5. (a) $H_0: p_1 = p_2$ versus $H_1: p_1 \neq p_2$ **(b)** $z_0 = -0.37$
(c) $-z_{0.025} = -1.96$; $z_{0.025} = 1.96$
(d) P-value $= 0.7114$. Because $-z_{0.025} < z_0 < z_{0.05}$ (or P-value $> \alpha$), we do not reject the null hypothesis. There is not sufficient evidence to conclude that $p_1 \neq p_2$.

6. (a) $H_0: p_1 = p_2$ versus $H_1: p_1 \neq p_2$ **(b)** $z_0 = -2.14$
(c) $-z_{0.025} = -1.96$; $z_{0.025} = 1.96$
(d) P-value $= 0.0324$. Because $z_0 < -z_{0.05}$ (or P-value $< \alpha$), we reject the null hypothesis. There is sufficient evidence to conclude that $p_1 \neq p_2$.

7. Lower bound: -0.075, upper bound: 0.015

8. Lower bound: -0.090, upper bound: 0.068

9. Lower bound: -0.063, upper bound: 0.043

10. Lower bound: -0.039, upper bound: 0.009

11. (a) Each sample can be thought of as a simple random sample; $n_1 \hat{p}_1(1 - \hat{p}_1) = 91 \geq 10$ and $n_2 \hat{p}_2(1 - \hat{p}_2) = 60 \geq 10$; and each sample is less than 5% of the population size.
$H_0: p_1 = p_2$; $H_1: p_1 > p_2$
Classical approach: $z_0 = 2.20 > z_{0.05} = 1.645$; reject H_0
P-value approach: P-value $= 0.0139 < \alpha = 0.05$; reject H_0
Yes. There is sufficient evidence at the $\alpha = 0.05$ level of significance to conclude that a higher proportion of subjects in the treatment group (taking Prevnar) experienced fever as a side effect than in the control (placebo) group.
(b) Lower bound: 0.01, upper bound: 0.07. We are 90% confident that the difference in the proportion of subjects who experience

a fever as a side effect between the experimental and control groups is between 0.01 and 0.07.

12. (a) Each sample can be thought of as a simple random sample; $n_1\hat{p}_1(1 - \hat{p}_1) = 95 \geq 10$ and $n_2\hat{p}_2(1 - \hat{p}_2) = 21 \geq 10$; and each sample is less than 5% of the population size.
H_0: $p_1 = p_2$, H_1: $p_1 \neq p_2$
Classical approach: $z_0 = -0.20$ is between $-z_{0.025} = -1.96$ and $z_{0.025} = 1.96$; do not reject H_0
P-value approach: P-value $= 0.8443 > \alpha = 0.05$; do not reject H_0
No. There is not sufficient evidence at the $\alpha = 0.05$ level of significance to conclude that a different proportion of subjects taking Prevnar experienced drowsiness as a side effect than in the control (placebo) group.
(b) Lower bound: -0.14, upper bound: 0.12. We are 99% confident that the difference in the proportion of subjects who experienced drowsiness as a side effect between the experimental and control group is between -0.14 and 0.12.

13. (a) Each sample is a simple random sample; $n_1\hat{p}_1(1 - \hat{p}_1) = 73 \geq 10$ and $n_2\hat{p}_2(1 - \hat{p}_2) = 76 \geq 10$; and each sample is less than 5% of the population size.
H_0: $p_1 = p_2$, H_1: $p_1 > p_2$
Classical approach: $z_0 = 0.98 < z_{0.10} = 1.28$; do not reject H_0
P-value approach: P-value $= 0.1635 > \alpha = 0.10$; do not reject H_0
There is not sufficient evidence at the $\alpha = 0.1$ level of significance to conclude that a higher proportion of individuals with at most an 8th-grade education consume too much cholesterol than of those with some college education.
(b) Lower bound: -0.04, upper bound: 0.11. We are 95% confident that the difference in the proportion of individuals with at most an 8th-grade education and individuals with some college education who consume too much cholesterol is between -0.04 and 0.11.

14. (a) Each sample is a simple random sample; $n_1\hat{p}_1(1 - \hat{p}_1) = 124 \geq 10$ and $n_2\hat{p}_2(1 - \hat{p}_2) = 120 \geq 10$; and each sample is less than 5% of the population size.
H_0: $p_1 = p_2$, H_1: $p_1 < p_2$
Classical approach: $z_0 = -3.69 < -z_{0.05} = -1.645$; reject H_0
P-value approach: P-value $< 0.0001 < \alpha = 0.05$; reject H_0
There is sufficient evidence at the $\alpha = 0.05$ level of significance to conclude that a lower proportion of females consumes too much saturated fat than of males.
(b) Lower bound: -0.17, upper bound: -0.05. We are 95% confident that the difference in the proportion of females and males who consume too much saturated fat is between 0.05 and 0.17. A lower proportion of females consumes too much saturated fat.

15. Provided we choose $\alpha \geq 0.01$, we can say the proportion of American adults who smoked at least one cigarette in the past week has decreased, because when testing the hypotheses H_0: $p_1 = p_2$ versus H_1: $p_1 < p_2$, where p_1 is the proportion of adults who smoked in 2005 and p_2 is the proportion of adults who smoked in 1990, we get a P-value $= 0.0041$ ($z_0 = -2.64$).

16. (a) Each sample is a simple random sample; $n_1\hat{p}_1(1 - \hat{p}_1) = 127 \geq 10$ and $n_2\hat{p}_2(1 - \hat{p}_2) = 108 \geq 10$; and each sample is less than 5% of the population size.
H_0: $p_{1999} = p_{1996}$, H_1: $p_{1999} < p_{1996}$
Classical approach: $z_0 = -3.84 < -z_{0.10} = -1.28$; reject H_0
P-value approach: P-value $= 0.0001 < \alpha = 0.10$; reject H_0
Yes. There is sufficient evidence at the $\alpha = 0.1$ level of significance to conclude that the proportion of adults who believe there is life of some form on other planets in the universe has decreased.
(b) Lower bound: -0.16, upper bound: -0.06. We are 90% confident that the difference in the proportion of adults in 1996 who believe there is life on other planets and the proportion in 1999 is between 0.06 and 0.16 and that a higher proportion in 1996 believed that there is life on other planets.

17. (a) Each sample is a simple random sample; $n_1\hat{p}_1(1 - \hat{p}_1) = 33 \geq 10$ and $n_2\hat{p}_2(1 - \hat{p}_2) = 115 \geq 10$; and each sample is less than 5% of the population size.
H_0: $p_1 = p_2$, H_1: $p_1 < p_2$
Classical approach: $z_0 = -6.74 < -z_{0.01} = -2.33$; reject H_0
P-value approach: P-value $< 0.0001 < \alpha = 0.01$; reject H_0

There is sufficient evidence at the $\alpha = 0.01$ level of significance to conclude that the proportion of children in the experimental group who contracted polio is less than the proportion of children in the control group who contracted polio.
(b) Lower bound: -0.0005, upper bound: -0.0003. We are 90% confident that the difference in the proportion of children who contract polio with the vaccine versus without the vaccine is between 0.0003 and 0.0005 and that a smaller percentage of children contracted polio in the experimental group.

18. (a) Each sample is a simple random sample, $n_1\hat{p}_1(1 - \hat{p}_1) = 248 \geq 10$, and $n_2\hat{p}_2(1 - \hat{p}_2) = 249 \geq 10$, and each sample is less than 5% of the population size.
H_0: $p_1 = p_2$, H_1: $p_1 \neq p_2$
Classical approach: $z_0 = 0.45$, which is between $-z_{0.025} = -1.96$ and $z_{0.025} = 1.96$; do not reject H_0
P-value approach: P-value $= 0.6528 > \alpha = 0.05$; do not reject H_0
There is not sufficient evidence at the $\alpha = 0.05$ level of significance to conclude that the proportion of American adults who believe in psychic or spiritual healing has changed between 2001 and 2005.
(b) Lower bound: -0.027, upper bound: 0.047. We are 90% confident that the difference in the proportion of adults who believe in psychic or spiritual healing is between -0.027 and 0.047.

19. (a) Each sample is a simple random sample, $n_1\hat{p}_1(1 - \hat{p}_1) = 48 \geq 10$ and $n_2\hat{p}_2(1 - \hat{p}_2) = 30 \geq 10$, and each sample is less than 5% of the population size.
H_0: $p_1 = p_2$, H_1: $p_1 > p_2$
Classical approach: $z_0 = 2.13 > z_{0.05} = 1.645$; reject H_0
P-value approach: P-value $= 0.0166 < \alpha = 0.05$; reject H_0
There is sufficient evidence at the $\alpha = 0.05$ level of significance to conclude that the proportion of individuals taking Clarinex and experiencing dry mouth is greater than that of those taking a placebo.
(b) No

20. (a) Each sample is a simple random sample, $n_1\hat{p}_1(1 - \hat{p}_1) = 446 \geq 10$ and $n_2\hat{p}_2(1 - \hat{p}_2) = 468 \geq 10$, and each sample is less than 5% of the population size.
H_0: $p_1 = p_2$, H_1: $p_1 < p_2$
Classical approach: $z_0 = -1.75 < -z_{0.05} = -1.645$; reject H_0
P-value approach: P-value $= 0.0401 < \alpha = 0.05$; reject H_0
There is sufficient evidence at the $\alpha = 0.05$ level of significance to conclude that a higher percentage of patients are cured with the new proposed treatment than with the current standard treatment.
(b) Answers will vary.

21. Answers will vary. There is not a statistically significant difference in the percentage of gun owners from the 1999 poll to the 2004 poll. When testing the hypotheses H_0: $p_1 = p_2$ versus H_1: $p_1 > p_2$, we get a P-value of 0.1586 ($Z_0 = 1.00 < z_{0.10} = 1.28$).

22. Answers will vary.

23. (a) $n = n_1 = n_2 = 1406$ **(b)** $n = n_1 = n_2 = 2135$

24. (a) $n = n_1 = n_2 = 2551$ **(b)** $n = n_1 = n_2 = 3383$

25. Since the P-value is less than 0.05, there is sufficient evidence at the $\alpha = 0.05$ level of significance to conclude that the drug is more effective in treating sleep apnea than the placebo.

Chapter Review Exercises (page 547)

1. Dependent   **2.** Independent   **3.** Independent
4. Dependent

5. (a)

Observation	1	2	3	4	5	6
$d_i = x_i - y_i$	-0.7	0.6	0	-0.1	-0.4	-0.4

(b) $d = -0.167$; $s_d = 0.450$
(c) Hypotheses: H_0: $\mu_d = 0$, H_1: $\mu_d < 0$
Classical approach: $t_0 = -0.907 > -t_{0.05} = -2.015$; do not reject the null hypothesis
P-value approach: P-value $= 0.2030 > \alpha = 0.05$; do not reject the null hypothesis
There is not sufficient evidence to conclude that the mean difference is less than zero.

(d) Lower bound: -0.79, upper bound: 0.45

6. (a)

Observation	1	2	3	4	5	6	7
$d_i = x_i - y_i$	0.2	-0.5	0.2	0.6	-0.6	-0.5	-0.8

(b) $d = -0.2$; $s_d = 0.526$
(c) Hypotheses: H_0: $\mu_d = 0$, H_1: $\mu_d \neq 0$
  Classical approach: $t_0 = -1.006$ is between $-t_{0.005} = -3.707$
  and $t_{0.005} = 3.707$; do not reject the null hypothesis
  P-value approach: P-value $= 0.3532 > \alpha = 0.01$; do not reject
  the null hypothesis
  There is not sufficient evidence to conclude that the mean difference is different from zero.
(d) Lower bound: -0.69, upper bound: 0.29

7. (a) Hypotheses: H_0: $\mu_1 = \mu_2$, H_1: $\mu_1 \neq \mu_2$
  Classical approach: $t_0 = 2.290 > t_{0.05} = 1.895$; reject the null
  hypothesis
  P-value approach: P-value $= 0.0351 < \alpha = 0.10$; reject the
  null hypothesis
  There is sufficient evidence at the $\alpha = 0.1$ level of significance
  to conclude that $\mu_1 \neq \mu_2$.
(b) Lower bound: 0.73, upper bound: 7.67

8. (a) Hypotheses: H_0: $\mu_1 = \mu_2$, H_1: $\mu_1 \neq \mu_2$
  Classical approach: $t_0 = -2.054$ is between $-t_{0.05} = -1.714$ and
  $t_{0.05} = 1.714$; do not reject the null hypothesis.
  P-value approach: P-value $= 0.0464 < \alpha = 0.05$; reject the
  null hypothesis
  Conclusion: Answers will vary depending on whether the classical approach or the P-value approach is used.
(b) Lower bound: -12.44, upper bound: 0.04

9. (a) Hypotheses: H_0: $\mu_1 = \mu_2$, H_1: $\mu_1 > \mu_2$
  Classical approach: $t_0 = 1.472 < t_{0.01} = 2.423$; do not reject the
  null hypothesis
  P-value approach: P-value $= 0.0726 > \alpha = 0.01$; do not reject
  the null hypothesis
  There is not sufficient evidence at the $\alpha = 0.01$ level of significance to conclude that the mean of population 1 is larger
  than the mean of population 2.
(b) Lower bound: -0.43, upper bound: 6.43

10. (a) Hypotheses: H_0: $\mu_1 = \mu_2$, H_1: $\mu_1 < \mu_2$
  Classical approach: $t_0 = -1.357 > -t_{0.05} = -1.895$; do not reject the null hypothesis
  P-value approach: P-value $= 0.0959 > \alpha = 0.05$; do not reject
  the null hypothesis
  There is not sufficient evidence at the $\alpha = 0.05$ level of significance to conclude that the mean of population 1 is less than
  the mean of population 2.
(b) Lower bound: -6.08, upper bound: 2.68

11. (a) H_0: $p_1 = p_2$ vs H_1: $p_1 \neq p_2$ **(b)** $z_0 = -1.68$
(c) $-z_{0.025} = -1.96$; $z_{0.025} = 1.96$
(d) P-value $= 0.0930$. Do not reject H_0. There is not sufficient evidence at the $\alpha = 0.05$ level of significance to conclude that
  $p_1 \neq p_2$ (because $-z_{0.025} < z_0 < z_{0.025}$ or P-value $> \alpha$).

12. (a) H_0: $p_1 = p_2$ vs H_1: $p_1 < p_2$ **(b)** $Z_0 = -0.62$
(c) $-z_{0.05} = -1.645$
(d) P-value $= 0.2611$. Do not reject H_0. There is not sufficient evidence at the $\alpha = 0.05$ level of significance to conclude that
  $p_1 < p_2$ (because $z_0 > -z_{0.025}$ or P-value $> \alpha$).

13. (a) The sampling method is dependent because the same individual is used for both measurements.
(b) Hypotheses: H_0: $\mu_d = 0$, H_1: $\mu_d \neq 0$
  Classical approach: $t_0 = 0.506$ is between $-t_{0.025} = -2.262$ and
  $t_{0.025} = 2.262$; do not reject the null hypothesis
  P-value approach: P-value $= 0.6209 > \alpha = 0.05$; do not reject
  the null hypothesis
  No. There is not sufficient evidence at the $\alpha = 0.05$ level of significance to conclude that an individual's arm span equals
  the individual's height.

14. (a) The testing method is independent since the dates selected for the Texas sample have no bearing on the dates chosen for the
  Illinois sample.

(b) Hypotheses: H_0: $\mu_{\text{Texas}} = \mu_{\text{Illinois}}$, H_1: $\mu_{\text{Texas}} \neq \mu_{\text{Illinois}}$
  Classical approach: $t_0 = 2.276 > t_{0.025} = 2.201$; reject the null
  hypothesis
  P-value approach: P-value $= 0.0387 < \alpha = 0.05$; reject the
  null hypothesis
  There is sufficient evidence at the $\alpha = 0.05$ level of significance
  to conclude that the acidity of the rain near Houston is different from its acidity near Chicago.

(c)

15. (a) The testing method is independent since the cars selected for the McDonald's sample had no bearing on the cars chosen for the
  Wendy's sample.
(b) Hypotheses: H_0: $\mu_{\text{McD}} = \mu_{\text{W}}$, H_1: $\mu_{\text{McD}} \neq \mu_{\text{W}}$
  Classical approach: $t_0 = -4.059 < -t_{0.05} = -1.706$; reject the
  null hypothesis
  P-value approach: P-value $= 0.0003 < \alpha = 0.10$; reject the
  null hypothesis
  Yes. There is sufficient evidence at the $\alpha = 0.10$ level of significance to conclude that the mean wait time is different at
  McDonald's and Wendy's drive-through windows.

(c)

16. (a) The sampling method is dependent because the same individuals are used in both samples. This is often referred to as a matched-pairs sample.
(b) Hypotheses: H_0: $\mu_d = 0$, H_1: $\mu_d > 0$
  Classical approach: $t_0 = 26.991 > t_{0.05} = 1.660$; reject the null
  hypothesis
  P-value approach: P-value $< 0.0001 < \alpha = 0.05$; reject the
  null hypothesis
  Yes. There is sufficient evidence at the $\alpha = 0.05$ level of significance to conclude that there is a positive BMI reduction 10
  months after gastric bypass surgery.
(c) Lower bound: 11.83, upper bound: 13.39

17. (a) Each sample is a simple random sample; $n_1\hat{p}_1(1 - \hat{p}_1) = 26 \geq 10$ and $n_2\hat{p}_2(1 - \hat{p}_2) = 45 \geq 10$; and each sample is less
  than 5% of the population size.
  Hypotheses: H_0: $p_{\text{exp}} = p_{\text{control}}$, H_1: $p_{\text{exp}} < p_{\text{control}}$
  Classical approach: $z_0 = -2.71 < -z_{0.01} = -2.33$; reject the
  null hypothesis
  P-value approach: P-value $= 0.0037 < \alpha = 0.01$; reject the
  null hypothesis
  Yes. There is sufficient evidence at the $\alpha = 0.01$ level of significance to conclude that a lower proportion of women in the
  experimental group experienced a bone fracture than in the control group.
(b) Lower bound: -0.057, upper bound: -0.009. We are 95% confident that the difference in the proportion of women who experienced a bone fracture between the experimental and control
  group is between -0.06 and -0.01.
(c) This is a completely randomized design. The treatment is drug.
  It has two levels: 5 mg of Actonel versus a placebo.

18. (a) Each sample is a simple random sample; $n_1\hat{p}_1(1-\hat{p}_1) = 66 \geq 10$ and $n_2\hat{p}_2(1-\hat{p}_2) = 39 \geq 10$; and each sample is less than 5% of the population size.
Hypotheses: $H_0: p_{exp} = p_{control}$, $H_1: p_{exp} > p_{control}$
Classical approach: $z_0 = 2.20 > z_{0.05} = 1.645$; reject the null hypothesis
P-value approach: P-value $= 0.0139 < \alpha = 0.05$; reject the null hypothesis
Yes. There is sufficient evidence at the $\alpha = 0.05$ level of significance to conclude that a higher proportion of subjects in the experimental group experienced a dry mouth than in the control group.
(b) Lower bound: 0.01, upper bound: 0.08. We are 90% confident that the difference in the proportion of subjects who experienced a dry mouth between the experimental and control group is between 0.01 and 0.08.

19. (a) $n = n_1 = n_2 = 2136$ **(b)** $n = n_1 = n_2 = 3383$
20. (a) $n = n_1 = n_2 = 762$ **(b)** $n = n_1 = n_2 = 1201$
21. Lower bound: -1.37, upper bound 2.17. Since the interval includes zero, we conclude that there is not sufficient evidence at the $\alpha = 0.05$ level of significance to conclude that arm span and height are not equal.
22. Lower bound: 0.055, upper bound 0.467
23. Lower bound: -123.48, upper bound -40.20. Answers will vary.
24. Answers will vary.

CHAPTER 12

12.1 Assess Your Understanding (page 557)

1. Answers will vary.
2. The χ^2 goodness-of-fit tests are always right tailed because the numerator in the test statistic is squared, making every test statistic other than a perfect fit positive. So we are measuring if $\chi_0^2 > \chi_\alpha^2$.
3. To perform a goodness-of-fit test, all expected frequencies must be greater than or equal to 1, and at least 80% of the expected frequencies must be greater than or equal to 5.
4. If the expected count of a category is less than 1, two (or more) of the categories can be combined so that a goodness-of-fit test can be used. Alternatively, the sample size can be increased.

5.

p_i	0.2	0.1	0.45	0.25
Expected counts	100	50	225	125

6.

p_i	0.15	0.3	0.35	0.20
Expected counts	105	210	245	140

7. (a) $\chi_0^2 = 2.72$ **(b)** df $= 3$ **(c)** $\chi_{0.05}^2 = 7.815$
(d) Do not reject H_0, since $\chi^2 < \chi_{0.05}^2$.
8. (a) $\chi_0^2 = 2.2$ **(b)** df $= 4$ **(c)** $\chi_{0.05}^2 = 9.488$
(d) Do not reject H_0, since $\chi^2 < \chi_{0.05}^2$.
9. (a) $\chi_0^2 = 12.56$ **(b)** df $= 4$ **(c)** $\chi_{0.05}^2 = 9.488$
(d) Reject H_0, since $\chi_0^2 > \chi_{0.05}^2$. There is sufficient evidence at the $\alpha = 0.05$ level of significance to reject that the random variable X is binomial with $n = 4$, $p = 0.8$.
10. (a) $\chi_0^2 = 11.987$ **(b)** df $= 4$ **(c)** $\chi_{0.05}^2 = 9.488$
(d) Reject H_0, since $\chi_0^2 > \chi_{0.05}^2$. There is sufficient evidence at the $\alpha = 0.05$ level of significance to reject that the random variable X is binomial with $n = 4$, $p = 0.3$.
11. Classical approach: $\chi_0^2 = 18.738 > \chi_{0.05}^2 = 11.071$; reject the null hypothesis.
P-value approach: P-value $< 0.005 < \alpha = 0.05$; reject the null hypothesis.
There is sufficient evidence at the $\alpha = 0.05$ level of significance to reject the null hypothesis that there are 13% brown, 14% yellow, 13% red, 20% orange, 24% blue, and 16% green candies in a bag of M&Ms.
12. Classical approach: $\chi_0^2 = 3.569 < \chi_{0.05}^2 = 11.071$; do not reject the null hypothesis.

P-value approach: P-value $> 0.10 > \alpha = 0.05$; do not reject the null hypothesis.
There is not sufficient evidence at the $\alpha = 0.05$ level of significance to reject the null hypothesis that there are 12% brown, 15% yellow, 12% red, 23% orange, 23% blue, and 15% green candies in a bag of peanut M&Ms.

13. (a) Answers will vary depending on the desired probability of making a Type I error.
(b) Classical approach: $\chi_0^2 = 21.693$; compare χ_0^2 to your chosen χ_α^2.
P-value approach: P-value is between 0.10 and 0.005.
(c) Answers will vary.

14. Classical approach: $\chi_0^2 = 5.092 < \chi_{0.05}^2 = 15.507$; do not reject the null hypothesis.
P-value approach: P-value $> 0.10 > \alpha = 0.05$; do not reject the null hypothesis.
There is not sufficient evidence at the $\alpha = 0.05$ level of significance to reject the null hypothesis that the distribution of the leading digits of the surface area of rivers follows Benford's law.

15. Classical approach: $\chi_0^2 = 1.268 < \chi_{0.05}^2 = 19.675$; do not reject the null hypothesis.
P-value approach: P-value $> 0.95 > \alpha = 0.05$; do not reject the null hypothesis.
There is not sufficient evidence at the $\alpha = 0.05$ level of significance to reject the null hypothesis that the distribution of birth month is uniform.

16. Classical approach: $\chi_0^2 = 19.03 > \chi_{0.05}^2 = 12.592$; reject the null hypothesis.
P-value approach: P-value $< 0.005 < \alpha = 0.05$; reject the null hypothesis.
There is sufficient evidence at the $\alpha = 0.05$ level of significance to reject the null hypothesis that bicycle fatalities are uniformly distributed over the days of the week.

17. Classical approach: $\chi_0^2 = 13.227 > \chi_{0.05}^2 = 12.592$; reject the null hypothesis.
P-value approach: P-value $< 0.05 < \alpha = 0.05$; reject the null hypothesis.
There is sufficient evidence at the $\alpha = 0.05$ level of significance to reject the null hypothesis that pedestrian deaths are uniformly distributed over the days of the week.

18. (a) Classical approach: $\chi_0^2 = 4.67 < \chi_{0.01}^2 = 15.086$; do not reject the null hypothesis.
P-value approach: P-value $> 0.10 > \alpha = 0.01$; do not reject the null hypothesis.
There is not sufficient evidence at the $\alpha = 0.01$ level of significance to conclude that the die is loaded.
(b) Answers will vary.

19. (a)

Grade	Relative Frequency	Observed Frequency	Expected Frequency
K-3	0.329	15	8.225
4-8	0.393	7	9.825
9-12	0.278	3	6.95

(b) Classical approach: $\chi_0^2 = 8.638 > \chi_{0.05}^2 = 5.991$; reject the null hypothesis.
P-value approach: P-value $< 0.025 < \alpha = 0.05$; reject the null hypothesis.
There is sufficient evidence at the $\alpha = 0.05$ level of significance to conclude that the grade distribution of home-schooled children in her district is different from the national proportions.

20. Classical approach: $\chi_0^2 = 1.119 < \chi_{0.05}^2 = 15.507$; do not reject the null hypothesis.
P-value approach: P-value $> 0.995 > \alpha = 0.05$; do not reject the null hypothesis.
There is not sufficient evidence at the $\alpha = 0.05$ level of significance to reject the null hypothesis that the first digit in Fibonacci numbers follows Benford's law.

21. (a) Answers will vary.
(b) Expect 20% of the numbers to be 1s, 20% to be 2s, 20% to be 3s, 20% to be 4s, and 20% to be 5s.
(c) Answers will vary.

22. (a) Answers will vary.
  (b) Expect 0.3277 of the numbers to be 0s, 0.4096 of the numbers to be 1s, 0.2048 to be 2s, 0.0512 to be 3s, 0.0064 to be 4s, and 0.0003 to be 5s.
  (c) Answers will vary.
23. (a) Expected number low birth weight: 8.52; expected number without low birth weight: 111.48.
  (b) H_0: $p = 0.071$ vs H_1: $p > 0.071$
    $\chi_0^2 = 1.225$; $\chi_{0.05}^2 = 3.841$; do not reject H_0, because $\chi_0^2 < \chi_{0.05}^2$.
  (c) Do not reject, because $z_0 < z_{0.05}$ (1.13 < 1.645).
24. (a) We would expect 103.2 of the 400 individuals sampled to live alone and 296.8 of the individuals not to be living alone.
  (b) Hypotheses: H_0: $p = 0.258$; H_1: $p > 0.258$
    $\chi_0^2 = 48.275 > \chi_{0.05}^2 = 3.841$; reject the null hypothesis.
    There is sufficient evidence at the $\alpha = 0.05$ level of significance to conclude that the proportion of Americans living by themselves is more than 0.258.
  (c) $z_0 = 6.948 > z_{0.05} = 1.645$ (P-value < 0.0001); reject the null hypothesis.
    There is sufficient evidence at the $\alpha = 0.05$ level of significance to conclude that the proportion of Americans living by themselves is more than 0.258.
25. $\sum (O - E) = 0$

12.2 Assess Your Understanding (page 572)

1. Answers may vary, but differences should include the following: chi-square test for independence compares two characteristics from a single population, whereas the chi-square test for homogeneity compares a single characteristic from two (or more) populations. Similarities should include the following: the procedures of the two tests and the assumptions of the two tests are the same.

2. Answers may vary.

3. (a) $\chi_0^2 = 1.701$
  (b) $\chi_{0.05}^2 = 5.991$. Since $\chi_0^2 < \chi_{0.05}^2$, do not reject H_0. There is evidence at the $\alpha = 0.05$ level of significance to support the belief that X and Y are independent. It appears that X and Y are not related.
  (c) P-value = 0.427

4. (a) $\chi_0^2 = 13.053$
  (b) $\chi_{0.05}^2 = 5.991$. Since $\chi_0^2 > \chi_{0.05}^2$, reject H_0. There is sufficient evidence at the $\alpha = 0.05$ level of significance to support the belief that X and Y are not independent. There is a relation between X and Y.
  (c) P-value = 0.001

5. $\chi_0^2 = 1.989$; $\chi_{0.01}^2 = 9.210$. Since $\chi_0^2 < \chi_{0.01}^2$, do not reject H_0. There is evidence at the $\alpha = 0.05$ level of significance to support the claim that the proportions are equal. We don't have enough evidence to conclude that at least one of the proportions is different from the others. P-value = 0.370

6. $\chi_0^2 = 3.533$; $\chi_{0.01}^2 = 9.210$. Since $\chi_0^2 < \chi_{0.01}^2$, do not reject H_0. There is evidence at the $\alpha = 0.01$ level of significance to support the claim that the proportions are equal. We don't have enough evidence to conclude that at least one of the proportions is different from the others. P-value = 0.171

7. (a)

	Both Biological/ Adoptive Parents	Single Parent	Parent and Stepparent	Nonparental Guardian
Had sexual intercourse	78.553	52.368	41.895	26.184
Did not have sexual intercourse	71.447	47.632	38.105	23.816

(b) (1) All expected frequencies are greater than or equal to 1, and (2) no more than 20% of the expected frequencies are less than 5.
(c) $\chi_0^2 = 10.357$
(d) $\chi_{0.05}^2 = 7.815$. Since $\chi_0^2 > \chi_{0.05}^2$, reject H_0. There is sufficient evidence at the $\alpha = 0.05$ level of significance to conclude that sexual activity and family structure are associated.
(e) The biggest difference between observed and expected occurs under the family structure in which both parents are present. Fewer females were sexually active than was expected when both parents were present. This means that having both parents present seems to have an impact on whether the child is sexually active.
(f)

	Both Biological/ Adoptive Parents	Single Parent	Parent and Stepparent	Nonparental Guardian
Had sexual intercourse	0.427	0.59	0.55	0.64
Did not have sexual intercourse	0.573	0.41	0.45	0.36

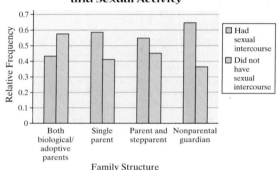

Family Structure and Sexual Activity

(g) P-value = 0.016

8. (a)

	Less Than 3 Months	3–5 Months	More Than 5 Months (or Never)
Intended	573.54	36.80	41.66
Unintended	73.01	4.68	5.3
Mistimed	179.45	11.51	13.04

(b) (1) All expected frequencies are greater than or equal to 1, and (2) no more than 20% of the expected frequencies are less than 5.
(c) $\chi_0^2 = 21.357$.
(d) $\chi_{0.05}^2 = 9.488$, P-value = 0.0005. Since $\chi_0^2 > \chi_{0.05}^2$, reject H_0. There is sufficient evidence at the $\alpha = 0.05$ level of significance to conclude that wantedness of pregnancy and months pregnant before prenatal care are associated.
(e) The biggest difference between observed and expected occurs between intended and less than 3 months. The observed is greater than the expected, which implies that intended pregnancies are likely to have early prenatal care.
(f)

	Less Than 3 Months	3–5 Months	More Than 5 Months (or Never)
Intended	0.910	0.040	0.051
Unintended	0.771	0.096	0.133
Mistimed	0.828	0.093	0.078

Prenatal Care by Wantedness of the Pregnancy

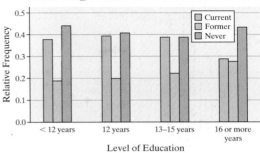

Smoking Status vs. Level of Education

9. (a) $\chi_0^2 = 32.926$; $\chi_{0.05}^2 = 16.919$; P-value $= 0.0001$. Since $\chi_0^2 > \chi_{0.05}^2$, reject H_0. There is sufficient evidence at the $\alpha = 0.05$ level of significance to conclude that level of education and area of country are associated.

(b) The cell corresponding to Midwest and not a high-school graduate contributed most to the test statistic. The expected was less than the observed.

(c)

| Area of Country | Level of Education | | | |
	Not High School Graduate	High School Graduate	Some College	Bachelor's or Higher
Northeast	0.146	0.212	0.162	0.215
Midwest	0.346	0.252	0.237	0.219
South	0.334	0.352	0.343	0.329
West	0.174	0.183	0.258	0.237
Total	1	1	1	1

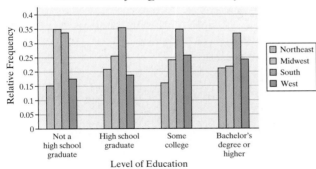

Education by Region of Country

10. (a) $\chi_0^2 = 7.803$; $\chi_{0.05}^2 = 12.592$. Since $\chi_0^2 < \chi_{0.05}^2$, do not reject H_0. There is not sufficient evidence at the $\alpha = 0.05$ level of significance to support the belief that smoking status and number of years of education are associated. It does not appear to be the case that number of years of education plays a role in determining smoking status.

(b) P-value $= 0.253$

(c)

Years of Education	Current	Former	Never	Total
<12	0.376	0.186	0.439	1
12	0.393	0.198	0.410	1
13–15	0.389	0.221	0.389	1
≥16	0.288	0.280	0.432	1

11. (a) $\chi_0^2 = 6.681$; $\chi_{0.05}^2 = 5.991$. Since $\chi_0^2 > \chi_{0.05}^2$, reject H_0. There is sufficient evidence at the $\alpha = 0.05$ level of significance to support the belief that age and opinion are associated. It appears to be the case that age plays a role in determining one's opinion regarding the legalization of marijuana. The 50 or older group is less likely to be in favor of the legalization of marijuana.

(b) P-value $= 0.035$

(c)

Opinion	18–29 Years Old	30–49 Years Old	50 Years or Older
For	0.768	0.752	0.684
Against	0.232	0.248	0.316
Total	1	1	1

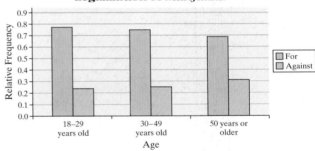

Legalization of Marijuana

12. (a) $\chi_0^2 = 0.036$; $\chi_{0.1}^2 = 2.706$. Since $\chi_0^2 < \chi_{0.1}^2$, do not reject H_0. There is not sufficient evidence at the $\alpha = 0.10$ level of significance to support the belief that gender and opinion are associated. It appears to be the case that gender is not associated with one's opinion regarding choice.

(b) P-value $= 0.849$

(c)

Gender	Pro-Life	Pro-Choice	Total
Men	0.496	0.504	1
Women	0.490	0.510	1

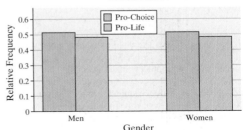

Abortion

13. (a) $\chi_0^2 = 0.946$; $\chi_{0.05}^2 = 7.815$. Since $\chi_0^2 < \chi_{0.05}^2$, do not reject H_0. There is not sufficient evidence at the $\alpha = 0.05$ level of significance to support the belief that gender and delinquencies are associated.
(b) P-value $= 0.814$
(c)

Gender	Delinquency			
	Person	Property	Drugs	Public Order
Female	0.198	0.188	0.152	0.214
Male	0.802	0.812	0.848	0.786
Total	1	1	1	1

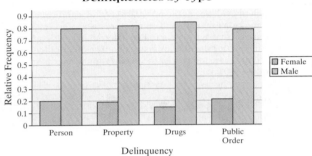

Delinquencies by Type

14. (a) $\chi_0^2 = 12.291$; $\chi_{0.1}^2 = 9.236$. Since $\chi_0^2 > \chi_{0.1}^2$, reject H_0. There is sufficient evidence at the $\alpha = 0.1$ level of significance to support the belief that gender and age are associated when it comes to visits to the emergency room.
(b) P-value $= 0.031$
(c)

Gender	Age					
	<15	15–24	25–44	45–64	65–74	≥75
Male	0.600	0.589	0.582	0.521	0.400	0.314
Female	0.400	0.411	0.418	0.479	0.600	0.686
Total	1	1	1	1	1	1

Emergency Room Visits

15. (a) H_0: $p_{18-29} = p_{30-49} = p_{50-64} = p_{65 \text{ and older}}$ versus
H_1: at least one of the proportions is not equal to the rest.
$\chi_0^2 = 6.000$; $\chi_{0.05}^2 = 7.815$. Since $\chi_0^2 < \chi_{0.05}^2$, do not reject H_0. There is not sufficient evidence at the $\alpha = 0.05$ level of significance to conclude that at least one proportion is different from the others. The evidence suggests that the proportion of individuals in each age group who smoked at least one cigarette in the past week is the same.
(b) P-value $= 0.112$

(c)

Smoking Status	Age			
	18–29	30–49	50–64	≥65
Smoked	0.3	0.2625	0.2875	0.15
Did not smoke	0.7	0.7375	0.7125	0.85
Total	1	1	1	1

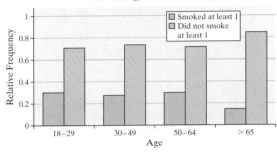

Smoking Status by Age

16. (a) H_0: $p_{\text{East}} = p_{\text{South}} = p_{\text{Midwest}} = p_{\text{West}}$ versus
H_1: at least one of the proportions is not equal to the rest.
$\chi_0^2 = 3.541$; $\chi_{0.1}^2 = 6.251$ Since $\chi_0^2 < \chi_{0.1}^2$ do not reject H_0. There is not sufficient evidence at the $\alpha = 0.1$ level of significance to conclude that at least one proportion is different from the others. The evidence suggests that the proportion of individuals in each region that is satisfied with the way things are going in the United States is the same.
(b) P-value $= 0.316$
(c)

Satisfaction	Region			
	East	South	Midwest	West
Satisfied	0.513	0.56	0.62	0.553
Dissatisfied	0.487	0.44	0.38	0.447
Total	1	1	1	

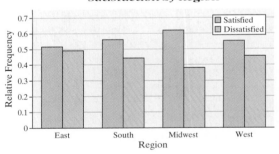

Satisfaction by Region

17. (a) H_0: $p_{\text{placebo}} = p_{50} = p_{100} = p_{200} = p_{\text{Naproxen}}$ versus
H_1: at least one of the proportions is not equal to the rest.
$\chi_0^2 = 49.703$; $\chi_{0.01}^2 = 13.277$. Since $\chi_0^2 > \chi_{0.01}^2$, reject H_0. There is sufficient evidence at the $\alpha = 0.01$ level of significance to support the belief that at least one proportion is different from the others. The evidence suggests that the subjects taking Naproxen experienced a higher incidence rate of ulcers than the other treatment groups.
(b) P-value < 0.001
(c)

Side Effect	Treatment				
	Placebo	Celebrex (50 mg)	Celebrex (100 mg)	Celebrex (200 mg)	Naproxen (500 mg)
Ulcer	0.023	0.034	0.031	0.059	0.162
No ulcer	0.977	0.966	0.969	0.941	0.838
Total	1	1	1	1	1

Side Effects

Why Did You Drop?

18. (a) H_0: $p_{Celebrex} = p_{placebo} = p_{Naproxen} = p_{Diclofenac} = p_{ibuprofen}$ versus H_1: at least one of the proportions is not equal to the rest. $\chi_0^2 = 4.673$; $\chi_{0.01}^2 = 13.277$. Since $\chi_0^2 < \chi_{0.01}^2$, do not reject H_0. There is not sufficient evidence at the $\alpha = 0.01$ level of significance to conclude that at least one proportion is different from the others. The evidence suggests that the proportions of individuals in each treatment group who experienced dizziness as a side effect are the same.
(b) P-value $= 0.323$
(c)

Side Effect	Drug				
	Celebrex	Placebo	Naproxen	Diclofenac	Ibuprofen
Dizziness	0.020	0.017	0.026	0.013	0.023
No Dizziness	0.980	0.983	0.974	0.987	0.977
Total	1	1	1	1	1

20. (a)

	Democrat	Republican
Grade school	7	2
High school	10	6
College	9	12

(b) $\chi_0^2 = 3.483$; $\chi_{0.1}^2 = 4.605$. Since $\chi_0^2 < \chi_{0.1}^2$, do not reject H_0. The evidence suggests that there is no association between level of education and political affiliation.
(c) P-value $= 0.175$

(d)

	Democrat	Republican	Total
Grade school	0.778	0.222	1
High school	0.625	0.375	1
College	0.429	0.571	1

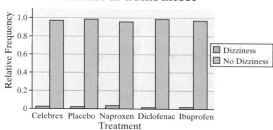

Dizziness as a Side Effect

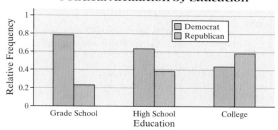

Political Affiliation by Education

19. (a)

	Course	Personal	Work
Female	13	5	3
Male	10	6	13

(b) $\chi_0^2 = 5.595$; $\chi_{0.1}^2 = 4.605$. Since $\chi_0^2 > \chi_{0.1}^2$, reject H_0. The evidence suggests that there is some relation between gender and drop reason. Females are more likely to drop because of the course, while males are more likely to drop because of work.
(c) P-value $= 0.061$

(d)

	Reason		
	Course	Personal	Work
Female	0.565	0.455	0.188
Male	0.435	0.545	0.813
Total	1	1	1

21. (a) The expected number of adult Americans who have smoked is 252. The expected number of adult Americans who have not smoked is 755.
(b) $\chi_0^2 = 9.527$
(c) H_0: $p_{2004} = p_{1990}$ versus H_1: $p_{2004} \neq p_{1990}$
Since $\chi_0^2 = 9.527 > \chi_\alpha^2 = 3.841$, reject the null hypothesis. There is evidence to conclude that in 2004 the proportion of adult Americans who smoke at least one cigarette a week is different from the proportion who smoked in 1990.
(d) $Z = 3.087$, $Z^2 = 9.5296$

12.3 Assess Your Understanding (page 589)

1. There are two requirements for conducting inference on the least squares regression model. First, for each particular value of the explanatory variable x, the corresponding responses in the population have a mean that depends linearly on x. Second, the response variables are normally distributed with mean $\mu_{y|x} = \beta_0 + \beta_1 x$ and standard deviation σ.

The first requirement is tested by examining a plot of the residuals against the explanatory variable. If the plot shows any discernible pattern, then the linear model is inappropriate. The normality requirement is checked by examining a normal probability plot of the residuals.

2. It is important to perform both graphical and analytical analysis when analyzing relations between two quantitative variables because the graph describes the type of relation that exists and the analytical description allows for statistical inference to be conducted.

3. The y-coordinates on the least-squares regression line represent the mean value of the response variable for any given value of the explanatory variable.

4. It is desirable to have the explanatory variables be widely dispersed so that the standard deviation of b_1, is smaller.

5. If the null hypothesis, H_0: $\beta_1 = 0$, is not rejected, the best guess for the value of the response variable for any value of the explanatory variable is the sample mean of y.

6. We do not conduct inference on the linear correlation coefficient because the hypothesis test on the slope and a hypothesis test on the linear correlation coefficient yield the same conclusion. Moreover, the requirements for conducting inference on the linear correlation coefficient are very hard to verify.

7. (a) $\beta_0 \approx b_0 = -2.3256$, $\beta_1 \approx b_1 = 2.0233$
   (b) $s_e = 0.5134$ is the point estimate for σ.
   (c) $s_{b_1} = 0.1238$
   (d) Since the P-value $< 0.001 < \alpha = 0.05$ (or $t_0 = 16.345 > t_{0.025} = 3.182$), we reject the null hypothesis and conclude that a linear relation exists between x and y.

8. (a) $\beta_0 \approx b_0 = -3.700$, $\beta_1 \approx b_1 = 1.100$
   (b) $s_e = 0.7303$ is the point estimate for σ.
   (c) $s_{b_1} = 0.1155$
   (d) Because the P-value $= 0.002 < \alpha = 0.05$ (or $t_0 = 9.526 > t_{0.025} = 3.182$), we reject the null hypothesis and conclude that a linear relation exists between x and y.

9. (a) $\beta_0 \approx b_0 = 1.200$, $\beta_1 \approx b_1 = 2.200$
   (b) $s_e = 0.8944$ is the point estimate for σ.
   (c) $s_{b_1} = 0.2828$
   (d) Because the P-value $= 0.004 < \alpha = 0.05$ (or $t_0 = 7.778 > t_{0.025} = 3.182$), we reject the null hypothesis and conclude that a linear relation exists between x and y.

10. (a) $\beta_0 \approx b_0 = 3.600$, $\beta_1 \approx b_1 = -1.800$
    (b) $s_e = 0.5164$ is the point estimate for σ.
    (c) $s_{b_1} = 0.1633$
    (d) Because the P-value $= 0.002 < \alpha = 0.05$ (or $t_0 = -11.023 < -t_{0.025} = -3.182$), we reject the null hypothesis and conclude that a linear relation exists between x and y.

11. (a) $\beta_0 \approx b_0 = 116.600$, $\beta_1 \approx b_1 = -0.7200$
    (b) $s_e = 3.2863$ is the point estimate for σ.
    (c) $s_{b_1} = 0.1039$
    (d) Because the P-value $= 0.006 < \alpha = 0.05$ (or $t_0 = -6.928 < -t_{0.025} = -3.182$), we reject the null hypothesis and conclude that a linear relation exists between x and y.

12. (a) $\beta_0 \approx b_0 = -3.30$, $\beta_1 \approx b_1 = 0.78$
    (b) $s_e = 1.7416$ is the point estimate for σ.
    (c) $s_{b_1} = 0.1102$
    (d) Because the P-value $= 0.006 < \alpha = 0.05$ (or $t_0 = 7.081 > t_{0.025} = 3.182$), we reject the null hypothesis and conclude that a linear relation exists between x and y.

13. (a) $\beta_0 \approx b_0 = 12.4932$, $\beta_1 \approx b_1 = 0.1827$
    (b) $s_e = 0.0954$
    (c) A normal probability plot of the residuals shows that they are approximately normally distributed.
    (d) $s_{b_1} = 0.02756$
    (e) Because the P-value $< 0.001 < \alpha = 0.01$ (or $t_0 = 6.63 > t_{0.005} = 3.250$), we reject the null hypothesis and conclude that a linear relation exists between a child's height and head circumference.
    (f) 95% confidence interval: Lower bound: 0.1203; Upper bound: 0.2450

(g) A good estimate of the child's head circumference would be 17.34 inches.

14. (a) $\beta_0 \approx b_0 = 1.1140$, $\beta_1 \approx b_1 = 1.3902$
    (b) $s_e = 0.4944$
    (c) A normal probability plot of the residuals shows that they are approximately normally distributed.
    (d) $s_{b_1} = 0.1501$
    (e) Because the P-value $< 0.001 < \alpha = 0.01$ (or $t_0 = 9.260 > t_{0.005} = 3.250$), we reject the null hypothesis and conclude that a linear relation exists between the right humerus and the right tibia of rats sent to space.
    (f) 99% confidence interval: Lower bound: 0.9024; Upper bound: 1.8780
    (g) The mean length of the right tibia on a rat whose right humerus is 25.93 mm is 37.162 mm.

15. (a) $\beta_0 \approx b_0 = 2675.6$, $\beta_1 \approx b_1 = 0.6764$
    (b) $s_e = 271.04$
    (c) A normal probability plot of the residuals shows that they are approximately normally distributed.
    (d) $s_{b_1} = 0.2055$
    (e) Because the P-value $= 0.011 < \alpha = 0.05$ (or $t_0 = 3.291 > t_{0.025} = 2.306$), we reject the null hypothesis and conclude that a linear relation exists between the 7-day strength and the 28-day strength of this type of concrete.
    (f) 95% confidence interval: Lower bound: 0.2025; Upper bound: 1.1503
    (g) The mean 28-day strength of this concrete if the 7-day strength is 3000 psi is 4704.8 psi.

16. (a) $\beta_0 \approx b_0 = 0.2088$, $\beta_1 \approx b_1 = 0.0575$
    (b) $s_e = 0.1121$
    (c) A normal probability plot of the residuals shows that they are approximately normally distributed.
    (d) $s_{b_1} = 0.0046$
    (e) Because the P-value $< 0.001 < \alpha = 0.10$ (or $t_0 = 12.502 > t_{0.05} = 1.796$), we reject the null hypothesis and conclude that a linear relation exists between the tar and nicotine levels in cigarettes.
    (f) 90% confidence interval: Lower bound: 0.0492; Upper bound: 0.0658
    (g) The mean amount of nicotine in a cigarette that has 12 mg of tar is 0.90 mg.

17. (a) $\beta_0 \approx b_0 = 0.5975$, $\beta_1 \approx b_1 = 0.9764$
    (b) $s_e = 2.3518$
    (c) A normal probability plot of the residuals shows that they are approximately normally distributed.
    (d) $s_{b_1} = 0.3379$
    (e) Because the P-value $= 0.018 < \alpha = 0.10$ (or $t_0 = 2.890 > t_{0.05} = 1.833$), we reject the null hypothesis and conclude that a linear relation exists between the rate of return of the S&P 500 index and the rate of return of UTX.
    (f) 90% confidence interval: Lower bound: 0.357; Upper bound: 1.598
    (g) When the mean rate of return of the S&P 500 index is 4.2%, the mean rate of return for United Technologies is 4.698%.

18. (a) $\beta_0 \approx b_0 = 607.7$, $\beta_1 \approx b_1 = 25.012$
    (b) $s_e = 64.8477$
    (c) A normal probability plot of the residuals shows that they are approximately normally distributed.
    (d) $s_{b_1} = 2.1898$
    (e) Because the P-value $< 0.001 < \alpha = 0.01$ (or $t_0 = 11.427 > t_{0.005} = 4.032$), we reject the null hypothesis and conclude that a linear relation exists between men's fat-free body mass and their energy expenditure during inactivity.
    (f) 99% confidence interval: Lower bound: 16.186; Upper bound: 33.838
    (g) A man with a fat-free body mass of 57.3 kg has a mean inactive energy expenditure level of 2040.9 kcal per day.

19. (a)

(b) $\hat{y} = 0.93 + 0.0821x$ **(c)** $s_e = 4.1512$

(d) A normal probability plot of the residuals shows that they are approximately normally distributed.

(e) $s_{b_1} = 0.0961$

(f) Because the P-value $= 0.41 > \alpha = 0.01$ (or $t_0 = 0.854 < t_{0.005} = 3.169$), we do not reject the null hypothesis and conclude that a linear relation does not exist between the number of calories per serving and the number of grams of sugar per serving in high-fiber cereal.

(g) 95% confidence interval: Lower bound: -0.1320; Upper bound: 0.2962

(h) Do not recommend using the least-squares regression line to predict the sugar content of the cereal because we did not reject the null hypothesis. A good guess to the sugar content would be $\bar{y} = 17.08$ grams.

20. (a)

(b) $\hat{y} = -45 + 3.3605x$; Because the P-value $= 0.0166 < \alpha = 0.05$ (or $t_0 = 3.130 > t_{0.025} = 2.365$), we reject the null hypothesis and conclude that a linear relation exists between the weight and height of baseball players.

(c) $\hat{y} = -63.4722 + 3.611x$; Because the P-value $= 0.1772 > \alpha = 0.05$ (or $t_0 = 1.529 < t_{0.025} = 2.447$), we do not reject the null hypothesis and conclude that a linear relation does not exist between baseball players' weight and height.

21. (a) $\hat{y} = 26.36$ when $x = 10$.

(b) $\mu_{y|10} = 26.36$

(c) The standard deviation of y when $x = 10$ is $s = 2.167$

(d) If the requirements for the least-squares regression model are satisfied, y will be normally distributed with $\mu_{y|10} = 26.36$ and $\sigma_{y|10} = 2.167$.

12.4 Assess Your Understanding (page 596)

1. Confidence intervals are used to measure the accuracy of the mean response of all the individuals in the population, whereas prediction intervals are used to measure the accuracy of a single individual's predicted value.

2. If the residuals are not normally distributed, confidence and prediction intervals can still be constructed if the sample is large ($n > 30$).

3. (a) $\hat{y} = 11.8$
(b) Lower bound: 10.8; Upper bound: 12.8
(c) $\hat{y} = 11.8$
(d) Lower bound: 9.9; Upper bound: 13.7
(e) The confidence interval is an interval estimate for the mean value of y at $x = 7$, whereas the prediction interval is an interval estimate for a single value of y at $x = 7$.

4. (a) $\hat{y} = 5.1$
(b) Lower bound: 4.0; Upper bound: 6.2
(c) $\hat{y} = 5.1$
(d) Lower bound: 2.5; Upper bound: 7.7
(e) The confidence interval is an interval estimate for the mean value of y at $x = 8$, whereas the prediction interval is an interval estimate for a single value of y at $x = 8$.

5. (a) $\hat{y} = 4.3$
(b) Lower bound: 2.5; Upper bound: 6.1
(c) $\hat{y} = 4.3$
(d) Lower bound: 0.9; Upper bound: 7.7

6. (a) $\hat{y} = 0.4$
(b) Lower bound: -0.5; Upper bound: 1.3
(c) $\hat{y} = 0.4$
(d) Lower bound: -1.1; Upper bound: 1.9

7. (a) $\hat{y} = 17.20$ inches
(b) 95% confidence interval: Lower bound: 17.12 inches; Upper bound: 17.28 inches
(c) $\hat{y} = 17.20$ inches
(d) 95% prediction interval: Lower bound: 16.97 inches; Upper bound: 17.43 inches
(e) The confidence interval is an interval estimate for the mean head circumference of all children who are 25.75 inches tall. The prediction interval is an interval estimate for the head circumference of a single child who is 25.75 inches tall.

8. (a) $\hat{y} = 37.023$ mm
(b) 95% confidence interval: Lower bound: 36.646 mm; Upper bound: 37.398 mm
(c) $\hat{y} = 37.023$ mm
(d) 95% prediction interval: Lower bound: 35.842 mm; Upper bound: 38.202 mm
(e) Although the predicted lengths of the right tibia in parts (a) and (c) are the same, the intervals are different because the distribution of the means, part (a), has less variability than the distribution of the individuals, part (c).

9. (a) $\hat{y} = 4400.4$ psi
(b) 95% confidence interval: Lower bound: 4147.8 psi; Upper bound: 4653.1 psi
(c) $\hat{y} = 4400.4$ psi
(d) 95% prediction interval: Lower bound: 3726.3 psi; Upper bound: 5074.6 psi
(e) The confidence interval is an interval estimate for the mean 28-day strength of all concrete cylinders that have a 7-day strength of 2550 psi. The prediction interval is an interval estimate for the 28-day strength of a single cylinder whose 7-day strength is 2550 psi.

10. (a) $\hat{y} = 0.90$ mg
(b) 95% confidence interval: Lower bound: 0.83 mg; Upper bound: 0.97 mg
(c) $\hat{y} = 0.90$ mg
(d) 95% prediction interval: Lower bound: 0.64 mg; Upper bound: 1.16 mg

(e) Although the predicted nicotine contents in parts (a) and (c) are the same, the intervals are different because the distribution of the mean nicotine level, part (a), has less variability than the distribution of the individual nicotine levels of a particular cigarette, part (c).

11. (a) $\hat{y} = 4.698\%$
   (b) 90% confidence interval: Lower bound: 2.234%; Upper bound: 7.163%
   (c) $\hat{y} = 4.698\%$
   (d) 90% prediction interval: Lower bound: -0.267%; Upper bound: 9.6664%
   (e) Although the predicted rate of return in parts (a) and (c) are the same, the intervals are different because the distribution of the mean rate of return, part (a), has less variability than the distribution of the individual rate of return of a particular month, part (c).

12. (a) $\hat{y} = 2040.9$ kcal
   (b) 99% confidence interval: Lower bound: 1932.3 kcal; Upper bound: 2149.5 kcal
   (c) $\hat{y} = 2040.9$ kcal
   (d) 99% prediction interval: Lower bound: 1757.7 kcal; Upper bound: 2324.0 kcal
   (e) Although the predicted energy expenditures in parts (a) and (c) are the same, the intervals are different because the distribution of the mean energy expenditure, part (a), has less variability than the distribution of the energy expenditure of a particular individual, part (c).

13. (a) It does not make sense to construct either a confidence interval or prediction interval based on the least squares regression equation because the evidence indicated that there was no linear relation between calories and sugar.
   (b) 95% confidence interval for the mean sugar content: Lower bound: 14.5 grams; Upper bound: 19.7 grams.

Chapter Review Exercises (page 599)

1. $\chi_0^2 = 0.578$; $\chi_{0.05}^2 = 5.991$; since $\chi_0^2 < \chi_{0.05}^2$, do not reject H_0. There is not sufficient evidence at the $\alpha = 0.05$ level of significance to conclude that the wheel is out of balance.

2. $\chi_0^2 = 4.94$; $\chi_{0.01}^2 = 23.209$; since $\chi_0^2 < \chi_{0.01}^2$, do not reject the null hypothesis. There is not sufficient evidence at the $\alpha = 0.01$ level of significance to conclude that the dice are not fair.

3. $\chi_0^2 = 9.709$; $\chi_{0.1}^2 = 9.236$; since $\chi_0^2 > \chi_{0.1}^2$, reject H_0. There is sufficient evidence at the $\alpha = 0.1$ level of significance to conclude that the distribution of educational attainment is different today than it was in 1994.

4. $\chi_0^2 = 29.223$; $\chi_{0.05}^2 = 5.991$; since $\chi_0^2 > \chi_{0.05}^2$, reject H_0. There is sufficient evidence at the $\alpha = 0.1$ level of significance to conclude that the distribution of school crime is different today than it was in 1992.

5. (a)

	Belief			
	Developed with God	Developed without God	Created in Present Form	No Opinion
February 2001	365.82	116.94	467.27	65.97
June 1993	366.18	117.06	467.73	66.03

   (b) All expected values are greater than 5.
   (c) $\chi_0^2 = 2.203$
   (d) $\chi_{0.05}^2 = 7.815$; do not reject H_0. The evidence indicates that it is reasonable to conclude that belief is independent of date.
   (e) No opinion in 2001 contributed the most.
   (f)

	Belief				
Date	Developed with God	Developed without God	Created in Present Form	No Opinion	Total
Feb 2001	0.370	0.120	0.450	0.060	1
June 1993	0.350	0.110	0.470	0.070	1

Evolution or Creation by Date

(g) P-value $= 0.531$

6. (a)

	Delivery Method	
	Vaginal	Caesarean
White	241.49	63.51
Black	47.51	12.49

   (b) All expected counts are greater than 5.
   (c) $\chi_0^2 = 0.031$
   (d) $\chi_{0.05}^2 = 3.841$; do not reject H_0. The evidence indicates that it is reasonable to conclude that race is independent of delivery method.
   (e) Black women who had caesarian deliveries contributed the most to the test statistic, but not enough to affect significance.
   (f)

	Delivery Method	
Race	Vaginal	Caesarian
White	0.837	0.829
Black	0.163	0.171
Total	1	1

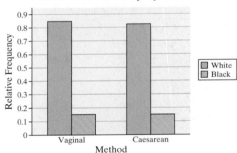

Method of Delivery by Race

(g) P-value $= 0.860$

7. $\chi_0^2 = 242.829$; $\chi_{0.01}^2 = 30.578$; reject H_0. The evidence indicates that it is reasonable to conclude that race is not independent of region.

8. $\chi_0^2 = 4.875$; $\chi_{0.05}^2 = 7.815$; do not reject H_0. The evidence indicates that it is reasonable to conclude that marital status is independent of gender.

9. $\chi_0^2 = 3.877$; $\chi_{0.05}^2 = 7.815$; do not reject H_0. The evidence indicates that it is reasonable to conclude that the proportion of Americans who had a common cold is the same for each region of the country.

10. $\chi_0^2 = 3.093$; $\chi_{0.05}^2 = 5.991$; do not reject H_0. The evidence indicates that it is reasonable to conclude that the proportion of Americans in the different age groups who believe Americans are willing to work hard is the same.

11. $\chi_0^2 = 126.168$; $\chi_{0.05}^2 = 7.815$; reject H_0. There is sufficient evidence to indicate that the proportion of Americans who are morning persons or night persons differs in at least one of the four age groups.

12. (a) Dustan Mohr had a better batting average both with runners in scoring position (0.196) and with no runners in scoring position (0.271).

(b) Darin Erstad had the better overall batting average (0.252).

13. (a) $\beta_0 \approx b_0 = 30.3848$; $\beta_1 \approx b_1 = -2.7977$;
$\hat{y} = 30.3848 - (2.7977)(3.8) = 19.75$, so the mean number of miles per gallon of cars with 3.8-liter engines is 19.8 miles per gallon.
(b) $s_e = 1.2692$
(c)

The normal probability plot indicates that the residuals are normally distributed.
(d) $s_b = 0.3859$
(e) There is evidence that a linear relation exists between engine displacement and city MPG (P-value < 0.001)
(f) 95% confidence interval about the slope of the true least-squares regression line: Lower bound: -3.6312, Upper bound: -1.9642.
(g) 90% confidence interval about the mean miles per gallon of all cars with a 3.8-liter engine: Lower bound: 19.065, Upper bound: 20.535.   **(h)** $\hat{y} = 19.8$
(h) 90% prediction interval about the miles per gallon of a particular car with a 3.8-liter engine: Lower bound: 17.552, Upper bound: 22.048.
(i) Although the predicted miles per gallon in parts (a) and (h) are the same, the intervals are different because the distribution of the means, part (a), has less variability than the distribution of the individuals, part (h).

14. (a) $\beta_0 \approx b_0 = -0.309$; $\beta_1 \approx b_1 = 0.2119$; $-0.309 + (0.2119) \cdot (80.2) = 16.685$; so the mean number of chirps when the temperature is 80.2°F is 16.69.   **(b)** $s_e = 0.9715$
(c)

The residuals are normally distributed.
(d) $s_b = 0.0387$
(e) There is evidence that a linear relation exists between temperature and the cricket's chirps (P-value < 0.001).
(f) 95% confidence interval about the slope of the true least-squares regression line: Lower bound: 0.1283, Upper bound: 0.2955.
(g) 90% confidence interval about the mean number of chirps at 80.2°F: Lower bound: 16.24, Upper bound: 17.13.
(h) When the temperature is 80.2°F, $\hat{y} = 16.69$ chirps.

(i) 90% prediction interval for the number of chirps of a particular cricket at 80.2°F: Lower bound: 14.91, Upper bound: 18.46.
(j) Although the predicted number of chirps in parts (a) and (h) are the same, the intervals are different because the distribution of the means, part (a), has less variability than the distribution of the individuals, part (h).

15. (a) $\beta_0 \approx b_0 = -399.2$; $\beta_1 \approx b_1 = 2.5315$; $-399.2 + (2.5315) \cdot (900) = 1879.15$; so the mean rent of a 900-square-foot apartment in Queens is $1879.15.
(b) $s_e = 229.547$
(c)

The residuals are normally distributed.
(d) $s_b = 0.2166$
(e) There is evidence that a linear relation exists between square footage of an apartment in Queens, NY and the monthly rent.
(f) 95% confidence interval about the slope of the true least-squares regression line: Lower bound: 2.0416, Upper bound: 3.0214.
(g) 90% confidence interval about the mean rent of 900 square foot apartments: Lower bound: $1752.20, Upper bound: $2006.00.
(h) When an apartment has 900 square feet, $\hat{y} = \$1879.15$.
(i) 90% prediction interval for the rent of a particular 900 square foot apartment: Lower bound: $1439.6, Upper bound: $2318.60.
(j) Although the predicted rents in parts (a) and (h) are the same, the intervals are different because the distribution of the means, part (a), has less variability than the distribution of the individuals, part (h).

16. (a) $\beta_0 \approx b_0 = 29.705$; $\beta_1 \approx b_1 = 2.635$; $29.705 + (2.635) \cdot (7) = 48.15$; so the mean height of a 7-year-old boy is 48.15 inches.
(b) $s_e = 2.45$
(c)

The residuals are normally distributed.
(d) $s_b = 0.2067$
(e) There is evidence that a linear relation exists between boys' age and height (P-value < 0.001).
(f) 95% confidence interval about the slope of the true least-squares regression line: Lower bound: 2.2008, Upper bound: 3.0694.
(g) 90% confidence interval about the mean height of 7-year-old boys: Lower bound: 47.11, Upper bound: 49.19 inches.

(h) The predicted height of a randomly chosen 7-year-old boy is 48.15 inches, $\hat{y} = 48.15$.

(i) 90% prediction interval for the height of a particular 7-year-old boy: Lower bound: 43.78, Upper bound: 52.52 inches.

(j) Although the predicted heights in parts (a) and (h) are the same, the intervals are different because the distribution of the means, part (a), has less variability than the distribution of the individuals, part (h).

17. (a) $\beta_0 \approx b_0 = 67.388$; $\beta_1 \approx b_1 = -0.263$

(b) $s_e = 11.1859$

(c)

The residuals are normally distributed.

(d) $s_b = 0.1680$

(e) There is not sufficient evidence at the $\alpha = 0.05$ level of significance to support the claim that a linear relation exists between a woman's age and grip strength (P-value $= 0.138$).

(f) Based on the answers to (d) and (e), a good guess of the grip strength of a 42-year-old female would be the mean strength of the population.

INDEX

Note: Pages beginning with "A-" locate entries in the Appendix; "C-" locate entries in the CD.

±, 405
Addition Rule for Disjoint Events, 238–49
 Benford's Law and, 239–41
 with contingency tables, 242–43
 General, 241–43
alpha. *See* Type I error
Alternative hypothesis, 454–57, 463
 definition of, 455
 structuring, 456
American Academy of Dermatology, 400
American Community Survey, 375
American Medical Association, 400
American Time Use Survey, 388
Analysis of variance (ANOVA), C-19–37
 definition of, C-19
 null hypothesis in, C-19–20
 one-way, C-19–37
  conceptual understanding of, C-24–25
  decision rule in, C-22
  equal population variances in, C-21
  normal distribution in, C-20–21
  requirements of, C-20–22
  robustness of, C-21
  testing hypotheses with, C-22–27
  using technology, C-23–24, C-33
 Type I error in, C-20
Anecdotal claims, 3
ANOVA table, C-27
Approach to problem solving, 6
Area
 under normal curve, 323–25, 345–47
  finding, 326
  interpreting, 325
  as probability, 324, 325, 331, 340–41
  as proportion, 324, 325, 331
  standard normal curve, 332–36
 as probability, 320–21
Arithmetic mean. *See* Mean(s)
Ars Conjectandi (Bernoulli), 224, 299
Associated variables, 179, 181
At-least probability, 253
Average, 106, 107. *See also* Mean(s)

Bar graph(s), 57–59, 61, 70, 569–70
 of conditional distribution, 569–70
 frequency and relative frequency, 57–58
 side-by-side, 58–59
 technology to draw, 58, 70
Before-after (pretest-posttest) experiments, 43
Behrens-Fisher problem, 521
Bell-shaped distribution, 80, 130

 Empirical Rule to describe, 131–32
Benford, Frank, 239, 560
Benford's Law, 239–41, 560
Bernoulli, Jacob, 224, 299
Bernoulli trial, 299
beta. *See* Type II error
Biased statistic, 127
Bias in nonsampling error, 33
Bimodal data set, 112
Binomial experiment
 criteria for, 298, 362
 identifying, 298–300
Binomial probability distribution, 298–314, 374
 binomial probability distribution function (pdf), 302–3
 binomial table to compute binomial probabilities, 303–5
 constructing, 300–302
 histograms of, 306–9, 318
 mean and standard deviation of
  binomial random variable, 305–6, 308–9
 normal approximation to, 362–67
 notation used in, 298
 using technology, 305, 313
Binomial probability distribution table, A-2–9
Binomial random variable, 298
 normal approximation to, 364–65
Binomial tables, 303–5
Bivariate data, 176. *See also* Relation between two variables
Blood pressure, 5
Box, George, 527
Boxplots, 161–64
 comparing two distributions using, 163–64
 constructing, 161–62
 distribution shape based on, 162–63
 technology to draw, 168

Callbacks, 34
Cardano, Fazio, 229
Cardano, Girolamo, 229
Categorical variable, 6–7
Causation, correlation vs., 184
Cavendish, Henry, 371
Cdf (cumulative distribution function), 305, 333
 inverse, 349
Cells, 243, 563
Census, 54
 definition of, 13
Census Bureau, 28
Center. *See* Mean(s)
Central Limit Theorem, 385, 387–88, 463
Central tendency, measures of, 107–23
 arithmetic mean. *See* Mean(s)

from grouped data, 142–43, 149
median, 78, 107, 110–11, 116, 159, 162
    computing, 110–11, 113, 123
    definition of, 110
    IQR and, 162
    shape of distribution identified using, 113–16
mode, 107, 111–13, 116
    bimodality, 112
    computing, 112
    definition of, 111
    multimodality, 112
    of qualitative data, 112–13
    of quantitative data, 112
trimmed mean, 122
Certainty, 225
Chart(s)
    Pareto, 58
    pie, 60–61, 70
Chebyshev, Pafnuty, 132, 133
Chebyshev's Inequality, 132–34
Chicago Tribune, 218
Chi-square (χ^2) distribution table, A-13
Chi-square distribution, 551
    characteristics of, C-8, C-14
    critical values for, C-7–9, C-13
    definition of, C-8
    hypothesis testing about population standard
        deviation and, C-14
Chi-square test, 180
    for homogeneity of proportions, 570–74
        definition of, 570
        steps in, 571–73
    for independence, 563–70
        definition of, 563
        expected counts in, 563–65
        expected frequencies in, 566
        steps in, 566–67
    using technology, 570, 580
Claim, 41
Class(es)
    data, 71, 73
    width of, 73
Classical approach to hypothesis testing, 463–64
    in chi-square test
        for homogeneity of proportions, 572
        for independence, 567, 568
    of difference between two population proportions,
        537, 538
    of difference of two means using independent samples,
        522, 524
    in goodness-of-fit test, 554–55, 556, 557–58
    in least-squares regression model, 586, 588
    about population mean
        known population standard deviation, 465–69
        unknown population standard deviation, 482, 483, 485
    about population proportion, 494–95, 496, 497
    about population standard deviation, C-15, C-16
    right-tailed, large sample, 467–68

to testing claims about matched-pairs data, 510
two-tailed, small sample, 468–69
Classical method, 227–31, 284
    empirical method compared to, 229–31
Closed question, 35
Cluster sampling, 26–27, 29
Coefficient of determination, 209–15
    computing, 211–12
        using technology, 212, 215
    definition of, 209
Column variable, 243, 563
Combinations, 274
    counting problems solved using, 270–73
    definition of, 271
    listing, 271
    of simple random samples, 272–73
    technology to compute, 272, 278–79
Complement of event, 244
Complement Rule, 244–45, 334
    at-least probabilities and, 253
Computational formula, 126
Conclusions, stating, 459–60
Conditional distribution, 569
    bar graph of, 569–70
    constructing, 569–70
Conditional probability, 255–65
    definition of, 256–57
    independence and, 261
    using the General Multiplication Rule, 258–61
Confidence, level of, 5, 406, 410
    margin of error and, 412–13
Confidence interval(s), 403–52
    definition of, 406, 595
    determining appropriate procedure for constructing,
        443–45
    for difference between two population proportions,
        520, 539–41
    about the difference of two means,
        526–28, 534
    hypothesis testing about population
        mean using, 473–74
    margin of error of, 406
    for mean response, 595–96, 597
    95%, 407–9, 410, 438
    about a population mean, 514
    about a population mean: known
        population standard deviation, 404–22
        constructing and interpreting, 405–12, 414–15
        margin of error in, 412–13
        outliers and, 415, 421
        point estimate of population mean, 404–5
        sample size and, 410, 413–15
        using technology, 412, 422
    about a population mean: unknown
        population standard deviation, 423–35
        constructing and interpreting, 426–30
        outliers and, 427, 428–29
        Student's *t*-distribution, 423–25

t-values, 425–26
    using technology, 428, 434–35
  about the population mean difference
      of matched-pairs data, 514–15, 520
  about a population proportion, 435–43
    constructing and interpreting, 436–38
    point estimate for population proportion, 435–36
    sample size determination, 439–40
    using technology, 437–38, 443
  about a population standard deviation, C-7–13
    constructing and interpreting, C-10–11
    critical values for chi-square distribution, C-7–9, C-13
    using technology, 443, C-11, C-13
  about the population variance, C-10–11
  for slope of regression line, 589–90
    constructing, 590
    definition, 589–90
  about standard deviation, C-10–11
Constants, 6
Consumer Reports, 48, 69, 141, 265, 314,
      353, 400, 434, 492, 533, 579
Contingency (two-way) table(s), 242–43,
      563. *See also* Chi-square test
  Addition Rule for Disjoint Events with, 242–43
  conditional distribution, 569–70
Continuity, correction for, 363
Continuous data, 8
  histograms of, 75–76
  quantitative, mean for, 143
  in tables, 73–75
Continuous distributions. *See* Normal probability
    distribution; Uniform probability distribution
Continuous random variable, 285–86
  probability density functions to find probabilities for,
    319–20
Control
  experiment vs. observational studies and, 15
Control group, 5, 15
Convenience sampling, 27–28
Correction factor, finite population, 399
Correction for continuity, 363
Correlation, causation vs., 184
Correlation coefficient
  linear, 179–81
    computing and interpreting, 182–83
    definition of, 180
    determination of linear relation and, 185
    properties of, 179–81
    technology to determine, 194–95
  table critical values for, A-14
Correlation matrix, 183
Counting problems, 265–79
  combinations for, 270–73, 274–75, 278–79
  Multiplication Rule for, 265–68
  permutations for, 268–70, 273–75, 278–79
  without repetition, 267
Counts, expected, 552–53, 563–65
Critical region (rejection region), 466

Critical value(s), 409, 466
  for chi-square distribution, C-7–9, C-13
  for *F*-distribution, C-34–37
Cryptosporidium, 214
Cumulative distribution function (cdf), 305, 333
  inverse, 349
Current Population Survey, 28
Current Population Survey: *Design and Methodology,*
    The, 28
Cutoff point, 225, 226

Darwin, Charles, 199
Data, 3–4. *See also* Qualitative data; Quantitative data
  bivariate, 176. *See also* Relation between two variables
  classes of, 71, 73
  collection of, 13–14. *See also*
    Experiment(s); Observational studies; Sampling
  continuous, 8
    histograms of, 75–76
    quantitative, mean for, 143
    in tables, 73–75
  discrete, 8, 72–73
  grouped, 142–49
    mean from, 142–43, 149
    standard deviation from, 144–46, 149
    variance from, 144–46
  polling, 14
  raw, 54, 73
  right skewed, 433
  time-series, 81
  univariate, 176
  variability in, 3–4
  variables vs., 8–9
Data organization and summary, 54–105.
    See also Numerically summarizing data
  graphical misrepresentations, 93–98
    by manipulating vertical scale, 94
  qualitative data, 8, 55–70
    bar graphs, 57–59, 61, 70
    pie charts, 60–61, 70
    tables, 55–56
  quantitative data, 8, 71–93
    dot plots, 80
    histograms, 72–73, 75–76, 92–93
    shape of distribution, 80–81
    split stems, 79
    stem-and-leaf plots, 76–79, 92–93
    tables, 71–72, 73–75
    time-series graphs, 81–82
Data sets, 54
  comparing, 59, 123–24
  small, 354
Decision rule, 466, 470
Degrees of freedom, 128, 426
de Moivre, Abraham, 322, 325, 362
Density function(s)
  exponential, 384

probability, 319–20
 normal, 347
 uniform, 321
Dependent events, 250
Dependent (response) variable, 14, 40, 41, 177–78
Dependent sampling, 508–10
 about two means. *See* Matched-pairs
  (dependent) design
Depreciation rate, 204
Depression, self-treatment for, 48
Descriptive statistics, 4
Designed experiment. *See also*
  Experiment(s): design of
 defined, 14, 40
 observational study vs., 14
Determination, coefficient of, 209–15
 computing, 211–12
  using technology, 212, 215
 definition of, 209
Deviation(s), 209–10
 explained, 210
 about the mean, 125
 total, 210
 unexplained, 210–11
Dice problem, 231
Die
 fair, 224
 loaded, 224
Dietary supplements, 48
Diophantus, 230
Discrete data, 8
 histograms of, 72–73
Discrete probability distributions, 284–317
 binomial, 298–314, 374
  binomial probability distribution function (pdf), 302–3
  binomial table to compute binomial probabilities, 303–5
  constructing, 300–302
  histograms of, 306–9, 318
  identifying binomial experiment, 298–300
  mean and standard deviation of
   binomial random variable, 305–6, 308–9
  normal approximation to, 362–67
  notation used in, 298
  using technology, 305, 313
 definition of, 286
 identifying, 286–87
 probability histograms of, 287–88
 rules for, 287
Discrete random variable, 285–97
 continuous random variables
   distinguished from, 285–86
 mean of, 288–92, 293, 297
  computing, 289
  defined, 289
  as an expected value, 291–92
  interpreting, 289–90
  using technology, 293, 297–98
 variance and standard deviation of, 292–93

Disjoint events, 238–41
 independent events vs., 250–51
Dispersion, measures of, 123–49, 159
 Chebyshev's Inequality, 132–34
 Empirical Rule, 131–32
 from grouped data, 142–49
 range, 124–25
  computing, 125
  definition of, 124
  interquartile (IQR), 159, 162
  technology to determine, 142
 standard deviation, 125, 129–31, 142, 180
  of binomial random variable, 305–6, 308–9
  confidence interval about, C-10–11
  of discrete random variables, 292–93
  from grouped data, 144–46, 149
  interpretations of, 130
  outlier distortion of, 155
  population, 129–30, 406, 582–83
  sample, 129–30, 583
  of sampling distribution of sample mean, 382
  technology to approximate, 146
  technology to determine, 142
  of two data sets, 130–31
 variance, 125–29, 130, 142
  of discrete random variables, 292–93
  from grouped data, 144–46
  population, 125–28, 144
  sample, 125, 127–29, 144
  technology to determine, 130, 142
Distribution. *See* Discrete probability distributions;
   Frequency distribution(s); Normal
    probability distribution
Distribution function, cumulative (cdf), 305, 333
 inverse, 349
Doctrine of Chance, The (Leibniz), 322
Dot plots, 80
Double-blind experiment, 5, 40
Dow Jones Industrial Average (DJIA), 94–95

EDA (exploratory data analysis), 159
Eléments de Géométrie (Legendre), 198
Elements of Sampling Theory and Methods
(Govindarajulu), 28
Empirical Method, 226–27, 229–31
 classical method compared to, 229–31
Empirical Rule, 131–32, 332, 343
 unusual results in binomial experiment and, 308–9
Equally likely outcomes, 227, 229
Error(s)
 input, 35
 margin of, 405, 438
  of confidence interval, 406
  definition of, 412
  level of confidence and, 412–13
  sample size and, 413–15, 439–40
 mean square due to (MSE), C-25

residual, 197–98
rounding, 129, 183
sampling, 33–39
  data checks to prevent, 35
  frame and, 34
  interviewer error, 34
  misrepresented answers, 35
  nonresponse and, 34
  nonsampling errors, 33
  order of questions, words, and responses, 36
  questionnaire design and, 35
  wording of questions, 35–36
standard, 423, 583–85
  computing, 583–84
  definition of, 583
  of the mean, 382
  for sampling distribution of population proportion, 494
Type I, 457–59, 464, 526
  in ANOVA, C-20
  probability of, 458–59, 464
Type II, 457–59
  probability of, 458–59
Estimator, biased, 127
Event(s), 224
  certain, 225
  complement of, 244
  dependent, 250
  disjoint, 238–41, 250–51
  impossible, 225
  independent, 249–51, 261
    Multiplication Rule for, 251–52, 261
  not so unusual, 225–26
  simple, 224
  unusual, 225–26
Excel, 18
  area under the normal curve using, 354
    scores corresponding, 354
  bar graph using, 58, 70
  binomial probabilities using, 305
  boxplots using, 168
  chi-square tests using, 580
  coefficient of determination using, 212, 215
  combinations using, 272, 279
  confidence intervals using, 599
    for population mean where
      population standard deviation is known, 422
    for population mean where
      population standard deviation is unknown, 435
    for population proportion, 443
    for population standard deviation, 443, C-13
  correlation coefficient using, 195
  difference of two population proportions using, 545
  factorials on, 279
  hypothesis testing using
    about population mean assuming
      known population standard deviation, 480

about population mean with
      unknown population standard deviation, 493
    about population proportion, 500
  least-squares regression line using, 202, 208
  least-squares regression model using, 594
  linear correlation coefficient using, 183
  mean and median using, 123
  normal probability plot using, 361
  one-way ANOVA using, C-23–24, C-33
  permutations using, 269–70, 279
  pie charts using, 70
  prediction intervals using, 599
  scatter diagrams using, 183, 195
  simulation using, 237
  standard error using, 584–85
  standard normal distribution using, 344
  time-series graphs on, 82
  two-sample t-tests, dependent samplingusing, 520
  two-sample t-tests, independent sampling using, 534
  z-score from a specified area to the left using, 337–38
Expected counts, 552–53, 563–65
  in chi-square test, 563–65
  in goodness-of-fit test, 552–53
Expected value, 291–92
Experiment(s), 54, 224, 227, 284
  design of, 40–47
    completely randomized design, 42–43
    matched-pairs design, 43–44
    simple random sampling and, 44
    steps in, 41
  double-blind, 5, 40
Experimental group, 5
Experimental units (subjects), 14, 40, 41
  in matched-pairs design, 43
Explained deviation, 210
Explanatory (predictor or independent) variable, 40, 177–78
Exploratory data analysis (EDA), 159
Exploratory Data Analysis (Tukey), 159
Exponential density function, 384
Exponential probability distribution, 420–21
Ex post facto studies. See Observational studies

Factor(s), 40, 41
Factorial notation, 269
Factorials, technology to compute, 278–79
Factorial symbol, 268
Fair die, 224
F-distribution, critical values for, C-34–37
Federal Trade Commission, 400
Fermat, Pierre de, 230, 231, 292
Fermat's Last Theorem, 230
Fibonacci sequence, 561–62
Finite population correction factor, 399
Fisher, Sir Ronald A., 40, 180, 553, C-13, C-20
Fisher's F-distribution, C-34–37
Five-number summary, 159–60

Frame, 16, 25, 34
Frequency, relative, 56
Frequency distribution(s). *See also*
       Relative frequency distribution
  based on boxplots, 162–63
  bell shaped, 80, 130
    Empirical Rule to describe, 131–32
  center of (average), 106. *See also* Mean(s)
  characteristics of, 106
  chi-square, 551
    characteristics of, C-8, C-14
    critical values for, C-7–9, C-13
    definition of, C-8
    hypothesis testing about population
       standard deviation and, C-14
  comparing, 163–64
  from continuous data, 73–74
  critical value for, 409
  of discrete data, 71–72
  frequency, 55–56
  mean of variable from, 142–43
  of qualitative data, 55–56
  relative, 56
  shape of, 80–81, 106
  spread of, 106
  symmetric, 130
  variance and standard deviation from, 145–46
F-test, 527–28
  computing, C-25
  idea behind, C-24

Gallup Organization, 15, 36
Galton, Sir Francis, 199
Gauss, Carl, 325
Gaussian distribution. *See* Normal probability distribution
General Addition Rule, 241–43
General Multiplication Rule, conditional probability using,
    258–61
Golden ratio, 561
Goodness-of-fit test, 551–62
  characteristics of chi-square distribution, 551
  definition of, 552
  expected counts, 552–53
  technology for, 558
  testing hypothesis using, 555–59
  test statistic for, 554
Gosset, William Sealey, 423
Graph(s), 106
  characteristics of good, 95
  dot plots, 80
  histograms
    binomial probability, 306–9, 318
    of continuous data, 75–76
    of discrete data, 72–73
    probability, 287–88
    technology to draw, 92–93
  of lines, C-2–3
  misleading or deceptive, 93–95

polar area diagram, 74
stem-and-leaf plots, 76–79
  constructing, 76–79
  split stems, 79
  using technology, 92–93
tables, 106
  binomial, 303–5
  continuous data in, 73–75
  discrete data in, 71–72
  open-ended, 73
  qualitative data in, 55–56
time-series, 81–82
Greek letters, use of, 108
Grouped data, 142–49
  mean from, 142–43, 149
  standard deviation from, 144–46, 149
  variance from, 144–46

Herbs, 48
Histogram(s)
  binomial probability, 306–9, 318
  of continuous data, 75–76
  of discrete data, 72–73
  of discrete probability distributions, 287–88
  technology to draw, 92–93
Homogeneity of proportions, chi-square test for, 570–74
  definition of, 570
  steps in, 571–73
Huygens, Christiaan, 292
Hypothesis/hypotheses
  alternative, 454–57, 463
    definition of, 455
    structuring, 456
  definition of, 455
  forming, 456–57
  null, 454–57, 463, 552, 555
    acceptance of, 459
    in ANOVA, C-19–20
    assumption of trueness of, 464
    definition of, 455
    structuring, 456
Hypothesis testing, 453–506. *See also*
       Inferences; One-tailed test(s);
       Two-tailed tests
  choosing method for, 500–502
  definition of, 455
  illustration of, 454
  large samples, 467–68, 470–71, 475
  logic of, 462–65
    classical approach, 463–64
    P-value approach, 464–65
  outcomes from, 457
  for a population mean assuming
      known population standard deviation, 462–80
    classical approach to, 465–69
    P-values approach to, 469–73
    technology in, 473, 479–80

using confidence intervals, 473–74
for a population mean in with
      unknown population standard deviation, 480–93
  classical approach to, 482, 483, 485
  outliers and, 482, 485
  P-value approach, 482, 483, 485
  technology in, 484, 493
  using large sample, 482–83
  using small sample, 484–86
for a population proportion, 493–500
  classical approach to, 494–95, 496, 497
  P-value approach to, 494–95, 496, 497
  right-tailed test, 495–96
  technology in, 497–98, 500
  two-tailed test, 496–97
for population standard deviation, C-13–19
  chi-square distribution and, C-14
  classical approach to, C-15, C-16
  left-tailed test, C-15–16
  P-value approach to, C-15, C-16
probability of Type I error, 458–59, 464
probability of Type II error, 458–59
small samples, 468–69, 471–72, 484–86
stating conclusions, 459–60
steps in, 454
Type I and Type II errors, 457–59

Impossible event, 225
Incentives, nonresponse and, 34
Independence
  chi-square test for, 563–70
    definition of, 563
    expected counts in, 563–65
    expected frequencies in, 566
    steps in, 566–67
  conditional probability and, 261
Independent events, 249–51, 261
  disjoint events vs., 250–51
  Multiplication Rule for, 251–52, 261
Independent (explanatory or predictor) variable, 40, 177–78
Independent samples, 508–10
  difference between two means from, 522–26
    confidence intervals regarding, 526–28, 534
    technology for, 525, 534
    testing claims regarding, 522–26
Independent trials, 299
Individual, population vs., 4
Inferences, 375, 507–604. *See also*
      Least-squares regression model
  reliability of, 375
  about two means
    dependent samples, 508–20
    independent samples, 522–26
  about two population proportions, 534–45
    confidence intervals, 520, 539
    sample size requirements for, 541–42
    testing, 535–39

using technology, 539, 545
Inferential statistics, 5, 41, 404, 453
Inflection points, 321–22
Input errors, 35
Integer distribution, 231
Intercept, 198, 201–2
Internet surveys, 27–28
Interquartile range (IQR), 159, 162
Interviewer error, 34
Inverse cumulative distribution function, 349

Journal of American Medical Association, 13–14

Kolmogorov, Andrei Nikolaevich, 260

Landon, Alfred M., 33
Laplace, Pierre Simon, 387
Law of Large Numbers, 223–24, 284, 381
Least-squares regression line, 195–208
  coefficient of determination, 209–15
  definition of, 198
  equation of, 198–99
  finding, 196–202, 208
    using technology, 202, 208
  interpreting the slope and *y*-intercept of, 201–2
  sum of squared residuals, 202–3
Least-squares regression model, 580–94
  confidence interval about slope of, 589–90
    constructing, 590
    definition of, 589–90
  confidence intervals for mean response, 595–96, 599
  definition of, 583
  example of, 581
  inference on the slope and intercept, 585–89
    using technology, 594
  normally distributed residuals in, 585
  prediction intervals for an individual response, 596–98, 599
  requirements of, 581–83
  robustness of, 587
  sampling distributions in, 581–82
  significance of, 580–94
  standard error of the estimate, 583–85
Left-tailed hypothesis testing. *See* One-tailed test(s)
Legendre, Adrien Marie, 198
Leibniz, Gottfried Wilhelm, 322
Level of confidence, 5, 406, 410
  margin of error and, 412–13
Life on the Mississippi (Twain), 208
Linear correlation coefficient, 179–84
  computing and interpreting, 182–83
  definition of, 180
  determination of linear relation and, 185
  properties of, 179–81
  using technology, 194–95
Linear equation, C-1, C-4–5

point-slope form of, C-3–4
slope and *y*-intercept of line from, C-5
slope-intercept form of, C-5
Linear relation, 185
Linear transformations, 122
Lines, C-1–6
graphing, C-2–3
horizontal, C-3–4
slope of, C-1–2
vertical, C-1
Literary Digest, 33
Loaded die, 224
Lower class limit, 73
Lurking variable, 3, 15, 177, 184

Margin of error, 405, 438
confidence interval and, 406
definition of, 412
level of confidence and, 412–13
sample size and, 413–15, 439–40
Matched-pairs (dependent) design, 43–44, 508
confidence intervals about population mean difference
of, 514–15, 520
testing claims about, 509–13
using technology, 513, 520
Mathematics, statistics vs., 4
Matrix, correlation, 183
Mean(s), 107–10, 116. *See also* Population mean;
Sample mean of binomial random variable,
305–6, 308–9
as center of gravity, 109–10
comparing three or more. *See* Analysis of variance
(ANOVA)
computing, 108–9
using technology, 113, 123
definition of, 107, 108
deviation about the, 125
of discrete random variable, 288–92, 293, 297
computing, 289
defined, 289
as an expected value, 291–92
interpreting, 289–90
using technology, 293, 297–98
from grouped data, 142–43, 149
least-squares regression model and, 582
outlier distortion of, 155
of sampling distribution of sample mean, 382
shape of distribution from, 113–16
standard error of the, 382
technology to approximate, 146
trimmed, 122
weighted, 144
Mean response, confidence intervals for, 595–96
Mean squares, C-25
due to error (MSE), C-25
due to treatment (MST), C-25
Mécanique céleste (Laplace), 387
Median, 78, 107, 110–11, 116, 159, 162

computing, 110–11
with even number of observations, 111
with odd number of observations, 110–11
using technology, 113, 123
definition of, 110
IQR and, 162
shape of distribution from, 113–16
Method of least squares. *See* Least-squares regression line
Midrange, 123
MINITAB, 18
area under standard normal curve using, 333
area under the normal curve using, 354
scores corresponding, 354
bar graph using, 70
binomial probabilities using, 313
boxplots using, 168
chi-square tests using, 570, 580
coefficient of determination using, 215
confidence intervals using, 598, 599
for population mean where
population standard deviation is known, 412, 422
for population mean where
population standard deviation is unknown, 435
for population proportion, 438, 443
for population standard deviation, 443, C-13
correlation coefficient using, 195
difference of two population proportions using, 539, 545
histogram for continuous data using, 76
hypothesis testing using
about population mean assuming
known population standard deviation, 473, 480
about population mean with
unknown population standard deviation, 493
about population proportion, 497–98, 500
least-squares regression line using, 202, 208
least-squares regression model using, 594
mean and median using, 123
normality assessed using, 357–58
normal probability plot using, 361
normal random variable using, 348–49
number summary using, 160
one-way ANOVA using, C-23–24, C-33
pie charts using, 70
prediction intervals using, 598, 599
scatter diagrams using, 195
simulation using, 237
standard normal distribution using, 344
stem-and-leaf plot using, 78
testing claims about matched-pairs data using, 513
time-series graphs on, 82
two-sample *t*-tests using
dependent sampling, 520
independent sampling, 534
Mode, 107, 111–13, 116
bimodality, 112
computing, 112
definition of, 111
multimodality, 112

of qualitative data, 112–13
of quantitative data, 112
Models, 324
Mood modifiers, herbal, 48
Multimodal data set, 112
Multimodal instruction, 531
Multiple linear regression model
  correlation matrix, 183
Multiplication Rule, 563
  for counting, 265–68
  for Independent Events, 251–52, 261
Multistage sampling, 28
Mutually exclusive (disjoint) events, 238–41
  independent events vs., 250–51

Necomb, Simon, 239
Negatively associated variables, 179, 181
Newcomb, Simon, 560
Newton, Isaac, 322
Neyman, Jerzy, 463
Nielsen Media Research, 28
Nightingale, Florence, 74
95% confidence interval, 407–9, 410, 438
Nominal variable, 13
Nonparametric procedures, 429
Nonresponse, 34
Nonsampling errors, 33
Normal curve
  area under, 323–25, 345–47
    finding, 326
    interpreting, 325
    as probability, 324, 325, 331, 340–41
    as proportion, 324, 325, 331
    standard normal curve, 332–36
  inflection points on, 321–22
Normal populations, sampling distribution of sample mean
    from, 377–83
Normal probability density function, 324, 347
Normal probability distribution, 318–71, 374
  applications of, 345–54
    area under a normal curve, 345–47
    value of normal random variable, 348–49
  area under, 323–25
    finding, 326
    interpreting, 325
    as proportion or probability, 324, 325, 331
  assessing normality, 354–61
    normal probability plots for, 354–61
  confidence interval construction and, 415
  graph of, 321–22
  least-squares regression model and, 585
  normal approximation to the binomial probability
    distribution, 362–67
  in one-way ANOVA, C-20–21
  properties of, 319–31
    statement of, 322–23
  relation between normal random
    variable and standard normal random variable, 325–27

standard, 326, 331–44
  area under the standard normal curve, 332–36, 340–41
  notation for probability of standard normal random
    variable, 340
  properties of, 331–32
  t-distribution compared to, 424
  technology to find, 344
  Z-scores for given area, 336–40
  technology to find, 353–54
  uniform probability distribution, 318, 319–21
  definition of, 319
Normal probability plot, 354–61
  drawing, 355–57
  linearity of, 355–57
  technology to assess, 357–58, 361
Normal random variable, 323–27
  standard, 325–27, 340–41
  standardizing, 326–27
  value of, 348–49
Normal score, 355
Not so unusual events, 225–26
Nouvelles méthodes pour la determination des orbites des comëtes (Legendre), 198
Null hypothesis, 454–57, 463, 552, 555
  acceptance of, 459
  in ANOVA, C-19–20
  assumption of trueness of, 464
  definition of, 455
  structuring, 456
Numerically summarizing data, 106–75
  boxplots, 161–64
    comparing two distributions using, 163–64
    constructing, 161–62
    distribution shape based on, 162–63
    technology to draw, 168
  five-number summary, 159–60
  measures of central tendency, 107–23
    arithmetic mean. See Mean(s)
    from grouped data, 142–43, 149
    median, 78, 107, 110–11, 113–16, 123, 159, 162
    midrange, 123
    mode, 107, 111–13, 116
    shape of distribution from mean and median, 113–16
    trimmed mean, 122
  measures of dispersion, 123–49, 159
    Chebyshev's Inequality, 132–34
    Empirical Rule, 131–32
    from grouped data, 142–49
    range, 124–25, 142, 159, 162
    standard deviation. See Standard deviation variance,
      125–29, 130, 142, 144–46, 292–93
  measures of position, 149–59
  outliers. See Outliers
  percentiles, 151–53, 346–47
  quartiles, 153–54, 155
  z-scores, 149–50, 326, 336–40, 355, 425

Observational studies, 13–16

defined, 14
experiment vs., 14–15
One-tailed test(s), 456, 466, 467–68, 470–71, 481–84, 495–96
  of difference between two means, 509, 510, 511
  of difference between two means: independent samples, 522
  of difference between two population proportions, 536, 537
  in least-squares regression model, 586–87
  about population proportion, 495–96
  about population standard deviation, C-15–16
One-way ANOVA, C-19–37
  conceptual understanding of, C-24–25
  decision rule in, C-22
  equal population variances in, C-21
  normal distribution in, C-20–21
  requirements of, C-20–22
  robustness of, C-21
  testing hypotheses with, C-22–27
  using technology, C-23–24, C-33
Open-ended tables, 73
Open question, 35
Ordinal variable, 13
Outcomes, 223
  equally likely, 227, 229
Outliers, 106, 155–56
  confidence intervals and
    for population mean where population standard deviation is known, 415, 421
    for population mean where population standard deviation is unknown, 427, 428–29
  hypothesis testing about population mean and, 482, 485
  quartiles to check, 155

Parameters, 107, 108
Pareto chart, 58
Pascal, Blaise, 230, 231, 292
Pearson, Egon, 463
Pearson, Karl, 180, 321, 423, 463, 553
Pearson product moment correlation coefficient.
  See Linear correlation coefficient
People Meter, 28
Percentile(s), 151–53
  with index as integer, 151–52
  with index not an integer, 152–53
  kth, 151–52
  quartiles, 153–54
  ranks by, 346–47
  of specific data value, 153
Permutations, 268–70, 274–75
  computing, 269–70
    using technology, 269–70, 278–79
  definition of, 268–69
  of distinct items, 274
  with nondistinct items, 273–75
Phone-in polling, 27–28
Pie charts, 60–61

constructing, 60–61
technology to draw, 70
Placebo, 5, 40
Point estimate
  definition of, 404
  of population mean, 404–5
  of population proportion, 435–36
  of two population means, 514
  of two population proportions, 536
Points, problem of, 231
Point-slope form of line, 197, C-3–4
Polar area diagram, 74
Polling, phone–in, 27–28
Polling data, 14
Pooled estimate of p, 536
Pooled t-statistic, 527–28
Pooling, 527
Population, 4
  with distribution unknown, 354
  mean of (μ), 107–9
  proportion of, 324
  underrepresentation of, 34
Population mean, 107–9, 142
  confidence interval about, 514
  confidence interval about, where the population standard deviation is known, 404–22
    constructing and interpreting, 405–12, 414–15
    margin of error in, 412–13
    outliers and, 415, 421
    point estimate of population mean, 404–5
    sample size and, 410, 413–15
    technology for, 412, 422
  confidence interval about, where the population standard deviation is unknown, 423–35
    constructing and interpreting, 426–30
    outliers and, 427, 428–29
    Student's t-distribution, 423–25
    technology for, 428, 434–35
    t-values, 425–26
  hypothesis testing about, 480–93
    known population standard deviation, 462–80
    unknown population standard deviation, 480–93
  point estimate of, 404–5
Population proportion, 324
  confidence intervals about, 435–43
    constructing and interpreting, 436–38
    point estimate for population proportion, 435–36
    sample size determination, 439–40
    technology for, 437–38, 443
  difference between two, 534–45
    confidence intervals, 520, 539–41
    sample size requirements for, 541–42
    testing, 535–39
    using technology, 539, 545
  hypothesis testing about, 493–500
    classical approach to, 494–95, 496, 497
    P-value approach to, 494–95, 496, 497
    right-tailed test, 495–96

technology in, 497–98, 500
  two-tailed test, 496–97
point estimate for, 435–36
sampling distribution of, 436
  standard error for, 494
Population size, sample size and, 395
Population standard deviation, 129–30
  confidence intervals about, C-7–13
    constructing and interpreting, C-10–11
    critical values for chi-square distribution, C-7–9, C-13
    using technology, 443, C-11, C-13
  hypothesis testing for, C-13–19
    chi-square distribution and, C-13
    classical approach to, C-15, C-16
    left-tailed test, C-15–16
    P-value approach to, C-15, C-16
  least-squares regression model and, 582–83
  margin of error of confidence interval and, 406
Population variance, 125–27, 144
  confidence intervals about, C-10–11
  hypothesis testing about, C-14–15
  in one-way ANOVA, C-21
Population *z*-score, 150
Position, measures of, 149–59
  outliers, 106, 155–56
  percentiles, 151–53
  quartiles, 153–54
  z-scores, 149–50
Positively associated variables, 179, 181
Practical significance, 538
  definition of, 474
  statistical significance vs., 474–75
Prediction intervals
  definition of, 595
  for an individual response, 596–98, 599
Predictor (independent or explanatory) variable, 40, 177–78
Pretest-posttest (before-after) experiments, 43
Probability(ies), 221–83
  Addition Rule for Disjoint Events, 238–49
    Benford's Law and, 239–41
    with contingency tables, 242–43
    General, 241–43
  area as, 320–21
  area under normal curve as, 324, 325, 331, 340–41
  at-least, 253
  classical, 227–31, 284
  Complement Rule, 244–45, 253, 334
  conditional, 255–65
    definition of, 256–57
    independence and, 261
    using the General Multiplication Rule, 258–61
  counting problems, 265–79
    combinations for, 270, 274–75, 278–79
    Multiplication Rule for, 265–68
    permutations for, 268–70, 273–75, 278–79
    without repetition, 267
  defined, 221, 223

Empirical Method to approximate, 226–27, 229–31
events and the sample space of probability experiment, 224
Multiplication Rule for Independent Events, 251–52, 261
relative frequency to approximate, 226
rules of, 223–37, 253
  understanding, 225–26
simulation to obtain, 231–32
  technology in, 237
subjective, 232
value of normal random variable corresponding to, 348–49
Probability density function (pdf), 319–20, 324
  normal, 324
Probability distribution, 374. *See also* Normal probability distribution
  binomial, 374
  exponential, 420–21
  testing claims regarding. *See* Contingency (two-way) table(s); Goodness-of-fit test
Probability experiment, 54, 224, 227, 284
  design of, 40–47
    completely randomized design, 42–43
    matched-pairs design, 43–44
    simple random sampling and, 44
    steps in, 41
  double-blind, 5, 40
Probability histograms of discrete probability distributions, 287–88
  binomial, 306–9
Probability model, 225, 284
  for random variables. *See* Discrete probability distributions
  from survey data, 227
Proportion(s). *See also* Population proportion; Sample proportion
  area under normal curve as, 324, 325, 331
  homogeneity of, 570–74
    definition of, 570
    steps in, 571–73
  value of normal random variable corresponding to, 348–49
P-value, definition of, 469
P-value approach to hypothesis testing, 464–65
  in chi-square test
    for homogeneity of proportions, 572
    for independence, 567, 568
  of difference between two means: independent samples, 522, 524
  of difference between two population proportions, 537, 538
  goodness-of-fit, 554–55, 556, 557–58
  in least-squares regression model, 586, 588
  of matched-pairs data, 510
  in one-way ANOVA, C-22
  about population mean
    known population standard deviation, 469–73
    unknown population standard deviation, 482, 483, 485

about population proportion, 494–95, 496, 497
about population standard deviation, C-15, C-16
right-tailed, large sample, 470–71
two-tailed, 471–72

Qualitative data, 8, 55–70
  bar graphs of, 57–59, 61, 70
  frequency distribution of, 55–56
    relative, 56
  mode of, 112–13
  pie charts of, 60–61, 70
  tables of, 55–56
Qualitative variable, 6–7
Quantitative data, 8, 71–93
  dot plots of, 80
  histograms of, 72–73, 75–76, 92–93
  mode of, 112
  shape of distribution of, 80–81
  stem-and-leaf plots of, 76–79, 92–93
    split stems, 79
  tables of, 71–72, 73–75
Quantitative variable, 6–7
Quartiles, 153–54
  checking for outliers using, 155
  using technology, 154
Questionnaire design, 35
Questions
  order of, 36
  wording of, 35–36
Questions and Answers in Attitude Surveys
    (Schuman and Presser), 36
Queuing theory, C-18

Randomization, 41
Randomized block design, 44
Random number generator, 18, 81
Random number table, A-1
Random sampling, 16–23, 25
  combinations of samples, 272–73
  confidence interval construction and, 414–15
  definition of, 16
  illustrating, 16
  obtaining sample, 16–23
Random variable(s), 374
  binomial, 298
    normal approximation to, 364–65
  continuous, 285–86
    probability density functions to find probabilities for, 319–20
  definition of, 285
  discrete, 285–97
    continuous random variables distinguished from, 285–86
    mean of, 288–92, 293, 297
    variance and standard deviation of, 292–93
  normal, 323–27
    standard, 325–27, 340–41

    standardizing, 326–27
    value of, 348–49
  probability models for. *See* Discrete probability distributions
  statistics as, 374
Range, 124–25, 142
  computing, 125
  definition of, 124–25
  interquartile (IQR), 159, 162
  technology to determine, 142
Ratio, golden, 561
Raw data, 54
  continuous, 73
Regression. *See* Least-squares regression model
Regression analysis, 199
Rejection region (critical region), 466
Relation between two variables, 176–219
  correlation versus causation, 184
  least-squares regression line, 195–208
    coefficient of determination, 209–15
    definition of, 198
    equation of, 198–99
    finding, 196–202, 208
    interpreting the slope and y-intercept of, 201–2
    sum of squared residuals, 202–3
  linear correlation coefficient, 179–84
    computing and interpreting, 182–83
    definition of, 180
    properties of, 179–80
    using technology, 194–95
  scatter diagrams, 177–79
    definition of, 177
    drawing, 177–78, 183, 194–95
Relative frequency(ies)
  probability using, 226
  of qualitative data, 56
Relative frequency distribution, 56
  from continuous data, 73–75
  of discrete data, 71–72
Reliability of inferences, 375
Replication, 41
Research objective, identification of, 4
Residual(s), 197–98
  normally distributed, 585
  sum of squared, 202–3
Response (dependent) variable, 14, 40, 41, 177–78
Rewards, nonresponse and, 34
Right skewed data, 433
Right-tailed hypothesis testing. *See* One-tailed test(s)
Rise, C-1
Robustness, 410, 467, 510
  of least-squares regression model, 587
  of one-way ANOVA, C-21
Roman letters, use of, 108
Roosevelt, Franklin D., 33
Rounding, 8
Rounding error, 129, 183
Rounding up vs. rounding off, 414

Row variable, 242–43, 563
Run, C-1

Sample(s)
  correlation coefficient, 180
  defined, 4
  matched-pairs (dependent), 508
    confidence intervals for, 514–15, 520
    testing claims about, 509–13
  mean of ($\bar{x}$) 107–9
  self-selected, 27
Sample mean, 107–9, 142
  sample size and, 381
  sampling distribution of, 375–92, 403, 407
    definition of, 375–76
    describing, 383
    mean and standard deviation of, 382
    from normal populations, 377–83
    from not-normal populations, 384–88
    shape of, 383
Sample proportion
  computing, 393
  definition of, 392
  sampling distribution of, 392–400
    describing, 392–95
    probabilities of, 396–97
  simulation to describe distribution of, 393–95
Sample size, 28
  confidence interval about population mean
    where population standard deviation is known and,
      410
  confidence interval about population proportion and,
      439–40
  for difference of two population proportions, 541–42
  distribution shape and, 385–87
  hypothesis testing and, 467–72, 475, 482–86
  margin of error and, 406, 413–15, 439–40
  population size and, 395
  sampling variability and, 381–82
Sample space, 224
Sample standard deviation, 129–30, 583
Sample variance, 125, 127–29, 144
Sample z-score, 150
Sampling, 16–39
  cluster, 26–27, 29
  convenience, 27–28
  dependent, 508–10
  errors in, 33–39
    data checks to prevent, 35
    frame and, 34
    interviewer error, 34
    misrepresented answers, 35
    nonresponse and, 34
    nonsampling errors, 33
    order of questions, words, and responses, 36
    questionnaire design and, 35
    wording of questions, 35–36

  goal in, 16
  independent, 508–10
  multistage, 28
  without replacement, 17
  sample size considerations, 28
  simple random, 16–23, 25
    combinations of, 272–73
    confidence interval construction and, 414–15
    definition of, 16
    illustrating, 16
    obtaining sample, 16–23
  stratified, 23–24, 25, 29
  survey, 14
  systematic, 25–26, 27, 29
Sampling distribution(s), 373–402
  concept of, 375–77
  of difference between two proportions, 535
  of difference of two means, 521–22
  in least-squares regression model, 581
  of population proportion, 436
    standard error for, 494
  of sample mean, 375–92, 403, 407
    definition of, 375–76
    describing, 383
    mean and standard deviation of, 382
    from normal populations, 377–83
    from not-normal populations, 384–88
    shape of, 383
  of sample proportion, 392–400
    describing, 392–95
    probabilities of, 396–97
Scatter diagrams, 177–79
  definition of, 177
  drawing, 177–78, 183, 194–95
Seed, 18
Self-selected samples, 27
Shape. *See* Normal probability distribution
Side-by-side bar graph, 58–59
Sigma (σ) 108
Significance, 538
  definition of, 463
  of least-squares regression model, 580–94
  practical, 538
    definition of, 474
    statistical vs., 474–75
  statistical, 538
    practical vs., 474–75
  Type I error and, 459
Simple events, 224
Simple random sample, 16–23, 25
  combinations of, 272–73
  confidence interval construction and, 414–15
  definition of, 16
  illustrating, 16
  obtaining, 16–23
Simpson's paradox, 604
Simulation, 231–32
  using technology, 237
Skewed distributions, 80

IQR for, 162
Skewness, 162
Slope, 198, 201–2
  defined, C-1
Slope-intercept form of equation of line, C-5
SPF (sun-protection factor), 353
Split stems, 79
Spread. *See* Standard deviation; Variance
Spreadsheets. *See* Statistical spreadsheets
Standard deviation, 125, 129–31, 142, 180
  of binomial random variable, 305–6, 308–9
  confidence interval about, C-10–11
  of discrete random variables, 292–93
  from grouped data, 144–46, 149
  interpretations of, 130
  outlier distortion of, 155
  population, 129–30
    confidence intervals about, C-7–13
    hypothesis testing for, C-13–19
    least-squares regression model and, 582–83
    margin of error of confidence interval and, 406
  sample, 129–30, 583
  of sampling distribution of sample mean, 382
  of two data sets, 130–31
  using technology, 142, 146
Standard error, 423, 583–85
  computing, 583–84
  definition of, 583
  of the mean, 382
  for sampling distribution of population proportion, 494
Standard normal distribution table, A-10–11
Standard normal probability distribution, 326, 331–44
  area under the standard normal curve, 332–36
    to left of z-score, 332–33, 335, 336
    as a probability, 340–41
    to right of z-score, 334–35, 336
    technology to find, 333–34, 337–38
    between two z-scores, 335, 336
    z-scores for given, 336–40
  notation for probability of standard normal random
      variable, 340
  properties of, 331–32
  t-distribution compared to, 424–25
  technology to find, 344
  z-scores for given area, 336–40
Standard normal random variables, 325–27, 340–41
Statistic, 107
  biased, 127
  as random variable, 373
  test, 466, 494
Statistical Abstract of the United States, 231
Statistical inference, 375
Statistical significance, 538
  practical significance vs., 474–75
Statistical spreadsheets. *See also* Excel
  area under standard normal curve using, 334
  pie charts on, 60
  Tally command, 56
Statistical thinking, 3–4

Statistics, 3–13
  definition of, 3–4
  descriptive, 4
  inferential, 5, 41
  mathematics vs., 4
  process of, 4–6
  variables in, 6–9
    data vs., 8–9
    discrete vs. continuous, 7–9
    nominal vs. ordinal, 13
    qualitative (categorical) vs. quantitative, 6–7
Status quo statement, 455, 456
Stem-and-leaf plots, 76–79
  constructing, 76–79
  split stems, 79
  using technology, 92–93
Strata, 23
Stratified random sampling, 23–24, 25, 29
Student's *t*. See *t*-distribution
Subject (experimental unit), 14, 40, 41
  in matched-pairs design, 43
Subjective probability, 232
Sum of squared residuals, 202–3
Sum of squares, C-27
Sunscreens, 353
Survey data, probability model from, 227
Surveys, 54, 438
  Internet, 27–28
Survey sampling, 14
Suzuki, Ichiro, 295
Symmetric distributions, 80, 130
Systematic sampling, 25–26, 27, 29

Tables, 106, A-1–14
  binomial, 303–5
  binomial probability distribution, A-2–9
  chi-square (χ^2) distribution, A-13
  continuous data in, 73–75
  critical values for correlation coefficient, A-14
  discrete data in, 71–72
  open-ended, 73
  qualitative data in, 55–56
  random number, A-1
  standard normal distribution, A-10–11
  t-distribution, A-12
t-distribution, 423–25
  finding values, 425–26
  hypothesis testing and, 480–81, 482
  properties of, 425, 480–81
  standard normal distribution compared to, 424–25
t-distribution table, A-12
Technology. *See also* Excel; MINITAB;
      TI-83/84 Plus graphing calculator
  ANOVA using
    one-way, C-23–24, C-33
  area under normal curve using, 346–47
  area under standard normal curve using, 333–34, 337–38
  binomial probabilities using, 305, 313

boxplots using, 168
chi-square tests using, 570, 580
coefficient of determination using, 215
combinations using, 272, 278–79
confidence intervals using, 598
   for population mean where population standard
      deviation is known, 412, 422
   for population mean where population standard
      deviation is unknown, 428, 434–35
   for population proportion, 437–38, 443
   for population standard deviation, 443, C-11, C-13
difference between two means using, 525, 534
difference between two population proportions using,
   539, 545
exact P-values using, 558
factorials using, 278–79
goodness-of-fit test using, 558
in hypothesis testing
   about population mean assuming known population
      standard deviation, 473, 479–80
   about population mean with unknown population
      standard deviation, 484, 493
   about population proportion, 497–98, 500
least-squares regression model using, 588–89, 594
linear correlation coefficient using, 183, 194–95
mean and standard deviation of discrete random
   variable using, 293, 297
mean using, 146
normal probability distribution using, 353–54
   standard, 344
normal probability plot using, 269–70, 361
normal random variable using, 348–49
number summary using, 160
permutations using, 269–70, 278–79
prediction intervals using, 598
scatter diagram using, 183
simple random sample using, 18–23
standard error using, 584
testing claims about matched-pairs
   data using, 513, 520
two-sample t-tests, independent sampling, 534
Test statistic, 466, 494
TI-83/84 Plus graphing calculator, 18–19, 22–23
area under normal curve using, 346–47, 353–54
   scores corresponding, 354
area under standard normal curve using, 333
binomial probabilities using, 313
boxplots using, 168
chi-square tests using, 580
coefficient of determination using, 215
combinations using, 272, 278–79
confidence intervals using, 599
   for population mean, 428
   for population mean where population standard
      deviation is known, 422
   for population mean where population standard
      deviation is unknown, 434–35
   for population proportion, 443

for population standard deviation, 443, C-13
correlation coefficient using, 195
difference between two population proportions using,
   545
factorials using, 278–79
in hypothesis testing
   about population mean assuming known population
      standard deviation, 479–80
   about population mean with unknown population
      standard deviation, 484, 493
   about population proportion, 500
least-squares regression line using, 202, 208
least-squares regression model using, 588–89, 594
mean and median on, 123
mean and standard deviation using
   approximation, 146
   from grouped data, 149
normal probability plot using, 361
one-way ANOVA using, C-23–24, C-33
permutations using, 269–70, 278–79
prediction intervals using, 599
quartiles using, 154
scatter diagrams using, 194
simulation using, 237
standard deviation of discrete random variable using,
   293, 297
standard normal distribution using, 344
time-series graphs on, 82
two-sample t-tests using
   dependent sampling, 520
   independent sampling, 534
variance using, 130
z-score from a specified area to the left using, 337–38
Time-series data, 81
Time-series graphs, 81–82
t-interval, 427
Total deviation, 209
Transformations, linear, 122
Treatment, 40
Tree diagram, 230
Trials, 298, 362
Trimmed mean, 122
t-statistic, 425–26
   interpretation of, 424
   pooled, 527–28
   two-sample, 526–28, 534
   Welch's approximate, 521–22
Tukey, John, 159, 163
Twain, Mark, 208
Two-tailed tests, 455, 466, 468–69, 470, 471–72, 481–82,
   496–97
   of difference between two means, 509, 510
   of difference between two population proportions,
     536, 537
   of difference of two means: independent samples, 522
   in least-squares regression model, 586–87
   about population proportion, 496–97
Two-way table. See Contingency (two-way) table(s)

Type I error, 457–59, 464, 526
  in ANOVA, C-20
  probability of, 458–59, 464
Type II error, 457–59
  probability of, 458–59

Underrepresentation of population, 34
Unexplained deviation, 209–10
Uniform density function, 321
Uniform probability distribution, 80, 318, 319–21
  definition of, 319
Unimodal instruction, 531
Univariate data, 176
Unusual events, 225–26
Upper class limit, 73

Value, expected, 291–92
Variable(s), 6–9. *See also* Random variable(s)
  associated, 179, 181
  column, 243, 563
  data vs., 8–9
  defined, 6
  dependent (response), 14, 40, 41, 177–78
  discrete vs. continuous, 7–9
  independent (explanatory or predictor), 40, 177–78
  lurking, 3, 15, 177, 184
  nominal vs. ordinal, 13
  relation between two. *See* Relation between
    two variables
  response, 14, 40, 41

row, 242–43, 563
Variance, 125–29, 130, 142
  of discrete random variables, 292–93
  from grouped data, 144–46
  population, 125–27, 144
  sample, 125, 127–29, 144
  technology to determine, 130, 142
Venn diagrams, 238
Vertical line, C-1
Voluntary response samples, 27

Weighted mean, 144
Welch, Bernard Lewis, 521
Welch's approximate t, 521–22
Wiles, Andrew, 230
Wunderlich, Carl Reinhold August, 488

Yakovlena, Vera, 260
y-intercept, 201–2, C-5

Z-interval, 410
z-score, 149–50, 425
  area under normal curve using, 326
  comparing, 150
  expected, 355
  for given area under standard normal probability
    distribution, 336–40
  population, 150
  sample, 150

CHAPTER 12 Additional Inferential Procedures

- Expected Counts (when testing for goodness of fit)

$$E_i = \mu_i = np_i \quad \text{for} \quad i = 1, 2, \ldots, k$$

- Expected Frequencies (when testing for independence or homogeneity of proportions)

$$\text{Expected frequency} = \frac{(\text{row total})(\text{column total})}{\text{table total}}$$

- Chi-Square Test Statistic

$$\chi^2 = \sum \frac{(\text{observed} - \text{expected})^2}{\text{expected}} = \sum \frac{(O_i - E_i)^2}{E_i}$$

$$i = 1, 2, \ldots, k$$

(1) All expected frequencies are greater than or equal to 1 and (2) no more than 20% of the expected frequencies are less than 5.

Use $k - 1$ degrees of freedom for goodness of fit.

Use $(r - 1)(c - 1)$ degrees of freedom when testing for independence or homogeneity of proportions (r is the number of rows, c is the number of columns).

- Standard Error of the Estimate

$$s_e = \sqrt{\frac{\sum (y_i - \hat{y}_i)^2}{n - 2}} = \sqrt{\frac{\sum \text{residuals}^2}{n - 2}}$$

- Standard error of b_1

$$s_{b_1} = \frac{s_e}{\sqrt{\sum (x_i - \overline{x})^2}}$$

- Test statistic for the Slope of the Least-Squares Regression Line

$$t_0 = \frac{b_1 - \beta_1}{s_e / \sqrt{\sum (x_i - \overline{x})^2}} = \frac{b_1 - \beta_1}{s_{b_1}}$$

- Confidence Interval for the Slope of the Regression Line
A $(1 - \alpha) \cdot 100\%$ confidence interval for the slope of the true regression line, β_1, is given by:

$$\text{Lower bound:} \quad b_1 - t_{\frac{\alpha}{2}} \cdot \frac{s_e}{\sqrt{\sum (x_i - \overline{x})^2}}$$

$$\text{Upper bound:} \quad b_1 + t_{\frac{\alpha}{2}} \cdot \frac{s_e}{\sqrt{\sum (x_i - \overline{x})^2}}$$

where $t_{\frac{\alpha}{2}}$ is computed with $n - 2$ degrees of freedom.

- Confidence Interval about the Mean Response of y, $\hat{y}$
A $(1 - \alpha) \cdot 100\%$ confidence interval for the mean response of y, $\hat{y}$, is given by

$$\text{Lower bound:} \quad \hat{y} - t_{\frac{\alpha}{2}} \cdot s_e \sqrt{\frac{1}{n} + \frac{(x^* - \overline{x})^2}{\sum (x_i - \overline{x})^2}}$$

$$\text{Upper bound:} \quad \hat{y} + t_{\frac{\alpha}{2}} \cdot s_e \sqrt{\frac{1}{n} + \frac{(x^* - \overline{x})^2}{\sum (x_i - \overline{x})^2}}$$

where x^* is the given value of the explanatory variable and $t_{\frac{\alpha}{2}}$ is the critical value with $n - 2$ degrees of freedom.

- Prediction Interval about an Individual Response, $\hat{y}$
A $(1 - \alpha) \cdot 100\%$ prediction interval for the individual response of y, $\hat{y}$, is given by

$$\text{Lower Bound:} \quad \hat{y} - t_{\frac{\alpha}{2}} \cdot s_e \sqrt{1 + \frac{1}{n} + \frac{(x^* - \overline{x})^2}{\sum (x_i - \overline{x})^2}}$$

$$\text{Upper Bound:} \quad \hat{y} + t_{\frac{\alpha}{2}} \cdot s_e \sqrt{1 + \frac{1}{n} + \frac{(x^* - \overline{x})^2}{\sum (x_i - \overline{x})^2}}$$

where x^* is the given value of the explanatory variable and $t_{\frac{\alpha}{2}}$ is the critical value with $n - 2$ degrees of freedom.

Table VI

Chi-Square (χ^2) Distribution
Area to the Right of Critical Value

Degrees of Freedom	0.995	0.99	0.975	0.95	0.90	0.10	0.05	0.025	0.01	0.005
1	—	—	0.001	0.004	0.016	2.706	3.841	5.024	6.635	7.879
2	0.010	0.020	0.051	0.103	0.211	4.605	5.991	7.378	9.210	10.597
3	0.072	0.115	0.216	0.352	0.584	6.251	7.815	9.348	11.345	12.838
4	0.207	0.297	0.484	0.711	1.064	7.779	9.488	11.143	13.277	14.860
5	0.412	0.554	0.831	1.145	1.610	9.236	11.071	12.833	15.086	16.750
6	0.676	0.872	1.237	1.635	2.204	10.645	12.592	14.449	16.812	18.548
7	0.989	1.239	1.690	2.167	2.833	12.017	14.067	16.013	18.475	20.278
8	1.344	1.646	2.180	2.733	3.490	13.362	15.507	17.535	20.090	21.955
9	1.735	2.088	2.700	3.325	4.168	14.684	16.919	19.023	21.666	23.589
10	2.156	2.558	3.247	3.940	4.865	15.987	18.307	20.483	23.209	25.188
11	2.603	3.053	3.816	4.575	5.578	17.275	19.675	21.920	24.725	26.757
12	3.074	3.571	4.404	5.226	6.304	18.549	21.026	23.337	26.217	28.299
13	3.565	4.107	5.009	5.892	7.042	19.812	22.362	24.736	27.688	29.819
14	4.075	4.660	5.629	6.571	7.790	21.064	23.685	26.119	29.141	31.319
15	4.601	5.229	6.262	7.261	8.547	22.307	24.996	27.488	30.578	32.801
16	5.142	5.812	6.908	7.962	9.312	23.542	26.296	28.845	32.000	34.267
17	5.697	6.408	7.564	8.672	10.085	24.769	27.587	30.191	33.409	35.718
18	6.265	7.015	8.231	9.390	10.865	25.989	28.869	31.526	34.805	37.156
19	6.844	7.633	8.907	10.117	11.651	27.204	30.144	32.852	36.191	38.582
20	7.434	8.260	9.591	10.851	12.443	28.412	31.410	34.170	37.566	39.997
21	8.034	8.897	10.283	11.591	13.240	29.615	32.671	35.479	38.932	41.401
22	8.643	9.542	10.982	12.338	14.042	30.813	33.924	36.781	40.289	42.796
23	9.260	10.196	11.689	13.091	14.848	32.007	35.172	38.076	41.638	44.181
24	9.886	10.856	12.401	13.848	15.659	33.196	36.415	39.364	42.980	45.559
25	10.520	11.524	13.120	14.611	16.473	34.382	37.652	40.646	44.314	46.928
26	11.160	12.198	13.844	15.379	17.292	35.563	38.885	41.923	45.642	48.290
27	11.808	12.879	14.573	16.151	18.114	36.741	40.113	43.194	46.963	49.645
28	12.461	13.565	15.308	16.928	18.939	37.916	41.337	44.461	48.278	50.993
29	13.121	14.257	16.047	17.708	19.768	39.087	42.557	45.722	49.588	52.336
30	13.787	14.954	16.791	18.493	20.599	40.256	43.773	46.979	50.892	53.672
40	20.707	22.164	24.433	26.509	29.051	51.805	55.758	59.342	63.691	66.766
50	27.991	29.707	32.357	34.764	37.689	63.167	67.505	71.420	76.154	79.490
60	35.534	37.485	40.482	43.188	46.459	74.397	79.082	83.298	88.379	91.952
70	43.275	45.442	48.758	51.739	55.329	85.527	90.531	95.023	100.425	104.215
80	51.172	53.540	57.153	60.391	64.278	96.578	101.879	106.629	112.329	116.321
90	59.196	61.754	65.647	69.126	73.291	107.565	113.145	118.136	124.116	128.299
100	67.328	70.065	74.222	77.929	82.358	118.498	124.342	129.561	135.807	140.169

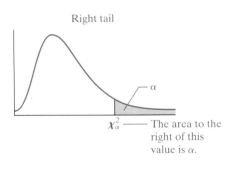

Right tail

χ_α^2 —— The area to the right of this value is α.

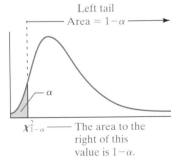

Left tail
Area = $1-\alpha$

$\chi_{1-\alpha}^2$ —— The area to the right of this value is $1-\alpha$.

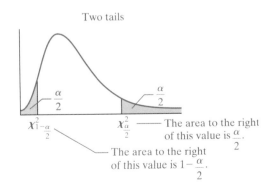

Two tails

$\chi_{1-\frac{\alpha}{2}}^2$

$\chi_{\frac{\alpha}{2}}^2$ —— The area to the right of this value is $\frac{\alpha}{2}$.

—— The area to the right of this value is $1-\frac{\alpha}{2}$.